In Synchrony with the Heavens

Volume One
The Call of the Muezzin

In Synchrony with the Heavens

Studies in Astronomical Timekeeping and Instrumentation in Medieval Islamic Civilization

Volume One

The Call of the Muezzin

Studies I-IX

By

David A. King

TUTA SUB AEGIDE PALLAS · 1683 ·

BRILL

LEIDEN · BOSTON
2014

This paperback was originally published in hardback under ISBN 90-04-12233-8

Cover illustration: *Prayer on the Rooftops* (1865) by Jean Léon Gérôme. Courtesy of the Kunsthalle, Hamburg. See fig. 2.1, page 861.

Library of Congress Cataloging-in-Publication Data
King, David A.
In synchrony with the heavens : studies in astronomical timekeeping and instrumentation in medieval Islamic civilization / by David A. King.
p. cm. – (Islamic philosophy, theology, and science, ISSN 0169-8729 ; v. 55)
Includes bibliographical references and indexes.
ISBN 9004122338 (v. 1 : acid-free paper) / 900414188X (v. 2 : acid-free paper)
1. Astronomy--Islamic countries--History. 2. Astronomy, Medieval. 3. Islam and science. 4. Time (Islamic law)
QB23 .K53 2004
520/.917/67

2004043523
CIP

This publication has been typeset in the multilingual "Brill" typeface. With over 5,100 characters covering Latin, IPA, Greek, and Cyrillic, this typeface is especially suitable for use in the humanities.
For more information, please see www.brill.com/brill-typeface.

ISBN 978-90-04-26179-2 (paperback, vol. 1)
ISBN 978-90-04-26180-8 (paperback, set, vol. 2 parts 1 & 2)
ISBN 978-90-04-26183-9 (paperback, set, vols. 1 & 2)

This book is printed on acid-free paper.

To the memory of Ḥabash (Baghdad), Ibn Yūnus (Cairo),
Abu ʻl-ʻUqūl (Taiz), al-Khalīlī (Damascus), and all the others,
in gratitude for the privilege of an encounter with their minds.

Individual parts of this work are dedicated to the teachers and colleagues
who helped me appreciate the achievements of these scholars
in their cultural contexts.

TABLE OF CONTENTS

Note: Vol. 2, entitled *Instruments of Mass Calculation* and containing Studies X-XV, is to be published by Brill Academic Publishers in 2004. This contains:

PREFACE

The obligation of Muslims to pray at specific times in a specific direction gave rise to a substantial literature during the period between *ca.* 750 and 1900. The prescriptions were interpreted at two different levels, namely, by the astronomers, who proposed mathematical solutions, and by the scholars of the religious law, who proposed non-mathematical solutions. This dichotomy between mathematical science and folk science, now to some extent documented, is not known to have led to any strife.

This particular aspect of the activity of the Muslim astronomers belongs to what I have called "Astronomy in the Service of Islam".[1] But it might also be considered as Islam's service to the astronomers:[2] it brought them work and perhaps occasionally recognition. But today, even in the Islamic world, their activities and achievements have been completely forgotten. Muslim readers in particular should appreciate that virtually all of the materials presented in this book have never been researched before in modern times: therefore, they should not be surprised to find that some of them have called for a reinterpretation of hypotheses proposed by previous generations who did not have access to these materials. In fact, only those Muslim astronomers whose works were used in the medieval Europe have received much recognition in modern times. It is my firm opinion that the history of Islamic astronomy merits study for its own sake, as part of the history of Islamic civilisation generally.[3] Furthermore, that history cannot be written from texts alone. In this study I have recourse to not a few texts, but my main sources are astronomical tables, most without any accompanying text, and also astronomical instruments.

Virtually none of the materials I present in this book were known in medieval Europe. Nevertheless, as the reader will learn, European astronomers from the 16th to the 20th century produced tables of the same kind as the earlier Islamic ones, without any inkling that they were not the first to do so.

The literature of the Muslim astronomers relating to timekeeping in general and the regulation of the prayer-times in particular is treated here first (**Parts I-II**), because at the time I worked on it over 20 years ago I had no conception that the legal scholars addressed the same topics at a different level. However, I have since included at least the arithmetical schemes for timekeeping proposed by the legal scholars and some of their comments on the definitions of the times of prayer (**III-IV**). I also investigate the social background of the Muslim astronomers charged with the regulation of the times of prayer (**V**) and a significant technical aspect of astronomical timekeeping and instrumentation, namely, universalism (**VI**). Some of the material in the scientific sources – in this case, a table for orienting massive ventilators

[1] This is the title of one of the volumes of reprints of some of my early publications (*Studies*, C).

[2] This point is made in Ragep, "Freeing Astronomy from Philosophy", p. 51.

[3] See the surveys in King, "Islamic Astronomy", and Saliba, "Islamic Astronomy and Astrology", as well as *idem*, "The Astrologer in Islamic Society". None of our generation has produced anything to really replace Nallino, "Islamic Astronomy" (1921).

on the rooftops of Cairene architecture in an astronomically-significant direction – has proven to be of prime importance for our understanding of other aspects of Islamic civilization – in this case, the layout of cities and religious architecture (**VIIa-b**). New evidence for the *Islamic* origin of the highly-sophisticated grids on some recently-discovered Islamic world-maps for finding the direction and distance to Mecca is also presented (**VIIc**). Furthermore, it is not without interest to compare the activities of Muslims involved with timekeeping with those of Christian monks in the European Middle Ages, some of whom also cared about timekeeping (**VIII**). And, finally for this volume, I present some reflections occasioned by the horrendous events of Autumn, 2001, and their aftermath (**IX**). In the second volume, I present a survey of Islamic instrumentation (**X**), and detailed studies of two varieties of instruments of Islamic origin known only from European sources (**XIIa-b**). Certain Islamic and medieval European instruments are based on an approximate formula for timekeeping, and I have here documented the history of that formula during a millennium (**XI**). I have appended various studies on individual astrolabes of particular historical significance (**XIIIa-d**) and the explicit and implicit geographical data on a large group of instruments (**XIV**). The volume concludes with an ordered checklist of early medieval Islamic and European instruments (**XV**).

As the title of this book implies, it contains a series of 'Studies', and since these deal with related subjects, there is a substantial amount of repetition, which anyone reading the whole text will surely find irksome, but I have chosen to preserve the original integrity of the individual 'Studies' and have used cross-referencing in the notes to the place where a given topic is treated most fully.

For convenience I use the term "medieval" for the era under discussion, which extends from the 9th to the 19th century, since the astronomical timekeeping that was practiced during that time in the Islamic world was neither "classical" nor "modern". However, the term appears only in the title of this book and very occasionally thereafter. I trust it will be clear to readers that the term "medieval" has no negative connotations, not least because much of the highly sophisticated material that I describe had no counterpart in any other civilization, neither before nor after. In the same way, "Islamic" refers to a civilization rather than a religion. "Islamic civilization" was the milieu in which Muslims, Jews and Christians could live and work together, and in which there was discussion of differences of opinion; their *convivencia* was not always a delight for all concerned, but generally it functioned better centuries ago than it does now.

In a sense, this work is a supplement to a work that does not exist yet: an overview of Islamic mathematical astronomy in general. In 1956, E. S. Kennedy published "A Survey of Islamic Astronomical Tables", listing some 125 astronomical handbooks with tables of the genre known as "*zīj*".[4] Of these, at that time, only those of al-Khwārizmī, al-Battānī, and al-Bīrūnī had been published. The first, which survives only in a Latin translation of an Andalusī recension, has been published optimally given the circumstances: Latin text, English translation and commentary. The second survives more or less in tact: the Arabic text has been published with a translation and commentary unfortunately in Latin. The third has been published in Arabic, and there is a Russian translation. Now these are important works, but they are only three out

[4] Kennedy, "*Zīj* Survey".

of some 125 which Kennedy was able to identify. In the 1970s, mainly during our Cairo days, Kennedy and I upped the number of known *zīj*es to over to 200. Also I had started to unearth vast numbers of tables that were not at all of the kind standard in *zīj*es.[5] Yet today, half a century after Kennedy started work on *zīj*es, not one more *zīj* has been published in its entirety. Now each *zīj* contains a couple of hundred pages of tables and text, so there are good reasons why only the hardy would get involved with a *zīj*. We were all delighted when Benno van Dalen undertook to prepare a much more detailed survey of the some 225 Islamic *zīj*es now known, and his Frankfurt-based project has made very satisfactory progress, so that a publication is in sight. In the meantime, however, the *Encyclopaedia of Islam* was progressing to the letter "Z", and I undertook to write the article on *zīj*es, with Julio Samsó contributing the early Eastern Islamic and Andalusī sections, and Bernard Goldstein the section on Hebrew *zīj*es. Such works contain:

- Tables for calendar conversion;
- Tables of trigonometric functions;
- Tables of stellar coordinates;
- Tables of geographical longitudes and latitudes;
- Tables of functions relating to spherical astronomy, such as solar declination and right and oblique ascensions; tables for timekeeping (not common in *zīj*es);
- Tables of solar, lunar and planetary mean-motions and equations; lunar and planetary latitudes; planetary stations; lunar and planetary visibility; equation of time;
- Tables for computing conjuctions and eclipses; and, last but not least,
- Tables for mathematical astrology.

Someone could (and should) write a book on each of these categories. Inevitably, what we wrote for the *Encyclopaedia* was far longer than what the Editors wanted or could accept. A short version was prepared for the *Encyclopaedia*, and was published with several illustrations. The longer version was published in *Suhayl* (Barcelona) in 2001, without any illustrations.[6] It was subtitled "An Interim Report" because Benno van Dalen has a much more detailed survey in store for us, and for the time being I refer the interested reader to his website.[7] My point is that the subject of this book, Islamic astronomical timekeeping, is just one part of a much greater entity, astronomy in medieval Islamic civilization.[8]

In two parallel studies, I have investigated the ways in which Muslims handled the determination of the sacred direction towards Mecca. In the first, I show how the Muslim astronomers treated the problem as one of mathematical geography, using formulae, tables and

[5] King, "Islamic Astronomical Tables".

[6] King & Samsó (with Goldstein), "Islamic Astronomical Handbooks and Tables".

[7] www.rz.uni-frankfurt.de/~dalen/index.htm (2003).

[8] See the overviews mentioned in n. 3 above. The interested reader will find over a hundred extracts from Islamic scientific manuscripts in *Cairo ENL Survey*, pp. 195-331, but should bear in mind that these are mainly from manuscripts post-dating *ca.* 1300. More representative pictorial overviews could be prepared from the collections in Turkey, Iran or Europe. UNESCO is about to publish a volume on illustrated scientifio manuscripts in Iranian collections, edited by Ziva Vesel. This is a step in the right direction, but it should be remembered that some of the most important scientific manuscripts are not illustrated at all. Nobody has taken seriously my list of Islamic scientific works which should be published in facsimile editions using the best manuscripts (King, "Islamic Scientific Manuscripts", pp. 128-134).

maps.[9] In the second, not yet published, I describe the ways in which the Muslim legal scholars developed a sacred folk geography based on astronomical horizon phenomena for the same ultimate purpose.[10] (I have not included the materials presented in the valuable 2000 study by Mònica Rius dealing in detail with materials relating to the qibla in al-Andalus and the Maghrib: the author is currently engaged in preparing an English version of her original Spanish text.[11]) We have also gained a good measure of control over medieval Islamic instruments: my catalogue of them, still in preparation, provides a useful research tool but is a long way from publication.[12] A task for the future is to confront the vast corpus of Islamic manuscripts dealing with different aspects of instrumentation.[13]

Around 1990 my assistant Kurt Maier in Frankfurt valiantly typed the text of **I** and **II**, but for all the symbols and formulae, into a computer. The formulae I typed in myself, with imposing curly brackets around multi-tiered numerators and denominators. I did not look at the new text of **I** and **II** for close to 10 years. In 1990 I published **III** as a separate article, and in 1996 I published **V**. Taking advantage of a sabbatical in the autumn of 1999 I spent some weeks on the entire text, inserting all the symbols and formulae (for the second time, for the first disappeared when I upgraded the software),[14] and cross-references and updating the notes

---

[9] King, *Mecca-Centred World-Maps*, pp. 47-127, is devoted to a survey of solutions to the qibla problem (see also n. 17).

[10] The study listed as King, *The Sacred Geography of Islam*, is not yet published, but a summary has appeared in the article "Makka. iv. As centre of the world" in EI_2, repr. in King, *Studies*, C-X.

[11] Rius, *La Alquibla en al-Andalus y al-Magrib* (in Spanish, with original texts edited in Arabic). An English version is planned.

[12] See King, "Medieval Instrument Catalogue", A-C, and also **VIII**. A table of contents from 1991 is to be found at www.uni-frankfurt.de/fb13/ign/instrument-catalogue.html.

[13] See already Sédillot-*père* and *fils*, *al-Marrākushī*, for a much-neglected medieval encyclopaedic work on instrumentation, and now Charette, *Mamluk Instrumentation*, on a remarkable 14th-century treatise describing over 100 instrument-types, and *idem* & King, *The Universal Astrolabe of Ibn al-Sarrāj: Innovation in Medieval Islamic Science* (forthcoming), on the most spectacular Islamic astronomical instrument also from the 14th century and the associated texts.

[14] It was naïve of me to think that formulae of all things would survive successive versions of software over several years. But I could not predict in advance the other computer problems I would have. Since many readers with a sense of humour have enjoyed my previous horror stories about computers (*Mecca-Centred World-Maps*, p. xxi, and *Ciphers of the Monks*, p. 19), I am confident that the following tales will also amuse them.

Many of my early problems were caused mainly by the fact that MSWord cannot handle large texts. Ever cautious, I had the text of this book saved in three separate files. I wonder how many modern people know that with software available in the mid 1990s it was not possible to insert more than a few dozen automatic cross-references in a text of such size before the computer would freeze: the other few hundred cross-references had to be inserted manually and fixed up at the stage of printing. Or that the English and German versions of MSWord 6 are not compatible and switching computers with different versions during a project can lead to unexpected consequences. But my favorite story this time round is what happened to my iMac bought in Germany in early 2000 with an English system and German software: over the months, the thing invariably froze at some stage of my work. I learned to live with this, not least because the same thing happens with my iBook purchased in the US. But when the time before the freezing of the iMac became intolerably short, I sought assistance from Apple, only to learn to my horror that I had purchased a computer without a serial number, one of that kind which "fell off some truck". So it was not guaranteed and Apple was not interested to help me with an unregistered computer. But my "dealer" had forgotten to send me a bill for the computer anyway, and his bill arrived only in December, 2000, in the very same week that I found out that the computer was not for real. I ignored the bill, and he ignored my telephone call to tell him that there was a severe problem with the computer. If and when his lawyer contacts me I shall send him this footnote.

The Frankfurt-based company Mac Org, the new Apple dealers in Frankfurt and the best I have ever worked with, kindly reinserted a new system and a new English version of Word 98. But the computer crashed upon

and bibliography, trying by means of subtle cross-references to suppress duplicated references, as well as making an attempt to improve the deathless prose, as far as this is possible in such a text. The word "table" occurs some 4,000 times in this book, and many people will not find descriptions of one table after another very exciting.[15] Also, when I prepared this new version,

---

opening (fortunately before I took it home), and it was ascertained that this English version of Word was defective, so I had to be content with a German version. At least this did not cause the computer to crash upon opening, but it had the annoying habit of always beginning spelling checks whenever files were opened, and then freezing. When one is dealing with hundreds of pages of coded text (such as a$f@$@ to produce oblique ascensions as α_ϕ) one does not need to be told that there are too many misspelled words for Word to control. It took me two days to turn off the spelling check (there are two places where this can be done, and the more obvious of the two was jammed because defective). Also the version I acquired had no subprogram for inserting formulae at all. Since my formulae were by now all one-liners (see the text to n. I-1:38) I conceded myself a victory against the system. It was about the same time that my Epson 760 Color Printer developed the singularly annoying habit of printing only the top half of the page for about one-third of every assignment.

In January 2001, some two weeks after converting the text to Word 95, I discovered that the three files on which the book was preserved had self-destructed: the texts of all three files were punctuated by large chunks of gibberish. I had to go back to the last backups on MSWord 6 and try to remember what changes I had made ... .

In August 2001, when both my iBook and iMac were defunct (constantly crashing) and I was spending some weeks in France, I ordered an English iMac with English keyboard from Apple in Ireland to be delivered in France ("delivery within three days"). Apple kindly sent me a French iMac with French keyboard, which I could have purchased locally had I wanted. It took a total of *four weeks* and several phone-calls a day, each time to different people of divers nationalities, each talking a multiplicity of languages, and most of them in Cork, Ireland, to get rid of the French computer and, only thereafter, to receive what I had ordered: the real English Apple arrived (from the Czech Republic) a week before I left France to return to Frankfurt. But just before I left, a French friend who is an Apple wizard came over to organize for me all new installations: he promptly changed the system to a more recent French one, also installing a French version of MSWord 2000. This combination actually seems to work, although frankly I prefer MSWord 5 because it was adequate for all sensible purposes and it did what I wanted, not *vice versa*. My Epson printer by this time had developed the habit of registering almost-new ink-cartridges as empty, and thirsting for new cartridges before it would print anything. Nevertheless I was able to squeeze the entire text out of it before returning to Frankfurt for the Winter Semester.

The final version of the book submitted to E. J. Brill in the Summer of 2002 was somehow achieved under the above-mentioned circumstances. But their printers could not receive the text by email, nor could they open the "superdisk" I sent them containing the text in four chunks. Then we found to our joint surprise that the powers that be do not make "superdisk" readers anymore, no doubt because like Eurochecks, they were so useful. Only in April, 2003, could the printers access the text.

My Epson Stylus Color 760 printer died of exhaustion in October, 2003, just when I had finished proof-reading the text and when I wanted to print a collection of chunks of corrected text and the indexes for Brill. I spent half a day travelling from my home in rural France to the city of Aubenas, where I purchased a C42 Plus. This turned out to be missing the CD to install it on my computer. The installation package from the Epson website is unusable. I spent another half day going to Nîmes to buy another printer.

A few days later, my iBook expired, taking with it half of the files saved on it, for which the back-ups were in Frankfurt. The materials for Volume 2 are all somehow retrievable somewhere, and interested readers will find new stories about this in my next publication.

The various chunks of Arabic text were all prepared in Cairo on a typewriter in the 1970s, then retyped on a computer in New York in the 1980s using al-Kaatib software. They all had to be retyped in the 1990s and early 2000s in various parts of the world using hardware unknown to me and software whose use is beyond me. This was achieved by Michael Kreutz in Bochum (**IV**), Sally Ragep in Norman, Oklahoma (**V**), Mónica Herrera in Frankfurt (**VIIb**), and François Charette in Frankfurt (**XIIa**). To dwell on the problems we encountered formatting these texts and making final changes would fill another page: my favourite was when the Arabic in the footnotes came out backwards. The formidable task of preparing the final printouts from the extremely volatile computer files was achieved by Mohammed Abu Zayed and Mónica Herrera in Frankfurt.

There were days at the turn of the Millennium when I yearned for the 1970s: at that time I was privileged to have the impeccable combination of a highly competent Egyptian secretary with two splendid Italian typewriters for English and Arabic.

[15] In 1973 I gave a lecture entitled "Medieval Islamic Prayer-Tables" at the American University of Beirut.

I had a much more broader respect for manuscripts than I had in my youth, so I have tried to add as much as is feasible about the provenance and date of the manuscripts. If such information is not given, the reader may assume that I am no longer prepared to hazard a guess from memory. I note that most colleagues do not care much about the origin of the manuscripts (or the provenance of the instruments) they are studying, being happy enough to have a copy or two from which to work. Nevertheless, the original environments of the manuscripts (and instruments) are extremely important. This study was much facilitated by the availability of some first-rate catalogues of Arabic scientific manuscripts, of which those by Wilhelm Ahlwardt (1893) for Berlin, McGulkin de Slane (1883-95) for Paris, and Henri-Paul-Joseph Renaud (1941) for El Escorial, are exemplary.

Apart from various studies by Glen Van Brummelen and François Charette,[16] the only other contributions to the subject of astronomical timekeeping over the 20-or-so years during which most of the text of this book (I mean, **I** and **II**) lay dormant are a series of derivative publications of my own – mostly surveys and encyclopaedia articles – which all contain references to, and promises concerning a mysterious "**SATMI**". The fact that **I** and **II** were not published years ago, as they should have been, and that **III**, **V**, **VIa**, and **VIb**, were published without **I** and **II** being accessible, has resulted in the rather cumbersome form in which the present work is presented. And the reader should also be aware that François Charette's *Mathematical Instrumentation in Fourteenth-Century Egypt and Syria*, published by E.J. Brill in 2003, could be considered as an *instant supplement* to the present work, since it deals with tables as well as instruments. Indeed, it features numerous tables and instruments that are not represented here at all. In the last few years, numerous Internet-sites have appeared dealing with Islamic prayer-times on the one hand, and historical instruments on the other: as far as I am aware, none add anything to the historical materials contained in this book: they do, however, offer the means to generate prayer-times for any date and any locality, and to purchase copies of medieval instruments. To François Charette too go my heart-felt thanks for numerous corrections to the penultimate version of this text, and no less for preparing all of the computer graphics.

None of my research over the last 20 years would have been possible without the infrastructure of the Institute for the History of Science in Frankfurt. I have to thank our Institute's librarian, Ryszard Dyga, and my CEO, Wolf-Dieter Wagner, for countless favours over many years, not least with regard to the acquisition of research materials – letters requesting this or that, library searches for this or that, ordering photos of this, preparing photocopies of that, ... Inevitably, this often had to be done in a hurry so that I could disappear for weeks on end and work in isolation, ...

Most of the photographs used to illustrate the two volumes of this book were kindly provided by the Deutsche Staatsbibliothek in Berlin; the Egyptian National Library in Cairo; the Chester Beatty Library in Dublin; the Museo di Storia della Scienza in Florence; the National Maritime

---

A young lady walked out a couple of minutes after I started: I later found out that she was interested in medieval Islamic furniture.

[16] See Van Brummelen, "al-Khalīlī's Auxiliary Tables"; and Charette, "Najm al-Dīn's Monumental Table", and *idem*, *Mamluk Instrumentation*.

Museum in Greenwich; the Süleymaniye and Topkapı Libraries in Istanbul; the British Museum and the British Library in London; the Bodleian Library and the Museum of History of Science in Oxford; the Bibliothèque Nationale de France in Paris; and Princeton University Library. It is a pleasure to record my gratitude to these splendid institutions. I also wish to thank Trudy Kamperveen and Tanja Cowall of Brill Academic Publications for their patience and understanding, and especially Ciska Palm of Palm Produkties, who somehow produced this book from a few computer files and a pile of photographs.

Whilst the materials presented here will be new to most Western readers, they will be completely new to Muslim readers unfamiliar with Western writings on the history of Islamic science.[17] (Alas virtually nothing of our research over the past 50 years is known in the Islamic world outside Turkey, Iran and India; neither is it is ever mentioned by those Muslim savants who write in the West on the "new" "Islamic science".) There is nothing on the subject of medieval Islamic astronomical timekeeping available for the modern Arabic reader other than a short article of mine in a general work on Islamic science (*Mawsūʿat taʾrīkh al-ʿulūm al-ʿarabiyya*) published in Beirut in 1997 and not widely available in the Muslim world anyway. In any case, no Muslim readers should be surprised that a study based on over 500 medieval manuscripts that no-one in modern times had ever looked at previously, should contain materials that are new and exciting, and which radically alter – I venture to say, improve – our picture of science in the medieval Islamic world.

[17] My book on the qibla (n. 9) was generously sponsored by the al-Furqan Islamic Heritage Foundation in London. The present work could not be sponsored by the Foundation because a colleague from C.N.R.S., Paris, reported to them that it contained not enough new material to merit their support. No other foundation or interested individual could be found to sponsor the publication, hence this book is somewhat more expensive than my book on the qibla, which was likewise mainly based on materials that had never been identified or studied previously.

PREFACE 2: SOME BRIEF REMARKS ON THE ISLAMIC SCIENTIFIC HERITAGE IN THE CONTEXT OF THE TRANSMISSION OF SCIENTIFIC IDEAS

Some out-dated notions wide-spread amongst the "informed public" and even amongst historians of science are that:

(1) The Muslims were fortunate enough to be the heirs to the sciences of Antiquity.

(2) They cultivated these sciences for a few centuries but never really achieved much that was original.

(3) They provided, mainly in Islamic Spain, a milieu in which eager Europeans emerging out of the Dark Ages could benefit from these Ancient Greek sciences once they had learned how to translate them from Arabic into Latin.

Islamic science, therefore, one might argue, is of no consequence *per se* for the development of global science and is important only insofar as it marks a rather obscure interlude between a more sophisticated Antiquity and a Europe that later became more civilized.

What happened in fact was something rather different. The Muslims did indeed inherit the sciences of Greek, Indian and Persian Antiquity. But within a few decades they had created out of this potpourri a new science, now written in Arabic and replete with new Muslim contributions, which flourished with innovations until the 15th century and continued thereafter without any further innovations of consequence until the 19th. Most of this activity took place in the Islamic East. Not much of this Islamic science was known in al-Andalus, and even less of it was available for any Europeans, however eager. In astronomy, the Europeans rescued from al-Andalus outdated and mutually inconsistent works such as the astronomical handbooks of al-Khwārīzm and al-Battānī (respectively representing the first Arabic redactions of the Indo-Persian and Greek traditions) and produced therefrom the unhappy mixture known as the *Toledan Tables*, influential all over Europe for a few centuries, including England. Indeed when I was a graduate student some 30 years ago I read a 1952 paper by Otto Neugebauer and Olaf Schmidt entitled "Indian Astronomy at Newminster in 1428". The main tables discussed were some tables of the tangent of the solar declination multiplied by a very curious coefficient. They were very much out of place in medieval England (see now **I-7.1.1** and **10.1**).

In 1997 my colleague, the eminent British historian of medieval science, Charles Burnett, published a book entitled *The Introduction of Arabic Learning into England*. It is an excellent book, not least because it deals with a subject on which Burnett is more qualified to write than anyone else. I take issue only with the title, my problem being the expression "Arabic learning" in the context of the ultimate destination, "medieval England". Since, as it happens, only a modest fraction of the grand total of Muslim scholarship was available in al-Andalus, and only a very small fraction of that ever made it to England in the Middle Ages, Burnett's title might be construed as a little misleading: in fact, apparently no Englishman in the Middle Ages ever came into contact with the full force of "Muslim scholarship" or "Arabic learning", for which he would have had to go to Baghdad or Cairo or Rayy, rather than, say, Toledo. When English scholars looked at Arabic scientific manuscripts in Oxford in the 17th century they were in for a few surprises: see Gerald Toomer's *Eastern Wisedome and Learning: The Study of Arabic in Seventeenth-Century England* (1996) and Gül Russell's *'Arabick' Interest in 17th-Century England* (1994). But by that time, it was already a bit late for actually exploiting the scientific bounty; the materials discovered tended to be more of historical interest. The real surprises came during the 19th and 20th centuries when historians from a multiplicity of national backgrounds investigated Islamic scientific manuscripts in libraries all over Europe and then in the Near East. Their investigations revealed a intellectual tradition of proportions that no medieval or Renaissance European could ever have imagined: anyone who might doubt this should look at the monumental bio-bibliographical writings of Heinrich Suter, Carl Brockelmann and Fuat Sezgin,[1] and, in the case of astronomy, also the early studies of J.-J. and L.-A. Sédillot, C. A. Nallino, H. Suter or E. S. Kennedy. These investigations are by no means over, and there is much more material to be unearthed in manuscript libraries and museums around the world. When investigating manuscripts that nobody has looked at for centuries it highly likely that one will unearth materials that call for a complete re-evaluation of some aspects of Muslim activity in a particular branch of knowledge. The task of the historian is to organize and present these materials. The task of the reader is to be prepared for a few surprises.

The present study introduces some of the materials from a very special trend in the Muslim scientific heritage. These materials have become known only in the past 30 years. Indeed, it has been my priviledge to look at most of them for the very first time. Inevitably, they modify the overall picture we have of Islamic science. And it so happens that the particular intellectual activity that inspired these materials is related to the religious obligation to pray at specific times. The material presented here makes nonsense of the popular modern notion that religion inevitably impedes scientific progress, for in this case, the requirements of the former actually inspired the progress of the latter for centuries. But the same material also enables us to witness the eventual drying up of initiative in Islamic science.[2] The reason for this phenomenon is seldom ever mentioned by historians of science who write on the decline of the scientific tradition in the Islamic world: within the medieval context, all of the problems had been solved, some many times over. Nowhere is this clearer than in the kind of sources that I describe in this book.

---

[1] As I was reading the proofs of this book, a monumental new publication – Boris A. Rosenfeld and Ekmeleddin İhsanoğlu, *Mathematicians, Astronomers and Other Scholars of Islamic Civilisation and their Works (7th-19th C.)*, Istanbul, 2003 – appeared. This is an English version of Galina P. Matvievskaya and Boris A. Rosenfeld, *Mathematicians and Astronomers of the Islamic Middle Ages and Their Works*, Moscow, 1983 (three volumes in Russian), supplemented with information from Ekmeleddin İhsanoğlu *et al.*, *Ottoman Astronomical / Mathematical / Geographical Literature*, Istanbul, 1997-2000 (altogether six volumes in Turkish). This new work is a mine of bio-bibliographical information on over 1,700 Muslim scientists and is of course an important contribution to our field. It is a useful addition to the standard bio-bibliographical works of Suter, Brockelmann and Sezgin, which should still be used alongside it. In the present volumes I have not included references to this new work.

[2] A useful essay by a scholar who is familiar with the scope and depth of Islamic science is Al-Hassan, "Decline of Islamic Science" (1994). It is significant that this essay deals with the decline that took place in and after the 16th century, not least because the available sources – available now to us, but not known to our predecessors or not previously studied – show that scientific initiative had not dried up before that time.

STATEMENT ON PREVIOUS PUBLICATION OF PARTS OF THIS WORK

Parts of this work have been published previously. They are included here because they are all essential to the overall picture provided by the ensemble of studies. They have been revised and brought into line with the other studies; in particular the references and bibliography have been updated.

Volume 1:
General remarks:

- **Parts I, II, IV, VIIc** and **IX** are published here for the first time.
- **Part VIII** is here published in English for the first time.
- The reader should please keep in mind that **Parts I-II** were written almost 30 years ago, **Parts IV** and **VIIb** some 20 years ago, **Parts III** and **VI** some 10 years ago. **Parts VIIc** and **IX** were written just a few months before this volume went to press.

Remarks to specific sections of this volume:

- **Parts I-II** have not been published previously.
- An earlier version of **Part III** was published in *Oriens* 32 (1990), pp. 191-249.
- **Part IV** has not been published previously.
- An earlier version of **Part V** was published in *Tradition, Transmission, Transformation: Proceedings of Two Conferences on Premodern Science Held at the University of Oklahoma,* F. Jamil Ragep & Sally P. Ragep, with Steven J. Livesey, eds., Leiden, New York & Cologne: E. J. Brill, 1996, pp. 285-346.
- Earlier versions of **Parts VIa** and **VIb** were published in *From Ancient Omens to Statistical Mechanics*: *Essays on the Exact Sciences Presented to Asger Aaboe*, J. Lennart Berggren & Bernard R. Goldstein, eds., *Acta Historica Scientiarum Naturalium et Medicinalium* (Copenhagen) 39 (1987), pp. 121-132; and *A Way Prepared*: *Essays on Islamic Culture in Honor of Richard Bayly Winder*, Farhad Kazemi & Robert B. McChesney, eds., New York: New York University Press, 1988, pp. 153-184, respectively.
- Earlier versions of **Parts VIIa** and **VIIb** were published in *Journal for the History of Astronomy* 26 (1995), pp. 253-274, and *Journal of the American Oriental Society* 104 (1984), pp. 97-133, respectively.
- **Part VIIc**, previously unpublished, supplements my recent book *World-Maps for Finding the Direction and Distance to Mecca: Innovation and Tradition in Islamic Science*, Leiden: E. J. Brill, and London: Al-Furqan Islamic Heritage Foundation, 1999.
- A German translation of an earlier version of **Part VIII** was published as "Aspekte angewandter Wissenschaften in Moscheen und Klöstern", *Berichte zur Wissenschafts-geschichte (Organ der Gesellschaft für Wissenschaftsgeschichte)* 18 (1995), pp. 85-95 and 137-149.
- **Part IX** has not been published previously.

Volume 2:
General remarks:
Parts XI, **XIIa-b**, **XIIIa** and **XV** have not been published previously.
Parts X and **XIIIc** are here published in English for the first time.

Remarks to specific sections of the second volume:

- An Italian translation of an earlier version of **Part X** was published as "Strumentazione astronomica nel mondo medievale islamico", in *Gli strumenti*, Gerard L'E. Turner, ed., Turin: Giulio Einaudi, 1991, pp. 154-189 and 581-585. The text has been considerably expanded.
- **Part XI** has not been published previously.
- The section of **Part XIIa** dealing with the Baghdad treatise has appeared separately as "A *Vetustissimus* Arabic Treatise on the *Quadrans Vetus*", in *Journal for the History of Astronomy* 33 (2002), pp. 237-255.
- A summary of **Part XIIb** entitled "14th-Century England or 9th-Century Baghdad? New Insights on the Origins of the Elusive Astronomical Instrument Called *Navicula de Venetiis*" was published in *Astronomy and Astrology from the Babylonians to Kepler – Essays Presented to Bernard R. Goldstein on the Occasion of his 65th Birthday*, Peter Barker, Alan C. Bowen, José Chabás, Gad Freudenthal and Tzvi Langermann, eds., a special issue of *Centaurus* 44 (2003).
- **Part XIIIa** has not been published before.
- **Part XIIIb** is modified from my article "Early Islamic Astronomical Instruments in Kuwaiti Collections", in Arlene Fullerton & Géza Fehérvári, eds., *Kuwait: Art and Architecture – Collection of Essays*, Kuwait (no publisher stated), 1995, pp. 76-96.
- A French version of **Part XIIIc** was part of my contribution "L'astronomie en Syrie à l'époque islamique", to the exhibition catalogue *Syrie – mémoire et civilisation*, Sophie Cluzan, Éric Delpont and Jeanne Mouliérac, eds., Paris: Flammarion (Institut du Monde Arabe), 1993, pp. 386-395 and 432-443.
- A modified version of **Part XIIId** was published as "A Remarkable Italian Astrolabe from *ca.* 1300 – Witness to an Ingenious Tradition of Non-Standard Astrolabes", in *MUSA MUSAEI: Studies on Scientific Instruments and Collections in Honour of Mara Miniati*, Marco Beretta, Paolo Galluzzi and Carlo Triarico, eds., Florence: Leo S. Olschki, 2003, pp. 29-52.
- **Part XIIIe** was first published as in *Suhayl – Journal for the History of the Exact and Natural Sciences in Islamic Civilisation* (Barcelona) 3 (2002/03), pp. 9-156.
- **Part XIV** was first published as "Bringing Astronomical Instruments back to Earth – The Geographical Data on Medieval Astrolabes to *ca.* 1100", in *Between Demonstration and Imagination: Essays in the History of Science and Philosophy Presented to John D. North*, Leiden: E. J. Brill, 1999, pp. 3-53. [This new version omits a significant error in the original, identified in the preamble.]
- **Part XV** has not been published previously.

MAIN BIBLIOGRAPHY AND BIBLIOGRAPHICAL ABBREVIATIONS

Notes: The reader will find here the main modern sources for the history of Islamic astronomy and astronomical instrumentation, including conference proceedings and exhibition catalogues. Many studies have been reprinted in various collected volumes, notably those published by Variorum; those individual studies related to astronomical timekeeping and related aspects of mathematical astronomy which are particularly relevant to this study are also listed here. The abbreviations introduced below are used throughout this book. Further bibliographical lists are given in **VIII** (comparison of aspects of Islamic and medieval European science) and **X** (instrumentation), and the works cited there are listed below without abbreviated titles. Earlier publications reprinted in this volume are indicated by diamonds. Detailed references to other works cited which are not directly related to our main subject are presented *ad loc.*

Aaboe Festschrift: *From Ancient Omens to Statistical Mechanics*: *Essays on the Exact Sciences Presented to Asger Aaboe*, J. Lennart Berggren and Bernard R. Goldstein, eds., *Acta Historica Scientiarum Naturalium et Medicinalium* (Copenhagen), 39 (1987).

Ackermann, "Instruments of Humphrey Cole": Silke Ackermann, "Scientific Instruments", in *eadem*, ed., *Humphrey Cole*, pp. 29-95.

— , "London BM *Quadrans vetus*": see EPACT.

— , "Sloane Astrolabe": see EPACT.

— , "Umarbeitung eines Astrolabs": *eadem*, "Mutabor: Die Umarbeitung eines mittelalterlichen Astrolabs im 17. Jahrhundert", in *Frankfurt IGN Festband*, pp. 193-209.

— , "Zinner Archive": *eadem*, "Dormant Treasures – The Zinner-Archive at Frankfurt University", *Nuncius – Annali di Storia della Scienza* (Florence) 16:2 (2001), pp. 711-722.

— , ed., *Humphrey Cole*: *eadem*, ed., *Humphrey Cole: Mint, Measurement and Maps in Elizabethan England*, (British Museum, Occasional Paper, no. 126), London: The British Museum, 1998.

– & Cherry, "Three Medieval English Quadrants": *eadem* and John Cherry, "Richard II, John Holland and Three Medieval Quadrants", *AS* 56 (1999), pp. 3-23.

Adıvar, Adnan, *La science chez les Turcs Ottomans*, Paris: G. P. Maisonneuve, 1939.

AHES: *Archive for History of Exact Sciences.*

Āghā Buzurg, *al-Dharīʿa*: Āghā Buzurg Muḥammad Muḥsin al-Tihrānī, *al-Dharīʿa ilā taṣānīf al-Shīʿa*, 25 vols., Najaf, later Tehran, 1965-1978.

Ahlwardt: see *Berlin Catalogue.*

AIHS: *Archives Internationales d'Histoire des Sciences.*

AIOS: *Arabische Instrumente in orientalistischen Studien*, Fuat Sezgin *et al*, eds., 6 vols., Frankfurt am Main: Institut für Geschichte der Arabisch-Islamischen Wissenschaften, 1990-91, repr. as *Islamic Astronomy and Mathematics*, vols. 85-90 (1998), with 6 further volumes *ibid.*, vols. 91-96 (1998). [Reprints of studies from the 19th and early 20th century: a monumental total of over 5,000 pages.]

Aleppo 1976 Symposium Proceedings: *Proceedings of the First International Symposium for the History of Arabic Science*, Ahmad Y. al-Hassan *et al.*, eds., 2 vols., Aleppo: Institute for the History of Arabic Science, 1978. Contains:

- D. A. King "Astronomical Timekeeping in Fourteenth-Century Syria", in II, pp. 75-84, repr. in *idem*, *Studies*, A-X.
- Fritz W. Zimmermann, "The Dustūr al-Munajjimīn of MS Paris, BN ar. no. 5968", in II, pp. 184-192.

Algiers BN Catalogue: E. Fagnan, *Catalogue général des manuscrits de la Bibliothèque Nationale d'Algérie* ... , first published Paris, 1893, 2nd edn., Algiers: Bibliothèque Nationale d'Algérie, 1995.

Ali: see al-Bīrūnī, *Geography.*

al-ʿĀmilī, *Aʿyān al-Shīʿa*: Muḥsin al-Amīn al-Ḥusaynī al-ʿĀmilī, *Aʿyān al-Shīʿa*, 56 vols., Damascus: Mabaʿat Ibn Zaydūn, 1935-1962.

Amsterdam NK 1990 Exhibition Catalogue: Anthony J. Turner, ed., *time*, Amsterdam: Tijd voor Tijd Foundation, 1990. [Catalogue of an exhibition at the Nieuwe Kirk, Amsterdam, in 1990.]

Anagnostakis, "Astrolabe Ecliptic Divisions": Christopher Anagnostakis, "How to Divide the Ecliptic on an Astrolabe", in *Aaboe Festschrift*, pp. 133-144.

Analema – Boletín de la Asociación de Amigos de los Relojes de Sol (Madrid).

Ankara (AÜDTCF) Dergisi and *Yayınlari*: *Ankara Üniversitesi Dil ve Tarih-Coğrafya Fakültesi Dergisi* and *Yayınları*.

Ansari & Ghori, "Two Treatises on Astronomical Instruments": S. M. Raza Ansari and S. A. Khan Ghori, "Two Treatises on Astronomical Instruments by ʿAbd al-Munʿim al-ʿĀmilī and Qāsim ʿAlī al-Qāyinī", *Delhi 1985 Colloquium Proceedings*, pp. 215-225.

Anthiaume & Sottas, "*Quadrans novus*": A. Anthiaume and J. Sottas, *L'astrolabe-quadrant du Musée des antiquités de Rouen – recherches sur les connaissances mathématiques, astronomiques et nautiques au moyen âge*, Paris: Librairie G. Thomas, 1910, repr. Paris: Librairie Alain Brieux, 1988.

Apian: see Röttel.

Araştırma: *Araştırma*, published by Ankara Üniversitesi Dil ve Tarih-Coğrafya Fakültesi, Felsefe Araştırmaları Enstitüsü.

Arcari & Costamagna, "Acquafredda Sundial", A-B: Barbara Arcari and Andrea Costamagna, "L'orologio a ore canoniche dell'Acquafredda", *L'Astronomia* (Rome) 211 (iuglio 2000), pp. 44-50 (A); and *eidem*, "Il quadrante solare dell'abbazia dell'Acquafredda", in *eidem*, *L'abbazia dell'Acquafredda* , Lenno (Como), (Comune di Lenno, Biblioteca Comunale Vittorio Antonini, Quaderno n. 3), 2002, pp. 115-121, and fig. 18 (B).

Archinard, "Cadrans solaires rectilignes": Margarida Archinard, "Les cadrans solaires rectilignes", *Nuncius* (Florence) 3:2 (1988), pp. 149-181.

— , "Geneva *Navicula*": *eadem*, "*Navicula de Venetiis* : une acquisition prestigieuse du Musée d'histoire des sciences [Genève]", *Genava*, N.S., 43 (1995), pp. 87-94, published separately with the same pagination, Geneva: Musée d'histoire des sciences, n.d. [1995].

— , "Unequal Hour Diagram": *eadem*, "The Diagram of Unequal Hours", *AS* 47 (1990), pp. 173-190.

Arnaldi & Schaldach, "Roman Cylinder Dial": Mario Arnaldi and Karl-Heinz Schaldach, "A Roman Cylinder Dial: Witness to a Forgotten Tradition", *JHA* 28 (1997), pp. 107-117.

Arvanitaki, "L'heure arabe": G. Arvanitaki, "Théorie de l'heure arabe", *Memoires de l'Institut Égyptien* 8 (1915), pp. 21-55.

AS: *Annals of Science*.

al-Aṣbaḥī, *al-Yawāqīt*: References in **IV** are to MS Cairo Dār al-Kutub *mīqāt* 948,1, fols. 1v-53r, of al-Aṣbaḥī's *Kitāb al-Yawāqīt fī ʿilm al-mawāqīt*.

ASP: *Arabic Science and Philosophy* (Cambridge University Press).

Assemani, *Globus arabicus*: Simon Assemani, *Globus caelestis cufico-arabicus Veliterni Musei Borgiani*, Padua, 1790, repr. in *IMA*, vol. 91 (1998), pp. 63-305, plus fold-outs.

Astrolabica: *Astrolabica* (Paris: Société Internationale de l'Astrolabe). [See especially no. 5: *Astrolabica N° 5: Etudes 1987-1989*, Anthony J. Turner, ed., Paris: Institut du Monde Arabe & Société Internationale de l'Astrolabe, 1989.]

Atkinson, "Points of the Compass": T. D. Atkinson, "Points of the Compass", in *ERE*, X, pp. 73-88.

Aʿyān al-Shīʿa: see al-ʿĀmilī, *Aʿyān al-Shīʿa*.

Azzawi, *History of Astronomy in Iraq*: ʿAbbās al-ʿAzzāwī, *Taʾrīkh ʿilm al-falak fī ʾl-ʿIrāq wa-ʿalāqātihi bi-ʾl-aqṭār al-islāmiyya wa-ʾl-ʿarabiyya fī ʾl-ʿuhūd al-tāliya li-ayyām al-ʿAbbāsiyyīn min sanat 656 H = 1258 M ilā sanat 1335 H = 1917 M* [= *A History of Astronomy in Iraq and its relations with Islamic and Arab regions in the period following the Abbasids, from 1258 A.D. to 1917 A.D.*], Baghdad: Iraqi Scientific Academy (al-Majmaʿ al-ʿilmī al-ʿIrāqī, 1958. [This Arabic work is independent of any earlier European bio-bibliographical works on the subject.]

Bagheri, "al-Kāshī on the Visible Horizon": Mohammad Bagheri, "A New Treatise by al-Kāshī on the Depression of the Visible Horizon", in *Bellagio 2000 Conference Proceedings*, pp. 357-368.

Bagvi, M. B. A. *Determination of the Direction of Qibla and the Islamic Timings*, Lahore: Sh. Muhammed Ashraf, 1972.

Baloch, "Measurement of Space and Time in Lower Sind": N. A. Baloch, "Measurement of Space and Time in the Lower Indus Valley of Sind", in *Islamabad 1979 Conference Proceedings*, pp. 168-196.

Barceló: see Labarta & Barceló.

Barmore, "Turkish Mosque Orientation": Frank E. Barmore, "Turkish Mosque Orientation and the Secular Variation of the Magnetic Declination", *Journal of Near Eastern Studies* 44 (1985), pp. 81-98.

Barthold, "Orientierung der ersten Moscheen": Wilhelm Barthold, "Die Orientierung der ersten muhammedanischen Moscheen", *Der Islam* 18 (1929), pp. 245-250.

Bashear, "Qibla Musharriqa": Suliman Bashear, "Qibla Musharriqa and Early Muslim Prayer in Churches", *The Muslim World* 81 (1991), pp. 267-282.

BEA: *Biographical Encyclopedia of Astronomers*, Thomas Hockey, ed., to be published by Kluwer Academic Publishers, Dordrecht.

Bennett, "Oxford *Navicula*": see EPACT.

Beer, "Astronomical Dating of Works of Art": Arthur Beer, "Astronomical Dating of Works of Art", *Vistas in Astronomy* 9 (1967), pp. 177-223.

Bellagio 2000 Conference Proceedings: *From China to Paris: 2000 Years Transmission of Mathematical Ideas*, (Boethius – Texte und Abhandlungen zur Geschichte der Mathematik und der Naturwissenschaften, Band 46), Yvonne Dold-Samplonius, Joseph W. Dauben, Menso Folkerts and Benno van Dalen, eds., Stuttgart: Franz Steiner, 2002. [Proceedings of a conference held at the Rockefeller Foundations's Research and Conference Center, Bellagio, during May, 2000.]

Benoit, "Cadran solaire monastique": J. L. Benoit, "Un cadran solaire monastique du XIIe siècle", *Le Cosmos* 59/1238 (1908), pp. 435-437.

Berggren, "Archimedes among the Ottomans": J. Lennart Berggren, "Archimedes among the Ottomans", in *Aaboe Festschrift*, pp. 101-109.

—, "Al-Bīrūnī on Plane Maps of the Sphere": *idem*, "Al-Bīrūnī on Plane Maps of the Sphere", *JHAS* 6 (1982), pp. 47-96.

—, "Four Analemmas for the Qibla": *idem*, "A Comparison of Four Analemmas for Determining the Azimuth of the Qibla", *JHAS* 4 (1980), pp. 69-80.

—, "Ḥabash's Analemma for Azimuth Curves": *idem*, "Ḥabash's Analemma for Representing Azimuth Circles on the Astrolabe", *ZGAIW* 7 (1991/92), pp. 23-30.

—, *Islamic Mathematics*: *idem*, *Episodes in the Mathematics of Medieval Islam*, New York, *etc.*: Springer, 1986.

—, "al-Kūhī on Astrolabe Construction": *idem*, "Abū Sahl al-Kūhī's Treatise on the Construction of the Astrolabe with Proof: Text, Translation and Commentary", *Physis* 31 (1994), pp. 141-252.

—, "Origins of the 'Method of the *Zīj*es'": *idem*, "The Origins of al-Bīrūnī's "Method of the *Zīj*es" in the Theory of Sundials", *Centaurus* 28 (1985), pp. 1-16.

—, "Spherical Trigonometry in the *Jāmiʿ Zīj*": *idem*, "Spherical Trigonometry in Kūshyār ibn Labbān's *Jāmiʿ Zīj*", in *Kennedy Festschrift*, pp. 15-33.

Berggren: see also *Aaboe Festschrift*.

Bergmann, *Innovationen im Quadrivium*: Werner Bergmann, *Innovationen im Quadrivium des 10. und 11. Jahrhunderts: Studien zur Einführung von Astrolab und Abakus im lateinischen Mittelalter*, Stuttgart: Franz Steiner, 1985.

Berlin Catalogue: Wilhelm Ahlwardt, *Die Handschriften-Verzeichnisse der Königlichen Bibliothek zu Berlin*, 5. Band: *Verzeichnis der arabischen Handschriften*, Berlin: A. Asher, 1893.

Biarne, "Le temps du moine": J. Biarne, "Le temps du moine d'après les premières règles monastiques d'Occident (IVe-VIe siècles)", in *Le temps chrétien de la fin de l'Antiquité au Moyen-Age, IIIe-XIIIe siècles*, (Colloques Internationaux du Centre National de la Recherche Scientifique, no. 604), Paris, 1984, pp. 99-128. [Not consulted.]

Bielefeld Catalogue: *Wissenschaftliche Instrumente und Sonnenuhren – Kunstgewerbesammlung der Stadt Bielefeld / Stiftung Huelsmann*, Dirk Syndram, ed., Munich: Callwey, 1989.

Bilfinger, *Die mittelalterlichen Horen*: Gustav Bilfinger, *Die mittelalterlichen Horen und die modernen Stunden – Ein Beitrag zur Kulturgeschichte*, Stuttgart: W. Kohlhammer, 1892.

al-Bīrūnī, *Astrolabe Construction*: al-Bīrūnī, *Istīʿāb al-wujūh al-mumkina fī ṣanʿat al-asṭurlāb*, al-Sayyid Muḥammad Akbar Jawādī al-Ḥusaynī, ed., with introductory notes by Ayatallāh Ḥasan Ḥasan Zāde al-Āmulī, Mashhad: Islamic Research Foundation, Astan Quds Razavi, 1380 HS [= 2001/02].

—, *Astrology*: R. Ramsay Wright, transl., *The Book of Instruction in the Elements of the Art of Astrology by ... al-Bīrūnī*, London: Luzac & Co., 1934, repr. Baghdad: al-Muthannā, n.d.

—, *Chronology*: C. Edward Sachau, *Chronologie orientalischer Völker von Alberuni*, Leipzig: F. A. Brockhaus & Otto Harrassowitz, 1923, repr. Baghdad: Maktabat al-Muthannā, n.d., translated as C. E. Sachau, *The Chronology of Ancient Nations – An English Version of the Arabic Text of the Athár-ul-bákiya of Albírúní, or 'Vestiges of the Past,' collected and reduced to writing by the author in A.H. 390-1, A.D. 1000*, London, 1879, repr. Frankfurt am Main: Minerva, 1984.

—, *India*: E. C. Sachau, *Alberuni's India*, Delhi, *etc.*: S. Chand & Co., 1964 reprint.

—, *Maqālīd*: Marie-Thérèse Debarnot, ed. and trans., *al-Bīrūnī: Kitāb Maqālīd ʿilm al-hayʾa. La trigonométrie sphérique chez les Arabes de l'est à la fin du Xe siècle,* Damascus: Institut Français de Damas (Publication no. 114), 1985.

—, *Qānūn*: al-Bīrūnī, *al-Qānūn al-Masʿūdī*, [Max Krause, ed.], 3 vols., Hyderabad, Deccan: Osmania Oriental Publications, 1954-1956.

—, *Shadows*: al Bīrūnī, *Ifrād al-maqāl fī amr al-ẓilāl*, no. 2 in *Rasāʾil al-Bīrūnī*, Hyderabad: Osmania Oriental Publications, 1948; with translation and commentary in Edward Stewart Kennedy, *The Exhaustive Treatise on Shadows by Abū al-Rayḥān Muḥammad b. Aḥmad al-Bīrūnī*, 2 vols., Aleppo: Institute for the History of Arabic Science, 1976.

— , *Taḥdīd*: al-Bīrūnī, *Kitāb Taḥdīd nihāyāt al-amākin*, ed. P. Bulgakov, *Majallat Maʿhad al-makhṭūṭāt al-ʿarabiyya* (Cairo) 8 (1962). [I have not seen the Ankara 1962 edn. by Muḥammad ibn Tāwīt al-Ṭanjī.] Translation in Jamil Ali, *The Determination of the Coordinates of Cities ... by al-Bīrūnī*, Beirut: American University of Beirut Press, 1967. Commentary in Edward Stewart Kennedy, *A Commentary upon Bīrūnī's Kitāb Taḥdīd [nihāyāt] al-amākin, An 11th Century Treatise on Mathematical Geography*, Beirut: American University of Beirut Press, 1973.

Blachère *et al.*, *Dictionnaire Arabe-Français-Anglais*: Régis Blachère *et al.*, *Dictionnaire Arabe-Français-Anglais*, 3 vols. to date, Paris: G.-P. Maisonneuve et Larose, 1967 to present.

Bonine, "The Morphogenesis of Iranian Cities": Michael E. Bonine, "The Morphogenesis of Iranian Cities", *Annals of the Association of American Geographers* 69 (1979), pp. 208-224.

— , "Sacred Direction and City Structure in Morocco": *idem*, "The Sacred Direction and City Structure: A Preliminary Analysis of the Islamic Cities of Morocco", *Muqarnas: An Annual on Islamic Art and Architecture* 7 (1990), pp. 50-72.

Borst, *Astrolab und Klosterreform*: Arno Borst, *Astrolab und Klosterreform an der Jahrtausendwende*, Heidelberg: Carl Winter, 1989. [Problematic.]

Bosworth, *Islamic Dynasties*: C. Edmund Bosworth, *The New Islamic Dynasties – A chronological and genealogical manual*, New York: Columbia University Press, 1996.

Bremner, "Shadow Table in Mul.Apin": Robert W. Bremner, "The Shadow Length Table in Mul.Apin", in *Graz 1991 Conference Proceedings*, pp. 367-382.

Brentjes, "The Republic of Letters in the Middle East": Sonja Brentjes, "The Interests of the Republic of Letters in the Middle East, 1550-1700", *Science in Context* 12:3 (1999), pp. 435-468.

Brentjes: see also *Frankfurt 2001 Conference Proceedings*.

Brice, *Atlas*: Brice, William C., ed., *An Historical Atlas of Islam*, Leiden: E. J. Brill, 1981.

Brice & Imber & Lorch, "Equatorial Semicircle": William Brice, Colin Imber and Richard P. Lorch, "The Dāʾire-yi Muʿaddel of Seydī ʿAlī Reʾīs", *Seminar on Early Islamic Science* (University of Manchester) 1 (1976). [Reviewed in King, *Studies*, B-XIII.]

Brieux, "Authenticité des astrolabes": Alain Brieux, "Les astrolabes: tests d'authenticité", *Art et curiosité*, sept. 1974 (unpaginated).

Brieux & Maddison, *Répertoire*: Alain Brieux and Francis R. Maddison, *Répertoire des facteurs d'astrolabes et de leurs œuvres*, in press (Paris: C.N.R.S.).

Brockelmann, *GAL*: Carl Brockelmann, *Geschichte der arabischen Litteratur*, 2 vols., 2nd edn., Leiden: E. J. Brill, 1943-49, and Supplementbände, 3 vols., Leiden: E. J. Brill, 1937-42.

Bruin, *al-Bīrūnī Newsletter*: A series of 38 mimeographed articles prepared between 1966 and 1971 by the late Professor Frans Bruin, then of the American University of Beirut. [For a complete list see the introduction to **XIIa**. Copies are available at the Department of History of Science and Medicine at Yale University; the University Library at the American University of Beirut; and the Institutes for History of Science in Frankfurt and Hamburg.] In this book the following are cited:

"The Astronomical Observatory of Ulugh Beg in Samarqand", *al-Bīrūnī Newsletter* 9 (1967).
"The Observational Instruments of al-ʿUrḍī ... ", *ibid.* 10 (1968).
"The Astronomical Observatory of Naṣīr al-Dīn al-Ṭūsī", *ibid.* 13 (1968).
"The Construction of Instruments ... by ... al-ʿĀmilī ... ", *ibid.* 15 (1968).
"The Fakhrī Sextant in Rayy", *ibid.* 19 (1969).
"The Instruments of Taqī al-Dīn ... ", *ibid.* 29 (1970).
"A Treatise on Small Instruments by ... al-Khāzinī ... ", *ibid.* 30 (1970).
"The Planispheric Astrolabe: Its History, Theory, and Use", *ibid.* 33, 34 and 35 (1970).

Brusa, "Navicelle": Giuseppe Brusa, "Le navicelle orarie di Venezia", *Annali dell'Istituto e Museo di Storia della Scienza di Firenze* 5 (1980), pp. 51-59.

Brussels 1984 Exhibition Catalogue: *La mesure du temps dans les collections Belges*, Brussels: Société Générale de Banque, 1984. [Catalogue of an exhibition at the Société Générale de Banque during 26.1.-7.4.1984.]

Brussels BR 2001 Exhibition Catalogue: *Des agrimensores romains aux arpenteurs du XVIe siècle*, Jan De Graeve and Jean Mosselmans, eds., Brussels: UGEB, 2001. [Catalogue of an exhibition organized by the Union des Géomètres-Experts immobiliers de Bruxelles held at the Bibliothèque Royale de Belgique during 22.11-21.12.2001.]

BSIS: *Bulletin of the Scientific Instrument Society*.

Buchner, "Antike Reiseuhren": Edmund Buchner, "Antike Reiseuhren", *Chiron* 1 (1971), pp. 457-482.

— , *Die Sonnenuhr des Augustus*: *idem*, *Die Sonnenuhr des Augustus*, Mainz: Philipp von Zabern, 1982.

Burnett, "ʿAbd al-Masīḥ of Winchester": Charles Burnett, "ʿAbd al-Masīḥ of Winchester", in *North Festschrift*, pp. 159-169.

—, "Antioch as a Link between Arabic and Latin Culture": *idem*, "Antioch as a Link between Arabic and Latin Culture in the Twelfth and Thirteenth Centuries", in *Occident et Proche-Orient : Contacts scientifiques au temps des Croisades, Actes du colloqie de Louvain-la-Neuve, 24 et 25 mars 1997*, Isabelle Draelants, Anne Tihon, Baudouin van den Abeele, eds., Turnhout (B): Brepols, 2000, pp. 1-78.

—, *The Introduction of Arabic Learning into England*: *idem*, *The Introduction of Arabic Learning into England*, (The Panizzi Lectures, 1996), London: The British Library, 1997.

—, "The Translation of Arabic Science into Latin": *idem*, "The Translation of Arabic Science into Latin: A Case of Alienation of Intellectual Property?", *Bulletin of the Royal Institute for Inter-Faith Studies* (Amman, Jordan) 4:1 (2002), pp. 145-157.

Burnett, ed, *Adelard of Bath*: *Adelard of Bath – An English Scientist of the Early Twelfth Century*, Charles Burnett, ed., (Warburg Institute Surveys and Texts, XIV), London: The Warburg Institute, University of London, 1987.

Cairo ENL Catalogue: David A. King, *A Catalogue of the Scientific Manuscripts in the Egyptian National Library* (in Arabic), 2 vols., Cairo: General Egyptian Book Organization, 1981 and 1986.

Cairo ENL Survey: *idem*, *A Survey of the Scientific Manuscripts in the Egyptian National Library* (in English), (American Research Center in Egypt, Catalogs, vol. 5), Winona Lake, Ind.: Eisenbrauns, 1987.

Caldo, "Astrolabi di Palermo": L. Caldo, "Astrolabi del Museo Nazionale di Palermo", *Pubblicazione dell' Osservatorio astronomico di Palermo, Memorie*, no. 65, 1936, repr. in *Coelum* 8 (June 1938), pp. 101-108, and 8 figs. [Not reprinted in *AIOS* or *IMA*.]

Calvo, *Ibn Bāṣo and his Universal Plate*: Emilia Calvo Labarta, *Abū ʿAlī al-Ḥusayn ibn Bāṣo (m. 716/1316): Risālat al-Ṣafīḥa al-ŷāmiʿa li-ŷamīʿ al-ʿurūḍ (Tratado sobre la lámina general para todas las latitudas)*, Madrid: Consejo Superior de Investigaciones Científicas and Instituto de Cooperación con el Mundo Árabe, 1993.

—, "Ibn Bāṣo's Universal Plate": *eadem*, "Ibn Bāṣo's Astrolabe [Plate] in the Maghrib and East", in *Vernet Festschrift*, II, pp. 755-767.

—, "Ibn Bāṣo's *Ṣafīḥa Mujayyaba*": *eadem*, "Transformation of Coordinates in Ibn Bāṣo's *al-Risāla fī 'l-ṣafīḥa al-mujayyaba dhāt al-awtār*", *JHAS* 12 (2001), pp. 3-21.

—, "La resolution graphique des questions astrologiques": *eadem*, "La resolution graphique des questions astrologiques à al-Andalus", *Actes du 3me Colloque Maghrébin sur l'Histoire des Mathématiques Arabes – Tipaza (Alger, Algérie) – 1-3 Décembre 1990*, Algiers: Association Algérienne d'Histoire des Mathématiques, 1998, pp. 31-44.

—, "The Use of Ibn Bāṣo's Universal Plate": *eadem*, "A Study of the Use of Ibn Bāṣo's Universal Astrolabe Plate", *AIHS* 50 (2000), pp. 264-295.

Cambridge DI 1998 Conference Proceedings: selected papers given at a conference "New perspectives on science in medieval Islam" held at the Dibner Institute, Cambridge, Mass., during Nov. 6-8, 1998, to appear.

Cambridge History of Iran, IV-VI: *The Cambridge History of Iran*, vol. IV: The Period from the Arab Invasion to the Saljuqs, (Richard N. Frye, ed.), vol. V: *The Saljuq and Mongol Periods* (John A. Boyle, ed.), and vol. VI: *The Timurid and Safavid Periods* (Peter Jackson and Laurence Lockhart, eds.), Cambridge: Cambridge University Press, 1975, 1968 and 1986. [See Kennedy, "Exact Sciences in Iran".]

Cambridge WMHS Catalogue: David J. Bryden, *The Whipple Museum of the History of Science, Catalogue 6: Sundials and Related Instruments*, Cambridge: The Whipple Museum of the History of Science, 1988.

Canobbio, "Fragment of a Spherical Astrolabe": Ernesto Canobbio, "An Important Fragment of a West Islamic Spherical Astrolabe", *Annali dell'Istituto e Museo di Storia della Scienza di Firenze* 1 (1976), pp. 37-41.

Cantor, *Geschichte der Mathematik*: Moritz Cantor, *Vorlesungen über die Geschichte der Mathematik* 2nd edn., 4 vols., repr. Leipzig: B. G. Teubner, 1913.

Carandell, "Analemma for the Qibla": Joan Carandell, "An Analemma for the Determination of the Azimuth of the Qibla in the *Risāla fī ʿilm al-ẓilāl* of Ibn al-Raqqām", *ZGAIW* 1 (1984), pp. 61-72.

—, *Ibn al-Raqqām sobre los cuadrantes solares*: *idem*, Risāla fī ʿilm al-ẓilāl *de Muḥammad Ibn al-Raqqām al-Andalusī: edicion, traducción y comentario*, Barcelona: Universidad de Barcelona, Instituto "Millás Vallicrosa" de Historia de la Ciencia Árabe, 1988.

Carboni, *Kitāb al-bulhān*: Stefano Carboni, *Il Kitāb al-bulhān di Oxford*, (*Eurasiatica* - Quaderni del Dipartimento di Studi Eurasiatici, Università degli Studi di Venezia, 6), Turin: Tirrenia, 1988.

Carra de Vaux, "L'astrolabe linéaire": Bernard Carra de Vaux, "L'astrolabe linéaire ou bâton d'al-Tousi", *Journal Asiatique*, 11. sér., 5 (1895), pp. 464-516, repr. in *AIOS*, III, pp. 181-233.

Casanova, "Cadran solaire syrien": Paul Casanova, "La montre du Sultan Nour ad Din (554 de l'Hégire = 1159-1160)", *Syria* 4 (1923), pp. 282-299, repr. in *AIOS*, IV, pp. 242-262.

Casulleras, "Andalusī Sundial": Josep Casulleras, "El último capítulo del *Kitāb al-Asrār fī natā'iŷ al-afkār*", in *Vernet Festschrift*, II, pp. 613-653.

Cátedra & Samsó, *Astrología de Enrique de Villena*: Pedro M. Cátedra (ed. and transl.) with Julio Samsó (comm.), *Tratado de astrología atribuido a Enrique de Villena*, Madrid: Río Tinto Minera, 1980.

Caussin, "Table Hakémite": A. P. Caussin de Perceval, "Le livre de la grande table Hakémite", in *Notices et extraits des manuscrits de la Bibliothèque nationale*, 7 (An XII [= 1804]), pp. 16-240, with separate pagination in the separatum.

Chabás, "Astronomy in 15th-Century Salamanca": José Chabás, "Astronomy in Salamanca in the Mid-Fifteenth Century: The Tabulae resolutae", *JHA* 29 (1998), pp. 167-175.

— , "Cahier d'un croisier": *idem*, "Le cahier d'astronomie d'un croisier du XVe siècle", *Nuncius* (Florence) 12:1 (1997), pp. 3-16.

— & Goldstein, *Zacut*: *idem* and Bernard R. Goldstein, *Astronomy in the Iberian Peninsula: Abrahan Zacut and the Transition from Manuscript to Print*, (*Transactions of the American Philosophical Society* 90:2), Philadelphia, Pa.: 2000.

Chapoutot-Remadi: see Souissi & Chapoutot-Remadi.

Charette, "Der geflügelte Quadrant": François Charette, "Der geflügelte Quadrant: Ein ungewöhnliches Sinusinstrument aus dem 14. Jh.", in *Saltzer Festschrift*, pp. 25-36.

— , "al-Khwārizmī on the Astrolabe": *idem* with Petra Schmidl, "Scientific Initiative in 9th-Century Baghdad: al-Khwārizmī on the Astrolabe and Related Treatises", to appear.

— , "The Library of Muṣṭafā Ṣidqī": *idem*, "A View of Scientific Life in Mid-Eighteenth Century Istanbul: The Library of Muṣṭafā Ṣidqī" [revised version of a paper presented at the DAVO Annual Congress (German Middle East Studies Association for Contemporary Research and Documentation), Hamburg, December 1999], forthcoming.

— , *Mamluk Instrumentation*: *idem*, *Mathematical Instrumentation in Fourteenth-Century Egypt and Syria – The Illustrated Treatise of Najm al-Dīn al-Miṣrī*, Leiden: E. J. Brill, 2003.

— , "Najm al-Dīn's Monumental Table": *idem*, "A Monumental Medieval Table for Solving the Problems of Spherical Astronomy for all Latitudes", *AIHS* 48 (1998), pp. 11-64.

— , *Orientalisme et histoire des sciences: l'historiographie européenne des sciences islamiques et hindoues, 1784-1900*, Master's thesis, Départment d'histoire, Université de Montréal, 1995.

— & King, "Survey of Islamic Tables for Astrolabe Construction": *idem* and David A. King, "A Survey of Medieval Islamic Tables for Astrolabe Construction", to appear.

— & King, *The Universal Astrolabe of Ibn al-Sarrāj*: *idem* and David A. King, *The Universal Astrolabe of Ibn al-Sarrāj*, to appear.

— & Schmidl, "Habash's Universal Plate": *idem* and Petra G. Schmidl, "A Universal Plate for Timekeeping with the Stars by Ḥabash al-Ḥāsib: Text, Translation and Preliminary Commentary", *Suhayl* 2 (2001), pp. 107-159.

Charette: see also *Frankfurt 2001 Conference Proceedings* and King & Charette.

Cheikho, "al-Mukḥula": Louis Cheikho (= Shaykhū), "al-Mukḥula" (in Arabic), *al-Machriq* [= *al-Mashriq*] (Beirut) 10 (1907) [repr. Munich: Kraus-Thomson Organization G.m.b.H., 1980], pp. 75-80, repr. in *IMA*, vol. 91 (1998), pp. 370-375.

Chidyaq, "Indicateur de qibla": Hana Chidyaq, "Un indicateur de qibla signé Muḥammad ʿIzz al-Ṣabbāgh", *Astrolabica* 5 (1989), pp. 27-36.

Christie's London 08.04.1998 Catalogue: *Christie's South Kensington – Exceptional Scientific and Engineering Works of Art, Instruments and Models – 8 April 1998*, London: Christie's, 1998.

Colin & Renaud, "Abū Miqraʿ": Georges Colin and Henri-Paul-Joseph Renaud, "Note sur le "*muwaqqit*" marocain Abū Muqriʿ – ou mieux Abū Miqraʿ – al-Baṭṭīwī (XIIIe s. J.C.)", *Hespéris* 25 (1933), pp. 94-96.

Comes, *Ecuatorios andalusíes* and "Ecuatorios andalusíes": Mercè Comes, *Ecuatorios andalusíes – Ibn al-Samḥ, al-Zarqālluh y Abū-l-Ṣalt*, Barcelona: Universidad de Barcelona, Facultad de Filología, *etc.*, 1991; and a summary in the chapter "Los ecuatorios andalusíes" in *Madrid MAN 1992 Exhibition Catalogue*, pp. 75-87.

The Compendium: *The Compendium – Journal of the North American Sundial Society*.

Cook, "Muslim Material on Comets and Meteors": David Cook, "A Survey of Muslim Material on Comets and Meteors", *JHA* 30 (1999), pp. 131-160.

Cotter, "Nautical Tables": Charles H. Cotter, "The Development of Nautical Astronomical Inspection Tables in the Period from 1770 to 1919", *Vistas in Astronomy* 20 (1976), pp. 245-247.

Crombie, "Griechisch-arabische Naturwissenschaften und Abendländisches Denken": Alistair C., Crombie, "Griechisch-arabische Naturwissenschaften und Abendländisches Denken", in Gereon Sievernich and Hendrik Budde, eds., *Europa und der Orient 800-1900*, Gütersloh and Munich: Bertelsmann Lexikon Verlag GmbH, 1989, pp. 102-131.

Curtze, "Urkunden zur Geschichte der Trigonometrie": Maximilian Curtze, "Urkunden zur Geschichte der Trigonometrie im christlichen Mittelalter", *Bibliotheca Mathematica*, 3. Folge, 1 (1900), pp. 321-416.

Al-Daffa & Stroyls, *Studies:* Ali A. Al-Daffa and John S. Stroyls, *Studies in the Exact Sciences in Medieval Islam*, New York: John Wiley & Sons, 1984. [To be used with caution.]

Dallal, "Al-Bīrūnī on Climates": Ahmad Dallal, "Al-Bīrūnī on Climates", *AIHS* 34(1984), pp. 3-18.

—, "Ibn al-Haytham's Universal Solution for the Qibla": *idem*, "Ibn al-Haytham's Universal Solution for Finding the Direction of the *Qibla* by Calculation", *ASP* 5 (1995), pp. 145-193.
— , "Islamic Science, Medicine, and Technology": *idem*, "Science, Medicine, and Technology – The Making of a Scientific Culture", in Esposito, ed., *The Oxford History of Islam*, pp. 154-213.
Damascus Catalogue: Ibrahim Khoury, *Fihris makhṭūṭāt Dār al-Kutub al-Ẓāhiriyya*, vol. VI: *'Ilm al-hay'a wa-mulḥaqātuhu*, Damascus: Majma' al-lugha al-'Arabiyya, 1389 H [= 1969].
Daniel, "Greenwich *Navicula*": Christopher Daniel, "The Fifth Navicula", *Clocks* (Hemel Hempstead, U.K.) 14 (1992), p. 37. [On the *navicula* now in Greenwich.]
Davidian, "al-Bīrūnī on the Time of Day": Davidian, Marie-Louise, "al-Bīrūnī on the Time of Day from Shadow Lengths", *JAOS* 80 (1960), pp. 330-335, repr. in Kennedy *et al.*, *Studies*, pp. 274-279.
Davidian: see also Kennedy & Davidian.
Debarnot, *al-Bīrūnī's* Maqālīd: see al-Bīrūnī, *Maqālīd.*
— , "Islamic Trigonometry": Marie-Thérèse Debarnot, "Trigonometry", in *EHAS*, II, pp. 495-538.
— , "*Zīj* of Ḥabash": *eadem*, "The *Zīj* of Ḥabash al-Ḥāsib: A Survey of MS Istanbul Yeni Cami 784/2", in *Kennedy Festschrift*, pp. 35-69.
Dekker, "Astrolabes and Dates and Dead Ends": Elly Dekker, "Of Astrolabes and Dates and Dead Ends", *AS* 49 (1992), pp. 175-184. [A critical study of Torode, "Dating Astrolabes".]
— , "Cartographic Grids from Iran": *eadem*, "Cartographic Grids from Iran: An Early Version of the Retro-Azimuthal Orthographic Projection", *The Cartographical Journal* (The British Cartographic Society, London) 37:2 (2000), pp. 109-116. [A discussion of the grids on the maps described in King, *Mecca-Centred World-Maps* – see now **VIIc**.]
—, "*Quadrans novus*": *eadem*, "An Unrecorded Medieval Astrolabe Quadrant from *c.* 1300", *Annals of Science* 52 (1995), pp. 1-47.
Delambre, *HAMA*: Jean-Baptiste Delambre, *Histoire de l'astronomie du moyen âge*, Paris, 1819, repr. New York: Johnson Reprints, 1965.
Delehar, "Illustrations": Peter Delehar, "Illustrations of Scientific Instruments in the *Gentleman's Magazine* 1746-1796", in *Turner Festschrift*, pp. 383-394.
Delphin, "L'astronomie au Maroc": Gäetan Delphin, "L'astronomie au Maroc", *Journal Asiatique*, 8e ser., 17 (1891), pp. 177-201.
De Smet: see Poulle & De Smet.
Destombes, "Astrolabe carolingien": Marcel Destombes, "Un astrolabe carolingien et l'origine de nos chiffres arabes", *AIHS* 58-59 (1962), pp. 1-45, repr. in *idem*, *Contributions*, pp. 153-200.
— , "Chiffres coufiques": *idem*, "Les chiffres coufiques des instruments astronomiques arabes", *Physis* 2:3 (1960), pp. 197-210, repr. in *idem*, *Contributions*, pp. 127-140.
— , *Contributions*: G. Schilder *et al.*, *M. Destombes (1905-1983): Contributions sélectionnées à l'Histoire de la Cartographie et des Instruments scientifiques,* Utrecht: Hes & Paris: A. G. Nizet, 1987.
— & Kennedy, "al-Ṣūfī on the Astrolabe": *idem* and Edward Stewart Kennedy, "Introduction to *Kitāb al-ʿAmal bil Asṭurlāb* [of al-Ṣūfī]", repr. in Kennedy *et al.*, *Studies*, pp. 405-447.
de Vries *et al.*, "Hafīr and Halazûn": Fer J. de Vries, William S. Maddux and Mac Oglesby, "Hafīr and Halazûn", *The Compendium – Journal of the North American Sundial Society* 6:2 (1999), pp. 23-26.
de Vries *et al.*, "Regiomontanus Sundials": Fer J. de Vries, Mac Oglesby, William S. Maddux and Warren Thom, "Regiomontanus, Apian and Capuchin Sundials", *The Compendium* 5:4 (1998), pp. 13-19.
al-Dharīʿa: see Āghā Buzurg, *al-Dharīʿa.*
D'Haenens, "La clepsydre de Villiers": Albert D'Haenens, "La clepsydre de Villiers (1267) – Comment on mesurait et vivait le temps dans une abbaye cistercienne au XIIIe siècle", in Harry Kühnel, ed., *Klösterliche Sachkultur des Spätmittelalters*, Vienna: Österreichische Akademie der Wissenschaften, 1980.
d'Hollander, *L'Astrolabe*: Raymond d'Hollander, *L'Astrolabe – Histoire, théorie et pratique*, Paris: Institut océanographique, 1999.
Dicks, "Ancient Astronomical Instruments": D. R. Dicks, "Ancient Astronomical Instruments", *Journal of the British Astronomical Association* 64 (1954), pp. 77-85.
Dizer, "Equatorial Semicircle": Muammer Dizer, "The *dāʾirat al-muʿaddal* in Kandilli Observatory ... ", *JHAS* 1 (1977), pp. 257-260.
Dizer: see also *Kandilli Instrument Handbook.*
Djebbar, *Une histoire de la science arabe*: Ahmed Djebbar, *Une histoire de la science arabe – Introduction à la connaissance du patrioine scientifique des pays d'Islam*, Paris: Seuil, 2001.
DMA: *Dictionary of the Middle Ages*, 13 vols., New York: Charles Scribner's Sons, 1982-89.
Dobrzycki: see Kremer & Dobrzycki.
Dold-Samplonius, "Measuring the Muqarnas": Yvonne Dold-Samplonius, "Practical Arabic Mathematics: Measuring the Muqarnas by al-Kāshī", *Centaurus* 35 (1992), pp. 193-242.

Dorn, "Drei arabische Instrumente": Bernhard Dorn, "Drei in der Kaiserlichen Öffentlichen Bibliothek zu St. Petersburg befindliche astronomische Instrumente mit arabischen Inschriften", *Mémoires de l'Académie Impériale des Sciences de St.-Pétersbourg,* VIIe série, 9:1 (1865), pp. 1-150 and 2 figs., repr. in *AIOS*, I, pp. 345-498.

Dozy, *Supplément aux dictionnaires arabes*: Reinhart Dozy, *Supplément aux dictionnaires arabes*, 2nd edn., 2 vols., Leiden: E. J. Brill & Paris: Maisonneuve Frères, 1927, repr. Beirut: Librarie du Liban, 1968.

Drecker, "Hermannus Contractus über das Astrolab": Joseph Drecker, "Hermannus Contractus über das Astrolab", *Isis* 16 (1931), pp. 200-219.

— , *Theorie der Sonnenuhren*: *idem*, *Die Theorie der Sonnenuhren*, vol. 1E of *Die Geschichte der Zeitmessung und der Uhren*, Ernst von Bassermann-Jordan, ed., Berlin & Leipzig: Vereinigung Wissenschaftlicher Verleger Walter de Gruyter & Co., 1925.

DSB: *Dictionary of Scientific Biography*, 14 vols. and 2 supp. vols., New York: Charles Scribner's Sons, 1970-80.

Dublin CB Arabic Catalogue: A. J. Arberry, *The Chester Beatty Library. A Handlist of the Arabic Manuscripts*, 7 vols., Dublin: Emery Walker (Ireland) Ltd. (vols. 1-2) and Hodges, Figgis & Co., Ltd. (vols. 3-7), 1955-64.

Dublin CB Persian Catalogue: A. J. Arberry, M. Minovi, and E. Blochet, *The Chester Beatty Library: A Catalogue of the Persian Manuscripts and Miniatures*, vol. I: MSS. 101-150, Dublin: Hodges Figgis & Co. Ltd., 1959.

Ducène, "*Kitāb Dalā'il al-qibla* d'Ibn al-Qāṣṣ": Jean-Charles Ducène, "Le *Kitāb Dalā'il al-qibla* d'Ibn al-Qāṣṣ: analyse des trois manuscrits et des emprunts d'Abū Ḥāmid al-Garnaṭī", *ZGAIW* 14 (2001), pp. 169-187.

EHAS: *Encyclopedia of the History of Arabic Science*, Roshdi Rashed, with Régis Morelon, eds., 3 vols., London: Routledge, 1996. [Also available in French as *Histoire de la science arabe*, 3 vols., Paris: Seuil, 1997, and in Arabic as *Mawsūʿat ta'rīkh al-ʿulūm al-ʿarabiyya*, 3 vols., Beirut: Markaz dirāsāt al-wiḥda al-ʿarabiyya & Muʿassasat ʿAbd al-Ḥamīd Shūmān, 1997.]

EI$_1$: *Encyclopaedia of Islam*, 1st edn., 4 vols., Leiden: E. J. Brill, 1913-1934, repr. in 9 vols., Leiden, *etc.*: E. J. Brill, 1987, and again in paper-back, 1993. [Also available in French and German.]

EI$_2$: *The Encyclopaedia of Islam*, new edn., 11 vols. and supplements, Leiden: E. J. Brill, 1960 to present, especially articles "Anwāʾ" (aspects of folk astronomy), "Asṭurlāb" (astrolabe), "ʿIlm al-hay'a" (here: astronomy generally), "ʿIlm al-ḥisāb" (arithmetic), "Ḳibla" (religious and astronomical aspects of the sacred direction), "Ḳuṭb" (celestial pole), "Layl and Nahār" (aspects of folk astronomy), "Makka. iv: as centre of the world" (sacred geography), "Maṭāliʿ" (right and oblique ascensions), "Maṭlaʿ" (astronomical risings and settings), "Mayl" (declination and obliquity), "Mīḳāt. ii. astronomical aspects" (astronomical timekeeping), "Minṭaḳa" (ecliptic, zodiac and obliquity), "Mizwala" (sundial), "Nudjūm" (stars); "Nudjūm, ʿIlm aḥkām al-" (astrology), "Rīḥ" (winds in the folk-astronomical tradition), "Rubʿ" (quadrant), "Samt" (direction), "Shakkāziyya" (universal astrolabic projections), "Taʿrīkh. 2. Era chronology in astronomical handbooks" (calendrics in *zīj*es), Ṭāsa" (magnetic compass), "Zīdj" (tables and astronomical handbooks) [a fuller version is in King & Samsó, "Islamic Astronomical Handbooks and Tables"]; and in the supplement: "Bādgīr" (ventilators). [Also available in French.]

EIr: *Encyclopaedia Iranica*, 10 vols. to date, London & New York: Routledge & Kegan Paul, 1983 to present.

Eisner, *Kalendarium of Nicholas of Lynn*: S. Eisner, ed., & *idem* and G. MacEoin, transl., *The Kalendarium of Nicholas of Lynn*, Athens, Ga.: University of Georgia Press (under the auspices of The Chaucer Library), 1980.

EJ: *Encylopaedia of Judaism*, 12 vols., New York & London: Funk & Wagnalls, 1901-1906.

Elwell-Sutton, L. P., *The Horoscope of Asadullāh Mīrzā – A Specimen of Nineteenth-Century Astrology*, Leiden: E. J. Brill, 1977.

Endreß, "Arabische Wissenschaftliche Literatur": Gerhard Endreß, "Die wissenschaftliche Literatur", Kapitel 8 in *Grundriß der Arabischen Philologie*, Band II: *Litteraturwissenschaft*, Helmut Gätje, ed., Wiesbaden: Dr. Ludwig Reichert, 1987, pp. 400-506.

EPACT: Website *Epact: Scientific Instruments of Medieval and Renaissance Europe*, located at www.mhs.ox.ac.uk/epact, including descriptions of individual instruments which can be accessed using the key-words ion the following artificial titles. See in particular Ackermann, "London BM *Quadrans Vetus*", and "Sloane Astrolabe"; Bennett, "Oxford *Navicula*"; Meliconi, "Brechte Plate", and "Oxford *Quadrans vetus*"; and A. J. Turner, "Florence *Navicula*", and "Florence *Quadrans vetus*".

ERE: *Encyclopaedia of Religion and Ethics*, James Hastings, ed., 12 vols., Edinburgh: T. & T. Clark, 1921.

Erlangen PMS SB: *Sitzungsberichte der Physikalisch-Medizinischen Sozietät zu Erlangen.*

Escorial Catalogue: Henri-Paul-Joseph Renaud, *Les manuscrits arabes de l'Escurial*, tome II, fascicule 3: *Sciences exactes et sciences occultes*, Paris: Paul Geuthner, 1941.

Esposito, ed., *The Oxford History of Islam*: *The Oxford History of Islam*, John L. Esposito, ed., Oxford & New York: Oxford University Press, 1999.

Evans, *Ancient Astronomy*: James Evans, *The History and Practice of Ancient Astronomy*, New York & Oxford: Oxford University Press, 1998.

Fantoni, *Orologi solari*: Girolamo Fantoni, *Orologi solari – Trattato completo di gnomonica*, Rome: Technimedia, 1988.

Fehérvári, "Quadrant of al-Mizzī": Géza Fehérvári, "An Eighth/Fourteenth-Century Quadrant of the Astrolabist al-Mizzī", *Bulletin of the School of Oriental and African Studies* (London) 36 (1973), pp. 115-117 and 2 pls.

Ferrari & Severino, *Meridiane islamiche*: Gianni Ferrari and Nicola Severino, *Appunti per uno studio delle meridiane islamiche*, privately distributed, 1997.

Field & Wright, "Gears from the Byzantines": Judith V. Field and Michael T. Wright, "Gears from the Byzantines: A Portable Sundial with Calendrical Gearing", *Annals of Science* 42 (1985), pp. 87-138, repr. in *eidem* and Hill, *Byzantine and Arabic Gearing*.

— & Hill & Wright, *Byzantine and Arabic Gearing*: *eidem* and Donald R. Hill, *Byzantine and Arabic Mathematical Gearing*, London: The Science Museum, 1985

Fierro & Samsó, eds., *The Formation of al-Andalus*, II: Maribel Fierro and Julio Samsó, eds., *The Formation of al-Andalus*, Part 2: Language, Religion, Culture and the Sciences, Aldershot, Brookfield, Vt.: Ashgate – Variorum, 1998.

Fletcher *et al.*, *Index of Mathematical Tables*: A. Fletcher, J. C. P. Miller, L. Rosenhead, and L. J. Comrie, *An Index of Mathematical Tables*, 2 vols., Oxford: Blackwell, 1962.

Forcada, "*Anwā*ʾ Books in al-Andalus": Miquel Forcada, "Books of *Anwā*ʾ in al-Andalus", in Fierro & Samsó, eds., *The Formation of al-Andalus*, II, pp. 305-327.

—, "Folk Astronomy of Aḥmad b. Fāris": *idem*, "Astrology and Folk Astronomy: the *Mukhtaṣar min al-Anwā*ʾ of Aḥmad b. Fāris", *Suhayl* 1 (2000), pp. 107-205.

—, "*Mīqāt* en los calendarios andalusíes", *idem*, "*Mīqāt* en los calendarios andalusíes", *al-Qanara* 11 (1990), pp. 59-69.

–, "The *Mukhtaṣar min al-Anwā*ʿ of Aḥmad ibn Fāris": "A New Andalusian Astronomical Source from the Fourth/Tenth Century: The *Mukhtaṣar min al-Anwā*ʿ of Aḥmad ibn Fāris", *in Vernet Festschrift*, II, 769-780.

—, "Les sources andalouses du Calendrier d'Ibn al-Bannā'", *idem*, "Les sources andalouses du Calendrier d'Ibn al-Bannā'", in *Actas del II Coloquio hispano-maroquí de ciencias históricas "Historia, ciencia y sociedad", Granada, 6-10 nov. 1989*, pp. 183-196.

Frank, *Zur Geschichte des Astrolabs*: Joseph Frank, *Zur Geschichte des Astrolabs*, Erlangen: Universitäts-Buchdruckerei, 1920, repr. in *AIOS*, IV, pp. 31-62.

—, "al-Khwārizmī über das Astrolab": *idem*, "Die Verwendung des Astrolabs nach al-Chwârizmî", *Abhandlungen zur Geschichte der Naturwissenschaften und der Medizin* 3 (1922), repr. in *AIOS*, IV, pp. 125-159, with a review by Carl Schoy repr. *ibid*, pp. 160-161.

—, "Zwei arabische Instrumente": *idem*, "Über zwei astronomische arabische Instrumente", *Zeitschrift für Instrumentenkunde* 41 (1921), pp. 193-200, repr. in *AIOS*, IV, pp. 63-70.

— & Meyerhof, "Mogulisches Astrolab": *idem* and Max Meyerhof, "Ein Astrolab aus dem indischen Mogulreiche", *Heidelberger Akten der von-Portheim-Stiftung* 13 (Heidelberg: Carl Winter), 1925, reprinted in *AIOS*, IV, pp. 307-356.

Frank: see also Wiedemann & Frank.

Frankfurt IGN Festband: *Ad radices – Festband zum fünfzigjährigen Bestehen des Instituts für Geschichte der Naturwissenschaften Frankfurt am Main*, Anton von Gotstedter, ed., Stuttgart: Franz Steiner, 1994.

Frankfurt 2001 Conference Proceedings: Sonja Brentjes, Benno van Dalen and François Charette, eds., *Certainty, Doubt, Error: Aspects of the Practice of Pre- and Early Modern Science. ...* , in *Early Science and Medicine* (E. J. Brill, Leiden) 7:3 (2002), pp. 173-310. [Proceedings of a conference held in Frankfurt am Main during 17-18.11.2001.]

Fück, *Arabische Studien*: Johann Fück, *Die arabischen Studien in Europa bis in den Anfang des 20. Jahrhunderts*, Leipzig: Otto Harrassowitz, 1955.

Fuller, "Universal Rectilinear Dials": A. W. Fuller, "Universal Rectilinear Dials", *The Mathematical Gazette* (London) 41 (1957), pp. 9-24.

Garbers, "Thābit ibn Qurra über Sonnenuhren": Karl Garbers, "Ein Werk Ṯābit b. Qurra's über ebene Sonnenuhren", *Quellen und Studien zur Geschichte der Mathematik, Astronomie und Physik*, Abt. A, Bd. 4 (1936).

García Avilés, *El tiempo y los astros*: Alejandro García Avilés, *El tiempo y los astros – Arte, ciencia y religion en la Alta Edad Media*, Murcia: Unversidad de Murcia, 2001.

García Franco, *Astrolabios en España*: Salvador García Franco, *Catálogo crítico de los astrolabios existentes en España*, Madrid, Instituto Histórico de Marina, 1945.

Gentleman's Magazine: see Delehar.

Ghori: see Ansari & Ghori.

Gibbs, *Greek and Roman Sundials*: Sharon L. Gibbs, *Greek and Roman Sundials*, New Haven & London: Yale University Press, 1976.

Gibbs: see also *Washington NMAH Catalogue*.

Gingerich, "Applications of Computers to the History of Astronomy": Owen Gingerich, "Applications of High-Speed Computers to the History of Astronomy", *Vistas in Astronomy* 9 (1967), pp. 229-236.

— , "Fake Astrolabes": *idem*, "Astronomical Scrapbook – Fake Astrolabes", *Sky & Telescope*, May, 1982, pp. 465-468.

— , "Paper Instruments with Moving Parts": *idem*, "Astronomical Instruments with Moving Parts", in *G. Turner Festschrift*, pp. 63-74.

— , "Sacrobosco Illustrated": *idem*, "Sacrobosco Illustrated", in *North Festschrift*, pp. 211-224.

— , "Zoomorphic Astrolabes": *idem*, "Zoomorphic Astrolabes and the Introduction of Arabic Star Names into Europe", in *Kennedy Festschrift*, pp. 89-104.

— & King & Saliba, "The ʿAbd al-Aʾimma Astrolabe Forgeries": *idem*, David A. King and George Saliba, "The ʿAbd al-Aʾimma Astrolabe Forgeries", *JHA* 3 (1972), pp. 188-198, repr. in King, *Studies*, B-VI.

Ginzel, *Handbuch der Chronologie*: F. K. Ginzel, *Handbuch der mathematischen und technischen Chronologie*, 3 vols., Leipzig: Hinrichs, 1906-14.

Girke, "Die frühesten islamischen Sternkataloge": Dorothea Girke, "Drei Beiträge zu den frühesten islamischen Sternkatalogen mit besonderer Rücksicht auf Hilfsfunktionen für die Zeitrechnung bei Nacht", Johann Wolfgang Goethe-Univer-sität, Frankfurt am Main, Institut für Geschichte der Naturwissenschaften, Preprint Series no. 8 (1988).

Glowatzki & Göttsche: *Die Tafeln des Regiomontanus*: E. Glowatzki and H. Göttsche: *Die Tafeln des Regiomontanus. Ein Jahrhundertwerk*, (Algorithmus. Studien zur Geschichte der Mathematik und der Naturwissenschaften), Munich: Institut für Geschichte der Naturwissenschaften, 1990.

Gockerell, "Telling Time without a Clock": Nina Gockerell, "Telling Time without a Clock", in *The Clockwork Universe: German Clocks and Automata 1550-1650*, Klaus Maurice and Otto Mayr, eds., New York: Neale Watson, 1980, pp. 131-143.

Goldstein, "Astronomical Instruments in Hebrew": Bernard R. Goldstein, "Description of Astronomical Instruments in Hebrew", in *Kennedy Festschrift*, pp. 105-141.

— , *The Astronomy of Levi ben Gerson*: *idem*, *The Astronomy of Levi ben Gerson (1288-1344)*, New York, *etc.*: Springer, 1985.

— , *Ibn al-Muthannâ on al-Khwārizmī*: *idem*, *Ibn al-Muthannâ's Commentary on the Astronomical Tables of al-Khwârizmî*, New Haven, Conn.: Yale University Press, 1967.

— , "A Medieval Table for Reckoning Time from Solar Altitude", *Scripta Mathematica* 27 (1963), pp. 61-66, repr. in Kennedy *et al.*, *Studies*, pp. 293-298.

— , *Studies*: *idem*, *Theory and Observation in Ancient and Medieval Astronomy*, London: Variorum, 1985.

Goldstein Festschrift: *Astronomy and Astrology from the Babylonians to Kepler – Essays Presented to Bernard R. Goldstein on the Occasion of his 65th Birthday*, Peter Barker, Alan C. Bowen, José Chabás, Gad Freudenthal and Tzvi Langermann, eds., a special issue of *Centaurus* 44 (2003).

Goldstein: see also *Aaboe Festschrift*; Chabás & Goldstein; and King & Samsó, "Astronomical Handbooks and Tables".

Goldziher, "Nachmittagzeit im Islam": Ignaz Goldziher, "Die Bedeutung der Nachmittagzeit im Islam", *Archiv für Religionswissenschaft* 9 (1906), pp. 293-302.

Gouk, *Ivory Sundials of Nuremberg*: Penelope Gouk, *The Ivory Sundials of Nuremberg 1500-1700*, Cambridge: Whipple Museum of the History of Science, 1988.

Gotha Catalogue: Wilhelm Pertsch, *Die arabischen Handschriften der Herzoglichen Bibliothek zu Gotha*, 5 vols., Gotha: Fried. Andr. Perthes, 1878-1892, reprinted Frankfurt: Institut für Geschichte der Arabisch-Islamischen Wissenschaften, 1987.

Graz 1991 Conference Proceedings: *Die Rolle der Astronomie in den Kulturen Mesopotamiens, Beiträge zum 3. Grazer Morgenländischen Symposion (23.-27. September 1991)*, Hannes D. Galter, ed., (Grazer Morgenländische Studien, Band 3), Graz, 1993.

Greenwich Astrolabe Booklet: *The Planispheric Astrolabe*, Greenwich: National Maritime Museum, 1976.

Greenwich Astrolabe Catalogue: Koenraad Van Cleempoel, ed., A catalogue of the astrolabes at the National Maritime Museum, Greenwich, to be published by Oxford University Press.

Greenwich Sundial Catalogue: Hester Higton, ed., *Sundials at Greenwich – A Catalogue of the Sundials, Nocturnals, and Horary Quadrants in the National Maritime Museum*, Oxford: Oxford University Press, 2002.

Gunella, "Abaco para transformar horas": Alessandro Gunella, "Un abaco para transformar horas iguales en disiguales y viceversa", *Analema* no. 32 (Mayo-Agosto 2001), pp. 12-13.

Gunella *et al.*, "Antecedentes del reloj de Rodrigo Zamorano": Alessandro Gunella, "Antecedentes del reloj de Rodrigo Zamorano", Giovanni Paltrinieri, "Sobre un instrumento de Oroncio Finé", and Manuel María

Valdés, "Antecedentes del reloj universal de Rodrigo Zamorano (1581)", *Analema* no. 24 (Septiembre-Diciembre 1998), pp. 3-4, 5-7, and 8-9.

Gunther, *Astrolabes*: Robert T. Gunther, *The Astrolabes of the World*, 2 vols., Oxford: Oxford University Press, 1932, repr. in 1 vol., London: The Holland Press, 1976.

— , *Early Science in Oxford*, II and V: *idem*, *Early Science in Oxford*, vol. II: *Astronomy*, Oxford: Clarendon Press, 1923, and vol. V: *Chaucer and Messahalla on the Astrolabe*, Oxford: University Press, 1929.

Haddad: see Kennedy & Pingree & Haddad.

Härtel, "Ludolfus Borchdorp de Brunswik": Helmar Härtel, "Ludolfus Borchdorp de Brunswik", *Braunschweigisches Jahrbuch* 68 (1987), pp. 113-120. [Establishes the identity of the maker of an astrolabe from Einbeck, *ca.* 1330.]

Hahn, *Medieval Mensuration*: Nan L. Britt Hahn, *Medieval Mensuration:* Quadrans vetus *and* Geometrie due sunt partes principales, *Transactions of the American Philosophical Society* 72:8 (1984).

Hakim Mohammed Said: see *Islamabad 1979 Conference Proceedings.*

Hamadanizadeh, "Medieval Islamic Interpolation Schemes": Javad Hamadanizadeh, "A Survey of Medieval Islamic Interpolation Schemes", in *Kennedy Festschrift*, pp. 143-152.

Hartner, "Astrolabe", A-B: Willy Hartner, "The Principle and Use of the Astrolabe", a chapter in *A Survey of Persian Art from prehistoric times to the present*, Arthur Upham Pope, ed., 9 vols., London & New York: Oxford University Press, 1938-39, III, pp. 2530-2554, reprinted with the original plates in *Astrolabica*, no. 1, Feb. 1978, Paris: Société Internationale de l'Astrolabe, and article "Asṭurlāb" in EI_2, both reprinted in *idem*, *Oriens-Occidens*, Hildesheim: Georg Olms, 1968, pp. 287-311 and pp. 312-318.

— , *Oriens-Occidens*, I-II: *idem*, *Oriens-Occidens – Ausgewählte Schriften zur Wissenschafts- und Kulturgeschichte – Festschrift zum 60. Geburtstag*, Hildesheim: Georg Olms, 1968 (I), and *Oriens-Occidens – Ausgewählte Schriften zur Wissenschafts- und Kulturgeschichte, Band II*, Yasukatsu Maeyama, ed., Hildesheim, *etc.*: Georg Olms, 1984 (II).

— & Ruska, "Katalog": Willy Hartner and Julius Ruska, "Katalog der orientalischen und lateinischen Originalhandschriften, Abschriften und Photokopien des Instituts für Geschichte der Medizin und der Naturwissenschaften in Berlin", *Quellen und Studien zur Geschichte der Naturwissenschaften* 7:2/3 (1939), pp. 1-149.

Hartner Festschrift: *ΠΡΙΣΜΑΤΑ – Naturwissenschaftsgeschichtliche Studien – Festschrift für Willy Hartner*, Yasukatsu Maeyama and Walter G. Saltzer, eds., Wiesbaden: Franz Steiner, 1977.

Al-Hassan, "Decline of Islamic Science": Ahmad Y. al-Hassan, "Factors behind the Decline of Islamic Science after the Sixteenth Century", in Sharifah Shifa al-Attas, ed., *Islām and the Challenge of Modernity: Historical and Contemporary Contexts, Proceedings of the Inaugural Symposium on Islām and the Challenge of Modernity ... at ... the International Institute of Islamic Thought and Civilization in Kuala Lumpur*, August 1-5, 1994, Kuala Lumpur: International Institute of Islamic Thought and Civilization, pp. 351-389.

Hastings: see *ERE*.

Hawkins & King, "Orientation of the Kaʿba": see King, *Studies*, C-XII.

Heidarzadeh, "Astronomy in Iran": Tofigh Heidarzadeh, "From the Maragha School to the Darolfonun - A Historical Review of Astronomy in Iran from the 13th to the 19th Century", paper presented at the Symposium on Science and Technology in the Turkish and Islamic World, Istanbul, 3-5 June, 1994. [Not printed in the *Proceedings*.]

Heinen, *Islamic Cosmology*: Anton Heinen, *Islamic Cosmology: A Study of as-Suyūṭī's* al-Hay'a al-sanīya fi-l-hay'a al-sunnīya, Beiruter Texte und Studien (Orient-Institut der Deutschen Morgenländischen Gesellschaft), Band 27, Beirut (for Franz Steiner Verlag, Wiesbaden), 1982.

Helmecke, "Berliner Astrolab des Muḥammad Zamān": Gisela Helmecke, "Das Berliner Astrolab des Muḥammad Zamān al-Mašhadī", *Forschungen und Berichte der Staatlichen Museen zu Berlin, Hauptstadt der DDR* (Berlin: Akademie-Verlag) 25 (1985), pp. 129-142 and pls. 31-32.

Higgins, "Classification of Sundials": Kathleen Higgins, "The Classification of Sundials", *Annals of Science* 9 (1953), pp. 342-358.

Higton, *Portable Sundials*: Hester K. Higton, *Sundials – An Illustrated History of Portable Dials*, London: Philip Wilson Publishers, Ltd., 2001.

Higton: see also *Greenwich Sundial Catalogue*.

Hill, "Al-Bīrūnī's Mechanical Calendar": Donald R. Hill, "Al-Bīrūnī's Mechanical Calendar", *AS* 42 (1985), pp. 139-163, repr. in *idem*, *Studies*, XIV, and also in Field & Hill & Wright, *Byzantine and Arabic Gearing.*

— , *Islamic Science and Engineering*: *idem*, *Islamic Science and Engineering*, Edinburgh: Edinburgh University Press, 1994.

— , *Studies*: *idem*, *Studies in Medieval Islamic Technology: From Philo to al-Jazarī – from Alexandria to Diyār Bakr*, David A. King, ed., Aldershot, Brookfield USA, Singapore and Sydney: Ashgate-Variorum, 1998.

Hill: see also Field & Wright & Hill.

G. F. Hill, *Arabic Numerals in Europe*: G. F. Hill, *The Development of Arabic Numerals in Europe Exhibited in Sixty-Four Tables*, Oxford: Clarendon Press, 1915.

Hinks, "Retro-Azimuthal Equidistant Projection": A. Hinks, "A Retro-Azimuthal Equidistant Projection of the Whole Sphere", *Geographical Journal* 73:2 (1929), pp. 245-247.

Hinz, *Islamische Masse und Gewichte*: Walter Hinz, *Islamische Masse und Gewichte*, Leiden: E. J. Brill, 1955.

HM: *Historia Mathematica*.

History of Cartography: *The History of Cartography*, J. B. Harley and David Woodward, eds., vol. I: *Cartography in Prehistoric, Ancient, and Medieval Europe and the Mediterranean*; vol. II, book 1: *Cartography in the Traditional Islamic and South Asian Societies*, Chicago, Ill., & London: The University of Chicago Press, 1987 and 1992. [See also the review of II:1 by Karen C. Pinto in *Harvard Middle Eastern and Islamic Review* 1:1 (1994), pp. 109-115.]

History of Oriental Astronomy: see *New Delhi 1985 Colloquium Proceedings*.

Hogendijk, "Al-Khwārizmī's Sine of the Hours": Jan P. Hogendijk, " Al-Khwārizmī's Tables of the "Sine of the Hours" and the Underlying Sine Table", *Historia scientiarum* 42 (1991), pp. 1-12.

—, "Mathematical Astrology": *idem*, "Mathematical Astrology in the Islamic Tradition" (dealing with houses, rays and progressions), in *Cambridge DI 1998 Conference Proceedings*, to appear.

—, "al-Muʿtaman ibn Hūd": *idem*, "al-Muʾtaman ibn Hūd. 11th Century King of Saragossa and Brilliant Mathematician", *HM* 22 (1995), pp. 1-18.

—, "al-Nayrīzī on the Qibla at Baghdad": *idem*, "Al-Nayrīzī's Mysterious Determination of the Azimuth of the Qibla at Baghdad", *Sciamvs* 1 (2000), pp. 49-70.

—, "Seasonal Hour Lines on Astrolabes and Sundials": *idem*, "The Contributions by Abū Naṣr ibn ʿIrāq and al-Ṣāghānī to the Theory of Seasonal Hour Lines on Astrolabes and Sundials", *ZGAIW* 14 (2001), pp. 1-30.

—, "Tables for Casting the Rays": *idem*, "The Mathematical Structure of Two Islamic Astrological Tables for Casting the Rays", *Centaurus* 32 (1989), pp. 171-202.

—, "Three Islamic Lunar Crescent Visibility Tables": *idem*, "Three Islamic Lunar Crescent Visibility Tables", *JHA* 19 (1988), pp. 29-44.

Hogendijk: see also Luckey, *Thesis*.

Hoyland, *Seeing Islam as Others Saw it*: Robert G. Hoyland, *Seeing Islam as Others Saw it: A Survey and Evaluation of Christian, Jewish and Zoroastrian Writings on Early Islam*, Princeton, NJ: The Darwin Press, 1997.

HM: *Historia Mathematica*.

HS: *History of Science*.

IAM: see *Islamic Astronomy and Mathematics*.

Ibish, ed., *Editing Islamic Manuscripts on Science*: *Editing Islamic Manuscripts on Science*, Yusuf Ibish, ed., London: Al-Furqan Islamic Heritage Foundation, 1999.

Ibn al-Nadīm, *Fihrist*: Ibn al-Nadīm, *Kitāb al-Fihrist*, ed. Gustav Flügel (1871), repr. Beirut: Khayats, 1964, transl. by Bayard Dodge as *The Fihrist of al-Nadīm*, 2 vols., New York and London: Columbia University Press, 1970.

Ibn Nujaym, *Rafʿ al-ghishāʿ*: References in **IV** are to MS Cairo Ṭalʾat *majāmīʿ* 200,5 (fols. 27r-30r, copied 1733/34) of his treatise *Rafʿ al-ghishāʾ ʿan waqtayi ʾl-ʿaṣr wa-ʾl-ʿishāʾ*.

Ibn Raḥīq, *al-Mawāqīt*: References in **IV** are to MS Berlin Ahlwardt 5664 of his treatise on folk astronomy.

Ibn al-Raqqām: see Carandell.

Id, "Analemma for Ascensions": Yusuf Id, "An Analemma Construction for Right and Oblique Ascensions", *The Mathematics Teacher* 62 (1969), pp. 669-672, repr. in Kennedy *et al.*, *Studies*, pp. 495-498.

İhsanoğlu, "Western Science in the Ottoman World": Ekmeleddin İhsanoğlu, "Introduction of Western Science to the Ottoman World: a Case Study of Modern Astronomy (1660-1860)", in *Istanbul 1987 Symposium Proceedings*, pp. 67-120.

— *et al.*, *Ottoman Astronomical Literature*: *idem et al.*, *Osmanlı astronomi literatürü tarihi – History of Astronomy Literature during the Ottoman Period* (in Turkish), 2 vols., (Studies and Sources on the History of Science, Series No. 7), Istanbul: Research Centre for Islamic History, Art and Culture (IRCICA), 1997.

—, *Ottoman Geographical Literature*: *eidem*, *Osmanlı coğrafya literatürü tarihi – History of Geographical Literature during the Ottoman Period* (in Turkish), 2 vols., (Studies and Sources on the History of Science, Series No. 9), Istanbul: Research Centre for Islamic History, Art and Culture (IRCICA), 2000.

—, *Ottoman Mathematical Literature*: *eidem*, *Osmanlı matematik literatürü tarihi – History of Astronomy Literature during the Ottoman Period* (in Turkish), 2 vols., (Studies and Sources on the History of Science, Series No. 8), Istanbul: Research Centre for Islamic History, Art and Culture (IRCICA), 1999.

İhsanoğlu: see also *Istanbul 1987 Symposium Proceedings*, and *Istanbul 1991 and 1994 Symposia Proceedings*.

Ilyas, *Astronomy of Islamic Times*: Mohammad Ilyas, *Astronomy of Islamic Times for the Twenty-first Century*, London & New York: Mansell, 1988.

— , *Islamic Calendar, Times & Qibla*: *idem*, *A Modern Guide to the Astronomical Calculations of Islamic Calendar, Times & Qibla*, Kuala Lumpur: Berita, 1984.

Imad-ad-Dean Ahmad, *Signs in the Heavens: A Muslim Astronomer's Perspective on Religion and Science*, Beltsville, Md.: Writers' Inc.-International, 1992.

Index Islamicus: *Index Islamicus,* numerous vols. to date, Cambridge: W. Heffer & Sons, starting in 1958, continued up to the present as *The Quarterly Index Islamicus*, London: Bowker-Saur, 4th edn. (including 1906-1999) available on CD-ROM from E. J. Brill, Leiden.

Ingolstadt Apian Exhibition: see Röttel.

Irani, "Arabic Numeral Forms": Rida A. K. Irani, "Arabic Numeral Forms", *Centaurus* 4 (1955), pp. 1-12, repr. in Kennedy *et al.*, *Studies*, pp. 710-721.

— , *The Jadwal al-Taqwīm of Ḥabash al-Ḥāsib*, unpublished Master's dissertation, American University of Beirut, 1956.

Islamabad 1979 Conference Proceedings: *History and Philosophy of Science – Proceedings of the International Congress of the History and Phlosophy of Science, Islamabad, 8-13 December, 1979*, Hakim Mohammed Said, ed., Nazimabad, Pakistan: Hamdard Foundation Press, n.d. (*ca.* 1980).

Islamic Geography: *Islamic Geography*, Fuat Sezgin, with Mazen Amawi, Carl Ehrig-Eggert and Eckhard Neubauer *et al.*, eds., 278 vols. to date, Frankfurt am Main: Institut für Geschichte der Arabisch-Islamischen Wissenschaften, 1992 to present. [Facsimile reprints of early writings, mainly 19th and early-20th century.]

Islamic Mathematics and Astronomy: *Islamic Mathematics and Astronomy*, *eidem*, eds., 112 vols. to date, Frankfurt am Main: Institut für Geschichte der Arabisch-Islamischen Wissenschaften, 1997 to present. [Reprints of texts and studies mainly from the 19th and early 20th centuries. Materials relating to al-Khwārizmī feature in vols. 1-7; al-Farghānī 9-10, 68; al-Battānī 11-13; Thābit ibn Qurra 21-22; Ibn Yūnus 25; al-Ṣūfī 26; Ibrāhīm ibn Sinān 27, 101; Abū Naṣr 28, 37; al-Bīrūnī 29-36, 74; Ibn al-Zarqālluh 39-40; al-Marrākushī 41-42; Ibn al-Bannā' 43-44; 'Umar al-Khayyām 45-46; al-Ṭūsī *et al.* 47-51; Ulugh Beg 52-55; al-Kāshī 56, 84; Ibn al-Haytham 57-58, 75; Abu 'l-Wafā' 60-61; al-Sijzī 66; Maimonides 67; Alfonso X 98-99; calendrics 64-65; astronomical instruments 85-96; Sayılı's *Observatory in Islam* 97; Nallino's *Ta'rīkh 'ilm al-falak* 100.]

Istanbul 1977 Symposium Proceedings: *Proceedings of the International Symposium on the Observatories in Islam (Istanbul, 1977)*, Muammer Dizer, ed., Istanbul: Milli Eğitim Basımevi, 1980.

Istanbul 1987 Symposium Proceedings: *Transfer of Modern Science & Technology to the Muslim World*, Ekmeleddin İhsanoğlu, ed., Istanbul: Reseach Centre for Islamic History, Art and Culture (IRCICA), 1982.

Istanbul 1991 and 1994 Symposia Proceedings: *Science in Islamic Civilisation*, Ekmeleddin İhsanoğlu and Feza Günergun, eds., Istanbul (IRCICA) 2000.

Istanbul: see also *Kandilli Instrument Handbook.*

Jachimowicz, "Islamic Cosmology": E. Jachimowicz, "Islamic Cosmology", in C. Blacker and M. Loewe, eds., *Ancient Cosmologies*, London: George Allen & Unwin Ltd., 1985, pp. 143-171.

Jackson, "Retro-Azimuthal Projections": J. E. Jackson, "Retro-Azimuthal Projections", *Survey Review* 19 (1968), pp. 319-328.

Janin, "Cadran solaire de Damas": Louis A. Janin, "Le cadran solaire de la Mosquée Umayyade à Damas", *Centaurus* 16 (1972), pp. 285-298, repr. in Kennedy & Ghanem, eds., *Ibn al-Shāṭir*, pp. 107-121.

— , "Un cadran solaire grec d'Afghanistan": *idem*, "Un cadran solaire grec à Aï Khanoum, Afghanistan", *L'Astronomie* (Paris) 92 (1978), pp. 357-362.

— , "Gnomonique tunisienne": *idem*, "Quelques aspects récents de la gnomonique tunisienne", *Revue de l'Occident musulman et de la Méditerranée* 23 (1977), pp. 202-221.

— , "Projection horizontale": *idem*, "Astrolabe et cadran solaire en projection horizontale", *Centaurus* 22 (1979), pp. 298-314.

— , "Ar-Rūdānī sur l'astrolabe sphérique": *idem*, "Un texte d'ar-Rūdānī sur l'astrolabe sphérique", *Annali dell'Istituto e Museo di Storia della Scienza di Firenze* 3 (1978), pp. 71-75. [See also Pellat, "L'astrolabe sphérique d'al-Rūdānī".]

Janin & King, "Cadran solaire d'Ibn Ṭūlūn": see King, *Studies*, B-XVI.

— & — , "Ibn al-Shāṭir's Compendium": see King, *Studies*, B-XII.

Janin & Rohr, "Deux astrolabes-quadrants turcs", *idem* and R. R. J. Rohr, "Deux astrolabes-quadrants turcs", *Centaurus* 19 (1975), pp. 108-124.

JAOS: *Journal of the American Oriental Society.*

Jaubert, "Capteurs de vents d'Égypte": Olivier Jaubert, "Capteurs de vents d'Égypte – Essai de typologie", *Annales islamologiques* (Institut Français d'Archéologie Orientale du Caire) 25 (1995), pp. 169-231.

Jensen, "Abū Naṣr's *Table of Minutes*": Claus Jensen, "Abū Nasr's Approach to Spherical Trigonometry as Developed in His Treatise *The Table of Minutes*", *Centaurus* 16 (1971), pp. 1-19.

— , "The Lunar Theories of al-Baghdādī": *idem*, "The Lunar Theories of al-Baghdādī", *AHES* 8 (1971-72), pp. 321-328.

JHA: *Journal for the History of Astronomy*.

JHAS: *Journal for the History of Arabic Science* (Institute for the History of Arabic Science, Aleppo).

Jiminez, "La qibla extraviada": Alfonso Jiminez, "La qibla extraviada", in *Cuadernos de Madīnat al-Zahrā'*, vol. III, Cordova, 1991, pp. 189-209.

JNES: *Journal of Near Eastern Studies*.

Jones, *Byzantine Astronomical Manual*: Alexander Jones, *An Eleventh-Century Manual of Arabo-Byzantine Astronomy*, (*Corpus des astronomes byzantins* III), Amsterdam: J. C. Gieben, 1987.

JRAS: *Journal of the Royal Asiatic Society*.

Juschkewitsch, *Mathematik im MA*: Adolf P. Juschkewitsch, *Geschichte der Mathematik im Mittelalter*, (first published in Russian in 1961), Leipzig: B. G. Teubner, 1964. [See also Youschkevitsch, *Mathématiques arabes*.]

Kamal al-Din, *Tables*: Mathematical tables and graphs by Husayn Kamal al-Din (then Professor in the Civil Engineering Dept., Riyadh University) in 3 volumes bearing the titles: 1 *Kebla* [*sic*] *Directions for the Whole World*, 2 *Praying Time Tables for the Whole World*, 3 *Curves of Praying Times for the Whole World*, Cairo: Dār al-Fikr al-'Arabī, for The Conference of Islamic Jurisprudence at the Islamic University of Imām Muḥammad ibn Sa'ūd, Saudi Arabia, n.d. [*ca*. 1970 ?]. [Copy inspected at Aligarh Muslim University in 1978, by which time the book was no longer available at the publishers in Cairo.]

Kandilli Instrument Handbook: Muammer Dizer, *Astronomi hazineleri*, (B.Ü. Yayınları no. 404), Istanbul: Boğaziçi Üniversitesi and Kandilli Rasathanesi, 1986.

Kaye, *The Observatories of Jai Singh*, A-B: George R. Kaye, *The Astronomical Observatories of Jai Singh*, (*Archaeological Survey of India*, New Imperial Series, vol. XL), and *A Guide to the Observatories at Delhi, Jaipur, etc*., Calcutta, 1918 and 1920, both repr. in *IAM*, vol. 93.

Kennedy, "al-Bīrūnī on Prayer-Times": Edward Stewart Kennedy, "al-Bīrūnī on the Muslim Times of Prayer", in *The Scholar and the Saint: Studies in Commemoration of Abū'l-Rayḥān al-Bīrūnī and Jalāl al-Dīn al-Rūmī*, Peter J. Chelkowski, ed., New York: New York University Press, 1975, pp. 83-94, repr. in *idem et al*., *Studies*, pp. 299-310.

— , *al-Bīrūnī's* Shadows: see al-Bīrūnī, *Shadows*.

— , "Comets in Islamic Astronomy and Astrology": *idem*, "Comets in Islamic Astronomy and Astrology", *JNES* 16 (1957), pp. 44-51, repr. in *idem et al*., *Studies*, pp. 311-318.

— , "The Digital Computer and the History of the Exact Sciences": *idem*, "The Digital Computer and the History of the Exact Sciences", *Centaurus* 12 (1967), pp. 107-113, repr. in *idem et al*., *Studies*, pp. 385-393.

— , "The Equatorium of Abū al-Ṣalt": *idem*, "The Equatorium of Abū al-Ṣalt", *Physis* 12 (1970), pp. 73-81, repr. in *idem et al*., *Studies*, pp. 481-491.

— ,"Exact Sciences in Iran", A-C: *idem*, "The Exact Sciences during the Abbasid Period", *Cambridge History of Iran*, IV, pp. 378-395 (A); "The Exact Sciences in Iran under the Seljuqs and Mongols", *ibid*., V, pp. 659-679 (B); "The Exact Sciences in Iran under the Timurids", *ibid*., VI, pp. 568-580.

— , "History of Trigonometry": *idem*, "Overview of the History of Trigonometry" (first published 1969), repr. in *idem et al*., *Studies*, pp. 3-29.

— , "Ibn al-A'lam's Tables": "The Astronomical Tables of Ibn al-A'lam", *JHAS* 1 (1977), pp. 13-23.

— , "Ibn al-Raqqām's Tables": *idem*, "The Astronomical Tables of Ibn al-Raqqām, a Scientist of Granada", *ZGAIW* 11 (1997), pp. 35-72.

— , "Islamic Mathematical Geography": *idem*, "Mathematical Geography", in *EHAS*, I, pp. 185-201.

— , "al-Kāshī on Instruments": *idem*, "al-Kāshī's Treatise on Astronomical Observational Instruments", *JNES* 20 (1961), pp. 98-108, repr. in *idem et al*., *Studies*, pp. 394-404.

— , *al-Kāshī's Equatorium*: *idem*, *The Planetary Equatorium of Jamshīd Ghiyāth al-Dīn al-Kāshī (d. 1429) ...* , Princeton N.J.,: Princeton University Press, 1960.

— , "Spherical Astronomy in al-Kāshī's *Khāqānī Zīj*": see *idem*, *Studies*, VII.

— , *Studies*: *idem*, *Astronomy and Astrology in the Medieval Islamic World*, (Variorum Collected Studies Series: CS600), Aldershot & Brookfield, Vt.: Ashgate-Variorum, 1998. Contains:

I "Ḥabash al-Ḥāsib on the Melon Astrolabe" (with Richard P. Lorch), first publication.

VII "Spherical Astronomy in Kāshī's *Khāqānī Zīj*", *ZGAIW* 2 (1990), pp. 1-46.

— , "Al-Ṣūfī on the Celestial Globe": *idem*, "Al-Ṣūfī on the Celestial Globe", *ZGAIW* 5 (1989), pp. 48-93.

— , "*Zīj* Survey": *idem*, "A Survey of Islamic Astronomical Tables", *Transactions of the American Philosophical Society*, N. S., 46:2 (1956), pp. 123-177, repr. with separate pagination, n.d. [*ca*. 1990].

— & Davidian, "Al-Qāyinī on Twilight": *idem* and Marie-Louise Davidian, "Al-Qāyinī on the Duration of Dawn and Twilight", *JNES* 20 (1961), pp. 145-153, repr. in *idem et al*., *Studies*, pp. 284-292.

— & Debarnot, "Two Mappings of al-Bīrūnī": *idem* and Marie-Thérèse Debarnot, "Two Mappings Proposed by al-Bīrūnī", *Zeitschrift für Geschichte der arabisch-islamischen Wissenschaften* (Frankfurt am Main) 1 (1984), pp. 144-147, repr. in *idem*, *Studies*, V.

— & Faris, "Yaḥyā b. Abī Manṣūr on Solar Eclipses": *idem* and Nazim Faris, "The Solar Eclipse Technique of Yaḥyā b. Abī Manṣūr", *JHA* 1 (1970), pp. 20-38, repr. in *idem et al.*, *Studies*, pp. 185-203.

— & Ghanem, *eds., Ibn al-Shāṭir*: *idem* and Imad Ghanem, eds., *The Life and Work of Ibn al-Shāṭir: an Arab Astronomer of the Fourteenth Century*, Aleppo: Institute for the History of Arabic Science, 1976.

— & Id, "Ḥabash's Analemma for the Qibla": *idem* & Yusuf Id, "A Letter of al-Bīrūnī: Ḥabash al-Ḥāsib's Analemma for the Qibla", *Historia Mathematica* 1 (1973), pp. 3-11, repr. in Kennedy *et al.*, *Studies*, pp. 621-629.

— & Kennedy, *Islamic Geographical Coordinates*: *idem* and Mary Helen Kennedy, *Geographical Coordinates of Localities from Islamic Sources*, Frankfurt am Main: Institut für Geschichte der Arabisch-Islamischen Wissenschaften, 1987.

— & King, "Ibn al-Majdī's Tables for Calculating Ephemerides": *idem* and David A. King, "Ibn al-Majdī's Tables for Calculating Ephemerides", *JHAS* 4 (1980), pp. 48-68, repr. in King, *Studies*, A-VI.

— —, "Indian Astronomy in Fourteenth-Century Fez": *eidem*, "Indian Astronomy in Fourteenth-Century Fez: The Versified *Zīj* of al-Qusunṭīnī [*sic*: read al-Qusunṭīnī]", *JHAS* 6 (1982), pp. 3-45, repr. in King, *Studies*, A-VIII.

— & Krikorian-Preisler, "Projecting the Rays": *idem* and H. Krikorian-Preisler, "The Astrological Doctrine of Projecting the Rays", *Al-Abhath* 25 (1972), pp. 3-15, repr. in Kennedy *et al.*, *Studies*, pp. 372-384.

— & Kunitzsch & Lorch, *Melon Astrolabe*: *idem*, Paul Kunitzsch and Richard P. Lorch, *The Melon-Shaped Astrolabe in Arabic Astronomy*, (*Boethius: Texte und Abhandlungen zur Geschichte der Mathematik und der Naturwissenschaften*, Band 43), Stuttgart: Franz Steiner, 1999.

— & Pingree, *Astrological History of Māshā'allāh*: *idem* and David Pingree, *The Astrological History of Māshā'allāh*, Cambridge, Ma.: Harvard University Press, 1971.

— & Pingree & Haddad, *al-Hāshimī on Astronomical Tables*: *idem*, David Pingree and Fuad I. Haddad, *The Book of Reasons behind Astronomical Tables* (Kitāb ʿilal al-zījāt) *by ʿAlī ibn Sulaymān al-Hāshimī*, Delmar, N.Y.: Scholars' Facsimiles & Reprints, 1981.

Kennedy & Ukashah, "The Chandelier Clock of Ibn Yūnis": *idem* and Walid Ukashah, "The Chandelier Clock of Ibn Yūnis", *Isis* 60 (1969), pp. 543-545, repr. in Kennedy *et al.*, *Studies*, pp. 499-501.

Kennedy *et al.*, *Studies*: E. S. Kennedy, Colleagues and Former Students, *Studies in the Islamic Exact Sciences*, David A. King and Mary Helen Kennedy, eds., Beirut: American University of Beirut, 1983. Contains:

- "Overview of the History of Trigonometry", *31st Yearbook of the National Council of Teachers of Mathematics*, Washington, D.C., 1969, pp. 333-359, repr. here pp. 3-29.
- "The Solar Eclipse Technique of Yahyā b. Abī Manṣūr" (with Nazim Faris), *JHA* 1 (1970), pp. 20-38, repr. here pp. 185-203.
- Mark Lesley, Mark, "Bīrūnī on Rising Times and Daylight Lengths", *Centaurus* 5 (1957), pp. 121-141, repr. here pp. 253-273.
- Marie-Louise Davidian, "al-Bīrūnī on the Time of Day from Shadow Lengths", *JAOS* 80 (1960), pp. 330-335, repr. here pp. 274-279.
- Nadi Nadir, "Abū al-Wafāʾ on the Solar Altitude", *The Mathematics Teacher* 53 (1960), pp. 460-463, repr. here pp. 280-283.
- "Al-Qāyinī on the Duration of Dawn and Twilight" (with Marie-Louise Davidian), *JNES* 20 (1961), pp. 145-153, repr. here, pp. 284-292.
- Bernard R. Goldstein, "A Medieval Table for Reckoning Time from Solar Altitude", *Scripta Mathematica* 27 (1963), pp. 61-66, repr. here pp. 293-298.
- "al-Bīrūnī on the Muslim Times of Prayer", in *The Scholar and the Saint ...* , P. J. Chelkowski, ed., New York, 1975, pp. 83-94, repr. here pp. 299-310.
- "Comets in Islamic Astronomy and Astrology", *JNES* 16 (1957), pp. 44-51, repr. here, pp. 311-318.
- "The Astrological Doctrine of Projecting the Rays" (with H. Krikorian-Preisler), *Al-Abhath* 25 (1972), pp. 3-15, repr. here pp. 372-384.
- "The Digital Computer and the History of the Exact Sciences", *Centaurus* 12 (1967), pp. 107-113, repr. here pp. 385-393.
- "al-Kāshī's Treatise on Astronomical Observational Instruments", *JNES* 20 (1961), pp. 98-108, repr. here pp. 394-404.
- "The Equatorium of Abū al-Ṣalt", *Physis* 12 (1970), pp. 73-81, repr. here pp. 481-491.
- "The Chandelier Clock of Ibn Yūnis" (with Walid Ukashah), *Isis* 60 (1969), pp. 543-545, repr. here, pp. 499-501.
- Yusuf Id, "An Analemma Construction for Right and Oblique Ascensions", *The Mathematics Teacher* 62 (1969), pp. 669-672, repr. here pp. 495-498.
- "A Letter of al-Bīrūnī: Ḥabash al-Ḥāsib's Analemma for the Qibla" (with Yusuf Id), *Historia Mathematica* 1 (1973), pp. 3-11, repr. here pp. 621-629.
- Rida A. K. Irani, "Arabic Numeral Forms", *Centaurus* 4 (1955), pp. 1-12, repr. here pp. 710-721.

Kennedy Festschrift: *From Deferent to Equant*: *Studies in the History of Science in the Ancient and Medieval Near East in Honor of E. S. Kennedy*, David A. King and George Saliba, eds., *Annals of the New York Academy of Sciences* 500 (1987). Contains:

- Marie-Thérèse Debarnot, "The *Zīj* of Ḥabash al-Ḥāsib: A Survey of MS Istanbul Yeni Cami 784/2", pp. 35-69.
- Javad Hamadanizadeh, "A Survey of Medieval Islamic Interpolation Schemes", pp. 143-152.
- D. A. King, "Some Early Islamic Tables for Lunar Crescent Visibility", pp. 185-225.
- G. Rosinska, "Tables of Decimal Trigonometric Functions from *ca.* 1450 to *ca.* 1550", pp. 419-426.

Kennedy: see also al-Bīrūnī, *Geography*, and *idem*, *Shadows*; Destombes & Kennedy; and King, *Studies*, A-VI and A-VIII (joint papers).

Kessler: "Mecca-Oriented Architecture in Cairo", A-C: Christel Kessler, "Mecca-Oriented Architecture and Urban Growth of Cairo", *Atti del Terzo Congresso di Studi Arabi e Islamici* (Ravello, 1966), Naples, 1967, p. 425 (A); "Funerary Architecture within the City", *Colloque International sur l'Histoire du Caire*, Cairo, 1969, pp. 257-267 (B); and "Mecca-Oriented Urban Architecture in Mamluk Cairo: The Madrasa-Mausoleum of Sultan Shaʿbān II, in *In Quest of an Islamic Humanism: Arab an Islamic Studies in Memory of Mohamed Nowaihi*, Cairo: American University of Cairo Press, 1984, pp. 97-108 (C).

Khalili Collection: see *London Khalili Collection* ... .

al-Khwārizmī: see Suter, *al-Khwārizmī*; Neugebauer, *al-Khwārizmī*; Goldstein, *Ibn al-Muthannā on al-Khwārizmī*; and also King, "al-Khwārizmī", and Rosenfeld *et al.*, eds., *Al-Khorezmi.*

ʿAbdallāh al-Khwārizmī, *Mafātīḥ al-ʿulūm*: Gerlof van Vloten, ed., *Mafâtîh al-olûm explicans vocabula technica scientiarum tam arabum quam peregrinorum auctore Abû Abdallah Mohammed ibn Ahmed ibn Jûsuf al-Kâtib al-Khowarezmi*, Leiden, 1895, also pirated edition: Cairo: Maṭbaʿat al-Sharq, n. d. [*ca.* 1960?], and Beirut, 1984 edn. [based on the 1895 edn.?].

Kimerling, "Adding the Qibla to Geographic Education": A. Kimerling, "Adding the Qibla to Geographic Education", *Journal of Geography* 101 (2002), pp. 20-26.

♦ King, "Architecture and Astronomy": David A. King, "Architecture and Astronomy: The Ventilators of Medieval Cairo and their Secrets", *JAOS* 104:1 (1984), pp. 97-133. [A new version is included here as **VIIb**.]

—, "Andalusī Astronomical Instruments": *idem*, "[Andalusī Astronomical Instruments]", in *Madrid & New York 1992 Exhibition Catalogue*, pp. 376-383.

— , "Andalusī Sundials": see *idem*, *Studies*, B-XV, also *idem*, "Cuadrantes solares andalusíes".

♦ — , "Aspekte": *idem*, "Aspekte angewandter Wissenschaften in Moscheen und Klöstern", *Berichte zur Wissenschaftsgeschichte* 18 (1995), pp. 85-95 and 137-149. [A translation of "Aspects of Science in Mosques and Monasteries", unpublished paper presented at the conference "Science and Religion in Islam, Judaism and Christendom" organised by Prof. David C. Lindberg, Madison, Wisc., April 15-17, 1993. [A new version is included here as **IX**.]

— , "Astronomical Alignments in Medieval Islamic Religious Architecture": see *idem*, *Studies*, C-XIII.

— , "Astronomical Instruments between East and West": *idem*, "Astronomical Instruments between East and West", in *Kommunikation zwischen Orient und Okzident – Alltag und Sachkultur*, Harry Kühnel, ed., (*Sitzungsberichte der Philosophisch-Historischen Klasse der Österreichischen Akademie der Wissenschaften*, Band 619, *Veröffentlichungen des Instituts für Realienkunde des Mittelalters und der Frühen Neuzeit*, Nr. 16), Vienna: Österreichische Akademie der Wissenschaften, 1994, pp. 143-198.

— , "Astronomical Timekeeping in Medieval Cairo": see *idem*, *Studies*, A-IX.

— , "Astronomical Timekeeping in Ottoman Turkey": see *idem*, *Studies*, A-XII.

— , "Astronomical Timekeeping in Medieval Syria": see *idem*, *Studies*, A-X.

— , "Astronomy in Fatimid Egypt": *idem*, "Aspects of Fatimid Astronomy: From Hard-Core Mathematical Astronomy to Architectural Orientations in Cairo", in Marianne Barrucand, ed., *L'Égypte Fatimide: son art et son histoire – Actes du colloqie organisé à Paris les 28, 29 et 30 mai 1998*, Paris: Presses de l'Université de Paris-Sorbonne, 1999, pp. 497-517.

— , "Astronomy of the Mamluks": see *idem*, *Studies*, A-III.

— , "Astronomy in the Maghrib": *idem*, "On the History of Astronomy in the Medieval Maghrib", in *Études Philosophiques et Sociologiques Dédiées à Jamal ed-Dine Alaoui*, Université Sidi Mohamed Ben Abdallah, Publications de la Faculté des Lettres et des Sciences Humaines Dhar El Mahraz – Fès, N° Spécial 14 (Département de Philosophie, Sociologie et Psychologie), Fez, 1998 [published 1999], pp. 27-61.

— , "Astronomie en Syrie": *idem*, "L'astronomie en Syrie à l'époque islamique", in *Paris IMA 1993-94 Exhibition Catalogue*, pp. 386-395 and 432-443.

— , *Astronomy in Yemen*: *idem*, *Mathematical Astronomy in Medieval Yemen – A Bio-Bibliographical Survey*, (Publications of the American Research Center in Egypt), Malibu, Ca.: Undena Publications, 1983.

— ,"al-Bazdawī on the Qibla in Tranoxania": *idem*, "al-Bazdawī on the Qibla in Early Islamic Tranoxania", *JHAS* 7 (1983/1986), pp. 3-38.

— , "Byzantine Astronomy": *idem*, an essay review of Pingree, *Astronomical Works of Gregory Chioniades*, and Jones, "Byzantine Astronomical Manual", in *Isis* 82 (1991), pp. 116-118.

— , "Catalogue of Medieval Astronomical Instruments", A-B: *idem*, "Medieval Astronomical Instruments: A Catalogue in Preparation", *Bulletin of the Scientific Instrument Society* no. 31 (Dec., 1991), pp. 3-7, and "1992 – A Good Year for Medieval Astronomical Instruments", *ibid.* no. 36 (March, 1993), pp. 17-18. [For a table of contents of the Frankfurt catalogue see www.uni-frankfurt.de/fb13/ign/instrument-catalogue.html.]

– , "Cuadrantes solares andalusíes": *idem*, "Los cuadrantes solares andalusíes", in *Madrid MAN 1992 Exhibition Catalogue*, pp. 89-102.

— , *Ciphers of the Monks*: *idem*, *The Ciphers of the Monks – A Forgotten Number Notation of the Middle Ages*, (*Boethius: Texte und Abhandlungen zur Geschichte der Mathematik und der Naturwissenschaften*, Band 44), Stuttgart: Franz Steiner, 2001.

—, "Earliest Muslim Geodetic Measurements": *idem*, "Too Many Cooks ... – A Newly-Rediscovered Account of the First Muslim Geodetic Measurements", *Suhayl* 1 (2000), pp. 207-241.
—, "Earliest Qibla Methods": see *idem*, *Studies*, C-XIV.
—, "Geography of Astrolabes": *idem*, "Bringing Astronomical Instruments Back to Earth: The Geographical Data on Medieval Astrolabes (to *ca.* 1100)", in *North Festschrift*, pp. 3-53.
—, *Ibn Yūnus*: *idem*, *The Astronomical Works of Ibn Yūnus*, unpublished doctoral dissertation, Yale University, 1972. [Available from ProQuest.com (formerly University Microfilms, Ann Arbor, Mich.), no. 7229740]
—, "Ibn Yūnus and the Pendulum": *idem*, "Ibn Yūnus and the Pendulum: A History of Errors", *AIHS* 29 (1979), pp. 35-52.
—, "Islamic Astronomical Tables": see *idem*, *Studies*, A-I.
—, "Islamic Astronomy", A-B: *idem*, "Islamic Astronomy", in *Astronomy before the Telescope*, Christopher B. F. Walker, ed., London: British Museum Press, 1996, repr. 1999, pp. 143-174 (A); and "Mathematical Astronomy in Islamic Civilisation", in Selin, *Astronomy across Cultures*, pp. 585-613 (B).
—, "Islamic Folk Astronomy", A-B: *idem*, "Folk Astronomy in the Service of Religion: The Case of Islam", in *Astronomies and Cultures*, Clive L. N. Ruggles & Nicholas J. Saunders, eds., Niwot, Co.: University Press of Colorado, 1993 [published 1994], pp. 124-138 (A), with a summary in *Archaeoastronomy in the 1990s*, Clive L. N. Ruggles, ed., Loughborough (U.K.): Group D Publications, Ltd., 1993, p. 346; and *idem*, "Applications of Folk Astronomy and Mathematical Astronomy to Aspects of Muslim Ritual", *The Arabist (Budapest Studies in Arabic)*, 13-14 (1995) (*Proceedings of the XIVth UEAI Congress, Budapest, 1988*, Alexander Fodor, ed., Part I), pp. 251-274 (B).
—, "Islamic Scientific Manuscripts": *idem*, "Some Remarks on Islamic Scientific Manuscripts and Instrument and Past, Present and Future Research", in John Cooper, ed., *The Significance of Islamic Manuscripts*, London: Al-Furqan Islamic Heritage Foundation, 1992, pp. 115-144 (also available in Arabic).
—, "Islamic Trigonometric Tables": *idem*, "A Brief Survey of Medieval Islamic Trigonometric Tables", paper delivered at the Colloquium "Innovation et transmission dans les mathématiques arabes", Paris, Institut du Monde Arabe, 25-27 Oct., 1990. [As yet unpublished.]
—, "Italian Astrolabe": *idem*, "A Remarkable Italian Astrolabe from *ca.* 1300 – Witness to an Ingenious Islamic Tradition of Non-Standard Astrolabes", in *Miniati Festschrift*, pp. 29-52.
—, "Al-Khalīlī's Auxiliary Tables": see *idem*, *Studies*, A-XI.
—, "Al-Khalīlī's Qibla Table": see *idem*, *Studies*, A-XIII.
—, "al-Khwārizmī": *idem*, "al-Khwārizmī and New Trends in Mathematical Astronomy in the Ninth Century", *Occasional Papers on the Near East* (New York University, Hagop Kevorkian Center for Near Eastern Studies), 2 (1983), 43 pp.
—, "Kuwait Astrolabes": *idem*, "Early Islamic Astronomical Instruments in Kuwaiti Collections", in Arlene Fullerton & Géza Fehérvári, eds., *Kuwait: Art and Architecture – Collection of Essays*, Kuwait (no publisher stated), 1995, pp. 76-96.
—, "Lunar Equation Table": *idem*, "A Double-Argument Table for the Lunar Equation Attributed to Ibn Yūnus", *Centaurus* 18 (1974), pp. 129-146, repr. in *idem*, *Studies*, A-V.
—, *Mecca-Centred World-Maps*: *idem*, *World-Maps for Finding the Direction and Distance to Mecca: Innovation and Tradition in Islamic Science*, Leiden: E. J. Brill, and London: Al-Furqan Islamic Heritage Foundation, 1999.
—, "A Medieval Arabic Report on Algebra before al-Khwārizmī": *idem*, "A Medieval Arabic Report on Algebra before al-Khwārizmī", *Al-Masāq*: *Studia Arabo-Islamica Mediterranea* 1 (1988), pp. 25-32.
—, "Medieval Islamic Multiplication Tables", A-B: see *idem*, *Studies*, A-XIV and XV.
—, "Monumental Syrian Astrolabe": *idem*, "The Monumental Syrian Astrolabe in the Maritime Museum, Istanbul", in *Sayılı Memorial Volume*, II, pp. 729-735 and 10 pls.
♦ —, "Muwaqqits": *idem*, "On the Role of the Muezzin and the *Muwaqqit* in Medieval Islamic Society", in *Oklahoma 1992 & 1993 Conference Proceedings*, pp. 285-346. [A new version is presented here as **V**.]
—, "*Navicula*": see *idem*, "Universal Horary Dials".
—, "Neglected Astrolabe": *idem*, "The Neglected Astrolabe", in *Mathematische Probleme in Mittelalter – Der lateinische und arabische Sprachbereich*, Menso Folkerts, ed., Wiesbaden: Harrassowitz, 1996, pp. 45-55.
—, "Nuremberg Astrolabes": *idem*, "Die Astrolabiensammlung des Germanischen Nationalmuseums", in *Nuremberg GNM 1992-93 Exhibition Catalogue*, I, pp. 101-114, and II, pp. 568-602, 640-643.
♦ —, "Orientation of Islamic Architecture": *idem*, "The Orientation of Medieval Islamic Religious Architecture and Cities", *JHA* 26 (1995), pp. 253-274. [A new version is presented here as **VIIa**.]
—, "Origin of the Astrolabe": see *idem*, *Studies*, B-III.
—, "*Quadrans vetus*": *idem*, "A *Vetustissimus* Arabic Treatise on the *Quadrans Vetus*", *JHA* 33 (2002), pp. 237-255
—, "Review of Higton": *idem*, "Towards a History from Antiquity to the Renaissance of Sundials and Other

Instruments for reckoning Times by the Sun and Stars", an essay review of *Greenwich Sundial Catalogue*, and Higton, *Portable Sundials*, to appear in *AS* • (200•).

— , "Review of Hill, *al-Jazarī*": *idem*, "Medieval Mechanical Devices", an essay review of Donald R. Hill's 1974 translation of al-Jazarī's *Book of Knowledge of Ingenious Mechanical Devices*, *HS* 13 (1975), pp. 284-289, repr. in *idem*, *Studies*, B-XX.

— , "Review of *London Khalili Collection Catalogue*": *idem*, "Cataloguing Islamic Astronomical Instruments", essay review in *Bibliotheca Orientalis* 57 (2000), cols. 247-258.

— , "Review of Sezgin, *GAS*, V": *idem*, "On the Sources for the Study of Early Islamic Mathematics", an essay review in *JAOS* 99 (1979), pp. 450-459.

— , "Review of Sezgin, GAS, VI": *idem*, "Early Islamic Astronomy", an essay review in *JHA* 12 (1981), pp. 55-59.

— , "Review of G. Turner, *Elizabethan Instruments*": review in *Journal of the History of Collections* (Oxford) 15:1 (2003), pp. 147-150.

— , "Review of Varisco & Smith, eds., *Al-Afḍal's Anthology*, in *Yemeni Update* 44 (2002) – see www.aiys.org/webdate/kngrev.html.

— , *The Sacred Geography of Islam*: *idem*, *The World about the Kaaba – The Sacred Folk Geography of Medieval Islam*, to be submitted to E. J. Brill.

♦ — , **SATMI**: This abbreviation has been used by the author since the mid 1970s to refer to the unpublished manuscript on which part of the present work (**I-II**) is based.

♦ — , "Shadow Schemes": *idem*, "A Survey of Medieval Islamic Shadow Schemes for Simple Timereckoning", *Oriens* 32 (1990), pp. 191-249. [A new version is presented here as **III**.]

— , "The *Shakkāzīya* Quadrant of al-Māridīnī": see *idem*, *Studies*, B-X.

— , "Star-Names on European Astrolabes": *idem*, "The Stars-Names on Three 14th-Century Astrolabes from Spain, France and Italy", in *Kunitzsch Festschrift*, pp. 307-333.

♦ — , "Strumentazione astronomica nel mondo medievale islamico", a chapter in *Gli strumenti*, Gerard L'E. Turner, ed., Turin: Giulio Einaudi, 1991, pp. 154-189 and pp. 581-585 (bibliography). [A new version of the English original is presented here as **X**.]

— , *Studies*, A: *idem*, *Islamic Mathematical Astronomy*, London: Variorum, 1986, 2nd rev. edn., Aldershot: Variorum, 1993. Includes reprints of:

I "On the Astronomical Tables of the Islamic Middle Ages", *Studia Copernicana* 13 (1975), pp. 37-56.
III "The Astronomy of the Mamluks", *Isis* 74 (1983), pp. 531-555.
V "A Double-Argument Table for the Lunar Equation Attributed to Ibn Yūnus" *Centaurus* 18 (1974), pp. 129-146.
VI "Ibn al-Majdī's Tables for Calculating Ephemerides" (with E. S. Kennedy), *JHAS* 4 (1980), pp. 48-68.
VII "Some Astronomical Observations from Thirteenth-Century Egypt" (with Owen Gingerich), *JHA* 13 (1982), pp. 121-128.
VIII "Indian Astronomy in Fourteenth-Century Fez: The Versified Zīj of al-Qusuntīnī [*sic*: read al-Qusanṭīnī]" (with E. S. Kennedy), *JHAS* 6 (1982), pp. 3-45.
IX "Ibn Yūnus' *Very Useful Tables* for Reckoning Time by the Sun", *AHES* 10 (1973), pp. 342-394.
X "Astronomical Timekeeping in Fourteenth-Century Syria", *Aleppo 1976 Symposium Proceedings*, II, pp. 75-84.
XI "Al-Khalīlī's Auxiliary Tables for Solving Problems of Spherical Astronomy", *JHA* 4 (1973), pp. 99-110.
XII "Astronomical Timekeeping in Ottoman Turkey", *Istanbul 1977 Symposium Proceedings*, pp. 245-269.
XIII "Al-Khalīlī's Qibla Table", *JNES* 34 (1975), pp. 81-122.
XIV "On Medieval Islamic Multiplication Tables", *HM* 1 (1974), pp. 317-323.
XV "Supplementary Notes on Medieval Islamic Multiplication Tables", *HM* 6 (1979), pp. 405-417.
XX "Medieval Mechanical Devices" (an essay review of Donald R. Hill's 1974 translation of al-Jazarī's *Book of Knowledge of Ingenious Mechanical Devices*), *HS* 13 (1975).

— , *Studies*, B: *idem*, *Islamic Astronomical Instruments*, London: Variorum, 1987, repr. Aldershot: Variorum, 1995. Includes reprints of:

II "The Medieval Yemeni Astrolabe in the Metropolitan Museum of Art in New York", *ZGAIW* 2 (1985), pp. 99-122, and 4 (1987/88), pp. 268-269 (corrections).
III "On the Origin of the Astrolabe According to the Medieval Arabic Sources", *JHAS* 5 (1981), pp. 43-83.
VI "The ʿAbd al-A'imma Astrolabe Forgeries" (with Owen Gingerich and George Saliba), *JHA* 3 (1972), pp. 188-198.
VII "On the Early History of the Universal Astrolabe in Islamic Astronomy and the Origin of the Term *Shakkāzīya* in Medieval Scientific Arabic", *JHAS* 3 (1979), pp. 244-257.
VIII "The Astrolabe of ʿAlī al-Wadāʿī", previously unpublished.
IX "The Astronomical Instruments of Ibn al-Sarrāj", previously-unpublished paper presented at the Second International Symposium for the History of Arabic Science, Aleppo, 1979.
X "An Analog Computer for Solving Problems of Spherical Astronomy: The *Shakkāzīya* Quadrant of Jamāl al-Dīn al-Māridīnī", *AIHS* 24 (1974), pp. 219-242.
XI "New Light on the *Zīj al-Ṣafā'iḥ* of Abū Jaʿfar al-Khāzin", *Centaurus* 23 (1980), pp. 105-117.
XII "Ibn al-Shāṭir's *Ṣandūq al-Yawāqīt*: An Astronomical Compendium" (with Louis Janin), *JHAS* 1 (1977), pp. 187-256.
XIV "Osmanische astronomische Handschriften und Instrumente", first published 1985.
XV "Three Sundials from Islamic Andalusia", *JHAS* 2 (1978), pp. 358-392.
XVI "Le cadran solaire de la mosquée d'Ibn Ṭūlūn au Caire" (with Louis Janin), *JHAS* 2 (1978), pp. 331-357.

XVII "The Sundial on the West Wall of the Madrasa of Sultan Qaytbay in Jersualem" (with Archibald G. Walls), *art and architecture research papers* (London) 15 (1979), pp. 16-21.

XVIII "A Fourteenth-Century Tunisian Sundial for Regulating the Times of Muslim Prayer", in *Hartner Festschrift*, pp. 187-202.

— , *Studies*, C: *idem*, *Astronomy in the Service of Islam*, Aldershot: Variorum, 1993. Includes reprints of:

I "Science in the Service of Religion: The Case of Islam", *impact of science on society* (U.N.E.S.C.O.), no. 159 (1991), pp. 245-262.

II "Some Early Islamic Tables for Lunar Crescent Visibility", in *Kennedy Festschrift*, pp. 185-225.

III "Ibn Yūnus on Lunar Crescent Visibility", *JHA* 19 (1988), pp. 155-168.

IV "Lunar Crescent Visibility Predictions in Medieval Islamic Ephemerides", in *Quest for Understanding – Arabic and Islamic Studies in Memory of Malcolm H. Kerr*, S. Seikaly, R. Baalbaki, P. Dodd, eds., Beirut: American University of Beirut, 1991, pp. 233-251.

♦ VI "Universal Solutions in Islamic Astronomy", in *Aaboe Festschrift*, pp. 121-132. [A new version is included here as **VIa**.]

♦ VII "Universal Solutions to Problems of Spherical Astronomy from Mamluk Egypt and Syria", in *The Way Prepared: Essays on Islamic Culture in Honor of R. Bayly Winder*, Farhad Kazemi and Robert B. McChesney, eds., New York: New York University Press, 1987, pp. 153-184. [A new version is included here as **VIb**.]

XII "On the Orientation of the Kaʿba" (with Gerald S. Hawkins), *JHA* 13 (1982), pp. 102-109.

XIII "Astronomical Alignments in Medieval Islamic Religious Architecture", *Annals of the New York Academy of Sciences* 385 (1982), pp. 303-312.

XIV "The Earliest Mathematical Methods and Tables for Finding the Direction of Mecca", *ZGAIW* 3 (1986), pp. 82-149, with corrections listed *ibid.* 4 (1987/88), p. 270.

— , "Tables for Lunar Crescent Visibility": see *idem*, *Studies*, C-II.

— , "Three Sundials from Islamic Andalusia": see *idem*, *Studies*, B-XV.

— , "The Toledo Astrolabe": *idem*, "An Astrolabe from 14th-Century Christian Spain with Inscriptions in Latin, Hebrew and Arabic – A Unique Testimonial to an Intercultural Encounter", *Suhayl* 3 (2002/03), pp. 9-156.

— , "Tunisian Sundial": see *idem*, *Studies*, B-XVIII.

— , "Universal Astrolabe": see *idem*, *Studies*, B-VII.

♦ — , "Universal Horary Dials", *idem*, "On Universal Horary Dials for Determining Time by the Sun and Stars", in **XIIb**. [A summary is in King, "*Navicula*": "14th-Century England or 9th-Century Baghdad? New Insights on the Origins of the Elusive Astronomical Instrument Called the *Navicula de Venetiis*", in *Goldstein Festschrift*.]

♦ — , "Universal Horary Quadrant": *idem*, "The Universal Horary Quadrant for Timekeeping by the Sun", in **IXa**. [A shorter version is "A *Vetustissimus* Arabic Treatise on the *Quadrans Vetus*", *JHA* 33 (2002), pp. 237-255.]

— , "Universal Quadrant of al-Māridīnī": see *idem*, *Studies*, B-X.

♦ — , "Universal Solutions in Islamic Astronomy": see *idem*, *Studies*, C-VI.

♦ — , "Universal Solutions from Mamluk Egypt and Syria": see *idem*, *Studies*, C-VII.

— , "Urbino Astrolabe": *idem*, "The Astrolabe Depicted in the Intarsia of the *Studiolo* of Archduke Federico in Urbino", in *La scienza del Ducato di Urbino – The Science of the Dukedom of Urbino*, Flavio Vetrano, ed., Urbino: Accademia Raffaello, 2001, pp. 101-139.

— , "Yemeni Astrolabe": see *idem*, *Studies*, B-II.

— , "The *Zīj al-Ṣafāʾiḥ*" : see *idem*, *Studies*, B-XI.

— & Charette, *The Universal Astrolabe of Ibn al-Sarrāj*: *idem* and François Charette, *The Universal Astrolabe of Ibn al-Sarrāj: Innovation in Medieval Islamic Science*, to appear.

King & Gingerich, "Astronomical Observations from 13th-Century Egypt": *idem* and Owen Gingerich, "Some Astronomical Observations from Thirteenth-Century Egypt", *JHA* 13 (1982), pp. 121-128, repr. in King, *Studies*, A-VII.

♦ – with Dorothea Girke, "An Approximate Trigonometric Formula for Astronomical Timekeeping and Related Tables, Sundials and Quadrants from *ca.* 500 to *ca.* 1900", *Johann Wolfgang Goethe-Universität Frankfurt am Main, Institut für Geschichte der Naturwissenschaften, Preprint Series*, 1 (1988). [A new version is included here as **XI**.]

— & Kennedy, "Ibn al-Majdī's Tables": *idem* and E. S. Kennedy, "Ibn al-Majdī's Tables for Calculating Ephemerides", *JHAS* 4 (1980), pp. 48-68, repr. in King, *Studies*, A-VI.

— & Maier, "London Catalan Astrolabe": *idem* and Kurt Maier, "The Medieval Catalan Astrolabe of the Society of Antiquaries, London", in *Vernet Festschrift*, II, pp. 673-718.

— & Samsó, "Islamic Astronomical Handbooks and Tables": *idem* and Julio Samsó, with a contribution by Bernard R. Goldstein, "Astronomical Handbooks and Tables from the Islamic World (750-1900): An Interim Report", *Suhayl* 2 (2001), pp. 9-105. [A summary is in the article "Zīdj" in EI_2 (illustrated).]

— & Turner, "Regiomontanus' Astrolabe": *idem* and Gerard L'E. Turner, "The Astrolabe Presented by Regiomontanus to Cardinal Bessarion in 1462", *Nuncius – Annali di Storia della Scienza* (Florence) 9 (1994), pp. 165-206, repr. in Turner, *Studies*.

— & Walls, "Jerusalem Sundial": see *idem*, *Studies*, B-XVII.

King: see also *Cairo ENL Catalogue* and *Survey*, and *Kennedy Festschrift*.

King-*père*, *Geared to the Stars*: Henry C. King with John R. Milburn, *Geared to the Stars*, Toronto: University of Toronto Press, 1978.

Klemm, *Georg Hartmann*: Hans Gunther Klemm, *Georg Hartmann aus Eggolsheim (1489-1564) – Leben und Werk eines fränkischen Mathematikers und Ingenieurs*, Forschheim (D): Ehrenbürg-Gymnasium (*Wissenschaftliche und künstlerische Beiträge*, Heft 8), 1990, 104 pp.

Knobel, *Ulughbeg's Catalogue of Stars*: E. B. Knobel, *Ulughbeg's Catalogue of Stars*, Washington, D.C.: Carnegie Institution, 1917.

Knorr, "*Quadrans Vetus*": Wilbur R. Knorr, "The Latin Sources of *Quadrans Vetus*, and What They Imply for Its Authorship and Date", in Edith Sylla and Michael McVaugh, eds., *Texts and Contexts in Ancient and Medieval Science – Studies on the Occasion of John E. Murdoch's Seventieth Birthday*, Leiden, *etc.*: E. J. Brill, 1997, pp. 23-67.

—, "Sacrobosco's *Quadrans*": *idem*, "Sacrobosco's *Quadrans*: Date and Sources", *Journal for the History of Astronomy* 28 (1997), pp. 187-222.

Kragten, *Greenwich Navicula*: Jan Kragten, *The Venetian Ship of Sibton Abbey*, Eindhoven (NL), published privately, 1992.

—, *Navicula*: *idem*, *The Little Ship of Venice* [*sic*] *(Navicula de Venetiis)*, Eindhoven (NL), published privately, 1989.

Krause, "Stambuler Handschriften": Max Krause, "Stambuler Handschriften islamischer Mathematiker", *Quellen und Studien zur Geschichte der Mathematik, Astronomie und Physik*, B 3:4 (1936), pp. 437-532.

Krause: see also al-Bīrūnī, *Qānūn*.

Kremer & Dobrzycki, "Peurbach and Marāgha Astronomy?": Richard L. Kremer and Jerzy Dobrzycki, "Peurbach and Marāgha Astronomy? The Ephemerides of Johannes Angelus and their Implications", *JHA* 27 (1996), pp. 187-237.

Kren, *Bibliography of Medieval Science and Technology*: Claudia Kren, *Medieval Science and Technology – A Selected, Annotated Bibliography*, New York and London: Garland Publishing, Inc., 1985.

—, "The Traveller's Dial": *eadem*, "The Traveller's Dial in the Late Middle Ages: The Chilinder", *Technology and Culture* 18 (1977), pp. 419-435;

Kunitzsch, "*Anwā*ʾ-Tradition": Paul Kunitzsch, "Über eine *anwā*ʾ-Tradition mit bisher unbekannten Sternnamen", *Sitzungsberichte der Bayerischen Akademie der Wissenschaften, Phil.-hist. Klasse*, Heft 5, (*Beiträge zur Lexikographie des Classischen Arabisch*, Nr. 4), Munich, 1983.

—, *Arabische Sternnamen in Europa*: *idem*, *Arabische Sternnamen in Europa*, Wiesbaden: Otto Harrassowitz, 1959.

—, "Fachausdrücke der Astrolabliteratur": *idem*, "Glossar der arabischen Fachausdrücke in der mittelalterlichen europäischen Astrolabliteratur", *Nachrichten der Akademie der Wissenschaften in Göttingen*, I. Philologisch-Historische Klasse, Jahrgang 1982, Nr. 11, pp. 455-571. [The separatum is paginated 1-117.]

—, "Ibn Ḥabīb's *Book on the Stars*": *idem*, "ʿAbd al-Malik ibn Ḥabīb's *Book on the Stars*", *ZGAIW* 9 (1994), pp. 161-194.

–, "Al-Khwārizmī as a Source for the *Sententie astrolabii*": *idem*, "Al-Khwārizmī as a Source for the *Sententie astrolabii*", in *Kennedy Festschrift*, pp. 227-236, repr. in *idem*, *Studies*, IX.

—, "Messahalla on the Astrolabe: *idem*, "On the Authenticity of the Treatise on the Composition and Use of the Astrolabe Ascribed to Messahalla", *AIHS* 31 (1981), pp. 42-62.

—, *Sternkatalog des Almagest*: *idem*, *Claudius Ptolemäus – Der Sternkatalog des Almagest – Die arabisch-mittelalterliche Tradition*, 3 vols., Wiesbaden: Otto Harrassowitz, 1986-1991.

—, *Sternnomenklatur der Araber*: *idem*, *Untersuchungen zur Sternnomenklatur der Araber*, Wiesbaden: Otto Harrassowitz, 1961.

—, *Sternverzeichnisse*: *idem*, *Typen von Sternverzeichnissen in astronomischen Handschriften des zehnten bis vierzehnten Jahrhunderts*, Wiesbaden: Otto Harrassowitz, 1966.

—, *Studies*: *idem*, *The Arabs and the Stars*, Northampton: Variorum, 1989.

—, "Al-Ṣūfī and the Astrolabe Stars": *idem*, "Al-Ṣūfī and the Astrolabe Stars", *ZGAIW* 6 (1990), pp. 151-166.

— & Dekker, "The Stars on the Carolingian Astrolabe": *idem* and Elly Dekker, "The Stars on the Rete of the So-Called "Carolingian Astrolabe"", in *Vernet Festschrift*, II, pp. 655-672.

Kunitzsch & Lorch, "*Matāliʿ al-samt*": *idem* and Richard Lorch, "Abū Naṣr and Ḥabash on *matāliʿ al-samt*", *ZGAIW* 9 (1994), pp. 43-82.

Kunitzsch Festschrift: *Sic itur ad astra. Studien zur Geschichte der Mathematik und Naturwissenschaften. Festschrift für den Arabisten Paul Kunitzsch zum 70. Geburtstag*, Menso Folkerts and Richard P. Lorch, eds., Wiesbaden: Otto Harrassowitz, 2000.

Kunitzsch: see also Kennedy & Kunitzsch & Lorch.

Kurz, *European Clocks*: Otto Kurz, *European Clocks and Watches in the Near East*, London: The Warburg Institute, and Leiden: E. J. Brill, 1975.

Labarta & Barceló, "Un nuevo fragmento de reloj de sol andalusí": Ana Labarta and Carmen Barceló, "Un nuevo fragmento de reloj de sol andalusí", *al-Qanṭara* (Madrid) 16 (1995), pp. 147-150.

— , "Ocho relojes de sol hispanomusulmanes", *eaedem*, "Ocho relojes de sol hispanomusulmanes", al-*Qanṭara* (Madrid) 9 (1988), pp. 231-247.

Lacroix, *Science in the Middle Ages*: Paul Lacroix, *Science and Literature in the Middle Ages and Renaissance*, translated from the French, first published in 1878, repr. New York: Frederick Ungar, 1964.

de La Lande, *Bibliographie Astronomique*: Jérôme de La Lande, *Bibliographie Astronomique ; avec L'Histoire de l'Astronomie jusqu'à 1802*, Paris: Imprimerie de la République, An XI = 1803.

Landes, *Revolution in Time*: David S. Landes, *Revolution in Time – Clocks and the Making of the Modern World*, Cambridge, Mass, and London: The Belknap Press of Harvard University Press, 1983

Lane, *Lexicon*: Edward W. Lane, *An Arabic-English Lexicon*, 8 pts., London: Williams and Norgate, 1863, repr. Beirut: Librairie du Liban, 1968.

Langermann, *Ibn al-Haytham's Cosmology*: Y. Tzvi Langermann, *Ibn al-Haytham's* On the Configuration of the World, (Series: Harvard Dissertations in the History of Science), New York & London: Garland Publishing, Inc., 1990.

— , *Studies*: *idem*, *The Jews and the Sciences in the Middle Ages*, Aldershot: Variorum, 1999.

Lee, "Astronomical Tables of Al Farsi": Samuel Lee, "Notice of the Astronomical Tables of Mohammed Abibekr Al Farsi ... ... ", *Transactions of the Cambridge Philosophical Society* 1 (1822), pp. 249-265.

Lehr, *De Geschiedenis van het Astronomisch Kunstuurwerk*: André Lehr, *De Geschiedenis van het Astronomisch Kunstuurwerk – Zijn techniek en muziek*, The Hague: Martinus Nijhoff, 1981.

Leipzig Catalogue: K. Vollers, *Katalog der islamischen ... Handschriften der Universitätsbibliothek zu Leipzig* ... , Leipzig: Otto Harrassowitz, 1906, reprinted Osnabrück: Biblio Verlag, 1975.

Le Strange, *Lands of the Eastern Caliphate*: Guy Le Strange, *The Lands of the Eastern Caliphate, Mesopotamia, Persia and Central Asia from the Moslem Conquest to the Time of Timur*, Cambridge: Cambridge Universty Press, 1905, repr. London: Frank Cass & Co. Ltd., 1966, and again in *Islamic Geography*, vol. 85 (1993).

Lesley, Mark, "Bīrūnī on Rising Times and Daylight Lengths", *Centaurus* 5 (1957), pp. 121-141, repr. in Kennedy *et al.*, *Studies*, pp. 253-273.

Levin, "Direction-Facing in Judaism and Islam": Daniel Z. Levin, "Which Way is Jerusalem? Which Way is Mecca? The Direction-Facing Problem in Religion and Geography", *Journal of Geography* (National Council for Geographic Education, Washington, D.C.) 101 (2002), pp. 27-37.

Lexikon des Mittelalters, 9 vols., Munich and Zurich: Artemis, 1980-99, especially articles "Astrologie", "Astronomie", "Instrumente", "Komputistik", "Mathematik".

Lindberg, David C., *The Beginnings of Western Science*, Chicago: University of Chicago Press, 1992.

Linton Collection Catalogue: *Scientific Instruments and Rare Books from the Collection of Leonard Linton, Point Lookout, N.Y.*, Paris: Alain Brieux, 1980.

Lippincott, "*Navicula*": Kristen Lippincott, "The Navicula Sundial", *Bulletin of the Scientific Instrument Society*, no. 35 (Dec., 1992), p. 22. [On the *navicula* now in Greenwich.]

Livingston, "Islamic Conical Sundial": John W. Livingston, "The *Mukḥula*, an Islamic Conical Sundial", *Centaurus* 16 (1972), pp. 299-308.

London BL Catalogue: (W. Cureton and C. Rieu), *Catalogus codicum manuscriptorum orientalium qui in Museo Britannico asservantur*, Part II: *Codices arabicos amplectens*, London: British Museum, 1871.

London BM Catalogue: F. A. B. Ward, *A Catalogue of European Scientific Instruments in the Department of Medieval and Later Antiquities of the British Museum*, London: British Museum Publications, 1981. [See also the essay review in Price, "Instruments in the British Museum".]

London Khalili Collection Catalogue: *The Nasser D. Khalili Collection of Islamic Art*, vol. XII: *Science, Tools and Magic*, by Francis Maddison and Emilie Savage-Smith, 2 pts., London: The Nour Foundation, in association with Azimuth Editions and Oxford University Press, 1997. [See King, "Cataloguing Islamic Instruments", for a review.]

London SM 1976 Exhibition Catalogue: Francis R. Maddison and Anthony J. Turner, *Catalogue of an Exhibition 'Science and Technology in Islam' ... , 1976*, [a catalogue of the exhibition The World of Islam, held at The Science Museum, London, in 1976], privately distributed [but unfortunately never published].

Lorch, "Astronomical Instruments of Jābir ibn Aflaḥ": see Richard P. Lorch, *Studies*, XVI.

— , "Al-Khāzinī's 'Sphere that Rotates by Itself'": see *idem*, *Studies*, XI.

— , "Mischastrolabien": Richard P. Lorch, "Mischastrolabien im arabisch-islamischen Kulturbereich", in *Frankfurt IGN Festschrift*, pp. 231-236.

— , "Rudolf of Bruges on the Astrolabe": *idem*, "The Treatise on the Astrolabe by Rudolf of Bruges", in *North Festschrift*, pp. 55-100.

— , "Al-Ṣaghānī on Projecting the Sphere": see *idem*, *Studies*, XVIII.

— , "Sine Quadrant": *idem*, "Some Early Applications of the Sine Quadrant", *Suhayl* 1 (2000), pp. 251-272.

— , "The *sphaera solida*": see *idem*, *Studies*, XII.

—, *Studies*: *idem*, *Arabic Mathematical Sciences – Instruments, Texts, Transmission*, Aldershot: Variorum, 1995. Contains *inter alia*:
XI "Al-Khāzinī's "Sphere that Rotates by Itself"", *JHAS* 4 (1980), pp. 287-329.
XII "The *sphaera solida* and Related Instruments", *Centaurus* 24 (1980), pp. 153-161.
XIII "Ḥabash al-Ḥāsib's Book on the Sphere and its Use" (with Paul Kunitzsch), *ZGAIW* 2 (1985), pp. 68-98.
XIV "The *Qibla*-Table Attributed to al-Khāzinī", *JHAS* 4 (1980), pp. 259-264.
XVI "The Astronomical Instruments of Jābir ibn Aflaḥ and the Torquetum", *Centaurus* 20 (1976), pp. 11-34.
XVII "A Note on the Horary Quadrant", *JHAS* 5 (1981), pp. 115-120.
XVIII "Al-Ṣaghānī's Treatise on Projecting the Sphere", in *Kennedy Festschrift*, pp. 237-252.

— , "Universal Horary Quadrant": see *idem*, *Studies*, XVII.

— & Kennedy, "Ḥabash al-Ḥāsib on the Melon Astrolabe": see Kennedy, *Studies*, I.

— & Kunitzsch, "Ḥabash on the Sphere": see *idem*, *Studies*, XIII.

Lorch: see also Kennedy & Kunitzsch & Lorch, and Kunitzsch & Lorch.

Luckey, "Das Analemma von Ptolemäus": Paul Luckey, "Das Analemma von Ptolemäus", *Astronomische Nachrichten*, Bd. 230, Nr. 5498 (1927), pp. 18-46.

—, "Beiträge", I/II: *idem*, "Beiträge zur Erforschung der islamischen Mathematik", *Orientalia*, N. S., 17 (1948), pp. 490-510, and 22 (1953), pp. 166-189, repr. in *IAM*, vols 96 and 56.

— , *Ibrāhīm ibn Sinān über die Sonnenuhren*: Jan Pieter Hogendijk, ed., *Die Schrift des Ibrāhīm b. Sinān b. Ṯābit über die Schatteninstrumente, übersetzt und erläutert von Paul Luckey (Phil. Diss. Tübingen, 1941)*, *Islamic Mathematics and Astronomy* vol. 101 (1999).

— , "Thābit ibn Qurra über die Sonnenuhren": *idem*, "Thābit b. Qurra's Buch über die ebenen Sonnenuhren", *ibid.*, Abt. B, Bd. 4 (1937-38), pp. 95-148, repr. in *IMA*, vol. 22 (1997), pp. 95-148.

— , "Zur Entstehung der Kugeldreiecksrechnung", *Deutsche Mathematik* 5 (1940), pp. 405-446, repr. in *IMA*, vol. 77 (1998), pp. 137-178.

MacKay, "The Qibla of Ottoman Mosques in Greece": Pierre MacKay, "The Qibla of Ottoman Mosques in Greece", paper presented at the Annual Meeting of the American Oriental Society, San Francisco, April, 1980.

Maddison, "Early Astronomical Instruments": Francis R. Maddison, "Early Astronomical and Mathematical Instruments ... ", *HS* 2 (1963), pp. 17-50.

— , "15th-Century Spherical Astrolabe": *idem*, "A Fifteenth-Century Spherical Astrolabe", *Physis* 4 (1962), pp. 101-109.

— , "Instruments arabes": *idem*, "Observatoires portatifs: les instruments arabes à usage pratique", in *HSA*, I, pp. 139-172 (not in the English version, *EHAS*).

— , "Locks": *idem*, "al-Jazarī's Combination Lock: Two Contemporary Examples", in *The Art of Syria and the Jazīra 1100-1250* (*Oxford Studies in Islamic Art* 1), Julian Raby, ed., Oxford: Oxford University Press, 1985, pp. 141-157.

— , "Medieval Instruments": *idem*, "Medieval Scientific Instruments and the Development of Navigational Instruments in the XVth and XVIth Centuries", *Revista da Universidade de Coimbra* 24 (1969), printed separately, Agrupamento de Estudos de Cartografia Antiga, XXX, Secção de Coimbra, 1969.

— , "Rojas Astrolabe Projection": *idem*, "Hugo Helt and the Rojas Astrolabe Projection", *Revista da Faculdade de Ciências* (Coimbra) 39 (1966), pp. 195-251 and 2 fold-outs.

Maddison: see also *London Khalili Catalogue*, *London SM Exhibition Catalogue*, and *Oxford Billmeir Supplement*.

Madrid 1980 Conference Proceedings: *Historia de la ciencia árabe*, Madrid: Real Academía de Ciencias Exactas, Fîsicas y Naturales, 1981.

Madrid MAN 1992 Exhibition Catalogue: *El legado científico andalusí*, Juan Vernet and Julio Samsó, eds., Madrid: Ministerio de Cultura, 1992. [A catalogue of an exhibition at the Museo Arqueológico Nacional, Madrid, April-June, 1992.]

Madrid & New York 1992 Exhibition Catalogue: Jerrilynn D. Dodds, ed., *Al-Andalus – The Art of Islamic Spain*, New York: The Metropolitan Museum of Art, 1992. [Catalogue of an exhibition held at the Alhambra, Granada, and then at The Metropolitan Museum of Art, New York, in 1992.]

Maistrov, *Nauchnye pribory*: L. E. Maistrov, *Nauchnye pribory*, Moscow: Nauk. 1968. [A catalogue in Russian of scientific instruments in collections in the former Soviet Union.]

Manchester Catalogue: Alois Mingana, *Catalogue of the Arabic Manuscripts in the John Rylands Library*, Manchester: Manchester University Press, 1934.

Maranyi, "Zwischen *ʿaṣr* and *maġrib* in Mekka": Miklos Maranyi, "Zwischen *ʿaṣr* and *maġrib* in Mekka – Ein Augenzeugenbericht von al-Laiṯ b. Saʿd aus dem Jahr 113/732", *Die Welt des Orients: wissenschaftliche Beiträge zur Kunde des Morgenlandes* (Göttingen: Vandenhoek & Ruprecht) 23 (1992), pp. 101-128.

al-Marrākushī, *A-Z of Astronomical Timekeeping*: Fuat Sezgin *et al.*, *Comprehensive Collection of Principles and Objectives in the Science of Timekeeping: Jāmiʿ al-mabādiʾ wa-'l-ghāyāt by al-Marrākushī*, (facsimile "edition" of MS Topkapı Ahmet III 3343), 2 vols., Frankfurt am Main: Institut für Geschichte der Arabisch-Islamischen Wissenschaften, 1984. [See also Sédillot-*père*, *Traité*, and Sédillot-*fils*, "Mémoire".]

Martí & Viladrich, "Tratados de astrolabio": Ramon Martí and Mercè Viladrich, "En torno a los tratados de uso del astrolabio hasta el siglo XIII en al-Andalus, la Marca Hispánica y Castilla", in Vernet, ed., *Nuevos estudios*, pp. 9-74.

Matvievskaya & Rosenfeld, *MAMS*: Galina P. Matvievskaya and Boris A. Rosenfeld, *Matematiki i astronomi musulmanskogo srednevekovya i ikh trudi*, 3 vols., Moscow: Nauk, 1983.

S. & K. Maurice, "Counting the Hours": Sigrid and Klaus Maurice, "Counting the Hours in Community Life of the 16th and 17th Centuries", in Maurice & Mayr, eds., *The Clockwork Universe*, pp. 144-156.

Maurice & Mayr, eds., *The Clockwork Universe*: *The Clockwork Universe – German Clocks and Automata 1550-1650*, Klaus Maurice and Otto Mayr, eds., New York: Neale Watson, 1980.

Mayer, *Islamic Astrolabists*: Leo A. Mayer, *Islamic Astrolabists and Their Works*, Geneva: Albert Kundig, 1956, with a supplement in *Aus der Welt der islamischen Kunst*, Richard Ettinghausen, ed., Berlin: Gebrüder Mann, 1959, pp. 293-296.

McCluskey, *Medieval Astronomy*: Stephen C. McCluskey, *Astronomies and Cultures in Early Medieval Europe*, Cambridge, *etc.*: Cambridge University Press, 1998.

McKay, "How Far to Mecca?": J. McKay, "How Far to Mecca?", *Mercator's World* 5:5 (2002), p. 58.

Meliconi, "Brechte Plate" and "Oxford *Quadrans vetus*": see EPACT.

Mercier, "Astronomical Tables in the 12th Century": Raymond Mercier, "Astronomical Tables in the Twelfth Century", in Burnett, ed, *Adelard of Bath*, pp. 87-118.

—, "English Orientalists and Islamic Astronomy": *idem*, "English Orientalists and Mathematical Astronomy", in Russell, ed., *'Arabick' Interest in 17th-Century England*, pp. 158-214.

—, "Ibn al-Aʿlam's Parameters": *idem*, "The Parameters of the *Zīj* of Ibn al-Aʿlam", *AIHS* 39 (1989), pp. 22-50.

Mestres, "Hyderabad MS of the *Zīj* of Ibn Isḥāq": Àngel Mestres, "Maghribī Astronomy in the 13th Century: a Description of Manuscript Hyderabad Andra Pradesh State Library 298", in *Vernet Festschrift*, I, pp. 383-443.

—, Zīj *of Ibn Isḥāq*: *idem*, *Materials andalusins en el Zīj d'Ibn Isḥāq al-Tūnisī* (edited text and tables, with introduction and commentary in English), doctoral thesis, University of Barcelona, 2000.

Meucci, *Globo celeste arabico*: Ferdinando Meucci, *Il globo celeste arabico del secolo XI nel Gabinetto degli strumenti antichi di astronomia, di fisica e di matematica del R. Istituto di studi superiori*, Florence: Successori Le Monnier, 1878, repr. in *AIOS*, II, pp. 126-144 plus pullout.

Meyer, *Istanbul Sundials*: Wolfgang Meyer, *İstanbul'daki güneş saatleri*, *Sandoz kültür yayınları* (Istanbul) 7 (1985). [See also *idem*, "Sundials of the Osmanic Era in Istanbul", in *Istanbul 1977 Symposium Proceedings*, pp. 193-202.]

Meyerhof: see Frank & Meyerhof.

Michel, "L'astrolabe linéaire": Henri Michel, "L'astrolabe linéaire d'al-Tusi", *Ciel et terre* 59 (1943), pp. 101-107, repr. in *IAM*, vol. 94.

—, *Images des sciences*: *idem*, *Images des sciences: les anciens instruments scientifiques vus par les artisites de leur temps*, Rhode-St-Genèse (B): Albert de Visscher, 1977.

—, "Construction des astrolabes persans": *idem*, "Méthodes de trace et d'exécution des astrolabes persans", *Ciel et terre* 57 (1941), pp. 481-496.

—, *Scientific Instruments in Art and History*: *idem*, *Scientific Instruments in Art and History*, London: •••, 1967.

—, *Traité de l'astrolabe*: *idem*, *Traité de l'astrolabe*, Paris: Gauthier-Villars, 1947, reprinted with a new preface by Francis R. Maddison, Paris: Librarie Alain Brieux, 1989.

— & Ben-Eli, "Cadran solaire polaire": *idem* and A. Ben-Eli, "Un cadran solaire remarquable", *Ciel et terre* 81 (1965), pp. 214-216.

Mielgo, "Analyzing Mean Motion Tables": Honorino Mielgo, "A Method of Analysis for Mean Motion Astronomical Tables", in *Vernet Festschrift*, I, pp. 159-180".

Mielgo: see also Samsó & Mielgo.

Milenio, "Cuadrante Vetustissimus": M. Milenio, "Cuadrante Vetustissimus", *Analema* no. 30 (Septiembre-Diciembre 2000), pp. 20-24.

—, "Relojes de altura Regiomontano": *idem*, "Sobre el diseño de la forma de los relojes de altura Regiomontano y Capuchino", *Analema* no. 27 (Septiembre-Diciembre 1999), pp. 11-14.

Millás Vallicrosa, *Assaig*: José M. Millás Vallicrosa, *Assaig d'història de les idees fisiques i matemàtiques a la Catalunya medieval*, vol. I, (Estudis Universitaris Catalans, sèrie monogràfica I), Barcelona: 1931.

—, *Azarquiel*: *idem*, *Estudios sobre Azarquiel*, Madrid & Granada: Consejo Superior de Investigaciones Científicas and Instituto Miguel Asín, 1943-50.

—, "Cuadrante con cursor": *idem*, "La introducción del cuadrante con cursor en Europa", *Isis* 17 (1932), pp. 218-258, repr. in *idem*, *Estudios*, I, pp. 65-110.

— , *Don Profeit Tibbón sobre l'assafea*: *idem*, *Don Profeit Tibbón*: *Tractat de l'assafea d'Azarquiel*, Barcelona: Universitat de Barcelona, 1933.

— , "Un ejemplar de azafea árabe de Azarquiel", *Al-Andalus* 9 (1944), pp. 11-119.

— , *Estudios* and *Nuevos estudios*: *idem*, *Estudios sobre historia de la ciencia española*, Barcelona: Consejo Superior de Investigaciones Científicas, Instituto "Luis Vives" de Filosofía, 1949, and *idem*, *Nuevos estudios sobre historia de la ciencia española*, Barcelona: Consejo Superior de Investigaciones Científicas, Instituto "Luis Vives" de Filosofía, and Asociacion para la Historia de la Ciencia Española, 1960, both repr., 2 vols., Madrid: Consejo Superior de Investigaciones Científicas, 1987.

— , "Estudios sobre Azarquiel: El tratado de la azafea", *Archeion* 14 (1932), pp. 393-419.

Millás-Vallicrosa Festschrift: *Homenaje a Millás-Vallicrosa*, 2 vols., Barcelona: Consejo Superior de Investigaciones Cientificas, 1959.

Millás Vendrell, "Apogeos planetarios en el ecuatorio de Azarquiel": Eduardo Millás Vendrell, "Las posiciones de los apogeos planetarios en el ecuatorio de Azarquiel", in Vernet, ed., *Nuevos estudios*, pp. 119-123.

Millás: see also Samsó & Millás.

Mills, "Canterbury Dial": Allan Mills, "The Canterbury Pendant: A Saxon Seasonal-Hour Altitude Dial", *Bulletin of the British Sundial Society* 95.2 (June 1995), pp. 39-44.

Miniati Festschrift: *MUSA MUSAEI: Studies on Scientific Instruments and Collections in Honour of Mara Miniati*, Marco Beretta, Paolo Galluzzi and Carlo Triarico, eds., Florence: Leo S. Olschki, 2003.

Monzawi, *Persian Manuscripts*: Aḥmad Monzawi, *A Catalogue of Persian Manuscripts - Fihrist-i nuskhahā-yi kha-i fārisī* (in Persian), 6 vols., Tehran: Regional Cultural Institute, 1348-1353 HS = 1969-1974.

Morelon, "Early Eastern Arabic Astronomy": Régis Morelon, "Eastern Arabic Astronomy between the Eighth and Eleventh Centuries", in *EHAS*, I, pp. 20-57.

— , *Thābit ibn Qurra*: *idem*, *Thābit ibn Qurra – oeuvres d'astronomie*, Paris: Les Belles Lettres, 1987.

Morelon: see also *EHAS*.

Morley, "Arabic Quadrant": William H. Morley, "Description of an Arabic Quadrant", *Journal of the Royal Asiatic Society* 17 (1860), pp. 322-330 and two figs., repr. in *AIOS*, I, pp. 326-336.

— , "Astrolabe of Shāh Ḥusayn": *idem*, *Description of a Planispheric Astrolabe Constructed for Sháh Sultán Husain Safawí ...* , London: Williams and Norgate, 1856, repr. as the introduction to Gunther, *Astrolabes*, I, and again in *AIOS*, I, pp. 249-325.

Mortillaro, "Astrolabio arabo": Vincenzo Mortillaro, *Illustrazione di un astrolabio arabo-sicula* [*sic*], Palermo: Stamperia Oretea, 1848, also appeared as "Illustrazione di un astrolabio arabo del nono secolo", in *idem*, *Opere*, IV, Palermo, 1848, pp. 110-135, repr. from the latter in *AIOS*, I, pp. 162-191.

Müller, *Tables des hauteurs du soleil*: Fréderic Christophe Müller, *Tables des hauteurs du soleil pour toute l'Allemagne et les pays qui la bornent au levant et au couchant*, Leipzig: Siegfried Lebereclt Crusius, 1791.

Munich Catalogue: Burkhard Stautz, *Die Astrolabiensammlungen des Deutschen Museums und des Bayerischen Nationalmuseums*, Munich: Oldenbourg, 1999.

Murdoch, John E., *Album of Science – Antiquity and the Middle Ages*, New York: Charles Scribner's Sons, 1984.

Murray, Alexander, *Reason and Society in the Middle Ages*, (first published 1978), Oxford: Clarendon Press, 1991 edn.

Nadir, Nadi, "Abū al-Wafā' on the Solar Altitude", *The Mathematics Teacher* 53 (1960), pp. 460-463, repr. in Kennedy *et al.*, *Studies*, pp., 280-283.

Naffah, "Cadran cylindrique": Christiane Naffah, "Un cadran cylindrique ottoman du XVIIIème siècle", *Astrolabica* 5 (1989), pp. 37-52.

Nallino, *al-Battānī*: Carlo Alfonso Nallino, ed., transl. and comm., *al-Battani sive Albatenii opus astronomicum*, 3 pts., Pubblicazioni del Real Osservatorio di Brera in Milano, no. XL, pts. I-III, Milan, 1899-1907, repr. in 1 vol., Hildesheim & New York: Georg Olms, 1977. [Pts. I-II were also reprinted Frankfurt am Main: Minerva, 1969; pt. III (the Arabic text) was reprinted in an undated, unlabelled pirate edition by [Muthannā, Baghdad], in the 1960s.]

— , "Islamic Astronomy": *idem*, "Sun, Moon, and Stars (Muhammadan)", in *ERE*, vol. XII (1921), pp. 88-101.

— , *Scritti*, V: *idem*, *Raccolta di scritti editi e inediti* (ed. Maria Nallino), vol. V: *Astrologia – Astronomia – Geografia*, Rome: Istituto per l'Oriente, 1944.

Nasr, Seyyed Hossein, *Islamic Science: An Illustrated Study*, London: World of Islam Festival Publishing Company Ltd., 1976, repr. Istanbul: "nsan Yayınları, 1989. [A Persian translation *ʿIlm dar Islām* was published in Tehran, 1366 HS [= 1988]. [This work is to be used with caution as far as any technical aspects of Islamic science are concerned.]

Necipoglu, *The Topkapı Scroll*: Gülrü Necipoglu, *The Topkapı Scroll – Geometry and Ornament in Islamic Architecture ...* , Santa Monica, Ca.: The Getty Center for the History of Art and the Humanities, 1995.

Neugebauer, *Abu Shaker's Chronography*: Otto Neugebauer, *Abu Shaker's "Chronography" – Treatise of the 13th Century on Chronological, Calendrical, and Astronomical Matters, written by a Christian Arab, preserved in Ethiopic*, Vienna: Österreichische Akademie der Wissenschaften, 1988.

— . "Astronomical and Calendrical Data in the Très Riches Heures": *idem*, "Astronomical and Calendrical Data in the Très Riches Heures", (originally published 1974), repr. in *idem*, *Astronomy and History*, pp. 507-520.

— , "Astronomische Papyri": *idem*, "Astronomische Papyri aus Wiener Sammlungen: II. Über griechische Wetterzeichen und Schattentafeln", *Sitzungsberichte der philosophisch-historischen Klasse der Österreichischen Akademie der Wissenschaften* 240:2 (1962), pp. 29-44.

— , "Early Greek Astronomy": *idem*, "On Some Aspects of Early Greek Astronomy", *Journal of the American Philosophical Society* 116 (1972), pp. 243-251, repr. in *idem*, *Essays*, pp. 361-369.

— , "The Early History of the Astrolabe": *idem*, "The Early History of the Astrolabe", *Isis* 40 (1949), pp. 240-256, repr. in *idem*, *Essays*, pp. 278-294.

— , *ESA*: *idem*, *The Exact Sciences in Antiquity*, reprint of the 1957 2nd edn., New York: Dover, 1969.

— , *Essays*: *idem*, *Astronomy and History – Selected Essays*, New York, *etc.*: Springer, 1983. Contains:

- Neugebauer & Olaf Schmidt, "Hindu Astronomy at Newminster in 1428", *AS* 8 (1952), pp. 221-228, repr. here pp. 425-432.

— , *Ethiopic Astronomy and Computus*: *idem*, *Ethiopic Astronomy and Computus*, (Österreichische Akademie der Wissenschaften, Philosophisch-Historische Klasse, Sitzungsberichte, 347. Band, Veröffentlichungen der Kommission für Geschichte der Mathematik, Naturwissenschaften und Medizin, Heft 22), Vienna, Verlag der Österreichischen Akademie der Wissenschaften, 1979.

— , *HAMA*: *idem*, *A History of Ancient Mathematical Astronomy*, Berlin, Heidelberg & New York: Springer, 1975.

— , *al-Khwārizmī*: *idem*, *The Astronomical Tables of al-Khwārizmī*, in *Kgl. Danske Vidensk. hist.-fil. Skrifter*, 4:2 (1962).

— , "Studies in Byzantine Astronomy": *idem*, "Commentary on the Astronomical Treatise Par. gr. 2425", *Mémoires de l'Académie Royale de Belgique* 49:4 (1969), pp. 3-45.

— & Schmidt, "Hindu Astronomy at Newminster": *idem* and Olaf Schmidt, "Hindu Astronomy at Newminster in 1428", *AS* 8 (1952), pp. 221-228, repr. in Neugebauer, *Essays*, pp. 425-432.

New Delhi 1985 Colloquium Proceedings: *History of Oriental Astronomy – Proceedings of an International Astronomical Colloquium No. 91, New Delhi, India, 13-16 November, 1985*, G. Swaru, A. K. Bag and K. Shukla, eds., Cambridge, *etc.*: Cambridge University Press, 1987.

Nicholas of Lynn, *Kalendarium*: *see Eisner.*

Nolte, "Die Armillarsphäre": Friedrich Nolte, "Die Armillarsphäre", *Abhandlungen zur Geschichte der Naturwissenschaften und der Medizin* (Erlangen: M. Mencke), Heft 2, 1922.

North, "Astrolabe": John D. North, "The Astrolabe", *Scientific American* 230 (1974), pp. 96-106, repr. in *idem*, *Essays*, no. 14.

— , *Chaucer's Universe*: *idem*, *Chaucer's Universe*, Oxford: Clarendon Press, 1988.

— , *Essays*, A-B:, *idem*, *Stars, Minds and Fate – Essays in Ancient and Medieval Cosmology* (A) and *The Universal Frame – Historical Essays in Astronomy, Natural Philosophy and Scientific Method* (B), London & Ronceverte, W.V.: Hambledon, 1989.

— , *Horoscopes and History*: *idem*, *Horoscopes and History*, (Warburg Institute Surveys and Texts, vol. 12), London: The Warburg Institute, 1986.

— , "Hour-Line Ritual": *idem*, "Astrolabes and the Hour-Line Ritual", *JHAS* 5 (1981), pp. 113-114, repr. in *idem*, *Essays*, A, no. 15.

— , "Meteoroscope": *idem*, "Werner, Apian, Blagrave and the Meteoroscope", *British Journal for the History of Science* 3 (1966-67), pp. 57-65 and 1 pl.

— , "*Opus quarundam rotarum mirabilium*", *Physis* 8 (1966), pp. 337-372, repr. in *idem*, *Essays*, I, pp. 135-169 (no. 11).]

— , "Review of King, *Mecca-Centred World-Maps*": *idem*, in *Bibliotheca Orientalis* (Leiden), 57 (2000), cols. 747-750.

— , *Richard of Wallingford*: *idem*, *Richard of Wallingford: An edition of his writings with introductions, English translation and commentary*, 3 vols., Oxford: The Clarendon Press, 1976.

North Festschrift: *Between Demonstration and Imagination: Essays in the History of Science and Philosophy Presented to John D. North*, Arjo Vanderjagt and Lodi Nauta, eds., Leiden: E. J. Brill, 1999.

Nuremberg GNM 1992-93 Exhibition Catalogue: *Focus Behaim-Globus*, Gerhard Bott, ed., 2 vols., Nuremberg: Germanisches Nationalmuseum, 1992.

Oestmann, "History of the Nocturnal": Günther Oestmann, "On the History of the Nocturnal", *Bulletin of the Scientific Instrument Society*, no. 69 (2001), pp. 5-9.

Oklahoma 1992 & 1993 Conference Proceedings: *Tradition, Transmission, Transformation: Proceedings of Two Conferences on Premodern Science Held at the University of Oklahoma,* F. Jamil Ragep and Sally P. Ragep, with Steven J. Livesey, eds., Leiden, New York & Cologne: E. J. Brill, 1996 [Conferences held during 6.-7.3.1992 and 25.-27.2.1993.]

Oxford BL Catalogue: J. Uri, *Bibliothecae Bodleianae codicum manuscriptorum orientalium* ... , Part I, Oxford: Clarendon Press, 1787.

Oxford MHS Billmeir Supplement Catalogue: [Francis R. Maddison], *A Supplement to a Catalogue of Scientific Instruments in the Collection of J. A. Billmeir, Esq., C. B. E.*, London: Frank Partridge & Sons, 1957

Padmākara Dube, "Rampur Astrolabes": Padmākara Dube, "Astrolabes in the State Library Rampur", *The Journal of the United Provinces Historical Society* (Lucknow) 4 (1929), pp. 1-11 and 7 pls., repr. in *IAM*, vol. 90.

Paltrinieri: see Gunella *et al.*

Paris BN Catalogue: (M. le Baron) McGulkin de Slane, *Bibliothèque nationale – Départment des manuscripts – Catalogue des manuscrits arabes*, Paris: Imprimerie nationale, 1883-1895.

Paris IMA 1993-94 Exhibition Catalogue: *Syrie – mémoire et civilisation*, Sophie Cluzan, Éric Delpont and Jeanne Mouliérac, eds., Paris: Flammarion (Institut du Monde Arabe), 1993. [Catalogue of an exhibition held at the Institut during 14.9.1993-28.2.1994.]

Paris IMA 1995 Exhibition Catalogue: *Itinéraire du savoir en Tunisie – Les temps forts de l'histoire tunisienne*, Hassen Annabi, Mounira Chapoutot-Remadi, Samia Kamarti, eds., Paris: CNRS Editions & Alif & Institut du Monde Arabe, 1995.

Paris IMA 2000 Exhibition Catalogue: *Les Andalousies de Damas à Cordoue*, Éric Delpont *et al.*, eds., Paris: Institut du Monde Arabe, 2000. [Catalogue of an exhibition held at the Institute during 28.11.2000-15.4.2001.]

Paris IMA Catalogue: Jeanne Mouliérac, "La Collection Marcel Destombes", in *Astrolabica No. 5 – Études 1987-1989*, Paris: Institut du Monde Arabe & Société Internationale de l'Astrolabe, 1989, pp. 77-126.

Paselk, "Instruments": Richard A. Paselk, "Early Scientific Instruments", Internet site at www.humboldt.edu/~rap1/EarlySciInstSite/Instruments (2001).

Pedersen, F. S., ed., *Astronomical Works of Petrus de Dacia and Petrus de S. Audomoro*: Fridericus Saaby Pedersen, ed., *Petri Philomenae de Dacia et Petri de S. Audomoro opera quadrivialia*, 2 pts., (*Corpus philosophorum danicorum medii aevi*, X.1-2), Copenhagen: Danske Sprog- og Litteraturselskab, 1983-1984.

—, *Toledan Tables*: *idem*, *The Toledan Tables: A review of the manuscripts and the textual versions with an edition*, 3 vols., Copenhagen: Dĕt Kongelige Danske Videnskabernes Selskab (*Historisk-filosofiske Skrifter* 24:1-3), 2002.

Pedersen, O., "Corpus Astronomicum": Olaf Pedersen, "The Corpus Astronomicum and the Traditions of Mediaeval Latin Astronomy", *Studia Copernicana* 13 (1975), pp. 57-96,

—, "Medieval European Astronomy": *idem*, "Astronomy", in David C. Lindberg, ed., *Science in the Middle Ages*, Chicago, Ill.: University of Chicago Press, 1978, pp. 303-337.

Pedersen,O.: see Ptolemy, *Almagest*.

Pellat, "L'astrolabe sphérique d'al-Rūdānī": Charles Pellat, "L'astrolabe sphérique d'al-Rūdānī", *Annali dell'Istituto e Museo di Storia della Scienza di Firenze* 26 (1974), pp. 7-82. [See also Janin, "Ar-Rūdānī sur l'astrolabe sphérique".]

—, *Le Calendrier de Cordoue*: *idem*, *Le Calendrier de Cordoue*, Leiden: E. J. Brill, 1961.

—, *Cinq calendriers égyptiens*: *idem*, *Cinq calendriers égyptiens*, (Textes arabes et études islamiques, tome XXVI), Cairo: Institut Français d'Archéologie Orientale du Caire, 1986.

Philibert, *La Qibla des mosquées maghrébines*: M. Philibert, *La Qibla et le Mih'râb. Différences constatées dans la direction des mosquées maghrébines. Raisons possibles, orientation par des procédés modernes*, Algiers, 1972 (privately-circulated photostat).

Physis: *Physis – Rivista internazionale di storia della scienza*.

Pinder-Wilson, "The Malcolm Celestial Globe": Ralph H. Pinder-Wilson, "The Malcolm Celestial Globe", *British Museum Yearbook I, The Classical Tradition*, London: British Museum, 1976, pp. 83-101.

Pingree, *Astronomical Works of Gregory Chioniades*: David Pingree, *The Astronomical Works of Gregory Chioniades*, vol. i: *The Zīj al-ʿAlāʾī*, 2 pts., (*Corpus des astronomes byzantins* II), Amsterdam: J. C. Gieben, 1985-86.

—, "Chionades": *idem*, "Gregory Chionades and Palaeologan Astronomy", *Dumbarton Oaks Papers* 18 (1964), pp. 135-160.

—, "Editing the *Zīj al-Sanjarī*": *idem*, "A Preliminary Assessment of the Problems of Editing the *Zīj al-Sanjarī* of al-Khāzinī", in Ibish, ed., *Editing Islamic Manuscripts on Science*.

—, "al-Fazārī": *idem*, "The Fragments of the Works of al-Fazārī", *JNES* 29 (1970), pp. 103-123.

—, "Greek Influence": *idem*, "The Greek Influence on Early Islamic Mathematical Astronomy", *JAOS* 93 (1973), pp. 32-43.

—, "Indian Influence": *idem*, "Indian Influence on Sasanian and Early Islamic Astronomy and Astrology", *The Journal of Oriental Research* (Madras) 34-35 (1964-66/1973), pp. 118-126.

—, "Sanskrit Astronomical Tables in the US": *idem*, "Sanskrit Astronomical Tables in the United States", *Transactions of the American Philosophical Society*, N.S., 58:3 (1968).

—, *The Thousands of Abū Maʿshar*: *idem*, *The Thousands of Abū Maʿshar*, London: The Warburg Institute, University of London, 1968.

— , "Yaʿqūb ibn Ṭāriq": *idem*, "The Fragments of the Works of Yaʿqūb ibn Ṭāriq", *JNES* 26 (1968), pp. 97-125.

Pingree: see also Kennedy & Pingree; and Kennedy, Pingree & Haddad.

Plofker, "Iterative Approximations": Kim Plofker, "Use and Transmission of Iterative Approximations in India and the Islamic World", in *Bellagio 2000 Conference Proceedings*, pp. 167-186.

— , "Spherical Astronomy in Medieval India": *eadem*, "The Astrolabe and Spherical Trigonometry in Medieval India", *JHA* 31 (2000), pp. 37-54.

Poulle, *Fusoris*: Emmanuel Poulle, *Un constructeur d'instruments astronomiques au 15e siècle – Jean Fusoris*, Paris: Honoré Champion, 1963.

— , "Heures inégales": *idem*, "L'Horlogerie a-t-elle tué les heures inégales", *Bibliothèque de l'Écoles des Chartes* (Paris & Geneva, Librairie Droz) 157 (1999), 137-156.

— , "Instruments astronomiques": Emmanuel Poulle, "Les instruments astronomiques du Moyen Age," Museum of the History of Science, Oxford, Selected Off-prints no. 7 (1969), repr. from *Le ruban rouge* 32 (mars 1967). du moyen âge", 1st edn. in *Le Ruban rouge,* no. 32 (mars 1967), pp. 18-29, repr. by the Museum of the History of Science, Oxford: Selected Off-prints, no. 7; 2nd edn. with different plates in *Astrolabica* 3, Paris: Société internationale de l'Astrolabe, 1983.

— , "*Quadrans novus*": *idem*, "Le quadrant nouveau médiéval", in 2 parts, *Journal des Savants*, avril-juin 1964, pp. 148-167, and juillet-septembre 1964, pp. 182-214.

— , "*Saphea*": *idem*, "Un instrument astronomique dans l'occident latin – la 'saphea'", *Studi Medievali* 10 (1969), pp. 491-510

— , "Sexagenarium": *idem*, "Théorie des planètes et trigonométrie au XVe siècle, d'après un équatoire inédit, le sexagenarium", *Journal des savants* (1966), pp. 129-161.

— , *Sources astronomiques*: *idem*, *Les sources astronomiques* (*textes, tables, instruments*), Turnhout (B): Brepols, 1981.

— , *Studies*: *idem*, *Astronomie planétaire au Moyen Âge latin*, Aldershot: Variorum, 1996.

— & De Smet, Tables astronomiques de Louvain: *idem* and Antoine De Smet, *Les tables astronomiques de Louvain de 1528*, Brussels: Editions Culture et Civilisation, 1976. [See also *idem*, *Studies*, IV.]

Price, *The Equatorie of the Planetis*: Derek J. de Solla Price (with R. M. Wilson), *The Equatorie of the Planetis edited from Peterhouse MS. 75.I*, Cambridge: Cambridge University Press, 1955.

— , "Fake Antique Scientific Instruments": *idem*, "Fake Antique Scientific Instruments", in *Actes du VIIIe Congrès International d'Histoire des Sciences, Florence, 3-9 Sept., 1956*, pp. 380-394 (separatum paginated 1-15).

— , "Fez Water Clocks": *idem*, "Medieval Water Clocks of the 14th Century in Fez, Morocco", *Proceedings of the Xth International Congress Congress of History of Science, (Ithaca, 1962)*, pp. 599-602.

— , *Gears from the Greeks*: *idem*, "Gears from the Greeks – The Antikythera Mechanism: A Calendar Computer from *ca.* 80 B.C.", *Transactions of The American Philosophical Society*, N.S., 64:7 (1974), repr. New York: Science History Publications, 1975.

— , "Instruments in the British Museum": *idem*, "Instruments in the British Museum", *Journal for the History of Astronomy* 13 (1982), pp. 129-132. [An essay review of *London BM Catalogue*.]

— , "*Navicula*": *idem*, "The Little Ship of Venice, a Middle English Instrument Tract", *Journal of the History of Medicine and Allied Sciences* 15 (1960), pp. 399-407.

— , "Portable Sundials in Antiquity": *idem*, "Portable Sundials in Antiquity, including an Account of a New Example from Aphrodisias", *Centaurus* 14 (1969), pp. 242-266.

— , "Precision Instruments to 1500": *idem*, "Precision Instruments to 1500", in Singer *et al.*, eds., *History of Technology*, III, pp. 582-619.

— , "Regiomontanus' Astrolabe"; "The First Scientific Instrument of the Renaissance", *Physis* 1 (1959), pp. 26-30.

— , *Science since Babylon*: *idem*, *Science since Babylon*, New Haven, Ct. & London: Yale University Press, 1976.

Price *et al.*, *Astrolabe Checklist*: Sharon Gibbs, Janice A. Henderson and Derek de Solla Price, *A Computerized Checklist of Astrolabes*, New Haven Conn.: Yale University, Department of History of Science and Medicine, 1973. [See also Price, "An International Checklist of Astrolabes", *AIHS* 8 (1955), pp. 243-263, and pp. 363-381.]

Princeton Catalogue: Rudolf Mach, *Catalogue of Arabic Manuscripts (Yahuda Section) in the Garrett Collection, Princeton University Library*, Princeton: Princeton University Press, 1977.

Ptolemy, *Almagest*: Gerald J. Toomer, *Ptolemy's Almagest, translated and annotated*, London: Gerald Duckworth & Co., 1984, and New York, *etc.*: Springer-Verlag, 1984; and Olaf Pedersen, *A Survey of the Almagest*, Odense: Odense University Press, 1974.

Poulle, *Fusoris*: Emmanuel Poulle, *Un constructeur d'instruments astronomiques au 15e siècle – Jean Fusoris*, Paris: Honoré Champion, 1963.

— , "Sexagenarium": *idem*, "Théorie des planètes et trigonométrie au XVe siècle, d'après un équatoire inédit, le sexagenarium", *Journal des savants* (1996), pp. 129-161.
— , *Studies*: *idem*, *Astronomie planétaire au Moyen Âge latin*, Aldershot (U.K.) & Brookfield, Vt.: Variorum, 1996.
Puig, "El astrolabio lineal en al-Andalus": Roser Puig, "Ibn Arqam al-Numayrī (m. 1259) y la introducción en al-Andalus del astrolabio lineal", in Vernet, ed., *Estudios*, B, pp. 101-103.
— , *Azafea de Azarquiel*: *eadem*, *Los tratados de construcción y uso de la azafea de Azarquiel*, Madrid: Instituto Hispano-Arabe de Cultura, 1987.
— , "Cuadrante de senos": *eadem*, "Una aportación andalusí a la difusión del cuadrante de senos", in *Yād-Nāma: in memoria di Alessandro Bausani*, Biancamaria Scarcia Amoretti and Lucia Rostagno, eds., 2 vols., Rome: Bardi, II, pp. 75-85.
— , "Ibn al-Zarqālluh's Orthographic Projection": *eadem*, "On the Eastern Sources of Ibn al-Zarqālluh's Orthographic Projection", in *Vernet Festschrift*, II, pp. 737-753.
— , "Instrumentos universales en al-Andalus": *eadem*, chapter "Instrumentos universales en al-Andalus", in *Madrid MAN 1992 Exhibition Catalogue*, pp. 67-73.
— , "Proyección ortogrâfica": *eadem*, "Proyección ortogrâfica en los instrumentos astronómicos del Oriente islámico a la Europa del Renacimiento", *Revista del Instituto Egipcio de Estudios Islámicos en Madrid* 26 (1993-94), pp. 249-261.
— , "La proyección ortográfica en el Libro de la açafeha": *eadem*, "La proyección ortográfica en el Libro de la açafeha alfonsí", in Samsó *et al.*, eds., *Alphonsine Astronomy*, A, pp. 125-138.
— , "*Ṣafīḥa shakkāzīya*": *eadem*, "Concerning the *ṣafīḥa shakkāzīya*", *ZGAIW* 2 (1985), pp. 123-139.
— , *al-Šakkāziyya*: *eadem*, *al-Šakkāziyya – Ibn al-Naqqāš al-Zarqālluh, edición, traducción y estudio*, Barcelona: Universidad de Barcelona, Instituto "Millás Vallicrosa" de Historia de la Ciencia Árabe, 1986.
— , "Al-Zarqālluh's Graphical Method": *eadem*, "Al-Zarqālluh's Graphical Method for Finding Lunar Distances", *Centaurus* 32 (1989), pp. 294-309.
al-Qanara: *al-Qanara – Revista de Estudios Árabes* (Madrid) 11
Rada, "Comets in Arabic Literature": W. S. Rada, "A Catalogue of Medieval Arabic and Islamic Observations of Comets during the Period AD 700-1600", *ZGAIW* 123 (1999/2000), pp. 71-91.
Ragep, "Freeing Astronomy from Philosophy": F. Jamil Ragep, "Freeing Astronomy from Philosophy: An Aspect of Islamic Influence on Science", *Osiris* 16 (2001), pp. 49-71.
— , "Ṭūsī and Copernicus": *idem*, "Ṭūsī and Copernicus: The Earth's Motion in Context", *Science in Context* 14 (2001), pp. 145-163.
— , *al-Ṭūsī's Memoir on Astronomy*: *idem*, *Naṣīr al-Dīn al-Ṭūsī's* Memoir on Astronomy (al-Tadhkira fī ʿilm al-hay'a), 2 vols., New York, *etc.*: Springer, 1993.
Ragep: see also *Oklahoma 1992 and 1993 Conference Proceedings.*
Rashed, "Méthodes d'interpolation": Roshdi Rashed, "al-Samaw'al, al-Bīrūnī et Brahmagupta: les méthodes d'interpolation", *ASP* 1 (1991), pp. 101-160, repr. in *idem*, *Studies*, XII.
— , *Studies*: *idem*, *Optique et mathématiques – Recherches sur l'histoire de la pensée scientifique en arabe*, Aldershot & Brookfield, Vt.: Variorum, 1992.
Rebstock, *Rechnen im islamischen Orient*: Ulrich Rebstock, *Rechnen im islamischen Orient – Die literarischen Spuren der praktischen Rechenkunst*, Darmstadt: Wissenschaftliche Buchgesellschaft, 1992.
Renaud, "Additions à Suter": Henri-Paul-Joseph Renaud, "Additions et corrections à Suter, 'Die Mathematiker und Astronomen der Araber'", *Isis* 18 (1932), pp. 166-183.
— , "Astronomie et astrologie marocaines": *idem*, "Astronomie et astrologie marocaines", *Hespéris* 29 (1942), pp. 41-63.
— , *Calendrier d'Ibn al-Bannā'*: *idem*, *Le Calendrier d'Ibn al-Bannā' de Marrakech (1256-1321 J.C.), Publications de l'Institut des Hautes-Etudes Marocaines*, tome XXXIV, Paris: Larose Editeurs, 1948.
— , "Déterminations marocaines de l'obliquité de l'écliptique": *idem*, "Déterminations marocaines de l'obliquité de l'écliptique", *Bulletin de l'Enseignement Public (Protectorat de la France au Maroc)* 28 (1941), pp. 321-336.
— , "Enseignement au Maroc": *idem*, "L'enseignement des sciences exactes et l'édition d'ouvrages scientifiques au Maroc avant l'occupatiuon européenne", *Hespéris* 14 (1932), pp. 78-89.
— , "Ibn al-Bannā'": *idem*, "Ibn al-Bannā' de Marrakesh: ṣūfī et mathématicien (XIIIe-XIVe s. J.C.)", *Hespéris* 25 (1933), pp. 13-42.
— , "Sur les lunes du Ramadan": *idem*, "Sur les lunes du Ramadan", *Hespéris* 32 (1945), pp. 51-68.
Renaud: see also Colin & Renaud, *Escorial Catalogue*, and Sarton, "Mosque Orientations".
Rico y Sinobas, ed., *Libros del saber*: D. Manuel Rico y Sinobas, ed., *Libros del saber de astronomía del Rey D. Alfonso X de Castille*, 5 vols., Madrid, 1873.
Ritter, "Griechisch-koptische Ziffern": Helmut Ritter, "Griechisch-koptische Ziffern in arabischen Manuskripten", *Rivista degli Studi Orientali* 16 (1936), pp. 212-214.

Rius, *La Alquibla en al-Andalus y al-Magrib*: Mònica Rius, *La Alquibla en al-Andalus y al-Magrib al-Aqṣà*, (*Anuari de Filologia* (Universitat de Barcelona) XXI (1998-99) B-3), Barcelona: Institut "Millás Vallicrosa" de Història de la Ciència Àrab, 2000. [An English version is in preparation.]

Riyadh 1985 Exhibition Catalogue: *The Unity of Islamic Art*, published in English and Arabic in Riyadh in 1985.

Rockford TM Catalogue, I-1 and I-3: *The Time Museum - Catalogue of the Collection,* Bruce Chandler, ed., vol. I: *Time Measuring Instruments,* Pt. 1: Anthony J. Turner, *Astrolabes, Astrolabe Related Instruments, and* Pt. 3: *idem, Water-Clocks, Sandglasses, Fire-clocks*, Rockford, Ill.: The Time Museum, 1985 and 1984.

Röttel, ed., *Peter Apian*: Karl Röttel, ed., *Peter Apian – Astronomie, Kosmographie und Mathematik am Beginn der Neuzeit, mit Austellungskatalog*, 2nd edn., (1st edn. published 1995), Eichstätt: Polygon-Verlag Buxheim, 1997.

Rohr, René R. J., "Astrolabium und Sonnenuhr im Dienste der Moschee", *Centaurus* 18 (1973), pp. 44-56.

— , *Sundials*: *idem, Sundials – History, Theory, and Practice*, (first published in French in 1965), Toronto: University of Toronto Press, 1975.

Rosenfeld & İhsanoğlu, *MAIC*: Boris A. Rosenfeld and Ekmeleddin İhsanoğlu, *Mathematicians, Astronomers and Other Scholars of Islamic Civilisation and their Works (7th-19th C.)*, Istanbul: Research Centre for Islamic History, Art and Culture (IRCICA), 2003.

— *et al.*, eds., *Al-Khorezmi*: *Al-Khorezmi*: *Muhammad ibn Musa al-Khorezmi* (in Russian), Boris A. Rosenfeld *et al.*, eds., Moscow: Nauk, 1983.

Rosenfeld: see also Matvievskaya & Rosenfeld.

Rosenthal, "Al-Asurlâbî and as-Samaw'al on Scientific Progress": Franz Rosenthal, "Al-Asurlâbî and as-Samaw'al on Scientific Progress", *Osiris* 9 (1950), pp. 555-564.

Rosenthal Festscrift: *Studies in Islam and the Ancient Near East Dedicated to Franz Rosenthal*, Jeannette Wakin, ed., a special issue of the *Journal of the American Oriental Society* (*JAOS* 104:1 (1984)).

Rosinska, "Early European Decimal Trigonometric Tables": G. Rosinska, "Tables of Decimal Trigonometric Functions from *ca.* 1450 to *ca.* 1550", in *Kennedy Festschrift*, pp. 419-426.

RSO: *Rivista degli Studi Orientali.*

Rubin, "The Kaʿba": Uri Rubin, "The Kaʿba – Aspects of its Ritual Functions and Position in Pre-Islamic and Early Islamic Times", *Jerusalem Studies on Arabic and Islam* 8 (1986), pp. 97-131.

— , "Morning and Evening Prayers in Early Islam": *idem*, "Morning and Evening Prayers in Early Islam", *Jerusalem Studies on Arabic and Islam* 10 (1987), pp. 40-64.

Ruggles & Saunders, eds., *Astronomies and Cultures*: Clive L. N. Ruggles and Nicholas J. Saunders, eds., *Astronomies and Cultures*, Niwot, Colorado: University Press of Colorado, 1993.

Russell, ed., *'Arabick' Interest in 17th-Century England*: *The 'Arabick' Interest of the Natural Philosophers in Seventeenth-Century England*, Gül A. Russell, ed., Leiden: E. J. Brill, 1994.

Ruska: see Hartner & Ruska.

Sabra, "Greek Science in Medieval Islam": Abdelhamid I. Sabra, "The Appropriation and Subsequent Naturalization of Greek Science in Medieval Islam: A Preliminary Statement", *HS* 25 (1987), pp. 223-243 repr. in *Oklahoma Proceedings*, pp. 3-27, and in *idem Studies*, I.

— , *Studies*: *idem*, *Optics, Astronomy and Logic – Studies in Arabic Science and Philosophy*, Aldershot: Variorum, 1994.

Sachau, *Alberuni's India*: see al-Bīrūnī, *India*.

Said: see Hakim Mohammed Said and *Islamabad 1979 Conference Proceedings*.

Saliba, "Arabic Planetary Theories after the 11th Century": George Saliba, "Arabic Planetary Theories after the Eleventh Century AD", in *EHAS*, I, pp. 58-127.

— , "Arabic Science and the Greek Legacy": *idem*, "Arabic Science and the Greek Legacy", in *Vernet Festschrift*, I, pp. 19-37.

— , "Artisans and Mathematicians in Medieval Islam": *idem*, "Artisans and Mathematicians in Medieval Islam" [an essay review of Necipoglo, *Topkapı Scroll*], *Journal of the American Oriental Society* 119 (1999), pp. 637-645.

— , "The Astronomical Tradition of Maragha": *idem*, "The Astronomical Tradition of Maragha: A Historical Survey and Prospects for Future Research", *ASP* 1 (1991), pp. 67-99, repr. in *idem*, *Studies*, pp. 258-290.

— , "Astronomy and Astrology in Medieval Arabic Thought": *idem*, "Astronomy and Astrology in Medieval Arabic Thought", in Roshdi Rashed and Joël Biard, eds., *Les doctrines de la science de l'antiquité à l'âge classique*, Leuwen / Louvain: Peeters, 1999, pp. 131-164.

— , "Astrolabe by Khafīf": *idem*, "A Sixteenth-Century Drawing of an Astrolabe made by Khafīf Ghulām ʿAlī b. ʿĪsā (c. 850 A.D.)", *Nuncius – Annali di Storia della Scienza* (Florence) 6 (1991), pp. 109-119.

— , "The Astrologer in Islamic Society": *idem*, "The Role of the Astrologer in Medieval Islamic Society", in *Bulletin d'études orientales* (Damascus: Institut Français de Damas), 44 (1992), pp. 45-67 and 6 figs.

— , "Easter Computation in *Zījes*": *idem*, "Easter Computation in Medieval [Islamic] Astronomical Handbooks", *Al-Abḥāth* 23 (1970), pp. 179-212, repr. in Kennedy *et al.*, *Studies*, pp. 677-709.

— , "Islamic Astronomy and Astrology": *idem*, "Astrology/Astronomy, Islamic" in *DMA*, vol. I, 1982, pp. 616-624, repr. in *idem*, *Studies*, no. 2.
— , "Al-Qūshjī's Reform of the Ptolemaic Model for Mercury": *idem*, "Al-Qūshjī's Reform of the Ptolemaic Model for Mercury", *ASP* 3 (1993), pp. 161-203.
— , "A 16th-Century Arabic Critique of Ptolemy": *idem*, "A Sixteenth-Century Arabic Critique of Ptolemy: The Work of Shams al-Dīn al-Khafrī", *JHA* 25 (1994), pp. 15-38.
— , *Studies*: *idem*, *A History of Arabic Astronomy – Planetary Theories during the Golden Age of Islam*, (New York University Studies in Near Eastern Civilization, XIX), New York: New York University Press, 1994.
— , "13th-Century Observational Notebook": *idem*, "The Observational Notebook of a Thirteenth-Century Astronomer", *Isis* 74 (1983), pp. 388-401, repr. in *idem*, *Studies*, no. 8.
Saliba: see also *Kennedy Festschrift* and *Washington NMAH Catalogue.*
Saltzer Festschrift: *Der Weg der Wahrheit – Aufsätze zur Einheit der Wissenschaftsgeschichte – Festgabe zum 60. Geburtstag von Walter G. Saltzer*, Peter Eisenhardt, Frank Linhard and Kaisar Petanides, eds., Hildesheim: Olms, 1999.
Samsó, "Astronomical Observations in the Maghrib": Julio Samsó, "Astronomical Observations in the Maghrib in the Fourteenth and Fifteenth Centuries", *Science in Context* 14 (2001), pp. 165-178.
— , "al-Bīrūnī in al-Andalus": *idem*, ""al-Bīrūnī" in al-Andalus", in *Vernet Festschrift*, II, pp. 583-612.
— , "Cuadrante *šakkāzī*": *idem*, "Una hipótesis sobre calculo por aproximación con el cuadrante šakkāzī", *Al-Andalus* 36 (1971), pp. 117-126.
— , *Ciencias en al-Andalus*: *idem*, *Las ciencias de los antiguos en al-Andalus*, Madrid: MAPFRE, 1992.
— , "Ecuatorio de Ibn al-Samḥ": *idem*, "Notas sobre el Ecuatorio de Ibn al-Samḥ", in Vernet, ed., *Nuevos estudios*, pp. 105-118.
— , *Estudios sobre Abū Naṣr*: *idem*, *Estudios sobre Abū Naṣr Manṣūr b. ʿAlī b. ʿIrāq*, Barcelona: Asociación para la Historia de la Ciencia Española, 1969.
— , "Instrumentos astronómicos": *idem*, "Instrumentos astronómicos", in [Juan Vernet, ed.], *Historia de la ciencia árabe*, Madrid: Real Academia de Ciencias Exactas, Fisicas y Naturales, 1981, pp. 97-126.
— , "Maghribī *Zīj*es": *idem*, "An Outline of the History of Maghribī *Zīj*es from the End of the Thirteenth Century", *JHA* 29 (1998), 93-102.
— , *Studies*: *idem*, *Islamic Astronomy and Medieval Spain*, Aldershot: Variorum, 1994.
— , "Zacut in the Islamic East": *idem*, "In Pursuit of Zacut's *Almanach Perpetuum* in the Islamic East", to appear in *ZGAIW*.
— , "*Zīj al-ṣafāʾiḥ*": *idem*, "Un nuevo dato sobre el *Zīŷ al-ṣafāʾiḥ* de Abū Ŷaʿfar al-Jāzin", *al-Qanara* 2 (1981), pp. 461-464.
— & Catalá, "Cuadrante *šakkāzī*": *idem* and M. A. Catalá, "Un instrumento astronómico de raigambre zarqālī: el cuadrante *šakkāzī* de Ibn Ṭībugā", *Memorias de la Real Academia de Buenas Letras de Barcelona* 13:1 (1971), pp. 5-31.
— & Mielgo, "Ibn Isḥāq and Ibn Muʿādh on the Qibla": Julio Samsó and Honorino Mielgo, "Ibn Isḥāq al-Tūnisī and Ibn Muʿādh al-Jayyānī on the Qibla" (not published previously), in Samsó, *Studies*, VI.
— & Millás, "Ibn al-Bannāʾ on Planetary Longitudes": *idem* and Eduardo Millás, "The Computation of Planetary Longitudes in the *Zīj* of Ibn al-Bannāʾ", *ASP* 8 (1998), pp. 259-286.
Samsó *et al.*, eds., *Alphonsine Astronomy*, A-B: *De Astronomia Alphonsi Regis – Proceedings of the Symposium on Alfonsine Astronomy held at Berkeley (August 1985) together with other papers on the same subject*, Mercè Comes, Roser Puig and Julio Samsó, eds., Barcelona: Universidad de Barcelona, Instituto "Millás Vallicrosa" de Historia de la Ciencia Árabe, 1987, and *"Ochava Espera" y "Astrofisica" – Textos y estudios sobre las fuentes árabes de la astronomía de Alfonso X*, Mercè Comes, Honorino Mielgo and Julio Samsó, eds., Barcelona: Universidad de Barcelona, Instituto "Millás Vallicrosa" de Historia de la Ciencia Árabe, *etc.*, 1990.
Samsó: see also Cátedra & Samsó; King & Samsó; *Madrid MAN 1992 Exhibition Catalogue*; *Santa Cruz 1985 Exhibition Catalogue*; and *Vernet Festschrift*; also Fierro & Samsó, eds., *The Formation of al-Andalus*, II.
Santa Cruz 1985 Exhibition Catalogue: Juan Vernet, Julio Samsó *et al.*, *Instrumentos astronomicos en la España medieval: Su influencia en Europa* (in Spanish and English), Santa Cruz de la Palma: Ministerio de Cultura, 1985.
Sarma, "The Astrolabe in Sanskrit": Sreeramula Rajeswara Sarma, "Yantrarāja: the Astrolabe in Sanskrit", *IJHS* 34 (1999), pp. 145-158.
— , "Astronomical Instruments in Mughal Miniatures": *idem*, "Astronomical Instruments in Mughal Miniatures", *Studien zur Indologie und Iranistik* 16-17 (1992), pp. 235-276.
— , "Lahore Astrolabists": *idem*, "The Lahore Family of Astrolabists and their Ouvrage", *Studies in History of Medicine and Science* (Hamdard University, Hamdard Nagar, New Delhi) 13 (1994), pp. 205-224.
— , "The *Ṣafīḥa Zarqāliyya* in India": *idem*, "The *Ṣafīḥa Zarqāliyya* in India", in *Vernet Festschrift*, II, pp. 719-735.
— , "Sulṭān, Sūri and the Astrolabe": *idem*, "Sulṭān, Sūri and the Astrolabe", *IJHS* 35 (2000), pp. 129-147.

Sarrus, "Astrolabe marocain": Frédéric Sarrus, "Description d'un astrolabe construit à Maroc en l'an 1208", *Mémoires de la Société d'histoire naturelle de Strasbourg* 4 (1853), pp. 1-32, and pls. 1-6, repr. in *AIOS*, I, pp. 193-230.

Sarton, *IHS*: George Sarton, *Introduction to the History of Science*, 5 vols. in 3, Baltimore: Williams and Wilkins & Co. for the Carnegie Institution, Washington, D.C., 1927-1948, repr. 1962.

— , "Mosque Orientations": *idem*, "Orientation of the mihrab in mosques", *Isis* 20 (1933), pp. 262-264, and in a reply by Henri-Paul-Joseph Renaud in *Isis* 34 (1942), p. 24.

SATMI: see King, **SATMI**.

Saunders, *All the Astrolabes*: Harold N. Saunders, *All the Astrolabes*, Oxford: Senecio, 1984.

Savage-Smith, *Islamicate Celestial Globes*: Emilie Savage-Smith, *Islamicate Celestial Globes – Their History, Construction, and Use*, Washington, D.C.: Smithsonian Institution Press, 1985. [See also the review by Paul Kunitzsch in *Der Islam* 64 (1987), pp. 367-370, with useful corrections.]

— , "Review": *eadem*, a review of King, *Mecca-Centred World-Maps*, in *BSIS* 66 (2000), pp. 32-35.

Savage-Smith: see also *London Khalili Collection Catalogue*.

Sayılı, "al-Khāzinī on Instruments": Aydın Sayılı, "Al-Khāzinī's Treatise on Astronomical Instruments", *Ankara (AÜDTCF) Dergisi* 14 (1956), pp. 18-19.

—, *The Observatory in Islam*: *idem*, *The Observatory in Islam*, (Publications of the Turkish Historical Society, Series VII: No. 38), Ankara: Türk Tarih Kurumu Basımevi, 1960, repr. New York: Arno, 1981.

Sayılı Memorial Volume: *Aydın Sayılı Özel Sayısı*, I-III, a special issue of *Erdem* (Ankara: Atatürk Kültür Merkezi), in three parts (9:25-27), Ankara: Türk Tarih Kurumu Basımevi, 1996-1997.

Schaldach, "Zinner-Archiv": Karl Heinz Schaldach, "Das Ernst Zinner-Archiv", in *Frankfurt IGN Festband*, pp. 25-28.

— & Rau, "Mittelalterliche Vertikalsonnenuhren": *idem* and Herbert Rau, "Vertikalsonnenuhren des 6. bis 14. Jahrhunderts", in *Frankfurt IGN Festband*, pp. 273-290.

Schaldach: see also Arnaldi & Schaldach.

Schechner, "Sundials": Sara Schechner, "The Material Culture of Astronomy in Daily Life: Sundials, Science and Social Change", *JHA* 32 (2001), pp. 189-222.

Schimmel, "The Sacred Geography of Islam": Annemarie Schimmel, "The Sacred Geography of Islam", in *Sacred Places and Profane Spaces: Essays in the Geographics of Judaism, Christianity, and Islam*, Jamie Scott and Paul Simpson-Housley, eds., New York, Westport, Ct., & London: Greenwood Press, 1991, pp. 163-175.

Schirmer, "Studien zur Astronomie der Araber": Oskar Schirmer, "Studien zur Astronomie der Araber", (includes II. Arabische Bestimmungen der Schiefe der Ekliptik), *Erlangen PMS SB* 58 (1926), pp. 33-88, repr. in *IAM*, vol. 22 (1997), pp. 1-56.

Schjellerup, *Description des étoiles fixes par al-Sûfi*: Hans C. F. C. Schjellerup, *Description des étoiles fixes ... par l'astronome persan Abd-al-Rahman al-Sûfi*, St. Petersburg: Académie Impériale des sciences, 1874, repr.s Frankfurt am Main: Institut für Geschichte der Arabisch-Islamischen Wissenschaften, 1986.

Schlachter, "Der Globus": A. Schlachter, "Der Globus, seine Entstehung und Verwendung in der Antike ... ", *Stoicheia, Studien zur Geschichte des antiken Weltbildes und der griechischen Wissenschaften* (Leipzig & Berlin: Teubner), Heft 8, 1927.

Schmalzl, *Geschichte des Quadranten*: Peter Schmalzl, *Zur Geschichte des Quadranten bei den Arabern*, Munich: Salesianische Offizin, 1929, repr. in *AIOS*, VI, pp. 189-331.

Schmidl, "Ein Astrolab aus dem 17. Jahrhundert": Petra Schmidl, "Ein Astrolab aus dem 17. Jahrhundert – prachtvoll und verfälscht", in *Frankfurt IGN Festband*, pp. 291-314.

—, "Early Sources on the Compass": *eadem*, "Two Early Arabic Sources on the Magnetic Compass", *Journal of Arabic and Islamic Studies* 1 (1996-97), pp. 81-132. [Electronic journal accessible on the Internet *via* http://www.uib.no/jais/jais.htm.]

— , *Islamische volksastronomische Abhandlungen*: *eadem*, *Volksastronomische Abhandlungen aus dem mittelalterlichen arabisch-islamischen Kulturraum*, doctoral dissertation in preparation (Institute for History of Science, Johann Wolfgang Goethe University, Frankfurt).

— , "Qibla Schemes in a Medieval Yemeni Text": *eadem*, "Finding the Direction to Mecca with Folk Astronomical Methods: Qibla Schemes in a Medieval Yemeni Text by al-Fārisī", in César Esteban and Juan Antonio Belmonte, eds., *Oxford VI and SEAC 99 – Astronomy and Cultural Divversity*, La Laguna (Tenerife): Organismo Autónomo de Museos del Cabildo de Tenerife, 1999, pp. 177-182.

—, "Qibla und Winde": *eadem*, "Zur Bestimmung der Qibla mittels der Winde", in *Saltzer Festschrift*, pp. 135-146.

Schmidl: see Charette & Schmidl.

Schmidt, "Daylight in Hindu Astronomy": Olaf Schmidt, "The Computation of the Length of Daylight in Hindu Astronomy", *Isis* 35 (1944), pp. 205-211.

Schmidt: see also Neugebauer & Schmidt.

Schoy, *Beiträge*: *Carl Schoy: Beiträge zur arabisch-islamischen Mathematik und Astronomie*, Fuat Sezgin *et al.*, eds., 2 vols., Frankfurt am Main: Institut für Geschichte der Arabisch-Islamischen Wissenschaften, 1988. Contains:

- "Studien über die Dämmerung", in I, pp. 89-94.
- "Ibn al-Haiṯam über die Qibla", in I, pp. 230-241.
- "Bestimmung der Ortsbreite nach Ibn Yūnus", in I, pp. 275-292
- *Schattentafeln*, in I, pp. 324-349.
- *Gnomonik der Araber*, in II, pp. 351-447.
- "Beiträge zur arabischen Trigonometrie", in II, pp. 448-483.
- "Spätarabische Sonnenuhren", in II, pp. 484-512.
- *Trigonometrie des al-Bīrūnī*, in II, pp. 629-746.

—, "Beiträge zur arabischen Trigonometrie": Carl Schoy, "Beiträge zur arabischen Trigonometrie (Originalstudien nach unedierten arabisch-astronomischen Manuscripten)", *Isis* 5 (1923), pp. 364-399, repr. in *idem*, *Beiträge*, II, pp. 448-483.

—, "Bestimmung der Ortsbreite nach Ibn Yūnus": *idem*, "Die Bestimmung der geographischen Breite eines Ortes ... nach dem arabichen Text der Ḥākimitischen Tafeln des Ibn Yūnus", *Annalen der Hydrographie und maritimen Meteorologie* 50 (1922), pp. 3-20, repr. in *idem*, *Beiträge*, I, pp. 275-292.

—, *Gnomonik der Araber*: *idem*, *Gnomonik der Araber*, in *Die Geschichte der Zeitmessung und der Uhren*, E. von Bassermann-Jordan, ed., Band 1F, Berlin & Leipzig: Walter de Gruyter, 1923, repr. in *idem*, *Beiträge*, II, pp. 351-447.

—, "Ibn al-Haitam über die Qibla": *idem*, "Abhandlung des al-Ḥasan ibn al-Ḥasan ibn al-Haiṯam (Alhazen) über die Bestimmung der Richtung der Qibla", *Zeitschrift der Deutschen Morgenländischen Gesellschaft* 75 (1921), pp. 242-253, repr. in *idem*, *Beiträge*, I, pp. 230-241, and in *Islamic Geography*, vol. 18 (1992), pp. 155-166.

—, *Schattentafeln*: *Über den Gnomonschatten und die Schattentafeln der arabischen Astronomie*, Hannover: Heinz Lafaire, 1923, repr. in *idem*, *Beiträge*, I, pp. 324-349.

—, "Spätarabische Sonnenuhren": *idem*, "Sonnenuhren der spätarabischen Astronomie", *Isis* 6 (1924), pp. 332-360, repr. in *idem*, *Beiträge*, II, pp. 484-512.

—, "Studien über die Dämmerung": *idem*, "Geschichtlich-astronomische Studien über die Dämmerung", *Naturwissenschaftliche Wochenschrift*, N.F., 14:14 (1915), pp. 209-214, repr. in *idem*, *Beiträge*, I, pp. 89-94.

—, *Trigonometrie des al-Bīrūnī*: *idem*, *Die trigonometrischen Lehren des persischen Astronomen Abu 'l-Raiḥān ... al-Bīrūnī dargestellt nach al-Qānūn al-Mas'ūdī*, Hannover: Heinz Lafaire, 1927, repr. in *idem*, *Beiträge*, II, pp. 629-746.

Schweinfurt 1993 Exhibition Catalogue: *450 Jahre Copernicus 'De revolutionibus' – Astronomische und mathematische Bücher aus Schweinfurter Bibliotheken*, Uwe Müller, ed., Schweinfurt: Stadtarchiv (Veröffentlichung Nr. 9), 1993.

Sciamvs: *Sciamvs – Sources and Commentaries in Exact Sciences* (Kyoto, Hokuto).

Sédillot-*fils*, *Mémoire*: Louis-Amélie Sédillot, "Mémoire sur les instruments astronomiques des Arabes", *Mémoires de l'Académie Royale des Inscriptions et Belles-Lettres de l'Institut de France* 1 (1844), pp. 1-229, repr. Frankfurt am Main: Institut für Geschichte der Arabisch-Islamischen Wissenschaften, 1989.

—, *Prolégomènes*: *idem*, *Prolégomènes des tables astronomiques d'Oloug-Beg – traduction et commentaire*, Paris: Firmin Didot Frères, 1853.

Sédillot-*père*, *Traité*: Jean-Jacques Sédillot, *Traité des instruments astronomiques des Arabes composé au treizième siècle par Aboul Hhassan Ali de Maroc ...* , 2 vols., Paris: Imprimerie Royale, 1834-1835, repr. in 1 vol., Frankfurt am Main: Institut für Geschichte der Arabisch-Islamischen Wissenschaften, 1985.

Seemann, "Die Instrumente der Sternwarte zu Maragha": Hugo J. Seemann, "Die Instrumente der Sternwarte zu Maragha nach den Mitteilungen von al-'Urḍī", *Erlangen PMS SB* 60 (1928), pp. 15-154, repr. in *AIOS*, VI, pp. 17-129 (see also the review by Josef Frank repr. *ibid.*, pp. 130-141).

—, *Das kugelförmige Astrolab*: *idem*, *Das kugelförmige Astrolab nach ... arabischen Quellen*, (*Abhandlungen zur Geschichte der Naturwissenschaften und der Medizin*, Heft 6), Erlangen, 1925, repr. in *AIOS*, IV, pp. 359-431, and again in *IAM*, vol 88.

Selin, ed., *Astrononomy across Cultures*: *Astrononomy across Cultures – The* [!] *History of Non-Western Astronomy*, Helaine Selin, ed., Dordrecht, *etc.*: Kluwer, 2000.

Serjeant, "S.W. Arabian Almanac": Robert B. Serjeant, "Star-Calendars and an Almanac from South-West Arabia", *Anthropos* 49 (1954), pp. 433-459, repr. in *idem*, *Farmers and Fishermen in Arabia: Studies in Customary Law and Practice*, G. Rex Smith, ed., Aldershot: Variorum, 1995, III.

Sezgin, *GAS*: Fuat Sezgin, *Geschichte des arabischen Schrifttums*, 12 vols. to date, Leiden: E. J. Brill, 1967 onwards, from 2000 onwards Frankfurt am Main: Institut für Geschichte der Arabisch-Islamischen Wissenschaften, especially V: *Mathematik*, 1974, VI: *Astronomie*, 1978, Band VII: *Astrologie, Meteorologie und Verwandtes*, 1979; X-XII: *Mathematische Geographie und Kartographie im Islam und ihr Fortleben im Abendland*, 2000.

— , "Jakobsstab": *idem*, "Zur Frage der Entdeckung des astronomischen Gerätes "Jakobsstab'" (in Arabic), *ZGAIW* 2 (1985), pp. A7-73. [To be used with caution.]

— *et al.*, *Scientific Manuscript*: *Manuscript of Arabic Mathematical and Astronomical Treatises reproduced from MS A.Y. 314 Istanbul Üniversitesi Kütüphanesi, Istanbul*, Fuat Sezgin *et al.*, eds., Frankfurt: IGAIW, 2001.

Sezgin: see also *AIOS*; *Islamic Astronomy and Mathematics*; *Islamic Geography*; and all publications of the Institut für Geschichte der Arabisch-Islamischen Wissenschaften, Frankfurt am Main.

Shabib, *al-Matījī on the Qibla*: ʿAmmār Saʿīd al-Shabīb, *Kitāb Dalāʾil al-Qibla – Abū ʿAlī al-Matījī*, unpublished doctoral dissertation, Universitat de Barcelona, 1992. [Uncritical edition, with useful appendixes.]

Sharma, *Sawai Jai Singh and his Astronomy*: Viruendra N. Sharma, *Sawai Jai Singh and his Astronomy*, Delhi, [1995].

Singer *et al.*, eds., *History of Technology*: Charles Singer *et al.*, eds., *A History of Technology*, 5 vols., Oxford: Clarendon Press, 1954-58.

Smart, *Spherical Astronomy*: W. M. Smart, *Text-Book on Spherical Astronomy*, 5th edn., Cambridge: Cambridge University Press, 1971.

Sotheby's London 25.02.93 Catalogue: *Sotheby's: Clocks, Watches, Mechanical Music & Instruments of Science & Technology, 25th February 1993*, London: Sotheby's, 1993. [Features the *navicula* now in Geneva.]

Souissi & Chapoutot-Remadi, "Les sciences en Tunisie": Mohamed Souissi and Mounira Chapoutot-Remadi, "Les mathématiques et l'astronomie entre Ifriqiya et Maghreb", in *Paris IMA 1995 Exhibition Catalogue*, pp. 64-71.

Stahlman, *Handy Tables*: William Duane Stahlman, *The Astronomical Tables of Codex Vaticanus Graecus 1291*, doctoral dissertation, Brown University, Providence, R.I., 1960. [Available from ProQuest.com (formerly University Microfilms, Ann Arbor, Mich.), no. 62-5761.]

Stautz, "Ein Astrolab aus dem Jahr 1420": Burkhard Stautz, "Ein Astrolab aus dem Jahr 1420", *NTM – Internationale Zeitschrift für Geschichte und Ethik der Naturwissenschaften, Technik und Medizin* (Basle: Birkhäuser), N.S., 5:3 (1997), pp. 142-159.

— , "Die früheste Formgebung der Astrolabien": *idem*, "Die früheste bekannte Formgebung der Astrolabien", in *Frankfurt IGN Festband*, pp. 315-328.

— , *Mathematisch-astronomische Darstellungen auf Instrumenten*: *idem*, *Untersuchungen von mathematisch-astronomischen Darstellungen auf mittelalterlichen Instrumenten islamischer und europäischer Herkunft*, Bassum (D): Verlag für Geschichte der Naturwissenschaften und der Technik, 1997.

— , "Max von Oppenheim's Instrumente", A-B: *idem*, "Sterne, Winde und Gebete – Islamische Instrumente", in *Faszination Orient – Max von Oppenheim – Forscher • Sammler • Diplomat*, Gabriele Teichmann and Gisela Völger, eds., Cologne: DuMont, 2001, pp. 372-381 (A); and the complete version in *Sterne, Winde und Gebete – Max von Oppenheims islamische Instrumente*, Hünfelden-Dauborn (D): Al-Daburnaha-Verlag, 2002 (B).

Stautz: see also *Munich Catalogue*.

Stebbins, "Portable Sun-Dial": Frederick A. Stebbins, "A Medieval Portable Sun-Dial", *Journal of the Royal Astronomical Society of Canada* (Toronto) 55:2 (1961), pp. 49-56.

Steingass, *Dictionary*: F. Steingass, *A Comprehensive Persian-English Dictionary*, Beirut: Librairie du Liban, 1975 (reprint of 1892 edn.).

Stern, "A Treatise on the Armillary Sphere by Dunas ibn Tamīm": Samuel M. Stern, "A Treatise on the Armillary Sphere by Dunas ibn Tamīm", in *Millás-Vallicrosa Festschrift*, II, pp. 373-382, repr. in *IAM*, vol. 96.

Stevens et al., eds., *The Oldest Latin Astrolabe*: Wesley M. Stevens, Guy Beaujouan and Anthony J. Turner, eds., *The Oldest Latin Astrolabe*, (*Physis* 32:2-3), Florence: Leo S. Olschki, 1995. [A promised new edition (Florence: Leo S. Olschki, 1998) does not seem to have materialized.]

Stevenson, *Terrestrial and Celestial Globes*: Edward Luther Stevenson, *Terrestrial and Celestial Globes. Their History and Construction*, 2 vols., New Haven, Conn., 1921.

Stevens, Wesley M., Guy Beaujouan and Anthony J. Turner, eds., *The Oldest Latin Astrolabe*, (*Physis* 32:2-3), Florence: Leo S. Olschki, 1995.

Stroyls: see Al-Daffa & Stroyls.

Swerdlow, "The Recovery of the Exact Sciences of Antiquity in the Renaissance": Noel Swerdlow, "The Recovery of the Exact Sciences of Antiquity: Mathematics, Astronomy, Geography", in Anthony Grafton, ed., *Rome Reborn – The Vatican Library and Renaissance Culture*, Washington, D.C.: Library of Congress, 1993, pp. 125-168.

Stöffler, *Elucidatio*: Johannes Stöffler, *Elucidatio fabricae ususque astrolabii*, Oppenheim, 1524 edn.

Storey, *Persian Literature*: Charles A. Storey, *Persian Literature – A Bio-bibliographical Survey*, 2 vols. (vol. I, part 1: *Qurʾanic Literature; History*, part 2: *Biography*, vol. II, part 1: *A. Mathematics. B. Weights and Measures. C. Astronomy and Astrology. D. Geography*, part 2: *E. Medicine*, part 3: *F. Encyclopaedias and*

Miscellanies. G. Arts and Crafts. H. Science. J. Occult Arts), London: Luzac & Co., Ltd., later The Royal Asiatic Society of Great Britain and Ireland, 1927-77.

Strasbourg 1995 Colloquium Proceedings: *La science dans le monde iranien à l'époque islamique, Actes du colloque tenu à l'Université des Sciences Humaines de Strasbourg 6-8 juin 1995*, Z. Vesel, H. Beikbaghban and B. Thierry de Crussol des Epesse, eds., (*La Bibliothèque iranienne* 50), Tehran: Institut Français de Recherche en Iran, 1998.

Suhayl: *Suhayl – Journal for the History of the Exact and Natural Sciences in Islamic Civilisation* (Barcelona).

Suter, *Beiträge*: *Heinrich Suter, Beiträge zur Geschichte der Mathematik und Astronomie im Islam*, Fuat Sezgin *et al.*, eds., 2 vols., Frankfurt am Main: Institut für Geschichte der Arabisch-Islamischen Wissenschaften, 1986.

— , *al-Khwārizmī*: *idem et al.*, eds., *Die astronomischen Tafeln des Muhammed ibn Musa al-Khwārizmī* ... , in *Kgl. Danske Vidensk. Skrifter*, 7. R., *Hist. og filos. Afd.*, 3:1 (1914).

— , *MAA*: *idem*, "Die Mathematiker und Astronomen der Araber und ihre Werke", *Abhandlungen zur Geschichte der mathematischen Wissenschaften* 10 (1900), and "Nachträge und Berichtigungen", *ibid.* 14 (1902), pp. 157-185, repr. Amsterdam: The Oriental Press, 1982, and again in *idem*, *Beiträge*, I, pp. 1-285 and 286-314.

Syndram: see *Bielefeld Catalogue.*

Tannery, "*Quadrans vetus*", A-B: Paul Tannery, "Le traité du quadrant de Maître Robert Anglès (Montpellier, XIIIe siècle). Texte latin et ancienne traduction grecque", *Bulletin des sciences mathématiques* (Paris) 32 (1897) (A), and *idem*, "Maitre Robert Anglès – Le traité du quadrant ... ", *ibid.* 34 (1899), pp. 145-150 (B), both repr. in *idem*, *Mémoires scientifiques publiés par J.-L. Heiberg*, V: *Sciences exactes au Moyen Âge (1887-1921)*, Toulouse: Édouard Privat & Paris: Gauthier-Villars, 1922, no. 8. pp. 118-197 (A) and 198-203 (B).

Tekeli, "The Astronomical Instruments of *Zīj-i-Shāhinshāhīya*": Sevim Tekeli, "The Astronomical Instruments of *Zīj-i-Shāhinshāhīya*", *Araştırma* 1 (1963), pp. 88-122.

— , "Equatorial Semicircle": *eadem*, "The 'Equatorial Armilla' of ʿIzz al-Dīn b. Muḥammad al-Wafā'ī and the Torquetum", *Ankara (AÜDTCF) Dergisi* 18 (1960), pp. 227-259.

— , "Ottoman Clocks": *eadem*, "The Clocks in the Ottoman Empire in the 16th Century", *Ankara (AÜDTCF) Yayınları* 171 (1966), pp. 121-339.

— , "The Instruments of Taqī al Dīn and Tycho Brahe": *eadem*, "The Comparison of the Instruments of Taqī al Dīn and Tycho Brahe", *Ankara (AÜDTGF) Dergisi* 16 (1958), pp. 377-386.

— , "Observational Instruments of Istanbul Observatory": *eadem*, "Observational Instruments of Istanbul Observatory", in *Istanbul 1977 Symposium Proceedings*, pp. 33-44.

— , "al-ʿUrḍī on Observations": *eadem*, ""The Article on the Quality of the Observations" of al-ʿUrḍī", *Araştırma* 8 (1970), pp. 57-169.

Thorndike, "Horologe of travelers": Lynn Thorndike, "On the cylinder called the horologe of travelers", *Isis* 13 (1929), pp. 51-52.

Tibbetts, *Arab Navigation*: Gerald R. Tibbetts, *Arab Navigation in the Indian Ocean before the Coming of the Portuguese* ... , (Oriental Translation Fund, N.S. XLII), London: The Royal Asiatic Society of Great Britain and Ireland, 1971, repr. 1981.

Tihon, "L'astronome Alim": Anne Tihon, "Sur l'identité de l'astronome Alim", *AIHS* 39 (1989), pp. 3-21, repr. in *eadem*, *Studies*, IV.

— , *Studies*: *eadem*, *Études d'astronomie byzantine*, Aldershot: Variorum, 1994.

Tobler, "Qibla Projections": Waldo Tobler, "Qibla, and Related, Map Projections", *Cartography and Geographic Information Science* 29:1 (2002), pp. 17-23.

Toomer, *Eastern Wisedome and Learning*: Gerald Toomer, *Eastern Wisedome and Learning: The Study of Arabic in Seventeenth-Century England*, Oxford: Clarendon Press, 1996.

— , "Solar Theory of al-Zarqāl": *idem*, "The Solar Theory of al-Zarqāl: A History of Errors", *Centaurus* 14 (1969), pp. 306-336.

— , "Toledan Tables": *idem*, "A Survey of the Toledan Tables", *Osiris* 15 (1968), pp. 5-174.

Toomer: see also Ptolemy, *Almagest.*

Torode, "Dating Astrolabes": Roland K. E. Torode, "A Mathematical System for Identifying the Stars of an Astrolabe and Finding its Age", *Astrolabica* 5 (1989), pp. 53-76. [To be used with caution – see Dekker, "Astrolabes and Dates and Dead Ends".]

Tunis 1988 Colloquium Proceedings: *Actes du 2e Colloque Maghrébin de l'Histoire des Mathématiques Arabes, Tunis, 1-3 Dec. 1988*, Tunis: Institut Supérieur de l'Éducation et de la Formation Continue, n.d. [*ca.* 1990].

Turkish Manuscripts Catalogue: *Türkiye yazmalarl toplu kataloğu, The Union Catalogue of Manuscripts in Turkey*, Istanbul: Edebiyat Fakültesi Basımevi, in progress.

Turner, A. J., *Early Scientific Instruments*: Anthony J. Turner, *Early Scientific Instruments – Europe 1400-1800*, London: Sotheby's, 1987.

—, "Florence *Navicula*": see EPACT.
—, "Florence *Quadrans vetus*": see EPACT.
—, "Sundials: History and Classification": *idem*, "Sundials: History and Classification", *HS* 27 (1989), pp. 303-318.
Turner, A. J.: see also *Amsterdam NK 1990 Exhibition Catalogue*, *London SM 1976 Exhibition Catalogue*, *Rockford TM Catalogue*, and Wynter & A. J. Turner.
Turner, G., *Antique Scientific Instruments*: Gerard L'E. Turner, *Antique Scientific Instruments*, Poole, Dorset: Blandford Press, 1980.
—, "Dating Astrolabes": *idem*, "A Critique of the Use of the First Point of Aries in Dating Astrolabes", in *Kunitzsch Festschrift*, pp. 548-554, repr. in *idem*, *Studies*.
—, *Elizabethan Instruments*: *idem*, *Elizabethan Instrument Makers – The Origins of the London Trade in Precision Instrument Making*, Oxford: Oxford University Press, 2000.
—, *Navicula*: *idem*, *The Late-Mediæval Navicula*, London: Trevor Philip & Sons Ltd., n.d. [*ca.* 1994]. [On the *navicula* now in Geneva.]
—, *Studies*: *Renaissance Astrolabes and their Makers*, Ashgate-Variorum, 2003.
—, "Whitwell's Addition to a 14th-Century Quadrant": *idem*, "Whitwell's Addition, *c.* 1595, to a Fourteenth-Century Quadrant", *Antiquities Journal* 85 (1985), pp. 454-455.
Turner Festschrift: *Making Instruments Count – Essays on Historical Scientific Instruments presented to Gerard L'Estrange Turner*, Robert G. Anderson, James A. Bennett, and Will F. Ryan, eds., Aldershot: Variorum, 1993.
Turner: see also King & Turner.
H. Turner, *Science in Medieval Islam*: Howard R. Turner, *Science in Medieval Islam – An Illustrated Introduction*, Austin: University of Texas Press, 1995.
Ünver, "Cadrans solaires en Turquie": A. Süheyl Ünver, "Sur les cadrans solaires horizontaux et verticaux en Turquie", *AIHS* 28-29 (1945), pp. 254-266.
—, *Istanbul Observatory*: *idem*, *"İstanbul rasathanesi* (in Turkish), Ankara: Türk Tarih Kurumu, 1969.
—, "Muwaqqits' Quarters": *idem*, "Osmanli türklerinde ilim tarihinde muvakkithaneler" (in Turkish), *Atatürk Konferanslari* (Ankara) 5 (1975), pp. 217-257.
Ullmann, Manfred, *Die Natur- und Geheimwissenschaften im Islam*, Leiden: E. J. Brill, 1972.
Valdés, "Antecedentes del astrolabio de Rojas": Manuel María Valdés, "Antecedentes del astrolabio de Rojas", *Analema* no. 32 (Mayo-Agosto 2001), pp. 14-18.
—, "El astrolabio de Rojas y la açafea": *idem*, "El astrolabio de Rojas y la açafea de Azarquiel", *Analema* no. 28 (Enero-Abril 2000), pp. 4-10.
Valdés: see also Gunella *et al.*
Validi, "Islam and Geography": Ahmet Zeki Validi, "Islam and the Science of Geography", *Islamic Culture – The Hyderabad Quarterly Review* 8 (1934), pp. 511-527.
Van Brummelen, "al-Khalīlī's Auxiliary Tables": Glen Van Brummelen, "The Numerical Structure of al-Khalīlī's Auxiliary Tables", *Physis* 28 (1991), pp. 667-697.
—, "Mathematical Tables in Ptolemy's *Almagest*": *idem*, "A Survey of the Mathematical Tables in Ptolemy's Almagest", in *Frankfurt IGN Festband*, pp. 155-170.
— & Butler, "Interdependence of Astronomical Tables": *idem* and Kenneth Butler, "Determining the Interdependence of Historical Astronomical Tables", *Journal of the American Statistical Association* 92 (1997), pp. 41-48.
van Cittert, *Utrecht Astrolabes*: P. H. van Cittert, *Astrolabes – A Critical Description of the Astrolabes, Noctilabes and Quadrants in the Care of the Utrecht University Museum*, Leiden: E. J. Brill, 1954.
Van Cleempoel, *Louvain Instruments*: Koenraad Van Cleempoel, *A Catalogue Raisonné of Scientific Instruments from the Louvain School, 1530 to 1600*, (*De diversis artibus*: Collections of studies from the Inernational Academy of the History of Science, vol 65 (N.S. 28)), Turnhout: Brepols, 2002.
Van Cleempoel: see also *Greenwich Astrolabe Catalogue*.
van Dalen, *Ancient and Mediaeval Astronomical Tables*: Benno van Dalen, *Ancient and Mediaeval Astronomical Tables: Mathematical Structure and Parameter Values*, Utrecht: Universiteit Utrecht, 1992.
—, "Statistical Method": *idem*, "A Statistical Method for Recovering Unknown Parameters from Medieval Astronomical Tables", *Centaurus* 32 (1989), pp. 85-145.
van Dalen: see also *Frankfurt 2001 Conference Proceedings*.
Van De Vyver, "Les premiers traductions latines de traités arabes sur l'astrolabe": A. Van De Vyver, "Les premiers traductions latines (Xe-XIe siècles) de traités arabes sur l'astrolabe", *Actes du 1er Congrès International de Géographie Historique*, 2 vols., Brussels, 1931, II, pp. 266-290, repr. in *AIOS*, VI, pp. 377-405.
Vardjavand, "Observatoire de Maraqe": P. Vardjavand, "Rapport sur les fouilles de l'observatoire de Maraqe", *Le monde Iranien et l'Islam* 3 (1975), pp. 119-123.

Vatican Borgia Catalogue: E. Tisserant, "Inventaire sommaire des manuscrits arabes du fonds Borgia à la Bibliothèque Vaticane", *Miscellanea Francesco Ehrle*, vol. V, (Vatican City: Biblioteca Apostolica Vaticana, 1924), pp. 1-34.

Varisco, "Islamic Folk Astronomy": Daniel M. Varisco, "Islamic Folk Astronomy", in Selin, ed., *Astronomy across Cultures*, pp. 615-650.

— , *Studies*: *idem*, *Medieval Folk Astronomy and Agriculture in Arabia and the Yemen*, (Variorum Collected Studies Series: CS585), Aldershot & Brookfield, Vt.: Ashgate – Variorum, 1997.

—, *Yemeni Almanac*: *idem*, *Medieval Agriculture and Islamic Science – The Almanac of a Yemeni Sultan*, Seattle, Wa.: University of Washington Press, 1993.

— , "Yemeni Almanac II": *idem*, "An Anonymous 14th-Century Almanac from Rasulid Yemen", *ZGAIW* 9 (1994), pp. 195-213.

— & Smith, eds., *al-Afḍal's Anthology*: *idem* and G. Rex Smith, eds., *al-Afḍal's Anthology*: *The Manuscript of al-Malik al-Afḍal al-'Abbās b. 'Alī b. Dā'ūd b. Yūsuf b. 'Umar b. 'Alī ibn Rasūl (d. 778/1377) – A Medieval Arabic Anthology from the Yemen*, [London?]: E. J. W. Gibb Memorial Trust, 1998.

Vernet, *Estudios*: Juan Vernet, *Estudios sobre historia de la ciencia medieval*, Barcelona: Universidad de Barcelona (Facultad de Filología) & Bellaterra: Universidad Autónoma de Barcelona (Facultad de Filosofía y Letras), 1979.

— , *Ibn al-Bannā*ʿ: *idem*, *Contribución al estudio de la labor astronómica de Ibn al-Bannā*ʿ, Tetuan: Editora Marroquî, 1952.

— , *Studies*: *idem*, *De ʿAbd al-Raḥmān I a Isabel II – Recopilación de estudios dispersos sobre Historia de la Ciencia y la Cultura Española ofrecida al autor por sus discípulos con ocasión de su LXV aniversario*, Barcelona: Universidad de Barcelona, Instituto "Millás Vallicrosa" de Historia de la Ciencia Árabe & Promociones y Publicaciones Universitarias, S.A., 1989.

— , "Las Tabulae Probatae": *idem*, "Las Tabulae Probatae" in *Millás-Vallicrosa Festschrift*, II, pp. 501-522, repr. in *idem*, *Estudios*, pp. 191-212.

Vernet, ed., *Historia de la ciencia árabe*: *idem*, ed., *Historia de la ciencia árabe*, Madrid: Real Academia de Ciencias Exactas, Fisicas y Naturales, 1981.

Vernet, ed., *Textos y estudios*, A-B: *idem*, ed., *Textos y estudios sobre astronomía española en el siglo XIII*, and *Nuevos estudios sobre astronomía española en el siglo de Alfonso* X, Barcelona: Instituto de Filología, Instutución "Milá y Fontanals", Consejo Superior de Investigaciones Científicas, 1981 and 1983.

Vernet Festschrift: *From Baghdad to Barcelona. Studies in the Islamic Exact Sciences in Honour of Prof. Juan Vernet*, Josep Casulleras and Julio Samsó, eds., (*Anuari de Filologia* (Universitat de Barcelona) XX (1996) B-2), 2 vols., Barcelona: Instituto "Millás Vallicrosa" de Historia de la Ciencia Árabe, 1996. Contains *inter alia*:

- George Saliba, "Arabic Science and the Greek Legacy", in I, pp. 19-37.
- Honorino Mielgo, "A Method of Analysis for Mean Motion Astronomical Tables", in I, pp. 159-180".
- Mercè Viladrich, "The *Mumtaḥan* Tradition in al-Andalus ... ", in I, pp. 253-265.
- Àngel Mestres, "Maghribī Astronomy in the 13th Century: a Description of Manuscript Hyderabad Andra Pradesh State Library 298", in I, pp. 383-443. [See now Mestres, Zīj *of Ibn Isḥāq*.]
- ""al-Bīrūnī" in al-Andalus", in II, pp. 583-612.
- Josep Casulleras, "El último capítulo del *Kitāb al-Asrār fī natāʾiŷ al-afkār*", in II, pp. 613-653.
- Paul Kunitzsch and Elly Dekker, "The Stars on the Rete of the So-Called "Carolingian Astrolabe"", in II, pp. 655-672.
- David A. King and Kurt Maier, "The Medieval Catalan Astrolabe of the Society of Antiquaries, London", in II, pp. 673-718.
- Sreeramula Rajeswara Sarma, "The *Ṣafīḥa Zarqāliyya* in India", in II, pp. 719-735.
- Roser Puig, "On the Eastern Sources of Ibn al-Zarqālluh's Orthographic Projection", in II, pp. 737-753.
- Emilia Calvo, "Ibn Bāṣo's Astrolabe [Plate] in the Maghrib and East", in II, pp. 755-767.
- Miquel Forcada, "A New Andalusian Astronomical Source from the Fourth/Tenth Century: The *Mukhtaṣar min al-Anwā*ʿ of Aḥmad ibn Fāris", *in* II, 769-780.

Viladrich, "Horary Quadrants": Mercè Viladrich, "Medieval Islamic Horary Quadrants for Specific Latitudes and Their Influence on the European Tradition", *Suhayl* 1 (2000), pp. 273-355.

— , *Ibn al-Samḥ on the Astrolabe*: *eadem*, Mercè Viladrich, *El* Kitāb al-ʿAmal bi-l-asṭurlāb (Llibre de l'ús de l'astrolabi) *d'Ibn al-Samḥ,* Barcelona: Institut d'Estudis Catalans (*Memòries de la Secció Històrico Arqueològica* 36), 1986.

— , "*Mumtaḥan* Influence in the *Calendar of Cordova*": *eadem*, "The *Mumtaḥan* Tradition in al-Andalus – Analysis of the Data from the *Calendar of Cordova* Related to the Entrance of the Sun in the Zodiacal Signs", in *Vernet Festschrift*, I, pp. 253-265.

— & Martí, "El *Libro dell Ataçir*": *eadem* and Ramon Martí, "Sobre el *Libro dell Ataçir* de los *Libros del Saber de Astronomía* de Alfonso X el Sabio", in Vernet, ed., *Nuevos estudios*, pp. 75-100.

Viladrich: see also Martí & Viladrich.

Villena, *Astrología*: see Cátedra & Samsó.

Villuendas, *Trigonometría de Ibn Muʿādh*: M. V. Villuendas, *La trigonometría europea en el siglo XI – estudio de la obra de Ibn Muʿāḏ* El Kitāb maŷhūlāt, Barcelona: Instituto de Historia de la Ciencia de la Real Academia de Buenas Letras, 1979.

von Braunmühl, *Geschichte der Trigonometrie*: Anton von Braunmühl, *Vorlesungen über Geschichte der Trigonometrie*, 2 pts., Leipzig: B. G. Teubner, 1900, repr. in 1 vol., Niederwalluf (D): Martin Sändig, 1971.

Walker, ed., *Astronomy before the Telescope*: *Astronomy before the Telescope*, Christopher B. F. Walker, ed., London: British Museum Press, 1996.

Washington NMAH Catalogue: Sharon Gibbs with George Saliba, *Planispheric Astrolabes from the National Museum of American History*, Washington, D.C.: Smithsonian Institution Press, 1984.

Washington 1992 Exhibition Catalogue: *Circa 1492 – Art in the Age of Exploration*, Jay A. Levenson, ed., Washington, D.C.: National Gallery of Art & New Haven & London: Yale University Press, 1991. {Catalogue of an exhibition held at the National Gallery of Art during 12.10.1991-12.01.1992.]

Waters, "Renaissance Navigation": D. W. Waters, "Science and the Techniques of Navigation in the Renaissance", *Maritime Monographs and Reports* (National Maritime Museum, Greenwich), no. 19 (1976), 39 pp.

Wensinck *et al.*, *Concordance*: Arent J. Wensinck *et al.*, *Concordance et indices de la tradition Musulmane*, 7 vols., Leiden: E. J. Brill, 1936-1969, paper-back reprint in 4 vols., n.d. (*ca.* 1992).

White, "Iconography of *Temperantia*": Lynn White, "The Iconography of *Temperantia* and the Virtuousness of Technology", first published in 1969, repr. in *idem*, *Medieval Religion and Technology: Collected Essays*, Berkeley, Los Angeles & London: University of California Press, 1978, pp. 181-204.

Wiedemann, *Aufsätze*: Eilhard Wiedemann, *Aufsätze zur arabischen Wissenschaftsgeschichte*, 2 vols., Hildesheim: Georg Olms, 1970 (originally published in *Erlangen PMS SB*, 1902-1928).

—, "Zur Geschichte des Kompasses bei den Arabern": *idem*, "Zur Geschichte des Kompasses bei den Arabern", *Verhandlungen der Deutschen Physikalischen Gesellschaft* 19/20 (1919), pp. 665-667, repr. in *idem*, *Schriften*, II, pp. 883-885, and also in *Islamic Geography*, vol. 15 (1992), pp. 271-273.

— , *Schriften*: Eilhard Wiedemann, *Gesammelte Schriften zur arabisch-islamischen Wissenschaftsgeschichte*, 3 vols., Frankfurt am Main: Institut für Geschichte der Arabisch-Islamischen Wissenschaften, 1984.

Wiedemann & Frank, "Gebetszeiten": *idem* and Josef Frank, "Die Gebetszeiten im Islam", *Erlangen PMS SB* 58 (1926), pp. 1-32, repr. in Wiedemann, *Aufsätze*, II, pp. 757-788.

— & — , "Zirkel zur Bestimmung der Gebetszeiten": *eidem*, "Zirkel zur Bestimmung der Gebetszeiten", *Erlangen PMS SB* 52 (1922), pp. 122-125, repr. in Wiedemann, *Aufsätze*, II, pp. 541-544.

— & Juynboll, "Avicenna über ein Beobachtungsinstrument": Wiedemann and W. Th. Juynboll, "Avicenna's Schrift über ein von ihm ersonnenes Beobachtungsinstrument", *Acta Orientalia* 5 (1927), pp. 81-167, repr. in *idem*, *Schriften*, II, pp. 1117-1203. [See also *ZGAIW* 2 (1985), pp. A47-73 (facsimile of the manuscript).]

— & Würschmidt, "Arabische kegelförmige Sonnenuhr": *idem* and Joseph Würschmidt, "Über eine arabische kegelförmige Sonnenuhr", *Archiv für die Geschichte der Naturwissenschaften und der Technik* 7 (1916), pp. 359-376, repr. in *idem*, *Aufsätze*, II, pp. 830-847.

Winter, "Safavid Science": Henry James Jacques Winter, "Persian Science in Safavid Times", in *Cambridge History of Iran*, VI, pp. 581-609.

Wolf, *Handbuch der Astronomie:* Rudolf Wolf, *Handbuch der Astronomie, ihrer Geschichte und Literatur*, 4 pts. in 2 vols., Zurich, 1890-91 edn., repr. Hildesheim & New York: Georg Olms, 1973.

Woepcke, "Arabisches Astrolab": Franz Woepcke, "Über ein in der königlichen Bibliothek zu Berlin befindliches arabisches Astrolabium", *Abhandlungen der Königlichen Akademie der Wissenschaften zu Berlin* (1855), pp. 1-31 (separatum), repr. in *idem*, *Études*, II, pp. 131-165, and again in *AIOS*, II, pp. 1-36.

— , *Études*: *idem*, *Études sur les mathématiques arabo-islamiques*, 2 vols., Frankfurt am Main: Institut für Geschichte der Arabisch-Islamischen Wissenschaften, 1986.

Worrell, "Qusṭā ibn Lūqā on the Celestial Globe": William H. Worrell, "Qusṭā ibn Lūqā on the Use of the Celestial Globe", *Isis* 35 (1944), pp. 285-293, repr. in *IAM*, vol. 76.

Wright: see al-Bīrūnī, *Astrology*.

Würschmidt, "Gedosi über den Quadranten": Joseph Würschmidt, "Die Schriften Gedosis über die Höhenparallelen und über die Sinustafel", *Erlangen PMS SB* 60 (1928), pp. 127-154, repr. in *AIOS*, VI, pp. 142-169.

— , "Osmanische Zeitrechnung": *idem*, "Die Zeitrechnung im Osmanischen Reich", *Deutsche Optische Wochenschrift*, 1917, pp. 88-100.

Würschmidt: see also Wiedemann & Würschmidt.

Wüstenfeld-Mahler, *Vergleichungs-Tabellen*: Bertold Spuler with Joachim Mayr, *Wüstenfeld-Mahler'sche Vergleichungs-Tabellen zur muslimischen und iranischen Zeitrechnung ...* , Wiesbaden: Franz Steiner, for the Deutsche Morgenländische Gesellschaft, 1961.

Wussing Festschrift: Sergei S. Demidov *et al.*, *Amphora – Festschrift für Hans Wussing zu seinem 65. Geburtstag*, Basle, *etc.*: Birkhäuser, 1992.

Wynter & A. J. Turner, *Scientific Instruments*: Harriet Wynter and Anthony J. Turner, *Scientific Instruments*, London: Studio Vista, 1975.

Youschkevitsch, *Mathématiques arabes*: Adolf P. Youschkevitsch, *Les mathématiques arabes (VIIIe-XVe siècles)*, Paris: Librairie Philosophique J. Vrin, 1976. [A partial translation of Juschkewitsch, *Mathematik im MA*.]

de Zambaur, *Manuel*: E. de Zambaur, *Manuel de généalogie et de chronologie de l'Islam*, first published 1955, repr. Bad Pyrmont: Heinz Lafaire, 1972.

Zenner, "Analemma": Armin Zenner, "Über das Analemma", *Jahresschrift der Deutschen Gesellschaft für Chronometrie* 39 (2000), pp. 211-226.

— , "Mondvolvellen": *idem*, "Der Mond – Volvellen und Aequatorien", *ibid.* 40 (2001), pp. 161-178.

— , "Sonnenuhr des Gerbert": *idem*, "Die Sonnenuhr des Gerbert von Aurilac", *ibid.* 38 (1999), pp. 150-158.

ZGAIW: *Zeitschrift für Geschichte der arabisch-islamischen Wissenschaften* (Institut für Geschichte der Arabisch-Islamischen Wissenschaften, Frankfurt am Main).

Zimmermann, "Dustūr al-Munajjimīn": Fritz W. Zimmermann, "The Dustūr al-Munajjimīn of MS Paris, BN ar. no. 5968", in *Aleppo 1976 Symposium Proceedings*, II, pp. 184-192.

Zinner, *Die ältesten Räderuhren*: Ernst Zinner, *Die ältesten Räderuhren und modernen Sonnenuhren*, (28. Bericht der Naturforschenden Gesellschaft, Bamberg), Bamberg, 1939.

— , *Alte Sonnenuhren*: *idem*, *Alte Sonnenuhren an europäischen Gebäuden*, Wiesbaden: Franz Steiner, 1964.

— , *Astronomische Instrumente*: *idem*, *Deutsche und niederländische astronomische Instrumente des 11.-18. Jahrhunderts*, 2nd edn., Munich: C. H. Beck, 1967, repr. 1972.

—, "Invento español": *idem*, "Un invento español en el siglo X", *Euclides – Revista mensuel de Ciencias Exactas, Fisicas, Quimicas, Naturales y sus Aplicaciones* (Madrid) 4:42 (1944), pp. 559-562.

— , "Regensburger Lehrgerät": *idem*, "Das mittelalterliche Lehrgerät für Sternkunde zu Regensburg und seine Beziehung zu Wilhelm von Hirsau", *Zeitschrift für Instrumentenkunde* 43 (1923), pp. 278-282.

— , *Verzeichnis*: *idem*, Verzeichnis der astronomischen Handschriften des deutschen Kulturgebiets, 1925. [See also Schaldach, "Zinner-Archiv", and Ackermann, "Zinner Archive".]

—, "Wissenschaftliche Instrumente": *idem*, "Wissenschaftliche Instrumente", unpublished, privately distributed, n.d. [after 1958].

Part I

A survey of tables for timekeeping by the sun and stars

To Ted Kennedy

ACKNOWLEDGEMENTS AND NOTES TO THIS VERSION

The present study (**I-II**) was made possible by the cooperation of the directors of numerous manuscript libraries in Europe, the Near East and the United States, who allowed me access to their collections of Arabic scientific manuscripts or provided me with microfilms of specific manuscripts. I should like to express my particular gratitude to the Süleymaniye Library in Istanbul and the Egyptian National Library in Cairo.

In 1972 I received a two-year research grant to catalogue the scientific manuscripts in the Egyptian National Library. When I arrived at the Library I did not know how many manuscripts there were or that, for security reasons, it would take me two years to obtain unlimited access to them. There was a sense in which I had to justify my existence. Whilst preparing my doctoral thesis on the astronomical works of Ibn Yūnus during 1969-72 I had written a description of the tables for timekeeping attributed to him in a Dublin manuscript, which was published in 1973. Even before that paper was published I realized that the tables were not all computed by Ibn Yūnus. It seemed like a good idea at the time to try to figure out just who had computed them. In the course of this pursuit, and especially when I found the tables of al-Khalīlī and Najm al-Dīn al-Miṣrī and realized their scope, I decided to write an article on the development of all such tables from the Islamic Middle Ages. Fortunately, my grant was extended for an additional five years, and the catalogue was indeed prepared and eventually published. Slowly but surely, my article on tables for timekeeping also grew and grew, and by the time I left Cairo for New York an extensive monograph was finally finished. But in New York my research interests changed, and I started to work on the qibla, on the notion of the world centred on the Kaʿba, and on mosque orientations. In fact, when I left New York for Frankfurt two monographs were completed, one from Cairo on timekeeping and one more recent from New York on the sacred geography of Islam. At least summaries of these were published in two articles ("Mīḳāt. ii" and "Makka. iv") in the *Encyclopedia of Islam*.

The research on medieval Islamic science conducted at the American Research Center in Egypt during the years 1972-79 was supported by the Smithsonian Institution and National Science Foundation, Washington, D.C. (1972-79), and by grants from the Penrose and Johnson Funds of the American Philosophical Society (1972-74) and the Ford Foundation (1977-79). The support of each of these five institutions is gratefully acknowledged. Professor Owen Gingerich of the Smithsonian Astrophysical Observatories and Harvard University generously sponsored this research. All of the various drafts of the text were typed by Nabila Hajj, who was the mainstay of the project from its inception right through to 1985, when the Arabic catalogue of the scientific manuscripts was finally published. Kurt Maier of Frankfurt typed much of the new text for the computer in the early 1990s. François Charette controlled the penultimate version for consistency but should not in any way be held responsible for a text penned by someone who at the time he wrote it was younger than Charette is now.

Two sets of tables have been investigated by the next generation of researchers and some of their results are noted in this book. First, Glen Van Brummelen attacked the tables of al-Khalīlī with statistical techniques that I could not hope to understand. Second, François Charette has produced masterly studies of the monumental triple-argument universal auxiliary table of Najm al-Dīn al-Miṣrī and the same author's remarkable treatise on instruments. He also helped me insert new materials by Najm al-Dīn in the final version of the present study.

This first part of the main study (**I**) is dedicated to the one man who influenced me most in my adult life. Ted Kennedy is a man for whom I have an undying feeling of gratitude. He is a man whom I can both love and respect and try – if in vain – in some ways to emulate. His achievements were to no small extent the result of the support of his wife, Mary Helen. I first met Ted Kennedy at Brown University in the autumn of 1969 when I was a graduate student at Yale. Asger Aaboe and Bernie Goldstein organized monthly expeditions there to gather with the Brown crowd, which then consisted of Otto Neugebauer, David Pingree, Abe Sachs and Gerald Toomer, with Ted Kennedy on sabbatical leave from the American University of Beirut. I had already encountered Ted Kennedy on paper, and I was fascinated by the fact that he was active in Beirut. It also seemed to me absurd that an Englishman, *manqué* Near East man, most at home in Dārfūr or Cairo, should be languishing in New Haven, Connecticut, when he too could be in Beirut. Sure enough, my wife and I spent the academic year 1970-71 there, getting to know the Kennedys and their world. I also worked on my dissertation, and enjoyed unlimited access to Ted's library, extensive microfilm collection and unpublished materials. Ted's generosity was boundless: one day George Saliba, the other of Ted's "terrible twins", discovered that Kodak had a special offer; we taped all of Ted's microfilms together and had a copy made for each of us, on several enormous spools.

In our early years in Cairo my wife and I went to Beirut regularly to see the Kennedys. During 1976-77 the Kennedys were in Cairo, living in an apartment next to ours. Ted and I braved the rigours of the Dār al-Kutub together; even with a sense of humour and a few provisions against the cold, this was not easy. When we moved to Frankfurt in 1985, the Kennedys were there too, thanks to the generosity of Professor Fuat Sezgin, for a couple of years. So our paths have crossed many times – in Providence, Beirut, Aynab, Aleppo, Cairo, Frankfurt, Barcelona, Princeton, New York City, and Doylestown, Pa., as well as various conference-sites such as Bucharest, Hamburg, Liège and Cambridge, Ma.

In November, 1999, I spoke at a dinner in Pittsburgh organised by a splinter group of the (U.S.) History of Science Society and attended by colleagues in the history of ancient and medieval science. I was waxing enthusiastically about the exciting materials available to historians of Islamic science. I happened to mention Ted Kennedy. A certain distinguished female colleague, versed in medieval European science of the Aristotelian variety and in American personalities, voiced the question: "You mean *the* Ted Kennedy?". The only answer was: "Of course!".

In this new version I have not added many new materials, since, after I stopped looking in the 1980s, I did not find any save those that more or less fell into my lap.[1] The few new

---

[1] But notice that few new materials of consequence appear to be documented in the monumental, newly-published volumes listed as İhsanoğlu *et al.*, *Ottoman Astronomical Literature*, and *Ottoman Mathematical Literature*. For details see n. 1:14 below.

materials are obvious from the table of contents in **Parts I-II** for they bear an asterisc appended to the section number. Several of these are due to Najm al-Dīn al-Miṣrī and have now been studied by François Charette.[2] Others have been found in two manuscripts which I saw for the first time in 2001, namely, Bursa Haraççioğlu 1177,4 and the precious Istanbul U.L. A314. On the other hand, much more could be written on aspects of timekeeping using astronomical instruments, a topic which I have pursued since arriving in Frankfurt and which has more recently been taken up by Charette. Also I did not add much more information about the manuscripts themselves. I wrote at least **Parts I-II** of this book when I was young and innocent; with increasing years I realize that the lack of information such as date and location of copying is unfortunate, but I am not inclined to add it now since I no longer trust my own notes and the majority of the manuscripts are not properly catalogued anyway. The reader may assume that the manuscripts were copied in the locations for which the tables were compiled; if they were copied elsewhere, unless they were universal, the tables would be useless anyway, apart from their inherent academic interest. It was in Cairo that a very high proportion of the tables were compiled, and it is not an accident that it was in Cairo that I wrote most of this book. Most researchers will find my English survey of the Cairo scientific manuscripts more accessible than the manuscripts themselves. However, the survey is full of references to a certain "**SATMI**". Here, finally, it is.

---

[2] See the paragraph associated with n. 16 to the main preface.

TABLE OF CONTENTS

CHAPTER 1

ON ISLAMIC TABLES FOR TIMEKEEPING

> "That cultural collaboration had ... become possible for the most different nations Al-Bîrûnî appreciates with full consciousness ... . For him Islam was culture rather than religion, and the Arabic language the language of Science rather than that of the Qur'ân." A. Z. Validi, "Islam and Geography" (1934), p. 519.

> "Wherever in the medieval world there were tables, real astronomy was practiced; where tables were lacking there were only dilettantes and dabblers." James Evans, *Ancient Astronomy* (1998), p. viii.

1.0 Introductory remarks

In this study I present a survey of all known examples of a category of Islamic astronomical tables[1] preserved in scientific manuscripts located mainly in libraries in Europe and the Near East. Since many of these precious sources are uncatalogued or not properly catalogued, and also since many of the tables are not clearly associated with an author or do not boast a clear title, the vast majority of the tables have not been investigated previously in modern times.[2]

The tables were intended as aids to the solution of the basic problems of spherical astronomy: given the altitude of the sun or any star, to find the time of day or night, the azimuth of the celestial body, and the longitude of the ascendant or horoscopus. Such tables, many of which contain several thousand entries, are generally not contained in the Islamic astronomical handbooks known as *zīj*es, the profusion and diversity of which were shown by Ted Kennedy already in 1950s.[3] The *zīj*es generally tend to have no tables for timekeeping beyond a set of

---

[1] In the context of the tables and Islamic science in general the term "medieval" refers to the period from *ca.* 750 to *ca.* 1900. The "renaissance" of the mathematical sciences took place in the Islamic world in the late 8th, 9th and 10th centuries. Muslim interest in these sciences was progressive until the 15th century, and thereafter and until the 19th century the Muslims continued to practice them but without further original input of consequence. This is nowhere more evident than in the tables discussed in this study. See also n. **II**-1:1.

[2] Prior to 1970 the only study of an Islamic table for astronomical timekeeping was Goldstein, "Medieval Table for Reckoning Time from Solar Altitude", which contains an analysis of the table described in **2.3.4**. King, "Astronomical Timekeeping in Medieval Cairo", includes an analysis of the tables from the Cairo corpus described in **2.1.1**, **4.7.1**, and **5.1.1**; *idem*, "al-Khwārizmī", pp. 7-11, a discussion of the earliest prayer-tables for Baghdad (**II-3.1**); *idem*, "Astronomical Timekeeping in Medieval Syria", and "Astronomical Timekeeping in Ottoman Turkey", and "Astronomy in the Maghrib", pp. 37-40, for three regional surveys. An overview is in the article "Mīḳāt, ii. Astronomical aspects" in EI_2, repr. in King, *Studies*, C-V. An overview in Arabic – the only one of its kind – is in the Arabic edition of *EHAS*, I, pp. 173-238. Several of the tables listed in **7.1** and the auxiliary tables discussed in **9.1**, **9.3**, **9.5** and **9.6** have been analyzed before: references are given *ad loc.* The existence of the tables described in **2.3.1** and **2.3.2** was noted in Suter, *MAA*, p. 50, and Kennedy, "*Zīj* Survey", p. 161, respectively.

[3] Kennedy, "*Zīj* Survey", published in 1956 and reprinted *ca.* 1990, contains a survey of about 120 Islamic *zīj*es and a partial analysis of several notable examples. The number of individual *zīj*es compiled by Muslim

standard spherical astronomical tables for the latitude for which the work was compiled: with a few exceptions I have not treated these standard tables here. I shall, however, have occasional recourse to the data-base of coordinates of the extensive geographical tables in Islamic *zīj*es prepared by Ted and Mary Helen Kennedy, which is useful for identifying the localities associated with latitudes underlying tables when these are not noted in the sources.[4]

Timekeeping – called *ʿilm al-mīqāt* in Arabic – was an important branch of Islamic astronomy (see **Fig. 1.0a**).[5] In addition to the obvious need for the professional astronomer to be able to determine the time of day or night accurately, the times of the five daily prayers in Islam are astronomically defined.[6] In a parallel study (**II**) I discuss numerous medieval tables for determining the prayer times in different localities, some of which supplement the tables for timekeeping considered in this study (here occasionally referred to as **I**).

Previous investigations by Otto Neugebauer have called attention to the simple arithmetical schemes known from Hellenistic, Byzantine and Ethiopic sources, which display midday shadows throughout the year or, less frequently, the shadow lengths for each seasonal hour of daylight.[7] In another study (**III**) I have analyzed over 50 Islamic shadow-schemes of one kind or another; these are characteristic of the tradition of non-mathematical folk astronomy which flourished in the Muslim world throughout medieval times.[8] Likewise, David Pingree has drawn attention to tables compiled by Indian astronomers displaying the length of daylight as a function of solar longitude.[9] The Islamic tables investigated in this study represent a far

---

astronomers between the 8th and the 19th century is closer to 200. King, "Islamic Astronomical Tables", first published in 1975, contains a preliminary account of various categories of Islamic astronomical tables which are *not* generally found in *zīj*es. See now the interim report in King & Samsó, "Islamic Astronomical Handbooks and Tables", with a summary in the article "Zīdj" in EI_2. New insights are to be anticipated from the new survey of *zīj*es currently in preparation by Benno van Dalen of Frankfurt. On the calendars used by Muslim astronomers see van Dalen's article "Taʾrīkh. 2. Era chronology in astronomical handbooks" in EI_2.

[4] Kennedy & Kennedy, *Islamic Geographical Coordinates*, presents *ca.* 15,000 sets of coordinates for *ca.* 2,500 localities from some 75 medieval works, organised according to locality, source, and increasing longitudes, and increasing latitudes. This important reference work is currently being updated by Mercè Comes in Barcelona.

For the purposes of the identification of localities that could be served by a given medieval value of a latitude, modern latitudes are often irrelevant and, more often than not, misleading.

[5] The best basic survey of Islamic astronomy remains that of Nallino, "Islamic Astronomy", published in 1921, which must be read in the light of the later contributions of E. S. Kennedy and his school. For surveys of "Islamic astronomy" and "astronomy in the service of Islam" written for a general audience, see Saliba, "Islamic Astronomy and Astrology"; and King, "Islamic Astronomy", A-B, and *idem*, "Science in the Service of ... Islam" (first published in 1991), repr. in *idem*, *Studies*, C-I. For the broader picture see Hill, *Islamic Science and Technology*, and Dallal, "Islamic Science, Medicine, and Technology". A useful survey of Islamic scientific literature in general is Endreß, "Wissenschaftliche Literatur". A more popular overview in French by a specialist is Djebbar, *Une histoire de la science arabe*. It has taken an interested layman to produce the first overview of Islamic science that could be used in schools or colleges in the English-speaking world: see H. Turner, *Science in Medieval Islam*.

[6] See further **II-1.1** on these definitions, and **IV** on the motivation behind those for the daylight prayers.

I use the expression "timekeeping" rather than, say "time-measurement" (as in *mésure du temps* or *mesura del tiempo*) or "time-reckoning" (as in *Zeitrechnung*), because, at least as far as the prayers were concerned, the times were pre-defined (*mawāqīt al-ṣalāt*), and Muslims have always endeavoured to "keep" those times.

[7] See Neugebauer, *HAMA*, II , pp. 736-747, for an overview.

[8] See King, "Islamic Folk Astronomy", and, most recently, Varisco, "Islamic Folk Astronomy".

[9] See Schmidt, "Daylight in Hindu Astronomy", for a detailed description of one such table, and Pingree,

Fig. 1.0a: A beautiful miniature of the astronomers of the Istanbul Observatory *ca.* 1577. The men are depicted with various instruments (see **X**), none of which are known to have survived the vicissitudes of time, but some of the manuscripts on the bookshelf behind the men holding the astrolabe are now in the University Library, Leiden. Several manuscripts in that collection bear the mark of ownership of Taqi ʿl-Dīn ibn Maʿrūf (see **Fig. II-5.1**), who is one of those two men. His compilations on astronomical timekeeping are surveyed in the present work (see **I-2.3.6** and **II-14.9**). [From MS Istanbul UL Yıldız F-1404, fol. 57r, courtesy of Istanbul University Library.]

"Sanskrit Astronomical Tables in the US", pp. 72-73, for references to several others. Most Islamic methods of determining time and azimuth are developments of methods originally adopted from Indian sources. See, for example, Davidian, "Al-Bīrūnī on the Time of Day"; E. S. Kennedy's commentary to al-Bīrūnī, *Shadows*; and, most recently, Lorch, "Sine Quadrant".

more sophisticated tradition of mathematical astronomy, and although they have no direct precursors in Greek astronomy, some of them in a sense represent a development of Ptolemaic tables for spherical astronomy, which included, for example, tables of ecliptic rising times for the seven "climates" of classical mathematical geography.[10] The influence of the climates in medieval astronomy and also astronomical instrumentation was far more widespread than has been recognized hitherto.[11] A significant part of the Islamic tradition which I investigate here was concerned with what I call "universal solutions", by which I mean solutions to problems of spherical astronomy which serve all latitudes; tables and instruments serving all of the climates are in a sense "universal" (see further **VIa-b**).

It has been possible to locate extensive corpuses of tables for timekeeping computed for each of Baghdad, Cairo, Taiz, Damascus, Jerusalem, Tunis, Alexandria and Istanbul. The earliest tables are those for Baghdad: some date from the 9th and 10th centuries, when Muslim astronomers had assimilated Indian and Ptolemaic astronomical ideas and were already making significant contributions of their own.[12] The principal centres of astronomical timekeeping from the 13th to the 15th century were Cairo and Damascus,[13] and from the 16th century onwards, Istanbul.[14] There was also impressive activity in Taiz and Tunis in the late 13th and 14th centuries. The growth of interest in timekeeping in Cairo lead to the establishment of the office of the *muwaqqit*, that is, the professional astronomer associated with a religious institution, one of whose functions it was to regulate the times of prayer. I have surveyed the activities of the *muwaqqit*s in another study (**V**). But as we shall see, the tables compiled by certain Syrian and Egyptian *muwaqqit*s between the 13th and 16th centuries relate not only to the times of prayer but also to astronomical timekeeping in general and they are particularly impressive from a mathematical point of view. The Yemeni and Tunisian tables form part of sophisticated traditions of mathematical astronomy in these regions which have only recently been

---

[10] In *Almagest*, II.8 (pp. 100-103) the individual and accumulated rising times of each 10° of the ecliptic are tabulated, and in the *Handy Tables* (Stahlman, *Handy Tables*, pp. 206-242) normed right ascensions and oblique ascensions are tabulated for each 1° of the ecliptic.

[11] See the EI_2 article "Iḳlīm" by A. Miquel, and on the importance of the climates in instrumentation see King, "Astronomical Instruments between East and West", esp. pp. 152, 168-169, and *idem*, "Geography of Astrolabes", pp. 6-9, as well as **VIa-b**.

The only Islamic table for timekeeping involving the climates is a 16th-century Egyptian table for twilight at each half-climate discussed in **II-8.2** (illustrated). A much earlier (Andalusī?) table for lunar crescent visibility computed for each of the climates is discussed in King, "Islamic Tables for Lunar Crescent Visibility", pp. 197-207, and also **VIa-7** (illustrated).

[12] On the foreign influences on early Islamic astronomy see Pingree, "Indian Influence", and *idem*, "Greek Influence", and the same author's article "*ʿIlm al-hayʾa*" in EI_2; and Saliba, "Arabic Science and the Greek Legacy". The only survey of early Islamic astronomy is Morelon, "Early Eastern Arabic Astronomy".

[13] Several Cairo and Damascus astronomers associated with timekeeping and their major works are listed in Suter, *MAA*; Brockelmann, *GAL*; and *Cairo ENL Survey*. For an overview of their activities see King, "Astronomy of the Mamluks".

[14] The Ottoman tradition in mathematical astronomy was until recently documented only in a few specialized studies by Aydın Sayılı and Sevim Tekeli and the art-historian Süheyl Ünver, and various symposium publications such as *Istanbul 1977 Symposium Proceedings* and *Istanbul 1991 and 1994 Symposia Proceedings*. An overview of Ottoman activity in timekeeping is in King, "Astronomical Timekeeping in Ottoman Turkey". We now have the six impressive volumes of references to hundreds of Ottoman scientific scholars and their works listed as İhsanoğlu *et al.*, *Ottoman Astronomical / Mathematical / Geographical Literature*.

investigated for the first time.[15/16] We shall be constantly confronted with the fact that astronomy in Islamic civilization did not progress uniformly, and that eventually, when all the relevant problems had been solved, some many times over, all initiative dried up. What any given astronomer might know in the late period was to some extent a matter of chance. It depended not least on his location, and the authorities locally recognized. The works of astronomers in the period from the 8th to the 11th period were not generally available, and some of them are known now thanks only to the labours of a series of orientalists.

The reader will observe that I have been able to locate virtually no material from al-Andalus[17] and rather little from medieval Iran, Central Asia and India. In al-Andalus at least, mathematical astronomy was apparently directed towards other ends. Yet there must have been some activity in this field: how else can one explain the appearance in the early 15th century of a table of solar altitude as a function of the date and the time of day computed for the latitude of Baez (see **10.1**)? Were corpuses of tables for timekeeping available for Rayy and Nishapur, Bukhara and Samarqand? I have no evidence that they were, but find it hard to believe that they were not. The rich collections of astronomical manuscripts preserved in Iran remain virtually untouched by modern scholarship,[18] but those manuscripts of Iranian provenance which I have examined elsewhere (including the former Uzbek S.S.R. and India) contain little material on timekeeping.[19] However, I do not doubt that further investigation of the vast manuscript sources available for the study of the history of Islamic astronomy will bring to light other examples of the various categories of tables described in this part of my study. Indeed, no-one will be happier than myself to see a supplement to the present work.[20]

Another question that must be posed is: were these tables ever used? I have no evidence (like a sentence here or there in a historical source) that they were. The fact that some of them were copied dozens of times over several centuries is all the indirect evidence that we have.

Furthermore, the Muslim astronomers are generally silent about the methods used to compile their tables, the interpolation schemes they employed,[21] and the nature of the tables for sexagesimal arithmetic[22] as well as of the trigonometric tables which they had at their disposal.[23] Some examples are shown in **Figs. 1.0b-f**. All that we have is the tables for

---

15 See King, *Astronomy in Yemen*, based on over 100 Yemeni astronomical manuscripts.

16 For a first attempt to survey the Maghribī tradition of astronomy see the second version of King, "Astronomy in the Maghrib", parts of which are already superceded by various studies of the Barcelona School, including Samsó, "Maghribī *Zījes*", and notably Mestres, Zīj *of Ibn Isḥāq*.

17 I use this term to indicate that part of the Iberian Peninsular under Muslim domination at any given time.

18 This is not to say that they are not catalogued, and much credit is due to Iranian scholars (especially al-ʿĀmilī, *Aʿyān al-Shīʿa* (56 vols., published 1935-1962), and Āghā Buzurg, *al-Dharīʿa* (25 vols., published 1965-1978), who have surveyed vast quantities of manuscripts mainly in Iranian collections. But tables of the kind studied here seldom bear titles that a cataloguer would pick up.

19 The standard bibliographical reference work on Iranian and Indian astronomy in Persian, listed as Storey, *PL*, II:1, contains no references to any works of consequence dealing with timekeeping. Likewise the *Aʿyān al-Shīʿa* and *al-Dharīʿa ilā taṣānīf al-Shīʿa* (see previous note), contain innumerable references to astronomical works in Persian.

20 See the text to n. 1:51.

21 On Islamic interpolation schemes see Hamadanizadeh, "Medieval Islamic Interpolation Schemes", and the references there cited, as well as Rashed, "Méthodes d'interpolation". On interpolation in tables of timekeeping see **II-5.8**. Some primitive schemes are discussed in **II-8.7** and **11.9**. The short notes in MS Cairo Zakiyya 917,4 would be worth investigating: see n. **II**-10:41.

22 See King, "Medieval Islamic Multiplication Tables", A-B.

23 On Islamic trigonometric tables see the ground-breaking studies of Carl Schoy listed as "Beiträge zur

Fig. 1.0b: An extract from a sexagesimal multiplication table displaying products m × n for m = 0;1, 0;2, ... , 59;59, 60;0 and n = 1 , 2, ... , 60. This extract shows products for m = 14;21, 14;22, ... , 14;40, and n = 1, 2, ... , 60. This manuscript was copied *ca.* 1700. Smaller multiplication tables with integral m are attested from the 10th century onwards. [From MS Paris BNF ar. 2552, fols. 44v-45r, courtesy of the Bibliothèque Nationale de France.]

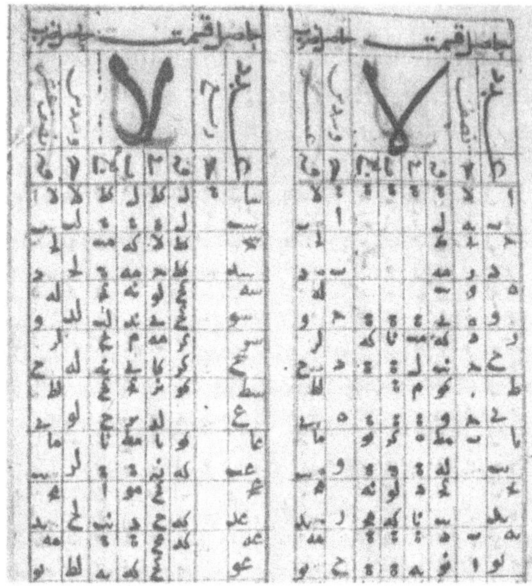

Fig. 1.0c: An extract from a table of sexagesimal quotients displaying quotients m ÷ n for m = 1, 2, ... , 60 and n = 1, 2, ... , 120. This extract shows quotients for m = 31 and 32. This table is unique of its kind. [From MS Cairo DMF 8,1, fols. 23v-24r, courtesy of the Egyptian National Library.]

Fig. 1.0d: An extract from the sine tables attributed to the 10th-century Cairene astronomer Ibn Yūnus. The table gives values of the sine to base 60 for each minute of each degree, with first differences for each second. Entries in the main table are to five sexagesimal digits, but they are based on the values for each 0;10°, which are taken from his major work, the *Ḥākimī Zīj*. Their attribution to Ibn Yūnus cannot be considered certain, given the penchant of Mamluk astronomers for "stretching" tables in this way (see **II-7.2**). The extract shows the entries for 22° and 23°. [From MS Berlin 5752, fols. 13r-14r, courtesy of the Deutsche Staatsbibliothek (Preußischer Kulturbesitz).]

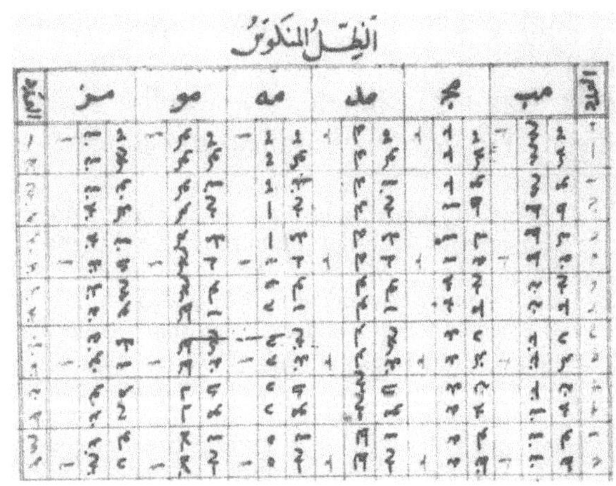

Fig. 1.0e: An extract from some trigonometric tables from Mamluk Egypt. This extract is from a table of cotangents to base 12 and values are given to three sexagesimal figures for each minute. This double-page serves arguments 42°-47°. The tables were copied by the Cairo muwaqqit Sayf al-Dīn Sāṭilmish *ca.* 1450. [From MS Cairo TM 80, fols. 23v-24r, courtesy of the Egyptian National Library.]

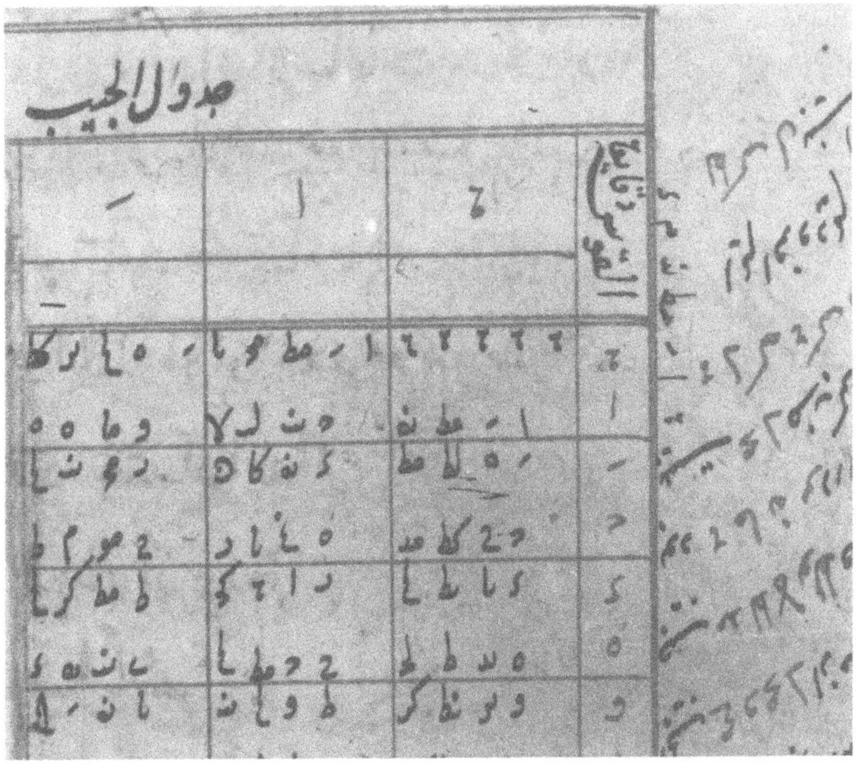

Fig. 1.0f: An extract from the sine tables compiled by Ulugh Beg in Samarqand *ca.* 1430. Entries are to five sexagesimal digits, with values for each minute of arc. This extract shows the first few entries in the table, serving 0-13 minutes over 0°, 1° and 2°. These tables were often copied separately from the *zīj*. [From an unidentified Cairo manuscript (one of those listed in *Cairo ENL Survey*, no. G49, 2.2.1), courtesy of the Egyptian National Library.]

timekeeping themselves, and detailed analysis of error patterns would doubtless furnish additional information on medieval computational techniques. The analysis of medieval tables has made vast strides in recent years.[24] See further **1.5** and **II-1.5**, also **II-10.10**.

Until a survey has been made of medieval and Renaissance European tables for spherical astronomy it will not be possible to ascertain whether any of the different categories of Islamic tables found their way to medieval Europe. There is, alas, no survey of medieval European astronomical tables in general, arranged either chronologically or regionally, nor has any

arabischen Trigonometrie", *Die trigonometrischen Lehren des al-Bîrûnî*, and *Schattentafeln* (on the cotangent and tangent functions), as well as numerous others reprinted in his *Beiträge*, and the following more recent studies: Kennedy, "Overview of the History of Trigonometry"; *idem*, "*Zīj* Survey", pp. 139-140; Berggren, *Islamic Mathematics*, pp. 127-156.

[24] On the first attempts the reader may consult Gingerich, "Applications of Computers to the History of Astronomy"; Kennedy, "The Digital Computer and the History of the Exact Sciences"; and King, "Astronomical Timekeeping in Medieval Cairo", p. 353.

On more recent activity see van Dalen, *Ancient and Mediaeval Astronomical Tables*, and *idem*, "Statistical Method for Recovering Parameters"; Van Brummelen, "Mathematical Tables in Ptolemy's *Almagest*"; *idem* & Butler, "Interdependence of Astronomical Tables"; Mielgo, "Analyzing Mean Motion Tables"; and King, *Mecca-Centred World-Maps*, pp. 163-168.

attempt been made to prepare one.[25] However, the very first medieval European manuscript that I looked at (in Gotha in 1973) and several others that I have seen in recent years (mainly during the 1990s) also contain tables for timekeeping, and it is clear that further research in this field would be worthwhile (see already **10.1-2**).

For inadequacies in the presentation of the material in **I-II**, I beg the indulgence of the reader. The problems associated with research on sources scattered between Princeton and Tashkent, and between Dublin and Sanaa, should be obvious. Not all of the libraries where the sources are housed have facilities for photographing, or even for photocopying manuscripts, and some of those that do have such facilities are administrated by bureaucrats, in-house or in-capital, who are too stupid to provide such services to researchers or who are so clever that they charge outrageous amounts for a single photo. I make no apologies for the organisation of the material presented: it seemed a good idea at the time to arrange the spherical astronomical tables according to their structure and the prayer-tables and the corpuses in which they are found chronologically according to provenance. Also, all this was done in the 1970s, when one worked with type-writers and revised texts using a technique known as "cut and paste". In this first part of my study the reader should bear in mind that the standard tables of spherical astronomy also occur in *zīj*es, and for these only unusual tables relating specifically to timekeeping have been included here.

This study owed its inception to the fact that the manuscripts in Berlin and Paris had long been catalogued in an excellent fashion, and that the descriptions of tables were sufficiently detailed to convey a good idea of the nature of the tables themselves. It owed its continuation and fruition to the fact that for several years I had unlimited access to manuscripts in Cairo and Istanbul. As I prepare the final version of this text I feel confident that most of the materials preserved in the libraries of those two cities, as well as the major European libraries, have been included. Libraries in the Maghrib, Iran and India have been exploited to a much lesser extent, and it may well be that new tables unknown to me now will one day be found there. Any future researcher would do well to start with some materials, mainly anonymous, listed by Ekmeleddin İhsanoğlu and his colleagues in Istanbul.[26] That researcher, if diligent, will find that some of these are indeed already treated in the present work, others maybe not. In any case, the present work should serve as a framework for controlling, classifying, and setting a context for any significant new sources.

---

[25] There are, of course, numerous valuable studies of particular corpuses of tables by scholars familiar with the manuscript sources. See, for example, Toomer, "Toledan Tables", and now F.S. Pedersen, *The Toledan Tables*; Poulle, *Studies*; Goldstein & Chabás, *Zacuto*. See also North, *Horoscopes and History*, on the way in which astronomical tables were used to compile horoscopes, and Chabás, "Cahier d'un croisier", on an astronomer's notebook.

[26] See İhsanoğlu *et al.*, *Ottoman Astronomical Literature*, II, pp. 729, 793-795, 801, 802, 804, 805, 836, 869-870, for various tables relating to timekeeping. For one table mentioned by İhsanoğlu *et al.* which is highly important and of a type not attested elsewhere see **8.5.1*** and **9.4*** below.

1.1 Notation used in the analysis

In this text notation in the form "(**1.2.3**)" stands for "(see Section 1.2.3)". Occasionally here in **I** there are also cross-references to **II** in the form "(**II-1.2.3**)", and *vice versa*.

In the mathematical analysis of tables I use the notation f(x,y) to indicate that the function f is tabulated with x as the horizontal argument and y as the vertical argument. The following notation is used freely in the sequel, On the auxiliary functions for timekeeping (B, C and G), and for calculating ascensions (e) and the azimuth (k and L) see further **Chs. 6-9**.

Latin

a	azimuth (usually measured from the prime vertical)
d	equation of half-daylight (= D – 90°)
B	an auxiliary function for timekeeping, called "absolute base", related to cos δ cos ϕ
C	an auxiliary function for timekeeping related to sin δ sin ϕ
D	semi diurnal arc (sun) or semi nocturnal arc (stars)
e	the auxiliary function tan δ tan ϕ
G	an auxiliary function for timekeeping related sec δ sec ϕ
h	instantaneous solar or stellar altitude
$\tilde{h}$	number of degrees in one seasonal day-hour (= $^1/_{12}$ 2D)
$\tilde{h}'$	number of degrees in one seasonal night-hour (= $^1/_{12}$ 2N)
h_a	solar altitude at the beginning of the afternoon prayer
h_0	altitude in the prime vertical
$-^{h}$	equinoctial hours (also used in the representation of functions, thus: $2D^h(\lambda) = {}^1/_{15}\, D(\lambda)$)
H	meridian altitude
H′	the difference between the Sines of the meridian and instantaneous altitudes (H′ = Sin h - Sin h)
k	an auxiliary function for azimuth calculations, related to sin h tan ϕ
L	an auxiliary function for azimuth calculations, related to sin ψ = sin δ / cos ϕ
L	terrestrial longitude
N	semi nocturnal arc
q	qibla, measured from the meridian
$-_{q}$	relates to the qibla
R	base for trigonometric functions (usually 60, although also 12 and 7 are used for Cotangents and Tangents)
$-^{sdh}$	seasonal day-hours
$-^{snh}$	seasonal night-hours
t	hour-angle (actually an arc in equatorial degrees, rather than an angle)
T	for the sun: time since sunrise or time until sunset; for a star: time since rising or time remaining until setting
z	horizonal shadow of a vertical gnomon (base given in parentheses as subscript)

z_a	shadow at the beginning of the afternoon prayer
z'	vertical shadow
Z	shadow at midday

Greek

α	right ascensions
α'	normed right ascensions ($\alpha' = \alpha + 90°$)
α_ϕ	oblique ascensions for latitude ϕ
α_r	oblique ascensions of the ascendant at daybreak
α_s	oblique ascensions of the ascendant at nightfall
δ	solar declination
δ^*	solar declination augmented by 90° (useful for finding $H = \delta^*-\phi$)
Δ	stellar declination
ε	obliquity of the ecliptic
θ	independent variable
λ	solar longitude or ecliptic longitude
λ'	see below
λ^*	see below
λ_H	longitude of the ascendant (horoscopus)
ρ	longitude of the co-ascendant (point of the ecliptic which rises with a star)
ϕ	terrestrial latitude
ψ	rising amplitude

Miscellaneous

$\bar{\theta}$	complement of θ
λ'	longitude measured from the nearer equinox
λ^*	longitude of the point on the ecliptic opposite the point with longitude λ ($\lambda^* = \lambda + 180°$)
[x]	integral part of x

In virtually all of the tables inspected the entries are written sexagesimally in standard Arabic alphabetical notation (*abjad*).[27] In one important table dating from the earliest period of Islamic astronomy and in most which date from recent centuries the entries are given in Arabic numerical notation – see **Figs. II-3.1** on the former and **I-2.2.4** for one example of the latter. In the analysis the term "digit" means "sexagesimal digit",[28] unless otherwise stated, so that, for example, a number written 23;50,21 represents $23 + {}^{50}/_{60} + {}^{21}/_{3600}$. In the middle of an Arabic text this number would be written "23 50 21 seconds". In the tables the number would be written "23 50 51", and the headings would indicate that the three figures are degrees (abbreviated *j* for *daraj*), minutes (*q* for *daqā'iq*) and seconds (*y* for *thawānī*).

[27] On this notation see Irani, "Arabic Numeral Forms", and Destombes, "Chiffres coufiques".
[28] See, for example, Kennedy, "*Zīj* Survey", p. 17, and Berggren, *Islamic Mathematics*, pp. 39-63.

The medieval trigonometric functions are to base R other than unity (see below). The functions are here denoted by the standard capital notation;[29] thus for a general arc θ:[30]

$$\text{Sin}\ \theta = \text{Sin}_R\ \theta = R \sin\theta\ ,$$
$$\text{Cos}\ \theta = \text{Sin}\ \bar{\theta} = R \cos\theta\ ,$$
$$\text{Vers}\ \theta = R - \text{Cos}\ \theta = R\,(1 - \cos\theta)\ ,\ etc.$$

The Cotangent and Tangent functions were generally, but not exclusively, used with the solar altitude as argument to represent the horizontal shadow of a vertical gnomon length n and the vertical shadow of a horizontal gnomon of the same length, respectively. Thus:

$$\text{Cot}\ h = \text{Cot}_n\ h = n \cot h\ ,$$
$$\text{Tan}\ h = \text{Tan}_n\ h = n \tan h\ .$$

I use Z for Cot H and z for Cot h and occasionally z′ for Tan h, with the base indicated in parentheses in the subscript, thus, for example, $Z_{(6;40)}$ stands for $\text{Cot}_{6;40}$ H. I also use z_a, *etc.*, for the shadow at the *ʿaṣr* prayer. The 'z', for Arabic *ẓill*, shadow", should serve as a reminder that the medieval Cotangent function is of a different nature from the medieval Sine function.

The base used for the Sine and related functions is generally 60, although some of the earliest sources use the Indian value 150.[31] The Cotangent and related functions are usually to base 12 or 7, although in some sources we find 60 (see further below). Most of the sources surveyed in this study use 12 (digits) or 7 (feet) for the gnomon length n. We shall also encounter occasional use of 6 or 6;40, intended as variants of 7;[32] as well as of 60, the last-mentioned intended to correspond to the tables of the Sine function. The use of base 5 is not to be compared with these, for the 5 here is simply 60÷12.[33] However, and this is of greater historical interest (if not significance), we shall also encounter the use of 1 and 10 in early sources, with entries expressed sexagesimally.[34] Very occasionally, we find 20 used as a base.[35]

In the analysis, it is to be assumed that all intervals of time are expressed in equatorial degrees unless otherwise indicated. These may be converted to equinoctial hours (denoted by superscript h) according to the relation:

$$T^{h} = T^{\circ} / 15 = 0;4\ T^{\circ}\ ,$$

or to seasonal day-hours (denoted by superscript sdh) according to:

$$T^{sdh} = 6\ T^{\circ} / D\ .$$

In certain Ottoman tables the time of day is given in equinoctial hours according to the convention that sunset is 12 o'clock.[36] If T represents the equinoctial hours that have elapsed

---

[29] On this see, for example, Kennedy, "*Zīj* Survey", pp. 139-140.

[30] The arguments of the medieval Sine and related functions were considered to be arcs of a base circle, as opposed to angles (a more modern notion in trigonometry).

[31] Base 150 is used, for example, in the Sine tables discussed in **4.1.1**, **4.2.2**, and also **7.1.1** and **7.2.1**.

[32] On these and other bases for the cotangent function see the text to n. **II**-1:17. For the use of 6 see **II-2.9** and **5.6**.

[33] See, for example, various tables discussed in **7.1.5** and **7-8**, and also **II-6.6-7**.

[34] For a Cotangent table to base 1 from the 9th century see **4.1.1**. For the use of 1 and 10 in the same 10th-century source see **4.2.3**. For the use of 10 (and 20) in a 14th-century source see **II-10.6**, especially n. 10:22. For the use of 100 in a 16th-century source see **7.1.12**.

[35] For 20 see **II-10.6**, especially nn. 10:22 and 39a.

[36] On this convention see Würschmidt, "Osmanische Zeitrechnung", and the last paragraphs of **II-1.2** and also **II-14.0**.

since sunrise on a day when daylight is 2D equinoctial hours long, then the time of day T′ according to this convention is:

$$T' = 12 - 2D + T \ .$$

1.2 The standard medieval formulae for timekeeping

For the convenience of the reader I list in this section the formulae that underlie the major tables described in the sequel. To simplify the formulae slightly I use modern notation for trigonometric functions, that is, base R = 1 rather than 60. Each of these formulae is numbered, and I refer to them in the analysis by these numbers preceded by **F**. The formulae underlying the minor tables discussed in the sequel are presented *ad loc.* Likewise approximate methods are discussed only in the analysis. As we shall see, the Muslim astronomers used an array of auxiliary functions to facilitate their computations: these are introduced in the immediate sequel in special script, which is thereafter abandoned but used again in **XIIb**. The formulae for which they provide simplified solutions are presented in rectangular frames.

The formulae derived below represent methods that would be outlined *in words* in the astronomical handbooks known as *zīj*es or in the instructions to the tables themselves.[37] At least from the 13th century onwards lists of formulae of the form a : b = c : d, with four quantities a,b,c,d arranged in columns were available: see **Fig. 1.2a** and **Table 10.3b**. Two examples are:

(1) the Sine of instantaneous altitude : the Cosine of the instantaneous altitude
= the length of the gnomon : the horizontal shadow,

that is, $\text{Sin } h : \text{Cos } h = n : z_n \ (= n : \text{Cot}_n \ h \ )$.

(8) the absolute base : the difference between the Sine of the meridian altitude and the Sine of the instantaneous altitude = unity : the Versed Sine of the hour-angle,

that is, $B : (\text{Sin } H - \text{Sin } h) = 1 : \text{Vers } t$.

It is perhaps appropriate that for technical reasons the formulae in this study are reproduced on a single line: in the first computer-generated version of this study they were more visually impressive, with large curly brackets and multi-level division bars, but with updated hard- and soft-ware these behemoths actually disappeared, their previous existence being indicated on my computer-screen by the message "**Fehler!**".[38] The present mode of rendition of the formulae should constantly remind the reader that the equivalent procedures in medieval Arabic texts

---

[37] On Islamic methods in spherical astronomy, see, for example, Neugebauer, *al-Khwārizmī* (al-Khwārizmī/al-Majrīṭī), and Goldstein, *Ibn al-Muthannā on al-Khwārizmī*; Nallino, *al-Battānī*; King, *Ibn Yūnus*; Samsó, *Estudios sobre Abū Naṣr*; Berggren, "Spherical Trigonometry in the *Jāmiʿ Zīj*" (Kūshyār); Schoy, *Die trigonometrischen Lehren des al-Bīrūnī*; Debarnot, *al-Bīrūnī's* Maqālīd; Villuendas, *Trigonometria de Ibn Muʿāḏ*; Sédillot-*père*, *Traité* (al-Marrākushī); Vernet, *Ibn al-Bannāʾ*; Kennedy, "Spherical Astronomy in Kāshī's *Khāqānī Zīj*"; and Sédillot-*fils*, *Prolégomènes* (Ulugh Beg).

On methods for timekeeping in particular see especially Nadir, "Abū al-Wafāʾ on the Solar Altitude"; and Davidian, "al-Bīrūnī on the Time of Day", and E. S. Kennedy's commentary on al-Bīrūnī, *Shadows*. See also the articles "Maṭāliʿ" [= ascensions] and "Mayl" [= declination] in EI_2.

For a modern discussion of spherical astronomy see Smart, *Spherical Astronomy*, esp. Ch. II.

[38] See n. 14 to the main preface.

Fig. 1.2a: A list of 32 sets of four quantities in spherical astronomy such that the ratio of the first to the second equals the ratio of the third to the fourth. Notice that all expressions are in words, without any recourse to symbols of any kind. This list is anonymous but was taken from the *magnum opus* of the late-13th-century Cairo astronomer Abū ʿAlī al-Marrākushī. [From MS Cairo DM 291,3, fols. 4v-5r, courtesy of the Egyptian National Library.]

involved sentences of several lines. The language of the astronomers of the late medieval period (13th-15th century) considerably facilitated the formulation of complicated mathematical procedures: the crisp, highly technical Arabic of al-Khalīlī (Damascus, *ca.* 1360) and al-Māridīnī (Cairo and/or Damascus, *ca.* 1375), let alone the garbled colloquial of Najm al-Dīn al-Miṣrī (Cairo, *ca.* 1325), is quite different in vocabulary and style from the Arabic of earlier scholars such as al-Battānī (Raqqa, *ca.* 910) and Ibn Yūnus (Cairo, *ca.* 990).[39]

All of the formulae are easily derived either from trivial considerations of the celestial sphere or from not-so-trivial orthogonal projections of the sphere, using the elegant procedure called the analemma that was known since classical Antiquity.[40] Certain Muslim astronomers preferred solutions to problems in spherical astronomy using the rules of spherical trigonometry,[41] but not those who specialised in timekeeping.

The celestial sphere

Fig. 1.2b shows a sphere of unit radius representing the celestial sphere about the observer at O on a fixed earth, with the zenith at Z and the local horizon NWSE. As a result of the rotation of the earth on its axis, the celestial sphere appears to the observer to rotate about

[39] The text of al-Battānī's *Zīj* is published in Nallino, *al-Battānī*, III. For an example of this late medieval scientific Arabic see the texts of al-Māridīnī's treatises on a quadrant of his own invention and on the use of his auxiliary tables published in King, "al-Māridīnī's Universal Quadrant", and al-Khalīlī's instructions on the use of his various tables translated in **II-10.4, 10.7** and **10.8**). For the Arabic of Najm al-Dīn see Charette, *Mamluk Instrumentation*.

[40] On the analemma, see, for example, Paul Luckey, "Das Analemma von Ptolemäus"; Id, "Analemma for Ascensions"; Kennedy & Id, "Ḥabash's Analemma for the Qibla"; Carandell, "Analemma for the Qibla"; E. S. Kennedy's commentary to al-Bīrūnī, *Shadows*; King, *Ibn Yūnus*, *passim*; Berggren, "Four Analemmas for the Qibla"; and, most recently, King, *Mecca-Centred World-Maps*, pp. 15-16.

[41] See, for example, Samsó, *Estudios sobre Abū Naṣr*; and al-Bīrūnī, *Maqālīd*.

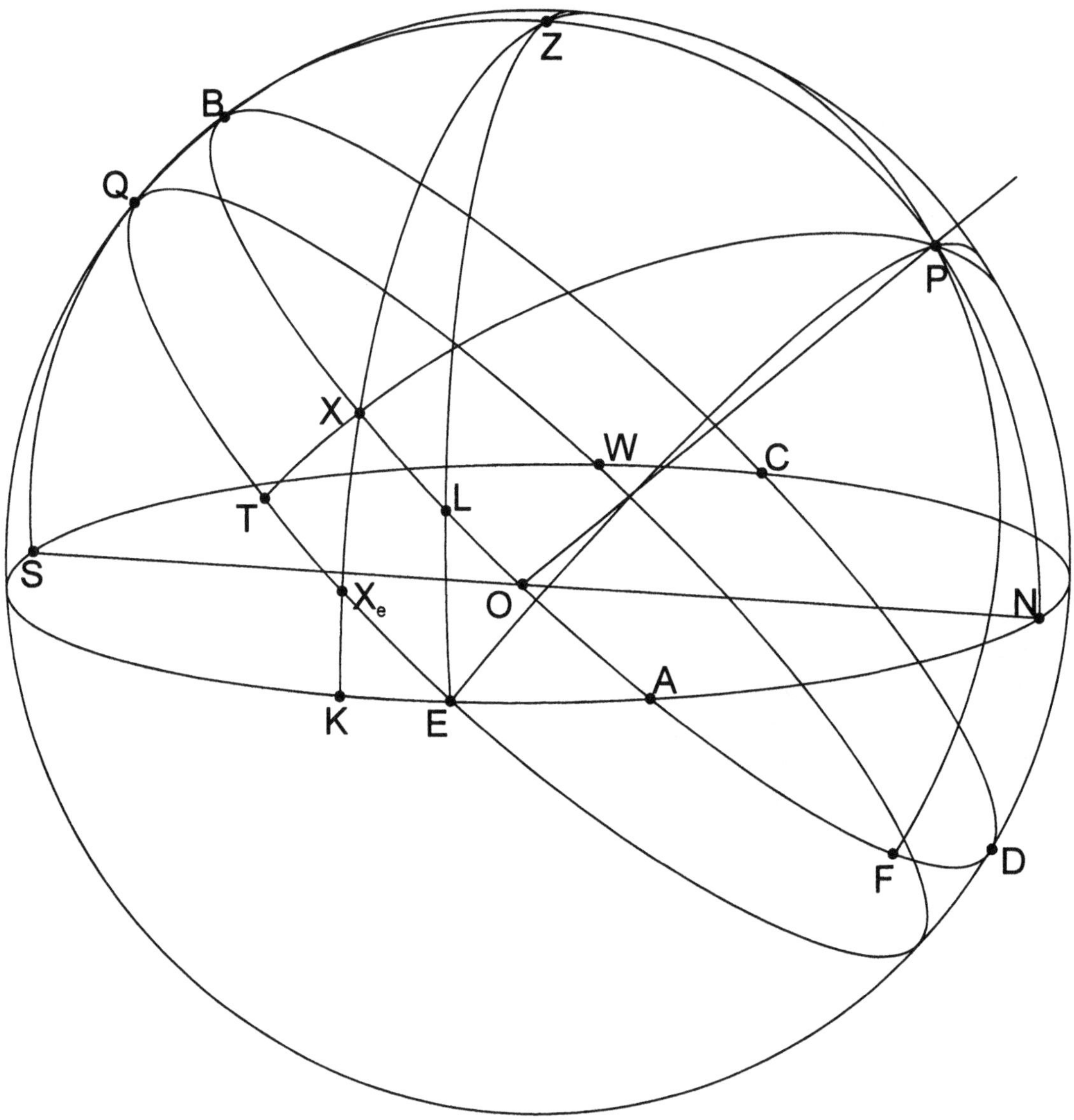

Fig. 1.2b. The celestial sphere. Compare **Fig. VIIc-4a**.

the axis OP once every 24 hours, where P is the celestial pole. The segment ZP defines the local meridian, intersecting the horizon at N and S, the north- and south-points, and the arc PN measures the local latitude ϕ (Arabic, *ʿarḍ al-balad*). The perpendicular plane EZW is the prime vertical, intersecting the horizon at E and W, the east- and west-points. The great circle EQW with its pole at P is the celestial equator.

The sun and stars participate in the apparent motion of the celestial sphere and are observed to traverse small circles parallel to the celestial equator, called declination circles or day-circles (*madārāt*). A given celestial body rises at A, culminates at B on the meridian, sets at C, and crosses the meridian below the horizon at D; its day-circle is ABCD.

The position of the day-circle relative to the celestial equator is determined by the declination of the celestial body, measured by arc BQ. I use Δ to denote declination in general or stellar declination in particular (*al-bu'd*) and δ only to denote solar declination (*al-mayl*). Observe that the meridian altitude H (*ghāyat al-irtifā'*), measured by arc SB, and the meridian altitude below the horizon H* are related to the local latitude and the declination by:

F1 $H = \bar{\phi} + \Delta$ and $H^* = \bar{\phi} - \Delta$.

Some Muslim astronomers tabulated $\delta^*(\lambda) = \delta(\lambda) + 90°$, with which $H(\lambda) = \delta^*(\lambda) - \phi$.

Two other quantities related to the latitude and declination are the rising amplitude ψ (*sa'at al-mashriq*), measured by arc EA on the horizon, and the altitude in the prime vertical h_0 (*al-irtifā' alladhī lā samt lahu*), measured by arc EL on the prime vertical or east-west colure, which is the great-circle ELZW perpendicular to the meridian.

The most important problem of spherical astronomy is, given the instantaneous altitude h (*al-irtifā'*) of a celestial body at X, to determine the time since rising T (*al-dā'ir*), measured by the arcs AX on the day-circle or FT on the celestial equator, or the hour-angle t (*faḍl al-dā'ir*), measured by the arcs XB or TQ. These times are clearly related by the simple formula:

F2 $T + t = D$,

where D is half the diurnal arc (*niṣf qaws al-nahār*), measured by arcs AB or FQ. The arc FE is the equation of half-daylight d (*ta'dīl niṣf qaws al-nahār*) and clearly:

F3 $D = 90° + d$.

When the sun or star is at G, the intersection of arcs AB and PE, the time since rising is d and the hour-angle is 90°. Another important problem of spherical astronomy is to determine the azimuth a of the celestial body (*samt*), measured by the arc EK on the horizon, where ZXK is the altitude circle of the celestial body.

Solar coordinates

As a result of the motion of the earth about the sun, the sun appears to move against the background of fixed stars on a great circle called the ecliptic which is inclined to the celestial equator at an angle $\varepsilon \approx 23^1/_2°$ called the obliquity (*al-mayl al-a'ẓam*). The sun moves along the ecliptic in a direction opposite to that of the apparent daily rotation and completes its journey around the ecliptic in a year.

When the sun is at one or other of the points of intersection of the ecliptic and celestial equator it rises at the east-point and sets at the west-point, and day is equal to night: these solar positions are the vernal and autumnal equinoxes. General solar positions on the ecliptic are determined by the sun's longitude λ (*darajat al-shams*), which is measured from the vernal equinox V. In **Fig. 1.2c** the longitude of the sun at X is measured by the arc VX.

The solar longitude at a given time can be determined either from the tables standard in *zīj*es, or from an ephemeris showing solar and other planetary positions for each day over a period of several years.[42] Another kind of table (called in Arabic *shabaka*) was common in tables for timekeeping: these show the solar longitude for each day of a period of four Syrian

[42] On the solar tables in *zīj*es see Kennedy, "*Zīj* Survey", p. 141. On Islamic ephemerides see the EI_2 article "Taḳwīm. i" by Michael Hofelich.

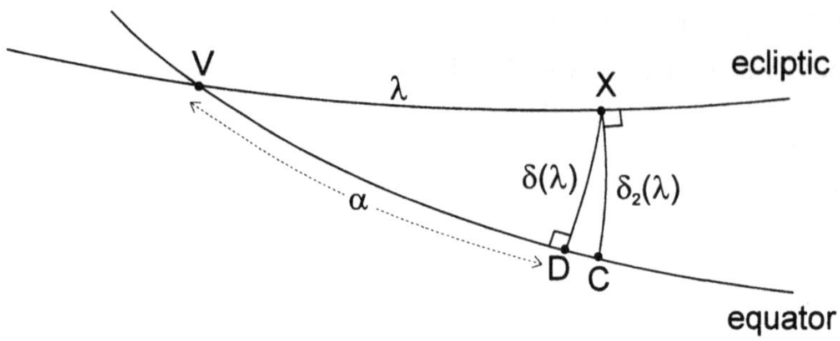

Fig. 1.2c.

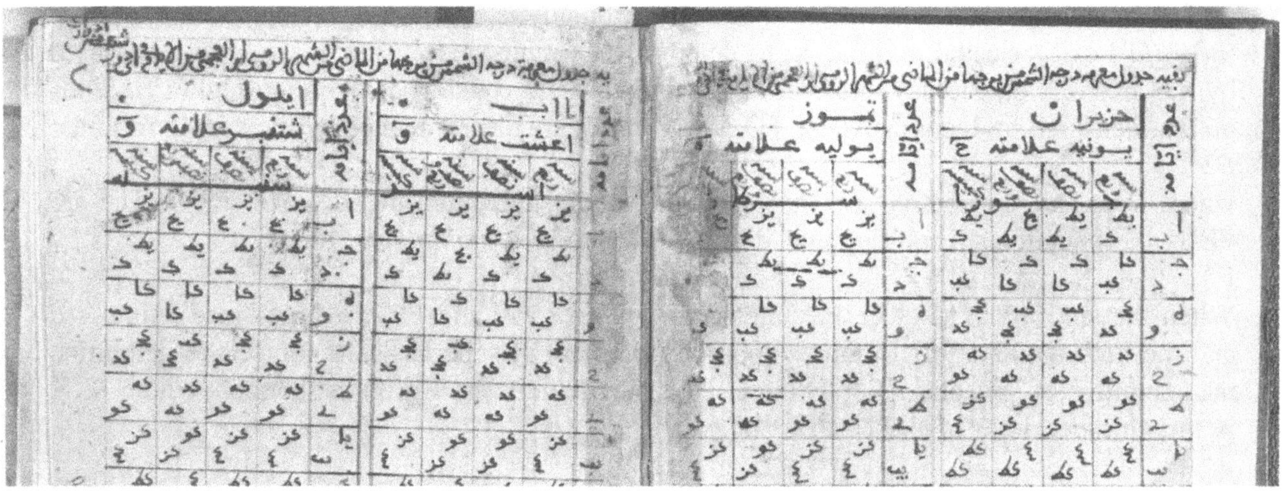

Fig. 1.2d: An extract from a Tunisian table showing the solar longitude as a function of the date in the Syrian (solar) calendar over a four-year cycle. This precedes the set of auxiliary tables for timekeeping described and illustrated in **9.7**. The tables were compiled about 1400 and the manuscript copied about 1600. [From MS Cairo DM 689, fols. 1v-2r, courtesy of the Egyptian National Library.]

or Coptic years (to cover the leap year) at a particular epoch.[43] For an example, see **Fig. 1.2d**.

As the sun moves around the ecliptic from the vernal equinox its declination to the north of the celestial equator increases and the length of daylight also increases, reaching a maximum when $\lambda = 90°$ and $\delta = +\varepsilon$ at the summer solstice. Likewise, as the sun moves past the autumnal equinox its declination to the south of the celestial equator increases and the length of daylight decreases, reaching a minimum when $\lambda = 270°$ and $\delta = -\varepsilon$ at the winter solstice. The declination is measured by the arc XD drawn perpendicular to the equator. The ecliptic is divided into twelve zodiacal signs of 30° each, six northerly and six southerly. Tables of $\delta(\lambda)$ are standard in *zīj*es.

[43] The only published example of a *shabaka* table is that of al-Marrākushī computed for Cairo, 992 Diocletian [= 1275/76], reproduced in Sédillot-*père*, *Traité*, I, pp. 136-137. See also nn. **I**-9:26 and **II**-10:19.

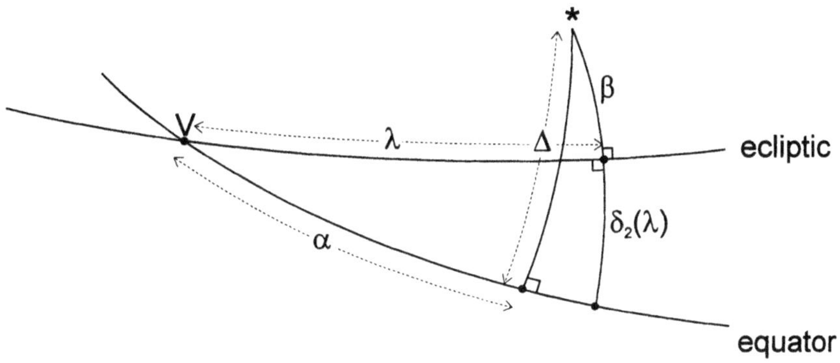

Fig. 1.2e.

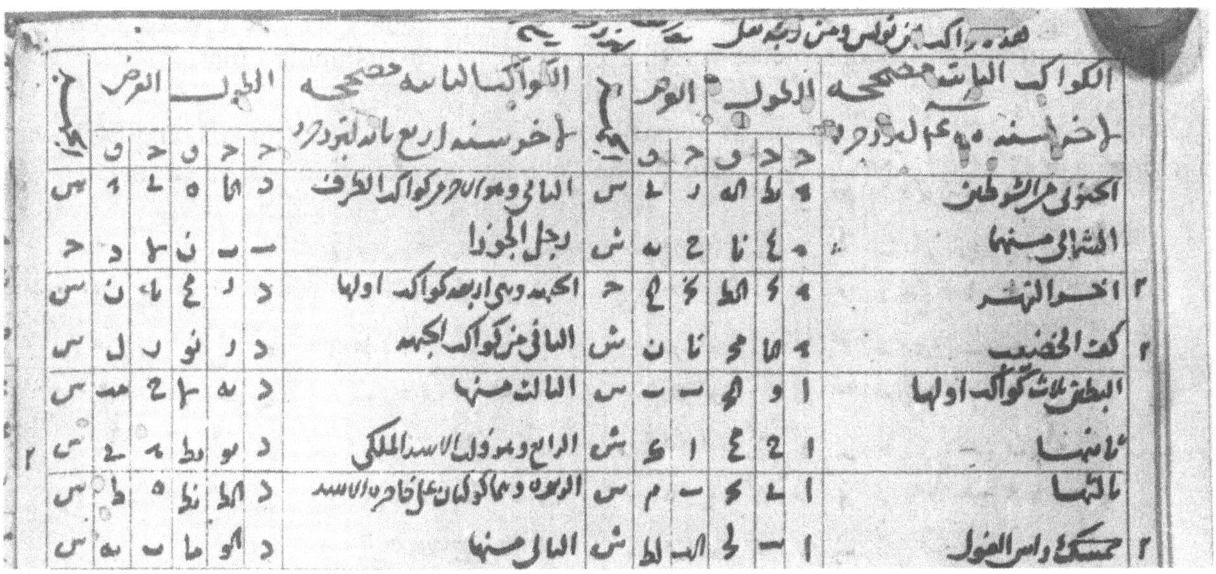

Fig. 1.2f: The star table of Ibn Yūnus, missing from the surviving fragments of his *zīj*, was discovered during in the 1970s in an Egyptian manuscript copied *ca.* 1475 by Aḥmad ibn Timurbāy. It displays longitudes and latitudes for 59 stars for epoch 400 Yazdigird, that is 1032 C.E., some 30 years after the death of Ibn Yūnus. (His solar, lunar and planetary tables run up to 1800 Yazdigird, that is, 2432 C.E., so he clearly was in the habit of planning ahead.) The table has been edited but has not yet been published. [From MS Cairo MM 188,1, fol. 72v, courtesy of the Egyptian National Library.]

The position of the sun relative to the celestial equator is measured by the declination perpendicular to the equator and a coordinate called right ascension α (*maṭāli*ʿ *darajat al-shams*) measured as VD along the equator from the vernal equinox V. Tables of $\alpha(\lambda)$ are also standard in *zīj*es. Another useful notion in spherical astronomy is the "second declination" δ_2 (*al-mayl al-thānī*), which is measured perpendicular to the ecliptic and represented by the arc XC. This quantity, occasionally tabulated in *zīj*es, is of use in the conversion of ecliptic and equatorial coordinates – see below.

Stellar coordinates

The positions of the stars in the heavens are conveniently measured with respect to the ecliptic or the celestial equator by two systems of orthogonal coordinates. In the first the coordinates are longitude and latitude, λ and β, and in the second they are right ascension and declination, α and Δ – see **Fig. 1.2e**. The phenomenon of precession causes stellar longitudes slowly to increase but does not affect the latitudes; both the right ascension and declination of the stars vary slowly with time. Islamic star catalogues display coordinates of prominent stars in one or other system, sometimes both, for a particular epoch: for an example see **Fig. 1.2f.**[44] *Zīj*es generally contain instructions on methods of converting from the one coordinate system to the other.[45] For calculating the time by night it is necessary to know the coördinates of the stars in the equatorial system.

Solar declination and right ascensions

We shall now derive δ, δ_2 and α as functions of λ by means of an analemma. Firstly we project the celestial sphere into the plane whose pole is the vernal equinox V – see **Fig. 1.2g**. AB and CD represent the projections of the celestial equator and the ecliptic so that:

$$\text{arc BD} = \text{arc AC} = \varepsilon \ .$$

O and X_1 represent the projections of V and X, with X_1Y perpendicular to AB, and EGF with midpoint at G represents the projection of the declination circle through X, so that:

$$\text{arc BF} = \delta \ , \ X_1Y = GO = \sin\delta \text{ and } GF = \cos\delta \ .$$

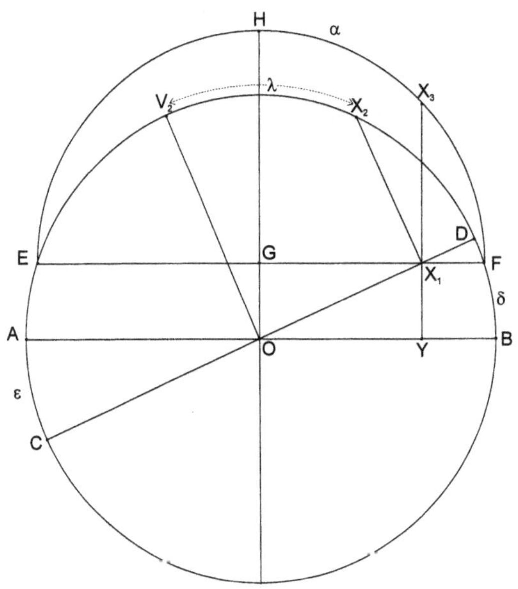

Fig. 1.2g.

[44] Kennedy, "*Zīj* Survey", p. 144a. The earliest Islamic star-catalogues from 9th- and 10th-century Baghdad are investigated in Girke, "Die frühesten islamischen Sternkataloge", alas not yet published. On the most significant star-catalogue of late Islamic astronomy see Knobel, *Ulughbeg's Catalogue of Stars*.

[45] The conversion of stellar coordinates from one system to the other can also be effected using an instrument such as the *ṣafīḥa* of the 11th-century Andalusī astronomer Ibn al-Zarqālluh (Azarquiel), which was specifically designed for that purpose. See the article "Shakkāziyya" in EI_2 and the references there cited, and also **X-5.2**.

Now we rotate the ecliptic into the working plane so that V and X move to V_2 and X_2. Clearly:

$$\text{arc } V_2X_2 = \lambda \quad \text{and} \quad OX_1 = \sin \lambda \ .$$

Next we rotate the declination circle about its axis EF into the semicircle EHF with midpoint at H on the working plane so that X moves to X_3. Clearly:

$$\text{arc } HX_3 = \alpha \text{ and } OY = GX_1 = \cos \delta \sin \alpha \ .$$

Now in ΔX_1OY we have:

$$\angle O = \varepsilon,\ OX_1 = \sin \lambda \ ,\ X_1Y = \sin \delta \text{ and } OY = \cos \delta \sin \alpha \ .$$

It follows immediately that:

F4 $$\delta = \text{arc sin } (\sin \varepsilon \sin \lambda)$$

and:

F5 $$\alpha = \text{arc sin } \{ \tan \delta \ / \ \tan \varepsilon \} \ .$$

Note that the second declination $\delta_2(\lambda)$ equals the declination corresponding to a longitude λ'' whose right ascension is λ, that is:

$$\delta_2(\lambda) = \delta(\lambda'') \text{ where } \alpha(\lambda'') = \lambda \ .$$

It follows from **F5** that:

$$\lambda = \text{arc sin } \{ \tan \delta(\lambda'') \ / \ \tan \varepsilon \} = \text{arc sin } \{ \tan \delta_2(\lambda) \ / \ \tan \varepsilon \} \ ,$$

whence:

F6 $$\delta_2(\lambda) = \text{arc tan } \{ \tan \varepsilon \sin \lambda \} \ .$$

Hour-angle and time since rising

To derive the time as a function of h, D and ϕ, we first project the celestial sphere onto the meridian plane – see **Fig. 1.2h**. The segments OP, OQ and A_1B represent the projections of the celestial axis, the celestial equator and the day-circle. Also X_1 and L_1 represent the projections of X and L. Observe that:

$$\text{arc } PN = \phi \ ,\ \text{arc } BQ = \Delta \ ,\ \text{arc } BS = H = \bar{\phi} + \Delta \ .$$

and:

$$OG_1 = \sin \Delta \ ,\ G_1B = \cos \Delta \ .$$

Now we draw the perpendiculars BB_2 and DD_2 from B and D to NS. Then:

$$BB_2 = \sin H \text{ and } DD_2 = \sin H^* \ .$$

Now the segment A_1D below the horizon measures the projection of the day-circle above the horizon for declination $-\Delta$. For the sun A_1D represents the projection of the day-circle for longitude $\lambda^* = \lambda + 180°$ (*naẓīr darajat al-shams*), so that:

$$H^*(\lambda) = H(\lambda^*) \ .$$

Next we rotate the celestial sphere about its axis ZO until X crosses the meridian at Y_1 and draw the perpendiculars X_1X_2 and Y_1Y_2 from X_1 and Y_1 to NS. Then since:

$$\text{arc } Y_1S = h \ ,$$

we have:

$$X_1X_2 = Y_1Y_2 = \sin h \text{ and } Y_2O = \cos h \ .$$

Similarly:

$$L_1O = \sin h_0 \ .$$

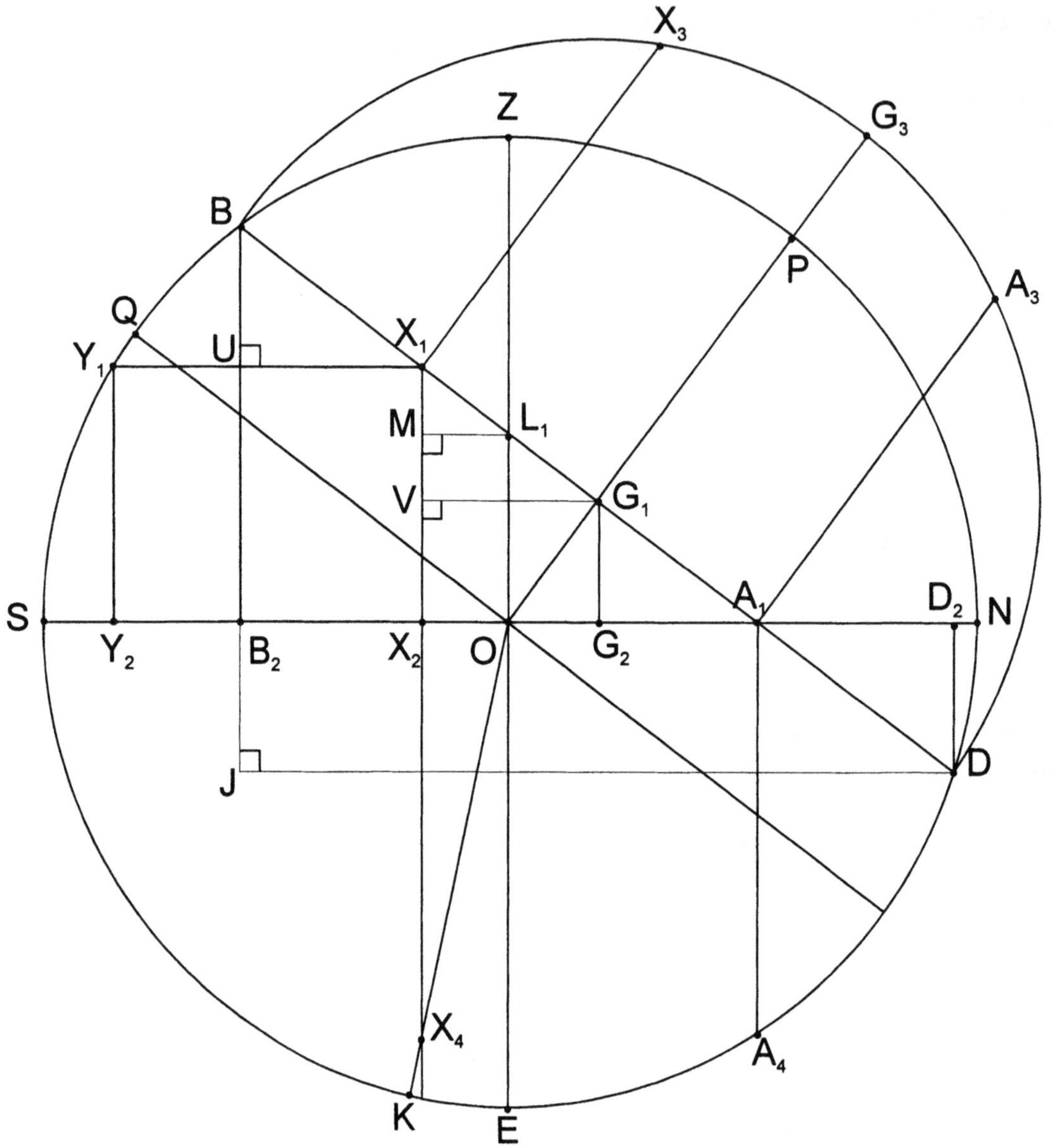

Fig. 1.2h. The celestial sphere reduced to two dimensions. Compare **Fig. VIIa-3.5**.

We now rotate the day-circle about its axis G_1B into the working plane so that A, G and X move to A_3, G_3 and X_3. Then:

$$\text{arc } A_3B = D, \quad \text{arc } A_3G_3 = d, \quad \text{arc } A_3X_3 = T, \quad \text{and} \quad \text{arc } X_3B = t\ ,$$

and since:

$$G_1B = \cos \Delta\ ,$$

we have:

$$A_1B = \text{vers } D \cos \Delta\ ,$$
$$A_1G_1 = \sin d \cos \Delta\ ,$$

$$A_1X_1 = \sin d \cos \Delta + \sin (T - d) \cos \Delta ,$$
$$G_1X_1 = \cos t \cos \Delta ,$$
$$X_1B = \text{vers } t \cos \Delta .$$

Now in ΔOA_1G_1, with G_1G_2 perpendicular to OA, we have:

$$\angle O = \phi,\ OG_1 = \sin \Delta ,\ A_1G_1 = \sin d \cos \Delta , \text{ and also } G_1G_2 = \sin \Delta \sin \phi ,$$

whence it follows that:

F7 $$d = \text{arc sin} (\tan \Delta \tan \phi) .$$

Further, since in ΔBB_2A_1:

$$\angle B = \phi,\ BB_2 = \sin H,\ A_1B = \text{vers } D \cos \Delta ,$$

and in ΔBJD, where J is defined by the perpendicular DJ onto BB_2 produced:

$$\angle B = \phi,\ BJ = BB_2 + D_2D = \sin H + \sin H^* ,$$
$$DB = 2\ G_1B = 2 \cos \Delta ,$$

it follows that:

$$\text{vers } D \cos \Delta / \sin H = 1 / \cos \phi = 2 \cos \Delta / (\sin H + \sin H^*)$$

F8 $$\text{vers } D / \sin H = 1 / (\cos \Delta \cos \phi) = 1 / [\tfrac{1}{2} (\sin H + \sin H^*)] .$$

The product $\cos \Delta \cos \phi$ and the ratio vers D / sin H are of considerable importance in medieval timekeeping. I denote:

$$\mathcal{B} = \cos \Delta \cos \phi \quad \text{and} \quad \mathcal{G} = 1 / \mathcal{B} = 1 / (\cos \Delta \cos \phi) .$$

Also of importance is the segment G_1G_2, the perpendicular from G_1 to SN, whose length we denote by $\mathcal{C}$. Clearly:

$$\mathcal{C} = \sin \Delta \sin \phi .$$

We can now derive four equivalent medieval formulae for t and T. Firstly, in ΔBUX_1, where X_1U is perpendicular to BB_2:

$$\angle B = \phi \text{ and } X_1B = \text{vers } t \cos \Delta ,$$

so that:

$$BU = \text{vers } t \cos \Delta \cos \phi .$$

But:

$$BU = BB_2 - X_1X_2 = \sin H - \sin h ,$$

so that:

F9 $$t = \text{arc vers } \{ (\sin H - \sin h) / (\cos \Delta \cos \phi) \}$$
$$= \text{arc vers } \{ (\sin H - \sin h) \cdot \text{vers } D / \sin H \} .$$

Note that using the auxiliary functions defined above, this reduces to:

$$t = \text{arc vers } \{ (\sin H - \sin h) / \mathcal{B} \} = \text{arc vers } \{ (\sin H - \sin h) \cdot \mathcal{G} \} .$$

Secondly, in $\Delta X_1X_2A_1$,

$$\angle X_1 = \phi \text{ and } A_1X_1 = A_1B - X_1B = \text{vers } D - \text{vers } t ,$$

so that:

$$X_1X_2 = (\text{vers } D - \text{vers } t) \cos \Delta \cos \phi .$$

But:

$$X_1X_2 = \sin h ,$$

so that:

F10 $$t = \text{arc vers} \{ \text{vers } D - \sin h / (\cos \Delta \cos \phi) \}$$
$$= \text{arc vers} \{ \text{vers } D - \sin h \text{ vers } D / \sin H \} .$$

Alternatively, since also:

$$A_1X_1 = \{ \sin d + \sin (T\text{-}d) \} \bullet \cos \Delta ,$$

we have:

F11 $$T = d + \text{arc sin} \{ \sin h / (\cos \Delta \cos \phi) - \sin d \}$$
$$= d + \text{arc sin} \{ \sin h \text{ vers } D / \sin H - \sin d \} .$$

Likewise in ΔX_1VG_1, where G_1V is perpendicular to X_1X_2:

$$\angle X_1 = \phi \text{ and } X_1G_1 = \cos t \cos \Delta ,$$

so that:

$$X_1V = \cos t \cos \Delta \cos \phi .$$

But:

$$X_1V = X_1X_2 - G_1G_2 = \sin h - \sin \Delta \sin \phi ,$$

so that:

F12 $$t = \text{arc cos} \{ [\sin h - \sin \Delta \sin \phi] / (\cos \Delta \cos \phi) \} .$$

This last is the modern formula for $t(h,\Delta,\phi)$. Notice that:

$$t = \text{arc cos} \{ [\sin h - \mathcal{C}] / \mathcal{B} \}$$

Finally, in ΔL_1OG_1 we have:

$$\angle L_1 = \phi,\ LO = \sin h_0, \text{ and } OG_1 = \sin \delta ,$$

whence:

F13 $$h_0 = \text{arc sin} \{ \sin \delta / \sin \phi \} .$$

Azimuth

To derive the azimuth a as a function of h, Δ and ϕ, we project the celestial sphere onto the horizon plane and rotate the projection about its axis SN into the semicircle SKEDN in the working plane. Points X and A move to X_4 on OK and A_4 on the semicircle SEN, respectively, and clearly:

$$\text{arc } EA_4 = \psi \text{ and } OA_1 = \sin \psi .$$

(For convenience we adopt the convention that $\psi \gtreqless O$ as $\delta \gtreqless 0$.) But in ΔOA_1G_1 we have also:

$$\angle O = \phi \text{ and } OG_1 = \sin \delta ,$$

whence:

F14 $$\psi = \text{arc sin} \{ \sin \delta / \cos \phi \} .$$

Furthermore, arc EK = a (now with the convention that $a \gtreqless 0$ as X is north or south of E), and since:

$$OX_4 = OY_2 = \cos h ,$$

we have:

$$OX_2 = \cos h \sin a .$$

Now in $\Delta X_1X_2A_1$:

$$\angle X_1 = \phi \text{ and } X_1X_2 = \sin h ,$$

so that:

$$A_1X_2 = \sin h \tan \phi \ .$$

This quantity is of importance in medieval azimuth calculations and I denote it by $\mathfrak{k}$. With

$$\mathfrak{k} = \sin h \tan \phi \ ,$$

we have further:

$$OX_2 = \cos h \sin a = A_1X_2 - OA_1 = \mathfrak{k} - \sin \psi \ ,$$

whence we derive the standard medieval formula:

F15 $$a = \text{arc sin} \ \{ \ [\ \sin h \tan \phi \ - \ \sin \Delta \ / \ \cos \phi \ ] \ / \ \cos h \ \} \ ,$$

or, in terms of the auxiliary function:

$$a \ = \ \text{arc sin} \ \{ \ [\ \mathfrak{k} - \sin \psi \ ] \ / \ \cos h \ \}$$

which is equivalent to the modern formula:

$$a = \text{arc sin} \ \{ \ [\ \sin h \sin \phi \ - \ \sin \delta \ ] \ / \ (\ \cos h \cos \phi \ ) \ \} \ .$$

Finally, note that in ΔX_1ML_1, where L_1M is perpendicular to X_1X_2:

$$\angle X_1 = \phi \ \text{and} \ X_1M = X_1X_2 - L_1O = \sin h - \sin h_0 \ ,$$

so that:

$$OX_2 = L_1M = (\sin h - \sin h_0) \bullet \tan \phi \ .$$

But:

$$OX_2 = \cos h \sin a \ ,$$

so that:

F16 $$a = \text{arc sin} \ \{ \ [\ \sin h - \sin h_0 \ ] \tan \phi \ / \ \cos h \ \} \ ,$$

as given in some medieval sources (see **5.3**).

Right and oblique ascensions

The rising times of segments of the ecliptic over the local horizon are measured by the corresponding segments of the celestial equator.[46] In **Fig. 1.2i** the arc VH of the ecliptic rises over the horizon at the equator (so that the celestial pole P is on the horizon) with arc VE of the celestial equator. The arc VE measures the right ascension α (*al-maṭāliʿ fi 'l-falak al-mustaqīm*) corresponding to the ecliptic arc VH. In **Fig. 1.2j** the arc VH rises over the horizon of a locality with latitude ϕ with arc VE of the celestial equator (so that the celestial pole P is at ϕ above the north point). This arc VE measures the oblique ascension α_ϕ (*al-maṭāliʿ al-baladiyya*) corresponding to the ecliptic arc VH. In this figure, if the longitude of H is λ and HF is drawn perpendicular to VE produced, then since VE = VF - EF, we have:

F18 $$\alpha_\phi(\lambda) = \alpha(\lambda) - d(\lambda,\phi) \ ,$$

where α and d are determined as in **F5** and **F7** above. Note also the important relation:

F19 $$\alpha_\phi(\lambda) - \alpha_\phi(\lambda^*) = 2D(\lambda) \ .$$

Horoscopus

The point of the ecliptic instantaneously rising over the local horizon (H in **Fig. 1.2h**) is called the ascendant or horoscopus (*al-ṭāliʿ*). The point where the ecliptic intersects the meridian is

---

[46] See the EI_2 article "Maṭāliʿ".

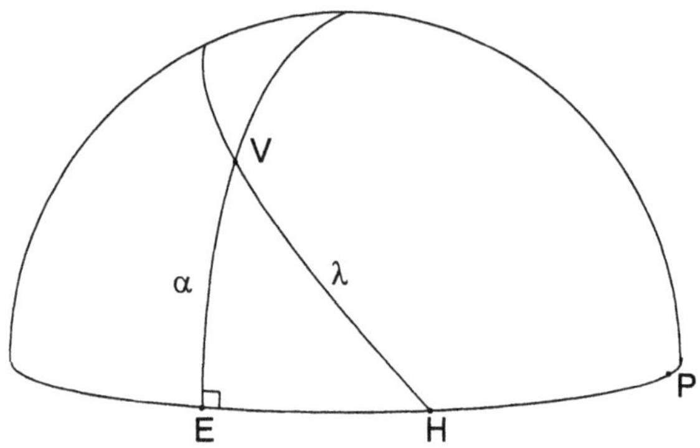

Fig. 1.2i.

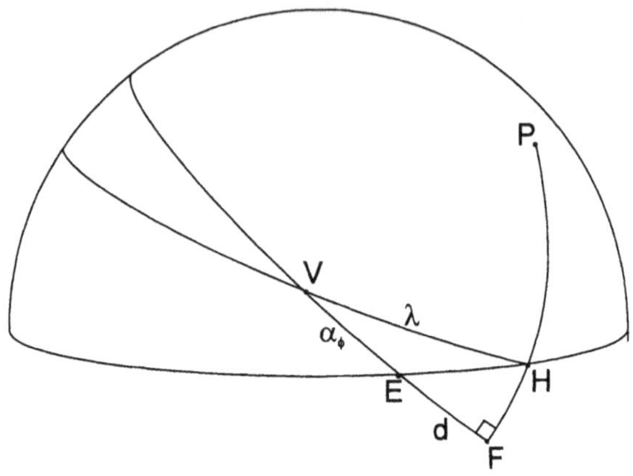

Fig. 1.2j.

called midheaven (*wasaṭ al-samāʾ*). The position of the ecliptic relative to the horizon and meridian is of prime importance in astrology, but also, needless to say, in timekeeping.

If we now denote the longitude of horoscopus H by λ_H when the sun at X has altitude h and longitude λ, it follows from a simple consideration of the celestial sphere that:

F20 $$\alpha_\phi(\lambda_H) = \alpha_\phi(\lambda) + T(h,\lambda) .$$

Likewise for a star with co-ascendant ρ and declination Δ:

F21 $$\alpha_\phi(\lambda_H) = \alpha_\phi(\rho) + T(h,\Delta) .$$

Note further that if λ_M denotes the longitude of upper midheaven M, then:

F22 $$\alpha_\phi(\lambda_H) = \alpha(\lambda_M) + 90° = \alpha'(\lambda_M) ,$$

where α' denotes normed right ascensions, measured from the winter solstice rather than the vernal equinox, and more commonly used in Islamic timekeeping than α. See further **IX**.

The sacred direction

Another problem discussed by the Muslim astronomers is the determination of the qibla or the direction of Mecca from a given locality.[47] The reader is referred to **VIIa-c** for more information on this subject.

Notes on technical terminology

The Arabic expressions for the various astronomical concepts are either mentioned already above or are presented *ad loc*. It may suffice here to add that the arguments in the various tables are invariably denoted as *ʿadad*, literally "number". The term *ḥiṣṣa*, pl. *ḥiṣaṣ*, literally "share" or "portion", means "duration", as in, for example, *ḥiṣṣat al-fajr*, "duration of morning twilight".[48] A special function may be indicated by the expression *maḥfūẓ*, literally, "that which is kept in mind".[49] On the kind of table denoted as *ṭaylasān* see the next Section.

[47] See already the EI_2 articles "Ḳibla. i. ritual and legal aspects" by A. J. Wensinck, and "Ḳibla. ii. astronomical aspects" and "Makka. iv. As centre of the world" by D. A. King, the last two repr. in *idem*, *Studies*, C-IX-X; and the survey of Muslim approaches to the qibla-problem in *idem*, *Mecca-Centred World-Maps*, pp. 47-127.

[48] See, for example, **I-8.0**, **8.2.1**, **8.4.2**, **9.4***, **9.6**, **9.9**, and **II-4.10**, **6.1**, **10.3**, **10.6**, **10.7**, **11.5**, **13.8***, **14.3** and **14.13**.

[49] As, for example, in **II-10.7** and **10.11**.

1.3 Classification of subject matter

In the sequel the reader should bear in mind that the tables of:

$\delta(\lambda)$, $\delta_2(\lambda)$ and $\alpha(\lambda)$ or $\alpha'(\lambda)$,

as well as tables of:

$H(\lambda)$, $d(\lambda)$, $D(\lambda)$ and $\alpha_\phi(\lambda)$ for some particular latitude,

were standard in Islamic *zīj*es.[50] In these tables values were usually given to two sexagesimal digits, occasionally three, for each degree of λ. As noted already, I do not include such common (not in the sense of vulgar) tables in this survey. I have endeavoured to arrange the analysis of other Islamic tables for astronomical timekeeping in such a way that new material can be easily classified and included in any future supplement.[51]

In **Ch. 2** I present examples of five kinds of tables for finding the time of day from the solar altitude and two for finding the time of night by the stars:

1. $T(h,\lambda)$ and $t(h,\lambda)$ for specific ϕ
2. $T(\lambda,h)$ and $t(\lambda,h)$ for specific ϕ
3. $T(H,h)$ and $t(H,h)$ for specific ϕ
4. $t(h)$ for various stars and specific ϕ
5. $T(H,h)$ for all ϕ
6. $T(H,h,D)$ for all ϕ
7. $T(\alpha',\lambda)$ for specific ϕ

In **Ch. 3** I present examples of four kinds of related tables for finding the longitude of the horoscopus from solar or stellar altitudes or from the time of day or night:

1. $\lambda_H(h,\lambda)$ for specific ϕ
2. $\lambda_H(h)$ for various stars and specific ϕ
3. $\lambda_H(T,\lambda)$ for specific ϕ
4. $\lambda_H(T)$ for various stars and specific ϕ

In **Ch. 4** we consider examples of eight different kinds of tables for finding the altitude of the sun or fixed stars, given the time, or the longitude of the horoscopus, or the azimuth:

1. $h(T)$ at the equinoxes and solstices for specific ϕ
2. $h(T,\lambda)$ for specific ϕ
3. $h(T,H)$ for all ϕ
4. $h(D,T)$ for specific ϕ
5. $h(\lambda_H)$ for various stars and specific ϕ
6. $h(T)$ for various stars and specific ϕ
7. $h(a,\lambda)$ for specific ϕ
8. $h_0(\lambda)$ for specific ϕ

Ch. 5 contains a discussion of examples of seven different kinds of tables for finding the azimuth of the sun or the stars:

1. $a(h,\lambda)$ for specific ϕ
2. $a(\lambda,h)$ for specific ϕ

---

[50] See Kennedy, "*Zīj* Survey", pp. 140-141, and King, *Ibn Yūnus*, II.3.
[51] See the text to n. 1:20.

3. a(H,h) for specific ϕ
4. a(T,λ) for specific ϕ
5. a(T) at the equinoxes for specific ϕ
6. $\psi(\lambda)$ for specific ϕ
7. $\psi(\Delta)$ for specific ϕ

In **Chs. 6, 7 and 8** I note the existence of numerous examples of minor auxiliary functions for timekeeping and azimuth calculations. **Ch. 9** concludes the study with a discussion of some 15 examples of more extensive auxiliary tables for solving problems of spherical astronomy for all latitudes. In **Ch. 10** I survey a few medieval European and Renaissance tables in the same spirit that have come to my attention.

1.4 The format of the tables

In general the Muslim astronomers took advantage of the symmetry of the functions they were tabulating, in order to avoid pointless duplication of entries. On the behaviour of the various functions see my previous study of the Cairo corpus, where various graphs are presented.[52]

For functions such as $\delta(\lambda)$, as well as $d(\lambda)$ and $\psi(\lambda)$ for particular latitudes, whose absolute values are symmetrical about the equinoxes and solstices, 90 entries suffice to display the function for each degree of λ. Tables of these functions were usually arranged in three columns of 30 entries, often with a single vertical argument running from 1 to 30, understood to represent 1°-30°, or 31°-60° or 61°-90° as necessary.

For functions such as $H(\lambda)$ and $D(\lambda)$ which are symmetrical about the solstices but not the equinoxes, two sets of 90 entries and an additional (non-zero) equinoctial value display the function for each degree of λ. Tables of these functions were usually arranged in six columns of 30 entries, the first three serving the northern signs and the second three the southern ones. A vertical argument running from 1 to 30 downwards serves the first and third quarters of the ecliptic, and a second one running from 0 to 29 upwards serves the second and fourth quadrants. The equinoctial value is thus omitted (but is generally obvious anyway). Alternatively, since both H and D increase monotonically between the winter and summer solstice, the tables might be arranged so that the first three columns serve the southern signs and the second three columns the northern ones. In this case, if the arguments run from 1 to 30 downwards and 0 to 29 upwards, the value of the function at the winter solstice will be omitted.

The functions $t(h,\lambda)$, $T(h,\lambda)$, $a(h,\lambda)$ and $h(a,\lambda)$ display the same symmetry as $H(\lambda)$ and $D(\lambda)$, and for a given value of the first argument a page or double page of tables serves to display the function for each degree of λ. See, for example, **Figs. 2.1.1a** and **II-5.6a**. If either of the first three functions are tabulated with λ as the horizontal argument, a given page or double page might serve each degree of λ from, say, the winter solstice to the summer solstice, but now the maximum vertical argument is H or [H] and hence varies with λ. See, for example, **Figs. 2.2.1-3**.

[52] See King, "Astronomical Timekeeping in Medieval Cairo" (n. 1:2), esp. pp. 361, 363, 365, 367 and 370.

The functions $\alpha(\lambda)$ and $\alpha_\phi(\lambda)$ are not symmetrical about the equinoxes or the solstices. Tables of these functions were usually arranged in 12 columns of 30 entries, with a single vertical argument running from 1 to 30 or from 0 to 29, with each column essentially serving a single zodiacal sign. The tables of $\lambda_H(h,\lambda)$ are of the same kind, and for a given value of h a double page of tables with 12 columns of 30 entries serves to display the function for each degree of λ. See, for example, **Fig. 3.1.1**.

The various functions f(H,h) discussed in the sequel are computed for a given range of H and each degree of h up to H. The resulting table, which is usually spread over several pages, is trapezoidal in shape. Such tables were generally called in medieval Arabic *ṭaylasān*, which means "shawl".[53] See, for example, **Figs. 2.3.1-2**.

For other large tables the format is devised to best suit the function tabulated, often with astounding ingenuity. See, for example, **Figs. 2.6.1** and **4.4.1**.

1.5 Recomputation of tables

> "One man's error is another man's data." Berman's corollary to Robert's axiom.
>
> "... (the) non-existence (of al-Khāzinī's qibla-table) does not prevent King from specifying its characteristics in considerable detail. This level of theoretical construction makes a reader seriously question imposing on data an order that might not be fully justified by the medieval texts (some of which are not even extant)." E. Savage-Smith, "Review of King, *Mecca-Centred World-Maps*", p. 34c.

Most of the tables discussed below were recomputed during the period 1970-74 with the electronic computers and myriads of punched cards at the American University of Beirut, Yale University or the American University in Cairo.[54] In certain cases, the computer was useful in facilitating the analysis of the tables and the determination of the underlying parameters. Very occasionally, one may be able to reconstruct a table that does not survive in its original form, or at least comment on its structure and the accuracy of its entries, using in both cases, other derivative tables that have survived.[55] Such techniques may inevitably be beyond the

---

[53] On this term see Goldstein, "Medieval Table for Reckoning Time", p. 61; and Albert Arazi, "Noms des vêtements d'après *al-Aḥādīṯ al-ḥisān fī faḍl al-ṭaylasān* d'al-Suyūṭī", *Arabica* 23 (1976), pp. 109-155. In Yedida Kalfon Stillman, *Arab Dress from the Dawn of Islam to Modern Times – A Short History*, Norman A. Stillman, ed., Leiden, *etc*.: E. J. Brill, 2000, pp. 18, 39, 44, 48, 51-52, 71, 73-74, 91, 96 and 179, there no mention of a *trapezoidal-shaped* shawl.

The term is uncommon in the medieval astronomical literature, apart from the titles of this kind of table. In fact, I know of only two other attestations, both by al-Bīrūnī (see **II-2.2**). In his treatise on the construction of astrolabes entitled *al-Istī'āb* (MS London BL Or. 5593, fol. 56v), al-Bīrūnī poses the question whether a particular variety of non-standard astrolabe "leads to the knowledge of the hours and the ascendants in the same way as the *ṭaylasān* table and different kinds of sundials do". (This statement is recorded in Wiedemann, *Aufsätze*, II, pp. 523-524; see also his n. 10 on p. 528.) Again, al-Bīrūnī in his *Chronology* (p. 132 of Sachau's translation) uses the term *ṭaylasān* to describe a table for calendar conversion. (This reference is already recorded in Kennedy, "*Zīj* Survey", p. 137, *ad* no. X201.)

[54] Computer time was made available in Beirut by the Mathematics Department, American University of Beirut; in New Haven by the Department of Near Eastern Languages and Literatures, Yale University; and in Cairo by the Smithsonian Institution, Washington, D.C.

[55] See, for example, Hogendijk, "al-Khwārizmī's Sine Table", and King, *Mecca-Centred World-Maps*, pp. 71-75.

comprehension of colleagues innocent of mathematics and not well-versed in the history of mathematical astronomy.

I have refrained from giving sample entries from the tables but occasionally comment on their accuracy. Several individual tables deserve more detailed analysis than I present in this survey, and now computer-assisted techniques are available than I cannot hope to understand, let alone to employ myself.[56]

---

[56] See various recent studies by Glen Van Brummelen and Benno van Dalen listed in n. 1:24.

CHAPTER 2

TABLES OF TIME AS A FUNCTION OF SOLAR AND STELLAR ALTITUDE

2.0 Introductory remarks

The functions tabulated in the tables described in this chapter are either the time since rising of the sun or star T (for altitudes in the east), or the hour-angle t. These are related by the simple formula (*cf.* **F2**):

$$T + t = D ,$$

where D is the semi diurnal arc. The Arabic terms for the time since rising and hour-angle are *al-dāʾir* and *faḍl al-dāʾir*, respectively.

In those tables for the sun in **2.1** and **2.2** the arguments are h and λ. The tables discussed in **2.3** and **2.5** have H and h as arguments, and in the case of those for the stars in **2.4** the argument is h, but a value of H underlies the table for each star. The tables of **2.5** display the time approximately in seasonal day-hours for all latitudes. The table described in **2.6** has three arguments – H, h and D – and serves both the sun and stars and works for any latitude. The tables for timekeeping by the stars discussed in **2.7** are of a different kind, giving the time as a function of the ascension of the star that is culminating and the solar longitude λ. Those in **2.8** take advantage of a convenient temporary celestial phenomenon. In most of the tables time is measured in equatorial degrees.

See **II** for the other, smaller tables which accompany such tables as these for Cairo, Mecca, Damietta, Jerusalem, Damascus, Aleppo, Anatolia and Istanbul.

2.1 Tables of time as a function of instantaneous solar altitude and solar longitude, for a specific latitude

In this section I note examples of tables of t(h,λ) or T(h,λ) for particular latitudes. Since both of these functions are dependent on δ, and since two points on the ecliptic equidistant from the solstices have the same declination, for a given altitude 181 values suffice to display either function for each integral degree of solar longitude ($\lambda = 0°$ and $\lambda' = 1°, 2°, \ldots, 90°, \delta \gtrless 0$). In the tables described below, full advantage is taken of the symmetry of the functions t(h,λ) and T(h,λ). For given h, generally six columns of 30 entries display the tabulated function. When for certain longitudes the sun does not attain the altitude in question, no entry is given in the table, and columns that would otherwise be devoid of entries are omitted. This format results in the omission of either the equinoctial value or a solstitial value. The format of the tables described in **2.2** obviates the need to prepare additional tables for either the equinoxes or the solstices (**2.1.1** and **5.1.1**). Each of the tables of t or T described in this section and in the next contains over 10,000 entries.

A few Muslim astronomers used tables of t(h,λ) or T(h,λ) to compile other tables for the duration of twilight,[1] based on the assumption that this phenomenon begins or ends when the sun is at a particular angle of depression below the horizon – see **Figs. 2.1.1a-b** and **II**, *passim*.

2.1.1 Ibn Yūnus / al-Maqsī / Ibn al-Kattānī / al-Bakhāniqī: Cairo

In the main corpus of tables for timekeeping that was used in medieval Cairo[2] the functions:

t(h,λ) and T(h,λ)

are tabulated for the domains:

h = 1°, 2°, ... , 83° and λ = 1°, 2°, ... , 90° and 181°, 182°, ... , 270° .

Entries are computed to two digits and are based on the parameters:

φ = 30;0° (Cairo-Fustat) and ε = 23;35° .

No entry is given for the equinoxes. These tables are rather accurately computed, although for higher altitudes there are errors that are sometimes as high as ±5 in the second digit, and the interpolation system used is inadequate.

The tables exist in numerous manuscript sources, which are discussed in greater detail in **II-4** and **5**. (See also **I-4.7.1** and **5.1.1** on certain other tables in the corpus.) The two sets of tables of T(h,λ) and t(h,λ) may be copied separately or both functions may be displayed on facing pages of the manuscript – see **Figs. 2.1.1a-b**. In some manuscripts the two functions are tabulated side by side with the solar azimuth, a triplet of values (T,t,a) being given for each pair of arguments (h,λ) – see **Figs. II-5.6a-b**.

In several manuscripts the tables are attributed to the late-10th-century Egyptian astronomer Ibn Yūnus,[3] who is certainly the compiler of the tables of a(h,λ) (**5.1.1**), but it seems that the tables of T(h,λ) in their present form were compiled by the late-13th-century Cairo astronomer al-Maqsī.[4] This is confirmed by the mid-14th-century Jerusalem astronomer al-Karakī (**2.2.1**) and also by his contemporary, the Cairo astronomer al-Bakhāniqī (see below).[5] Nevertheless, I suspect that al-Maqsī based his tables of T(h,λ) on an earlier set of tables by Ibn Yūnus which has not survived (see, however, **II-5.3-4**). About the middle of the 14th century the Egyptian astronomer Ibn al-Kattānī[6] prepared a set of tables of t(h,λ) using al-Maqsī's tables of T(h,λ): Ibn al-Kattānī's tables exist in a unique copy in his own hand, namely MS Istanbul Kılıç Ali Paşa 684. Not long thereafter, the astronomer al-Bakhāniqī rearranged the tables so that al-

---

[1] See the article "Shafaḳ" on twilight in EI_2.

[2] This corpus was analyzed for the first time in King, "Ibn Yūnus' *Very Useful Tables* for Reckoning Time by the Sun", *AHES* 10 (1973), pp. 342-394, repr. in *idem*, *Studies*, A-IX (hereafter "Astronomical Timekeeping in Medieval Cairo"). However, that early study was based on a set of manuscripts in which the tables are mainly attributed to Ibn Yūnus (see next note). The corpus is investigated anew in **II-4-5**.

[3] On Ibn Yūnus see the article in *DSB*. The introduction and first few chapters of his major work are published with French translation in Caussin, "Table Hakémite", and his treatment of spherical astronomy is analyzed in King, *Ibn Yūnus*. On his influence in later Egyptian astronomy see *idem*, "Astronomy in Fatimid Egypt", and *idem*, "Astronomy of the Mamluks".

[4] On al-Maqsī (Suter, *MAA*, no. 383) see *Cairo ENL Survey*, no. C15, and King, "Astronomy of the Mamluks", pp. 540, 541 and 547.

[5] On al-Bakhāniqī see Brockelmann, *GAL*, SII, pp. 158 and 1019; *Cairo ENL Survey*, no. C28; and King, *Astronomy in Yemen*, no. 13.

[6] On Ibn al-Kattānī (Suter, *MAA*, no. 411) see *Cairo ENL Survey*, no. C32.

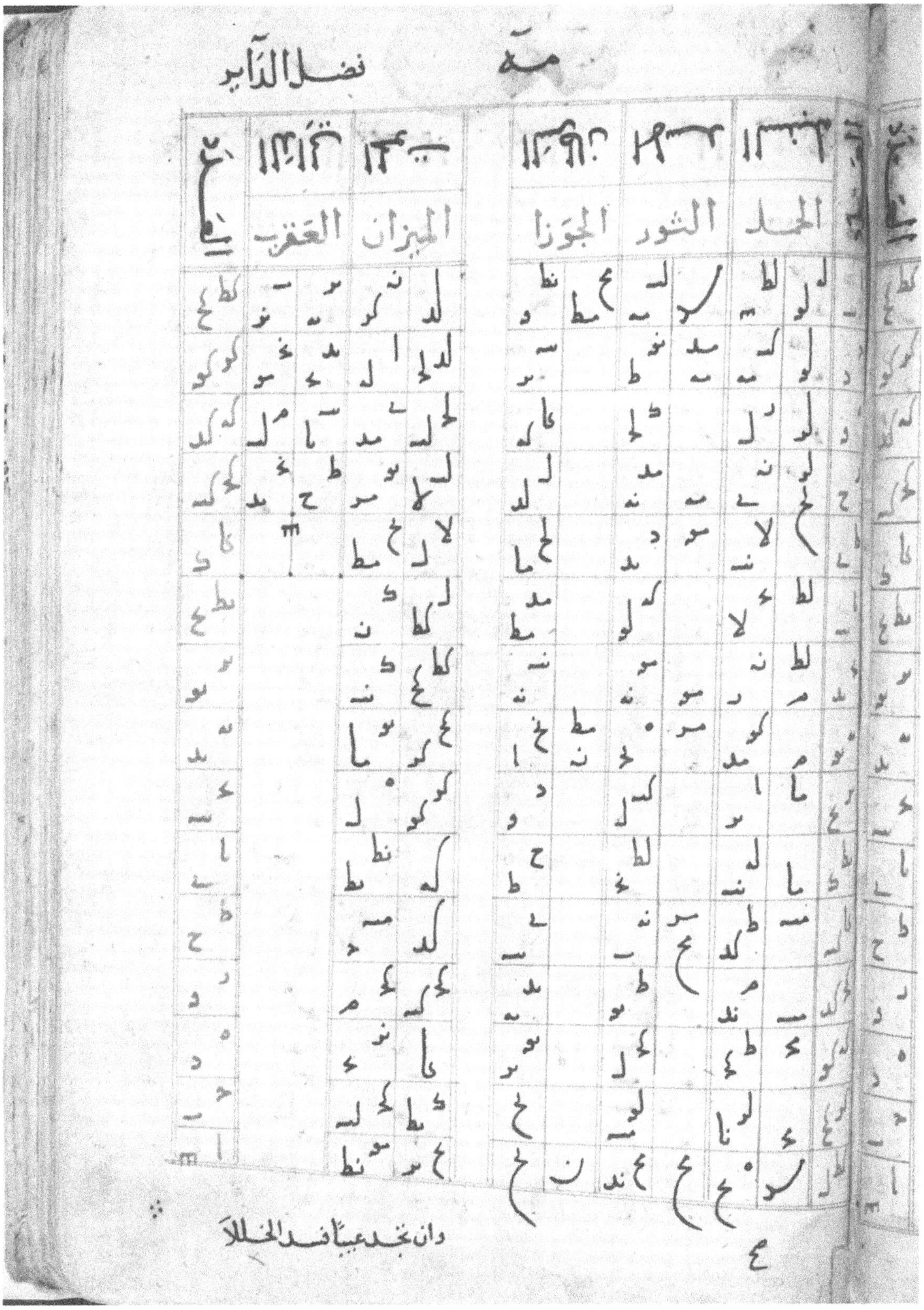
مه
فضل الدائر
الحمل الثور الجوزا
الميزان العقرب
وان تجد عيبا فسد الخللا

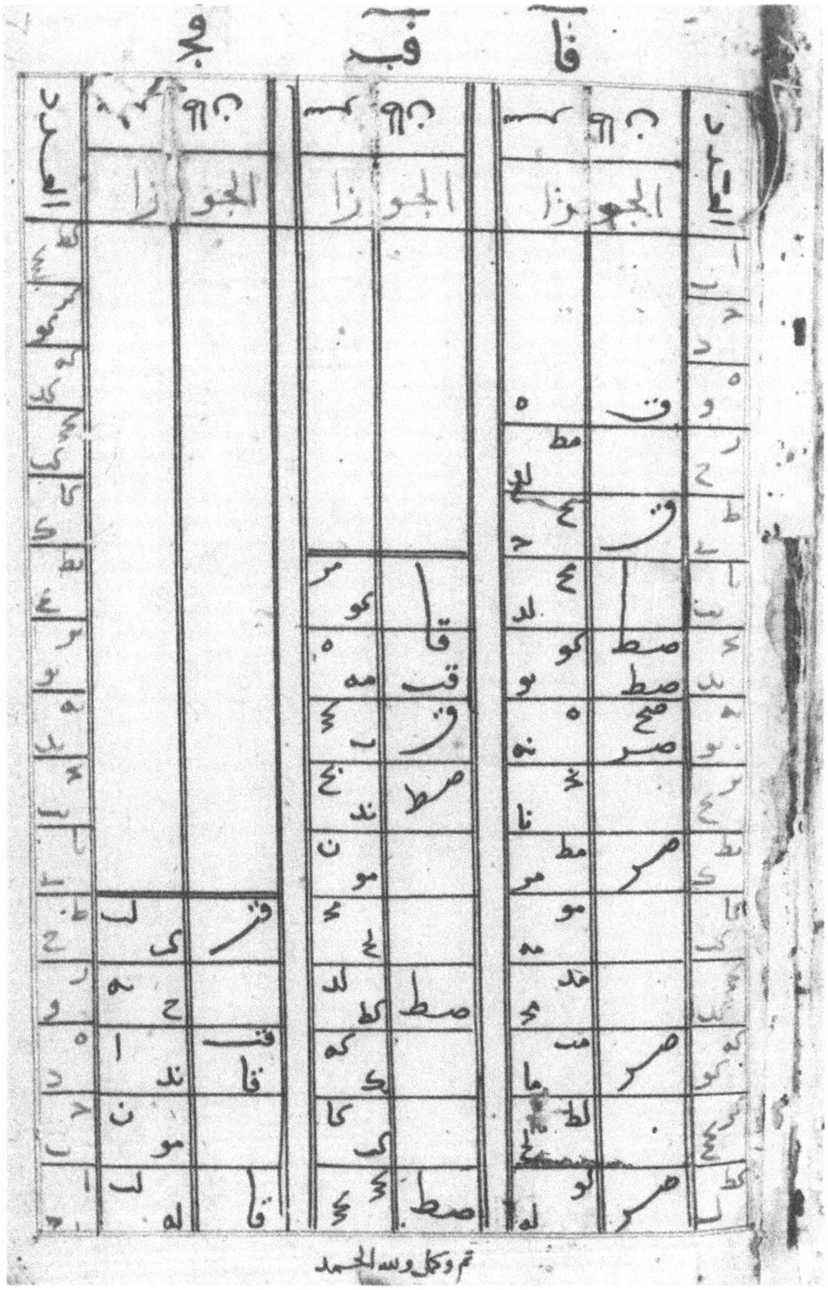

Fig. 2.1.1b: An extract from the tables of the time since sunrise for altitudes 81°-83°. See also **Figs. II-5.4a** and **5.6b**. [From MS Cairo DM 444, courtesy of the Egyptian National Library.]

←

Fig. 2.1.1a: An extract from the tables of the hour-angle serving altitude 45°. [MS Dublin CB 3673, courtesy of the Chester Beatty Library.]

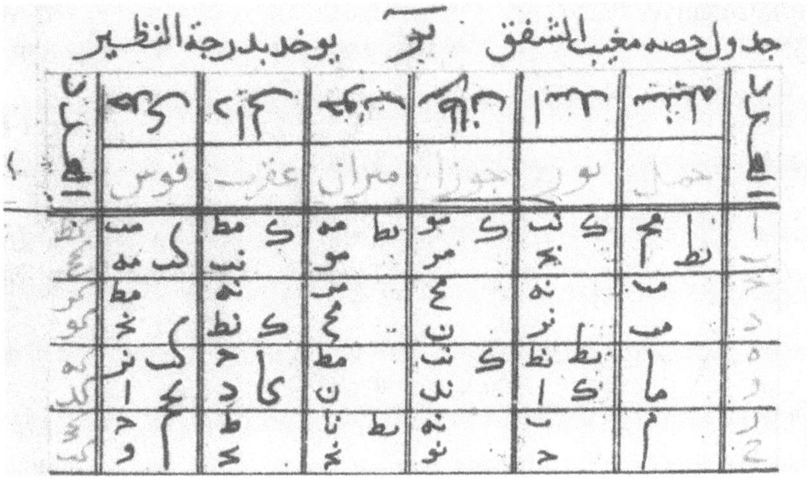

Fig. 2.1.1c: The table of time since sunrise for solar altitude 16° with the instruction that the table serves to find the duration of evening twilight if one enters with the opposite point of the ecliptic from the longitude of the sun. [From MS Cairo DM 444, courtesy of the Egyptian National Library.]

Maqsī's values of T(h,λ), Ibn al-Kattānī's values of t(h,λ) and Ibn Yūnus' values of a(h,λ) were displayed side by side.

The following list indicates the manuscript sources for the tables of T, t and a for Cairo and their attribution. (The tables in MS Cairo TR 354 falsely attributed to Ibn Yūnus are computed for Alexandria – see **2.1.6**.) The notation T/t means that T and t are tabulated on facing pages, and the notation T/t/a means that T, t and a are tabulated together, with triplets of values for each pair of arguments. Otherwise the functions listed are tabulated separately. An asterisk indicates that tables are preceded by an introduction on their use prepared by al-Maqsī, and a double asterisk indicates that this introduction is accompanied by some notes by al-Bakhāniqī describing the way in which he had rearranged the tables (see further **II-5.6**).

Berlin Ahlwardt 5753	T/t, and a	Ibn Yūnus
Cairo TR 191	t	Ibn Yūnus
Dublin CB 3673	T/t and a	Ibn Yūnus*
Escorial ár. 924,7	a	Ibn Yūnus
Cairo DM 108	T/t/a	Ibn Yūnus**
Cairo Azhar *falak* 4382	t and a	Ibn Yūnus
Cairo MM 137	a	Ibn Yūnus
Dublin CB 4078	T/t	anon.
Gotha A 1410	a	Ibn Yūnus
Cairo DM 53	T/t/a	al-Maqsī**
Cairo DM 45	T/t/a	anon.
Gotha A1402	T	al-Maqsī
Istanbul Kılıç Ali Paşa 684	t	Ibn al-Kattānī
Leipzig 817	T/t	anon.
Istanbul Nuruosmaniye 2903	T/t/a	anon.*

Istanbul Nuruosmaniye 2925	T/t/a	al-Maqsī**
Cairo TR 354	T/t/a (ϕ=31°)	Ibn Yūnus
Cairo MM 64	a	anon.
Cairo DM 444	T	al-Maqsī*
Cairo DM 690	T/t/a	anon.
Cairo DM 777	T/t	anon.
Cairo DM 616	T/t/a	al-Maqsī**
Cairo DM 739	T/t/a	al-Maqsī**
Cairo DM 776,1	T	anon.
Cairo DM 778	t	anon.
Cairo DM 786	T/t/a	anon.
Cairo DM 1101	a	anon.
Cairo DM 1108,9	a	anon. (**5.2.1**)
Cairo DM 1109	T/t/a	Ibn Yūnus
Cairo DM 1224	T/t/a	Ibn Yūnus
Cairo K 4044	t and a	al-Maqsī (t)/Ibn Yūnus (a)

There is no indication in any of the sources investigated of the way in which the tables of T(h,λ) were compiled. The simplest method would be to compile tables of t(h,λ) first. A formula for t(h,λ) that is not given in Ibn Yūnus' major work, the *Ḥākimī Zīj*, but which is outlined in MS Paris BNF ar. 2513, fol. 62v, of a recension of the 13th-century Egyptian *Muṣṭalaḥ Zīj* based mainly on his writings (**6.7.1**), is the following (*cf.* **F10**):

$$\text{Vers } t(h,\lambda) = \text{Vers } D(\lambda) - \text{Sin } h \bullet G(\lambda) = \text{Vers } D(\lambda) - p(h,\lambda),$$

where p is an auxiliary function not used elsewhere and G is defined by (see **6.0**):

$$G(\lambda) = G_1(\phi) \bullet G_2(\lambda) = R \,/\, \text{Cos } \phi \bullet R \,/\, \text{Cos } \delta(\lambda) \;.$$

A table of $G_2(\lambda)$ accompanies the theoretical discussion in MS Paris BNF ar. 2513 (**6.7.1**). In the *Ḥākimī Zīj* itself, Ibn Yūnus describes the computation of T(h,λ) rather than t(h,λ).[7] His method is the following. Firstly form the product:

$$p(h,\lambda) = \text{Sin } h \bullet G(\lambda) \;,$$

where G(λ) is defined by any one of three equivalent expressions (*cf.* **F8**):

$$G(\lambda) = R^2 \,/\, (\text{ Cos } \delta(\lambda) \text{ Cos } \phi \text{ }) = \text{Vers } D(\lambda) \,/\, \text{Sin } H(\lambda)$$
$$= R \,/\, \{ \tfrac{1}{2} \; [\text{ Sin } H(\lambda) - \text{Sin } H(\lambda^*) \text{] } \}$$

and then (*cf.* **F11**):

$$T(h,\lambda) = d(\lambda) - \text{arc Sin } \{ \text{ } p(h,\lambda) - \text{Sin } d(\lambda) \text{ } \} \;,$$

where d is the equation of half-daylight (= D - 90°). In the *Ḥākimī Zīj* Ibn Yūnus tabulated both d(λ) and Sin d(λ) to three digits for each degree of λ, giving no reason for tabulating the latter.

In passing we note that since:

$$H(\lambda) \;=\; \bar{\phi} + \delta(\lambda) \quad \text{and} \quad H(\lambda^*) \;=\; \bar{\phi} - \delta(\lambda) \;,$$

the identity:

$$\cos \delta \cos \phi \;=\; \tfrac{1}{2} \;\{ \cos (\phi\text{-}\delta) - \cos (\phi+\delta) \}$$

[7] *Cf.* King, *Ibn Yūnus*, III.15.3.

underlies the equivalence of the first and the third of the above expressions for G(λ). This was noted already by Jean-Baptiste Delambre in his *Histoire de l'astronomie du moyen âge*, published in Paris in 1820.[8] His pronouncement:

> "Ebn Jounis a changé depuis cosϕ cosδ en $^1/_2$ {cos(ϕ-δ) - cos(ϕ+δ)}, et c'est le premier exemple qu'on trouve de cette pratique connue sous le nom de prostaphérèse,"

led to the notion that Ibn Yūnus was:[9]

> "the first to propound the prostapherical formula ... which before the invention of logarithms was of great value to astronomers as it transformed the complicated multiplication of trigonometric functions expressed in sexagesimal fractions into an addition."

This has percolated through both the scholarly and popular literature on Islamic mathematics.[10] The fact that Ibn Yūnus did not propound *any* prostaphaeretical formulae does not, however, detract from his remarkable achievements in mathematical astronomy.

2.1.2 Abu 'l-ʿUqūl: Taiz

MSS Berlin Ahlwardt 5720 (Mq. 733,3), copied in Mukhkha in 1209 H [= 1795], and Milan Ambrosiana C84 are two fragments of an extensive corpus of spherical astronomical tables for the latitude of Taiz, computed by the Yemeni astronomer Abu 'l-ʿUqūl about 1300.[11] The corpus is entitled *Mirʾāt al-zamān*, "Mirror of Time", and more information on it is con-tained in **II-12.1**. A manuscript in a private collection in Sanaa of an astronomical miscellany by the Yemeni ruler al-Sulṭān al-Afḍal, compiled about 1375 (**II-12.4**),[12] contains other tables from this corpus, no longer extant in other two sources. Abu 'l-ʿUqūl also compiled a *zīj* called the *Mukhtār Zīj*, extant in MS London BL Or. 3624 and based mainly on the work of Ibn Yūnus.

The Berlin manuscript (fols. 128v-166r) contains a complete set of tables of the functions:

$$T(h,\lambda) \text{ and } t(h,\lambda)$$

computed to two digits for the domains:

$$h = 1°, 2°, \ldots, 90° \text{ and } \lambda = 1°, 2°, \ldots, 90° \text{ and } 181°, 182°, \ldots, 270°$$

and based on the parameters:

$$\phi = 13;40° \text{ (Taiz) and } \varepsilon = 23;35°.$$

The tables are less accurately computed than those for Cairo and Damascus (**2.1.1** and **2.1.4**). An extract is shown in **Fig. 2.1.2**. On the first page of the tables it is stated that they are for the town of Zabid, which is incorrect. On fols. 88r-92v of the Milan manuscript there is an incomplete set of tables of T(h,λ) only, for arguments h = 52°, 53°, ... , 75°. In both the Berlin and Milan copies there are also tables of $\lambda_H(h,\lambda)$ (**3.1.1**), which were doubtless compiled using

---

8 Delambre, *HAMA*, pp. 112 and 164.

9 Quoted from Heinrich Suter's article "Ibn Yūnus" in EI_1. See also, for example, Sezgin, *GAS*, V, p. 342.

10 For some comments on this see Al-Daffa & Stroyls, *Studies*, pp. 27-28. (This work is otherwise to be used with extreme caution.) See also King, *Ibn Yūnus*, pp. 7 and 149; *idem*, "Astronomical Timekeeping in Egypt", p. 360; and *idem*, "Astronomy in Fatimid Egypt", p. 508.

11 On Abu 'l-ʿUqūl see King, *Astronomy in Yemen*, no. 9. In Varisco, *Yemeni Almanac*, p. 13, he is identified as Muḥammad ibn Aḥmad al-Ṭabarī, the first teacher appointed by al-Malik al-Muʾayyad Dāʾūd, the brother and successor of al-Ashraf, to the Muʾayyadiyya *madrasa* in Taiz.

12 On al-Afḍal see King, *Astronomy in Yemen*, no. 18. The manuscript is now available in facsimile in Varisco & Smith, eds., *al-Afḍal's Anthology*.

Fig. 2.1.2: The tables of Abu 'l-ʿUqūl displaying the time since sunrise and the hour-angle for altitude 35°. [From MS Berlin Ahlwardt 5720, courtesy of the Deutsche Staatsbibliothek (Preußischer Kulturbesitz).]

these tables of T(h,λ) and t(h,λ). However, I know of no auxiliary tables of any kind which Abu 'l-ʿUqūl might have used to compile his main tables for timekeeping.

2.1.3 al-Mizzī: Damascus

The early-14th-century Syrian astronomer al-Mizzī[13] is known to have compiled a set of tables of t(h,λ) for the latitude of Damascus, taken as 33;27°. These tables do not exist in the known manuscript sources, but are mentioned in MS Leipzig 808, fol. 2v, by al-Karakī (**2.2.1**). MS Cairo MM 62, dating from *ca.* 1400, contains a set of prayer-tables compiled by al-Mizzī for parameters ϕ = 33;27° and ε = 23;33°, having the same format as the prayer-tables in the main Cairo corpus. See further **II-9.2**.

2.1.4 al-Khalīlī: Damascus

Shams al-Dīn al-Khalīlī, who worked in Damascus about 1360,[14] compiled a set of tables of t(h,λ) based on the parameters newly derived by his better-known colleague Ibn al-Shāṭir,[15] namely:

$$\phi = 33;30° \text{ (Damascus) and } \varepsilon = 23;31° \ .$$

These tables, which form part of an extensive corpus compiled by al-Khalīlī, are extant in

[13] On al-Mizzī see Suter, *MAA*, no. 406; Mayer, *Islamic Astrolabists*, pp. 61-62; *Cairo ENL Survey*, no. C34; and King, "Astronomical Timekeeping in Syria", pp. 78-79.

[14] On al-Khalīlī (Suter, *MAA*, no. 418) see my article in *DSB*, supplement.

[15] On Ibn al-Shāṭir (Suter, *MAA*, no. 416, and Kennedy, "*Zīj* Survey", no. 11) see my article in *DSB* and

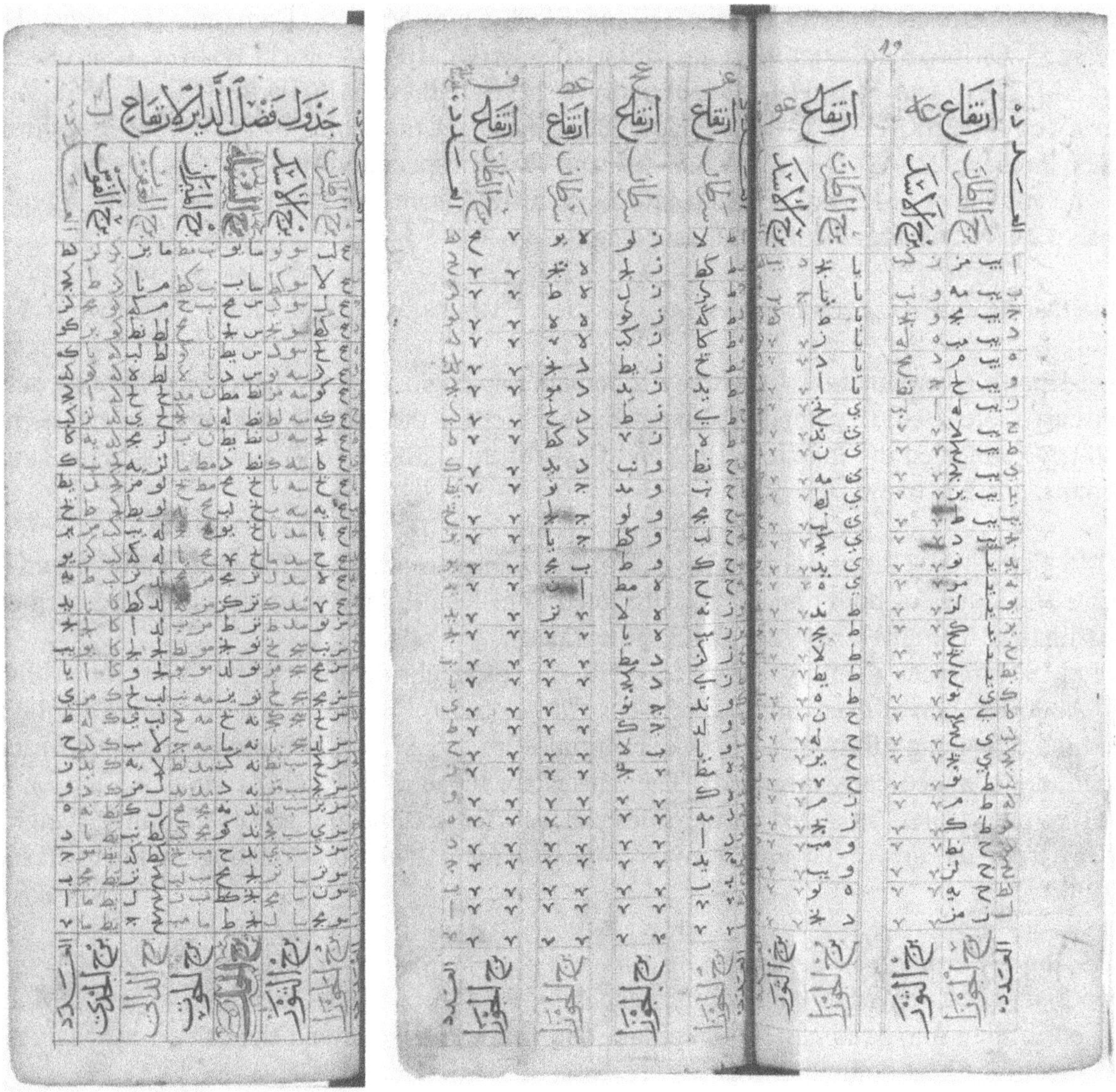

Fig. 2.1.4a-b: al-Khalīlī's hour-angle tables for altitudes 30° (a) and 75°-80° (b). Note the different format from the tables of the Cairo corpus shown in **Fig. 2.1.1**. [From MS Paris BNF ar. 2558, courtesy of the Bibliothèque Nationale de France.]

---

the references there cited, to which add Kennedy & Ghanem, eds., *Ibn al-Shāṭir*, and King, "Astronomy of the Mamluks", pp. 538-539 and 547-548. Several more recent papers reprinted in Saliba, *Studies*, deal with his contributions to theoretical astronomy. At the time I wrote the *DSB* article it was not known that Muslim astronomers concerned themselves with non-Ptolemaic models for several centuries after the time of Ibn al-Shāṭir. Several new sources from later centuries, albeit none yet from Syria, have been located and described in numerous papers by George Saliba: see, for example, his "The Astronomical Tradition of Maragha" (1991), "Arabic Planetary Theories after the 11th Century" (1996), "Al-Qūshjī's Reform of the Ptolemaic Model for Mercury" (1993), "A 16th-Century Arabic Critique of Ptolemy" (1994), and "Arabic Planetary Theories after the 11th Century" (1996).

several copies including MSS Paris BNF ar. 2558, Cairo MM 71, Damascus Ẓāhiriyya 3116 and 9233, Oxford Seld. Supp. 100, Oxford Marsh 39 (Uri 1042), fols. 113v-142v, Cairo K 8525 and ṬM 228,5, and Mosul al-Muḥammadiyya 129. Of these, MS Paris BNF ar. 2558 is the finest copy, dated 811 H [= 1408] and containing all of the tables of the corpus. See further **4.8.3**, **5.6.6**, **6.1.2**, **6.3.2**, **6.5.1**, **8.3.2**, **9.5** and **II-10**, especially **10.5**.

The format of al-Khalīlī's hour-angle table differs from that used in the earlier Egyptian tables (**2.1.1**), the argument domains being:

$$h = 1°, 2°, \dots , 80° \quad \text{and} \quad \lambda = 271°, 272°, \dots , 359°, 0°, 1°, \dots , 90° \ .$$

Note that no entry is given for the winter solstice. Extracts from these tables are shown in **Figs. 2.1.4a-b**.

al-Khalīlī's computational accuracy is remarkably high. He may have used his first set of auxiliary tables (**9.4**) to compile the hour-angle tables for the latitude of Damascus, but he may also have tabulated the functions B(λ) and C(λ) for this latitude (**6.0**) and then used the simple formula (*cf.* **F12**):

$$t(h,\lambda) = \text{arc Cos } \{ \text{ R} \bullet [\text{ Sin h} - C(\lambda) \ ] \ / \ B(\lambda) \ \} \ .$$

In MS Cairo MM 71 triplets of entries (t,T,a) for Damascus are given for each pair of arguments (h,λ). The values of both t and T are attributed to al-Khalīlī, and the values of a are attributed to Shihāb al-Dīn al-Ḥalabī: see **5.1.2** (illustrated).

In MS Damascus Ẓāhiriyya 7387, fols. 57v-60r, of an abridgement of the *Zīj* of Ibn al-Shāṭir by ʿAbd al-Raḥīm al-Qazwīnī of Damascus (*fl. ca.* 1610),[16] there is a set of tables displaying the time remaining till moonset for lunar altitudes h = 7°, 8°, ... , 16° and each degree of lunar longitude (symmetrically arranged). The entries are those of al-Khalīlī's tables of T(h,λ), and al-Qazwīnī's tables, as stated in the title, can be used to find the arc of visibility (*qaws al-ru'ya*) from the arc of tarrying (*qaws al-makth*) for calculations relating to lunar crescent visibility.[17]

2.1.5 Ibn al-Rashīdī (?): Jerusalem

Two disordered copies of tables for timekeeping for Cairo, MSS Cairo DM 45 and DM 153, copied in the same hand *ca.* 1650, contain odd folios from one and the same set of tables of the function t(h,λ) computed for:

$$\phi = 32;0° \text{ (Jerusalem) and } \varepsilon = 23;35° \ ,$$

with the same format as the Cairo tables. The altitude arguments 22°-23° (MS DM 153, fols. 11r-11v), 24°-25° (MS DM 45, fols. 27r-27v), and 34°-35° (MS DM 153, fols. 13r-13v) head the fragments of the tables which have found their way into these two manuscripts. The six pages of tables are copied in the same hand. There is no indication of the parameters underlying the tables, which were determined by inspection.

The entries in these fragments are reasonably accurately computed and are identical to the corresponding entries in al-Karakī's tables of t(λ,h) – note the change in format – for Jerusalem (**2.2.1**). Since al-Karakī states in his introduction to his tables that (the 14th-century Egyptian astronomer) Ibn al-Rashīdī also compiled a set of tables of t(h,λ) (latitude unspecified), I am

---

16 On al-Qazwīnī see *Cairo ENL Survey*, no. D38.

17 On these concepts see my article "Ru'yat al-hilāl" [= lunar crescent visibility] in EI_2.

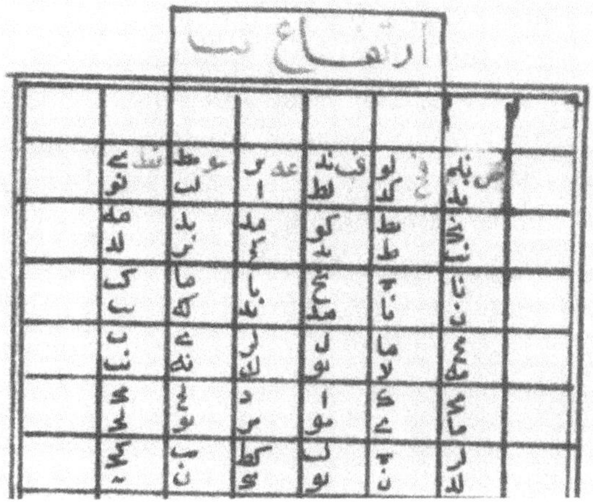

Fig. 2.1.5: An extract for solar altitude 12° from an anonymous corpus of hour-angle tables for Jerusalem. The corpus was copied *ca.* 1900, possibly from a manuscript in the Khālidiyya Library in Jerusalem. [From MS Cairo ṬM 81,1, courtesy of the Egyptian National Library.]

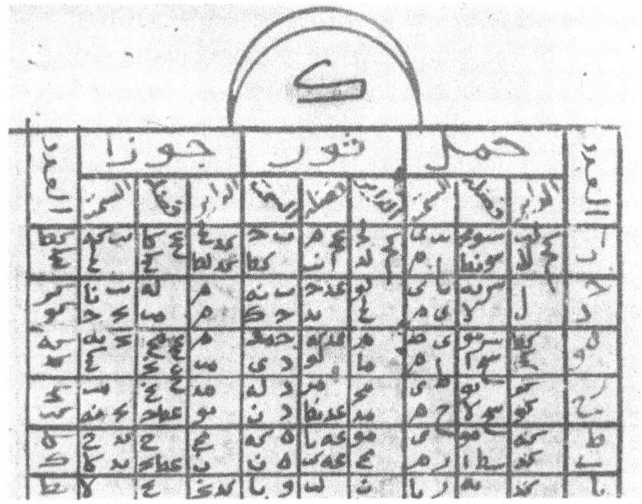

Fig. 2.1.6: An extract from the anonymous tables of the time since sunrise, the hour-angle and the azimuth for Alexandria showing the entries for solar altitude 20°. [From MS Cairo TR 354, courtesy of the Egyptian National Library.]

inclined to think that these fragments are from Ibn al-Rashīdī's hour-angle tables.[18] Ibn al-Rashīdī also made some corrections to Ibn Yūnus' azimuth tables where these had been incorrectly copied (**5.1.1**).

In MS ṬM 81,1, copied *ca.* 1900, there is a complete set of tables of the function t(h,λ) with values in degrees and minutes, for the same parameters:

$$\phi = 32;0° \text{ (Jerusalem) and } \varepsilon = 23;35° .$$

These tables are copied without the solar longitude arguments, and without the degrees of the entries, except at the head of each column. I have not been able to check that the entries in various sub-tables are the same as those in the fragments mentioned above, but it seems unlikely that there would be two different sets in existence. See also **I-2.2.6** and **II-6.12** and **II-11.12**.

2.1.6 Anonymous: Alexandria

The apparently unique source MS Cairo TR 354, copied *ca.* 1700,[19] contains a set of tables of the functions:

$$T(h,\lambda),\ t(h,\lambda) \text{ and } a(h,\lambda) ,$$

computed to two digits for the domains:

$$h = 1°, 2°, \ldots , 82°, \text{ and } \lambda = 1°, 2°, \ldots , 90° \text{ and } 181°, 182°, \ldots , 270°$$

and the parameters:

$$\phi = 31;0° \text{ (Alexandria) and } \varepsilon = 23;35° .$$

These tables, which contain over 30,000 entries, are falsely attributed to Ibn Yūnus on the title folio and they are less carefully computed than the corresponding tables for Cairo (**2.1.1** and **5.1.1**). Values of the three functions are displayed side by side for each pair of arguments –

[18] On Ibn al-Rashīdī, whose full name was Shams al-Dīn Abū ʿAbdallāh Muḥammad ibn Burhān al-Dīn Ibrāhīm al-Rashīdī, see *Cairo ENL Survey*, no. C39, and **II-4.1.3** and **6.11**.

[19] See *Cairo ENL Survey*, no. C143.

see **Fig. 2.1.6** – as in al-Bakhāniqī's edition of the Cairo corpus – see **Figs. II-5.6a-b**. The errors in the sum of corresponding values of t and T are related to those in a table of D(λ) for these parameters in MSS Oxford Marsh 676 (Uri 944 = 995) and Princeton Yahuda 861,1, originally compiled by Najm al-Dīn al-Miṣrī (**2.6.1**). I know of no tables of auxiliary functions based on latitude 31;0° which might have been used to compile the larger set. Several manuscripts exist of a corpus of prayer-tables based on the same parameters: see further **II-8.5**. However, the identity of the compiler of these tables remains obscure.

MS Cairo ṬM 216, copied in 1003 H [= 1594/95], contains a different set of tables of the functions T(h,λ) and t(h,λ) based on the parameters ϕ = 31;0° and ε = 23;35° (?). No compiler is associated with this set either.

2.1.7 Aḥmad Efendī: Istanbul

MS Istanbul Kandilli 196 contains a set of tables of the function t(h,λ) computed to two digits for the domains:

$$h = 1°, 2°, \ldots, 72° \text{ and } \lambda = 1°, 2°, \ldots, 90° \text{ and } 181°, 182°, \ldots, 270° .$$

The underlying parameters are found by inspection to be:

$$\phi = 41;0° \text{ (Istanbul) and } \varepsilon = 23;30° .$$

The tables are attributed to "Aḥmed Efendī known as al-Miṣrī al-Islāmbūlī", which double epithet suggests that he was an Egyptian who had taken up permanent residence in Istanbul,[20] and they are dated 1095 H [= 1684]. They are rather accurately computed.

2.1.8 al-Kutubī: Cairo

MSS Cairo DM 1104, ṬJ 811,8, DM 149, ṬM 88 and ṬM 142, DM 812 and 1103 contain a set of tables for timekeeping prepared in the year 1150 H [= 1737/38] by ʿAbd al-Laṭīf al-Dimashqī, also known as al-Kutubī.[21] The tables, which were prepared for the latitude of Cairo, display the times before midday and after midday when the sun has a given altitude and longitude, for each degree of both arguments – see **Fig. 2.1.8**. The entries are given in equinoctial hours and minutes, according to the Ottoman convention that sunset is 12 o'clock (**1.1**). The format of the tables is identical with that of the earlier tables of T(h,λ) and t(h,λ) in the main Cairo corpus (**2.1.1**), from which these tables were no doubt derived. If we denote the two times tabulated by al-Kutubī as t_1 and t_2, then:

$$t_1(h,\lambda) = {}^1/_{15}\,\{180° - D(\lambda) - t(h,\lambda)\} \text{ and}$$
$$t_2(h,\lambda) = {}^1/_{15}\,\{180° - D(\lambda) + t(h,\lambda)\} = {}^1/_{15}\,\{180° - T(h,\lambda)\} .$$

al-Kutubī states that he also included a correction for refraction at the horizon. See further **II-7.14**.

2.2 Tables of time as a function of solar longitude and instantaneous altitude for a given latitude

These tables differ from those discussed in **2.1** only in that the arguments h and λ are interchanged. For each degree of λ a given table displays entries for each degree of h up to

[20] On Aḥmad Efendī see *Cairo ENL Survey*, nos. H28. He does not appear to be listed in İhsanoğlu *et al.*, *Ottoman Astronomical Literature*.

[21] On al-Kutubī see *Cairo ENL Survey*, no. D76; and İhsanoğlu *et al.*, *Ottoman Astronomical Literature*, I, pp. 427-429, no. 281.

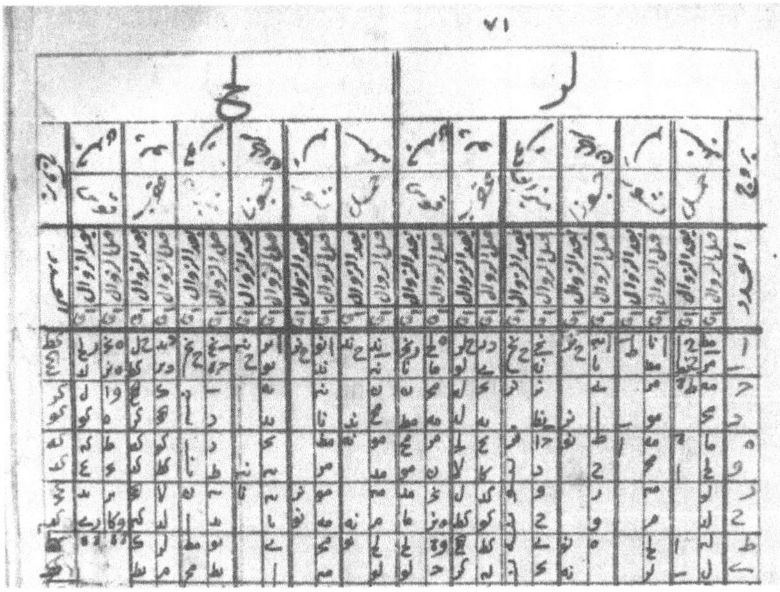

Fig. 2.1.8: al-Kutubī's tables for solar altitudes 37°-38°. These are based on the earlier tables of the Cairo corpus, although the author gives no indication of this. [From MS Cairo ṬJ 811,8, fol. 71r, courtesy of the Egyptian National Library.]

the maximum for the solar longitude in question. that is, H(λ) or the smaller integral value [H(λ)]. This arrangement has the advantage that entries can be included both for the equinoxes and for the solstices.

2.2.1 al-Karakī: Jerusalem

MS Leipzig UB 808, fols. 3r-93r, copied in 805 H [= 1402], contains a set of tables for timekeeping computed for the latitude of Jerusalem. The tables are introduced in the name of the 14th-century astronomer al-Karakī[22] and display, separately in two different hands, the functions:

$$T(\lambda,h) \text{ and } t(\lambda,h)$$

with entries computed to two digits for the domains:

$$\lambda = 270°, 271°, \ldots , 359°, 0°, 1°, \ldots , 90° \text{ and } h = 1°, 2°, \ldots , [H(\lambda)], H(\lambda) ,$$

and based on the parameters:

$$\phi = 32;0° \text{ (Jerusalem) and } \varepsilon = 23;35° .$$

The values are rather accurately computed. **Figs. 2.2.1a-b** show sample sub-tables from the Leipzig manuscript. See further **II-9.4** on the Jerusalem corpus.

In the introduction to his tables (fol. 2v) al-Karakī states that he wished to emulate al-Maqsī, who had computed T(h,λ) for ϕ = 30° (**2.1.1**), his own teacher al-Mizzī, who had computed t(h,λ) for ϕ = 33;27° (**2.1.2**), and Ibn al-Rashīdī, who had computed t(h,λ) for an unspecified latitude. None of these scholars had prepared tables showing both T and t, and al-Karakī states that there he saw his opportunity to join their ranks by compiling tables of both functions for

[22] On al-Karakī see *Cairo ENL Survey*, no. C35.

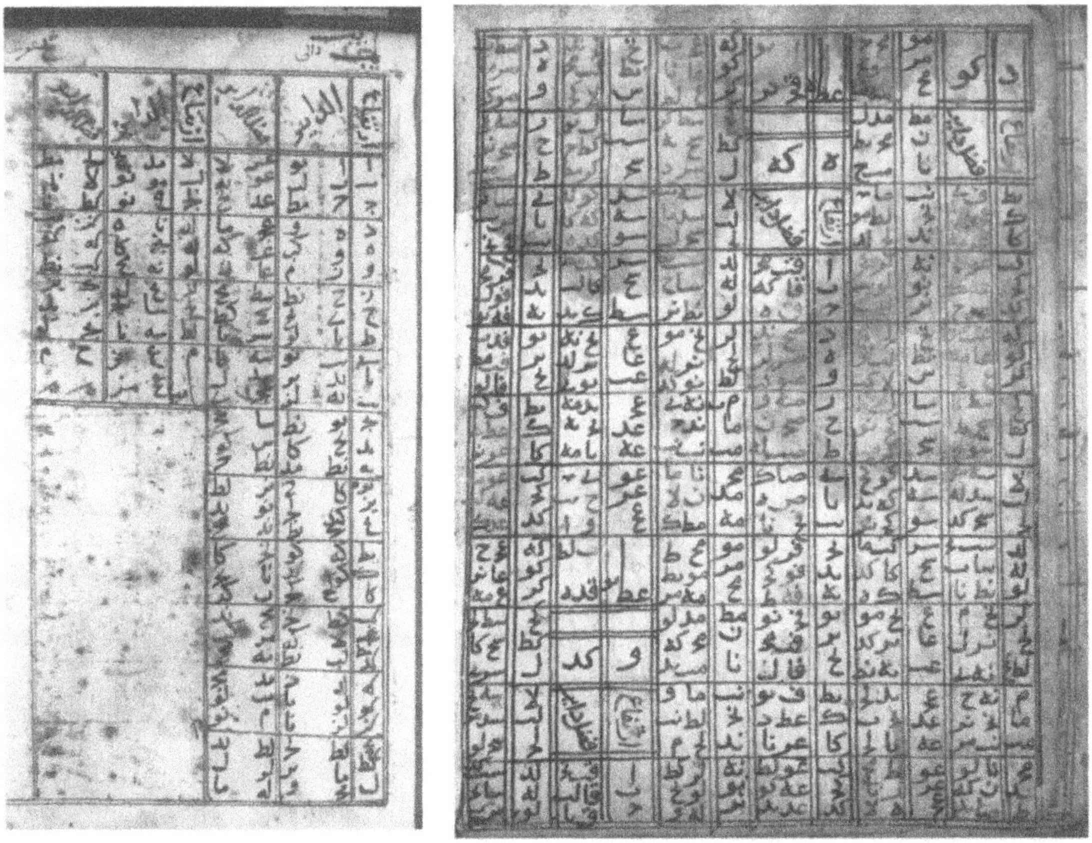

Fig. 2.2.1a-b: The tables of the time since sunrise for solar longitude Aquarius 12° / Scorpio 18° (a) and the hour-angle for longitudes 4°-6° of a certain sign (b) in the Jerusalem corpus. [From MS Leipzig UB 808, fol. 23v and 119v, courtesy of the Universitätsbibliothek.]

Jerusalem. With due respect to al-Karakī, it seems to me probable that Ibn al-Rashīdī's tables of t(h,λ) were computed for Jerusalem and that al-Karakī simply changed the format and added the values of T(λ,h). The entries for t(λ,h) in the Leipzig manuscript are identical with the corresponding entries in the tables of t(h,λ) for Jerusalem in MSS Cairo DM 45 and Cairo DM 153, which I suspect were computed by al-Rashīdī (**2.1.5**).

In fols. 94v-123v of the Leipzig manuscript there is another set of hour-angle tables in a different hand – see **Fig. 2.2.1b**. It is stated at the beginning and at the end of the tables that they are for latitude 32;40° (Ramla), but in fact the tables are simply those of al-Karakī for latitude 32;0°.

In MS Princeton Yahuda 861,1, copied *ca.* 1600, amidst a set of anonymous prayer-tables for latitude 32°, there is an odd table of T(h) and t(h) computed for the equinoxes, which apart from copyist's errors has the same entries as al-Karakī's tables of T(λ,h) and t(λ,h) for λ = 0°. Likewise the entries in the twilight tables for latitude 32° in MS Princeton Yahuda 861,1 are related to the entries for solar altitudes 20° and 16° in al-Karakī's set. See further **II-8.1** and **9.5.**

2.2.2 Anonymous: Edirne

MS Oxford arab. e. 93, fols. 3v-25r, contains a set of tables of t(λ,h) and T(λ,h) for the domains:

λ = 270°, 271°, ... , 90° and h = 1°, 2°, ... , [H(λ)], H(λ)

based on the parameters:

φ = 41;30° (Edirne) and ε = 23;30°.

An extract is shown in **Fig. 2.2.2**. The manuscript is undated and the tables are anonymous and without introduction, but the city of Edirne is mentioned at the head of some of the tables. I have not investigated their accuracy.

2.2.3 Ṣāliḥ Efendī: Istanbul

The tables for timekeeping compiled in the late 18th century by Ṣāliḥ Efendī Miʿmārī "the architect",[23] display values of the functions T(λ,h) and t(λ,h) computed to three digits with remarkable accuracy for the domains:

λ = 270°, 271°, ... , 90° and h = 1°, 2°, ... , [H(λ)], H(λ)

and based on the parameters:

φ = 41;0° (Istanbul) and ε = 23;28,54°.

An extract is shown in **Fig. 2.2.3**. The distinctive value of ε used by Ṣāliḥ Efendī is due to his predecessor Taqi 'l-Dīn, who worked in Istanbul two centuries previously (see, for example, **6.4.8**).

MSS Princeton Yahuda 353, Istanbul Aşir Efendi 224, Istanbul Kandilli 219, Vienna 2379 (Mixt. 989), Cairo K 18199, Cairo ṬM 151 and 215, and perhaps also Baghdad Awqāf 324/ 12230, are complete copies of these tables, which contain over 80,000 entries. An extract from the main tables for timekeeping in the superior Princeton manuscript is shown in **Fig. 2.2.3**. A given sub-table displays entries for one particular solar longitude, with the times for solar altitudes in the east and west expressed in equatorial degrees, equinoctial hours and seasonal day-hours. The times in equinoctial hours are also given (and now labelled *al-muwāfiqa*) according to the Ottoman convention that sunset is 12 o'clock (**1.2** and **II-14.0**). Various functions relating to timekeeping are displayed around each sub-table. Those relating to the afternoon prayers and twilight are also tabulated separately at the end of the main tables. For more details see **II-14.13**.

The method by which the values of t(λ,h) were computed is clear from some of the auxiliary functions displayed on each sub-table. For each value of λ Ṣāliḥ Efendī tabulates B(λ) and C(λ), as well as, for each degree of h, the function b(λ,h). These three functions are labelled *al-aṣl al-muṭlaq*, *buʿd al-quṭr* and *al-aṣl al-muʿaddal*, respectively (see further **6.0**). The hour-angle is then given by the simple formula (*cf.* **F12**):

$$\text{Cos } t(\lambda,h) = R \cdot b(\lambda,h) / B(\lambda) .$$

Numerous other sources contain smaller sets of tables for Istanbul that are based on those of Ṣāliḥ Efendī. For example, MSS Istanbul Bağdatli Vehbi Efendi 990, Istanbul Lala Ismail 287, Istanbul Husrev Paşa 232, Istanbul Kiliç Ali Paşa 677, Istanbul Kandilli 220, Istanbul

---

[23] On Ṣāliḥ Efendī see *ibid.*, no. H36; and İhsanoğlu *et al.*, *Ottoman Astronomical Literature*, II, pp. 453-458, no. 303.

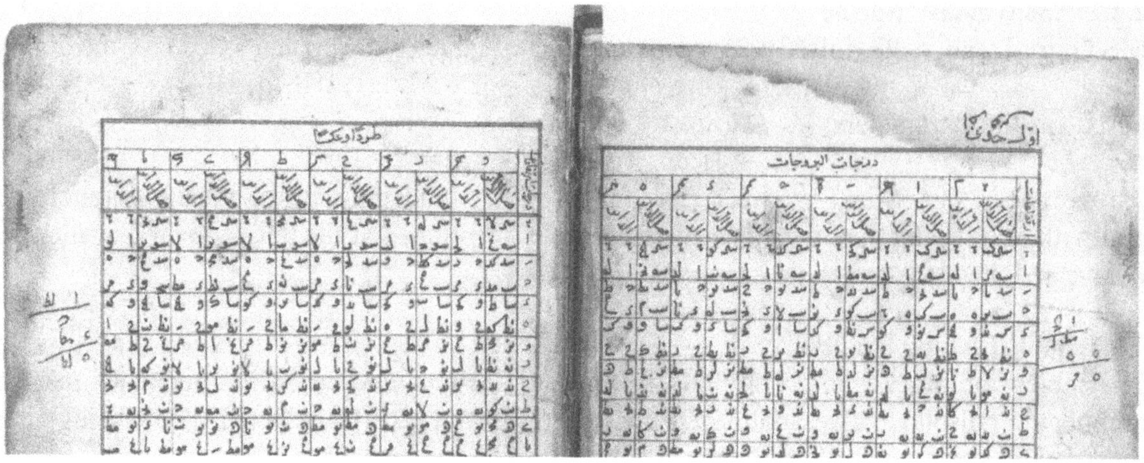

Fig. 2.2.2: The tables for Capricorn 0°-11° / Sagittarius 30°-19° in the Edirne corpus. [From MS Oxford arab. e. 93, fols. 3v-4r, courtesy of the Bodleian Library.]

Fig. 2.2.3: The tables for Taurus 14° / Leo 16° in the corpus of Ṣāliḥ Efendī. [From MS Princeton Yahuda 353, courtesy of the Special Collections, Princeton University Library.]

Kandilli 440 and 441, and Cairo ṬM 120 contain tables displaying the times before and after midday when the sun has the given longitude and altitude, with entries in equinoctial hours, minutes and seconds, expressed according to the Ottoman convention. Likewise the tables in MSS Istanbul Esat Efendi 1979, Istanbul UL T1963 and T1964 display these same times in equinoctial hours and minutes. Most of these sets of tables are anonymous and most of them are labelled in Turkish *irtifā cedveli* < *irtifāʿ jadwali*, "altitude tables". In MS Istanbul Husrev Paşa 232, also apparently MS Baghdad Awqāf 325/12248, the tables are attributed to the *muwaqqit* Muḥammad Ṣādiq Jihāngīrī. An abridged version of Ṣāliḥ Efendī's tables appears on an Ottoman quadrant.[24]

---

[24] See *Khalili Collection Catalogue*, II, p. 268, no. 157, and King, "Review", col. 256.

2.2.4 Ḥusayn Ḥusnī: Mecca

MS Cairo K 4002 contains a set of tables for timekeeping for Mecca prepared in 1239 H [= 1823/24] by the Ottoman astronomer Ḥusayn Ḥusnī, also known for his translation of the "*zīj*" of the French astronomer Lalande.[25] The function tabulated is:

$$T'(\lambda,h)$$

for arguments:

$$\lambda = 270°, 271°, \ldots , 359°, 0°, \ldots , 90° \quad \text{and} \quad h = 0°, 1°, \ldots , [H(\lambda)] .$$

The times are expressed according to the Ottoman convention and values are given in equinoctial hours, minutes and seconds for solar altitudes in both east and west: see the extract in **Fig. 2.2.4**. The entries are in Hindu-Arabic numerals. The underlying latitude is stated to be ϕ = 21;45° (Mecca). I have not investigated the accuracy of these tables. al-Ḥusnī's value of the latitude of Mecca is not used in any other Islamic astronomical work known to me. The accurate value is 21;26°, so that Ḥusnī's value was no improvement over the standard medieval values 21°, 21;20°, 21;30° and 21;40°.[26] On the prayer-tables accompanying Ḥusnī's main tables for timekeeping see **II-12.9**.

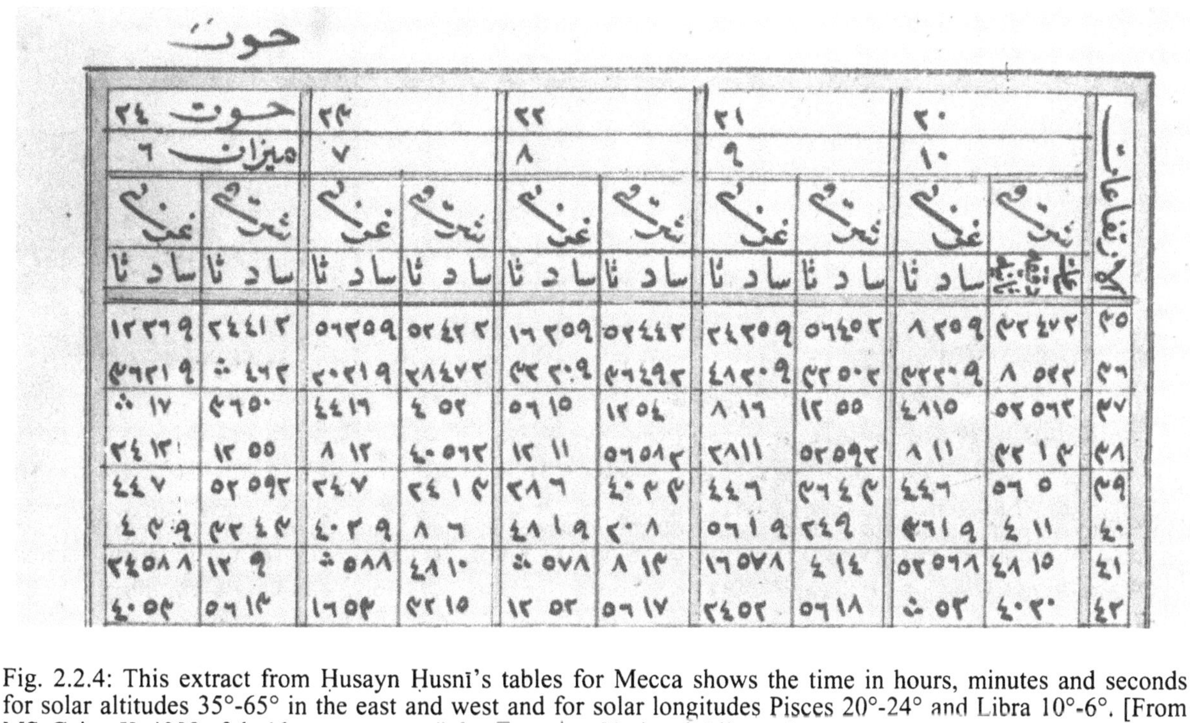

Fig. 2.2.4: This extract from Ḥusayn Ḥusnī's tables for Mecca shows the time in hours, minutes and seconds for solar altitudes 35°-65° in the east and west and for solar longitudes Pisces 20°-24° and Libra 10°-6°. [From MS Cairo K 4002, fol. 10v, courtesy of the Egyptian National Library.]

[25] On Ḥusayn Ḥusnī see Azzawi, *History of Astronomy in Iraq*, pp. 268, 284 and 285; *Cairo ENL Survey*, no. H47; and Ihsanoğlu *et al.*, *Ottoman Astronomical Literature*, II, pp. 581-584, no. 419.

[26] On Islamic values for the latitude of Mecca see King, "Earliest Muslim Geodetic Measurements", pp. 225-226.

2.2.5 al-Ṭanṭāwī: Damascus

MSS Cairo TR 129 and DM 1007 and an unnumbered manuscript in *al-Maktaba al-ʿArabiyya* bookshop in Damascus in 1973 contain a set of tables of the function:

$$[T(\lambda,h)]^{h}$$

attributed to the Damascus astronomer al-Ṭanṭāwī (d. 1889).[27] The values are computed for Damascus, are expressed in equinoctial hours and minutes and are probably derived from al-Khalīlī's tables of t(h,λ) or T(h,λ) (**2.1.4**), which, however, have a different format. Other tables to which al-Ṭanṭāwī put his name were likewise computed by his predecessor of five centuries.

2.2.6 Anonymous: Jerusalem

MS Cairo ṬM 81,1, copied *ca.* 1900, contains some very late tables displaying values of the function:

$$T(\lambda,h)$$

expressed in hours and minutes, with entries for altitude in both east and west – see **Fig. 2.2.6**. They are based on the parameter λ = 32;0° (Jerusalem). I have not tried to establish whether or not these tables were derived from those described in **2.1.5** or **2.2.1**. See further **II-11.12**.

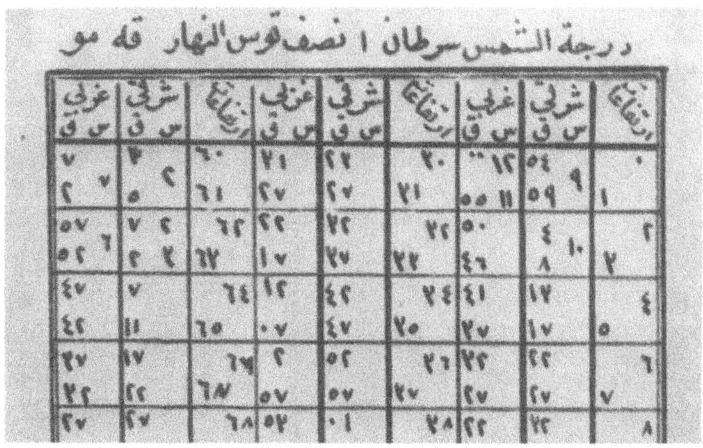

Fig. 2.2.6: An extract from some tables for Jerusalem preserved in a late manuscript. These are probably based on those illustrated in **Fig. 2.1.5**. [From MS Cairo ṬM 81,1, fols. 1v-2r, courtesy of the Egyptian National Library.]

2.3 Tables of time as a function of meridian altitude and instantaneous altitude, for a specific latitude

The tables listed in this section display the function T(H,h) and can in principle be used for either the sun or the stars. Since h ≤ H the tables are trapezoidal in shape, and in medieval

[27] On al-Ṭanṭāwī see *Cairo ENL Survey*, no. D123; and İhsanoğlu *et al.*, *Ottoman Astronomical Literature*, II, pp. 646-649, no. 496.

Arabic these tables are generally refered to by the term *ṭaylasān* (**1.4**). In such tables only the entries for $\bar{\phi}-\varepsilon \leq H \leq \bar{\phi}+\varepsilon$ can be used for the sun. The entries for general arguments H can be used to find the time elapsed since the rising of a star with declination Δ such that $\Delta = H - \bar{\phi}$. Tables of T(H,h) computed for a given latitude for timekeeping by the sun require about one-third as many entries as tables of T(h,λ), if entries are given for each degree of argument in both cases.

No indication is given in the sources of the method of computation of the entries in these tables, but it seems reasonable to assume that one would first compute the value of D corresponding to H and then apply the standard formula (*cf.* **F9**):[28]

$$\text{Vers } t(H,h) = [\ \text{Sin } H - \text{Sin } h\ ] \cdot \text{Vers } D \ /\ \text{Sin } H\ .$$

Given D and t the computation of T is trivial. Note that the function T(H,h) is independent of the obliquity ε.

Other tables of this kind are attested in some copies of the *Toledan Tables* and in a medieval Hebrew source. In the former, values are given for each degree of both arguments, and the meridian altitude runs up to about 80°. In the latter, the horizontal arguments are periods of days of the year (during which the meridian altitude is assumed constant) and the vertical argument is the solar altitude in 5°-intervals. See further **10.1**.

2.3.1 ʿAlī ibn Amājūr: Baghdad

Some of the earliest known Islamic tables for reckoning time from solar or stellar altitudes are due to the Baghdad astronomer ʿAlī ibn Amājūr (*fl. ca.* 950).[29] One of these is contained in MS Paris BNF ar. 2486, fols. 239r-255r, a unique manuscript of the *Zīj* of al-Baghdādī copied by the author for himself in 684 H [= 1285].[30] (On the other table attributed to Ibn Amājūr see **2.5.1**.) The function tabulated here is:

$$T(H,h)\ ,$$

with values computed to two digits for the domains:

$$H = 21°, 22°, \ldots , 84°, \text{ and } h = 1°, 2°, \ldots , H\ ,$$

and based on the parameter:

$$\phi = 33;25° \text{ (Baghdad)}\ .$$

The entries in this table, which number about 3,300, are computed with remarkable accuracy. The lower and upper limits for H indicate that the table was intended for timekeeping by the stars as well as by the sun. An extract from this table is shown in **Fig. 2.3.1**.

2.3.2 al-Ṭūsī: Maragha

Another such table is contained in some copies of the *Īlkhānī Zīj*, prepared under the celebrated scholar Naṣīr al-Dīn al-Ṭūsī (d. 1274)[31] at the observatory in Marāgha in N.W. Iran. I have

[28] See, for example, Nallino, *al-Battānī*, I, pp. 189-191, King, *Ibn Yūnus*, III.15.3 (pp. 149-150), and Nadir, "Abu 'l-Wafāʾ on the Solar Altitude", on this formula as used by al-Battānī (*ca.* 910), Ibn Yūnus (*ca.* 980), and Abu 'l-Wafāʾ (*ca.* 1000), respectively. See also Goldstein, *Ibn al-Muthannā on al-Khwārizmī*, pp. 208-209; and Davidian, "al-Bīrūnī on the Time of Day", p. 333.

[29] On Ibn Amājūr see Suter, *MAA*, no. 99, and Sezgin, *GAS*, V, p. 282, where this *ṭaylasān* table is mentioned.

[30] On al-Baghdādī and his *zīj* see Kennedy, "*Zīj* Survey", no. 3; Kennedy, "Comets in Islamic Astronomy and Astrology", p. 48; and Jensen, "The Lunar Theories of al-Baghdādī", p. 322.

[31] On al-Ṭūsī and the *Īlkhānī Zīj* see Suter, *MAA*, no. 368; Kennedy, "*Zīj* Survey", no. 6; and Storey, *PL*,

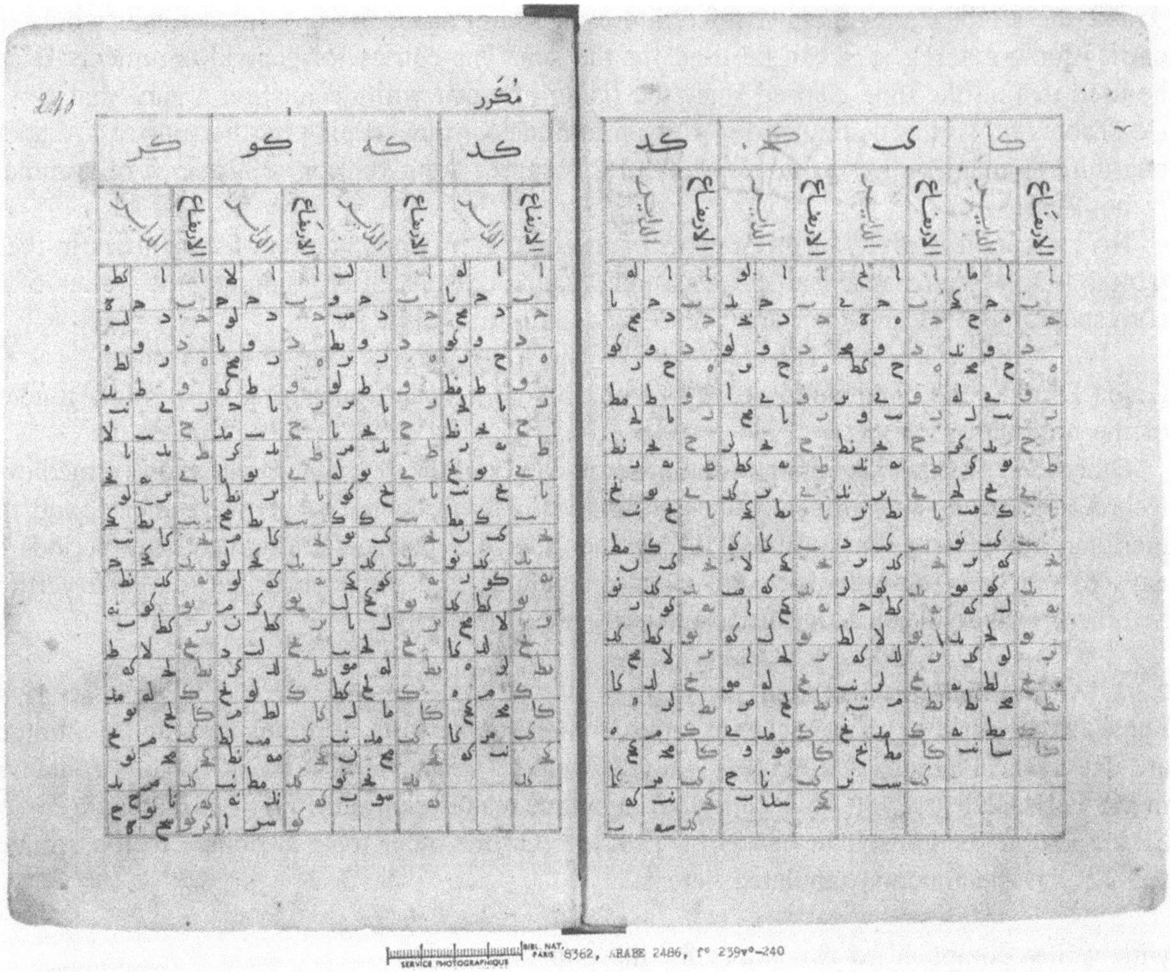

Fig. 2.3.1: The sub-tables for solar altitudes 21°-27° in the Baghdad corpus of ʿAlī ibn Amājūr. Note that the scribe copied the column for 24° twice, realising his error just before he finished the second set and marking it *mukarrar*, "repeated". [From MS Paris BNF ar. 2486, 239v-240r, courtesy of the Bibliothèque Nationale de France.]

examined MS Cairo DMF 1, 136 fols., copied at Maragha in 692 H [= 1293], as well as MSS Florence Medici 269, fols. 150r-152v, Istanbul UL P1418,2, fols. 234r-237v, and Oxford Hunt. 143, fols. 155v-160v, of this table. There the function:

$$T(H,h)$$

is tabulated to three digits for the domains:

$$H = 29;10°, 30°, 31°, \ldots , 75°, 76;10°, \text{ and } h = 1°, 2°, \ldots , H ,$$

based on the parameter:

$$\phi = 37;20° \text{ (Maragha).}$$

II:1, pp. 58-60; and, most recently, Ragep, *al-Ṭūsī's Memoir on Astronomy*, and the same author's article "al-Ṭūsī. 3. As scientist" in EI_2. The existence of the *Īlkhānī ṭaylasān* table was noted in Kennedy, *op. cit.*, p. 161.

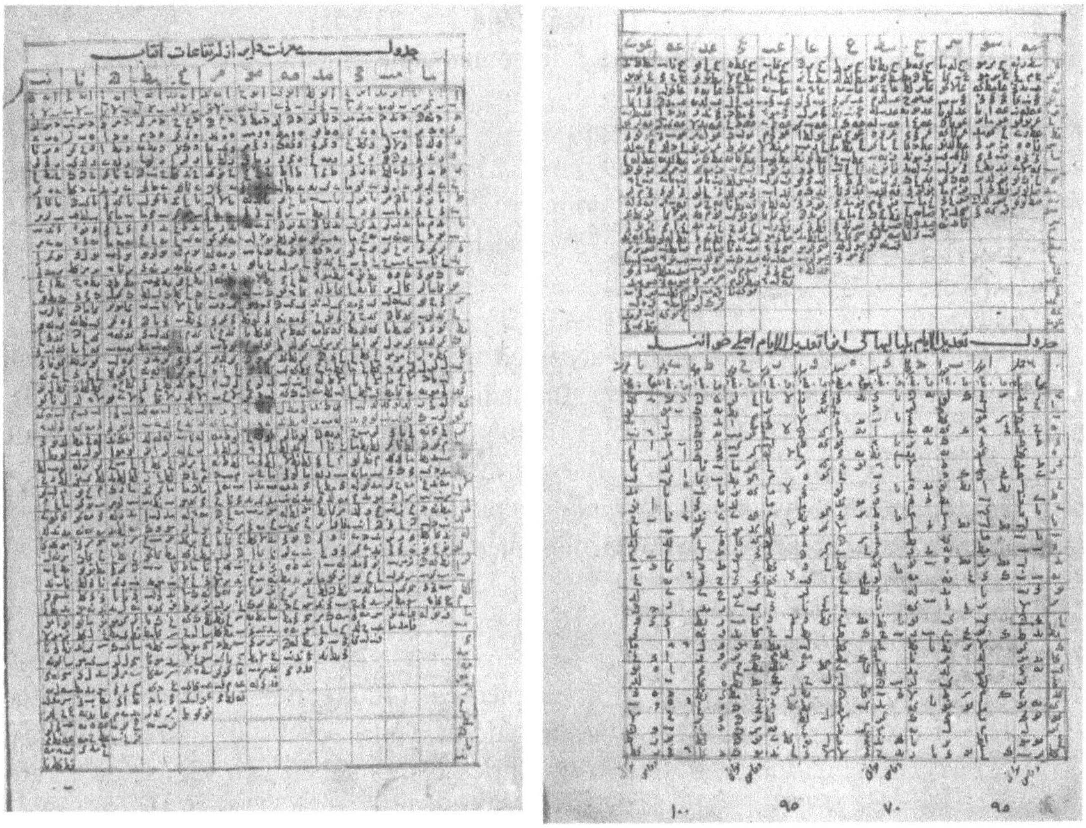

Fig. 2.3.2a-b: Some of the entries in a splendid late 13th-century copy of al-Ṭūsī's tables for meridian altitudes 41°-64° (a) and 65°-75° and then 76;10° (b). After this *ṭaylasān* table there is a table for the equation of time. From MS Cairo DMF 1, courtesy of the Egyptian National Library.]

Two extracts are displayed in **Figs. 2.3.2a-b**. The minimum and maximum values of H indicate that the value assumed for ε was the *Īlkhānī* value 23;30°, although as noted above the values of T(H,h) are independent of this parameter. Since the second digits in the entries are often in error, the third ones are meaningless. Values are also given in equinoctial hours, minutes and seconds, and these are generally accurate to the nearest minute.

2.3.3 Anonymous: Shiraz

MS Paris BNF supp. pers. 1488 is a unique copy from *ca.* 1400 of the Persian *Zīj-i Ashrafī*, compiled in Shiraz by Sanjar al-Kamālī in the year 702 H [= 1302/03].[32] This *zīj* contains several tables relevant to the present study. One of these is a table (fols. 205r-207r) displaying:

$$T(H,h)$$

with entries to two sexagesimal digits for each degree of both arguments. Values of D(H) are also given. The underlying parameters are:

---

[32] On Sanjar al-Kamālī and the *Ashrafī Zīj* see Kennedy, "*Zīj* Survey", no. 4, and Storey, *PL*, II:1, p. 64.

$$\phi = 29;30° \text{ (Shiraz) (and } \varepsilon = 23;35°) \text{ .}$$

See also **2.5.1**, **5.8.3**, **6.4.5**, **7.2.2** and **8.1.2** for other tables in the *Zīj-i Ashrafī*.

2.3.4 Anonymous: unspecified locality (Iran)

An anonymous table in MS Leiden Or. 199, fols. 21v-27v, copied *ca.* 1400, displays the same function computed to two digits for the domains:

$$H = 30°, 31°, \ldots , 78°, \text{ and } h = 1°, 2°, \ldots , H \text{ ,}$$

and:

$$\phi = 36;0° \text{ .}$$

The limits for H suggest that the author accepted the Indian value 24;0° for ε. The table is preceded by a short introduction in Persian. Immediately before this on fol. 21r of the Leiden manuscript there is a table of $\alpha_\phi(\lambda)$ for $\phi = 37;40°$ (locality unspecified). The *ṭaylasān* table, which contains many errors difficult to attribute to copyists, has been studied previously by Bernard Goldstein,[33] and its structure is now explained (**XI-4.2**). In brief, the entries were computed using the approximate formula presented in **2.5** and then converted to equatorial degrees for latitude 36°.

2.3.5 Anonymous: Tunis

MS Berlin Ahlwardt 5724 (Wetzstein 1150), copied *ca.* 1700, contains an anonymous corpus of tables for timekeeping computed for the latitude of Tunis *ca.* 1400.[34] One of these (fols. 16r-28v) displays the hour-angle t(H,h), with entries computed to two digits for arguments:

$$H = 29;25°, 30°, 31°, \ldots , 76°, 76;35°, \text{ and } h = 1°, 2°, \ldots , H \text{ .}$$

The underlying latitude is 37;0° (Tunis), and the limits for H imply that the value of ε was taken as 23;35°. The entries are rather accurately computed, and the format of the tables differs from that used in those discussed in **2.3.1-4** above – see the extract in **Fig. 2.3.5**. For other tables in the Tunis corpus see **2.4.1**, **8.1.3**, **8.3.3** and **8.4.1** and also **II-13.2**.

2.3.6 Taqi 'l-Dīn: Istanbul

MS Istanbul Kandilli 208 is a collection of works by the 16th-century Ottoman astronomer Taqi 'l-Dīn copied in his own hand.[35] Fols. 96r-100v contain a *ṭaylasān* table displaying t(H,h) and T(H,h) with values computed to two digits for each degree of both arguments: see **Fig. 2.3.6** for an extract. The underlying parameters (ε underlies the argument H) are:

$$\phi = 41° \text{ (Istanbul) (and } \varepsilon = 23;30°).$$

2.3.7 Anonymous: Edirne (?)

MS Cairo KhMT 2,1 (fol. 1r), copied *ca.* 1750, contains part of an anonymous *ṭaylasān* table displaying the function t(H,h) with values computed to two digits for each degree of both

---

[33] See Goldstein, "Medieval Table for Reckoning Time", the first modern study of a large medieval table for timekeeping.

[34] On the manuscript see *Berlin Catalogue*, Ahlwardt no. 5724. *Cf.* King, "Astronomy in the Maghrib", p. 38, for brief notes on this manuscript and others reflecting the level of astronomy in medieval Tunis.

[35] On Taqi 'l-Dīn (Suter, *MAA*, no. 471 and *Cairo ENL Survey*, no. H12) see Sayılı, The *Observatory in Islam*, pp. 289-305, the article "Taḳī al-Dīn" in EI_2, and now also İhsanoğlu *et al.*, *Ottoman Astronomical Literature*, I, pp. 199-217, no. 96.

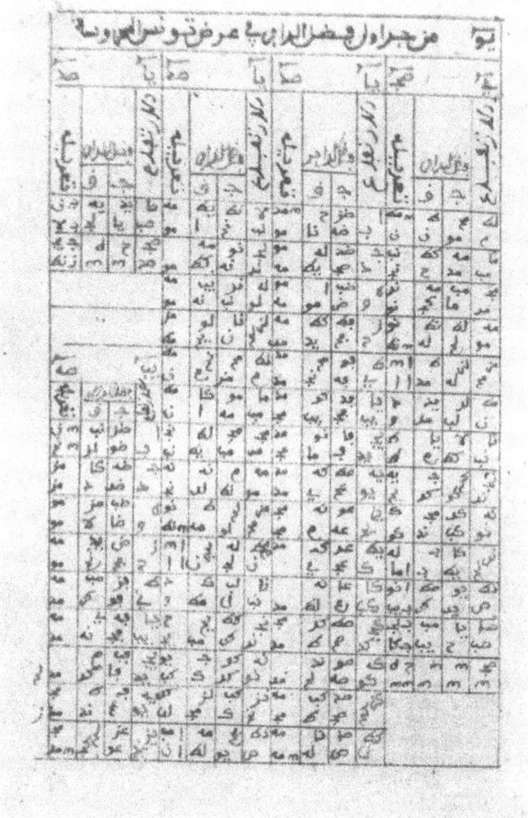

Fig. 2.3.5: The sub-tables for solar meridian altitudes 63°-64° in the Tunis corpus. [From MS Berlin Ahlwardt 5724, fols. 24r, courtesy of the Deutsche Staatsbibliothek (Preußischer Kulturbesitz).]

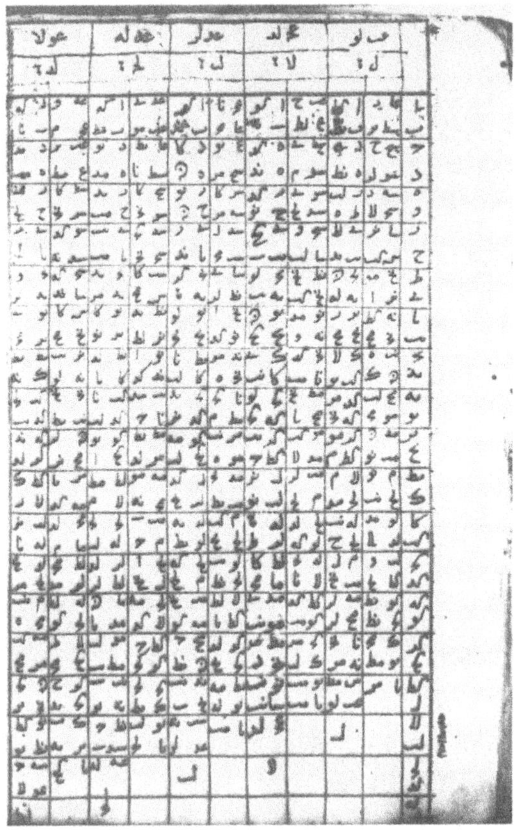

Fig. 2.3.6: The sub-tables for solar meridian altitudes 30°-34° from the *ṭaylasān* table of Taqi 'l-Dīn. [From MS Istanbul Kandilli 208, courtesy of Kandilli Observatory.]

arguments. The underlying parameters are ϕ = 42° and ε = 23;39°, and the table may have been intended to serve Edirne.

2.4 Tables of time as a function of stellar altitude, for a specific latitude

Such tables are clearly analogous to those described in **2.3** since for each star there is an underlying value of H. See also **6.16**.

2.4.1 Anonymous: Tunis

MS Berlin Ahlwardt 5724, fols. 55r-56r, of the anonymous corpus for Tunis (**2.3.5**) contains some simple tables displaying the hour-angle from stellar altitudes. For some 21 stars the function t(h) is tabulated for each degree of h up to the maximum for the star in question, computed for the latitude of Tunis (36;40° or maybe 37;0°). Entries are given to one digit only – see **Fig. 2.4.1**. The table may have been computed using the auxiliary table for latitude 36;40° on fol. 42v of the Berlin manuscript (on which see **6.8.2**).

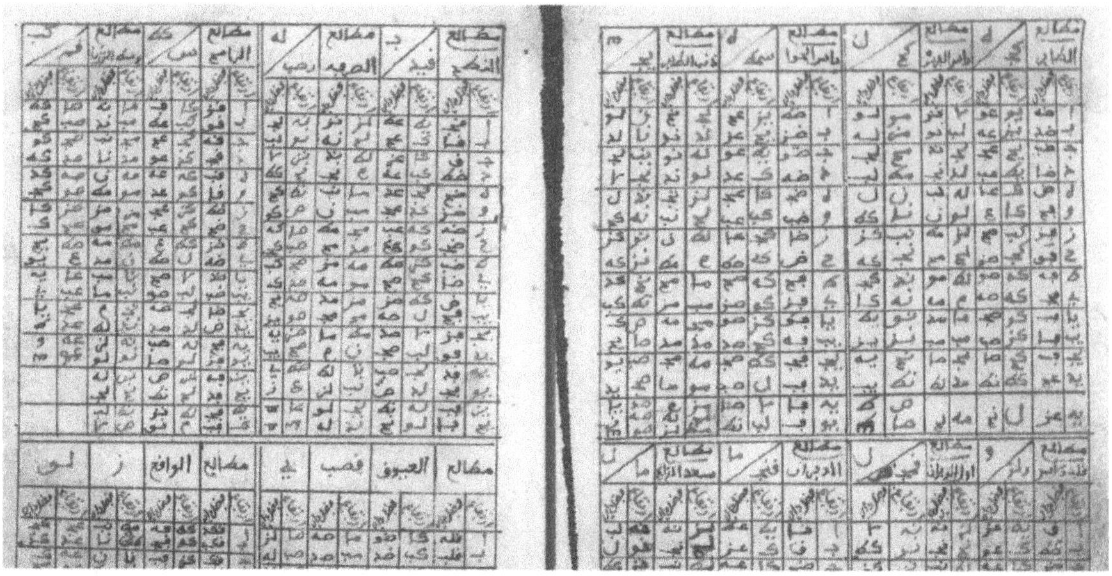

Fig. 2.4.1: Tables from the Tunis corpus for 14 stars. For each star the ascensions at culmination are also given. [From MS Berlin Ahlwardt 5724. fols. 55v-56r, courtesy of the Deutsche Staatsbibliothek (Preußischer Kulturbesitz).]

2.5 Tables of time since sunrise as a function of meridian altitude and instantaneous altitude, for all latitudes

The tables discussed in this section are also usually called by the name *ṭaylasān* (**2.3**) and are based on an approximate formula:

$$T(H,h) \approx {}^{1}/_{15}\ \text{arc Sin}\ \{\ R\ \text{Sin}\ h\ /\ \text{Sin}\ H\ \}$$ ḥ .

This is a good approximation to the actual value of T(H,h). The expression is accurate for the equinoxes (and, of course, when $h = 0°$ or H). I have computed the error incurred by using the equivalent expression in degrees, namely:

$$T(H,h) \approx D\ /\ 90\ \bullet\ \text{arc Sin}\ \{\ R\ \text{Sin}\ h\ /\ \text{Sin}\ H\ \}\ ,$$

in the domain:

$$\bar{\phi} - \varepsilon \leq H \leq \bar{\phi} + \varepsilon\ ,$$

corresponding to solar altitudes for various values of ϕ. The approximate values are less than the accurate values, but the maximum errors for latitudes 15° (Yemen) and 30° (Cairo) are only about 1° and $1^{1}/_{2}°$, respectively. The tables could also be used for timekeeping by the stars.

This approximate formula for reckoning time from solar altitude was used in the early 9th century by astronomers such as al-Khwārizmī and Ḥabash.[36] Certain later Muslim astronomers

[36] Ibn al-Muthannā (10th century?) attributes this method to al-Khwārizmī (Goldstein, *Ibn al-Muthannā on al-Khwārizmī*, pp. 81-82 and 207-208), and states that in addition al-Khwārizmī proposed an exact formula (*ibid.*, pp. 83). However, the approximate method is not included in the Latin translation of the recension of al-Khwārizmī's *Zīj* by al-Majrīṭī (*ca.* 1000) (Kennedy, "*Zīj* Survey", no. 21).

The duration of twilight is determined by an equivalent method in MS Berlin Ahlwardt 5750, fol. 153r, of a recension of the *Zīj* of Ḥabash (Kennedy, *op. cit.*, no. 15) but the attribution of the twilight theory in the Berlin

used an equivalent formula to compile tables of the duration of twilight: see **II-3.3** and **3.7** for tables of this kind for Baghdad and Alamut (?). Others tabulated the solar altitude as a function of time using this formula: see **4.3**. The approximation is further attested in a Byzantine astronomical treatise,[37] and also underlies a table in an English manuscript of the *Toledan Tables* (**No. 10.1**). See also **II-1.4** and **XI**.

It is also worth noting that certain Muslim astronomers in early Abbasid times as well as later astronomers concerned with timekeeping in Cairo and Damascus used an approximation ultimately of Indian origin[38] involving the increase of a gnomon shadow over its midday minimum, $\Delta z = \mathrm{Cot}_n\ h - \mathrm{Cot}_n\ H$, namely:

$$T(h,H) \approx \{\ 6n\ /\ (\Delta z + n)\ \}^{sdh}$$

for various values of the base n. No tables based on this formula are known to me.

2.5.1 ʿAlī ibn Amājūr (Baghdad)

MS Paris BNF supp. pers. 1488 of the early 14th-century Persian *Ashrafī Zīj* contains the table of T(H,h) for Shiraz discussed in **2.3.3** above, preceded by another table (fols. 201v-204v) of the function $T(H,h)^{sdh}$ with entries computed to two digits for the domains:

$$H = 1°, 2°, \ldots, 90° \quad \text{and} \quad h = 1°, 2°, \ldots, H\ .$$

A note by this table states that it is due to ʿAlī ibn Amājūr and that he included it in his *zīj* called *Zīj al-Ṭaylasān* (**2.3.1**): see **Fig. 2.5.1** for an extract. There is no mention of the fact that the table is intended to serve all latitudes.

2.5.2 Anonymous (Baghdad / Damascus)

The anonymous MS Paris BNF ar. 2514, copied in 612 H [= 1215], bears the title *Kitāb fīhi maʿrifat Zīj al-Ṭaylasān al-laylī wa-'l-nahārī fī sāʿāt khāṣṣa^{tan}*. "Book containing instructions for the *Ṭaylasān Zīj* for timekeeping by day and night in hours".[39] The first part of this work (fols. 1v-26r), which may be all that the title refers to, consists of a set of tables of:

$$T(H,h)^{sdh},\ T(H,h)^{h},\ \mathrm{Cot}_{12}\ h,\ \mathrm{Cot}_{6;40}\ h,\ h_a(H)$$

for each degree of H up to 90° and/or each degree of h (≤ H). An extract is illustrated in **Fig. 2.5.2**.

The values of the first function differ from those attributed to Ibn Amājūr in the *Ashrafī Zīj* (**2.5.1**). Values of the second function are given only for 33° ≤ H ≤ 80° (the pages for 15° ≤ H ≤ 32° are missing from the manuscript), and these relate to timekeeping by day to a specific latitude, whereas the values of the first function are intended to serve any latitude. The entries for the third and fourth functions are independent of H and the same values are tabulated for each value of H. The entries for the fifth function are independent of h, and only one value is tabulated for each value of H. The value of ϕ underlying the table is about 33;30°, which would serve either Baghdad or Damascus. A smaller table (fols. 26v-27r) displays functions:

---

manuscript to Ḥabash is not certain, since it is not contained in MS Istanbul Yeni Cami 784 of his *Zīj* (*ibid.*, no. 16, and Debarnot, "Zīj of Ḥabash"). See further **XI-5.0**.

[37] See Neugebauer, "Studies in Byzantine Astronomy", pp. 11-12, and Jones, *Byzantine Astronomical Manual*, pp. 154-155.

[38] On this formula see Pingree, "Indian Influence", pp. 121-122; Davidian, "Bīrūnī on the Time of Day", pp. 331-332 and 334; and now **II-1.4**, **III-1.3**, and **XI-1.2**.

[39] *Paris Catalogue*, pp. 446-447.

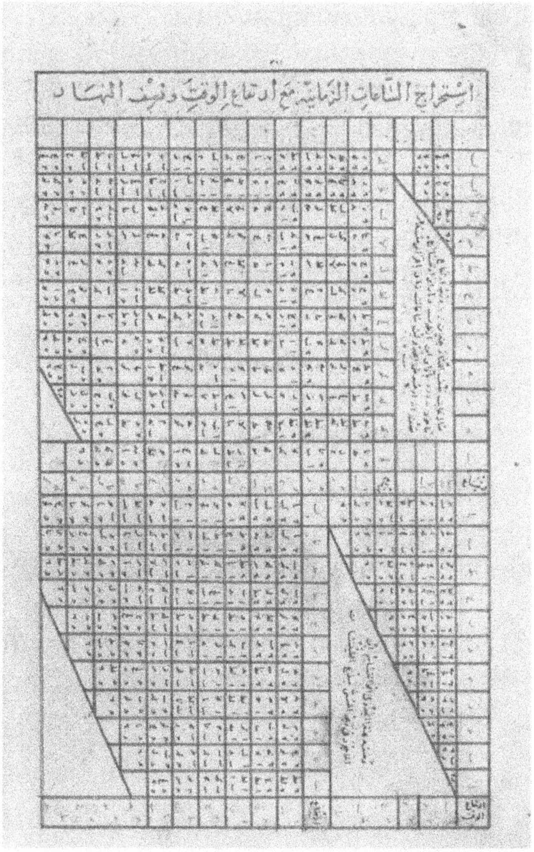

Fig. 2.5.1: Part of ʿAlī ibn Amājūr's *ṭaylasān* table. [From MS Paris BNF supp. pers. 1488, fol. 201v, courtesy of the Bibliothèque Nationale de France.]

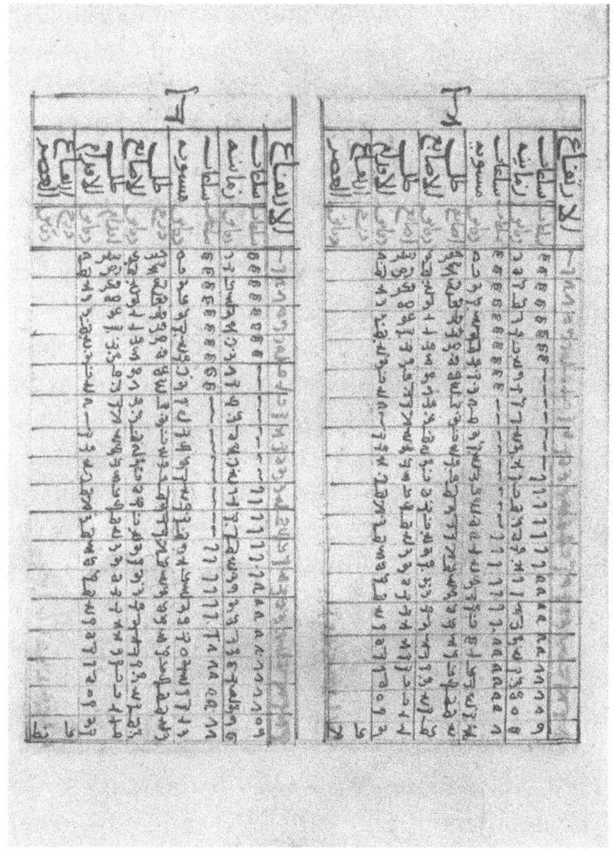

Fig. 2.5.2: The part of an anonymous *ṭaylasān* table serving solar meridian altitudes 33°-34°. [From MS Paris BNF ar. 2514, fol. 3r, courtesy of the Bibliothèque Nationale de France.]

$$f_1(h),\ f_2(h),\ \mathrm{Cot}_{12}\ h,\ \mathrm{Cot}_{6;40}\ h,\ h_a(H) \text{ and } \mathrm{Sin}\ h$$

to two digits for each degree of h (*al-irtifāʿ*). The first two functions are labelled *sāʿāt muʿwajja*, seasonal hours, and *sāʿāt mustawiya*, equinoctial hours, and appear to display the time since sunrise in hours and minutes for altitude h when the meridian altitude is 90°. The remainder of the tables in this manuscript relate to timekeeping to the stars (**3.2.1**).

2.5.3 Anonymous (Yemen)

The anonymous Yemeni *Zīj* preserved in MS Paris BNF ar. 2523, which was compiled *ca.* 1375,[40] contains a table of T(H,h) (fols. 97v-100v). Entries are computed to two digits for the domains:

$$H = 46°, 47°, \ldots, 90°, \text{ and } h = 1°, 2°, \ldots, H\ .$$

Note that the table is adequate for use in the Yemen where the minimum value of H for the sun is about 50°. See also **4.3.3** for a related table preserved in the same manuscript.

2.6 Tables of the time since rising of the sun or fixed stars as a function of their altitude, for all latitudes

To prepare a table which will give the time since rising of a celestial body for all latitudes, one can either tabulate T(H,h) for each degree of latitude or generate the function T in terms of h and pairs of arguments such as (δ,ϕ), (δ,d) or (H,D). The last pair has the advantage that – in modern terms – both arguments are always positive.

2.6.1 Najm al-Dīn al-Miṣrī (Cairo)

The apparently unique sources MSS Cairo MM 132 and Oxford Marsh 676 (Uri 944 = 995), now preserved in Cairo and Oxford, are the two halves of a single manuscript copied *ca.* 1325 which contains a table for timekeeping serving all latitudes. With a total of over 420,000 entries this is the largest table known from the medieval period.

A note on the flyleaf of the Oxford manuscript attributes the tables to Najm al-Dīn Abū ʿAbdallāh Muḥammad ibn Muḥammad al-Miṣrī, an Egyptian astronomer of the early 14th century before the 1970s virtually unknown to the scholarly literature. Thanks to the labours of François Charette, he is now well established as one of the most important astronomers of the Mamluk period.[41] No compiler is mentioned in the Cairo manuscript, but the title page is missing. Najm al-Dīn also wrote a treatise on spherical astronomy, extant in MS Milan Ambrosiana 227a (C49), fols. 85v-97r, copied in 1386, and the tables and treatises are clearly the work of the same person. It may be that the tables in MS Cairo MM 72, which were copied in 747 H [= 1346/47], are also by Najm al-Dīn (**4.4.1**).

The main function tabulated is:

$$T(H,h,D)$$

for the domains:

$$H = 2°, 3°, \dots , 73° \text{ (Cairo MS) and } 74°, 75°, \dots , 90° \text{ (Oxford MS)} ,$$
$$h = 1°, 2°, \dots , (H\text{-}1°) \quad \text{and} \quad D = D^*(H), [D^*(H)+1°], \dots , 180° .$$

For each value of H there are several pages of tables, in which the arguments h and D are entered horizontally and vertically, respectively. The lower limit D* for D is discussed below. **Fig. 2.6.1** shows an extract from the Cairo manuscript.

Clearly the two arguments H and D depend on the local latitude ϕ and the declination of the celestial body in question. There is no table of δ(λ) with which one could find H(λ) for the sun, but the work does contain a star table giving the equatorial coordinates of about 370 stars (fols. 148v-151r of the Oxford MS). On the other hand, tables of D(ϕ,λ) for each degree of ϕ from 1° to 90° and each degree of λ are provided (fols. 123r-142v of the Oxford manuscript), as well as tables of d(ϕ,Δ) (fols. 152r-166v) for each degree of ϕ and each degree of Δ from 1° to $\bar{\phi}$ (see further **7.0**). With such tables the appropriate arguments H and D can be found very easily for given ϕ and δ or Δ. The main table then gives immediately the time

---

[40] On this *zīj* see King, *Astronomy in Yemen*, no. 17.

[41] On Najm al-Dīn al-Miṣrī (Suter, *MAA*, no. 460, where two different individuals with this name are listed together; *Cairo ENL Survey*, no. C16; King, "Astronomy of the Mamluks", pp. 540-541) see now Charette, "Najm al-Dīn's Monumental Table", and *idem*, *Mamluk Instrumentation*. I have not updated my text to incorporate Charette's findings.

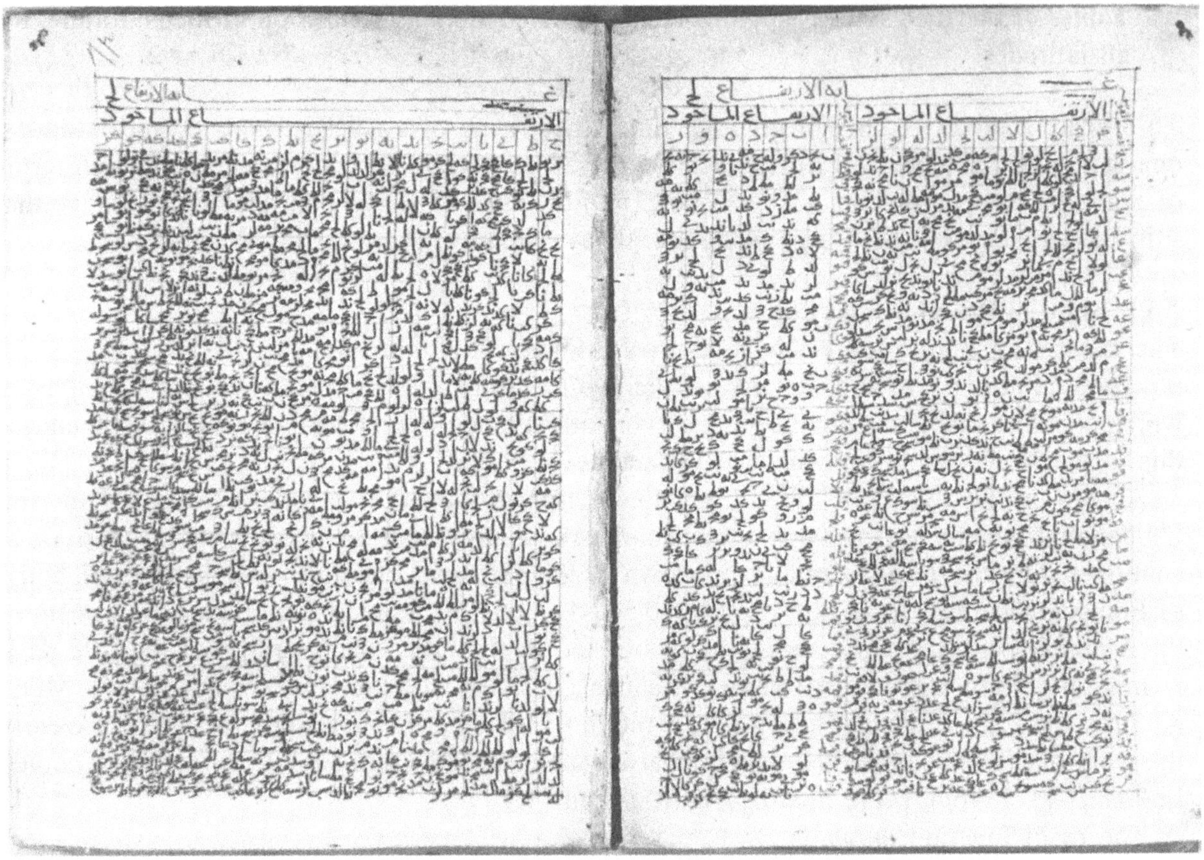

Fig. 2.6.1: An extract from Najm al-Dīn's monumental table serving meridian altitude 38°. Values are shown for altitudes 27°-37° and half arc of visibility 63°-87°, and then altitudes 1°-27° and half arc 108°-152°. [From MS Cairo MM 132, fols. 82v-83r, courtesy of the Egyptian National Library.]

since sunrise or since the rising of the star in question. For given H, it is not difficult to show that D varies between a certain minimum, defined by $\phi = -\delta = {}^1/_2\ \bar{H}$, and 180°. The lower limit D* for D in each table is simply the smallest integer greater than this minimum, *i.e.*:

$$D^*(H) = [\min D(H,\phi)] + 1 \ .$$

The compilation of the table is in fact extremely straightforward. The formula for finding T (= D - t) from H, h and D outlined in Ch. 12 of Najm al-Dīn's treatise on spherical astronomy (on which see **II-2.5**) is the following (*cf.* **F8** and **F10**):

$$\text{Vers}\ t(H,h,D) = \text{Vers}\ D - \text{Sin}\ h \cdot \text{Vers}\ D\ /\ \text{Sin}\ H \ .$$

Note that this is the equivalent to:

$$\text{Vers}\ t = \text{Vers}\ D\ \{\ 1\ -\ \text{Sin}\ h\ /\ \text{Sin}\ H\ \}\ ,$$

and that if one had first compiled a table of the function:

$$H''(H,h) = 1\ -\ \text{Sin}\ h\ /\ \text{Sin}\ H\ ,$$

the table of t(H,h,D) could be generated with considerable facility. Ibn al-Mushrif (**6.13.1**) tabulated the related function:

$$H'(H,h) = \text{Sin}\ H\ -\ \text{Sin}\ h\ ,$$

but no Islamic tables of H″ have been located.

In a sample of 100 random entries in Najm al-Dīn's main table about 55 showed errors of ±3 or less in the second digit, 35 showed errors between ±4 and ±9, and there were 10 errors between ±10 and ±15. It should be borne in mind that 0;15° corresponds to one minute of time, and I make bold to say that Najm al-Dīn's table shows the time for any solar or stellar altitude with *the vast majority of entries accurate to the nearest minute.*

Only in 1982 with the discovery of an introduction by him in MS Dublin CB Persian 102,1, did it become clear that Najm al-Dīn also intended this table as a universal auxiliary table: see further **9.3***. For a complete discussion of all aspects of this table, the reader is referred to the publications of François Charette.

2.7 Tables of the time of night as a function of the normed right ascensions of culminating stars and the solar longitude, for a specific latitude

2.7.1 Shihāb al-Dīn al-Ḥalabī: Damascus

MS Cairo K 8525 is a copy of part of al-Khalīlī's corpus of tables for Damascus (**2.1.4**) in the hand of the Damascus astronomer Shihāb al-Dīn al-Ḥalabī (d. 1455).[42] Certain additional tables are stated to have been computed by al-Ḥalabī himself, including a table of the oblique ascensions of the ascendant at daybreak (based on a value 20° for the solar depression rather than al-Khalīlī's 19°), and a complete set of azimuth tables (**5.1.2**), as well as a table (fols. 14r-20r) displaying the time remaining until daybreak as a function of solar longitude for seven culminating stars which vary for each zodiacal sign – see **Fig. 2.7.1** for an extract. Values are given in equatorial degrees and minutes and are based on the relation:

$$f(\lambda) = \alpha' - \alpha_r(\lambda) ,$$

where α' is the normed oblique ascension of the particular star (= $\alpha_\phi(\lambda_H)$, where λ_H is the longitude of the horoscopus when the star is culminating) and $\alpha_r(\lambda)$ is the "ascensions at daybreak" (= $\alpha_\phi(\lambda_H)$, where λ_H is the longitude of the horoscopus at daybreak). These values are easily derived from the entries for α' in the star catalogue in MS Cairo K 8525 and al-Ḥalabī's table of $\alpha_r(\lambda)$ in the same manuscript, which is derived from al-Khalīlī's tables of $\alpha_\phi(\lambda)$ and $T(h,\lambda)$ for Damascus using $\alpha_r(\lambda) = \alpha_\phi(\lambda) - T(h,\lambda^*)$ with $h_r = -20°$.

2.7.2 Muḥammad ibn Kātib Sinān: Istanbul

MSS Istanbul Ayasofya 2710 and Istanbul Topkapı T 3046 (Ahmet III 3515) are the only known copies of an extensive set of tables for timekeeping by the stars by the late-15th-/ early-16th-century Ottoman astronomer Muḥammad ibn (Kātib) Sinān.[43] According to the introduction, the compiler was a *muwaqqit* at *al-ʿataba al-ʿulya 'l-sulṭāniyya*, "the Sublime Porte", in Istanbul. The work is entitled *Mīzān al-kawākib*, "Balance of the Stars", and is dedicated to the Sultan Sulaymān ibn Sālim (*reg.* 1520-1566).

The main tables are preceded by a star catalogue displaying the declination and normed right ascension ($\alpha' = \alpha + 90°$) for about 510 stars, apparently for Ramaḍān, 916 H [= December,

[42] On al-Ḥalabī (Suter, *MAA*, no. 434) see *Cairo ENL Survey*, no. C69.

[43] On Muḥammad ibn Kātib Sinān (Suter, *MAA*, no. 455) see *Cairo ENL Survey*, no. H8; and İhsanoğlu *et al.*, *Ottoman Astronomical Literature*, I, pp. 84-90, no. 46.

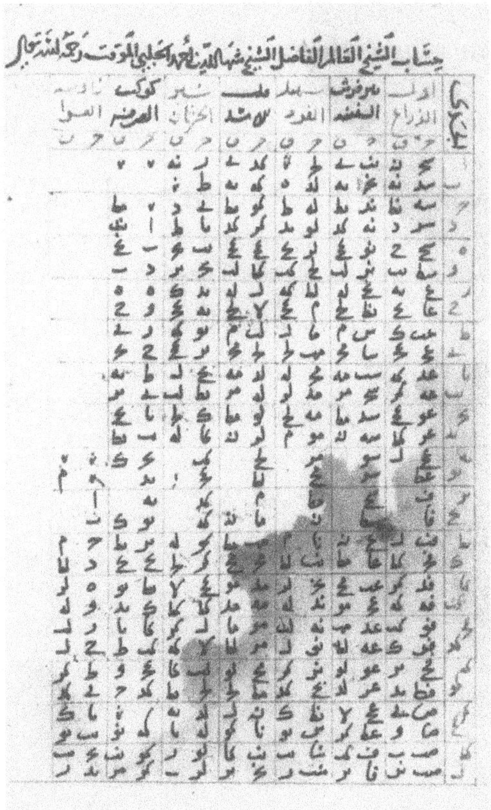

Fig. 2.7.1: The part of al-Ḥalabī's table serving solar longitudes in each degree of Capricorn and Aquarius. [From MS Cairo K 8525, fols. 14v-15r, courtesy of the Egyptian National Library.]

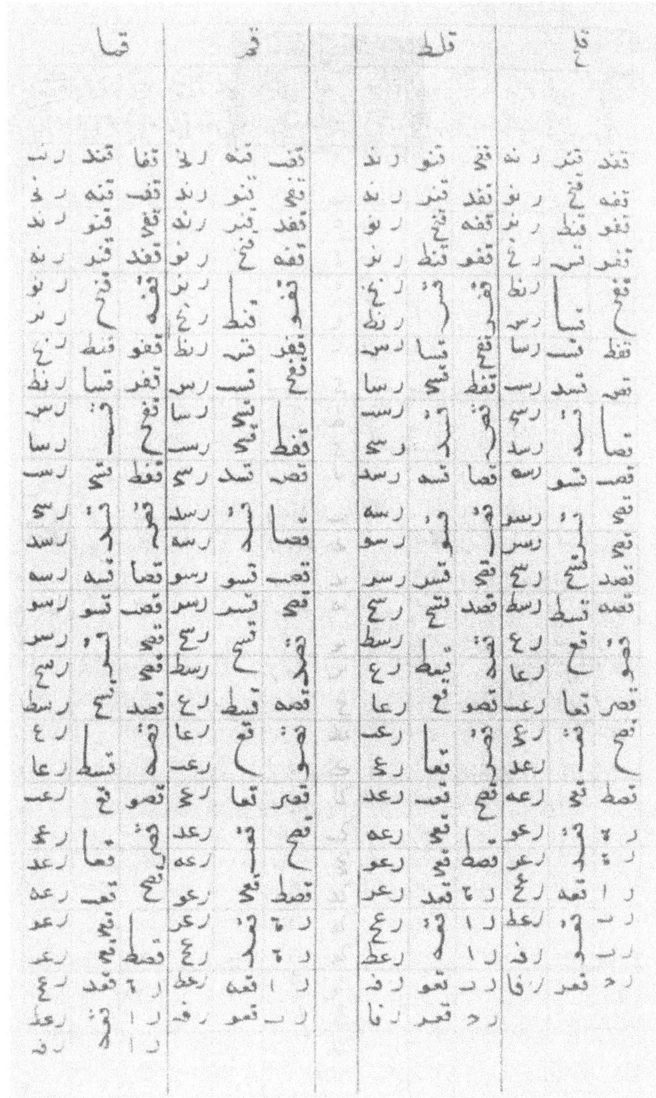

Fig. 2.7.2: The section of Ibn Kātib Sinān's enormous table showing the time of night for each degree of the ascensions of culminating stars from 138° to 141°, entered horizontally, for each degree of solar longitude in Aquarius, entered vertically. [From MS Istanbul Topkapı T 3046 (Ahmet III 3515), courtesy of the Topkapı Library.]

1510]. All of the right ascensions have been increased by 2;0° in the margin of the star catalogue. The main tables themselves occupy about 500 pages in each of the manuscripts and contain about 240,000 entries. The horizontal argument is intended to be the normed right ascension α', and the vertical argument is the solar longitude λ; the argument increments are 1° for both. Several dozen consecutive pages serve each 30° of solar longitude. Four functions are tabulated side by side for each pair of arguments:

t_1 the time since sunset,
t_2 the time remaining until sunrise,

t_3 the time remaining until daybreak, and
t_4 the time remaining until midday.

Entries are given in equatorial degrees to one digit only. See the extract in **Fig. 2.7.2**. To use the table one simply feeds in the normed right ascensions of the star which is culminating instantaneously and the solar longitude and reads off the appropriate time. The four functions tabulated are defined by the simple relations:

$$t_1(\alpha',\lambda) = \alpha' - \alpha_\phi(\lambda)$$
$$t_2(\alpha',\lambda) = 2N(\lambda) - t_1$$
$$t_3(\alpha',\lambda) = t_2 - r(\lambda)$$
$$t_4(\alpha',\lambda) = t_2 + D(\lambda)$$

where r is the duration of morning twilight. For each zodiacal sign (I-XII) of solar longitude, an appropriate range of arguments is given for the ascensions, namely:

I: 204°-259°; II: 235-20; III: 274-47; IV: 315-84; V: 330-115; VI: 5-140;
VII: 5-168; VIII: 33-196; IX: 50-249; X: 78-293; XI: 100-321; XII: 190-345

The entries in the tables correspond roughly to the latitude $\phi = 41;0°$ (Istanbul), and in MS Istanbul Ayasofya 2708 of Muḥammad ibn Kātib Sinān's treatise on timekeeping he uses parameters 20° and 16° for morning and evening twilight, although in MS Istanbul Ayasofya 2590 of his Turkish introduction to al-Khalīlī's universal auxiliary tables (**9.5**), written in 897 H [= 1491], he uses al-Khalīlī's parameters 19° and 17°. I have not investigated the accuracy of the tables.

2.8 Tables of the time of night as a function of the longitude of the horoscopus, for a specific latitude

The star *al-kaff al-khaḍīb* (= β Cassiopeiae)[44] was used by Ottoman and Safavid astronomers for timekeeping by night because its right ascension was close to zero, that is, it culminated with the vernal equinox. These tables represent a special case of the theory underlying those treated in **2.7**.

2.8.1 Anonymous: Istanbul

MS Cairo ZK 782,5 (fols. 25v-28r), copied *ca.* 1700, contains an isolated table of a function for timekeeping, which according to information in the margin is computed for the star *al-kaff al-khaḍīb* using coordinates $\alpha' = 86;50°$ and $\Delta = 56;37°$ for the year 950 H [= 1543/44] – see the extract in **Fig. 2.8.1**. The accompanying instructions are in Turkish. The tabulated function is computed to two digits for each degree of ecliptic longitude. It displays discontinuities at arguments 106° and 286° and in the three argument domains 0°-105°, 106°-285° and 286°-360° is labelled *al-māḍī mina 'l-ṭulū'*, "time since rising", *al-māḍī mina 'l-ghurūb*, "time since setting", and again *al-māḍī mina 'l-ṭulū'*. The function tabulated is simply:

$$f(\lambda) = \alpha' - \alpha_\phi(\lambda) \quad \text{for } \lambda < 106° \text{ and } \lambda > 286° \quad \text{or}$$
$$\alpha' - [\alpha_\phi(\lambda) - 180°] \text{ for } 106° \leq \lambda \leq 285° ,$$

for the latitude of Istanbul, taken as 41°.

[44] See Kunitzsch, *Sternnomenklatur der Araber*, p. 72, nos. 136a-c. This star was considered as one of the hands associated with the female face identified as *al-thurayyā*, the Pleiades (*ibid.*, pp. 114-115, no. 306).

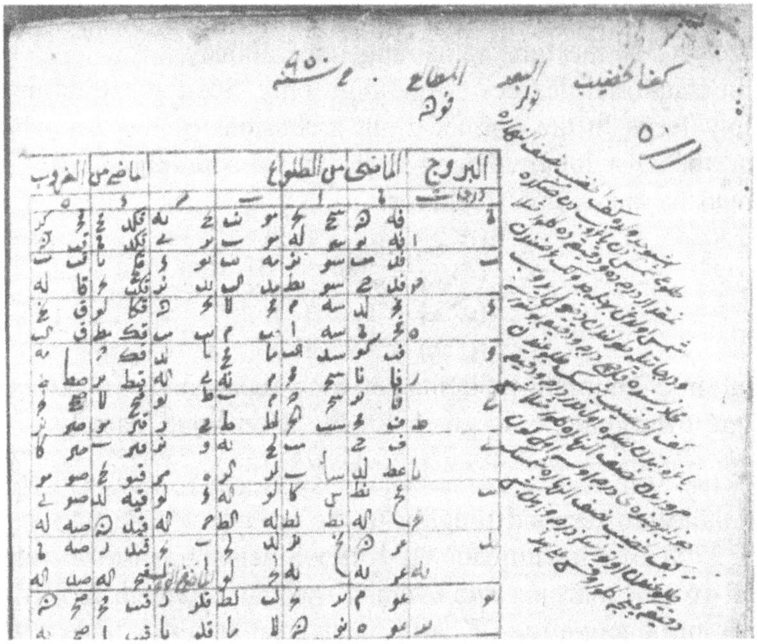

Fig. 2.8.1: An extract from a table for timekeeping by the star *al-kaff al-khaḍīb*, chosen because of its fortuitous position relative to the vernal equinox in the 16th century. [From MS Cairo ZK 782,5, fols. 25v-26r, courtesy of the Egyptian National Library.]

2.8.2 Muḥammad Zamān Mashhadī: Meshed

MS Cairo ṬFF 14 (111 fols.), penned *ca.* 1700, is a unique copy of a Persian astronomical work entitled *Tuḥfa-yi Sulaymānī* [*sic*] compiled by the astronomer and instrument-maker Muḥammad Zamān ibn Sharaf al-Dīn al-Mashhadī.[45] The treatise was compiled in 1078 H [= 1667/68] and it contains some spherical astronomical tables computed for the latitude of Meshed, taken as 37°. On fol. 42v there is a table entitled *jadwal wuṣūl kaff al-khaḍīb bi-niṣf al-nahār li-ṭūl Mashhad-i muqaddas bi-taʾrīkh gh-ʿ-ḥ*, "table of the culmination of *al-kaff al-khaḍīb* for the longitude of Meshed in the year 1078 (H)". Entries are given in hours and minutes for each degree of solar longitude. In the star catalogue on fol. 102r the right ascension of the star is given as 359;0°. See **II-3.14** on the prayer-tables in this work.

2.8.3 Anonymous: unspecified locality

In MS Aleppo Awqāf 948, p. 20, copied about 1800, there is a similar table following a larger set of tables of the astrological houses for an unspecified latitude,[46] probably that of Aleppo, where the manuscript is now preserved.

---

[45] On Muḥammad Zamān al-Mashhadī (**I-2.8.2**) see Mayer, *Islamic Astrolabists*, pp. 78-79 and 87; *Cairo ENL Survey*, no. G76; and King, *Mecca-Centred World-Maps*, pp. 172-175.

[46] See n. 3:2.

CHAPTER 3

TABLES OF THE LONGITUDE OF THE HOROSCOPUS AS A FUNCTION OF SOLAR AND STELLAR ALTITUDE

3.0 Introductory remarks

Tables displaying the longitude of the horoscopus are related to the categories described in **Ch. 2** by virtue of the simple formulae (*cf.* **F20**):

$$\alpha_\phi\{\lambda_H(h,\lambda)\} = \alpha_\phi(\lambda) + T(h,\lambda)$$

for the sun and:

$$\alpha_\phi\{\lambda_H(h)\} = \alpha_\phi(\rho) + T(h)$$

for a fixed star. It should be remembered that tables of oblique ascensions $\alpha_\phi(\lambda)$ for different latitudes were standard equipment of the Muslim astronomers.[1]

The horoscopus is of prime importance in astrology as well as in timekeeping and the Muslim astronomers also compiled extensive tables displaying the longitudes of the astrological houses and the projection of the rays.[2] These tables are not discussed in the present study.[3] See **10.1** on a table displaying the function $\lambda_H(T,\lambda)$ for a particular latitude found in an early-15th-century manuscript of the *Alphonsine Tables* written in Germany.

3.1 Tables of the longitude of the horoscopus as a function of the instantaneous solar altitude and solar longitude, for a specific latitude

3.1.1 Abu 'l-ʿUqūl: Taiz

In the corpus of tables of Taiz by Abu 'l-ʿUqūl (**2.1.2**) there is a set displaying the function:

$$\lambda_H(h,\lambda)$$

for:

$$h = 1°, 2°, \dots, 90° \text{ and } \lambda = 1°, 2°, \dots, 360°$$

and based on parameters:

$$\phi = 13;40° \text{ (Taiz) and } \varepsilon = 23;35° .$$

Entries are given to degrees and minutes for both eastern and western altitudes. An extract is shown in **Fig. 3.1.1**. There are over 40,000 entries in the entire set, which is complete in both the Berlin and Milan copies. They were probably computed by Abu 'l-ʿUqūl using his tables of $T(h,\lambda)$ and the table of $\alpha_\phi(\lambda)$ for these parameters in the *Mukhtār Zīj* (MS London BL Or. 3624, fols. 178v-179r). With such tables at hand the compilation of a table of $\lambda_H(h,\lambda)$

[1] See the article "Maṭāliʿ" in EI_2.2

[2] On these concepts see now Hogendijk, "Mathematical Astrology".

[3] For examples of such tables see those of al-Majrīṭī (*fl. ca.* 1000 – see the article in *DSB* by Juan Vernet) published in Suter, *al-Khwārizmī*, pp. 195-229; analyzed in Kennedy & Krikorian-Preisler, "Projecting the Rays"; and explained in Hogendijk, "Tables for Casting the Rays".

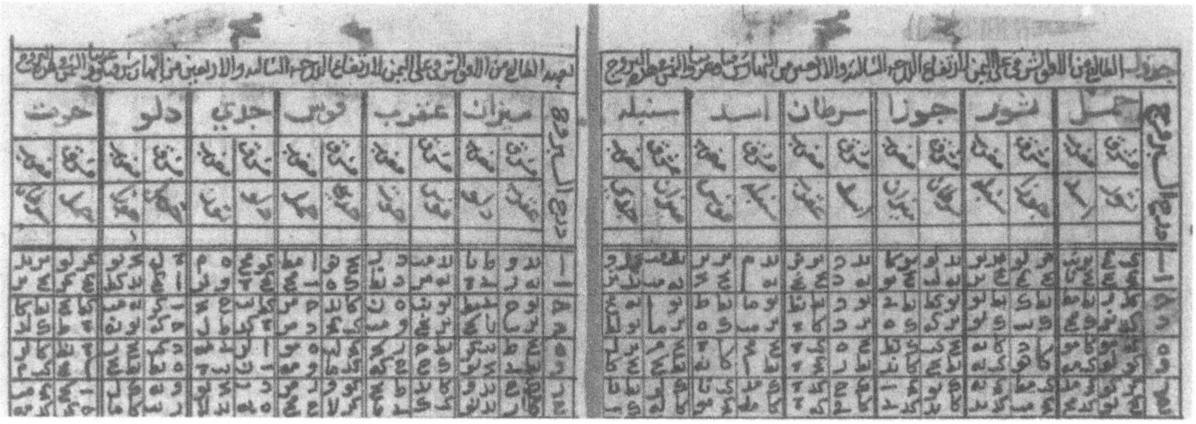

Fig. 3.1.1: The part of Abu ʻl-ʼUqūl's enormous table displaying the longitude of the horoscopus for solar altitude 43° in the east and west. [From MS Berlin Ahlwardt 5720, courtesy of the Deutsche Staatsbibliothek (Preußischer Kulturbesitz).]

is tedious but very straightforward. Given also a table of inverse oblique ascensions $\lambda(\alpha_\phi)$, the compilation of the larger table is made even easier.

3.2 Tables of the longitude of the horoscopus as a function of stellar altitude, for a specific latitude

3.2.1 Anonymous: Qandahar

At the end of MS Berlin Ahlwardt 5751 (Mq. 101,1), copied *ca.* 1300, of the *Zīj* of Kūshyār ibn Labbān, who worked in Iran *ca.* 1000,[4] there are various tables of considerable historical interest that have been taken from other early Islamic sources. All of these tables merit detailed study. One such is a star catalogue (pp. 188-189) displaying values of twelve functions for 25 stars: see **Fig. 3.2.1a**. These functions are:

λ	*al-ṭūl*	longitude
β	*al-ʿarḍ*	latitude
$\delta_2(\lambda)$	*al-mayl*	second declination
Δ	*al-buʿd ʿan muʿaddil al-nahār*	declination
$\bar{H}$	*al-inḥirāf ʿan samt al-raʾs*	zenith distance when culminating
H	*al-irtifāʿ*	altitude when culminating
D	*niṣf qaws al-nahār*	half arc of visibility
$\alpha_\phi(\rho)$	*maṭāliʿ ajzāʾ al-ṭulūʿ*	ascensions of the co-ascendant
ρ	*daraj al-ṭulūʿ*	co-ascendant (degree of the ecliptic which rises with the star)
σ	*daraj al-ghurūb*	co-descendant

[4] On Kūshyār (Suter, *MAA*, no. 192) see Kennedy, "*Zīj* Survey", no. 9, and Sezgin, *GAS*, V, pp. 343-345, VI, pp. 246-249 and VII, pp. 182-183. On the theoretical treatment of spherical astronomy in his *Zīj* see Berggren, "Spherical Trigonometry in the *Jāmiʿ Zīj*".

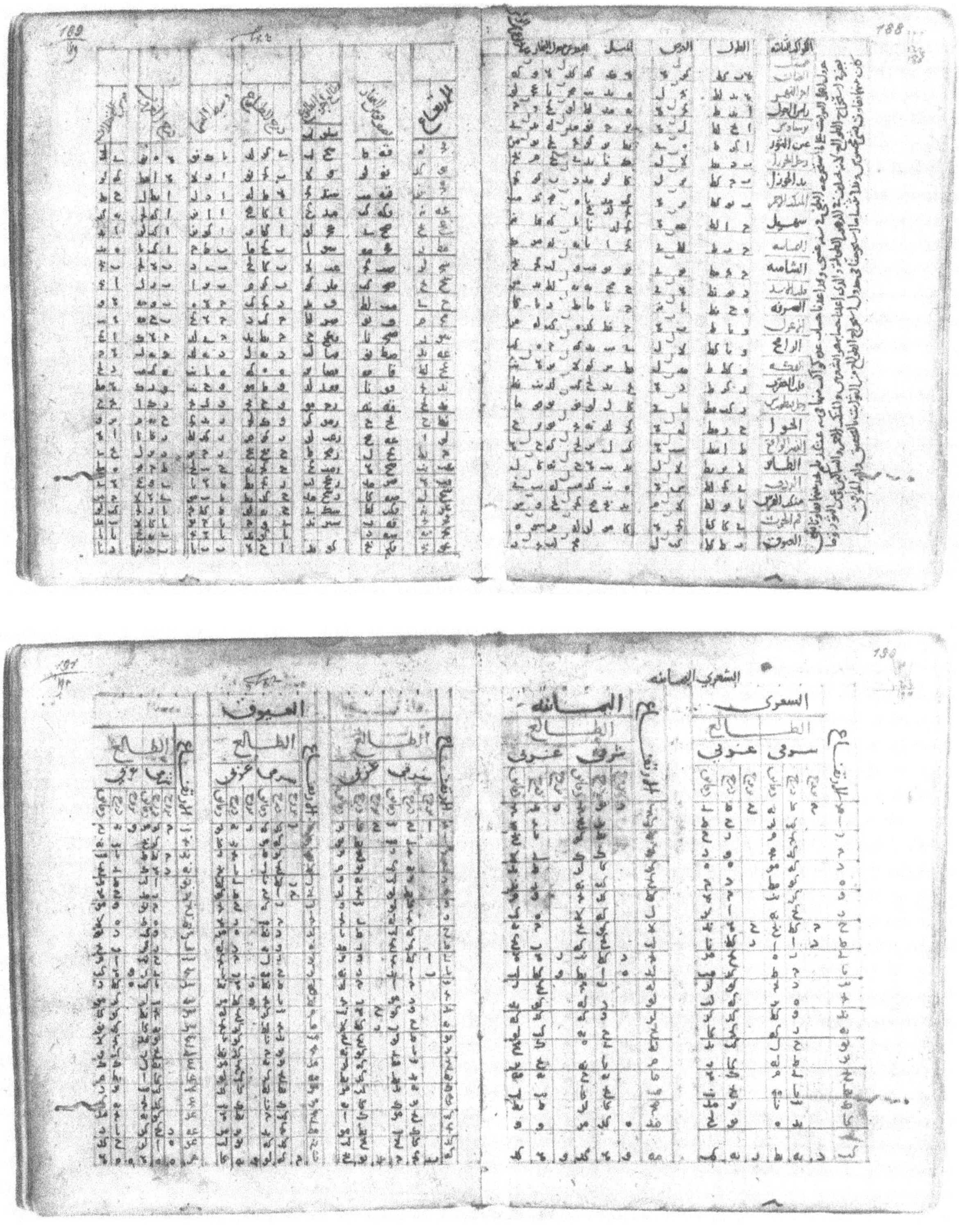

Figs. 3.2.1a-b: The star catalogue with additional information for timekeeping by the stars (a), and the sub-tables for timekeeping by Sirius and Capella (b). [From MS Berlin Ahlwardt 5751, pp. 188-189 and 190-191, courtesy of the Deutsche Staatsbibliothek (Preußischer Kulturbesitz).]

τ	*wasaṭ al-samā'*	co-culminant (degree of the ecliptic which culminates with the star)
ζ	*qaws al-ikhtilāf*	unidentified function

The underlying latitude is found by inspection to be 30;25°. A note by the side of these tables states:

> "(This is) a table for operations (of timekeeping) with the fixed stars according to the results of al-Balkhī in the year 1310. We recomputed (the entries for) several stars for the year 1424 and did not find any difference which would affect the determination of the (co)ascendants, because they (?) do not vary over long periods. We recomputed values for Sirius and Procyon, Betelgeuze, Spica and Arcturus, and Aldebaran (*al-shi'rayn wa-'l-mankib al-ayman wa-'l-simākayn wa-'ayn al-thawr*),[5] and for these there was no perceptible difference. We used what we derived in the table for finding the ascendants accurately from the fixed stars. God grants success."

The dates 1310 and 1424, which are written in the *abjad* notation in the text, must relate to the calendar of Alexander, and thus correspond to 1000 and 1125 in the Julian calendar. Also from the table itself (the longitude of Regulus is 136;39°) a date of *ca.* 1050 is suggested.[6] At the time of writing this in the 1970s I was not aware of any astronomer named al-Balkhī who was active in Southern al-'Irāq or Iran *ca.* 1000, although 'Alī ibn Aḥmad al-Nasawī compiled a *zīj* in central Iran in the early 11th century. The latitude 30;25° was used for Qandahar (now in Afghanistan) in a recension of the geographical tables of al-Khwārizmī (**7.1.1**),[7] and we can be certain that the table was intended for use there.

The next four pages in the Berlin manuscript (pp. 190-194) contain the only remaining part of these ascendant tables: see **Fig. 3.2.1b**. The function tabulated is $\lambda_H(h)$ for each degree of eastern and western altitudes for Sirius, Capella, Vega and Arcturus. Entries are given in zodiacal signs, degrees and minutes, and the underlying latitude appears to be 30;25°. Both star catalogue and ascendant tables seem to be unrelated to the auxiliary tables that immediately precede them (on pp. 188-189) in the Berlin manuscript (**6.16.1**), although these auxiliary tables do indeed serve to compute $\lambda_H(h)$ for different stars. See **5.8.1** on another table from this collection in the same manuscript. Both tables are featured again in **IX**.

3.2.2 Anonymous: Baghdad / Damascus

MS Paris BNF ar. 2514, fols. 28v-48v, (**2.5.2**), copied in the year 612 H [= 1215], contains some anonymous tables of the functions $\lambda_H(h)$ and $\alpha_\phi\{\lambda_H(h)\}$ computed for 21 stars. Values are given in signs, degrees and minutes for each degree of eastern and western altitude up to the maximum for each star: see **Fig. 3.2.2** for an extract. The underlying latitude is about 33;30°, which would serve either Baghdad or Damascus.

---

5 On these six stars see Kunitzsch, *Sternnomenklatur der Araber*, pp. 111-112, nos. 289-290; p. 57, no. 158; p. 105, nos. 269-270, and p. 51, no. 69, respectively.

6 For example Ibn Yūnus has Regulus at Leo 16° *ca.* 1000, and the rate of precession is 1° in *ca.* 70 years.

7 Kennedy & Kennedy, *Islamic Geographical Coordinates*, pp. 261-262.

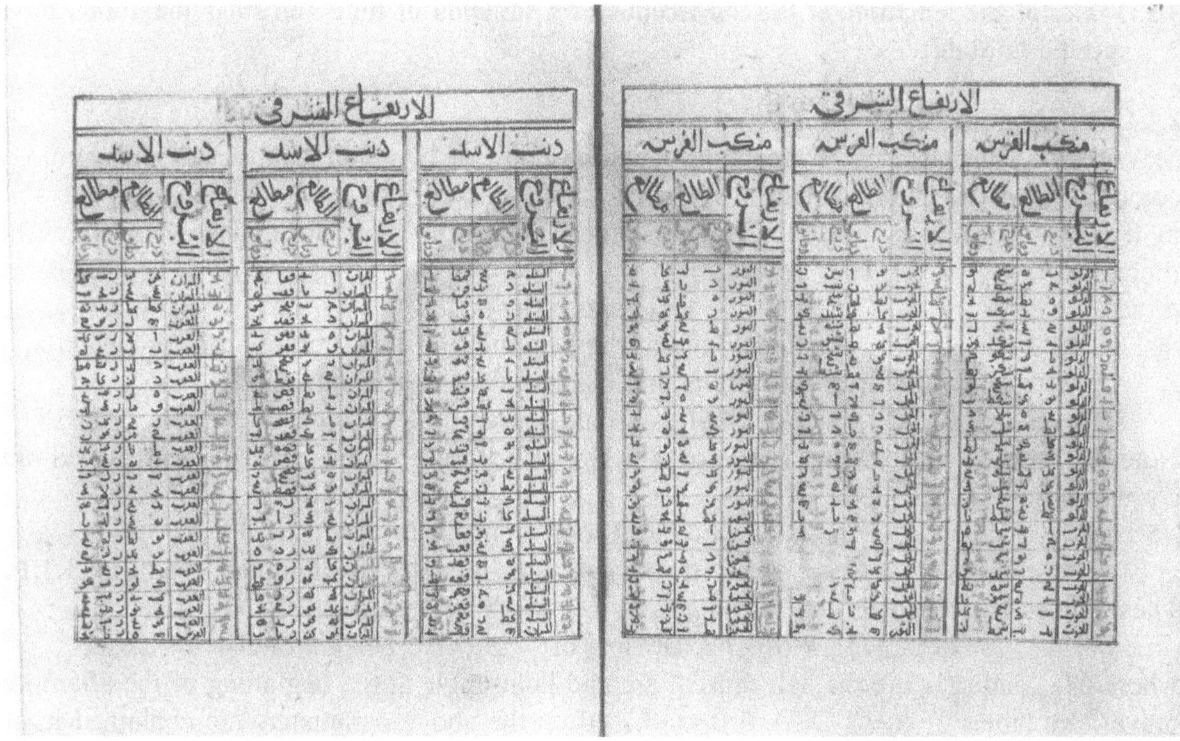

Fig. 3.2.2: An extract from the anonymous tables for timekeeping by the stars, serving *mankib al-faras* and *dhanab al-asad*. [From MS Paris BNF ar. 2514, fols. 31v-32r, courtesy of the Bibliothèque Nationale de France.]

3.2.3 Ibn Dāʿir: Sanaa

MS Berlin Ahlwardt 5720 of Abu 'l-ʿUqūl's corpus of tables for Taiz (**2.1.2**) contains three tables (fols. 121v, 124r-127r) computed for the latitude of Sanaa and taken from another source. The tables are all for timekeeping by the stars, and it is stated (fol. 121v) that the stellar coordinates were based on the observations of Ibn Yūnus, but the name of the compiler of the tables and the date for which he computed the stellar coordinates have been deliberately erased. However, it is just possible to read the name ʿAbdallāh ibn Ṣalāḥ Dāʿir,[8] which also appears elsewhere (fol. 121r) in reference to some instructions on how to use tables of stellar altitude at the seasonal hours. Ibn Dāʿir (d. 1013 [= 1604] is otherwise known as the author of a geographical treatise compiled for the Ottoman Sultan Murad III. See **4.5.1** on these altitude tables and **3.4.1** on the other tables for Sanaa in this set, and also **II-12.6**.

The first of the three tables (fols. 121v, 124r-126r), which is also the largest, displays the longitude of the horoscopus as a function of the altitude of 16 stars. Entries are given in signs, degrees and minutes and for each degree of altitude up to the maximum for each star, and there are two sub-tables for altitudes in the east and west.

---

[8] On Ibn Dāʿir see King, *Astronomy in Yemen*, no. 31, and now Ihsanoğlu *et al.*, *Ottoman Geographical Literature*, I, pp. 76-78, no. 39. The word Dāʿir means "profligate": see the article "Zuʿʿār" by Thierry Bianquis in EI_2.

3.3 Tables of the longitude of the horoscopus as a function of time and solar longitude, for a specific latitude

3.3.1 al-Sulṭān al-Ashraf: Sanaa

MS Oxford Hunt. 233 is an apparently unique copy of a monumental work on astrology compiled by the Yemeni Sultan al-Ashraf (*fl. ca.* 1295).[9] The work contains considerable material on timekeeping, including several tables. One of the latter (fols. 63v-69r) displays the functions:

$$H(\lambda) \text{ and } h_a(\lambda) \ ,$$

that is, the solar altitudes at midday and at the beginning of the afternoon prayer (when $z_a = Z + n$). Values are given to two digits and are based on the parameters:

$$\phi = 14;30° \text{ (Sanaa) and } \varepsilon = 24° \ .$$

Note the use of the Indian value of ε! Alongside these values are also displayed the corresponding values of:

$$\lambda_H(\lambda)$$

at both these times. Values are given in zodiacal signs, degrees and minutes – see **Fig. 3.3.1a**. These tables are based on the formulae:

$$\alpha_\phi\{\lambda_H(\lambda)\} = \alpha_\phi(\lambda) + D(\lambda) \quad \text{or} \quad \alpha_\phi(\lambda) + D(\lambda) + t_a(\lambda) \ ,$$

where $D(\lambda)$ and $t_a(\lambda)$ are the half diurnal arc and hour-angle at the beginning of the afternoon prayer. No tables of $\alpha_\phi(\lambda)$, $D(\lambda)$, or $t_a(\lambda)$ based on the above parameters are contained in al-Ashraf's treatise or in any other known Yemeni source.

Elsewhere in al-Ashraf's treatise (fols. 80v-92r) there are tables of the function:

$$\lambda_H(T,\lambda) \ ,$$

with values expressed in zodiacal signs, degrees and minutes for the domains:

$$T = 1, 2, \ldots, 12^{sdh} \text{ and } \lambda = 1°, 2°, \ldots, 360° \ .$$

The function is called *ṭāliʿ sāʿāt (al-)nahār*, "the ascendant of the hours of daylight" – see **Fig. 3.1.1b**.

See **4.2.3** on another table in MS Oxford Hunt. 233 and **4.1.4** and **5.4.4** on tables in a second work by al-Ashraf, and also **II-12.2**.

3.3.2 Abu 'l-ʿUqūl: Taiz

In the 14th-century Yemeni miscellany preserved in a manuscript in a private collection in Sanaa (**2.1.2**) there is an incomplete set of tables of the function:

$$\lambda_H(T,\lambda)$$

computed to signs, degrees and minutes, for the domains:

$$T = 1, 2, \ldots, 12^{sdh} \text{ and } \lambda = 1°, 2°, \ldots, 360° .$$

The underlying parameters are:

$$\phi = 13;37° \text{ (Taiz) and } \varepsilon = 23;35° \ .$$

[9] On the astronomical works of al-Ashraf (Suter, *MAA*, no. 394 and Mayer, *Islamic Astrolabists*, pp. 83-84) see King, *Astronomy in Yemen*, no. 8.

Fig. 3.3.1a: An extract from al-Ashraf's tables giving the altitude of the sun at midday and the beginning of the *ʿaṣr* and the corresponding ascendant. This sub-table serves solar longitudes in Aries. [From MS Oxford Hunt. 233, fol. 63v, courtesy of the Bodleian Library.]

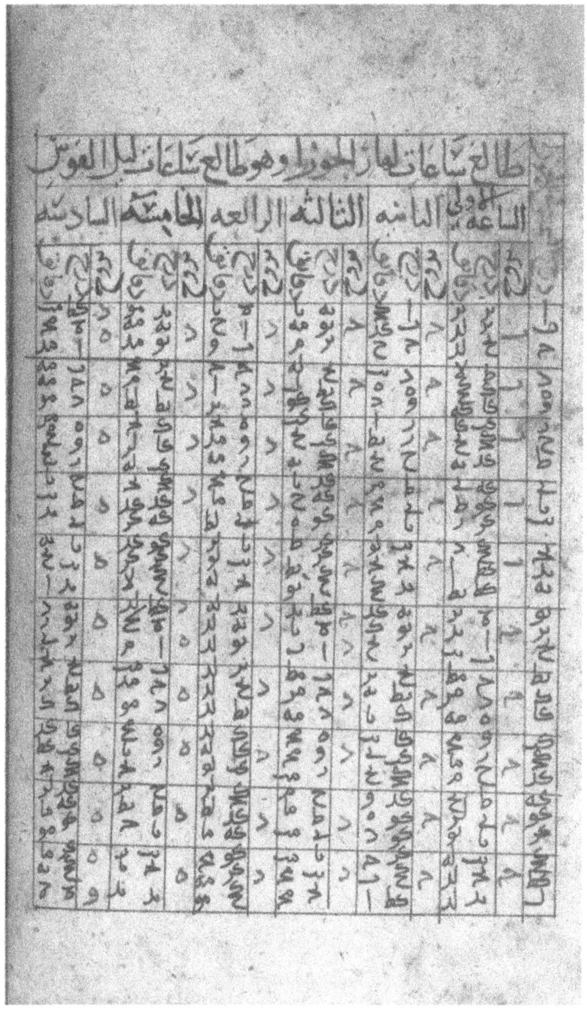

Fig. 3.3.1b: An extract from al-Ashraf's tables showing the ascendants at the hours as a function of solar longitude, here Gemini. The same entries are for the night hours for solar longitudes in Sagittarius. [From MS Oxford Hunt. 233, fol. 82v, courtesy of the Bodleian Library.]

An extract is shown in **Fig. 3.3.2**. The tables are unattributed, but the smaller table described in **4.2.5** is based on the same distinctive latitude and is attributed to Abu 'l-ʿUqūl. It seems that this latitude for Taiz was derived by Abu 'l-ʿUqūl after he had computed his corpus of tables for latitude 13;40° (**2.1.2**). A table of $\alpha_\phi(\lambda)$ for the same parameters is preserved in the anonymous late-14th-century Yemeni *zīj* MS Paris BNF ar. 2523, fol. 95v (**2.5.3**). Likewise the tables for marking astrolabes and sundials for Taiz compiled by the Rasulid Sultan al-Ashraf about 1295 which are extant in MS Cairo TR 105 (**4.1.4**) are also based on ϕ = 13;37°, but the obliquity used is the *Īlkhānī* value 23;30° (**2.3.2**).

The function $\lambda_H(T,\lambda)$ is called *ṭawāliʿ al-sāʿāt al-zamāniyya*, "the ascendants of the seasonal

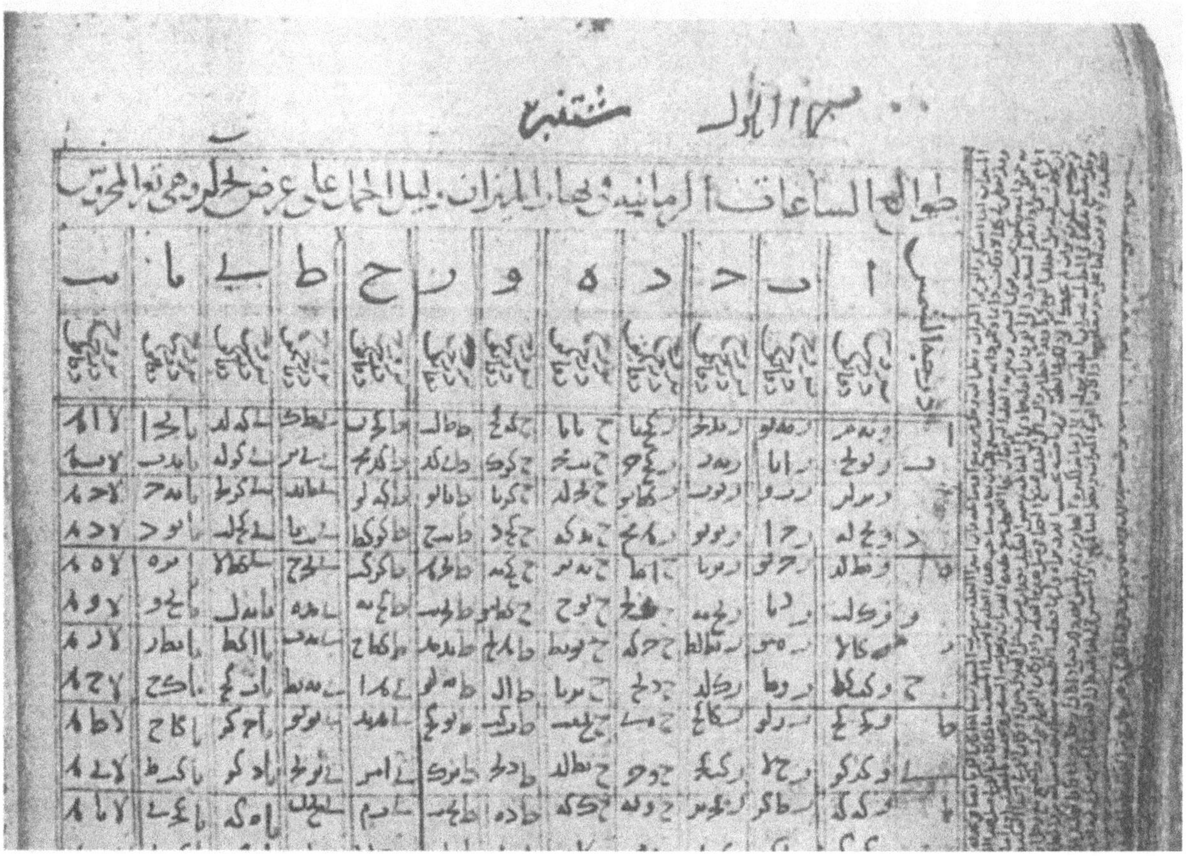

Fig. 3.3.2: An extract from the anonymous tables of the horoscopus at the hours of day and night for the latitude of Taiz. [From a manuscript in a private library in Sanaa; taken from the facsimile in Varisco & Smith, eds., *al-Afḍal's Anthology*, p. 114, with permission of Professor Daniel Varisco.]

hours", and the existence of a similar table in a 15th-century German manuscript has been noted in **3.0** (see also **10.1**). Implicit in the headings of the Yemeni table is the fact that the horoscopus is the same at the nth seasonal hour of the day for solar longitude λ as it is at the nth seasonal hour of the night for solar longitude $\lambda^* = \lambda + 180°$. A given sub-table on one page serves one zodiacal sign, but there are only five sub-tables in the manuscript now. The compilation of such a table is quite straightforward, since if $T = n^{sdh}$,

$$\alpha_\phi\{\lambda_H(T,\lambda)\} \;=\; \alpha_\phi(\lambda) + n/12 \bullet \{2D(\lambda,\phi)\}\ .$$

3.3.3 Anonymous: Anatolia

MS Istanbul Nuruosmaniye 2782, completed in Sivas in 774 H [= 1371/72], contains an anonymous set of calendrical and astrological tables perhaps of Seljuk origin. MS Cambridge Browne O.1, copied *ca.* 1400, is the only known copy of the *Zīj-i mufrad* by the 11th(?)-century astronomer Muḥammad ibn Ayyub al-Ṭabarī. Both of these contain a remarkable table (fols. 53v-54r of the Nuruosmaniye manuscript and fol. 179r of the Cambridge one); the captions are in Persian in the former and in Arabic in the latter (see **Fig. IV-5.3**). The table displays the function:

$$\lambda_H(T_n,\lambda)$$

for the domains:

$$\lambda = 0°, 6°, \ldots , 354° \quad \text{and} \quad T_n(\lambda) = n/8 \bullet \{2D(\lambda)\} \ (n = 1, 2, \ldots , 7)$$

as well as at daybreak and nightfall. The functions are called simply *ṭāliʿ-i awqāt*, "the ascendant at different times of day". The underlying latitude is found by inspection to be about $\phi = 38°$, and it is not possible to determine the value used for ε; this latitude perhaps corresponds to Konya. Entries are given in signs and degrees and it is stated that they were derived using an astrolabe. The underlying formula is trivial, namely:

$$\alpha_\phi \{\lambda_H(T,\lambda)\} = \alpha_\phi(\lambda) + T_n(\lambda) .$$

Using an astrolabe fitted with a plate for latitude ϕ one sets the ecliptic longitude λ on the rete over the local horizon on the plate and can immediately read off $\alpha_\phi(\lambda)$ and $D(\lambda)$. One then rotates the rete by the amount $T_n(\lambda)$ in the direction of the daily rotation and simply reads off the longitude of the new point where the ecliptic intersects the horizon, which is $\lambda_H(T_n,\lambda)$. See also **II-14.2** and **IV-5.3** for further information on this unusual table.

3.4 Tables of the longitude of the horoscopus as a function of the time since rising of the stars, for a specific latitude

3.4.1 Ibn Dāʿir: Sanaa

One of the anonymous tables for the latitude of Sanaa in MS Berlin Ahlwardt 5720 (**3.2.3**) displays the longitude of the horoscopus for each "seasonal hour" of visibility of 18 fixed stars (fol. 127r) – see **Fig. 3.4.1**. These "hours" are simply twelfths of the arcs of visibility of each particular star. The compilation of such a table is trivial since, if λ_n be the longitude of the horoscopus at the nth "hour" and 2N the arc of visibility of a star whose co-ascendant is ρ, then:

$$\alpha_\phi(\lambda_n) = \alpha_\phi(\rho) + n/12 \bullet (2N) .$$

See also **4.6.1** below on a related table by Ibn Dāʿir, and further **II-12.6**.

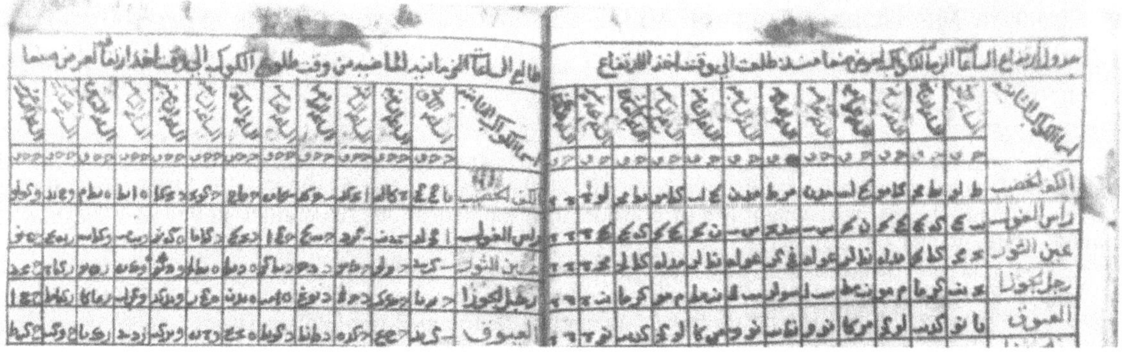

Fig. 3.4.1: The table by Ibn Dāʿir on the right displays the horoscopus at each "seasonal hour" of visibility for 18 stars. The one on the left shows the corresponding altitudes (**4.6.1**). [From MS Berlin Ahlwardt 5720, fols 126v-127r, courtesy of the Deutsche Staatsbibliothek (Preußischer Kulturbesitz).]

CHAPTER 4

TABLES OF SOLAR AND STELLAR ALTITUDE

4.0 Introductory remarks

The extensive stellar and solar altitude tables located in various Islamic sources are closely related to the tables described in **Chs. 2** and **3**. The numerous smaller solar altitude and azimuth tables for marking solstitial shadow traces on sundials belong to another category of Islamic astronomical astronomical tables which merit a separate study,[1] and I have intentionally limited examples of such altitude tables cited below to the tables of Ḥabash (Pseudo-al-Khwārizmī), al-Battānī, al-Marrākushī, al-Maqsī and the Yemeni Sultan al-Ashraf. Tables of solar meridian latitudes for specific latitudes were standard in *zīj*es, and already in the 9th century we find a small table of H(λ,φ) for each of the climates (see **Fig. VIa-1.1**).

4.1 Tables of solar altitude at the solstices as a function of time, for a specific latitude

Tables of the solar altitude for each seasonal hour at the solstices, and more especially the corresponding gnomon shadow lengths and azimuths, were used by Muslim astronomers for constructing the points of intersection of the hour lines with the solstitial shadow traces on horizontal sundials.[2]

4.1.1 Pseudo-al-Khwārizmī / Ḥabash: various latitudes

MS Istanbul Ayasofya 4830, fols. 231v-235r, copied in Damascus in 626 H [= 1228/29], contains a treatise on sundials attributed to the early-9th-century Baghdad astronomer al-Khwārizmī (**7.1.1**).[3] In fact, the treatise is more likely due to his late contemporary Ḥabash al-Ḥāsib (**9.1**).[4] The work consists mainly of tables for constructing horizontal sundials with markings for each seasonal hour. An extract is shown in **Fig. 4.1.1a**. A related set of tables, appended to the treatise on astrolabes and sundials by the late-10th-century scholar al-Sijzī,[5] is extant in MS Istanbul Topkapı A3342 (disordered, correctly: fols. 123r-153v then 114r-122v), copied also in Damascus in 634 H [= 1236/37]. An extract is shown in **Fig. 4.1.1b**. Each of these sources merits detailed study.

---

[1] See King, "Islamic Astronomical Tables", pp. 51-53, and *idem* & Samsó, Islamic Astronomical Tables", pp. 92-94, for preliminary discussions of the various kinds of tables used by the Muslim astronomers for marking curves on sundials, many of which are of considerable sophistication. A detailed survey of all extant Islamic sundial tables would be worthwhile.

[2] See my article "Mizwala" in EI_2, repr. in King, *Studies*, C-VIII, also **X-7**.

[3] On al-Khwārizmī see n. 7:5. On these tables see already King, "al-Khwārizmī", pp. 7-11. An edition of the tables and a Russian translation of the accompanying text is in Rosenfeld *et al.*, eds., *Al-Khorezmi*, pp. 221-234.

[4] For the arguments see King, *Mecca-Centred World-Maps*, p. 350. In Charette, *Mamluk Instrumentation*, p. 181, on the other hand, it is argued that the tables are indeed by al-Khwārizmī.

[5] On al-Sijzī (Suter, *MAA*, no. 185) see Sezgin, *GAS*, V, pp. 329-334, VI, pp. 224-226, and VII, pp. 177-182.

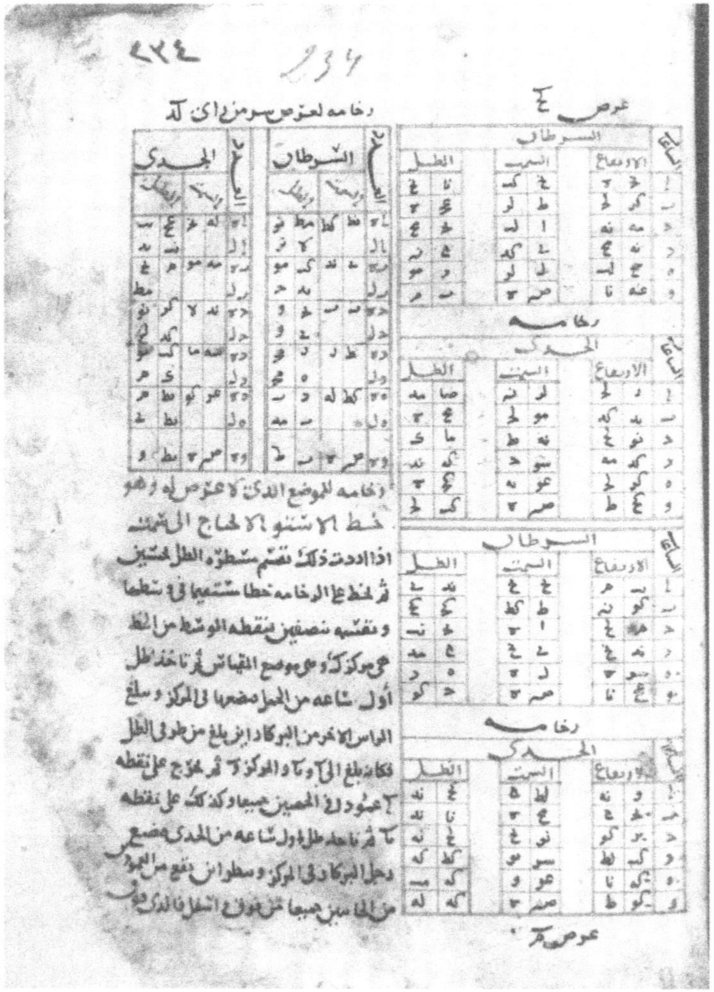

Fig. 4.1.1a: On the right we see the sub-tables for latitudes 38° and 40° from the treatise on sundial construction attributed to al-Khwārizmī in the unique manuscript, but more probably due to Ḥabash. Part of the reason for the latter attribution is that many of Ḥabash's works include tables for Samarra, with latitude 34°, here seen on the left. See also **Fig. VIa-10.1**. [From MS Istanbul Ayasofya 4830, fol. 234r, courtesy of the Süleymaniye Library.]

Fig. 4.1.1c: A set of basic tables probably to be associated with Ḥabash. Values are given of the sine to base 150 (*al-jayb*) and the cotangent to base 1 (*al-ẓill*), as well as the excess of daylight for an unspecified latitude – actually 34° – labelled "every 15° is one hour" (*kull y-h sāʿa*) and the solar declination (*al-mayl*). The argument labelled "altitude" (*al-irtifāʿ*) can serve all of the functions: a general arc for the sine, the solar altitude for the cotangent, and the solar longitude for the other two. [From MS Istanbul Ayasofya 4830, fol. 188v, courtesy of the Süleymaniye Library.]

Fig. 4.1.1b: Some sub-tables in the Topkapı manuscript, serving latitudes 21°, 28;30°, (33°, 40°, 35° and 30°). [From MS Istanbul Topkapı A3342, courtesy of the Topkapı Library.]

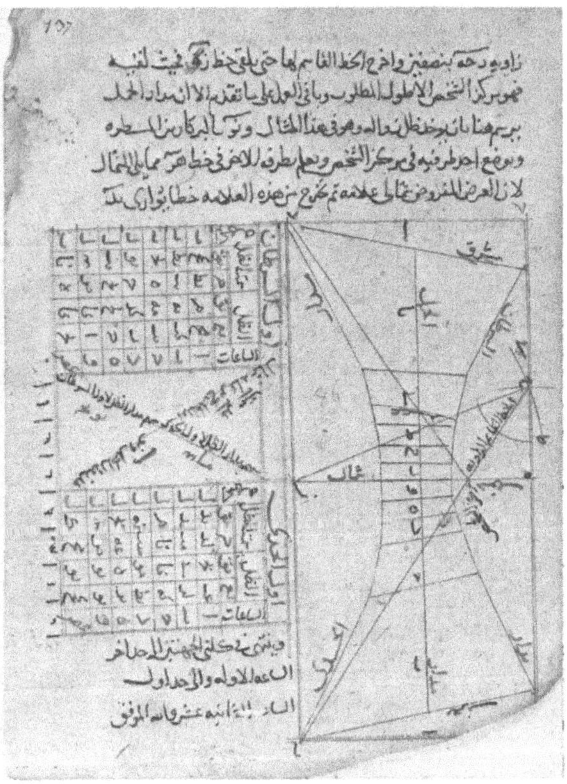

Fig. 4.1.3: The tables for constructing a horizontal sundial for latitude 30° in the treatise of Marrākushī. [From MS Paris BNF 2507, fol. 137r, courtesy of the Bibliothèque Nationale de Paris.]

Fig. 4.1.4: This extract from al-Ashraf's tables for sundial construction serves latitude 21°. [From MS Cairo TR 105, fols. 132v-133r, courtesy of the Egyptian National Library.]

The tables in the Ayasofya manuscript display the functions:

$$h(T), a(T) \text{ and } z_{(12)}(T),$$

giving values for arguments T = 1, 2, ... , 5^{sdh} at the summer and winter solstices for:

$$\phi = 0°, 15°, 18°, 20°, 24°, 27°, 30°, 33°, 34°, 35°, 38° \text{ and } 40°.$$

Entries are given to two digits and the underlying value of ε is 23;51°. For φ = 34° (Samarra) values are given for each $0;30^{sdh}$ and for φ = 33° (Baghdad) values of the second and third functions are given each $0;10^{sdh}$ in a separate table. The special attention paid to the latitude of Samarra is a further indication that the tables are by Ḥabash.[6]

The tables on fols. 119v-122v of the Topkapı manuscript are in some disorder. Those for:

$$\phi = 21°, 24°, 27°, 30°, 33°, 35°, 38° \text{ and } 40°$$

are based on ε = 23;51°, and those for:

$$\phi = 28;30° \text{ and } 36°$$

are based on ε = 23;35° and ε = 23;30°, respectively. Further investigation of the two manuscripts might establish whether any of these tables is due to al-Sijzī.

In the pages preceding the treatise attributed to al-Khwārizmī in the Ayasofya manuscript there are some other Abbasid treatises on astronomy containing a few simple tables of functions, such as (fol. 188v):

$$H(\lambda) \text{ for } \phi = 33;0° \text{ and } 34°\ ,$$

based on ε = 23;35°, called *al-mayl al-mumtaḥan*, by which may be meant "the obliquity of the *Mumtaḥan Zīj*" (**9.1**).[7] In any case, this is one of the two values used by Ḥabash (see again **9.1**).[8] Another set (fol. 188v – see **Fig. 4.1.1c**) displays the functions:

$$Sin_{150}\ \theta,\ Cot_1\ \theta,\ 2d(\lambda)\ (\phi = 34° \text{ and } \varepsilon = 23;35°), \text{ and } \delta(\lambda)\ (\varepsilon = 23;35°).$$

Values are given to two digits for each degree of argument.[9] The use of base 1 is remarkable.

Note that one of the tables discussed above displays h(T) for each $0;10^{sdh}$ of T for latitude 33° (Baghdad). In this connection it is worth mentioning that the 10th-century bibliographer Ibn al-Nadīm attributes to Yaḥyā ibn Abī Manṣūr,[10] a leading astronomer of early-9th-century Baghdad, a work entitled *Maqāla fī ʿamal irtifāʿ suds sāʿa li-ʿarḍ Madīnat al-salām*, "Treatise on the determination of the (solar) altitude for each one-sixth of an hour for the latitude of Baghdad". This work is no longer extant, but we can assume from the title that it must have been a set of tables displaying the solar altitude as a function of time of day, with argument increments of 0;10 hours, which might have been equinoctial or seasonal. The other argument was probably solar longitude λ or half daylight D. On the other hand, the work refered to by Ibn al-Nadīm may be identical with the table I have described above. However, a few spherical

---

[6] See King, *Mecca-Centred World-Maps*, p. 350, for a list of his works in which this latitude is featured, usually to the exclusion of all others.

[7] On Yaḥyā ibn Abī Manṣūr and the *Mumtaḥan Zīj* see Kennedy, "*Zīj* Survey", no. 51; Vernet, "Las Tabulae Probatae", the same author's article in *DSB*; and Sezgin, *GAS*, V, pp. 227-228, and VI, 136-137. A distorted facsimile of the Escorial manuscript was published in Frankfurt (IGAIW) in 1986.

[8] See, in particular, n. 9:5.

[9] On the sine tables see n. 7:10 to **7.1.1**.

[10] See Sezgin, *GAS*, VI, p. 137, no 2. The only other early Islamic work mentioned by Ibn al-Nadīm which relates specifically to timekeeping is a treatise by Ibn Karnib (see Sezgin, *GAS*, V, p. 275), who lived in Baghdad probably in the latter half of the 9th century, on the determination of the time of day from the solar altitude.

astronomical tables for Baghdad are contained in the recension of Yaḥyā's *Mumtaḥan Zīj* preserved in MS Escorial ár. 927: these are based on parameters ϕ = 33;21° and ε = 23;33°.

4.1.2 al-Battānī: Raqqa

MS Escorial ár. 908, fol. 240r, of the *Zīj* of the celebrated early-10th-century Syrian astronomer al-Battānī contains two small tables of the functions h(T) and a(T) for arguments T = 1, 2, ..., 12^{sdh} and T = 1, 2, ..., 6^{sdh} at the summer and winter solstices.[11] The underlying parameters are:

ϕ = 36;0° (Raqqa) and ε = 23;35°

and the entries are to two and three digits respectively. See also **5.4.2** on another table associated with al-Battānī.

4.1.3 al-Maqsī and al-Marrākushī: various latitudes

The Egyptian astronomer al-Maqsī (**2.1.1**) was a contemporary of al-Marrākushī (**4.2.4**), but the relationship between the two scholars is as yet unclear. Both compiled tables for marking the shadow traces on sundials oriented in various planes, but al-Maqsī's tables, of which I have examined MSS Cairo Azhar *falak* 5528, Dublin CB 4090 and Cairo DM 103, are more extensive. His tables for horizontal sundials, to which I restrict attention here, display the solar altitude and azimuth, as well as the length of gnomon shadow, that is:

$h(T)$, $a(T)$ and $z_{(12)}(T)$,

computed to two digits for T = 1, 2, ... , 6^{sdh} at the summer and winter solstices for:

ϕ = 0° (Equator), 21;0° (Mecca), 24;0° (Medina), 30;0° (Cairo) and 32;0° (Jerusalem).

For Cairo the same functions are also tabulated for each equinoctial hour at the solstices. al-Marrākushī's treatise (I.492 and 494) also contains tables of a(T) and $z_{(12)}(T)$ for latitudes 0° and 30°. An extract is shown in **Fig. 4.1.3**.

4.1.4 al-Sulṭān al-Ashraf: Yemen and Hejaz

MS Cairo TR 105 is a copy of the treatise on the astrolabe, horizontal sundial and magnetic compass by the late-13th-century Yemeni ruler Sultan al-Ashraf (**3.3.1**), apparently in the author's hand. The treatise includes tables of the functions:

$z_{(12)}(T)$ and $a(T)$

for each seasonal day-hour at the solstices and equinoxes, computed for latitudes:

13;0°, 13;37°, 14;0°, 14;30°, 15°, 21° (Mecca), and 24° (Medina) .

Note the use of the distinctive parameter 13;37°, which is intended for Taiz (**3.3.2**). Extracts are shown in **Fig. 4.1.4**.

4.1.5 Anonymous: Istanbul

In MS Cairo ṬM 255,10 (fol. 91r), copied *ca.* 1700, following a collection of prayer-tables and treatises on instruments, there is an isolated table displaying values of h(t) to two digits

[11] On al-Battānī see Willy Hartner's fine article in *DSB*. His *Zīj* is published in Nallino, *al-Battānī*, where the sundial tables occur in II, p. 188 (see also p. 296).

for each 5° of t at the equinoxes and both solstices, computed for $\phi = 41°$ (Istanbul). There is also a table displaying values of h(T) for each seasonal hour at the solstices.

4.2 Tables of solar altitude as a function of time since sunrise and solar longitude, for a specific latitude

The following tables display the solar altitude at the seasonal hours corresponding to a wider range of values of λ than in the tables discussed in **4.1**. On the computation of these tables see **4.2.2** below.

4.2.1 Anonymous: Baghdad

In MS Berlin Ahlwardt 5793 (Landberg 56,2), fol. 97v, copied 783 H [= 1381], at the end of two treatises on the construction and use of the astrolabe by al-Khwārizmī (**4.1.1**), there is a table of a function $f(\lambda,T)$ computed to two digits (labelled *daqāʾiq* and *thawānī*, minutes and seconds) for each zodiacal sign and T = 1, 2, ... , 6^{sdh}. The function is labelled *jayb al-sāʿāt*, that is, "sine of the hours", and its structure, which escaped me at first encounter, has been explained by Jan Hogendijk.[12] See **I-4.3.1** and especially **XI-9.3**.

4.2.2 Anonymous: Baghdad

In MS Istanbul Ayasofya 4830, fol. 197r, at the end of a short anonymous treatise of Abbasid origin on the construction of an horary quadrant (fols. 196v-197r, copied in Damascus in 626 H [= 1228/29]), there is a table displaying the solar altitude to the nearest degree (the second sexagesimal digit of each entry is always zero) at each seasonal hour from 1 to 6 for each zodiacal sign.[13] From the entries for T = 6 it is immediately obvious that the table is based on the parameters:

$$\phi = 33° \text{ (Baghdad) and } \varepsilon = 24° \text{ (or } 23;51°) .$$

From the text we learn that the values $h_i(T)$ for the signs i = 1, 2, ... , 12 are derived using the approximate formula:

$$\mathrm{Sin}_R\, h_i(T) \approx \mathrm{Sin}_R\,(15T) \cdot \mathrm{Sin}_R\, H_i \,/\, R ,$$

where H_i is the meridian altitude for the ith sign and R = 150. See further **XI-4.1** (illustrated).

4.2.3 Saʿīd ibn Khafīf al-Samarqandī or Ibn al-Ādāmī: Baghdad

MS Paris BNF ar. 2506,1 (fols. 1r-62r) is a unique copy from the late 15th century of a treatise on sundial theory by either Ibn al-Ādamī or Saʿīd ibn Khafīf al-Samarqandī, both of whom lived in the 10th century.[14] On fols. 31v-33v there are some tables for facilitating the

[12] On the treatises see now Charette & Schmidl, "al-Khwārizmī on the Astrolabe". On this table see already King, "al-Khwārizmī", pp. 25 and 29, and Hogendijk, "al-Khwārizmī's Sine of the Hours".

[13] On this treatise and table see *ibid.*, pp. 30-31.

[14] On Ibn al-Ādamī see Sezgin, *GAS*, VI, pp. 179-180, and on al-Samarqandī see *ibid.*, VI, pp. 216-217, and *Cairo ENL Survey*, no. B58. The title page and the first few folios of this manuscript are in the hand of the late-15th-century Egyptian astronomer Ibn Abi ʾl-Fatḥ al-Ṣūfī (**9.10**), and the main part is in an older hand. On the title folio al-Ṣūfī wrote that the work was by al-Ḥusayn ibn Muḥammad al-Ādamī but that in another manuscript that he had seen, the work was attributed to Saʿīd ibn Khafīf al-Samarqandī.

construction of vertical sundials. The functions tabulated are:

$$a(T,\lambda) \text{ and } h(T,\lambda)$$

for the domains:

$$T = 0;30,\ 1;0,\ \ldots,\ 12^{sdh} \quad \text{and} \quad \lambda = 0°,\ 30°,\ 60°,\ 90°,\ 210°,\ 240°,\ 270° \ .$$

The azimuths are measured "clockwise" from the north point (not south as stated). Values are given to three sexagesimal digits and are based on parameters:

$$\phi = 33° \text{ (Baghdad) and } \varepsilon = 24° \ .$$

Another set of tables entitled *jadwal al-inḥirāfāt*, "table of inclinations", follows. Three functions are tabulated, namely:

$$\mathrm{Sin}_{10}\ \theta,\ \mathrm{Cot}_{10}\ \theta \text{ and } \mathrm{Cot}_1\ \theta\ ,$$

and values are given to three digits for each degree of argument. The fact that the author uses base 10 is extremely interesting.

4.2.4 al-Marrākushī: Cairo

The extensive treatise on spherical astronomy and instruments by the Cairo astronomer of Maghribi origin Abū ʿAlī al-Marrākushī, entitled *Jāmiʿ al-mabādiʾ wa-'l-ghāyāt fī ʿilm al-mīqāt*, which means "An A to Z of Astronomical Timekeeping", was compiled in Cairo in the late 13th century. The first part of this work dealing with spherical astronomy and sundials (hereafter, I) was translated into French by Jean-Jacques Sédillot (*père*), and a summary of some of the material on other instruments in the second part (hereafter, II) was published by Louis-Amélie Sédillot (*fils*).[15] The treatise was rather popular amongst later astronomers in Egypt, Syria and Turkey, and exists in several manuscript copies: I have consulted MSS Istanbul Selim Ağa 866, and Paris BNF ar. 2507-2508, 14th century, of which the latter was used by the Sédillots. For more on the spherical astronomy and tables in al-Marrākushī's *opus* see **II-2.7** and **6.7**.

One of the various tables presented by al-Marrākushī (I.454) displays the function:

$$h(T,\lambda)$$

with entries to two sexagesimal digits for the domains:

$$T = 1,\ 2,\ \ldots,\ 6^{sdh} \text{ and } \lambda' = 0°,\ 30°,\ 60°,\ 90° \ (\delta \gtrless 0)$$

for the parameters $\phi = 30;0°$ (Cairo) and $\varepsilon = 23;35°$.

To determine $h(T,\lambda)$ al-Marrākushī prescribes the following equivalent methods (I.266-267, *cf.* **F10**):

$$h = \text{arc Sin } \{ [\text{Vers } D - \text{Vers } t] \cdot \text{Sin } H \ / \ \text{Vers } D \}$$
$$= \text{arc Sin } \{ \text{Sin } H(\lambda) \ - \ \text{Vers } t \cdot B(\lambda) \} \ ,$$

and presents a numerical example for $T = 1^{sdh}$ and $\lambda = 90°$ using each method. He apparently did not tabulate $B(\lambda)$ for Cairo (**6.4.2** and **6.4.3**).

Elsewhere in the same work (I.428-429 and 436-437), al-Marrākushī tabulates $z(T,\lambda)$ and

---

[15] On al-Marrākushī see Suter, *MAA*, no. 363; *Cairo ENL Survey*, no. C17; and the article in EI_2. Sédillot-*père*, *Traité*, contains a translation of the first half of al-Marrākushī's treatise on spherical astronomy and sundials. Sédillot-*fils*, *Mémoire*, contains a summary of the second half dealing with instruments. A Topkapı manuscript of the *Jāmiʿ al-mabādiʾ wa-'l-ghāyāt* was published by photo-offset at the Institut für Geschichte der Arabisch-Islamischen Wissenschaften, Frankfurt, in 1984.

$z'(T,\lambda)$ to base 12 for each seasonal day-hour and each 10° interval of λ, as well as (II.76) the function $h(T,\lambda)$, giving values to two sexagesimal digits for the domains:

$$T = 1, 2, \ldots, 6^h \ (1^h = 15°) \quad \text{and} \quad \lambda' = 0°, 30°, 60°, 90° \ (\delta \gtrsim 0°)$$

and for the same parameters. The corresponding solar altitudes in the prime vertical and in the azimuth of the qibla are also given.

4.2.5 al-Sulṭān al-Ashraf: Sanaa

In MS Oxford Hunt. 233, fols. 73r-78v, of the astrological treatise of the Yemeni Sultan al-Ashraf (**3.3.1** and also **4.1.4**), there is a set of tables of the function:

$$h(T,\lambda)$$

with values computed to two digits for the domains:

$$T = 1, 2, \ldots, 6^{sdh} \quad \text{and} \quad \lambda = 1°, 2°, \ldots, 360° \ .$$

The underlying parameters are:

$$\phi = 14;30° \ (\text{Sanaa}) \quad \text{and} \quad \varepsilon = 24;0° \ ,$$

and the function is called *irtifāʿ sāʿāt (al-)nahār*, "the (solar) altitude for each hour of daylight".

4.2.6 Abu 'l-ʿUqūl: Taiz

In the astronomical miscellany of the Yemeni Sultan al-Afḍal preserved in a manuscript in a private collection in Sanaa (**2.1.2**), there is a set of tables of the function:

$$h(\lambda,T)$$

computed to two digits for the same domains, but based on the parameters:

$$\phi = 13;37° \ (\text{Taiz}) \quad \text{and} \quad \varepsilon = 23;35° \ .$$

It is stated in the margin that this table is taken from the *Zīj* of Abu 'l-ʿUqūl, but the latitude 13;40° is used for Taiz in both his main tables for timekeeping and in the *Mukhtār Zīj*. This may mean that Abu 'l-ʿUqūl prepared two *zīj*es. (See also **3.1.1** and **3.3.2**.)

4.2.7 Najm al-Dīn al-Miṣrī: 4th climate (?)

For the many tables of solar altitude in the treatise of the Cairene astronomer Najm al-Dīn (**2.6.1**) on instruments I refer the reader to the work of François Charette.[16] Some of these tables, intended for the construction of horary markings on different kinds of instruments, are computed for latitude 36°, which is not for Aleppo, as I thought in the days when I (falsely) attributed the (anonymous) treatise to the Aleppo scholar Ibn al-Sarrāj (**VIb-5**) . Rather it seems that this latitude was for the 4th climate. It remains a mystery why Najm al-Dīn did not compute all of his tables for the latitude of Cairo.

4.3 Tables of solar altitude as a function of time since sunrise and meridian altitude, for all latitudes

A number of sources contain tables of the function $h(T,H)$, where T is measured in seasonal hours, based on the approximate formula:

$$h(T,H) \approx \text{arc Sin } \{ \text{ Sin } H \cdot \text{Sin } (15\ T) \ / \ R \ \} \ .$$

[16] Charette, *Mamluk Instrumentation.*

These tables are intended to be valid for all latitudes and the underlying formula is a transformation of the one discussed in **2.5** above. On a similar table in a 14th-century English astronomical manuscript of both the *Toledan* and *Alphonsine Tables* see **10.1**.

4.3.1 Anonymous (Baghdad)

The anonymous tables in MS Berlin Ahlwardt 5793 (Landberg 56,2), fols. 93r-95v, copied 783 H [= 1381] (**4.2.1**), appended to al-Khwārizmī's two treatises on the construction and use of the astrolabe, clearly date from the earliest days of mathematical astronomy in Baghdad and can indeed be shown to be by al-Khwārizmī. Following a table of normed right ascensions $\alpha'(\lambda)$ (fol. 93v, see **7.1.1**, n. 7:11) and some tables for latitude 33° (fol. 94r, see **II-3.1**) there is a table[17] of the function:

$$h(T,H)$$

with entries to one digit (fol. 94v), computed for the domains:

$$T = 1, 2, \ldots, 6^{sdh} \text{ and } H = 25°, 26°, \ldots, 90° .$$

There is also a table (fols. 95r-95v) of a function:

$$f(T,H) ,$$

labelled *jayb al-sāʿāt*, "Sine of the hours", and with entries to two digits for the same domains. See already **4.2.1** and especially **XI-9.3**.

4.3.1* Muḥyi 'l-Dīn al-Maghribī (?) (Maragha)

Another table of the function:

$$h(T,H)$$

is contained in MS Medina Aref Hikmet *mīqāt* 1, fols. 196r-196v, of the *Zīj* for Maragha by the 13th-century astronomer Muḥyi 'l-Din al-Maghribī (**5.6.3**). This particular copy was completed in 691 H [= 1292], and the table is not contained in any of the other manuscripts of al-Maghribi's various *zīj*es (he wrote at least three).

4.3.2 al-Marrākushī (Cairo)

In al-Marrākushī's treatise on spherical astronomy (**4.2.4**) there is a table (I.457) of the function:

$$h(T,H)$$

with the entries to two digits for:

$$T = 1, 2, \ldots, 5^{sdh} \quad \text{and} \quad H = 5°, 10°, \ldots, 90° .$$

(Clearly h = H when $T = 6^{sdh}$.) The solar altitude h_a at the beginning of the interval for the afternoon prayer is also given for each value of H. Elsewhere in the same work (I.253) al-Marrākushī tabulates $z'_{(12)}(T,H)$ to two digits for the same domains and also $T = 6^{sdh}$).

4.3.3 Anonymous (Yemen)

The anonymous late-14th-century Yemeni *Zīj* preserved in MS Paris BNF ar. 2523 (**2.5.3**) contains a table (fol. 86r) of the function h(T,H) with values computed to two digits for the domains $T = 1, 2, \ldots, 11^{sdh}$ and $H = 51°, 52°, \ldots, 90°$. See **Fig. 4.3.3**. Note that the table

[17] On this table see King, "al-Khwārizmī", pp. 7 and 10-11.

Fig. 4.3.3: A table of the tangent of the solar declination (**7.1.10**) and the anonymous Yemeni table serving solar meridian altitudes 11°-90°. [From MS Paris BNF ar. 2523, fol. 86r, courtesy of the Bibliothèque Nationale de France.]

is adequate for use in the Yemen, where the minimum value of the solar altitude is about 50°. See also **II-12.3** on this table.

4.3.4 Anonymous (Yemen)

A manuscript that was in the private collection of Sayyid Aḥmad ʿAbd al-Qādir al-Ahdal in Zabid in the 1980s contains a copy of the 16th-century Yemeni *zīj* compiled by Muḥammad al-Daylamī and entitled *Zād al-musāfir*.[18] This contains two similar tables but for vertical arguments H = 51°, 52°, ... , 90° and H = 40°, 41°, ... , 90°. The entries generally differ from those in MS Paris BNF ar. 2523 mentioned in **4.3.3**.

[18] On this work see further King, *Astronomy in Yemen*, no. 28.

4.3.5 al-Ṣāliḥī (?) (Damascus)

In MS Baghdad Awqāf 2966/6294, fol. 136r, of an anonymous *Zīj* containing tables due to Ibn al-Shāṭir (Damascus, *ca.* 1350), Ulugh Beg (Samarqand, *ca.* 1440) and ʿAbd al-Raḥmān ibn Banafshā al-Ṣāliḥī (Damascus, *ca.* 1500),[19] there is a table of the same function for arguments T = 1, 2, ... , 5^{sdh} and H = 1°, 2°, ... , 90°. In MS Leiden Or. 65, fol. 110r, of a *zīj* attributed to Ibn al-Shāṭir which purports to be entitled *Tuḥfat al-nāẓir* but which seems to be merely an early modification of his *Zīj*, there is a similar, if not identical, table. I have not come across such a table in the various manuscripts of Ibn al-Shāṭir's *Zīj*.

4.3.6 Ḥusayn Quṣʿa (?) (Tunis)

In MS Princeton Yahuda 147c of the *Zīj* for Tunis compiled by Ḥusayn Quṣʿa[20] *ca.* 1650 and based mainly on the *Zīj* of Ulugh Beg, there is a page ruled for a table with identical title and format to the one in MS Baghdad Awqāf 2966/6294 (see the previous section), but no entries have been copied.

4.3.7 Zacuto (?) (Salamanca) / Anonymous (Yemen)

In MS Milan Ambrosiana C82, a Yemeni copy from 1086 H [= 1675] of the Maghribi version of the perpetual almanac of the late-15th-century Jewish astronomer Abraham Zacuto of Salamanca[21] there are various additional tables (fols. 142r-148r), one of which is attributed to the 13th-century Tunisian astronomer Ibn Isḥāq (**6.9.1***). One anonymous table (fol. 144v) displays values of the function h(T,H) to one digit for arguments T = 1, 2, ... , 5^{sdh} and H = 1, 2, ..., 90°. This particular table is not contained in MS Hyderabad Āṣafiyya 298 of the Syrian recension of Ibn Isḥāq's *Zīj*. See further **6.9.1*** and **7.1.5***.

4.4 Tables of solar altitude as a function of the semi diurnal arc and the time since sunrise

4.4.1 Najm al-Dīn al-Miṣrī (?): Cairo

The unique source MS Cairo MM 72, copied in 747 H [= 1346/47], contains an anonymous set of tables of the function:

$$h(D,T)$$

with entries computed to two digits for the domains:

$$D = 104;36°, 104°, 103°, \ldots , 76°, 75;24° \quad \text{and} \quad T = 1°, 2°, \ldots , [2D]$$

and based on the parameters:

$$\phi = 30;0° \text{ (Cairo)} \quad \text{and} \quad \varepsilon = 23;35° .$$

The entries, which number over 5,500, are rather accurately computed. Note that 104;36° and 75;24° are the maximum and minimum values of D for these parameters. A given pair of facing

[19] On Ibn al-Shāṭir see n. 2:15. On Ulugh Beg see *Cairo ENL Survey*, no. G49, the articles by T. N. Kari-Niazov in *DSB* and by Beatrice F. Manz in EI_2. On al-Ṣāliḥī (Suter, *MAA*, no. 454) see *Cairo ENL Survey*, no. C87.

[20] This author is not listed in Suter, *MAA*; Brockelmann, *GAL*; or Kennedy, "*Zīj* Survey". See *Cairo ENL Survey*, no. F47 (also no. F53); Samsó, "Maghribī *Zījes*", p. 98; and İhsanoğlu *et al.*, *Ottoman Astronomical Literature*, I, p. 314.

[21] On this work see now Chabás & Goldstein, *Zacut*, where there is no mention of any tables for timekeeping in the original version. On this table see now Samsó, "Zacut in the Islamic East".

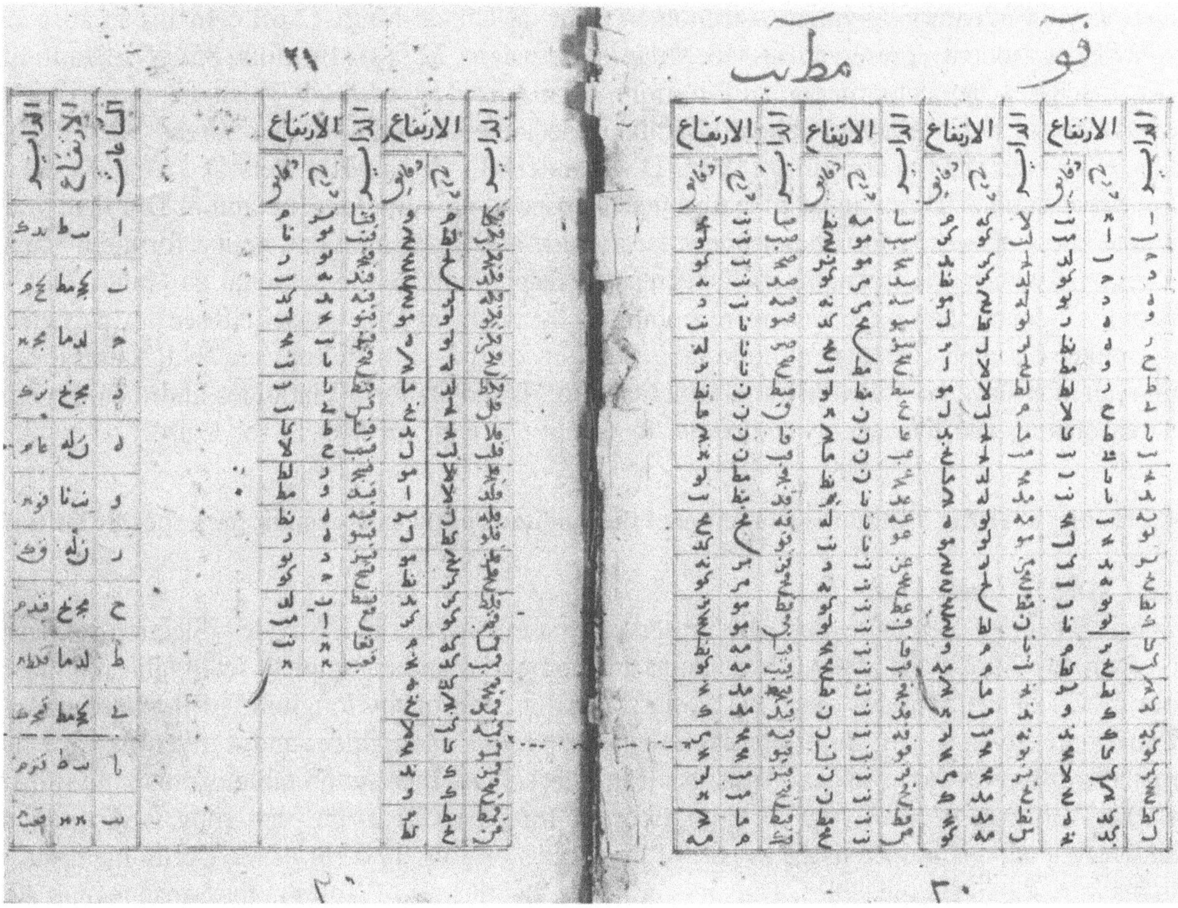

Fig. 4.4.1: An extract from the anonymous Mamluk tables of solar altitude serving half arc of daylight 86° (hour-angle at the *ʿaṣr* 49;12°). [From MS Cairo MM 72, fols. 20v-21r, courtesy of the Egyptian National Library.]

pages in the manuscript serves a particular value of D, and the two unlabelled numbers at the head of each table are in fact the argument D and the associated time from midday to the beginning of the interval for the afternoon prayer t_a. By the side of each sub-table are the solar altitudes at each seasonal hour for the given value of D. **Fig. 4.4.1** shows a pair of facing pages of these tables.

The title folio and all but the last page of the introduction to these tables are missing from the manuscript. The copyist was Ibn al-Kattānī, who compiled the hour-angle tables for Cairo in MS Istanbul Kılıç Ali Paşa 684, also penned in his elegant hand (**2.1.1** and **II-4.1.4**).[22] I have previously suggested that these tables might have been computed by Ibn Yūnus,[23] but the possibility that the tables in MS Cairo MM 72 were computed by Ibn al-Kattānī himself cannot be entirely excluded. However, it now seems more likely that they were computed by

[22] On Ibn al-Kattānī (Suter, *MAA*, no. 411) see *Cairo ENL Survey*, no. C32.
[23] King, "Astronomical Timekeeping in Medieval Cairo", pp. 365 and 391.

the early-14th-century Egyptian astronomer Najm al-Dīn al-Miṣrī (**2.6.1**). In his treatise on spherical astronomy preserved in MS Milan Ambrosiana 227a (C49), fols. 85v-97r, Najm al-Dīn discusses the determination of h from D and T. The method which he prescribes is equivalent to the following standard formula used, for example, by Ibn Yūnus[24] (*cf.* **F10**):

$$h(D,T) = \text{arc Sin } \{ [\text{ Vers } D - \text{Vers } (D\text{-}T)] \bullet \text{Sin } H / \text{Vers } D \} .$$

For each value of D one must first find the corresponding value of H. Najm al-Dīn then adds a note about a method for finding the solar altitude at the seasonal day-hours: for the nth hour $T = n \bullet \tilde{h}$ where $\tilde{h}$ is the number of degrees corresponding to one seasonal day-hour. This is strong evidence that the tables were compiled by Najm al-Dīn himself. I see no particular advantage to tables of solar altitude as a function of the semi diurnal arc, but Najm al-Dīn may already have compiled his enormous table of T(H,h,D) for all latitudes and then thought it reasonable, and had energy and time, to compile a table of h(D,T) for Cairo.

4.5 Tables of stellar altitude as a function of the longitude of the horoscopus, for a specific latitude

4.5.1 Abu 'l-ʿUqūl: Taiz

The 14th-century Yemeni miscellany preserved in a manuscript in a private collection in Sanaa (see already **2.1.2**, **3.3.1** and **4.2.3**) contains an anonymous and incomplete set of tables (4 pages out of 12) of the altitudes of up to 12 stars as a function of the longitude of the horoscopus. Entries are given to two digits for each degree of ecliptic longitude, and a given page of the manuscript serves two zodiacal signs: see **Fig. 4.5.1**. Now there is no table of this kind in MS Berlin Ahlwardt 5720 of Abu 'l-ʿUqūl's corpus for Taiz (**2.1.2**) but that copy is incomplete and such a table is indeed listed as part of the corpus on fols. 17v-18r of the Berlin manuscript in his introduction to the tables. Thus it is probable that this set of tables in the Sanaa manuscript is due to Abu 'l-ʿUqūl and is based on the value 13;40° for the latitude of Taiz.

Abu 'l-ʿUqūl also compiled an extensive set of tables displaying the altitudes of various prominent stars at daybreak as functions of the longitude of the horoscopus. These are extant in the Berlin manuscript mentioned above: see **Fig. 4.5.2** and also **II-12.1**.

4.6 Tables of stellar altitude as a function of time of night, for a specific latitude

4.6.1 Ibn Dāʿir: Sanaa

On fol. 126v of the Berlin copy of Ibn Dāʿir's tables for Sanaa (**3.2.3** and **3.4.1**) there is a table displaying the altitudes of 18 fixed stars at each "seasonal hour" of visibility, with entries to two digits – see **Fig. 3.4.1**.

---

[24] King, *Ibn Yūnus*, III.15.4e.

Fig. 4.5.1: An extract from Abu ʻl-ʻUqūl's tables of the altitudes of various stars as a function of the horoscopus, here the first few degrees the sign of Leo. [From Varisco & Smith, eds., *al-Afḍal's Anthology*, p. 173, with kind permission of Professor Daniel Varisco.]

Fig. 4.5.2: An extract from Abu 'l-ʿUqūl's tables of stellar altitudes in the east and west serving each degree of the horoscopus in Aquarius. [From MS Berlin Ahlwardt 5720, courtesy of the Deutsche Staatsbibliothek (Preußischer Kulturbesitz).]

4.7 Tables of solar altitudes in certain azimuths as a function of solar longitude, for a specific latitude

The azimuths for which a knowledge of the corresponding solar altitude is of some importance in Islamic astronomy are defined by: (i) the meridian; (ii) the prime vertical; and (iii) the qibla or azimuth of Mecca. In Cairo a fourth direction was of interest to the astronomers because of the layout of the architecture of the city, namely, (iv) the direction for orienting the ventilators on the roofs of religious and secular buildings, these being a common feature of the medieval city.

Tables of the solar meridian altitude $H(\lambda)$ for a particular latitude are standard in *zīj*es, and their compilation is trivial since (*cf.* **F1**):

$$H(\lambda,\phi) \;=\; \bar{\phi} + \delta(\lambda)\ .$$

Numerous Islamic sources contain tables of the function:

$$\delta'(\lambda) \;=\; 90° + \delta(\lambda)\ ,$$

to be used for finding the meridian altitude for any latitude. Clearly:

$$H(\lambda,\phi) \;=\; \delta'(\lambda) - \phi\ .$$

Tables of $h_0(\lambda)$, the solar altitude in the prime vertical, are less common. Those located thus far are discussed in **4.8** below.

Finally, tables of $h_q(\lambda)$, the solar altitude in the azimuth of Mecca, are attested for a wider variety of localities. These are discussed in **II-3.6, 3.9, 4.7, 10.6**, *etc.*[25]

To find $h(a,\lambda)$ for a particular azimuth, one can either use interpolation in tables of $a(h,\lambda)$ if these were available (**5.1**) or calculate the altitude directly. The method of Ibn Yūnus is as

[25] On Ibn Yūnus' tables of $h_q(\lambda)$ for Cairo see already Schoy, *Schattentafeln*, p. 42; King, *Ibn Yūnus*, III.28.3, and *idem*, "Astronomical Timekeeping in Medieval Cairo", p. 368. See also **II-4.7**.

follows.[26] First compute the altitude h_e of the sun at an equinox when it has azimuth a, thus:

$$h_e = \text{arc Sin } \{ \text{ R Sin a } / \sqrt{ [(\text{R Sin } \phi / \text{Cos } \phi)^2 + \text{Sin}^2 \text{ a }] } \} .$$

Secondly compute a "correction arc" C for the solstitial points, thus:

$$C = \text{arc Sin } \{ \text{ Cos } h_e \text{ Sin } \varepsilon / \text{Sin } \phi \} ,$$

and then a correction arc c(λ) for general solar longitudes, thus:

$$c(\lambda) = \text{arc Sin } \{ \text{ Sin C Sin } \lambda / \text{R} \} .$$

Finally the solar altitude is given by:

$$h(a,\lambda) = h_e(a) + c(\lambda) \qquad (c \gtrless 0 \text{ as } \delta \gtrless 0) .$$

Only one set of tables of h(a,λ) for general azimuths has come to light.

4.7.1 Ibn Yūnus: Cairo

In MS Oxford Hunt. 331, fols. 27v-42r, of the *Ḥākimī Zīj* Ibn Yūnus (**2.1.1**) presents a set of tables of the function:

$$h(a,\lambda)$$

for the domains:

a = 30°, 40°, 42°, 45°, 50°, 54°, 60°, 66°, 70°, 75° and λ = 1°, 2°, ... , 360°.

Values are carefully computed to two digits and are based on his parameters φ = 30;0° (Cairo) and ε = 23;35°.[27] The tables for a = 30° and 60° are also found in MSS Dublin CB 3673 and Cairo DM 153 of the Cairo corpus of tables for timekeeping (see further **II-4.6**).

Ibn Yūnus states in the *Zīj* that these tables are for finding the meridian. The azimuths are supposed to have been chosen so that the meridian could be determined using a simple geometrical construction once the direction corresponding to the azimuth a has been found by observation of the solar altitude. There are, of course, easier methods for finding the meridian, several of which are outlined elsewhere in the *Ḥākimī Zīj*.

Another table in the main Cairo corpus displays the altitude of the sun in the azimuth in which ventilators were oriented in popular Egyptian practice. This azimuth is that of the rising sun at the winter solstice, that is, about 27;30° S. of E. for Cairo, and it is understood that the base of the back of the ventilator should be aligned in this direction so that (a) the ventilator be aligned with the building on which it is place, and (b) the front be open to the favorable winds from the north. The table, which exists in numerous manuscripts, *e.g.* MS Dublin CB 3673, fol. 84v, may have been computed by Ibn Yūnus, but in MS Istanbul Nuruosmaniye 2925 one manifestation of it is attributed to Ibn al-Rashīdī (**2.1.5**).[28] See further **II-4.8** and **VIIb-3-4**.

4.8 Tables of solar altitude in the prime vertical, for a specific latitude

The solar altitude in the prime vertical, $h_0(\lambda)$, is called in Arabic *al-irtifāʿ alladhī lā samt lahu* or *al-irtifāʿ al-ʿadīm al-samt*, that is, "the altitude with no azimuth". It is determined by the simple formula (*cf.* **F13**):

---

[26] See King, *Ibn Yūnus*, III.23.3, on the similar methods of Ibn Yūnus and al-Bīrūnī.

[27] See also *ibid.*, III.22.2, and *idem*, "Astronomical Timekeeping in Medieval Cairo", pp. 364-365.

[28] See also *ibid.*, pp. 371-373.

$$h_0(\lambda) = \text{arc Sin } \{ \text{ R Sin } \delta(\lambda) \text{ / Sin } \phi \text{ } \} \text{ ,}$$

and like the solar rising azimuth (**5.5** and **8.3**) is easily determined from Sin $\delta(\lambda)$ (**6.1**). The following tables of $h_0(\lambda)$ for particular latitudes have been located.

4.8.1 Ibn Yūnus: Cairo

In MS Leiden Or. 143, p. 360, of the *Ḥākimī Zīj*, Ibn Yūnus (**2.1.1**) tabulated $h_0(\lambda)$ to three digits for each degree of λ and parameters ϕ = 30;0° (Cairo) and ε = 23;35°.

4.8.2 al-Mizzī: Damascus

The prayer-tables of al-Mizzī preserved in MS Cairo MM 62 (**2.1.3**) display $h_0(\lambda)$ to two digits for each degree of λ and parameters ϕ = 33;27° (Damascus) and ε = 23;33°.

4.8.3 al-Khalīlī: Damascus

al-Mizzī's younger colleague al-Khalīlī (**2.1.4**) recomputed $h_0(\lambda)$ for parameters ϕ = 33;30° (Damascus) and ε = 23;31°. His values are also given to two digits for each degree of λ, and the table is contained, for example, in MS Paris BNF ar. 2558, fol. 51v.

4.8.4 Anonymous: Cairo

In several manuscripts of the main Cairo corpus (**2.1.1**) there occur tables of $h_0(\lambda)$ to two digits, ostensibly computed for Ibn Yūnus' parameters (**4.8.1**) but containing numerous garbled entries. Thus, for example, the 14th-century copy MS Dublin CB 3673, fol. 12v, contains a table where the entry for λ = 30° (which should be ε = 23;35° for ϕ = 30°) is 24;34°, and the entry for λ = 90° (which should be 53;9°) is 53;46°. Another such table in the same copy of the corpus (fol. 3v) has corresponding entries 24;34° and 53;9°. Again, the 16th-century Cairo astronomer al-Minūfī (**II-7.1**) recomputed some of the simpler tables in the corpus for ε = 23;30° Thus, for example, he prepared a new table of $\psi(\lambda)$ (**5.6.9**), but merely took over a corrupt table of $h_0(\lambda)$: his table of this function in MS Cairo DM 107 is the same as that in MS Dublin CB 3673, fol. 3v. Finally, the values of $h_0(\lambda)$ for λ = 30° and 90° in the prayer-tables of the 16th-century Egyptian astronomer al-Fawānīsī (**II-7.5**), preserved, for example, in MS Oxford Seld. Supp. 99, have been distorted by further copyist's errors to 23;47° and 53;21°.

4.8.5 al-Asyūṭī: Assiut

The late-15th-century Egyptian astronomer al-Asyūṭī (**II-6.16**) in his prayer-tables for Assiut extant in MS Cairo DM 188 tabulated $h_0(\lambda)$ to two digits for each degree of λ and parameters ϕ = 27;0° (Assiut) and ε = 23;35°.

4.8.6 al-Maḥallī: Damietta

In a manuscript formerly in the Institut für Geschichte der Medizin und der Naturwissenschaften in Berlin and apparently no longer extant (see **5.1.4**), there was a table of $h_0(\lambda)$ computed for ϕ = 31;25° (Damietta). This table is not contained in MS Cairo DM 106, an incomplete copy of the prayer-tables of Quṭb al-Dīn al-Maḥallī (**II-8.6** and also **I-5.1.4**) for parameters ϕ = 31;25° and ε = 23;35°, but may have originally been part of this set.

4.8.7 al-Dīsṭī: Lattakia

MSS Aleppo Awqāf 911 and Leiden Or. 2808(2) of a set of prayer-tables computed *ca.* 1700 for Lattakia by al-Dīsṭī (**II-11.8**) contain a table of $\psi(\lambda)$ with values to two digits for each degree of λ and parameters ϕ = 34;30° (Lattakia) and ε = 23;35°.

4.8.8 Anonymous: Alexandria

At the end of the tables for the latitude of Alexandria preserved in MS Cairo TR 354 (**2.1.6**) there is a single table of $h_0(\lambda)$, with entries to two digits for each degree of λ. The underlying parameters are ϕ = 31;0° (Alexandria) and ε = 23;30°. The value of the obliquity differs from that used for the main tables.

MS Cairo DM 1207 of the corpus of prayer-tables for Alexandria (**II-8.5**) contains a different table of $h_0(\lambda)$ with values to two digits for each degree of λ based on parameters ϕ = 31;0° (Alexandria) and ε = 23;35°.

4.8.9 Yūsuf Kilārjī: Crete

In MS Cairo DM 834 of a set of prayer-tables for Crete compiled by the early-18th-century Egyptian astronomer Yūsuf Kilārjī (**II-8.8**) there is a table of $h_0(\lambda)$ computed to two digits for each degree of λ and based on the parameters ϕ = 35;30° (Crete) and ε = 23;35°.

4.8.10 Anonymous: Nablus

In MS Cairo ṬM 81 at the end of some late timekeeping tables for Jerusalem (**2.2.6**) there are three tables computed for latitude ϕ = 32;10° (Nablus). The functions tabulated are $N(\lambda)$, $d(\lambda)$ and $h_0(\lambda)$.

CHAPTER 5

TABLES OF SOLAR AZIMUTH

5.0 Introductory remarks

The calculation of the azimuth of the sun and stars is a standard topic of Islamic spherical astronomy, and certain Muslim astronomers compiled tables of the solar azimuth as a function of the altitude. In **5.1** I describe tables having h and λ as arguments, and in **5.2** a table having arguments H and h. The tables of azimuth as a function of time discussed in **5.3** are samples of a category important in Islamic sundial theory.

The function a(h,λ) displays the same symmetry as t(h,λ) and T(h,λ) (**2.1**). It is generally defined in *zīj*es[1] in terms of two auxiliary functions k(h,ϕ) and Sin ψ(λ,ϕ), with which (*cf.* **F15**):

$$a(h,\lambda,\phi) = \text{arc Sin } \{ \text{ R} \bullet [\text{ k}(h,\phi) \ - \ \sin \psi(\lambda,\phi) \ ] \ / \cos h \ \} \ .$$

Certain Muslim astronomers tabulated k(h) and Sin ψ(λ) for specific latitudes, and at least one table of the secant function:

$$G_1(h) = R \ / \ \text{Cos } h \ ,$$

especially intended for azimuth calculations is attested in the Islamic sources. al-Khalīlī tabulated k(h) and Sin ψ(λ) for each degree of latitude, and another auxiliary function for deriving the azimuth from their difference. The determination of the azimuth with these three functions tabulated is very simple indeed: see further **Ch. 8** and especially **Sections 8.5** and **9.4a**.

In most Islamic sources the azimuth (Arabic, *samt*) is reckoned from the prime vertical. The azimuth of Mecca (Arabic, *inḥirāf al-qibla*), on the other hand, is generally reckoned from the meridian.

5.1 Tables of solar azimuth as a function of solar altitude and solar longitude, for a specific latitude

5.1.1 Ibn Yūnus: Cairo

Ibn Yūnus (**2.1.1**) compiled a set of tables called *Kitāb al-Samt*, "Azimuth Tables", displaying values of the function:

$$a(h,\lambda)$$

to two digits for the domains:

$$h = 1°, 2°, \ldots , 83° \quad \text{and} \quad \lambda = 1°, 2°, \ldots , 90° \text{ and } 181°, 182°, \ldots , 270°$$

and based on the parameters:

$$\phi = 30° \text{ (Cairo-Fustat) and } \varepsilon = 23;35° \ .$$

---

[1] See, for example, Nallino, *al-Battānī*, I, pp. 183-184; Luckey, "Beiträge", I, pp. 500-501 (al-Māhānī); and King, *Ibn Yūnus*, III.20.

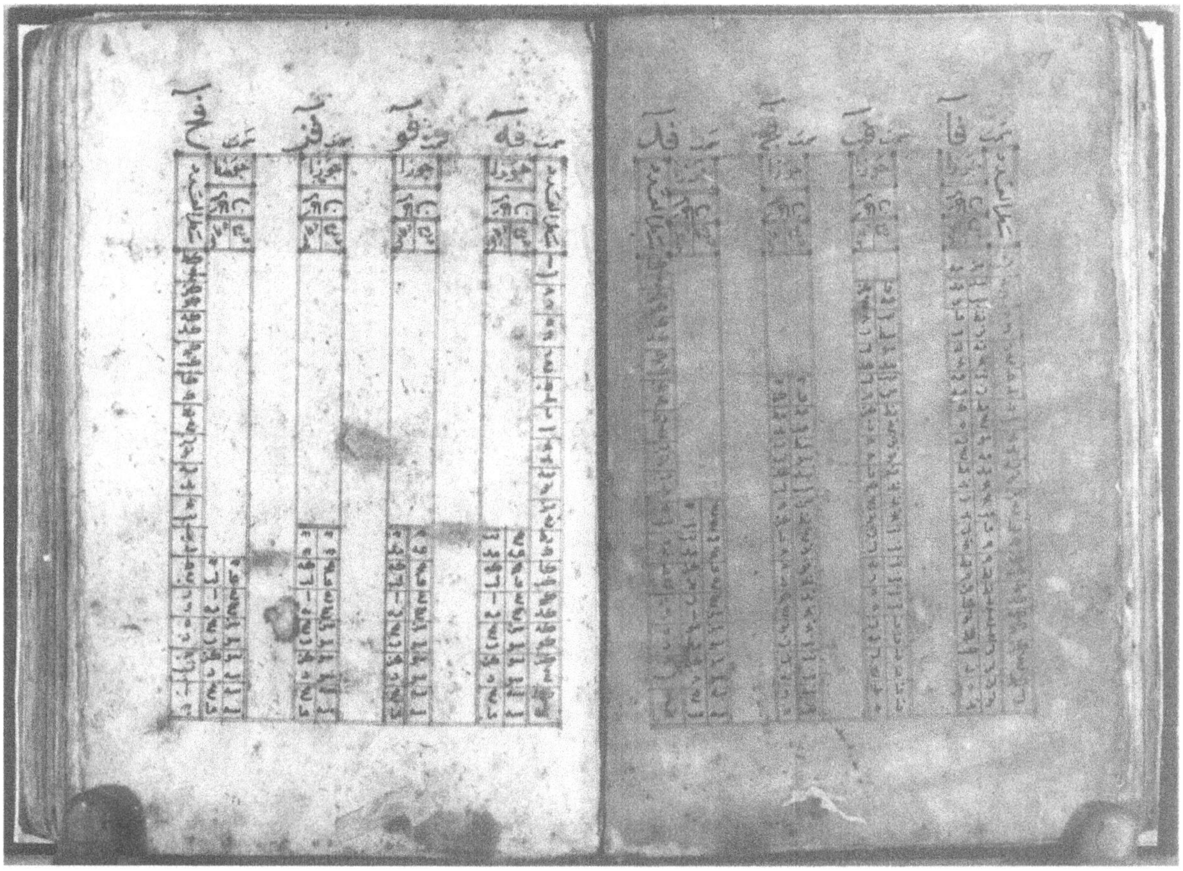

Fig. 5.1.1a: The additional values of the solar azimuth for solar altitudes from 84° to 88° in some early Mamluk copies of Ibn Yūnus' tables defy explanation. [From MS Dublin Chester Beatty 3673, fols. 86v-87r, courtesy of the Chester Beatty Library.]

These tables, extant in MSS Berlin Ahlwardt 5753 (incomplete), Dublin CB 3673, Escorial ár. 924,7, Cairo DM 108, Cairo Azhar *falak* 4382, Cairo MM 137, Gotha A 1410, Cairo DM 53, Istanbul Nuruosmaniye 2903 and 2925, Cairo MM 64, Cairo DM 690, DM 616, DM 739, DM 786, DM 1101, DM 1224 and Cairo K 4044, are very accurately computed. Tables for h = 30° and 35° are found in the *Ḥākimī Zīj*, MSS Leiden Or. 143, pp. 389-390, and Oxford Hunt. 331, fols. 124v-127v.[2] On the related tables of a(λ,h) in MS Cairo DM 1108,9 see **5.2.1** below.

It seems that the entries in Ibn Yūnus' tables for altitudes above 80° were either not complete or were not contained in any manuscripts that were available in the 14th century. Additional entries for altitudes 81° to 89° were added by some anonymous incompetent who was oblivious of the fact that the maximum solar altitude in Cairo is about $83^1/_2$°. Thus, for example, MS

[2] On Ibn Yūnus' azimuth tables see also King, *Ibn Yūnus*, III.22.2 (*Ḥākimī Zīj*); and *idem*, "Astronomical Timekeeping in Medieval Cairo", pp. 362-364 (*Kitāb al-Samt*). The tables for altitude 30° in the two sources are reproduced *ibid.*, pp. 354-355.

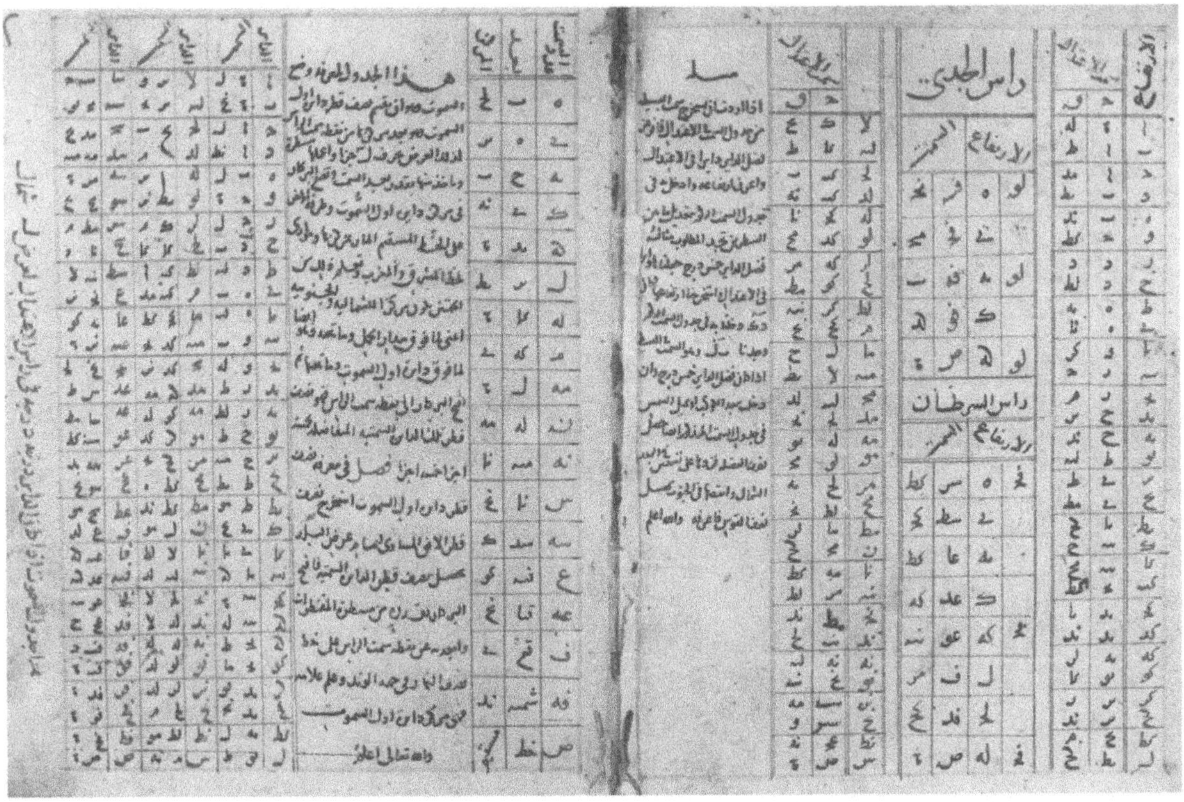

Fig. 5.1.1b: Ibn Yūnus' table of the azimuth as a function of the solar altitude at the equinoxes and solstices. On the left is an auxiliary table for sundial construction, of the kind for which another detailed survey would be worthwhile. [From Cairo MM 57, fols. 1v-2r, courtesy of the Egyptian National Library.]

Gotha A 1410 contains entries for h running from 1° to 89°, MS Dublin CB 3673 entries to 88° (see **Fig. 5.1.1a**), MS Cairo MM 137 entries to 86°. The entries for 81° to 83° were recomputed properly by Ibn al-Rāshidī (**2.1.5**), and were incorporated in al-Bakhāniqī's edition of the Cairo corpus (**2.1.1**). Thus, for example, MS Cairo Azhar *falak* 4382 contains a corrected set of azimuth tables with entries to 83° for h, and in MSS Cairo DM 108 and Cairo DM 53 these corrected entries are tabulated alongside the corresponding values of T(h,λ) and t(h,λ). Further information on these tables is given in **II-4.4** and **II-5.2**.

Ibn Yūnus also tabulated the function a(h) at the equinoxes for h = 1°, 2°, ... , 60° (= $\bar{\phi}$) with additional values for the solstices. This table is extant in MSS Cairo Azhar *falak* 4382, fol. 71r, and Cairo MM 57, fol. 1v – see **Fig. 5.1.1b**. MS Vatican Borg. ar. 217,2 (fols. 6r-7r) of al-Khaṭā'ī's auxiliary tables (**6.15.1**) contains a table of this same function for the equinoxes with minor variants from Ibn Yūnus' table. Furthermore, al-Khaṭā'ī reiterates Ibn Yūnus' remark in the *Ḥākimī Zīj* to the effect that such a table also displays d(Δ).[3] Notice that if δ = 0 the formula for a(h) is (*cf.* **F15**):

$$a(h) = \text{arc Sin} \{ \text{Tan } h \text{ Tan } \phi \text{ / R} \} ,$$

[3] *Cf.* King, *Ibn Yūnus*, III.15.1b, and *idem*, "Astronomical Timekeeping in Medieval Cairo", pp. 389-390.

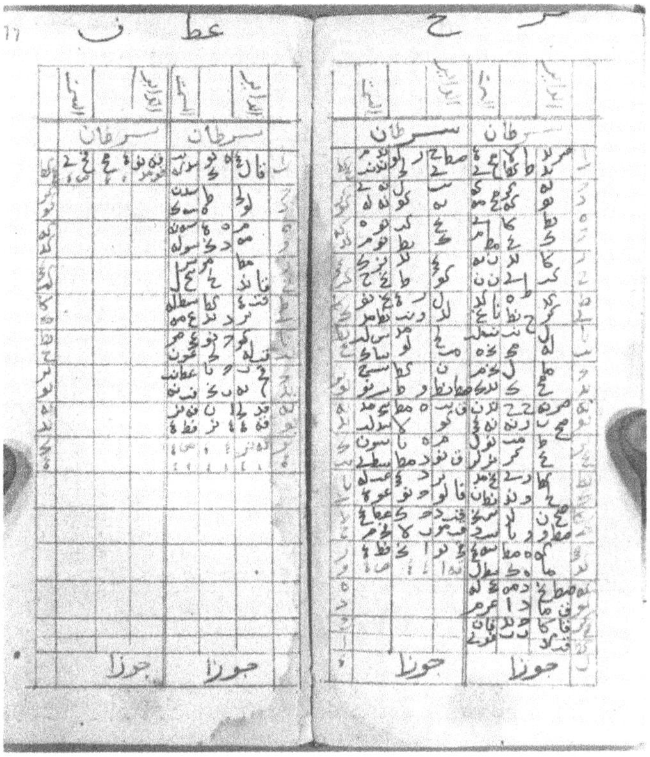

Fig. 5.1.2: The last sub-tables in the corpus for Damascus, in which al-Ḥalabī's values for the azimuth are tabulated alongside al-Khalīlī's values of the time since sunrise and time to midday. [From MS Cairo MM 71, fols. 65v-66r, courtesy of the Egyptian National Library.]

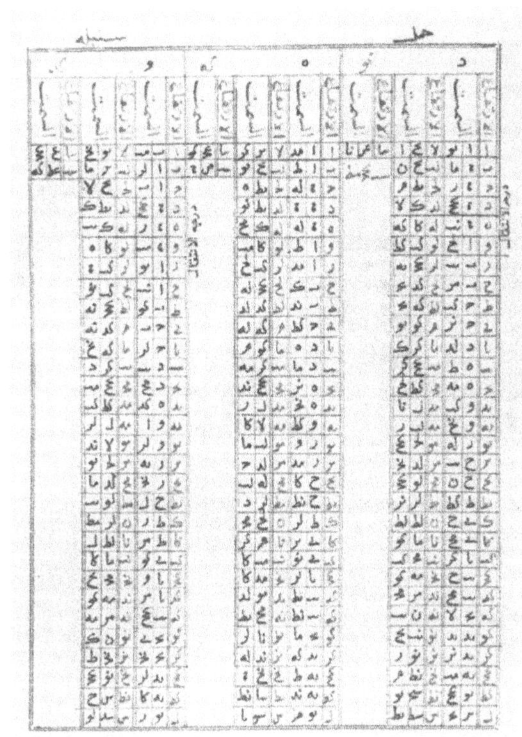

Fig. 5.2.1: An extract from the anonymous solar azimuth tables for Cairo serving solar longitudes Aries 4°-6° / Virgo 24°-24°. [From MS Cairo DM 1108,9, courtesy of the Egyptian National Library.]

and that d(Δ) is defined by (*cf.* **F7**):

$$d(\Delta) = \text{arc Sin } \{ \text{ Tan } \Delta \text{ Tan } \phi \ / \ R \ \} \ .$$

On Ibn Yūnus' auxiliary tables for azimuth calculations see **8.1.1** and **8.3.1**.

5.1.2 Shihāb al-Dīn al-Ḥalabī: Damascus

al-Khalīlī's tables of t(h,λ) and T(h,λ) for Damascus (**2.1.3**) were supplemented by the 15th-century astronomer Shihāb al-Dīn al-Ḥalabī (**2.8.1**) with a complete set of tables of a(h,λ) based on al-Khalīlī's parameters. These are contained in MS Cairo MM 71 where they are tabulated side by side with al-Khalīlī's functions (see **Fig. 5.1.2**), and in MSS Damascus Ẓāhiriyya 9227 and Cairo K 8525 where they are tabulated separately. See further **II-11.2**.

5.1.3 Anonymous: Alexandria

The tables for timekeeping in MS Cairo TR 354 computed for the latitude of Alexandria which are falsely attributed to Ibn Yūnus (**2.1.6**) also give values of a(h,λ) based on the same parameters as those of t(h,λ) and T(h,λ). See further **II-8.5**.

5.1.4 Anonymous: Damietta (?)

In a manuscript of late Egyptian provenance which in 1939 was in the Institut für Geschichte der Naturwissenschaften in Berlin, according to a description by Willy Hartner, there was a set of solar azimuth tables for an unspecified latitude occupying about 25 pages (fols. 259r-271r).[4] Since these tables were followed by two other spherical astronomical tables for the latitude of the Nile Delta town of Damietta, namely $\phi = 31;25°$, it may be that the azimuth tables were also computed for this parameter. The other functions tabulated for the latitude of Damietta were the solar altitude in the prime vertical $h_0(\lambda)$ (**4.8.7**) and the (? solar altitude in the ?) azimuth of the qibla $h_q(\lambda)$. Elsewhere in the former Berlin manuscript (fols. 58v-61v) Hartner noted various other spherical astronomical tables for latitude 31;25° attributed to the mid-17th-century astronomer Quṭb al-Dīn al-Maḥallī. A more complete set of prayer-tables for Damietta by this individual is contained in MS Cairo DM 106: see **II-8.6**.

The possibility that the main azimuth tables were simply those of Ibn Yūnus for Cairo (**5.1.1**) cannot be excluded. In the same manuscript (fols. 71r-81r) Hartner noted some 20 pages of "*Tafeln des Vorübergangs*" for arguments 2 to 22 (degrees?) attributed to an individual named ʿAbd al-Rāziq al-Munāwī: perhaps these were an incomplete set of tables displaying the time since sunrise or the hour-angle.

5.2 Tables of solar azimuth as a function of solar longitude and solar altitude, for a specific latitude

5.2.1 Anonymous: Cairo

MS Cairo DM 1108,9 (fols. 35v-59r, copied 1053 H [=1642/43]) contains an anonymous set of tables of the function:

$$a(\lambda,h)$$

for Cairo. The tables begin with those for $\lambda = 1°$ and go up to $\lambda = 90°$ and then run from $\lambda = 181°$ up to $\lambda = 270°$, so that no values are given for the equinoxes – see **Fig. 5.2.1**. The anonymous compiler overlooked the fact that the advantage in having separate pages of tables of a(h) for different longitudes enables the equinoctial values to be displayed conveniently (**2.2**) and he also overlooked Ibn Yūnus' special table of a(h) at the equinoxes. The entries in the tables in MS Cairo DM 1108,9 vary little from the corresponding entries in Ibn Yūnus' tables of $a(h,\lambda)$, from which they were no doubt lifted.

5.3 Tables of solar azimuth as a function of meridian altitude and instantaneous altitude

The simplest method for calculating a(H,h) involves first finding the Sine of the altitude in the prime vertical (**4.8**) corresponding to each value of H, thus (*cf.* **F13**):

$$\text{Sin } h_0(H) = \text{R Sin } (H - \bar{\phi}) / \text{Sin } \phi \ .$$

Then the azimuth is given by (*cf.* **F16**):

$$a(H,h) = \text{arc Sin } \{ [\text{Sin } h - \text{Sin } h_0(H)] \cdot \text{Tan } \phi / \text{Cos } h \} \ .$$

---

[4] *Cf.* Hartner & Ruska, "Katalog", pp. 55-56.

Note that a(H,h) is independent of ε. Only one Islamic table of this function has come to my attention.

5.3.1 al-Khalīlī (Damascus)

MS Dublin CB 4091 of al-Khalīlī's set of minor auxiliary tables for timekeeping by the sun (**9.4**) also contains a set of tables of the function:

a(H,h)

computed to two digits for the domains:

H = 33°, 34°, ... , 80° and h = 1°, 2°, ... , H.

The underlying latitude is 33;30° (Damascus) and number of entries is about 2,700. Horizontal differences are also tabulated and labelled positive or negative by the Arabic letters *d* (for *zā'id*, increasing) or *ṣ* (for *nāqiṣ*, decreasing). See further **II-10.3** on al-Khalīlī's minor corpus.

5.4 Tables of solar azimuth as a function of time and solar longitude, for a specific latitude

The solar azimuth tables noted in this section were probably compiled from the corresponding solar altitude tables listed in **4.1**. The following list is not exhaustive.

5.4.1 Pseudo-al-Khwārizmī (Ḥabash): various latitudes

See **4.1.1** on the tables of a(T) for each seasonal hour at the solstices computed for various latitudes attributed to al-Khwārizmī but most probably due to Ḥabash.

5.4.2 al-Battānī: Raqqa

See **4.1.2** on the small table of the same kind computed for the longitude of Raqqa and contained in MS Escorial ár. 908 of the *Zīj* of al-Battānī.

5.4.3 Ibn al-Ādamī or Saʿīd ibn Khafīf al-Samarqandī: Baghdad

See **4.2.3** on the tables of a(T,λ) for each seasonal hour and each zodiacal sign computed for Baghdad in the treatise on sundials by Ibn al-Adamī or al-Samarqandī.

5.4.4 al-Maqsī: Cairo

See **4.1.3** on the tables of al-Maqsī for Cairo.

5.4.5 al-Marrākushī: Cairo

See **4.2.4** on al-Marrākushī's tables of the same kind, also for Cairo.

5.4.6 al-Sulṭān al-Ashraf

See **4.1.4** on the tables of a(T) for each seasonal hour at the equinoxes and solstices computed for various latitudes in the Yemen and Hejaz by the Sultan al-Ashraf.

5.5 Tables of solar azimuth at the equinoxes as a function of time, for a specific latitude

Tables of solar azimuth at the equinoxes are useful in sundial construction. The azimuth is related to the time since sunrise by the simple formula:

$$a(T,\phi) = \text{arc Tan } \{ \text{ Sin } \phi \text{ Tan } T / R \} .$$

5.5.1 Anonymous: Cairo

MS Cairo MM 58,1, fol. 2r, copied *ca.* 1450, of an anonymous set of spherical astronomical tables for Cairo contains a table of the function a(T) at the equinoxes for T = 1°, 2°, ... , 90° based on ϕ = 30;0° (Cairo) – see **Fig. 5.1.1b**. Note that for this latitude:

$$a(T) = \text{arc Tan } \{ \tfrac{1}{2} \text{ Tan } T \} .$$

However, the table was probably not compiled using this formula. In the instructions the anonymous compiler notes that one can find the solar altitude at the equinox h(T) using a special table and then use the accompanying table in MS Cairo MM 58 of a(h) at the equinoxes (**5.1.1**) to compute a(T). He gives an example in which the value of h(T) is the same as that for h(T,90°) in the tables of h(T,D) preserved in MS Cairo MM 72 (**4.6.1**) and interpolates in the table of a(h) to derive the value of a(T) in his table.

5.5.2 Ibn Abi 'l-Fatḥ al-Ṣūfī: all latitudes

A more extensive table occurs in MSS Istanbul Hamidiye 874,4, copied 1104 H [= 1692/93], Istanbul Nuruosmaniye 2904, and Cairo DM 196 of a short treatise on sundials written in 878 H [= 1473] by the Egyptian astronomer Ibn Abi 'l-Fatḥ al-Ṣūfī (**9.10**). The treatise is entitled *Nuzhat al-nāẓir fī waḍʿ khuṭūṭ faḍl al-dāʾir*, "Diversion for those Concerned with Marking Hour-angle Curves on Sundials" and the table is called *al-jadwal al-mushtarak*, "the combined (?) table". Like al-Māridīnī's *shabaka* (**9.6**), al-Ṣūfī's table resembles a sexagesimal multiplication table in appearance.[5]

The function tabulated is found by inspection to be:

$$f(x,y) = \text{arc Tan } \{ x \cdot \text{Tan } y / R \} ,$$

and the table gives values to two digits for the domains:

$$x = 1, 2, \ldots , 60 \quad \text{and} \quad y = 1°, 2°, \ldots , 90° .$$

It is intended to display the solar azimuth at the equinox as a function of time, since:

$$a(T,\phi) = f(\text{Sin } \phi,T) ,$$

although the arguments are labelled *jayb al-ʿarḍ*, "Sine of the latitude", and *faḍl al-dāʾir*, "hour-angle". In MS Cairo MM 61 an incomplete anonymous copy of this table has been appended to an anonymous set of prayer-tables for Cairo, and in MS Cairo MM 67,1 the same table reappears with a set of instructions in Turkish, again without any mention of al-Ṣūfī. The entries in al-Ṣūfī's table for x = 30, which correspond to ϕ = 30°, differ from those in the anonymous table of a(T) for Cairo mentioned in **5.5.1** and are generally more accurate.

[5] See n. 1:22.

5.6 Tables of solar rising amplitude, for a specific latitude

The solar rising amplitude (*saʿat al-mashriq* in Arabic) is determined by the equivalent expressions (*cf.* **F4** and **F14**):

$$\psi(\lambda) = \text{arc Sin} \{ \text{R} \sin \delta(\lambda) / \text{Cos } \phi \} = \text{arc Sin} \{ \text{Sin } \varepsilon \text{ Sin } \lambda / \text{Cos } \phi \}$$
$$= \text{arc Sin} \{ \text{Sin} [\max \psi(\lambda)] \cdot \text{Sin } \lambda / \text{R} \} .$$

Several tables of $\psi(\lambda)$ for specific latitudes are preserved, occasionally accompanying tables of Sin $\delta(\lambda)$ or Sin $\psi(\lambda)$ (**6.1** and **8.3**).

5.6.1 Anonymous: Baghdad

MS Alexandria 5577J, penned *ca.* 1400 (?), is a copy of the *Zīj* for the Yemen entitled *Taysīr al-maṭālib ...*, "Facilitating the Problems ... ", by the late-13th-century astronomer al-Kawāshī.[6] This work is of considerable interest because it contains material derived from a *zīj* by Ibn Yūnus (**2.1.1**) other than the *Ḥākimī Zīj* and an earlier *zīj* for Baghdad, as yet unidentified. al-Kawāshī also records some observations which he made in Egypt between 1273 and 1284.[7] MS London BL Or. 9116 is another copy of this work which lacks the title folio.

Alongside various simple spherical astronomical tables for the latitudes of Aden and Taiz, al-Kawāshī presents a table of $\psi(\lambda)$ for Baghdad. There are no other tables for Baghdad in his *Zīj*. Values of $\psi(\lambda)$ are given to three digits for each degree of λ and are based on the parameters ϕ = 33;0° (Baghdad) and ε = 23;33°. The same table occurs in MS Berlin Ahlwardt 5773/5776,1 (Mq. 98,16+18), fol. 47r, of a fragment of an unidentified *zīj*, probably of Egyptian origin (see also **5.6.4**). The value used for the obliquity and for the latitude of Baghdad suggests a 9th-century source (see **9.1**).

5.6.2 Ibn Yūnus: Cairo

In MS Leiden Or. 143, p. 356, of the *Ḥākimī Zīj* of Ibn Yūnus (**2.1.1**) there is a table of $\psi(\lambda)$ to three digits for the parameters ϕ = 30;0° (Cairo) and ε = 23;35°. Ibn Yūnus' values are extremely accurately computed. He also tabulated Sin $\psi(\lambda)$ for Cairo and Baghdad: see **8.3.1**. See also **5.6.9** on some later and less accurate tables of $\psi(\lambda)$ for Cairo.

5.6.3 Muḥyi 'l-Dīn al-Maghribī: Maragha

The mid-13th-century astronomer Muḥyi 'l-Dīn al-Maghribī[8] tabulated $\psi(\lambda)$ for the latitude of Maragha, where he worked (**4.3.1***). The table is preserved in MSS Dublin CB 3665, fol. 96v, and Meshed Shrine Library 332(103), fol. 99r, of his *Zīj*, and values are given to three digits for each 1° of λ and are based on the parameters ϕ = 37;20,30° (Maragha) and ε = 23;30°.

5.6.4 al-Marrākushī: Cairo

In MS Paris BNF ar. 2508, fol. 29r, al-Marrākushī (**4.2.4**) presents a table of $\psi(\lambda)$ for parameters

---

6 The existence of this work is noted in Brockelmann, *GAL*, I, p. 625. On al-Kawāshī and his *zīj* see King, *Astronomy in Yemen*, no. 7.

7 Published in King & Gingerich, "Astronomical Observations from 13th-Century Egypt".

8 On al-Maghribī see Suter, *MAA*, no. 376; and Kennedy, "*Zīj* Survey", nos. 41 and 108. Another *zīj* for Maragha by al-Maghribī is preserved in Cairo: see *Cairo ENL Catalogue*, I, *sub* MM 188, *etc.*

ϕ = 30;0° (Cairo) and ε = 23;35°. The values are given to two digits for each 5° of λ and are intended to serve the construction of the markings on a horizontal dial (an orthogonal projection of the celestial sphere onto the plane of the horizon). They are not based on the corresponding values of Ibn Yūnus (**5.6.2**). Yet another table based on the same parameters but now with values to two digits for each 1° of λ is found in MS Berlin Ahlwardt 5773/5776,1 (Mq. 98,16+18), fol. 25v, of an Egyptian *zīj* (**5.6.1**).

5.6.5 al-Mizzī: Damascus

The prayer-tables of al-Mizzī, preserved in MS Cairo MM 62 (**2.1.3**), display $\psi(\lambda)$ for Damascus (fol. 7v). Values are carefully computed to two digits for each 1° of λ and are based on al-Mizzī's distinctive parameters ϕ = 33;27° (Damascus) and ε = 23;33°.

5.6.6 al-Khalīlī: Damascus

Several copies of the later prayer-tables for Damascus compiled by al-Khalīlī (**2.1.4**), for example MS Paris BNF ar. 2558, fol. 51v, contain a table of $\psi(\lambda)$ with values carefully computed to two digits for each degree of λ and parameters ϕ = 33;30° (Damascus) and ε = 23;31°. al-Khalīlī also tabulated Sin $\psi(\lambda)$ (**6.1.2** and **8.3.2**).

5.6.7 Anonymous: Tunis

MS Berlin Ahlwardt 5724, fols. 29v-30r, of the main corpus of tables for Tunis (**2.3.5**) contains tables of both $\psi(\lambda)$ and Sin $\psi(\lambda)$ carefully computed to two digits for each degree of λ and based on the parameters ϕ = 37;0° (Tunis) and ε = 23;35°. See also **8.3.3** on the latter.

5.6.8 Anonymous: Isfahan

MS Istanbul Topkapı B 411, an encyclopedia compiled *ca.* 1413 for Iskandar Sulṭān ibn ʿUmar Shaykh of Isfahan, contains an anonymous collection of astronomical and astrological tables. Amongst these tables is a set of prayer-tables for parameters ϕ = 32;25° (Isfahan) and ε = 23;30°, including a table displaying $\psi(\lambda)$ for each 1° of λ. See further **II-3.15** on the other tables in this set and also **VIIc-4c** on the latitude used for Isfahan, which was apparently derived by al-Khāzimī in the 11th century.

5.6.9 Miscellaneous: Cairo

In numerous sets of late Egyptian prayer-tables for Cairo dating from after the 16th century (**4.8.4**), such as MS Cairo MM 15 of the prayer-tables of al-Lādhiqī, or MS Alexandria 4441J of the prayer-tables of Ibn Abī Rāya, or MS Cairo MM 81 of the prayer-tables of al-Ikhṣāṣī, the function $\psi(\lambda)$ is tabulated to two digits for the parameters of Ibn Yūnus (**5.6.2**). However, most of these tables are corrupt and the errors seem to represent an attempt to recompute the function rather than to be mere copyists' errors. These tables are characterized by the entry 23;29° for λ = 60°. Note that:

$$\psi = \varepsilon \text{ for } \lambda = 60° \text{ and } \phi = 30° \ ,$$

so that the entry should be 23;35°. Also the entry for λ = 90° is 27;25° rather than the accurate value 27;31°. See **4.8.4** on some related tables of the solar altitude in the prime vertical.

5.6.10 al-Asyūṭī: Assiut

In MS Cairo DM 188 of the prayer-tables for Assiut prepared by al-Asyūṭī (**4.8.5**), there is a table of $\psi(\lambda)$ with values to two digits for each 1° of λ and parameters ϕ = 27;0° (Assiut) and ε = 23;35°.

5.6.11 al-Dīsṭī: Lattakia

MS Aleppo Awqāf 911 of al-Dīsṭī's prayer-tables for Lattakia (**4.8.7**) contains a table of $\psi(\lambda)$ with values to two digits for each 1° of λ and parameters ϕ = 34;30° (Lattakia) and ε = 23;30°.

5.6.12 Anonymous: Alexandria

MS Cairo DM 1207 of the corpus of tables for Alexandria (**4.8.8**) contains a table of $\psi(\lambda)$ for parameters ϕ = 31;0° (Alexandria) and ε = 23;35°.

5.6.13 Yūsuf Kilārjī: Crete

MS Cairo DM 834 of Yūsuf Kilārjī's prayer-tables for Crete (**4.8.9**) contains a table of $\psi(\lambda)$ for parameters ϕ = 35;30° (Crete) and ε = 23;35°.

5.7 Tables of rising amplitude as a function of declination, for a specific latitude

Only one Islamic table for finding the rising amplitude of non-circumpolar stars has come to light. The underlying formula is (*cf.* **F14**):

$$\psi(\Delta) = \text{arc Sin } \{ \text{ R Sin } \Delta \text{ / Cos } \phi \} .$$

Note that al-Khalīlī's auxiliary function $f_\phi(\theta)$ (**9.5**) displays Sin $\psi(\Delta,\phi)$ for each 1° of both arguments.

5.7.1 Anonymous: Cairo

MS Cairo MM 43, penned *ca.* 1450, is an Egyptian copy of al-Khalīlī's universal auxiliary tables (**6.4.3** and **9.5**), contains an anonymous table of the function $\psi(\Delta)$ computed for ϕ = 30° (Cairo). Values are given to two digits for each 1° of Δ up to 60° (= $\bar{\phi}$).

5.8 Tables of maximum solar rising amplitude as a function of latitude

In *Almagest* VI.12 Ptolemy presented a table of the maximum solar rising amplitude for the seven climates. The values for each zodiacal sign are given in degrees and minutes and are arranged in concentric circles for each climate, the so-called "horizon diagrams".[9] The underlying value of ε is 23;51,20°. I have not consulted any of the Arabic versions of the *Almagest* in which this diagram is surely reproduced, but in three sets of Islamic sources we find similar tables.

---

[9] See Ptolemy, *Almagest*, transl. Gerald Toomer, p. 320; and Neugebauer, *HAMA*, I, pp. 37-39. On the climates see n. 1:11.

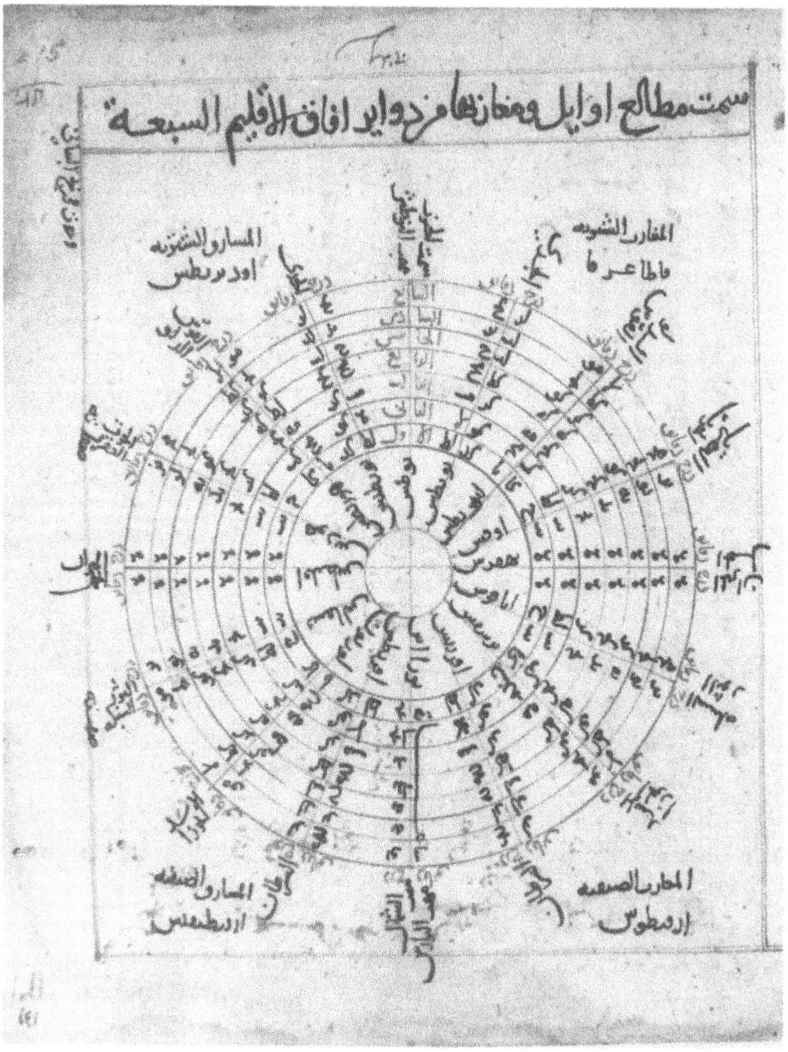

Fig. 5.8.1: A table of solar rising amplitudes in the Ptolemaic tradition. Values are given for each pair of zodiacal signs with the same amplitude (and declination), radially for the first climate out to the seventh. The (false) impression given is of directions around the horizon, not least because the 16 winds are displayed in the inner ring. [From MS Berlin Ahlwardt 5751, p. 215, courtesy of the Deutsche Staatsbibliothek (Preußischer Kulturbesitz).]

5.8.1 al-Battānī (Raqqa)

In MS Escorial ár. 908, fol. 196v, of the *Zīj* of al-Battānī[10] (**4.1.2**), there is a horizon diagram displaying solar rising amplitudes for the seven climates as in the *Almagest* but with values based on $\varepsilon = 23;35°$. al-Battānī's table is reproduced in MS Berlin Ahlwardt 5751 (Mq. 101,1), p. 215, copied *ca.* 1300, amidst various tables appended to a copy of the *Zīj* of Kūshyār – see **3.2.1** and **Fig. 5.8.1** – and in MS Milan Ambrosiana C86, fol. 167v, of a small collection of extracts (on fols. 167r-168v) from al-Battānī's *Zīj* copied in the Yemen.

[10] Nallino, *al-Battānī*, II, p. 92.

5.8.2 Abu 'l-Wafā' (Baghdad)

MS Paris BNF ar. 2494, from the 12th century, is an incomplete copy of the *Zīj* by Abu 'l-Wafā' al-Būzajānī (*fl.* Baghdad *ca.* 970) entitled *al-Majisṭī* after Ptolemy's *magnum opus.*[11] Unfortunately all of Abu 'l-Wafā''s tables are missing from this manuscript. On fol. 28r he states that he computed a table of the maximum solar rising amplitude for each 0;15° of latitude up to 60°.

5.8.3 Sanjar al-Kamālī (Shiraz)

In MS Paris BNF supp. pers. 1488, fol. 100r, of the *Ashrafī Zīj* (**2.3.3**) there is a table of max $\psi(\phi)$ with entries to *four* sexagesimal digits for each degree of ϕ from 1° to 60°, based on ε = 23;35°.

---

[11] On Abu 'l-Wafā' (Suter, *MAA*, no. 167) see Sezgin, *GAS*, VI, pp. 222-224, and the references there cited, also the article by A. P. Youschkevitch in *DSB*.

CHAPTER 6

TABLES OF AUXILIARY FUNCTIONS FOR TIMEKEEPING

6.0 Introductory remarks

The computation of time from solar altitude (*cf.* **F9-12**) is considerably facilitated if the function:

$$B(\lambda) \;=\; \mathrm{Cos}\;\delta(\lambda)\;\mathrm{Cos}\;\phi\;/\;R$$

is tabulated first. This function B is called in Arabic *al-aṣl al-muṭlaq*, "the absolute base", or simply *al-aṣl*, "the base". Tables of the function:

$$G(\lambda) = R^2 \;/\; [\mathrm{Cos}\;\delta(\lambda)\;\mathrm{Cos}\;\phi]$$

would be yet more useful, but Muslim astronomers generally prefered to use B(λ). To compute B(λ) it is necessary either to compute Cos δ(λ) or Sin H(λ). Given a table of Sin H(λ), B(λ) can be found using (*cf.* **F8**):

$$B(\lambda) = {}^{1}/_{2}\;\{\;\mathrm{Sin}\;H(\lambda) + \mathrm{Sin}\;H(\lambda^*)\;\}$$

This is clearly valid since:

$$H(\lambda) \;=\; \bar{\phi} + \delta(\lambda) \quad \text{and} \quad H(\lambda^*) \;=\; \bar{\phi} - \delta(\lambda)$$

and (in modern notation):

$$\cos\delta\;\cos\phi = {}^{1}/_{2}\;\;\{\;\sin\;(\;\bar{\phi} + \delta\;) + \sin\;(\;\bar{\phi} - \delta\;)\;\}.$$

See also **2.1.1**.

Another important function of Islamic timekeeping is:

$$C(\lambda) = \mathrm{Sin}\;\delta(\lambda)\;\mathrm{Sin}\;\phi\;/\;R\;,$$

called in Arabic *buʿd al-quṭr*, literally "the distance of the diameter". It measures the height of the day-circle centre above the horizon. To compute C(λ) one can either use Sin δ(λ) or B(λ) together with Sin H(λ) since:

$$C(\lambda) = B(\lambda)\;-\;\mathrm{Sin}\;H(\lambda^*)\;.$$

Given tables of B(λ) or G(λ) and C(λ) or Sin H(λ), the determination of the hour-angle reduces to an application of the equivalent formulae:

$$\mathrm{Vers}\;t(h,\lambda) = [\mathrm{Sin}\;H(\lambda)\;-\;\mathrm{Sin}\;h]\;\bullet\;R/B(\lambda) = [\mathrm{Sin}\;H(\lambda)\;-\;\mathrm{Sin}\;h]\;\bullet\;G(\lambda)$$
$$\mathrm{Cos}\;\;t(h,\lambda) = [\mathrm{Sin}\;h\;-\;C(\lambda)]\;\bullet\;R/B(\lambda) = [\mathrm{Sin}\;H(\lambda)\;-\;C(\lambda)]\;\bullet\;G(\lambda)\;.$$

If we now define the function:

$$H'(H,h) = \mathrm{Sin}\;H\;-\;\mathrm{Sin}\;h\;,$$

called in Arabic *faḍl al-jaybayn*, "the difference between the two sines", the first formula for the hour-angle reduces to:

$$\mathrm{Vers}\;t \;=\; H'\bullet R/B \;=\; H'\bullet G\;.$$

Likewise if we define the function:

$$b(h,\lambda) = \mathrm{Sin}\;h\;-\;C(\lambda)\;,$$

called in Arabic *al-aṣl al-muʿaddal*, "the modified base", the second formula for the hour-angle reduces to:

$$\text{Cos } t = b \cdot R/B = b \cdot G .$$

Similarly the functions:

$B(\Delta) = \text{Cos } \Delta \text{ Cos } \phi / R$, $C(\Delta) = \text{Sin } \Delta \text{ Sin } \phi / R$ and $G(\Delta) = R^2 / [\text{Cos } \Delta \text{ Cos } \phi]$ are useful in computing time from stellar altitudes.

I now present a description of various Islamic tables of Sin δ, Cos δ and Sin H (**6.1**. to **6.3**), B, C, G and related functions (**6.4** to **6.11**), b and H′ (**6.12** and **6.13**), and the inverse trigonometric functions necessary to find the hour-angle once one has used the functions B, C and G or Sin H (**6.14** and **6.15**). Finally I discuss certain star catalogues which present numerical information for each star to facilitate reckoning time by night (**6.16**).

6.1 Tables of the sine of the solar declination

The function Sin δ(λ) (Arabic, *jayb al-mayl*) is easily computed using:

$$\text{Sin } \delta(\lambda) = \text{Sin } \varepsilon \text{ Sin } \lambda / R .$$

At least three simple tables of the quantities:

$$n/R \cdot \text{Sin } \varepsilon \quad \text{and} \quad n/R \cdot \text{Cos } \varepsilon \qquad (n = 1, 2, \dots , R)$$

are attested in the Islamic sources, two computed by Ibn Yūnus (MS Leiden Or. 143, pp. 236-237)[1] and Abu 'l-Wafā' (MS Escorial ár. 927, fol. 51r),[2] and another compiled several centuries later by al-Ḥamzāwī (MS Cairo TR 119, pp. 602-603).[3] Such tables for Sin ε can be used to facilitate multiplication by Sin ε in order to generate a table of Sin δ(λ).

In turn, a table of Sin δ(λ) can be used to prepare: (1) tables of δ(λ), which are standard in *zīj*es; (2) tables of the auxiliary function C(λ) for particular latitudes, which are fairly common in corpuses of tables for timekeeping (**6.10**); and (3) tables of $h_0(\lambda)$ and ψ(λ) (**4.8**, **5.5** and **8.3**). Such a table can also be used to facilitate the transformation of ecliptic to equatorial coordinates by one of the standard Islamic methods.[4] The following tables of Sin δ(λ) have come to my attention.

6.1.1 Anonymous (Baghdad?)

In MS Berlin Ahlwardt 5750 (Wetzstein 90), copied *ca.* 1300, of the anonymous recension of one of the *zīj*es of Ḥabash (**9.1** and **9.2**) there is a set of tables of:

$$\delta(\lambda), \text{ Sin } \delta(\lambda) \text{ and Cos } \delta(\lambda) ,$$

computed to three digits for each degree of λ and based on the parameter ε = 23;35°. The identity of the compiler of these tables is uncertain.

6.1.2 al-Khalīlī (Damascus)

In MS Cairo DM 184, copied in 1171 H [= 1757/58], of the prayer-tables of the Syrian astronomer al-Manāshīrī (b. 1573, d. 1630),[5] the function Sin δ(λ) is tabulated alongside

[1] See King, *Ibn Yūnus*, III.11.4c; and *idem*, "Medieval Islamic Multiplication Tables", B, pp. 409 and 411.

[2] These tables are mentioned in Kennedy, "*Zīj* Survey", p. 145. On Abu 'l-Wafā' see n. 5:11.

[3] On al-Jamālī Yūsuf ibn Qurqmās al-Ḥamzāwī see *Cairo ENL Survey*, no. C91. On these tables of his see King, "Islamic Multiplication Tables", B, pp. 409 and 410.

[4] See, for example, King, *Ibn Yūnus*, III.39.1a-b.

[5] On al-Manāshīrī see Brockelmann, *GAL*, I, p. 427; *Cairo ENL Survey*, no. D8; and İhsanoğlu *et al.*, *Ottoman Astronomical Literature*, I, pp. pp. 274-275, no. 138.

numerous other spherical astronomical tables for Damascus attributed to al-Khalīlī (**2.1.4**). This particular table, which is based on parameter $\varepsilon = 23;31°$, is also contained in MS Damascus Ẓāhiriyya 9233, p. 156, of the main Damascus corpus for timekeeping, there introduced in the name of al-Ṭanṭāwī (**2.2.5**). See also **6.3.2** and **8.3.2** and **II-11.10** and **11.13**.

6.1.3 Riḍwān Efendī (Cairo)

In MS Istanbul S. Esad Efendi Medresesi 119,1 of the prayer-tables of the late-17th-century Egyptian astronomer Riḍwān Efendī,[6] virtually all of which were lifted from earlier sources, there is a set of tables of:

$$\text{Sin } \delta(\lambda), \text{ Cos } \delta(\lambda) \text{ and Tan } \delta(\lambda)$$

computed to three digits for each degree of λ, and based on the distinctive parameter of Ulugh Beg of Samarqand, namely $\varepsilon = 23;30,17°$. Although Riḍwān Efendī lifted various tables for $\varepsilon = 23;35°$ from earlier sources, these for the new obliquity were probably computed by himself. On the other tables see **6.2.6** and **7.1.1**, and on the corpus associated with Riḍwān Efendī see **II-7.10**.

6.1.4 Taqi 'l-Dīn (Istanbul)

In MS Istanbul Nuruosmaniye 2930, fol. 23r, of the *zīj* entitled *Sidrat muntaha 'l-afkār*, "The Lotus Tree in the Seventh Heaven of Reflection", by Taqi 'l-Dīn (**2.3.6**), there is a set of tables of the functions:

$$\delta(\lambda), \text{ Sin } \delta(\lambda) \text{ and Tan } \delta(\lambda)$$

computed to four sexagesimal digits and based on the parameter $\varepsilon = 23;28,54°$ derived from his "new Murād Khān observations". See also **6.4.8** below.

6.2 Tables of the cosine of the solar declination

To compute Cos $\delta(\lambda)$ for given λ the easiest procedure is to compute $\delta(\lambda)$ first and then use a Sine table to find Cos $\delta(\lambda)$. In the Islamic sources we find tables of both Cos $\delta(\lambda)$ and Cos $\delta(\bar{\lambda})$. The former can be used to compute $B(\lambda)$ for a particular latitude (**6.4**) and the latter is of limited use in certain operations in spherical astronomy, such as the conversion from ecliptic to equatorial coordinates.[7] The following tables of Cos $\delta(\lambda)$ or Cos $\delta(\bar{\lambda})$ are known.

6.2.1 Ḥabash (Baghdad / Damascus)

One of the auxiliary functions in the *Jadwal al-taqwīm* of Ḥabash (**9.1**) is simply Cos $\delta(\bar{\lambda})$. Values are given to three digits for each degree of λ, computed for $\varepsilon = 23;35°$. Ḥabash's table reappears in MS Paris BNF supp. pers. 1488, fol. 207v, of the early-14th-century Persian *Ashrafī Zīj* (**2.3.3**) where the function is referred to as *taʿdīl al-buʿd al-awwal ʿan muʿaddil al-nahār*

---

[6] On Riḍwān Efendī (**I-6.1.3**) see Dorn, "Drei arabische Instrumente", pp. 32-34; Brockelmann, *GAL*, II, p. 471, and SII, p. 487, Azzawi, *History of Astronomy in Iraq*, pp. 319 320; Mayer, *Islamic Astrolabists*, pp. 81-82, Kennedy, "*Zīj* Survey", no. X209; and İhsanoğlu *et al.*, *Ottoman Astronomical Literature*, I, pp. 377-384, no. 246. See also **II-7.10**.

[7] *Cf.* Jensen, "Abū Naṣr's *Table of Minutes*", pp. 6-7 and 14.

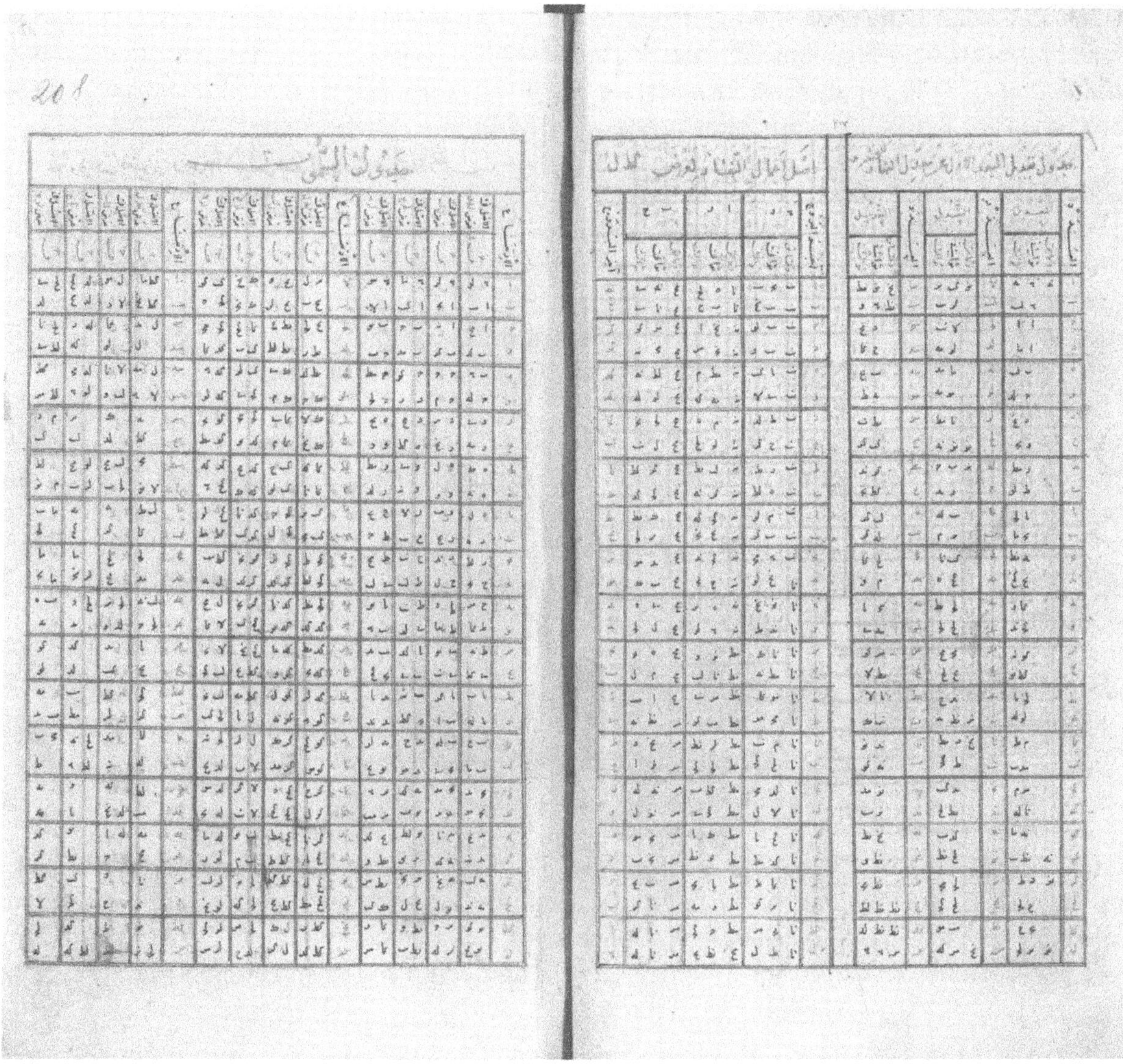

Fig. 6.2.1: Ḥabash's table of the cosine of the declination of the complement of the solar longitude, together with a table of the absolute base for latitude 29;30° (**6.4.5**), and tables of the auxiliary azimuth function for various latitudes in Iran (**8.1.2**). [From MS Paris BNF supp. pers. 1488, fol. 207v, courtesy of the Bibliothèque Nationale de France.]

and is attributed to *x-kh-s-r-w*, where *x* denotes a carrier which could serve as a *ḥā*ʾ, *tā*ʾ or *thā*ʾ, *nūn* or *yā*ʾ, the combination corrupted from *ḥ-b-sh* for Ḥabash! – see **Fig. 6.2.1**.

6.2.2 Anonymous (Baghdad ?)

In MS Berlin Ahlwardt 5750 of the anonymous recension of a *zīj* of Ḥabash (**6.1.1** and **6.2.1**) there are tables of Cos δ(λ) and Cos $\delta(\bar{\lambda})$ (**9.2**), both computed to three digits for each degree of λ and based on ε = 23;35°.

6.2.3 Abū Naṣr (Khwarizm)

One of the auxiliary functions tabulated by the late-10th-century scholar Abū Naṣr in his *Jadwal al-daqā'iq* (**9.3**) is simply Cos $\delta(\bar{\lambda})$. Values are given to four digits for each degree of λ and are based on the Ptolemaic parameter ε = 23;51,20°.

6.2.4 Anonymous (Cairo)

In one copy of al-Bakhāniqī's edition of the Cairo corpus (**2.1.1**), namely MS Istanbul Nuruosmaniye 2925, the main tables for timekeeping are followed by tables of Cos $\delta(\lambda)$ (called *jayb tamām al-mayl*) and $C(\lambda)$ (**6.10.1**). Entries are given to two digits and the underlying value of ε is 23;35°. The same pair of tables occurs in MS Alexandria 4441J of the prayer-tables for Cairo of Ibn Abī Rāya (*fl. ca.* 1710 – see **II-7.9**), with many copyist's errors.

6.2.5 Anonymous (Tunis)

MS Cairo DM 689 of the anonymous Tunisian auxiliary tables for timekeeping (**9.7**) contains a set of tables of Sin $H(\lambda)$ computed for various latitudes (**6.3.3**). One of these is for latitude 0° (called *jayb al-ghāya fī kull balad lā ʿarḍ lahu*) and the tabulated function is thus Sin [90°-$\delta(\lambda)$] = Cos $\delta(\lambda)$. Values are given to two digits for each degree of λ and are based on ε = 23;35°.

6.2.6 Riḍwān Efendī (Cairo)

MS Istanbul S. Esad Efendi Medresesi 119, fol. 59v, of the prayer-tables of Riḍwān Efendī (**6.1.3**) contains a table of Cos $\delta(\lambda)$ for Cairo to three digits based on ε = 23;30,17°.

6.2.7 ʿAbdallāh al-Ḥalabī (?) (Aleppo)

MS Aleppo Awqāf 943 contains a set of prayer-tables for Aleppo attributed to ʿAbdallāh al-Ḥalabī (**II-11.14**), who was *muwaqqit* at the Umayyad mosque in Aleppo about 1750. One of the tables presented by al-Ḥalabī, but not necessarily computed by him, is of the function Cos $\delta(\lambda)$, with values displayed for each degree of λ and computed to three digits for ε = 23;30°.

6.3 Tables of the sine of the meridian altitude, for a specific latitude

Tables of $H(\lambda)$ for a particular latitude are standard in *zīj*es. The underlying formula is (*cf.* **F1**):

$$H(\lambda) \;=\; \bar{\phi} + \delta(\lambda) \;.$$

Tables of Sin $H(\lambda)$ (Arabic, *jayb al-ghāya*) are less common and can be used either to compute the function $B(\lambda)$ (**6.4**) or to find the hour-angle $t(h,\lambda)$ using the formula (*cf.* **F9**):

$$\text{Vers } t(h,\lambda) = [\ \text{Sin } H(\lambda) - \text{Sin } h\ ] \cdot R \,/\, B(\lambda)\ .$$

6.3.1 al-Khaṭā'ī (Cairo)

In MS Vatican Borg. ar. 217,2, of the auxiliary tables of al-Khaṭā'ī (**6.15.1**) the function Sin $H(\lambda)$ is tabulated to three digits for each degree of λ and parameters ϕ = 30;0° (Cairo) and ε = 23;35°. See also **6.4.2**.

6.3.2 al-Khalīlī: Damascus

In MS Cairo DM 184 of the prayer-tables of al-Munāshirī (**6.1.2**) the function Sin H(λ) is tabulated for ϕ = 33;30° (Damascus) and ε = 23;31°. These tables are attributed to al-Khalīlī (**2.1.4**).

6.3.3 Anonymous: Tunis

In MS Cairo DM 689 of the anonymous Tunisian auxiliary tables (**9.7**) Sin H(λ) is tabulated to two digits for each degree of λ and a total of 23 latitudes:

21;40° (Mecca), 25°(Medina), 0° (**6.2.5**), 30°, 30;30°, ... , 38°,
as well as 33;40° [Fez], 36;40° [Tunis] and 37;10° (locality ?) .

The latitude 37;10° is curious, for no locality to the north of Tunis comes to mind, and it may be intended as an alternative value for that city (for which 37;0° is used elsewhere). The underlying value of ε is 23;35°. The table for ϕ = 36;40° is also contained in MS Berlin Ahlwardt 5724 (**2.3.5**).

6.3.4 Anonymous: Tlemcen

On the table of Sin H(λ) for parameters ϕ = 35;0° (Tlemcen) and ε = 23;35° in MS London BL Or. 411,2 of an anonymous commentary on the astronomical poem of al-Jādarī see **6.15.2.**

6.3.5 Anonymous: Aleppo

One of the functions tabulated by ʿAbdallāh al-Ḥalabī (**6.2.7**) in his prayer-tables for parameters ϕ = 36;0° (Aleppo) and ε = 23;30° is Sin H(λ). al-Ḥalabī also displays values of the difference between Sin H(λ) and the Sines of the solar altitudes at various times of prayer – see **II-11.14.**

6.4 Tables of the absolute base for timekeeping by the sun, for a specific latitude

Several sources listed below contain tables of B(λ) for particular latitudes. The function is invariably called *al-aṣl al-muṭlaq* or simply *al-aṣl.* Note that in al-Khalīlī's first set of auxiliary tables (**9.4** and also **9.7** and **9.11**) the function $^1/_2$B(ϕ,λ) is tabulated for each degree of both arguments.

6.4.1 al-Nasawī: Rayy

MSS Cairo ṬFF 11, copied *ca.* 1400, and London BL Add. 27261, copied in 814 H [= 1411], contain a Persian treatise entitled *Rawḍat al-munajjimīn*, "The Garden of the Astronomers", compiled by Shāhmardān Rāzī,[8] a student of the astronomer-mathematician Abu 'l-Ḥasan ʿAlī ibn Aḥmad al-Nasawī who worked in Rayy in the late 10th century, but whose major astronomical work is lost.[9] One of the tables in the *Rawḍa* (fol. 78v of the Cairo manuscript) is specifically attributed to al-Nasawī: see **Fig. 6.4.1**. The function tabulated is simply B(λ)

[8] On Shāhmardān Rāzī see Storey, *PL*, II:1, p. 45, no. 81; Sezgin, *GAS*, VI, p. 246; and *Cairo ENL Survey*, no. B91.

[9] On al-Nasawī (Suter, *MAA*, no. 214) see Sezgin, *GAS*, V, pp. 345-348, VI, pp. 245-246, and VII, pp. 182, 402, and 410-411, also III, p. 311, and the article by A. S. Saidan in *DSB*.

Fig. 6.4.1: This table of the auxiliary function called the "base" (*aṣl*) or later the "absolute base" (*al-aṣl al-muṭlaq*), computed in the 10th century by a scholar well versed in mathematical astronomy as well as pure mathematics, rather than by a scholar specializing in astronomical timekeeping, shows that Muslim interest in practical astronomy started early. [From MS Cairo ṬFF 11, fol. 78v, courtesy of the Egyptian National Library.]

and is referred to as *aṣl* in the text. The values are computed to three digits for each degree of λ with first differences (*tafāḍul*) to two digits. The latitude 35;34,40° is mentioned in the accompanying text, and is attested for Rayy, being associated with the observations of al-Khujandī in the year 994.[10] However, the text further states that the Cosine of this latitude is 48;47,52, whereas the accurate value is 48;47,58,31. In fact, the value in the text was derived

[10] On the derivation of this value of the latitude of Rayy see Schirmer, "Studien zur Astronomie der Araber", pp. 63-79.

by *linear interpolation* (!) between the entries for 35° and 36° given to three (or perhaps four) significant digits. If we, not unreasonably, assume that this erroneous value of the Cosine was actually used to compile the table, Benno van Dalen has shown that the values used for Cos $\delta(\lambda)$ were extremely accurate for ε = 23;35,44 plus or minus 1 second. This unfortunately does not correspond to the obliquity associated with the first derivation of the latitude 35;34,40°, namely, 23;32,21°. We note in passing that 23;35,43,57,41 is the arc Sine of 24;1: this may somehow be relevant to explaining the makeup of this table.

Other spherical astronomical tables in the *Rawḍa*, for example, those for $\alpha_\phi(\lambda)$ (fol. 59v of the Cairo manuscript), are computed for ϕ = 36°, a popular medieval value for Rayy and also for the 4th climate.[11]

6.4.1* al-Baghdādī (?): Baghdad

In MS Paris BNF ar. 2486, fol. 236r, of the *Zīj* of al-Baghdādī (**2.3.1**), there is a table of $B(\lambda)$ computed to three digits for each degree of λ and based on the parameters ϕ = 33;25° (Baghdad) and ε = 23;35°.

6.4.1 Early anonymous (N. Iran)**

MS Istanbul U.B. A 314, copied *ca.* 1600 (?), is a precious collection of early Islamic treatises on mathematics, astronomy and astronomical instruments which has only in 2001 attracted the attention it merits, and a facsimile is now available in published form.[12] I refrain from commenting on the treatises at this point, confident that the next generation of scholars will occupy themselves with this manuscript, for there is something in it for everyone. In fact, there is one table for spherical astronomy in the collection, and it cannot be attributed or dated, but I suspect that it is very early. Certainly it is very corrupt. It occurs on fol. 47v, quite out of place, and shows values to three digits of three functions labelled *al-aṣl*, *jayb taʿdīl al-nahār*, and *taʿdīl al-nahār*, for each degree of the argument, which is labelled *buʿd muʿaddal al-nahār*. The "title " of the table translates: "Multiply the Cosine of the local latitude by the Cosine of the declination (lit., distance from the celestial equator) and divide it by 60 and the result is the base (*al-aṣl*)." The three functions:

$$B(\Delta), \quad \text{Sin } d(\Delta) \quad \text{and} \quad d(\Delta)$$

appear to be computed for a specific latitude of about 38°, which would serve a locality in Northern Iran. The treatise on planetary and spherical astronomy immediately preceding this table mentions Isfahan several times, so it is unrelated. There are no entries for the second and third functions beyond Δ = 55°, a feature which shows how corrupt the table is (see further **7.2.0***).

6.4.2 al-Khaṭāʾī: Cairo

In the auxiliary tables of al-Khaṭāʾī in MS Vatican Borg. ar. 217,2 (**6.15.1**), there is a table of $B(\lambda)$ to three digits for each degree of λ, based on the parameters ϕ = 30;0° (Cairo) and ε = 23;35°. A table of Sin $H(\lambda)$ to three digits accompanies this one of $B(\lambda)$ (**6.3.1**) and al-

[11] Kennedy & Kennedy, *Islamic Geographical Coordinates*, p. 284, and King, "Geography of Astrolabes", pp. 6-9.

[12] See Sezgin *et al.*, *Scientific Manuscript.* On the importance of this see the introduction by Fuat Sezgin (pp. v-x) and also the remarks in **VIIc** below.

Khaṭā'ī points out the fact that the latter can be computed from the former using the relation noted in **6.0**. Indeed the errors in the table of B(λ) can be explained by the fact that the values of Sin H(λ) were derived from H(λ) by linear interpolation in an accompanying table of Sin θ which displays the function to three digits for each integral value of θ.

In MS Berlin Ahlwardt 5710, fol. 24r, of the prayer-tables of Riḍwān Efendī (**6.1.3**) the same table of B(λ) occurs with a few additional copyists' mistakes, alongside a more accurately computed table of C(λ). Note that C(λ) is easier to compute than B(λ) for φ = 30° (see further **6.10.1**).

6.4.3 Anonymous: Cairo

In MS Cairo MM 43, penned *ca.* 1450, an Egyptian copy of al-Khalīlī's universal auxiliary table (**9.5**), numerous other spherical astronomical functions are tabulated for the latitude of Cairo. There is a table of B(λ) computed to two digits for each 0;15° of λ, based on the same parameters as the one described in the previous section. Likewise in MS Cairo MM 241 of the auxiliary tables of Ibn al-Mushrif (**9.8**), there is a table of B(λ) for Cairo computed to two digits for each degree of λ.

6.4.4 Anonymous: Mecca

In MS Cairo MM 68, copied *ca.* 1500, of an anonymous set of prayer-tables for Mecca (**II-6.10**) the function B(Δ) is tabulated to two digits for each 0;5° of Δ up to Δ = ε = 23;35°. The underlying value of φ is 21;0°. See **6.6** below.

6.4.5 Sanjar al-Kamālī: Shiraz

In MS Paris BNF supp. pers. 1488, fol. 207v, of the *Ashrafī Zīj* (**2.3.3**) there is a table of B(λ) for each degree of λ computed to three digits for the parameters φ = 29;30° (Shiraz) and ε = 23;35°. The function is referred to as *aṣl aʿmāl al-nahār*, "the basic function for operations (of timekeeping) by day" – see **Fig. 6.2.1**.

6.4.6 Anonymous: Anatolia

MS Istanbul Süleymaniye 1037/32 (fols. 282v-285v) of an anonymous set of Seljuk prayer-tables for an unspecified locality in Anatolia (**3.3.3**) contains additional tables of B(λ) to two digits for each degree of λ. The underlying parameters are φ = 38;30° and ε = 23;35°.

6.4.7 Anonymous: Tunis

In MS Berlin Ahlwardt 5724, fol. 42r (**2.3.5**), and also in MS Princeton Yahuda 147c, p. 113, of the Tunisian *Quṣʿī Zīj* of Ḥusayn Quṣʿa (**4.3.6**) there is a table of B(λ) computed to two digits for each degree of λ and based on the parameters φ = 36;40° (Tunis) and ε = 23;35°. In the Berlin copy the function is referred to as *aṣl faḍl dāʾir al-shams*, "the base for (computing) the hour-angle of the sun". Both sources contain a table of Sin H(λ) for Tunis with which that of B(λ) might have been compiled (**6.3.3**).

6.4.8 Taqi 'l-Dīn: Istanbul

MS Istanbul Esat Efendi 1976, fol. 25r, of the *Zīj* of Taqi 'l-Dīn entitled *Jarīdat al-durar* (**6.1.4**) contains a table of B(λ) computed to four significant *decimal* digits. The latitude served is φ

= 40;58° (Istanbul), which is stated to underlie a table of $\alpha_\phi(\lambda)$ computed to four sexagesimal digits elsewhere in this particular manuscript. In his earlier works Taqi al-Dīn used 41;15° for the latitude of Istanbul (*e.g.*, in the geographical coordinates attributed to him in MS Istanbul Kandilli 441 and in his treatise *Rayḥānat al-rūḥ* on sundials in MS Vatican ar. 1424). The underlying value of ε is 23;28,54°, the distinctive parameter attributed to Taqi 'l-Dīn in other sources. See also **6.10.6** and **7.1.12** on other tables by the Istanbul astronomer.

6.4.9 Anonymous: Istanbul

MS Istanbul UL T1824,1 (fols. 3r-9r) of a set of anonymous prayer-tables for Istanbul contains tables of B(λ) computed to two digits for each degree of λ for the parameters ϕ = 41;15° (Istanbul) and ε = 23;30°. See also **6.5.2** and **6.10.7** and **II-12.8** for more information on these prayer-tables.

6.4.10 al-Dīsṭī: Lattakia

MSS Aleppo Awqāf 911 and Leiden Or. 2808(2) of al-Dīsṭī's tables for Lattakia (**4.8.7**) include tables of B(λ) and C(λ) with values to three digits for each degree of λ based on the parameters ϕ = 34;30° (Lattakia) and ε = 23;30°. See also **6.10.8**.

6.4.11 al-Jannād: various localities in the Western Maghrib

MS Cairo TR 338,2, penned *ca.* 1850, is the only known copy of a set of prayer-tables for various latitudes in the Western Maghrib (*i.e.*, what is now Morocco) compiled *ca.* 1775 by Abū ʿAbdallāh al-Jannād, an individual whose name is new to the modern literature. Details are given in **II-13.5**. Amongst the tables are some for B(λ) and C(λ) with entries to two digits for each degree of λ. The latitudes underlying al-Jannād's tables are 31° (Marrakesh, *etc.*); 34° (Meknes, Fez, *etc.*); 30° (Sijilmasa, *etc.*).

6.4.12 al-Maḥallī: Damietta

In the prayer-tables for Damietta compiled by Quṭb al-Dīn al-Maḥallī about 1650 (**5.1.4**) there are tables of B(λ) and C(λ) computed to two digits for each degree of λ and based on ϕ = 31;25° (Damietta) and ε = 23;35°. Of the various known copies of al-Maḥallī's prayer-tables only MS Cairo DM 1014, fol. 11r, contains these auxiliary tables.

6.4.13 Anonymous: Alexandria

In an anonymous set of prayer-tables preserved in MS Cairo DM 1207 (**4.8.8**) there are tables of B(λ) and C(λ) computed to two digits for each degree of λ and based on ϕ = 31;0° (Alexandria) and ε = 23;35°. The possiblity that these tables were intended for Maḥalla rather than Alexandria cannot be excluded.

6.4.14 Yūsuf Kilārjī: Crete

In MS Cairo DM 834 of the prayer-tables for Crete compiled by Yūsuf Kilārjī (**4.8.9**) there are tables of B(λ) and C(λ) computed to three digits for each degree of λ and based on ϕ = 35;30° (Crete) and ε = 23;35°.

6.4.15 Anonymous: Tlemcen

A table of the function $^1/_2$ B(λ) computed for parameters φ = 35;0° (Tlemcen) and ε = 23;35° (?) is contained in the anonymous tables for Tlemcen in MS London BL Or. 411,2. See further **6.15.2** on the main tables in this source.

6.4.16 Anonymous: Sfax

MS Cairo K 7584,1, copied in 1257 H [= 1841/42] in an elegant Maghribi hand, contains an extensive fragment of an anonymous treatise on spherical astronomy compiled in Sfax (see further **II-13.8**). The work was written after the 15th century since the obliquity used is that of Ulugh Beg and before 1257 H [= 1841/42] when the manuscript was copied. The work contains a few tables of various spherical astronomical functions including B(λ) and C(λ) for parameters:

φ = 34;48° (Sfax) and ε = 23;30,17° .

See further **6.10.13** and **8.4.2** on other tables in this source.

6.4.17 Ṣāliḥ Efendī: Istanbul

In the tables of Ṣāliḥ Efendī (**2.2.3**) the function B(λ) is tabulated to three digits for each 1° of λ. The underlying parameters are φ = 41;0° (Istanbul) and ε = 23;28,54°.

6.4.18 Anonymous: Istanbul

In several late copies of the simplified version of Ṣāliḥ Efendī's tables for timekeeping (**2.2.3**), *e.g.*, MS Istanbul Kandilli 441, there is a small set of tables appended which is entitled *jayb āfāqī*, literally, "universal Sine". In fact the tables display the four functions:

δ(λ), C(λ), B(λ) and d(λ) ,

each computed to two sexagesimal digits for each 1° of λ. The underlying parameters appear to be φ = 41;0° (Istanbul) and ε = 23;28°. The tables are arranged in three columns of thirty entries beginning with λ = 0°, so that there is no entry for λ = 90°. See also **6.10.15** on the table of C(λ).

6.4.19 ʿAbdallāh al-Ḥalabī (?): Aleppo

MS Aleppo Awqāf 943 of ʿAbdallāh al-Ḥalabī's prayer-tables for Aleppo (**6.2.7**) contains a table of B(λ) computed to three digits for each degree of λ and based on parameters φ = 35;50° (Aleppo) and ε = 23;30°.

6.4.20 al-ʿAlamī: Fez

MS Cairo TR 132,2 (pp. 2-139, *ca.* 1850) is the only known copy of a commentary by ʿAbd al-Salām al-ʿAlamī on an earlier work by ʿAbd al-ʿAzīz ibn ʿAbd al-Salām al-Warjānī entitled *Taḥrīr al-mawāqīt*, roughly "The Correct Times of Prayer", and compiled in 1142 H [= 1729/30]. Both individuals are new to the modern literature. The manuscript is datable to *ca.* 1875 and the treatment of spherical astronomy in the commentary is rather sophisticated. In particular the treatise contains tables (pp. 95 and 92) of the functions B(λ) and C(λ) computed to two digits for parameters:

φ = 34;10° (Fez) and ε = 23;29° .

See now **II-13.8*** for prayer-tables based on this latitude.

6.5 Tables for marking the absolute base on trigonometric quadrants

The purpose of the tables described below is to facilitate the marking of B(λ) on the instrument called the sine quadrant (Arabic, *al-rubʿ al-mujayyab*), a quadrant fitted with a trigonometric grid that was popular amongst Muslim astronomers from the 9th to the 19th century.[13] Complicated trigonometric problems can be solved with the sine quadrant without any calculation whatsoever. The grid consists of two orthogonal sets of equi-spaced lines drawn parallel to the axes of the grid and dividing each axis into 60 parts. The circumference is divided into 90 equal degree intervals. A thread attached to the centre of the quadrant bears a movable marker (Arabic, *murī* or *mūrī*). On the use of the instrument to compute the hour-angle from solar or stellar altitudes **6.16.4** below.

6.5.1 al-Khalīlī: Damascus

Tables of B(λ) for the latitude of Damascus do not occur in the known manuscripts of tables due to al-Khalīlī, although in his smaller set of auxiliary tables (**9.4**) he tabulates $^1/_2$ B(λ) for each degree of latitude and also ϕ = 33;30°.

However, MSS Oxford Seld. Supp. 100, fol. 16v, and Paris BNF ar. 2558, fol. 33r, contain two tables by al-Khalīlī, presented without instructions regarding their use. The two tabulated functions, having λ as argument, are called respectively *qaws al-aṣl bi-'l-mūrī*, "arc of the base (for a quadrant) with a marker" and *qaws al-aṣl bi-ghayr mūrī*", arc of the base (for a quadrant) without a marker". I denote these functions by b_1 and b_2 and find by inspection that:

$$b_1(\lambda) = \text{arc Sin} \{ B(\lambda) \} \quad \text{and} \quad b_2(\lambda) = \text{arc Tan} \{ B(\lambda) \}$$

for al-Khalīlī's values of ε and ϕ for Damascus.

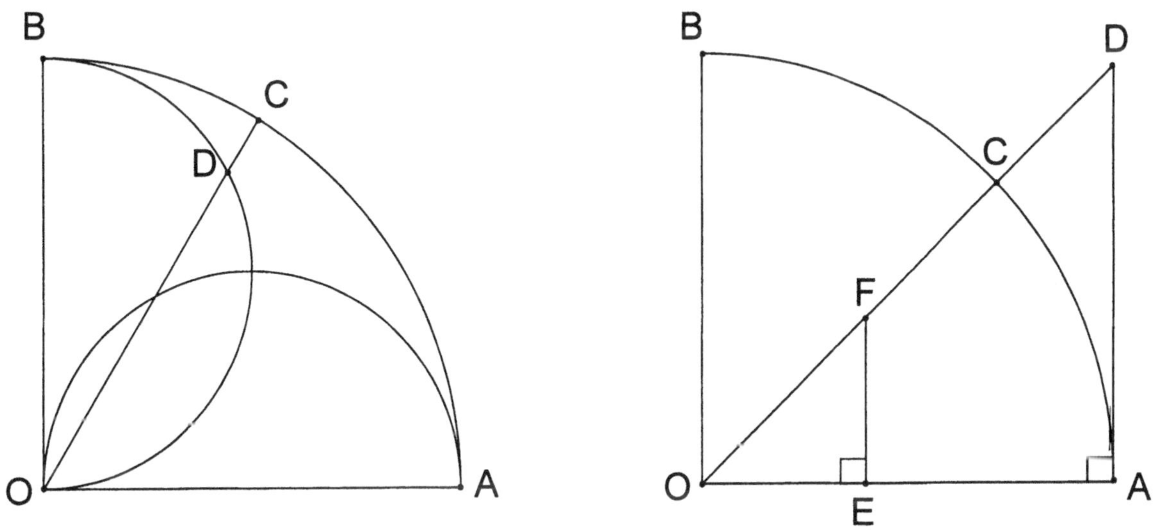

Fig. 6.5.1a-b: Diagrams illustrating the use of al-Khalīlī's tables for marking the absolute base on a sine quadrant.

[13] On the sine quadrant, see, for example, Schmalzl, *Geschichte des Quadranten*; Würschmidt, "Gedosi über den Quadranten"; and King, "al-Khwārizmī", pp. 28-29. The relatively complicated operation of finding the qibla with a sine quadrant is discussed in detail in King, "al-Khalīlī's Qibla Table", pp. 111-115.

The first function $b_1(\lambda)$ is intended to be used with a sine quadrant (radius R = 60) marked with two special semi-circles (Arabic, *dā'irat al-tajyīb*) for finding Sines, drawn with the two axes of the quadrant as diameters. In **Fig. 6.5.1a** if the thread OC cuts off the arc AC equal to $b_1(\lambda)$ then OD equals Sin $b_1(\lambda)$, that is, $B(\lambda)$. Thus the marker can be moved to D and the quadrant is ready to be used for computations of time from solar altitude which involve a division by $B(\lambda)$.

The second function $b_2(\lambda)$ is intended to be used with a Sine quadrant bearing the usual grid and thread, but no marker. The use of this instrument is described by al-Khalīlī in MS Cairo MM 201, fols. 11v-12r, which also contains the table of $b_2(\lambda)$ (fol. 12r). His treatise on this quadrant is of considerable interest but I here restrict attention to the use of the table of $b_2(\lambda)$. In **Fig. 6.5.1b** the thread OC is moved to the position such that the arc AC equals $b_2(\lambda)$. To divide any quantity, say x, by $B(\lambda)$, simply measure EF equal x and then, because AD equals Tan $b_2(\lambda)$, that is, $B(\lambda)$, OE equals Rx / $B(\lambda)$. Such a simple operation is clearly extremely useful in timekeeping problems.

6.5.2 Anonymous: Istanbul

In MS Istanbul UL T1824,1 of a set of prayer-tables for Istanbul (**6.4.9**) the function called *qaws al-aṣl*, "arc of the base", is in fact $b_1(\lambda)$ as defined above. Values of $B(\lambda)$ and $b_1(\lambda)$ are given to two digits for each degree of λ and the underlying parameters are $\phi = 41;15°$ (Istanbul) and $\varepsilon = 23;30°$. There are no instructions accompanying the tables in the manuscript and I assume that the purpose behind the table of $b_1(\lambda)$ was to facilitate marking $B(\lambda)$ on a sine quadrant.

6.5.2* Anonymous: Maghrib

See **II-13.8*** for some tables of the same kind, probably for Fez.

6.6 Tables of the absolute base for timekeeping by the stars

The existence of a table of $B(\Delta)$ for the latitude of Mecca has been noted in **6.4.4**. Since the argument runs only as far as $\Delta = \varepsilon$ the table was clearly intended to be used for timekeeping by the sun. No other tables of $B(\Delta)$ for particular latitudes which would be suitable for timekeeping by the stars have been located. See, however, **6.16.4** below. Note also that al-Ṣūfī's second auxiliary function (**9.10**) actually defines $B(\phi,\Delta)$ for each degree of both arguments, and that al-Māridīnī's second auxiliary function M_2 (**9.6**) can be used to find $B(\phi,\Delta)$ with facility, namely, by simply using $\bar{\phi}$ and $\bar{\Delta}$ as arguments.

6.7 Tables of the reciprocal of the absolute base and related functions for timekeeping by the sun

It is rather curious that we find tables of the function $B(\lambda)$ rather than its reciprocal $G(\lambda)$ in the Islamic sources. As yet I have not come across any tables of G specifically intended for timekeeping by the sun – see, however, **6.8**. The following tables of functions related to $G(\lambda)$ have been located.

6.7.1 Anonymous: Cairo

In the *Ḥākimī Zīj* Ibn Yūnus suggests the method outlined in **2.1.1** above for finding t from h and λ. The function G he calls *al-nisba*, "the ratio", since:

$$G(\lambda) = R^2 / [\text{Cos } \delta(\lambda) \text{ Cos } \phi] = \text{Vers } D(\lambda) / \text{Sin } H(\lambda) .$$

I have not yet located a table of $G(\lambda)$ for the latitude of Cairo in any of the manuscripts containing material ultimately due to Ibn Yūnus.

In one such source, however, containing a recension of the anonymous 13th-century Egyptian *Muṣṭalaḥ Zīj*,[14] MS Paris BNF ar. 2513, fols. 62r and 62v, copied in the 13th century, there is a table of at least one related function. The instructions in fact mention two related functions and indicate that the first, called *nisbat al-iʿtidāl*, "the ratio at the equinoxes" ($\delta = 0°$), is:

$$G_1(\theta) = R / \text{Cos } \theta .$$

However, the tabulated function is not $G_1(\theta)$ and I am unable to explain it. The table bears no title and the argument column is not visible on the microfilm which I have used. The 16 entries in the table are as follows:

13;11, 9	15;23,35	21;43,30	33;30,26
13;14,17	16;24,43	24; 0,30	42; 0,34
13;34,19	17;51,51	26;22,30	49;30,17
14;20,34	19;16, 0	28;45, 0	57; 0,53

This first table may be unrelated to spherical astronomy.

The second table accompanying these instructions is inappropriately entitled *sahm faḍl al-dāʾir*, "Versed Sine of the hour-angle". The function tabulated is in fact:

$$G_2(\lambda) = R / \text{Cos } \delta(\lambda) .$$

Values are given to three digits but are badly garbled. It is not possible to determine the value of ε underlying the table. The instructions indicate that the "ratio" may be found from these tables using either:

$$G(\lambda,\phi) = G_1(\phi) \cdot G_1(\delta) \quad \text{or} \quad G(\lambda,\phi) = G_1(\phi) \cdot G_2(\lambda) .$$

6.7.2 al-Marrākushī: Cairo

In the treatise of al-Marrākushī (**4.2.4**), there is a set of values (I.264) of a function $G_3(\delta)$, as follows:

δ	8°	13°	16°	18°	20°	22°	23;35°
G_3	9	10	11	12	13	14	15

This function is simply:

$$G_3(\delta) = R^3 / [\text{Cos } \delta \text{ Cos } \phi] - R$$

for latitude 30° and its use is explained in **9.12**, where I describe a table of G_3 for all declinations and latitudes.

[14] On the *Muṣṭalaḥ Zīj* see Kennedy, "Zīj Survey", no. 47; *Cairo ENL Survey*, no. C12; King, "Lunar Equation Table", p. 135; and *idem*, "Astronomy of the Mamluks", pp. 535-536. See also **Fig. I-9.8a**.

6.8 Tables of the reciprocal of the absolute base for timekeeping by the sun and stars

The reader will by now have gained the impression that the Muslim astronomers tabulated virtually every function that was of any use in computing the hour-angle from solar or stellar altitudes. Yet I have not come across any Islamic tables of the function:

$$G = \text{Vers } D \ / \ \text{Sin } H = R^2 \ / \ [\ \text{Cos } \Delta \ \text{Cos } \phi\ ] \ ,$$

either for a specific latitude or for all latitudes, such as might have been used to facilitate the computation of a *ṭaylasān* table of T(H,h) (**2.3**) or a universal table of T(H,h,D) (**2.6**). Note that using such a table of G(H) for a particular latitude or G(H,D) for all latitudes, along with a table of the function:

$$H'(H,h) \ = \ \text{Sin } H \ - \ \text{Sin } h \ ,$$

such as compiled by Ibn al-Mushrif (**6.13.1**), the computation of the hour-angle reduces to (*cf.* **F9**):

$$\text{Vers } t \ = \ H' \cdot G \ .$$

On the other hand, two tables of the function G(Δ) have come to light and are described below. The first was compiled for Cairo, where, however, as far as we know, no *ṭaylasān* tables were prepared. The second was compiled for Tunis and may have been used to compute a set of tables for reckoning time by stellar altitudes. Note also that al-Wafāʾī's first auxiliary function (**9.9**) defines G(Δ,φ) for each degree of Δ and some 13 different values of φ.

6.8.1 Anonymous: Cairo

In the Egyptian copy of al-Khalīlī's *Universal Table* MS Cairo MM 43 (**6.4.3**) there is a table entitled *jadwal al-dāʾir*, "table for finding time". The function tabulated is simply G(Δ) computed to two digits for:

$$\phi = 30° \text{ (Cairo)} \quad \text{and} \quad \Delta = 1°, 2°, \ldots , 60° \ (= \bar{\phi}).$$

A page of anonymous tables in MS Oxford Seld. Supp. 99, fol. 63v, at the end of the prayer-tables for Cairo by al-Fawānīsī (**4.8.4**) contains two incomplete tables entitled *jadwal al-dāʾir min al-falak* and *jadwal al-dāʾir min al-sahm*. The first of these displays G(Δ) for latitude 30°, computed to two digits for each degree of argument (I have not been able to compare the entries with those in the Cairo manuscript) and the second displays arc Vers (x) (see further **6.14.2**).

6.8.2 Anonymous: Tunis

MS Berlin Ahlwardt 5724, fol. 42v, of the anonymous corpus of tables for Tunis based on φ = 36;40° (**2.3.5**) contains a table of G(Δ) computed to three digits for this value of φ and arguments Δ = 1°, 2°, ... , 60°. The table is entitled *jadwal aṣl faḍl dāʾir al-shams wa-'l-kawākib*, "table of the base (for calculating) the hour-angle of the sun and stars" – see **Fig. 6.8.2**. This table may have been used to compile the tables of t(h) for various stars that occur on fols. 55r-56r of the Berlin manuscript (**2.4.1**).

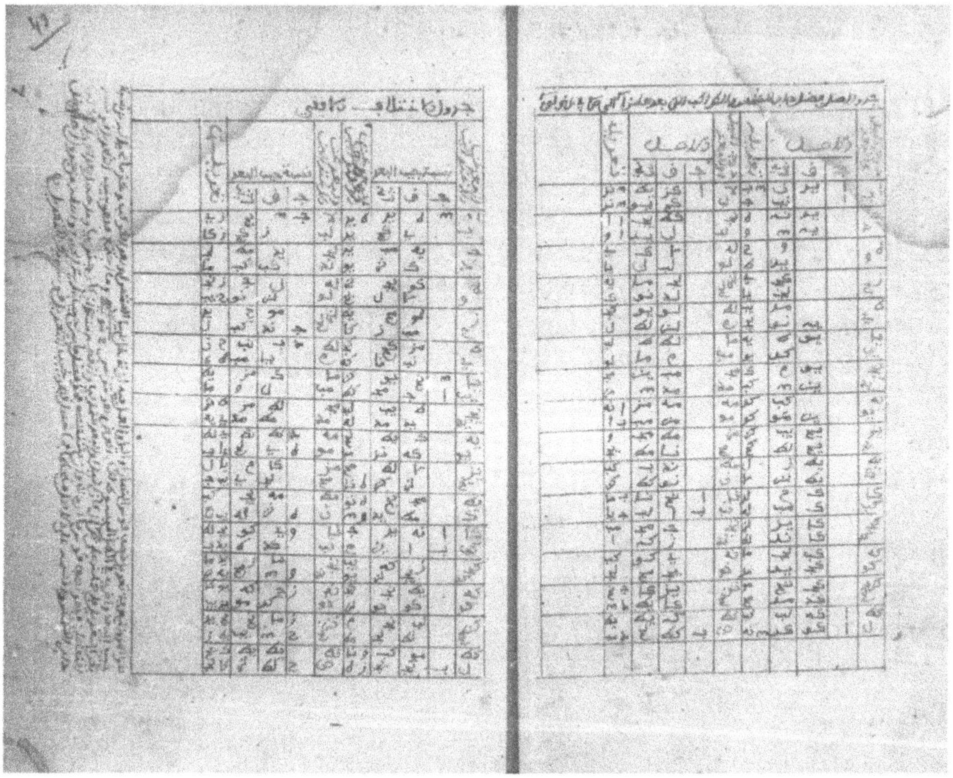

Fig. 6.8.2: The table of the reciprocal of the absolute base in the Tunis corpus, together with a table of the sine of the declination labelled as if it were an auxiliary azimuth function (see **7.1.5**). [From MS Berlin Ahlwardt 5724, fols. 42v-43r, courtesy of the Deutsche Staatsbibliothek (Preußischer Kulturbesitz).]

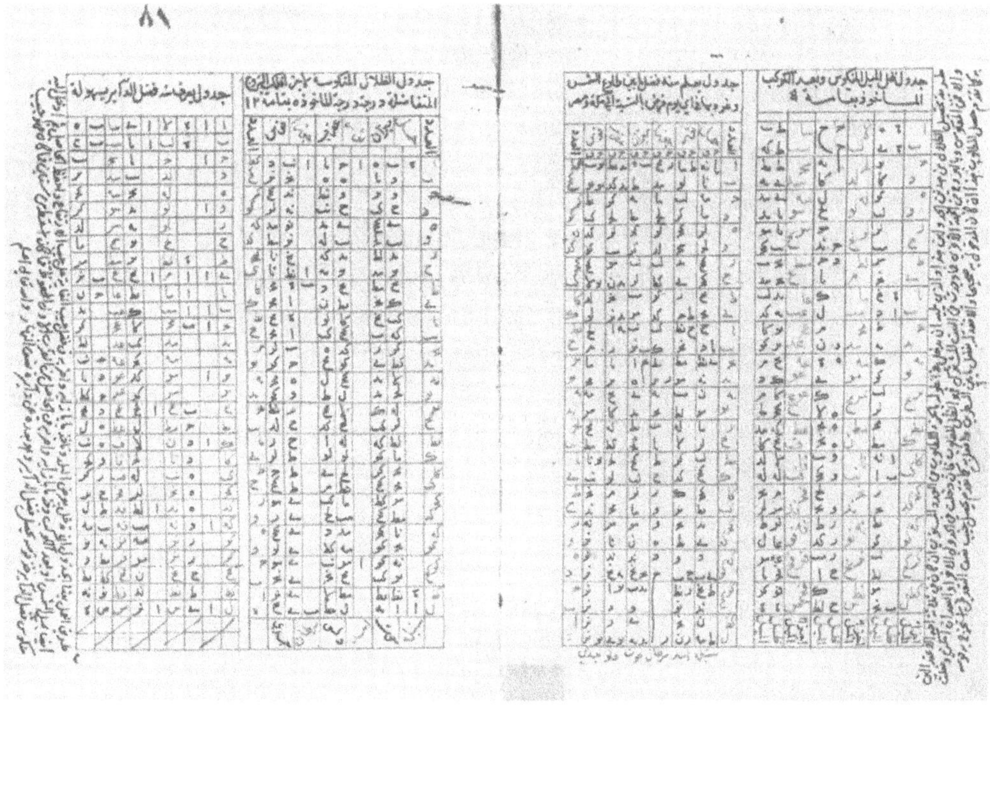

6.9 Tables of the secant function for timekeeping

The Muslim astronomers did not generally exploit the standard trigonometric functions other than the sine and (co-)tangent.[15] However, a table of the cosecant function is found in one version of the *Zīj* of the early-9th-century astronomer Ḥabash and others are found in the sets of tables used by the Muslim astronomers for marking curves on astrolabes and quadrants (see further **9.2**). Also, a few simple tables of the secant function have recently come to light but they are not referred to by the standard, if uncommon, Arabic term *quṭr al-ẓill*, "hypotenuse of the (horizontal) shadow" (*i.e.*, the length of the hypotenuse of the triangle formed by a vertical gnomon and its shadow on a horizontal plane). Rather, they are considered as auxiliary tables for timekeeping. See also **8.4** for tables of the secant function intended for azimuth calculations.

I denote the secant function as found in the Arabic sources by G_1, where:

$$G_1(\theta) = R / \text{Cos}\ \theta .$$

It can be used to find the auxiliary function G (**6.7** and **6.8**) using the relation:

$$G(\Delta,\phi) = G_1(\Delta) \cdot G_1(\phi) .$$

Thus, for example, in the text of the *Zīj* of Ibn al-Bannāʾ (**6.9.4**) the formula outlined for finding t from h, Δ and ϕ is equivalent to (*cf.* **F9**):

$$\text{Vers}\ t(h,\Delta,\phi) = (\ \text{Sin}\ H - \text{Sin}\ h\ ) \cdot G_1(\phi) \cdot G_1(\Delta) .$$

Note that each of the four tables of G_1 listed below contains the same significant error in the entry for $\theta = 60°$. (Compare the Cosecant table of Ḥabash discussed in **9.1**.) All of these tables merit a new investigation.

6.9.1 Ibn al-Zarqālluh (Toledo)

The Andalusī astronomer Ibn al-Zarqālluh (*fl.* Toledo, *ca.* 1060) compiled a set of auxiliary tables for generating planetary ephemerides known as the *Almanach of Azarquiel*. These have been published by José-Maria Millás Vallicrosa, using the unique Arabic copy MS Munich ar. 853 and a Catalan version of the *Almanach*.[16] One of the spherical astronomical tables in the set (p. 226) is entitled *jadwal fuḍūl khaṭṭ al-madār ʿalā 'l-irtifāʿ*, literally, "table of the excesses of the line of the day-circle over the altitude". (This makes no sense either in Arabic or English and the Arabic is doubtless corrupt.) The function is in fact $G_1(\theta) = R / \text{Cos}\ (\theta)$, and values are given to three digits for each 3° of θ and also 88° and 89°. In view of the importance of this table and since Millás' version can hardly be considered an accurate representation at least of the Arabic version, I reproduce the entire table from MS Munich ar. 853, fol. 44v,

←
Fig. 6.9.3: A table of the reciprocal of the absolute base (far left) amidst some anonymous tables for Cairo. Two other tables display the tangent of the declination and the tangent of the solar declination (both omitted from **7.1**!). Another displays for each degree of solar longitude the difference between the lengths of daylight in Cairo and Mecca, which should be investigated in the light of the acompanying treatise. [From MS Cairo MM 204,7, fols. 80v-81r, courtesy of the Egyptian National Library.]

---

[15] On Islamic trigonometric tables in general see n. 1.23.

[16] Millás, *Azarquiel*, pp. 72-237. More recent studies on Ibn al-Zarqālluh include Toomer, "Solar Theory of al-Zarqāl", p. 331; and Julio Samsó and Eduardo Millás, "Ibn al-Bannāʾ, Ibn Isḥāq and Ibn al-Zarqālluh's Solar Theory", in Samsó, *Studies*, X. See also the article "al-Zarqālī" by Juan Vernet in *DSB*.

The main part of the table is as follows. (The errors in the third digit are indicated in square brackets.)

3°	1;0, 5[a]	[0]	27	1; 7,21	[+1]	51	1;35,27	[+7]
6	1;0,20	[0]	30	1; 9;17	[0]	54	1;42, 2	[-3]
9	1;0,45	[0]	33	1;11,32	[0]	57	1;50,10	[0]
12	1;1,21	[+1]	36	1;14,10	[0]	60	2; 1, 2	[!!!][b]
15	1;2, 5	[-2]	39	1;17,11	[-1]	63	2;12[c],10[d]	[0]
18	1;3, 6	[+1]	42	1;20,44	[0]	66	2;27,31	[0]
21	1;4,16	[0]	45	1;24,51	[0]	69	2;47,27	[+1]
24	1;5;40	[-1]	48	1;29,40	[0]	72	3;14,10	[0]

[a] Text: 30, Millás: 5. [b] See below. [c] Text: 17, Millás: 12. [d] Text and Millás: 2

The remaining entries in the table in the Munich manuscript, compared with Millás' entries and the accurate values of Sec θ, are as follows:

Text		Millás		Accurate	
75°	3;51,50	75°	3;51;50	75°	3;51;49
78	4;48,21	78	4;48,21	78	4;48,35
81	5;23,33(!)	81	6;23,33	81	6;23,33
84	6;14, 1(!)	84	9;34, 1	84	9;34, 0
87	7; 6,30(!)	87	19; 6,30	87	19; 6,25
88	8;18,42(!)	90(!)	29,18,42	88	28;39,11
89	9;17,42(!)	- (!)	56;17,40	89	57;17,56
90	-; 0,47(!!)			90	infinite

The last few entries in the Munich manuscript are thus badly garbled. Of considerable interest is the entry for θ = 60°, namely 2;1,2 rather than the accurate 2;0,0. It is difficult to account for this error in view of the fact that most of the other entries are fairly carefully computed. The tables of $G_1(\theta)$ discussed in **6.9.3-5** have the entry 2;1 for θ = 60°.

6.9.1* Ibn Isḥāq (Tunis)

MS Hyderabad Āṣafiyya 298 of the *Zīj* of Ibn Isḥāq, located only in 1978, is an extremely important new source for the history of Islamic astronomy.1 It contains a table of the function $R^2/\text{Cos}(\theta)$ labelled *nisbat al-madārāt* (table no. 90 in the manuscript). Values are given to two significant sexagesimal digits, and the entry for argument 60 is 1;1, in error for [2];1. See also **7.1.5*** and **II-13.1**.

6.9.2 Anonymous (Cairo)

The table in MS Paris BNF ar. 2513 of the *Muṣṭalaḥ Zīj* which purports to be of the function $G_1(\theta)$ has already been discussed in **6.7.1**.

[17] In the 1980s all one that was known about Ibn Isḥāq was what was recorded in Suter, *MAA*, no. 356, and King, *Astronomy in Yemen*, p. 83. His *zīj* is not listed in Kennedy, "*Zīj* Survey". The Zabid manuscrip of the *zīj* of the Yemenī astronomer al-Daylamī (**4.3.4**) contains some planetary tables due to Ibn Isḥāq. MS Milan Ambrosiana C82 of the perpetual almanac of Zacuto (**4.3.7**) contains some spherical astronomical tables due to him. On the Hyderabad manuscript see King, "Astronomy in the Maghrib", p. 32; Samsó, "Maghribī *Zīj*es", p. 93; and, more especially, Mestres, "Hyderabad MS of the *Zīj* of Ibn Isḥāq", and *idem*, Zīj *of Ibn Isḥāq*. In the new edition of Mestres (*ibid.*, pp. 275 and 279-281), a value 2;0 occurs at argument 61°, also in the recomputation! The values in the table for 59°-61° are 1;57, 1;1 (*sic*) and 2;5.

6.9.3 Anonymous (Cairo)

In the anonymous set of spherical astronomical tables for Cairo contained in MS Cairo MM 204,7, copied in 1052 H [= 1642/43] and described in **II-4.2**, there is one entitled *jadwal yuʿraf minhu faḍl al-dāʾir bi-suhūla*, "table for finding the hour-angle easily" (fol. 81r): see **Fig. 6.9.3**. The function tabulated is simply $G_1(\theta)$, computed to two digits for each degree of θ, and the entry for θ = 60° is 2;1. The tables are appended to an anonymous treatise on finding the times of sunrise and sunset in localities other than one's own (fols. 77v-80r).

6.9.4 Ibn al-Bannāʾ (Marrakesh)

In the *Zīj* of Ibn al-Bannāʾ,[18] compiled in Marrakesh about 1300, there is a small table entitled *nisbat al-madārāt*, literally, "the ratio of the day-circles". I have consulted MSS Dublin CB 4087, fol. 46r, Escorial ár. 909,1, fol. 36v, and fol. 34v of an unnumbered manuscript in the Museo Naval, Madrid. The function tabulated is $G_1(\theta)$, computed to two digits for each 3° of θ, and the entry for θ = 60° is 2;1. See also **6.9.5** below.

6.9.5 Ibn al-Raqqām (Tunis)

In MS Istanbul Kandilli 249, copied *ca.* 1800, of the *Shāmil Zīj* as well as in the unnumbered manuscript in the Museo Naval, Madrid (fol. 41r), of the *Qawīm Zīj*, both compiled by the early-14th-century Tunisian astronomer Ibn al-Raqqām,[19] there is a table entitled *nisbat al-madārāt* (**6.9.4**) displaying $G_1(\theta)$ to two digits for each degree of θ. The entry for θ = 60° is again 2;1.

6.10 Tables of the altitude of the solar day-circle centre and related functions

Several sources listed below contain tables of:

$$C(\lambda) = \text{Sin}\ \delta(\lambda)\ \text{Sin}\ \phi\ /\ R$$

for particular latitudes. This function is called *buʿd al-quṭr* in Arabic, literally, "the distance of the diameter", and is easily derived from Sin δ(λ) (**6.1**).

6.10.1 Anonymous: Cairo

In MS Berlin Ahlwardt 5710, fol. 24r, of the prayer-tables of Riḍwān Efendī (**6.1.3**), there is a table of C(λ) to three digits for each degree of λ. The underlying parameters are φ = 30;0° (Cairo) and ε = 23;35°. This table, which is very carefully computed, was doubtless lifted from some earlier source, like most of Riḍwān's tables. His table of the function B(λ) is the same

[18] On Ibn al-Bannāʾ see Suter, *MAA*, no. 399; and *Cairo ENL Survey*, no. F23; as well as Renaud, "Ibn al-Bannāʾ". His *zīj*, not listed in Kennedy, "*Zīj* Survey", is entitled *Minhāj al-ṭālib* and is extant in several manuscripts. The text of the introduction to the *zīj* and a Spanish translation were published in Vernet, *Ibn al-Bannāʾ*, and the planetary tables have recently been studied in Samsó & Millás, "Ibn al-Bannāʾ on Planetary Longitudes".

[19] On Ibn al-Raqqām see Suter, *MAA*, nos. 388 and 417; and *Cairo ENL Survey*, no. F22. His two *zīj*es are not listed in Kennedy, "*Zīj* Survey", but see now *idem*, "Ibn al-Raqqām's Tables". See also Carandell, *Ibn al-Raqqām sobre los cuadrantes solares*.

as that of al-Khaṭāʾī (**6.4.2**), but there is no corresponding table of C(λ) in MS Vatican Borg. ar. 217,2 of al-Khaṭāʾī's set. Note that for $\phi = 30°$:

$$C(\lambda) = {}^1/_2 \text{ Sin } \delta(\lambda) .$$

6.10.2 Anonymous: Cairo

MS Istanbul Nuruosmaniye 2925 of al-Bakhāniqī's edition of the Cairo corpus (**2.1.1** and **6.2.4**) contains a table of C(λ) for Cairo. The function is tabulated to two digits for the same parameters as the table in **6.10.1**. MS Alexandria 4441J of the prayer-tables for Cairo of Ibn Abī Rāya (**6.2.4**) contains a corrupt version of this same table.

6.10.3 Anonymous: Cairo

In MS Paris BNF ar. 2513, fol. 53v, of the anonymous recension of the Egyptian *Muṣṭalaḥ Zīj* (**6.7.1** and **II-6.6**), there is a table of the function:

$$\text{arc Sin } \{C(\lambda)\}$$

entitled *irtifāʿ madār al-shams*, "altitude of the (centre of the) solar day-circle". The entries are given to two digits and are based on the parameters ϕ = 30;0° (Cairo) and ε = 23;35°. The function measures the altitude in degrees of a point on the celestial sphere at the same distance above the horizon as the centre of the solar day-circle. This corresponds to the altitude of the sun when the hour-angle is 90° (for $\delta > 0$ only). I can think of no good reason for tabulating this function except to facilitate marking C(λ) on a Sine quadrant – see **6.5** above.

6.10.4 Anonymous: Anatolia

MS Istanbul Süleymaniye 1037/32 (**6.4.6**) contains a table of C(λ) to two digits for each degree of λ, based on the parameters ϕ = 38;30° (Anatolia ?) and ε = 23;35°.

6.10.5 Anonymous: Tunis

In MS Princeton Yahuda 147c, p. 113, of the Tunisian *Quṣʿī Zīj* of Ḥusayn Quṣʿa (**6.4.7**) there is a table of C(λ) for each degree of λ computed to two digits for the parameters ϕ = 36;40° (Tunis) and ε = 23;35°.

6.10.6 Taqi 'l-Dīn: Istanbul

In MS Istanbul Esat Efendi 1976, fol. 25r, of the *Zīj* of Taqi 'l-Dīn (**6.4.8**), C(λ) is tabulated to four significant *decimal* digits of λ and the entries are apparently based on the parameters ϕ = 40;58° (Istanbul) and ε = 23;28,54°.

6.10.7 Anonymous: Istanbul

In MS Istanbul UL T1824,1 (**6.4.9**) C(λ) is tabulated to two digits for each degree of λ and parameters ϕ = 41;15° (Istanbul) and ε = 23;30°.

6.10.8 al-Dīsṭī: Lattakia

MSS Aleppo Awqāf 911 and Leiden Or. 2808(2) of al-Dīsṭī's prayer-tables for Lattakia (**6.4.10**) contains a table of C(λ) with values to two digits for each degree of λ for the parameters ϕ = 34;30° (Lattakia) and ε = 23;30°.

6.10.9 al-Jannād: various localities in the Western Maghrib

MS Cairo TR 338,2 of al-Jannād's prayer-tables for the Western Maghrib (**6.4.11**) contain tables of C(λ) for latitudes φ = 30°, 31° and 34°.

6.10.10 al-Maḥallī: Damietta

Quṭb al-Dīn al-Maḥallī (**6.4.12**) tabulated C(λ) to two digits for each degree of λ and parameters φ = 31;25° (Damietta) and ε = 23;35°.

6.10.11 Anonymous: Alexandria

The anonymous set of prayer-tables preserved in MS Cairo DM 1207 (**6.4.13**) contains a table of C(λ) computed to two digits for each degree of λ and based on the parameter φ = 31;0° (Alexandria).

6.10.12 Yūsuf Kilārjī: Crete

Yūsuf Kilārjī (**6.4.14**) tabulated C(λ) to three digits for each degree of λ for the parameters φ = 35;30° (Crete) and ε = 23;35°.

6.10.13 Anonymous: Sfax

MS Cairo K 7584,1 (**6.4.16**) contains a table of C(λ) on the parameters φ = 34;48° (Sfax) and ε = 23;30,17°.

6.10.14 Ṣāliḥ Efendī: Istanbul

Ṣāliḥ Efendī (**6.4.17**) also tabulated C(λ), giving values to three sexagesimal digits for the parameters φ = 41;0° (Istanbul) and ε = 23;28,54°.

6.10.15 Anonymous: Istanbul

MS Istanbul Kandilli 441 (**6.4.18**) contains a table of C(λ) apparently based on parameters φ = 41;0° (Istanbul) and ε = 23;28°.

6.10.16 ʿAbdallāh al-Ḥalabī (?): Aleppo

MS Aleppo Awqāf 943 of the prayer-tables attributed to ʿAbdallāh al-Ḥalabī (**6.4.19**) contains a table of C(λ) for parameters φ = 35;50° (Aleppo) and ε = 23;30°.

6.10.17 al-ʿAlamī: Fez

MS Cairo TR 132,2 of the treatise by ʿAbd al-Salām al-Ḥusnī al-ʿAlamī (**6.4.20**) contains a table of C(λ) computed to two digits for each degree of λ and based on parameters φ = 34;10° (Fez) and ε = 23;29°.

6.11 Tables of the height of the centre of the arc of visibility for timekeeping by the stars

No tables of C(Δ) for particular latitudes have been located in the manuscript sources. (See, however, **6.16.4** below where C is tabulated for specific stars.) Note that both al-Ṣūfī's first auxiliary function (**9.10**) and al-Māridīnī's second auxiliary function M_2 (**9.6**) define $C(\phi,\Delta)$ for each degree of both arguments.

6.12 Tables of the modified base

The function:

$$b(h,\lambda) = \text{Sin } h - C(\lambda) \ ,$$

called *al-aṣl al-muʿaddal*, "the modified base", is useful in timekeeping because the hour-angle $t(h,\lambda)$ is defined in terms of b by the simple relation (**6.0**):

$$\text{Cos } t(h,\lambda) = R \cdot b(h,\lambda) \ / \ B(\lambda) = b(h,\lambda) \cdot G(\lambda) \ .$$

In only one set of hour-angle tables has the compiler left us the set of tables of $b(h,\lambda)$ which he used to compute the hour-angle.

6.12.1 Ṣāliḥ Efendī: Istanbul

In the tables of Ṣāliḥ Efendī for Istanbul (**2.2.3** and **Fig. 2.2.3**), the function $b(h,\lambda)$ is displayed to three digits for each degree of λ and each value of h up to $[H(\lambda)]$. Ṣāliḥ Efendī also tabulated $B(\lambda)$ and $C(\lambda)$ to three digits for each degree of λ (**6.4.17** and **6.10.14**), but not $G(\lambda)$.

6.13 Tables of the difference between the sines of the meridian altitude and the instantaneous altitude

The function:

$$H'(H,h) = \text{Sin } H - \text{Sin } h$$

is useful in timekeeping because the hour-angle $t(H,h)$ is defined in terms of H' by the simple relation (**6.0** and **6.8**):

$$\text{Vers } t(H,h) = H'(H,h) \cdot G(H) \ .$$

Only one table of this function has come to light.

6.13.1 Ibn al-Mushrif (Cairo)

The early-15th-century Egyptian astronomer Ibn al-Mushrif (**9.8**) tabulated $H'(H,h)$ to two digits for the domains:

$$H = 1°, 2°, \ldots , 90° \quad \text{and} \quad h = 1°, 2°, \ldots , (H\text{-}1) \ .$$

These tables are contained in MS Cairo MM 241 of his auxiliary tables.

6.14 Tables of the inverse versed sine function

Tables of inverse trigonometric functions are comparatively rare in the Islamic sources. al-Marrākushī (**4.2.4**) tabulated the inverse Sine and Cotangent (base 12) (I.121-124 and 168-169), and a later Egyptian copy of al-Khalīlī's *Universal Table*, MS Cairo MM 43 (**9.5**), contains tables of the inverse Sine and Tangent (base 12). Again, a 14th-century copy of the main Cairo corpus, MS Dublin CB 3673, fol. 8r (**2.1.1**), and also MS London BL Or. 3624, fol. 169v, of the *Mukhtār Zīj* of Abu 'l-ʿUqūl (**2.1.2**) contain tables of the inverse Sine function, the latter probably taken from an earlier source. Tables of inverse spherical astronomical functions are likewise rare. MS Cairo MM 43, mentioned above, contains a table of the function λ(δ) with values to two digits for each degree of δ, and MS Cairo MM 58, fol. 7r, contains a similar table with argument increment 0;15°.

In timekeeping the most useful tables of inverse trigonometric functions would be for the Cosine and Versed Sine. No tables of the former have come to my attention, but notice that al-Khalīlī's auxiliary function K and al-Ṣūfī's function S (**9.5** and **9.10**) perform the final operation for finding t(h,D). Several tables of the inversed Versed Sine have been located, and the auxiliary functions V and V′ tabulated by al-Khaṭāʾī, al-Khalīlī and Ibn al-Mushrif (**6.15.1**, **9.4** and **9.8**) likewise perform the final operation for finding t(h,H).

6.14.1 al-Wafāʾī (Cairo)

In his auxiliary tables preserved in MSS Istanbul Nuruosmaniye 2921,2 (fols. 22r-26v) and Vatican Borg. ar. 217,1 (fols. 1v-5v), the 15th-century Egyptian astronomer al-Wafāʾī (**9.9**) tabulated arc Vers (x) to two digits for arguments:

x = 0;5, 0;10, ... , 0;30, 0;40, ... , 2;0, 2;15, ... , 3;0, 3;20, ... ,
4;0, 4;30, ... , 16, 17, ... , 120.

The table is entitled *jadwal faḍl al-dāʾir*, "hour-angle table".

6.14.2 Anonymous (Cairo)

Several other Egyptian sources contain tables of arc Vers (x) to two digits for x = 1, 2, ... , 120. Such tables are contained in, for example, MS Oxford Seld. Supp. 99, fol. 63v, of al-Fawānīsī's prayer-tables for Cairo (**6.8.1**), where the function is called *al-dāʾir min al-sahm*, "time from the Versed Sine", and MS Cairo MM 203 of an anonymous set of auxiliary tables (**9.12**), where it is called *qaws al-sahm*, "the arc of the Versed Sine". I have not compared the entries in these two tables.

6.14.3 Anonymous (Damascus)

MS Istanbul Hamidiye 1453 (fols. 232v-266v) of a 15th-century Turkish copy of al-Khalīlī's *Universal Table* (**9.5**) contains (fol. 266r) tables of arc Sin (x) and arc Vers (x), the latter being referred to as *faḍl al-dāʾir min qibal al-sahm*, "the hour-angle as a function of the Versed Sine". See also **7.0**.

6.14.4 Anonymous (Aleppo)

MS Aleppo Awqāf 943, copied in Aleppo *ca.* 1850, contains an anonymous set of tables of the function:

$$V'(x,y) = \text{arc Vers}\,\{\, R \cdot y \,/\, x \,\}$$

for arguments x = 35, 36, ... , 60 and y = 1, 2, ... , 60. The function is referred to as *faḍl al-dāʾir*, "the hour-angle", and the two arguments are labelled *al-aṣl*, "the base", and *mā bayn al-jaybayn*, "the difference between the two Sines". See **9.8** and also **6.2.7** on other tables in this Aleppo manuscript.

Fig. 6.14.3: Tables of the inverse Sine and Versed Sine functions and a table of the maximum half daylight for all latitudes (based on obliquity 23;35°) appended at the end of a copy of al-Khalīlī's universal auxiliary tables (see **Fig. I-9.5a-b**), of which the last few sub-tables for his third function are shown on the right. From MS Istanbul Ḥamidıye 1453, fols. 265v-266r, courtesy of the Süleymaniye Library, Istanbul.]

6.15 Tables of the hour-angle as a function of the absolute base

I now describe a table for finding the hour-angle which has the absolute base B as one argument. This is devised for a particular latitude. Note that the auxiliary tables of al-Khalīlī for timekeeping by the sun (**9.4** and also **9.7** and **9.11**) have $B' = \frac{1}{2}$ B as one argument but serve

Fig. 6.15.1. An extract from the auxiliary tables of al-Khaṭāʾī, showing the surviving part of the table of V′. [From MS Vatican Borg. ar. 217,2, courtesy of the Biblioteca Apostolica Vaticana.]

all latitudes. Likewise the later auxiliary tables of Ibn al-Mushrif and al-Ṣūfī (**9.8** and **9.10**) have B as one argument and serve all latitudes.

6.15.1 al-Khaṭāʾī: Cairo

MS Vatican Borg. ar. 217,2 (fols. 6r-7r), copied *ca.* 1500, contains part of a set of auxiliary tables intended for use in Cairo, attributed to Muḥammad ibn al-Amīr Fakhr al-Dīn ʿUthmān al-Khaṭāʾī, an individual whose name is new to the history of Islamic science. His father was a Mamluk prince who flourished around 1450.[20] (The other tables in this manuscript are by al-Wafāʾī: see **9.9**.) The main function tabulated by al-Khaṭāʾī is discussed below. We first consider the smaller tables of the standard functions which accompany the main set. Those having solar longitude as argument have the same format as the tables for timekeeping in the main Cairo corpus (**2.1.1** and **5.1.1**) and the underlying parameters are the same, namely, $\phi = 30;0°$ and $\varepsilon = 23;35°$.

(a) A table of Sin θ to three digits for each degree of θ.

(b) A table of Sin H(λ) to three digits (**6.3.1**).

(c) A table of B(λ) to three digits (**6.4.2**).

(d) A table of the solar altitude at the beginning of the afternoon prayer $h_a(\lambda)$. The entries are not the same as those in the main corpus of tables for timekeeping compiled for Cairo (**II-4.9**).

(e) A table of the solar azimuth as a function of the altitude, computed for the equinoxes. Values of a(h) are given to two digits for each degree of h from 1° to 60° ($= \bar{\phi}$) and but for minor variants the entries are the same as those of Ibn Yūnus (**5.1.1**).

(f) A table of a function f(λ) which I am unable to identify. The table bears no title and is not referred to in the extant fragment of the instructions. The equinoctial value of f is noted in the margin. The following sample entries from the table may eventually facilitate the identification of the function tabulated:

λ	0°	30	60	90	180	210	240	270
f(λ)	54,45	58;24	59;49	59;59*	54;45	48;23	42;28	39;57

* read 60;0 (?)

At first sight the entries for $\delta > 0$ resemble the values of Cos $\delta(\bar{\lambda})$, but closer investigation reveals that this is not the function tabulated.

The entries for Sin H(λ) and B(λ) are generally in error by several digits in the third place. Notes at the side of the tables indicate how to derive B(λ) from Sin H(λ), and indeed the errors in these tables can be explained by the fact that the values of Sin H were found by linear interpolation in the Sine table. al-Khaṭāʾī also points out that the table of a(h) displays d(Δ) (**5.1.1**), adding that Ibn Yūnus had compiled azimuth tables for the latitude of Cairo. Otherwise there are no references to any of his predecessors, but unfortunately – as already noted – the instructions to his tables in the Vatican manuscript are incomplete.

The main set of tables is likewise incomplete. The function tabulated is:

$$V'(x,y) = \text{arc Vers} \{ R \cdot y / x \},$$

and values are given to two digits for the domains:

$$x = 47;37, 47;57, \ldots , 51;57 \quad \text{and} \quad y = 1, 2, \ldots , Y(x),$$

where Y(x) is a certain maximum defined below. The instructions for finding t(h,λ) indicate that one should first find B(λ) and:

$$H'(h,\lambda) = \text{Sin } H(\lambda) - \text{Sin } h$$

[20] de Zambaur, *Manuel*, p. 105.

using tables (a), (b) and (c) above. Then these arguments are to be entered in the main table, which is called *jadwal faḍl al-dāʾir*, "hour-angle table": see **Fig. 6.15.1**. The result is indeed the hour-angle, since:

$$t(h,\lambda) = V\ \{\ B(\lambda),\ H'(h,\lambda)\ \}\ .$$

The argument x runs in intervals of 0;20 between the limits of B(λ) for Cairo and the argument y runs in unit intervals up to Y(x), which is the greatest integer less than the value of Sin H corresponding to the value of δ underlying the argument x. The table originally contained about 825 entries, but in the Vatican manuscript only the page for y ≥ 31 is to be found now. The entries are rather accurately computed.

Being devised for latitude 30° al-Khaṭāʾī's tables are less useful than al-Khalīlī's auxiliary tables for timekeeping by the sun (**9.4**) which serve all latitudes. The possibility that al-Khalīlī knew of al-Khaṭāʾī's tables can be ruled out because the Damascene astronomer preceded the Cairene astronomer by close to a century. Nevertheless, the existence of al-Khaṭāʾī's tables implies that there were no tables of t(h,λ) and/or T(h,λ) currently available for Cairo (see, however, **2.1.1**). Ibn al-Mushrif's auxiliary tables (**9.8**) are an extension of the kind of tables compiled by al-Khaṭāʾī, for timekeeping by both the sun and the stars and intended to serve all latitudes.

6.15.2 Anonymous: Tlemcen

MS London BL Or. 411,2 is the only known copy of an anonymous commentary on the astronomical poem of the late-14th-century Maghribi scholar al-Jādarī.[21] The commentary was written in Tlemcen at the end of the 14th century and it contains several spherical astronomical tables computed for parameters:

$$\phi = 35;0°\ \text{(Tlemcen) and}\ \varepsilon = 23;35°\ .$$

Details are given in **II-13.6**. In passing, we note that the commentary contains interesting historical accounts of trepidation,[22] twilight and the obliquity of the ecliptic.

Amongst the spherical astronomical tables are three displaying the functions:

$$\text{Sin H},\ {}^1/_2\ \text{B and k}$$

(**6.3.4**, **6.4.15** and **8.1.4**), as well as another displaying the function:

$$t\ \{\ {}^1/_2\ B,\ [\text{Sin H - Sin h}]\ \}$$

for the same latitude. I have not been able to compare these tables with the corresponding ones in the Tunisian corpus of auxiliary tables (**9.7**), from which they were probably lifted.

6.16 Tables of auxiliary functions for timekeeping by the stars

Most *zīj*es contain a catalogue displaying the coordinates of prominent stars for a particular epoch, either in the ecliptic or equatorial system or both.[23] A smaller number of *zīj*es and tables

---

[21] On al-Jādarī see Suter, *MAA*, no. 424a, Renaud, "Additions à Suter", no. 424a, and *Cairo ENL Survey*, no. F26. The anonymous commentary has been previously attributed to Ibn al-Ḥabbāk (Suter, *MAA*, no. 435 and *Cairo ENL Survey*, no. F28), which cannot be correct because he is mentioned in the text.

[22] For a survey of trepidation in the Islamic sources see King & Samsó, "Islamic Astronomical Handbooks and Tables", Section 3.7. See also the text to n. **II**-12:20.

[23] See n. 1:44.

for timekeeping give additional numerical information about particular stars.[24] Thus, for example, al-Battānī in his *Zīj*, after his main catalogue which gives the coordinates of 533 stars, tabulates the meridian altitude, half-arc of visibility of five stars, as well as the longitudes of the points of the ecliptic which rise, culminate and set with each star, all computed for the latitude of Raqqa (36°).[25] The two Yemeni astronomers Abu 'l-ʿUqūl (*ca.* 1300) and Ibn *al-M-s-r-b* (?) (*ca.* 1325)[26] tabulated the same functions for 30 stars, computed respectively for the latitude of Taiz (13;40°) and Zabid (14;40°) (in both the Berlin and Sanaa manuscripts). Our present concern is with a category of star catalogue which displays information particularly useful for reckoning time by the stars, of which the following have come to my attention.

6.16.1 Anonymous: Qandahar (?)

Amongst the various tables copied at the end of MS Berlin Ahlwardt 5751 (Mq. 101,1), copied *ca.* 1300, of the *Zīj* of Kūshyār ibn Labbān (**3.2.1**), there is an anonymous set of auxiliary tables (pp. 186-187) for finding the ascendant at night (*al-ṭāliʿ bi-'l-layl*) from stellar altitudes – see **Fig. 6.16.1**. These tables appear not to be related to the tables of the ascendant at night that follow immediately in the Berlin manuscript. For each of 34 stars five quantities are tabulated to two sexagesimal digits, labelled as follows:

1 *taḍrub jayb irtifāʿ al-waqt fīhi abad*[an]
multiply the Sine of the instantaneous altitude by this;

2 *taqsim mā yakhruj mina 'l-ḍarb ʿalayhi abad*[an]
divide the product of the multiplication by this;

3 *tanquṣ al-maqsūm minhu abad*[an] *wa-tajʿal al-bāqī qaws*[an] *maʿkūsa*[tan]
subtract the quotient from this and take the inverse Versed Sine;

4 *tanquṣ al-qaws al-maʿkūsa minhu aw tuzād ʿalayhi*
subtract or add the arc from or to this; and

5 *yuzād mā baqiya aw balugha mina 'l-qaws ʿalayhi abad*[an],
add the result to this.

The quantities mentioned relate to the following functions:

1 and 3: Vers D, 2: Sin H, 4: D, and 5: $\alpha_\phi(\rho)$

and the instructions conform to the two formulae (*cf.* **F9** and **F21**):

$$T(h) = D - \text{arc Vers} \{ \text{Vers } D - \text{Sin } h \text{ Vers } D / \text{Sin } H \}$$

$$\text{and } \alpha_\phi(\lambda_H) = \alpha_\phi(\rho) + T(h).$$

The underlying latitude in these tables is about 30° (this needs to be investigated further) and they were probably intended for use in some city in the South of Greater Iran, possibly Qandahar (see **3.2.1**). We shall return to these tables in **IX**.

---

[24] See Girke, "Die frühesten islamischen Sternkataloge" (cited in n. 1:44). A prime example of such tables are those presented by Ḥabash al-Ḥāsib for his instrument for reckoning time by night for any latitude (**XIIb-12**), on which the various quantities are registered graphically: see Charette & Schmidl, "Ḥabash's Universal Plate", pp. 138-146.

[25] *Cf.* Kennedy, "*Zīj* Survey", p. 155.

[26] King, *Astronomy in Yemen*, no. 12. Perhaps this is Ibn al-Mushrif, who cannot, however, be identical with the 15th-century Cairene astronomer with the same name (see **I-9.8** and **II-6.15**).

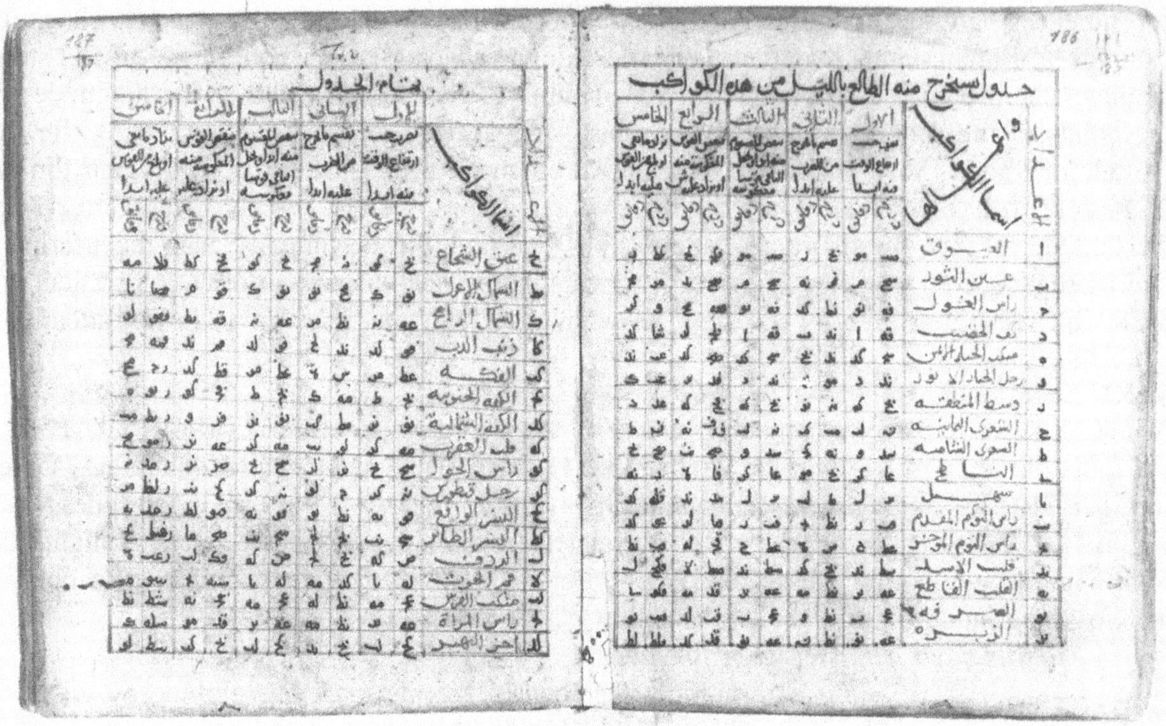

Fig. 6.16.1: Some auxiliary tables for timekeeping by the stars, apparently for the latitude of Qandahar. [From MS Berlin Ahlwardt 5751, pp. 186-187, courtesy of the Deutsche Staatsbibliothek (Preußischer Kulturbesitz).]

6.16.2 Anonymous: unspecified locality in Iran or al-ʿIrāq

In MS Berlin Ahlwardt 5750, fols. 62v-63r, of the anonymous recension of the *Zīj* of Ḥabash (**9.1** and **9.2**), there is a table displaying the following functions for thirty stars:

λ	*darajat al-kawkab fī 'l-ṭūl*
β	*al-ʿurūḍ*
Δ	*buʿduhā ʿan muʿaddil al-nahār*
α'	*mamarr al-kawkab fī wasaṭ al-samāʾ*
H	*irtifāʿ niṣf nahārihā*
$\alpha_\phi(\rho)$	*maṭāliʿ al-daraja al-ṭāliʿa maʿahu*
D	*niṣf qaws nahārihā*
ρ	*al-daraja al-ṭāliʿa maʿahu*
Cos Δ	*anṣāf al-aqṭār (i.e. radii)*

The last of tabulated functions is of interest to the present study, being the radius of the day-circle, that is, the "day-radius" of Indian astronomy. The hour-angle t corresponding to the instantaneous altitude h of a particular star can be found using (*cf.* **F9**):

$$t = \text{arc Vers } \{ [\text{Sin } H - \text{Sin } h] \cdot R^2 / [\text{Cos } \Delta \text{ Cos } \phi] \} .$$

Note that it is H rather than Sin H which is tabulated, and also that it would have been more useful (**6.0**) to tabulate the product:

$$B(\Delta,\phi) = \text{Cos } \Delta \text{ Cos } \phi ,$$

rather than simply Cos Δ.

The underlying latitude is 33;25° (Baghdad), which is used elsewhere in the Berlin manuscript, as in the table of oblique ascensions. By virtue of the epoch 304 – whether Hijra (?) [= 916/17] or Yazdigird (?) [= 936] – the star catalogue cannot be due to Ḥabash himself. The star catalogue in MS Istanbul Yeni Cami 784,2, of Ḥabash's *Zīj* (see fol. 192r) is simply that of the *Mumtaḥan Zīj*, compiled for the year 214 H [= 829/30] (**6.16.3**).[27]

6.16.3 Anonymous: Baghdad

MS Escorial ár. 927, copied *ca.* 1300, contains an anonymous recension of the early-9th-century *Mumtaḥan Zīj* (**4.1.1**). It contains a star catalogue dated 380 Yazdigird [= 1012] (fols. 96v-97r), derived by addition of a correction for precession from the *Mumtaḥan* star catalogue dated 214 H [= 829/30], which immediately precedes it in the manuscript (fol. 96r): see **Fig. 6.16.3**. In this later catalogue the following information is given for each of 18 stars:

λ	*mawḍiʿ min falak al-burūj*
β	*al-ʿarḍ*
Δ	*al-abʿād ʿan muʿaddil al-nahār*
μ	*darajat al-mamarr fī wasaṭ al-samāʾ*
D	*niṣf qaws al-nahār*
Vers D	*jayb al-nahār*
H	*irtifāʿ niṣf al-nahār*
Sin H	*jayb irtifāʿ niṣf al-nahār*
$\alpha_\phi(\rho)$	*maṭāliʿ al-daraja al-ṭāliʿa maʿ al-kawkab*
ρ	*al-daraja al-ṭāliʿa maʿ al-kawkab*

The underlying latitude is 33;21° (Baghdad), a parameter used in some other spherical astronomical tables in MS Escorial ár. 927 (**4.1.1**).[28]

6.16.4 Anonymous: Fez and/or Meknes

In 1891 Gäetan Delphin published photographs of an unsigned Moroccan calculating disc (*ṣafīḥa*) dated 1782.[29] This instrument incorporates two small tables displaying the values of the functions:

$$C, B \text{ and } \alpha$$

for 16 prominent stars. Values are given to one sexagesimal digit and the underlying latitude is either 33° as stated on the side of the instrument bearing the tables or 34° as stated on the other side of the instrument. Between the two tables there is engraved a trigonometric grid of the kind known in medieval Arabic as *al-rubʿ al-mujayyab*, complete with semi-circles for finding sines and cosines (**6.5**). With such a grid one could determine the hour-angle from an observation of the altitude of any of the stars using the formula (**6.12**):

---

[27] Debarnot, "*Zīj* of Ḥabash al-Ḥāsib", p. 57. These tables are edited in an unpublished paper by Dorothea Girke (see n. 1:44). See now Charette & Schmidl. "Ḥabash's Universal Plate", p. 139, n. 113.

[28] There are no geographical tables in the Escorial manuscript. According to Kennedy & Kennedy, *Islamic Geographical Coordinates*, pp. 55-56, only the *zīj*es of al-Maghribī (Kennedy, "*Zīj* Survey", nos. 41 and 108 – see note 5:8) out of all the numerous Islamic sources investigated have this value for Baghdad. See also King, "Earliest Muslim Geodetic Measurements" (cited in n. 2:26), pp. 221 and 225-227.

[29] See Delphin, "L'astronomie au Maroc".

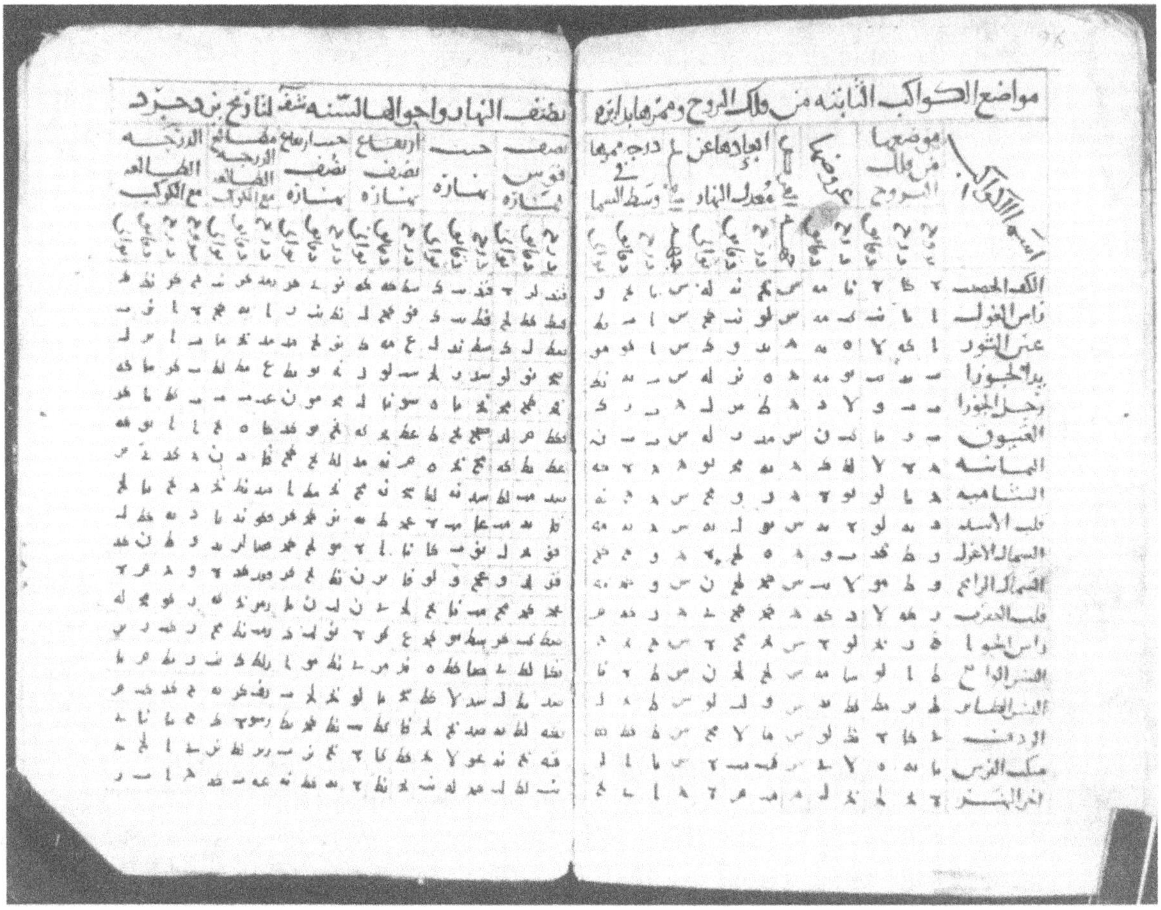

Fig. 6.16.3: The star-table in a recension of the *Mumtaḥan Zīj*, showing various functions useful for timekeeping in addition to the basic ecliptic coordinates of the stars and their declinations. [From MS Escorial ár. 927, fols. 96v-97r, courtesy of the Biblioteca de El Escorial, taken from the 1986 Frankfurt facsimile.]

$$\text{Cos } t = R [\text{Sin } h - C] / B .$$

On the other side of the instrument there is an alidade and altitude scale for making such measurements of stellar altitude, as well as a table purporting to display the rising times of the signs of the zodiac for each degree of latitude from 1° to 30° (the entries do not make much sense to me).

CHAPTER 7

TABLES OF AUXILIARY FUNCTIONS FOR CALCULATING THE LENGTH OF DAYLIGHT AND RIGHT AND OBLIQUE ASCENSIONS

7.0 Introductory remarks

As noted in **1.3**, most *zīj*es contain tables of:

$$\delta(\lambda) \text{ and } \alpha(\lambda) \text{ or } \alpha'(\lambda)$$

and of:

$$d(\lambda) \text{ or } D(\lambda) \text{ and } \alpha_\phi(\lambda)$$

for some particular latitude. These functions are related by simple formulae such as (*cf.* **F5**, **F7** and **F18**):

$$\alpha(\lambda) = \text{arc Sin } \{ \text{ R Tan } \delta(\lambda) \ / \text{ Tan } \varepsilon \ \}$$
$$d(\lambda,\phi) = D(\lambda,\phi) - 90^\circ = \text{arc Sin } \{ \text{ Tan } \delta(\lambda) \text{ Tan } \phi \ / \text{ R } \}$$
$$\alpha_\phi(\lambda) = \alpha(\lambda) - d(\lambda,\phi) \ .$$

It is immediately apparent that the function Tan $\delta(\lambda)$ is important in the determination of $\alpha(\lambda)$ and $d(\lambda)$. In **7.1** I list all known Islamic tables of this function. In **7.2** I note the existence of a table of proportional parts of Tan ϕ for various latitudes. Another Islamic formula for $d(\lambda,\phi)$ (which follows from **F5** and **F7**) is:

$$d(\lambda,\phi) = \text{arc Sin } \{ \text{ Sin [max } d(\phi)] \cdot \text{Sin } \alpha(\lambda) \ / \text{ R } \} \ .$$

Various Islamic sources contain tables of the functions:

$$\text{Sin [max } d(\phi)] \quad \text{and} \quad \text{Sin } \alpha(\lambda) \ / \text{ R} \ ,$$

which I discuss in **7.3** below. Some of the Cairo and Damascus astronomers who specialized in timekeeping derived approximate formulae for the standard functions of spherical astronomy (see further **II-1.4**): in **7.4** I note the only example known to me of a table based on such an approximation.

The purpose of these auxiliary tables was to facilitate the preparation of tables of d, D or α_ϕ for various latitudes. A table of max $d(\phi)$ to two digits for each degree of latitude is appended to one copy of al-Khalīlī's universal auxiliary tables but being based on $\varepsilon = 23;35^\circ$ is clearly taken from an earlier source: see **Fig. 6.14.3**. His predecessor al-Marrākushī (**4.2.4**, etc.) had tabulated max $d(\phi)$ to three digits for each degree of ϕ from 1° to 66°.[1] In the extensive tables of Najm al-Dīn al-Miṣrī (**2.6.1**) there are tables of $D(\phi,\lambda)$ and $d(\phi,\Delta)$ for each degree of λ and Δ and each degree of ϕ; curiously, alongside these two tables with together over 25,000 entries, there are no associated tables of tan $\delta(\lambda)$ or tan Δ.[2] In the *Ḥākimī Zīj* of Ibn Yūnus there are tables of $\alpha_\phi(\lambda)$ for each degree of both arguments (max $\phi = 48°$),[3] and such extensive tables

[1] al-Marrākushī, *A-Z of Astronomical Timekeeping*, I, pp. 115, and Sédillot-*père*, *Traité*, I, p. 246.
[2] See now Charette, "Najm al-Dīn's Monumental Table", pp. 22-23.
[3] On these see King, *Ibn Yūnus*, III.14.5.

of oblique ascensions are also contained in the later *zīj*es of al-Ṭūsī, al-Kāshī and Ulugh Beg.[4] Ibn Yūnus' entries are computed to degrees and minutes, al-Ṭūsī's and al-Kāshī's to seconds, and Ulugh Beg's to thirds.

7.1 Tables of the Tangent of the declination for calculating the equation of half-daylight for any latitude and right ascensions

As noted in **7.0**, the function Tan $\delta(\lambda)$ is of use in computing either $d(\lambda)$ or $\alpha(\lambda)$. Most of the tables of Tan $\delta(\lambda)$ listed below have been mentioned in previous studies. Note that tables of the auxiliary functions Sin $d(\lambda)$ and Sin $\alpha(\lambda)$, which are easily derived from tables of Tan $\delta(\lambda)$, also occur in the Islamic sources (see **7.1.2** and **7.1.5**; and **7.2**, **9.2** and **9.3**, respectively). See also **Fig. I-6.9.3** for some other tables of Tan Δ and Tan $\delta(\lambda)$, which by an oversight have not been included in this section.

7.1.1 al-Khwārizmī (Baghdad)

The *Zīj* of the early-9th-century Baghdad astronomer al-Khwārizmī (**4.1.1**) is no longer extant in its original form.[5] However, we do possess a Latin translation by Adelard of Bath based on a greatly modified recension of the *Zīj* by the Andalusī astronomer al-Majrīṭī (*ca.* 1000).[6] There is no table of Tan $\delta(\lambda)$ in this version, but we know from statements by Ibn al-Muthannā (10th century ?) and al-Bīrūnī (*ca.* 1025) that al-Khwārizmī's *Zīj* contained a table entitled *fuḍūl al-maṭāliʿ li-'l-arḍ kullihā*, "ascensional differences for the whole earth".[7] Now in the *Toledan Tables*, a corpus of tables derived mainly from the *Zīj*es of al-Khwārizmī and al-Battānī,[8] as well as in a 15th-century English manuscript,[9] there is a table of the function:

$$e(\lambda) = 150 \tan \delta(\lambda) / 12 .$$

Values are given to three digits and are based on al-Khwārizmī's parameter $\varepsilon = 23;51°$. See further **10.1**. With this function $d(\lambda,\phi)$ can be found using:

$$d(\lambda,\phi) = \text{arc Sin}_{150} \{ e(\lambda) \cdot \text{Tan}_{12} \phi \} .$$

It is well established that al-Khwārizmī's Sine table[10] was based on the Indian parameter 150

[4] See Kennedy, "*Zīj* Survey", pp. 161, 164, and 166, respectively.

[5] On al-Khwārizmī see Gerald Toomer's article in *DSB*, and on various Abbasid works attributed to him, but not all by him, King, "al-Khwārizmī".

[6] See Suter, *al-Khwārizmī*, and Neugebauer, *al-Khwārizmī*, for text, tables, translation, and analysis, and also Goldstein, *Ibn al-Muthannā on al-Khwārizmī*.

[7] See Goldstein, *op. cit.*, pp. 78-81 and 204-206, and Lesley, "Bīrūnī on Rising Times and Daylight Lengths", pp. 125-127, on these sources.

[8] See Toomer, "Toledan Tables", p. 33, and now Pedersen, *The Toledan Tables*, III, pp. 986-991.

[9] See Neugebauer & Schmidt, "Hindu Astronomy at Newminster", p. 226/430.

[10] al-Khwarizmī's original Sine table has been skilfully reconstructed in Hogendijk, "al-Khwārizmī's Sine of the Hours" (see **4.2.1**).

Another Sine table based on R = 150 survives in the *Toledan Tables* and the same 15th-century English manuscript (*cf.* Toomer, "Toledan Tables", pp. 27-28; Neugebauer & Schmidt, "Hindu Astronomy at Newminster", p. 226/430; North, *Richard of Wallingford*, II, pp. 12-14, and *idem*, *Horoscopes and History*, p. 14). A similar table is found in MS London BL Or. 3624, fol. 170r, of the *Mukhtār Zīj* of the 13th-century Yemeni astronomer Abu 'l-ʿUqūl (see n. 2:11); it is entitled *jadwal jayb al-Hind*, "table of Indian sines", and contains numerous variants from the table published by Neugebauer and Schmidt. Again, in MS Istanbul Ayasofya 4830, copied in Damascus in 626 H [= 1228/29], fol. 188v, amidst various treatises dating from 9th-century al-ʿIrāq (see further **4.1.1**), there is another Sine table based on R = 150. The table from the Latin translation of the

and that he used the value for ε from the *Handy Tables*, namely 23;51° and it seems fairly certain that the above-mentioned tables are ultimately due to al-Khwārizmī.[11] The factor 150/12 essentially converts the product of two tangents (the second to base 12) into a sine (to base 150). See also **7.1.9** on what is probably another such table from 9th-century Baghdad.

7.1.2 Anonymous (Baghdad)

MS Alexandria 5577J of the *Zīj* of the Yemeni astronomer al-Kawāshī (**5.6.1**) contains tables of:

$$\delta(\lambda),\ \delta_2(\lambda) \text{ and } \mathrm{Tan}_{60}\ \delta(\lambda)\ ,$$

each computed to three digits and based on the parameter ε = 23;33°. The same parameter underlies al-Kawāshī's tables of H(λ) and $\alpha_\phi(\lambda)$ for the latitudes of Aden (11°) and Taiz (14;40°), but I suspect that the three tables mentioned above were taken from an earlier ʿIrāqī source. The entry for λ = 90° in the table of Tan δ(λ) is 26;9,53, whereas the accurate value of Tan 23;33° is 26;9,4. See also **7.1.7** on other ʿIrāqī tables of Tan δ(λ) in another Yemeni source and **7.1.9** on a table in a Mamluk Egyptian source also based on obliquity 23;33° and with entries to three digits.

7.1.3 Ibn Yūnus (Cairo)

In MS Leiden Or. 143, p. 271 of the *Ḥākimī Zīj* of Ibn Yūnus (**2.1.1**) the function:

$$e(\lambda) = \mathrm{Tan}_{60}\ \delta(\lambda)$$

is tabulated to four sexagesimal digits for ε = 23;35°.[12] The values are rather accurate: the errors in the fourth digit are less than ±4 in one third of the entries and less than ±30 in the remainder. The entry for λ = 90° is 26;11,33,23 whereas Tan 23;35° is accurately 26;11,33,16 to three sexagesimal fractions. Ibn Yūnus states that he compiled this table by dividing R Sin δ(λ) by Cos δ(λ) for each degree of λ. He refers to the function e(λ) by the same name that al-Khwārizmī used for it and does not explicitly state that it is the Tangent of the declination. His rules for finding α(λ) and δ(λ,φ) using the table are equivalent to the formulae:

$$\alpha(\lambda) = \text{arc Sin } \{\ R \cdot e(\lambda)\ /\ e(90°)\ \}$$
$$d(\lambda,\phi) = \text{arc Sin } \{\ e(\lambda) \cdot \text{Sin } \phi\ /\ \text{Cos } \phi\ \}\ .$$

In MS Leiden Or. 143, pp. 270-271 and 347 of the *Zīj*, Ibn Yūnus tabulates the related auxiliary functions:

$$\text{Sin } \alpha(\lambda) \text{ and Sin } d(\lambda)\ (\phi = 30°)\ ,$$

Zīj of al-Majrīṭī published in Suter, *al-Khwārizmī*, pp. 169-170, which is to base 60, is not original to al-Khwārizmī's *Zīj*. On these tables see also King, "al-Khwārizmī", pp. 2 and 34, n. 12.

[11] In MS Berlin Ahlwardt 5793 (Landberg 56), fols. 93v-95v, appended to a unique copy of the treatise on the construction and use of the astrolabe by al-Khwārizmī (see King, "al-Khwārizmī", p. 23), there are some tables which may also be by him. Of particular interest is a table of normed right ascensions α′(λ) with values to 3 digits for each 3° of λ from 273° to 360°. This may be al-Khwarizmī's original table, for as shown in Neugebauer & Schmidt, *op. cit.*, p. 224, the values in the table of α(λ) in al-Majrīṭī's *Zīj* do not correspond to values of α(λ) which underlie certain calculations in a text relating to the use of al-Khwārizmī's table of e(λ). Neugebauer and Schmidt noted three values of α(λ) in this text as follows:

0;54,53,40 (λ = 1°), 1;49,33,20 (λ = 2°), and 27;53 (λ = 30°).

The corresponding values which I derive (using linear interpolation for λ = 1° and 2°) from the table of α′(λ) are 0;54,54, 1;49,48, and 27;53,26, so that this table of α′(λ) is probably related to al-Khwārizmī.

[12] See further King, *Ibn Yūnus*, III.13.3d, on this table.

as well as the standard functions:

$$\alpha(\lambda) \text{ and } d(\lambda)\ (\phi = 30°)\ .$$

His tables of:

$$\text{Tan } \delta(\lambda),\ \text{Sin } \alpha(\lambda) \text{ and } \alpha(\lambda)$$

are also found in MS London BL Or. 3624, fol. 173v-176r, of the *Mukhtār Zīj* of Abu 'l-ʿUqūl (**2.1.2** and also **7.2.1**).

7.1.4 Kūshyār ibn Labbān (Iran, locality uncertain)

On fol. 304r of MS Istanbul Yeni Cami 784, fols. 230r-376v, of the *Zīj* of Kūshyār ibn Labbān (**3.2.1**), there is a table[13] entitled *ẓill al-mayl al-awwal*, "tangent of the first declination", displaying the function:

$$e(\lambda) = \text{Tan}_{60}\ \delta(\lambda)\ .$$

Entries are given to three digits for each degree of λ and are based on ε = 23;35°. The table is considerably less accurate than that of Ibn Yūnus (**7.1.2**). In particular the value of e(90°) is given as 26;11,41 (accurately 26;11,33) and this was probably derived by linear interpolation in a table of Tangents. In the Tangent tables in the Istanbul manuscript (fol. 262v) we find the accurate entries 25;28,7 and 26;42,39 for arguments 23° and 24°, whence by linear interpolation:

$$\text{Tan } 23;35° = 26;11,41,30 = 26;11,41 \text{ (by truncation)}\ .$$

Kūshyār's table also occurs in MS Paris BNF ar. 5968, fol. 41v, of the anonymous *Dustūr al-munajjimīn*, a *zīj* compiled probably in N. Syria (rather than Alamut in N. W. Iran as has been maintained) in the 12th century and based mainly on the *zīj*es of al-Battānī, Kūshyār, ʿAbū Jaʿfar al-Khāzin and al-Bīrūnī.[14]

In MS Gotha A1402 of the main corpus of tables for timekeeping used in medieval Cairo (**2.1.1**), but in no other such source known to me, there is a table of $\text{Tan}_{60}\ \delta(\lambda)$ with entries based on ε = 23;35° and having 26;11,41 for λ = 90°. I have not been able to ascertain whether this table is related to that of Kūshyār. It is perhaps worth noting that the 13th-century Egyptian astronomer al-Maqsī also used 26;41 for Tan 23;35° in his treatise on sundials (**4.1.3**) rather than the 26;11,40 used by his contemporary al-Marrākushī (**7.1.7**).

7.1.5 Ibn al-Zarqālluh (Toledo)

Amongst the spherical astronomical tables in the *Almanac* of Azarquiel (**6.9.1**) there is one entitled *jadwal fuḍūl al-maṭāliʿ*, "table of the excesses of the ascensions" (p. 225) and the function tabulated is:

$$e(\lambda) = {}^{1}/_{12}\ \text{Tan}_{60}\ \delta(\lambda)\ .$$

Values are given to three digits and it is stated that the underlying value of ε is 23;33°. This value of ε is the entry for argument λ = 90° in the accompanying table of δ(λ) in the *Almanac*

13 The existence of this table was noted in Kennedy, "*Zīj* Survey", p. 156. There is more work to be done on the two *zīj*es of Kūshyār (see n. 3:4); in particular, an investigation of the tables relating to spherical astronomy in the various available manuscripts might establish the location of his activities. In the *DSB* article on him it is stated, after a medieval source, that he was active in Baghdad, but somehow I doubt this.

14 On this work, not listed in Kennedy, "*Zīj* Survey", see Zimmermann, "Dustūr al-Munajjimīn".

(p. 174). The entry for $\lambda = 90°$ in the table of $e(\lambda)$ is 2;10,46 (accurately, 2;10,45). The same table with minor variants occurs in MS Istanbul Kandilli 249 of the *Zīj* of the early-14th-century Tunisian astronomer Ibn al-Raqqām (**6.9.4**), whose declination table, however, is based on the distinctive value 23;32,40° for ε. Here the function is called *nisab juyūb al-faḍlāt li-ʿamal al-maṭāliʿ*, "the ratios of the Sines (*i.e.*, Sines and Cosines) of the excesses (?) for finding ascensions".

A related table in the *Almanac* of Ibn al-Zarqālluh (p. 226) is entitled *jadwal anṣāf fuḍūl ẓuhūrāt al-kawākib*, "table of the half excesses of visibility of stars", and displays values of the function:

$$e(\Delta) = {}^{1}/_{12}\ \mathrm{Tan}_{60}\ \Delta$$

to three digits for each 3° of Δ (and also $\Delta = 88°$ and 89°). It is of interest that the entry for 45° is 4;59,59 rather than 5;0,0: this suggests that the table was not compiled directly from a tangent or cotangent table. (See also **6.9.1**.)

Another table of the same function $e(\Delta)$ displaying values to three digits for each degree of argument from 1° to 60°, and also having the entry 4;59,59 for argument 45°, is contained in MS Berlin Ahlwardt 5724, fol. 43r, of an anonymous corpus of tables for Tunis (**2.3.5**). In the Berlin copy the table is entitled *jadwal al-ikhtilāf al-ufuqī*, "table of variation due to the horizon" and the function is labelled *nisbat jayb al-buʿd*, "the ratio of the Sine of the declination" – see **Fig. 6.8.2**.

The advantage to be gained from dividing $\mathrm{Tan}_{60}\ \delta(\lambda)$ or $\mathrm{Tan}_{60}\ \Delta$ by 12 is that using only tables of Cotangents to base 12 and Sines to base 60 one can compute $d(\lambda)$ or $d(\Delta)$ for any latitude with facility. The formulae are:

$$d(\lambda,\phi) = \mathrm{arc\ Sin}_{60}\ \{\ \mathrm{Tan}_{60}\ \delta(\lambda)\ /\ 12\ \cdot\ \mathrm{Cot}_{12}\ \bar{\phi}\ \}$$
$$d(\Delta,\phi) = \mathrm{arc\ Sin}_{60}\ \{\ \mathrm{Tan}_{60}\ \Delta\ /\ 12\ \cdot\ \mathrm{Cot}_{12}\ \bar{\phi}\ \}\ .$$

7.1.5* Ibn Isḥāq (Tunis)

MS Hyderabad Āṣafiyya 298 of the *Zīj* of the 13th-century Tunisian astronomer Ibn Isḥāq (**6.9.1***) contains the same table of the function labelled *al-ikhtilāf al-ufuqī* and *jayb nisbat al-buʿd*, "the sine of the ratio of the distance" (table no. 65 in this manuscript), as well as of the function ${}^{1}/_{12}\ \mathrm{Tan}_{60}\ \delta(\lambda)$ labelled *al-fuḍūl* (with maximum value 2;10,46) (table no. 91).[15]

7.1.6 al-Baghdādī (Baghdad)

In MS Paris BNF ar. 2486, fol. 235r, of the late-13th-century *Zīj* of al-Baghdādī (**2.3.1**), there is a table of:

$$e(\lambda) = \mathrm{Tan}_{60}\ \delta(\lambda)$$

with entries to four digits for each degree of λ.[16] The values are based on ε – 23;35° and are slightly less accurate than those of Ibn Yūnus (**7.1.3**). The entry for $\lambda = 90°$, however, is 26;11,33,16, which is more accurate than Ibn Yūnus' value. I do not discount the possibility that this table was computed by ʿAlī ibn Amājūr in the 10th century (**2.3.1** and **6.4.1***). In MS

[15] See Mestres, Zīj *of Ibn Isḥāq*, pp. 278 and 282.
[16] See Lesley, "Bīrūnī on Rising Times and Daylight Lengths", p. 127.

Paris BNF ar. 2486, fol. 235v, there is a table of the function Sin d(λ), perhaps based on the above-mentioned table of e(λ) and computed for ϕ = 33;25° (Baghdad).

7.1.7 al-Fahhād (?) / al-Marrākushī / al-Fārisī (Cairo / Yemen)

In al-Marrākushī's treatise on spherical astronomy and instrumentation (**4.2.4**), there are tables (I.183-185) of the functions:

$$\delta(\lambda),\ \delta_2(\lambda) \text{ and } \mathrm{Tan}_{60}\ \delta(\lambda)\ (\varepsilon = 23;35°)\ .$$

The first two functions are given to two digits and the third to three digits, for each integral value of λ. Now in the roughly contemporaneous *Muẓaffarī Zīj* of the Yemeni astronomer Muḥammad ibn Abī Bakr al-Fārisī,[17] *e.g.*, MS Cambridge Gg. 3.27, fols. 104v-105r, copied *ca.* 1400, there are tables of the same three functions, each to three digits. al-Marrākushī's values of δ(λ) and $\delta_2(\lambda)$, which contain several errors, can be derived from the corresponding *Muẓaffarī* values by truncation, and the values of Tan δ(λ) in the two sources are identical, but for copyists' errors. The *Muẓaffarī Zīj* is based mainly on the *ʿAlāʾī Zīj* of the mid-12th-century astronomer al-Fahhād.[18] A 15th-century Byzantine astronomical manuscript containing material derived mainly from the *ʿAlāʾī Zīj* likewise contains a table of Tan δ(λ).[19] I have not been able to ascertain whether this Byzantine table is the same as that of al-Marrākushī and al-Fārisī (see also **7.2.2**), but it seems probable that the Islamic tables are due to al-Fahhād or one of his sources. They are characterized by the entry 26;11,40 for argument 90° rather than the 26;11,41 found in the table of Kūshyār (**7.1.4**).

al-Marrākushī also tabulated (I.209-210) the two functions:

$$\mathrm{Tan}_{60}\ \delta(\lambda)\ /\ 12 \quad \text{and } \mathrm{Tan}_{60}\ \Delta\ /\ 12$$

to three and two digits, respectively, for each degree of arguments λ and Δ. He calls the functions *faḍla*, literally "excess". The first table is based on obliquity 23;35° and thus differs from the earlier table of Ibn al-Zarqālluh (**7.1.5**). In MS Cairo MM 43, fol. 42r, of an Egyptian copy of al-Khalīlī's *Universal Table* (**9.5**), two tables of the same functions occur, with numerous variants from al-Marrākushī's tables as published by Sédillot. (See further **7.1.8**.)

7.1.8 Anonymous (Cairo)

In MS Paris BNF ar. 2513, fol. 81r, of a recension of the 13th-century Egyptian *Muṣṭalaḥ Zīj* (**6.7.1**) there is a table of:

$$e(\lambda) \;=\; \mathrm{Tan}_{60}\ \delta(\lambda)\ /\ 12$$

based on ε = 23;35° and with entries computed to three digits. The entries differ from those of al-Marrākushī (**7.1.7**) but the same table occurs in MS Cairo MM 43, fol. 42r, in a set of anonymous spherical astronomical tables for Cairo which also contains al-Khalīlī's universal auxiliary tables (**9.5**); in MS Damascus Ẓāhiriyya 3116, fol. 61r, where it is appended to a

---

[17] On al-Fārisī and the *Muẓaffarī Zīj* see Suter, *MAA*, nos. 349 and 349N; Kennedy, "*Zīj* Survey", no. 54, and King, *Astronomy in Yemen*, no. 6. An 1822 study Lee, "Astronomical Tables of Al Farsi", is still most useful. See also n. 7:26 below and **II-2.2** and **12.0**.

[18] On al-Fahhad see Kennedy, "*Zīj* Survey", nos. 23 and 84, *inter alia*, and, most recently, Pingree, *Astronomical Works of Gregory Chioniades*.

[19] See Neugebauer, "Studies in Byzantine Astronomy", p. 36b (entries for fols. 43lr-433r).

set of al-Khalīlī's hour-angle tables for Damascus (**2.1.4**); and in MS Cairo TR 275, p. 73, copied in 858 H [= 1454], of the *zīj* entitled *al-Lumʿa* by the early-15th-century Egyptian astronomer al-Kawm al-Rīshī[20] (but not in other copies of this *zīj*); and in the manuscript in a private collection in Sanaa of the astronomical miscellany of the 14th-century Yemeni ruler al-Sulṭān al-Afḍal (**2.1.2**). A significant entry in all of these sources is 1;50,41 for λ = 60°: al-Marrākushī has the accurate value 1;50,48.

7.1.9 Anonymous (Cairo)

In MS Paris BNF ar. 2520, fol. 69v, copied *ca.* 1400, of another recension of the *Muṣṭalaḥ Zīj* there is a table of:

$$e(\lambda) = \text{Tan}_{60}\ \delta(\lambda)$$

with entries computed to three digits – see **Fig. 9.1a**. The underlying value of ε is 23;33° and I suspect that the table was taken from earlier source, not least because it precedes a copy of the auxiliary tables of Ḥabash (**9.1**). The entry for argument 90° is 26;9,3, whereas the accurate value of Tan 23;33° is 26;29,4. See also **7.1.2** above.

7.1.10 Anonymous (Yemen)

In the anonymous late-14th-century Yemeni *zīj* MS Paris BNF ar. 2523 (**2.5.3** and **4.3.3**) there is a table (fol. 85v) entitled *jadwal ẓill al-mayl al-awwal al-mankūs*, "table of the Tangent of the first declination", displaying the function:

$$e(\lambda) = \text{Tan}_{60}\ \delta(\lambda)$$

to three digits for each degree of λ and apparently based on ε = 23;35° – see **Fig. 4.3.3**. The entries are rather inaccurate: for example, the table gives e(90°) = Tan ε = 26;12,0°, whereas the accurate value of Tan 23;35° is 26;11,33.

7.1.11 Riḍwān Efendī (Cairo)

In MS Istanbul S. Esad Efendi Medresesi 119, fol. 59v, of the prayer-tables of Riḍwān Efendī for Cairo (**6.1.3**), the function:

$$e(\lambda) = \text{Tan}_{60}\ \delta(\lambda)$$

is tabulated alongside:

$$\delta(\lambda),\ \text{Sin}\ \delta(\lambda)\ \text{and}\ \text{Cos}\ \delta(\lambda)\ .$$

Values are given to three digits for each degree of λ and are based on Ulugh Beg's parameter ε = 23;30,17°. I see no reason to doubt that these tables are due to Riḍwān.

7.1.12 Taqi 'l-Dīn (Istanbul)

In MS Istanbul Nuruosmaniye 2930, fol. 23r, of the *zīj* entitled *Sidrat muntaha 'l-afkār* of Taqi 'l-Dīn (**6.1.4**), there is a table of the same function with entries to four *sexagesimal* digits for each degree of λ from 1° to 89°. The underlying parameter is ε = 23;28,54°.

Again, in MS Istanbul Esat Efendi 1976, fol. 25v, of Taqi 'l-Dīn's later *zīj* called *Jarīdat al-durar* (**6.4.8**), the function:

$$e(\lambda) = \text{Tan}_{100}\ \delta(\lambda)$$

---

[20] On al-Kawm al-Rīshī (Suter, *MAA*, no. 428) see *Cairo ENL Survey*, no. D41.

is tabulated to four significant *decimal* digits for each degree of λ. This table and accompanying one of δ(λ) correspond closely to recomputation with the parameter ε = 23;28,54°. Taqi 'l-Dīn's Sine and Tangent tables in MS Istanbul Esat Efendi 1976 are also given decimally, but to base 10 rather than base 1.[21]

7.2 Tables of auxiliary functions for calculating the equation of half daylight for any latitude

As noted in **7.0**, a widely-used Islamic formula for d(λ,φ) is the following:

$$d(\lambda,\phi) = \text{arc Sin } \{ \text{ Sin } [\max d(\phi)] \cdot \text{Sin } \alpha(\lambda) \ / \ R \ \} ,$$

and tables of the two auxiliary functions:

$$\text{Sin } [\max d(\phi)] \quad \text{and Sin } \alpha(\lambda) \ / \ R$$

are found in various *zīj*es. They are generally labelled *jayb ta'dīl al-nahār al-kullī*, "Sine of the maximum equation of (half-)daylight", and *daqā'iq al-nisab*, which means "minutes of the interpolation factor". Note that Abū Naṣr's auxiliary functions f_2 and f_3 are equivalent to these (**9.3**). Other combinations of functions with less limited practical application are found in our sources, such as the tables of α(λ) and Sin α(λ) (for each 10° of λ) and of max d(φ) (for each 0;30° of φ) in the *Zīj* of al-Battānī.[22] Tables of max d(φ) are attested already in the *Handy Tables*.[23] In **7.4** I remark on the similarity between a Byzantine table for finding d(λ,φ) approximately and the Islamic tables of Sin α(λ) / R. Note that the tables in **7.2.0***, which came to my attention only in December, 2001, have Δ as the argument and are for a fixed φ.

7.2.0* Early anonymous (N. Iran)
MS Istanbul U.L. A 314 (**6.4.1****) contains tables of

$$\text{Sin } d(\Delta) \quad \text{and} \quad d(\Delta)$$

with values to three digits for each degree of argument up to 55°. The tables appear to be computed for a latitude of about 38°, but this copy is extremely corrupt. It seems that we are dealing with an early production for a locality in Northern Iran.

7.2.1 Abu 'l-'Uqūl (?) (Taiz)
Tables of the three functions:

$$\max d(\phi) \ , \text{ Sin } [\max d(\phi)] \quad \text{and Sin } \alpha(\lambda) \ /R$$

occur in MS London BL Or. 3624, fol. 185v, of the *Mukhtār Zīj* of the Yemeni astronomer Abu 'l-'Uqūl (**2.1.2**). Values are given to two digits and are based on ε = 23;35°. The argument domains are:

$$\phi = 1°, 2°, \ldots , 60° \text{ and } \lambda = 1°, 2°, \ldots , 90°.$$

A marginal note states that to find the third function one should take the Sine of α(λ) and divide it by 150 or 60, according to which base the Sines are computed. A table of "Indian Sines" to base 150 as well as another to base 60 are found elsewhere in the *Zīj* (*cf.* n. 7:10 to **7.1.1**).

---

[21] See n. 1:22.
[22] Nallino, *al-Battānī*, II, pp. 58-59; also Neugebauer, *HAMA*, II, pp. 980-982.
[23] See, for example, Stahlman, *Handy Tables*, p. 264.

The entries in the tables are rather accurately computed, but, unlike many of the other standard spherical astronomical tables in the *Zīj*, are not taken from the *Ḥākimī Zīj* of Ibn Yūnus (**7.1.3**). Indeed it would appear that Abu 'l-ʿUqūl took these auxiliary tables from the same source from which he took the table of "Indian Sines", the identity of which is not yet established. The attribution of these auxiliary tables to Ibn Yūnus is put in question by the fact that the values of max d(ϕ) do not generally correspond to his values for λ = 90° in his tables of oblique ascensions, and also, for example, by the fact that the auxiliary tables have Sin [max d(45°)] (= Tan ε) = 23;11, whereas Ibn Yūnus' value for Tan ε is 23;11,33,23 (≈ 23;12). The formula for d(λ,ϕ) noted in **7.2** was used by Ibn Yūnus, but it was also used by certain of his predecessors such as al-Battānī and anyway is equivalent to the method proposed by Ptolemy in the *Almagest*.[24]

7.2.2 al-Khāzinī (Merw) / al-Fārisī (Aden)

In MS Vatican ar. 761, fol. 152r, of the *Sanjarī Zīj* of al-Khāzinī,[25] compiled in Marw *ca.* 1120, there is a set of tables of the same three functions also based on ε = 23;35° but less accurately computed. Certain significant errors in these tables are also found in the set in MS Cambridge Gg. 3.27, fol. 108r, of the *Muẓaffarī Zīj* of al-Fārisī, compiled in the Yemen about 1260 (**7.1.6**). Since al-Fārisī did not know of the *Sanjarī Zīj*,[26] both sets of tables must have a common source. Auxiliary tables of this kind are found in an early-15th-century Byzantine astronomical manuscript that has been investigated successively by Otto Neugebauer and Alexander Jones: this manuscript includes considerable material from the *Sanjarī Zīj*.[27] Tables of max d(ϕ) and Sin [max d(ϕ)] for ϕ = 16°, 17°, ... , 45° based on ε = 23;35° are also contained in MS Paris BNF ar. 5968, fol. 171r, of the anonymous *Dustūr al-munajjimīn* (**7.1.4**).

In MS Paris BNF supp. pers. 1488, fol. 109v, of the *Ashrafī Zīj* (**2.3.3**) there is a table entitled *taʿādīl al-nahār al-kulliyya al-muʿtabara*, "universal checked equations of daylight", also computed for ε = 23;35°. Three functions are displayed for each degree of ϕ from 1° to 60°, namely:

max d(ϕ), Sin [max d(ϕ)], and f(ϕ) ,

where f(ϕ) is a function labelled *taʿdīl ajzāʾ al-sāʿāt*, "equation of the degrees of the hours", whose nature and purpose escapes me. I can find no reference to the function in the text immediately preceding these tables. Many years ago, Jan Hogendijk explained its purpose to me but both of us have forgotten what it was.

A table of the function Sin α(λ) /R accompanies these three, and the set is followed by tables of oblique ascensions for each degree of λ and each degree of ϕ from 1° to 42°, with separate tables for 0° and 29;30° (Shiraz).

---

24 *Cf.* King, *Ibn Yūnus*, III.14.1d; Nallino, *al-Battānī*, I, p. 188; Neugebauer, *al-Khwārizmī*, p. 51; and *Almagest*, II.7 (pp. 92-94).

25 On al-Khāzinī and the *Sanjarī Zīj* see Kennedy, "*Zīj* Survey", no. 27, and the article by Robert E. Hall in *DSB*. The existence of these tables is noted by Kennedy on p. 159. More recent studies include King, *Mecca-Centred World-Maps*, pp. 71-75 and 564-585, and Pingree, "Editing the *Zīj al-Sanjarī*", pp. 105-113.

26 al-Fārisī lists some 28 *zīj*es known to him (see Lee, "Notice" (cited in n. 7:17), and also Kennedy, "*Zīj* Survey", no. 54), and the *Sanjarī* is not one of them.

27 Neugebauer, "Studies in Byzantine Astronomy", p. 30 (Appendix 12).

7.2.3 Anonymous (locality uncertain)

In MS Baghdad Awqāf 2966/6294, fol. 9v, of an anonymous *Zīj* containing tables mainly by Ibn al-Shāṭir, al-Ṣāliḥī and Ulugh Beg (**4.3.5**), there is a table of $\delta(\lambda)$ based on Ibn al-Shāṭir's parameter, $\varepsilon = 23;31°$, and a set of the same three auxiliary functions (for $\phi = 1°, 2°, \ldots, 66°$) based on $\varepsilon = 23;35°$. The entries are significantly different from those in the tables noted in **7.2.1** and **7.2.2**. Likewise, in MS Istanbul Hafid Efendi 181, fol. 93v, of the redaction of the *Īlkhānī Zīj* for Damascus prepared by Shihāb al-Dīn al-Ḥalabī (**5.1.2**) the auxiliary functions appear again, also based on $\varepsilon = 23;35°$. In MS Damascus Ẓāhiriyya 7387 of a recension of the *Zīj* of Ibn al-Shāṭir by al-Qazwīnī the same set based on $\varepsilon = 23;35°$ occurs once more. However, there are no such tables in MS Oxford Seld. A30 of the original *Zīj* of Ibn al-Shāṭir or in MS Cairo TR 275 of the Egyptian recension by al-Kawm al-Rīshī (**7.1.8**). The same set of tables occurs in the late Tunisian source MS Cairo K 7584,1 (**6.4.16**), in which other spherical astronomical tables are based on Ulugh Beg's parameter $\varepsilon = 23;30,17°$.

7.2.4 Ḥusayn Quṣʿa (Tunis)

MS Princeton Yahuda 147c of the *Quṣʿī Zīj* of the Tunisian astronomer Ḥusayn Quṣʿa (**4.3.6**) contains a page ruled and titled for the three auxiliary functions but there are no entries. It is stated in the title that $\varepsilon = 23;30°$.

7.3 Tables of proportional parts of the tangent of the local latitude, for calculating the equation of half-daylight

7.3.1 Ibn al-Mushrif (Cairo)

In MS Cairo MM 241 of the auxiliary tables of the Egyptian astronomer Ibn al-Mushrif (**9.8**) there is one set of tables of a function f(x,y). The arguments x and y are labelled *ẓill ʿarḍ* and *ẓill mayl*, that is, tangents of the latitude and declination. The horizontal arguments are:

$$21°,\ 24°,\ 30°,\ 31°,\ 32° \text{ and } 33;24°\ ,$$

and the vertical arguments are 1°, 2°, ... , 60. The function tabulated is:

$$f(\phi,n) \;=\; n/R \cdot \mathrm{Tan}_R\ \phi \qquad (R = 60)$$

where ϕ and n are the horizontal and vertical arguments, so that the table can be used to find the quantity:

$$\mathrm{Tan}\ \Delta \,/\, R \;\cdot\; \mathrm{Tan}\ \phi$$

which is Sin $d(\Delta,\phi)$. The latitudes correspond to Mecca, Medina, Cairo, Alexandria, Jerusalem and Damascus. The value for the latitude of Damascus, 33;24°, is attested in only one other known Islamic source, namely, the fragment of a *zīj* preserved in MS Utrecht Or. 1442.[28]

The values in the table are rather carelessly copied, which suggests that they were taken from an earlier source. The underlying values of Tan ϕ are respectively:

$$23;1,55,\quad 26;42,49,\quad 34;38,23,\quad 36;3,6,\quad 37;29,32 \quad\text{and}\quad 39;33,46\ ,$$

all of which are accurate. Ibn al-Mushrif notes that this table can also be used to find the auxiliary azimuth function, k(h): see further **8.2.1**.

---

28 Kennedy & Kennedy, *Islamic Geographical Coordinates*, p. 473 (*sub* UTT). See also n. **II**-10:34.

7.4 Tables of auxiliary functions for calculating the arc of daylight approximately

In this section I describe a table for computing the arc of daylight for any latitude, underlying which is the assumption that $d(\Delta,\phi)$ varies linearly with each of Δ and ϕ. It is worth mentioning that in various Byzantine manuscripts of the *Handy Tables* there is a table of a function $\xi(\lambda)$, tabulated to one digit for each degree of λ, with which the equation of daylight is given approximately using:

$$d(\lambda,\phi) \approx \xi(\lambda) \cdot \max d(\phi) .$$

William Stahlman has suggested that this table is a Byzantine addition to the *Handy Tables*.[29] By virtue of the fact that:

$$\xi(\lambda) \approx \mathrm{Sin}\ \alpha(\lambda) / R$$

and that tables of $\mathrm{Sin}\ \alpha(\lambda)/R$ (to two digits) were used by certain Muslim astronomers to determine $d(\lambda,\phi)$ (**7.2**), it may be that this particular table is ultimately of Islamic provenance. On the other hand, if the table is due to Theon then it may have provided the inspiration for the more elaborate Islamic tables.

7.4.1 Anonymous (Cairo)

In MS Paris BNF ar. 2513, fol. 61r, of the recension of the 13th-century Egyptian *Muṣṭalaḥ Zīj* (**6.7.1**) there is a table entitled *jadwal li-maʿrifat qaws al-nahār fī kull balad*, "table for finding the arc of daylight in any locality". The arguments are labelled *al-mayl*, "declination" and the entries *al-faḍla*, "equation of daylight"; the tabulated function is simply:

$$f(\delta) = 0;11\ \delta$$

for $\delta = 1°, 2°, \dots, 24°$. There are no instructions accompanying the table, but what is intended is that the equation of daylight in equinoctial hours d′ is very approximately given by:

$$d'(\delta,\phi) = 2\ d(\delta,\phi) / 15 \approx f(\delta) \cdot \phi / R .$$

This same approximation is mentioned in MS Istanbul Hamidiye 1453, fol. 222v, of an anonymous Egyptian treatise on timekeeping, which I suspect is by Najm al-Dīn al-Miṣrī (**2.6.1** and **II-6.5**). Another such approximation, namely:

$$\max d(\phi) \approx {}^{1}/_{2}\ \phi ,$$

is recorded on the title folio of MS Cairo DM 188, a copy of al-Asyūṭī's prayer-tables for Assiut (**II-6.16**).

---

[29] See Stahlman, *Handy Tables*, pp. 265 and 117-122, for table and commentary; also Neugebauer, *HAMA*, II, pp. 980-982.

CHAPTER 8

TABLES OF AUXILIARY FUNCTIONS FOR AZIMUTH CALCULATIONS

8.0 Introductory remarks

As noted above (**5.1**), the standard Islamic method for finding the solar azimuth from the solar altitude involves the use of two auxiliary functions (*cf.* **F15**). The first of these is called in Arabic *ikhtilāf al-ufq*, literally "difference of the horizon" or *ḥiṣṣat al-samt*, "azimuth component", and is defined by:

$$k(h,\phi) = \text{Sin } h \text{ Sin } \phi \;/\; \text{Cos } \phi \;.$$

For lack of simple, meaningful English equivalent, I shall refer to this function as the "auxiliary azimuth function". The second auxiliary function is called in Arabic *jayb saʿat al-mashriq* or simply *jayb al-saʿa*, which means "Sine of the rising amplitude". This is determined by:

$$L(\delta,\phi) = \text{Sin } \psi(\delta,\phi) = R \text{ Sin } \delta \;/\; \text{Cos } \phi \;.$$

With these two functions the azimuth is defined by:

$$a(h,\delta,\phi) = \text{arc Cos } \{ \; R \; [\; k(h,\phi) - L(\delta,\phi) \;] \; / \; \text{Cos } h \; \} \;.$$

In **8.1** and **8.3** I describe the tables of k(h) and L(δ,φ) for specific latitudes which have been located in the Islamic sources, and in **8.5** al-Khalīlī's tables of both functions for each latitude. The quantity:

$$m(h,\delta,\phi) = k(h,\phi) - L(\delta,\phi)$$

is usually called *taʿdīl al-samt*, "equation of the azimuth", in late medieval Arabic. At least one Muslim astronomer compiled a table of the secant function $G_1(h)$ (see already **6.9**) specifically to facilitate azimuth calculations. With this the determination of the azimuth reduces to:

$$a(h,\delta,\phi) = \text{arc Sin } \{ \; m(h,\delta,\phi) \cdot G_1(h) \; \}.$$

I discuss this secant table in **8.4**. al-Khalīlī further tabulated a double-argument auxiliary function with the equation of the azimuth as one argument and the solar altitude as the other (**8.5.1***): the function gives the azimuth without further ado.

8.1 Tables of the auxiliary azimuth function, for a specific latitude

Tables of k(h) for particular latitudes are found in the following sources. Note that al-Khalīlī's auxiliary function g_ϕ (**9.5**) defines k for each degree of both h and φ, thus:

$$g_\phi(h) = \text{Sin } h \text{ Sin } \phi \;/\; \text{Cos } \phi = k(h,\phi) \;.$$

8.1.1 Ibn Yūnus (Cairo and Baghdad)

Ibn Yūnus (**2.1.1**) compiled tables of k(h) for latitudes 30;0° (Cairo-Fustat) and 33;25° (Baghdad).[1] These tables occur in the *Ḥākimī Zīj*, MS Leiden Or. 143, pp. 381 and 357,

[1] On these tables and those in **8.3.1** see further King, *Ibn Yūnus*, III.18.2 and 19.2. The table for Baghdad is published already in Schoy, *Gnomonik der Araber*, p. 81.

respectively. Values are given to three digits for each degree of argument and are extremely accurately computed. Ibn Yūnus doubtless used his table for Cairo to compile his extensive azimuth tables (**5.1.1**).

8.1.2 Sanjar al-Kamālī: Shiraz, Rayy, and two other cities in Iran

In MS Paris BNF supp. pers. 1488, fol. 208r, of the *Ashrafī Zīj* (**2.3.1**) there are tables of k(h) for the latitudes of four cities in Iran – see **Fig. 6.2.1**. They bear the title *jadwal al-sumūt*, "table of azimuths", but the tabulated function is referred to as *aṣl*, "base". Values are given to two digits for each degree of h and the underlying latitudes are stated to be:

$$29;30°, \quad 30;22°, \quad 33;18° \quad \text{and} \quad 36;0° \ .$$

The first latitude is that of Shiraz and is used in other tables in the *Zīj* (see, for example, **6.4.5**), but the second and third values do not occur in the *Ashrafī* geographical tables, or in any other known medieval geographical tables,[2] and so the tables of k(h) may have been taken from an earlier source. The fourth value is standard for Rayy and the 4th climate.[3]

8.1.3 Anonymous: Tunis

In MS Berlin Ahlwardt 5724, fols. 30v-31r, of the anonymous corpus of tables for Tunis (**2.3.4**) there is a table of k(h) with entries to two digits for $\phi = 37;0°$ (Tunis). See also **8.3.3** and **8.4.1** on related tables in the same source.

8.1.4 Anonymous: Tlemcen

One of the anonymous tables for Tlemcen in MS London BL Or. 411,2 (**6.4.15**) displays the function k(h) with entries to two digits for $\phi = 35;0°$ (Tlemcen). There is no accompanying table of Sin $\psi(\lambda)$ or even $\psi(\lambda)$.

8.2 Tables of proportional parts of the tangent of the local latitude, for calculating the auxiliary azimuth function

8.2.1 Ibn al-Mushrif (Cairo)

Ibn al-Mushrif, in his instructions to his auxiliary tables in MS Cairo MM 241 (**7.3.1**), notes that his tables of the function:

$$f(\phi,n) \; = \; n/R \cdot \mathrm{Tan}_R\, \phi \qquad (R = 60)$$

for various latitudes can also be used to find the auxiliary azimuth function (*ḥiṣṣat al-samt*). One simply feeds in Sin h as the vertical argument, since:

$$k(h,\phi) \; = \; f(\phi,\mathrm{Sin}\ h) \ .$$

[2] In other words, they are not found in Kennedy & Kennedy, *Islamic Geographical Coordinates*. For the geographical table in the *Ashrafī Zīj* see now King, *Mecca-Centred World-Maps*, pp. 74-75 and 564-585.

[3] See Kennedy & Kennedy, *Islamic Geographical Coordinates*, p. 284, and King, "Geography of Astrolabes", pp. 6-8.

8.3 Tables of the Sine of the solar rising amplitude, for a specific latitude

The function $L(\delta,\phi)$ = Sin $\psi(\lambda)$ is related to Sin $\delta(\lambda)$ (**6.1**) by the relations:

$$\text{Sin } \psi(\lambda) = R \text{ Sin } \delta(\lambda) / \text{Cos } \phi = \text{Sin } \varepsilon \text{ Sin } \lambda / \text{Cos } \phi = \text{Sin [max } \psi(\phi)] \text{ Sin } \lambda / R .$$

Thus Sin $\psi(\lambda)$ can be computed from Sin $\delta(\lambda)$ (**6.1**) or Sin [max $\psi(\phi)$]. Note that Abū Naṣr's auxiliary function $f_1(\phi)$ (**9.3**) defines Sin [max $\psi(\phi)$], and that al-Khalīlī's auxiliary function f_ϕ (**9.5**) defines the Sine of the rising amplitude for each degree of declination and latitude, thus:

$$f_\phi(\Delta) = R \text{ Sin } \Delta / \text{Cos } \phi = \text{Sin } \psi(\Delta,\phi) .$$

Tables of $L(\lambda)$ = Sin $\psi(\lambda)$ for particular latitudes are located in the sources indicated below. Related tables of $\psi(\lambda)$ and max $\psi(\lambda)$ occur in various Islamic sources (**5.4** and **5.8**).

8.3.1 Ibn Yūnus: Cairo and Baghdad

In MS Leiden Or. 143, pp. 356-357, of the *Ḥākimī Zīj*, Ibn Yūnus tabulates Sin $\psi(\lambda)$ for the latitudes of Cairo and Baghdad (**8.1.1**).[4] He also tabulates $\psi(\lambda)$ for Cairo (**5.6.2**). The underlying value of ε is 23;35° and entries are given to three digits for each degree of λ.

8.3.2 al-Khalīlī: Damascus

MS Cairo DM 184 of al-Manāshīrī's prayer-tables for Damascus (**6.1.2**), which he attributes to al-Khalīlī, contains a table of Sin $\psi(\lambda)$ with entries computed to two digits for the parameters:

$$\phi = 33;30° \text{ and } \varepsilon = 23;31° .$$

The same table is contained in MS Damascus Ẓāhiriyya 9233, p. 156, of the main Damascus corpus, where it is introduced in the name of al-Ṭanṭāwī (see also **6.1.2**). al-Khalīlī also tabulated $\psi(\lambda)$ for these parameters (**5.6.6**).

8.3.3 Anonymous: Tunis

In MS Berlin Ahlwardt 5724, fol. 30r (**8.1.3**), there is a table of Sin $\psi(\lambda)$ with entries to two digits for the parameters ϕ = 37;0° (Tunis) and ε = 23;35°. There is also a table of $\psi(\lambda)$ for these parameters (**5.6.7**).

8.4 Tables of the secant function for azimuth calculations

On tables of the secant function intended to facilitate the computation of the hour-angle see **6.9** above.

8.4.1 Anonymous (Tunis)

In MS Berlin Ahlwardt 5724, fols. 30v-31r, alongside the table of k(h) for Tunis (**8.1.3**), there is a table of the function:

$$G_1(h) = R / \text{Cos } h$$

with entries to two digits for each degree of argument. The two tables are labelled jointly *jadwal*

[4] The table for Baghdad is published already in Schoy, *Gnomonik der Araber*, p. 81.

Fig. 8.4.2: The very special auxiliary function, actually just the secant, for calculating the rising amplitude, found in the anonymous Sfax treatise. That such a table just "pops up" in such a work points to the fact that a large proportion of our potential sources are lost for all time. [From MS Cairo K 7584,1, fol. 55r, courtesy of the Egyptian National Library.]

al-samt li-'l-shams wa-'l-kawākib, "table for solar and stellar azimuth" and there is no reference to the use of the function G_1 in timekeeping calculations.

8.4.2 Anonymous (Sfax)

MS Cairo K 7584,1, fol. 55r, of the anonymous astronomical compilation for Sfax (**6.4.16**) contains a table of the function:

$$G(\phi) = R / \text{Cos } \phi$$

labelled *ḥiṣaṣ*, literally, "shares" or "portions",[5] with values to three digits for each degree of ϕ. The table, shown in **Fig. 8.4.2**, occurs in a section dealing with the computation of ψ and the text prescribes the formula:

$$\text{Sin } \psi(\delta,\phi) = \text{Sin } \delta \cdot G(\phi) .$$

[5] See n. 1:48.

8.5 Auxiliary tables for finding the solar azimuth for any latitude

Only in January, 2001, did the following tables come to my attention. I mention them here only in passing: a more detailed account is in **9.4*** and **II-10.3***.

8.5.1* al-Khalīlī

MS Bursa Haraççioğlu 1177,4 (fols. 72r-90r), copied *ca.* 1450, is a unique copy of a set of tables by Shams al-Dīn al-Khalīlī, not identified as the compiler, which contain sub-tables for the following functions:

$$L(\lambda) = \text{Sin}\ \psi(\lambda) \text{ and } k(h)$$

for each 1° of ϕ from 1° to 49°, and each 1° of the arguments λ and h. The value of ε underlying the first table is about 23;30°, possibly 23;31°. With these the instructions describe how to determine the *taʿdīl al-samt*. The third sub-table displays the function:

$$K(x,y) = \text{arc Cos}\ \{\ x \bullet R\ /\ \text{Cos}\ h\ \}$$

for each unit value of x and each 1° of h. These universal auxiliary tables represent the ultimate solution to the problem of determining the solar azimuth a(h,λ,ϕ). See further **9.4***, where they are considered in the light of al-Khalīlī's other tables.

CHAPTER 9

TABLES OF AUXILIARY FUNCTIONS FOR SOLVING THE PROBLEMS OF SPHERICAL ASTRONOMY FOR ALL LATITUDES

9.0 Introductory remarks

In the 9th century Ḥabash compiled a set of tables of functions which had no specific astronomical significance, but which were so conceived that combinations of them would lead to the solution of certain problems in spherical astronomy. In the 10th century al-Nayrīzī and Abū Naṣr compiled other sets of such functions. These three sets of tables contain a few hundred entries and they are not ideally suited for use in the solution of problems in timekeeping. However, in the early 14th century, the Cairo astronomer compiled a monumental table for timekeeping, with close to half a million entries, and proposed its use as a universal auxiliary table for solving all of the standard problems of spherical astronomy, including the determination of the qibla and the coversion of ecliptic and equatorial coordinates. Such a table, before the age of printing, was doomed from the outset, and it is small wonder that it survives only in a single copy in the hand of its author. In the mid 14th century, the Damascus astronomer al-Khalīlī compiled his more subtle, and eventually more useful, universal auxiliary tables. These contain over 13,000 entries and were likewise devised to solve the standard problems of timekeeping for all latitudes, with no calculation beyond addition, subtraction and interpolation.

In the sequel I briefly describe the auxiliary tables of Ḥabash, al-Nayrīzī, Abū Naṣr and al-Khalīlī (**9.1**, **9.2**, **9.3** and **9.5**), all of which are known from previous studies, and I also present a description of several other sets of auxiliary tables which have come to light during recent research. Most of these, like al-Khalīlī's tables, were intended to be used for problems in timekeeping. Most of the functions displayed in these tables are extensions of the auxiliary functions discussed in **Chs. 6** and **8** to serve more than one latitude. Only the auxiliary tables of al-Khalīlī, Ibn al-Mushrif and al-Wafāʾī (**9.5**, **9.8** and **9.9**) take advantage of the fact that the problems of determining the hour-angle and azimuth from the altitude, declination and latitude, are mathematically equivalent.

9.1 The "Rectification Table" of Ḥabash al-Ḥāsib

Ḥabash, called al-Ḥāsib, meaning "the astronomer" rather than "the calculator", lived for over a hundred years and was most active in Baghdad, Samarra and Damascus in the first half of the 9th century.[1] He was the author of several *zīj*es, one of which survives in a later recension

[1] On Ḥabash (Suter, *MAA*, no. 22) see Kennedy, "*Zīj* Survey", nos. 15 and 16; Sezgin, *GAS*, V, pp. 275-276, and VI, pp. 173-175; King, *Mecca-Centred World-Maps*, pp. 40-41 and 345-349; and also nn. 2-4 below.

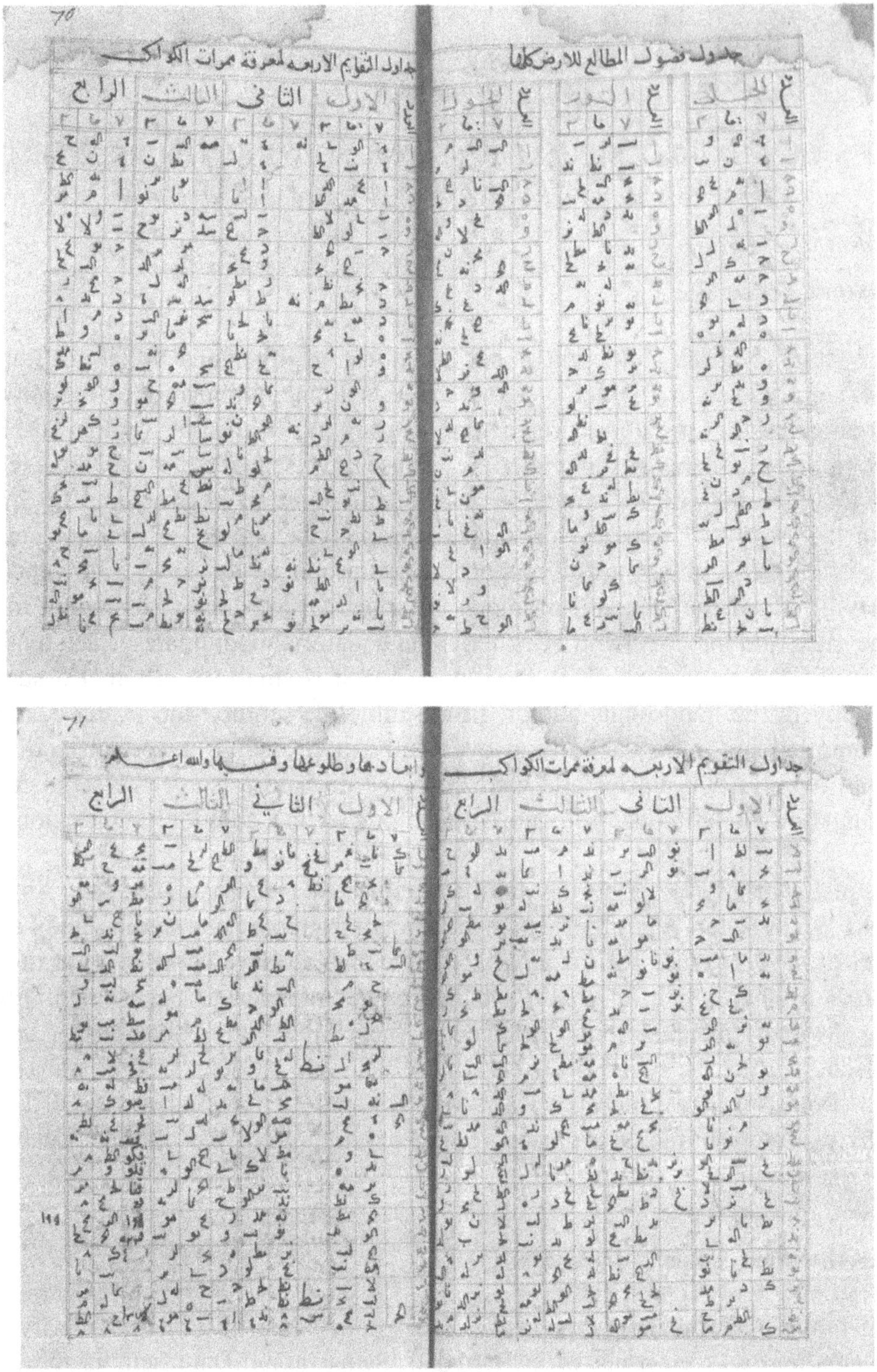

Fig. 9.1a-b: The auxiliary tables of Habash as they occur in a Mamluk *zīj*. On the right of the first double page is a table of the tangent of the solar declination (**7.1.9**). [From MS Paris BNF ar. 2520, fols. 69v-71r, courtesy of the Bibliothèque Nationale de France.]

in MS Istanbul Yeni Cami 784, fols. 69v-229r, from the 13th century. Another anonymous recension of a different *zīj* by Ḥabash is contained in MS Berlin Ahlwardt 5750 (Wetzstein 90), copied *ca.* 1300. The first source has been studied in detail by Marie-Thérèse Debarnot.[2]

The Istanbul manuscript contains two related sets of auxiliary tables, both apparently compiled by Ḥabash himself and both entitled *Jadwal al-taqwīm*, "Rectification Table".[3] The two sets of tables in the Istanbul manuscript have been analyzed in detail in a valuable study by Rida A. K. Irani, alas unpublished.[4] The main difference between the two sets is that one is based on $\varepsilon = 23;35°$ and the other on $\varepsilon = 23;33°$. Ibn Yūnus (**2.1.1**) noted with disapproval that in his "*Mumtaḥan Zīj* called *al-Qānūn*" Ḥabash used $\varepsilon = 23;35°$ in the solar declination tables and $\varepsilon = 23;33°$ in the auxiliary tables.[5]

A set of Ḥabash's tables for $\varepsilon = 23;35°$ is also contained in a treatise preserved in MS Bankipore 2468,8 (fols. 50v-66r), copied 631 H [= 1233/34], which consists of a discussion by Abū Naṣr (**9.3**) of Ḥabash's tables and procedures.[6] Yet other sources exist for the study of Ḥabash's tables.[7] For example, there are various related auxiliary tables in MS Berlin Ahlwardt 5750 of the recension of Ḥabash's *Zīj* (**9.2**), and another copy of the auxiliary tables has been located in MS Paris BNF ar. 2520 of a recension of the Egyptian *Muṣṭalaḥ Zīj* – see **6.7.1** and **Fig. 9.1a-b**. Again, a table of one of Ḥabash's functions occurs in MS Paris BNF supp. pers. 1488 of the Persian *Ashrafī Zīj* – see **6.2.1** and **Fig. 6.2.1**.

Ḥabash's auxiliary functions are the following (R = 60):[8]

$$F_1(\lambda) = \delta_2(\lambda) = \text{arc Tan } \{ \text{ Tan } \varepsilon \text{ Sin } \lambda \text{ / R } \}$$
$$F_2(\lambda) = \text{Cos } \delta(\bar{\lambda}) = \text{Cos } \{ \text{ arc Sin } [\text{ Cos } \lambda \text{ Sin } \varepsilon \text{ / R }] \}$$
$$F_3(\lambda) = \text{R Cos } \lambda \text{ / } F_2(\lambda) = \text{R Cos } \lambda \text{ / Cos } \delta(\bar{\lambda})$$
$$F_{4a}(\theta) = \text{Tan } \theta \text{ Sin } \varepsilon \text{ / R}$$
$$F_{4b}(\theta) = \text{Tan } \theta \ .$$

They are of use in solving particular problems of spherical astronomy, though not those directly relating to timekeeping. Values are given to three digits for each degree of argument and I have not investigated their accuracy. Maybe before I retire I shall organize a seminar to prepare a publishable version of Ḥabash's work.

MS Leiden Or. 468 (280 fols., copied *ca.* 1400), contains the introduction and first of five *maqāla*s of a work entitled *Kanz al-yawāqīt fī 'stīʿāb al-mawāqīt* and compiled by an anony-

---

[2] See Debarnot, "*Zīj* of Ḥabash".

[3] The use of the word *taqwīm* here is curious: in medieval Arabic scientific terminology it generally means "the operation of finding the longitudes of the sun, moon, and planets", or "an ephemeris (in which the planetary longitudes are displayed for intervals of a few days)". See n. 1:42.

[4] Irani, "*Jadwal al-Taqwīm* of Ḥabash al-Ḥāsib". I have a photocopy of a carbon copy.

[5] MS Leiden UB Or. 143, p. 223, lines 10-13, translated in Schoy, "Bestimmung der Ortsbreite nach Ibn Yūnus", p. 11.

[6] See Sezgin, *GAS*, VI, p. 243.

[7] Note that: (i) The tables of F3 in MSS Istanbul YC 784 (fols. 226v-227r) and Paris ar. 2520 need further investigation since if they are indeed intended to display the function associated with F_3 the errors are larger than in the rest of the table. See Irani, *op. cit.*, p. 75. (ii) The table of F_2 in MS Berlin 5750 (fols. 85r-87v) contains significant variants from those in MSS Istanbul YC 784 (fols. 147r-148v) and Bankipore 2468,8 and may not be due to Ḥabash.

[8] As noted in Irani, *op. cit.*, p. 51, the functions tabulated in MS Istanbul YC 784 (fols. 226v-227r) (see previous note) are to base 1. Those in MS Paris ar. 2520, however, are to base 60. Abū Naṣr confirms that Ḥabash used bases 60 and 1 in his two sets of auxiliary tables (*ibid.*, p. 128).

mous Egyptian author *ca.* 1350.[9] The work consists mainly of material culled from the *Muṣṭalaḥ Zīj* and the *Kitāb al-Mabādiʾ wa-ʾl-ghāyāt* of al-Marrākushī (**II:6.6-7**), both compiled in Cairo in the late 13th century. Besides various spherical astronomical tables also found in al-Marrākushī's treatise, the *Kanz al-yawāqīt* contains a set of auxiliary tables not found in any other known sources. The tables (fol. 147v-150r) are entitled *jadwal al-nisab*, "table of ratios" and display values of four functions for each half-degree of argument θ from 0;30° to 90° to three sexagesimal digits. The entries in the tables are garbled but the tabulated functions are recognizable as sexagesimal multiples of:

$$\cos \delta(\bar{\theta}) \ , \quad \cos \theta \ / \ \cos \delta(\bar{\theta}) \ , \quad \tan \varepsilon \tan \theta \quad \text{and} \quad \tan \theta \ .$$

I have not compared the entries in this Egyptian source with those found in MS Paris ar. 2520 of the Egyptian *Muṣṭalaḥ Zīj* mentioned above.

9.2 The auxiliary tables of al-Nayrīzī and an anonymous set

In MS Berlin Ahlwardt 5750 of the anonymous recension of one of the *zīj*es of Ḥabash (**9.1**) there are two sets of auxiliary tables, the first (fols. 82r-84v) entitled *al-jadāwil al-jāmiʿa*, "the universal tables", and the second (fols. 85r-87v) entitled *jadwal al-nisab*, "table of ratios". Various operations with the first set are outlined in the text of the *Zīj*. These tables have not been discussed before in any detail,[10] and they deserve a more thorough investigation than is possible here. Indeed our knowledge of early Islamic mathematical astronomy in general will be greatly improved when each of MSS Istanbul Yeni Cami 784 (fols. 69v-229r) and Berlin Ahlwardt 5750 of Ḥabash's *Zīj*, as well as MS Escorial ár. 927 of the anonymous recension of the *Mumtaḥan Zīj*, have been subjected to a critical comparative analysis.[11]

The first set is clearly related to *al-jadwal al-jāmiʿ*, "the universal table", of the late-9th-century astronomer al-Nayrīzī.[12] Abū Naṣr (**9.3**) mentions that al-Nayrīzī incorporated Ḥabash's *jadwal al-taqwīm* into his *Zīj* and added some more functions of his own.[13] This *zīj* is alas no longer extant. The first set of auxiliary tables in the Berlin manuscript displays ten functions, with entries computed to three digits for each degree of argument. Five of these are simply as follows (R = 60):

$$\text{Sin } \theta \text{ and Vers } \theta \ ,$$
$$\delta(\lambda), \text{ Sin } \delta(\lambda) \text{ and Cos } \delta(\lambda) \ (\varepsilon = 23;35^\circ) \ .$$

The remainder are labelled "Ḥabash's second, third and fourth (functions), and al-Nayrīzī's first and second (functions)". The first three are indeed Ḥabash's functions F_2 (computed for $\varepsilon = 23;33^\circ$), F_3 and F_{4a} (computed for $\varepsilon = 23;35^\circ$), and the remaining two are simply F_2 and F_3 (also for $\varepsilon = 23;35^\circ$), to base 150 rather than 60.

---

[9] King, *Mecca-Centred World-Maps*, pp. 80-83 and 114, esp. n. 82 on p. 82.

[10] The existence of some of these tables was first noted in Schoy, "Beiträge zur arabischen Trigonometrie", p. 392. See also Irani, *op. cit.*, pp. 32-33.

[11] On these three sources see Kennedy, "*Zīj* Survey", nos. 15, 16, and 51, and the abstracts of contents on pp. 145-147 and 151-154. On the *Mumtaḥan Zīj* see also n. 4:7.

[12] On al-Nayrīzī (Suter, *MAA*, no. 88) see Kennedy, "*Zīj* Survey", nos. 46 and 75, and Sezgin, *GAS*, V, pp. 283-285, and VI, pp. 177-182.

[13] See Irani, *op. cit.*, p. 19.

This curious assortment of tables is followed immediately by the second set displaying another ten functions. The entries are also computed to three digits for each degree of argument, and the functions tabulated are the following (R = 60):

$$g_1(\lambda) = \text{Cos } \delta(\bar{\lambda}) \qquad [= F_2(\lambda)]$$
$$g_2(\lambda) = \text{Sin } \alpha(\bar{\lambda}) \qquad [= g_{10}(\bar{\lambda})]$$
$$g_3(\lambda) = \alpha(\bar{\lambda})$$
$$g_4(\phi) = \text{Tan } \phi \text{ Tan } \varepsilon \,/\, R$$
$$g_5(\theta) = \text{Tan } \theta$$
$$g_6(\lambda) = \lambda(\alpha)$$
$$g_7(\lambda) = \delta(\lambda)$$
$$g_8(\theta) = R \,/\, \text{Sin } \theta$$
$$g_9(\theta) = \text{Tan } \theta \,/\, R$$
$$g_{10}(\lambda) = \text{Sin } \alpha(\lambda) \qquad [= g_2(\bar{\lambda})]$$

Note the following points. First, the tables of g_1, g_2, g_3, g_4, g_6, g_7 and g_{10} are based on ε = 23;35° but the entries in the table of g_{10} do not correspond precisely to those in the table of g_2. Second, the table of g_1 differs from both of Ḥabash's tables of F_2. Third, the value of Tan ε underlying the table of g_4 is 26;11,34 (the accurate value is 26;11,33). (See **7.1** on other values of this parameter.) Fourth, $g_4(\phi)$ is tabulated only up to ϕ = 60°. For the remaining arguments up to 90° a different function, which I am unable to interpret, is tabulated. Sample entries are:

61°	0;47,15	.		85	4;49,22
62	0;49,13	.		86	6;14,34
63	0;51,53	75	1;37,45	87	8;19,20
64	0;53,41	.		88	12;31,30
65	0;56,11	.		89	26;30, 0
.		80	2;33,33	90	50; 0, 0
.		.			
70	1;11,18	.			

Fifth, the function f_8 is simply the modern cosecant: tables of this function were also used in Islamic instrument construction.[14]

Here again we have at first sight a rather motley set of auxiliary tables. But we do well not to forget that we are dealing with tables compiled almost over a thousand years ago. The anonymous compiler has included a function equivalent to Ḥabash's F_2 and Abū Naṣr's f_5 as well as others equivalent to Abū Naṣr's f_2 and f_3. I have not investigated the accuracy of either set of auxiliary tables in the Berlin manuscript.

---

[14] The cosecant table is published in Schoy, "Beiträge zur arabischen Trigonometrie", p. 15, from defective photographs (Schoy was able to publish only 60 entries rather than the 90 that are found in the manuscript).

For an example of a cosecant table for use in marking curves on astrolabes and quadrants see the Ottoman source MS Princeton UL Yahuda 373, fol. 74r. Here the cosecant and cotangent are tabulated side by side to three sexagesimal digits for each degree of arc: on one possible application of such a pair of tables see King, "al-Māridīnī's Universal Quadrant", App. A.

9.3 Abū Naṣr's "Table of Minutes"

Abū Naṣr ibn ʿIrāq[15] worked in Khwārizm *ca.* 1000 and was a teacher of the illustrious al-Bīrūnī.[16] He was the author of numerous works of considerable merit, two of which concern the present study. The first of these is a commentary on the auxiliary tables of Ḥabash and al-Nayrīzī (**9.1-2**), and the second is a treatise written to al-Bīrūnī in which he introduces a set of auxiliary tables of his own. Both of these texts were published in Hyderabad in 1948 using MS Bankipore 2468,14 (fols. 86v-93r), copied 631 H [= 1233/34], and Abū Naṣr's auxiliary tables have been analyzed in a valuable study by Claus Jensen.[17] MS Oxford Bodleian Thurston 3, fols. 111r-114r, copied 675 H [= 1276/77] is another copy of Abū Naṣr's treatise on his own auxiliary tables – see **Fig. 9.3**.

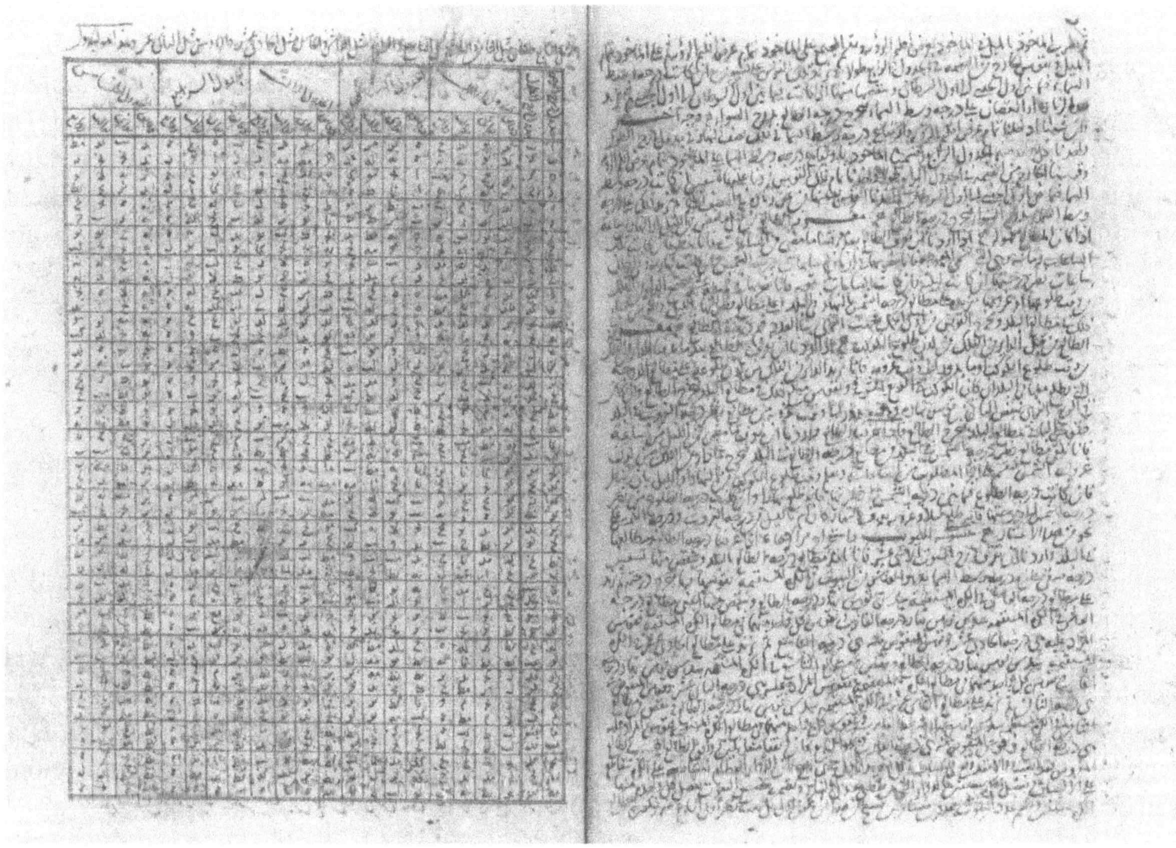

Fig. 9.3: The end of Abū Naṣr's treatise on his auxiliary tables and the first tables. [From MS Oxford Bodleian Thurston 3, fols. 113v-114r, courtesy of the Bodleian Library.]

---

[15] On Abū Naṣr (Suter, *MAA*, no. 186) see the article "Manṣūr ibn ʿAlī ibn ʿIrāq" by Julio Samsó in *DSB*, and Sezgin, *GAS*, V, pp. 338-341, and VI, pp. 242-245, and more especially Samsó, *Estudios sobre Abū Naṣr*.
[16] On al-Bīrūnī see the splendid article by Ted Kennedy in *DSB*, and also **II-2.2**.
[17] See Jensen, "Abū Naṣr's *Table of Minutes*".

The underlying trigonometric functions used by Abū Naṣr are to base 1 rather than 60, but values are, of course, given sexagesimally. This explains the title *Jadwal al-daqāʾiq*, "Table of Minutes (*i.e.*, sexagesimal fractions)", which he gave to his auxiliary tables. The functions tabulated are the following:

$$f_1(\phi) = \sin \varepsilon \,/\, \cos \phi$$
$$f_2(\phi) = \tan \varepsilon \tan \phi$$
$$f_3(\lambda) = \tan \delta(\lambda) \,/\, \tan \varepsilon$$
$$f_4(\theta) = \sin \theta$$
$$f_5(\lambda) = \cos \delta(\bar{\lambda}) \,.$$

Abū Naṣr used Ptolemy's value 23;51,20° for ε rather than the value 23;35° which was generally accepted in his time. His tables contain 225 entries, rather carelessly computed to four sexagesimal digits.[18]

Abū Naṣr's $f_4(\theta)$ is simply the sine function, and the other functions $f_1(\phi)$, $f_2(\phi)$ and $f_3(\lambda)$ are respectively:

$$\sin\,[\max\, \psi(\phi)] \,, \quad \sin\,[\max\, d(\phi)] \quad \text{and} \quad \sin \alpha(\lambda) \,.$$

Tables of max $\psi(\phi)$, max $d(\phi)$ and its Sine, and Sin $\alpha(\lambda)$ occur in several other Islamic sources (**7.2** and **9.2**). Note that Abū Naṣr's f_5 is equivalent to Ḥabash's F_2.

Abū Naṣr's functions are not ideally suited to solving the problems of timekeeping. For example, to find $T(h,\lambda,\phi)$ he suggests a method which involves first finding $\delta(\lambda)$ and $\sin d(\lambda,\phi)$. Thus we begin by using the relations:

$$\delta(\lambda) = f_4^{-1}\,\{\, f_4(\lambda) \cdot f_4(\varepsilon) \,\} \quad \text{and} \quad \sin d(\lambda,\phi) = f_2(\phi) \cdot f_3(\lambda)$$

and then apply:

$$T(h,\lambda,\phi) = f_4^{-1}\,(\sin d) + f_4^{-1}\,\{\, [\, f_4(h) \cdot \sin d + f_4(h) \,] \,/\, f_4(\,\bar{\phi}+\delta\,) - \sin d \,\} \,.$$

It is not difficult to show that this last formula is equivalent to **F11**.

Abū Naṣr does not describe the computation of the corresponding azimuth using his tables. The procedure would be clumsy, although the value of $\sin \psi(\lambda)$ is conveniently given by:

$$\sin \psi(\lambda) = f_1(\phi) \cdot f_4(\lambda) \,.$$

9.3* Najm al-Dīn al-Miṣrī's universal auxiliary tables

Najm al-Dīn al-Miṣrī (**I-2.6.1**) intended his monumental triple argument table of a function $F(x,y,z)$ to be used as a universal auxiliary table. A commentary on the use of this table for this purpose is in MS Dublin 102,1, copied *ca.* 1325, and has been studied in detail by François Charette.[19] I refrain here from a detailed commentary, not least in order to encourage the reader to consult Charette's study, which is to be regarded as a supplement to the present work. Suffice it to say that Najm al-Dīn describes not only the use of the table to find the time since rising and the azimuth, and special cases thereof such as the determination of the length of daylight and rising amplitude, but also to find the qibla using the accurate formula and to covert between ecliptic and equatorial coordinates. Also, that Najm al-Dīn shows as much originality in his

[18] On the accuracy of these tables see King, "Al-Khalīlī's Auxiliary Tables", p. 107, n. 7.

[19] Charette, "Najm al-Dīn's Monumental Table".

Figs. 9.4a-b: Two extracts from al-Khalīlī's universal auxiliary tables for timekeeping by the sun. The first (a) shows a sub-table for the (half) *aṣl* function for latitudes 31°, 32°, 33°, 33;30° (Damascus), and 34°, with the solar longitude entered vertically. The second (b) shows a sub-table for the hour-angle as a function of the same *aṣl* function entered horizontally, here with arguments 29 and 30, and the difference of the Sines of the meridian and instantaneous altitudes entered vertically. [From MS Dublin CB 4091, fols. 163r and 166r, courtesy of the Chester Beatty Library.]

instructions as in his tirelessness at compiling his enormous table. However, all this seems to have taken its toll on his mind.

9.4 al-Khalīlī's auxiliary tables for timekeeping by the sun

In the unique source MS Dublin CB 4091, copied 833 H [= 1429], there is a set of auxiliary tables by the Damascus astronomer al-Khalīlī (**2.1.4**) for finding $t(h,\lambda,\phi)$, which is quite different from his *Universal Table* (**9.5**). The principal functions tabulated are the following:

$$B'(\phi,\lambda) = \text{Cos}\ \delta(\lambda)\ \text{Cos}\ \phi\ /\ 2R = {}^{1}/_{2}\ B(\phi,\lambda) \qquad (\varepsilon = 23;31°)$$

for the domains:

$$\phi = 1°, 2°, \ldots, 49° \text{ as well as } 33;30° \text{ (Damascus) and } \lambda = 1°, 2°, \ldots, 90°,$$

and:

$$V'(x,y) = \text{arc Vers}\ \{\ R \cdot y\ /\ 2x\ \}$$

for the domains:

$$x = 30, 29, \ldots, 19 \text{ and } y = 0;10, 0;20, \ldots, 60;0\ .$$

These two sets of tables – see **Figs. 9.4a-b** – contain respectively 4,500 and 4,680 entries, rather accurately computed to two digits. Horizontal differences are shown in the second set. The first function is called *al-aṣl*, "the base" (*cf.* **6.0**) but a marginal note at the beginning of the table points out that it is in fact *niṣf al-aṣl*, "half the base". The second function is called *faḍl al-dā'ir āfāqī min al-aṣl wa-min faḍl jayb al-ghāya ʿan jayb al-irtifāʿ*, "the hour-angle for all latitudes as a function of the base and the excess of the Sine of the meridian altitude over the Sine of the (instantaneous) altitude".

al-Khalīlī's tables are complete with instructions and are preceded by two small tables of the function:

$$\delta'(\lambda) = 90° + \delta(\lambda)\ (\varepsilon = 23;31°)$$

and the Sine function, with entries to two digits for each degree of argument. To find $t(h,\lambda,\phi)$ al-Khalīlī proposes first using the table of $\delta'(\lambda)$ to find $H(\lambda,\phi)$, thus:

$$H(\lambda,\phi) = \delta'(\lambda) - \phi\ ,$$

and then using the Sine table to find Sin H and Sin h in order to establish their difference:

$$H'(h,\lambda,\phi) = \text{Sin}\ H(\lambda,\phi) - \text{Sin}\ h\ (> 0)\ .$$

Then, using the two main sets of auxiliary tables one should find $B'(\phi,\lambda)$, with which:

$$t(h,\lambda,\phi) = V'\ \{\ B'(\phi,\lambda)\ ,\ H'(h,\lambda,\phi)\ \}\ .$$

al-Khalīlī's tables are clearly an extension of those of al-Khaṭā'ī (**6.15.1**), which were devised for a specific latitude, but the two sets may have been compiled independently. Although only one copy of these original auxiliary tables of al-Khalīlī has come to light, as opposed to the several extant copies of his *Universal Table*, it appears that they became known in both Tunis and Cairo: see further **9.7** and **9.11** and also **II-10.11.**

9.4* al-Khalīlī's universal auxiliary tables for finding the solar azimuth

The following tables are mentioned by al-Khalīlī is the introduction to his main universal auxiliary tables (**9.5**) but the unique surviving copy came to my attention only in January, 2001.

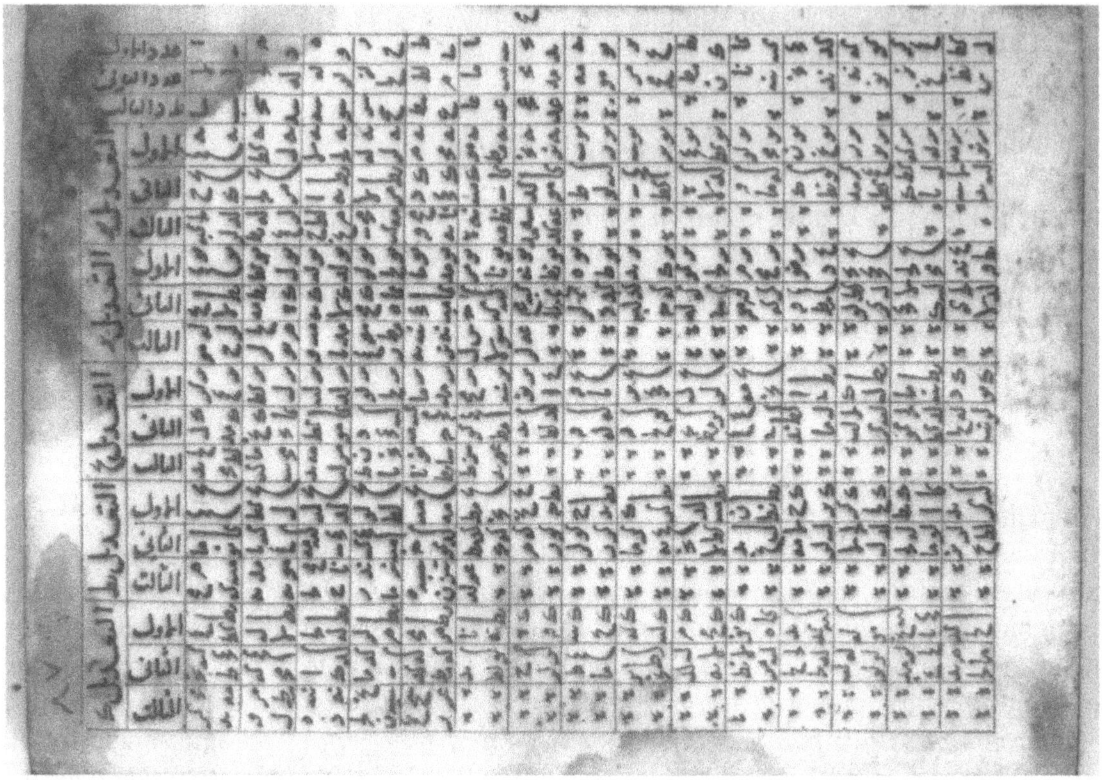

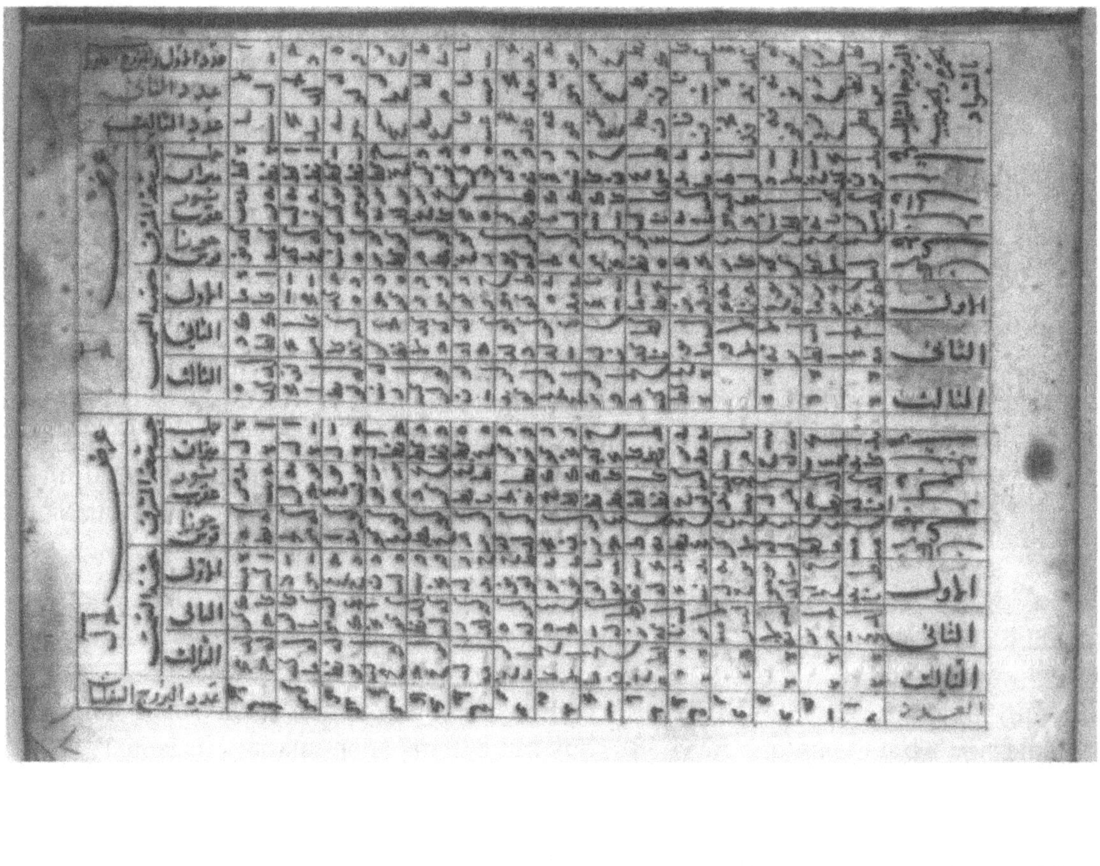

Fig. 9.4*a-b: Extracts from al-Khalīlī's tables of f' and g_{ϕ} for latitudes 33° and 33;30° (a) and his table of K for arguments 16, 17, ..., 20 (b). See also **Fig. II-10.3*a-b**. [From MS Bursa Haraççioğlu 1177,4, fols. 80v-81r and 86v-87r, courtesy of the Genel Kütüphanesi.]

I have mentioned them briefly in **8.5.1***, and they will be treated again in the context of al-Khalīlī's other tables in **II-10.3***.

MS Bursa Haraççioğlu 1177,4 (fols. 72r-90r), copied *ca.* 1450, is a unique copy of a set of tables entitled *Jadwal al-samt li-kulli 'rtifāʿ fī ʿurūḍ al-aqālīm al-sabʿa*, "table of the azimuth for all altitudes in the latitudes of the seven climates (*i.e.*, for all latitudes).[20] No compiler is mentioned, but it can only be Shams al-Dīn al-Khalīlī (**9.4** and **9.5**). The set contain sub-tables for the following functions:

$$f'_{\phi}(\lambda) = R \text{ Sin } \delta(\lambda) \;/\; \text{Cos } \phi \quad \text{and} \quad g_{\phi}(h) = \text{Sin } h \text{ Tan } \phi \;/\; R$$

for the domains:

$$\lambda \text{ and } h = 1°, 2°, \ldots, 90° ,$$
$$\phi = 1°, 2°, \ldots, 48°, \text{ as well as } 33;30° \text{ (Damascus)} ,$$

and:

$$K(x,h) = \text{arc Sin } \{ R \cdot x \;/\; \text{Cos } h \}$$

for the domains:

$$x = 1, 2, \ldots, 59 \text{ and } h = 0°, 1°, \ldots, n(x) ,$$

where $n(x)$ is the largest integer such that $x \leq \text{Cos } n(x)$. The functions f'_{ϕ} and g_{ϕ} are labelled *jayb saʿat al-mashriq* and *ḥiṣṣat al-samt* by al-Khalīlī, and it is clear that they represent Sin $\psi(\lambda)$ and $k(h)$ – see **5.0** and **8.5**. The value of ε underlying the first table is about 23;30°, possibly 23;31°.

The instructions describe how to determine the *taʿdīl al-samt* using the first two tables, and then how to find the azimuth from the third table. These universal auxiliary tables represent the ultimate solution to the problem of determining the solar azimuth $a(h,\lambda,\phi)$. It is clear that all al-Khalīlī had to do to compile his splendid tables described in the next section was to replace the table of $f'_{\phi}(\lambda)$ with a new one of $f_{\phi}(\theta)$ and rewrite the instructions!

9.5 al-Khalīlī's "Universal Table"

After compiling his two sets of auxiliary tables for timekeeping by the sun (**9.4**) and for finding the solar azimuth (**9.4***), al-Khalīlī prepared another set which can be used for solving all of the standard problems of spherical astronomy for any latitude.[21] These tables are appropriately called *al-Jadwal al-āfāqī*, "The Universal Table". They are preserved in several sources, including MSS Paris BNF ar. 2558, Berlin Ahlwardt 5754/5/6 (Wetzstein 1138) and 5739 (Wetzstein 1144), Escorial ár. 931,8, fols. 171r-211v, Istanbul Ayasofya 2590, Istanbul Hamidiye 1453 (fols. 232v-266v), Istanbul Serez 1914, Oxford Marsh 95 (Uri 961), Cairo MM 43, Cairo MM 98, Cairo DM 758, and Princeton Yahuda 861,2. See also **II-10.7**.

The functions tabulated by al-Khalīlī are:

$$f_{\phi}(\theta) = R \text{ Sin } \theta \;/\; \text{Cos } \phi \quad \text{and} \quad g_{\phi}(\theta) = \text{Sin } \theta \text{ Tan } \phi \;/\; R$$

for the domains:

$$\theta = 1°, 2°, \ldots, 90° ,$$
$$\phi = 1°, 2°, \ldots, 55° \text{ as well as } 21;30° \text{ (Mecca) and } 33;30° \text{ (Damascus)} ,$$

[20] I owe the reference to this manuscript to İhsanoğlu *et al.*, *Ottoman Astronomical Literature*, II, pp. 805. My thanks are due to Dr. Sonja Brentjes of Berlin for showing me a microfilm thereof.

[21] See King, "al-Khalīlī's Auxiliary Tables", for the first analysis.

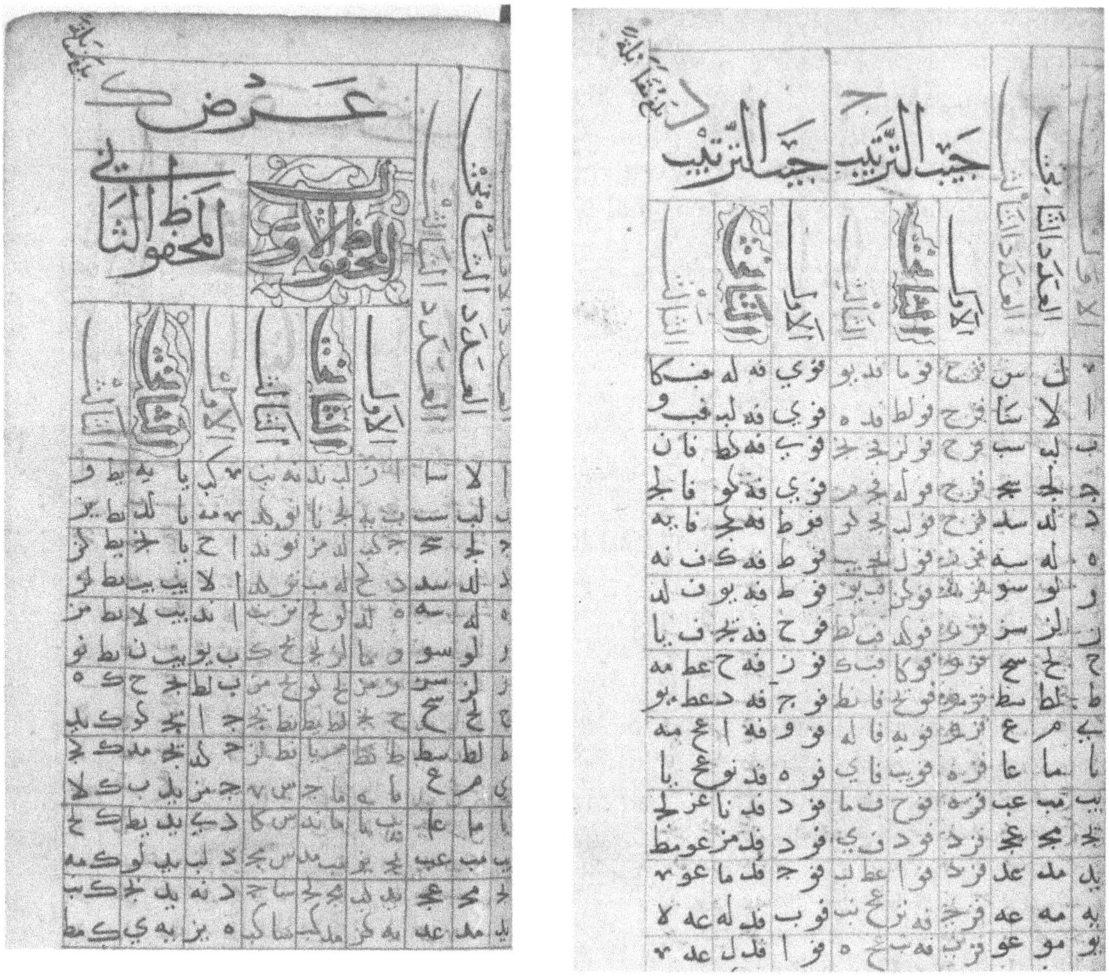

Fig. 9.5a-b: The sub-tables of al-Khalīlī's first and second functions for latitude 20° (a) and his third function for arguments 3-4 (b). See also **Fig. I-6.14.3**. [From MS Paris BNF ar. 2558, courtesy of the Bibliothèque Nationale de France.]

and:

$$K(x,y) = \text{arc Cos } \{ R \cdot x / \text{Cos } y \}$$

for the domains:

$$x = 1, 2, \ldots , 59 \text{ and } y = 0°, 1°, \ldots , n(x) ,$$

where n(x) is the largest integer such that $x \leq \text{Cos } n(x)$. al-Khalīlī calls f_ϕ and g_ϕ *al-maḥfūẓ al-awwal* and *al-thānī*, "the first and second functions", and he refers to K by the same expression used for the argument x, namely *jayb al-tartīb,* "the auxiliary Sine". His tables contain over 13,000 entries, rather accurately computed – see **Fig. 9.5a-b**. The values for the second and third function were, of course, simply taken over from his universal auxiliary tables for finding the solar azimuth (**9.4***).

To find $D(\Delta,\phi)$, $t(h,\Delta,\phi)$ and $a(h,\Delta,\phi)$ using his tables al-Khalīlī outlines the following rules:

$$D(\Delta,\phi) = 180° - K \{ g_\phi(\Delta) , \Delta \} \ (\Delta>0) \quad \text{or} \quad K \{ g_\phi(\Delta) , \Delta \} \ (\Delta<0)$$

$$t(h,\Delta,\phi) = K \{ [f_\phi(h) - g_\phi(\Delta)], \Delta \} \quad \text{and} \quad a(h,\Delta,\phi) = K \{ [g_\phi(h) - f_\phi(\Delta)], h \} .$$

The functions are devised to take advantage of the equivalence of the problems of determining the hour-angle and azimuth. Note that al-Khalīlī's f_ϕ and g_ϕ are simply extensions of the auxiliary functions k and Sin ψ used in Islamic azimuth calculations (**8.0**), since:

$$f_\phi(\Delta) = \text{Sin}\ \psi(\Delta,\phi) \quad \text{and} \quad g_\phi(h) = k(h,\phi) .$$

al-Khalīlī's *Universal Table* was used for several centuries in Syria and also in Cairo, the Maghrib, and Turkey. MSS Cairo MM 43, Cairo MM 98, Cairo DM 758 and Princeton Yahuda 861,2 are Egyptian copies of these tables, and it is reasonable to assume that both al-Wafā'ī and al-Ṣūfī (**9.9** and **9.10**) had seen the *Universal Table*. MS London BL Add. 9599,31 is a late Maghribi copy of al-Khalīlī's table. MS Istanbul Hamidiye 1453 (fols. 232v-266v) is a copy of the tables prepared *ca.* 1475 in Edirne, and MS Istanbul Ayasofya 2590 is another Turkish copy of the tables preceded by a Turkish translation of al-Khalīlī's instructions, prepared in 896 H [= 1491] by the Ottoman astronomer Muḥammad ibn Kātib Sinān (**2.7.2**). In an unnumbered manuscript formerly (*ca.* 1970) in the private collection of the late Professor Buhairi of the American University of Beirut, there are tables of f_ϕ and g_ϕ for $\phi = 41°$ (Istanbul) and of G for x = 41.[22] The hapless Ottoman copyist who put these three tables together was unaware that the argument x is in no way related to the latitude (**II-10.11** and **14.6**).

9.6 al-Māridīnī's auxiliary tables

The late-14th-century astronomer Jamāl al-Dīn al-Māridīnī, who appears to have worked in both Damascus and Cairo,[23] compiled an auxiliary table which is extant in MS Paris BNF ar. 2525,1 (fols. 1v-16v), copied *ca.* 1450, and also in the much later copy MS Cairo K 4026. al-Māridīnī called his table *al-shabaka*, which means "net" or "grid" and I have analyzed it in detail in a previous publication.[24] The *shabaka* displays values of three trivial functions, arranged in a single 90 × 90 table containing 8,100 entries. No indication is given in the instructions that more than one function is tabulated and in the sequel I use M(x,y) to denote all three functions, which are in fact:

$$M_1(x,y) = R \cdot y / x \qquad \text{for } 0 \le x \le 59,\ y < x$$
$$M_2(x,y) = \text{Sin}\ x° \cdot \text{Sin}\ y° / R \qquad \text{for } 0° \le x \le 90°,\ x < y \le 90°$$
$$M_3(x,y) = 24\ y / x \qquad \text{for } 60 \le x \le 90,\ y < x$$

The function M_3 is intended to be used for calculating conjunctions and does not concern the present study.

To derive the hour-angle using his table al-Māridīnī suggests finding the three quantities:

$$B(\phi,\Delta) = M(\bar{\phi},\bar{\Delta})\ (\phi > \Delta) \quad \text{or} \quad M(\bar{\Delta},\bar{\phi})\ (\phi < \Delta)$$
$$C(\phi,\Delta) = M(\phi,\Delta)\ (\phi < \Delta) \quad \text{or} \quad M(\Delta,\phi)\ (\phi > \Delta)$$
$$b(h,\Delta,\phi) = \text{Sin}\ h \ \square\ C(\phi,\Delta)\ (\text{as } \Delta \gtrless 0) ,$$

---

[22] I am indebted to the late Prof. M. Buhairi of the American University of Beirut for showing me this manuscript from his personal collection.

[23] On al-Māridīnī (Suter, *MAA*, no. 421) see *Cairo ENL Survey*, no. C47; and King, "al-Māridīnī's Universal Quadrant", p. 219, n. 2.

[24] *Ibid.*, pp. 231-240. The term *shabaka* is not to be confused with the same term used for a table displaying the solar longitude over a four-year period: see n. 9:26.

which he calls respectively *al-aṣl al-muṭlaq*, *jayb irtifāʿ quṭr al-madār and al-aṣl al-muʿaddal* (see further **6.0**). With these t is defined by:

$$\text{Cos } t(h,\Delta,\phi) = M\{\ B(\phi,\Delta)\ ,\ b(h,\Delta,\phi)\ \}\ .$$

To find the azimuth a(h,Δ,φ) one of al-Māridīnī's methods involves first finding the arc:

$$\xi = \text{arc Sin (Tan } \phi)$$

and then the three quantities:

$$k(h,\phi) = M(\xi,h)\ (\xi < h) \quad \text{or} \quad M(h,\xi)\ (\xi > h)$$

$$\text{Sin } \psi(\Delta,\phi) = M(\text{Cos } \phi,\text{Sin } \Delta)$$

$$m(h,\Delta,\phi) = k(h,\phi)\ \square\ \text{Sin } \psi(\Delta,\phi)\ (\text{as } \Delta \gtreqless 0)$$

which he calls respectively *ḥiṣṣat al-samt*; *jayb al-saʿa* and *taʿdīl al-samt.* al-Māridīnī also suggests the following alternative method. First form the three quantities:

$$p(h,\phi) = M(h,\phi)\ (\phi > h) \quad \text{or} \quad M(\phi,h)\ (\phi < h)$$

$$q(h,\Delta,\phi) = p(h,\phi) - \text{Sin } \Delta$$

$$r(h,\phi) = M(\bar{\phi},\bar{h})\ (\phi > h) \quad \text{or} \quad M(\bar{h},\bar{\phi})\ (\phi < h)$$

and then:

$$\text{Sin } a(h,\Delta,\phi) = M\{\ r(h,\phi),\ q(h,\Delta,\phi)\ \}\ .$$

9.7 Anonymous Tunisian tables for timekeeping by the sun

MS Cairo DM 689, copied *ca.* 1600, contains an extensive set of auxiliary tables copied in an elegant Maghribi hand.[25] The tables conclude with a star catalogue dated 801 H [= 1398] and they appear to have been compiled in Tunis. The title folio, instructions and first few tables are missing from the manuscript, which begins with the last page of a set of tables of the *shabaka* variety[26] displaying the solar longitude for each day of a period of four Syrian years (see **Fig. 1.2d**). The tables for timekeeping follow this – see the extracts in **Figs. 9.7a-b**.

The main functions tabulated are the same as those in al-Khalīlī's auxiliary tables for timekeeping by the sun (**9.4**), namely B′(φ,λ) and V′(x,y). In fact these Tunisian tables are merely an extension of al-Khalīlī's tables and the corresponding entries in both sets are the same. The Tunisian tables of B′ are computed for each degree of φ from 1° to 48° and also 21;40° (Mecca) and various non-integral latitudes between 30° and 40° some of which were intended to serve specific localities in Ifriqiyya and the Maghrib and perhaps also Sicily. al-Khalīlī had a separate table for latitude 33;30° (Damascus) and the Tunisian set has separate tables for latitudes:

30;30°, 31;30°, 32;30°, 33;30°, 33;40°, 34;30°, 35;30°,
36;30°, 36;40°, 37;10°, 37;30°, 38;30°, 39;30° .

The Tunisian tables of V′(x,y) are simply those of al-Khalīlī rearranged so that the horizontal argument is increasing.

In MS Cairo DM 689 the tables of B′ are preceded by a set of tables of the function Sin H(λ), called *jayb al-ghāya*, computed to two digits for each degree of λ and the latitudes of

[25] *Cairo ENL Survey*, no. F30.
[26] See n. 1:43.

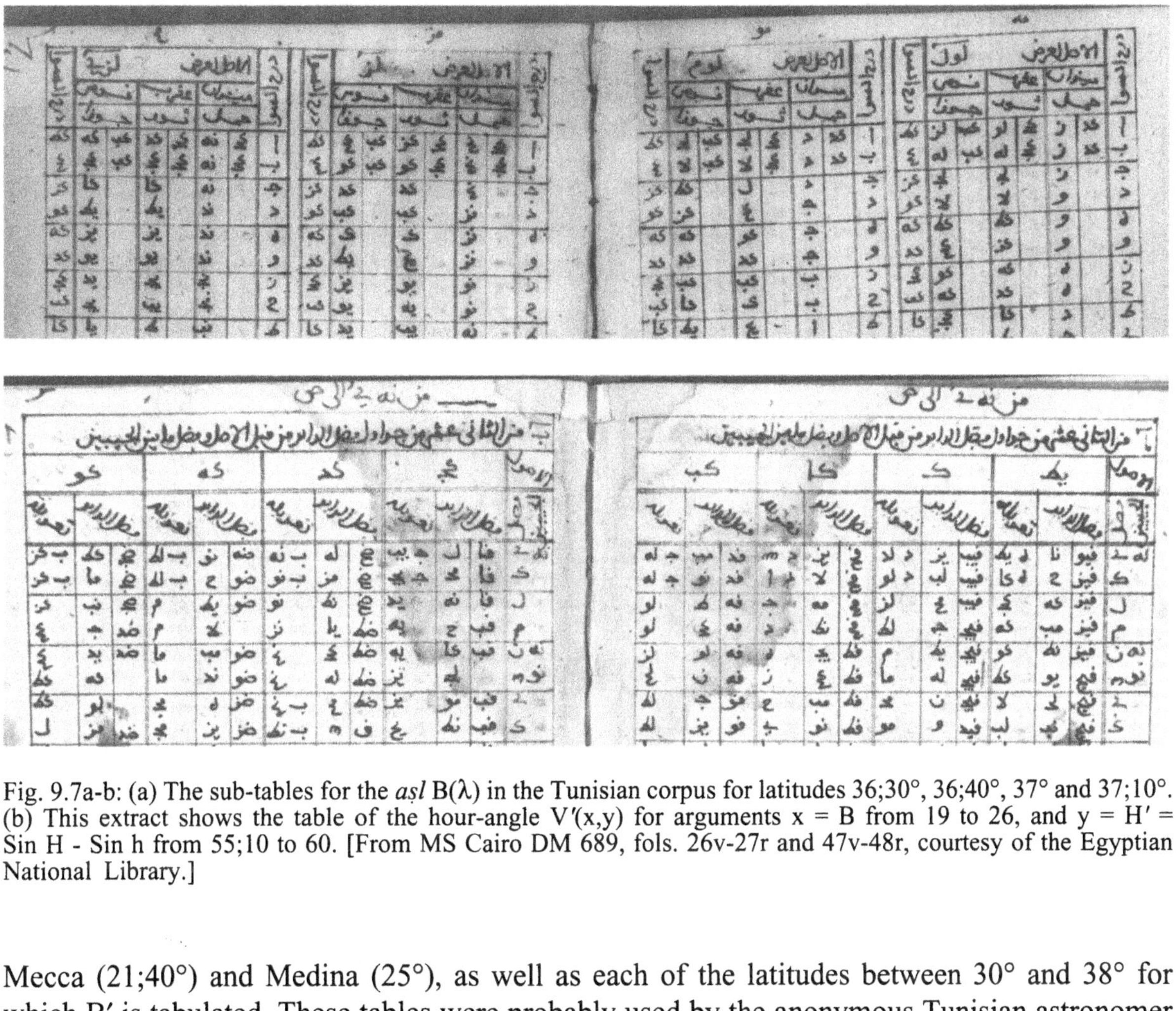

Fig. 9.7a-b: (a) The sub-tables for the *asl* B(λ) in the Tunisian corpus for latitudes 36;30°, 36;40°, 37° and 37;10°. (b) This extract shows the table of the hour-angle V′(x,y) for arguments x = B from 19 to 26, and y = H′ = Sin H - Sin h from 55;10 to 60. [From MS Cairo DM 689, fols. 26v-27r and 47v-48r, courtesy of the Egyptian National Library.]

Mecca (21;40°) and Medina (25°), as well as each of the latitudes between 30° and 38° for which B′ is tabulated. These tables were probably used by the anonymous Tunisian astronomer to compile his tables of B′, since:

$$B'(\phi,\lambda) = {}^1/_2 \; \{ \text{Sin } H(\phi,\lambda) + \text{Sin } H(\phi,\lambda^*) \} \; .$$

The underlying value of ε is 23;35°, whereas al-Khalīlī used 23;31°. However, the change in ε hardly affects the values of B′(λ) given to two digits. The Tunisian tables also contain a table of Sin H(λ) for φ = 0°, which is simply Cos δ(λ). See further **6.2.5** and **6.3.3**, and **II-13.4**.

9.8 Ibn al-Mushrif's auxiliary tables

MS Cairo MM 241 is an apparently unique copy from *ca.* 1450 of a set of auxiliary tables by Abū Bakr ibn Ismāʿīl, known as Ibn al-Mushrif (see already **6.13.1**, **7.3.1** and **8.2.1**).[27] The work is entitled *Nūr al-aḥdhāq li-maʿrifat aʿmāl al-falak fī sāʾir al-āfāq*, "The Light of the

[27] On Ibn al-Mushrif see *Cairo ENL Survey*, no. C43. He was still active in 848 H [= 1444/45]: see **II-5.7**, so he cannot be identical with the individual with the same name mentioned in King, *Astronomy in Yemen*, no. 12, and **6.16**.

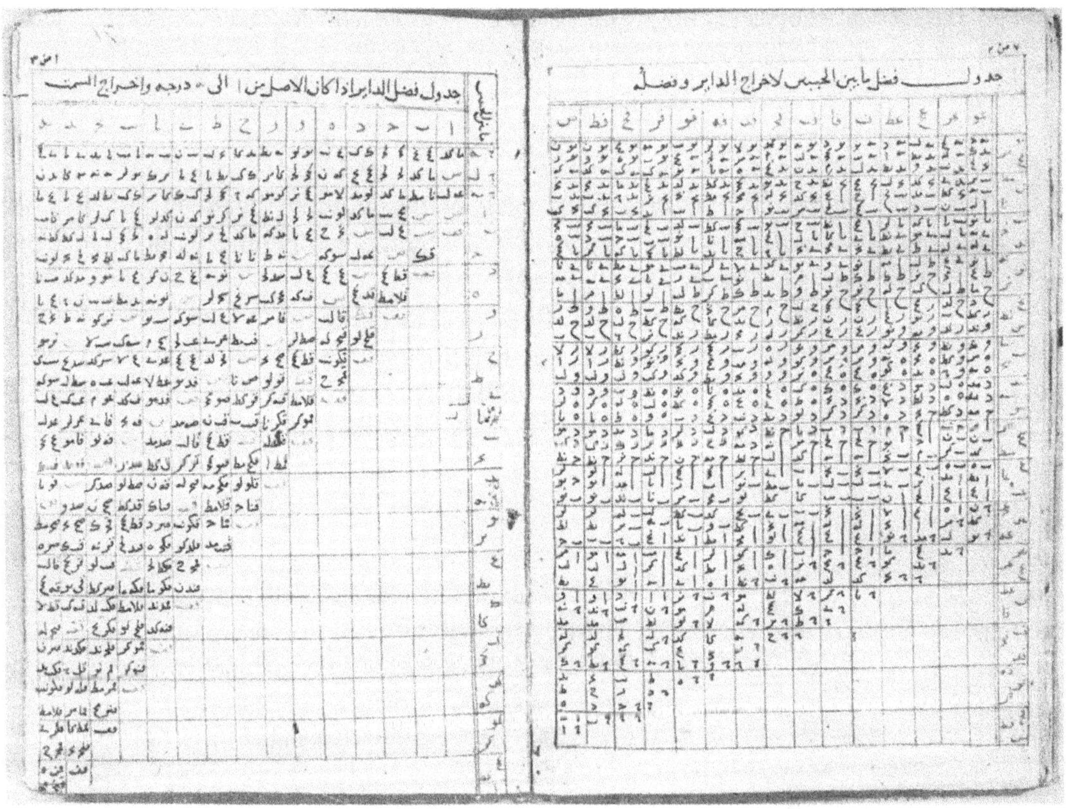

Intelligent for Performing Astronomical Calculations for all Latitudes", and appears to have been compiled in Cairo in 1435. The title-folio of this manuscript is shown in **Fig. 9.8a**. The three main functions tabulated are:

$$B(\phi,\Delta) = \text{Cos}\ \phi\ \text{Cos}\ \Delta\ /\ R$$

for the domains:

$$\phi = 0°, 1°, \ldots, 89° \text{ and } \Delta = 0°, 1°, \ldots, 90° ;$$

$$H'(H,h) = \text{Sin}\ H - \text{Sin}\ h$$

for the domains:

$$H = 1°, 2°, \ldots, 90° \quad \text{and} \quad h = 1, 2°, \ldots, H - 1° ;$$

and:

$$V'(x,y) = \text{arc Vers}\ \{\ R \cdot y\ /\ x\ \}$$

for the domains:

$$x = 1, 2, \ldots, 60 \quad \text{and} \quad y = 0;15, 0;30, 0;45, 1, 2, \ldots, 60 .$$

These three functions are called *al-aṣl*, "the base", *faḍl mā bayn al-jaybayn*, "the difference between the two Sines", and *faḍl al-dā'ir idhā kān al-aṣl kadhā daraja wa-ikhrāj al-samt*, which means "hour-angle or azimuth for base so-and-so". An extract is shown in **Fig. 9.8b**. The tables are complete with instructions and to find the hour-angle t(h,Δ,φ) one simply finds B(φ,Δ) and H′(H,h) and uses these as arguments in the third table, thus:

$$t(h,\Delta,\phi) = V'\ \{\ B(\phi,\Delta)\ ,\ H'(H,h)\ \} .$$

The method for finding the azimuth a(h,Δ,φ) is to find B(φ,h) and:

$$H'\ \{\ (\ \bar{\phi} + h\ ), \Delta\ \} = \text{Sin}\ (\ \bar{\phi} + h\ ) - \text{Sin}\ \Delta$$

using the first two tables, and then the azimuth (measured from the north point) is given by:

$$a(h,\Delta,\phi) = V'\{\ B(\phi,h), H'\{\ (\ \bar{\phi} + h\ ), \Delta\ \}\ \} .$$

Various other minor functions are tabulated by Ibn al-Mushrif including δ(λ) for ε = 23;35° and B(λ) for φ = 30° (**6.4.3**) and the inverse Versed Sine function (**6.14**), as well as a table for finding the equation of half daylight or the auxiliary azimuth function (**7.3.1** and **8.2.1**). One small table, which is not referred to in the instructions and whose purpose eludes me, is entitled *jadwal taḥwīl juyūb al-tartīb*, "Table for converting the auxiliary Sines". The argument is *al-ʿarḍ*, "latitude", and the values of the function f(φ) are labelled *al-taʿdīl*, "equation" or "interpolation factor". The tabulated function displays discontinuities between φ = 30° and 31° and between φ = 31° and 32°. Sample entries are:

←
Figs. 9.8a-b: (a) The title-folio of the unique copy of Ibn al-Mushrif's tables. The owner was the copyist Abu 'l-Yumn Muḥammad ibn Muḥammad ibn Muḥammad ibn ʿArab (*Cairo ENL Catalogue*, I, pp. 714 and 761). There is a note in a different hand listing the planetary apogees for epoch 740 H [= 1339/40] from the *Muṣṭalaḥ Zīj*, the most popular *zīj* in medieval Egypt. A further note in the lower left corner is in the hand of Ibn Abi 'l-Fatḥ al-Ṣūfī, the leading astronomer in Egypt *ca.* 1500: it deals with what one might call "advanced timekeeping". (b) The last part of the table of the second function and the beginning of the table of the third function in Ibn al-Mushrif's set. By the nature of the functions both tables are of the *ṭaylasān* variety. [From MS Cairo MM 241, fols. 1r and 11v-12r, courtesy of the Egyptian National Library.]

ϕ	$f(\phi)$						
		.		30	; 0,35,24	.	
1°	; 0, 0, 0	.		31	1; 0,37,11	.	
2	; 0, 1, 27	20	; 0,20,36	32	1; 1,16,48	40	1; 7,49,50
3	; 0, 2, 21	.		33	1; 1,24,24	.	
.		.		.		.	
.		29	; 0,33,37	35	1; 3,25,19	45	1;13,29, 4
.							
10	; 0, 9, 16						

The tabulated function is certainly neither Sec ϕ nor arc Vers (x) / x nor Vers ϕ.

MS Cairo DM 512, copied *ca.* 1080 H [≈ 1670], contains an extract from Ibn al-Mushrif's tables by Sūdūn al-Bashtakī.[28] The author presents only the tables of V′(x,y), with his own introduction. MS Cairo MM 209,1 (fols. 1r-37r) contains tables of normed right ascension to three digits for each minute of ecliptic longitude attributed to "Abū Bakr ibn al-Mushrif", copied in 873 H [= 1468/69] from a copy by the author dated 848 H [= 1444/45].

9.9 al-Wafā'ī's auxiliary tables

al-Wafā'ī was an Egyptian astronomer who died *ca.* 1470.[29] MSS Vatican Borg. ar. 217,1 (fols. 1v-5v), copied *ca.* 1500, and Istanbul Nuruosmaniye 2921,2 (fols. 22r-26v) contain a small set of auxiliary tables attributed to him entitled *Kifāyat al-waqt li-maʿrifat al-dā'ir wa-faḍlihi wa-'l-samt*, which might be rendered "All you need to find the time, hour-angle and azimuth".

The first function tabulated is:

$$G(\phi,\theta) = R^3 / [\ \mathrm{Cos}\ \phi\ \mathrm{Cos}\ \theta\ ]$$

with values to three digits for horizontal arguments (latitudes):

ϕ = 12° (Hadramawt, Aden), 15° (Zabid, Sanaa), 18° (Ethiopia), 21° (Aswan, Jidda), 24° (Qift, Yathrib (= Medina)), 27° (Akhmim, Yenbo), 30° (Fayyum, Aqaba), 32° (Jerusalem, Hebron), 33;30° (Damascus, Tripoli), 36° (Mardin, Tunis), 39° (Malatya), 42° (Edirne) and 45° (Constantinople)

and vertical argument:

$$\theta = 0°, 1°, \ldots , 89° ,$$

and the second function is simply:

$$V(x) = \mathrm{arc\ Vers}(x) ,$$

with values to degrees and minutes for:

x = 0;5, 0;10, ... , 0;30, 0;40, ... , 2;0, 2;15, ... , 3;0, 3;20, ... , 4;0, 4;30, ... , 16, 17, 18, ... , 120 .

The two tables of these functions are labelled *jadwal al-ḥiṣaṣ* and *jadwal faḍl al-dā'ir*, that is "argument table" and "hour-angle table". These titles generally refer to tables giving the times of prayer and tables in which the hour-angle is given as a function of solar altitude, respectively. An extract is shown in **Fig. 9.9a-b**.

---

28 On Sūdūn see *Cairo ENL Survey*, no. C81.

29 On al-Wafā'ī see Suter, *MAA*, no. 437, and *Cairo ENL Survey*, no. C98.

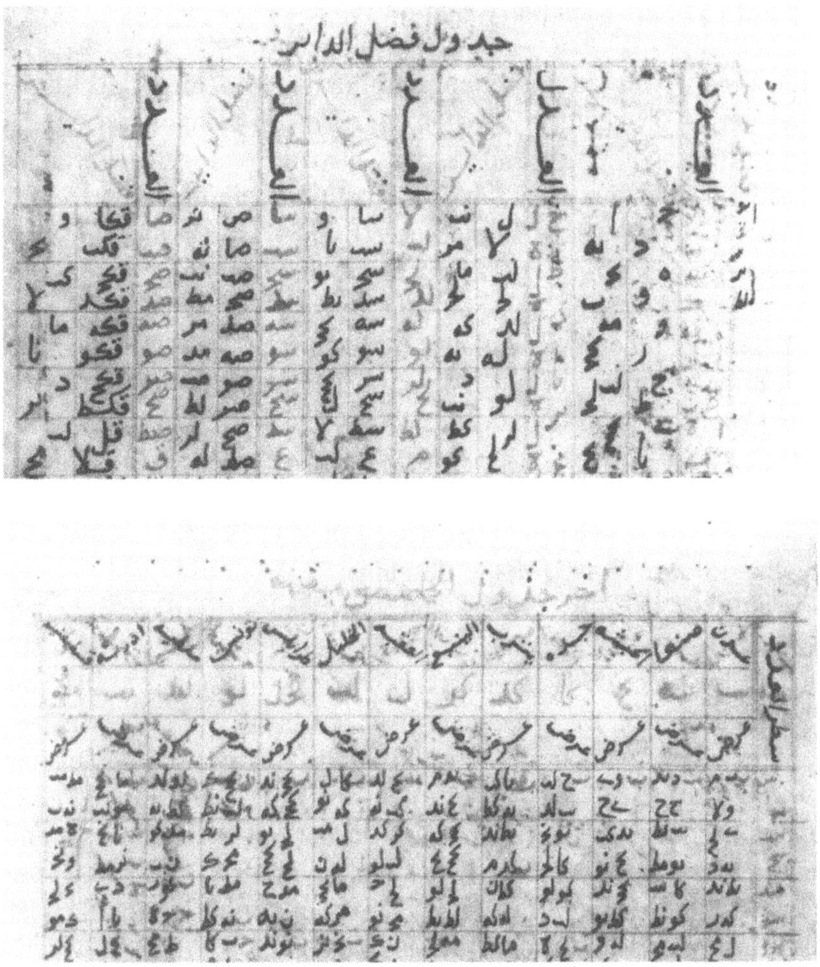

Fig. 9.9a-b: Extracts from the auxiliary tables of al-Wafā'ī. They are followed by part of the incomplete auxiliary tables of al-Khaṭā'ī (**6.15.1**). [From MS Vatican Borg. ar. 217,1, courtesy of the Biblioteca Apostolica Vaticana.]

al-Wafā'ī's tables, which contain 1,320 entries, are rather accurately computed and are complete with instructions. To find t from h, Δ and φ one first computes:

$$H'(h,\Delta,\phi) = \text{Sin } H(\Delta,\phi) - \text{Sin } h = \text{Sin } (\bar{\phi} + \Delta) - \text{Sin } h$$

and then uses:

$$t(h,\Delta,\phi) = V \{ H'(h,\Delta,\phi) \cdot G(\phi,\Delta) \} .$$

Similarly, to find a(h,Δ,φ) one computes:

$$H'(\Delta,h,\phi) = \text{Sin } (\bar{\phi} + h) - \text{Sin } \Delta$$

and then uses:

$$a(h,\Delta,\phi) = V \{ H'(\Delta,h,\phi) \cdot G(\phi,h) \} ,$$

where a is measured from the north point. al-Wafā'ī's use of the functions G and H to determine t or a with equal facility is ingenious and it fully exploits the equivalence of the underlying mathematical problems. He also demonstrates how to compute the duration of twilight and the direction of Mecca with his tables.

9.10 Ibn Abi 'l-Fatḥ al-Ṣūfī's auxiliary tables

Ibn Abi 'l-Fatḥ al-Ṣūfī was an Egyptian astronomer who died *ca.* 1495.[30] He prepared a recension of the early-15th-century *Zīj* of Ulugh Beg of Samarqand, adapting the planetary tables for the longitudes of Cairo and he was also the author of several short treatises on quadrants and sundials. His auxiliary tables are contained in the apparently unique source MS Oxford Seld. Supp. 101 (Uri 1040), copied in 918 H [= 1513], and they are entitled *Kitāb al-Jawāhir fī maʿrifat al-samt wa-faḍl al-dāʾir*, "Book of Jewels for Finding the Azimuth and the Hour-angle".

The first and second functions tabulated by al-Ṣūfī are the following:

$$C(\xi,\eta) = \text{Sin}\ \xi\ \text{Sin}\ \eta\ /\ R \quad \text{and} \quad B(\xi,\eta) = \text{Cos}\ \xi\ \text{Cos}\ \eta\ /\ R\ ,$$

computed to two digits for the domains:

$$\xi = 0°, 1°, \ldots, 89°, \ \eta = 0°, 1°, \ldots, 90° \text{ and also } \eta = 33;30° \text{ and } 23;35° \ (= \varepsilon)\ .$$

The third is:

$$S(x,y) = \text{arc Cos}\ \{\ R \cdot x\ /\ y\ \}\ ,$$

computed likewise to two digits for the domains:

$$x = 0;30,\ 1;0,\ \ldots,\ 59;30 \quad \text{and} \quad y = [x] + 1,\ [x] + 2,\ \ldots,\ 60\ .$$

The first two functions are called *ufq* and *aṣl* respectively, that is, "horizon" and "base" (see further **6.0**). The third function is referred to by the term applied to the argument x, namely, *al-muqanṭara*, "almucantar", for what reason is not clear.

al-Ṣūfī's tables, which contain about 15,000 entries, are more sophisticated than those of al-Māridīnī (**9.6**) but are hardly an improvement over the *Universal Table* of al-Khalīlī (**9.5**). In fact, it seems that they owe their inspiration to both al-Māridīnī's set and al-Khalīlī's first set of auxiliary tables (**9.4**): compare al-Ṣūfī's function C with al-Māridīnī's M_1 and his functions B and S with al-Khalīlī's B′ and V′. The link with al-Khalīlī is also apparent from the fact that al-Ṣūfī, working in Cairo, tabulates B and C for $\phi = 33;30°$, which serves Damascus. Likewise it may be that al-Ṣūfī noticed that al-Māridīnī's function M_1 would be more useful if one could feed in the non-integral value of ε as one argument in order to facilitate the computation of $\delta(\lambda)$: his tabulating C for $\eta = 23;35°$ satisfies this need. al-Ṣūfī clearly had not learned much from the investigations of the Damascus and Samarqand astronomers on the secular change of ε: he preferred to use the standard value, 23;35°. He begins his work with a table for finding the solar longitude for the period 877-1004 H [= 1472-1596], based on Ibn Yūnus' solar parameters derived five centuries previously. al-Ṣūfī's auxiliary tables can bc used for computations involving the stars as well as the sun, whereas al-Khalīlī's first set of auxiliary tables are for timekeeping by the sun only. al-Ṣūfī notes that his two functions B and C are related by:

$$B(\xi,\eta) = C(\bar{\xi},\bar{\eta})$$

but states that he preferred to tabulate both functions rather than only one. To find the hour-angle $t(h,\Delta,\phi)$ his instructions prescribe the simple rule:

[30] On Ibn Abi 'l Fatḥ al-Ṣūfī see Suter, *MAA*, no. 447; Kennedy, "*Zīj* Survey", no. 37; *Cairo ENL Survey*, no. C98; and İhsanoğlu *et al.*, *Ottoman Astronomical Literature*, I, pp. 116-126, no. 58 (where this work is not mentioned).

$$t(h,\Delta,\phi) = S \{ [\text{Sin } h - C(\phi,\Delta)] \cdot B(\phi,\Delta) \},$$

but to find the corresponding azimuth $a(h,\Delta,\phi)$, as in the case of al-Māridīnī's tables, the procedure is more complicated. First form:

$$m(h,\Delta,\phi) = \text{Cos } [S \{ C(\phi,h), \text{Cos } \phi \}] + \text{Sin } \psi(\Delta,\phi)$$

and then:

$$a(h,\Delta,\phi) = 90° - S \{ m(h,\Delta,\phi) , \text{Cos } h \} .$$

Note that Cos $\{S(x,y)\}$ is simply R • x / y and that al-Ṣūfī's tables are not suited to determine Sin $\psi(\Delta,\phi)$ easily.

9.11 Two sets of anonymous Egyptian auxiliary tables for timekeeping by the sun

MS Istanbul S. Esad Efendi Medresesi 119,2 (fols. 82v-97v), copied *ca.* 1700, contains a set of auxiliary tables bound at the end of a set of prayer-tables for Cairo by Riḍwān Efendī (**6.1.3**). The same auxiliary tables occur in MS Cairo DM 644,1 from *ca.* 1700, where they are followed by a star catalogue dated 1061 H [= 1651]. In both sources the work is anonymous. The title is *Fatḥ al-Karīm al-Bāqī fī maʿrifat al-dāʾir wa-faḍlihi āfāqī*, "The Victory of God, the Noble and Eternal, for Finding the Time since Sunrise and the Hour-angle for all Latitudes".[31]

The main functions tabulated are:

$$B'(\phi,\lambda) = {}^1/_2\, B(\phi,\lambda) = {}^1/_2 \text{ Cos } \delta(\lambda) \text{ Cos } \phi \;/\; R$$

for the domains:

$\lambda = 1°, 2°, \ldots , 90°$ and $\phi = 1°, 2°, \ldots , 27°, 28°, 30°, 32°, 33;30°, 36°, 38°, \ldots , 48°$,

and:

$$V'(x,y) = \text{arc Vers } \{ R \cdot y / 2x \}$$

for the domains:

$$x = 19, 20, \ldots , 30 \quad \text{and} \quad y = 0;5, 0;10, \ldots , 60;0 .$$

The first auxiliary function is labelled *al-aṣl*, "(half) the base", and the arguments in the table of the second function are called *al-aṣl al-muʿaddal*, "the modified base", and *faḍl al-jaybayn*, "the difference between the two Sines". The entries in the tables, which are given to two digits, were lifted from al-Khalīlī's auxiliary tables for timekeeping by the sun (**9.4**). Note that in al-Khalīlī's set B′ is tabulated for some 50 values of ϕ and the increment in the argument y for the tables of V′ is 0;10 rather than 0;5. Also, in the later Egyptian set no horizontal differences are given in the tables of V′ but the simplified format makes them slightly easier to use.

MS Cairo MM 203,3 (fols. 153r-160v), copied *ca.* 1700, contains an anonymous set of tables, probably of Egyptian provenance, entitled *Daqāʾiq al-raqāʾiq fī maʿrifat faḍl al-dāʾir li-sāʾir al-āfāq*, which means something like "Nice Numbers for Finding the Hour-angle for any Latitude". The set consists of simple tables of the inverse Versed Sine (**6.14.2**) and Sine functions and solar declination (based on $\varepsilon = 23;35°$), followed by a table of a function:

$$G_3(\phi,\delta)$$

tabulated to *one* significant digit for the domains:

$$\phi = 1°, 2°, \ldots , 60° \quad \text{and} \quad \delta = 1°, 2°, \ldots , 24° .$$

---

[31] A third copy in Diyarbakır, also anonymous, is mentioned in İhsanoğlu *et al.*, *Ottoman Astronomical Literature*, II, p. 743.

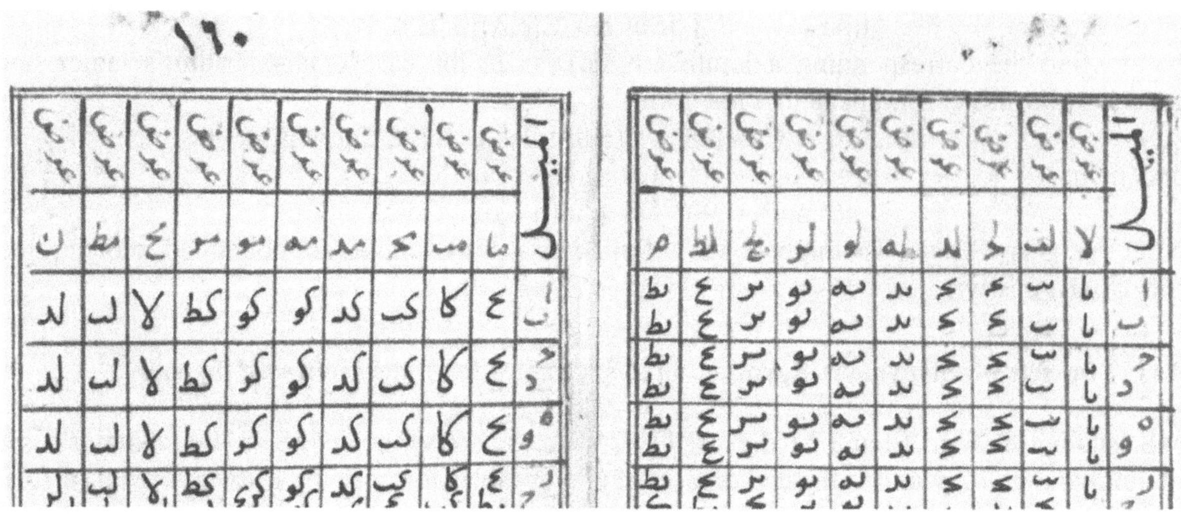

Fig. 9.11: An extract from some anonymous Egyptian auxiliary tables, serving latitudes 31°-50°. [From MS Cairo MM 203, fols. 159v-160r, courtesy of the Egyptian National Library.]

See the extract in **Fig. 9.11**. The methods outlined in the instructions for finding D and t by means of this function are equivalent to the formulae:

$$\text{Vers } D(H,\phi) = \text{Sin } H \ + \ \text{Sin } H \cdot G_3(\phi,\delta) \ / \ R \ ,$$

$$\text{Vers } t(h,H,\phi) = [\ \text{Sin } H - \text{Sin } h\ ] \ + \ [\ \text{Sin } H - \text{Sin } h\ ] \cdot G_3(\phi,\delta) \ .$$

It is clear that for these formulae to be valid we must have:

$$G_3 = R^3 \ / \ [\ \text{Cos } \delta \text{ Cos } \phi\ ] \ - \ R \ ,$$

which is confirmed by inspection of the table. The inspiration for tabulating this function was probably the little table of G_3 for the latitude of Cairo due to al-Marrākushī (**6.7.2**).

9.12 Saʿīd Beg Zāde's auxiliary tables for timekeeping

It is not inappropriate that I should conclude this study with a brief description of the late Ottoman set of auxiliary tables for timekeeping which shows the influence of Western mathematics on traditional Islamic astronomy. MS Istanbul Kandilli 226 is a copy of the auxiliary tables of Saʿīd (or Sayyid ?) Beg Zāde for computing the prayer-times in Turkish time for any latitude.[32] The main tables are based on the principle that to find the hour-angle in equinoctial hours and minutes corresponding to a particular prayer time t_i one uses three functions:

$$H_i''(H),\ G''(\lambda,\phi) \text{ and } V''(x) \ ,$$

thus:

$$t_i = V''\{\ H_i''(H) + G''(\lambda,\phi)\ \}.$$

Likewise, to find the hour-angle corresponding to a particular solar altitude one uses other tables to find a function $H''(H,h)$, and then uses:

---

32 Apparently not mentioned *ibid.*

$$t_i = V''\{ H_i''(H,h) + G''(\lambda,\phi) \} .$$

These functions are the following (compare the auxiliary tables of al-Wafāʾī discussed in **9.9** above !):

$$H''(H,h) = \log [\text{Sin } H - \text{Sin } h] = \log H'(H,h) ,$$

$$G''(\lambda,\phi) = \log \{ R^2 / [\text{Cos } \delta(\lambda) \text{ Cos } \phi] \} = \log G(\lambda,\phi),$$

$$V''(x) = {}^1/_{15} \text{ arc Vers (alog } x).$$

Both the trigonometric functions and the logarithms are to base 60 and are written sexagesimally! Further details are given in **II-14.16**.

MS Istanbul UL T4203 contains another set of auxiliary tables for computing the hour-angle, also using logarithms but not the sexagesimal system. The tables were computed by an individual identified simply as Jamāl,[33] and I have not investigated their underlying structure.

---

[33] *Ditto.*

CHAPTER 10

EUROPEAN TABLES FOR TIMEKEEPING

10.1 Some medieval European tables for timekeeping

"Über das Wie und Warum der Benutzung dieser Tabellen können wir nur spekulieren."
D. A. King, in *Schweinfurt 1993 Exhibition Catalogue*, p. 352.

I have not looked systematically for tables for timekeeping in medieval European manuscripts or early printed works; however, I am confident that these are far more widespread than has hitherto been thought. I mention here just a few examples from the published literature and from European manuscripts which have come to my attention by chance. One feature of these tables is clear: medieval European astronomers liked to tabulate the solar altitude as a function of time. The reason is not completely clear, although it had to do with the regulation of clocks.[1] Muslim astronomers, on the other hand, preferred tabulating something that one wants to know, namely, the time, as a function of something that one can measure, namely, solar or stellar altitude.

We begin with the *Toledan Tables*, a motley corpus of tables derived mainly from the *Zījes* of al-Khwārizmī and al-Battānī, which has been surveyed by Gerald Toomer, with a detailed new study by Fritz Pedersen.[2] In this corpus, as well as in a 15th-century English manuscript, there is a table of the function $e(\lambda) = {}^{1}/_{12}$ 150 tan $\delta(\lambda)$ (**7.1.1**), labelled *tabula differencie ascensionum universe terre*.[3] A table of h(T,H) for all latitudes with entries tabulated to two digits for each 0;15° of T (up to 6^{sdh}) and each 0;30° of h from 10° to 90°, with a total of over 3,500 entries (**4.3**), in a 14th-century English astronomical manuscript of both the *Toledan* and *Alphonsine Tables*, has also been noted by Toomer.[4] Universal tables of T(H,h) and h(H,T) based on the standard approximate formula are also found in two Byzantine manuscripts.[5]

The *Alphonsine Tables* generally replaced the *Toledan Tables* and were used in different recensions all over Europe until the 16th century. An isolated table displaying the function $\lambda_H(T,\lambda)$ for a particular latitude (**3.0**) is found in an early-15th-century manuscript of the *Alphonsine Tables* written in Germany: it is entitled *Tabel der ofsteygenden Czeychn* and the entries are given in zodiacal signs and degrees for each hour and each degree of solar longitude.[6]

---

[1] Geoffrey Chaucer referred to them as "tables ... for the governaunce of a clokke": see North, *Chaucer's Universe*, p. 87, and also Eisner, ed., *Kalendarium of Nicholas of Lynn*, pp. 29ff.

[2] See Toomer, "Toledan Tables", and now Pedersen, *The Toledan Tables*.

[3] *Ibid.*, p. 33, and Neugebauer & Schmidt, "Hindu Astronomy at Newminster", p. 226. See also North, *Richard of Wallingford*, II, pp. 12-14, and *idem*, *Horoscopes and History*, p. 14. See now Pedersen, *Toledan Tables*, III, pp. 986-991.

[4] Toomer, "Toledan Tables", p. 155. See now Pedersen, *op. cit.*, III, pp. 1134-1138.

[5] Jones, *Byzantine Astronomical Manual*, p. 174.

[6] I have examined MS Gotha Forschungsbibliothek Membr. I. 109, apparently dating from 1428, in which the tables occur on fols. 22r-24v.

ENERO

días	horas	X	XIIII	XVII	XXII	XXV	XXVII	XXX
	VII	grados						
	VIII	VII	VII	VIII	VIII	IX	X	XI
	IX	XVI	XVI	XVII	XVII	XVIII	XVIII	
	X	XXIIII	XXIIII	XXV	XXV	XXVI	XXVII	XXVIII
	XI	XXIX	XXX	XXXI	XXXII	XXXIII	XXXIIII	XXXV
	XII	XXXI	XXXII	XXXIII	XXXIIII	XXXV	XXXVI	XXXVII
	horas	grados	gra.	gra.	gra.	gra.	gra.	gra.

JUNIO

días		X	XXII
horas	h. V	g. VII	VI
grados	VI	g. XIX	XVIII
	VII	XXX	XXIX
	VIII	XXXXI	XXXX
	IX	LII	LI
	X	LXIIII	LXIII
	XI	LXXII	LXXI
	XII	LXXVI	LXXV

Fig. 10.1a: Extracts from the table of solar altitude at the hours in the treatise on astrology attributed to Enrique de Villena. The format of the table is dictated by the changes in meridian altitude of 1°. [From Cátedra & Samsó, *Astrología de Enrique de Villena*, pp. 171 and 173.]

A table of $h_i(H)$ for the equatorial hours and each degree of H was computed for the latitude of Baeza (38°) by pseudo-Enrique de Villena *ca.* 1430: see **Fig. 10.1a.**[7] The entries display error patterns which could be taken as implying that they were based on the approximate formula; however, the table may have been compiled using an instrument such as an astrolabe, for even the values at the equinoxes contain errors of as much as 2°. A less extensive table of this kind, with entries for each 5° of h, is found in a medieval Hebrew manuscript.[8]

Tables of $h_i(\lambda)$ ($\Delta\lambda = 30°$) and/or the associated vertical or horiontal shadows abound in the medieval European sources, but few have been published, and even fewer investigated.[9] There is, for example, a table of solar altitudes h_i for latitude 48° in the treatise on the astrolabe attributed to Hermannus Contractus (1013-1054).[10] The author states that he derived the entries, which are given to degrees, using an astrolabe and they do indeed correspond quite well to computation with the exact formula. The available version of the table of vertical shadows $z'_i(\lambda)$

[7] See Cátedra & Samsó, *Astrología de Enrique de Villena*, pp. 171-176, and the commentary on pp. 56-57.

[8] MS Munich heb. 343, which was brought to my attention by Professor Bernard Goldstein in the 1970s.

[9] Zinner, *Astronomische Instrumente*, pp. 159 and 50-51, also lists tables for Oxford, London, Rome, Venice, Nuremberg, Augsburg and Regensburg, which I have not investigated. See North, *Richard of Wallingford*, I, p. 18, on a table for Oxford by John Maudith (14th century), and *idem*, *Chaucer*, p. 114, on various other tables of this kind.

[10] On Hermannus see the article by Claudia Kren in *DSB*. For the table see Gunther, *Early Science in Oxford*, II, p. 419 (but it is not contained in the treatise translated in Joseph Drecker, "Hermannus Contractus über das Astrolab", *Isis* 16 (1931), pp. 200-219).

= $\mathrm{Tan}_{12}\ h_i(\lambda)$ ($\Delta\lambda = 30°$) for latitude 47;46° (Vienna) compiled by John of Gmunden (*fl. ca.* 1430) for marking the hour-curves on a cylindrical sundial, is full of errors.[11]

In the *Kalendarium* of Nicholas of Lynn (1386) there are tables of the solar altitude h_i and shadows z (base 6!) for the seasonal hours on each day of the year for the latitude of Oxford, taken as 51;50°.[12] Petrus de Dacia, canon in Roskilde, Denmark, *ca.* 1300, presented tables of meridian altitudes H and lengths of day and night 2D and 2N for each day of the year, computed for latitude 48;50°, serving Paris, and similar to others by William of St Cloud in his *Kalendarium* edited in 1296.[13] In one manuscript of the treatise on astrolabe construction by Jean Fusoris (Paris *ca.* 1400) there is a table, also for Paris, of the solar altitude $h_i(\lambda)$ at each equinoctial hour for each degree of λ and a separate table for each 15 days of the year specifically for constructing a cylindrical sundial.[14]

This activity was clearly continued by Renaissance astronomers, but there is precious little documentation as yet. A manuscript in the Stadtsarchiv in Schweinfurt contains a mixture of tables in the hands of two 16th-century German astronomers, Johannes Hommel and Johannes Praetorius.[15] The former contributed two tables for a latitude of 51° (location unspecified), displaying of h(T,λ) and a(T,λ) for each equinoctial hour and each 1° of λ. The latter contributed tables of h(T,λ) for latitudes 50° (locality unspecified) and 49;26° (Nuremberg is specifically mentioned), for $\Delta\lambda = 3°$ and 5°, respectively, as well as a set of tables for coordinate conversion for latitude 50° (copied in Cracow) after the model of the tables appended to Ptolemy's *Analemma*.[16] Also in the hand of Hommel we find a table of h(φ,λ) for each 1° of φ from 39° to 63° and each 1° of λ. No other such universal table is known from the medieval and Renaissance European sources.

In Europe, between the late 18th and the early 20th century, there appeared a series of extensive tables for timekeeping for purposes of navigation, of which I present scant details.[17] Already in 1770 the French astronomer César-François Cassini (= Cassini III) prepared in his *Almanach perpetuel pour trouver l'heure par tous les degrés de hauteur du soleil* a "skeleton set" of tables for latitudes between 34° and 51°. Joseph-Jérôme Lalande indicated the usefulness of such tables in the second edition of his *Traité de l'astronomie* (1771), and referred to the tables computed by Cassini. We can take a closer look at the tables published in Leipzig in 1791 by one Friedrich Christoph Müller. These display the two times of day in the morning and afternoon for each day of the year when the sun has a given altitude. Entries are given

---

[11] On John of Gmunden see the article by Kurt Vogel in *DSB*. The table is published in Claudia Kren, "The Traveller's Dial", p. 431.

[12] Eisner, *Kalendarium of Nicholas of Lynn*, pp. 68-69 (Jan.), 74-75 (Feb.), *etc.*, and the commentary on p. 26. On the latitude see *ibid.*, p. 17. See also North, *Chaucer*, pp. 115-116, and on the problems of the edition see *ibid.*, p. 115, n. 54. These tables include values for twilight: see also Pedersen, *Toledan Tables*, III, pp. 1132-1133.

[13] Pedersen, F. S., ed., *Astronomical Works of Petrus de Dacia and Petrus de S. Audomoro*, I, pp. 39 and 320-322.

[14] Poulle, *Fusoris*, pp. 184-185 and 186. On this manuscript see also King, *Ciphers of the Monks*, p. 155, and for more on Fusoris see *ibid.*, pp. 397-398.

[15] On the manuscript (AvS Ha 21) see *Schweinfurt 1993 Exhibition Catalogue*, pp. 351-353, no. 169.

[16] See n. 1:40.

[17] The information in this paragraph is taken from Cotter, "Nautical Tables". Alas the details of the individual tables are not standardized. The subject could use a new study.

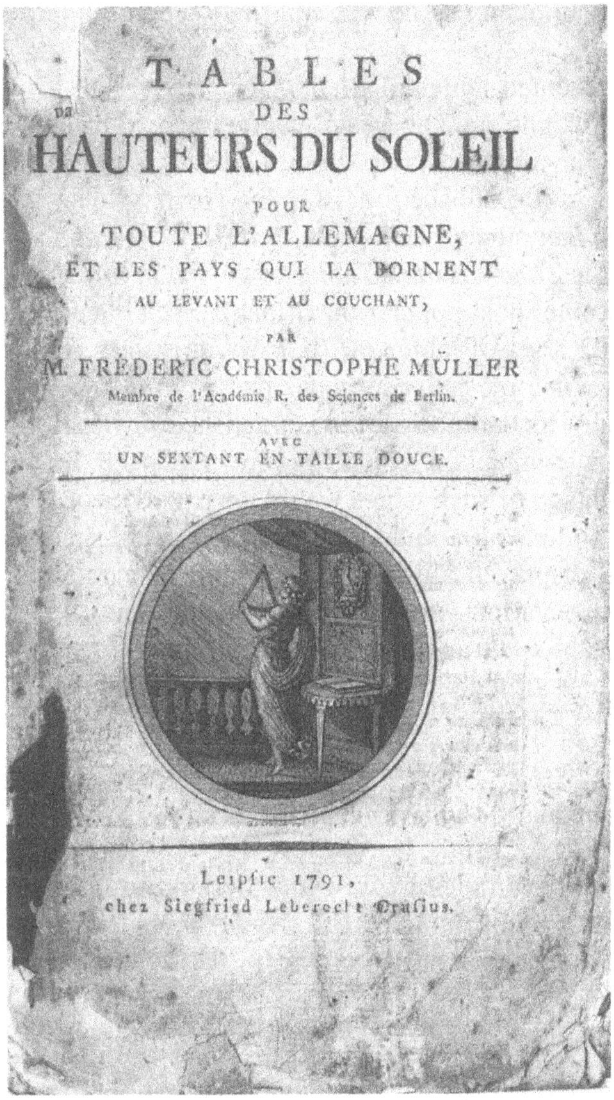
TABLES DES HAUTEURS DU SOLEIL POUR TOUTE L'ALLEMAGNE, ET LES PAYS QUI LA BORNENT AU LEVANT ET AU COUCHANT, PAR M. FRÉDERIC CHRISTOPHE MÜLLER Membre de l'Académie R. des Sciences de Berlin. AVEC UN SEXTANT EN TAILLE DOUCE.

Leipsic 1791, chez Siegfried Lebrecht Crusius.

Elévation du Pole, 49 dégrés.

HAUTEUR DU SOLEIL, 20 DÉGRÉS.

Avant-midi.

	Janv.	Fevr.	Mars	Avr.	Mai	Juin	Juil.	Aout	Sept.	Oct.	Nov.	Dec.
1			8 49	7 41	6 53	6 23	6 18	6 40	7 24	8 22	9 42	
2		10 8	8 46	7 39	6 52	6 22	6 19	6 41	7 26	8 24	9 45	
3		10 4	8 44	7 37	6 50	6 22	6 19	6 42	7 27	8 26	9 48	
4		10 1	8 41	7 36	6 49	6 21	6 19	6 44	7 29	8 29	9 51	
5		9 57	8 39	7 34	6 48	6 21	6 19	6 45	7 30	8 31	9 54	
6		9 54	8 36	7 32	6 47	6 20	6 20	6 46	7 32	8 33	9 58	
7		9 51	8 34	7 30	6 46	6 20	6 20	6 47	7 34	8 36	10 1	
8		9 48	8 32	7 28	6 44	6 19	6 20	6 49	7 36	8 38	10 5	
9		9 45	8 29	7 27	6 43	6 19	6 21	6 50	7 37	8 41	10 9	
10		9 42	8 27	7 25	6 42	6 19	6 21	6 51	7 39	[illegible] 43		
11		9 39	8 25	7 23	6 41	6 18	6 22	6 52	7 41	8 45		
12		9 36	8 23	7 21	6 40	6 18	6 23	6 53	7 43	8 48		
13		9 33	8 21	7 20	6 39	6 18	6 23	6 55	7 45	8 50		
14		9 30	8 18	7 18	6 38	6 18	6 24	6 56	7 47	8 53		
15		9 27	8 16	7 16	6 37	6 18	6 24	6 57	7 49	8 55		
16		9 24	8 14	7 15	6 36	6 17	6 25	6 59	7 51	8 58		
17		9 21	8 12	7 13	6 35	6 17	6 26	7 1	7 53	9 1		
18		9 18	8 9	7 12	6 34	6 17	6 26	7 2	7 55	9 3		
19		9 15	8 7	7 10	6 33	6 17	6 27	7 3	7 57	9 6		
20		9 12	8 5	7 9	6 32	6 17	3 28	7 5	7 59	9 8		
21		9 10	8 3	7 7	6 31	6 17	6 29	7 6	8 1	9 11		
22		9 7	8 1	7 6	6 30	6 17	6 30	7 8	8 3	9 13		
23		9 5	7 59	7 4	6 29	6 17	6 31	7 9	8 5	9 16		
24		9 2	7 57	7 3	6 28	6 17	6 32	7 11	8 7	9 18		
25		8 59	7 55	7 2	6 27	6 17	6 33	7 12	8 9	9 21		
26		8 57	7 53	7 0	6 27	6 17	6 34	7 14	8 11	9 24		
27		8 54	7 51	6 58	6 26	6 18	6 35	7 15	8 13	9 27		
28		8 52	7 49	6 57	6 25	6 18	6 36	7 17	8 15	9 30		
29		8 50	7 47	6 55	6 24	6 18	6 37	7 18	8 18	9 33		
30			7 45	6 54	6 24	6 18	6 38	7 20	8 20	9 37		
31			7 43		6 23		6 39	7 22		9 39		

Figs. 10.1b: The title-page and an extract from the tables of F. C. Müller, 1791, showing time as a function of solar altitude throughout the year for a range of latitudes in Germany. The corpus bears the title "tables of solar altitudes", which is mathematically inexact, and curiously corresponds precisely to the Turkish expression *irtifāʿ jadwali* (*irtifā cedveli*) for tables of the same kind. Müller's introduction makes no mention of contemporaneous Turkish tables, and the West was to remain innocent of any knowledge of them for almost another two centuries. [From Müller, *Tables des hauteurs du soleil.*]

to hours and minutes for each degree of altitude, and there are tables for each degree of latitude from 47° to 54°, intended to serve all of Germany and the countries surrounding it. See **Fig. 10.1b** for an extract from the French edition, apparently published simultaneously with the original German version.[18] In his preface the author complained about the difficulties of

[18] It is a pleasure to thank Dr. Živa Vesel of Paris for sending me a copy of the French edition of these tables, listed in the bibliography as Müller, *Tables des hauteurs du soleil.*

compiling the tables and finding a publisher, and mentioned that he was considering compiling similar tables for the solar azimuth.

In his *Abrégé de navigation* (1793) Lalande presented tables of t(h,δ,φ) for each value of h between 2° and 48°, each degree of δ between 0° and 24°, and each degree of φ from 0° to 60°. The tables displayed the first differences and an auxiliary table was provided to facilitate interpolation. The first English tables of this kind were published in 1827 by Thomas Lynn: they were not widely used except by officers of the Honourable East India Company, to whom Lynn was engaged as an examiner in navigation. The *Time-Azimuth-Tables* of the naval officer John Burdwood published from 1852 onwards became most popular on British vessels. Percy Davis of the Royal Naval Office published in 1902 a set of tables of t(h,δ,φ) with values for each degree of each argument. In 1913 a set of tables of stellar altitudes and azimuths, for declinations between -30° and +30°, and specifically for latitude 55°, was published in Berlin for the German Navy; within the next few years companion volumes were published for latitudes 50° and 70°. (It is not clear what the advantage of such tables would have had beyond enabling a vessel to be steered along one of three parallels of latitude.) A set of tables published by the United States military authorities in 1919 displayed altitude and azimuth in parallel columns (no further information available). In fact, various astronomers from the late 19th century to the mid 20th century compiled tables of the hour-angle and the solar azimuth, both as functions of solar altitude and declination, or solar altitude as a function of hour-angle and declination, all for extensive ranges of latitudes.

10.2 European auxiliary tables for spherical trigonometry and astronomy

Although previous studies of Islamic auxiliary tables have failed to draw attention to European tables of the same kind, the tradition of compiling auxiliary tables for solving problems of spherical trigonometry – inevitably relating mainly to spherical astronomy – was continued in Europe.[19]

Regiomontanus' table for solving spherical triangles in his *Tabula primi mobilis* prepared in 1463 displays values of:

$$x(a,b) = \text{arc sin } \{ \sin a \cdot \sin b \}$$

for each degree of both arguments. Values are given in degrees, minutes and seconds, and are generally accurate in the last digit. First horizontal and vertical differences are also tabulated. As Anton von Braunmühl has shown, linear interpolation applied horizontally and vertically yields results in which the seconds are not at all to be trusted (his example shows an error of about 30″). This table is useful for solving right-angled spherical triangles with any of the rules involving only sines and cosines.

[19] The information in this section is summarized from Cantor, *Geschichte der Mathematik*, II, pp. 273-275; von Braunmühl, *Geschichte der Trigonometrie*, I, pp. 122-124 and 231-236, II, p. 231; Wolf, *Handbuch der Astronomie*, I, p. 438; and Fletcher *et al.*, *Index of Mathematical Tables*, esp. I, pp. 224-227 (altitude tables, azimuth tables and hour-angle tables). The tables of F. C. Müller are not mentioned in any of these works. As far as I know, the Renaissance tables have never been studied in any depth. Glowatzki & Göttsche, *Die Tafeln des Regiomontanus*, deals only with standard trigonometric tables, but not only those of Regiomontanus.

Regiomontanus' tables called *Tabulae directionum* for converting ecliptic to equatorial coordinates were first published in 1475 and again in a new edition by Erasmus Reinhold in 1554. For each degree (?) of ecliptic longitude values of the following functions are tabulated:

$$\alpha' , \quad \delta_2, \quad \sin \gamma \quad \text{and} \quad \cot \gamma ,$$

where δ_2 is the "second declination" (measured perpendicular to the ecliptic). With these Δ and α are found using:

$$\sin \Delta = \sin \{ \beta + \delta_2 \} \sin \gamma \quad \text{and} \quad \sin \{ \alpha' - \alpha \} = \tan \Delta \cot \gamma .$$

The auxiliary tables of Ḥabash also include tables of δ_2 and Sin γ for the same purpose.

The year 1604 saw the publication in Venice of Magini's *Tabulae primi mobilis, quas directorium vulgo dicunt*. Five years later an extensive commentary was published in Bologna, which was one of the most widely used books among contemporary astronomers. Magini's tables display four main functions:

$$x(a,b) = \text{arc sin} \{ \sin a \cdot \sin b \}$$
$$y(a,b) = \text{arc sin} \{ \sin a / \sin b \}$$
$$z(a,b) = \text{arc tan} \{ \sin a \cdot \tan b \}$$
$$w(a,b) = \text{arc sin} \{ \tan a \cdot \tan b \} ,$$

again with values for each degree of both arguments. In the 1609 edition values are given for each 0;10°. The first function is that of Regiomontanus, and all four serve the direct solution of all of the standard formulae of spherical astronomy with the exception of the cosine formula, which requires two applications. In addition, Magini's *Magnus canon trigonometricus* contains tables of the following functions for each minute of argument:

$$\sin x, \text{ arc sin } \{(\sin x)/10\}, \text{ vers } x, \text{ arc sin } \{(\text{vers } x)/10\}, \tan x,$$
$$\text{arc sin } \{(\tan x)/10\}, \text{ arc sin } (\tan x), \sec x \text{ and arc sin } \{(\sec x)/10\}.$$

Although Delambre noted that Magini's tables were soon rendered superfluous by the invention of logarithms, von Braunmühl was more considerate, writing that:

> "Magini with his *primum mobile* had given Italy a work the like of which was not attested in the history of trigonometry, neither previously [*sic* !!] nor for decades thereafter."

Of course, neither scholar was aware of any Islamic tables of this kind.

In 1849 Adolf Heegemann prepared new tables displaying the functions:

$$x(a,b) = \text{arc sin} \{ \sin a \cdot \sin b \}$$
$$z(a,b) = \text{arc tan} \{ \sin a \cdot \tan b \} ,$$

apparently without knowing that Regiomontanus and Magini had produced similar tables several centuries before. It was only in the 19th century with Inman's *Nautical Tables* and Raper's *Practice of Navigation* that tables for finding an angle A of a general spherical triangle became available. These displayed sin $^1/_2$ A, computed using:

$$\sin^2 (^1/_2 A) = {}^1/_2 \text{ vers } A = \sin (s-b) \sin (s-c) / (\sin b \sin c) ,$$

where $s = {}^1/_2 (a + b + c)$. We should point out that, using this same notation, it may be said that the tables of such Muslim scholars as Najm al-Dīn al-Miṣrī and al-Khalīlī serve to determine the angle A(a,b,c) in terms of the sides of a general spherical triangle for problems of spherical astronomy. When:

$$a = 90° - h , \quad b = 90° - \delta \quad \text{and} \quad c = 90° - \phi ,$$

A is the hour-angle, and, when a and b are switched, A is the azimuth.

It is not necessary to assume a knowledge of the spherical cosine formula when an equivalent solution occurs in an ancient or medieval text. Thus, for example, it is often asserted that al-Battānī (*ca.* 910) knew the formula and used it to derive the hour-angle from an instantaneous celestial altitude.[20] This is simply not the case. al-Battānī merely used an Indian procedure for deriving the hour-angle by projection methods. His formula can, of course, be shown to be *equivalent* to an application of the spherical cosine rule. As far as we know, no Muslim scholar wrote on the cosine formula *per se*, and we cannot even assume that even Najm al-Dīn al-Miṣrī or al-Khalīlī knew of it in its generality.[21] All that can be asserted is that both of them and many other medieval Muslim astrnomers were fully aware of the mathematical equivalence of the two major problems of spherical astronomy (determination of the hour-angle and the azimuth from an instantaneous celestial altitude) which in modern terms can most readily be solved by an application of the formula. As far as the available sources indicate, most of the astronomers who produced the tables I have described preferred projection methods to spherical trigonometry anyway.

There is no evidence that Regiomontanus or any of the later European astronomers were familiar with the Islamic auxiliary tables. Nor did the orientalists at Oxford in the 16th and 17th centuries, for all their interest in the Islamic astronomical texts and tables preserved in Oxford, encounter any Islamic tables for timekeeping.[22] Nevertheless, the auxiliary tables of the Muslim astronomers, particularly the splendid universal tables of Najm al-Dīn al-Miṣrī and al-Khalīlī, as well as some of the more impressive of tables of other Muslim astronomers from the 9th to the 16th century, would surely have been of interest to any serious European astronomer, navigator or naval officer from the 18th to the 20th century. For during that time, numerous sets of tables of trigonometric functions with two or three arguments were compiled.[23]

Certainly the tables of the Muslim astronomers merit appropriate consideration in any new history of astronomy or history of trigonometry. But most people would think that these histories have already been written, and it is unlikely that newly-discovered historical sources, such as a few hundred medieval tables, will change old attitudes.

---

[20] Similar confusion about the use of the cosine formula for deriving the azimuth from the altitude is found in Sezgin, *GAS*, V, pp. 261 and 288, and VI, p. 185 (based on earlier secondary sources).

[21] Similar confusion relates to the attribution of the so-called prosthaphaeresis formula of trigonometry to Ibn Yūnus: see the remarks at the end of **2.1.1**.

[22] See Mercier, "English Orientalists and Islamic Astronomy". As I have shown in a study listed as "Ibn Yūnus and the Pendulum", the misconception that Ibn Yūnus discovered the principle of the pendulum, an association now widely spread on the Internet, actually stems from Edward Bernard in 17th-century Oxford.

[23] Fletcher *et al.*, *Index of Mathematical Tables*, esp. I, pp. 169-195 (tables of trigonometric functions and miscellaneous functions associated with the circle and sphere), 207-229 (trigonometric functions of two or three arguments).

Part II

A survey of tables for regulating the times of prayer

To Bernard Goldstein

ACKNOWLEDGEMENTS AND NOTES TO THIS VERSION

This study is dedicated to the best teacher and thesis advisor a graduate student could ever have wished for. Bernard Goldstein was just that, a man whose generosity, humour and breadth of scholarship I have always admired and felt fortunate to have experienced and enjoyed. It was he who taught me how to read Arabic scientific manuscripts and what to do with the numbers one finds in them: in his seminar we read the Arabic version of Ptolemy's *Planetary Hypotheses*, in which he had been the first to discover Ptolemy's inter(n)esting spheres.

One day Bernie told me to get hold of microfilms of the Leiden, Oxford and Paris manuscripts of the *Zīj* of Ibn Yūnus and redo Delambre: this would serve as a doctoral thesis. But I should also take a look at a certain Dublin manuscript: it had been catalogued by A. J. Arberry as another copy of the *Zīj*. The Dublin manuscript turned out to be unrelated to the *Zīj*, rather it was a copy of the Cairo corpus of tables for timekeeping, and the reader will imagine my surprise and delight as I drooled over the microfilm for the first time. When I returned to Yale in 1971 after a year in Beirut with Ted Kennedy, I showed Bernie not only the first draft of my thesis on Ibn Yūnus, but also my analysis of the corpus of tables for timekeeping attributed to him in the Dublin manuscript. Bernie corrected the first draft of my thesis over a weekend. "Should I include the timekeeping stuff in the thesis?", I asked. "No", he replied, fully aware of the syndrome of the graduate student who cannot finish a thesis. He was right, of course, and as soon as the thesis was submitted, I turned back to the timekeeping. Thus Bernie is directly responsible for the genesis of this study, albeit unintentionally. (Now this study is published, and the thesis not.)

When my wife and I went off to Cairo in 1972, Bernie gave me a compass and a book entitled *How to Survive in the Outdoors*, both of which I have always cherished. I came out of Egypt after "seven fat years" rather than the formidable "forty years" which others are said to have taken. I used Bernie's compass off and on, not to get back to the U.S. but to measure orientations of various medieval mosques; thus he also unintentionally inspired another major area of my research (see **VIIa**). In 2001 he further gave me a very useful little Eskimo man, carved out of bone, and with a flat head: this one should turn upside down and address when one has problems with colleagues.

At the special session of the Annual Meeting of the (U.S.) History of Science Society held in Bernie's honour in Pittsburg in November, 1999, four colleagues presented state-of-the-art reports on their joint research with him, in ancient Greek, medieval European and Renaissance astronomy, and, last but not least, things Keplerian. There were no presentations on Islamic or Hebrew astronomy, which was a pity because Bernie's contributions to both have been of the same order of magnitude. However, the *Festschrift* dedicated to him in 2003 contains contributions relating to each of these six widely-different various areas of the history of astronomy and is appropriately entitled *Astronomy and Astrology from the Babylonians to Kepler*. In each of these areas Bernie Goldstein has made ground-breaking discoveries, and not only because he too had the best teacher.

TABLE OF CONTENTS

CHAPTER 1

ON ISLAMIC PRAYER-TABLES

> "But now Islam has appeared in the eastern and western parts of the world and has spread between Andalusia in the West and parts of China and central India in the East, and between Abyssinia and Nubia in the south and the Turks and Slavs in the North. It has united all the different nations in one bond of love, a handiwork which can be made by God only." al-Bīrūnī (Afghanistan, *ca.* 1025), *Taḥdīd*, p. 225, transl. by J. Ali, p. 190-191.

> "The historian of the Islamic exact sciences is frequently confronted with an *embarras de richesse* – hundreds of manuscript sources which have never been studied in modern times." E. S. Kennedy, "Islamic Mathematical Geography" (1996), p. 185.

1.0 Introduction

The times of the five daily prayers of Islam are astronomically defined and for a given locality vary throughout the year. Given a clear sky and unrestricted view of the local horizon, it is possible to regulate the daytime prayers using a gnomon and the nighttime prayers by observation of twilight phenomena. The legal scholars of medieval Islam, who spoke with authority, advocated the methods of folk astronomy, devoid of any mathematics beyond simple arithmetic, in their discussions of this subject (see **III**). However, the astronomers of medieval Islam developed a new branch of science out of the religious obligation to pray at specific times, which they called *ʿilm al-mīqāt*, the science of astronomical timekeeping. In the early centuries of Islamic astronomy, this topic was treated almost as in passing in *zīj*es and treatises on instruments. But in Egypt in the 13th century a new professional class emerged, astronomers who specialized in this discipline and who were associated with a particular religious institution. These men were called *muwaqqit*s, literally, "timekeepers"; astronomers who also worked in this discipline independently bore the epithet *mīqātī* (see **V**). The writings of these men consist mainly of treatises on the determination of the prayer-times either by direct calculation or with the aid of analogue computers such as the astrolabe and astrolabic quadrant or calculating devices such as the trigonometric quadrant. But they also produced an enormous amount of tables of one sort or another as aids to their task. Muslim astronomers were also concerned with the determination of the qibla, that is the direction of Mecca, which the Muslim should face in prayer, and also the determination of the visibility of the lunar crescent, and hence the regulation of the lunar Muslim calendar with its holy months of fasting and religious festivals.

We have already seen the practical aids to their work which the astronomers and *muwaqqit*s had at their disposal in the form of tables for timekeeping by the sun and stars (**I**). The most tangible and colourful evidence of the activity of the *muwaqqit*s and their role in Islamic societies is to be found in the mathematical tables which they prepared displaying the prayer-times throughout the year for a particular locality. It is clearly convenient to have the prayer-times tabulated, either for each degree of solar longitude or for each day of the year, particularly in localities where clouds are frequent or where the local horizon is obscured by mountains or tall buildings. However, such tables must be used in conjunction with some kind of instrument for measuring the passage of time, be it an astrolabe, a quadrant or a sundial, or, from the 17th/18th/19th century onwards (depending on the location), a clock. The times of the muezzin's call to prayer in the modern Islamic world are still regulated by tables, compiled by modern methods, and displayed in almanacs, calendars and pocket-diaries (see **V-13**).

The purpose of this study is to present a survey of the medieval prayer-tables which have been located in recent years in various libraries around the world.[1] The vast majority have never been studied previously, indeed, they were not known to exist. These tables are of interest not only by virtue of their mathematical sophistication and accuracy and for the light they cast on medieval computational techniques, but also for the limited but significant information they reveal on one aspect of devotional life in medieval Islam. I discuss mainly tables that were used in Egypt and Syria, since these were the centres of *ʿilm al-mīqāt* in medieval times. However, I have also included descriptions of all medieval prayer-tables from other areas known to me in 1975 (and in 1999 I know of no others). I make no claim to have exhausted the available material even for the study of Egyptian and Syrian prayer-tables, although I suspect that any fresh tables from these two areas that may come to light will not differ greatly from those described in the sequel. However, new Egyptian and Syrian manuscripts may clarify the picture which I present of the transmission of prayer-tables in the medieval period and help to establish the identity of those who compiled the tables, where this is still in doubt. The reader will observe that many of the tables that I discuss exist in what appear to be unique manuscripts. However, the main corpuses of tables for Cairo and Damascus and also Istanbul, exist in dozen manuscripts.

In an early publication I analyzed in some detail the corpus of tables for timekeeping that was used in medieval Cairo.[2] In each of the seven manuscripts that formed the basis of that analysis the tables are attributed to the 10th-century astronomer Ibn Yūnus. Fresh manuscript material enables me now to present new information on the main Cairo corpus (**Chs. 4** and **5**), showing that some of the tables were in fact not computed by Ibn Yūnus but by later Egyptian astronomers. There are still many questions surrounding our knowledge of astronomy in Egypt before about 1400, particularly with regard to the activities of such individuals as Najm al-Dīn al-Miṣrī and al-Marrākushī (**Ch. 6**). Other later Egyptian tables for timekeeping (**Chs. 7-8**) can generally be shown to be closely related to the main Cairo corpus. The

[1] I use the terms "medieval" to cover the entire period from the 8th to the 19th century: see n. **I**-1:1. I use the expression "prayer-table" to denote any set of spherical astronomical tables displaying functions relating to twilight or the afternoon prayer. Many medieval prayer-tables also contain tables of standard spherical astronomical functions: these I have discussed. Many *zīj*es (see n. **I**-1:3) contain similar tables of spherical astronomical functions: these I have generally not discussed. In this study I also treat tables for finding the local qibla by means of the sun, but not tables displaying the qibla for a whole range of latitudes and longitudes, except for that of al-Khalīlī (**10.8**), the only such table in a corpus of tables for timekeeeping.

[2] See n. 4:1 below.

impressive developments in astronomical timekeeping in 14th-century Syria and their later influence (**Chs. 9-11**) seem to owe at least part of their inspiration to the main Cairo corpus. Likewise, the tables compiled in the Yemen, in Tunis and in Istanbul (**Chs. 12-14**) are related to the Egyptian and Syrian traditions. Various treatises on timekeeping and on the use of instruments (**Ch. 2**) cast light on the mathematical techniques that were used by the astronomers of medieval Islam for this purpose.

I have grouped together various sets of prayer-tables for timekeeping compiled in al-ʿIrāq and Iran that have been located in the manuscript sources (**Ch. 3**), but these are inadequate to convey any clear picture of the development of tables for timekeeping in these two areas. Notice that even al-Bīrūnī, the leading scientist of Islam, when writing on timekeeping at the beginning of the 11th century (**2.2**), does not mention the existence of any such tables. Likewise, I have found no tables of Andalusī origin. I suspect that new material located during future research might demand a reappraisal of the development of *mīqāt* particularly in al-ʿIrāq and Iran, as well as al-Andalus. Prayer-tables of the modern kind (**V-13**) are, together with the determination of lunar crescent visibility at the beginning and end of Ramaḍān, and the orientation of the *miḥrābs* in mosques towards Mecca, the sole vestiges of a tradition of mathematical astronomy which knew no rival for 750 years.

As was the case with the tables surveyed in **I**, the tables discussed in this study had, with very few exeptions, never been investigated previously before *ca.* 1970. I know no reference in the Arabic literary sources to tables for regulating the times of prayer, and there is not a trace of such tables in the standard bibliographical works of Carl Brockelmann, Fuat Sezgin and Charles Storey.[3] Furthermore, tables for regulating the prayer-times are seldom contained in the Islamic astronomical handbooks known as *zīj*es.[4] Most of the manuscripts on which this study is based were either miscatalogued or not catalogued at all, and in general the tables were discovered only by foraging through large numbers of manuscripts. Catalogues such as that of Wilhelm Ahlwardt for the Berlin manuscripts give sufficiently detailed descriptions of each manuscript, and even individual tables are mentioned.

The information that I present on Islamic prayer-tables represent the results of only a preliminary survey of the available material. Certainly many of the sources that I describe in the sequel deserve further investigation.

1.1 On the definition of the times of Muslim prayer

The definition of the intervals during the five daily prayers should be performed have been traditionally traced to certain verses of the *Qurʾān* and various *ḥadīth* attributed to the Prophet Muḥammad. The *ḥadīth* dealing with the prayer-times have been subjected to a preliminary analysis by Arent J. Wensinck,[5] and some of the various definitions found in the later astronomical sources have been discussed in a valuable study by Eilhard Wiedemann and

---

[3] See nn. **I**-1:13 and 19.

[4] See n. **I**-1:3.

[5] See the article "Mīḳāt, i. Legal Aspects" by Arent J. Wensinck in EI_2.

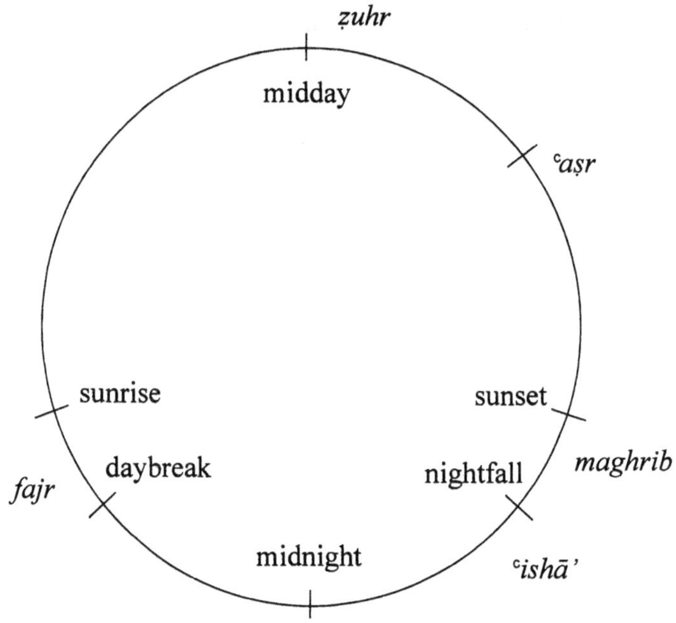

Fig. 1.1a: The Islamic day, showing the times for the beginnings of the five prayers at sunset, nightfall, daybreak, midday and mid-afternoon.

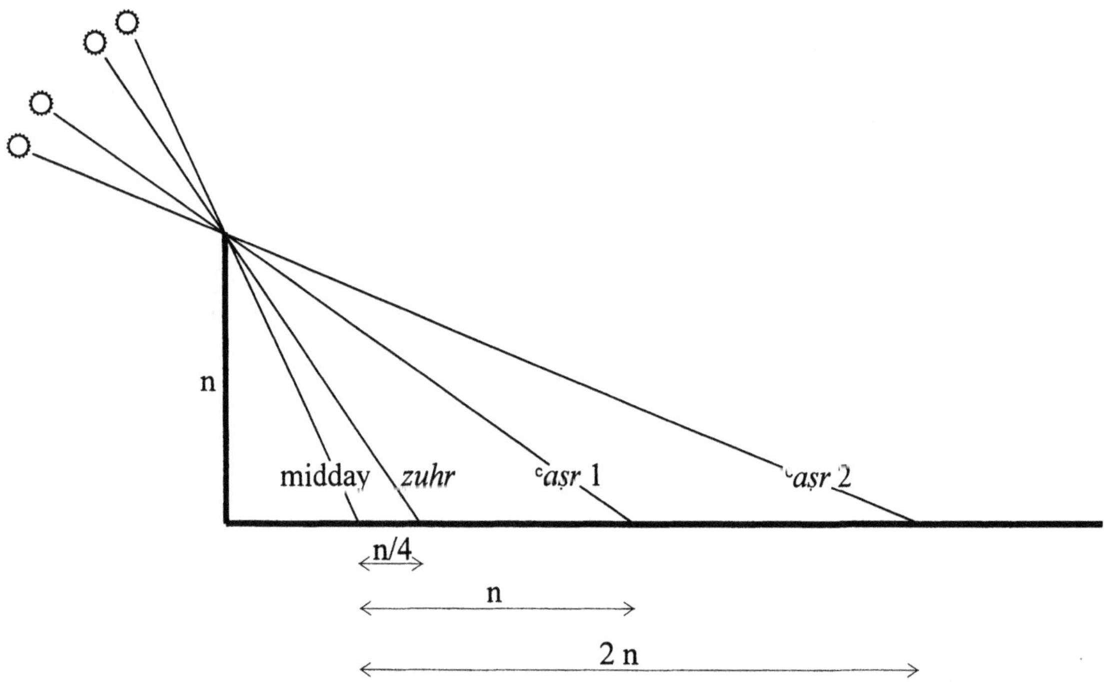

Fig. 1.1b: The standard definition of the beginning of the interval for the *ʿaṣr* is the time when the shadow increase equals the gnomon-length. In some circles, notably amongst the Ḥanafīs, the end of the permitted interval is defined when the shadow increase is double that amount. In Andalusī and Maghribī practice, the interval for the *ẓuhr* prayer begins when the shadow increase is one-quarter of the gnomon-length.

Joseph Frank.[6] In the 1970s Ted Kennedy published a translation and commentary of al-Bīrūnī's book *On Shadows*, probably the most important single work on timekeeping compiled in the Middle Ages, albeit without tables.[7]

The definitions which were used by the Muslim astronomers, and which underlie most of the tables discussed in this study, are as follows: The Muslim day begins at sunset, and the interval during which the first prayer (*al-maghrib*) is to be performed lasts from sunset to nightfall. The interval for the second prayer (*al-ʿishāʾ*) begins at nightfall and lasts until daybreak. The third prayer (*al-fajr*) is to be performed during the interval between daybreak and sunrise. The permitted time for the fourth prayer (*al-ẓuhr*) begins when the sun has crossed the meridian and ends when the interval for the fifth prayer (*al-ʿaṣr*) begins; namely, when the shadow of an object equals its meridian shadow increased by the length of the object. The interval for the fifth prayer may last until the shadow increases again by the length of the object or until sunset. In the medieval period there were occasional variations to this practice, such as, for example, the definition of the *ẓuhr* in al-Andalus and the Maghrib, whereby the prayer began not at midday but when the shadow increase over its midday minimum was one-quarter of the length of the gnomon. Also in some communities there was a prayer at mid-morning called the *ḍuḥā*. Indeed there is sufficient material available to explain the origin of the curious definitions for the *ẓuhr* and *ʿaṣr*, which are not specifically mentioned either in the *Qurʾān* or the *ḥadīth*. I have investigated the origin of these definitions in **IV**.

The determination of the prayer-times either by observation or by computation was a relatively simple problem for a competent medieval astronomer. The midday and afternoon prayer-times can be determined with a facility using a gnomon. The corresponding solar altitude can be readily computed so that one could also use an astrolabe or a quadrant to know when the time for prayer had arrived. To determine the times from sunrise to midday, and from midday to the beginning of the afternoon prayer (when the sun has a predetermined altitude), are standard problems of medieval spherical astronomy. The times of the prayers at nightfall and daybreak can be likewise be determined either by observation or by calculation. To compute the duration of twilight one must assume that the transitions between twilight and total darkness, that is, daybreak and nightfall, occur when the sun has a certain angle of depression below the horizon. The determination of the duration of twilight is then a simple extension of the problem of finding the time to sunrise or since sunset when the sun has a particular altitude, which is again a standard problem of medieval spherical astronomy. Thus from an astronomical or mathematical point of view the determination of the prayer-times is a straightforward application of certain standard procedures in spherical astronomy.

Underlying several of the tables amongst those I shall be describing is a value for the qibla or local direction of Mecca. the reader is referred to **VIIa-c** for an introduction to this topic.

[6] See Wiedemann & Frank, "Gebetszeiten".

[7] Listed under al-Bīrūnī, *Shadows*. For a summary of the parts relevant to this study see Kennedy, "al-Bīrūnī on Prayer-Times".

1.2 On the development of astronomical timekeeping in Islam

In early Islam the times of prayer were regulated by observation. In the case of the daytime prayers, the *ẓuhr* and *ʿaṣr*, a gnomon erected in the mosque could be used to determine the times with facility. Such a gnomon exists to this day in the beautiful Mosque of Janad in the Yemen (see **Fig III-1.2b**); the mosque dates from the 7th century, but I do know when this gnomon, a (concrete?) block of rectangular cross-section about 2 metres high and about 15 cm thick, was erected. In the case of the nighttime prayers, the moment of sunset can generally be ascertained from the vantage of a minaret, and the twilight phenomena are likewise readily observable.

In early Islam, time in general was reckoned either in seasonal hours of the day or night or with respect to the times of prayers. The time of day could be calculated from shadow length by simple formulae ultimately of Indian origin, as recorded in the treatises of al-Bīrūnī and Ibn Raḥīq (**2.2-3**). The Bedouin of the Arabian peninsula have possessed, since pre-Islamic times, a detailed astronomical folklore relating the changing night sky to meteorological and agricultural patterns; in particular, timekeeping by night was effected by observing the lunar mansions.[8] The time of night in early Islam could also be measured, for example, by a simple candle-clocks: one such clock described in the medieval sources consists of a set of twelve oil lamps which are filled in such a way that one goes out at each hour of the night.[9]

From the 9th century onwards the times of the *ẓuhr* and *ʿaṣr* were sometimes regulated by sundials.[10] Also, individual prayer-times could be determined using special curves marked on astrolabes and quadrants.[11] However, the passion of the early Muslim astronomers for the compilation of tables found a ready outlet in astronomical timekeeping. Already in the mid 10th century ʿAlī ibn Amājūr in Baghdad compiled two extensive tables displaying the time since sunrise as a function of instantaneous solar altitude and meridian altitude. One of these was based on an accurate formula and computed for the latitude of Baghdad, and the other was based on an approximate formula and was intended to serve all latitudes (**3.2**). Ibn Amājur and various other Muslim astronomers, even from the 9th century, likewise tabulated certain functions relating to the times of prayer, such as the solar altitude at the beginning of the *ʿaṣr* as a function of its meridian altitude, and the duration of twilight assuming that daybreak and

---

[8] See, for example, the articles "Anwāʾ" and "Layl and Nahār" by Charles Pellat and "Manāzil" by Paul Kunitzsch in EI_2, and now Varisco, "Islamic Folk Astronomy".

[9] This clock is described in Kennedy & Ukashah, "The Chandelier Clock of Ibn Yūnis". The author is the ʿIrāqī craftsman Yūnus al-Asṭurlābī, not the celebrated Egyptian astronomer: see King, *Ibn Yūnus*, I.2.10 and III.15.1c, and *idem*, "Review of Hill, *al-Jazarī*", n. 2 on p. 286.

[10] On Islamic gnomonics see Schoy, *Gnomonik der Araber*, and, more recently, the survey in the article "Mizwala" in EI_2, repr. in King, *Studies*, C-VIII, and also **X-7**. Many new insights are to be gained from Charette, *Mamluk Instrumentation*.

[11] On the astrolabe in Islam see the articles "Asṭurlāb" by Willy Hartner in EI_2, also listed under Hartner, "Astrolabe". On the quadrant see the ground-breaking study Schmalzl, *Geschichte des Quadranten*, and my article "Rubʿ" in EI_2. Many new insights are to be gained from Charette, *op. cit.* Several Islamic instruments displaying the curves for the various times of prayer are illustrated in Gunther, *Astrolabes*, I, and Michel, *Traité de l'astrolabe*. See further **X-4**.

nightfall occur when the sun is at a fixed arc of depression below the horizon (**3.1** and **3.3**). Later some astronomers in Iran followed the tradition of Ibn Amājūr in compiling tables of the time since sunrise for the latitudes of Maragha and Shiraz, *inter alia* (**3.8**, **3.11** and **3.12**).

The corpus of tables for Cairo associated with Ibn Yūnus (**Ch. 4**) marks a new trend in astronomical timekeeping in Islam. The major part of the corpus, which contains some 200 tables and over 30,000 entries, consists of tables for reckoning time and solar azimuth from solar altitude. Other tables relate to the beginning of the afternoon prayer, the duration of morning and evening twilight, and the time when the sun is in the azimuth of Mecca. Numerous simple spherical astronomical functions are also tabulated, such as the solar meridian altitude and the rising amplitude. All of the functions are displayed for each integral degree of solar longitude, and the times are usually given in equatorial degrees and minutes. Not all the tables in this corpus were computed by Ibn Yūnus (**Ch. 5**), but the corpus set the pattern for most subsequent prayer-tables in Egypt and Syria. In certain tables prepared by later Egyptian *muwaqqit*s (**Chs. 7-8**) we find the values of the functions relating to the prayer-times in the Cairo corpus tabulated for each minute of solar longitude, with values derived from these by interpolation. The extensive tables for regulating time by the sun and stars for all latitudes compiled by Najm al-Dīn al-Miṣrī in the early 14th century (**6.2**) are rather different in character from those of the Cairo corpus, but they were not widely used. In another set of tables, apparently dating from the 15th century, we find prayer-tables like those in the Cairo corpus computed for each degree of terrestrial latitude (**8.1**). The Cairo corpus itself was used in Egypt as late as the 19th century alongside some of its later manifestations.

In Syria, tables for timekeeping were computed independently by three 14th-century astronomers of exceptional competence, namely, al-Mizzī, Ibn al-Shāṭir and al-Khalīlī (**Chs. 9-10**). al-Mizzī, who studied astronomy in Cairo and then moved to Damascus, computed a set of hour-angle tables and prayer-tables for his city which were virtually identical in their conception to those of the Cairo corpus. Ibn al-Shāṭir computed some prayer-tables for an unspecified locality near Damascus, but this was his colleague al-Khalīlī who was responsible for the extensive corpus of tables for Damascus that was used in Syria until the beginning of the 20th century. The main corpuses of tables that were used in the Yemen and in Tunis also owe their inspiration to the Egyptian and Syrian traditions. The achievements of the *muwaqqit*s in Egypt and Syria after about 1500 are unimpressive, but this is partly explained by the fact that the tables which were their major concern had already been computed by competent astronomers in earlier centuries.

In the earliest Islamic prayer-tables, the functions relating to the times of prayer are tabulated with the solar longitude as argument. In later tables of this kind, the argument was the date in a calendar based on the solar year, such as the Syrian or Coptic calendars. Additional calendrical tables could enable the user to find the corresponding date in a solar-based calendar to use in such prayer-tables. Tables displaying the prayer-times for a given Muslim year were also compiled.

In medieval prayer-tables it was usually the permitted intervals for the prayers that were tabulated, these being expressed in equatorial degrees and minutes. In Ottoman times it generally became the practise to tabulate the times of the prayers in equinoctical hours and minutes, according to the Ottoman convention that sunset is 12 o'clock.[12] The rationale for this system derives from the fact that the Muslim day begins at sunset and also that the Muslim astronomers generally used equinoctial hours rather than seasonal hours. However, this system is hardly convenient when used with mechanical clocks, and when Western clocks were introduced in the Ottoman Empire they were at first generally used with the Turkish time system, the hands of the clock being adjusted to 12 o'clock every evening at sunset. This custom still prevails in the Arabian peninsula, although Yemeni friends of mine have claimed, and not during a *qāt* session, that their watches register 12 o'clock at sunset throughout the year. Modern prayer-tables are in a sense an extension of the Ottoman-type tables, now displaying the times in "Western time": see some examples in **Fig. V-13.1**. Also, since they are intended to be used with clocks, the time given for midday usually incorporates the equation of time.[13]

1.3 Notation used in the analysis

In the text notation in the form "(**1.2.3**)" stands for "(see Section 1.2.3)". Occasionally in **Part II** there are also cross-references to **Part I** in the form "(**I-1.2.3**)", and *vice versa*.

In the mathematical analysis of tables the notation f(x,y) is used for a function f tabulated so that x is the horizontal argument and y is the vertical argument. I use the following mathematical notation:

Latin

a	azimuth, measured from the prime vertical
a′	earliest permitted time for the afternoon prayer (Ottoman convention)
$^{-}{}_{a}$	relates to the earliest time for the afternoon prayer
b′	time for the end of the afternoon prayer (Ottoman convention)
$^{-}{}_{b}$	relates to the end of the interval for the afternoon prayer
$b_{1,2}$	auxiliary functions related to B
B	the "absolute base", an auxiliary function of Islamic timekeeping (see **I-6.0**)
C	the "altitude of the day-circle centre", an auxiliary function of Islamic timekeeping (see **I-6.0**)
d	half-excess of daylight, or equation of half daylight (d = D-90°)
∂	("uncial d") time of the *ḍuḥā*
$^{-}{}_{\partial}$	relates to the time of the *ḍuḥā* (only in Maghribi and Ottoman sources)
D	semi diurnal arc
e	the auxiliary function tan δ tan ϕ
G	an auxiliary function for timekeeping related to sec δ sec ϕ
h	altitude
h_0	solar altitude in the prime vertical
h_a	solar altitude at the beginning of the interval for the afternoon prayer
h_b	solar altitude at the end of the interval for the afternoon prayer
h_q	solar altitude in the azimuth of the qibla

[12] On this see Würschmidt, "Osmanische Zeitrechnung", and the text to n. **I**-1:36.

[13] On this notion in medieval astronomy see the article "Taʿdīl al-zamān" in *EI*$_2$.

h_{q*} solar altitude in the direction perpendicular to the qibla
h_r solar depression at daybreak
h_s solar depression at nightfall
$h_{\tilde{\upsilon}}$ solar altitude in the azimuth of the ventilator (*i.e.,* the azimuth of the rising sun at the winter solstice)
h_z solar altitude at the beginning of the noon prayer (usually equals H)
-^{h} equinoctial hours (also used in the representation of functions, thus: $2D^h(\lambda) = D(\lambda) / 15$)
$\tilde{h}$ number of equatorial degrees corresponding to one seasonal day-hour
$1/\tilde{h}$ factor for converting equatorial degrees to seasonal day-hours
$\tilde{h}'$ number of equatorial degrees corresponding to one seasonal night-hour
H solar meridian altitude
H′ the difference between the Sines of the meridian and instantaneous altitudes (H′ = Sin h - Sin h)
-_H relates to the ascendant (horoscopus)
i′ time of the *imsāk* (Ottoman convention)
j′ time of the call to prayer in Ramaḍān (Ottoman convention)
k an auxiliary function for azimuth calculations (see **I-8.0**)
L terrestrial longitude
m′ time of midday (Ottoman convention)
-_M relates to upper midheaven
-^{m} equinoctial minutes
n length of darkness or length of gnomon
2N nocturnal arc
q azimuth of the qibla, measured from the meridian
-_q relates to the qibla
q* the direction perpendicular to the qibla
-_{q*} relates to the direction perpendicular to the qibla
r duration of morning twilight
r′ time of daybreak (Ottoman convention)
-_r relates to daybreak
R sexagesimal base (60)
R′ time of sunrise (Ottoman convention)
s duration of twilight
s′ time of nightfall (Ottoman convention)
-_s relates to nightfall
S midday shadow
-^{sdh} seasonal day-hours
-^{snh} seasonal night-hours
t hour-angle (actually an arc measured in equatorial degrees rather than an angle; measured from the meridian)
t_a time from midday to the beginning of the interval for the afternoon prayer (see also a′ for the time of the *ʿaṣr* according to the Ottoman convention)

$\underline{t}_{a/b}$	refers to a time between the beginning and end of the ʿaṣr (see **14.10**)
t_b	time from midday to the end of the interval for the afternoon prayer (see also b′)
t_q	hour-angle when the sun is in the azimuth of the qibla
t_{q*}	hour-angle when the sun is in the direction perpendicular to the qibla
t_z	hour-angle at the beginning of the midday prayer (see h_z)
$-_t$	relates to the *taʾhīb* (Maghribi sources only)
T	time since sunrise or remaining until sunset (measured from horizon)
T_a	time from the beginning of the interval for the afternoon prayer to sunset
T_q	time from sunrise to the moment when the sun is in the azimuth of the qibla
z	horizonal shadow of a vertical gnomon (base given in parentheses as subscript)
z_a	shadow at the beginning of the afternoon prayer
z′	vertical shadow
$-_z$	relates to the *ẓuhr*
Z	horizontal shadow at midday
Greek	
α	right ascension
α'	normed right ascension ($\alpha' = \alpha - 90°$)
α_a	oblique ascension of the ascendant at the ʿ*aṣr*
α_r	oblique ascension of the ascendant at daybreak
α_s	oblique ascension of the ascendant at nightfall
α_σ	ascension of the ascendant at the time of the *salām*
α_ϕ	oblique ascension for latitude ϕ
δ	solar declination
$\delta*$	solar declination augmented by 90° (useful since $H = \delta* - \phi$)
δ_2	second declination (see **F6** in **I-1.3**)
ΔD	correction to D for horizontal refraction
Δz	increase of the shadow over its midday minimum
$\Delta\alpha$	rising times of the zodiacal signs
ε	obliquity of the ecliptic
ι'	time of the prayer at the ʿ*īd* (Ottoman convention)
θ	independent variable
λ	solar longitude or ecliptic longitude
λ'	elongation of sun from nearer equinox
$\lambda*$	longitude of point of ecliptic opposite point with longitude λ ($\lambda* = \lambda + 180°$)
λ_H	longitude of the ascendant (horoscopus)
λ_M	longitude of upper midheaven
σ	time from sunset to the *salām*
$-_\sigma$	relates to the time of the *salām*
τ	time from sunset to the time of the *ṭafy* in Ramaḍān
τ_a	time beforc the beginning of the afternoon prayer
τ_z	time from the beginning of the midday prayer (see h_z) to the beginning of the afternoon prayer

$-_\tau$	relates to the time of the *ṭafy*
ϕ	terrestrial latitude
ψ	rising amplitude

Miscellaneous

$\approx$	approximately equals
$-^*$	opposite point of ecliptic
$\bar{\theta}$	complement of θ to 90°
$-'$	for times reckoned according to the Ottoman convention
; ,	sexagesimal semi-colon and comma

In all of the tables consulted in the manuscripts, the entries are, unless otherwise stated, given to two sexagesimal digits and are expressed in standard medieval Arabic alphabetical notation.[14] Most intervals of time are expressed in equatorial degrees and minutes, where:

$$360^\circ = 24^h \text{ and } 1^\circ = 4^m .$$

In the few tables in which times are given in seasonal day or night-hours the reader should bear in mind that:

$$12^{sdh} = (2D)^\circ \quad \text{and} \quad 12^{snh} = (2N)^\circ .$$

In various Ottoman tables other numerical notation and units are used, and the times:

$$m,\ r',\ R',\ s' \text{ and } a'$$

are given according to the convention whereby sunset is reckoned as 12 o'clock.[15]

The capital notation for medieval trigonometric functions, now standard in modern literature on the history of the exact sciences,[16] indicates that the function is computed to a base R, usually 60 for the Sine and Cosine of a general arc θ, thus, for example:

$$\text{Sin}_R\ \theta = R \sin \theta .$$

The Cotangent or Tangent is usually defined with an argument that is usually a celestial altitude, since these functions measure, respectively, the horizontal shadow of a vertical gnomon length n or the vertical shadow of a horizontal gnmon. Thus for any celestial altitude h:

$$\text{Cot } h = n \cot h .$$

The most common bases for these functions are 12 (digits) and 7 (feet). However, other bases were used, such as 1, 6, $6^1/_2$, $6^2/_3$, 10, 20 and 60,[17] and one "artificial" base of 5 (= 60/12).[18] I use Z for Cot H and z for Cot h and occasionally z′ for Tan h, with the base indicated in parentheses in the subscript, thus, for example, $Z_{(6;40)}$ stands for $\text{Cot}_{6;40}$ H. I also use z_a, *etc.*, for the shadow at the *ʿaṣr* prayer. The 'z', for Arabic *ẓill*, shadow", is intended as a reminder that the medieval Cotangent function is of a different nature from the medieval Sine function.

I assume that the reader has some familiarity with spherical astronomy and is familiar with **I**.[19] Note that the hour-angle $t(h,\lambda,\phi)$ corresponding to an instantaneous solar altitude h, solar

---

14 On this notation see the text to n. **I**-1:27.

15 See the text to n. **I**-1:12.

16 See the text to n. **I**-1:28.

17 See the text to n. **I**-1:31-35. On the reasoning behind the use of these bases see the text to n. **III**-0:17.

18 See, for example, the various tables discussed in **I-7.1.5** and **7-8**.

19 A brief account of the standard medieval formulae and references to the published literature on spherical astronomy in medieval Islam is given in **I-1.2**. A discussion of the determination of the prayer-times by modern methods is contained in Arvanitaki, "L'heure arabe" (1915), and, more recently, Ilyas, *Astronomy of Islamic Times*, and *Islamic Calendar, Times & Qibla*.

longitude λ and local latitude ϕ, is symmetrical with respect to the solstices and so the equinoctial value and 180 values suffice to define it for each integral degree of solar longitude. This also holds for functions defining the permitted intervals for the prayers. In virtually all the tables consulted, full advantage is taken of the symmetry of the functions tabulated. I denote the longitude argument for such functions by λ' ($\delta \gtreqless 0°$), where λ' measures the solar elongation from nearer equinox. Generally early tables begin with entries for Aries and Virgo and later ones begin with entries for Capricorn and Sagittarius.

1.4 On some approximate formulae for timekeeping

Some of the Muslim astrononomers whose writings and tables we shall be considering used approximate formulae, as a result of which computation is much facilitated. I mention just two of these here, since they recur frequently.

First, an Indian "arithmetical" formula for finding the time in seasonal hours ($T < 6$) from the increase in the shadow length Δz for a gnomon length n is equivalent to the following:

$$T \approx 6 \cdot n / (\Delta z + n) .$$

Note that when $\Delta z = 0$ then $T = 6$, and when $\Delta z \to \infty$ then $T = 0$. For intermediate hours the results are respectable to a first approximation. As we shall see, this formula underlies the standard definitions for the *ʿaṣr* as well as the Andalusī and Maghribī definitions of the *ẓuhr* (see further **IV-2.4**). See **2.3**, **2.4**, **2.5**, **2.9**, and **9.3** (Ibn al-Shāṭir!) for the formula in the present study, and also **III** and **IV**.

Second, an approximation based on trigonometric rather than simple arithmetical considerations. The time in seasonal day-hours T (< 6) is given in terms of the instantaneous solar altitude h and the meridian altitude H thus:

$$T \approx {}^{1}/_{15} \cdot \text{arc Sin} \{ R \text{ Sin } h / \text{Sin } H \} .$$

Here $T = 0$ when $h = 0°$ and $T = 6$ when $h = H$. The formula is in fact accurate when the sun is at the equinoxes. For intermediate solar longitudes and latitudes between 15° and 40° this formula too gives results that are reasonable – actually more reasonable than one would expect – to a first approximation. We find this formula in numerous of the sources studied here, from the 9th century onwards: see **I-2.5** and **4.3**, **II-2.4**, **2.5**, **2.9**, **3.2** and **3.3**, and further **XI**. It underlies the universal horary quadrant that was a standard feature of astrolabes, both Islamic and European, although those who used it in Europe were apparently aware that the formula is no longer reliable at European latitudes.

In addition, various approximate rules for quick calculation of various spherical astronomical functions are proposed by certain Muslim astronomers: see **2.4**, **2.4a** (Andalusī source), **2.5**, **2,9**, **3.1** (early ʿAbbāsid source!), **3.11**, **6.1**, **6.3**, **6.12**, and **9.3** (Ibn al-Shāṭir!). It seems that those who proposed such rules looked at tables of a function $f(\lambda)$, noted the changes between the values at the equinoxes f_0 and solstices, and then determined the factors m and n (+ or -) in simple fractions which would best fit in a simple relation of the kind:

$$f(\lambda) = f_0 + m \cdot \delta(\lambda) \text{ for } \delta > 0 \quad \text{and} \quad f_0 + n \cdot \delta(\lambda) \text{ for } \delta < 0 .$$

Sometimes other variables, such as d or $\bar{H}$ were used instead of δ. I am not aware of any tables based on such approximations.

1.5 On the recomputation of medieval prayer-tables

Given the algebraic expression defining a particular function and the underlying parameters it is possible to compile extensive tables of the function with an electronic computer (see already **I-1.5**). The use of the computer in the analysis of medieval Islamic tables was pioneered by Ted Kennedy, and is now being pursued by Benno van Dalen, Glen Van Brummelen and François Charette.[20] Once the underlying latitude is established, it is often useful to consult the lists of coordinates taken from Islamic sources prepared by Ted and Mary Helen Kennedy.[21]

In 1970-71 I recomputed the main Cairo corpus in the Computer Centre at the American University of Beirut and in 1971-72 recomputed similar sets for other latitudes at the Yale University Computer Center. At the Computer Centre of the American University in Cairo, programmes were prepared in 1972-73 to recompute prayer-tables of each of the various types described in the sequel for any latitude. A given set for a particular latitude can be recomputed in a matter of seconds once the underlying parameters have been established.[22]

The tables discussed in the present study contain a total of several million entries. Some of them merit publication *in toto*, others are of little scientific or historical interest. In this study I present sample entries only from tables that I consider to be of singular interest for one reason or another. The values agree with recomputation except where the error in the second digit is shown in square brackets, computed according to the convention:

error = (value in text) - (recomputed value) .

Rather than give sample entries, I have generally considered it more important to identify the structure and underlying parameters of all the tables, to comment on their accuracy, as well as to present illustrations of as many tables as possible.[23]

---

[20] See n. **I**-1:24.

[21] See n. **I**-1:4.

[22] Computer time was made available by the Department of Mathematics, The American University of Beirut; the Department of Near Eastern Languages and Literatures, Yale University; and the Smithsonian Institution, respectively.

[23] Ideally, sample entries should be given for tables forming part of a clearly defined corpus of prayer-tables for a given locality. (A model for such an endeavour for standard astronomical tables is Toomer, "Toledan Tables".) In most cases such entries (say for each 30°) are adequate to assess the accuracy of the tables and to identify related material. In King, "Astronomical Timekeeping in Medieval Cairo", pp. 380-384, I present sample entries from the tables in the Cairo corpus, and in an unpublished study *Shams al-Dīn al-Khalīlī and the Culmination of the Islamic Science of Astronomical Timekeeping*, yet more extensive samples from the Damascus corpus. The latter work is unpublished not least because the prospective publisher, the Third World Academy of Sciences, some years ago understandably baulked at the prospect of printing masses of tables as well as photos of the most important manuscripts (including MSS Paris ar. 2558 and Dublin CB 4091). The text, without the tables, is now incorporated in **Ch. 10** of the present study.

CHAPTER 2

SELECTED MEDIEVAL ARABIC TREATISES ON TIMEKEEPING

2.0 Introductory remarks

An enormous amount of Arabic scientific literature deals with astronomical timekeeping. Also, the topic of spherical astronomy is dealt with in the majority of Islamic *zīj*es. In order to display the importance of this literature for the documentation of the history of the development of mathematical methods, and the history of astronomical inquiry into such phenomena as twilight and refraction at the horizon, I have selected a few of the more historically interesting of the available treatises and have attempted to survey their contents. I have deliberately included one treatise in the folk astronomical tradition (**2.3**). There are others, also of widely varying levels of competence on the part of their original authors, to which the interested reader may have recourse: Kūshyār ibn Labbān on spherical trigonometry;[1] Abū Naṣr's treatises on spherical astronomy;[2] al-Bīrūnī's treatise on spherical trigonometry applied to problems of spherical astronomy;[3] al-Kāshī's treatment of spherical astronomy in his *Khāqānī Zīj*;[4] and Abū Miqraʿ's treatment of timekeeping by folk astronomical methods.[5] Also the simple techniques for reckoning time of night by the lunar mansions have been studied recently by Miquel Forcada.[6] In addition there are several accounts in the published literature of the solution of such problems using astronomical instruments, such as al-Khwārizmī on the astrolabe;[7] Ḥabash on the celestial globe;[8] Ibn al-Zarqālluh on the universal plate (*ṣafīḥa shakkāziyya*);[9] Ibn al-Shāṭir on an astronomical "compendium" (multi-purpose instrument) of his own design;[10] and Jamāl al-Māridīnī on the universal quadrant.[11] See further **X**. As this book goes to press, I am gratified that Julio Samsó has published an account of a medieval Maghribī treatise preserved in MSS London B.L. 411 and Cairo K 4311 which deals critically with parameters for twilight (and also trepidation) as discussed by various earlier authors.[12]

---

[1] See Berggren, "Spherical Trigonometry in the *Jāmiʿ Zīj*".
[2] See Samsó, *Estudios sobre Abū Naṣr.*
[3] See the study of Marie-Thérèse Debarnot listed under al-Bīrūnī, *Maqālīd.*
[4] See Kennedy, "Spherical Astronomy in al-Kāshī's *Khāqānī Zīj*".
[5] See Colin & Renaud, "Abū Miqraʿ".
[6] See Forcada, "*Mīqāt* en los calendarios andalusíes".
[7] See already Frank, "al-Khwārizmī über das Astrolab". A new edition, translation and commentary have been prepared by François Charette and Petra Schmidl.
[8] Lorch & Kunitzsch, "Ḥabash on the Sphere".
[9] See n. **I**-1:45.
[10] See Janin & King, "Ibn al-Shāṭir's Compendium".
[11] See King, "al-Māridīnī's Universal Quadrant" (n. **I**-9:23).
[12] Samsó, "Astronomical Observations in the Maghrib", pp. 174-175. On the unidentified author and his work see *Cairo ENL Survey*, no. F26, and on some of his tables for timekeeping see **13.6**.

2.1 A poem on the times of prayer attributed to the Imām al-Shāfiʿī and Ibn Yūnus

In MS Cairo DM 181, fols. 46v-48r, copied *ca.* 1800, there is a short poem on the prayer-times in *ṭawīl* metre attributed to Ibn Yūnus (**4.1**): see Fig. **III-9.7b**. The Egyptian astronomer was renowned as a poet and several of his poems are preserved in various anthologies and biographical lexicons.[13] However, in MSS Berlin Ahlwardt 5820 (Wetzstein 175), fol. 65r, 5700 (Landberg 953), fol. 11r, copied in 944 H [= 1537/38] and 1231 H [= 1816], respectively, a similar poem is attributed to the famous legal scholar al-Shāfiʿī (d. 819/20). This poem is also attributed to al-Shāfiʿī in various later Egyptian works on timekeeping, such as those of ʿAbd al-Raḥmān al-Tājūrī and Yūsuf Kilārjī (**8.8**). al-Shāfiʿī was renowned for his poetry, but this particular poem is not included in a modern anthology of his poetical works.[14] The poem in both Berlin copies is extremely corrupt and is clearly an improvisation of the original poem by someone who was not very gifted in either astronomy or poetry. I present a prose translation of the poem from MS Cairo DM 181, incorporating a few corrections necessary to the sense of the poem, some of which are confirmed by the readings in the second Berlin manuscript (Ahlwardt 5700):[15]

> "Knowing the prayer-times is a prescribed duty for discerning Muslims. This is summarized in the *Qurʾān*, my friend, and was explained by Aḥmad [*i.e.*, the Prophet Muḥammad, referred to as Aḥmad in the *Qurʾān*], the most outstanding of men. Perform the midday prayer whenever you observe the shadow starting to increase. Add a length (of the gnomon) to the shadow: this gives you (the shadow) at the time of the afternoon prayer. At sunset get up and perform the evening prayer: this is the only permissible time. Perform the night prayer when, looking at the sky, you see the upper part of the evening twilight fade away and disappear. As for the end of this prayer time, if you wish you can wait until one-third of the night; indeed, it is better that you wait. But do not wait until whiteness appears (on the eastern horizon): it will last for a time in the sky. Bear in mind that there are two stages of daybreak according to our doctrine: distinguish between them carefully – you are the one who decides this. The first daybreak looks like a wolf's tail rising in the sky: this is the false dawn. The later one is the true dawn: you see it illuminate the sky like a fire. The end of this prayer-time is sunrise: at that moment the best time for prayer is over. There is no virtue in a person who is neglectful of the prayer-times and he has no knowledge of Him who is to be worshipped."

Ibn Yūnus is the more likely author because in the time of al-Shāfiʿī the standard definitions mentioned here were not yet definitively formulated (**IV-3**).

---

[13] On Ibn Yūnus' poetry see King, *Ibn Yūnus*, I.1, and the references there cited.

[14] On al-Shāfiʿī's poetry see Sezgin, *GAS*, III, p. 647.

[15] The text of the poem is published in my contribution to the *Aleppo 1976 Symposium Proceedings*, I (Arabic section), pp. 393-394, but is not included in the English version "Astronomical Timekeeping in Medieval Syria", repr. in King, *Studies*, A-X.

2.2 al-Bīrūnī on tables for timekeeping

al-Bīrūnī was the most outstanding scientist of the Islamic Middle Ages.[16] He was a prolific writer and did original work in fields as varied as ethnography, geology and astronomy. Our present concern is with his *Ifrād al-maqāl fī amr al-ẓilāl*, "Unique Treatise on Matters Relating to Shadows", in which he devotes a whole chapter to the opinions of the religious leaders about the times of prayer and the obligation to determine them properly. The Arabic text of this work was published in Hyderabad in 1948 and E. S. Kennedy published a translation and commentary in 1976.[17] al-Bīrūnī's treatise is a mine of information concerning the prayer-times and time-keeping in general. However, al-Bīrūnī does not make a single reference to tables for regulating the prayer-times, although in his treatise on the construction of the astrolabe entitled *al-Istīʿāb* he does mention *ṭaylasān* tables for time-keeping.[18] Thus this study is in a sense a supplement to Kennedy's study of al-Bīrūnī's *Shadows*.

2.3 Ibn Raḥīq's treatise on timekeeping

MS Berlin Ahlwardt 5664 (Landberg 108), achieved *ca.* 1350, is an apparently unique copy of a treatise on folk astronomy by an individual named Abū ʿAbdallāh Muḥammad ibn Raḥīq ibn ʿAbd al-Karīm, otherwise unknown to the literature.[19] The treatise was compiled in Mecca in the early 11th century and it contains information of a very simple kind on the calendar, the lunar mansions and timekeeping. Some of the material on the latter topic is said to be taken from "the large book" of Abū ʿAlī ʿAraqa, the muezzin of the mosque of ʿAmr (*al-jāmiʿ al-ʿatīq*) in Fusṭāṭ. The treatise is an important source for our knowledge of the practices of the muezzins of early Islam, before sophisticated mathematical methods were applied to time-keeping. A detailed study of this and two other Rasulid Yemeni treatises on the same subjects – by Ibrāhīm ibn ʿAlī al-Aṣbaḥī[20] and Muḥammad ibn Abī Bakr al-Fārisī[21] – is currently being conducted by Petra Schmidl.[22]

Ibn Raḥīq begins his treatise by asserting that the times of prayer should be determined by observation with one's own eyes (*al-ʿiyān wa-'l-raṣad*), not by using an astrolabe or "any of this astronomy nonsense" (Arabic, *tanjīm*: the word usually means "astrology"). He states that he will follow the *ḥadīth* about the Archangel Gabriel recorded in the canonical collections of Muslim and al-Bukhārī (**IV-1.2**), rather than the writings of those take their learning from "the infidels and the *Sindhind*". (By these he means the Greek and Indian traditions in Islamic astronomy, respectively.)

With regard to the *ẓuhr* prayer Ibn Raḥīq recommends the use of an "Indian sundial" (*al-rukhāma al-mabniya* (?) *al-hindiyya*) with which one can watch for the moment when the

[16] On al-Bīrūnī see the splendid article by E. S. Kennedy in *DSB*. His work on timekeeping is listed as al-Bīrūnī, *Shadows*.
[17] See n. 1:7.
[18] See n. **I**-1:53.
[19] On Ibn Raḥīq and his works see King, *Astronomy in Yemen*, no. 2, and also **III-3.2** and **IV-1.5**.
[20] On al-Aṣbaḥī see King, *Astronomy in Yemen*, no. 5; *Cairo ENL Survey*, no. E5; and **IV-1.5**, *etc.*
[21] On al-Fārisī see n. **I**-7:17, and also **III-4.2** and **IV-1.5**.
[22] See Schmidl, *Islamische Volksastronomische Abhandlungen*, and also *eadem*, "Qibla und Winde".

shadow cast by the sun begins to increase.[23] Ibn Raḥīq records that ʿAraqa had reported that Aḥmad al-Rāzī, the muezzin at the Sacred Mosque in Mecca, had observed the sun over a period of seven years and had found that the shadow length for the Coptic months, expressed in terms of the height of a man (*qāma*), were as follows (beginning with the 10th month):

X: $^1/_3 \cdot {}^1/_5$	XI: 0 [a]	XII: $^1/_6$
I: $^1/_3 + {}^1/_2 \cdot {}^1/_6$ [b]	II: $^1/_2$	III: $^2/_3 + {}^1/_4$
IV: $1 + {}^1/_6 \cdot {}^1/_8$	V: $1 - {}^1/_2 \cdot {}^1/_8$	VI: $^1/_3 + {}^1/_4$
VII: $^1/_4 + {}^1/_8$	VIII: $^1/_6$	IX: 0

[a] Arabic: *lā shayʾ* [b] text has: $1/33 + {}^1/_2 = {}^1/_6$

These fractions, written in words in the text, can be simplied thus:

$^1/_{18}$ 0 $^1/_6$ $^5/_{12}$ $^1/_2$ $^{11}/_{12}$ $1^1/_{48}$ $^{15}/_{16}$ $^7/_{12}$ $^3/_8$ $^1/_6$ 0

Ibn Raḥīq states that the beginning of the *ʿaṣr* prayer should be when the shadow is $6^1/_2$ feet (*qadam*, pl. *aqdām*) more than the midday shadow, which indicates that he favored $6^1/_2$ units for the length of the gnomon. He records the following maximum shadow lengths reported by Abū ʿAlī ʿAraqa:

Mecca	$7 - {}^1/_5$	(= $6^4/_5$)
Medina	$7 + {}^1/_2$	
Sirrayn (in Yemen)	$7 - {}^1/_3$	(= $6^2/_3$)
Alexandria, Baghdad, Crete	10	
Jerusalem	9 + 2/3	
Cairo-Fusṭāṭ (*Miṣr*)	9	
Ifrīqiyya (*i.e.*, Tunis), Rayy	11	
Qum	13	
Kufa	$8 + {}^1/_2$	
Armenia	6	(read 16?)

ʿAraqa also measured the shadow at Sanaa on Kayhak 25 (year unspecified), in the company of the *ḥadīth* scholar Aḥmad ibn Ṣāliḥ, and found it to be $4^1/_2$ feet.

In a section on measuring time of day approximately from shadow lengths Ibn Raḥīq presents a method for finding the seasonal hours from the excess of the instantaneous shadow over the midday shadow, which he asserts is valid for the latitudes of al-ʿIrāq, Syria, the Maghrib and Egypt. When these excesses are:

39, $19^1/_2$, $6^1/_2$, $3^1/_4$, 1+

according to Ibn Raḥīq, the time is:

1, 2, 3, 4, 5 or 11, 10, 9, 8, 7

seasonal hours after sunrise. The figure for 5 and 7 seasonal hours is expressed as *qadam rājiḥ*, which means "just more than one foot". No explanation is given for these figures. Consider, however, the following approximation for the time T in seasonal day-hours as a function of the excess Δz of the instantaneous shadow of a gnomon length n over the midday shadow, namely:

$$T \approx 6 \cdot n / [\, n + \Delta z \,] .$$

[23] It is not completely clear what is meant here. The name "Indian circle" (*al-dāʾira al-hindiyya*) is usually given in the Arabic sources to a simple device and procedure – a gnomon and a a geometric construction based on observations with it – for determining the local meridian. On this see Kennedy, *al-Bīrūnī's* Shadows, II, pp. 80-90, and, most recently, King, *Mecca-Centred World-Maps*, pp. 93 and 94, and also n. 24 on p. 134.

This formula, which is found in several Islamic sources (**1.5**), is ultimately of Indian origin (**III-1.3** and **IV-2.4**). If we substitute T = 3, 4 and 5^{sdh}, we obtain respectively:

$$x = 6^1/_2 , \quad 3^1/_4 \quad \text{and} \quad 1^3/_{10} ,$$

as in the text. Further, if x is large compared with n and we use the cruder approximation:

$$T \approx 6 \cdot n / x ,$$

we obtain for T = 1^{sdh} and 2^{sdh} Ibn Raḥīq's values:

$$x = 39 \quad \text{and} \quad 19^1/_2 .$$

Thus each of the five values is explained. Ibn Raḥīq adds a note that the 9th hour marks the beginning of the *ʿaṣr* prayer according to the fourth caliph ʿAlī ibn Abī Ṭālib (**III-2.2** and **IV-2.2**).

Here we have, perhaps, the explanation of the origin of the definition of the *ʿaṣr* which was not stated in the *ḥadīth* but which was adopted already in the 8th century and has persisted to this day. If we postulate that the *ʿaṣr* divides the period from midday to sunset into two equal intervals, then, according to the approximate Indian formula quoted above, the length of the shadow at the *ʿaṣr* is equal to the midday shadow increased by the length of the gnomon. If this is indeed the origin of the definition of the *ʿaṣr*, Ibn Raḥīq has underestimated the debt of Islam to at least the Indian tradition in early Islamic astronomy. Note also that the shadow length at the beginning of the 10th hour, according to the Indian formula though not according to Ibn Raḥīq, is in excess of the midday shadow by twice the length of the gnomon. This is Abū Ḥanīfa's definition of the *ʿaṣr* (**1.1** and **IV-3**).

Ibn Raḥīq states that the evening prayer (*al-maghrib*) should begin when redness appears in the western sky after sunset. He distinguishes between the true and false dawn for the beginning of the morning prayer (*al-fajr*).

Our author quotes from a work of Muhammad ibn Surāqa al-ʿĀmirī (d. 1019) information about the determination of the qibla in various parts of the Muslim world. The following extract is typical of the information contained in an extensive corpus of material dealing with what I call sacred geography and which I have surveyed elsewhere.[24] It is traditional knowledge, independent of the scientific tradition of mathematical geography and cartography in medieval Islam.

> "The people of Alexandria and Egypt up to Kairouan and Sus (*al-Sūs al-aqṣā*) and the Black Sea and all places in the same direction (with regard to Mecca) should face the same direction as (when one is standing in front of the Kaʿba) between its western corner to the draindripe (at the middle of the north-western wall). In any of these places, if one should stand with the three main stars of the Great Bear (*banāt naʿsh*) at one's right shoulder when they are setting and at one's left shoulder when they are rising, with the celestial pole behind one's back and the west wind (*al-dabūr*) on one's right, then one is facing the direction of the Kaʿba."

A large part of Ibn Raḥīq's treatise consists of simple information about each of the lunar mansions throughout the year, such as is found in several later Yemeni works (**12**). The following is an extract:

---

[24] This kind of qibla determination is surveyed in my unpublished study *The Sacred Geography of Islam*, summarized in the article "Makka. iv. As centre of the world" in EI_2, repr. in King, *Studies*, C-X. See also Schmidl, *Islamische Volksastronomische Abhandlungen* (n. 2:22).

"The lunar mansion *al-Iklīl* rises at daybreak on Hātūr 17, which is Tishrīn II 13, November 13 and Ādharmāh 10. Morning twilight (?: *al-ẓill*) is when the next mansion *al-Qalb* (is rising over the horizon) and the sun is in (the next mansion) *al-Shawla*. The beginning of the night is when al-*Haqʿa* is rising. The mansions culminating at dawn and nightfall are *al-Jabha* and *al-Muʾakhkhar*. This mansion is (completely) visible on the 12th day of its rising and contains three bright stars which (form) the crown of the Scorpion. The midday shadow at Mecca – may God exalt her – is six and one third feet."

2.4 Ibn al-Ḥammāmī's treatise on timekeeping

MS Istanbul Haci Mahmud Efendi 5713, penned in the 12th century, is a unique copy of three Egyptian astronomical treatises of considerable historical interest. These are: (a) a treatise on timekeeping by an individual called Ibn al-Ḥammāmī (fols. 1r-10v), (b) a treatise on the astrolabic quadrant by the same author (fols. 10v-25v), and (c) a treatise on the trigonometric quadrant by Fatḥ al-Dīn al-Qaysī (fols. 25v-34v). The treatise on the astrolabic quadrant predates the earliest previously-known Arabic treatises on quadrants (by al-Mizzī) by at least a century.[25] I have no information on either of the two authors.[26]

Our present concern is with the first of these treatises, which contains various approximate rules for the standard functions of timekeeping, such as those given by the later Egyptian astronomers Najm al-Dīn al-Miṣrī, al-Bakhāniqī and al-Fawānīsī (**3.3**, **3.11** and **6.12**).

Ibn al-Ḥammāmī first records the following approximate changes in δ and α_ϕ for each 30° of λ measured from the equinox:

$$\delta: 12° \quad 8° \quad 4° \quad \text{and} \quad \alpha_\phi: 21° \quad 24° \quad 30° \quad 34° \quad 36° \quad 35°,$$

and the following midday shadows for the zodiacal signs:

7 4 2 1;21 2 4 7 10 14 16;16 14 10

To find the time T in seasonal hours since sunrise from the instantaneous and midday shadows z and Z he prescribes the rule:

$$T = 72 / [(z + n) - Z] \qquad (n = 12) .$$

To find the shadow z_1 and z_2 at the (second optimum?) time for the midday prayer (*ẓill ikhtiyār al-ẓuhr*) and the end of the midday prayer (*ẓill ākhirihi*), the author prescribes:

$$z_1 = Z + {}^1/_4\, n \quad \text{and} \quad z_2 = Z + n \qquad (n = 12) .$$

Likewise the shadows z_3 and z_4 at the beginning and end of the afternoon prayer (*ẓill awwal waqt al-ʿaṣr* and *ẓill ākhirihi*) are given as:

$$z_3 = z_2 + \varepsilon \quad \text{and} \quad z_4 = Z + 2n ,$$

where ε is referred to as *adnā ziyāda*, "the last (perceptible) increase". I know of no other Egyptian treatise in which the times of the midday and afternoon prayers are defined in this way (**IV-4**).

---

[25] On quadrants see my EI_2 article "Rubʿ" and **X-6**.

[26] Their full names are Abu 'l-Ḥasan ʿAlī Ibn Muḥammad ibn al-Ḥammāmī and Fatḥ al-Dīn Abū ʿUthmān ibn Hibatallāh ibn Abi 'l-Ḥawāfir. The copyist of the manuscript was ʿAbd al-Raḥmān ibn Muḥammad ibn ʿAbd al-Muhayman al-Shāfiʿī, and the date of copying is not clearly legible. The day was Wednesday, Ramaḍān 1, and the year ??1 Hijra. The Süleymaniye card index has 504 H, which is untenable.

In the course of his discussion of the determination of the standard functions of timekeeping Ibn al-Ḥammāmī presents several approximate rules. As an example I quote the rule for finding t_a and T_a:

> "To find the time between the midday and afternoon prayers and between the afternoon prayer and sunset, add one-half of one-sixth of the solar declination to 52° for the northern signs and subtract one-quarter of the declination and one-sixth of the declination for the southern signs. The result will be the time between the midday and afternoon prayers. Subtract this from half the diurnal arc and the remainder will be the time between the afternoon prayer and sunset. God knows best."

In all there are five such approximate rules:

(a)	$h_a \approx 32;30° + {}^2/_5\ \delta$	$(\delta \gtrless 0)$
(b)	$h_q \approx 47° + \delta + {}^2/_5\ \delta$	$(\delta \gtrless 0)$
(c)	$t_a \approx 52° + {}^1/_2 \cdot {}^1/_6\ \delta$	$(\delta > 0)$
	$t_a \approx 52° - {}^1/_4\ \delta - {}^1/_6\ \delta$	$(\delta < 0)$
(d)	$s \approx 18;30° + {}^1/_8\ \delta$	$(\delta > 0)$
	$s \approx 18;30° + {}^1/_2 \cdot {}^1/_8\ \|\delta\|$	$(\delta < 0)$
(e)	$r \approx 23;30° + {}^1/_5\ \delta$	$(\delta > 0)$
	$r \approx 23;30° + {}^1/_2 \cdot {}^1/_6\ \delta$	$(\delta < 0)$

Rule (b) corresponds slightly more closely to the paramater q = 52° than to 53° (though see below) and rules (d) and (e) imply that Ibn al-Ḥammāmī accepted the parameters 20° and 16° for morning and evening twilight, which is confirmed on fols. 22r and 22v of this Istanbul copy of his treatise on the astrolabic quadrant. I consider it likely that Ibn al-Ḥammāmī derived these simple rules by inspecting some tables of these five functions. Those in the main Cairo corpus give the following values for the equinoxes and the solstices:

	h_a	h_q q=52°	h_q q=53°	t_a	s h_s=-16°	r h_r=-20°
EQ (λ=1°)	32;33	47;23	46;43	51;54	18;35	23;18
SS	41;58	80; 2	79;48	53;48	22;13	28;28
WS	23; 0	13;39	12;32	41;53	20; 4	24;54

This having been said, it should be pointed out that Ibn al-Ḥammāmī predates al-Maqsī and so was more likely to have access to earlier tables that may not have been preserved for us. See, for example, **5.3**.

Our author concludes his treatise by describing some simple operations with an astrolabe, including how to use the alidade to indicate the direction of the qibla, stated as 37° south of east (q = 53°).

2.4a Ibn Bāṣo's rules for the prayer-times

The Andalusī astronomer Ibn Bāṣo (d. 1316) is best known for his ingenious universal plate, which became a feature of many later Maghribi and even a few Eastern Islamic astrolabes.[27]

[27] See Calvo, *Ibn Bāṣo and his Universal Plate*, where the definitions are discussed on pp. 78-80.

In his treatise on this plate he advocates approximate rules for the times of the *ẓuhr* and *ʿaṣr*, as follows:

$$h_z \approx (H - 10°) - {}^1/_{10}\,(H - 30°) \quad \text{for } H > 30° \quad \text{and} \quad h_z \approx H - 10° \quad \text{for } H < 30°$$

$$h_a \approx {}^1/_2\,H + {}^1/_{10}\,(80° - H) \quad \text{for } H < 80° \quad \text{and} \quad h_a \approx {}^1/_2\,H \text{ for } H > 80°$$

The second rule is curious because for the latitude of, say, Cordova, the solar altitude does not reach 80°. I have not encountered these elsewhere, although certain Egyptian treatises (**2.9** and **6.1**) have a similar rule for h_a with underlying latitude 30°.

2.5 Najm al-Dīn al-Miṣrī's treatise on timekeeping

MS Istanbul Hamidiye 1453, fols. 219r-228v and 228v-230, copied in Damascus in 869 H [= 1464/65], contains two related treatises on timekeeping, the second being attributed to Najm al-Dīn al-Miṣrī, who worked in Cairo in the early 14th century (see **I-2.6.1** and **I-9.3***, and also **II-6.5** on his extensive tables for timekeeping). The first treatise is entitled *Taḥrīr al-maqāla fī maʿrifat al-awqāt bi-ghayr āla*, and the second *Ikhtiṣār al-maqāla* ... , meaning respectively "Redaction of" and "Abridgement of" of a "Treatise on Determining the Times of Prayer without Instruments". A third treatise on timekeeping by Najm al-Dīn al-Miṣrī is preserved in MS Milan Ambrosiana 227a (C49), fols. 85v-97r, and represents a much more sophisticated level of spherical astronomy. Neither of the first two treatises is related to the third, since they both deal only with approximate methods, and not always the same ones at that. However, I see no problem in accepting Najm al-Dīn as the author of all three.

In the first treatise Najm al-Dīn first records the following approximate changes in δ, α and α_ϕ for each 30° of λ measured from the equinox:

$$\delta: \; 12° - 8° - 4°$$

$$\alpha: \; 28° - 30° - 32°$$

$$\alpha_\phi: \; 21° - 24° - 30° - 35° - 35° - 35°$$

To find the shadow lengths (base 12) corresponding to a particular altitude he considers the four cases:

$$\text{(a)} \;\; h < 27°\,, \quad \text{(b)} \;\; 27° < h < 45°\,, \quad \text{(c)} \;\; 45° < h < 63° \quad \text{and} \quad \text{(d)} \;\; h > 63°$$

stating that the corresponding shadow lengths are respectively:

$$\text{(a)} \;\; h\,/\,4^1/_2\,, \quad \text{(b)} \;\; (h\text{-}27)/3 + 6\,, \quad \text{(c)} \;\; (63\text{-}h)/3 + 6\,, \quad \text{and} \quad \text{(d)} \;\; (90\text{-}h)\,/\,4^1/_2\,,$$

where the first two are vertical (*mankūs*) and the last two horizontal (*mabsūṭ*). He also notes that the horizontal and vertical shadows z and z′ are related by:

$$z \cdot z' = 144 \quad (= 12^2)\,.$$

Najm al-Dīn then discusses the altitudes corresponding to (vertical) shadow lengths in feet (*aqdām*) and mentions the values:

z	1	2	3	4	5	6	6;40
h	9°	17	24	31	37	42	45

For timekeeping by the sun he notes that the length of half-daylight in equinoctial hours D′ is given approximately by:

$$D' \approx {}^{11}/_{60} \cdot \delta \cdot \phi\,,$$

a formula which underlies a simple table in MS Paris BNF ar. 2513 of the contemporary

Muṣṭalaḥ Zīj (**6.6** and **I-7.4.1**), and then prescribes the following rules for finding the time in seasonal hours since sunrise, T, from the instantaneous and midday shadows, z and Z:

$$T \approx 40 / [\ (6;40 + z) - Z\ ] \qquad \text{(shadows in feet)}$$
$$T \approx 40 / [\ (12 + z) - Z\ ] \qquad \text{(shadows in digits)}$$

The author also presents the following approximate rules for finding the duration of morning and evening twilight r and s in seasonal hours when the solar longitude is λ:

$$r \approx 72 / [\ (33+12) - Z^*\ ] \quad \text{and} \quad s \approx 72 / [\ (42+12) - Z^*\ ]\ ,$$

where Z^* is the horizontal midday shadow ($n = 12$) cast by the sun when it has longitude $\lambda^* = \lambda + 180°$. Note that $\mathrm{Cot}_{12}\, 20° \approx 33$ and $\mathrm{Cot}_{12}\, 16° \approx 42$, so that these rules are simply extensions of the above rule for reckoning time by day.

In the second treatise Najm al-Din records the following approximate changes in δ, α and α_ϕ for each 30° of λ measured from the equinox:

δ: 12° - 8° - 4° α: 28° - 30° - 32° and α_ϕ: 21° - 24° - 30° - 34° - 36° - 35°

and the following midday shadows ($n = 12$) for the zodiacal signs beginning with Aries:

7 4 2 1;20 2 4 7 10 14 16;15 14 10 .

Note that the solar meridian altitude in Cairo at the equinox and solstices are 60°, 83;35°, 36;25° and the corresponding shadows are 6;56, 1;21, and 16;16.

The remainder of the treatise deals with the determination of the standard functions of timekeeping. Najm al-Dīn mentions the following approximate formulae:

(a) $h_a \approx 32;32° \pm {}^2/_5\, |\delta| \qquad (\delta \gtrless 0)$

(b) $d \approx {}^1/_2\, \delta + {}^1/_4\, \delta$

(c) $t_a \approx 52° + {}^1/_3 \cdot {}^1/_5\, d \qquad (\delta > 0)$

$t_a \approx 52° - {}^1/_3\, |d| \qquad (\delta < 0)$

(d) $\psi \approx \delta + {}^1/_6\, \delta$

(e) $h_q \approx 47° + \delta + {}^1/_6\, \delta + {}^1/_4\, \delta \qquad (\delta \gtrless 0)$

(f) $h_0 \approx 2\delta \qquad (2\delta < 15°)$

$h_0 \approx 30° + (4\delta - 30°) + {}^1/_4\, (4\delta - 30°) \qquad (2\delta > 15°)$

(g) $\alpha_s \approx \alpha_\phi(\lambda^*) + 20°$

(h) $\alpha_r \approx \alpha_\phi(\lambda) - 25°$

Note that in (a) the value 32;32° was probably taken from a table of $h_a(\lambda)$ in which it was the first entry (for Aries 1°): the tables of this function in the main Cairo corpus (**4.9**) have 32;33° for the first entry but the corresponding table in the *Muṣṭalaḥ Zīj* (**6.6**) and in the auxiliary tables of al-Khaṭā'ī (**6.8**) have 32;32°. Rules (b) and (c) are rather poor approximations, but may be garbled. The values 20° and 25° in (g) and (h) are approximations for the duration of twilight at the equinoxes and the summer and winter solstices, which are approximately:

$18^1/_2°$, $22^1/_4°$, 20 and $23^1/_3°$, $28^1/_2°$, 25

respectively. Rules (d) and (e) are also used by al-Bakhāniqī and al-Fawānīsī (**2.9** and **7.5**).

The treatise concludes with an approximate rule for finding the solar azimuth a_a at the time of the afternoon prayer, namely:

$$a_a \approx 21° + \delta + {}^1/_8\, \delta \qquad (\delta \gtrless 0)$$

(measured south from the east point). The azimuths at the equinoxes and solstices are in fact close to:

EQ: 21°S SS: 6°N WS: 48°S ,

so the rule is a fair approximation. (No tables of $a_a(\lambda)$ are known to have been prepared for Cairo, but the function is discussed in several later treatises on sundial theory.)

In passing I note the existence of a related anonymous treatise preserved in MS Escorial ár. 961,3 (fols. 13v-16v), copied in 863 H [= 1459]. This consists of a preface and 21 very short chapters, each dealing with one of the standard functions of timekeeping. The approximations for d, ψ, α_s and α_r given in this source are the same as those of Najm al-Dīn, and the other approximations are clearly related to his:

(a) $h_a \approx 32° + {}^2/_5\,\delta$

(b) $t_a \approx 52° + {}^1/_3 \cdot {}^1/_5\,d$

(c) $h_q \approx 47° + {}^1/_6\,\delta + {}^1/_4\,\delta$

(d) $h_0 \approx 2\delta$ $(\delta < 15°)$

$h_0 \approx 30° + {}^1/_4\,(2\delta - 30°)$ $(\delta > 15°)$

(e) $a_a \approx 21° + {}^1/_8\,\delta$.

The work attributed to Najm al-Dīn al-Miṣrī in MS Milan Ambrosiana 227a (C49), fols. 85v-97r, is entitled *al-Risāla al-ḥisābiyya fī 'l-aʿmāl al-āfāqiyya*, which means "Treatise on Computations in Spherical Astronomy for all Latitudes". This short treatise of 31 chapters contains material of considerably greater sophistication than his two other treatises. All of the theoretical procedures advocated are exact, and are expressed in the terse technical Arabic standard in later Egyptian and Syrian works on astronomy. I shall refer to the section on reckoning time from celestial altitude elsewhere (**5.2**), and here restrict attention to the only historical reference in the work, which occurs in Ch. 11 on the determination of twilight. Najm al-Dīn advocates the parameter 18° for both morning and evening phenomena but states that if one is calculating "for Cairo in particular according to the opinions of the celebrated Ibn Yūnus", one should use 20° and 16°. This reference confirms that Ibn Yūnus wrote more on twilight than has been preserved in MS Leiden Or. 143 of his *Ḥākimī Zīj* (**4.5** and **5.1**) and indicates that the Egyptian tradition of using the parameters 20° and 16° which we shall encounter frequently was much older than the 13th century.

2.6 al-Dīrīnī's poem on timekeeping

ʿAbd al-ʿAzīz ibn Aḥmad al-Dīrīnī was an Egyptian dervish who lived from about 1215 to 1297. He was the author of numerous works, one of which deals with astronomical timekeeping.[28] This particular compilation is a poem with intermittent prose passages and simple tables and diagrams (see **Fig. 2.6a**). I have examined MS Istanbul Hamidiye 1453, fols. 85v-102v, copied in Edirne in 869 H [=1464/65], and MS Istanbul Ayasofya 2711 of this poem, which is entitled *al-Yawāqīt fī ʿilm al-mawāqīt*, "Gems for Timekeeping".

al-Dīrīnī's discussion of calendars, lunar mansions and the motions of the sun and moon do not directly concern the present study. He notes the following midday shadows for the twelve signs beginning with Cancer, which appear to be based on n = 6;40:

$^2/_3$ 1 2 $3^1/_2$ $5\,^1/_4 + {}^1/_8$ $7^1/_4$ $8^1/_8$ $7^1/_4$ ...

[28] On al-Dīrīnī see Brockelmann, *GAL*, I, pp. 588-589, and SI, pp. 810-811; and *Cairo ENL Survey*, no. C14.

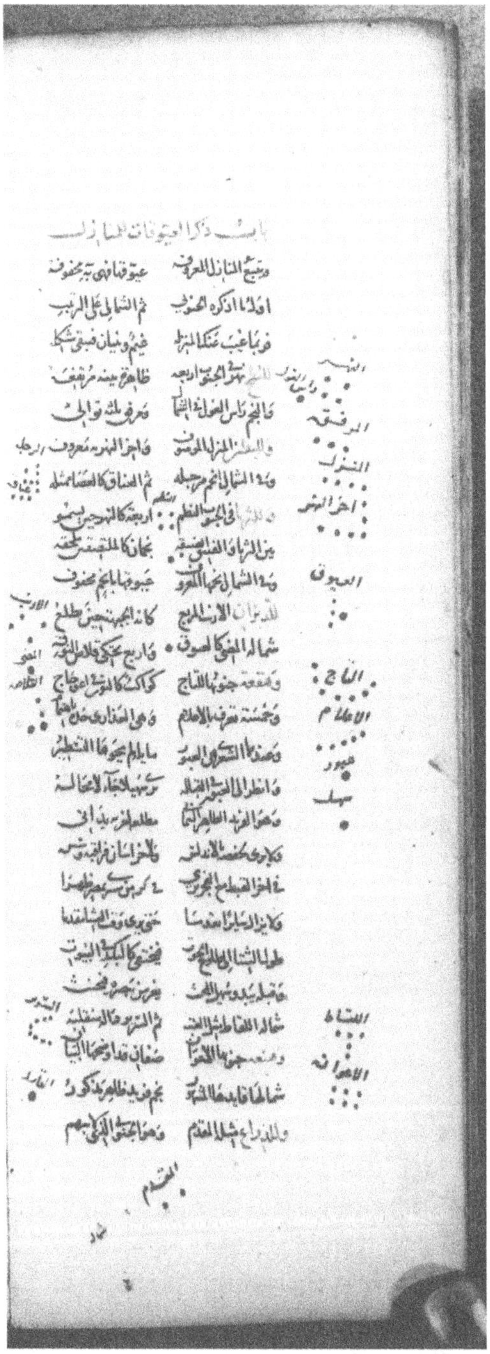

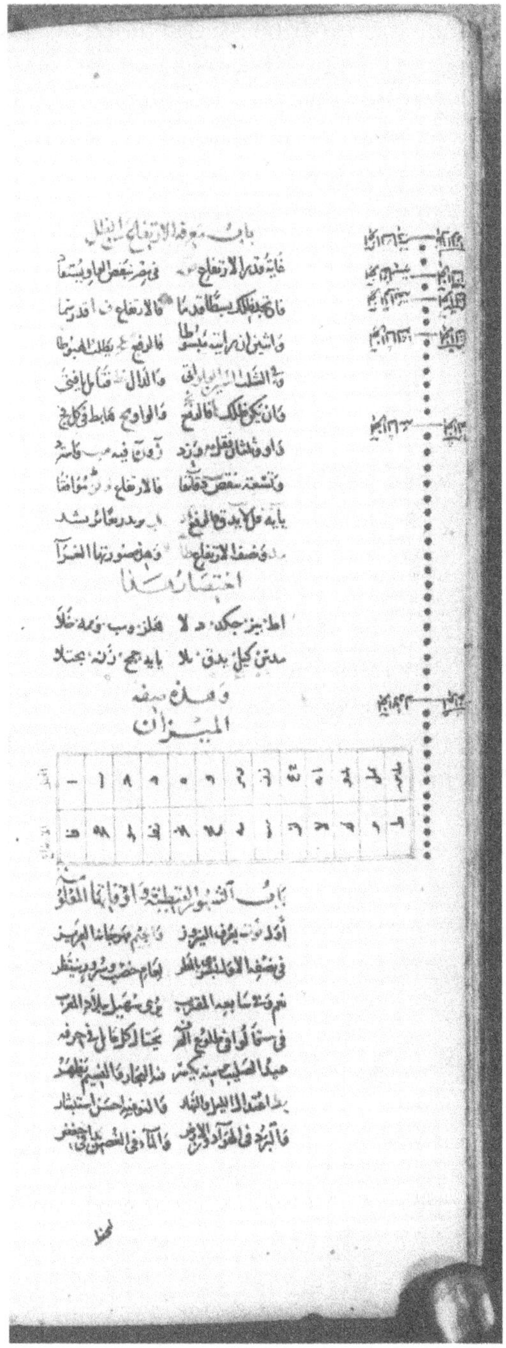

Figs. 2.6a-b: Illustrations of the bright stars (*ʿayyūqāt*) in each of the 28 lunar mansions (a), and a curious table purporting to show the solar altitude for each integral shadow length (to base 6;40!) (b). [From MS Istanbul Hamidiye 1453, fols. 87v-88r and 97v-98r, courtesy of the Süleymaniye Library, Istanbul.]

He also notes the daily change in shadow length for each sign. In one of the two manuscripts there is a table displaying the midday shadows for each of the Coptic year. al-Dīrīnī's values for the solar meridian altitude and for the lengths of daylight in equatorial degrees are:

84°	80	72	60	48	40	36	40	...
210°	205	195	180	165	155	150	155	...

His values for the change in oblique ascensions for each 30° of ecliptic longitude are:

21°	24	30	34	36	35	..

A table of solar altitude corresponding to particular horizontal shadow is also presented (see **Fig. 2.6b**). The first six entries:

z	1	2	3	4	5	6
h	81°	73	66	59	53	48

correspond to those in the first treatise of Najm al-Dīn al-Miṣrī (**2.5**); the remaining entries are somewhat confused. The reader should bear in mind that they are written in *abjad* notation in the manuscripts:

z (Hamidiye)	6;47	7;50	8;48	11;15	14; 6	22;30	44;*60*
z (Ayasofya)	6;*80*	7;50	8;48	11;15	14;*100*	22;30	44;*60*
h	45°	42°	37°	31°	24°	17°	9°

It is difficult to explain these numbers. However, it is clear that z(45°) was intended to be 6;40 and the *mīm* representing 40 has been miscopied as *mīm-zāy* 47 and (by a further stretch of the imagination) *fā'* 80. The values of h in the second part of the table are the complements of those in the first part, and for these complementary values we should have shadows z and z′ such that:

$$z \cdot z' = (6;40)^2 = (6^2/_3)^2 = 400 / 9 .$$

In order to complete the table one must divide this product by 1, 2, 3, 4, 5 and 6 to obtain the shadow lengths corresponding to 9°, 17°, 24°, 31° and 42°. The values that one obtains this way are:

z	7;24	8;53	11; 7	14;49	22;13	44;27
h	42°	37	31	24	7	9

I suggest that whoever compiled the table in al-Dīrīnī's treatise was incapable of expressing the quotient 400/9 as a sexagesimal and wrote z(9°) = 44;60. This explains the values z(17°) = 22;30 and z(31°) = 11;15. To find z(24°) our author divided 44;60 by 3 and somehow derived 14;100 (the *qāf* for 100 was confused for *wāw*, 6 in the Ayasofya copy) and hence 7;50 for z(42°). To find z(37°) he divided 44;60 by 5 and was content to a settle for $8^4/_5$ = 8;48 as the quotient. The above explanation is somewhat far-fetched, but how else can one explain the absurd entries in the table?

2.7 The spherical astronomy in al-Marrākushī's A-Z

Abū ʿAli al-Marrākushī was the author of a substantial *summa* of spherical astronomy and instruments entitled *Jāmiʿ al-mabādi' wa-'l-ghāyāt*. Part of this work was translated by Jean-Jacques Sédillot (*père*) in 1834-35 as *Traité des instruments astronomiques des Arabes*, although in fact the part that he translated deals only with spherical astronomy and sundials.

A summary of part of the remainder of al-Marrākushī's treatise was published by Louis-Amélie Sédillot (*fils*) in 1844 as *Mémoire sur les instruments des Arabes*, but considerable material of interest was omitted.[29] It has not been adequately stressed in previous studies that al-Marrākushī's treatise was compiled in Cairo; al-Marrākushī is invariable listed amongst Moroccan astronomers since, as his name indicates, his family originated in Marrakesh. In his treatise he mentions that he travelled extensively in al-Andalus, North Africa and the Near East, but in all of his tables numerical examples are computed for Cairo. Sédillot-*père* misdated al-Marrākushī to about 1230 in spite of the fact that his solar longitude tables and star catalogue are computed for 1275/76 and 1282, respectively.

al-Marrākushī's treatise was very popular amongst later Muslim astronomers, if only in the central lands of Islam: Egypt, Syria and Turkey.[30] It exists in several manuscript copies and numerous manuscripts that I have examined in Cairo contain spherical astronomical tables extracted from his treatise. It is rather curious that al-Marrākushī makes no reference to Ibn Yūnus, in spite of the fact that he recomputed several tables already computed by Ibn Yūnus three centuries previously, and for the same parameters, namely:

$$\phi = 30;0^\circ \text{ and } \varepsilon = 23;35^\circ .$$

Likewise al-Marrākushī does not refer to al-Maqsī (**3.5**), who must have been working in Cairo at the same time, and who compiled tables for sundial construction similar to, but more extensive than his own. al-Maqsī in the introduction to his tables does not mention al-Marrākushī either. On the other hand, al-Khalīlī (**10.8**) mentions that he favoured al-Marrākushī's method for determining the qibla, and his colleague Ibn al-Shāṭir (**7.3**) surely found his inspiration for his list of 184 standard spherical astronomical formulae in al-Marrākushī's list of 62 of the same (see below).

Whilst considerable uncertainty clouds our knowledge of al-Marrākushī's sources, his work in Cairo, and his later influence, we are at least fortunate in having the studies of the Sédillots on his major work. In the following brief survey of his spherical astronomy and related tables I have used the sudies of *père* and *fils*, hereafter labelled I and II respectively and referred to by page numbers, and also the 14th-century manuscript used by the Sédillots, namely MS Paris BNF ar. 2507-2508, as well as MS Istanbul Selim Ağa 866.[31] We should bear in mind that the work is a monumental encyclopaedia on the subject, to which the following extracts can hardly do justice. See also **6.7** on the extensive tables in the work.

Simple problems in spherical astronomy

al-Marrākushī uses the approximation Sin $\varepsilon \approx 24$ (I.180) and the value Tan $\varepsilon = 26;11,40$ (I.185). al-Maqsī, for comparison, used 24 and 26;11,41. al-Marrākushī gives the following approximation for finding $\delta(\lambda)$:

---

[29] On al-Marrākushī (**I-4.2.4**, *etc.*) see Suter, *MAA*, no. 363; *Cairo ENL Survey*, no. C17; my article in *EI_2*; and now Charette, *Mamluk Instrumentation*, pp. 9-13. Sédillot-*père*, *Traité*, contains a translation of the first half of al-Marrākushī's treatise on spherical astronomy and sundials. Sédillot-*fils*, *Mémoire*, contains a summary of the second half dealing with instruments.

[30] That is, as far as I know, it was not known in the either in the Maghrib or in Iran and points further east. In *London Khalili Catalogue*, pp. 191-192, no 118, a manuscript copied by a Tabrīzī is presented supposedly from "Iran or India", and this thesis of mine countered. However the manuscript is in a typical Ottoman hand.

[31] The latter was published in facsimile, after a fashion, by the Institut für Geschichte der Arabisch-Islamischen Wissenschaften, Frankfurt, in 1984.

$$\text{Sin } \delta(\lambda) = \text{Sin } \varepsilon \text{ Sin } \lambda \ / \ R \approx 24 \text{ Sin } \lambda \ / \ 60 = {}^2/_5 \text{ Sin } \lambda \ .$$

Later in his treatise he gives an example of the computation of the solar rising amplitude, choosing the summer solstice (I.303). He derives 27;31° and also suggests the approximation:

$$\text{Sin } \psi = R \text{ Sin } \delta \ / \ \text{Cos } \phi = 1;9,16 \text{ Sin } \delta \approx 1^3/_{20} \cdot \text{Sin } \delta$$

for $\phi = 30°$. In a similar example for the solar altitude in the prime vertical at the summer solstice (I.306), he computes Sin $h_0 = 48$, from which I derive 53;8° for h_0 rather than the more accurate value 53;9°. He also notes that for $\phi = 30°$:

$$\text{Sin } h_0 \ = \ R \text{ Sin } \delta \ / \ \text{Sin } \phi \ = \ 2 \text{ Sin } \delta \ .$$

Timekeeping

I have discussed elsewhere (**I-6.7.2**) al-Marrākushī's small auxiliary table for timekeeping operations (I.264). It is rather curious that he does not tabulate the functions:

$$B(\lambda) = \text{Cos } \delta(\lambda) \text{ Cos } \phi \ / \ R \quad \text{or} \quad B(\Delta) = \text{Cos } \Delta \text{ Cos } \phi \ / \ R \ ,$$

which he calls the *aṣl*, "base", and uses freely throughout his discussion of the determination of time from solar and stellar altitudes (I.258ff.). I have also discussed (**I-7.1.7**) al-Marrākushī's auxiliary tables for finding the half-excess of daylight (I.209). The method he expounds is repeated by al-Bakhāniqī (see **5.6**, note (k) to the translation).

al-Marrākushī advocates the parameters 20° and 16° for morning and evening twilight (I.295-296). He states that observations which he had made in various latitudes between 20° and 45° led him to the conclusion that these parameters, or "what differs from them by one degree", by which he surely means 19° and 17°, were the most suitable.

Standard formulae of spherical astronomy

al-Marrākushī presents a list of some 62 proportions, which represent the standard formulae of spherical astronomy (pp. 352-359). As an example I quote no. 55 in his list, which reads (in words):

1	2	3	4
Cos h	Cos δ	Sin t	Cos a

and indicates that:

$$\text{Cos h} \ / \ \text{Cos } \delta \ = \ \text{Sin t} \ / \ \text{Cos a} \ .$$

Less than a century later Ibn al-Shāṭir compiled a list of 184 such proportions in his *Zīj* (MS Oxford Seld. A30) and his colleague al-Khalīlī 24 proportions in his minor auxiliary tables in MS Dublin CB 4091 (**10.3b**).

My investigations thus far show that there is no material in the *Jāmiʿ al-mabādiʾ wa-'l-ghāyāt* which was taken directly from Ibn Yūnus' more sophisticated treatment of spherical astronomy in the *Ḥākimī Zīj*. al-Marrākushī makes no reference to Ibn Yūnus or any other Egyptian astronomers, and does not mention the tables for timekeeping which were available in Cairo in his time. Some of his individual tables are found in several later collections of tables, such as those in MSS Cairo MM 43, copied *ca.* 1450, and the manuscript in a private collection in Sanaa, copied *ca.* 1375, which were used in Cairo and the Yemen respectively (see **II-6.14** and **12.4**), but further research is necessary before we can ascertain the extent of his influence in later Islamic astronomy.

2.8 al-Abharī's treatise on instruments

MS Escorial ár. 965, copied in Syria in 719 H [= 1319/20], contains a treatise on instruments entitled *Lawāmiʿ al-wasāʾil fī maṭāliʿ al-rasāʾil* by Amīn al-Dīn ʿAbd al-Raḥmān ibn ʿUmar al-Abharī.[32] The author pays tribute in his introduction to the scholar-prince Abu 'l-Fidāʾ Ismāʿīl (*reg.* 1310-1332)[33] and (his father?) Nūr al-Dīn Abu 'l- Ḥasan ʿAlī ibn Maḥmūd, and from this and other internal evidence it is clear that the Escorial manuscript was copied not long after the treatise was finished. MS Gotha A1414 (date?), which I have not examined, is a later copy of the same treatise.[34]

al-Abharī discusses at great length (the Escorial manuscript contains 170 folios) the problem that can be solved using four instruments: (1) the universal plate known as *shakkāziyya*;[35] (2) the celestial globe (*al-kura dhāt al-kursī*); (3) a trigonometric grid (*al-dustūr wa-rubʿ al-mujayyab*); and (4) what appears to be the standard astrolabe (here called *al-āfāqiyya al-asṭurlābiyya*). He gives no information on any previous writings on these instruments, but does present a set of planetary parameters that are not attested in any known *zīj*. Of relevance to the present discussion are the parameters 18° and 16° advocated for morning and evening twilight (fols. 49v-50r and 171r-171v of the Escorial manuscript), although 18° is given for both phenomena elsewhere (fols. 83v-84r). Also of interest is the statement (fol. 62v) that the ʿ*aṣr* prayer begins when $z_a = Z + n$ and ends when $z_b = Z + 2n$, where the gnomon length n is either 12 or 7 (fol. 134v).

2.9 al-Bakhāniqī's treatise on timekeeping

The mid-14th-century Egyptian astronomer al-Bakhāniqī is known to us as the editor of a recension of the main Cairo corpus (**4.1.5** and **5.6**). He also wrote a treatise on spherical astronomy and the use of the quadrant, extant in MS Berlin Ahlwardt 5860 (Sprenger 1835), copied *ca.* 1400. This treatise contains some approximate formulae for determining certain of the standard functions of *mīqāt* for the latitude of Cairo. al-Bakhāniqī does not say how he derived these formulae and does not refer to any of his predecessors, such as Najm al-Dīn al-Miṣrī (**2.5**), who had presented similar formulae. The principal formulae proposed by al-Bakhāniqī, outlined in words in the text, are the following:

(a) $d \approx \frac{1}{2} (\delta + \frac{1}{4} \delta)$

(b) $h_a \approx \frac{1}{2} H + \frac{1}{10} [\max H - H]$

(c) $t_a \approx 52° + 2\ \delta/\varepsilon$ $(\delta > 0)$

$t_a \approx 52° - 10\ |\delta|/\varepsilon$ $(\delta < 0)$

(d) $\psi \approx \delta + \frac{1}{6} \delta$

---

[32] On Amīn al-Dīn al-Abharī (not to be confused with the well-known Athīr al-Dīn al-Abharī) see Suter, *MAA*, nos. 369 and 393 (confused).

[33] On Abu 'l-Fidāʾ see Suter, *MAA*, no. 392, and the articles in *DSB* by Juan Vernet and in EI_2 by H. A. R. Gibb.

[34] *Gotha Catalogue*, pp. 64-65.

[35] See the article "Shakkāziyya" in EI_2, and **X-5.2**.

(e)	$h_0 \approx 2\delta$	$(\delta < 15°)$
	$h_0 \approx 2\delta + {}^1/_4\,(2\,\delta - 30°)$	$(\delta > 15°)$
(f)	$h_q \approx 47° + \delta + {}^1/_4\,\delta + {}^1/_6\,\delta$	
(g)	$s \approx 18;30° + {}^1/_6\,\delta$	$(\delta > 0)$
	$s \approx 18;30° + {}^1/_2 \cdot {}^1/_6\,\lvert\delta\rvert$	$(\delta < 0)$
(h)	$r \approx 23;30° + {}^1/_6\,\delta$	$(\delta > 0)$
	$r \approx 23;30° + {}^1/_2 \cdot {}^1/_6\,\lvert\delta\rvert$	$(\delta < 0)$

Other rules are given for determining ascensions, the times of rising and setting of the moon, and the rising of the lunar mansions.

The formulae for twilight (*cf.* Ibn al-Ḥammāmī's formulae in **2.4**) are based on parameters 20° and 16° for morning and evening. However in the Berlin copy (fol. 25v) of his treatise on the quadrant al-Bakhāniqī prescribes the parameters 19° and 17°, which also underlie the twilight tables in his edition of the Cairo corpus **4.10** and **5.6**.

Another approximate formula propounded by al-Bakhāniqī is for finding the time of day in seasonal hours, T, from the length of the shadow, z, cast by a gnomon of length n. The formula appears to be garbled in the text (fols. 9v-10r of the Berlin copy), and may be represented algebraically thus:

$$T \approx \{\, 36 \,/\, [\,(z+n) - Z\,]\,\}^{\mathrm{sdh}}\,,$$

where the gnomon length n is not specified. Elsewhere in this same treatise al-Bakhāniqī uses base 12 for the Cotangent function, for which the corresponding formula would be:

$$T \approx \{\, 6 \cdot n \,/\, [\,(z+n) - Z\,]\,\}^{\mathrm{sdh}} \quad (n = 12)\,,$$

It is difficult to explain the figure 36 in the numerator rather than 72. However, in al-Bakhāniqī's introduction to his edition of the Cairo corpus (**5.5**) in MS Cairo DM 108 he mentions Cotangents to base 6. The use of this base is uncommon in the Islamic sources,[36] but in light of this evidence it seems that al-Bakhāniqī's approximate formula for the time of day was intended to be:

$$T \approx \{\, 6 \cdot n \,/\, [\,(z+n) - z'\,]\,\}^{\mathrm{sdh}} \quad (n = 6)\,.$$

2.10 Sibṭ al-Māridīnī on the differences of opinion of the religious scholars and the *muwaqqit*s

Sibṭ al-Māridīnī was one of the leading astronomers in Cairo towards the end of the 15th century; he was a prolific author, mainly of treatises on spherical astronomy and instrumentation.[37] One of his works (see also **5.8** on another) exists in an apparently unique copy, MS Princeton Garret Hitti 531 (1960). This is a discussion of the validity of the times of prayer determined by the *muwaqqit*s, and it is disappointing for the paucity of information which it presents on our subject. Nevertheless I present a free translation of the entire treatise.

> "Questions and answers by the scholar Sibṭ al-Māridīnī concerning the determination of the prayer-times, and whether or not the opinion of the astronomers agrees with the opinion of the legal scholars, and other similar problems.

---

36 See n. **I**-1:32.

37 On Sibṭ al-Māridīnī see Suter, *MAA*, no. 445; *Cairo ENL Survey*, no. C97; and İhsanoğlu, *et al.*, *Ottoman Astronomical Literature*, II, p. 1014 (index).

In the Name of God, the Merciful and Compassionate. The lawyers, who are the *imāms* of religion and the scholars of the Muslims – may God grant them all success for their obedience – have written in their books, and commented thereon with their legal opinions, concerning those daily phenomena which can actually be seen, such as: sunset, when someone who is fasting can break the fast; the disappearance of the red twilight glow, when the time for the evening prayer begins; and the first visibility of the lunar crescent, when the fast of Ramaḍān begins.
They also dealt with similar problems like the determination of midday, on which the determination of all the prayer-times depends, noting that when the sun is in the east the shadow it casts will be in the west, and that as the sun's altitude increases the length of this shadow decreases, and stops decreasing when the sun culminates; and that as soon as the sun starts to sink from the meridian, the midday shadow starts to increase, and this phenomenon marks the beginning of the time of the midday prayer. If the sanctity of this prayer-time is neglected, and this increase has not taken place, the prayer is not correct. Now it has become apparent that most muezzins for a long time have not been taking heed of this. They say: "We take heed of what the professional time-keepers have done, and of what they have prescribed in their treatises on quadrants and astrolabes; this is what we go by." What they mean is that they consider the actual time of midday as the beginning of the prayer-time and do not wait to observe the increase in shadow length prescribed by the legal scholars. It is clear that the opinion of the muezzins is less correct than that of the legal scholars, and it is the latter opinion which should be used as a basis for the determination of the prayer-time."

2.11 A late Syrian treatise on fifteen times of day with religious significance

MS Aleppo Awqāf 970, copied *ca.* 1850, contains a short treatise of a few folios compiled in Aleppo and deals with 15 times of day with religious significance. This treatise is of considerable interest for late Ottoman practice (**IV-5.4** and **6.1**). I have come across no other treatise in which these 15 times are astronomically defined. The times, which include an indication of when the performance of the prayer is no longer approved (*karāha*), are as follows:

(1) the *ẓuhr*;
(2)-(5) the first and second *ʿaṣr* and the times of *karāha* for both *ʿaṣrs*;
(6)-(7) sunset and the time of *karāha* for the *maghrib* prayer;
(8)-(9) the first and second *ʿishāʾ*s;
(10) the *imsāk*;
(11)-(12) the false dawn and the true dawn;
(13) sunrise;
(14) the small *ḍaḥwa*; and
(15) the large *ḍaḥwa*.

The author states that the time of *karāha* for the *ʿaṣr* is when the solar altitude is $4^1/_2°$. The time of the *karāha* for the *maghrib* is when the stars become visible (*waqt iḥtibāk al-nujūm*), namely, when sun is 10° below the horizon (h = -10°). The first and second *ʿishāʾ* occur at the disappearance of the red and the white twilight glow, respectively, that is, when h = -17° and -19°. The false dawn occurs when h = -20°, altered in the text to -21°, and the true dawn when h = -19°. The *imsāk* is 20 minutes before the true dawn. The small *ḍaḥwa*, the time before which prayer is forbidden and "the time for the prayer at the *ʿīd* and the *ḍuḥā*", is when h = +4$^1/_2$°. The large *ḍaḥwa*, which is "the time after which the statement of intention to fast is not valid and after which the *ḍuḥā* prayer should not (be performed)", is halfway between the true dawn and sunset, when the sun is at the mid-point of the "legal arc of daylight" (*qaws al-nahār al-sharʿī*).

The treatise includes some remarks about the effect of refraction at the horizon and concludes with some tables for facilitating the determination of the hour-angle at these times: see further **11.15**.

CHAPTER 3

ʿIRĀQĪ AND IRANIAN TABLES FOR TIMEKEEPING

3.0 Introductory remarks

Islamic astronomical timekeeping and indeed Islamic mathematical astronomy in general started in 8th-century Baghdad. However, the number of Abbasid astronomical sources which survive is small indeed compared with the original output of the early Muslim astronomers. Ted Kennedy's survey of the Islamic astronomical handbooks known as *zīj*es indicates that at least two dozen *zīj*es, that is a substantial fraction of all significant known Islamic *zīj*es, were compiled in Abbasid al-ʿIrāq, Syria and Iran.[1] Very little material on timekeeping survives from these regions in Abbasid times:[2] the few tables I have found are discussed below. These, although of considerable historical interest, hardly constitute a corpus of tables that could have been used by muezzins. The Abbasid tables occur mainly in *zīj*es, in contrast to later tables for timekeeping from other parts of the Muslim world. We can, however, detect a tradition, which apparently started in Baghdad with the astronomer ʿAlī ibn Amājūr (**3.2**), of compiling tables of the time of day as a function of solar altitude (and solar meridian altitude) and of time of night as a function of stellar altitudes, and which influenced later astronomers at least in Iran and the Yemen (*cf.* **3.8**, **3.10-11** and **12.3**). The writings of al-Bīrūnī[3] illustrate the development of the mathematical methods of astronomical timekeeping, from early Abbasid sources when they relied on Indian techniques down to his own time when he and his predecessors had developed spherical trigonometry. Certain of the tables which have survived from al-ʿIrāq and Iran are based on the approximate methods adopted from Indian astronomy (**3.2-4**). There is every reason to suppose that further research on Arabic scientific manuscripts, particularly those preserved in Iran and Turkey, will bring to light other examples of tables for timekeeping from the early period of Islamic astronomy.

After the fall of Baghdad in 1258, no scientific works of any consequence are known to have been compiled in al-ʿIrāq. The few astronomical manuscripts preserved in modern Iraqi libraries are virtually untouched by modern scholarship, but the manuscript catalogues seem to confirm this lack of scientific endeavour.[4] I have found no complete sets of ʿIrāqī tables for timekeeping postdating the 13th century, although prayer-tables of the Ottoman kind were doubtless prepared for such important centres as Mosul, Baghdad and Basra.

---

[1] King & Samsó, "Islamic Astronomical Handbooks and Tables", pp. 31-44.

[2] The standard bio-bibliographical source for early Islamic astronomy is Sezgin, *GAS*, VI. Sezgin does not mention astronomical timekeeping at all.

[3] I think here especially of his works *Ifrād al-maqāl fī amr al-ẓilāl* and *al-Maqālīd fī ʿilm al-hayʾa*, listed in the bibliography as al-Bīrūnī, *Shadows*, and *Maqālīd*.

[4] Apart from Suter, *MAA*, and Brockelmann, *GAL*, the Arabic work Azzawi, *History of Astronomy in Iraq*, is the only bio-bibliographical source for this area and period. Unfortunately the author was unaware of any other modern scholarship in the history of Islamic astronomy, but occasionally he mentions treatises and manuscripts not listed in other modern sources.

On the other hand, creative interest in astronomy continued in Iran and Central Asia, both under the Timurids and later under the Safavids.[5] But the tables for timekeeping that we find are also few and far between. Nevertheless, there is every hope that such materials will be uncovered when the rich manuscript collections in Iran are exploited for their scientific contents.

3.1 al-Khwārizmī's tables for Baghdad

The anonymous tables in MS Berlin Ahlwardt 5793 (Landberg 56), fols. 93v-95v, copied *ca.* 1450, are appended to al-Farghānī's treatise on the construction of the astrolabe (fols. 1r-77r)[6] and two treatises on the construction and use of the astrolabe by al-Khwārizmī (fols. 77v-93r).[7] In view of the nature of these tables, I deem likely that they were compiled in the time of al-Farghānī and al-Khwārizmī, namely the early 9th century, and the most probable author is al-Khwārizmī himself.

Firstly, there is a table (fol. 93v) of normed right ascensions, $\alpha'(\lambda)$, with entries to three digits for $\lambda = 273°, 276°, \ldots, 360°$, accurately computed but for a few scribal errors for $\varepsilon = 23;33°$, the first *Mumtaḥan* value of this parameter.[8] The second table (fol. 94r – see **Fig. 3.1**) displays the shadows to base 12 at the beginning of the midday prayer, and at the beginning and end of the afternoon prayer, that is:

$$z_z, \ z_a \ \text{and} \ z_b \ .$$

Entries are given to one digit for each 6° of λ' ($\delta \gtrsim 0°$). The latitude is stated as 13°, which is a scribal error for 33°. The values in the table correspond *roughly but not precisely* to the standard definitions of later Islamic practice:

$$z_{z(12)} = Z_{(12)} \ , \ z_{a(12)} = Z_{(12)} + 12 \ \text{and} \ z_{b(12)} = Z_{(12)} + 24 \ .$$

In particular, it seems that the values of z_a and z_b were computed using functions of H different from these. The table merits detailed investigation.

In al-Khwārizmī's treatise on the astrolabe the following definitions are given for the solar altitude at the *ẓuhr* and *ʿaṣr* (fol. 89v of the Berlin manuscript):

$$h_z \approx H - 7° \quad \text{and} \quad h_a \approx {}^1/_2\, H + {}^1/_{10}\, \bar{H} \ .$$

---

[5] See the survey articles Kennedy, "Seljuq and Mongol Science", and "Timurid Science", and Winter, "Safavid Science", and on Safavid astronomy see more recently King, *Mecca-Centred World-Maps*, pp. 128-138. The standard Western bio-bibliographical reference work on Persian manuscripts is Storey, *PL*, II:1, although much can now be added from Iranian reference works, such as those of Āghā Buzurg, *al-Dharīʿa*, and Monzawi, *Persian Manuscripts*.

[6] On al-Farghānī see the article in *DSB* by A. I. Sabra, and Sezgin, *GAS*, VI, pp. 149-151. On his astrolabe tables see King & Samsó, "Islamic Astronomical Handbooks and Tables", Section 4.11. His astrolabe treatise is currently being edited by Richard Lorch (Munich).

[7] On al-Khwārizmī (**I-7.1.1**) see the article in *DSB* by G. Toomer; Sezgin, *GAS*, VI, pp. 140-143; and on various Abbasid works attributed to him, but not all by him, King, "al-Khwārizmī". One of his treatises, on the use of the astrolabe, is translated in Frank, "al-Khwārizmī über das Astrolab". On the other, on the construction of the astrolabe, see King, "al-Khwārizmī", pp. 23-27. An edition of both treatises with English translation and commentary has been prepared by François Charette and Petra Schmidl.

[8] See **I-5.6.1** and **9.1**.

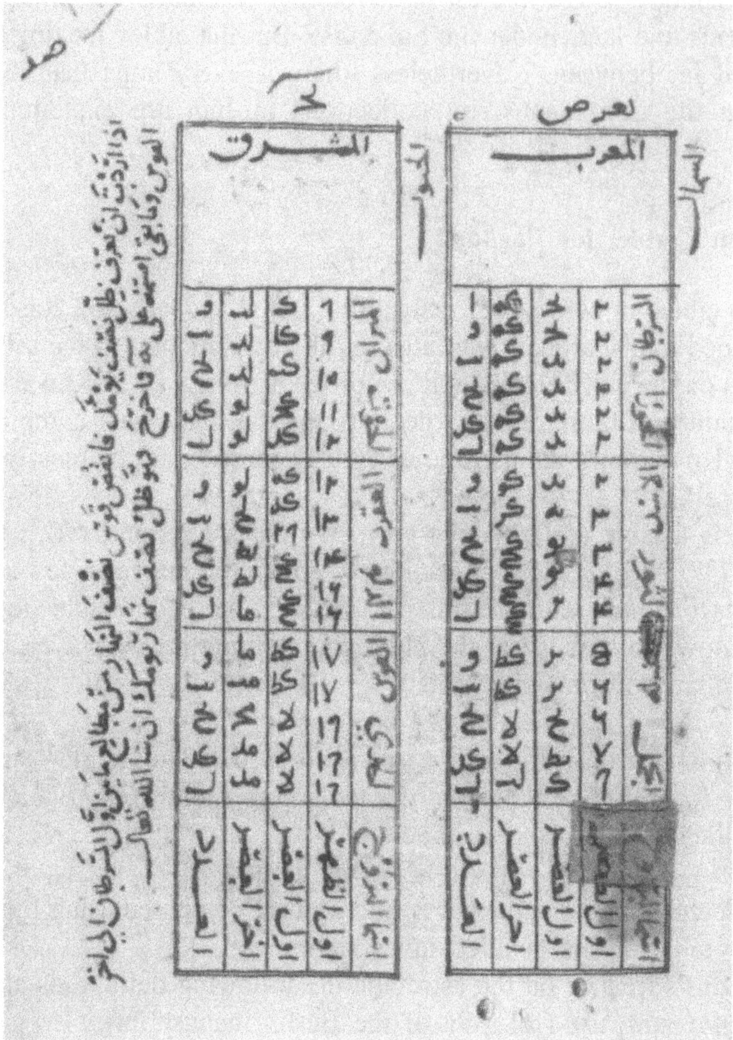

Fig. 3.1: al-Khwārizmī's prayer-tables for Baghdad. Note the sporadic use of Hindu-Arabic numerals in the table: this is most unusual in tables copied in the Mamluk period. It may indicate that al-Khwārizmī used such notation in his original table. From MS Berlin Ahlwardt 5793 fol. 94r, courtesy of the Deutsche Staatsbibliothek (Preußischer Kulturbesitz).]

It remains to explain these formula, the first of which is not attested in any other source known to me, and the second of which is similar to that used by later astronomers in both Egypt and Syria (see, for example, **2.9**).

al-Khwārizmī's definition of the *ẓuhr* is intended to correspond to a time one seasonal hour after midday as defined by shadow lengths using the Indian formula:

$$T \approx \{\ 6 \cdot n\ /\ (\Delta z + n)\ \}^{\mathrm{sdh}},$$

discussed above (**1.4**): see further **IV-4.5**.

The second of al-Khwārizmī's formulae is more a careful approximate solution to the condition $\Delta z = n$. If we add $n = 12$ to the midday shadows for Baghdad noted above we obtain:

30;29 19;48 13;54

and the corresponding altitudes are:

$$21\tfrac{1}{2}^\circ \qquad 31^\circ \qquad 41^\circ \,.$$

We notice that these altitudes are a little more than one-half the meridian altitudes. The rationale behind al-Khwārizmī's formula in which a fraction of $\bar{H}$ is added to $\tfrac{1}{2}H$ to obtain h_a can be understood from the following:

	h_a	$\tfrac{1}{2}$ H	$h_a - \tfrac{1}{2}$ H	$\bar{H}$	$\tfrac{1}{10}\bar{H}$
WS	$21\tfrac{1}{2}^\circ$	$16\tfrac{1}{2}^\circ$	5°	57°	6°
EQ	31	$28\tfrac{1}{2}$	$2\tfrac{1}{2}$	33	3
SS	41	$40\tfrac{1}{2}$	$\tfrac{1}{2}$	9	1

The shadow lengths given in the tables for the *ẓuhr* and *ʿaṣr* are not based on solar altitudes derived by these formulae of al-Khwārizmī.

The tables which follow the prayer-tables (fols. 94v-95v of the Berlin manuscript) display the solar altitude as a function of the seasonal hours and the meridian altitude, as well as a curious (and superfluous) auxiliary function for finding the sine of the altitude from the same arguments. See further **I-4.2.1** and **4.3.1** and especially **XI-9.3**.

In the margin by the side of the prayer-table is the following note, with no further explanation:

> "If you want to know the shadow at midday (*ẓill niṣf yawmika*) subtract the arc of half daylight (*qaws niṣf al-nahār*) from the rising time (of the arc of the ecliptic) from the beginning of Cancer to the end of Sagittarius (*maṭāliʿ mā bayn awwal al-sarṭān ilā ākhir al-qaws*) and divide the remainder by 15: the result will be the shadow at midday, if God – may He be exalted – wills."

Now the rising time of the arc between $\lambda = 90°$ and $\lambda = 270°$ is either:

$$180^\circ \quad \text{or} \quad 2 \max D(\phi)\,,$$

according as the ascensions are measured in *sphaera recta* or in *sphaera obliqua*. Thus this method may be represented by the formula:

$$Z = [\, 2 \max D(\phi) - D(\lambda) \,] \,/\, 15\,.$$

This method requires further investigation.

3.2 ʿAlī ibn Amājūr's *ṭaylasān* tables

The most impressive known achievement of any Abbasid astronomer in astronomical timekeeping is the *ṭaylasān* table of Abu 'l-Qāsim ʿAlī ibn Amājūr, who worked in Baghdad in the mid 10th century.[9] This displays the function T(H,h) for each degree of both arguments such that $21^\circ \le H \le 84^\circ$ and $1^\circ \le h \le H$. Values are given in equatorial degrees to two digits and are based on the parameter $\phi = 33;25^\circ$ (Baghdad). They are extremely carefully computed, being either accurate or in error by ±1 in the second digit, rarely more. Since for the sun $33;0^\circ = \bar{\phi} - \varepsilon \le H \le \bar{\phi} + \varepsilon = 80;10^\circ$, ʿAlī ibn Amājūr clearly intended his table to be used for timekeeping by the stars as well. See further **I-2.3.1**. Ibn Amājūr's table is contained in MS Paris BNF ar.

[9] On Ibn Amājūr (n. **I**-2:29) see Sezgin, *GAS*, VI, pp. 177-178, where the table in MS Paris BNF ar. 2486 is listed.

2486, fols. 239v-255r, of the *Zīj* of al-Baghdādī (copied by the author in 1285), where it is expressly attributed to the earlier scholar and called *al-zīj al-ṭaylasān* (fol. 239r), the word *zīj* here apparently meaning a large table rather than an astronomical handbook.

In MS Paris BNF supp. pers. 1488, copied *ca.* 1500, of the early-14th-century Persian *Ashrafī Zīj*,[10] there is a similar table computed for an unspecified locality, actually Shiraz (**3.10**), preceded by yet another *ṭaylasān* table (fols. 201v-204v), of which it is stated: "this table was compiled by Abu 'l-Qāsim al-Mājūr in his *zīj* called *al-Zīj al-Ṭaylasān*". The title reads *istikhraj al-sāʿāt al-zamāniyya min irtifāʿ al-waqt wa-niṣf al-nahār*, "for finding the temporal hours from the instantaneous and meridian altitudes". The function tabulated is T(H,h) in seasonal hours and minutes for degree of both arguments such that $1° \leq h \leq H \leq 90°$. The table is in fact intended to display the time since rising for any latitude and the tabulated function is based on the approximate formula:

$$T(H,h) \approx \text{arc Sin} \{ R \text{ Sin } h \,/\, \text{Sin } H \}^{\text{sdh}} .$$

See further **I-2.5.1** on the table and **II-1.4** on the formula, also **XI**.

Thus Ibn Amājūr's work contained at least two extensive tables, one for Baghdad and the other for all latitudes. I suspect that several of the other spherical astronomical tables in the *Zīj* of al-Baghdādī (**3.2**) are also due to Ibn Amājūr. We are most fortunate to have these *ṭaylasān* tables of his, since the other five *zīj*es compiled by him and his sons have vanished almost without trace. The little evidence we have about the Amājūr family ranks them close to the top of the list of Muslim astronomers.

3.3 The prayer-tables for Baghdad in the *Zīj* of al-Baghdādī

The unique MS Paris BNF ar. 2486, copied 684 H [= 1285], of the *Zīj* of al-Baghdādī (**3.2**) contains numerous spherical astronomical tables, but I here restrict comments to those found on fols. 117r-122v. Several functions are tabulated to *four* digits for the parameters $\phi = 33;25°$ (Baghdad) and $\varepsilon = 23;35°$, namely:

$$d, \; 2D^h, \; \tilde{h} \; \text{and} \; H .$$

These are followed by a set of prayer-tables – see **Fig. 3.3** – for the same parameters, with values given to one or two digits for each day of the Syrian year. The corresponding position of the sun (*mawḍiʿ al-shams*) in the zodiacal signs (*al-aḥwāl*) is given (Tishrīn I 1 corresponds to Libra 16°) and the functions tabulated are as follows:

(1) Cot_7 H is given to two digits.

(2) A function is labelled *ṭulūʿ al-fajr*, "daybreak", is actually the number of seasonal night-hours and minutes from sunset to daybreak (see below). To find the duration of evening twilight one should subtract the values from 12 (see below).

(3) Values of functions H, h_a and h_b are given to one digit.

al-Baghdādī's twilight table is based on the parameter 17°. The function tabulated is:

$$r(\lambda) = 12^{\text{snh}} - s(\lambda)$$

[10] On the *Ashrafī Zīj* see n. **I**-2:32.

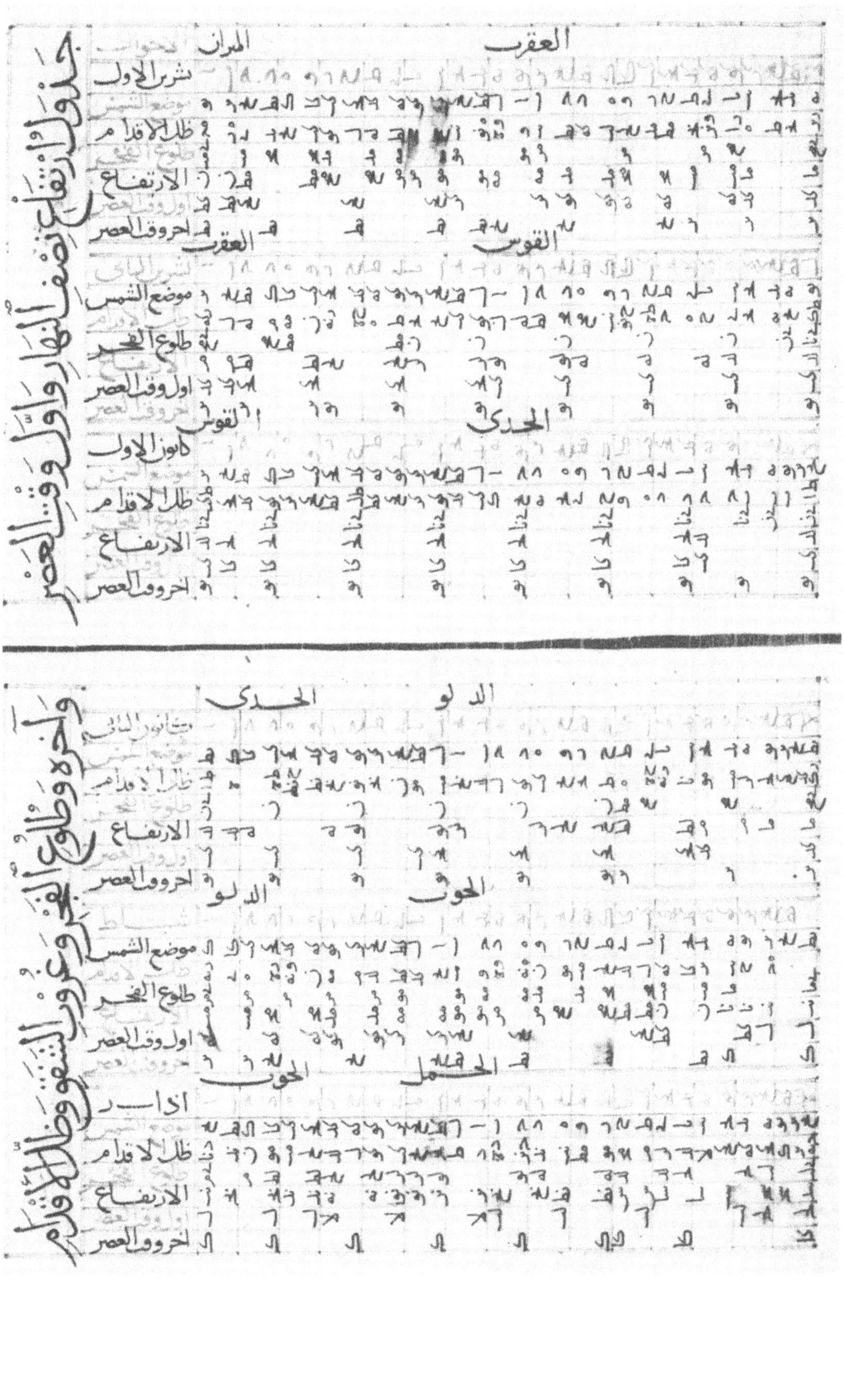

Fig. 3.3: An extract from the prayer-tables in the *Zīj* of al-Baghdādī. This double page serves the month of Tishrīn I (October). [From MS Paris BNF ar. 2486, fols. 120v-121r, courtesy of the Bibliothèque Nationale de France.]

where s(λ) is defined by the approximate formula:

$$s(\lambda) \approx {}^{1}/_{15} \text{ arc Sin } \{ R \cdot \text{Sin } 17° / \text{Sin } H(\lambda*) \}^{snh} .$$

This same formula, though with parameter 18° is prescribed in MS Berlin Ahlwardt 5750, fol. 153v, of a recension of the *Zīj* of Ḥabash (**I-9.1**). Likewise, the twilight tables in the *Dustūr al-munajjimīn* (**3.7**) are based on this formula but with parameter 16°. The formula is derived from approximate formula for T(h,H) in seasonal day-hours that underlies, for example, the universal table of T(h,H) prepared by ʿAlī ibn Amājūr (**3.2**). The last of al-Baghdādī's tables displays the functions:

$$h_a(H) \text{ and } h_b(H)$$

to two digits for each degree of H up to 90°.

3.4 Anonymous timekeeping tables for Baghdad (?)

MS Paris BNF ar. 2514 contains an anonymous set of tables for timekeeping, copied or maybe compiled in 612 H [= 1215]. The work consists of two parts, for timekeeping by day and night, respectively. The first set of tables display the time since rising in seasonal hours as a function of meridian altitude and instantaneous altitude. The underlying formula is approximate and the entries can be used for all latitudes; however, for meridian altitudes such that $\bar{\phi} - \varepsilon \leq H \leq \bar{\phi} + \varepsilon$, the time since rising for the sun is given in equinoctial hours for the latitude $\phi \approx 33;30°$, which could serve Baghdad or Damascus. For each value of H the value of h_a is also given. On this table see further **I-2.5.2**.

The remainder of the tables (fols. 28v-48v) display the longitude of the horoscopus and its oblique ascensions as a function of the altitudes of various stars. See further **I-3.2.2**. On some similar tables for latitude 30;25° (Qandahar) see also **I-3.2.1**.

3.5 al-Qāyinī's twilight tables for Qayin

A pair of tables for the duration of morning and evening twilight was computed by an individual named ʿAlī ibn ʿAbdallāh al-Qāyinī who lived in the city of Qayin in Eastern Iran, probably *ca.* 1000.[11] These tables are preserved in the unique MS Bankipore 2468,23 (fols. 114v-115r), copied in 632 H [= 1234/35], from which the text was published in Hyderabad in 1947; they have been studied by Marie-Louise Davidian and Ted Kennedy.[12]

al-Qāyinī states in his introduction that he has tabulated the functions $r(\lambda_H)$ and $s(\lambda_H + 180°)$, where $\lambda_H(\lambda)$ is the longitude of the horoscopus at daybreak in Qayin when the solar longitude is λ. He also states that the latitude of Qayin is 33;55°, a value not known from other sources,[13] and that he is using the parameter 17° for both morning and evening twilight. In the introduction it is noted that:

$$r(\lambda_H) = s(\lambda_H + 180°) = s(\lambda_H*) ,$$

[11] On al-Qā'inī see Sezgin, *GAS*, V, p. 337, and VI, p. 242.
[12] Kennedy & Davidian, "Al-Qāyinī on Twilight".
[13] Kennedy & Kennedy, *Islamic Geographical Coordinates*, p. 258.

and this relation also holds for the tabulated functions, values of which are given in degrees and minutes for each 5° of argument. However, the tabulated functions do not correspond to the accurate values of these functions for these parameters. Kennedy and Davidian demonstrated that the entries corresponded more closely to the lengths of twilight as a function of solar longitude, but then we are confronted with the fact that if λ is considered constant for a given day then $r(\lambda) = s(\lambda)$.

To compute the function $r(\lambda_H)$ using an accurate formula we first determine the arc of the ecliptic $\Delta\lambda$ which rises during morning twilight. The simplest formula is suggested by al-Qāyinī, namely:

$$\Delta\lambda = \text{arc Sin } \{ r \cdot \text{Sin } 17°/ \text{ Cos } \nu \}$$

where $\nu(\lambda_H)$ is the function known as *ʿarḍ iqlīm al-ruʾya*, "the latitude of visible climate", which measures the angle between the ecliptic and the horizon.[14] With this $r(\lambda_H)$ is given by:

$$r(\lambda_H) = \alpha_\phi(\lambda_H + \Delta\lambda) - \alpha_\phi(\lambda_H) \ .$$

No tables of $\nu(\lambda)$ or $\alpha_\phi(\lambda)$ specifically for latitude 33;55° are attested in the known manuscript sources.

Like Kennedy and Davidian I am unable to explain al-Qāyinī's table. I have considered the possibility that he used a table for the "latitude of visible climate" which was computed for the latitudes of one of the climates rather than for Qayin, and also that the quantity $\Delta\lambda$ was computed by an approximate formula. But neither of these possibilities seems to account for the peculiarities of al-Qāyinī's functions.

3.6 al-Bazdawī's qibla table for Transoxania

Abu 'l-Yusr al-Bazdawī was a Ḥanafī legal scholar active in Samarqand in the 11th century.[15] He wrote a treatise on the qibla in early Islamic Transoxania which is extant in a unique manuscript preserved at one time in Sohag, from which MS Cairo B 19385 was copied in 1355 H [= 1936].[16] al-Bazdawī promises a table displaying the solar altitude in the azimuth of the qibla h_q which will serve Samarqand, Bukhara and Nasw. The table is not contained in the Cairo manuscript, but al-Bazdawī mentions three values from it which are the following:

EQ: 40° SS: 66° W: 11° .

For $\phi = 40°$, which al-Bazdawī gives for Samarqand, these correspond roughly to $q = 45°$, which is a happy compromise between the directions of due west, used for the qibla by the Ḥanafīs in Samarqand, and due south, used there by the Shāfiʿīs.[17]

[14] On this concept see Kennedy, "*Zīj* Survey", p. 145b. See also nn. 3:29 and 7:15.

[15] On al-Bazdawī see Brockelmann, *GAL*, I, p. 460, and SI, pp. 637-638, and Sezgin, *GAS*, I, pp. 412-428.

[16] Text, translation, and commentary are contained in King, "al-Bazdawī on the Qibla in Tranoxania". That study is based on the Cairo manuscript because the original in Sohag had disappeared when I tried to locate it in the 1970s.

[17] *Op. cit.*, and King, *Mecca-Centred World-Maps*, p. 125.

3.7 Anonymous twilight tables for the fourth climate

The anonymous *zīj* preserved in MS Paris BNF ar. 5968, copied *ca.* 1250, and known as the *Dustūr al-munajjimīn* is based mainly on the *zījes* of al-Battānī, Kushyār, Abū Jaʿfar al-Khāzin and al-Bīrūnī.[18] There are two tables for morning and evening twilight in this *zīj* (fols. 187v-188), which are based on the approximate formulae noted in **3.3**. The tables are anonymous, and someone has written a non-commital *li-baʿḍihim*, "by one of the astronomers", by each of the titles – see **Fig. 3.7**. The underlying parameter is found by inspection to be 16°, although the parameters mentioned in the accompanying text are firstly 18° and then 17° (fols. 187r). Corresponding entries in the two tables add up to twelve seasonal night-hours: the table for evening twilight measures the duration of twilight and the other measures the time from sunset

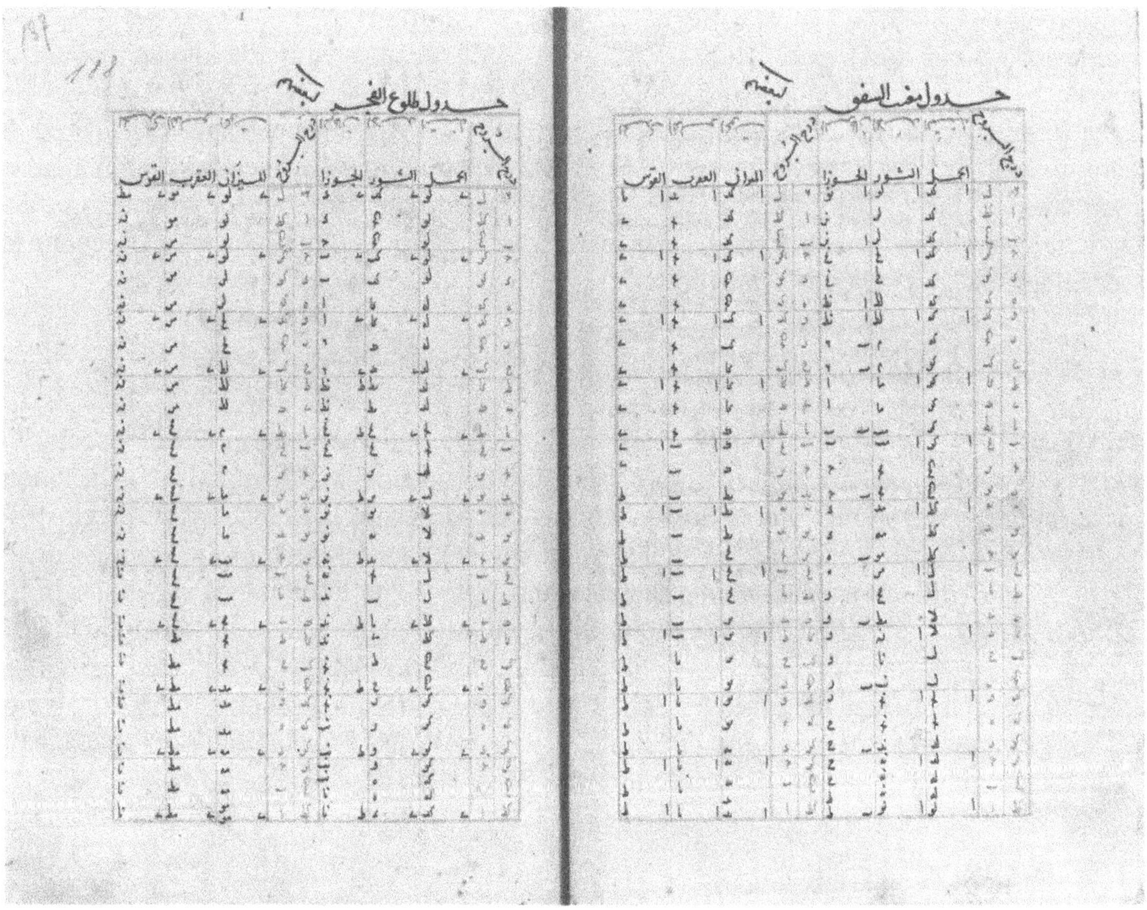

Fig. 3.7: The tables for the duration of twilight in the the *Dustūr al-munajjimīn*. The format is the same as that of the tables in the Cairo corpus, and it may be that there is Egyptian influence behind these Iranian tables. [From MS Paris BNF ar. 5968. fols. 187v-188r, courtesy of the Bibliothèque Nationale de France.]

[18] This *zīj* is not listed in Kennedy, "*Zīj* Survey", but the introduction has been studied in Zimmermann, "Dustūr al-munajjimīn" (cited in n. **I**-7:14). See also Sezgin, *GAS*, VI, pp. 63-64.

until daybreak. The underlying latitude is as stated, namely, $\phi = 36;21°$. This is the value used in several Islamic sources for the Ismāʿīlī stronghold of Alamut (and also Nishapur), although it was probably derived by calculation as the latitude of the 4th climate rather than derived from any observation.[19] Elsewhere in the *Dustūr* (fols. 161r-163r) there are tables of the functions:

$$H,\ d,\ D,\ 2D^h \text{ and } \tilde{h},$$

with values to two digits for parameters $\phi = 36;21°$ and $\varepsilon = 23;35°$. See further **XI-5.2**.

3.8 The *Īlkhānī ṭaylasān* table for Maragha

The *Īlkhānī Zīj* of Naṣīr al-Dīn al-Ṭūsī and his collaborators at the Observatory in Maragha *ca.* 1260[20] contains a *ṭaylasān* table displaying the times since sunrise as a function of the solar meridian altitude and instantaneous altitude for $\phi = 37;20°$ (Maragha). The table is not contained in all of the numerous available copies of the *Zīj* since its use is restricted to one latitude, but it is contained, for example, in MS Florence Medici 269, fols. 150r-152v, and Oxford Hunt. 143, fols. 155v-160v, of the *Zīj*. See further **I-2.3.2**.

3.9 al-Maghribī's qibla table for Maragha

Muḥyi 'l-Dīn al-Maghribī was one of the astronomers who assisted Naṣīr al-Dīn al-Ṭūsī at the Observatory in Maragha (**I-4.3.1*** and **5.6.3**).[21] His own *Zīj* purports to be more representative of the work done at Maragha than al-Ṭūsī's *Īlkhānī Zīj*, but neither *Zīj* has been fully studied yet: a brief survey has been made of the *Īlkhānī Zīj*, but the *Zīj* of Muḥyi 'l-Din al-Maghribī has hardly been studied at all. Whilst the *Īlkhānī Zīj* exists in several copies, only three copies are known of al-Maghribī's *Zīj*, namely, MSS Dublin CB 3665, Meshed Shrine Library 332(103), and Medina Aref Hikmet *mīqāt* 1.

Certain features of al-Maghribī's *Zīj* indicate his debt to Ibn Yūnus (**5.1**). First, his Sine tables are based on values lifted from the *Ḥākimī Zīj*. (Likewise, his Tangent table is based on values lifted from al-Bīrūnī's *Qānūn*.) Second, other than Ibn Yūnus he is the only known compiler of a *zīj* who tabulates both the solar rising amplitude and the solar altitude in the azimuth of the qibla (MSS Dublin CB 3665, fols. 96v-97r and Meshed Shrine Library 332(103), fols. 99r-99v).

al-Maghribī tabulated $\psi(\lambda)$ to three digits for each degree of λ, and the latitude of Maragha, taken as 37;20,30° (**I-5.6.3**). Ibn Yūnus had also tabulated the function to three digits for the latitude of Cairo (**5.1**). al-Maghribī's table of $h_q(\lambda)$ gives values to two digits for each degree of λ, as did Ibn Yūnus' table for Cairo (see also **5.1**). At the head of al-Maghribī's table of $h_q(\lambda)$ in both sources it is stated that the underlying parameter is $q = 16;59°$. In al-Maghribī's

---

[19] Kennedy & Kennedy, *Islamic Geographical Coordinates*, pp. 15 and 245-246; also King, "Geography of Astrolabes", pp. 6-9.

[20] On al-Ṭūsī and the *Īlkhānī Zīj* see n. I-2:31. The existence of the *Īlkhānī ṭaylasān* table was noted in Kennedy, "*Zīj* Survey", p. 161.

[21] On Muḥyi 'l-Dīn al-Maghribī see n. 5:8.

geographical tables (MSS Dublin CB 3665, fols. 57v-59r, and Meshed Shrine Library 332(103), fols. 61r-61v) we find the coordinates:

Mecca	ϕ: 21;40°	L: 77;0°
Maragha	37;20,(30)°	82;0°

Accurately computed for these coordinates q is 16;49°. However, in MS Istanbul Ayasofya 2984, fols. 85-86v, of the early-14th-century Persian *Zīj* of al-Wābiknawī,[22] the calculation is presented on which the value 16;59° is based. The value of q which underlies the table of $h_q(\lambda)$, however, appears to be 16;49° rather than 16;59° and the value used for ε is 23;30°. al-Maghribī has the three values: 51;27°, 75;38°, 27;15° for h_q at the equinoxes and summer and winter solstices. Recomputation with the parameter 16;49° reproduces precisely these values. Recomputation with the parameter 16;59° yields 51;25°, 75;37°, 27;13°. al-Maghribī's table of h_q differs in format from Ibn Yūnus' table in the *Ḥākimī Zīj*. The latter is arranged in 12 columns of 30 entries for each degree of λ, so that no advantage is taken of the symmetry of the function. In the main Cairo corpus $h_q(\lambda)$ is displayed in six columns of 30 entries and no value is given for the equinoxes. In al-Maghribī's table the function is displayed in six columns of 30 entries beginning with Aries, but the vertical argument runs from 0 to 29 instead of Ibn Yūnus' 1 to 30. At the bottom of the third column (for Gemini) and the sixth column (for Sagittarius) an extra entry is added for argument 30. Thus values are given for the equinoxes and both solstices. al-Maghribī's entries are rather accurately computed.

The Medina manuscript, but not the other two copies of the *Zīj,* contains a table displaying the solar altitude at each seasonal hour for each degree of solar meridian altitude. This table is based on an approximate formula – see further **I-4.3.1***.

3.10 The *ṭaylasān* table for Shiraz in the *Ashrafī Zīj*

MS Paris BNF supp. pers. 1488, copied *ca.* 1550, is an unique copy of the *Ashrafī Zīj,* compiled in 702 H [= 1302/03] by the Iranian astronomer Sanjar al-Kamālī also known as Sayf al-Munajjim (**3.2**). It contains two *ṭaylasān* tables, the first being the universal table of Ibn Amājūr (**3.2**) and the second being computed for ϕ = 29;30° (Shiraz). Various other tables in the *Ashrafī Zīj* relate to spherical astronomy and have been discussed in **I-2.3.3**, *etc.* – see **Fig. I-6.2.1** for some of these. However, there are no tables for regulating the times of prayer.

3.11 An anonymous *ṭaylasān* table for Northern Iran

MS Leiden Or. 199, fols. 21v-27v, contains an anonymous *ṭaylasān* table of the function t(H,h) for latitude 36°. This table, which contains many errors difficult to attribute to copyists, has been studied by Bernard R. Goldstein.[23] See further **I-2.3.4** and, on the structure, **XI-4.2** (illustrated).

---

[22] Kennedy, "*Zīj* Survey", no. 35 (not in Storey, *PL*, II:1); and King, *Mecca-Centred World-Maps*, n. 84 on p. 158.

[23] See Goldstein, "Medieval Table for Reckoning Time", the first modern study of a medieval Islamic table for timekeeping.

3.12 Anonymous timekeeping tables for Baghdad

MS Cambridge Add. 3527 (date?) contains a set of anonymous timekeeping tables for Baghdad, bound in some confusion. The underlying latitude is ϕ = 33;25° (Baghdad). The functions tabulated for each degree of argument λ are α_ϕ (fols. 120v-121r), $2D^h$ (fol. 121v), and α' (fols. 158v-159). In view of the format of the tables, which have the main vertical argument running from 0 to 29° rather than the more sensible 1° to 30°, I estimate that they are rather late, perhaps from the 16th century. There is also a table for converting hours and minutes to equatorial degrees for each minute from 1^m to 15^h59^m (fols. 122v, 157r-158r) and another displaying the astrological houses for latitude 33;25°.[24] Two other tables in the set, each copied in different hands, are spurious, namely:

(1) a table displaying the astrological houses for latitude 32° (fols. 166v-172r); and
(2) a table displaying the time of daybreak for Baghdad according to the Ottoman convention for each day of the Persian year (fol. 165v) – see the next section.

3.13 An anonymous table for twilight at Baghdad

The anomalous table in MS Cambridge Add. 3527 (**3.12**), fol. 165v, displays the duration of twilight (*sāʿāt ṣubḥ wa-shafaq*) for the latitude of Baghdad, stated to be 33;25°. Values are given in equinoctial hours and minutes for each day of the Persian year starting with the month of Farwadīn and ending with the five intercalary days. The maximum value of $1;41^h$ corresponds nicely to a solar depression of 17° and the minimum of $1;17^h$ equally well to 16°, so the table invites further investigation.

3.14 Muḥammad Zamān al-Mashhadī's prayer-tables for Meshed

MS Cairo ṬFF 14, copied *ca.* 1700, is the only copy known to me of an astronomical work entitled *Tuhfa-yi Sulaymani* compiled in 1078 H [= 1667/68] by Muḥammad Zamān ibn Sharaf al-Dīn Ḥusayn al-Mashhadī, well-known as a maker of astronomical instruments.[25] The treatise contains some tables of various spherical astronomical functions (fols. 103r-105v) such as:

$$\alpha(\lambda),\ \alpha'(\lambda),\ \text{and}\ \alpha_\phi(\lambda)$$

based on the parameter ϕ = 37;0° (Meshed), a standard value.[26] There is also a table displaying the time of culmination of the star *al-kaff al-khaḍīb* (= β Cas) as a function of solar longitude (fol. 42v) – see further **I-2.8.2**. Two prayer-tables concern the present study. They are entitled *jadwal-i irtifāʿ-i aftāb dar samt-i qibla-yi Mashad-i muqaddas bi-taqwim-i shams birikrand* (fol. 102v) – see **Fig. 3.14** – and *jadwal-i sāʿāt mā bayn ṭulūʿ-i ṣubḥ wa shams wa ghurūb-i shams wa maghīb-i shafaq* (fol. 111r) and display the functions:

$$h_q(\lambda) \quad \text{and} \quad r^h(\lambda)\ \text{or}\ s^h(\lambda)\ .$$

[24] On the houses see n. **I**-3:2.
[25] On Muḥammad Zamān see n. **I**-2:45.
[26] See Kennedy & Kennedy, *Islamic Geographical Coordinates*, p. 221, and King, *Mecca-Centred World-Maps*, pp. 172-175.

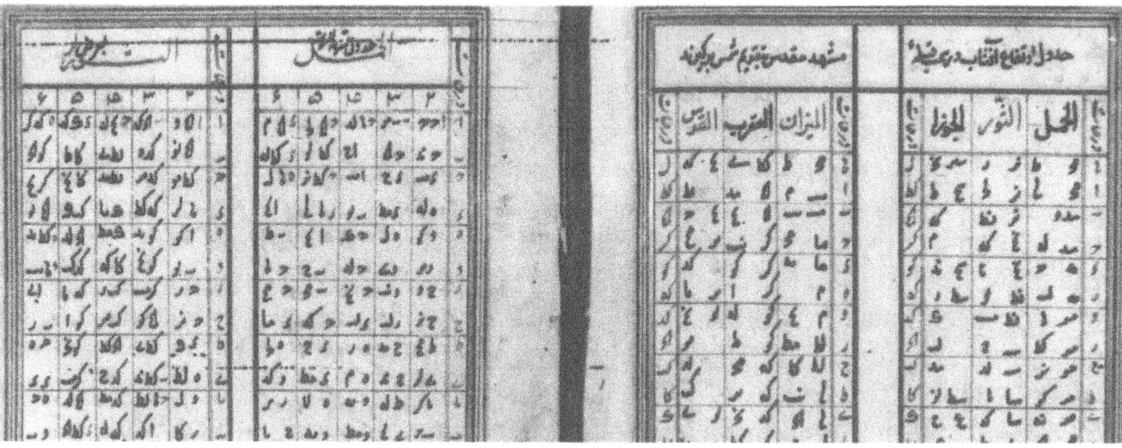

Fig. 3.14: Muḥammad Zamān's table of the solar altitude in the azimuth of the qibla at Meshed, followed by the first and second tables in a set for each of the zodiacal signs for the equalisation of the astrological houses. [From MS Cairo ṬFF 14, fols. 102v-103r, courtesy of the Egyptian National Library.]

The format is the same as that of the twilight tables in the Cairo corpus (**4.10**). The underlying latitude here is stated on other tables to be 37°. The qibla used by Muḥammad Zamān for Meshed in the gazetteers engraved on one of his astrolabes is 45;23° (accurately for his coordinates 45;9°); on another he has 55;28°(!), and on another no value at all (!).[27]

3.15 Anonymous prayer-tables for Isfahan

MS Istanbul Topkapı B 411, an encyclopedia compiled *ca.* 1413 for Isakandar Sulṭān ibn ʿUmar Shaykh of Isfahan, contains an extensive section on astronomy with numerous tables and illustrated constellation figures.[28] Amidst the tables is a set of prayer-tables for Isfahan. Values of the following functions are displayed to two digits for each degree of λ' ($\delta \gtrless 0$):

$d, \psi, d, D, \tilde{h}, 2D^h, H, Z_{(12)}, Z_{(7)}, h_a, z_{a(12)}, z_{a(7)}, t_a^h, t_a, 12/2D, h_b, z_{b(12)}, (2N\text{-}r)^h$ and r^h .

The underlying parameters are:

$$\phi = 32;25° \text{ and } \varepsilon = 23;30° .$$

Although no locality is specified, elswhere in this compilation there are tables of the function called *ʿarḍ-i iqlīm-i ruʾya*, "latitude of visible climate"[29] for:

$$\phi = 29;30° \text{ (Shiraz) and } 32;25° \text{ (Isfahan).}$$

The times defined by h_a and h_b are called *awwal al-ʿaṣr* and *ajzāʾ-yi ikhtiyārāt*, the latter suggesting an astrological association but simply an unfortunate error for *ākhir al-ikhtiyār*, "the end of the most favourable time". The names of all the functions tabulated are given in Arabic although the remainder of the astronomical section is in Persian.

---

[27] *Ibid.*, p. 504.

[28] See *ibid.*, pp. 143-145, on the world-map in this manuscript.

[29] See n. 3:14.

3.16 Prayer-tables for Isfahan on instruments by ʿAbd al-Aʾimma

The celebrated Iranian astrolabist ʿAbd al-Aʾimma, who worked in Isfahan about 1700 and is best known for his elegant astrolabes,[30] engraved a set of shadow tables on a sundial and qibla-indicator which he made and which is now preserved in the Time Museum, Rockford, Illinois. Another unsigned instrument of the same kind is in the Museum of the History of Science, Oxford: see **Fig. 3.16.**[31] The instrument is fitted with a pin-gnomon at the centre a set of

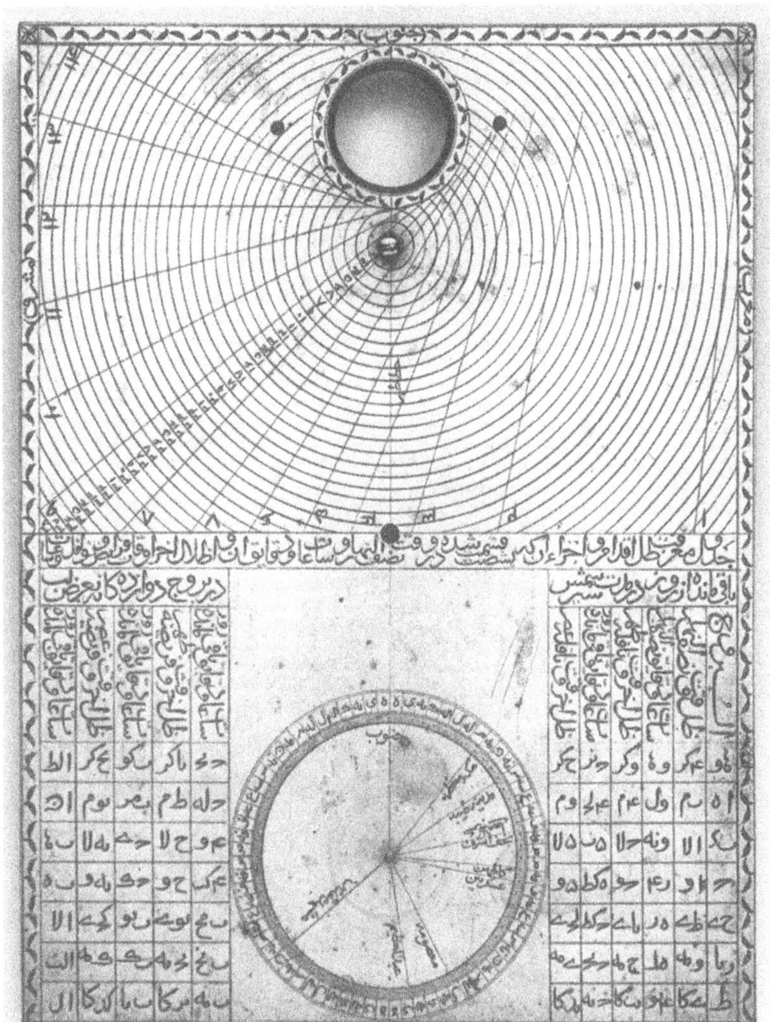

Fig. 3.16: The prayer-tables on the compass *cum* qibla-indicator attributable to ʿAbd al-Aʾimma. [Photo courtesy of Museum of the History of Science, Oxford.]

[30] On ʿAbd al-Aʾimma see Mayer, *Islamic Astrolabists*, pp. 23-26, and supplement, p. 293; Gingerich & King & Saliba, "The ʿAbd al-Aʾimma Astrolabe Forgeries"; and also King, *Mecca-Centred World-Maps*, pp. 175-177.

[31] On these two instruments see *ibid*., pp. 118-121. On a fake see *Khalili Collection Catalogue*, II, p. 274 (no. 162).

quadrants of a circle radii 1, 2, ... , 40 where 12 is the length of the gnomon. These curves can be used to regulate the *ẓuhr* and *ʿaṣr* using the tables, which display values of the functions:

$$z_z \ (= Z), \quad z_a \quad \text{and} \quad T_a \ (\text{in hours})$$

to two sexagimals for each zodiacal sign. The underlying latitude is $\phi = 32°$ (Isfahan).[32] Given the paucity of Safavid manuscript sources on timekeeping these tables are of particular historical interest. It is unlikely that ʿAbd al-A'imma computed them himself, for, as far as we know, neither he nor any of his colleagues – astronomers and instrument-makers – from 17th-century Isfahan were not into compiling tables.[33]

3.17 Anonymous Ottoman-type prayer-tables for Yarqand

I recall seeing a set of such tables in a manuscript at the Oriental Institute in Tashkent. My notes thereon have disappeared.

---

[32] On the qibla values for various cities also engraved on these instruments see *ibid.*, pp. 118-121, 177, 518-519, and 545.

[33] *Ibid.*, pp. 128-138.

CHAPTER 4

THE MAIN CAIRO CORPUS OF TABLES FOR TIMEKEEPING

4.0 Introductory remarks

The Cairo corpus exists in a number of a manuscript sources, usually containing tables from three main categories:

(1) tables of solar azimuth as a function of solar altitude and longitude;
(2) tables of time as a function of solar altitude and longitude;
(3) tables of standard spherical astronomical functions of solar longitude and various other functions for regulating the prayer-times.

All of these tables are computed for the latitude of Cairo taken as 30;0° and obliquity 23;35°. No two manuscripts have been found yet containing precisely the same tables.

In the first category the function tabulated is a(h,λ), and the tables contain over 10,000 entries. In the second the functions tabulated are either t(h,λ) or T(h,λ) or both. These are closely related and the computation of one given the other is trivial, since their sum is D(λ). Each set contains over 10,000 entries. The third category consists of a total of about 30 tables, each containing 90 or 180 entries. Most of these can be derived very easily from basic tables of δ(λ) (as is the case with, say, H, whence z_a and h_a can be found), or d(λ) (as is the case with, say, D, 2N, $\tilde{h}$, $1/\tilde{h}$, $\tilde{h}$), or directly from the tables of T(h,λ) (as in the case with r and s). Likewise, given h_a or h_q, it is not difficult to find the corresponding hour-angles and time since sunrise, t_a and T_a, and t_q and T_q, using the tables of t(h,λ) and T(h,λ) or D(λ).

Most of the sources which I used in my first study of the Cairo corpus[1] mention the 10th-century astronomer Ibn Yūnus as the compiler, but several sources identified thereafter include instructions by the the late-13th-century astronomer al-Maqsī and others contain some notes by the mid-14th-century astronomer al-Bakhāniqī. In numerous late manuscripts in which only the tables of standard spherical astronomical functions and the prayer-tables are found, these tables are attributed to the 16th-century astronomers al-Lādhiqī and al-Ikhṣāṣī (**7.8**).

The location in the manuscripts of the various categories of main tables, as well as the authors to whom they are attributed, is shown below (see already **I-2.1.1**). (The tables in MS Cairo TR 354 falsely attributed to Ibn Yūnus are computed for Alexandria – see **8.5**.) The notation T/t means that T and t are tabulated on facing pages, and the notation T/t/a means that T, t and a are tabulated together, with triplets of values for each pair of arguments. Otherwise the functions listed are tabulated separately. An asterisk indicates that tables are preceded by an introduction on their use prepared by al-Maqsī (**5.4**), and a double asterisk indicates that this introduction is accompanied by some notes by al-Bakhāniqī describing the way in which he had rearranged the tables (see further **5.6**).

[1] See King, "Astronomical Timekeeping in Medieval Cairo", cited in n. **I**-2:2. That study was based mainly on MSS Dublin CB 3673 and Cairo DM 108, also Berlin Ahlwardt 5753, Cairo Azhar *falak* 4382, Cairo TR 191, and Escorial ár. 924,7.

Berlin Ahlwardt 5753	T/t and a	Ibn Yūnus
Cairo TR 191	t	Ibn Yūnus
Dublin CB 3673	T/t and a	Ibn Yūnus*
Escorial ár. 924,7	a	Ibn Yūnus
Cairo DM 108	T/t/a	Ibn Yūnus**
Cairo Azhar *falak* 4382	t and a	Ibn Yūnus
Cairo MM 137	a	Ibn Yūnus
Dublin CB 4078	T/t	anon.
Gotha A 1410	a	Ibn Yūnus
Cairo DM 53	T/t/a	al-Maqsī**
Cairo DM 45	T/t/a	anon.
Gotha A1402	T	al-Maqsī
Istanbul Kılıç Ali Paşa 684	t	Ibn al-Kattānī
Leipzig 817	T/t	anon.
Istanbul Nuruosmaniye 2903	T/t/a	anon.*
Istanbul Nuruosmaniye 2925	T/t/a	al-Maqsī**
Cairo TR 354	T/t/a (ϕ=31°)	Ibn Yūnus
Cairo MM 64	a	anon.
Cairo DM 444	T	al-Maqsī*
Cairo DM 690	T/t/a	anon.
Cairo DM 777	T/t	anon.
Cairo DM 616	T/t/a	al-Maqsī**
Cairo DM 739	T/t/a	al-Maqsī**
Cairo DM 776,1	T	anon.
Cairo DM 778	t	anon.
Cairo DM 786	T/t/a	anon.
Cairo DM 1101	a	anon.
Cairo DM 1108,9	a	anon. (**I-5.2.1**)
Cairo DM 1109	T/t/a	Ibn Yūnus
Cairo DM 1224	T/t/a	Ibn Yūnus
Cairo K 4044	t and a	al-Maqsī (t)/ Ibn Yūnus (a)

Having already analyzed the main Cairo corpus from a mathematical point of view in my previous study, I propose to consider each of the tables afresh (**4.3** to **4.9**), paying particular attention to the question of authorship. I begin with a few brief remarks (in **4.1**) about each of the individuals known to have contributed to the corpus. I then (in **4.2**) present brief descriptions of several new manuscripts of the corpus to supplement those in my original analysis.

My main purpose in **4.3-9** is to describe the tables in two manuscripts of the corpus, MS Dublin CB 3673, copied in the 14th century, and MS Cairo DM 108, copied at the beginning of the 19th century, in the light of the several dozen related manuscripts that I have located.

In **4.3** I discuss the tables of standard functions for spherical astronomy. In **4.4** I discuss the tables of a(h,λ) and show that they were indeed compiled by Ibn Yūnus. Certain manuscripts of the azimuth tables contain some nonsensical values added by an incompetent anonymous and in others these have been corrected by the early-14th-century astronomer Ibn al-Rashīdī. In **4.5** I discuss the tables of t(h,λ) and T(h,λ) and present the evidence which makes it clear that these tables in their present form were not compiled by Ibn Yūnus. Three centuries after his time, al-Maqsī claimed to have prepared the tables of T(h,λ), but I suspect that he had at his disposal some tables for timekeeping by Ibn Yūnus which have not survived in their original form. In the 14th century Ibn al-Kattānī prepared a set of tables of t(h,λ) using al-Maqsī's tables of T(h,λ). In **4.6-11** I attempt to cast some light on the highly complicated problem of the authorship of the minor tables in the corpus. Most of the mathematically-significant tables were either computed by Ibn Yūnus or owe their inspiration to him. The development of the main Cairo corpus from the 10th to the 14th century is then discussed in **Ch. 5**.

4.1 The main contributors

4.1.1 Ibn Yūnus

Abu 'l-Ḥasan ʿAlī ibn 'Abd al-Raḥmān known as Ibn Yūnus worked in Fusṭāṭ in the last quarter of the 10th century. Much of the activity described in this book owes its original inspiration to him. According to his biographers, Ibn Yūnus had a comic appearance: in the only illustration of him, in which he appears with his patron al-Ḥākim, he looks reasonably normal – see **Fig. I-4.1.1**. He is already well known to the history of Islamic science as the author of an extensive *zīj* called the *Ḥākimī Zīj*, which is one of the finest works of its genre.[2] His extensive and highly sophisticated treatment of spherical astronomy in the *Ḥākimi Zīj* includes numerous tables (**5.1**). Indeed, virtually all of the theory underlying the mathematically-significant tables in the main Cairo corpus and also some of the individual tables are contained in the *Ḥākimi Zīj*. Ibn Yūnus compiled more than one *zīj*: for example, the *Mukhtār Zīj* by the Yemeni astronomer Abu 'l-ʿUqūl (**12.1**) is based on a *zīj* of Ibn Yūnus other than the *Ḥākimī*. Ibn Yūnus also appears to have compiled extensive tables of the Sine and Cotangent functions for each minute of argument,[3] as well as a large set of double-argument tables for the lunar equation.[4] In **5.3** I assess the evidence that there is a "missing link" between his known works and the main tables of the corpus in the form of a lost work of his dealing with practical aspects of timekeeping.

4.1.2 al-Maqsī

Shihāb al-Dīn Aḥmad ibn ʿUmar al-Maqsī was one of the leading astronomers in Cairo about 1275.[5] In my previous study of the Cairo corpus, on the basis of my examination of the set of tables in MS Cairo DM 53, introduced in al-Maqsī's name, I suggested that he had merely plagiarized some of Ibn Yūnus' tables. I have since been able to consult other such sets, for example, MSS Gotha A1402, Istanbul Nuruosmaniye 2925 and Cairo DM 616, and al-Maqsī's

---

[2] On Ibn Yūnus see n. **I**-2:3.

[3] King, *Ibn Yūnus*, pp. 85-89 and 109. See also *ibid.*, pp. 96-99, on tables of the solar declination for each minute of solar longitude, with differences for each second, attributed to him.

[4] King, "Lunar Equation Table".

[5] On al-Maqsī see n. **I**-2:4.

Fig. 4.1.1: A miniature showing a rather angry-looking Caliph al-Ḥākim apparently giving Ibn Yūnus a hard time. The astronomer is holding his *zīj*, marked with Hindu-Arabic numerals on the spine. Actually the *Ḥākimī Zīj* would have filled several volumes of this size. [From MS Istanbul Topkapı Revan Köşku 1638 (400 folios) of *al-Qānūn fī 'l-dunyā*, an astrological work of the prolific Ibn Zunbul (İhsanoğlu *et al.*, *Ottoman Astronomical Literature*, I, pp. 183-184, no. 84); photo courtesy of the Topkapı Library.]

contribution is reconsidered in **5.4** below. I have also studied the only other known work attributed to al-Maqsī, a treatise on sundial theory extant in MSS Cairo Azhar *falak* 5528, Dublin CB 4090 and Cairo DM 103. This consists mainly of tables, which are to be used for marking shadow traces and curves for the seasonal hours on sundials erected in various planes.[6] There are over 100 tables each containing about 30 entries and this illustrates that al-Maqsī was not afraid of extensive calculation. These tables are related to the more accurate but less extensive ones of his contemporary Abū ʿAlī al-Marrākushī (**6.7**) although neither individual refers to the other. The treatment of the standard problems of spherical astronomy in al-Maqsī's introduction to his sundial tables is not impressive, and it is of interest that in this introduction al-Maqsī used some material from the *Ḥākimī Zīj* without mentioning his illustrious predecessor.

[6] On al-Maqsī's sundial tables see King, "Islamic Astronomical Tables", pp. 51-53, and *idem*, "Astronomy of the Mamluks", pp. 547-548, and also **I-4.1.3** and **X-7.1**.

4.1.3 Ibn al-Rashīdī

Shams al-Dīn Muḥammad ibn Ibrāhīm known as Ibn al-Rashīdī appears to have worked in Cairo about 1300,[7] after al-Maqsī and before al-Bakhāniqī. He was the compiler of a set of prayer-tables for Mecca (**6.11**) and is known to us also by the references of al-Karakī (**9.4**), who was probably his late contemporary, and of al-Bakhāniqī (**5.6**). al-Karakī states that Ibn al-Rashīdī compiled some hour-angle tables, but does not specify the underlying latitude. I am inclined to think that al-Karaki's own tables for Jerusalem were in fact based on a set of tables for latitude 32° computed by Ibn al-Rashīdī. al-Bakhāniqī mentions that Ibn al-Rashīdī had recomputed some of the values in Ibn Yūnus' azimuth tables where they had been inaccurately copied. On Ibn al-Rashīdī's contribution to the Cairo corpus see **4.8** and **5.2**.

4.1.4 Ibn al-Kattānī

Muḥammad ibn Muḥammad ibn Muḥammad ibn ʿAbd al-Qawī al-Qurashī known as Ibn al-Kattānī was an astronomer and instrument-maker and also a fine penman.[8] MSS Cairo MM 72, Istanbul Kılıç Ali Paşa 684, and Dublin CB 3673,1 are copied in his elegant hand and are dated 747 H [= 1346/47], 768 H [= 1361] and 772 H [= 1371], respectively (see **Figs. 4.2b**, **4.7a** and **I-4.1.1**). The first of these manuscripts contains a set of tables of solar altitude for Cairo: the title folio is missing and no compiler is mentioned, although I suspect that these tables are due to Najm al-Dīn al-Miṣrī (**6.5**). The second manuscript contains a set of tables for timekeeping for Cairo which Ibn al-Kattānī says he computed himself. These include a complete set of tables of t(h,λ) and various additional tables for regulating the times of prayer. Ibn al-Kattānī's name is not mentioned in any other known source in connection with the authorship of part of the Cairo corpus. The third manuscript is a copy of Ibn Yūnus' tables of a(h,λ) (**5.2**) which contains additional garbled entries for solar altitude up to 88°. Ibn al-Kattānī's integrity is put in question by the fact that he claimed to have computed t(h,λ) but was incapable of copying tables of a(h,λ) properly. Other inconsistencies in his tables in MS Istanbul Kılıç Ali Paşa 684 confirm that Ibn al-Kattānī was a better copyist than astronomer. I discuss his tables in **5.5**.

4.1.5 al-Bakhāniqī

Shams al-Dīn Aḥmad ibn Muḥammad al-Bakhāniqī worked in Cairo and also in the Yemen in the 14th century.[9] He was at some time associated with the Azhar Mosque since he is also called al-Azharī. MS Istanbul Topkapı A3343 of the *summa* of al-Marrākushī (**2.7** and **6.7**) bears his mark of ownership dated 776 H [= 1374/75]. However, he must have been active for rather a long time because one of his works was compiled for a Yemeni *wazīr* and can be dated *ca.* 1325. This particular compilation, extant only in MS Dublin CB 4092, was an extension of the tables of the early 9th-century Baghdad astronomer al-Farghānī for constructing the various markings on astrolabe plates.[10] al-Farghānī's table enabled the user to construct

[7] On Ibn al-Rashīdī (**I-2.1.5**), whose full name was Shams al-Dīn Abū ʿAbdallāh Muḥammad ibn Burhān al-Dīn Ibrāhīm al-Rashīdī, see n. **I**-2:18.

[8] On Ibn al-Kattānī see n. **I**-2:6.

[9] On al-Bakhāniqī see n. **I**-2:5.

[10] On al-Farghānī and his astrolabe tables see the article in *DSB* by A. I. Sabra and King, "Islamic

the altitude and azimuth circles on plates for each degree of latitude from 15° to 50° and are easily derived from a single very useful little table of an auxiliary trigonometric function; al-Bakhāniqī simply completed the set for each degree of latitude from 0° to 90°. Another work by al-Bakhāniqī is a treatise on the prayer-times in which he outlines a set of approximations for the standard functions of timekeeping: I have summarized this in **2.9**.

al-Bakhāniqī was responsible for rearranging the tables in the Cairo corpus so that triplets of values (T,t,a) are displayed side by side as functions of (h,λ). I assess his other contributions to the corpus in **5.6**.

4.2 The manuscript sources

The manuscripts in which the main tables of the Cairo corpus are to be found date from the 14th to the 19th century.[11] No two copies contain the same tables. No single manuscript contains reliable information on the identity of the compiler of the individual tables. In my previous analysis I relied on MSS Berlin Ahlwardt 5753, Cairo TR 191, Dublin CB 3673, Escorial ár. 924,7, Cairo DM 108, Cairo Azhar *falak* 4382 and Cairo MM 137; all of these happened to be preserved in collections which had been catalogued and all happened to be attributed to Ibn Yūnus, which explains how I located them first and why I had no problem attributing all of the tables to him. I also noted the existence of MS Cairo DM 53, in which part of the corpus is attributed to al-Maqsī, of al-Bakhāniqī's notes in MS Cairo DM 108, and of various copies of al-Lādhiqī's prayer-tables, but I assumed that each of these authors had merely plagiarized tables of Ibn Yūnus. The new sources listed below help explain more precisely the development of the corpus between the 10th and the 14th centuries and although my doubts about the integrity of al-Maqsī, al-Bakhāniqī and al-Lādhiqī still appears to be correct (**5.4, 5.6** and **7.8**), one should bear in mind that in the world of the Cairo astronomers in the 13th and 14th centuries it must have been difficult to simply "plagiarize" tables.[12] The discovery of other early copies of the corpus might well cast new light on its development and on the identity of the compilers of some of the tables, where this is still in doubt.

MS Berlin Ahlwardt 5753 (Landberg 574):[13]
This is a disordered fragment of the corpus copied *ca.* 1400. The title on fol. 1r reads *Kitāb fīhi 'l-Samt li-'bn Yūnus al-ḥāsib*, "Azimuth Tables by Ibn Yūnus, the astronomer" (see **Fig. 4.2a**). There are several notices of possession on the title folio, one dated 840 H [= 1436]. The tables on fols. 1v-20r are in considerable confusion. Each is headed only by the degree of solar altitude. For altitudes 1° through 19° (fols. 1v-10v) the function tabulated is a(h,λ);

Astronomical Tables", pp. 53-55; *idem* & Samsó, "Islamic Astronomical Handbooks and Tables", pp. 91-92; and **X-4.2**.

[11] On the Cairo manuscripts see *Cairo ENL Catalogue* (in Arabic) I, in which they are listed by accession numbers, and II, especially under 3.1.2 (azimuth tables of Ibn Yūnus), 3.1.2 (tables of al-Maqsī), and 3.1.10 (edition of al-Bakhāniqī); as well as *Cairo ENL Survey* (in English), no. B59/3.1.1 (Ibn Yūnus), C15/3.1.2 (al-Maqsī), and C28 (al-Bakhāniqī). For Istanbul manuscripts some catalogues have been prepared after I worked on the manuscripts, but I have not seen them. For manuscripts from other catalogued collections references to the catalogues are given *ad loc.*

[12] See also nn. 4:31 and 10:2.

[13] *Berlin Catalogue*, pp. 206-207.

Fig. 4.2a: The title folio of the Berlin manuscript, in which the azimuth tables are attributed to Ibn Yūnus. The tables that follow are, however, a disordered mixture of tables of a (which correspond to the title), then t and T as functions of h and λ. See also **Figs. II-5a-b**. This was the only manuscript of the corpus known to Carl Schoy when he worked on Ibn Yūnus in the early years of the 20th century. [From MS Berlin Ahlwardt 5753, fol. 1r, courtesy of the Deutsche Staats-bibliothek, Berlin.]

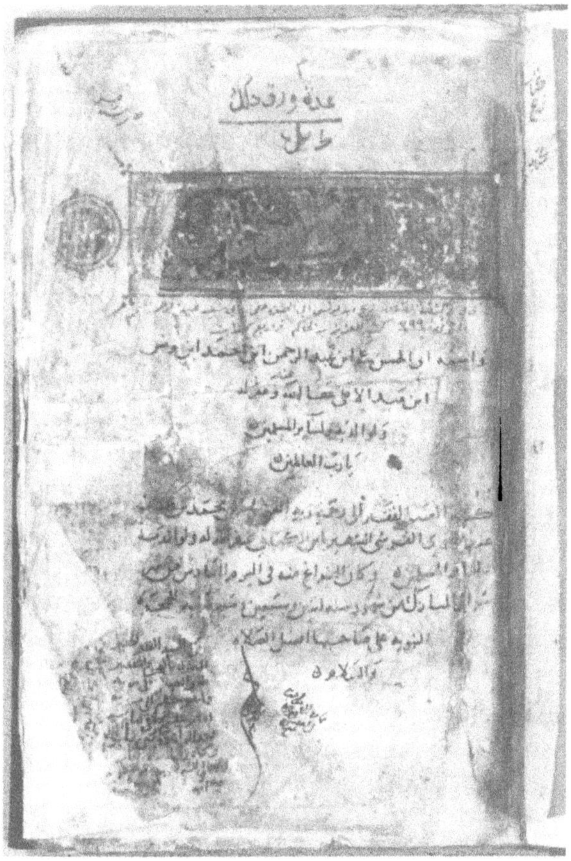

Fig. 4.2b: The title-folio of the Dublin manuscript. The decorated title, illegible in this photograph, announces only the azimuth tables of Ibn Yūnus, and the double tables of the time since sunrise and time before midday which follow are in fact in a different hand. Compare **Figs. I-2.1.1a** and **I-5.1.1a**. This did not stop me in the early 1970s from blithely assuming that the whole lot were by Ibn Yūnus. I had ordered a microfilm of the manuscript because Bernard Goldstein had told me it had been catalogued as a "new" copy of the *Ḥākimī Zīj*. [From MS Dublin CB 3673, fol. 121r, courtesy of the Chester Beatty Library.]

for 8° and 9° (fols. 11r-11v) the function is t(h,λ); and for 52° to 83° (fols. 12v-20r) we are dealing with T(h,λ)! On fol. 20v there is a table of $t_a(\lambda)$, which was published by Carl Schoy,[14] and on fol. 21r, a table of $1/\tilde{h}(\lambda)$. An additional folio at the end of the manuscript contains an illegible title and some instructions on a calculation in spherical astronomy.

MS Cairo TR 191:
This is a copy of part of the corpus written in 1215 H [= 1800/01] by ʿAbd al-Bāriʾ al-ʿAshmāwī (also the copyist of MS Cairo DM 108 – see below). The title reads *Jadāwil faḍl al-dāʾir min*

[14] Schoy, *Gnomonik der Araber*, p. 53.

qibal al-irtifāʿ ḥisāb ... Ibn Yūnus, and most of the manuscript consists of a complete set of tables of t(h,λ). These are followed by tables of the functions:

$$H,\ s,\ n,\ T_q,\ z_{a(12)},\ r,\ 2N,\ Z_{(12)},\ t_q,\ \sigma,\ \alpha_\phi,\ \alpha',\ \alpha_s \text{ and } \alpha_r\ .$$

The twilight tables are based on parameters 19° and 17°, the qibla tables on q = 53° and the *salām* is assumed to be 1° before daybreak. The manuscript concludes with an undated catalogue of the equatorial coordinates of 60 stars, copied by al-ʿAshmāwī in 1217 H [= 1802/03].

MS Dublin CB 3673:[15]

This manuscript (annoyingly foliated 121r-1v as for a Latin manuscript) is a virtually complete copy of the Cairo corpus, incorrectly catalogued as the *Ḥākimī Zīj*. The title on fol. 121r reads: *Kitāb al-Samt li-bn Yūnus*, but this refers only to the first part of the tables: see **Fig. 4.2b**. This folio also contains some notices of possession, a floriated *ṭughrā* signature, and the date of copying, 772 H [= 1371]. The tables are copied in the distinctive elegant hand of Muhammad ibn Muhammad ibn Muḥammad ibn ʿAbd al-Qawī al-Qurashī known as Ibn al-Kattānī – see the notes to MS Cairo MM 72 below. This part of the manuscript contains a complete set of tables of a(h,λ), with additional entries for altitudes 84° to 88° (fols. 121v-96r) (see **Fig. 5.1.1a**), and these are followed by tables of

$$h_q\ (q = 52°),\quad h_0,\quad d \quad\text{and}\quad h_v\ .$$

(The entries in the table of h_v decrease to 0;41° at λ = 270°. If Ibn al-Rashīdī compiled this table, why is it presented here together with the uncorrected azimuth tables that al-Bakhāniqī states Ibn al-Rashīdī had corrected?)

The remaining part of this manuscript contains tables of T(h,λ) and t(h,λ) (fols. 82v-15r) and various other functions for timekeeping and the prayers, also copied in an elegant hand, but not that of Ibn al-Kattānī: see **Figs. 2.1.1a** and **II-4.7a**. This part of the manuscript has no separate title page but begins with the instructions on the use of the tables (fols. 83r-83v), in which the name of al-Maqsī has been altered to that of Ibn Yūnus. The minor functions tabulated (fols. 15v-3r) are:

$$t_a,\ d,\ H,\ h_q\ (q{=}53°),\ T_a,\ \tilde{h},\ \psi,\ h_0,\ D,\ t_a,\ \delta,\ r\ (h_r{=}20°),\ s\ (h_s{=}16°),$$
$$h_{a=60°},\ h_{a=30°},\ t_q\ (q{=}53°),\ \text{arc Sin}\ (x),\ r\ (h_r{=}19°),\ s\ (h_s{=}17°),\ 2N,\ n,\ 2N,\ r\ (h_r{=}20°),$$
$$d,\ H,\ \delta,\ \psi,\ h_0,\ 1/\tilde{h}$$

Fol. 82r has been filled with odd items such as notices of possession, verses of poetry, and a list of Graeco-Coptic numerals, given with their Arabic *abjad* equivalents.[16] The manuscript ends with some astrological writings in a different hand (fols. 2r-2v).

MS Escorial ár. 924,7:[17]

This is an incomplete copy of Ibn Yūnus' azimuth tables dating from the 14th (?) century. The title on fol. 30r reads: *Kitāb al-Samt li-bn Yūnus wa-Kitāb al-Ẓill maḥlūl daqīqa daqīqa*, "Azimuth Tables of Ibn Yūnus, and Cotangent Tables computed for each Minute of Argument". Below this is the curious note: *fī ʿilm al-rumūz li-fiqh al-majmaʿ*, "on the science of secret signs for the under-standing of the community (?)". Fols. 30v-63v contain azimuth tables for solar

15 *Dublin CB Arabic Catalogue*, p. 73. The date of copying is misread as 672 H.
16 On these see, for example, Ritter, "Griechisch-koptische Ziffern". See also n. 4:21.
17 *Escorial Catalogue*, pp. 32-33.

altitudes from 1° to 78°. The remainder of the azimuth tables, and the Cotangent tables, are missing from the manuscript.

MS Cairo DM 108:
This is a fairly complete copy of the main Cairo corpus as edited by al-Bakhāniqī. It was copied by ʿAbd al-Bāriʾ (ibn) Naṣr al-ʿAshmawi (fol. 74v) in 1218 H [= 1803] (see also MS Cairo TR 191 above) (see **Fig. II-4.11a**). The title folio gives the title as *Kitāb Ghāyat al-intifāʿ fī maʿrifat al-dāʾir wa-faḍlihi wa-'l-samt min qibal al-irtifāʿ* and identifies the author as Ibn Yūnus. The manuscript contains al-Maqsī's introduction followed by al-Bakhāniqī's notes (see **Fig. II-5.6c**), a complete set of tables of the functions (T,t,a) (see **Fig. II-5.6b**), followed by various prayer-tables (see below), as well as some calendrical tables tables not found in the other main copies of the corpus. The latter part of al-Maqsī's introductions, al-Bakhāniqī's notes, and the first five folios of tables are in the older and more elegant hand of the copyist of MS Cairo DM 777 (see below). The parameters for twilight in this fragment are 20° and 16°.

Following the main set of tables are some calendrical tables for tables for finding the date in the Hijra calendar for the years 1521-1680 in the Coptic calendar [= 1219-1383 H], and tables of the following functions:

$$H, D, t_a, h_a, T_a, h_q, t_q, h_{q*}, 2N, n, s, r, \sigma, \Delta D, \tilde{h}, 2D^h, \tilde{h}, \tau \text{ and } t_{q*}\ .$$

The twilight tables are based on parameters 19° and 17°, the qibla tables on q = 53° and q* = 37°, and the *salām* is taken to be 1° before daybreak. See **Fig. II-4.11a**.

MS Cairo Azhar falak 4382:
This is a carefully-copied manuscript of tables of t(h,λ) (fols. 1v-35r) and a(h,λ) (fols. 36v-70v), and also of a(h) for certain altitudes at the equinoxes and solstices (fol. 71r) and ψ(λ) together with $h_0(\lambda)$ (fol. 71v). All of the tables are in the same 14th-century hand. The hour-angle tables are preceded by a page on which the title *Kitāb Faḍl al-dāʾir* is mentioned but no author is identified. The azimuth tables, which contain entries for arguments h = 1° to 83°, are preceded by a page on which the title *Kitāb al-Samt* is mentioned and Ibn Yūnus named as author.

MS Cairo MM 137:
This manuscript, which was copied about 1500, contains a complete set of Ibn Yūnus' azimuth tables bearing the title *Kitāb al-Samt*. The altitude argument runs beyond 83° to 86° (**4.4**). Information about the solar altitude for which the azimuth changes direction is given in the margins of the sub-tables.

MS Dublin CB 4078 (new):
This is a complete set of carefully-copied tables of T(h,λ) and t(h,λ) for Cairo dating from *ca.* 1450. The title appears to read *Kitāb al-Dāʾir wa-faḍluhu*, and there is no mention of the compiler or date of copying.

MS Gotha A 1410 (new):[18]
This consists mainly of a complete set of Ibn Yūnus' azimuth tables with entries for altitudes

[18] *Gotha Catalogue*, III, p. 62.

up to 89°. The copyist was ʿAlī ibn Muḥammad ibn ʿAlī al-Dalāmī[19] and the date of completion 841 H [= 1437/38]. An undated notice of possession names the owner as Jalāl al-Dīn Muḥammad al-Ramlī, *muwaqqit* at the the Umayyad Mosque (in Damascus). There are additional tables of h_q (q = 53°) and h_v (with entry 0;41° for $\lambda = 270°$). The remaining tables were originally from a different manuscript and consist of an incomplete anonymous set of double-argument lunar equation tables, attributed elsewhere to Ibn Yūnus.[20]

MS Cairo MM 184 (new):
This contains tables (fols. 7v-10r, copied *ca.* 1700) of the following functions:

$$h_q,\ \delta*,\ h_v,\ B \text{ and } C\ ,$$

as well as a table of the difference between sunrise and sunset at Mecca and Cairo, such as is contained in MS Cairo MM 204. The table of h_q is based on q = 53° and the entry for h_v for $\lambda = 270°$ is 0;41°.

MS Cairo DM 190,2 (new):
The tables (fols. 25r-33r) listed below were copied *ca.* 1550 and begin with a page of notes on the values of ΔD for each sign (max. 62, min. 32), the (Coptic) numbers for the *warrāqīn*,[21] and a statement that the sun moves 18 miles on its sphere each minute. The tables display the following functions:

$$D,\ t_a,\ h_a,\ T_a,\ s,\ n,\ r,\ \delta,\ h_0,\ \alpha_s,\ \alpha_\sigma,\ \alpha_\phi \text{ and } \alpha'\ .$$

MS Cairo DM 954,2:
This manuscript, copied *ca.* 1475, contains an incomplete set of tables from the corpus (fols. 2r-10v) with no accompanying title-page. The functions tabulated are:

$$t_q,\ h_{q*},\ t_{q*},\ \tau,\ \sigma,\ h_v,\ Z_{(12)},\ h_{a=60°},\ \tilde{h},\ z_{a(12)},\ h_{a=30°},$$
$$\alpha',\ (D{+}s),\ \delta,\ d,\ H,\ 2N,\ \tilde{h} \text{ and } h_q\ .$$

The tables relating to the qibla are based on q = 53° and the table of h_v has 0;41° at $\lambda = 270°$.

MS Cairo DM 53 (new):
This contains a complete set of tables of T(h,λ), t(h,λ) and a(h,λ) with a triplet of entries for each pair of arguments. The colophon indicates that the copyist was the Azhar *muwaqqit* ʿAbd al-Fattāḥ ibn Muḥammad al-Dulajī and that the manuscript was completed in the Nile Valley town of *Jirjā al-ʿarīḍa* (Girga) in 1117 H [= 1705]. The first folio bears the title *Kitāb al-Dāʾir wa-faḍluhu wa-'l-samt* and the tables are preceded by al-Maqsī's introduction and al-Bakhāniqī's notes. In the introduction the parameters advocated for twilight are 19° and 17°. At the end of the manuscript are some absurd instructions on the use of the tables for latitudes other than 30°.

MS Cairo DM 45:
This contains a collection of spherical astronomical tables in different hands and in considerable disorder. The various parts of the manuscript date from about 1700. The first function tabulated, as indicated on the title folio, is the oblique ascensions of the ascendant at the beginning of

[19] On al-Dalāmī and other manuscripts copied by him see *Cairo ENL Catalogue*, I, p. 704.
[20] See King, "Lunar Equation Table" (n. 4:4), p. 132.
[21] See already n. 4:16.

the afternoon prayer "according to the new observation", *i.e.* based on obliquity 23;30° (**7.1**). The tables which follow are for finding the entry of the sun into each zodiacal sign and are due to Riḍwān Efendī, as stated (**7.10**). The next tables display the functions (T,t,a) for ϕ = 30°, a triplet of entries being given for each pair of arguments (h,λ) but only for altitude arguments 21° to 57°. These are followed by tables of H for ϕ = 30° (ε = 23;35°), as well as for eight different values of ϕ between 32° and 45° (ε = 23;30°). Of particular interest is an odd folio of tables (fol. 27) of an unspecified function f(h,λ) tabulated for arguments h = 24° and 25° and each degree of solar longitude: the function is in fact t(h,λ) for ϕ = 32° (**6.7**). The manuscript also contains an incomplete sexagesimal multiplication table.

MS Cairo MM 204:
This manuscript, copied in 1052 H [= 1642/43], contains an unusual set of tables displaying the following functions (fols. 80v-81r and 86r):

$$\mathrm{Tan}_5\ \Delta,\ \mathrm{Tan}_5\ \delta(\lambda),\ G(\theta),\ h_a(H)\ \text{and}\ \mathrm{Cot}_{12}\ h$$

for each degree of argument. The table of G is labelled "table for finding the hour-angle with ease": see further **I-6.9.3** (illustrated). Another table in this set displays the difference between sunrise or sunset at Mecca and Cairo for each degree of λ, such as is also found in MS Cairo MM 184 (see above).

MS Cairo MM 58,2:
This was copied 1450 by Ibn Abi 'l-Fatḥ al-Ṣūfī and contains tables (fols. 2v-12v) of the following functions:

$$\alpha_\phi\ (\phi=31°,\ \text{Alexandria}),\ \alpha_\sigma,\ \alpha,\ \alpha_s,\ n,\ h_a,$$
$$\lambda(\delta)\ (\Delta\delta = 0;15°),\ h_v,\ h_a(H),\ \text{and}\ 2D^h\ (\Delta\lambda = 0;6°)\ ,$$

as well as some astrological tables. The table of h_v has 0;41° at λ = 270°.

MS Gotha A1402 (new):[21a]
This manuscript (date ?) bears the title *Risāla fī 'l-falak li-'l-Ṣūfī*, "Treatise on astronomy by al-Ṣūfī (that is, al-Maqsī)", and contains a complete set of tables of T(h,λ) preceded by an introduction introduced in the name of al-Maqsī (written al-Maqdisī, as if he or his forbears were from Jerusalem, a curious mistake for a Cairene copyist, who must have known the suburb of al-Maqs) and followed by an incomplete and disordered set of various other tables for timekeeping. These additional tables are numbered and are bound in the manuscript now as follows:

$$1\text{: } t_a;\quad 2\text{: } D;\quad 3\text{: } T_q;\quad 4\text{: } 1/\tilde{h};\quad 5\text{: } h_q;\quad 23\text{: Tan } \delta;$$
$$24\text{: } \mathrm{Tan}_{60}\ \theta;\quad 9/10\text{: d and } \delta;\quad 11\text{: } h_q\ (= \text{no. } 5);\quad 21\text{: Vers } \theta;\quad \text{and}\quad 22\text{: Sin } \theta\ .$$

It is clear that there were originally some two dozen additional tables. The first four appear to be original to al-Maqsī's set (**5.4**). All of the qibla tables are based on q = 53°. The table of $\delta(\lambda)$ with values to three digits for each degree of λ is based on that of Ibn Yūnus in the *Ḥākimī Zīj*. The table of Tan $\delta(\lambda)$, also giving values to three digits, is not identical to that of al-Marrākushī, but the entry 26;11,41 for λ = 90° is used by al-Maqsī in his treatise on sundials.[22] The three trigonometric tables give values to three digits for each degree of arc. In the introduction the parameters advocated for twilight are 20° and 16°.

[21a] *Gotha Catalogue*, III, pp. 56-57.
[22] See n. 4:6.

MS Gotha A1411 (new):[22a]
This consists of an anonymous almanac (fols. 1r-6v) followed by four pages of tables from the corpus in a different hand (fols. 7r-8v). As well as a table of h_q (q = 53°) and α_ϕ for Cairo (the values are those of al-Marrākushī), these extra pages contain a table of values of the azimuth of the qibla for each degree of latitude and longitude difference from Mecca. This table, which is based on an approximate formula and is attributed to Yūsuf ibn al-Damīrī, is of a kind found in other sources that I have discussed elswhere (**4.7** below). However, no such qibla tables are attested in any of the other known copies of the Cairo corpus.

MS Istanbul Kılıç Ali Paşa 684 (new):[23]
This is an apparently unique copy of Ibn al-Kattānī's set of tables, beautifully copied in his own hand and dated 768 H [= 1361]. It is a valuable source of information on the authorship of some of the tables in the Cairo corpus. See further **5.5**.

MS Leipzig 817 (new):[24]
This 16th-century manuscript bears no title, but contains an incomplete set of tables of t(h,λ) for Cairo (fols. 1v-32v, h = 1°, 2°, ... , 68°), followed by an incomplete copy of the instructions (fols. 33r-35r, the first page is missing), and a complete set of tables of T(h,λ) for Cairo (fols. 35v-67r). Neither tables nor text, which are all written in a single hand, bear the name of any compiler or copyist. The parameters for twilight mentioned in the introduction are 19° and 17°.

MS Istanbul Nuruosmaniye 2903 (new):
This bears no title and begins with an anonymous introduction (due to al-Maqsī). The parameters advocated for twilight are 20° and 16° but the copyist added in the margin that 19° and 17° were preferable. The remainder of the manuscript consists of a complete set of tables of (T,t,a) as functions of (h,λ), copied in a different hand by Khalīl al-Ḥarīrī in 1089 H [= 1678].

MS Istanbul Nuruosmaniye 2925 (new):
This undated manuscript begins with tables of the functions $2D^h$ and $\tilde{h}$ and continues with al-Maqsī's introduction and al-Bakhāniqī's notes. The main tables display (T,t,a) as functions of (h,λ) and are complete. These are followed by tables of:

$$h_v,\ \text{Cos}\ \delta,\ B,\ h_q,\ t_q,\ h_{q*},\ t_{q*}\ \text{and}\ Z_{(12)}\ .$$

The table of h_v (with 0;41° for λ = 270°) is stated to have been computed by Ibn al-Rashīdī.

MS Paris BNF ar. 2520,2:
See **8.1**.

MS Cairo MM 43:
Various tables from the Cairo and Damascus corpora, copied in the hand of [ʿAlī ibn Muḥammad al-Dalāmī] *ca*. 1450. See **6.14** and **10.2**.

22a *Gotha Catalogue*, III, p. 63.
23 Overlooked in Rosenfeld & İhsanoğlu, *MAIC*, no. 735.
24 *Leipzig Catalogue*, p. 264.

MS Cairo MM 64:
A beautifully-executed copy of Ibn Yūnus' azimuth tables, dating from about 1400. The title folio is missing, and there is no colophon. The manuscript begins with the last page of a set of tables displaying the solar longitude for each day of a four-year period in the Coptic calendar. For each month of the Coptic year four columns display the solar longitude in signs, degrees and minutes for each of the four years, labelled *sana rubʿ*, *sana niṣf*, *sana niṣf wa-rubʿ*, *sana kabīs*. Only the tables for Baʾūna, Abīb and Misrā are contained in the manuscript now. The entry for Baʾūna 1 (first year) is Gemini 12;16°. The altitude argument in the azimuth tables runs up to 89° and there is a marginal note which attempts to rationalize this (**4.4**). For each altitude above 36° the meridian altitude corresponding to the last longitude argument for which there is an entry for the azimuth is given to three digits and is called *ghāyat irtifāʿ hātayn al-darajatayn ruʾūs al-jadwal,* "the meridian altitude for these two solar longitudes (whose signs are given) at the head of the table".

MS Cairo DM 444:
This manuscript contains a complete set of tables of T(h,λ) copied in 858 H [= 1454] in the elegant hand of Abu 'l-Yumn Muḥammad ibn Muḥammad ibn ʿArab al-Shabībī (or al-Shayyibī).[25] It bears a notice of possession (fol. 3r) by the Cairo *muwaqqit* ʿAbd al-Raḥmān al-Ṭūlūnī[26] dated 1046 H [= 1636/37]. The title is given as *Kitāb Ghāyat al-intifāʿ fī maʿrifat al-dāʾir min al-falak min qibal al-irtifāʿ li-ʿarḍ lām shamāl li-'l-Maqsī*, and the tables are preceded by al-Maqsī's introduction (with parameters 20° and 16° advocated for twilight). al-Maqsī's four additional tables are not contained in this manuscript; indeed the tables of T conclude with the words: *tamma wa-kamala wa-l-'llāh al-ḥamd*, "it is completed and fulfilled and praise be to God". The tables of T(h,λ) end on the recto side of the last folio and the verso is blank except for a notice of sale dated 997 H [= 1588/89]. The tables for 17° and 19° bear instructions for finding the duration of twilight. See **Figs. I-2.1.1b-c** and **II-5.6a**.

MS Cairo DM 690:
This consists mainly of two incomplete sets of tables of (T,t,a) bound in disorder. The handwriting in both cases is neat and can be dated to about 1500. The last folio of tables (for altitude arguments 40°-42°) is copied in the later hand of Ḥasan al-Ṭablāwī *ca.* 1700.[27] There is no title folio. The tables for altitudes 19° and 17° bear instructions for finding the duration of morning and evening twilight.

MS Cairo DM 777:
This contains an incomplete set of tables of T(h,λ) and t(h,λ) on facing pages, for values of h from 2° to 46° and then 76° to 83°. There is no title folio and no compiler is mentioned. The tables are copied in the elegant hand of the copyist of MS Cairo MM 241 of Ibn al-Mushrif's auxiliary tables (**6.9**) and can thus be dated *ca.* 1500; see also the notes on MS Cairo DM 616 below.

---

[25] On other manuscripts copied by Abu 'l-Yumn see *Cairo ENL Catalogue*, I, p. 714.

[26] On ʿAbd al-Raḥmān al-Ṭūlūnī see *Cairo ENL Survey*, no. D16, and on his library see **V-10**.

[27] Close to 20 scientific manuscripts copied by him *ca.* 1700 are listed in *Cairo ENL Catalogue*, I, p. 688 (also n. 4:29). See also *Cairo ENL Survey*, no. D56, on some tables for constructing sundials apparently computed by him, as well as İhsanoğlu *et al.*, *Ottoman Astronomical Literature*, I, p. 347, no. 212.

MS Cairo DM 616:
This consists of four parts, the first two forming a complete set of tables of (T,t,a) preceded by al-Maqsī's introduction and al-Bakhāniqī's notes. The title page and tables for altitude arguments 7° and above are written in the same hand and dated (10)98 H [= 1686/87]. The title is given as *Kitāb Kifāyat al-waqt fī maʿrifat al-dāʾir wa-faḍlihi wa-'l-samt* and is attributed to Abu 'l-ʿAbbās Aḥmad ibn ʿUmar al-Ṣūfī, that is, al-Maqsī. Part of the introduction and the tables for altitude arguments 1° to 7° (fols. 2r-9v) are in an older hand datable to about 1500, which is in fact the same hand as that of MSS Cairo MM 241 and Cairo DM 777 (see above). The third part of the manuscript, copied about 1600], consists of a set of prayer-tables displaying values of the functions:

$$H, D, h_a, t_a, T_a, d, n, 2N, \sigma, \alpha_\sigma, s \text{ and } r ,$$

side by side for each degree of λ starting with $\lambda = 1°$. The fourth part of the manuscript contains some tables for finding the longitude of the horoscopus and upper midheaven from the oblique ascensions of the horoscopus.

MS Cairo DM 651,5 (new):
This source, copied *ca.* 1700, contains (fols. 59r-64v) a set of tables of various functions from the corpus tabulated together, namely:

$$H, D \text{ and } N, 2D \text{ and } 2N, t_a \text{ and } T_a, s \text{ and } r, h_a \text{ and } z_{a(12)} ,$$

as well as tables of α_ϕ and $\tilde{h}$ and a table of geographical coordinates.

MS Cairo DM 739:
This is in the hand of the copyist of the title folio of MS Cairo DM 616 (see above) and can hence be dated to about 1685. It bears the title *Kitāb Kifāyat al-waqt fī maʿrifat al-dāʾir wa-faḍlihi wa-'l-samt*, attributed to Abu 'l-ʿAbbas Aḥmad ibn ʿUmar al-Ṣūfī, that is, al-Maqsī. It contains al-Maqsī's introduction followed by al-Bakhāniqī's notes and a complete set of tables of the functions (T,t,a). These tables end on the recto side of the last folio (fol. 64). On the verso there are displayed values of $a(h,\lambda)$ only, for altitude arguments 84° to 88°, and values of $a(h)$ at the solstices.

MS Cairo DM 776, fols. 1v and 10r-43r:
This contains a complete set of tables of $T(h,\lambda)$, apparently copied in 981 H [= 1573/74]. The title folio, which is in the hand of an owner Muhammad Sinār / Sannār (?) al-Aḥmadī,[28] gives this date of copying and the title as *al-Dāʾir li-'rtifāʿ al-shams*. No compiler is mentioned and there is no introduction. The tables for arguments 16° and 20° are marked as suitable for twilight by the original copyist, and are marked *b-ṭ-l* for *baṭal(a)*, "is no good", in a different hand. The tables for 17° and 19° are marked in another different hand as suitable for twilight according to "the new observations", a phrase which usually refers to the Samarqand observations of Ulugh Beg, but here may refer to the opinion of the Cairo editor of the Samarqand *zīj*, namely, Ibn Abi 'l-Fatḥ al-Ṣūfī.

[28] He is to be identified with the Muḥammad Sinār / Sannār who was a muwaqqit at the Aḥmadī (from Sayyid Aḥmad Badawī) Mosque in Tanta, on whom see *Cairo ENL Survey*, no. D108; and also İhsanoğlu *et al.*, *Ottoman Astronomical Literature*, II, pp. 625-626, no. 460.

MS Cairo DM 778:
This was copied by Ḥasan ibn ʿAlī al-Ghamrī and appears to date from *ca.* 1500. It contains a disordered incomplete set of hour-angle tables followed by various prayer-tables. The title is given as *Kitāb Faḍl al-dāʾir* and no compiler is mentioned. Following the main set there are tables of:

$$D, t_a, T_a, s, h_a, r \text{ and } \sigma .$$

The twilight tables are based on parameters 19° and 17° and the *salām* is assumed to be 1° before daybreak. These prayer-tables are copied in the same hand as the hour-angle tables, but the colophon occurs at the end of the latter.

MS Cairo DM 786:
An incomplete copy of the tables of (T,t,a) for altitude arguments 45° to 83°. The manuscript bears neither title nor colophon but is copied in the distinctive and untidy hand of Ḥasan al-Tablāwī[29] and can thus be dated *ca.* 1700.

MS Cairo DM 1101:
A late copy of Ibn Yūnus' azimuth tables, perhaps copied *ca.* 1785, and entitled *Jadwal al-samt al-maḥlūl daqīqa daqīqa*. Entries are given for altitude arguments up to 89°.

MS Cairo DM 1108,9 (fols. 35v-59r):
An unique copy, dated 1053 H [= 1643/44], of a set of azimuth tables for Cairo arranged so that for each degree of solar longitude the solar altitude is entered vertically. The altitude arguments are arranged in columns of up to 30 entries, and the meridian altitude is given to three digits below the last integral altitude argument for certain degrees of solar longitude. The last southern azimuth in each solar longitude sub-table is written in red ink and the corresponding solar altitude argument is labelled *darajat al-intiqāl.*

MS Cairo DM 1109:
A complete set of tables of (T,t,a) copied in 1128 H [= 1716] by Ḥasan ibn ʿAbdallāh, nicknamed *al-Qiṭṭ al-miskīn*, perhaps meaning "Poor Pussy". There is no introduction and the title *al-Dāʾir wa-faḍluhu wa-ʾl-samt ḥisāb Ibn Yūnus al-Miṣrī li-ʿarḍ Miṣr* is a later addition. On fol. 1r there is a list of pilgrim stations on the road from Cairo to Mecca with latitudes and qiblas (see **8.4** and **10.9**).

MS Cairo DM 1224:
An incomplete copy of the tables of the functions (T,t,a) bearing the title *Jadāwil al-dāʾir wa-faḍl al-dāʾir wa-ʾl-samt min qibal al-irtifāʿ* and attributed to Ibn Yūnus. There are some simple instruments on the use of the tables but no other introduction. The manuscript was copied by ʿAbd al-Bāriʾ al-ʿAshmāwī (see MSS Cairo TR 191 and DM 108) and can hence be dated *ca.* 1800. The tables end with those for altitude argument 51°.

MS Cairo K 4044:
This copy of the main tables in the corpus is dated 1308 H [= 1890/91]. The work is divided into two parts, the first being entitled *Kitāb sharḥ ʿalā faḍl al-dāʾir li-l shaykh Abi ʾl-ʿAbbās*

[29] See n. 4:27.

al-Maqsī. This refers to al-Maqsī's introduction which precedes a complete set of tables of T(h,λ), not t(h,λ) as one might have expected from the title. The second part of the manuscript is written in a different hand but also dated 1308 H. It is entitled *Kitāb Maḥlūl al-samt li-bn Yūnus li-ʿarḍ lām shamāl* and contains a set of azimuth tables for altitude arguments up to 83°. The last folio of the manuscript contains tables of the solar azimuth at the equinoxes and solstices and the functions ψ(λ) and h_0(λ).

MS Princeton Yahuda 861,1:
See **8.1**.
MS Cairo ṬJ 367,5, fols. 35r-40r (new):
Copied *ca.* 1450, this contains part of an introduction and part of a solar longitude table, followed by tables of the functions:

r, H, h_q, h_0, ψ, d and δ

from the corpus. The work bears the title *Natījat al-afkār fī aʿmāl al-layl wa-'l-nahār* and is attributed to the 15th-century Egyptian astronomer ʿIzz al-Dīn al-Wafāʾī. However, the prayer tables are not mentioned in the introduction (see further **5.7**). The same introduction and solar tables, followed by a much more extensive set of prayer-tables for different latitudes, is contained in MS Princeton Yahuda 861,1 (**8.1**).

4.3 Tables of standard spherical astronomical functions

Some copies of the corpus contain a number of spherical astronomical functions of the kind standard in Islamic *zījes*. The notes which follow give a clearer picture of those tables than was possible in my original analysis.

(a) Solar declination, δ(λ)

MSS Dublin CB 3673, Gotha A1402 and Cairo DM 153 are the only copies of the corpus in which tables of δ(λ) occur, but in most manuscripts there are tables of the related function H(λ) – see (c) below. The entries in these tables of δ(λ) are given to three digits for each degree of λ′ and contain several variants from the corresponding entries in the *Ḥākimī Zīj,* which are more accurate.

(b) Half excess of daylight, d(λ)

The tables of d(λ) to two digits in the Chester Beatty manuscript are *not* based on the corresponding table in the *Ḥākimī Zīj*, which has entries to three digits: about one-sixth of the entries in the table in the former source are in error by ±1 in the second digit. See further (d) below on the related tables of D(λ).

(c) Meridian altitude, H(λ)

The errors in a table of H(λ) will reflect those in the underlying values of δ(λ) since H = $\bar{\phi}$ + δ. Furthermore, the values for δ > 0 and δ < 0 can be compared to distinguish computational errors from copyists' errors. Now the table of H in MS Dublin CB 3673, fol. 14v, contains only one error of the former kind: for λ = 15° we find H = 65;58°, and for λ = 195° we find

H = 54;2°, which means that δ(15°) was taken as 5;58° instead of the accurate value 5;57°. If this table were based on Ibn Yūnus' declination table in the *Ḥākimī Zīj* the entries would all be correct, since of his entries for integral values of λ only seven are in error by only ±1 in the *third* sexagesimal digit. However, it is possible that Ibn Yūnus' accurately-computed entry 5;56,37° for δ(15°) was misread as 5;57,37° and rounded to 5;58°. In MS Dublin CB 3673, fols. 4r and 11r, however, it is clearly 5;56,37°.

I do not think that al-Maqsī could have computed δ(λ) accurately to two digits. al-Marrākushī's declination table (computed to two digits) contains some nine errors of -1 in the second digit, but it seems that he compiled this table by truncating values to three digits found in an earlier ʿIrāqī source.

(d) Semi diurnal arc, D(λ)

The tables of D(λ) to two digits in MSS Dublin CB 3673 and Cairo DM 108 were compiled from the table of δ(λ) to two digits in the Dublin manuscript, using the simple relation D = 90° + d (d ≷ 0 as δ ≷ 0). As noted in (b) above this table of δ(λ) is *not* derived from Ibn Yūnus' carefully-computed table in the *Ḥākimī Zīj*, which gives entries to three digits. al-Bakhāniqī does not state that al-Maqsī computed either d or D but al-Maqsī does refer to a table of D(λ) in his introduction to his tables of T(h,λ). I suspect that al-Maqsī did compute these two tables himself and that he used his table of D(λ) to compute T(h,λ) from some hour-angle tables by Ibn Yūnus. Perhaps al-Bakhāniqī had his own table of D(λ) which he had used to make minor alterations in the tables of t(h,λ) starting from al-Maqsī's tables of T(h,λ). See further (e) below.

(e) Nocturnal arc, 2N(λ)

The tables of 2N(λ) in MS Dublin CB 3673, fols. 5v and 6v, are based on different tables of D(λ). The values of 2N on fol. 6v in the Dublin manuscript are fairly closely related to those of D on fol. 12r. The values of 2N on fol. 5v are based on a set of values of D or d which is significantly different from those on fol. 12r and also from the more accurate set of values of d in MS Paris BNF ar. 2513, fol. 53v, of the *Muṣṭalaḥ Zīj*. The table of length of darkness n(λ) (see **4.10**) on fol. 6r of the Dublin copy is based on the table of 2N(λ) on fol. 6v.

(f) Length of seasonal day-hours in degrees, $\tilde{h}(\lambda)$

This is derived by dividing 2D by 12. In the *Ḥākimī Zīj* Ibn Yūnus computes it to four digits and the third digit of his entries is generally accurate. In MS Dublin CB 3673, fol. 13r, the table of $\tilde{h}(\lambda)$ to two digits contains 5 errors of ±1 in the second digit, so that it was not derived from the *Ḥākimī* values by rounding.

(g) Factor for converting equatorial degrees to seasonal day-hours, $1/\tilde{h}(\lambda)$

The table of $1/\tilde{h}(\lambda)$ to two digits in MS Dublin CB 3673, fol. 3r, contains some 40 errors of ±1 in the second digit. It was probably prepared by al-Maqsī using rounded values of $\tilde{h}$. In the introductions to MSS Dublin CB 3673 and Cairo DM 108, both of which sources contain the table, al-Maqsī states that he has included tables of $t_a(\lambda)$ and $1/\tilde{h}(\lambda)$ "at the end of the book" (**5.4**). In the Dublin copy a marginal note by the side of the table of $1/\tilde{h}(\lambda)$ states "this is the end of the book", but in the Cairo copy there are four tables following the one of $1/\tilde{h}(\lambda)$. In MS Berlin Ahlwardt 5753, fols. 20v and 21r, however, immediately following an incomplete set of tables of $T(h,\lambda)$, there are others of $t_a(\lambda)$ and $1/\tilde{h}(\lambda)$ which conclude the work.

(h) Length of daylight in equatorial hours, $2D^h(\lambda)$

The table of $2D^h(\lambda) = 2D(\lambda)/15$ to two digits occurs only in MS Cairo DM 108, fol. 75v, of the main sources. There are four errors of ±1 in the second digits of the entries, and that for arguments which differ from those for which $\tilde{h}(\lambda) = 2D(\lambda)/12$ is in error. The function is not tabulated in the *Ḥākimī Zīj and* is not referred to in al-Maqsī's introduction to the corpus. al-Bakhāniqī states that it was he himself who added this table (**5.6**).

(i) Normed right ascensions, $\alpha'(\lambda)$

My original implication that the tables of $\alpha'(\lambda)$ in MSS Hartford, Conn., Theological Seminary 621 and Paris BNF ar. 2553 of the prayer-tables of al-Lādhiqī (**7.8**) were taken from the *Ḥākimī Zīj* is incorrect. In his *Zīj* Ibn Yūnus computes $\alpha(\lambda)$ carefully to three digits. The table of $\alpha'(\lambda)$ to two digits in MS Leiden Or. 143 of the *Zīj* is not derived from this: it is either spurious or taken from an earlier work of his. On the other hand, the table of $\alpha'(\lambda)$ to three digits in MS London BL Or. 3624 of the Yemeni *Mukhtar Zīj* is based on the *Ḥākimī* table of $\alpha(\lambda)$, and the table with entries to two digits in MS Cairo MM 43 appears to be based on this. Likewise, al-Marrākushī (**6.7**) tabulated $\alpha'(\lambda)$ carefully to three digits: his values differ from those of Ibn Yūnus and are in general slightly less accurate. The values of $\alpha'(\lambda)$ in the Hartford and Paris BNF ar. 2553 copies of al-Lādhiqī's tables can *not* be derived from either of Ibn Yūnus' or al-Marrākushī's three-digit values by rounding and they are not identical with those to two digits in MS Leiden Or. 143.

(j) Oblique ascensions for Cairo, $\alpha_\phi(\lambda)$ for $\phi = 30°$

The tables of $\alpha_\phi(\lambda)$ to two digits in the Hartford and Paris copies of al-Lādhiqī's prayer-tables contain numerous errors, which fact immediately distinguishes them from the corresponding table in the *Ḥākimī Zīj,* since it contains but two small errors. In fact, these tables are al-Marrākushī's. On the other hand, the table of $\alpha_\phi(\lambda)$ in the *Natīja* attributed to al-Wafā'ī (**8.1**), which contains a few copyist's errors, appears to have been taken from the *Ḥākimī Zīj.*

(k) Solar altitude in the prime vertical, $h_0(\lambda)$

The *Ḥākimī Zīj* contains a table of $h_0(\lambda)$ carefully computed to three sexagimal digits. In MS Dublin CB 3673, fol. 85v, of the corpus contains this same table with numerous copyist's errors, but it is two other tables of $h_0(\lambda)$ to two digits in the Chester Beatty manuscript, also occurring in other copies of the corpus, which merit comment. Firstly note that:

$$h_0(\lambda) = \text{arc Sin } \{ \text{ R Sin } \delta(\lambda) \text{ / Sin } \phi \} = \text{arc Sin } \{ 2 \text{ Sin } \delta(\lambda) \} \quad \text{for} \quad \phi = 30° .$$

Since also:

$$\text{Sin } \delta = \text{Sin } \varepsilon \text{ Sin } \lambda \text{ / R} ,$$

we have:

$$h_0 = \varepsilon \text{ for } \lambda = 30° .$$

Now in the table of $h_0(\lambda)$ in MS Dublin CB 3673, fol. 3v, the entry for $\lambda = 30°$ is 24;34°, which is clearly absurd. The entry for $\lambda = 29°$ has been miscopied as 23;49° instead of 22;49,(32)°: the 24 in 24;34° appears to have been "fixed up" and the 34 can be explained as a copyist's error for 35°. Nevertheless, the entry for $\lambda = 31°$ is 24;21°. (Even in the table to three digits in MS Dublin CB 3673, fol. 85v, the entry for 30° is miscopied as 23;45,0°.) These same errors occur in the Hartford and Paris copies of al-Lādhiqī's prayer-tables (**7.8**), but in MS Princeton Yahuda 861,1 of the *Natīja* attributed to al-Wafā'ī (**8.1**) they have been corrected. In the table of $h_0(\lambda)$ on fol. 12v of the Dublin manuscript, the entries in the third column have been further garbled, apparently by someone who believed that the maximum value of h_0 was 53;46° rather than the accurate value 53;9°. Again, in the prayer-tables of al-Minūfī (**7.1**) most of the standard functions are recomputed for $\varepsilon = 23;30°$: his table of h_0 is an exception, since he simply plagiarized an earlier source. The entries in his table for λ = 30° and 90° are again 24;34° and 53;9°. Finally, in the prayer-tables of al-Fawānīsī (**7.5**), the values of h_0 corresponding to $\lambda = 30°$ and 90° are respectively 23;47° and 53;21°, both inaccurate.

(l) Solar rising azimuth, $\psi(\lambda)$

This function is defined by:

$$\psi(\lambda) \;=\; \text{arc Sin } \{ \text{ R Sin } \delta(\lambda) \;/\; \text{Cos } \phi \ \} \ ,$$

and so for $\phi = 30°$ is slightly more difficult to compute than $h_0(\lambda)$ (see (k) above). In the *Ḥākimī Zīj* Ibn Yūnus computes $\psi(\lambda)$ carefully to thirds. His table occurs again in MS Dublin CB 3673, fol. 84r, and those on fols. 12v and 3v of the same manuscript, computed to two digits, are slightly less accurate than can be obtained from his values by rounding. Nevertheless the errors are much smaller than those in the associated tables of h_0.

(m) Arc Sines

The table of arc Sin (x) in MS Dublin CB 3673, fol. 8r, contains some errors which are identical to those in the corresponding entries in the tables in MS London BL Or. 3624, fol. 169v, of the Yemeni *Mukhtār Zīj*, a work containing numerous tables of Ibn Yūnus and other errors which are not. The errors in the table in the Chester Beatty manuscript also differ from these in al-Marrākushī's table of arc Sines (**6.7**). It is possible to derive arc Sines accurately to two sexagesimal digits from Ibn Yūnus' extensive Sine tables in MS Berlin Ahlwardt 5752 (Landberg 1038), copied 895 H [= 1490], but we do know that Ibn Yūnus used Sine tables less accurate than those in the *Ḥākimī Zīj* in his early work.

The minor tables in the corpus can now be subdivided according to their provenance as follows:

$h_0(\lambda)$ and $\psi(\lambda)$ (3 digits)	Ibn Yūnus
$H(\lambda)$	based on Ibn Yūnus
$\delta(\lambda)$, $d(\lambda)$, $D(\lambda,)$, $2D^h(\lambda)$, $2N(\lambda)$, $\tilde{h}(\lambda)$, $\alpha'(\lambda)$	anonymous
$h_0(\lambda)$ and $\psi(\lambda)$ (2 digits)	anonymous
$1/\tilde{h}(\lambda)$	al-Maqsī
$2D^h(\lambda)$	al-Bakhāniqī

$\alpha_\phi(\lambda)$ ($\phi = 30°$) al-Marrākushī
arc Sin (x) anonymous

The tables of $h_0(\lambda)$ and $\psi(\lambda)$ to three digits were appended to Ibn Yūnus' *Kitāb al-Samt* (**5.2**). The table of $1/\bar{h}(\lambda)$ was appended to al-Maqsī's tables of time since sunrise (**5.4**). al-Bakhāniqī appears to have compiled the table of $2D^h(\lambda)$ and added those of $\alpha'(\lambda)$ and $\alpha_\phi(\lambda)$ ($\phi = 30°$) to his edition of the corpus (**5.6**). The identity of the person who compiled the remaining tables remains obscure.

4.4 Tables of solar azimuth

These, called *Kitāb al-Samt* when copied separately, display the solar azimuth $a(h,\lambda)$ measured from the prime vertical, for each integral degree of solar altitude h up to the maximum 83° and each degree of solar longitude λ' ($\delta \gtrless 0$).

In each of MS Berlin Ahlwardt 5753, Dublin CB 3673, Escorial ár. 924,7, Cairo Azhar *falak* 4382, Cairo MM 137 and Gotha A 1410, the title *Kitāb al-Samt* is given and Ibn Yūnus is named as the compiler (see **Figs. 4.2a-b**). al-Bakhāniqī and al-Khaṭāʾī (**5.6** and **6.8**) also say that Ibn Yūnus compiled the azimuth tables. I see no reason to doubt this. In the *Ḥākimī Zīj* there are tables of $a(\lambda)$ for altitudes 30° and 35° with values for each degree of λ in 12 columns of 30 entries (**5.1**): no reason is given for choosing these two altitudes. The two tables in the *Zīj* are slightly more accurately computed than the corresponding tables in the *Kitāb al-Samt*.[30]

It seems that Ibn Yūnus' azimuth tables for altitudes 80° to 83° were either not completed or were lost at some stage in their transmission. Some anonymous with a very limited knowledge of spherical astronomy computed a set for 80° to 89° (extant in MS Gotha A 1410; MSS Dublin CB 3673 and Cairo MM 137 have entries up to 88° and 86°: see **Fig. I-5.1.1a**). A corrected set for 80° to 83° was compiled by Ibn al-Rashīdī (**4.1.3**) and these azimuth tables were used by al-Bakhāniqī (MSS Cairo DM 108 and DM 53) and were also copied separately (MS Cairo Azhar *falak* 4382). I must withdraw my previous statement that al-Maqsī "plagiarized" Ibn Yūnus' azimuth tables,[31] since now it is clear that al-Bakhāniqī put together the tables of T, t and a, and the reference to azimuth tables in the introduction to MS Dublin CB 3673 is probably not original to al-Maqsī's introduction. Ibn Yūnus tabulated certain other functions along with the azimuth – see **5.2**.

In the margin of the tables in some copies of the corpus, *e.g.*, MSS Dublin CB 3673 and Escorial ár. 924,7, for altitudes 1° to 53° it is stated for which solar longitudes the azimuth changes from north to south for that particular altitude. In MS Berlin Ahlwardt 5753 the change in direction is more crudely indicated. In the margin of the tables for altitude 54° in the Chester Beatty manuscript we find the comment that "there are no more azimuths in the north for this altitude and those above it". When, for certain longitudes, the sun does not attain a given altitude, the entries in the table are left blank. For altitudes of about 60° and above, the entries for more than one altitude can be tabulated on one folio. In the Escorial manuscript, for a given

[30] Both are edited in King, "Astronomical Timekeeping in Medieval Cairo", pp. 354-355.
[31] *Ibid.*, p. 364. See also the text to nn. 4:12 and 10:2.

altitude, the meridian altitude is given for the longitude whose azimuth is closest to 90°. Thus, in the tables for altitude 37°, the last significant entry in the right-hand side of the table is an azimuth of 87;3° for Sagittarius 17° and Capricorn 13°. The next and final entry is given as 0;0°, which does not mean due east or west, but that the sun does not attain the altitude 37° for Sagittarius 18° or Capricorn 12°. The meridian altitude is accurately stated as 37;3,22° for the solar longitudes having azimuth 87;3°. The purpose of giving this detail is not fully clear to me. However, in the unique MS Cairo DM 1108,9, copied in 1053 H [= 1643/44], the solar azimuth is displayed as a function a(λ,h), with values a(h) for each value of λ, and the values of H(λ) are given for each λ. It seems unlikely that Ibn Yūnus' original tables were in this form, not least because he tabulated a(h,λ) for two altitudes in the *Ḥākimī Zīj* (**5.1**).

By virtue of the format of the tables no values are shown for the equinoxes. However, in some copies of the corpus, such as MS Cairo Azhar *falak* 4382, fol. 71r, there is a table displaying a(h) at the equinoxes for each degree of h from 1° to 60° (= $\bar{\phi}$) and also for each 0;5° of h between 36;5° and 36;25° (= $\bar{\phi}$-ε) at the winter solstice and again for each 0;5° of h between 83;5° and 83;30° and also for 83;33° and 83;35° (= $\bar{\phi}$+ε) at the summer solstice. al-Bakhāniqī (**5.6**) states that this table was computed by Ibn al-Rashīdī (**4.1.3**).

4.5 Tables of time since sunrise and the hour-angle

These, often called *Kitāb al-Dā'ir* and *Kitāb Faḍl al-dā'ir* when tabulated separately, display T(h,λ) and t(h,λ) for each integral degree of h up to the maximum 83° and each degree of λ' (δ ≷ 0). There are no such tables in the *Ḥākimī Zīj*. In MSS Dublin CB 3673, Dublin CB 4078, Leipzig 817 and Cairo DM 777, the tables of both functions for a given altitude generally occupy two facing pages of the manuscript. In MSS Cairo TR 191 and Cairo Azhar *falak* 4382 only t is presented, and in MS Gotha A1402 only T. In MS Berlin Ahlwardt 5753 the functions are tabulated separately. In MSS Cairo DM 108, Cairo DM 53, Istanbul Nuruosmaniye 2903 and 2925, Cairo DM 690 and DM 616, triplets of values (T,t,a) are given for each pair of arguments (h,λ), where a is the solar azimuth. See **Figs. I-2.1.1a-b**, **2.1.1c**, **II-5-4b** and **5.6a**.

These tables of T and t are less accurately computed than the azimuth tables. In general the entries are less accurate for higher altitudes, as a result of the nature of the tabulated functions. When the three functions are tabulated side by side, for altitudes over 80° the number of longitude arguments for which entries are given for T and t differs slightly from the number for which entries are given for a: see **Fig. 5.6b**.

The values of T and t are related by the values of D(λ) tabulated elsewhere in the corpus (**4.3d**). al-Maqsī (**5.4**) claims to have computed the tables of T(h,λ) himself. Each of al-Karakī, Ibn al-Kattānī and al-Bakhāniqī (**9.4, 5.5** and **5.6**) attribute these tables to him. Ibn al-Kattānī claims to have computed the tables of t(h,λ) from al-Maqsī's tables of T(h,λ) and al-Bakhāniqī claims to have done the same, adding that his entries "may differ in some cases from previous calculations".

I am now of the opinion that al-Maqsī did compile the tables of T(h,λ) and Ibn al-Kattānī compiled the tables of t(h,λ). The fact remains that in order to tabulate T(h,λ) one must first compute t(h,λ): I suspect that al-Maqsī used an earlier set of tables of either t(h,λ) or T(h,λ) compiled by Ibn Yūnus which are no longer extant in their original form and were not available to astronomers after al-Maqsī: see further **5.2** and **5.3**.

4.6 Tables of solar altitude in specific azimuths

Two tables of the function h(a,λ) for a = 30° and 60° and each degree of λ′ (δ ≷ 0) occur in MSS Dublin CB 3673, Cairo DM 954,2 of the Cairo corpus. The second table alone is also found in MS Cairo DM 153,2, of the corpus, and the first table alone is found in MS Cairo DM 620 of al-Qaymarī's tables for the afternoon prayer (**7.3**). The functions are labelled *al-irtifāʿ idhā kāna 'l-samt lām/sīn*. In MS Cairo DM 153 Ibn Yūnus is specifically identified as the compiler and indeed the same tables occur along with others based on different azimuths in the *Ḥākimī Zīj* (**5.1**), where Ibn Yūnus describes the way in which they can be used to determine the meridian. The tables in each of the above-mentioned copies of the corpus are arranged symmetrically in the same way as the azimuth and time tables, whereas those in the *Zīj* give values of h for each degree of λ in 12 columns of 30 entries.

The Cairo corpus also contains tables of solar altitude in the azimuths of the qibla and the rising sun at the winter solstice: see **4.7** and **4.8**.

4.7 Tables of the solar altitude and the time of day when the sun is in the azimuth of the qibla

The tables in the corpus relating to the qibla are based on the assumption that the qibla at Cairo-Fusṭāṭ is 53° E of S. This value is rounded from the 53;0,6° and 53;0,17° derived by Ibn Yūnus in his *Zīj*, and it was used at least by the astronomers for the qibla in Cairo throughout the medieval period. (Other directions were used by different interest groups, such as the azimuth of the rising sun at the winter solstice: see **4.8**.) Ibn Yūnus' discussion of the qibla in his *Zīj* concludes with a table of the solar altitude in the azimuth of the qibla, $h_q(\lambda)$, based on the parameter q = 52°, computed "some time ago". The value 53° is based on a different value of the longitude difference between Cairo and Mecca, but no further information is provided about the way in which this longitude difference was measured or what coordinates were used to derive the value q = 52°.

The table of $h_q(\lambda)$ in MS Dublin CB 3673, fol. 86v, of the corpus is based on the parameter q = 52° and is arranged with values for each degree of λ, that is, in 12 columns of 30 entries. There are numerous corrupt readings, as can be seen by checking entries symmetrically-placed with respect to the solstices or by comparing the table with the more accurate one in the *Ḥākimī Zīj*. This table is not contained in any other copies of the corpus, although al-Marrākushī (**6.7**) tabulated $h_q(\lambda)$ for each 30° of λ′ and six of the seven values, which are based on q = 52°, are the same as the corresponding entries in the *Ḥākimī Zīj*. The Dublin manuscript (fol. 14r), and several other copies of the corpus contain a table of $h_q(\lambda)$ for q = 53°, symmetrically arranged like the azimuth and time tables, but less accurately computed: see **Fig. 4.7a**.

A number of sources, including the Dublin manuscript, contain a table of $t_q(\lambda)$, the hour-angle when the sun is in the direction of Mecca, also symmetrically arranged. This is based on q = 53° and is not carefully computed. Using the values of solar altitudes for this azimuth

Fig. 4.7a: The table of the hour-angle when the sun is in the azimuth of the qibla at Cairo. On the left is a table for the duration of evening twilight, based on a value of 17° for the angle of solar depression below the horizon. [From MS Dublin CB 3673, courtesy of the Chester Beatty Library.]

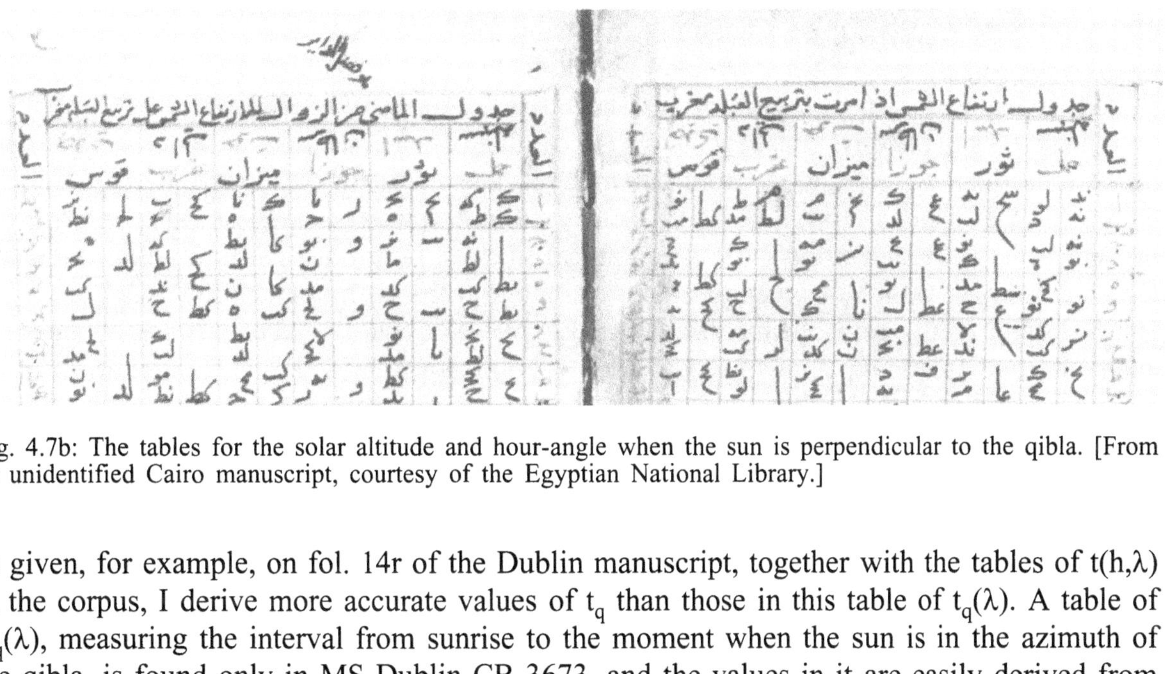

Fig. 4.7b: The tables for the solar altitude and hour-angle when the sun is perpendicular to the qibla. [From an unidentified Cairo manuscript, courtesy of the Egyptian National Library.]

as given, for example, on fol. 14r of the Dublin manuscript, together with the tables of $t(h,\lambda)$ in the corpus, I derive more accurate values of t_q than those in this table of $t_q(\lambda)$. A table of $T_q(\lambda)$, measuring the interval from sunrise to the moment when the sun is in the azimuth of the qibla, is found only in MS Dublin CB 3673, and the values in it are easily derived from the tables of $t_q(\lambda)$ and $D(\lambda)$.

In a few sources, including MS Cairo DM 108, there are tables for the solar altitude $h_{q*}(\lambda)$ and the corresponding hour-angle $t_{q*}(\lambda)$ when the sun is in the direction perpendicular to the qibla (*ʿalā tarbīʿ al-qibla*). These tables are based on $q = 53°$ and are less carefully computed than those of h_q and t_q: see **Fig. 4.7b**.

The table of h_q ($q = 52°$), which occurs in only one source, is certainly due to Ibn Yūnus. The tables of h_q and h_{q*} ($q = 53°$) are also due to him, if my interpretation of al-Bakhāniqī's remarks (**5.6**) is correct. The table of T_q appeared already in al-Maqsī's set compiled in the late 13th century (see **5.4**) and was probably based on an earlier table of t_q. The table of t_{q*} appears to be a later addition.

One copy of the corpus, MS Gotha A1411, contains a table displaying the qibla as a function of the latitude and longitude differences from Mecca. Values of $q(\Delta\phi,\Delta L)$ are given for each degree of each argument from 1° up to 20° and are based on the approximate formula:

$$\text{Tan } q \quad = \quad \text{R Sin } \Delta L \ / \ \text{Sin } \Delta\phi \ .$$

The table is attributed to Yūsuf ibn al-Damīrī, an individual as yet unidentified.[32] However, the same table exists in several other manuscripts, all of Egyptian provenance, and in one of these the compiler is named as Yūsuf al-Damīrī and in another as the 9th-century Baghdad astronomer al-Khwārizmī. I have discussed this table and related approximate methods for determining the qibla in a separate study.[33]

4.8 A table for orienting ventilators

Several copies of the corpus contain a table displaying the solar altitude, h_v, in the azimuth of the ventilator (*al-bādahanj*), for each 1° of λ′ (Δ ≷ 0). The table is based on an azimuth of 27;30° (S of E), that is, the azimuth of the rising sun at the winter solstice. It is understood that the back of the base of the ventilator should be aligned in this direction and that the front should be exposed to the winds from the north east. See **Figs. VIIb-3.1** and **4.1**.

We know from medieval travel accounts that ventilators were a common feature on the roofs of the multi-storey houses of al-Qāhira and Fusṭāṭ; those few which survive in Cairo are late Ottoman constructions. The table of $h_v(\lambda)$ in the Cairo corpus, as well as other information on ventilators in medieval Egyptian astronomical sources, inform us that in Mamluk and early Ottoman times it was the custom to erect ventilators in this astronomically-defined direction. But the qibla of the first Muslim settlers in Egypt was towards winter sunrise, so that, for example, the Mosque of ʿAmr in Fusṭāṭ faces this direction. In addition, the entire medieval city of al-Qāhira is, by hydrographical convenience and human design accompanied with a good deal of luck, aligned in this direction. The ventilators were simply aligned with the roughly-orthogonal street-plan of the medieval city, and they were in fact open to a direction which is not entirely appropriate for gathering favourable winds.

The entries in the table of $h_v(\lambda)$, which are given to two digits, are not as accurately computed as those in Ibn Yūnus' tables of $h(a,\lambda)$ (**4.4**). Many of the entries are in error by about -0;5°, which would suggest that the value used for $h_e(a)$ (**4.6** and **5.2**) was in error by this amount. In certain sources, e.g., MSS Cairo MM 58 and Istanbul Nuruosmaniye 2925, the entries in the sixth column decrease to 0;41° rather than to 0;0° as one would expect, since the sun rises in this azimuth when λ = 270°. We can be certain that the correction to 0;41° represents an attempt to account for the displacement of the visible horizon from the true horizon as a result of refraction (see further **4.11**). Yet not all later astronomers included this adjustment: the

---

[32] In 2001 a set of tables of double argument planetary (but no lunar) equation tables compiled by him was located in MS Istanbul Selim Ağa 728,2, fols. 38r-181r, copied in the 14th century. These merit detailed investigation.

[33] King, "Earliest Qibla Methods", pp. 133-141, 110-112, and especially p. 112.

corresponding table of h_v in MS Cairo DM 190 of al-Lādhiqī's prayer-tables, for example, has entries which decrease to 0;0°. The difference would hardly affect the efficiency of the ventilators, which were not happily oriented perpendicular to winter sunrise anyway.

al-Bakhāniqī (**5.6**) implies that this table was appended to Ibn Yūnus' azimuth tables, but in MS Istanbul Nuruosmaniye 2925 it is stated that the table (now with entry 0;41° at the winter solstice) was computed by Ibn al-Rashīdī (**5.2**). In MSS Dublin CB 3673 and Gotha A 1410, however, the table of h_v (again with entry 0;41°) is appended to sets of azimuth tables uncorrected by Ibn al-Rashīdī, which suggests an earlier compiler, perhaps Ibn Yūnus. Certainly there were ventilators in Cairo-Fusṭāṭ already in the 12th century, if not in the 10th. About the year 1200 the ʿIrāqī scholar ʿAbd al-Laṭīf al-Baghdādī visited Cairo and wrote: "one hardly sees any houses without ventilators". On the significance of the ventilators and their orientation for understanding the layout of the medieval city see **VIIb**.

4.9 Tables for determining the time of the afternoon prayer

The tables in the corpus relating to the afternoon prayer are based on the standard definition that the permitted time for the prayer commences when the gnomon shadow exceeds its midday minimum by the length of the gnomon, and ends at sunset. The most commonly-tabulated functions are:

$$h_a(\lambda),\quad t_a(\lambda)\quad \text{and}\quad T_a(\lambda)\ ,$$

labelled respectively *irtifāʿ awwal waqt al-ʿaṣr*, *al-dāʾir bayna 'l-ẓuhr wa-'l-ʿaṣr* and *mā bayn al-ʿaṣr wa-'l-ghurūb* and representing the solar altitude and the hour-angle at the beginning of the prescribed interval and the time remaining until sunset: see **Fig. 5.4b**. In MS Cairo TR 191 only there are tables of:

$$z_{a(60)}(\lambda)\quad \text{and}\quad z_{a(12)}(\lambda)$$

instead, which measure the shadow at the time for the prayer. Clearly, at least one of these quantities, easily determined from the corresponding shadow at midday, Z, by addition of one gnomon length, would need to be calculated in the process of deriving h_a.

al-Maqsī claims to have computed the table of $t_a(\lambda)$ himself (**5.4**). The entries in the table for $h_a(\lambda)$ in the corpus differ slighly from those in the corresponding tables in the *Muṣṭalaḥ Zīj* and in the auxiliary tables of al-Khaṭāʾī (**6.6** and **6.8**).

4.10 Tables for twilight

The tables of $r(\lambda)$ and $s(\lambda)$, measuring the duration of morning and evening twilight, are labelled respectively *ḥiṣṣat al-fajr wa-huwa min ṭulūʿ al-fajr ilā ṭulūʿ al-shams* and *ḥiṣṣat al-shafaq wa-huwa min al-ghurūb li-maghīb al-shafaq al-aḥmar*. These titles indicate only that morning twilight lasts from daybreak to sunrise and evening twilight from sunset to the disappearance of the red twilight glow. The underlying values of the solar depression are 19° for morning twilight and 17° for evening twilight and the two tables are derived from the tables of $T(h,\lambda)$ in the corpus (**4.3**) using:

$$r(\lambda) = T(19°,\lambda^*)\quad \text{and}\quad s(\lambda) = T(17°,\lambda^*)\ .$$

MS Dublin CB 3673 contains these tables – see **Fig. 4.7a** – and also another pair based on the parameters 20° and 16°, both marked *ḍaʿīf*, "weak", in a later hand, suggesting that someone objected to these parameters.

Ibn Yūnus discussed the determination of twilight in the *Ḥākimī Zīj*[34] but did not present any tables for twilight. An interesting question that cannot yet be answered definitively, is whether he himself computed any tables for twilight, using either the parameter 18° mentioned in the *Ḥākimī Zīj* or the parameters 20° and 16° attributed to him by Najm al-Dīn al-Miṣrī (**2.5**) three centuries after his time. al-Maqsī (**5.4**) did not include any twilight tables in his set but advocated the parameters 20° and 16° in his introduction. Ibn al-Kattānī (**5.5**) included twilight tables based on parameters 19° and 16° in his set and also additional tables for morning twilight based on 20°, which parameter he says was used by recent scholars (presumably he was referring to al-Marrākushī and al-Maqsī), and 18°, which he says was used by earlier scholars (presumably referring to Ibn Yūnus). The parameters 19° and 17° were favoured by most later *muwaqqit*s in Egypt and Syria from the 15th century onwards. Various tables for twilight purporting to be adjusted for refraction at the horizon occur in late Egyptian prayer-tables in the tradition of al-Minūfī (**7.1**).

The table in the corpus displaying the duration of darkness, n(λ), labelled *jawf al-layl*, was compiled using the simple relation:

$$n(\lambda) = 2N(\lambda) - s(\lambda) - r(\lambda) \qquad (h_r = -19°, \ h_s = -17°).$$

As indicated by the title of the table in MS Dublin CB 3673, darkness lasts from the beginning of the prayer at nightfall to the "true dawn" (*wa-huwa mā bayn al-ʿishāʾ al-ākhira wa-ṭulūʿ al-fajr al-ṣādiq*).

The tables of r(λ), s(λ) and n(λ) are the main twilight tables in the corpus, but in various late copies such as MSS Cairo DM 108 and Princeton Yahuda 861,1 (**8.1**), as well as in the prayer-tables of al-Lādhiqī (**7.8**), other tables derived from these are presented either to facilitate timekeeping relative to the times of nightfall and daybreak or to serve the determination of various twilight-related religious institutions. These I now describe.

Firstly, tables of $\alpha_s(\lambda)$, labelled *maṭāliʿ al-shafaq*, "the (oblique) ascensions of (the ascendant at) nightfall", occur in various late Egyptian prayer-tables, notably those of al-Lādhiqī, and are based on the relation:

$$\alpha_s(\lambda) = \alpha_\phi(\lambda_H) = \alpha_\phi(\lambda^*) + s(\lambda) \ ,$$

where H denotes the ascendant at nightfall and the parameter for s is 17°. Tables of $\alpha_r(\lambda)$, labelled *maṭāliʿ al-fajr*, "the (oblique) ascensions of (the ascendant at) daybreak", are less common in late Egyptian prayer-tables, the function $\alpha_\sigma(\lambda)$ (see below) generally being preferred. The underlying relation is:

$$\alpha_r(\lambda) = \alpha_\phi(\lambda_H) = \alpha_\phi(\lambda) - r(\lambda) \ ,$$

where H denotes the ascendant at daybreak and the parameter for r is 19°. I have no clues as to the identity of the individual who first compiled these tables. However, the earliest reference to the two functions in the sources known to me is in the treatise by Najm al-Dīn al-Miṣrī on spherical astronomy (MS Milan Ambrosiana 227a (C49), Ch. 24: see **2.5**), written *ca.* 1325. He prescribes procedures that are equivalent to the above formulae.

[34] See King, *Ibn Yūnus*, III.16.

A group of tables occuring in MS Cairo DM 108 of the corpus, as well as in the prayer-tables of al-Lādhiqī, is associated with two activities called *salām* and *ṭafy*. It was the custom in medieval Egypt to light candles (*qanādīl*) on the minarets of mosques during the nights of Ramaḍān: the term *ṭafy* refers to the extinction of these candles 20 minutes before daybreak, which may have been the signal to the faithful that the day's fasting should begin. (In Egypt today the beginning of the fast is called the *imsāk* and is also 20 minutes before daybreak.) The *salām* was a special call of the muezzin, a few minutes before daybreak, invoking blessings on the Prophet Muḥammad. It is clear that the *muwaqqit*s in medieval Cairo were not of one mind concerning the precise time of the *salām*. From the sources investigated below it appears that the *salām* was variously timed to begin 1°, 1;30° or 2° (that is, 4, 6 or 8 minutes) before daybreak: see **Fig. 4.10** for a table based on a parameter of 1°.

The table in MS Cairo DM 108 and in most copies of al-Lādhiqī's prayer-tables which is entitled "time from sunset to the *salām*, two (equatorial) degrees before daybreak" is based on the relation:

$$\sigma(\lambda) \quad = \quad 2N(\lambda) - r(\lambda) - 2^\circ \qquad\qquad (h_r = -19^\circ)$$

In MS Paris ar. 2553 of al-Lādhiqī's tables the corresponding entries are 1° more, so that the *salām* is assumed to be at 1° before daybreak. In a later anonymous set of this function preserved in MS Cairo DM 57, computed to three digits for each 0;6° of λ by linear interpolation (**7.2**), the entries are based on those in MS Paris ar. 2553. Whilst this means that the *salām* is 1° before daybreak, the title of the tables states that it is 2° before. On the other hand, a note by the title states that the *shaykh* who compiled the tables was in error because the time-difference is actually 1° not 2°.

A related table is found only in al-Lādhiqī's prayer-tables. It is entitled "(oblique) ascensions of (the ascendant at the time of) the *salām*, two (equatorial) degrees before daybreak". The title also states that if one subtracts three degrees from the entries the result will be the "ascensions of the *ṭafy*". An entry is given for each degree of λ in 12 columns of 30 entries. If H is the ascendant at daybreak, then:

$$\alpha_\sigma(\lambda) \quad = \quad \alpha_\phi(\lambda_H) \quad = \quad \alpha_\phi(\lambda) - r(\lambda) - 2^\circ \qquad\qquad (h_r = -19^\circ)\ .$$

In MS Istanbul Reisülküttap Mustafa Efendi 582 the table of α_σ is based on the assumption that the *salām* is 1° before daybreak. In later related tables of this function in MSS Cairo DM 157, datable *ca.* 850, and DM 158, datable *ca.* 1600, values are given for each 0;3° or 0;6° of λ, respectively, generated by linear interpolation. In these two sources the entries are larger than those in the Hartford copy by 0;30° and 1°, respectively.

Tables of the function τ(λ), *al-dā'ir li-'l-ṭafy fī Ramaḍān ʿalā anna baynahu wa-bayn al-fajr hā' daraj*, "time from sunset to the *ṭafy* in Ramaḍān, 5° before daybreak", occur only in MSS Princeton Yahuda 861,1 and Cairo DM 108 amongst the sources investigated. In the former the function tabulated is:

$$\tau(\lambda) \quad = \quad 2N(\lambda) - r(\lambda) - 5^\circ \qquad\qquad (h_r = -19^\circ)\ ,$$

as one might expect. In the latter the function tabulated is in fact:

$$\tau'(\lambda) \quad = \quad 2N(\lambda) - r(\lambda) - 2\Delta D(\lambda) \qquad\qquad (h_r = -19^\circ)$$

where the values of ΔD, a correction for refraction at the horizon, are those attributed elsewhere to Ibn Yūnus (**4.11**). Note that this formula includes a correction for both horizons. This table

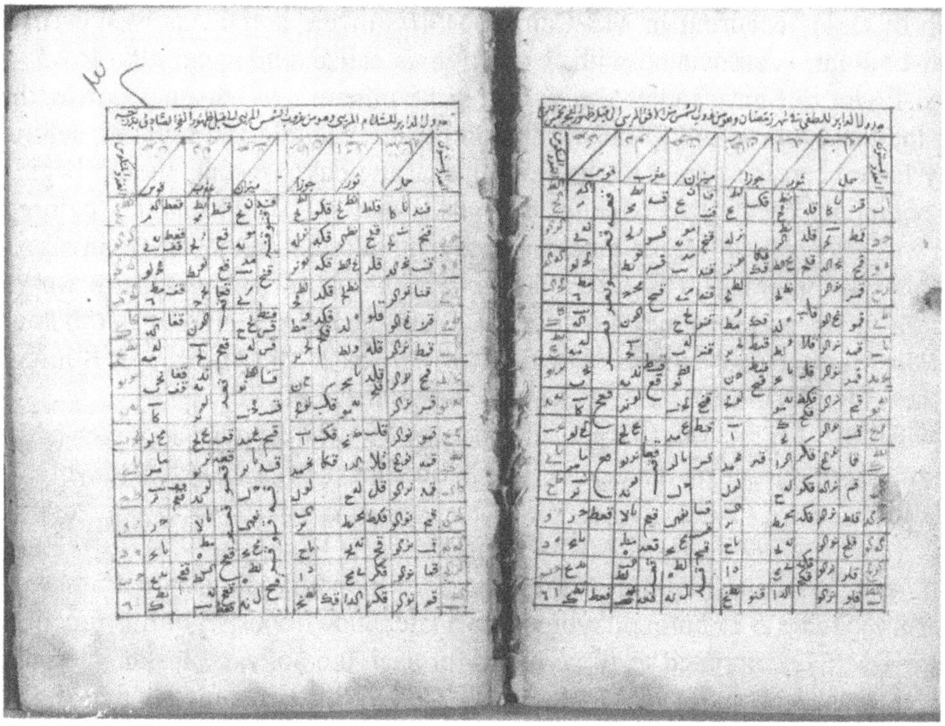

Fig. 4.10: On the right is a table displaying the time from sunset until the moment for extinguishing the candles in the lanterns on the minarets (*al-ṭafy*) in Ramaḍān, taken as 5°, that is, 20 minutes, before daybreak. On the left is a table displaying the corresponding time to the benediction on the Prophet (*al-salām*), taken as 1°, that is, 4 minutes, before daybreak. The entries in the second table are, of course, 4° more than the corresponding entries in the first. Both tables are adjusted using the "difference minutes" for the time of sunset over the visible (*marīʿ*) horizon. [From an unidentified Cairo manuscript of the Cairo corpus, courtesy of the Egyptian National Library.]

is the only one in MS Cairo DM 108 which is adjusted, silently at that, for refraction at the horizon.[35]

Tables of the function $\alpha_\tau(\lambda)$, labelled *maṭāliʿ al-ṭafy*, "the (oblique) ascensions (of the ascendant at the time) of the *ṭafy*", occur in MS Cairo DM 153 and Princeton Yahuda 861,1 of the sources investigated, as well as in MS Istanbul Reisülküttap Mustafa Efendi 582 of al-Lādhiqī's prayer-tables. Now the table of α_τ in MS Princeton Yahuda 861,1 is not based on the table of $\tau(\lambda)$ in MS Cairo DM 108. Rather, the tabulated function is:

$$\alpha_\tau(\lambda) = \alpha_\phi(\lambda) - r(\lambda) - 5° \qquad (h_r = -19°)\ ,$$

as one would expect. The table in this Istanbul manuscript is in fact based on:

$$\alpha_\tau(\lambda) = \alpha_\phi(\lambda) - r(\lambda) - 5° - \Delta D(\lambda) \qquad (h_r = -19)\ ,$$

[35] In "Astronomical Timekeeping in Medieval Cairo", p. 371, I noted that the entries would correspond to the relation one would expect (namely, that used in the Princeton manuscript) only if the value used for h_r was about -20;15°. The structure of this anomalous table is now explained.

using the values of ΔD attributed elsewhere to Ibn Yūnus. This formula includes a correction ΔD only for the western horizon, but there is no indication of any correction at all in the title of the table. Furthermore, the accompanying table of $\alpha_\sigma(\lambda)$ in the same manuscript is not corrected for horizontal refraction.

I have not found any references to functions relating to the *salām* and *ṭafy* in the Egyptian astronomical literature. In MS Princeton Yahuda 861,1 (**8.1** and **9.5**) there is a table giving the time from sunset to the *salām* for Jerusalem, but the underlying time is daybreak rather than a few minutes before. Shihāb al-Dīn al-Ḥalabī, writing in Damascus, stated that daybreak (*al-fajr*) is the time of the prayer-call (*al-ādhān*) in Ramaḍān and of the *salām* (written *s-l-m*) in the other months of the year (**11.2**). In MS Damascus Ẓāhiriyya 7564 (**11.11**) there are some cryptic notes about the time for the *salām* as practiced in Aleppo.

4.11 A table of corrections for refraction at the horizon

In MS Cairo DM 108 of the corpus, dating from the early 19th century, there is a table entitled *daqā'iq al-ikhtilāf*, literally, "minutes of difference", here denoted by $\Delta D(\lambda)$: see **Fig. 4.11a**. The function is tabulated to two sexagesimal digits for each degree of λ. In some manuscripts of al-Lādhiqī's prayer-tables (**7.8**), such as MS Cairo DM 190, values of the same function to one digit are given for each zodiacal sign, along with the remark that it measures the time in minutes (*daqā'iq*) of an equatorial degree taken by the sun to move between the true horizon (*al-ufq al-ḥaqīqī*) and the visible horizon (*al-ufq al-mar'ī*).

This table in MS Cairo DM 108 also occurs in the prayer-tables of the early-16th-century astronomer Muḥammad al-Minūfī (MS Cairo DM 107, but not MS Cairo DM 177). al-Minūfī modified each of the tables associated with horizon phenomena in the prayer-tables of the Cairo corpus to account for the effect of refraction at the horizon (**7.1**). The same table occurs in MS Alexandria 4441J of the prayer-tables of the 18th-century *muwaqqit* Ibn Abī Rāya (**7.9**). Although al-Minūfī's table of $\Delta D(\lambda)$ is the earliest table of the function in any of the known sources, the underlying theory, at least that part of it which is in reasonable conformity with physical reality, appears, as I shall now attempt to show, to date from the 10th century.

The entries in the tables of $\Delta D(\lambda)$ in MSS Cairo DM 108 and DM 107, and Alexandria 4441J were calculated by linear interpolation between the values at the equinoxes and the solstices. The significant entries, in minutes of an equatorial degree, are the following:

EQ: 47 SS: 62 WS: 32 .

The attribution of these values to Ibn Yūnus is claimed by al-Minūfī. In the introduction to his prayer-tables (*e.g.*, MS Cairo DM 107, fol. 7v), he states that Ibn Yūnus had found the difference minutes to be 47 at the equinoxes, increasing by 5 for each sign up to the summer solstice and decreasing by 5 for each sign to the winter solstice. We may presume that al-Minūfī had access to a work by Ibn Yūnus that is no longer extant, for none of his known works contains any mention of refraction. In an earlier, 15th-century Egyptian source, these same values are associated with the 14th-century Damascus astronomer Ibn al-Shāṭir (**5.7**), but this association is less likely.

Now only part of the theory described by al-Minūfī and reflected in the tables makes sense.

Fig. 4.11a: A table for the time taken by the sun to move between the true and visible horizons. [From MS Cairo DM 108, fol. 74v, courtesy of the Egyptian National Library.]

It seems reasonable to assume that Ibn Yūnus would have considered the visible horizon to be at a certain arc of depression below the true horizon. If this is so, the above equinoctial and solstitial values are easily shown to be mutually inconsistent. On the other hand, if the three values are the results of observations made at the equinoxes and solstices, then only the equinoctial value is reasonable. But it seems much more likely that the solstitial values were simply computed from the equinoctial value, which was possibly derived by observation. In any case, it is worthwhile to investigate the three values further.

Consider **Fig. 4.11b**, which represents a portion of the western sky about the west-point in plane cross-section. The sun sets over the horizon at an angle $\bar{\phi}$. AB represents the true horizon and A′B′ the visible horizon at an arc of depression Δh below the true horizon. The problem is to find to a first approximation how long the sun takes to move between these two horizons. XX′ represents the path of the centre of the sun at the equinoxes and YY′ and ZZ′ its path at the summer and winter solstices, respectively. Clearly:

$$XX' = YY' = ZZ' = \Delta h \sec \phi \ .$$

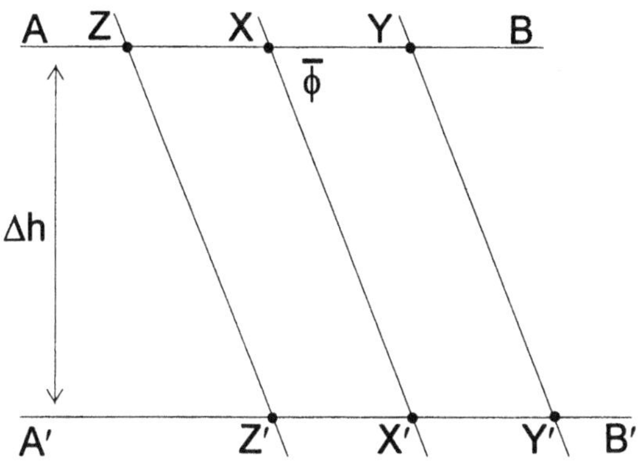

Fig. 4.11b: A plane projection of the visible and true horizons (AB and A′B′), and the solar day-circles at the equinoxes (XX′), the summer solstice (YY′) and the winter solstice (ZZ′). Not to scale.

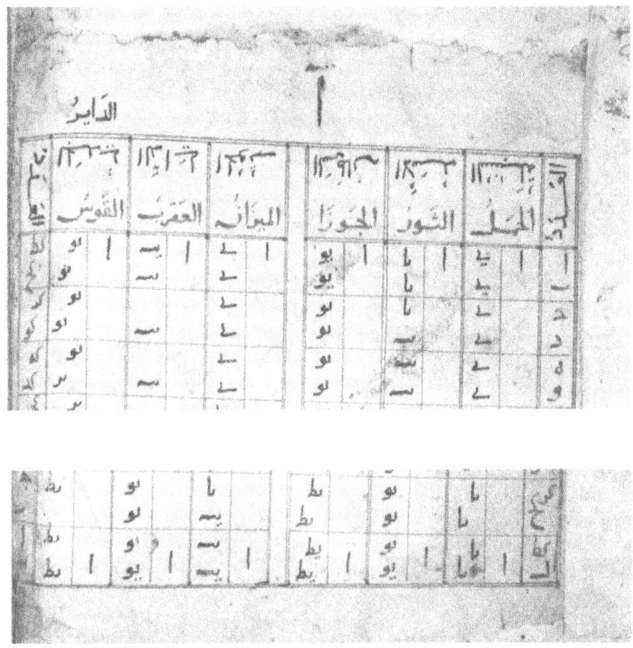

Fig. 4.11c: The table of the time since sunrise for altitude 1° from the corpus. [From MS Dublin CB 3673, courtesy of the Chester Beatty Library.]

Now the time taken by the sun to move from X to X′ is measured by the corresponding arc of the celestial equator. We assume, to a first approximation, that the difference minutes at the equinoxes, ΔD_0, equal $\Delta h \sec \phi$. Since Ibn Yūnus, according to al-Minūfī, took these as 0;47° for latitude 30°, we have:

$$\Delta D_0 \;=\; \Delta h \sec 30° \;=\; 0;47° \,,$$

whence:

$$\Delta h \;\approx\; 0;40° \;=\; {}^2/_3° \,.$$

This, then, is an approximate value of the vertical displacement of the horizon due to refraction corresponding to the value for ΔD_0 attributed to Ibn Yūnus. Note, however, that the value 0;41° is used in the table of h_v (**4.8**) to represent the apparent altitude of the sun when its centre is on the true horizon. The modern value for this displacement is 0°35′.[36]

The solstitial values attributed to Ibn Yūnus are, in the light of modern knowledge, less reasonable. A distance of 0;47° on either of the solstitial day-circles corresponds approximately to an arc of:

$$0;47° \sec \varepsilon \;=\; 0;51°$$

on the celestial equator and it is clear that *the time taken by the sun to move between the two horizons is the same at both solstices*. This time is some 0;4° more than the corresponding time at the equinoxes, according to the above calculation. Note too that, if we assume a value

[36] For a modern discussion see Smart, *Spherical Astronomy* (n. 1:37), pp. 69-70, *etc.*

of 0;40° for Δh, the corresponding times ΔD can be found from tables of $T(h,\lambda)$ in the same way as the duration of twilight (**4.10**). Thus:

$$\Delta D(\lambda) = T(^2/_3{}^\circ,\lambda^*) \approx {}^2/_3\ T(1^\circ,\lambda^*) .$$

From the table of $T(h,\lambda)$ for 1° (see **Fig. 4.11c**) we can read off the time taken by the sun to move 1° below the horizon, namely:

EQ: 70 SS/WS: 79 ,

from which it follows that the time taken by the sun to move $^2/_3{}^\circ$ below the horizon is approximately:

EQ: 47 SS/WS: 53 .

Since the variation in ΔD is relatively small, it is quite reasonable to interpolate linearly between the equinoctial and solstitial values of ΔD for intermediate solar longitudes.

Ibn Yūnus was a master of spherical astronomy, and I deem it unlikely that he would make such an elementary error as to give radically different values of ΔD at the solstices. Rather, I suspect that he originally stated that *the correction was 47 minutes at the equinoxes and increased about 5 minutes to maxima at the solstices*, which would be virtually correct. The assertion of al-Minūfī that Ibn Yūnus had stated that *the correction was 47 minutes at the equinoxes and changed by 5 minutes for each sign with a maximum at the summer solstice and a minimum at the winter solstice* is then simply garbled. My hypothesis has no substantiation in the manuscripts examined thus far and my investigations of the various Arabic manuscript sources dealing with refraction at the horizon contribute little to our understanding of the problem as treated by Ibn Yūnus and (mis)construed by his successors.

The theory of refraction at the horizon associated with Ibn Yūnus as interpreted by al-Minūfī was adopted by the latter's son ʿAbd al-Qādir al-Minūfī to serve all latitudes (**8.2**), using the approximation:

$$\Delta D(\phi,\lambda) \approx \phi / 30^\circ \cdot \Delta D(30^\circ,\lambda) .$$

Four sources have been identified in which values based on this formula are presented, namely, for Mecca, Crete, Damascus and Istanbul (**6.10**, **8.8**, **11.13** and **14.9**). No Muslim astronomer is known to have noted that the underlying equinoctial and solstitial values were mutually inconsistent.

4.12 The titles and instructions

The "instructions" accompanying the table in MS Dublin CB 3673 and the slightly modified version in MS Cairo DM 108 contain virtually no information about the way in which the tables were intended to be used in practice, nor do they reveal how the tables were computed. In the Dublin copy the title *Kitāb Ghāyat al-intifāʿ* ... , which may be (and has been) rendered "Very Useful Tables ... ", is given to the whole corpus. Also al-Kattānī called his tables *Kitāb Nihāyat al-intifāʿ* ... , which may be rendered "Extremely Useful Tables ... ". Otherwise the various copies of the tables of $T(h,\lambda)$, $t(h,\lambda)$ and $a(h,\lambda)$ bear the titles *Kitab al-Dāʾir*, *Kitāb Faḍl al-dāʾir* and *Kitāb al-Samt*, or various combinations, or they bear no titles at all.

The introduction begins with pious statements about God as the Creator of the Universe. The Qurʾānic verse X.5/5:

> "(God) made the sun as a shining light and made the moon a luminary and measured out its mansions"

is continued thus (fol. 83r:5):

> "for perfecting prognostications about what He orders and what He effects."

Then the verse XXXVII.6/5:

> "(God) adorned the lower heavens with the stars"

is continued (MS CB 3673, fol. 83r:5-6):

> "and the times are known by what rises and sets."

Since, states the author, astronomical timekeeping (*ʿilm al-mawāqīt*) was ordained by God, he thought he would prepare these table to be the "limit of usefulness" (*ghāyat al-intifāʿ*). The instructions proper are translated in full below, using MSS fols. 83r-83v of the Dublin manuscript and fols. 2v-3r of the Cairo one. The paragraph numbers are my own. §9 does not occur in copies of the introduction that are not followed by the azimuth tables. §10a appears to be original to al-Maqsī's introduction, whereas §10b, whose highly sophisticated contents merit special commentary, was probably inserted by someone more competent that al-Maqsī. The instructions read:

> 1 To determine the time since sunrise up to any instant you wish, enter the solar longitude in the table of the instantaneous altitude and you will find the required time, if the solar altitude is in the east. If it is in west, subtract the value in the table from the diurnal arc and the remainder will be the time up to that instant. God – may He be exalted – knows best. Example: Suppose the sun is in 21° of Taurus and the instantaneous altitude is 30°. We find 35;38°[a] in the table: this is the time from sunrise to the instant in question. If the altitude is in the west, subtract 35;38° from the diurnal arc, which is 201;46°[b]: the remainder is 166;8°, which is the required time.
>
> 2 To determine how much of the day has passed and how many hours of daylight are left, divide the time since sunrise by the number of degrees corresponding to a seasonal day-hour for the day in question, or by 15 if you want equinoctial hours. The result will be the number of hours of daylight that have passed. Subtract the seasonal hours from 12 and the equinoctial hours from the number of hours of daylight in question: the remainder will be the number of hours of daylight left. Example: Suppose the sun is in 11° of Aries and the instantaneous altitude is 45° in the east. We find the time since sunrise for this altitude as before: it is 53;19°[c]. We divide this by the number of degrees corresponding to a seasonal day-hour, which is 15;24°[d], and the result is [3][e];27. These are the hours of daylight that have passed; subtract them from 12 and the remainder, which is the number of hours of daylight left, is 8;33. If we divide the time since sunrise by 12, the quotient is [3;33][f] and this is the number of equinoctial hours that have elapsed. We subtract these from the number of equinoctial hours of daylight on the day in question, which is 12;20, and the remainder is 8;47. This is the number of hours of daylight remaining. *I have included a table at the end of the book in which the entry corresponding to the solar longitude is multiplied by the time since sunrise to give the seasonal day-hours passed since sunrise.*
>
> 3 To determine (the altitude of the sun for) the seasonal and equinoctial hours, enter the solar longitude in the tables of the time since sunrise for each altitude and turn

through table after table until you find a time as large as the degrees corresponding to the seasonal or equinoctial hours. Interpolate if necessary and the result with be the required altitude. Example: Suppose the sun is in the last degree of the sign of Gemini and we want the altitude at the end of the first seasonal hour. We enter the solar longitude in the tables, turning through table after table, and find opposite the longitude in the table for (altitude argument) 13° a time of 16;24°[g]. We subtract this from the number of degrees in one seasonal hour, which is 17;26°[h], and the remainder is 1;2°. We divide this by the difference between the time [in the tables for altitude 13° and for altitude 14°, which difference is 1;15°. The ratio of one to the other is 0;50, which we add to the first altitude][i] and so the required altitude is 13;50°. If we want the altitude after one equatorial hour, we enter the solar longitude in the tables and we find opposite it 13;56°. We subtract this from 15° and the remainder is 1;4°, which we divide by the difference between the time for altitude 11° and that for 12°, which is 1;15°. The ratio of the one to the other is 0;51, which we add to the first altitude, obtaining 11;51° for the required altitude.

4 To determine the ascendant at the eastern horizon, add the time since sunrise to the oblique ascensions of the solar longitude and the result will be (the oblique ascensions of) the ascendant at that time. Example: Suppose the sun is in 18° of Scorpio and the instantaneous altitude is 25°. We take the time since sunrise for this altitude and find it to be 33;20°[j]. We add this to the oblique ascensions of the sun, namely, 235;52°[k], and the sum is 269;12°. We take the longitude corresponding to this: the result is 16;18°, which is the degree of Sagittarius that is rising.

5 To determine the hour-angle at the beginning of the afternoon prayer, you should know that *there is a table for this included at the end of the book.* You enter the solar longitude as argument and read off the required value.

6 *To determine the semi diurnal arc, you enter the solar longitude in the appropriate table* and read the required value.

7 To determine the duration of morning twilight, which is the remainder of the nocturnal arc to sunrise, you enter the longitude directly opposite the solar longitude on the ecliptic in the table of the time since sunrise for altitude 19° and the value you find will be the required time.

8 To determine the time from sunrise to the moment when the sun is in the azimuth of the qibla, enter the solar longitude in the appropriate table and read off the required value.

9 To determine the azimuth, proceed as with the tables of time. The direction of the azimuth is south if the sun is in the southern zodiacal signs, or if it is the northern signs and the altitude is greater than the altitude in the prime vertical; otherwise, it is north. God knows best.

10a Numerous interesting facts about the operations of timekeeping by day and night will be revealed to anyone who examines these tables closely. However, in this book we have chosen to keep the discussion short.

10b Certain interesting facts will be revealed to anyone who examines these tables closely. For example, if the solar altitude equals the declination, then both the azimuth and the time since sunrise are equal to the complement of the other, Further, if you take the declination as altitude and you know the complement of the time since sunrise for the solar longitude, then, if the latitude is 30°, it will be the half excess of daylight. Numerous other facts and points of interest relating to night and day also become evident. God – may He be exalted – knows best."

[a] As on fol. 54v. [b] As on fol. 12r. [c] As on fol. 39v. [d] Fol. 13r has the more accurate value 15;25°. [e] Text: 4 (*sic*). [f] Text: 4;13 (*sic*). [g] As on fol. 71v. [h] As on fol. 13r. [i] [ ] missing in the MS. [j] As on fol. 59v. [k] This is accurate, and is the value given in the *Ḥākimī Zīj*. MS Hartford TS 621, p. 21, has 235;51°.

These instructions (with the exception, perhaps, of §10b) were written by al-Maqsī, not by Ibn Yūnus as I stated in my first study of the Cairo corpus.[37] They were intended to precede his tables of T(h,λ) discussed in **5.4**. This explains why there is no mention here of the hour-angle t(h,λ) and most of the minor functions tabulated in the corpus. Nevertheless the style differs markedly from that of al-Maqsī's introduction to his tables for sundial construction.[38]

In the new source MS Gotha A1402 the instructions are in al-Maqsī's name and are followed by his tables of T(h,λ). In MS Cairo DM 53 they also bear al-Maqsī's name and are followed by al-Bakhāniqī's arrangement of the tables of (T,t,a). In MS Cairo DM 108 the introduction ends with al-Maqsī's name. However, in MS Dublin CB 3673, which was the major source of my first study of the Cairo corpus, the instructions are introduced in the name of Ibn Yūnus. It is now apparent that the earlier scholar's name was inserted by the copyist: it is preceded by a badly-written phrase with Arabic letters improperly pointed, which I originally read as *anwaʾ al-nās* and, with considerable stretching of the imagination, interpreted as meaning something like "most proficient of astrologers". However, the phrase must be a corruption of Abu 'l-ʿAbbās, the *kunya* of al-Maqsī. The copyist of the Chester Beatty manuscript was of the opinion that Ibn Yūnus deserved more credit that al-Maqsī for the tables which he was copying: that copy does indeed contain more tables by Ibn Yūnus than any other known source. He omitted al-Maqsī's name and inserted that of Ibn Yūnus, not realizing, perhaps, that Abu 'l-ʿAbbās was al-Maqsī's *kunya* rather than that of Ibn Yūnus. This copyist also included the remarks of §10b above, which imply that the author had before him tables of *both* a(h,λ) and T(h,λ). I do not think that either al-Maqsī or al-Bakhāniqī would have noticed these subtleties. Rather, they are reminiscent of Ibn Yūnus' remark in Ch. 15 of the *Ḥākimī Zīj* (MS Leiden Or. 143, p. 335:17-20):[39]

> "If for a particular locality there are no (tables of) oblique ascensions available but there is a table of the azimuth computed for the first point of Aries giving values for each degree of solar altitude up to 24°, then enter the solar declination as argument in this table and the corresponding value in the table will be the half excess of daylight."

In al-Maqsī's introduction the parameters advocated for morning and evening twilight were probably originally 20° and 16°. This is the case in MSS Cairo DM 108 and Gotha A1402,

[37] King, "Astronomical Timekeeping in Medieval Cairo", pp. 376-377.
[38] See n. 4:6.
[39] King, *Ibn Yūnus*, III.15.1(b).

but in MS Dublin CB 3673 only the value for morning twilight is mentioned, and in MS Cairo DM 53 the parameters 19° and 17° are proposed. On the other hand, MS Dublin CB 3673 is the only copy of the corpus containing twilight tables based on parameters 20° and 16° (although tables based on the more common parameters 19° and 17° are also included in this manuscript). See further **4.10**.

In the introduction to the tables in the Chester Beatty manuscript separate tables of:

$$\tilde{h}(\lambda),\ t_a(\lambda),\ D(\lambda) \text{ and } T_q(\lambda)$$

are specifically mentioned. In the introduction in MS Cairo DM 108 only separate tables of:

$$\tilde{h}(\lambda),\ D(\lambda) \text{ and } T_q(\lambda)$$

are specifically mentioned. In both sources there is mention of the tables of a(h,λ), but this seems to be a later addition to al-Maqsī's original introduction. In **5.4** I discuss al-Maqsī's introduction and tables further.

CHAPTER 5

THE DEVELOPMENT OF THE MAIN CAIRO CORPUS

5.0 Introductory remarks

It is not yet possible to give a completely clear picture of the development of the main Cairo corpus described in **Ch. 4**. New manuscripts of tables from the corpus could add considerably to our knowledge and it is not too much to hope that some will turn up in future investigations of uncataloged manuscript collections.

5.1 The tables for spherical astronomy in the *Ḥākimī Zīj*

The celebrated Fatimid astronomer Ibn Yūnus (**I-2.1.1** and **II-4.1.1**) tabulated numerous spherical astronomical functions in his *Ḥākimī Zīj*.[1] The title-folio of the Leiden manuscript is shown in **Fig. 5.1**. It is written in the hand of Ibn Abi 'l-Fatḥ al-Ṣūfī, and bears a notice of ownership by Taqi 'l-Dīn ibn Maʿrūf, indicating that it once formed part of his library at the short-lived observatory in Istanbul.[2] No other *zīj* contains so many tables of this kind; Ibn Yūnus thus began a trend which was to distinguish the later traditions of astronomical timekeeping in Egypt and Syria. The tables in the *Zīj* which have values computed to minutes, seconds or thirds [here indicated by [1], [2] and [3]) for each 1° of λ (or h, where relevant), occasionally for each 0;10° (here indicated by *). The underlying parameters, where relevant, are:

$$\phi = 30;0° \quad \text{and} \quad \varepsilon = 23;35° .$$

The tables display the following functions:

δ(λ)* [2]; α(λ) [2]; α′(λ) [1]; Sin α(λ) [3]; Tan δ(λ) [3];
$\alpha_\phi(\lambda)$ [1] (for each 1° of φ from 1° to 48°);
d(λ) [3] and Sin d(λ) [3];
ψ(λ) [2] and Sin ψ(λ) [2]; $h_0(\lambda)$ [2]; k(h) [2];
a(λ) [1] for two different altitudes; h(λ) [1] for ten different azimuths;
and $h_q(\lambda)$ [1] (q = 52°).

Of particular importance for the development of the main Cairo corpus was the inclusion of tables of a(h,λ) for two values of h (**I-5.1.1**), and h(h,λ) for ten values of a (**I-4.7.1**); such tables are not known from earlier sources. The tables of:

---

[1] On Ibn Yūnus see n. **I**-2:3. King, *Ibn Yūnus*, deals specifically with the spherical astronomy in the *Ḥākimī Zīj*.

[2] See also King, "Astronomy in Fatimid Egypt", p. 499, and also Ünver, *Istanbul Observatory* (in Turkish), figs. 41-42, for the marks of ownership on other manuscripts from Taqi 'l-Dīn's library.

Fig. 5.1: The title folio of the *Ḥākimī Zīj*, containing biographical information on the author in the hand of the late-15th-century Cairene astronomer Ibn Abi 'l-Fatḥ al-Ṣūfī (**I-9.10**). On the right is a notice of possession of the Istanbul astronomer Taqi 'l-Dīn (see **Fig. I-1.0a**). [From MS Leiden Or. 143, fol. 1r, courtesy of the Universiteitsbibliotheek.]

$$\psi(\lambda),\ h_0(\lambda);\ k(h),\ \text{and}\ h_q(\lambda)$$

are also the earliest known tables of their kind. Also significant is the inclusion of the various auxiliary functions, notably:

Sin $\alpha(\lambda)$ (**I-7.2**); Tan $\delta(\lambda)$ (**I-7.1.3**); Sin $d(\lambda)$;
$k(h)$ (**I-8.1.1**); and Sin $\psi(\lambda)$ (**I-8.3.1**) .

There are also tables of Sin $\psi(\lambda)$ [3] and k(h) [2] for $\phi = 33;25°$, serving Baghdad and perhaps an indication of the source of part of his inspiration.

5.2 Ibn Yūnus' solar azimuth tables

Ibn Yūnus compiled a work called *Kitāb al-Samt* containing a complete set of tables of $a(h,\lambda)$. Subsequently the entries for $h \geq 80°$ were lost and they were recomputed by an anonymous incompetent who included tables for $80° \leq h \leq 89°$ even though the maximum solar altitude at Cairo is only about $83^1/_2°$. In the 14th century the tables for $80° \leq h \leq 83°$ were recomputed more carefully by Ibn al-Rashīdī (**4.4**). al-Bakhāniqī used the better set for his edition of the corpus (**5.6**), extant in MSS Cairo DM 108, Cairo DM 53, Istanbul Nuruosmaniye 2903 and 2925, Cairo DM 690 and Cairo DM 616, but the uncorrected tables also survive in MSS Dublin CB 3673, Cairo MM 137 and Gotha A 1410 of the *Kitāb al-Samt*. In these three sources, entries are given up to 88°, 86° and 89°, respectively: see **Fig. I-5.1.1a**. The corrected tables of the *Kitāb al-Samt* were also copied separately, as in MS Cairo Azhar *falak* 4382.

Some copies of the *Kitāb al-Samt* have tables of h_v, h_q (q = 52° or 53°), h_{q*} (q = 53°) and a(h) ($\lambda' = 0°$ and 90°) appended, as well as ψ and h_0. MS Dublin CB 3673 contains a set of uncorrected tables of $a(h,\lambda)$ as well as of h_v, h_q (q = 52°), ψ and h_0. MS Gotha A 1410 contains the uncorrected azimuth tables and also tables of h_q (q = 53°) and h_v. MS Cairo Azhar *falak* 4382 contains the azimuth tables and a(h) ($\lambda' = 0°$ and 90°). MSS Berlin Ahlwardt 5753 and Escorial ár. 924,7 of the azimuth tables are incomplete. The tables of h_q (q = 52°), and ψ and h_0 (both to three digits), were taken from the *Ḥākimī Zīj*. The rather carelessly computed table of h_v usually has the entry 0;41° for the winter solstice rather than zero, to correct for refraction at the horizon. The remark in MS Istanbul Nuruosmaniye 2925 that the table of h_v (which has the entry 0;41°) is due to Ibn al-Rashīdī may mean either that he computed the table or that he introduced the modification for the winter solstice. If I interpret al-Bakhāniqī correctly as asserting that the table of h_{q*} (q = 53°) was due to Ibn Yūnus, then it is reasonable to suppose that Ibn Yūnus also computed h_q (q = 53°), although these tables for q = 53 ° are less accurate than his table for q = 52°.

5.3 A lost work of Ibn Yūnus?

I conjecture that Ibn Yūnus also compiled a small set of prayer-tables for the latitude of Cairo-Fusṭāṭ, rather like those for Mecca preserved in MS Cairo MM 68 (**6.10**), and containing tables of such functions as:

$$H, D, h_a, t_a, r \text{ and } s \,.$$

Perhaps it was in such a work that he proposed the equinoctial and solstitial values for the correction ΔD due to refraction at the horizon. MS Cairo MM 68 contains a set of these corrections for Mecca.

al-Bakhāniqī (**5.6**) appears to state that at the end of a copy of Ibn Yūnus' azimuth tables used by him there had been added (*uḍyafa*) tables of the functions:

$$t_q,\ H,\ h_a,\ z_a,\ t_a,\ T_a,\ \tilde{h},\ r \text{ and } s\ .$$

These might have been Ibn Yūnus' prayer-tables and I should expect that the table of t_q would be based on the parameter q = 53° and that the parameters underlying the tables of r and s would be 20° and 16°.

The use of the Arabic passive *uḍyafa*, "it was added," is a little curious in this context, and the verb should be feminine, *i.e. uḍyafat*, since it refers to the noun *jadāwil*, "tables". Note that the same Arabic word written *'ḍyf* can also be read *uḍīfu*, "I am adding", which does not fit syntactically, but the word could be a corruption of *aḍiftu*, "I added", which makes more grammatical sense. This would mean that it was it was al-Bakhāniqī who compiled the prayer-tables, which seems very unlikely. He lived in the mid-14th century when prayer-tables for Damascus had already been compiled by individuals such as al-Mizzī (**9.2**), apparently using the model of the Egyptian tables.

As noted above (**2.5**), Najm al-Dīn al-Miṣrī attributes the twilight parameters 20° and 16° to Ibn Yūnus. He is clearly quoting a work compiled after the *Ḥākimī Zīj*, since there the parameter 18° is given for both morning and evening. Likewise ʿAbd al-Qādir al-Minūfī (**8.2**) states that Ibn Yūnus presented his values for the effect of refraction at the horizon in a work "other than his *Zīj*". This suggests that Ibn Yūnus compiled either a treatise in which these topics were discussed, or, more likely, a small set of tables of r(λ) and s(λ) based on parameters 20° and 16 ° and instructions on how to correct these for the effect of refraction at the horizon.

I have located three different tables of the function $h_a(\lambda)$ for Cairo (*cf.* **4.9**) and consider it likely that Ibn Yūnus compiled one of them. I also think the table of $t_a(\lambda)$ in the corpus is due to Ibn Yūnus. The supposition that Ibn Yūnus tabulated r(λ) and s(λ) explains why al-Maqsī merely indicated the way in which these functions could be derived from the tables of T(h,λ). I suspect that al-Maqsī's table of $T_q(\lambda)$ (q = 53°) is based on a table of $t_q(\lambda)$ computed by Ibn Yūnus.

In the *Ḥākimī Zīj* Ibn Yūnus presents several methods for finding T and t, of which one is suitable for tabulating T(h,λ). He prescribes:

$$T(h,\lambda) \ = \ d(\lambda) + \text{arc Sin} \{ \text{Sin } h \cdot G(\lambda) - \text{Sin } d(\lambda) \}\ ,$$

where:

$$G(\lambda) \ = \ R^2 / \{ \text{Cos } \delta(\lambda) \text{ Cos } \phi \}\ .$$

Note also that:

$$t(h,\lambda) = \text{arc Cos} \{ \text{Sin } h \cdot G(\lambda) - \text{Sin } d(\lambda) \}\ ,$$

although this is not explicitly stated by Ibn Yūnus. However, he did carefully tabulate both d(λ) and Sin d(λ). He also discussed the important function G(λ) at great length, but did not tabulate it in the *Ḥākimī Zīj*. A table of:

$$G'(\lambda) \ = \ R / \text{Cos } \delta(\lambda)$$

occurs in MS Paris BNF ar. 2513, fol. 62v, of the 13th-century Egyptian *Muṣṭalaḥ Zīj* (**6.6**) and the entries are badly garbled, which suggests successive careless copying from an earlier source. A simpler formula for tabulating t(h,λ) not explicitly stated in the *Ḥākimī Zīj* is:

$$\text{Cos } t(h,\lambda) \ = \ \text{Sin } h \cdot G(\lambda) - \text{Sin } d(\lambda)\ .$$

This follows immediately from another formula that Ibn Yūnus does use,[3] namely:

[3] King, *Ibn Yūnus*, III.15.3c and e.

$$\text{Vers } t(h,\lambda) = \text{Vers } D(\lambda) - \text{Sin } h \cdot G(\lambda) .$$

Again $t(h,\lambda)$ can be found from $a(h,\lambda)$ by using another formula expounded by Ibn Yūnus,[4] namely:

$$t = \text{arc Sin } \{ \text{Cos } h \cdot \text{Cos } a \, / \, \text{Cos } \delta \} .$$

It seems very unlikely that Ibn Yūnus would have used this indirect method of finding $t(h,\lambda)$, even though he had tabulated $a(h,\lambda)$ and possibly also the reciprocal of Cos $\delta(\lambda)$. al-Bakhāniqī states that he added to the corpus a table called *al-dā'ir al-āfāqī*, of which a reasonable translation would be "table for finding time for all latitudes", but he says that the table was for finding *the solar longitude* by a method of Ibn Yūnus. Suppose the illustrious astronomer had compiled a set of tables of the function:

$$p(h,\lambda) = \text{Sin } h \cdot G(\lambda).$$

To do this it would be reasonable to first tabulate:

$$n \cdot G(\lambda) , \qquad n = 1, 2, \ldots , R \ (= 60)$$

so that if Sin h = x;y then:

$$p(h,\lambda) = x \cdot G(\lambda) + y \cdot G(\lambda) \, / \, R ,$$

and the multiplication is transformed into an addition.[5]

Again, Najm al-Dīn's enormous table of T(H,h,D) for all latitudes was compiled in Cairo not long after al-Maqsī's table of $T(h,\lambda)$. No *ṭaylasān* tables of T(H,h) are known to have been compiled specifically for the latitude of Cairo, but it is appropriate to mention that Ibn Yūnus had access to the *Zīj*es of the Banū Amājūr, one of which contained such a table for Baghdad (**3.2**). My current hypothesis is that Ibn Yūnus compiled a set of tables for timekeeping which inspired (a) Najm al-Dīn al-Miṣrī to compile his universal table for timekeeping; (b) al-Maqsī to compile his table of $T(h,\lambda)$ for Cairo; and, in addition, (c) Abu 'l-ʿUqūl to compile his tables of $T(h,\lambda)$ for Taiz.

5.4 al-Maqsī's tables of time since sunrise

The tables for timekeeping called *Kitāb al-Dā'ir* compiled by al-Maqsī (**4.5**) consisted of a complete set of tables of $T(h,\lambda)$ and four additional tables of:

$$D(\lambda),\ t_a(\lambda),\ T_q(\lambda)\ (q = 53°) \text{ and } 1/\bar{h}(\lambda) .$$

The tables were preceded by an introduction in which each of these functions was specifically mentioned (**4.12**). MS Gotha A1402 (**4.1**) is the closest manuscript to al-Maqsī's original set. It contains his introduction and a complete set of tables of $T(h,\lambda)$ followed by the four smaller tables in the order noted above. The parameters advocated for twilight in the introduction are 20° and 16°. The other tables which follow the one of $1/\bar{h}(\lambda)$ are to be regarded as later additions to al-Maqsī's set. In MS Berlin Ahlwardt 5753 (**4.1**) there are tables of $t_a(\lambda)$ and $1/\bar{h}(\lambda)$ immediately following a set of tables of $T(h,\lambda)$: see **Figs 5.4a-b**. However, in all other sources investigated $T(h,\lambda)$ is tabulated either together with $t(h,\lambda)$, or together with $t(h,\lambda)$ and $a(h,\lambda)$ in al-Bakhāniqī's arrangement of the corpus.

---

[4] *Ibid.*, III.15.3d.

[5] See n. **I**-1:22.

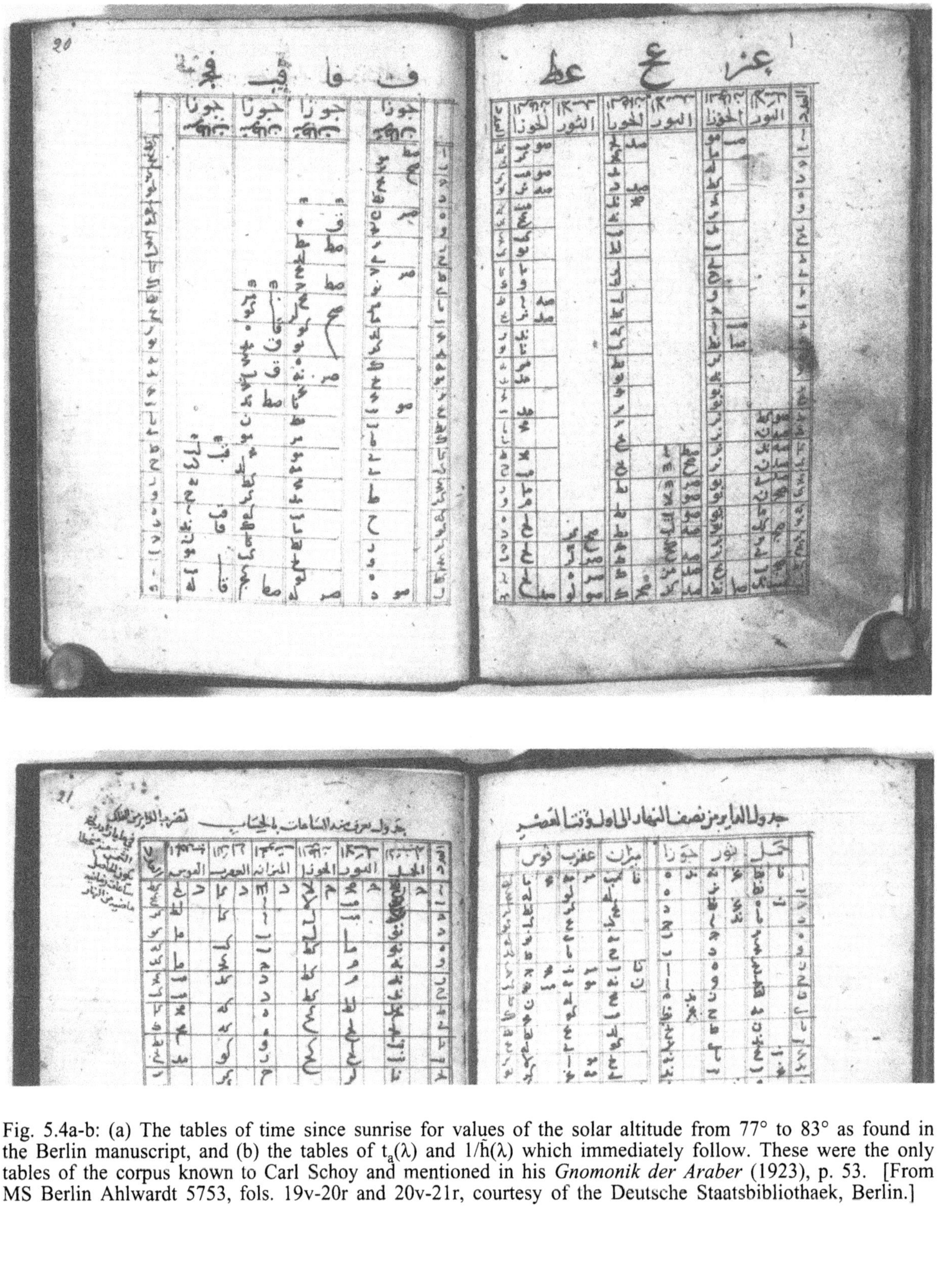

Fig. 5.4a-b: (a) The tables of time since sunrise for values of the solar altitude from 77° to 83° as found in the Berlin manuscript, and (b) the tables of $t_a(\lambda)$ and $1/\bar{h}(\lambda)$ which immediately follow. These were the only tables of the corpus known to Carl Schoy and mentioned in his *Gnomonik der Araber* (1923), p. 53. [From MS Berlin Ahlwardt 5753, fols. 19v-20r and 20v-21r, courtesy of the Deutsche Staatsbibliothaek, Berlin.]

al-Maqsī uses the phrase *raʾaytu an uḥarrir*, "I thought I would compile / edit / prepare" when introducing his tables. He makes no reference to any earlier tables of the same kind. In the introduction to his sundial tables he uses the verb *ḥasaba*, "to compute", and mentions various of his predecessors and contemporaries, with the notable exception of al-Marrākushī who compiled similar tables (**2.7**). It seems to me probable that al-Maqsī compiled his tables of T(h,λ) by using his own table of D(λ) and some earlier hour-angle tables. I suspect also that his table of $t_a(\lambda)$ is not original and that his table of $T_q(\lambda)$ was derived from an earlier table of $t_q(\lambda)$: I think Ibn Yūnus was compiler of both the tables of $t_a(\lambda)$ and $t_q(\lambda)$. Finally, al-Maqsī's table of $1/\tilde{h}(\lambda)$, which is rather carelessly computed, was probably compiled from rounded values of D(λ).

5.5 Ibn al-Kattānī's hour-angle tables and prayer-tables

MS Istanbul Kılıç Ali Paşa 684 is a unique copy of an extensive set of tables compiled by Ibn al-Kattānī (**4.1.4**). The manuscript was executed in Cairo in the author's own elegant hand and is dated 768 H [= 1366]. MS Cairo MM 72, preserved in Cairo, contains a set of solar altitude tables copied in 747 H [= 1346/47] by Ibn al-Kattānī but not necessarily computed by him (**6.9**, illustrated).

The Kılıç Ali Paşa manuscript is of considerable importance for the present study. The work it contains is entitled *Kitāb Nihāyat al-intifāʿ fī maʿrifat faḍl al-dāʾir min al-falak min qibal al-irtifāʿ*, "Extremely Useful Book for Finding the Hour-angle from the (Solar) Altitude", and was patronized by Burhān al-Dīn Ibrāhīm al-Battanūnī.[6] In his introduction to the tables Ibn al-Kattānī states that when he saw the tables of time since sunrise (*jadwal al-dāʾir min al-falak*) of al-Maqsī, he got upset about the fact that the format of the tables caused the upward arguments to run only from 0 to 29 and he wanted to make both vertical arguments run to 30°. He gives an example for λ = 360° and h = 60° and computes T = 90°, remarking that al-Maqsī gives the value 85;1°, and there is a difference of 4;59°. Ibn al-Kattānī's naïveté is well illustrated by the above: al-Maqsī's value 85;1°, the first entry in his tables of T(h,λ) for h = 60°, is for λ = 1° rather than the equinox. Ibn al-Kattānī goes on to say that most of the astronomical works he knew of dealt with the hour-angle rather than the time since sunrise and this fact was what inspired him to prepare the present work. The major part of the tables which follow consists of a complete set of tables of the hour-angle t(h,λ) derived from al-Maqsī's tables of T(h,λ). The entries are the same as those in the hour-angle tables in the main Cairo corpus (**4.5**) except that there are 31 rows of entries on each page rather than 30, both arguments running from 0° to 30°. Thus Ibn al-Kattānī has contrived to include the values for the equinoxes in the hour-angle tables: of course, the other entries in the top row merely duplicate the corresponding entries in the bottom row.

The remainder of the tables of Ibn al-Kattānī consists mainly of a set of prayer-tables, having basically the same entries as the Cairo corpus, but Ibn al-Kattānī's distinctive format. The functions which he tabulates are:

---

[6] I have not identified this individual.

h_q and t_q ($q = 53°$), H, h_a, D, $\tilde{h}$, t_a, T_a, r (see below), s, ψ, δ*, α′, α_ϕ, α_r, Δα and α* . Three different tables are given for r, the duration of morning twilight, based on parameters 20°, "according to recent scholars" (*raʾy al-mutaʾakhkhirīn*), 18°, "according to earlier scholars" (*raʾy al-mutaqaddimīn*), and 19°, "according to the compiler (of these tables)" (*raʾy wāḍiʿihi*). In each of these three tables the entries in the first row other than the equinoctial value are *not* the same as the corresponding entries in the last row: that is, Ibn al-Kattānī – may God forgive him – has interpolated between the values for, say, arguments 1^s 30° (= 2^s 0°!) and 2^s 1°, to find the value for 2^s 0°. One table is given for s, the duration of morning twilight, and the underlying parameter, which is not stated, is 16°; it does not display the absurdities of the tables for evening twilight. The table of ψ, the solar rising amplitude, is computed to three sexagesimal digits and is taken originally from the *Ḥākimī Zīj*, although this is not stated. It contains numerous copyist's errors.
The remaining tables consist of a solar longitude table giving positions in signs, degrees and minutes for each day of the Coptic year 1081 Diocletian [= 1364] (the entry for Tūt 1 is Virgo 13;55°) and two tables displaying the names of stars culminating at daybreak and culminating at nightfall for each degree of solar longitude.

Ibn al-Kattānī seems to have been the first to compile a complete set of hour-angle tables for Cairo, but it is very curious that his name is nowhere associated with the numerous other copies of the hour-angle tables which I have inspected. I doubt that any of his prayer-tables are original. Only in MS Istanbul Kılıç Ali Paşa 684 are there tables for Cairo with Ibn al-Kattānī's format, and later Egyptian astronomers, with good reason, prefered the original format of Ibn Yūnus. I know of only one other table for Cairo which displays the equinoctial value: in the margin of MS Gotha A 1410, fol. 36r, by the side of a table of h_q ($q = 53°$) the equinoctial value 46;13° is given. This is the same value given by Ibn al-Kattānī.

5.6 al-Bakhāniqī's edition of the corpus

Of prime importance for our investigations is a note by al-Bakhāniqī (**5.1**) which he added to al-Maqsī's introduction (**5.4**) when preparing his own edition of the corpus. The distinguishing feature and virtually sole innovation of this edition is that the values of the time since sunrise, hour-angle and azimuth are tabulated together, a triplet of entries (T,t,a) being given for each pair of arguments (h,λ): see **Fig. 5.6a-b**. al-Bakhāniqī's note is contained, for example, in MSS Cairo DM 108, Cairo DM 53, Istanbul Nuruosmaniye 2925, Cairo DM 616 and Cairo DM 739. The version in MS Cairo DM 53 is incomplete and rather corrupt. I present an annotated translation of the version in MS Cairo DM 108 (see **Fig. 5.6c**). (The texts in the other manuscripts offer no variants of consequence.) The most important sections, which are omitted in MS Cairo DM 53, are in italics in the translation.

> "The wretched slave of God Aḥmad ibn Muḥammad ibn Aḥmad al-Azharī known as al-Bakhāniqī said: I have arranged this book so that under each zodiacal sign there are three tables. The first of these gives the time since sunrise by al-Maqsī – may God have mercy on him – and the second gives the hour-angle. *The values of this latter function may differ in some cases from previous calculations:*[a] I calculated them by using the corresponding values (of the time since sunrise and half diurnal arc).[b] The

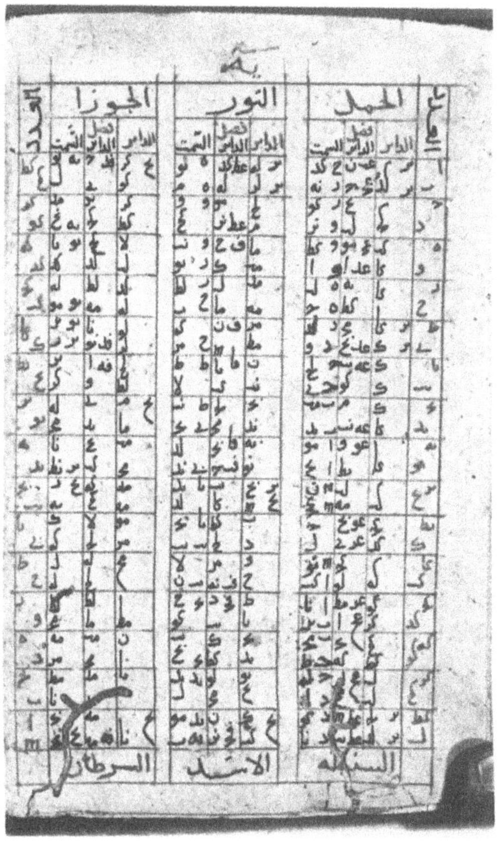

Fig. 5.6a: In al-Bakhāniqī's edition of the main tables of the Cairo corpus, the time since sunrise, the time till midday and the azimuth are tabulated side by side, as a function of the solar altitude, here 15°. [From MS Cairo DM 444, courtesy of the Egyptian National Library.]

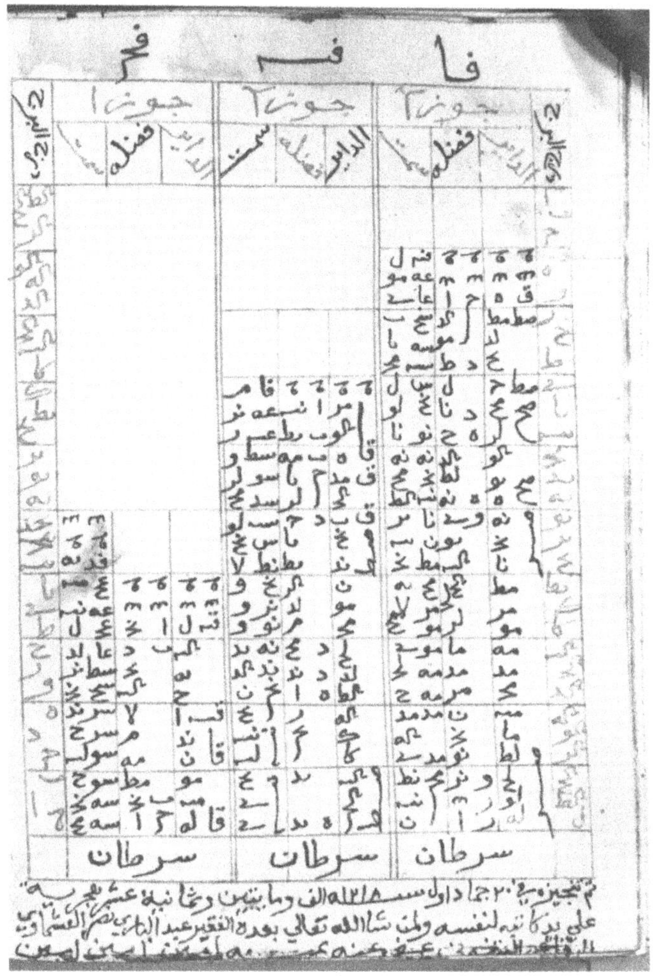

Fig. 5.6b: The last page of tables showing the problems with the triplets of entries for t and T, as compared with those for a: see **4.5**. [From MS Cairo DM 108, fol. 65v, courtesy of the Egyptian National Library.]

third function is the azimuth for the altitude in question computed by the Shaykh Shams al-Dīn Abū ʿAbdallāh Muḥammad ibn Ibrāhīm ibn ʿAbdallāh al-Rashīdī – may God have mercy upon him[c] – correcting the errors introduced by repeated copying of the azimuth tables of Abu 'l-Ḥasan[d] ibn Yūnus – may God have mercy on him. *He [e] had preceded these with two tables on a single page, one giving the solar rising azimuth and the other the solar altitude in the prime vertical. (He also included) tables of the solar altitude in the azimuth of the ventilator and in the azimuths of the qibla (S.E.) and the perpendicular direction (S.W.), on the basis that its azimuth (measured from the prime vertical) was 37°, and an additional complementary [f] table for the azimuth at the equinoxes and solstices. To these there was added [g] at the end of the book the tables which al-Maqsī had at the end of his book, [h] as well as the hour-angle when the sun is in the azimuth of the qibla, the solar meridian altitude, the solar*

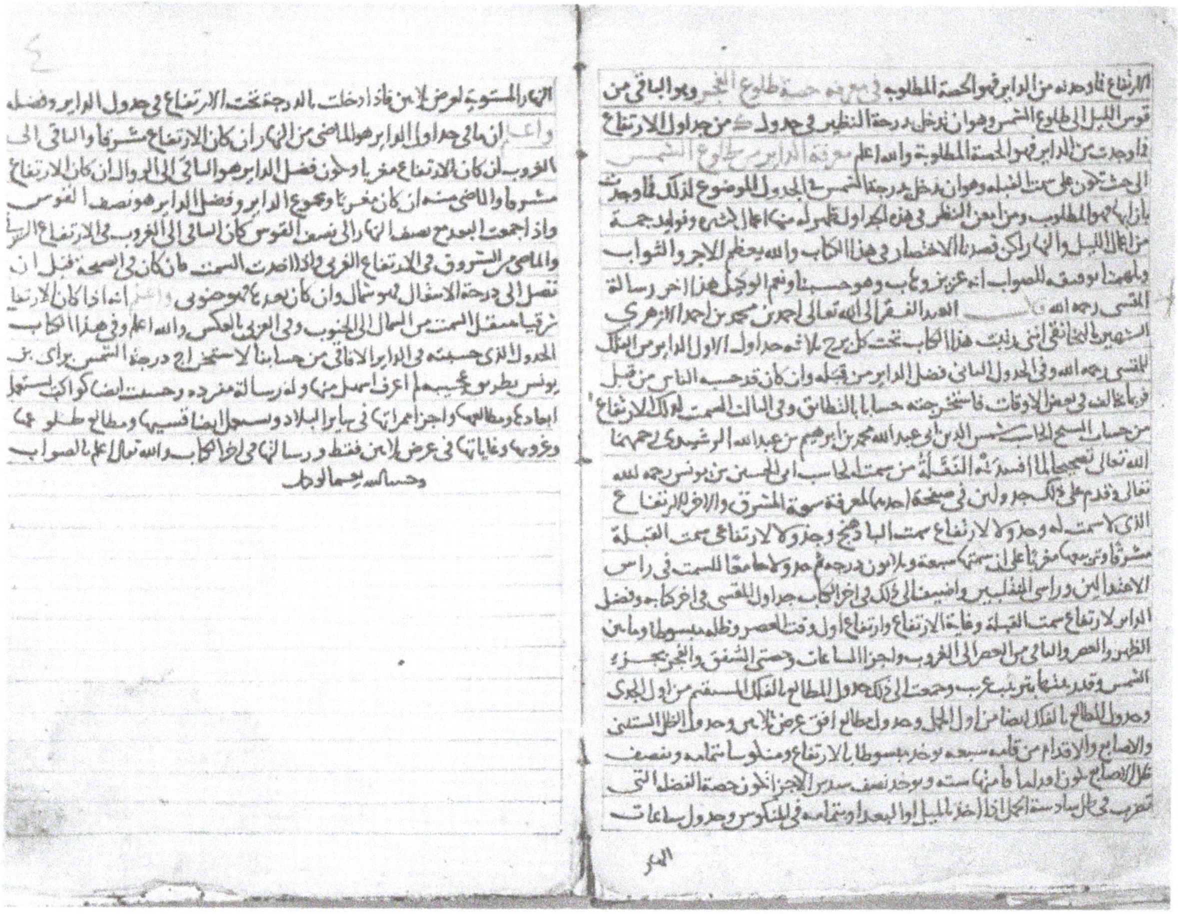

Fig. 5.6c: The text of al-Bakhāniqī's remarks on the corpus. [From MS Cairo DM 108, fols. 3v-4r, courtesy of the Egyptian National Library.]

altitude at the beginning of the afternoon prayer and its Cotangent, the time from midday to the beginning of the afternoon prayer and the time remaining till sunset, and also the number of equatorial degrees in the seasonal day-hours, and the duration of morning and evening twilight, (all) as functions of solar longitude.

I arranged these tables in a satisfactory fashion,[i] *and added to them others giving the normed right ascensions, regular right ascensions and oblique ascensions for latitude 30°, as well as tables of Cotangents to base 60, 12 and 7.*[j] If you enter these with the altitude you obtain the horizontal shadow of a vertical gnomon, and if you enter with the complement you obtain the vertical shadow of a horizontal gnomon. If you take one-half of the Cotangents to base 12 you obtain *Cotangents to base 6*, and if you take one-half of one-sixth of the entries you get the function which, with the declination as argument, is multiplied by the same function, that is the Cotangent, of the equinoctial meridian altitude or the Tangent of its complement (to find the Sine of of the half excess of daylight).[k] *(I also added) a table of the number of equatorial hours of daylight for latitude 30°.*

If you enter the solar longitude under the appropriate solar altitude in the tables of the time since sunrise and hour-angle you get for eastern altitudes the time since sunrise and the time till midday, and for western altitudes the time till sunset and the time since midday. The sum of the time since sunrise and the hour-angle is the semi diurnal arc. If you add the time from midday to the semi diurnal arc you get the time to sunset for eastern altitudes and the time since sunrise for western altitudes. As for the azimuth, it will be northern if (the solar longitude) on the given page is before the longitude at which the azimuth changes direction (from north to south), and it will be southern if the solar longitude is after this particular longitude. For altitudes in the east the azimuth moves from north to south and for altitudes in the west the opposite is the case. But God knows better.

In this book (I have also included) the universal table which I computed for finding the time: this is for obtaining the (hour-angle) [text: solar longitude] *according to the admirable method of Ibn Yūnus. I know of no easier method, and he had a treatise (on the subject) which was unrivalled.*[l] *I also computed a star catalogue, in which the declinations, right ascensions, and degrees of transit can be used in all localities, but the arcs of visibility, oblique ascensions of their risings and settings, as well as their meridian altitudes, are for the latitude 30° only. The text (explaining their use) is at the end of this book.*[m] God knows the truth best and, knowing all our works, is our best Judge."

Notes:

(a) This sentence implies that al-Bakhāniqī had seen some tables of $t(h,\lambda)$ computed for Cairo but preferred to prepare a new set.

(b) The expression *bi-'l-taṭābuq* means using the relation:

$$T(h,\lambda) = D(\lambda) - t(h,\lambda)$$

to compute the entries.

(c) The text has "them both" rather than "him", and the phrase should probably occur after the name of Ibn Yūnus, and refer to both him and al-Rashīdī. On the other hand, it may refer to al-Rashīdī and his father, but his grandfather's name is also mentioned.

(d) The text has incorrectly Abu 'l-Ḥusayn.

(e) This could refer to either Ibn Yūnus or Ibn al-Rashīdī – the latter seems the more likely.

(f) The Arabic term *jāmiʿ* seems to indicate that this supplementary table together with the main azimuth tables display the solar azimuth for *all* longitudes.

(g) I read the Arabic as *uḍayfa*, but since the subject is plural the word should be *uḍyafat*. Note that al-Bakhāniqī did not use the word *aḍāfa*, which would refer to Ibn al-Rashīdī.

(h) These are the tables of the functions D, t_a, T_q and $1/\tilde{h}$ – see below.

(i) This seems a reasonable rendering of *rattabtuhā tartīban*. The word *ʿajīb* generally means "wonderful / admirable / strange".

(j) Called *sittīnī*, *iṣbaʿ* and *qadam*, "sexagesimal, digit, and foot", respectively. These expressions are standard.

(k) This procedure is equivalent to the formula:

$$\text{Sin } d = {}^{1}/_{12} \cdot \text{Tan}_{60}\,\Delta\,\text{Cot}_{12}\,\bar{\phi}\,,$$

Δ being the declination of a star or the sun, and $\bar{\phi}$ being the co-latitude. See **I-7.0** and **7.1**.

(l) The text has *darajat al-shams* "solar longitude" rather than "hour-angle", a curious error. al-Bakhāniqī may have been referring to the function:

$$G(\lambda) \;=\; R^2 \,/\, \{\, \mathrm{Cos}\ \delta(\lambda)\ \mathrm{Cos}\ \phi \,\}$$

discussed by Ibn Yūnus in Ch. 15 of the *Ḥākimī Zīj* (see also **I-6.7.1**).[7] The term *risāla* means "treatise" or "method" in medieval scientific Arabic: it probably refers to the passage in the *Ḥākimī Zīj*.

(m) al-Bakhāniqī's star catalogue and notes thereon have not been identified in any of the known manuscripts of the corpus.

Thus it appears that al-Bakhāniqī rearranged the tables of Ibn Yūnus and al-Maqsī as follows. Taking Ibn Yūnus' tables of a(h,λ) corrected by Ibn al-Rashīdī (**5.2**), al-Maqsī's tables of T(h,λ) (**5.3**), and Ibn al-Kattānī's tables of t(h,λ) (**5.4**), he then tabulated the values (T,t,a) as triplets for each pair of arguments (h,λ). This arrangement of the tables is found for example, in MSS Cairo DM 108, Cairo DM 53, Istanbul Nuruosmaniye 2925, Cairo DM 616, and Cairo DM 739. If I understand al-Bakhāniqī properly he is saying that Ibn al-Rashīdī added the tables of:

$$\psi,\ h_0,\ h_v,\ h_q,\ h_{q*}\ \text{and}\ a(h)\ \text{for}\ \lambda = 0°,\ 90°,\ 270°$$

before the azimuth tables, and that after the azimuth tables he had put the tables which al-Maqsī had at the end of his book, namely, tables of:

$$D,\ t_a,\ T_q\ (q = 53°),\ \text{and}\ 1/\tilde{h}\ ,$$

as well as the tables of:

$$t_q,\ H,\ h_a\ \text{and}\ z_a,\ t_a\ \text{and}\ T_a,\ 2D^h,\ r\ \text{and}\ s\ .$$

But if Ibn al-Rashīdī had indeed added all these tables, why is he not generally acknowledged as the author of all the minor tables in the corpus? His name is specifically mentioned in the tables of the corpus only in relation to the two tables of $h_v(\lambda)$ and a(h) at the equinoxes and solstices.

To al-Maqsī's introduction al-Bakhāniqī added the note translated above, describing what he had done to the tables. The five manuscripts listed above contain this modified introduction as well as the tables of (T,t,a), but MS Cairo DM 108 also contains a variety of other tables. In al-Bakhāniqī's note in that copy (DM 108), but not in MS Cairo DM 53, he states that he included tables of α, α' and α_ϕ (ϕ = 30°). These are not contained in MS Cairo DM 108, but in later sets of prayer-tables for Cairo we find, for example, Ibn Yūnus' tables of α' (MS Cairo MM 43) and α_ϕ (ϕ – 30°) (MS Princeton Yahuda 861,1), and al-Marrākushī's tables of α_ϕ (ϕ = 30°) (MSS Hartford TS 621 and Paris BNF ar. 2507). al-Bakhāniqī further asserts that he included tables of the Cotangent function to bases 60, 12, and 7: these are likewise not contained in MS Cairo DM 108. He also mentions Cotangents to base 6 (*cf.* **2.9**).[8] Finally, al-Bakhāniqī states that he himself added a table of $\tilde{h}$ to the corpus (extant in MS Cairo DM 108) as well as a star catalogue and a table for finding the hour-angle (if my correction "hour-angle" for

[7] King, *Ibn Yūnus*, III.15.3-4.
[8] See n. **I**-1:37.

"solar longitude is correct), which may have been some kind of auxiliary table (both not extant in the known sources).

5.7 al-Wafāʾī's prayer-tables and notes on refraction at the horizon

MS Cairo ṬJ 367,5, fols. 35r-40r, copied about 1450, contains a short introduction and a solar longitude table followed by tables of the functions:

$$r, H, h_q, h_0, \psi, d, \text{and } \delta$$

from the corpus. The work bears the title *Natījat al-afkār fī aʿmāl al-layl wa-'l-nahār* and is attributed to the 15th-century Egyptian astronomer al-Wafāʾī (**1-9.9**). The solar table (fol. 37v) is not complete and the prayer-tables which follow (fols. 38r-40r) are in the same hand but are not mentioned in al-Wafāʾī's introduction. If the title, which mentions timekeeping by day and night, is original, it seems probable that the prayer-tables were included in the work by al-Wafāʾī himself. If it is indeed the case then it was al-Wafāʾī who set the trend followed by later Egyptian astronomers such as al-Lādhiqī, al-Ikhṣāṣī, Ibn Abī Rāya and Riḍwān Efendī (**7.8-10**), of presenting updated tables for calendar conversion and finding the solar longitude for a given date, and including selected tables from the already available corpus.

MS Cairo DM 157,2, copied *ca.* 1450, contains a set of tables of the function $\alpha_\sigma(\lambda)$, the oblique ascension at the time of the *salām*, computed for each six minutes of argument with entries given to minutes. These tables are specifically attributed to al-Wafāʾī. A similar set of tables for the time of nightfall is attributed in MS Cairo DM 1218, copied *ca.* 1500, to Muḥammad al-Ghazāwī, a *muwaqqit* in the Muqassam (?) Mosque in Cairo, whose name is new to the modern literature. MS Cairo MM 209,1 contains a set of normed right ascensions, $\alpha'(\lambda)$ (ε = 23;35°), with values for each minute of argument attributed to Ibn al-Mushrif (**I-9.8** and **II-6.15**), copied in 873 H [= 1468/69] by Ibn Abi 'l-Fatḥ al-Ṣūfī (**6.15**) from a copy by the author dated 848 H [= 1444/45]. Apart from the earlier ascension tables of Baylak al-Qipjāqī (**6.4**), all other Islamic tables of this kind, which are a particularly Egyptian phenomenon, and in which values for each degree of solar longitude are stretched to each six, three, or even one minute of argument, date from the period of Ibrāhīm ibn Qāyitbāy and ʿAbd al-Qādir al-Minūfī, namely the late 16th and early 17th centuries. See further **7.2**.

MSS Cairo MM 181,2 (fols. 4v-5r, copied 1030 H [= 1620/21], TR 303,2 (fol. 33v, copied 1266 H [= 1849/50]), and MM 155,2 (fol. 23v, *ca.* 1700, defective) are copies of a short treatise (*nubdha*) attributed to al-Wafāʾī (at least in the first source) concerning the *daqāʾiq al-ikhtilāf*, that is, the the time taken by the sun to cross between the visible and true horizons (**4.11**). The first two sources are displayed in **Fig. 5.7**. In the treatise it is stated that the values:

EQ: 47 SS: 62 WS: 32

were derived by thc Damascus astronomer Ibn al-Shaṭīr (**9.3**). This is curious for the following reasons:

(1) the Minūfīs, father and son, attributed these values to Ibn Yūnus (**7.1** and **8.2**);

(2) Ibn al-Shaṭīr is known to have written that he had observed the half arcs of visibility of the sun and stars and found that they were greater than the arcs found by calculation by an amount exceeding $^2/_3$° (**9.3**), but is not otherwise known to have written more

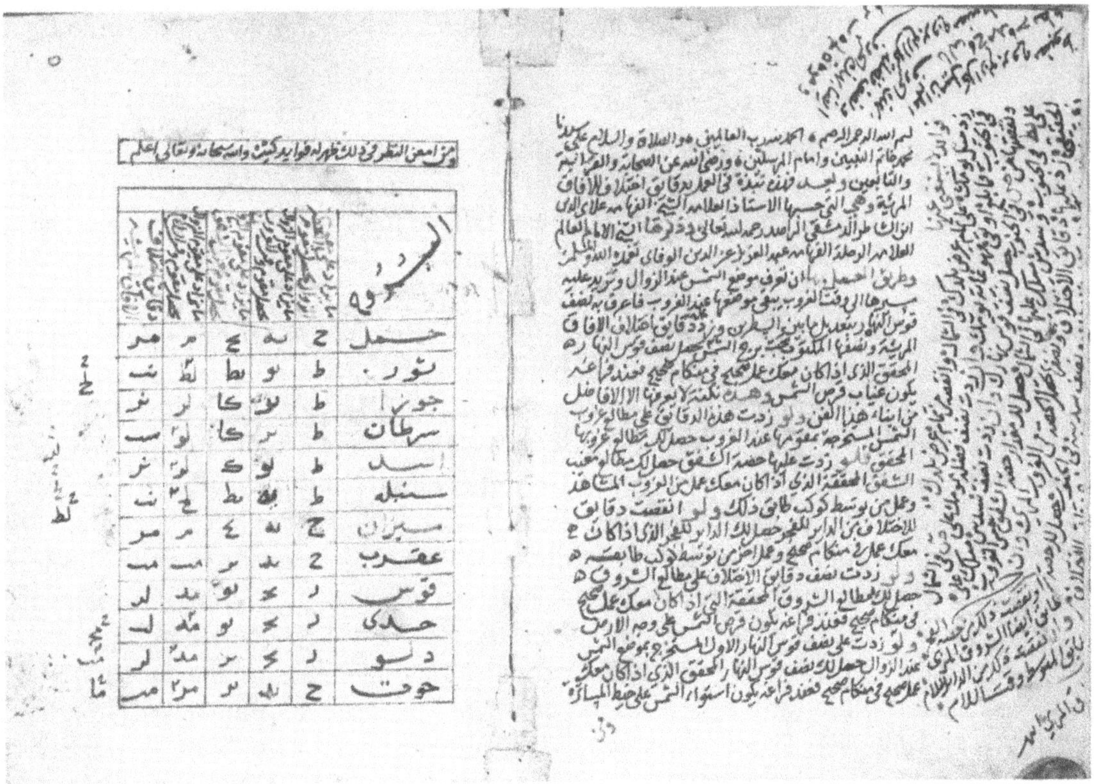

↑
Fig. 5.7a: Notes attributed to al-Wafāʾī on the time taken by the sun to travel between the true and visible horizons. [From MS Cairo MM 181,2 (fols. 4v-5r), courtesy of the Egyptian National Library.]

→
Fig. 5.7b: The same notes abridged and without the attribution to al-Wafāʾī, but note that in both these Cairo manuscripts the theory is attributed to the illustrious Ibn al-Shāṭir. [From MS Cairo TR 303,2 (fol. 33v), courtesy of the Egyptian National Library.]

on the subject. There is no reference to the subject in Ibn al-Shaṭir's prayer-tables or in his introduction to them (see again **9.3**).

(3) If Ibn al-Shaṭir had indeed computed a set of values of the "difference minutes" for the zodiacal signs he would surely have computed them for Damascus, rather than Cairo. If he had computed them for Cairo, he would surely also have computed them for Damascus. But the later Syrian tables for ΔD(λ) are based on the values for Cairo adapted for Damascus (**11.13**).

(4) If al-Wafā'ī was indeed responsible for these notes, I find it curious that he did not refer to the "difference minutes" in any of his other known works, not least, in his treatise and tables for timekeeping discussed above.

The treatise explains how the "difference minutes" are to be used, and concludes with a table displaying for each zodiacal sign the following quantities, all in minutes of arc:

(1-4) the amounts to be added to the solar longitude at midday to find the solar longitude at the time of the *ʿaṣr* prayer, sunset, nightfall, and the time of the *salām*.

(5) The time taken by the sun to move between the two horizons, labelled *daqāʾiq ikhtilāf al-āfāq al-marʾiyya*.

Similar tables are contained in most of the copies of al-Lādhiqī's prayer-tables for Cairo (**7.8**).

5.8 Sibṭ al-Māridīnī's notes on interpolation in timekeeping tables

Sibṭ al-Māridīnī (**2.10**), in his treatise on sexagesimal arithmetic entitled *Raqāʾiq al-ḥaqāʾiq fī ḥisāb al-daraj wa-'l-daqāʾiq*, "The Finest Truths concerning Sexagesimal Arithmetic", discusses interpolation[9] in tables for timekeeping. The relevant passage (I have used MS Gotha A1390, fol. 33v) translates as follows:

> "The conclusion of this work deals with interpolation (*taʿdīl mā bayn al-saṭrayn*). The tables of the semi diurnal arc, the half excess of daylight, solar rising amplitude, ascensions, and the like, are generally computed for each integral degree of argument although they may also be computed for each minute. Certain tables are generally computes for each minute, such as the (Co)tangent, Sine and Versed Sine. Entries in these are usually given for argument differences of a minute. Likewise the tables of the inverse (Co)tangent, Sine and Versed Sine also have entries for each minute of argument. In tables of this sort there is no need to interpolate... ."

It is of interest that he mentions tables of spherical astronomical functions and trigonometric functions computed for each minute of argument: see below.

Sibṭ al-Māridīnī then gives examples of how to use linear interpolation in a table of, say, D(λ) for non-integral λ. He also shows how to interpolate in double-argument tables, quoting an example from "his teacher", that is, the celebrated Ibn al-Majdī (1359-1447).[10] To find the solar azimuth when the altitude is 19;24° and the solar longitude is 53;20° he interpolates in

[9] See n. **I**-1:21.

[10] On Ibn al-Majdī see Suter, *MAA*, no. 432; *Cairo ENL Survey*, no. C62; and also King & Kennedy, "Ibn al-Majdī's Tables", pp. 48-49.

the azimuth tables for altitudes 19° and 20° as follows. The relevant entries which he cites (all accurate) are:

	h = 19°	20°
λ = 53°	11; 2°	10;31°
54°	11;20	10;49

(as in Ibn Yūnus' tables), and our author computes what he calls *al-taʿdīl al-awwal* and *al-thānī*, "the first and second interpolations":

$$a(19°, 53;20°) = 11;8° \text{ and } a(20°, 53;20°) = 10;37° ,$$

from which:

$$a(19;24°, 53;20°) = 10;56° .$$

He states that the accurate value of a is 10;55,36°, which is just one second too high, and calls this kind of interpolation *taʿdīl al-tajyīb*. He makes no mention of the author of the tables.

Sibṭ al-Māridīnī's remark that certain tables for timekeeping had entries for each minute of argument (solar longitude) is confirmed by the tables of Ibn al-Mushrif, Ibrāhīm ibn Qāyitbāy and al-Wafāʾī (**5.7**). The Sine and Cotangent tables attributed to Ibn Yūnus give values also to each minute, but to five sexagesimal digits.[11] By the 15th century such tables were standard in the major centres of astronomy in the Islamic world, the most popular ones being the Sine and Cotangent tables from the *Zīj* of Ulugh Beg.[12] However, there was also an independent Egyptian tradition of tables of the standard trigonometric functions with entries to three digits for each minute of argument, of which several manuscripts survive: see **Fig. I-1.0e** for an example.[13]

---

[11] King, *Ibn Yūnus*, III.10.5, and *idem*, *Mecca-Centred World-Maps*, p. 166.
[12] Schoy, *Trigometrie des al-Bīrūnī*, pp. 92-100; see also King, *Mecca-Centred World-Maps*, p. 165.
[13] These are listed in *Cairo ENL Survey*, no. C137.

CHAPTER 6

OTHER EARLY EGYPTIAN TABLES FOR TIMEKEEPING

6.0 Introductory remarks

Virtually no Egyptian astronomical works survive from the period between Ibn Yūnus and the middle of the 13th century. (This accounts, for example, for the importance of the treatises by Ibn al-Ḥammāmī discussed in **2.4**.) For the history of astronomy in Egypt during that period we have little more than biographical references to various individuals, and al-Maqrīzī's account of the observatory that was built in Cairo in the 12th century.[1] From late 13th-century Cairo we have the A-Z on spherical astronomy and instruments by Abū ʿAlī al-Marrākushī, the treatise on sundial construction by al-Maqsī, and a *zīj* called *al-Zīj al-Muṣṭalaḥ*. al-Marrākushī's treatise enjoyed considerable popularity in later centuries in Egypt, Syria, and Turkey. I discuss his spherical astronomical tables in **6.7**. In his encyclopedia *Irshād al-qāṣid*, the early-14th-century scholar Ibn al-Akfānī states that the two books on *mīqāt* that were used (in Egypt) in his time were a short work *Nafāʾis al-yawāqīt* and the longer work *Jāmiʿ al-mabādiʾ wa-ʾl-ghāyāt* of Abū ʿAlī al-Marrākushī.[2] He does not mention the author of the first work, but it may be that the *Durar al-yawāqīt* described in **3.2** is identical with, or at least related to the *Nafāʾis al-yawāqīt*. Ibn al-Akfānī also mentions that the closest astronomical observations to his time were those conducted under Hulagu by Naṣīr al-Dīn al-Ṭūsī and his associates, and that in Egypt the yearly ephemerides were prepared from a *zīj* which had been compiled from a number of other *zīj*es and which was called the *Muṣṭalaḥ Zīj*. In **6.6** I discuss the spherical astronomical tables found in the two extant versions of this work. Various Egyptian astronomers compiled auxiliary tables for facilitating the solution of problems in spherical astronomy. The historical context of the monumental table of Najm al-Dīn al-Miṣrī (**6.5**) is still something of a mystery, even though François Charette has produced a detailed study of the table itself and gathered all the known material relating to its author. It seems that the tables of al-Khaṭāʾī (**6.8**) do not, after all, and as one might have expected, precede those of al-Khalīlī of Damascus (**10.3**). al-Khalīlī's tables were also used in Egypt (**6.14**). All of the other Egyptian auxiliary tables postdate those of al-Khalīlī (**6.15**). Egyptian astronomers, in particular Ibn al-Rashīdī, compiled timekeeping tables for the latitudes of Mecca and Jerusalem (**6.11** and **6.12**).

6.1 A 13th-century muezzin's manual

MS Oxford Bodley Or. 133 is a beautifully-illustrated manuscript comprising mainly an astrological work *Kitāb al-Bulhān* (*sic* for *al-Burhān* ?) by an otherwise unknown ʿAbd al-Ḥasan

[1] This activity is described in Sayılı, *The Observatory in Islam*, pp. 167-175.
[2] Ibn al-Akfānī, *Irshād al-qāṣid*, p. 5. See also Charette, *Mamluk Instrumentation*, p. 12.

ibn Aḥmad ibn ʿAlī ibn al-Ḥasan al-Iṣfahānī and two other shorter works in poetry and calligraphy.[3] A treatise on astronomical timekeeping, written in a different hand but datable to the 14th century, has been bound in the manuscript (fols. 94v-130r) after the astrological work. It is entitled *Kitāb al-Durar wa-'l-yawāqīt fī ʿilm al-raṣad wa-'l-mawāqīt*, "Book of Pearls and Sapphires on Astronomical Observations and Timekeeping". On the title pages (fols. 94v-95r) the author's name has been partially deleted, but the following letters are still legible: *Abī A... ... ibn ...d ibn Aḥmad*, al-.ā... al-... (perhaps, *al-Shāfiʿī al-madhhab*, referring to his adherance to the Shāfiʿī legal school) *al-Miṣrī al-waṭn* (the Egyptian). This name, however, does not correspond exactly with the author's name which appears on fol. 97v, namely: [Abu] 'l-ʿAbbās Aḥmad ibn Abī ʿAbdallāh Muḥammad ibn Aḥmad. The copying of the manuscript was completed in 734 H [= 1334], and the name of the copyist was Abu 'l-Ḥasan ʿAlī ibn (?) Muḥammad ibn (?) Baṣīr (?). The original work can be dated to the early 13th century (see below).

The treatise is a complete *muwaqqit*'s manual for timekeeping by the lunar mansions and regulating the prayer-times approximately (see also **V-5**). It may be that Ibn al-Akfānī, writing in the early 14th century, was referring to this work when he stated that the book on simple timekeeping which was currently most widely used in Egypt was called *Nafā'is al-yawāqīt* (**6.0**). The treatise begins with a set of prayer-tables for Cairo (fols. 98r-104r) of a kind not attested in any other Egyptian source: see **Fig. 6.1**. For each day of the Coptic year and corresponding date in the Syrian calendar the lunar mansion rising at daybreak is indicated and following functions are tabulated:

λ	*darajat al-shams*
H	*ghāyat al-irtifāʿ*
$Z_{(12)}$	*ẓill al-zawāl*
h_a	*irtifāʿ al-ʿaṣr*
2D	*qaws al-nahār*
2N	*qaws al-layl*
h_q	*samt al-qibla*
$z_{q(12)}$ (= $\mathrm{Cot}_{12}\ h_q$)	*ẓill al-samt*
r	*ḥiṣṣat al-fajr*.

The values of H, 2D and 2N are given in degrees and minutes, and the remaining functions are displayed to the nearest degree or digit. From the values of H the parameters are found to be:

$$\phi = 30° \text{ (Cairo) and } \varepsilon = 23;35° .$$

All of the functions are approximate. The values given for H for each 30° of λ starting from the winter solstice are:

36;25° 40;8 48;4 60;0 70;36 79;56 83;35 ,

but starting from the summer solstice are:

83;35° 80;0 72;6 60;0 48;4 40;20 36;25 .

The values of δ vary in each of the four quadrants of the ecliptic! In the text the author states

[3] On the astrological work see Ullmann, *Die Natur- und Geheimwissenschaften im Islam*, pp. 344-345; Sezgin, *GAS*, VII, pp. 24-25; and the detailed study of the art-historical aspects in Carboni, *Kitāb al bulhān*

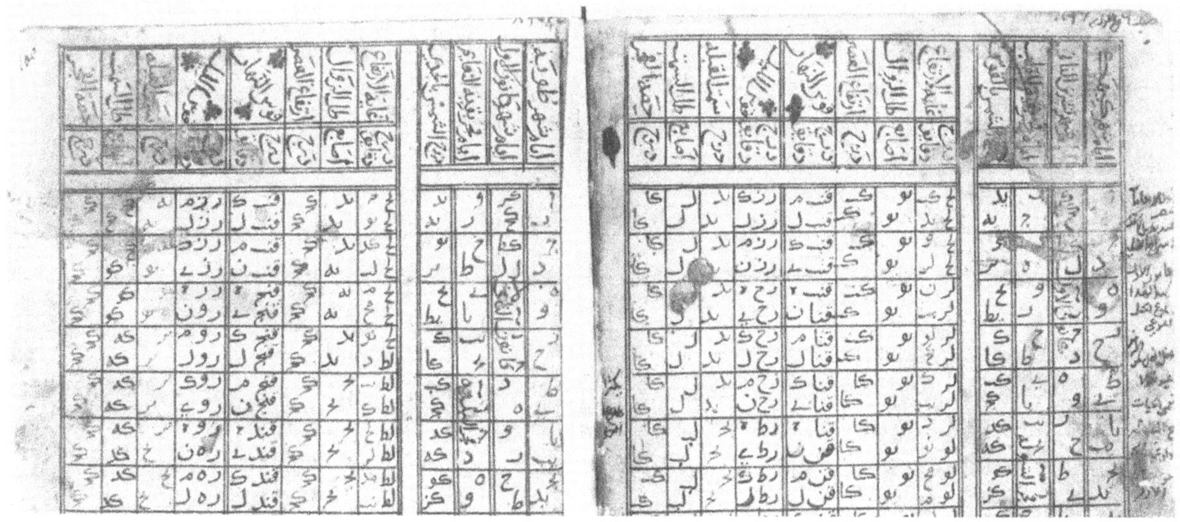

Fig. 6.1: The prayer-tables for the months of Kayhak and Ṭūba in the *Kitāb al-Durar wa-'l-yawāqīt*. [From MS Oxford Bodley Or. 133, fols. 99v-100r, courtesy of the Bodleian Library.]

that δ varies by 12°, 8°, and 4° for each 30° of λ measured from the equinox, but he has not used these figures in his tables. The scheme underlying the table of 2D, which is outlined elsewhere in the text, is that the function increases by 0;30° for each degree of Aries, 0;20° for each degree of Taurus, and 0;10° for each degree of Gemini, *etc*. This explains, for example, the extremal values 150;0° and 210;0° (accurately, 150;48° and 209;12°). The function $Z_{(12)}$ is made to assume values:

16 13 10 7 4 2 1;21

for each 30° of solar longitude starting from the winter solstice (entries less than 2 are given to two digits). The underlying scheme is clearly very crude, although the (accurate) value 1;21 for the summer solstice was probably copied from another source. The accurately computed shadow lengths are:

16;16 14;26 10;38 6;56 4;0 2;03 1;21

The values of h_a for each 30° of λ starting at the winter solstice are:

21 25 26 31 37 40 42 ,

compared with the accurate values:

23 24 28 33 37 40 42 .

In the text the author advocates the approximate rule (*cf.* **2.9**):

$$h_a \approx \frac{1}{2} H + \frac{1}{10} (83° - H) ,$$

which gives values:

23 24 27 32 37 40 42 .

The tabulated function h_q assumes values:

13 19 35 45 58 71 80

for each 30° of λ starting at the winter solstice. The corresponding values in Ibn Yūnus' table of h_q ($q = 52°$) and the other table of h_q in the Cairo corpus ($q = 53°$) round to:

14	19	31	47	63	75	80
13	18	30	46	62	75	80 ,

respectively. Using the approximate formulae for h_q found in the treatises of Najm al-Dīn al-Miṣrī and al-Bakhāniqī (**2.5** and **2.9**), namely:

$$h_q \approx 47° + 1;25\ \delta \quad \text{and} \quad h_q \approx 47° + 1;24\ \delta:$$

we obtain values:

14	18	31	47	63	76	80 .

Again, using linear interpolation between the extremal values 13 and 80, we obtain:

13	24	35	47	58	69	80 .

Thus the tabulated values correspond to none of these approximations.

No other known Egyptian prayer-tables display the function $z_{q(12)}$ (= $\text{Cot}_{12}\ h_q$), using which one can determine the qibla from the gnomon shadow.

The tabulated values of r are:

22	21	21	24	25	25	27

for each 30° of λ starting at the winter solstice. It is difficult to explain these values. For parameters 18°, 19°, and 20°, the corresponding accurate values are respectively:

22	22	21	21	22	24	25,
24	23	22	22	23	26	27
25	24	23	23	25	27	28 .

Note that the compiler realized that r displays a secondary maximum at the winter solstice.

Most of the remainder of the treatise and the other tables are for timekeeping by the lunar mansions. Some of the tables resemble those found in a later Yemeni almanac (**12.5**). Thus there are tables displaying the names of the lunar mansions rising, culminating and setting at nightfall, and at one-third, one-half, two-thirds and three-quarters of the night, as well as at daybreak, for each 13 days of the Coptic year. Other tables display the 13th divisions of the lunar mansions rising at daybreak and the solar longitude for each day of the Coptic year (the entry in the latter for Tūt 1 is Virgo 14°). The longitudes of 23 prominent stars are given; these correspond to the early 13th century (*e.g.* the longitude of Aldebaran is 2^s 29;37°), enabling us to date the compilation of the treatise to the early 13th century.

6.2 Prayer-tables attributed to Ibn Qudāma

MS Damascus Ẓāhiriyya 10732, fols. 3r-3v, copied *ca.* 1600, contains some simple prayer-tables for an unspecified latitude with values of the functions:

$$D,\ t_a,\ T_a,\ s,\ r,\ n \text{ and } 2N$$

to one degree for each zodiacal sign. The tables occur on a single folio that may be related to the text that immediately precedes and follows, namely, simple material on the lunar mansions and shadow lengths attributed to the late-12th-/early-13th-century legal scholar and theologian of Jerusalem, Ibn Qudāma.[4] Whether or not this is the case, the values in the table can be derived from those in the main Cairo corpus.

[4] On Ibn Qudāma see the article "Ibn Ḳudāma" in *EI*$_2$ by George Makdisi. On the shadow scheme associated with him see **III-8.1**.

6.3 Prayer-tables in an anonymous Egyptian almanac

MS Cairo DM 187 (74 fols., copied *ca.* 1850) is the only known copy of an anonymous almanac based on the Coptic calendar of the same kind as the *Calendar of Cordova*,[5] and accompanied by various simple calendrical and astrological tables. This almanac merits detailed investigation. I am unable to ascertain the date of its compilation, but have noted a list of planetary apogees amongst the tables, as follows:

Saturn	Jupiter	Mars	Sun (and Venus)	Mercury
245°173°	125°	83°	202°	

These correspond very closely to those in the 9th-century Abbasid *Mumtaḥan Zīj*, which are as follows for the year 199 Yazdigird [= 831]:[6]

244;30°	172;32	128;33	82;39	201;0

On the other hand, some prayer-tables contained in this work – see **Fig. 6.3** – do not, I think, predate the 14th century. These display values of the following functions for each five days of the Coptic month:

D (*al-ẓuhr*), t_a (*al-ʿaṣr*), T_a (*al-maghrib*), s (*al-ʿishāʾ*), n (*jawf al-layl*), r (*al-fajr*)

Values are given in degrees and also in hours (misspelled *saʿāt* !) and minutes, and correspond to those in the main Cairo corpus. The twilight tables are based on 19° and 17°, which are not attested in Egyptian sources before the 14th century.

At the end of the almanac there are various notes, including one expounding some approximate formulae for timekeeping, of which the first three are not attested elsewhere. These are as follows:

$$t_a \approx {}^4/_7\, D\,,\quad s \approx {}^1/_8\,(2N)\,,\quad r \approx {}^1/_7\,(2N) \text{ and } h_a \approx {}^1/_2\, H + {}^1/_2 \cdot {}^1/_6\,[90° - {}^1/_2\, H]\,.$$

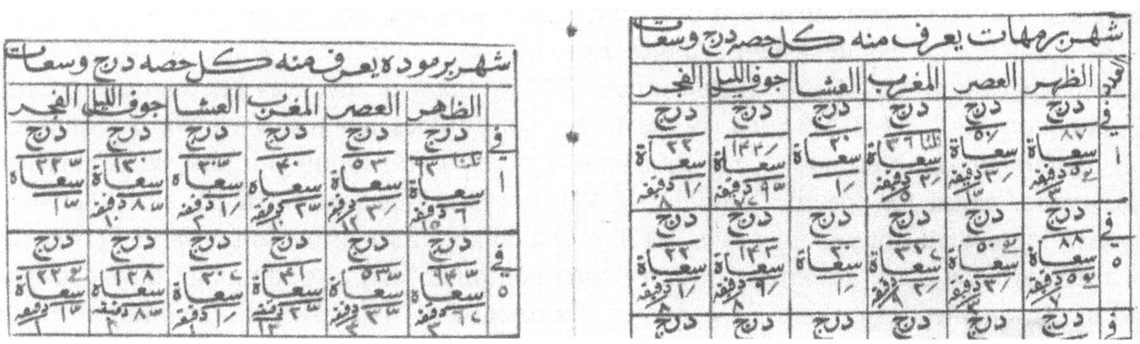

Fig. 6.3: The part of the prayer-tables in an anonymous Egyptian almanac serving the months of Baramhāt and Barmūda. [From MS Cairo DM 187, courtesy of the Egyptian National Library.]

[5] This is published with translation in Charles Pellat, *Calendrier de Cordoue*. On the shadow schemes in that work and also those in MS Cairo DM 187 see **III-5.1** and **9.7a**.

[6] On the *Mumtaḥan Zīj* see n. **I**-4:7. The apogees given are extracted from Caussin, *Table Hakémite* (cited in n. **I**-2:3), pp. 234 ff.

6.4 Baylak al-Qipjāqī's tables of ascensions

MS Cairo MM 41, copied in 1191 H [= 1777/78], contains a set of tables of the right ascensions for each minute of argument computed to three digits, attributed to the well-known scholar Baylak ibn ʿAbdallāh al-Qipjāqī[7] and stated to have been compiled in 674 H [= 1274/75]. The manuscript contains 41 folios of tables. There is an introduction by Sayf (?) al-Dīn Ibn al-Mushrif (**I-9.10**).

6.5 Najm al-Dīn al-Miṣrī's universal table for timekeeping

The tables of either of the two functions t(h,λ) or T(h,λ) which were compiled for particular latitudes each contain over 10,000 entries. To compile a table displaying either function for each degree of latitude would be an enormous undertaking. Nevertheless, around the year 1325 the Egyptian astronomer Najm al-Dīn al-Miṣrī (**2.5**) compiled a table displaying for any latitude the time since rising of either the sun or any non-circular star as a function of the solar or stellar altitude. This remarkable table is the largest known table from the entire medieval period and contains over 440,000 entries.

MSS Cairo MM 132 and Oxford Marsh 676 (Uri 944 = 995) are two halves of a unique copy of Najm al-Dīn's tables, copied *ca.* 1325 probably by the author. The arguments which one feeds into the main table to find the time since rising of any celestial body for any latitude are H, h, and D: for details see **I-2.6.1**. The table can also be used for solving all problems of spherical astronomy for any latitude: for details see **I-9.3***. Other tables are included to facilitate the determination of H and D for any latitude, starting from the solar longitude λ or stellar declination Δ. Solar tables based on Ibn Yūnus' parameters, and are a star catalogue giving the equatorial coordinates of 367 stars, are also provided.

The existence of Najm al-Dīn's tables raised certain questions. What, if any, was his relationship with al-Maqsī and al-Marrākushī? What, again if any, was his relationship with the *Muṣṭalaḥ Zīj* ? What were his sources? One surely does not just sit down to compile a table containing over close to half a million entries which works for all latitudes, without having seen smaller tables for specific latitudes. If Najm al-Dīn preceded al-Maqsī, as I believe he did, then there were not yet any tables for T(h,λ) available for Cairo. No tables of T(H,h), the so-called *ṭaylasān zīj*es, are known to have been compiled for Cairo, although they had been used by astronomers elsewhere for centuries. I suspect that Najm al-Dīn's inspiration for his main table might have come from some tables of Ibn Yūnus which are no longer extant. Suppose, for example, that Ibn Yūnus had compiled a table of:

$$G(H,D) = \text{Vers } D \,/\, \text{Sin } H \,,$$

for each degree of both arguments. The values of T(H,h,D) can then be generated using the simple formulae:

[7] On Baylak see the article by Martin Plessner in *DSB*, and *Cairo ENL Survey*, no. C9. The unique manuscript of a treatise on mechanical clocks by Riḍwān al-Khurasānī (MS Istanbul Köprülü 949 – see *Cairo ENL Survey*, no. C1) was copied by Baylak in 658 H [= 1260]; this treatise is now published – see Hill, *Studies*, p. xix, n. 4.

Vers t(H,h,D) = (Sin H - Sin h) • G(H,D) and T(H,h,D) = D - t(H,h,D) .
It seems to me that certain errors in the tables of h(D,T) for Cairo (**6.9**) which I believe were also computed by Najm al-Dīn point to the fact that they are based on an auxiliary table computed by someone else. For example, the entries for T = 104° and 105° in the table for D = max D = 104;36° are respectively 83;30° and 83;31° when they should be 83;34° and 83;35° (= max H).

It would not surprise me if further research on Egyptian astronomical manuscripts revealed the existence of a table of G(H,D) computed by Ibn Yūnus. Tables of related functions compiled by later Egyptian astronomers have been located in the manuscript sources: see **I-6.7-9**.

What did surprise me in 1982 was the discovery in MS Dublin CB 102,1 of Najm al-Dīn's instructions on the use of his table as a universal auxiliary table. Only in 1998 was an account of this published, namely by François Charette:[8] see the summary in **I-9.3***.

6.6 The spherical astronomical tables in the *Muṣṭalaḥ Zīj*

The *Muṣṭalaḥ* ("Technical") *Zīj* appears to have been the most popular *Zīj* from the 13th to the 15th centuries (**6.0**). The 17th-century Turkish bibliographer Ḥājjī Khalīfa attributes it to an individual named Muḥammad ibn Muḥammad al-Fāriqī,[9] on whom I have no further information, and he notes that it is based on the *Zīj* of Ibn Yūnus. Two recensions of this work are known to me, MSS Paris BNF ar. 2513 from the 13th century and Paris ar. 2520 from the 14th century, and both are incorrectly attributed to Ibn Yūnus on their title folios. In the former, the *zīj* is entitled *Kitāb al-Taʿlīm fī waḍʿ al-taqwīm*, "Book of Instruction on Finding Planetary Positions", and in the latter *Kitāb al-Zīj al-Muṣṭalaḥ*. The two copies are by no means identical, but their relationship to each other is firmly established by the similarity of the introductory text and the planetary tables, if not the spherical astronomical tables.

The planetary tables in the two Paris manuscripts are a motley collection lifted from various sources, including the *Zīj* of Ibn al-Aʿlam,[10] the *Zīj*es of Ibn Yūnus, the *Zīj* of al-Kammād,[11] and a *zīj* whose name is variously written as *Shāhī, Shāwī,* and *Shāmī,* and which may be the *Shāhī Zīj* of Ḥusām al-Dīn Sālār.[12] There are also tables in both manuscripts taken from Abbasid sources such as *Mumtaḥan Zīj* (solar eclipse table, MSS ar. 2513, fol. 48r, and ar. 2520, fols. 58v-59r),[13] the *Zīj* of Ḥabash (auxiliary tables for spherical astronomy, MS ar. 2520, fols. 70r-71r),[14] and another early *zīj,* perhaps of Andalusī origin (lunar crescent visibility table for the seven climates based on Indian theory, MS ar. 2513, fol. 71v).[15] MS ar. 2520 seems to be closer to the original than MS ar. 2513. In the former the mean motion tables run from

---

[8] Charette, "Najm al-Dīn's Monumental Table".
[9] On the *Muṣṭalaḥ Zīj* see n. **I**-6:14.
[10] On Ibn al-Aʿlam see Sezgin, *GAS*, VI, pp. 215-216, and now Kennedy, "Ibn al-Aʿlam's Tables"; Tihon, "L'astronome Alim"; and Mercier, "Ibn al-Aʿlam's Parameters".
[11] On al-Kammād see Suter, *MAA*, no. 487 and Kennedy, "*Zīj* Survey", nos. 5, 66, 72.
[12] On Ḥusām al-Dīn Salār see Suter, *MAA*, no. 482, and Kennedy, "*Zīj* Survey", no. 32.
[13] On these tables see Kennedy & Faris, "Yaḥyā b. Abī Manṣūr on Solar Eclipses".
[14] On Ḥabash see n. **I**-9:1.
[15] King, "Tables for Lunar Crescent Visibility", pp. 197-207, and **VIa-7**.

630 H [= 1232] and in the latter from 750 H [= 1349]. In MS ar. 2513, fol. 11r (but not in MS ar. 2520) there is a numerical example of the determination of the solar longitude for the year 682 H [= 1283]. I suspect that further investigation of these and related sources will reveal a connection with the scholar Ibn al-Lubūdī, who lived in the mid 13th century and is known to have moved from Damascus to Cairo.[16] He is also known to have compiled two *zīj*es, one of which was based on the *Shāhī Zīj*. At this stage of the research it is perhaps not amiss to mention an anonymous Yemeni *zīj* preserved in MS Paris BNF ar. 2523, copied in the late 14th century, which was compiled in Taiz in 775 H [= 1374] and contains planetary tables from the *Ḥākimī, Shāhī* and *Muntaḥal* (?) *Zīj*es.[17] Thus the combination of *Ḥākimī* and *Shāhī* tables is attested in both Egypt and the Yemen, but we also know that the Yemeni astronomers of the 14th century were familiar with the *Muṣṭalaḥ Zīj*.[18]

The potential value of each of these manuscripts for furthering our knowledge of Islamic astronomy is obvious, and all of them deserve detailed study. My present purpose is to briefly note certain of the spherical astronomical tables which occur in the two copies of the *Muṣṭalaḥ Zīj* and relate to our subject. Most of them are not found in any other Egyptian sources currently known to me.

Geographical coordinates

The latitude of Cairo is taken as 30;0° in the spherical astronomical tables in the *Muṣṭalaḥ Zīj*. However, MSS Paris ar. 2513 and 2520 contain two different sets of geographical tables, the former also with values of q,[19] as is shown by the following entries:

Source	2513, fols. 87v-88v			2520, fols. 82v-83r	
Locality	ϕ	L	q	ϕ	L
Cairo	30;0°	55;0°	52;30°	*29;55°*	54;55°
Damascus	33;27	60;0	39;50	33;30	60;0
Jerusalem	32;0	57;50	43;30	32;10	57;20
Mecca	21;20	67;0	-	21;30	67;0

Note that the distinctive value 29;55° for the latitude of Cairo was also used by al-Marrākushī (**6.7**), and that the distinctive value 33;27° for the latitude of Damascus was also used by al-Mizzī (**9.2**). Accurately computed for the coordinates displayed in MS ar. 2513, the three qibla values are respectively 54;2°, 28;51°, and 39;42° using the exact formula and 54;4°, 30;8°, and 40;43° using the standard approximate formula. Thus the stated qibla values are wretchedly inaccurate, or based on other coordinates.[20] al-Marrākushī (**6.7**) states that the qibla value 52;30° for Cairo was widely accepted there, and it may be that it was conceived as a happy compromise between Ibn Yūnus' first and second values, 52° and 53°.

[16] On Ibn al-Lubūdī see n. 9:10.

[17] See n. 6:9.

[18] King, *Astronomy in Yemen*, pp. 36-37.

[19] Overlooked in King, *Mecca-Centred World-Maps*, pp. 76-84, where, however, several other wretched Egyptian tables of this kind are discussed.

[20] *Ibid.*, p. 80. See also **9.1**.

Obliquity of the ecliptic

In MS ar. 2520, fol. 72r, in what purports to be the text of the *Zīj* (there is no corresponding section in MS ar. 2513) it is stated that "*al-Sharīf*", which is the common designation of the 10th-century Baghdad astronomer Ibn al-Aʿlam (see above) and is used to refer to him elsewhere in MS ar. 2520, found the solstitial meridian altitudes of the sun to be 80;10° and 33;6° and derived the value 23;33° for the obliquity. The corresponding latitude is 33;27°, and so the observation must have been made at Damascus (see **9.2**). However, Ibn al-Aʿlam is not known to have visited Damascus. Ibn Yūnus (MS Leiden Or. 143, p. 223) attributes the value 23;34,2° for the obliquity to him.[21]

In the various tables in MSS ar. 2513 and 2520 (see below) the four values $\varepsilon = 23;30°$, 23;33°, 23,35° and 23;51,20° are used. In general we can associate the first with the *Īlkhānī Zīj* (and perhaps the *Shāhī Zīj*), the second with the *Mumtaḥan Zīj* and the *Zīj* of Ḥabash (and Ibn al-Aʿlam?), the third with Ḥabash again as well as al-Battānī and Ibn Yūnus, and the fourth with Ptolemy.

Trigonometric and spherical astronomical tables

In MS ar. 2520 we find the following tables of this kind:

(1) Tables of Sin θ, $\text{Tan}_{60}\ \theta$, $\text{Cot}_{12}\ h$, and $\text{Cot}_7\ h$ (fols. 67v-69r). Values are given to three digits for each degree of argument.

(2) A table of $\delta(\lambda)$ with values to three digits for each 0;6° of λ (fols. 50v-51v). The table is based on $\varepsilon = 23;30°$, and it is stated that it was taken from the *Īlkhānī Zīj*.
Tables of $H(\lambda)$ and $2D^h(\lambda)$ for $\phi = 30°$ with values to three digits for each degree of λ (fols. 63v-64r), the first based on $\varepsilon = 23;30°$ and the second on $\varepsilon = 23;35°$.
A table of $\text{Tan}_{60}\ \delta(\lambda)$, labelled *fuḍūl al-maṭāliʿ li-'l-arḍ kullihā*, "ascensional differences for all the earth", with values to three digits for each degree of λ. The underlying value of ε is 23;33,0°. See further **I-7.1.9**.

(3) Tables of $\alpha(\lambda)$, $\alpha'(\lambda)$, and $\alpha_\phi(\lambda)$ ($\lambda = 30°$) with values to two digits for each degree of λ (fols. 60v-63r). These tables are based on $\varepsilon = 23;35°$; the first two appear to be derived from Ibn Yūnus' table of $\alpha(\lambda)$ to three digits, and the third is the same as his table in the *Ḥākimī Zīj*.

(4) Anonymous auxiliary tables based on $\varepsilon = 23;33°$, in fact due to Ḥabash (fols. 70r-71r) – see **I-9.1**.

MS ar. 2513 contains the following spherical astronomical tables:

(1) A table of $\text{Tan}_{60}\ \theta$ to three digits for each degree of argument, which differs from the one in MS ar. 2520.

(2) A table of $\delta(\lambda)$ with values to three digits for each degree of λ (fol. 35r), based on $\varepsilon = 23;33,0°$. But for minor variants this is the same as the table in the Mosul recension of the *Mumtaḥan Zīj* preserved in MS Escorial ár. 927.
A table of $H(\lambda)$ for $\phi = 30°$, with values to three digits for each degree of λ (fol. 49r), based on $\varepsilon = 23;35,0°$. The underlying values of $\delta(\lambda)$ are those of the *Ḥākimī Zīj*.

[21] See King, "Earliest Muslim Geodetic Measurements", pp. 226-227, on Ibn al-Aʿlam's determination of the latitude of Baghdad as 33;21°.

A table of $\delta^*(\lambda) = 90° + \delta(\lambda)$ to two digits (fol. 55r), which is based on $\varepsilon = 23;35°$, but on values of $\delta(\lambda)$ which are less accurate than those of Ibn Yūnus.

A table of $\text{Tan}_5\ \delta(\lambda)$ to three digits based on $\varepsilon = 23;35°$ (fol. 81r). This table differs from that of al-Marrākushī (**6.7**) but is the same as that in MS Cairo MM 43 , fol. 42r, of a set of anonymous spherical astronomical tables for Cairo. See further **I-7.1.8** on this table and **II-6.14** on the others in MS Cairo MM 43.

(3) A table of $d(\lambda)$ for $\phi = 30°$, with values to two digits for each degree of λ (fol. 53v), based on the values to three digits in the *Ḥākimī Zīj*.

A table of the maximum half-excess of daylight in equinoctial hours, max $d^h(\phi)$, for each degree of ϕ from 1° to 66°, with values to three digits (fol. 57v). The table is based on Ptolemy's value of the obliquity, $\varepsilon = 23;51,20°$.

Tables of $\alpha'(\lambda)$ and $\alpha_\phi(\lambda)$ ($\phi = 30°$) (fols. 46v-47r), as in MS ar. 2520.

(4) A table of $h_a(H)$ to two digits for each degree of H (fol. 89v).

Also, a table of $h_a(\lambda)$ with entries to two digits for each degree of λ (fol. 54r). The entries differ slightly from those in the main corpus and the auxiliary tables of al-Khaṭāʾī (**4.9**).

(5) Auxiliary tables for timekeeping by the sun, based on $\phi = 30°$ and $\varepsilon = 23;35°$ (fols. 53v and 62v). The functions tabulated are:

$$\text{arc Sin } \{ \text{ Sin } \delta(\lambda) \sin \phi \ / \ R \ \} \text{ and } R \ / \ \text{Cos } \delta(\lambda) \ .$$

See further **I-6.7.1** and **I-6.10.3** on these.

6.7 al-Marrākushī's spherical astronomical tables for Cairo

al-Marrākushī's *summa* on spherical astronomy and instruments was the most important single work on this subject to be compiled in medieval times (see already **2.7** and **6.0**). It was also the most influential, if only in Egypt, Syria and Turkey. I here describe only the tables presented by al-Marrākushī and discuss his remarks on the qibla at Cairo and on the twilight.

The treatise begins with a discussion of calendars and trigonometry, with approximate tables, as well as a table displaying the solar longitude for the year 1275/76 and a star catalogue for 1282. al-Marrākushī was of the opinion that the obliquity of the ecliptic oscillated between 23;53° and 23;33° and that there was an associated oscillatory motion of the equinoxes on the ecliptic. For the present discussion it suffices to note that in all of his tables which are based on a value of the obliquity he used $\varepsilon = 23;35°$.

al-Marrākushī tabulated $\delta(\lambda)$ to two digits for each degree of λ (I.183). Nine of his ninety entries are in error by -1 in the second digit, which suggests he rounded the entries in a table of $\delta(\lambda)$ to three digits. He used the approximate value Sin ε = Sin 23;35 $\approx$ 24 instead of the more accurate value 24;0,18 in his numerical example. He also tabulated $\delta_2(\lambda)$, the second declination, in similar fashion (I.184). al-Marrākushī then tabulated Tan $\delta(\lambda)$ for $\varepsilon = 23;35°$ to three digits for each degree of λ (I.185) – see **I-7.1.7**. His entries are again considerably less accurate than those in the corresponding table in the *Ḥākimī Zīj*. In particular he uses Tan $\delta(90°)$ = Tan 23;35° = 26;11;40, a value which can be obtained from interpolating between correct values of Tan 23° and Tan 24°. Ibn Yūnus has the more accurate value 26;11,33,(23).

al-Marrākushī also tabulated $\lambda(\delta)$ to two digits for each 0;15° of δ up to 23;35° (I.186/187).

Two rather unusual tables (see further **I-7.1.7**) presented by al-Marrākushī (I.209-210) display the functions:

$$f(\lambda) = {}^1/_{12}\ \mathrm{Tan}_{60}\ \delta(\lambda) \quad \text{and} \quad g(\Delta) = {}^1/_{12}\ \mathrm{Tan}_{60}\ \Delta$$

to three and two digits respectively for each degree of arguments λ and Δ. The purpose of these tables is to determine the half excess of daylight using only another of Cotangents to base 12, thus:

$$\mathrm{Sin}\ d(\lambda) = f(\lambda) \cdot \mathrm{Cot}_{12}\ \bar{\phi} \quad \text{and} \quad \mathrm{Sin}\ d(\Delta) = g(\Delta) \cdot \mathrm{Cot}_{12}\ \bar{\phi}\ .$$

The use of these functions reflects a strong sentimental attachment to the base 12, which was characteristic of the *muwaqqits* of medieval Egypt.

al-Marrākushī's next tables in the treatise display $d(\phi,\lambda)$ for each 6° of both arguments (I.214-215). His entries for $\phi = 30°$ are slightly different from the corresponding entries in the main Cairo corpus. His tables of $\alpha'(\lambda)$ to three digits for each degree of λ (I.222-225) are less accurate than those which can be derived from the corresponding values of $\alpha(\lambda)$ in the *Ḥākimī Zīj*. He also tabulates $\alpha_\phi(\lambda)$ to degrees and minutes for each 6° of both arguments ϕ and λ (I.230-235), whereas Ibn Yūnus tabulated them for each 1° of both arguments. al-Marrākushī does, however, include a table of $\alpha_\phi(\lambda)$ for $\phi = 30°$ with entries for each degree of λ in degrees and minutes (I.238-240), which again are less accurate than those in the *Ḥākimī Zīj*. Another table of al-Marrākushī's displays the length of the longest day in hours, minutes, and seconds for each degree of ϕ up to 66° (I.246).

al-Marrākushī presents a number of tables scattered throughout the remainder of his treatise which relate the solar altitude and shadow to the time of day. All of the entries in these tables are to two digits, and the functions of time have $T = 1, 2, \ldots, 5$ or 6^{sdh} as arguments and sometimes also the time of the afternoon prayer. The tables which are independent of terrestrial latitude are of the functions:

$h(T,H)$	$H = 5°, 10°, \ldots, 90°$	(I.457)
$z_{(12)}(T,H)$	$H = 5°, 10°, \ldots, 90°$	(I.253)
$z'_{(12)}(T,x)$	$x = Z'_{(12)} = 1, 2, \ldots, 12$	(I.254)
$z_{(12)}(T,x),\ z'_{(12)}(T,x)$	$x = Z_{(12)} = 1, 2, \ldots, 36$	(I.256-257)
$h_a(H),\ \mathrm{Sin}\ h_a(H),\ z_a'{}_{(12)}(H)$	$H = 5°, 10°, \ldots, 90°$	(I.271)

The tables specifically for $\phi = 30°$ are of the functions:

$h(T,\lambda)$	$\lambda = 270°, 300°, \ldots, 90°$	(I.454)
$z_{(12)}(T,\lambda),\ z'_{(12)}(T,\lambda)$	$\lambda = 270°, 280°, \ldots, 90°$	(I.428-429, 436-437)
$h_0(\lambda),\ \psi(\lambda)$	$\lambda = 30°, 60°, 90°$	(II)
$h_q(\lambda)$	$\lambda = 270°, 300°, \ldots, 90°$	(II)

Finally there is a table displaying $h(T,\lambda)$ for each *equinoctial* hour from 1 to 6 and each 30° of λ. Each of these small tables for timekeeping is of limited practical use. See further **I-4.2.4**. The remaining tables in the treatise are specifically intended for use with sundials, and do not concern the present study.

Geographical coordinates and qibla

In his lists of geographical coordinates (I.202-204, 315-317) al-Marrākushī has the following entries for Mecca and Cairo:

Mecca	φ: 21;0°	L: 77;0°
Cairo	29;55°	64;50° (I.316) / 64;54° (I.320)

The latitude of Cairo he claims to have measured himself; however, this value is also attested in earlier sources.[22] Also, in most of his numerical examples he uses the standard value 30;0° (*e.g.*, I.260, 303, 306, 433) rather than 29;55° (*e.g.* I.320, 339). He computes the qibla for Cairo using coordinates (I.320-322):

Mecca	φ: 21°	L: 77°
Cairo	30°	65°

to be 37;3° measured from the prime vertical (q = 52;57°). Ibn Yūnus' value 53,0,17°, derived from the same latitudes and longitude difference, is much closer to the accurately-computed value 53;0,23°. al-Marrākushī then computes the qibla at Cairo for his own coordinates (with 64;50° for the longitude of Cairo), and finds it to be 36;18° or 36;19° (q = 53;42° or 53;41°). The first of these values is in fact accurately computed. He notes that if one uses 21;40° for the latitude of Mecca the azimuth (measured from the prime vertical) will be smaller, adding that the qibla value 37;30° (q = 52,30°) (Sédillot has 35;30° here, which must be a copyist's error) which is "widely accepted in Cairo" is "quite incorrect". This value q = 52;30° occurs again in the geographical tables in MS Paris BNF ar. 2513 of the *Muṣṭalaḥ Zīj* (**6.6**), and was probably intended as a happy compromise between Ibn Yūnus' first and second values, 52° and 53°. Finally, it is worth noting that al-Marrākushī's table of $h_q(\lambda)$ (see above) is based on parameter q = 52°, although this is not stated. Further, six of al-Marrākushī's seven values of h_q are the same as those of Ibn Yūnus in the *Ḥākimī Zīj*.

As I have already noted (**2.7**), al-Marrākushī advocated the parameter 20° and 16° for morning and evening twilight. He presents several numerical examples, and computes r to be 23;16°, 28;29°, and 24;54° (Sédillot has incorrectly 24;15°) at the equinoxes, summer and winter solstices, and s to be 18;33°, 22;13°, and 20;4°. Now in MS Dublin CB 3673 of the main Cairo corpus there are tables of r(λ) and s(λ) based on these parameters (**4.10**). No values are given for the equinoxes because of the format of the tables but the solstitial values given, each of which is accurately computed, are respectively:

r: 28;28° and 24;54° and s: 22;13° and 20;4° .

6.8 al-Khaṭā'ī's auxiliary tables for timekeeping by the sun

Muhammad ibn al-Amīr Fakhr al-Dīn ʿUthmān al-Khaṭāʾī is an individual whose name is new to the modern literature; his father was a Mamluk prince who flourished around 1450 (**I-6.15.1**). His auxiliary tables, which contained about 825 entries, are partially extant in MS Vatican Borg. ar. 217,2 (fols. 6r-7v), copied *ca.* 1500. They are less general in their application than other later Egyptian and Syrian tables, being specifically intended for finding the hour-angle from the solar altitude at the latitude of Cairo. The functions tabulated are related to those of al-Khalīlī's minor set of auxiliary tables, but the two sets may have been compiled inde-

[22] See Kennedy & Kennedy, *Islamic Geographical Coordinates*, p. 111, and now King, "Geography of Astrolabes", p. 9 and n. 23.

pendently. al-Khaṭāʾī also tabulated $h_a(\lambda)$, and a(h) at the equinoxes. His table of the first function is more accurate than the ones in the main corpus and the *Muṣṭalaḥ Zīj* (**4.4** and **6.6**). See further **I-6.15.1**.

6.9 Anonymous solar altitude tables for Cairo

MS Cairo MM 72, copied in 747 H [= 1346/47] by Ibn al-Kattānī (**5.5**), contains a curious set of tables preceded by an introduction of which unfortunately only the last page remains. The function tabulated is not explained in the surviving portion of the introduction but in fact it is the solar altitude for values of the time since sunrise and different values of the semi diurnal arc, that is, h(D,T). Values of this function to two digits are given for the domains:

$$D = 104;36°, 104°, 103°, \dots, 76°, 75;24° \quad \text{and} \quad T = 1°, 2°, \dots, [2D] ,$$

where [2D] is the largest integer less than 2D. The underlying parameters are:

$$\phi = 30;0° \text{ (Cairo) and } \varepsilon = 23;35° ,$$

as in the main Cairo corpus. (Note that 104;36° and 75;24° are the extremal values of D for these parameters.) For each double page of tables serving a value of D, a value of the function $t_a(D)$ is given alongside the argument D at the head of the tables. The altitude of the sun at each seasonal and equinoctial hour of daylight is also displayed for each value of D at the side of the main tables. I suspect that these tables may be by Najm al-Dīn al-Miṣrī (**6.5**): more information is in **I-4.4.1** (illustrated).

6.10 Anonymous prayer-tables for Mecca

MS Cairo MM 68, copied *ca*. 1500, contains an anonymous set of prayer-tables computed for latitude 21°, that is, for Mecca.[23] The work bears no title and no date, but I do not doubt that it was compiled in Egypt. Establishing the identity of the compiler of these tables would probably solve some of the remaining problems associated with the main Cairo corpus. The format of the tables in this Cairo manuscript is the same as that of the main corpus, and the functions tabulated are:

$$H, D, h_a, t_a, T_a, 2N, (2n - r), r, s, \alpha_s \text{ and } \alpha_\phi .$$

The underlying value of ε is 23;35° and the tables are rather accurately computed.

The tables for latitude 21° in the corpus of prayer-tables for all latitudes preserved in MS Princeton Yahuda 861,1 (**8.1**) are quite different from these, being very carelessly computed. The table for α_ϕ in the Cairo manuscript is identical with that of Ibn Yūnus for latitude 21° in MS Leiden Or. 143 of the *Ḥākimī Zīj*, containing the same slight errors. The tables for twilight, that is, of the functions r and s, are based on parameters 19° and 17°, and do not display the error made by al-Mizzī in his tables for Damascus (**9.2**). MS Cairo MM 68 also contains a table of a function called *al-aṣl al-muṭlaq*, "the absolute base", defined by:

$$B(\delta) \quad = \quad \text{Cos}\ \delta\ \text{Cos}\ \phi\ /\ R .$$

Values are given to two digits for each 0;5° of argument up to δ = ε. (See further **I-6.4.4**.)

[23] *Cairo ENL Survey*, no. C40.

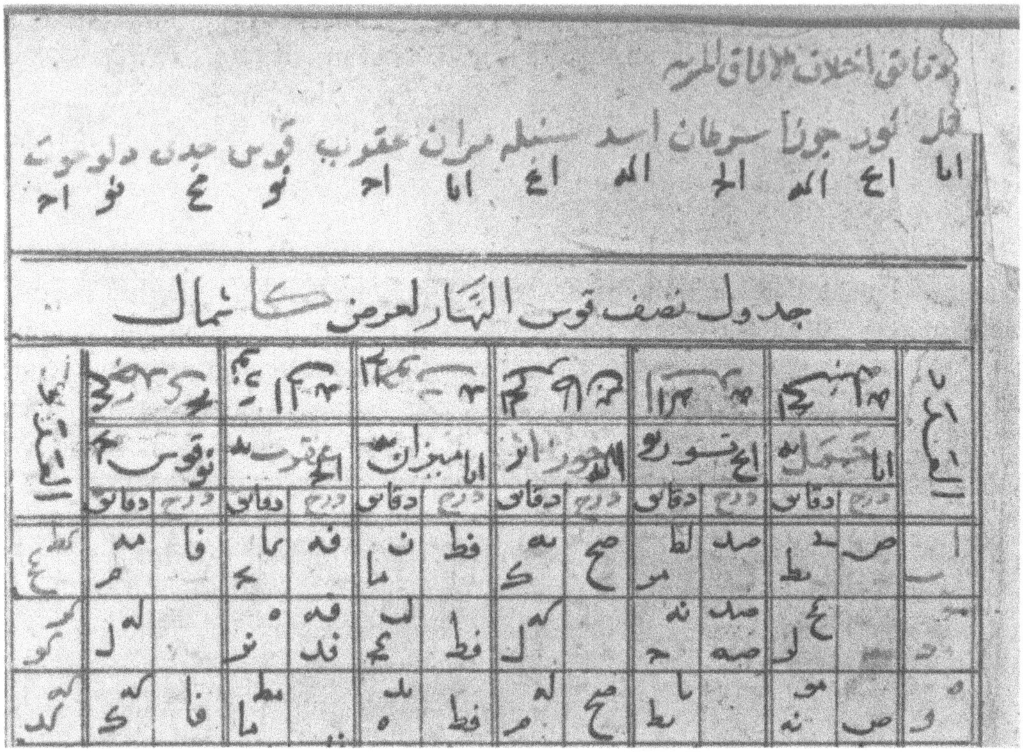

Fig. 6.10: The values above the table of the half arc of daylight represent the time taken by the sun to move between the true and the visible horizons. [From MS Cairo MM 68, courtesy of the Egyptian National Library.]

At the top of the table of D(λ) is a set of numbers for each of the signs (see **Fig. 6.10**) . They are labelled *daqāʾiq ikhtilāf al-āfāq al-marʾiyya*, "the "difference minutes" of the visible horizons", and it is clear that they are intended to represent the effect of refraction at the horizon on the arc of daylight, that is 2(ΔD), twice the "difference minutes" of the Cairo astronomers (**4.11**). The values given for the twelve signs starting with Aries are as follows:

1;11° 1;18 1;25 1;33 1;25 1;18 1;11 1;3 0;56 0;48 0;56 1;3

Thus the equinoctial value is intended to be 1;11° and the values for the summer and winter solstices are respectively 0;22° more and 0;23° less. The equinoctial value corresponds to an angle of depression below the horizon of approximately:

$$(^1/_2 \cdot 1;11°) \cos 21° = 0;33° ,$$

but this is not what is intended. Consider the values of ΔD for $\phi = 21°$ which are given by al-Minūfī (**8.2**). These yield the following rounded values for 2(ΔD), "according to al-Minūfī" (A) and "according to earlier scholars" (B):

A: 1;6 1;13 1;20 1;27 1;20 1;13 1;6 0;59 0;52 0;45 0;52 0;59
B: 1;6 1;14 1;21 1;28 1;21 1;14 1;6 1;0 0;53 0;46 0;53 1;0

These values are close to those given in the Cairo manuscript. Actually, however, if we multiply Ibn Yūnus' values by a factor of 2 • 22;40°/30°, rather than 2 • 21°/30°, which is the factor underlying al-Minūfī's values, we obtain the set:

1;11 1;19 1;26 1;34 1;26 1;19 1;11 1;3 0;56 0;48 0;56 1;3 .

Now these correspond very closely to those for Mecca in the Cairo manuscript. It remains only to explain the number 22;40°. Perhaps the values of 2(ΔD) were originally intended for ϕ = 21;40°,[24] another widely-used medieval value for the latitude of Mecca, and a careless error was made in the computation.

6.11 Ibn al-Rashīdī's prayer-tables for Mecca

MS Cairo Sh 89,4 (fols. 29v-32v), copied in 1025 H [= 1616], is the only known copy of the introduction to a set of prayer-tables for latitude 21°, that is, Mecca, by Ibn al-Rashīdī (**4.1**).[25] The tables are no longer contained in the manuscript but they are described in the text. The latitude 21° is specifically mentioned, as well as the parameters 19° ands 17° for twilight (*cf.* **6.10**). Ibn al-Rashīdī mentions ten tables of the following functions:

$$T(h) \text{ and } t(h) \text{ for } \lambda = 0°, H, D, h_a, t_a, h_b, T_a, s, r, \alpha_\phi .$$

The first table has its counterpart in the tables for Jerusalem in MS Princeton Yahuda 861,1 which I suspect were also computed by Ibn al-Rashīdī (**8.1** and **9.5**).

Added in proof: My notes to these tables indicate that MS Leiden UB Or. 2805, which I have not consulted, may contain the same tables.

6.12 Anonymous timekeeping tables for Jerusalem

Two disordered copies of timekeeping tables for Cairo, MSS Cairo DM 45 and Cairo DM 153, contain odd folios copied in the same hand *ca.* 1650 from a set of tables of the hour-angle t(h,λ) computed for

$$\phi = 32;0° \text{ (Jerusalem) and } \varepsilon = 23;35° .$$

The tables have the same format as the Cairo corpus and are probably of Egyptian provenance.[26] The entries are identical with the corresponding entries in al-Karakī's tables of t(λ,h) (**9.4**). Since al-Karakī states in his introduction that the 14th-century Egyptian astronomer Ibn al-Rashīdī had compiled a set of tables of t(h,λ) for an unspecified latitude I suspect these tables for Jerusalem are by Ibn al-Rashīdī. See also **I-2.1.5**.

6.13 al-Bakhāniqī's tables for the afternoon prayer at different latitudes

MS Cairo MM 33,2 (fols. 6v-7r), penned *ca.* 1600, is the only copy known to me of a table for the afternoon prayer attributed to al-Bakhāniqī, the editor of the main Cairo corpus (**4.1.5** and **5.6**). The table is entitled *jadwal maʿrifat faḍl dāʾir al-ʿaṣr li-ruʾūs al-burūj*, "table for finding the hour-angle at the beginning of the afternoon prayer at the first points of the zodiacal signs". The function tabulated is:

$$t_a(\phi,\lambda)$$

---

[24] King, "Earliest Muslim Geodetic Measurements", pp. 225-226.
[25] See *Cairo ENL Survey*, no. C39/3.1.5.
[26] *Ibid.*

for argument domains:

$\phi = 10°, 11°, \ldots, 50°$ and $\lambda = 90°, 60°, 30°, 0°, 210°, 240°, 270°$,

and entries are given in degrees and minutes. See also **VIb-7** (illustrated).

6.14 Four Egyptian copies of al-Khalīlī's universal auxiliary tables

MS Cairo MM 43 is an Egyptian manuscript, perhaps dating from around 1600, which contains al-Khalīlī's universal auxiliary tables (**10.5**), complete with his introduction but without mention of his name. The functions f_ϕ and g_ϕ are tabulated for values of ϕ up to 50° rather than 55° and the function G is tabulated for values up to 55 rather than 59. These auxiliary tables are preceded by tables of:

$\delta(\lambda)$ ($\varepsilon = 23;31°$) and $h_a(H)$

to two digits for each degree of argument, and are followed by a mixed bag of anonymous spherical astronomical tables. The following functions are tabulated to two digits for each integral unit of argument unless otherwise stated:

(1) $B(\lambda)$ ($\phi = 30°$ and $\varepsilon = 23;55°$), $\Delta\lambda = 0;15°$
(2) max $D(\phi)$ ($\varepsilon \approx 23;30°$), to hours, minutes and seconds, $\phi = 1°, 2°, \ldots, 66°$
(3) $\alpha_\phi(\lambda)$ ($\varepsilon = 23;35°$), $\phi = 6°, 12°, \ldots, 66°$, $\Delta\lambda = 6°$
(4) arc Sin (x)
(5) $\alpha(\lambda)$
(6) arc Tan_{12} (x)
(7) $H(h_a)$, $h_a = 1°, 2°, \ldots, 45°$ (for use on quadrants)
(8) $\psi(\Delta)$ ($\phi = 30°$), $\Delta = 1°, 2°, \ldots, 60°$
(9) $\lambda(\delta)$, $\delta = 1°, 2°, \ldots, 23°, 23;15°, 23;30°, 23;35°$
(10) arc Cot_{12} (x) (garbled)
(11) $h_a(H)$, $\Delta H = 0;30°$
(12) Cot_{12} h, $\Delta h = 0;30°$
(13) $^1/_{12}$ Tan Δ
(14) $^1/_{12}$ Tan $\delta(\lambda)$ ($\varepsilon = 23;35°$), 3 digits
(15) Vers θ, 3 digits
(16) Sin θ, 3 digits
(17) $\delta(\lambda)$ ($\varepsilon = 23;35°$)
(18) $\alpha_\phi(\lambda)$ ($\varepsilon = 23;35°$, $\phi = 21°, 24°, 36°$)
(19) $\alpha(\lambda)$ ($\varepsilon = 23;35°$)
(20) $\Delta\alpha_n(\phi)$ ($\varepsilon = 23;35°$), $\phi = 1°, 2°, \ldots, 67°$, $n = 1, 2, \ldots, 12$
(21) $G(\Delta)$ ($\phi = 30°$), $\Delta = 1°, 2°, \ldots, 60°$
(22) $d(\Delta)$ ($\phi = 30°$), $\Delta = 1°, 2°, \ldots, 60°$.

All of the tables are anonymous except no. 20 which is attributed to Ibn Yūnus. There is also a list of geographical coordinates and qiblas[27] and tables for marking curves on astrolabes. On tables nos. 1, 8, 13-14, and 21 see further **I-6.4.3**, **5.6.1**, **7.1.7** and **6.8.1**, respectively.

[27] Overlooked in King, *Mecca-Centred World-Maps*, pp. 76-86.

Three other Egyptian copies of al-Khalīlī's universal auxiliary tables are in MSS Cairo MM 98,2, copied *ca.* 1700, Princeton Yahuda 861,2, and Cairo DM 758, copied *ca.* 1650. In the first the tables are anonymous. In the second the auxiliary tables have been bound in the midst of the tables of the *Natīja al-kubrā* (**8.1**): here they are incorrectly entitled *Faḍl dā'ir al-Khalīlī*, "al-Khalīlī's hour-angle (tables)". In the third the tables also bear the title *Kitāb Faḍl al-dā'ir* and are attributed to al-Khalīlī. Between the introduction and the tables themselves is an anomalous table of $h_v(\lambda)$ for Cairo attributed to Ibn al-Rashīdī (**4.8**), with a remark that for other latitudes the azimuth of the *bādahanj* is also the rising point of Capricorn (!): see **Fig. VIIb-4.1**.

Finally, MSS Cairo DM 644,1, copied *ca.* 1700, and Istanbul S. Esad Efendi Medresesi 119,2, date ?, are two copies of an anonymous Egyptian set of auxiliary tables entitled *Fatḥ al-Karīm al-Bāqī fī maʿrifat al-dā'ir wa-faḍlihi āfāqī*, "The Victory of God, the Noble and Eternal, for Finding the Time Since Sunrise and Hour-Angle for all Latitudes". These tables were lifted from al-Khalīlī's auxiliary tables for timekeeping by the sun (**10.3**). See further **I-9.11**.

6.15 The auxiliary tables of Ibn al-Mushrif, al-Māridīnī, al-Wafā'ī, and al-Ṣūfī

At least five Egyptian astronomers compiled sets of auxiliary tables for solving the standard problems of spherical astronomy.[28] Tables of this kind containing a few hundred entries were first devised by Ḥabash and Abū Naṣr in the 9th and 10th centuries, but the most outstanding example is the set containing over 13,000 entries compiled by al-Khalīlī in the 14th century (**10.7**). Ḥabash's tables were known in Egypt: they are referred to by Ibn Yūnus and were incorporated into one recension of the 13th century *Muṣṭalaḥ Zīj* (**6.6**). al-Khalīlī's universal auxiliary tables, as well as his minor set, were also known in Egypt (**6.14**). I have surveyed the development of Islamic auxiliary tables, including the Egyptian examples noted below, in **I-9**.

Jamāl al-Dīn ʿAbdallāh ibn Khalīl al-Māridīnī (d. 1406) was one of the leading astronomers in Cairo and/or Damascus in his time (**I-9.6**). His auxiliary tables, which contain 8,100 entries, offer little practical advantage to the user because of the simple nature of the functions tabulated. They are contained in MS Paris BNF ar. 2525,1 (fol. 1v-16v), copied *ca.* 1450, where they precede an incomplete set of prayer-tables for all latitudes (**8.1**), and also MS Cairo K 4026 (late). See further **I-9.6**.

Abū Bakr ibn Ismāʿīl known as Ibn al-Mushrif worked in Cairo in the first half of the 15th century, and his name is new to literature.[29] MS Cairo MM 241, penned *ca.* 1450, is a unique copy of a set of auxiliary tables compiled by him containing over 15,000 entries. The tables take advantage of the mathematical equivalence of the problems of determining the hour-angle and the azimuth from solar or stellar altitude. See further **I-9.8**.

ʿIzz al-Dīn ʿAbd al-ʿAzīz ibn Muḥammad al-Wafā'ī (d. *ca.* 1470) was one of the leading astronomers in Egypt in the mid-15th century (**I-9.9** and **II-5.7**). His auxiliary tables are extant

[28] See already King, "Universal Solutions from Mamluk Egypt and Syria", of which a new version is in **VIb**.

[29] On Ibn al-Mushrif (**I-9.8**) see *Cairo ENL Survey*, no. C43. The individual with the same name mentioned in King, *Astronomy in Yemen*, no. 12, is a Yemeni astronomer from the early 14th century.

in MSS Vatican Borg. ar. 217,1 (fols. 1v-5v) and Istanbul Nuruosmaniye 2921,2 (fols. 22r-26v), and contain about 1,500 entries. They are based on the function G(λ) discussed at length by Ibn Yūnus, and, like those of Ibn al-Mushrif and al-Khalīlī, exploit the mathematical equivalence of the problems of determining the hour-angle and azimuth from solar or stellar altitude. See further **I-9.9**.

Ibn Abi 'l-Fatḥ al-Ṣūfī (d. *ca.* 1495) was another of the leading astronomers in Cairo in the late 15th century.[30] He prepared an extensive *zīj* based on those of Ibn al-Shāṭir (*ca.* 1350) and Ulugh Beg (*ca.* 1430), adapting the tables for the geographical coordinates of Cairo. Nevertheless in his own solar tables which precede his auxiliary tables in MS Oxford Seld. Supp. 101 (Uri 1040) he prefered to use the parameters of Ibn Yūnus derived five centuries previously. (The title-folio of MS Leiden Or. 143 of the *Ḥākimī Zīj* bears biographical information on Ibn Yūnus in the hand of al-Ṣūfī: see **Fig. 5.1**) al-Ṣūfī's auxiliary tables contain over 15,000 entries and are more suitable for finding the hour-angle from solar or stellar altitudes for any latitude than for finding the azimuth. See further **I-9.10**.

6.16 al-Asyūṭī's prayer-tables for Assiut

MS Cairo DM 188,1 (fols. 1r-10r), copied *ca.* 1700, contains a set of prayer-tables entitled *Nuzhat al-abṣār fī aʿmāl al-layl wa-'l-nahār li-ʿarḍ kāf-zāy*, "Delights of the Eyes for Astronomical Operations at Latitude 27°", that is, for the city of Assiut. On the title page it is stated that the tables were prepared by Muḥammad (ibn ʿAbd al-Qādir ibn Dallāl) al-Wafā'ī (al-Asyūṭī),[31] a student of Muḥammad ibn (Abī) al-Fatḥ al-Ṣūfī (**6.15**).

The functions tabulated are the following, with identical format to the tables in the Cairo corpus:

$$\delta, H, d, D, \tilde{h}, 2N, \psi, h_0, Z_{(12)}, z_{a(12)}, t_a, T_a, r, s, (2N\text{-}r), (r{+}D) \text{ and } \alpha_\phi .$$

al-Asyūṭī was not a bad student: his tables are rather carefully computed and are more accurate than the tables for latitude 27° in the *Natīja* attributed to al-Wafā'ī (**8.1**). However, al-Asyūṭī preferred to use 23;35° for the obliquity rather than the more up-to-date value 23;30,(17)° of Ulugh Beg. His tables of r and s are based on parameters 19° and 17° respectively, and the former is more accurately computed than the latter. His tables of α_ϕ ($\phi = 27°$) and δ are less accurate than those in the *Ḥākimī Zīj* (**5.1**). On the title folio of MS Cairo DM 188 there is a note stating the approximation:

$$\max d(\phi) \approx {}^{1}/_{2}\, \phi .$$

al-Asyūṭī would doubtless have disapproved of this: his own tables give the accurate value 12;51° for max d = d(90°) ($\phi = 27°$), and the value one can derive with this approximate formula, namely 13;30°, is in error by almost $^{2}/_{3}$°.

[30] On Ibn Abi 'l-Fatḥ al-Ṣūfī see n. **I**-9:30.

[31] On al-Asyūṭī see *Cairo ENL Survey*, no. C101.

6.17 Ibn Ṭūghān's table for determining the times of moonrise and moonset

MS Cairo MM 167,4 (fols. 158v-160r), copied in 989 H [= 1581], contains a short introduction and table of two pages by Jamāl al-Dīn Yūsuf ibn Ṭūghān al-Qiṭṭajī, otherwise known to us as the author of a treatise on astrology.[32] The table, displayed in **Fig. 6.17**, enables the user to determine the time between sunset and moonset from the nocturnal arc, entered horizontally, and the days elapsed since conjunction, entered vertically. Entries are given in degrees and minutes for each 6° of nocturnal arc and each day from 1 up to 16. If more than 14 days of the month have passed the instructions tell us to subtract 14 from the number of days passed since conjunction and enter the remainder to find the time passed since sunset when the moon will rise. The author acknowledges that the underlying procedure is approximate, and in fact the entries f(2N,n) are determined by simple relation:

$$f = \eta \cdot 2N/180 \cdot n ,$$

where η is the mean daily relative motion of the sun and moon, 12;53°.

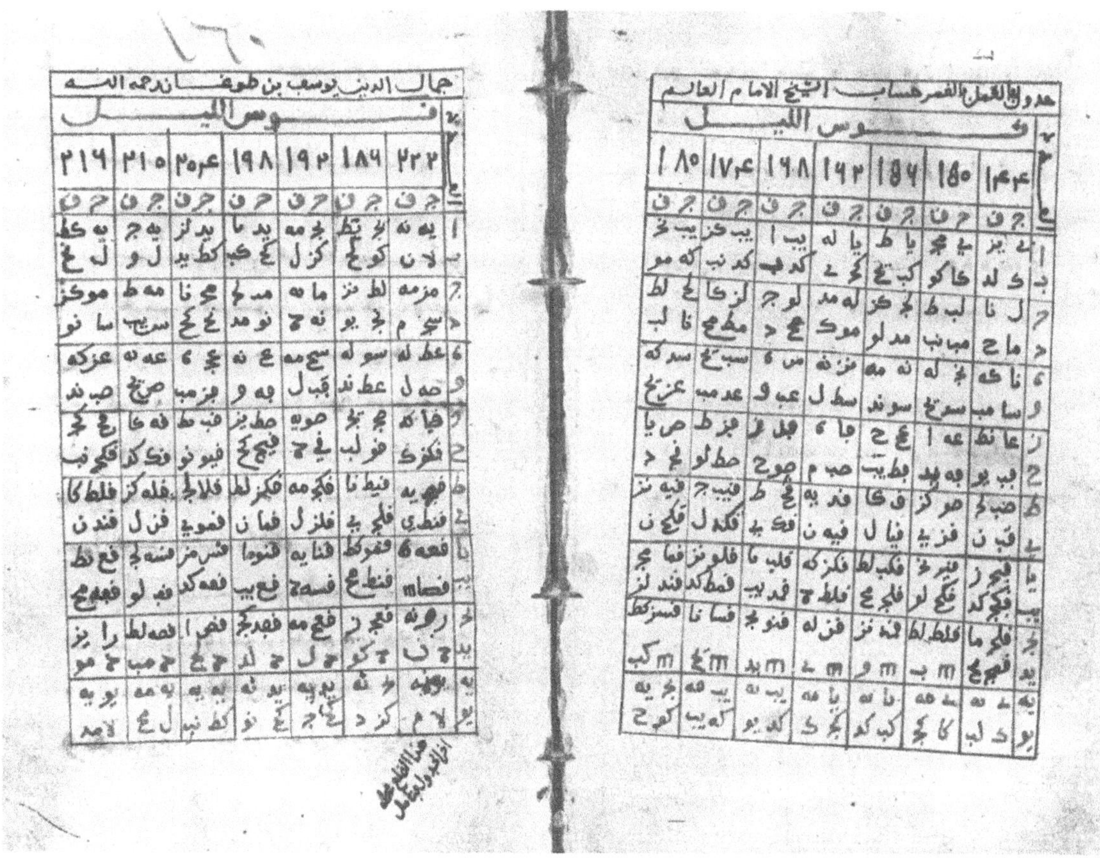

Fig. 6.17: A unique table for finding the times of moonrise and moonset. Note that the first column on the left-hand page should be the last column, as noted by the copyist. [From MS Cairo MM 167,4, fols. 159v-160r, courtesy of the Egyptian National Library.]

[32] On Ibn Ṭūghān see *Cairo ENL Survey*, no. C92; and İhsanoğlu *et al.*, *Ottoman Astronomical Literature*, II, pp. 228-229, no. 108.

CHAPTER 7

LATE MODIFICATIONS TO THE MAIN CAIRO CORPUS

7.0 Introductory remarks

From a mathematical point of view it was difficult for the astronomers of medieval Egypt to improve in any way upon the tables in the main Cairo corpus, these being in general very carefully computed, and the changes in the obliquity of the ecliptic over the centuries had an effect on the prayer-times that was negligible. Nevertheless, one Egyptian astronomer endeavoured to modify at least the prayer-tables in the corpus for a more up-to-date value of the obliquity (**7.1**). Certain *muwaqqit*s with time on their hands succumbed to the temptation to stretch some of the tables in the corpus to display values for each minute of solar longitude instead of each degree (**7.2**). Others rearranged the format of the tables and added information about timekeeping by the stars (**7.5**). Yet others were content to prepare some simple tables for finding the solar longitude for a date in the Hijra calendar and supplement these with prayer-tables lifted from the corpus (**7.7-9**).

From a practical point of view it was clearly more convenient to display the times of prayer in almanacs with values arranged for each day of the year (**7.4**). For those who did not like tables at all, the times of prayers for each day of the year were written out in words (**7.6**, **7.7** and **7.10**). The Egyptian astronomers also modified the tables of the corpus so that the entries were expressed in hours and minutes according to the Ottoman convention that sunset is 12 o'clock (see **7.11** and **7.12-13**; on the convention see **14.0**).

7.1 al-Minūfī's prayer-tables

Shams al-Dīn Muḥammad ibn Nāṣir al-Dīn al-Minūfī was a *muwaqqit* at the Ghawriyya *madrasa* in Cairo[1] who lived in the mid 16th century and compiled a set of prayer-tables in which he sought to modify the tables of the main Cairo corpus. Complete copies of these are MSS Cairo DM 470 and 467, the former, in 50 fols., penned *ca.* 1570 by the author's son ʿAbd al-Qādir (**7.2** and **8.2**), the second in 1144 H [= 1731/32]. The tables are entitled *ʿAnwān al-muhimmāt fī taḥrīr al-awqāt*, "The Epitome of Important Matters in Fixing Prayer-Times".

In the introduction the author claims that the tables are computed according to the "new observations" and that he has reworked the tables of "the *Natīja*" according to the new value of the obliquity, meaning the value 23;30(,17)° found by the astronomers of Ulugh Beg in

[1] Muhammad al-Minufī is not mentioned in the modern bio-bibliographical sources before *Cairo ENL Survey*, no. C120; and Ihsanoğlu *et al.*, *Ottoman Astronomical Literature*, I, pp. 187-188, no. 88 (listing mainly Cairo manuscripts). The suggestion in King, "Astronomical Timekeeping in Medieval Cairo", p. 373, n. 68, that he might be identical with Muhammad ibn Abi 'l-Fatḥ al-Ṣūfī (see **6.15**) is incorrect.

Samarqand. In fact, only a few of the simpler tables have been recomputed. It would be interesting to know precisely what al-Minūfī meant by *Natīja*: perhaps the complete set of prayer-tables for Cairo appended to those of T, t and a in the edition of the main corpus by al-Bakhāniqī (**5.6**)?

The other modification which al-Minūfī introduced was to incorporate corrections for refraction at the horizon in all tables involving a period of time measured from sunrise or sunset. In the introduction to his tables he mentions that because of the nature of light rays, an observer can see more than half of the celestial sphere. According to Ibn al-Haytham, he says, a person standing at the equator $3^1/_2$ cubits above the surface of the earth sees 0;4,26 more than half of the celestial sphere. I have been unable to trace this statement in any treatise associated with the illustrious 11th-century physicist (see, however, **9.3**, also **14.9**). The remark makes little sense, especially in the light of what follows. al-Minūfī proceeds to state that Ibn Yūnus fixed the interval between the moment when the centre of the sun is on the true horizon and the moment when its centre is on the apparent horizon as 0;47° at the equinoxes, increasing 0;5° for each sign to the summer solstice and decreasing similarly to the winter solstice. These are the corrections ΔD which are tabulated in certain late copies of the Cairo corpus (**4.11**), including al-Minūfī's prayer-tables (see below). al-Minūfī's son ʿAbd al-Qādir wrote at length about this function ΔD and tabulated it for all latitudes (**8.2**).

al-Minūfī tabulates most of the functions found in the main corpus, modified as follows. For those functions which involve horizon phenomena the function is tabulated for the sun on the true horizon and then again for the sun on the apparent horizon, the second table having the correction ΔD taken into account. Another feature that is not found in the main corpus is the instructions presented in marginal notes concerning the exact time of day for which to determine the solar longitude to use as argument in the tables. Thus, for example, there is a table of t_a, the time from midday to the afternoon prayer: the instructions are to enter the true longitude of the sun at the time of the afternoon prayer as argument. In the table of T_a, the time from the afternoon prayer to sunset, the instructions indicate that the true longitude of the sun at sunset should now be the argument and the value corrected with the "difference minutes" and the "minutes of the solar radius" to find the time when the sun has disappeared over the apparent horizon. This table is followed by another in which the "difference minutes" are taken into account, and one need only correct for the solar radius to obtain the value required. There is no table of the "minutes of solar motion" between midday and the other prayer-times to facilitate the computation of the argument, such as is found in a note by al-Wafāʾī (**5.7**) and al-Lādhiqī's tables (**7.8**).

al-Minūfī's work contains tables of the following functions (those adjusted for the difference minutes are asterisked):

$$D, D^*, t_a, h_a, T_a, T_a^*, H, h_v, d, \delta, T_q, t_q, Z_{(12)}, h_q, \psi, h_0, 2D^h, 2D^{*h}, \tilde{h}, 2D^{*h},$$
$$s, r, (r{+}D), (D{+}s), 2N, 2N^*, n, n^*, \sigma, \sigma^*, \tau, \Delta D, \alpha_s, \alpha_\tau, \alpha', \alpha \text{ and } \alpha_f .$$

The work concludes with a star table giving equatorial coordinates for the year 961 H [= 1551].

The functions d (and hence D, 2N, $2D^h$, $\tilde{h}$, *etc.*), δ (and hence H), ψ, h_a and t_a, and α' and α_ϕ (ϕ = 30°) have been recomputed for ε = 23;30°. The functions h_q, T_q, t_q, h_0, r and s, and h_v, have been taken from the main corpus, being based on ε = 23;35°. Functions such as n (= 2N-(r+s)) are based on values computed with different parameters.

al-Minūfī's knowledge of spherical astronomy was such that he reproduced the garbled entries:

$$\psi(60°) = 23;29° \text{ and } h_0(30°) = 23;34°$$

from earlier tables. Both these values should be the same as ε for latitude 30° (*cf.* **4.3**), that is, 23;30° for $\psi(60°)$ and 23;35° for $h_0(30°)$. Furthermore, he did not realize that Ibn Yūnus' values for the difference minutes were mutually inconsistent. al-Minūfī's efforts to improve the corpus were hardly successful, but he was the only Egyptian *muwaqqit* to attempt to radically modify it.

Another work on timekeeping by al-Minūfī is a treatise entitled *Naẓm al-yawāqīt fī taḥrīr aʿmāl al-mawāqīt,* "Organizing the Gems concerning the Correction of the Operations of Timekeeping", and is extant in MS Cairo MM 235,1 (fols. 1r-7r), copied *ca.* 1600 by his son ʿAbd al-Qādir. In this al-Minūfī mentions the "difference minutes" and the way in which they should be used, and cites the *Tanqīḥ al-manāẓir*, the well-known work on optics by the late-13th-/early-14th-century Baghdad scholar Kamāl al-Dīn al-Fārisī,[2] rather than Ibn Yūnus or Ibn al-Haytham. See further **8.2**.

7.2 The prayer-tables of ʿAbd al-Qādir al-Minūfī and Ibrāhīm ibn Qāyitbāy

ʿAbd al-Qādir al-Minūfī[3] was the son of Muḥammad al-Minūfī who modified the main Cairo corpus (**7.1**). He wrote on the subject of refraction of the horizon (**8.2**) and also compiled some prayer-tables in which the tables of various functions, mainly those tabulated by his father, are stretched by linear interpolation to display values for smaller argument increments. In this he was following an earlier tradition of al-Wafāʾī and Ibn al-Mushrif (**5.7**).

MS Cairo DM 1101 was copied in the hand of al-Minūfī-Jr. in 974 H [= 1566/67] and contains tables of the functions:

$$D + \Delta D,\ 2D^h,\ t_a \text{ and } s$$

with values to three digits for each 0;3° of λ. MS Cairo MM 66, copied in the previous year by al-Minūfī-Jr., contains tables of $\alpha'(\lambda)$ with values to three digits for each minute of argument, and MS Cairo MM 45, copied *ca.* 1625 by Ibrāhīm ibn Qāyitbāy (see below) contains similar tables with entries to five digits (lifted from Ulugh Beg), both attributed to ʿAbd al-Qādir al-Minūfī.

Ibrāhīm ibn Qāyitbāy was a student of Shams al-Dīn al-Minūfī who worked in Cairo *ca.* 1625.[4] Several manuscripts preserved in Cairo contain prayer-tables computed and also copied by him, similar to those of ʿAbd al-Qādir al-Minūfī. The functions tabulated in the various manuscripts are as follows:

t_a (DM 34 and DM 152), α_s (DM 153,1 and DM 740,2), ΔD (DM 682),
$T_a + \Delta D$ (DM 152) and $D + \Delta D$ (DM 33)

See **Fig. 7.2b** for a sample of the first of these, and **Figs. 7.2a** and **c** for their title-folios. Some

[2] On Kamāl al-Dīn see the article by Roshdi Rashed in *DSB*. See also n. 8:9.

[3] On ʿAbd al-Qādir al-Minūfī see Suter, MAA, no. 479 (confused); *Cairo ENL Survey*, no. C121; and İhsanoğlu *et al.*, *Ottoman Astronomical Literature*, I, pp. 217-220, no. 97.

[4] Ibrāhīm ibn Qāʾitbāy is not mentioned in the modern sources besides *Cairo ENL Survey*, no. C122, and İhsanoğlu *et al.*, *Ottoman Astronomical Literature*, I, pp. 264-265, no. 130 (lists only Cairo manuscripts).

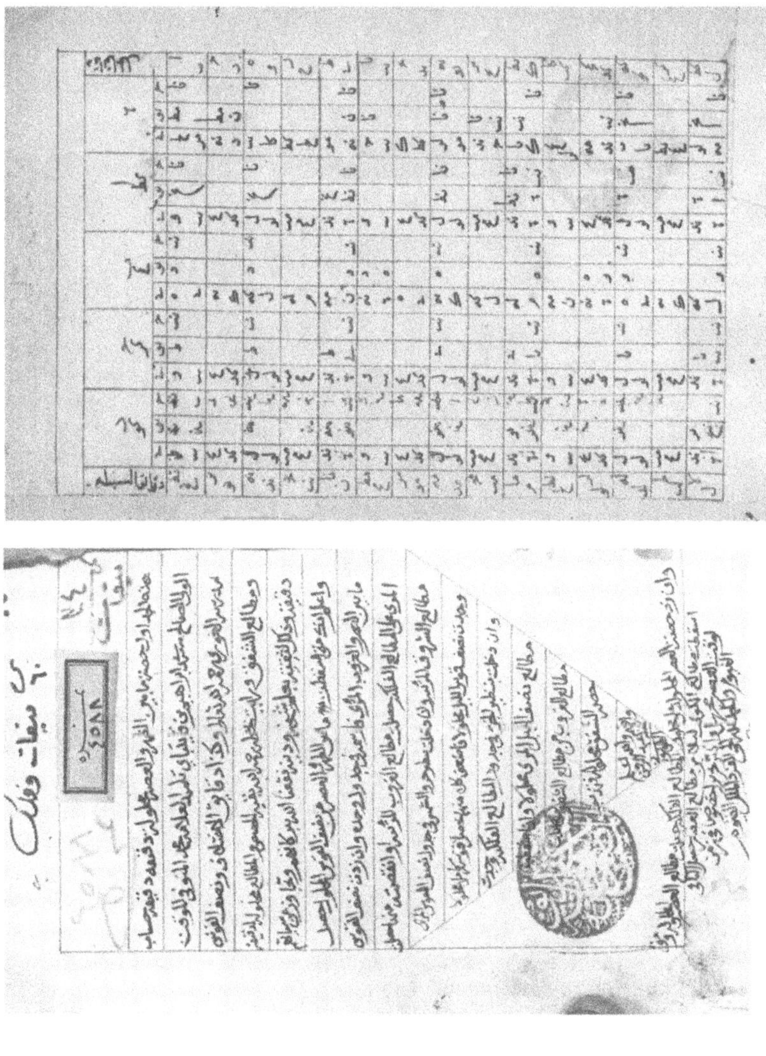

Fig. 7.2a-b: The title-folio and an extract from the "extended" tables of the hour-angle at the beginning of the *'aṣr*. This manuscript belonged to the celebrated 17th-century astronomer and muwaqqit 'Abd al-Raḥmān al-Ṭūlūnī: see **V-10**. [From MS Cairo DM 34, courtesy of the Egyptian National Library.]

Fig. 7.2c: The title-folio of Ibrāhīm ibn Qāyitbāy's "extended" tables of the "difference minutes". This manuscript also belonged to al-Ṭūlūnī. [From MS Cairo DM 682, courtesy of the Egyptian National Library.]

other anonymous prayer-tables of the same kind are preserved in various other Cairo manuscripts. For example, MSS Cairo DM 57,1 and 151,2, originally one codex of seven folios copied *ca.* 1600, contain a set of tables of σ (*al-dā'ir li-'l-salām*) for each 0;6° of λ, based on the assumption that the *salām* is 2° before daybreak, and MSS DM 545, 158 and 423, copied about the same time, contain tables of α_σ (*maṭāliʿ al-salām*) for each 0;3° of λ based on the assumption that the *salām* is 1° before daybreak.

7.3 al-Qaymarī's tables for the afternoon prayer

MS Cairo DM 620,9 (fols. 58r-59r + 64r-72v), copied *ca.* 1450, contains a set of tables entitled *jadwal al-bāqī li-'l-ʿaṣr*, "table of time remaining until the afternoon prayer", and attributed to an individual called al-Qaymarī, on whom I have no further information. The function tabulated is simply:

$$\tau_a(h,\lambda) = t_a(\lambda) - t(h,\lambda)$$

and values are given for each integral degree of h such that $h_a(\lambda) \leq h \leq H(\lambda)$ for each degree of λ. An extract is shown in **Fig. 7.3**. See **11.3** on two similar sets of tables for Damascus.

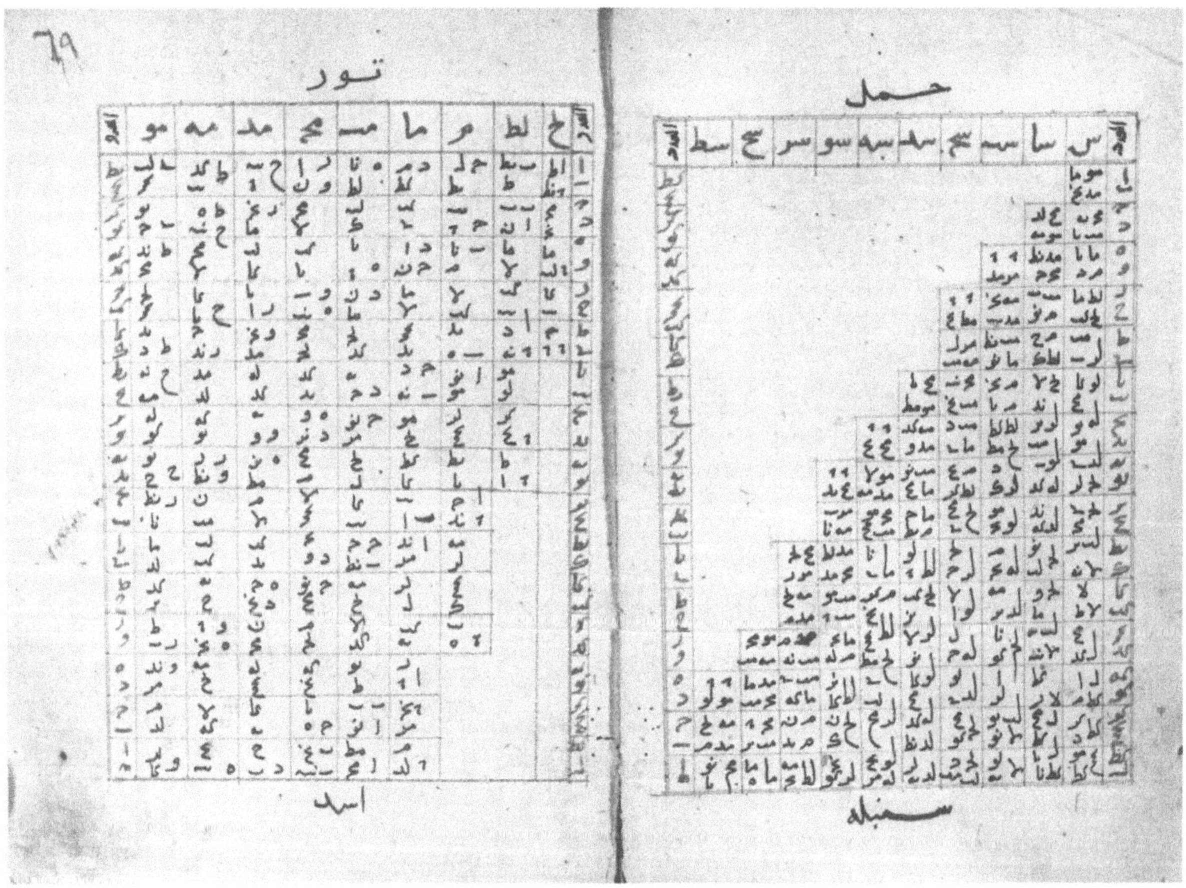

Fig. 7.3: An extract from al-Qaymarī's tables, serving Aries and Taurus / Virgo and Leo. [From MS Cairo DM 620,9, fols. 68v-69r, courtesy of the Egyptian National Library.]

These tables are followed (fols. 72r and 72v) by others of $\psi(\lambda)$ and $h_0(\lambda)$ (the entries, which are given to two digits, are not corrupt as in other such tables in the Cairo corpus – see **4.3**) and $h(a,\lambda)$ for $a = 30°$ (**4.6**).

7.4 Ibn Abi 'l-Khayr al-Ḥusnī's prayer-tables

MSS Cairo TR 114 (74 pp.) and ZK 154 (91 fols.), both from *ca.* 1700, are two copies of a treatise entitled *Itḥāf al-ḥabīb bi-maʿrifat al-tawqīʿāt wa-'l-awqāt wa-'l-qibla bi-'l-taqrīb* by the 16th-century Egyptian astronomer Muḥammad ibn Abi 'l-Khayr al-Ḥusnī.[5] In this work al-Ḥusnī has included the main prayer-tables of the main corpus in the text of his treatise. For each day of the Coptic year he gives the values of the functions:

$$D,\ t_a,\ T_a,\ s,\ n \text{ and } r\ ,$$

with values given to two digits and written in *abjad* notation. See **8.3** on some sophisticated tables for twilight by al-Ḥusnī.

7.5 al-Fawānīsī's prayer-tables for Cairo

MSS Paris BNF ar. 2545, copied *ca.* 1500 in an elegant hand, and Oxford Seld. Supp. 99 are copies of an extensive set of prayer-tables entitled *Natījat al-afkār fī aʿmāl al-layl wa-'l-nahār*, "The Result of Thoughts about the Operations of Timekeeping by Night and Day", the same title used by al-Lādhiqī (**7.8**). These were compiled by a late-16th-century Egyptian astronomer named Muḥammad ibn ʿUmar ibn Ṣiddīq ibn ʿUmar al-Bakrī al-Fawānīsī.[6] It is not difficult to show that very little of consequence in this at first sight imposing work is original.

In the introduction to the tables the author notes certain approximate rules for determining the standard functions of *mīqāt*, but makes no mention of the fact that similar rules had been formulated by his predecessors Najm al-Dīn al-Miṣrī and al-Bakhāniqī – see **2.5** and **2.9** above. Each of al-Fawānīsī's rules is intended for latitude 30°, except those for functions h_a and H, of which he incorrectly states that they work well for latitudes between 0° and 40°. His rules, outlined in words in the text, are the following:

(a)	$d \approx \frac{1}{2}(\delta + \frac{1}{4}\delta)$	
(b)	$\psi \approx \delta + \frac{1}{6}\delta$	
(c)	$h_q \approx 47° + \delta + \frac{1}{4}\delta + \frac{1}{6}\delta$	
(d)	$T_q \approx 57° + \delta + \frac{1}{2}\delta + \frac{1}{6}\delta$	
(e)	$t_a \approx 52° + \frac{1}{2} \cdot \frac{1}{6}\delta$	$(\delta > 0)$
	$t_a \approx 52° - \lfloor \lvert\delta\rvert/4 + \lvert\delta\rvert/6 \rfloor$	$(\delta < 0)$
(f)	$h_a \approx \frac{1}{2} H + \frac{1}{10}[\max H - H]$	
(g)	$h_b \approx \frac{1}{2} h_a + \frac{1}{10}[\max H - h_a]$	

[5] On Ibn Abi 'l-Khayr al-Ḥusnī see Suter, *MAA*, no. 511; Brockelmann, *GAL*, SII, p. 485; *Cairo ENL Survey*, no. C124; and İhsanoğlu *et al.*, *Ottoman Astronomical Literature*, pp. 255-262, no. 127.

[6] On al-Fawānīsī see Suter, *MAA*, no. 475; Brockelmann, *GAL*, II, p. 469, and SII, p. 485; *Cairo ENL Survey*, no. D10; and İhsanoğlu *et al.*, *Ottoman Astronomical Literature*, I, pp. 237-238, no. 111.

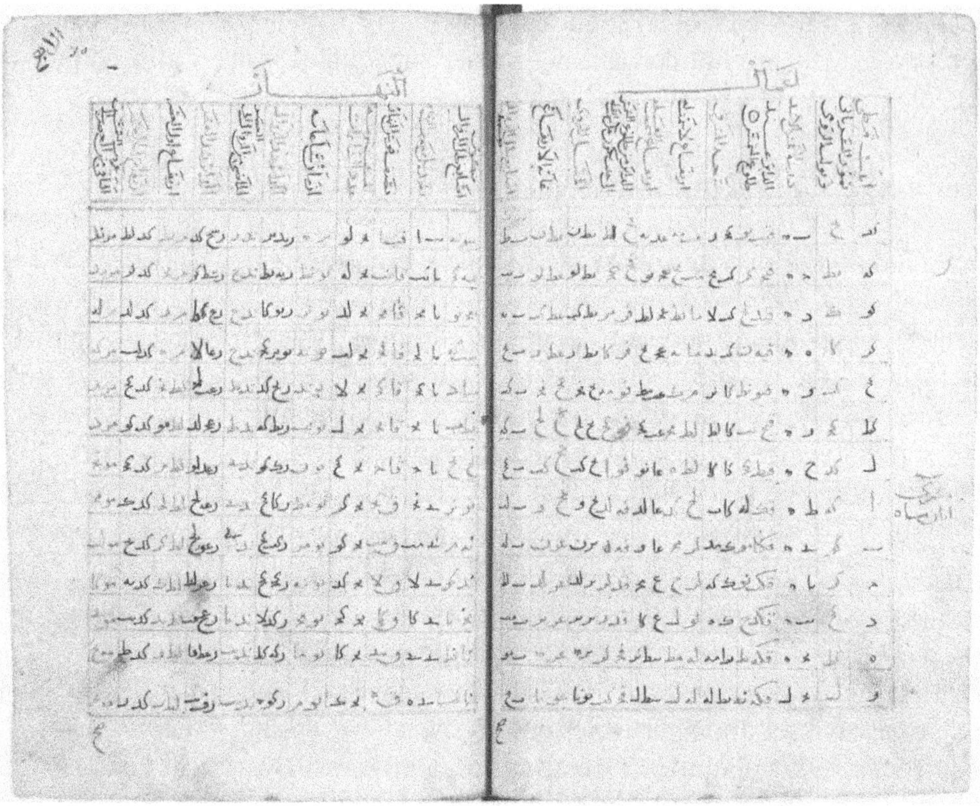

(h) $r \approx 21;0° + {}^1/_8\,\delta + {}^1/_2 \cdot {}^1/_6\,\delta$ $(\delta > 0)$
$r \approx 21;0° + {}^1/_2 \cdot {}^1/_8\,|\delta|$ $(\delta < 0)$
(i) $s \approx 18;30° + {}^1/_8\,\delta$ $(\delta > 0)$
$s \approx 18;30° + {}^1/_2 \cdot {}^1/_8\,|\delta|$ $(\delta < 0)$

al-Fawānīsī states that his formula for r is based on the parameter 18°, rather than 20° which was the commonly-accepted value in his day. On the other hand, he notes, his formula for s is based on the commonly-used parameter 16° rather than 18°, which had been used before his time. Some of the same approximations are given in a short anonymous treatise on time-keeping extant in MS Escorial ár. 961,3 (fols. 13v-16v), copied in 863 H [= 1459], that is, before the time of al-Fawānīsī (see **2.5**). In an another anonymous source, MS Cairo DM 181,3, (fols. 43r-46v), copied *ca.* 1800, his rule for t_a appears together with two rules for the duration of twilight (for $\delta > 0$ only?) which differ from both his rule and that of al-Bakhāniqī, namely:

$$r \approx 22° + {}^1/_5\,\delta \quad \text{and} \quad s \approx 20° + {}^1/_2 \cdot {}^1/_8\,\delta + {}^1/_6\,\delta\ .$$

These are clearly based on parameters 19° and 17°, respectively.

The tables which follow this introduction are not based on these approximate formulae. For each day of the Coptic year al-Fawānīsī tabulates the corresponding date in the Syrian calendar, the solar longitude (usually to the nearest degree, occasionally to the nearest half-degree), and then some 19 functions relating to spherical astronomy and the prayer-times – see the extract in **Fig. 7.5a**. The entries for a given day are written across a pair of facing pages: such an opening of the manuscript serves 13 days, grouped according to which of the 28 lunar mansions rises at daybreak. For Tūt 1 the solar longitude is given as Virgo 13° and this date is the third day during which the 11th mansion al-Kharāthān rises at dawn. The functions tabulated by al-Fawānīsī for each day are the following:

α_ϕ	"the amount of rotation of the celestial sphere (from the equinox) at sunrise", *i.e.*, oblique ascension (*al-dā'ir ʿinda ṭulūʿ al-shams*)
ψ	solar rising amplitude (*saʿat al-mashriq*)
h_0	solar amplitude in the prime vertical (*al-irtifāʿ alladhī lā samt lahu*)
h_q	solar altitude in the azimuth of the qibla (*irtifāʿ samt al-qibla*)
T_q	time from sunrise to the moment when the sun is in the azimuth of the qibla (*al-dā'ir min ṭulūʿ al-shams ilā ḥayth takūn al-shams ʿalā samt al-qibla*)
δ	solar declination (*al-mayl al-juz'ī*)
H	solar meridian longitude (*ghāyat al-irtifāʿ*)
$Z_{(12)}$	horizontal meridian shadow to base 12 (*aṣābiʿ ẓill al-zawāl mabsūṭ*)
$Z'_{(12)}$	vertical meridian shadow to base 12 (*aṣābiʿ ẓill al-zawāl mankūs*)
d	equation of half daylight (*taʿdīl niṣf al-nahār*)
D	semi diurnal arc (*niṣf qaws al-nahār*)

←
Fig. 7.5a: An extract from al-Fawānīsī's prayer-tables for day-time, displaying 19 different functions for each day of the Coptic year, here Āb 24 to Misrā 6. [From MS Paris BNF ar. 2545, fols. 29v-30r, courtesy of the Bibliothèque Nationale de France.]

Fig. 7.5b: An extract from al-Fawānīsī's prayer-tables for night-time, serving the end of Abīb and the beginning of Misrā. The tables have a very distinctive format. [From MS Paris BNF ar. 2545, fols. 30v-31r, courtesy of the Bibliothèque Nationale de France.]

$2D^h$ number of equinoctial hours of daylight (*ʿadad sāʿātihi al-mustawiya*)
$\tilde{h}$ number of equatorial degrees in one seasonal hour of daylight (*azmān sāʿātihi*)
α' "the amount of rotation at midday", *i.e.*, the normed right ascensions (*al-dāʾir ʿind al-zawāl*)
t_a time from midday to the afternoon prayer (*al-māḍī min al-zawāl ila ʾl-ʿaṣr*)
α_a "the amount of rotation at the afternoon prayer", *i.e.*, the oblique ascensions of the horoscopus at that time (*al-dāʾir ʿind al-ʿaṣr*)
h_a solar altitude or the beginning of the afternoon prayer (*irtifāʿ awwal al-aṣr*)
h_b solar altitude at the end of the afternoon prayer (*irtifāʿ ākhir al-ʿaṣr*)
T_a time remaining at the beginning of the afternoon prayer until sunset (*al-bāqī min al-ʿaṣr ilā ghurūb al-shams*)

With the following reservations, the values are essentially those of the main Cairo corpus:

(a) The values of h_0 are extremely corrupt. The entries for solar longitude 30° and 90° are 23;47° and 53;21° rather than the correct values 23;35° and 53;9° – see **4.3**. The entry 53;21° is probably to be explained as a scribal error for 53;9°: the Arabic notations for 21 (*kāf-alif*) and 9 (*ṭāʾ*) are quite similar.

(b) The values of h_q are likewise rather corrupt. They are based on $q = 52°$ and are ultimately derived from Ibn Yūnus' table of h_q in the *Ḥākimī Zīj* (**5.1**). However, a deliberate attempt has been made to adjust the entries near the summer solstice so that the reading for $\lambda = 90°$ is 80;39° rather than 80;2°. Nevertheless al-Fawānīsī's values of T_q can be derived by subtracting the values of t_q ($q = 53°$) from those of D in the main corpus!

(c) There are no tables of the functions $\mathrm{Tan}_{12}\, h_a(\lambda)$ or $\alpha_a(\lambda)$ in the known manuscripts of the corpus. The former can be found directly from h_a or by using:

$$\mathrm{Tan}_{12}\, h_a \cdot z_{a(12)} = 144 \ ,$$

and the latter is defined by:

$$\alpha_a = \alpha' + t_a \ .$$

Some tables of α_a for Cairo apparently based upon $\varepsilon = 23;30°$ and computed by Riḍwān Efendī are contained in MS Cairo DM 45 – see **7.10**.

(d) There are no tables of the function $h_b(\lambda)$ in the known manuscripts of the corpus. al-Fawānīsī's values are fairly accurate, but the computation of h_b from H is straightforward. The only known tables of the corresponding time $t_b(\lambda)$ computed for Cairo are in the later source MS Istanbul Kandilli 424 – see **7.11**.

For the nights corresponding to each day of the Coptic year al-Fawānīsī tabulates eight functions, namely:

(1) $\alpha_\phi(\lambda^*)$ "the amount of rotation at sunset", *i.e.*, the oblique ascensions of the ascendant at sunset (*al-dāʾir ʿinda ʾl-ghurūb*)
(2) $\alpha_s(\lambda) + 2°$ "the amount of rotation at nightfall with two degrees added to the ascensions", *i.e.*, the oblique ascensions of the ascendant at nightfall plus 2° (*al-dāʾir ʿinda ghiyāb al-shafaq bi-ziyādat darajatayn ʿala ʾl-maṭāliʿ*)
(3) $2N(\lambda)$ nocturnal arc (*qaws al-layl bi-kamālihi*)
(4) $2N^h(\lambda)$ length of night in equinoctial hours (*ʿadad al-sāʿāt al-mustawiya*)
(5) $\tilde{h}'(\lambda)$ length of the seasonal night-hours in degrees (*azmān al-sāʿāt*)

(6) $2N(\lambda) - r(\lambda)$ time from sunset to daybreak (*al-māḍī min ghurūb al-shams ilā ṭulūʿ al-fajr*)

(7) $\alpha'(\lambda) + 180°$ "the amount of rotation at midnight", *i.e.*, the oblique ascensions of the ascendant at midnight (*al-dāʾir ʿind niṣf al-layl*)

(8) $\alpha_r(\lambda)$ "the amount of rotation at daybreak", *i.e.*, the oblique ascensions of the ascendant at daybreak (*al-dāʾir ʿinda ʾl-fajr*)

Approximate values of $r(\lambda)$ and $s(\lambda)$ are given at the top of each pair of facing pages. These are expressed to the nearest 0;30° and are based on angles of depression of 18° and 16°, respectively. They are intended to be used for all the nights served by the given pair of pages. The 2nd, 6th, and 8th functions are based on these approximate values of r and s. al-Fawānīsī did not copy the tables of α_r and α_s in the corpus because these are based on parameters 19° and 17°. He gives no explanation why he adds 2° to the values of $\alpha_s(\lambda)$: perhaps he was not really convinced that 16° was a better parameter for the evening twilight than 18°. He also names the stars which are culminating at nightfall, midnight, and daybreak, or gives what is perhaps intended to be their distance from the meridian at these times, for each night of the year. (See **5.5** for a different set of solar tables compiled by Ibn al-Kattānī.)

The anonymous tables in MS Oxford Hyde 32 (Nic. 284), fols. 69v-81r, are related to those of al-Fawānīsī. The following functions are tabulated for each day of the Coptic year:

$$\lambda,\ Z_{(12)},\ H,\ h_a,\ \alpha',\ D,\ t_a,\ T_a \text{ and } 2N\ .$$

For the corresponding nights the altitudes of three prominent stars in the eastern and western sky at daybreak are given, as well as the values of various functions relating to nightfall and daybreak.

7.6 Two almanacs for the years 1015 and 1044 H

MS Paris BNF ar. 2571 contains an almanac for Cairo for the year 1015 H [= 1606/07], by Aḥmad ibn Muḥammad al-Ḥusnī,[7] entitled *Taqwīm al-sana al-qamariyya*, literally, "Ephemeris for the Lunar Year". al-Ḥusnī begins his almanac by paying tribute to his teacher, the Ottoman astronomer Jamāl al-Dīn Abū Dāʾūd Sulaymān (ibn Ḥamza), otherwise known as Ibn Bakhshīsh,[8] and also in passing mentions the Ottoman Sultān Aḥmad Khān (reg. 1603-1617).

The almanac contains calendrical information and remarks on the nature of the lunar crescent at first visibility for each month of the year. al-Ḥusnī states that the lunar crescent whose appearance marks the beginning of the year will be seen on a Sunday evening (*laylat al-ithnayn* or "night of Monday"). The first day of the year would thus be a Monday, which would correspond to April 30, 1606 (Julian calendar): modern tables give Muḥarram 1 as Sunday, April 29 for that year. For the first and 15th day of the lunar month al-Ḥusnī also gives the values of:

$$D,\ t_a,\ T_a,\ s,\ n \text{ and } r$$

written out in words, and expressed in equatorial degrees, with parts of a degree expressed as simple fractions. In Ramaḍān the time τ is also shown, based on the assumption that the

[7] This al-Ḥusnī is not mentioned in the modern sources.

[8] On Ibn Bakhshīsh see Azzawi, *History of Astronomy in Iraq*, pp. 181; *Cairo ENL Survey*, no. C117; and İhsanoğlu *et al.*, *Ottoman Astronomical Literature*, I, pp. 194-198, no. 95.

ṭafy occurs 5° before daybreak. The solar longitude on Muḥarram 1 is taken as 48°, which is reasonable.

The entries in al-Ḥusnī's prayer-tables for a given day are generally mutually inconsistent. The values of the functions t_a, s and r are rounded from the corresponding entries in the main Cairo corpus, but those for D, T_a and n have been haphazardly adjusted by what was probably intended to be a correction for horizon phenomena. al-Ḥusnī's values of the *niṣf qaws al-nahār al-sharʿī*, "semi diurnal arc according to religious law", are in excess of the geometric semi diurnal arc by approximately the following amounts:

VE: 0;3° SS: 0;55 AE: 0;51 WS: 0;43

He gives no explanation of these corrections.

MS Cairo ṬM 87 contains a similar almanac for the year 1044 H [= 1634/35] attributed to Muḥammad al-Kutubī,[9] which I have not investigated.

7.7 al-ʿAjmāwī's prayer-tables for Cairo

ʿAbd al-Qādir ibn Aḥmad al-ʿAjmāwī al-Azharī[10] was a *muwaqqit* in the *madrasa* of Sulṭān Ḥasan in Cairo who compiled a set of prayer-tables entitled *Nuzhat al-nāẓir fī maʿrifat mā bayn al-awqāt min al-dawāʾir*, "The Delights of Knowing the Intervals between the Prayer-times". MS Paris BNF ar. 2578,2, dated 1024 H [= 1616], and also Cairo ZK 287, 203 fols., penned 1056 H [= 1646/47], are copies of this work, containing the prayer-times *written out in words* for each day of the Coptic year. An extract is displayed in **Fig. 7.7**. Values are given for the functions:

$$D, t_a, T_a, s, n \text{ and } r ,$$

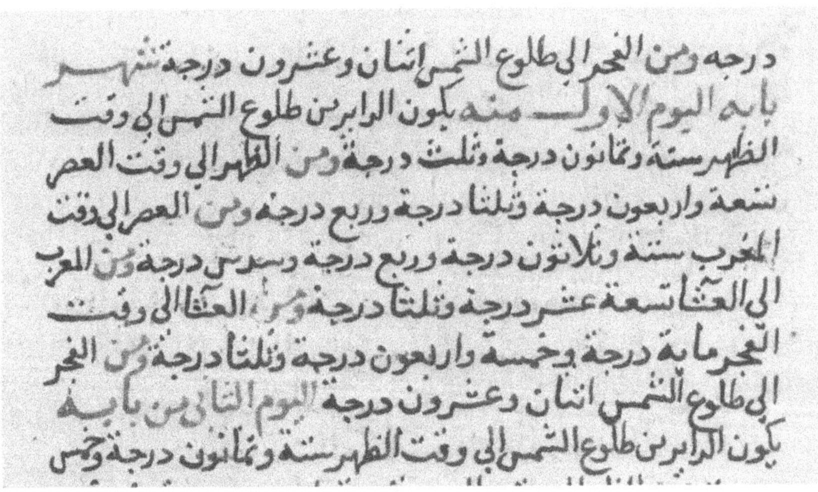

درجه ومن الفجر الى طلوع الشمس اثنان وعشرون درجة شهر
بابه اليوم الاول منه يكون الدائر من طلوع الشمس الى وقت
الظهر ستة وثمانون درجة وثلث درجة ومن الظهر الى وقت العصر
تسعة واربعون درجة وثلثا درجة وربع درجة ومن العصر الى وقت
المغرب ستة وثلاثون درجة وربع درجة وسدس درجة ومن المغرب
الى العشا تسعة عشر درجة وثلثا درجة ومن العشا الى وقت
الفجر مائة درجة وخمسة واربعون درجة وثلثا درجة ومن الفجر
الى طلوع الشمس اثنان وعشرون درجة اليوم الثاني من بابه
يكون الدائر من طلوع الشمس الى وقت الظهر ستة وثمانون درجة وخمس

Fig. 7.7: An extract from al-ʿAjmāwī's prayer-tables written as text. [From MS Paris BNF ar. 2578, fol. 46r, courtesy of the Bibliothèque Nationale de France.]

---

[9] On Muḥammad al-Kutubī see *Cairo ENL Survey*, no. D27; and İhsanoğlu *et al.*, *Ottoman Astronomical Literature*, I, p. 276, no. 141.

[10] On al-ʿAjmāwī see Brockelmann, *GAL*, SII, p. 1018, and *Cairo ENL Survey*, D11.

and are those of the main Cairo corpus with the equatorial minutes rounded so that they can be expressed as simple fractions.

7.8 The prayer-tables of al-Lādhiqī and al-Ikhṣāṣī

A very popular set of Egyptian prayer-tables, extant in numerous manuscript sources, is attributed to an individual named Shams al-Dīn Muḥammad ibn Muḥammad al-Lādhiqī.[11] The title of the tables is usually *Natījat al-afkār fī aʿmāl al-layl wa-'l-nahār*, "The Result of Reflection on Matters of Night and Day", although the title *Bughyat al-nafs* ... , "The Desire of the Soul ... ", also occurs and relates to the same work.

I have examined several manuscripts of al-Lādhiqī's prayer-tables, including MSS Cairo DM 198, Cairo DM 61, Hartford Theological Seminary 621, Cairo DM 190 and Paris BNF ar. 2553 (other manuscripts are listed below). Each contains an introduction on calendars, consisting of ten sections and a conclusion, all of no great interest. al-Lādhiqī mentions, however, that he compiled his solar tables for the longitude of Cairo, 54;55°, a value also used by al-Marrākushī (**6.7**). The first source, MS Cairo DM 198, contains no tables at all. The second and third contain a table of solar positions for Cairo for each day of the Syrian year. This table and other simple calendrical tables, followed by various others from the main corpus, namely:

$$H, d, D, 2N, \alpha', \psi, h_0, h_a, t_a, T_a, r, s, n, \sigma, \alpha_\sigma \text{ and } \alpha_s .$$

The fifth source, MS Cairo DM 190, contains these and also tables of h_q and h_v, the first based on $q = 53°$ and the latter with entry 0;0° at $\lambda = 270°$.

A very simple table occuring in most of the copies of al-Lādhiqī's work which I have consulted gives the corrections to be added to the solar position at noon in order to find the solar longitude at:

(1) the beginning of the afternoon prayer;
(2) sunset;
(3) nightfall;
(4) the time of the *salām*, and
(5) sunrise.

Values of these corrections in minutes are given for each zodiacal sign, but their use is only fully exploited by al-Minūfī (**7.1**). In al-Lādhiqī's tables the difference minutes ΔD (according to the theory associated with Ibn Yūnus) and the mean solar motion in a day are also tabulated. An earlier table of this kind is attributed to al-Wafā'ī (**5.7**).

Other manuscripts of al-Lādhiqī's tables that have come to my attention are MSS Gotha A1399, Istanbul (Millet Genel) Ali Emiri 2743, Istanbul Haci Beşir Ağa 674,2, Istanbul Reisülküttap Mustafa Efendi 582, Cairo DM 307, Cairo MM 14, 15, 16, Cairo MM 25, Cairo MM 221, 222, 223, 224, 225, and Istanbul UL A 4082,1 (fols. 1r-20v). There is no mention of Ibn Yūnus, or of al-Maqsī, in any of the manuscripts consulted. It seems that the prayer-tables were not "plagiarized" by al-Lādhiqī himself, but were simply used widely together with

[11] On al-Lādhiqī see Suter, *MAA*, no. 519; Brockelmann, *GAL*, SII, p. 1023; *Cairo ENL Survey*, no. C132; and İhsanoğlu *et al.*, *Ottoman Astronomical Literature*, I, pp. 229-230, no. 109, and p. 760.

his table for solar longitude by later *muwaqqits*. In MS Berlin Ahlwardt 5765 [= Wetzstein 1149, fols. 85v-88v] his introduction is followed by a set of prayer-tables for the latitude of Nablus (**11.7**). I have no information on the life of al-Lādhiqī, but his name indicates that his family was at one time associated with Lattakia in Syria.

A similar work, with a slightly different introduction and different arrangement of the tables is attributed to Muḥammad ibn Haykal al-Ikhṣāṣī, a *muwaqqit* at the Azhar Mosque in Cairo.[12] al-Ikhṣāṣī's tables are preserved in MSS Cairo DM 60, 480, 739, and MM 81.

7.9 Ibn Abī Rāya's prayer-tables

MS Alexandria 4441J is the only copy known to me of a set of prayer-tables for Cairo attributed to an individual called Ibn Abī Rāya, a *muwaqqit* at the Mu'ayyad Mosque in Cairo at the beginning of the 18th century.[13] The work is entitled *Zuhrat al-afkār fī aʿmāl al-layl wa-'l-nahār*, "The Beauty of Reflection on the Operations of Timekeeping by Night and Day", and begins with an introduction of 13 chapters. A table of solar longitude for each day of the Coptic year (with Virgo 16;12° for Tūt 1) is is said to be computed for longitude 54;55° (**7.8**).

The other simple calendrical, astrological and solar tables in this work do not concern the present study, save to mention one table displaying the lunar mansion corresponding to the position of the sun, such as is found in MS Istanbul Bağdatli Vehbi Efendi 887 of the tables of al-Ḥakīm al-Lādhiqī for Lattakia (**11.9**). This particular table is said to be computed for the year 1117 H [= 1705/06] by Ibn Abī Rāya.

The main set of prayer-tables is preceded by individual tables of ψ (computed for $\varepsilon = 23;30°$ but having 23;29° for $\lambda = 60°$) and h_0 (computed for $\varepsilon = 23;35°$ but having 23;54° for $\lambda = 30°$) (see **4.3** on this problem), as well as garbled tables of Cos $\delta(\lambda)$ (called *jayb tamām al-mayl*) and $C(\lambda)$ (called *buʿd al-quṭr li-'l-jayb* (*sic*)) (see **I-6.2** and **6.4**). The main tables display 16 functions side by side for each degree of λ beginning with $\lambda = 1°$. The first and third of these, D and D′, are called respectively *niṣf qaws al-nahār al-sharqī* and *al-gharbī*, semi diurnal arc in the east and in the west. The former is simply the semi diurnal arc and the latter the semi diurnal arc corrected for refraction at the horizon. The last function tabulated, which I denote by [D(λ) - min D], is called *ziyādat al-nahār*, and measures the excess of the length of daylight over its minimum at the winter solstice; I have not found this elsewhere. The underlying value of ε is 23;35°, and the tables are lifted from those in the main corpus rather than those of al-Minūfī (**7.1**). The 16 functions tabulated are:

$$D, H, D', h_a, t_a, T_a, \mathfrak{s}, n, r, 2N, h_q, \Delta D, Z_{(12)}, (2N\text{-}r), (D\text{+}2N\text{-}r) \text{ and } [D - \min D] .$$

The work concludes with a set of tables of:

$$\alpha', \alpha_s, \alpha_r, \alpha_\phi \text{ and } \alpha^* ,$$

as well as a star catalogue displaying the equatorial coordinates of about 285 stars for an unspecified date.

---

[12] On al-Ikhṣāṣī see *Cairo ENL Survey*, no. D3; and İhsanoğlu *et al.*, *Ottoman Astronomical Literature*, I, pp. 348-349, no. 215.

[13] Ibn Abī Rāya is new to the modern literature.

7.10 Riḍwān Efendī's prayer-tables

The celebrated astronomer and author Riḍwān Efendī al-Falakī,[14] who died in 1710, compiled a lengthy volume on timekeeping called *Dustūr uṣūl ʿilm al-mīqāt wa-natījat al-naẓr fī taḥrīr al-awqāt*, "Statutes of the Principles of Timekeeping and the Results of Reconsidering the Prayer-times". I have consulted MSS Cairo Azhar *falak* 8216, Berlin Ahlwardt 5710, Istanbul S. Esad Efendi Medresesi 119,1, Istanbul Nuruosmaniye 2781, Cairo TR 116 and TR 276 of this work. Apparently, the only original part is a set of tables for determining the times of the sun's entry into each of the signs: the treatment of the problems of spherical astronomy in the introduction is standard and all the prayer-tables which follow have been lifted from earlier sources.

These manuscripts contain all of the prayer-tables from the main corpus, and Riḍwān Efendī has made no effort to recompute even the simplest of them for the value of the obliquity used in his *Zīj*, namely, 23;30,17°. This *Zīj al-Riḍwānī*, or *al-Zīj al-Mufīd*, of which I have examined the copy preserved in MS Princeton Garret Hitti 1004, is based entirely on that of Ulugh Beg, which in Cairo was available in Arabic in the recension of Ibn Abi 'l-Fatḥ al-Ṣūfī (**6.15**). MS Cairo DM 45 contains some tables computed by Riḍwān Efendī giving the oblique ascensions, α_a, at the beginning of the interval for the afternoon prayer, "according to the new observation". The "new" observation of Ulugh Beg was almost three centuries prior to Riḍwān's activity in Cairo.

7.11 Miscellaneous anonymous prayer-tables for Cairo

MS Cairo DM 200 (7 fols., *ca.* 1750) contains a set of tables of the functions:

$$D,\ t_a,\ T_a,\ s,\ n,\ r, \text{ and the solar longitude } \lambda\ ,$$

tabulated side by side for each day of the Syrian calendar (equivalent dates in the Coptic calendar are also given). Values are given to two digits. MS Cairo DM 531,1 contains sets of sub-tables displaying values of the first six functions to one digit for each five days of the Coptic calendar.

MS Cairo DM 46 (17 fols., *ca.* 1750) contains a motley collection of prayer-tables for Cairo, some with entries to two digits taken from the main corpus, and others recomputed for Ulugh Beg's values of the obliquity, 23;30,17°, with entries to three digits. Four of the tables display the "latitude of visible climate", ν, and its complement, $\bar{\nu}$, which is the altitude of the pole of the ecliptic, as well as $\mathrm{Cos}_{60}\ \nu$ and $\mathrm{Tan}_{60}\ \nu$, with values to three digits for each degree of λ.[15] The tables that were not lifted from the corpus display functions:

$$h_b,\ T_a,\ T_b,\ (T_a\text{+}D),\ (T_b\text{+}D),\ \tau_{a/b},\ 2N,\ n,\ (2N\text{-}r) \text{ and } h_q\ ,$$

and the tables that were lifted from the corpus display functions:

$$2N,\ n,\ s,\ r,\ t_q,\ t_{q*},\ T_q,\ h_{a=30^\circ} \text{ and } h_{a=60^\circ}\ .$$

No author is associated with any of them.

---

[14] On Riḍwān Efendī see n. **I**-6:6.

[15] See n. 3:14 on this concept.

MS Istanbul Kandilli 424, copied in 1164 H (= 1750/51), contains (fols. 1v-4r) an anonymous set of tables displaying the longitude of the astrological houses for Cairo as a function of the ascensions of the horoscopus, computed "according to the Ptolemaic method",[16] and (fols. 5r-6v) an anonymous set of Ottoman-type prayer-tables for Cairo. I know of no other copies of either set. The prayer-tables display seven functions side by side for each degree of solar longitude beginning with Aries, and the times are given in equinoctial hours and minutes according to the Ottoman convention (**14.0**). The tables are entitled *jadwal sāʿāt ʿaqrab al-sāʿāt*, literally, "tables of the hours of the clock-hand". The functions tabulated are the following:

s′	*ʿishāʾ*	the time of nightfall
j′	*ādhān Ramaḍān*	the time of the call to prayer in Ramaḍān
r′	*fajr*	daybreak, the time of morning prayer
R′	*shurūq*	the time of sunrise
m′	*ẓuhr*	the time of the midday prayer
a′	*ʿaṣr awwal*	the time of the first afternoon prayer
b′	*ʿaṣr thānī*	the time of the second afternoon prayer.

A note by the title states that the tables are computed for the setting of the upper limb of the sun over the apparent horizon and the crossing of the entire solar disc across the meridian. It is stated that these modifications take into consideration the correction for the difference between the true and the apparent horizons and the radius of the sun, and that the tables are based on the parameters of Ulugh Beg (*i.e.*, $\varepsilon = 23;30,17°$).

Comparison of the tables with a set of Ottoman-type prayer-tables computed for $\phi = 30;0°$ and $\varepsilon = 23;30,17°$ and not modified for horizon phenomena reveals the following:

a) The tables of r′ and s′ are based on parameters 19° and 17°, and the time j′ is always 20^m before r′.

b) The times m′, a′, and b′ are modified by a correction of approximately:

EQ: 4^m SS: 6^m WS: 3^m ,

and the times r′ (and hence j′) and R′ are modified by double these amounts. The radically different amounts of this correction for the solstices suggest that the compiler used the corrections for refraction associated with Ibn Yūnus (**4.11**). Indeed if we add 0;15° (an approximate value of the time for half the solar disc to set over the horizon) to Ibn Yūnus' values we obtain

EQ: $0;47° + 0;15° = 1;2° \approx 4^m$
SS: $1;2° + 0;15° = 1;17° \approx 5^m$
WS: $0;32° + 0;15° = 0;47° \approx 3^m$.

(c) It is difficult to confirm whether or not the compiler of the tables introduced a correction of about $+0;15° \approx +1^m$ to the time m′ in order to ensure that the whole solar disc had crossed the meridian at the time of the midday-prayer.

I have not investigated whether or not most of the entries in these tables can be derived from the prayer-tables of al-Minūfī (**7.1**). No earlier tables of t_b for Cairo are attested yet, but the function h_b was tabulated by al-Fawānīsī (**7.5**).

[16] On the houses see n. **I**-3:2.

7.12 al-Khāwānikī's prayer-tables

Some manuscripts of al-Kutubī's tables (**7.14**), such as MSS Cairo ṬM 88 and ṬM 142, contain a set of prayer-tables for Cairo computed by the 18th-century Egyptian astronomer Ramaḍān ibn Ṣāliḥ al-Khāwānikī.[17] The following functions are tabulated in hours and minutes according to the Ottoman convention, with values for each day of the Coptic year:

s′, i′, r′, R′, m′, a′ and b′ .

The time of the *imsāk*, i′, is 20^m before daybreak, r′. The table for b′ occurs only in the first source. I have not been able to check whether these tables are identical to those in MS Istanbul Kandilli 424 (**7.11**).

7.13 Ḥasan al-Jabartī's prayer-tables

MS Cairo TR 346, penned *ca.* 1875, is the only copy known to me of a set of prayer-tables for Cairo attributed to Ḥasan al-Jabartī (d. 1774), father of the celebrated historian ʿAbd al-Raḥmān al-Jabartī.[18] The tables are entitled *Mawqiʿ ʿaqrab al-sāʿa murattaba ʿala 'l-shuhūr al-Qibṭiyya*, "The Place where the Watch-hand Falls (at the Times of Prayer), arranged according to the Coptic Months", and display the times:

s′, i′, r′, R′, m′ and t_a

for each day of the Coptic year. The time of the *imsāk*, i′, is 20^m before daybreak, r′.

7.14 al-Kutubī's timekeeping tables

MSS Cairo DM 149, ṬJ 811,8, ṬM 88 and ṬM 142, DM 812, DM 1103 and 1104 contain a set of tables for timekeeping prepared in the year 1150 H [= 1737/38] by ʿAbd al-Laṭīf al-Dimashqī, known as al-Kutubī.[19] These display for the latitude of Cairo the times in hours and minutes according to the Ottoman convention, before and after midday when the sun has a given altitude and longitude, for each degree of both arguments: see further **I-2.1.8** and the extract illustrated there The entries are doubtless based on the entries in the tables of t(h,λ) and T(h,λ) in the main corpus (**4.5**). al-Kutubī states in his introduction that in his calculations he has taken into consideration the apparent radius of the solar disc and the "difference minutes", as well as the increase in solar longitude from midday to the time in question.

---

[17] On al-Khawānikī see Brockelmann, *GAL*, II, pp. 471-472, and SII, p. 48; Azzawi, *History of Astronomy in Iraq*, pp. 320-321; *Cairo ENL Survey*, no. D78; and İhsanoğlu *et al.*, *Ottoman Astronomical Literature*, I, pp. 418-426, no. 279.

[18] On Hasan al-Jabartī see Brockelmann, *GAL*, II, p. 472, and SII, p. 487; *Cairo ENL Survey*, no. D91; and İhsanoğlu *et al.*, *Ottoman Astronomical Literature*, II, pp. 472-479, no. 314. On his son see the article "al-Djabartī" by David Ayalon in EI_2.

[19] On al-Kutubī (**I-2.1.8**) see *Cairo ENL Survey*, no. D76; and İhsanoğlu *et al.*, *Ottoman Astronomical Literature*, I, pp. 427-429, no. 281.

CHAPTER 8

OTHER LATE EGYPTIAN TABLES FOR TIMEKEEPING

8.0 Introductory remarks

In the sequel I discuss a set of late Egyptian prayer-tables for for each degree of latitude between Mecca and Istanbul (**8.1**), as well as sets of such tables specifically for use in Alexandria, Damietta, Rosetta and Crete (**8.5-8**). These are all clearly inspired by the tables in the main Cairo corpus. I also discuss ʿAbd al-Qādir al-Minūfī's writings on refraction at the horizon and his tables purporting to display the time taken by the sun to pass from the true horizon to the visible horizon for all latitudes (**8.2**), and Ibn Abi 'l-Khayr al-Ḥusnī's tables for computing the duration of twilight for all latitudes (**8.3**). al-Nabatītī's prayer-tables for the pilgrim route between Cairo and Mecca (**8.4**) have unfortunately not survived. The much later tables of Muḥammad Ḥifẓī serve all localities of consequence in Egypt and the Hejaz (**8.10**). These, and also the set for Cairo by ʿAbd al-Majīd Efendī printed in 1895 (**8.9**), bring our discussion of one thousand years of astronomical timekeeping in Egypt to a close.

8.1 Anonymous prayer-tables for all latitudes

MS Princeton Yahuda 861,1, penned *ca.* 1600, contains a set of prayer-tables computed for each integral degree of latitude between 21° (Mecca) and 41° (Istanbul). The tables are entitled *al-Natīja al-kubrā*, which may be rendered "Universal Prayer-tables". (The term *natīja* means "calendar" or "prayer-tables" in late medieval scientific Arabic.) On the title-folio the tables are attributed to ʿIzz al-Dīn al-Wafāʾī, a *muwaqqit* at the Muʾayyad Mosque in Cairo who died about 1470.[1] This attribution is highly doubtful, for reasons which will become clear below, but at least some of the tables are due to him. The manuscript is carelessly copied in an untidy hand, and is bound in some disorder. According to the title folio al-Wafāʾī's tables also included a set of auxiliary tables, but these are those of al-Khalīlī – see **10.7** – and are described as such in their sub-title. al-Wafāʾī did compile some auxiliary tables of his own, which are extant in another source – see **6.15**.

The manuscript begins with a short introduction on the use of the table displaying the solar longitude for each day of the Coptic year. This is stated to have been computed for longitude 55° by al-Wafāʾī (the entry for Tūt 1 is Virgo 14;39°). A star catalogue showing the right ascension of 72 stars is likewise attributed to al-Wafāʾī. Neither solar longitude table nor star catalogue is dated. There follow two sets of prayer-tables, the first for Cairo, and the second for all latitudes, including that of Cairo.

---

[1] On al-Wafāʾī see n. **I**-9:29.

In the first set of prayer-tables the functions:

δ, H, D, h_a, t_a, T_a, 2N, σ, α_σ, s and r

are tabulated for latitude 30°. The entries are those of the main Cairo corpus, but the format differs. For each pair of zodiacal signs these functions are displayed in columns across two facing pages of manuscript. The use of this format was perhaps inspired by the earlier prayer-tables of al-Khalīlī – see **10.6**. However, this first set of tables begins with Aries and Virgo, whereas al-Khalīlī's tables begin with Capricorn and Sagittarius.

In the second set most of the eight functions:

D, H, h_a, t_a, r, s, n and α_ϕ

are given for latitudes (no localities are mentioned):

21° (Mecca), 24° (Medina), 30° (Cairo), 31° (Alexandria), 32° (Jerusalem), 33;30° (Damascus and Bagdad), 34° (Tripoli and Homs) and 36° (Aleppo).

These are ostensibly based on obliquity 23;35°. *For other integral latitudes between 21° and 41° the functions D and H are given.* The function D is also tabulated for latitude 15° (Yemen). Other standard functions are given for various latitudes, without any pattern: for example, s is given for latitude 22° but not r, and for latitude 27° t_a but not h_a. It is not difficult to show that virtually all the 150 tables were lifted from other sources. My analysis reveals the following:

(a) The tables of D(λ) for all latitudes except 30° are the same as those of Najm al-Dīn al-Miṣrī in MS Oxford Marsh 676 (Uri 944 = 995) (**6.5**), where D(λ) is tabulated for each value of ϕ from 1° to 89°. This also explains the presence of the odd table of D(λ) for latitude 15°.

(b) The entries in the tables for 21° are quite different from the ones in MS Cairo MM 68 (**6.10**), being carelessly computed. The tables of t_a and D in the Princeton manuscript also occur in MS Istanbul Köprülü 1619 of an anonymous set of prayer-tables for Cairo.

(c) The 23 different functions tabulated for latitude 30° are taken from the main Cairo corpus.

(d) The 15 different functions for latitude 32° are taken from an independent source. There are tables of h_q and T_q for this latitude, based on a particular value of q: the only other tables for the qibla in the *Natīja* are those for latitude 30°. Also, the tables of r and s are based on parameters 20° and 16° rather than 19° and 17° as in the tables for other latitudes. I suspect that these tables were originally computed by Ibn al-Rashīdī or al-Karakī – see **9.5** for a further discussion.

(e) The tables for latitude 33;30° are those of al-Khalīlī (**10.6**) and are based on a value of 23;31° for the obliquity.

(f) The tables for latitude 34° are the same as those for Tripoli in MS Damascus Ẓāhiriyya 4893 (**11.6**).

Certain of the tables are particularly wretched: in the one giving h_a(λ) for latitude 39°, for example, it appears that there are three different functions tabulated rather than one! Some of these errors are due to copyists, but the fact that most of them are original is confirmed by examination of a second source for these tables.

MS Paris BNF ar. 2525,2 (fols. 17r-36v), copied *ca.* 1450, contains part of a beautifully-copied set of anonymous prayer-tables for various latitudes. This section of the manuscript is unrelated to those which precede and follow and is clearly incomplete. It begins with a table of α_ϕ for latitude 25°, and continues with tables of:

$$D, H \text{ and } \alpha_\phi$$

for latitudes 26° to 29° (except that for 27° t_a is tabulated rather than H). There follow tables of 24 different functions computed for the latitude of Cairo. The tables of D and H or t_a for latitudes 26° to 29° are virtually the same as those in the Princeton manuscript, being carelessly computed. On the other hand, those of α_ϕ in the Paris manuscript are fairly accurately computed: except for copyist's errors (particularly for ϕ = 25°), they are the same as the corresponding tables in the *Ḥākimī Zīj*. All of the tables for latitude 30° are the same as those in the main Cairo corpus.

Other functions tabulated in the Princeton manuscript are the following:

(a) "meridian altitude for all latitudes": this is simply $\delta^*(\lambda) = 90° + \delta(\lambda)$ tabulated for two digits for each degree of λ. Note that $H(\lambda,\phi) = [90° + \delta(\lambda)] - \phi$.

(b) "altitude for the afternoon prayer for all latitudes": this is simply $h_a(H)$ to two digits for each degree of H from 1° to 90°.

(c) "rising times of the signs for different latitudes": the rising times $\Delta\alpha$ for each pair of zodiacal signs are given to two digits for each degree of ϕ from 1° to 60°. The table is based on ε = 23;35°, and contains numerous errors. In MS Cairo MM 58 (**I-5.5.1**) there is a similar table based on ε = 23;51,20° (Ptolemy's value), which is attributed to Ibn Yūnus: the same table occurs in the Yemeni astronomical miscellany in the manuscript in a private collection in Sanaa (**12.4**), without the false attribution.

(d) "time since sunrise and hour-angle at the equinox for latitude 31°": entries are given for both functions for each degree of solar altitude from 1° to 58°. The table is based on latitude 32°, not 31° as stated.

There is also an anonymous table of the latitudes of the stations on the Syrian pilgrim route to Mecca. The entries are corrupt but were clearly lifted from the table in al-Khalīlī's Damascus corpus (**10.9**).

The celebrated Egyptian historian al-Jabartī (d. *ca.* 1825)[2] attributes two works entitled *al-Natīja al-kubrā* and *al-Natīja al-ṣughrā* to the astronomer Riḍwān Efendī (**7.9**), adding that these works were well known and in common use in his time. These two titles could refer to sets of prayer-tables (*natīja*) for all latitudes (*kubrā*, literally, "larger") and for one latitude (*ṣughrā*, literally, "smaller"). No universal prayer-tables attributed to Riḍwān survive in the manuscript sources. His prayer-tables for Cairo survive in numerous copies, several of which are in his own hand (**7.10**). The possibility that Riḍwan was the compiler of the wretched tables in MS Princeton Yahuda 861,1 cannot be excluded, but I consider it unlikely.

8.2 ʿAbd al-Qādir al-Minūfī's tables of corrections for refraction at the horizon at all latitudes

MS Istanbul Nuruosmaniye 2929,4 (fols. 22r-23v) contains two sets of tables compiled by ʿAbd al-Qādir al-Minūfī (see **7.2** on his father) in 975 H [= 1567] displaying the correction for

[2] Quoted in Dorn, "Drei arabische Instrumente", pp. 32-33.

refraction at the horizon for all latitudes (*daqāʾiq ikhtilāf mā bayn al-ufq al-ḥaqīqī wa-'l-marʾī li-jamīʿ al-ʿurūḍ*).[3] The first set (A) is "according to the opinion of ʿAbd al-Qādir ibn Muḥammad al-Minūfī", and the second set (B) is "according to the opinion of earlier astronomers" but also computed by al-Minūfī. Both tables display the function $\Delta D(\lambda,\phi)$ for arguments:

$$\lambda = 270°, 300°, 330°, 0°, 30°, 60°, 90° \quad \text{and} \quad \phi = 1°, 2°, \ldots, 60°.$$

Entries are given to four digits in A and three in B. No indication is given in the brief introduction to the tables of precisely what al-Minūfī thought he was doing when he compiled these tables. The two sets are based on the assumption that ΔD varies linearly with ϕ, so that the entries are merely proportional parts of the entries for $\phi = 60°$. The entries for $\phi = 30°$ in the two sets are:

	$\lambda = 270°$	$\lambda = 0°$	$\lambda = 90°$
A:	0;32,54,0	0;47,59,30	1;3,5,0
B:	0;32,0	0;47,0,0	1;2,0,0

Thus al-Minūfī's new tables (set A) represent an extension of his puerile attempt to "improve" the values attributed to Ibn Yūnus in other sources.

The notion that $\Delta D \propto \phi$ is not without importance for the development of astronomical timekeeping in Islam. We find in other sources sets of values of ΔD for the latitudes of Mecca, Crete, Damascus, and Istanbul, apparently based or intended to be based on this assumption. See further **6.10**, **8.8**, **11.13** and **14.9**.

MS Dublin CB 4067 contains the preface to a work entitled *Ḥadaqat an-nāẓir fī 'khtilāf al-manāẓir*, "The Pupil of the Eye of the Observer of Parallax (!)", by ʿAbd al-Qādir al-Minūfī. In his preface the author mentions that the work contains an introduction, seven chapters, and a conclusion, but the preface is followed by several pages of religious invocations and repetitions of the name of the Prophet. al-Minūfī Jr. states that he wishes to discuss the problem of the true and visible celestial circles using rigorous geometrical proofs. His father had told him that the scholar Aḥmad (*sic*) ibn Yūnus had fixed the difference between the two horizons (*mā bayna 'l-ufuqayn*) as:

EQ: 47^m SS: 62^m WS: 32^m

and that ʿAlāʾ al-Dīn Ibn al-Shāṭir (**9.3**) fixed them as 2^m more than Ibn Yūnus, *i.e.*,

EQ: 49^m SS: 64^m WS: 34^m .

al-Minūfī Jr. could not understand how Ibn al-Shāṭir had arrived at these values. He states that the only earlier astronomer who discussed this matter was Abu 'l-Ḥasan (*sic*) ibn al-Haytham,[4] who stated that a person of height $3^1/_2$ cubits could see 0;4,26 more than half of the celestial sphere. al-Minūfī Jr. could not understand what this meant, but he knew that Ibn al-Haytham wrote two large volumes on parallax (!) which he had not seen and in which perhaps Ibn al-

[3] On ʿAbd al-Qādir al-Minūfī see Suter, MAA, no. 479, and Brockelmann, GAL, SII, p. 486 (both confused); and *Cairo ENL Survey*, no. C121.

[4] On Ibn al-Haytham see the article in *DSB* by Abdelhamid I. Sabra. Prof. Sabra assures me that this subject is not discussed in Ibn al-Haytham's *Optics*. I have examined a manuscript in Oxford of a short treatise by Ibn al-Haytham on the fact that the amount of the celestial sphere that is visible is more than half, but it does not contain the numerical value given by al-Minūfī. However, several other manuscripts of two versions of this are listed in Sezgin, *GAS*, VI, p. 260.

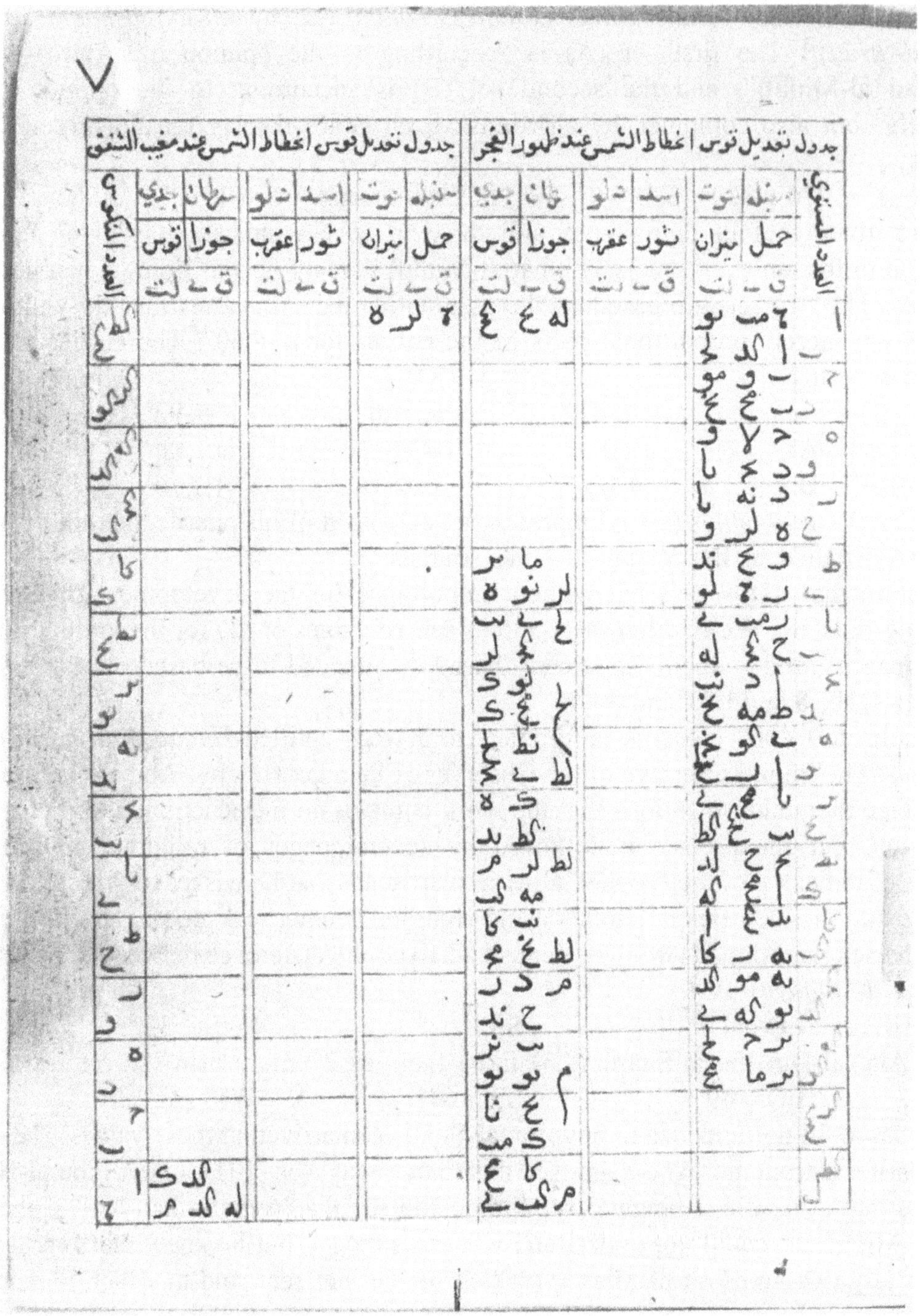

Fig. 8.2: Some mysterious tables in ʿAbd al-Qādir al-Minūfī's treatise *Rafʿ al-khilāf.* They purport to display the "correction" to the arc of depression of the sun at daybreak and nightfall as a result of the effect of refraction at the horizon, but most of the entries have been left blank, and it is not clear that our author knew what he was doing. [From MS Cairo MM 123, fol. 7r, courtesy of the Egyptian National Library.]

Haytham had explained the matter. However, he adds that Samaw'al Yaḥyā ibn ʿAbbās al-Maghribī, in Part I of Book V of his work entitled *Kitāb al-Tahdhīb fī ṣināʿat al-nujūm*, had made a table showing the amount of the celestial sphere visible for altitudes above the surface up to 12,000 cubits, apparently using Ibn al-Haytham's method.[5] al-Minūfī Jr. mentions that Abu 'l-Ḥasan (*sic*) ibn ʿAlī ibn ʿUmar al-Marrākushī (**2.7**) was asked in Rajab 694 H [= May/June 1295] whether one could find the amount of the celestial sphere one could see standing on the surface of the earth, and that he replied in the affirmative and had given geometrical and numerical demonstrations which ʿAbd al-Qādir approved. Unfortunately, our author's preface concludes without giving any more historical information.

Another treatise by ʿAbd al-Qādir al-Minūfī is preserved in MS Cairo MM 123, copied *ca.* 1600. This is entitled *Rafʿ al-khilāf fī ʿamal daqāʾiq al-ikhtilāf,* "The Removal of the Difference of Opinion on Operations with the Difference Minutes", and in it he mentions both Ibn Yūnus and Ibn al-Haytham again. It is not clear to me that al-Minūfī Jr. really knew what he was talking about, but certainly the entire tradition of which he is at the tail end merits detailed investigation. Some incomplete tables are found in this manuscript, whose purpose escapes me: see **Fig. 8.2**.

8.3 Ibn Abi 'l-Khayr al-Ḥusnī's twilight tables for all latitudes

MS Cairo DM 1108,3 (fols. 5v-10v, copied in 1052 H [= 1642/43]) is a unique copy of a treatise entitled *Kashf al-karabāt fī taḥqīq masāʾil yaḥtāj ilayhi ṭālib ʿilm al-awqāt*, which means something like "The Exposure of Apprehensions concerning the Proper Handling of Methods by the Student of Timekeeping", by Abu 'l-Khayr al-Ḥusnī (see already **7.4**). Part of the introduction is devoted to a discussion of the "difference minutes" in which the values:

47 52 57 62 57 52 47 42 37 32 37 42

for each sign beginning with Aries are attributed to Ibn al-Shāṭir (see **9.3**). The remaining part of the introduction is devoted to a discussion of the determination of the duration of twilight by computation, without reference to any observations.

The work concludes with a set of tables displaying the duration of morning and evening twilight r and s for each half-climate at the solstices and equinoxes. al-Ḥusnī first gives values of:

$$\phi \text{ and } \mathrm{Tan}_{60}\,\phi$$

for each half-climate and then for the solstices gives values of:

$$\mathrm{Sin}\,d,\ d,\ D,\ D^h\ .$$

[5] On Samaw'al al-Maghribī (Suter, *MAA*, no. 302, and *Cairo ENL Survey*, no. G9), an extremely important scholar of 12th-century Iran, see the article in *DSB* by Adel Anbouba, where his astronomical works are unfortunately overlooked. On the introduction to his astronomical treatise see Rosenthal, "Al-Asṭurlâbî and as-Samaw'al on Scientific Progress". After this study was completed (the first time round), I obtained microfilms of the Oxford and Leiden manuscripts of Samaw'al's astronomical works and found that there is indeed a table such as was described by al-Minūfī. al-Samaw'al's astronomical works merit a separate publication, which will contain much that is new to the history of Islamic science. See already Sezgin, *GAS*, VI, pp. 65-66. Parts of the texts were investigated in my seminar on Arabic scientific manuscripts at Frankfurt during 2000-01. See also the text to n. **V**-10:12.

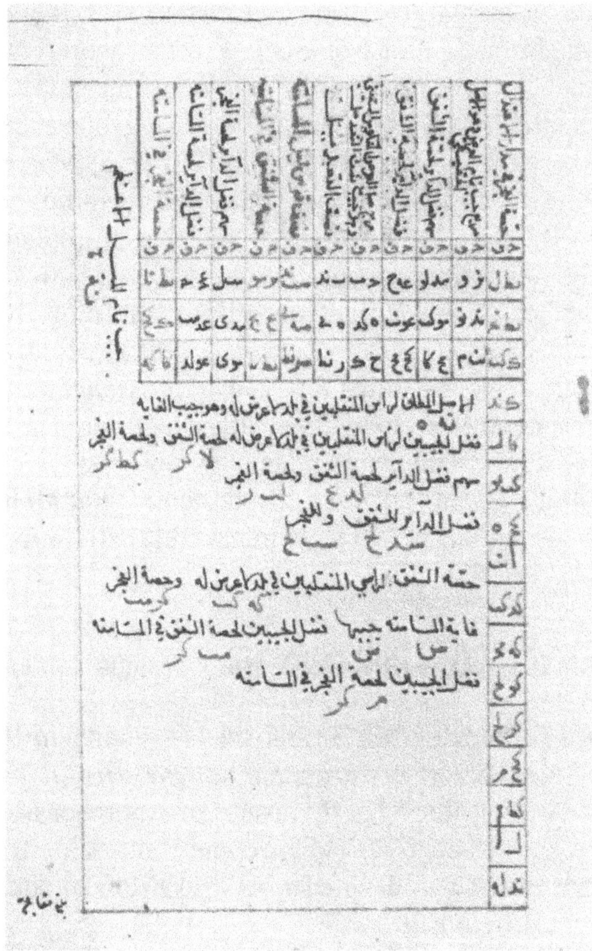

Then for each of the solstices and the equinoxes, as well as for the solar longitude for which the sun passes through the zenith (*i.e.*, such that $\delta(\lambda) = \phi$), he gives values of the quantities:

Sin H*, [Sin H* - 19°], [Sin H* - 17°], B, Vers t_r, Vers t_s, r and s ,

where t_r and t_s are the hour-angles at daybreak and nightfall, thus illustrating the steps by which r and s were calculated. The tables are shown in **Fig. 8.3**. See further **VIb-19** and also **II-11.14** on the tables of ʿAbdallāh al-Ḥalabī for Aleppo.

8.4 al-Nabatītī's prayer-tables for the pilgrimage to Mecca

MS Gotha A1412, which was lost from the Thüringsche Landesbibliothek, Gotha, during World War II, contained a set of four pages of prayer-tables compiled by ʿAbd al-Munʿim al-Nabatītī for the year 1041 H [= 1631/32].[6] His prayer-tables, of which no other copy has come to my attention, were intended to be used by pilgrims bound for Mecca, probably on the route from Cairo. The Gotha catalogue records the introduction, which translates as follows:

> "I computed these useful tables, according to the latest observations, for finding the times of prayer (*ḥiṣaṣ al-awqāt*) by day and night at the stations on the pilgrim road during the lunar year 1041."

Several other Arabic sources contain lists of geographical coordinates and qiblas for the stations on the pilgrim roads from Cairo or Damascus to Mecca.[7] As an example, I cite here the tables in MS Paris BNF ar. 2560, fol. 164v, which apparently form part of the *zīj* entitled *al-Rawḍ al-ʿāṭir* by the Damascus astronomer Ibn Zurayq (*fl. ca.* 1400).[8] These display the latitudes of some 32 stations on the pilgrim route from Damascus to Mecca, and the latitudes, longitudes, and qiblas of 15 stations on the pilgrim route from Cairo to Mecca. For the few stations listed in both tables the latitudes given are generally different. See also **10.9**.

8.5 Anonymous timekeeping tables for Alexandria

MS Cairo TR 354, copied *ca.* 1700, contains about 140 pages of tables, including a complete set of tables of the functions:

$$t(h,\lambda),\ T(h,\lambda),\ \text{and}\ a(h,\lambda)$$

←
Fig. 8.3: The tables of Abu 'l-Khayr al-Ḥusnī for determining the length of twilight for all latitudes. The tables are of particular interest in that the compiler presents all of the auxiliary functions which he used. [From MS Cairo DM 1108,3, fols. 9r-10v, courtesy of the Egyptian National Library.]

6 *Gotha Catalogue*, III, p. 63. On ʿAbd al-Munʿim al-Nabatītī see İhsanoğlu *et al.*, *Ottoman Astronomical Literature*, I, pp. 308-309, no. 167, citing the Gotha manuscript and a short treatise on timekeeping extant in Berlin. I have seen a copy of an almanac by ʿAbd al-Munʿim al-Nabatītī for the year 1073 H [= 1662/63], consisting mainly of simple calendrical tables, but I forget where. He is not to be confused with ʿAbd al-Qādir al-Nabatītī (*Cairo ENL Survey*, no. D28, and İhsanoğlu *et al.*, *op. cit.*, I, p. 279, no. 143), author of a treatise on the agricultural and meteorological aspects (*tawqīʿāt*) of the Syrian calendar, extant in MS Cairo MM 108,1, or the latter's son, ʿAlī ibn ʿAbd al-Qādir al-Nabatītī, author of a treatise on the algebra of inheritance and others on simple timekeeping (*Cairo ENL Survey*, no. D28, and İhsanoğlu *et al.*, *op. cit.*, I, pp. 286-287, no. 151).

7 Some examples are mentioned in King, *Mecca-Centred World-Maps*, pp. 84-86.

8 On Ibn Zurayq see n. 11:4.

computed to two digits for each degree of both arguments. The underlying parameters are found by inspection to be:

$$\phi = 31° \text{ (Alexandria) and } \varepsilon = 23;35° .$$

The maximum solar altitude argument is thus 82° rather than 83° as in the main Cairo corpus. The three functions are tabulated side by side as in al-Bakhāniqī's edition of the Cairo corpus and in MS Cairo MM 71 of al-Khalīlī's and Shihāb al-Dīn al-Ḥalabī's tables for Damascus (**5.6**, **10.5** and **11.2**). Following these extensive tables is a single table of h_0 based on the parameters $\phi = 31°$ and $\varepsilon = 23;30°$.

On the title-page it is stated that the tables are due to Ibn Yūnus, but the person who wrote this knew so little about the illustrious Egyptian astronomer that he called him Ibn Yūnus al-Mawṣilī (*i.e.* from Mosul), confusing him with an ʿIrāqī scientist also named Kamāl al-Dīn ibn Yūnus.[9] It is easily shown that the tables are not due to Ibn Yūnus the Egyptian.

Firstly, the tables are generally less accurately computed than those in the Cairo corpus. Secondly, the values of t and T in MS Cairo TR 354 are related by those in the table of $D(\lambda)$ for $\phi = 31°$ found in MS Princeton Yahuda 861,1 of the prayer-tables for all latitudes, which are ultimately due to Najm al-Dīn al-Miṣrī (**8.1** and **6.5**). Again, the tables of $T(h,\lambda)$ for altitudes 19° and 17° in MS Cairo TR 354 show suspicious resemblance to the tables of $r(\lambda)$ and $s(\lambda)$ for $\phi = 31°$ in MS Princeton Yahuda 861,1.

MS Cairo ṬM 216,3 (fols. 14r-18v, 23r-87r, copied *ca.* 1800), contains a different set of tables of:

$$T(h,\lambda) \text{ and } t(h,\lambda)$$

for parameters:

$$\phi = 31° \quad \text{and} \quad \varepsilon = ?.$$

The tables are incomplete and bound in disorder. Tables of the two functions are copied on facing pages.

MS Cairo DM 1207,1 (fols. 1r-57v) is a copy of a set of prayer-tables for latitude 31° prepared in 1168 H [= 1754/55] by Ḥusayn, a *muwaqqit* in Rosetta, from a copy in the hand of the author Riḍwān, dated 1090 H [= 1679/80]. The tables are not related to those in MS Cairo TR 354 discussed above. The work begins with a brief introduction with calendrical, solar, and astrological tables, followed by a set of geographical tables (fol. 18r) in which we find the following entries for various cities in the Delta:

	L	φ
Cairo	54;55°	30; 0°
Damietta	53;50	31;25
Maḥalla	54;30	31; 0
Alexandria	51;54	30;58
Rosetta	52;50	31;20
(Mecca	67; 0	21; 0)

The prayer-tables (fols. 21v-35r) display the following functions:

δ, d, ψ, h_0, h_q, $Z_{(12)}$, H, D, D″, h_a, t_a, T_a, s, r, 2N, n, B, C, h_q, α', α_ϕ, α_s and α_r .

[9] On Kamāl al-Dīn ibn Yūnus see n. 7:2 above. Similar confusion also occurred in the 20th century: see King, "Ibn Yūnus and the Pendulum", pp. 43-47.

Values appear to be based on $\phi = 31°$ (ε uncertain), although the table of $Z_{(12)}$ and the second table of h_q are lifted from the Cairo corpus and are based on $\phi = 30°$ (with $q = 37°$). The tables for twilight are based on parameters 19° and 17° but the tables are not the same as the corresponding ones in MSS Cairo TR 354 and Cairo ṬM 216. The first table of h_q is based on parameters $\phi = 31°$, and I was not able to investigate the underlying value of q. The function D″ represents the time between midday and when the sun has set over the visible horizon and appears to be computed using:

$$D'' = D + \Delta D + 0;16° ,$$

where ΔD are the "difference minutes" for Cairo (**4.11**). The work concludes (fols. 37v-57r) with an undated star catalogue.

MS Cairo ZK 324, copied *ca.* 1800, is another set of prayer-tables for latitude 31° containing the same introduction, calendrical tables, and most of the prayer-tables that are in MS Cairo DM 1207. MS Cairo DM 1211, copied in 1101 H [= 1689/90] in the hand of Riḍwān Efendī, contains some of these prayer-tables, namely:

$$Z_{(12)}, H, h_a, t_a, T_a, s , r, 2N, n , \alpha', \alpha_\phi \text{ and } \alpha_s ,$$

with a note by the first that it is for $\phi = 30°$ not 31°.

MS Cairo ṬM 216, fols. 19r-22v, copied *ca.* 1700, contains another set of prayer-tables for $\phi = 31°$ displaying functions:

$$D, t_a, T_a, h_a, s, r \text{ and } n .$$

Clearly Riḍwān Efendī had some input in certain of these tables, but I am not inclined to attribute the main tables of (T,t,a) for latitude 31° to him without new evidence.

8.6 Quṭb al-Dīn al-Maḥallī's prayer-tables for Damietta

MS Cairo ṬM 241,1 (fols. 1r-33r), copied 1245 H [= 1829/30], and several other inferior copies preserved in the Egyptian National Library, contains a set of prayer-tables for the latitude of Damietta compiled by the Egyptian astronomer Quṭb al-Dīn al-Maḥallī in 1088 H [= 1677/78].[10] The following functions are displayed to two digits for each degree of λ:

$$H, \delta, d, D, h_a, t_a, T_a, s, r, n, \alpha_\phi, \alpha_T \text{ and } \alpha' ,$$

and based on the parameters:

$$\phi = 31;25° \text{ and } \varepsilon = 23;35° ,$$

with 19° and 17° for twilight. Two different conventions are used for the argument λ. In the tables of T_a, α_ϕ, α_T and α', the arguments are as in the main Cairo corpus. In those of H, δ, d, D, t_a, s, r and n, the vertical arguments are still 1 to 30 downwards and 0 to 29 upwards, but Aries 1° corresponds to $\lambda = 0°$. Thus the first entry in the table is for the vernal equinox and no entries are given for the solstices. The upward argument is absurd, and implies that Virgo 29° corresponds to the autumnal equinox. The hapless soul who arranged these tables in this way so that the first entry was for $\lambda = 0°$ rather than $\lambda = 1°$ inadvertently rearranged the table of h_a twice, since the first entry is for $\lambda = 359°$! The entries for δ and d (and also

[10] On Quṭb al-Dīn al-Maḥallī see *Cairo ENL Survey*, no. D37; and İhsanoğlu *et al.*, *Ottoman Astronomical Literature*, I, pp. 312-313, no. 173 (lists other new sources).

H and D) are fairly accurately computed, but the table of α_ϕ is carelessly computed. The tables of t_a and T_a, which have two different formats, show corresponding error patterns, but some of the entries in each are garbled.

In an Egyptian manuscript which in 1939 was in the Institut für Geschichte der Medizin und der Naturwissenschaften in Berlin, according to a description by Willy Hartner,[11] there were some tables giving the solar azimuth as a function of the altitude for an unspecified latitude (**I-5.1.4**), as well as a table of h_0 for $\phi = 31;25°$ (not found in MS Cairo DM 106 – see **I-4.8.7**) and another relating to the azimuth of Mecca, probably of the function h_q. Perhaps the azimuth tables were computed for latitude 31;25°, which was the standard value for Damietta. Elsewhere in the former Berlin manuscript there were some other spherical astronomical tables for this same latitude, as well as a table of the corrections for refraction at the horizon – Hartner's *Minuten des Abstandes,* this last table being attributed to an individual identified as al-Hindī.[12]

8.7 Anonymous prayer-tables for Rosetta

MS Cairo DM 1207,2 (fols. 57v-58v, copied *ca.* 1850) – see **Fig. 8.7** – contains a set of tables of the functions:

2N, D and d ,

computed for latitude $\phi = 31;20°$ (Rosetta). The tables are copied in an atrocious hand and are very carelessly computed. Consider, for example, the following values of d compared with the correctly computed values d′ (for $\varepsilon = 23;35°$):

λ	15°	30	45	60	75	90
d	3;51°	7;43	10;17	12;51	14; 8	15;25
d′	3;38°	7; 8	10;21	13; 0	14;47	15;25

Note that the value given for argument 30° is about one-half that given for 90°, and that the values for 15°, 45°, and 75° were probably derived from those for 30°, 60°, and 90° by linear interpolation. See also **11.9** for another example of simplistic interpolation of this kind.[13]

8.8 Yūsuf Kilārjī's prayer-tables for Crete

al-Jamālī Yūsuf ibn Yūsuf, Kilārjī (*major domo*) of Ḥasan Āghā, was a student of Riḍwān Efendī who became *rūznāmejī* in Cairo and later worked in Mecca and the Yemen; he died in 1153 H [= 1741/42].[14] One of his works was a set of prayer-tables for Crete, extant in the unique copy MS Cairo DM 834, 70 fols., copied in 1120 H [= 1708/09], perhaps by the author. In his introduction Yūsuf Kilārjī states that he prepared these tables because there were none

---

[11] Hartner & Ruska, "Katalog", pp. 55-56.

[12] Possibly to be identified with Shams al-Dīn Muḥammad al-Hunayd(ī), on whom see *Cairo ENL Survey*, nos. C133-134 (though see also no. D41 on a "real" (?) Indian working in Cairo, albeit in mathematics rather than astronomy); and İhsanoğlu *et al.*, *Ottoman Astronomical Literature*, I, p. 116, no. 55.

[13] See n. **I-1**:21.

[14] On Yūsuf Kilārjī see Dorn, "Drei arabische Instrumente", p. 35; *Cairo ENL Survey*, no. D61; and İhsanoğlu *et al.*, *Ottoman Astronomical Literature*, I, pp. 412-415, no. 275.

Fig. 8.7: Some messy prayer-tables for Rosetta, displaying the duration of night and half the "true" length of daylight. [From MS Cairo DM 1207,2, courtesy of the Egyptian National Library.]

available for Crete. He mentions the traditions relating to refraction at the horizon attributed to Ibn al-Haytham and Ibn Yūnus and refers to a universal table displaying $\Delta D(\phi)$, although without mentioning al-Minūfī (**8.2**). His introduction ends with the poem on the prayer-times which he attributes to al-Shāfiʿī (**2.1**). After some tables for calendar conversion and solar longitude and velocity Yūsuf Kilārjī presents tables with the format of the main Cairo corpus of the following functions, mostly with values to three digits:

$$\delta,\ d,\ \psi,\ h_0,\ B,\ C,\ h_q,\ H,\ D,\ \Delta D,\ (D+\ \Delta D),\ h_a,\ t_a,\ T_a,\ Z_{(12)},\ Z_{(7)},\ z_{a(7)},\ z_{b(7)},$$
$$(r+D),\ s,\ r,\ 2N,\ n,\ D^h,\ 2D^h,\ \tilde{h},\ 2N^h,\ t_a^{\ h},\ T_a^{\ h},\ s^h,\ r^h \text{ and } (2N\text{-}r)^h\ .$$

These are followed by tables with entries for each five days for the functions:

$$s^h,\ (2n\text{-}r)^h,\ r^h,\ 2N^h,\ (r+D)^h,\ t_a^{\ h},\ T_a^{\ h} \text{ and } 2D^h\ .$$

The work concludes with tables of the functions:

$$\alpha_\phi,\ \alpha_\phi(\lambda^*),\ \alpha_s \text{ and } \alpha_r\ ,$$

and a star catalogue for 1120 H [= 1708/09] as well as an unusual kind of table displaying the solar longitude for which various stars become visible in the west. The parameters underlying the prayer-tables are:

$$\phi = 35;30° \text{ (Crete) and } \varepsilon = 23;30,17°\ .$$

The maxima and minima for ΔD are 1;13,12,0 and 0;37,52,0. The corresponding values of ΔD for ϕ = 35;30° according to the theory of ʿAbd al-Qādir al-Minūfī (**8.2**) are 1;13,22 and 0;37,52°. The parameters used for twilight are 19° and 17°. The azimuth of the qibla underlying the table of h_q has not been investigated. See also **14.10** on some Ottoman-type prayer-tables for the latitude of Khania in Crete.

8.9 ʿAbd al-Majīd Efendī's prayer-tables for Cairo and al-Rashīdī's table for finding the qibla

MS Cairo Azhar *falak* 18082 consists of a set of prayer-tables for Cairo printed there in 1312 H [= 1894/95], and compiled by ʿAbd al-Majīd Efendī.[15] These display the times of daybreak, sunrise, midday, the afternoon prayer, sunset and nightfall in *ifranjī*, that is, European time. At the end of this work there is a list of geographical coordinates and qiblas of several cities, giving the value 134;32,58° (from the north) as the qibla for Cairo. It is stated that these values are taken from a book by Ismāʿīl Pāshā al-Falakī.[16] This qibla is about 1½° north of the actual (*i.e.*, modern) qibla for Cairo, but is about 7½° south of the qibla favoured by the medieval Egyptian astronomers (q = 53°) (**4.7**).

Following this list of coordinates is a discussion of the use of a compass for finding the qibla. Then ʿAbd al-Majīd Efendī presents a table for determining the qibla which he states was compiled by Muḥammad al-Rashīdī al-Falakī.[17] This purports to give the time of each day of the Coptic year when the sun indicates the azimuth of Mecca. The times are given according to the Ottoman convention that sunset is 12 o'clock (**14.0**), and the underlying azimuth of Mecca which gives the closest fit to the entries in the table (maximum: 4;19^h, mean: 3;40^h, and minimum: 3;0^h) is about 54°, measured from the meridian. Thus it appears that al-Rashīdī simply converted the table of t_q in the Cairo corpus to hours and minutes, Turkish time.

8.10 Muḥammad Ḥifẓī's prayer-tables for Egypt and the Hejaz

MSS Cairo K 3792-3802, copied *ca.* 1900, consist of eleven small booklets of prayer-tables compiled by Muḥammad Efendī Ḥifẓī, an official of the Egyptian Customs in ʿAbbāsiyya, Cairo: see **Fig. 8.10a-b**.[18] The tables display the times:

r′ (*fajr*), R′ (*shurūq*), ι′ (*ʿīd*), z′ (*zawāl*), m′ (*ẓuhr*),
a′ (*ʿaṣr awwal*), b′ (*ʿaṣr thānī*), S′ (*maghrib ifranjī*), s′ (*ʿishāʾ*)

Generally, separate tables are given for "Arab" (*ʿarabī*), *i.e.*, "Turkish", and local mean time (*ifranjī*). Where the time for the *ʿīd* prayer is given it is usually between 20 and 30^m after sunrise. The localities served by these tables are Cairo (ϕ = 30°); Assiut (27°); Qusayr (26;7°); Sohag

15 On ʿAbd al-Majīd al-Falakī see Azzawi, *History of Astronomy in Iraq*, p. 325. He does not seem to be listed in İhsanoğlu *et al.*, *Ottoman Astronomical Literature*.

16 On Ismāʿīl Pāshā al-Falakī see Azzawi, *op. cit.*, pp. 328-329. He does not seem to be listed in İhsanoğlu *et al.*, *Ottoman Astronomical Literature*.

17 This al-Rashīdī is apparently not mentioned in the modern sources.

18 See *Cairo ENL Survey*, no. D194.

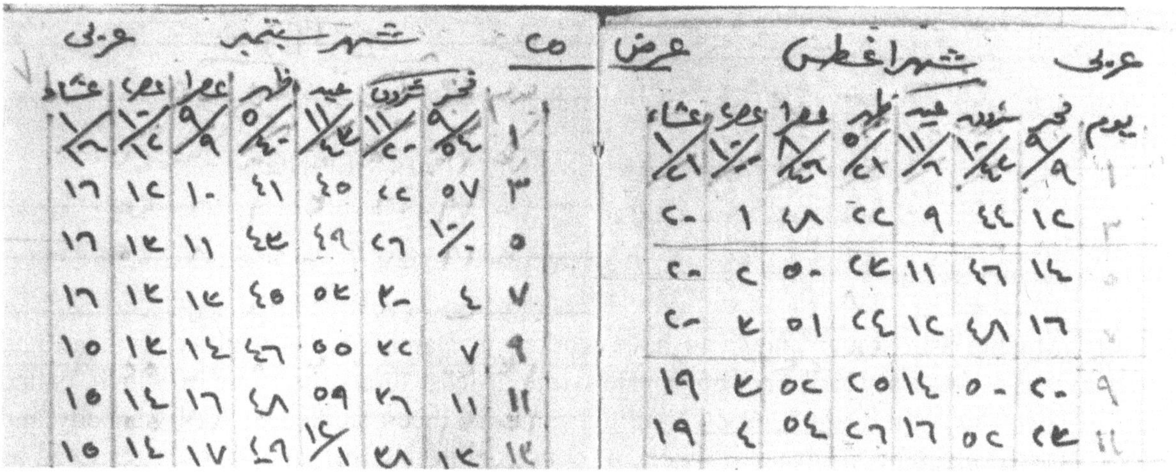

Fig. 8.10a: An extract from Muḥammad Efendī Ḥifẓī's prayer-tables for latitude 25° and Turkish time (*'arabī*), serving August and September. The entries are in minutes, with the appropriate hour indicated above a diagonal slash in the first entry of each column. [From MSS Cairo K 3792-3802, courtesy of the Egyptian National Library.]

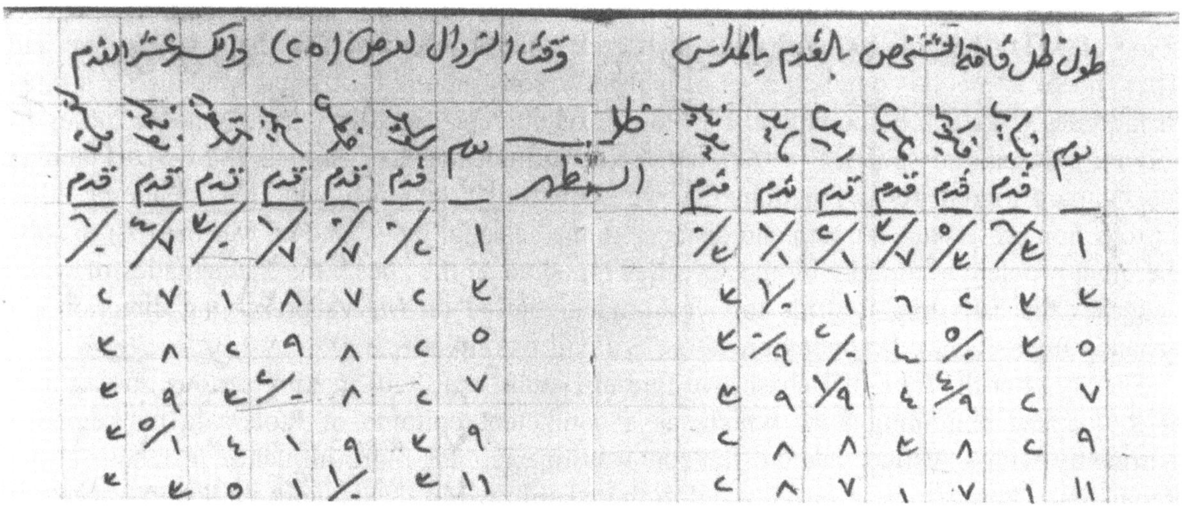

Fig. 8.10b: These tables display the shadow at the *ẓuhr* for every two days of the year, also for latitude 25°. The entries are in feet and decimal fractions thereof. [Same source.]

and Akhmim (26;30°); Qus (26°); Edfu and Esna (25°); Medina (both 24° and 24;10°); *Markaz al-Durr* (?) (23°); Korosko (*Markaz al-sidr*) (22;30°); Wadi Halfa (22°); Mecca (both 21;28° and 21;30°); and Jabal al-Ṭūr (in the Sinai) (28;5°). Some other funtions are tabulated for certain latitudes.

CHAPTER 9

EARLY SYRIAN TABLES FOR TIMEKEEPING

9.0 Introductory remarks[1]

The first Muslim astronomer known to have worked in Damascus (as well as Baghdad and Samarra) was Ḥabash al-Ḥāsib (*ca.* 850).[2] He wrote at least three zījes, one of which, called *al-Zīj al-Dimashqī*, "the Damascus *Zīj*", survives and has been studied by Ted Kennedy and Marie-Thérèse Debarnot.[3] Ḥabash did work of considerable interest, and is to be regarded as the most original astronomer of the early Islamic period, as well as one of the most prolific. His early work in mathematical astronomy shows considerable Indian influence, but this is supplanted by Greek influence and then – he lived for 100 years – we find him solving higly complicated problems which went beyond the interests of earlier traditions. His work was in a sense experimental for the Islamic world was not yet ready for some of his achievements, some of which had far less influence in later Islamic astronomy than they shoud have had. Thus, for example, he compiled a set of tables for constructing horizontal sundials for a range of latitudes (**I-4.1.1**); he compiled the first set of universal auxiliary trigonometric tables for solving problems of spherical astronomy for all latitudes (**I-9.1**); he was the first to tabulate the standard trigonometric functions that we know today; he conceived an astrolabe based on a projection preserving azimuth and distance to the celestial pole;[4] and he was the first to solve the qibla problem accurately by the so-called "method of the *zīj*es", the mathematics of which underlies the cartographic grids for world-maps centred on Mecca preserving direction and distance to the centre, known from three Safavid examples.[5]

Shortly after the time of Ḥabash, we find al-Battānī (*ca.* 910) making serious observations at Raqqa and compiling a *zīj* which, as a competent epitome of Ptolemaic mathematical astronomy with a distinct Islamic flavour, was to exert far more influence in medieval and Renaissance Europe than it actually did in the Islamic world.[6] This work marks the successful appropriation and synthesis of Ptolemaic planetary astronomy and Greek and Indian spherical astronomy and trigonometry, a process to which Ḥabash had been the main contributor.

The only astronomers know to have prepared *zīj*es in Syria between the time of Ḥabash and al-Battānī and the 14th century, when Damascus became the leading centre of astronomy in the Islamic (and indeed in the entire) world are:

---

[1] I have attempted an overview of astronomy in Syria in *Paris IMA 1993-94 Catalogue*, pp. 386-395. See also my "Astronomy of the Mamluks".

[2] On Ḥabash see n. **I**-9:1.

[3] Kennedy, "*Zīj* Survey", pp. 151-154, and Debarnot, "*Zīj* of Ḥabash".

[4] See Kennedy & Lorch & Kunitzsch, *Melon Astrolabe*.

[5] See King, *Mecca-Centred World-Maps*, esp. pp. 61-64 and 241, and also **VIIc**.

[6] On al-Battānī see n. **I**-4:11.

(1) Ibn al-Dahhān (*ca.* 1170) who worked in the service of the Ayyubid ruler Ṣalāḥ al-Dīn (Saladin);[7]
(2) Mu'ayyad al-Dīn al-ʿUrḍī (*ca.* 1250), who later assisted in the observations at Maragha in Iran under the direction of Naṣīr al-Dīn al-Ṭūsī;[8]
(3) Muḥyi 'l-Dīn al-Maghribī (*ca.* 1250), who also went on to work in Maragha;[9] and
(4) Ibn al-Lubūdī (*ca.* 1250), who worked in both Damascus and Cairo.[10]

The *zīj*es of Ibn al-Dahhān, al-ʿUrḍī and Ibn al-Lubūdī are no longer extant, and the *zīj* which al-Maghribī compiled in Damascus, called *Tāj al-azyāj*, "Crown of the *Zīj*es", survives but has not been studied in detail. A fragment from the *Zīj* of Ibn al-Dahhān preserved in a Yemeni manuscript is discussed in **9.1**. The recovery of al-ʿUrḍī's *Zīj* would contribute to our knowledge of the development of Islamic planetary theory, since al-ʿUrḍī's work influenced Ibn al-Shāṭir (see below). Likewise a detailed study of the *Zīj* of al-Maghribī and his later *zīj*es prepared at Maragha, would also contribute towards clarifying our knowledge of astronomy in Syria and its influence on the Maragha school. Important for our understanding of the influence of Andalusī and Maghribī influence in Syria is the recension for Damascus of the *Zīj* of the 13th-century Tunisian astronomer Ibn Isḥāq (**13.1**).

In the 14th century a number of astronomers flourished in Syria, mostly associated with the Umayyad Mosque in Damascus.[11] Ibn al-Sarraj, who worked in Aleppo apparently without institutional backing, specialized in devising new astronomical instruments.[12] Foremost amongst Damascus scholars were Ibn al-Shāṭir, who made substantial contributions to planetary theory,[13] and al-Mizzī[14] and al-Khalīlī,[15] whose interests were in astronomical timekeeping. Each of these 14th-century astromomers devised different kinds of quadrants and other devices for solving problems in spherical astronomy.[16] There was especial interest in instruments which would provide solutions for all latitudes (**VIb**). Both Ibn al-Shāṭir and al-Mizzī had studied astronomy in Alexandria before taking up their positions at the Umayyad Mosque, and the Egyptian influence is apparent in their interests and achievements.

al-Khalīlī's extensive tables for timekeeping (**Ch. 10**), which were used in one form or another from the 14th to the 19th century, replaced an earlier set compiled by al-Mizzī (**9.2**). Ibn al-Shāṭir also prepared some prayer-tables (**9.3**). Another student of al-Mizzī was al-Karakī,

---

[7] On Ibn al-Dahhān see n. 9:17.

[8] On al-ʿUrḍī see Suter, *MAA*, pp. 147 and 154; Kennedy, "*Zīj* Survey", no. 42; *Cairo ENL Survey*, no. C170; and Saliba, "13th-Century Observational Notebook". Saliba has also edited the text of his treatise on theoretical astronomy (Beirut, 1990).

[9] On al-Maghribī see n. **I**-5:8.

[10] On Ibn al-Lubūdī see Suter, *MAA*, no. 365, Kennedy, "*Zīj* Survey", nos. 86 and 87, and Azzawi, *History of Astronomy in Iraq*, pp. 155-156; and King, "Astronomy of the Mamluks", pp. 533-534.

[11] For a survey of their achievements in astronomical timekeeping see King, "Astronomical Timekeeping in Medieval Syria".

[12] On Ibn al-Sarrāj (Suter, no. 508 (confused), and *Cairo ENL Survey*, no. C26) and his instruments see King, "The Astronomical Instruments of Ibn al-Sarrāj", published for the first time in *idem*, *Studies*, B-IX, and also **X-5.2**. See also Charette & King, *The Universal Astrolabe of Ibn al-Sarrāj* (forthcoming).

[13] See n. **I**-2:15.

[14] On al-Mizzī see n. 9:21.

[15] On al-Khalīlī see n. 10:1.

[16] On some of these see Schmalzl, *Geschichte des Quadranten*, and **VIb-8**, as well as Charette, *Mamluk Instrumentation*.

who worked in Jerusalem and adapted the tables of Ibn al-Rashīdī to form the main Jerusalem corpus (**9.4**).

Damascus was razed to the ground by the Mongols under Timurlang in 1401. There was little creative astronomical activity of any consequence in Syria thereafter: the centre of Islamic planetary astronomy moved to Samarqand, and the centre of astronomical timekeeping moved back to Cairo for a century or two before it finally moved to Istanbul.

9.1 A fragment from the *Zīj* of Ibn al-Dahhān

Fakhr al-Dīn Abū Shujāʿ Muḥammad ibn ʿAlī known as Ibn al-Dahhān, who worked in Damascus *ca.* 565 H [≈ *ca.* 1170], was an astronomer associated with the celebrated ruler Ṣalāḥ al-Dīn (Saladin).[17] He compiled a *zīj*, which is unfortunately no longer extant. However, a fragment survives in the Sanaa miscellany of al-Afḍal (**12.4**), and consists of a set of tables giving the values of the following functions at the summer solstice:

$$d,\ d^h,\ H,\ Z_{(12)} \text{ and } z_{a(12)}$$

for some 120 cities.[18]: see **Fig. 9.1**. Values are given to two digits. The underlying value of ε is 23;35°. The relation between the last two functions is simply:

$$z_a = \text{Cot } h_a = \text{Cot } H + 12 = Z + 12.$$

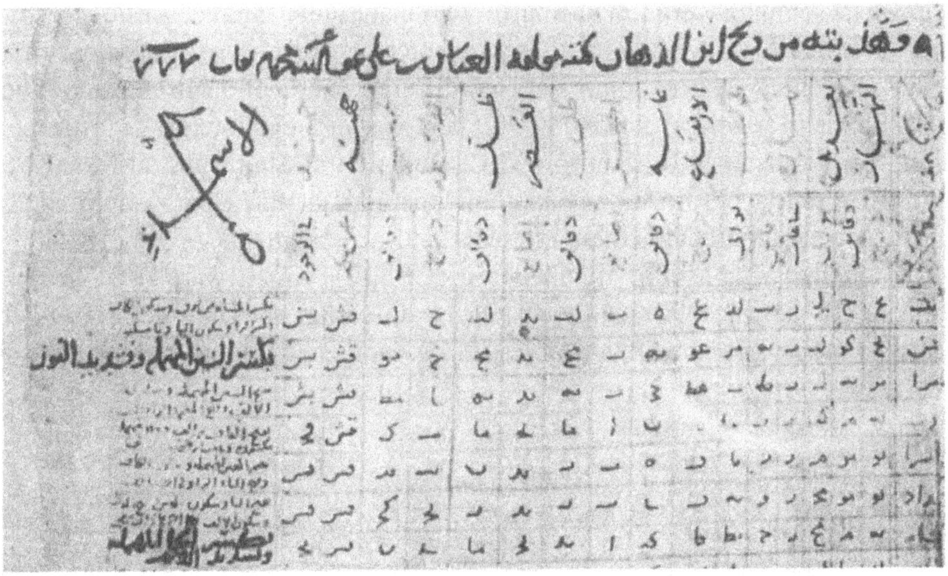

Fig. 9.1: An extract from the beginning of Ibn al-Dahhān's tables. Notice that the place-names are not visible on this page. A reconstruction of the entire table and the geographical coordinates underlying it would be possible and even worthwhile; but where are the volunteers? [From Varisco & Smith, eds., *al-Afḍal's Anthology*, p. 151, courtesy of Professor Daniel Varisco.]

[17] On Ibn al-Dahhān see Suter, *MAA*, no. 310; Kennedy, "*Zīj* Survey", no. 89; and King, *Astronomy in Yemen*, p. 37. Add to II n. 10:17: This Ibn al-Dahhān is not to be confused with ʿImād al-Dīn ibn al-Dahhān, chief of the astronomers in Mosul *ca.* 1285, for a reference to whom see Cook, "Muslim Material on Comets and Meteors", p. 147, and Rada, "Comets in Arabic Literature", p. 45.

[18] See now the facsimile in Varisco & Smith, eds., *al-Afḍal's Anthology*, pp. 151-154.

The azimuth of the qibla measured from the meridian is also given,[19] as well as the vowelling of the Arabic names of the cities. Not all of the city names are visible on the microfilm which I have used, but there seems to be no entry for Mecca (as one can see from the qibla-values, if not the names). Sample significant entries for other cities (values of ϕ are calculated from those of H) are:

Locality	d	max H	[ϕ]	q
Cairo	14;3°	84;20°	[29;15°]	54;39°
Jerusalem	16;5	80;35	[33;0]	41;21
Damascus	16;40	80;5	[33;30]	28;29

Ibn al-Dahhān is known to have visited Cairo, but his value for the latitude of that city, confirmed by the value of d (so that it is not a scribal error for 29;55°), is too small by $^3/_4$°.[20] Likewise his value for the latitude of Jerusalem is too large by about 1°. (However, a later qibla table compiled for Jerusalem is based on the parameter 41;30°, rather than on values which can be derived using a more accurate latitude for Jerusalem (see further **9.5**).) Of the Islamic geographical tables investigated by the Kennedys (**1.3**), only the table in al-Bīrūnī's *Qānūn* has the same latitudes for these three cities as Ibn al-Dahhān. However, Ibn al-Dahhān's qibla values do not correspond to those that can be computed from al-Biruni's coordinates using either the exact or the standard approximate formulae.

9.2 al-Mizzī's hour-angle tables and prayer-tables for Damascus

The astronomer Shams al-Dīn Muḥammad ibn Aḥmad ibn ʿAbd al-Raḥīm al-Mizzī (1291-1343), whose family originated from a suburb of Damascus, studied under Ibn al-Akfānī in Alexandria, and then moved back to Damascus where he eventually worked in the Umayyad Mosque.[21] He was a fine instrument-maker, and several instruments of his construction and treatises on their use have survived.[22] He also compiled some tables for timekeeping. The main set appears to be no longer extant, but, according to his student al-Karakī (**9.4**), it consisted of tables of the hour-angle t(h,λ) computed for latitude 33;27°, that is, Damascus: see **I-2.1.3**. al-Mizzī's minor set of tables survives in MS Cairo MM 62, copied *ca.* 1400. The tables are not accompanied by any introduction, and the functions tabulated are:

$$\delta,\ H,\ Z_{(7)},\ d,\ D,\ \tilde{h},\ 2D^h,\ 2N,\ h_0,\ \psi,\ \alpha,\ \alpha_\phi,\ h_a,\ t_a,\ T_a,\ r \text{ and } s\ .$$

The entries are arranged with the format of the main Cairo corpus (see **Fig. 9.2**), and are based on al-Mizzī's distinctive parameters:

$$\phi = 33;27° \text{ and } \varepsilon = 23;33°\ ,$$

on which see below. They are very accurately computed, although al-Mizzī used an interpolation system in the fourth columns of his tables of r and s which failed to take account of the actual behaviour of these functions for solar longitudes in Libra and Pisces. This means

[19] This table was advertently overlooked in King, *Mecca-Centred World-Maps*.
[20] On the unhappy value 29;15° see King, "Geography of Astrolabes", p. 9 and n. 23.
[21] On al-Mizzī see n. **I**-2:13.
[22] Mayer, *Islamic Astrolabists*, pp. 61-62. See also *Paris IMA 1993-94 Exhibition Catalogue*, p. 438, no. 333, on one of his quadrants.

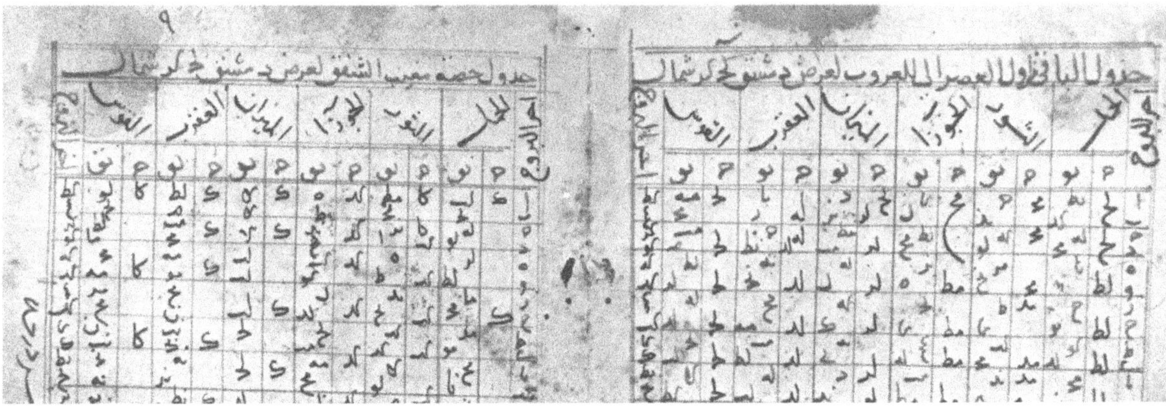

Fig. 9.2: An extract from al-Mizzī's prayer-tables showing the sub-tables for the time from the beginning of the *ʿaṣr* to sunset and the duration of evening twilight (based on a solar depression of 17°). [From MS Cairo MM 62, fols. 8v-9r, courtesy of the Egyptian National Library.]

that the corresponding entries in his hour-angle tables would also be in error. The 15th-century astronomer Jamāl al-Dīn al-Māridīnī[23] states that al-Mizzī used parameters 20° and 16° for morning and evening twilight, but the tables of r and s in this Cairo manuscript are based on 19° and 17°.

MS Florence Laurenziana 24 is the only known copy of a recension of the *Īlkhānī Zīj* for Mosul, prepared by Dāʾūd al-Mawṣilī.[24] On fols. 1r-1v there is an incomplete set of tables that are unrelated to the *zīj* of al-Mawṣilī. On fol. 1r there is a simple table for finding Easter and another of $\delta(\lambda)$ with entries to two digits for each degree of λ and based on $\varepsilon = 23;33°$. On fol. 1v there is the first page of a set of prayer-tables for an unspecific latitude. Entries are given to two digits for each degree of λ from 1° from 30°. The functions tabulated are:

$$H,\ D,\ \psi,\ h_0,\ h_a,\ t_a,\ 2D^h,\ \tilde{h},\ s \text{ and } r\ ,$$

and the values are those of al-Mizzī for Damascus.

The latitude 33;27° for Damascus is attested in MS Paris BNF ar. 2513, fols. 87v-88v, of the Egyptian 13th-century *Muṣṭalaḥ Zīj*, and from MS Paris BNF ar. 2520, fol. 72r, of another rescension of the same work we learn that both parameters were derived by observations made by an individual referred to as *al-Sharīf*, to be identified with the 10th-century Baghdad astronomer Ibn al-Aʿlam (**6.6**). Ibn al-Shāṭir also used these parameters in his early work: they are found, for example, in Chs. 18 and 114 of his treatise on the "complete quadrant" (MS Damascus Ẓāhiriyya 3098). In the year 765 H [= 1363/64] Ibn al-Shāṭir derived a new pair of parameters, for which al-Khalīlī computed a new corpus of tables for Damascus (**Ch. 10**) to replace those of al-Mizzī.

---

[23] I forget where I read this, but surely in an instrument treatise by al-Māridīnī (**6.15**).

[24] On the *Īlkhānī Zīj* see Kennedy, "*Zīj* Survey", no. 6. Dāʾūd al-Mawṣilī, not mentioned in the modern literature, states that he was asked to prepare the recension by *al-ṭabīb* (the medic) ʿĪsā and that it took him twenty days to translate the introduction into Arabic.

9.3 Ibn al-Shāṭir's prayer-tables for Tripoli (?)

The Damascus astronomer Ibn al-Shāṭir (1306-1375/80), author of two *zīj*es and an extensive treatise on theoretical astronomy in which he explained his new models for the sun, moon, and planets,[25] also compiled a set of prayer-tables for latitude 34°, possibly intended for the new Mamluk city of Tripoli (accurate latitude 34°27′). The evidence for this is contained in a short treatise attributed to him, preserved in MS Leiden Or. 1001, fols. 108r-113r, which was originally intended as an introduction to the tables. The tables themselves are extant in a different manuscript (see below).

Ibn al-Shāṭir begins his introduction by stating that one of his friends asked him to prepare a set of prayer-tables for latitude 34°. The functions:

$$H, Z_{(7)}, D, 2N, h_a, t_a, T_a, r, s, \text{ and } t_q$$

are specifically mentioned (fol. 110v:16-19). Ibn al-Shāṭir's description of the method of finding the value of a given function for a particular solar longitude indicates that the functions were tabulated side by side. Thus they would resemble the tables of al-Khalīlī in appearance, rather than those of Ibn Yūnus and al-Mizzī. He also notes how to compute:

$$\delta, d, \tilde{h}, 2D^h \text{ and } z_{a(7)}$$

from the entries in the tables, indicating that he did not actually tabulate these functions.

MS Cairo DM 1170,2 (fols. 11r-22v), copied about 1700, is the only known copy of an anonymous set of prayer-tables for latitude 34° which answer the description of Ibn al-Shāṭir's introduction in the Leiden manuscript: see **Fig. 9.3**. The tables on fols. 14v-19r are preceded by an introduction, of which the beginning is missing, and by calendrical and solar longitude tables (fols. 11r-14r), and they are followed by a star catalogue for the end of the year 1033 H [= 1624] and an almanac (fols. 19v-22v). According to a note in the margin of the star catalogue (fol. 21r) the entries in this, and probably also the other introductory tables, were adopted by Muḥammad ibn al-Ḥulwānī[26] for 1033 H [= 1623/24] from another set compiled by ʿAlāʾ al-Dīn ibn Nāṣir al-Din al-Ṭarābulusī, a *muwaqqit* at the Umayyad Mosque in Damascus, for the end of 984 H [= 1577]. Values of the following functions:

$$H, Z_{(7)}, D, 2D^h, h_a, t_a, T_a, 2N, s, r, n \text{ and } t_q ,$$

computed to two digits, are displayed side by side for each degree of λ beginning with Capricorn 1°. The underlying parameters are:

$$\phi = 34;0° \text{ and } \varepsilon = 23;31°.$$

I have not been able to investigate the parameters used for twilight or the qibla. Additional tables display the functions:

$$d, \alpha_\phi \text{ and } \alpha', s \text{ and } r$$

for the same ϕ and ε. It seems to me most likely that this collection of tables are the prayer-tables of Ibn al-Shāṭir.

Added in June, 2002: Whilst finalizing the illustrations for this study I realized that in the 1970s I had not included mention of the tables for latitude 34°, specifically intended for Tripoli,

[25] On Ibn al-Shāṭir see the references cited in n. **I**-2:15.

[26] On Muḥammad ibn Ibrāhīm al-Ḥulwānī al-Ḥusnī al-Shāfiʿī see *Cairo ENL Survey*, no. D22; and İhsanoğlu *et al.*, *Ottoman Astronomical Literature*, I, pp. 280-281, no. 146.

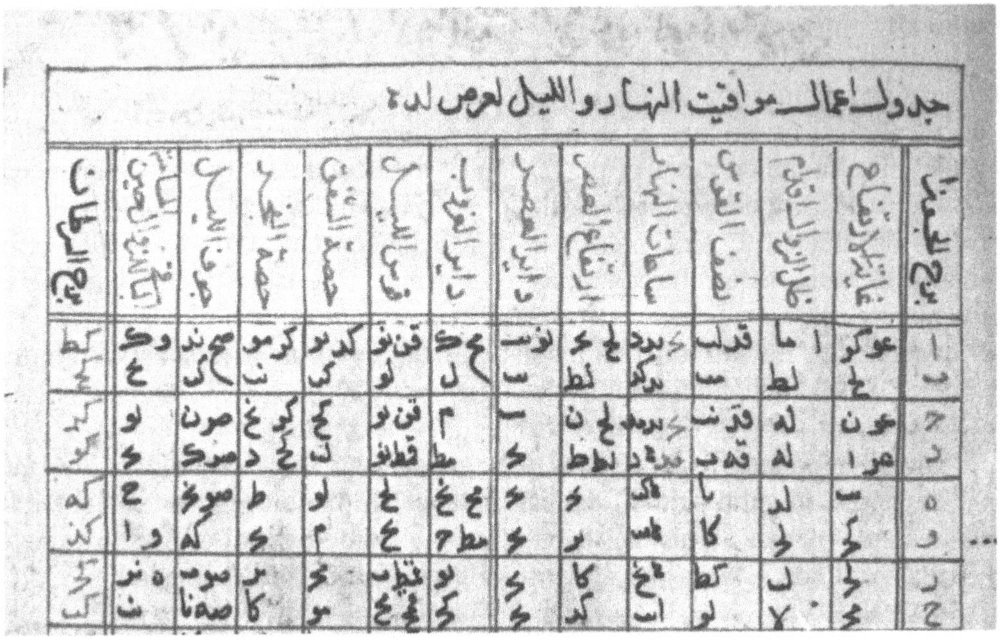

Fig. 9.3a: An extract serving solar longitudes in Taurus and Cancer from a set of prayer-tables for latitude 34° probably due to Ibn al-Shāṭir. [From MS Cairo DM 1170,2, courtesy of the Egyptian National Library.]

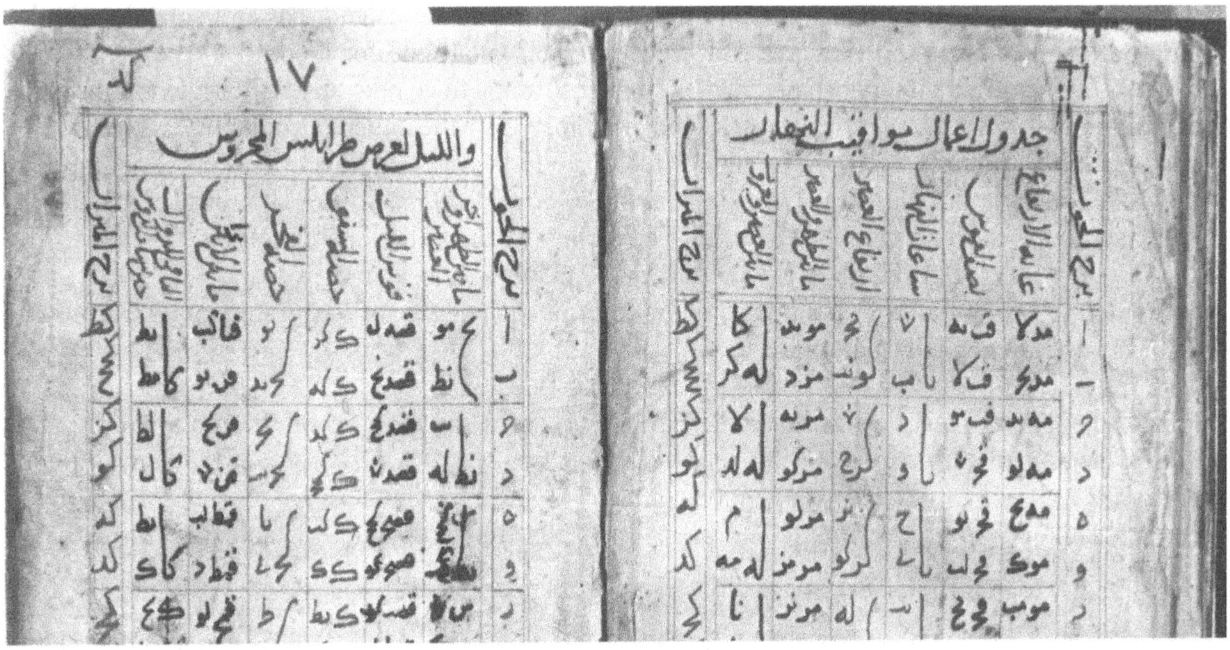

Fig. 9.3b: An extract for solar longitudes in Pisces from an an anonymous set of prayer-tables for Tripoli (latitude 34;0°). [From MS Cairo ṬM 228,3, fols. 16v-17r, courtesy of the Egyptian National Library.]

contained in MS Cairo ṬM 228,3, fols. 9r-9v + 16r-21r, copied *ca.* 1500.[27] I cannot compare these now with those in the Cairo DM manuscript. Suffice it here to note that the functions tabulated are:

$$H, D, 2D^h, h_a, t_a, T_a, t_b, 2N, s, r, n, \text{and } t_q ,$$

and to present an extract in **Fig. 9.3b**. The functions t_b and n are referred to as *mā bayn al-ẓuhr wa-ākhar al-ʿaṣr* and *mā bayn al-ādhānayn*, respectively. Clearly, the two sets, and possibly also those in **11.6**, merit a new investigation.

The treatise in the Leiden manuscript also contains a description of how to compute the solar longitude from the date, the time of day from shadow lengths, the solar altitude at the beginning of the afternoon prayer, and the times of rising and setting of the moon: all these methods are approximate though this is not stated. For example, to find the time elapsed since sunrise when the shadow is z (measured to base 7), the author suggests:

$$T \approx \{ 42 / [(z + 7) - Z_{(7)}] \}^{sdh} .$$

Equivalent versions of this Indian approximate formula were used by various Islamic astronomers from the 8th century onwards,[28] but one is suprised to see an astronomer of the calibre of Ibn al-Shāṭir advocating it. Likewise, the method outlined by him for finding the altitude at the beginning of the afternoon prayer involves the approximation:

$$h_a \approx {}^1/_2 [H + {}^1/_2 \cdot {}^1/_6 \bar{H}] ,$$

and the reader is warned against using:

$$h_a \approx {}^1/_2 H + {}^1/_2 \cdot {}^1/_6 \bar{H} .$$

Approximate methods of this kind are also given in Ibn al-Shāṭir's treatise on the complete quadrant (MS Damascus Ẓāhiriyya 3098, *e.g.* Chs. 114 and 173). For similar approximations in other sources, see the references in **1.4**.

In his treatise on planetary theory entitled *Nihāyat al-su'l* (MS Oxford Marsh 139, fol. 7r), Ibn al-Shāṭir mentions that the person who stated that the visible portion of the heaven is 0;4,26 more than the invisible portion was in error. (Shams al-Dīn al-Minūfī (**7.1**) associates that value with the illustrious Ibn al-Haytham: see also **14.9**.) He adds that he had observed the half arcs of visibility of the sun and stars and found that they were greater than the arcs found by calculation by an amount exceeding $^2/_3°$. In what purports to be a set of notes due to the Egyptian astronomer ʿIzz al-Dīn al-Wafāʾī the values of the "difference minutes" for Cairo (4.11) are attributed to Ibn al-Shāṭir (5.7, also 8.3), but I feel that this attribution is somewhat suspect.

In a study of Ibn al-Shāṭir's contributions in astronomical timekeeping it would be amiss not to mention the magnificent sundial which he constructed for the main minaret of the Umayyad Mosque. The remains of this instrument are today preserved in the garden of the Archeological Museum in Damascus, but in its original position there is a faithful replica made by the late-19th-century *muwaqqit* al-Ṭanṭāwī (**11.13**). The sundial serves to regulate time with respect to each of the five times of prayer.[29]

---

27 *Cairo ENL Survey*, no. C142, *sub* 3.1.17.

28 See **1.4** and also Arabic **Text 2.9** to **IV**.

29 See Janin, "Cadran solaire de Damas"; King in *Paris IMA 1993-94 Exhibition Catalogue*, p. 439, no. 334, and also **X-7.2**.

9.4 al-Karakī's timekeeping tables for Jerusalem

MS Leipzig UB 808, fols. 3r-93r, penned in 805 H [= 1402], is an apparently unique copy of a set of tables for timekeeping compiled by Zayn al-Dīn Abū Bakr al-Karakī[30] for the latitude of Jerusalem. He appears to have been a student of al-Mizzī (**9.2**) and can thus be dated to the mid 14th century. The introduction to his tables on fol. 2v is of considerable relevance to the present study and I present a translation:

> "In the Name of God, the Merciful and Compassionate. The *shaykh* and *imām* Zayn al-Dīn Abū Bakr ibn Muḥammad ibn Ayyūb al-Tamīmī known as al-Karakī, *muwaqqit* in Sacred Jerusalem – may God have mercy on him – said the following in the first section (of his book), which concerned the compilation of tables of time since sunrise and the hour-angle for latitude 32° north. When I saw that the virtuous *shaykh*, scholar, and calculator Jamāl al-Din Abu 'l-ʿAbbās Aḥmad ibn ʿUmar ibn Ismāʿīl ibn Muḥammad ibn Abī Bakr al-Ṣūfī al-Maqsī – may God have mercy on him – had compiled tables of the time since sunrise for latitude 30°, which require further calculation to find the hour-angle, and that our teacher the virtuous *shaykh* and meticulous scholar Shams al-Dīn Abū ʿAbdallāh Muḥammad ibn Aḥmad ibn ʿAbd al-Rahīm al-Mizzī – may God Almighty have mercy on him – had compiled tables of the hour-angle for latitude 33;27°, which also require further calculation to find the time since sunrise, and that the *shaykh* Shams al-Dīn Ibn al-Rashīdī – may God have mercy on him – had put the altitude at the head of each table and the hour-angle opposite the solar longitude, I wanted to participate with them in such compilations. So I put the significant (functions) together to facilitate the work of the observer, and tabulated the time since sunrise and the hour-angle opposite the altitude with the solar longitude at the head of the tables, beginning at the first point of the ascending zodiacal signs (*i.e.*, the winter solstice). The same tables can be used for the descending signs in the opposite direction (since the last table is for the summer solstice). I did this, asking help from God and placing my trust in Him ... ."

al-Karakī's tables do indeed display the functions:

$$T(\lambda,h) \text{ and } t(\lambda,h)$$

for each degree of both arguments. Note that values are given for both solstices and the equinoxes. For each value of λ, D(λ) is also tabulated. The underlying parameters are:

$$\phi = 32;0° \text{ and } \varepsilon = 23;35° .$$

The entries, which are given to two digits, are rather accurately computed. Note that al-Karakī does not actually say that he computed both of the functions himself. Indeed, it may be that he only computed the time since sunrise from someone else's hour-angle tables. On fols. 94v-123v of the Leipzig manuscript, in a different hand, there is another set of hour-angle tables arranged in the same way as al-Karakī's tables, and also showing D(λ). The title and colophon indicate that they are for latitude 31;40°. In fact they are for 32;0° and the entries are the same as those in al-Karakī's set. A note in the incomplete introduction to this second set on fol. 94r

[30] The Leipzig manuscript of al-Karakī's tables is mentioned in Brockelmann, *GAL*, SII, p. 156, after *Leipzig Catalogue*, p. 261.

mentions the completion of the minaret on a mosque in Ramla in 797 H [= 1395]. The value 31;40° is used for Ramla in several Islamic geographical tables.[31]

al-Karakī was clearly impressed with al-Maqsī's tables of $T(h,\lambda)$ for Cairo (**5.4**), and al-Mizzī's tables of $t(h,\lambda)$ for Damascus (**9.2**). It is a pity that he does not state the latitude for which Ibn al-Rashīdī compiled his hour-angle tables. His name is not otherwise associated with any of the various hour-angle tables that I have found in the manuscript sources. It seems to me likely that Ibn al-Rashīdī compiled a set of tables of $t(h,\lambda)$ for Jerusalem, which al-Karakī then used to compile his tables with a different format. Now fragments of tables of $t(h,\lambda)$ for latitude 32° exist in MSS Cairo DM 45 and DM 153 (**6.12**) with identical corresponding entries to al-Karakī's tables. If my supposition is correct, al-Karakī's tables for Jerusalem are merely an adaption of Ibn al-Rashīdī's tables, which were probably compiled in Cairo. See also the next section.

9.5 Anonymous prayer-tables for Jerusalem

The tables for latitude 32° in MS Princeton Yahuda 861,1 of the *Natīja al-kubrā* attributed to al-Wafā'ī (**8.1**) differ from those for other latitudes in such a way that it is immediately obvious that they were lifted from various independent sources. The following functions are tabulated for this latitude:

$$D,\ 2D^h,\ 2N,\ H,\ h_a,\ t_a,\ h_q,\ T_q,\ r,\ s,\ n,\ \alpha_\phi,\ \alpha_\sigma \text{ and } \sigma\ .$$

The entries in certain of the tables are reasonably accurately computed and in others are very carelessly computed. My investigation of the tables reveals the following.

(1) The table of D, which is carelessly computed, is the same as the corresponding table in MS Oxford Marsh 676 (Uri 944 = 995) due to Najm al-Dīn al-Miṣrī (**6.5**).

(2) The table of $2D^h$, the length of daylight in equinoctial hours, is a particularly wretched specimen. Some of the entries, which are badly garbled, are in hours and minutes and others are in hours and degrees.

(3) The table of 2N, which is rather accurately computed, is based on values of D other than those in (1). In fact it is based on the values of D used by al-Karakī to compile his tables of $T(\lambda,h)$ from (Ibn al-Rashīdī's ?) tables of $T(h,\lambda)$ (**9.4**).

(4-6) The tables of H, h_a and t_a are likewise rather accurately computed.

(7-8) The tables of h_q and T_q are based on a value of q for some particular locality, although neither parameter nor locality is specified for either function. The latitude 32° is the only latitude other than 30° for which h_q and T_q are tabulated in the Princeton manuscript. However, in the title of the first function tabulated for latitude 32°, namely $D(\lambda)$, the cities of Gaza, Ramla and Jerusalem are specifically mentioned, and by inspection I find the underlying value of q to be 41;30° (measured from the meridian). The two tables are reasonable accurately computed. Now in the geographical tables in MS Princeton Yahuda 861,1 the following entries for localities in Palestine are given:

---

[31] Kennedy & Kennedy, *Islamic Geographical Coordinates*, pp. 280-281.

	L	ϕ	q	q (acc.)	q (approx.)
Gaza	56;30°	32;0°	52;40°	42;59°	43;41°
Ramla	57;20	32;20	45;0	39;37	40;31
Jerusalem	58;35	32;10	45;51	35;58	37;5
Nablus	58;15	32;20	37;0	36;41	37;45
(Mecca	67;0	21;0	-	-	-

In each of several earlier Egyptian and Syrian sources I have consulted,[32] the qibla values have been carelessly computed, and in some cases miscopied to boot. However, in none of these sources do we find the qibla value 41;30°. It was probably derived for Jerusalem. Indeed it was perhaps based on the value 33;0° for the latitude of Jerusalem, as found, for example, in the tables of Ibn al-Dahhān (**9.1**). Notice that if we take

$$L = 56;30° \text{ and } \phi = 33;0 \quad \text{with} \quad L_M = 67;0° \text{ and } \phi_M = 21;30° ,$$

the accurate and approximate values of q are *41;37*° and 42;26°, and if we take:

$$L = 57;0° \text{ and } \phi = 33;0° \quad \text{with} \quad L_M = 67;0° \text{ and } \phi_M = 21;40° ,$$

then the accurate and approximate values of q are 40;31° and *41;28*°. Ibn al-Dahhān himself gives 41;21° for the qibla at Jerusalem.

(9-11) The tables for twilight are based on parameters 20° and 16° for morning and evening and are reasonably accurately computed. The entries are the same as the corresponding ones in al-Karakī's tables of T(λ,h) for h = 20° and 16°. All of the other twilight tables in the Princeton manuscript are based on parameters 19° and 17°.

(12) The table of α_ϕ is carelessly computed, but is based on a different set of values of d than was used to compile the table of D (see (1) above).

(13) The table of α_s is based on a more accurate set of values of α_ϕ than (12), using:

$$\alpha_s(\lambda) = \alpha_\phi(\lambda + 180°) + s(\lambda) .$$

(14) The table of σ(λ), displaying the time from sunset to the *salām*, is ostensibly based on:

$$\sigma(\lambda) = 2N(\lambda) - r(\lambda) ,$$

so that the *salām* was to be performed precisely at daybreak rather than a few minutes before as was the case in Egypt (**4.10**). However, the last four columns of entries are badly garbled.

(15) Finally, amongst the tables for latitude 31° in the Princeton manuscript there is one displaying the functions t(h) and T(h) for each degree of h at the equinoxes. The underlying latitude is 32° not 31° as stated, and the tables originally formed part of the more extensive set of tables compiled by al-Karakī. The entries are badly garbled,

From these investigations it is clear that there existed a set of carefully-computed prayer-tables for Jerusalem displaying at least the functions:

$$D, 2N, H, h_a, t_a, h_q, T_q, r, s, n \text{ and } \alpha_\phi ,$$

thus complementing al-Karakī's extensive set of tables for timekeeping.

---

[32] Namely, the geographical tables in MSS Paris BNF ar. 2513, fol. 89r, and Paris BNF ar. 2520, fol. 82v, of the *Muṣṭalaḥ Zīj*; MS Oxford Seld. A30, fols. 155r-157v, of the *Zīj* of Ibn al-Shāṭir; MSS Paris BNF ar. 5968, fols. 162v-163r, of the anonymous *Dastūr al-munajjimīn*; and Gotha A1403 of the derivative *zīj* of Ibn Zurayq. See further King, *Mecca-Centred World-Maps*, pp. 76-84, on the unhappy state of such tables.

CHAPTER 10

THE DAMASCUS CORPUS OF AL-KHALĪLĪ

10.0 Introductory remarks

An impressive corpus of astronomical tables was used in Damascus from the 14th to the 19th century. The tables in their entirety contain over 40,000 entries. Their compiler was a Damascene astronomer named Shams al-Dīn al-Khalīlī, a man recognized only in the 1970s for his remarkable competence in computation.[1] al-Khalīlī's corpus is more clearly-defined than the Cairo corpus, by which it was obviously inspired. We know who compiled the tables, and we are even provileged to have an earlier work of his which shows how his thinking had progressed. And the question why we had to wait to the late 20th century to have a clear picture of his achievements is simple to answer: his tables were rehashed and modified over the centuries and numerous of the manuscripts bear the names of later astronomers who simply appropriated them.[2]

Damascus was home to a vigorous tradition of astronomy in the 14th century, centred on the Umayyad Mosque in the heart of the city.[3] This tradition was but one of many colourful aspects of life under the Mamluks, who ruled Egypt, Syria, Palestine and the Hejaz from the mid-13th to the early 16th century. A team of astronomers called *muwaqqit*s (that is, professional timekeepers), which included al-Khalīlī, was employed at the Mosque. Their main duty was to determine the times of prayer so that the muezzins could perform the call to worship from the minarets at the appropriate time. In addition, they specialized in the regulation of the lunar calendar and the determination of the qibla. The interests of some of the *muwaqqit*s extended far beyond these somewhat mundane tasks (**V**).

The chief *muwaqqit* at the Umayyad Mosque during the 1360s and 1370s was ʿAlāʾ al-Dīn ʿAlī ibn Ibrāhīm called Ibn al-Shāṭir, known today for his innovative and successful research on geometric models for the sun, moon and planets (**9.3**). An earlier Damascus astronomer was al-Mizzī, who started the business of compiling prayer-tables for the city (**9.2**). As we shall see, both al-Mizzī and Ibn al-Shāṭir influenced the work of al-Khalīlī. Both had themselves

[1] Before the 1970s al-Khalīlī was known only from the bio-bibliographical references in Suter, *MAA*, no. 418, Brockelmann, *GAL*, II, pp. 156-157, SII, p. 157, and Sarton, *IHS*, III:2, pp. 1526-27. My first studies on his tables were "al-Khalīlī's Auxiliary Tables" (1973) and "al-Khalīlī's Qibla Table" (1975). A summary of his achievements and a critical list of the available manuscripts of his works appeared in *DSB*, XV (Supp. vol. I), pp. 259-261. See also *Cairo ENL Survey*, no. C37.

[2] "Appropriated" is a nicer word than "plagiarized": see n. 4:12 above. The Damascus of the astronomers, like Cairo, was a "small world" and it was not possible for astronomers to pass off as their own tables that were computed by their illustrious predecessor.

[3] This activity is surveyed in King, "Astronomy of the Mamluks", and *idem*, "Astronomie en Syrie". On timekeeping in particular see my "Astronomical Timekeeping in Medieval Syria".

been influenced by developments in Egypt in the 13th century (**6.0**). al-Khalīlī recomputed al-Mizzī's tables for the new parameters (local latitude and obliquity of the ecliptic) derived by Ibn al-Shāṭir in 765 H [= 1363/64], namely:

$$\phi = 33;30° \quad \text{and} \quad \varepsilon = 23;31°.$$

His corpus of tables for solar timekeeping and the regulation of the times of prayer for Damascus continued to be used in one form or another until the 19th century. In addition, al-Khalīlī prepared a set of auxiliary tables for solving the standard problems of spherical astronomy for all latitudes, as well as a universal qibla table.

The work of the early-14th-century Aleppo astronomer Shihāb al-Dīn Aḥmad ibn Abī Bakr called Ibn al-Sarrāj, which represents the high point of Islamic instrumentation, was also known to al-Khalīlī.[4] Indeed, the tables of al-Khalīlī and the instruments of Ibn al-Sarrāj are singularly important developments in the particularly Islamic tradition of providing solutions to the problems of spherical astronomy for all latitudes, which I have documented elsewhere (**VIa-b**). However, al-Khalīlī makes no reference to any of his predecessors except Abū ʿAlī al-Marrākushī (**10.8**).

In later centuries al-Khalīlī's universal auxiliary tables (**10.7**) were also used in Egypt, Tunisia and Turkey. As far as we know, al-Khalīlī had no influence on the development of astronomy or trigonometry in Europe. There in the 15th and 16th centuries auxiliary trigonometric tables were developed by a series of astronomers of whom the most important were Regiomontanus and Magini (**I-10.1-2**). Yet despite the sophistication of these European tables they are not as useful for practical purposes as those developed by al-Khalīlī. His tables would have been of great interest to European astronomers of the from the 15th century right through to the early 20th century, had they been known to them.

10.1 Shams al-Dīn al-Khalīlī

Biographical information on Shams al-Dīn Abū ʿAbdallāh Muḥammad ibn Muḥammad ibn Muḥammad al-Khalīlī is scant indeed, although the epithet al-Khalīlī indicates that his family came from Hebron. His name is given in the above form in MS Berlin Wetzstein 1138 (Ahlwardt 5754-6), fol. 1r, and is attested in the same form in numerous other manuscripts. On fol. 1v of the same Berlin manuscript he is also called al-Ḥanafī, indicating that he adhered to the predominant Ḥanafī legal school. His minor auxiliary tables were completed in 763 H [= 1361] (MS Dublin CB 4091, fol. 1r) and his main corpus some years later. In the earlier work he is referred to as a *muwaqqit* at the Yılbughā Mosque, a new mosque in Damascus established by the governor Sayf al-Dīn Yılbughā in 747 H [= 1346].[5] But in copies of his main corpus he is referred to as a muezzin (MS Paris BNF ar. 2558, fols. 1r and 64r) or a *muwaqqit* (MS Berlin Wetzstein 1138 (Ahlwardt 5754-6), fol. 1r) at the Umayyad Mosque,[6] which as we have seen had become the main centre of astronomy in the 14th-century Islamic

---

[4] On Ibn al-Sarrāj see n. 9:12.

[5] On the Yılbugha Mosque see Henri Sauvaire, "Description de Damas: Sur les Grandes-Mosquées", *Journal Asiatique*, mars-avril 1896, pp. 185-285, *etc.*, esp. pp. 236-237.

[6] On the Umayyad Mosque see *op. cit.*, pp. 185-230.

world. The present author is not aware of any references to al-Khalīlī in the standard Mamluk biographical sources. al-Khalīlī had a nephew who was also an astronomer. Sharaf al-Dīn Abū ʿImrān Mūsā ibn Muḥammad ibn ʿUthmān al-Khalīlī studied under his maternal uncle and is known from several short treatises on timekeeping.[7]

10.2 The manuscript sources

al-Khalīlī's minor auxiliary tables and solar azimuth tables survive in the unique MS Dublin CB 4091, copied in 833 H [= 1429]. Likewise, his auziliary tables for finding the solar azimuth for all latitudes are known only from MS Bursa Haraççioğlu 1177,4, copied *ca.* 1425. The various tables in the main Damascus corpus survive in numerous copies, in most of which they are correctly attributed to al-Khalīlī. MS Paris BNF ar. 2558, the best copy by far, is not only the most complete but also the earliest, having been penned in the year 811 H [= 1411]. See **Fig. 10.2**. It even bears evidence of having been proofread by the copyist. The title folio gives a very good indication of the contents (see **10.4**). A table of contents of this and various other early copies of the Damascus corpus is given below.

Although only one complete manuscript survives, there is little doubt that the entire corpus was frequently copied. Various parts were also copied separately, as the surviving manuscripts attest. The distribution of the tables in the sources is as follows:

Fig. 10.2: The title-folio of the magnificent Paris copy of al-Khalīlī's corpus. [From MS Paris ar. 2558, courtesy of the Bibliothèque Nationale de France.]

[7] On Sharaf al-Dīn al-Khalīlī see Suter, no. 427, and *Cairo ENL Survey*, no. C38.

Location of the various tables of the main corpus in the available manuscripts

I al-Khalīlī's prayer tables
II al-Khalīlī's hour-angle tables
III al-Ḥalabī's azimuth tables
IV al-Khalīlī's universal auxiliary tables
V al-Khalīlī's qibla table

MS	I	II	III	IV	V
Berlin Ahlwardt 5754-6	*	-	-	*	*
Berlin Ahlwardt 5758	*	-	-	-	-
Berlin Ahlwardt 5759/5771	*	-	-	-	-
Berlin Ahlwardt 5760/5772	*	-	-	-	-
Berlin Ahlwardt 5739	-	-	-	*	-
Beirut PC	-	-	-	*	-
Cairo MM 71	-	*	*	-	-
Cairo MM 43	-	-	-	*	-
Cairo DM 184	*	-	-	-	-
Cairo MM 98	-	-	-	*	-
Cairo ṬM 255,7	*	-	-	-	-
Cairo DM 758	-	-	-	*	-
Cairo K 8525	*	*	*	-	-
Cairo ṬM 228,5	*	*	-	*	-
Cairo ṬM 173	*	*	-	-	-
Cairo TR 129	*	*	-	-	-
Cairo DM 1007	*	*	-	-	-
Damascus 3116	-	*	-	-	-
Damascus 10387 (?)	*	-	-	-	-
Damascus 9227	-	-	*	-	-
Damascus 9233	*	*	-	-	-
Damascus 4893	*	-	-	-	-
Escorial ár. 931,8	-	-	-	*	-
Gotha A1412	*	-	-	-	-
Istanbul Ayasofya 2590	-	-	-	*	-
Istanbul Esat Efendi 1990	*	-	-	-	-
Istanbul Hamidiye 1453,3	-	-	-	*	-
Istanbul Serez 1914	-	-	-	*	-
Leipzig 814	*	-	-	-	-
London BL Add. 9599,31	-	-	-	*	-
Mosul al-Muḥammadiyya 129	-	*	-	-	-
Oxford Seld. Supp. 100	*	*	-	-	-
Oxford Marsh 39 (Uri 1042)	-	*	-	-	-
Oxford Marsh 95 (Uri 961)	*	-	-	-	-
Paris BNF ar. 2558	*	*	-	*	*
Paris BNF ar. 2560,11	-	-		-	*
Princeton Yahuda 861,2	*	-	-	*	-

Descriptions of the most important manuscripts

As in the case of the manuscripts of the Cairo corpus (**4.2**), it seems worthwhile to present descriptions of the most significant copies of the Damascus corpus.[8]

MS Dublin CB 4091:[9]

Title folio:	(MS 4090): Statement that the copyist was Shihāb al-Dīn al-Biqāʿī al-Shāfiʿī (?) (not identified). Notices of possession by Muḥammad ibn Muḥammad ibn Aḥmad al-..... (illegible); Jalāl al-Dīn Muḥammad ibn Muḥammad al-Ramlī, *muwaqqit* at the Umayyad Mosque; Muḥammad al-*Qwʾf* (?, unidentified); and the Egyptian astronomer ʿAlī al-Khashshāb al-Falakī al-Dimyāṭī (*Cairo ENL Survey*, no. D120).
1r:	(MS 4091): Title (see **10.3a**); introduction to the auxiliary tables; notice of possession by al-Khashshāb (see above and also fol. 17r).
1v:	Table of "universal meridian altitudes", table of Sines, table of solar altitude at the *ʿaṣr*.
2r-6v:	Tables of the half base.
7r-16v:	Tables for finding the hour-angle. Colophon.
17r:	Blank except for notice of possession by al-Khashshāb.
17v:	Introduction to list of formulae and solar azimuth tables.
17v:	Table of formulae for spherical astronomy.
18r-24r:	Tables of solar azimuth.
24r:	Colophon dated 24 Ṣafar 833 H [= 22 November 1429].

MS Bursa Haraççioğlu 1177,4, fols. 72r-90r:[10]

Notes: Only a part of this manuscript, with folios numbered 72-92 in fairly recent Syro-Egyptian-type Arabic numerals, is available to me for study. The hand is an elegant *naskhī*, dateable between about 1400 and 1450.

72r:	The title reads: *jadwal al-samt li-kulli ʾrtifāʿ fī ʿurūḍ al-aqālīm al-sabʿa*, that is, "Table of the Azimuth for all Altitudes in the Latitudes of the Seven Climates".
72v:	Instructions, without indication of the author.
73r-85r:	Tables of the first and second auxiliary functions.
85v-90r:	Tables of ther third auxiliary function.
90v:	A note in a similar, likewise elegant hand: *naẓara fīhi al-ʿabd al-faqīr ila ʾllāh taʿālā Muḥammad ibn Muḥammad al-Tīzīnī al-Mīqātī sanat 867 wa-lam yaktub minhu shay* (*sic* for *shayʾan*), that is: "(This manuscript was) inspected by ... Muḥammad ibn Muḥammad al-Tīzīnī al-Mīqātī in the year 867 (Hijra) [= 1462/63], and he did not write anything of it (??)."

---

[8] These were prepared in 1993 in the process of the compilation of my detailed study mentioned in n. 1:23 above. I prefer them to the descriptions presented in **4.2**.

[9] The manuscript (MS 4091) is part of a collection of treatises in the same hand (MSS 4090-92). The Arabic foliation runs from 159 to 181 skipping a folio between 165 and 166. The modern foliation is also incorrect. The notes here follow the correct foliation.

[10] I am most grateful to my colleague Dr. Sonja Brentjes for obtaining a microfilm of this manuscript.

MS Paris BNF ar. 2558:[11]
Notes: The folios are numbered on the verso, and there are two marked '2'. Throughout the manuscript there appear the words *balagha* (*sic*) *muqābalatuhu*, indicating that it was proofread. No changes or corrections have been inserted, however.

1r:	Title folio, illustrated in **Fig. 10.2**. The full title of the corpus is given (see the beginning of **10.4** for a translation). The manuscript was prepared for the treasury (*khizāna*) of the chief chamberlain (*amīr ḥājib*) of Damascus, Qarabughā al-ʿAlā'ī. Reader's notice of Buṭrus ibn Dīb, *turjumān* of the "Sultan" of France (is this how the manuscript came to Paris?). The copyist is named as Aḥmad ibn ʿAlī al-Khabbāz (on fols. 64r and 105v) and the date of copying as Jumādā I 11, 811 H [= October 2, 1411] on fol. 64r, and Jumādā II of the same year on fol. 105v. The handwriting is a most elegant *naskhī*.
1v:	al-Khalīlī's introduction to his calendrical and solar tables (duplicated in mirror image on fol. 2r).
2v-3r:	al-Tīzīnī's introduction to his calendrical and solar tables.
3v-8r:	al-Tīzīnī's calendrical tables.
8v-9r:	al-Tīzīnī's solar tables.
9v-14r:	Prayer-tables for Damascus.
14v-18r:	Tables of ascensions.
18v-20r:	Tables of stellar coordinates.
20v-50r:	Hour-angle tables for Damascus.
50v:	Table for finding solar motion in fractions of a day.
51r:	Tables of Cotangents, and solar rising amplitude and altitude in the prime vertical.
51v:	Tables of terrestrial longitudes, latitudes and qiblas.
52r:	Tables of latitudes of localities on the Syrian pilgrim route.
52v-53r:	al-Khalīlī's introduction to his qibla table.
53r:	Tables for marking the base on sine quadrants.
53v-61r:	al-Khalīlī's qibla table.
61v-64r:	al-Khalīlī's introduction to his universal auxiliary tables.
64r:	Copyist's colophon (see also fol. 105v).
64r:	Table of solar declination and solar altitude at the *ʿaṣr*.
64v-92r:	al-Khalīlī's universal auxiliary tables (first and second functions).
92v-104r:	al-Khalīlī's universal auxiliary tables (third function).
104v-105v:	Table of normed right ascensions.
105v:	Colophon (see also fol. 64r). Notices of possession by Ibrāhīm ibn Aḥmad ibn Yaḥyā ibn ʿAbd al-Ṭāhir (!?) dated 915 H [= 1509/10], and Muḥammad *R-ḥ-ʾ-l-h* (?) al-*Ḥ-m-ʾ-r-y* (?) *al-muʾadhdhin* (the muezzin), both unidentified.

MS Berlin Wetzstein 1138 (Ahlwardt 5754-6):[12]
Notes: The MS is in three different hands: A = 1v-55r + 64r-94r (elegant *naskhī*, *ca.* 1500); B = 56v-61v (later); and C = 62v-64r (very late).

[11] *Paris Catalogue*, p. 460; also *Paris IMA 1993-94 Exhibition Catalogue*, p. 440. no. 335.
[12] *Berlin Catalogue*, pp. 207-208.

No title folio.

1v-4r:	al-Khalīlī's introduction to his universal auxiliary tables.
4r-5v:	Sharaf al-Dīn al-Khalīlī's notes on the use of his uncle's tables.
6r:	Title for al-Khalīlī's universal auxiliary tables (*jadwal faḍl al-dāʾir al-āfāqī*).
6v-46v:	The universal auxiliary tables (including a sub-table for $\phi = 21;30°$). For each latitude various localities are named. The tables run to latitude 55° with a page ruled for 56° without entries.
47r:	al-Khalīlī's introduction to his qibla table.
47v-55r:	The qibla table.
55v:	Blank.
56r:	Scribbles.
56v-58v:	al-Khalīlī's introduction to his universal auxiliary tables (incomplete), in hand B.
59r:	Cotangent table to base 10 (argument 1°-90°).
59v:	Sine table.
60r-61r:	Notes on Cotangents.
61v:	Table of solar declination and solar altitude at the *ʿaṣr* (attributed to al-Khalīlī).
62r:	Blank.
62v-94r:	Special tables for the *ʿaṣr* (hand C, hand A resumes on fol. 64r).
94v:	Blank, no colophon.

MS Escorial ár. 931,8, fols. 171r-211v [13]

171r:	Title folio (*Kitāb al-Dāʾir wa-faḍl al-dāʾir wa-ʾl-samt*) attributed to al-Khalīlī.
171v-209r:	al-Khalīlī's universal auxiliary tables (no sub-table for $\phi = 21;30°$, latitude argument only up to 50°).
209v:	Table of solar declination ($\varepsilon = 23;35°$!).

No colophon. Dated 933 H [= 1526].

MS Cairo MM 43:

Notes: The manuscript contains some 50 unnumbered folios and the tables are unattributed. It is of Egyptian provenance and the copyist is recognizable as ʿAlī ibn Muḥammad al-Dalāmī and thus datable to *ca.* 1450. The tables listed here are followed by a variety of other tables for trigonometry, spherical astronomy and instrument construction (some universal, some specifically for the latitude of Cairo). There is also a table of geographical coordinates.

1r:	Miscellaneous scribblings. No title.
1v-5r:	(al-Khalīlī's) introduction to his universal auxiliary tables. Definitions of the functions f and g (fol. 5r).
5v:	Tables of solar declination ($\varepsilon = 23;31°$) and universal *ʿaṣr*.
6r-24r:	(al-Khalīlī's) tables of the first and second functions ($\phi = 1°$-$50°$ including 21;30° and 33;30°) and the third function.

[13] *Escorial Catalogue*, pp. 42-43.

MS Cairo ṬM 228,5:
Notes: All of the tables are anonymous. The manuscript is undated, but is copied in a legible *naskhī* script which is estimated *ca.* 1500. The folios have been bound in disorder, but the universal auxiliary tables and the hour-angle tables are complete.

1r-3r:	No title. (al-Khalīlī's) introduction to the universal auxiliary tables.
3r-3v:	Notes on the determination of the qibla and the use of the sine quadrant by an unidentified astronomer Zayn al-Dīn ʿUmar al-Zuhrī (*Cairo ENL Survey,* no. C79).
4r:	Tables of solar declination (ε = 23;31°) and universal *ʿaṣr*.
4v-7v:	(al-Khalīlī's) tables of the first and second functions (ϕ = 1°-7° only).
8r:	(al-Khalīlī's) hour-angle tables (h = 79°-80° only).
8v:	Notes on the determination of the qibla.
9r-9v:	Prayer-tables for Tripoli (latitude 34°) (incomplete, only 6 functions for Capricorn and 6 for Aquarius): see fols. 16r-21r and **9.3**.
10r-13r:	(al-Tīzīnī's) tables for converting from the Hijra to the Syrian calendar (incomplete).
13v-14r:	(al-Tīzīnī's) table of solar longitude.
14v-15r:	Table of solar mean motion.
15v:	Blank.
16r-21r:	Continuation of prayer-tables for Tripoli (12 functions for the remainder of Aquarius up to Gemini): see contents of fols. 9r-9v above.
21v-25v:	Tables of ascensions (for Damascus).
26v:	Cotangent tables (bases 12 and 7).
27r-28r:	Table of solar altitude at the *ʿaṣr* (for each 0;15° of H).
28r-60v:	(al-Khalīlī's) hour-angle tables (h = 1°-78° only).
61r-85r:	Remainder of (al-Khalīlī's) first two functions (ϕ = 8°-54° including ϕ = 21;30° and 33;30°).
85v-86r:	Geographical tables as follows: (a) list of localities, mainly in Egypt, with their coordinates and qiblas; (b) list of localities giving latitudes only; and (c) list of localities on the Egyptian pilgrim route with coordinates and qiblas.
86v-98r:	Tables of (al-Khalīlī's) third function (complete).
98v-101r:	Miscellaneous notes on calendars, scribblings, notice of possession by Muḥammad Ṣalāḥ al-Dīn (1)117 H [= 1705/06] (fol. 98v).

10.3 The minor corpus

The set of tables contained in the unique source MS Dublin CB 4091 was compiled before the main Damascus corpus. It consists of three parts, which are considered in turn (see already **I-9.4** and **I-5.3.1** on the tables, here considered in more detail). al-Khalīlī refers to these tables in the introduction to his universal auxiliary tables (**10.7**).

a) Auxiliary tables for solar timekeeping

In MS Dublin CB 4091, fols. 2r-16v, a pair of auxiliary tables by al-Khalīlī for finding $t(h,\lambda,\phi)$ survive. These are quite different from his universal auxiliary table. The principal functions tabulated are the following:

$$B'(\lambda,\phi) = \text{Cos}\ \delta(\lambda)\ \text{Cos}\ \phi\ /\ 2R = \ ^1/_2\ B(\lambda,\phi) \quad (\varepsilon = 23;31°)\ ,$$

computed for the domains:

$$\phi = 1°,\ 2°,\ \dots\ ,\ 49° \text{ and } 33;30°, \text{ and } \lambda = 1°,\ 2°,\ \dots,\ 90°,$$

and:

$$V'(x,y) = \text{arc Vers}\ \{\ R\ y\ /\ 2\ x\ \}\ ,$$

computed for the domains:

$$x = 30,\ 29,\ \dots,\ 19 \quad \text{and} \quad y = 0;10,\ 0;20,\ \dots\ ,\ 60;0.$$

These two sets of tables contain respectively 4,500 and 4,680 entries, rather accurately computed to two digits. Horizontal differences are shown in the second set. The first function is called *al-aṣl*, "the base" (*cf.* **I-6.0**) but a marginal note at the beginning of the table points out that it is actually *niṣf al-aṣl*, "half the base". The second function is called *faḍl al-dā'ir āfāqī min al-aṣl wa-min faḍl jayb al-ghāya ʿan jayb al-irtifāʿ*, "the hour-angle for all latitudes as a function of the base and the excess of the Sine of the meridian altitude over the Sine of the instantaneous altitude". The instructions translate as follows:

> "The Book entitled "A Compilation on Timekeeping in the Different Regions" (*Mawḍiʿ al-awqāt fī 'l-aqālīm al-muqassamāt*) by the scholar Shams al-Dīn Muḥammad al-Khalīlī *al-muwaqqit.*
>
> In the Name of God the Merciful and Compassionate. I have no success but through God, and upon Him I have placed my trust. The celebrated scholar Shams al-Dīn Muḥammad al-Khalīlī, the *muwaqqit* at the Mosque of al-Sayfī (= Sayf al-Dīn) Yılbughā – may God have mercy upon him – said:
>
> Praise be to God who takes away (our) fears and brings about miracles. May (God) bless and save Muḥammad and his family, the virtuous and honorable, with a blessing by which I beg for what is good and which will erase for me what is bad.
>
> I consider the hour-angle to be one of the most important concepts (*abwāb*) in astronomy because it is the basis for finding the times of the obligatory prayers. So I wanted to prepare a table for finding it for (all) inhabited regions. I ask God – who knows the secrets and what is hidden – that He should protect me from all kinds of mistakes, and that He should make it useful to all those who know about timekeeping (lit., the number of hours and degrees). Verily He responds to prayers and gives what is asked (of Him). I called (this work) "Compilation on Timekeeping in the Different Regions" (*Mawḍiʿ* ... ... *al-maqsūmāt*). God is sufficient for me and He is the best guardian; (He it is who is) the Creator of the Earth and of the Heavens.
>
> Chapter on the determination of the hour-angle, half the arcs of daylight and night, the time of the *ʿaṣr* and the time remaining until sunset, and the duration of morning and evening twilight, in localities between the equator and latitude 49°.
>
> You find the base from the appropriate table, entering with the longitude of the sun and the local latitude. Then find the meridian altitude from the appropriate table and find its Sine from the Sine table using proportional parts of the first differences (*bi-*

nisbat al-tafāḍul). Next find the difference between the Sines of the meridian altitude and the instantaneous altitude by subtracting the Sine of the altitude from the Sine of the meridian altitude. Enter with the base (as horizontal argument) in the universal hour-angle table until you come opposite the difference between the two Sines (as vertical argument), using linear interpolation if (the difference) is not equal to the argument. The result will be the hour-angle if the base is integral. If it is not integral, round the fraction to unity and determine the hour-angle by interpolating linearly for the difference between the two Sines. Then find the first difference from the appropriate box and recall the fraction which you rounded: if it was one-half or one-third or more or less, add one-half or one-third or more or less of the first difference to the hour-angle.

Example. The base is 24;15. We round (the fraction) to unity and obtain 25. The difference between the two Sines is 29;30. With this we find the hour-angle, which is 65;48. We find that the first difference is 1;32, and the amount we added to the fraction to make it whole is 0;45. We add one-half and one-quarter of the difference, namely 1;9, to the hour-angle and the result is 66;57°, which is the required hour-angle.

If we enter with the base (as horizontal argument) until we come opposite [the Sine of] the meridian altitude or the Sine of the meridian altitude of the opposite degree of the ecliptic, the result will be the half arc of daylight or of night (, respectively). (To determine) the time from midday to the *ʿaṣr*, find the altitude at the *ʿaṣr* from the special table and then find the corresponding hour-angle. The time remaining until sunset is the excess of half the arc of visibility over the time from midday to the *ʿaṣr*. (To determine) the durations of morning and evening twilight (*al-ḥiṣṣatān*), enter with the base (as horizontal argument) until we come opposite the excess of the Sine of the meridian altitude of the opposite point of the ecliptic over the Sine of 17° for evening twilight and 19° for morning twilight. The resulting hour-angles are the times between evening and morning twilight and midnight: subtract either of them from the half nocturnal arc and the remainder will be the duration of twilight. God knows best. The calculation was completed in the Holy Month of Muḥarram in the year 763 H [= November, 1361]."

The main tables are preceded by two small tables, the first of which is for the function:

$$\delta^*(\lambda) = 90° + \delta(\lambda) \quad (\varepsilon = 23;31°)$$

labelled *al ghāyāt āfāqiyya baʿd isqāṭ al-ʿarḍ minhā*, "meridian altitudes for all latitudes when the latitude is subtracted". The second is for the Sine function, with entries to two digits for each degree of argument. To find $t(h,\lambda,\phi)$ al-Khalīlī proposes first using the table of $\delta^*(\lambda)$ to find $H(\lambda,\phi)$, thus:

$$H(\lambda,\phi) = \delta^*(\lambda) - \phi$$

and then using the Sine table to find Sin H and Sin h to determine their difference:

$$H'(h,\lambda,\phi) = \text{Sin } H(\lambda,\phi) - \text{Sin } h.$$

Using the two main sets of auxiliary tables one should then be able to find $B'(\lambda,\phi)$ and finally:

$$t(h,\lambda,\phi) = V' \{ B'(\lambda,\phi), H'(h,\lambda,\phi) \} .$$

al-Khalīlī does not mention that his tables can also be used to determine the solar azimuth, but there is an obscure reference in the introduction to his major tables which indicates that he was aware of this possibility. The procedure would be to form the quantity:

$$H''(h,\lambda,\phi) = \text{Sin}(\bar{\phi} + h) - \text{Sin}\,\delta(\lambda),$$

with which the azimuth (measured from the north point) is given by:

$$a(h,\lambda,\phi) = V'\{B'(\lambda,\phi), H''(h,\lambda,\phi)\}.$$

Although only one copy of these original auxiliary tables has come to light, as opposed to the several extant copies of the universal table, it is clear that their circulation was not limited. They were known even in Tunis and Cairo (see **10.11**).

b) Formulae for timekeeping

MS Dublin CB 4091, fol. 17v, contains a list of 24 formulae for spherical astronomy. They are expressed in terms of a set of four quantities written in words. As explained in the introduction (see the second paragraph of the translation in (c) below), for the tetrad (a,b,c,d) the relation a/b = c/d is valid. See **Table 10.3b**.

Tabulation of formulae in this fashion was not initiated by al-Khalīlī, for a list of 62 formulae appears in the treatise of Abū ʿAlī al-Marrākushī (**6.7**) and a list of 184 (many, of course, redundant because they are trivially equivalent to others) appears in the *Zīj* of Ibn al-Shāṭir.[14] See **Fig. I-1.2a** for a similar list. This method of recording formulae was no doubt considered useful for memorization, for although algebraic notation would facilitate the tabulation, the symbols would still have to be defined in words.

Let us consider first a simple example, the fourth entry in the table. It indicates that:

horizontal shadow ÷ length of gnomon = length of gnomon ÷ vertical shadow

that is:

$$\text{Cot}_n\, h \div n = n \div \text{Tan}_n\, h,$$

or, in modern terms:

$$\cot h = 1 / \tan h.$$

More complicated examples are nos. 5, 8, 7, and 6. The first defines a quantity called the "base" (*al-aṣl*), here denoted by B (**I-6.0**):

Cosine of latitude reduced ÷ base = unity ÷ Cosine of latitude reduced ,

where the term "reduced" (*munḥaṭṭ^an^*) means "divided by 60"), so that:

$$B = \text{Cos}\,\delta\ \text{Cos}\,\phi / R^2.$$

The second presents an equivalent definition of this quantity, namely:

base ÷ Sine of meridian altitude = unity ÷ Versed Sine of semi diurnal arc ,

that is:

$$B = \text{Sin}\,H / \text{Vers}\,D.$$

With this quantity the third relation defines another quantity called the "auxiliary Sine" (*jayb al-tartīb*), here denoted by j:

base ÷ Sine of altitude = unity ÷ auxiliary Sine ,

whence:

$$j = \text{Sin}\,h / B = \text{Sin}h \cdot \text{Vers}\,D / \text{Sin}\,H.$$

---

[14] Sédillot-*père*, *Traité*, I, pp. 351-359, and Kennedy, "*Zīj* Survey", p. 163a, respectively. See now Charette, *Mamluk Instrumentation*, pp. 357-358, on Najm al-Dīn al-Misrī's list of 30 such formulae.

Table 10.3b
List of formulae for spherical astronomy
(MS Dublin CB 4091, fol. 17v)

1. Sine of rising amplitude	Sine of declination	60	Cosine of latitude
2. [Cosine][a] of altitude	[Sine][b] of altitude	Length of gnomon	Horizontal shadow
3. Cosine of altitude	[Sine][c] of altitude	Length of gnomon	Vertical shadow
4. Horizontal shadow	Length of gnomon	Length of gnomon	Vertical shadow
5. Cosine of latitude reduced	Base	Unity	Cosine of declination reduced
6. Base	Difference between Sines of meridian and instantaneous altitudes	Unity	Versed Sine of hour-angle
7. Base	Sine of altitude	Unity	Auxiliary Sine (*jayb al-tartīb*)
8. Base	Sine of meridian altitude	Unity	Versed Sine of half diurnal arc
9. Hypotenuse of horizontal shadow	Length of gnomon	60	Sine of altitude
10. Hypotenuse of vertical shadow	Length of gnomon	60	Cosine of altitude
11. Sine of ecliptic distance from equinox	Sine of declination	60	Sine of obliquity
12. Sine of latitude	Tangent of complement of rising amplitude	60	Tangent of half diurnal arc
13. Sine of altitude	Component (*ḥiṣṣa*) of the azimuth	Cosine of latitude	Sine of latitude
14. Cosine of azimuth	Component (*ḥiṣṣa*) of hour-angle	Cosine of declination	Cosine of altitude
15. Sine of altitude in prime vertical	Sine of declination	Total Sine	Sine of latitude
16. Sine of altitude in prime vertical	Sine of rising amplitude	Cosine of latitude	Sine of latitude
17. Tangent of solar declination (base 60)	Sine of half excess	Unity	Tangent of latitude (base 60)
18. Sine of half excess	60	Tangent of declination	Tangent of complement of latitude
19. Tangent of half excess	Tangent of latitude	Sine of declination	Cosine of rising amplitude
20. Sine of declination	Cosine of declination	"The excess"	5
21. Tangent of declination	12	"The excess"	5
22. Cotangent of latitude	Sine of half excess	Unity	"The excess"
23. Sine of ecliptic distance from equinox	Tangent of second declination	60	Tangent of obliquity
24. Cosine of declination	Cosine of rising amplitude	60	Sine of half diurnal arc

Notes: [a] text Sine (!); [b] text Cosine (!); [c] text Cosine (!)

The fourth relation allows a determination of the hour-angle from the difference between the Sines of the meridian and instantaneous altitudes:

$$\text{base} \cdot \text{Sine of altitude} = \text{unity} \cdot \text{auxiliary Sine} ,$$

whence:

$$\text{Vers } t = [\text{Sin } H - \text{sin } h] = R^2 [\text{Sin } H - \text{Sin } h] / [\text{Cos } \delta \text{ Cos } \phi]$$

$$= [\text{Sin } H - \text{Sin } h] \cdot \text{Vers } D / \text{Sin } H = \text{Vers } D - \text{Sin } h \cdot \text{Vers } D / \text{Sin } H ,$$

or, in the simplest possible terms:

$$\text{Vers } t = \text{Vers } D - j .$$

Rules such as these, and the accompanying technical terminology, were standard knowledge amongst contemporaneous Muslim astronomers.

c) Solar azimuth tables for Damascus

In MS Dublin CB 4091, fols. 18r-24r, there is a table of about 2,700 entries for the solar azimuth at Damascus. The function tabulated is a(H,h) computed to two digits for the domains:

$$H = 33°, 34°, \ldots , 80° \text{ and } h = 1°, 2°, \ldots , H .$$

The underlying latitude is 33;30° as stated at the end of the tables, and the values are independent of the obliquity. Horizontal differences are also tabulated and labelled positive or negative by the Arabic letters *d* (for *zā'id*, increasing) or *ṣ* (for *nāqiṣ*, decreasing). The use of the tables is explained in the short introduction (Dublin CB 4091, fol. 17v):

> "In the Name of God the Merciful and Compassionate. O my Lord, make (things) easy and help, O Generous One! The celebrated scholar (*al-shaykh al-imām al-ʿallāma*) Shams al-Dīn Muḥammad al-Khalīlī – may God have mercy upon him – said: Praise be to God, Lord of the Worlds. Blessings and Salvation for Muhammad, the most excellent of God's creation, and his Family and all of his Companions.
>
> These are four tables which contain all (important) astronomical operations. The way to use them is to look for the quantity you want. If it is one of the outer values multiply one of the inner entries by the other and divide the product by the outer entry which is known: the result will be the unknown outer quantity. If the unknown quantity is one of the inner two, multiply one of the outer quantities by the other and divide it by the inner quantity which is known: the result will be the unknown inner quantity. Section on the use of the table of the azimuth. You enter with the instantaneous altitude and the maximum altitude for the solar longitude (text has incorrectly: *ghāyat jadwal juz' al-shams*) and what you find is the azimuth. The direction of the azimuth from meridian altitude 33° to the limit of 57° is always southerly. The direction for the other meridian altitudes is also southerly if the instantaneous altitude is greater than the altitude in the prime vertical, and northerly if it is less.
>
> Section. If the meridian altitude is non-integral take a proportion corresponding to its fractional part of the first difference (*fa-khudh min al-tafāḍul bi-nisbatihi*) and add it to the azimuth of the differences which are marked positive (with the letter *d* for *zā'id*, increasing) and subtract it from the azimuth if they are marked negative (with the letter *ṣ* for *nāqiṣ*, decreasing): the result will be the azimuth. Example. The altitude is 24° and the meridian altitude 40;20°. You find (the entry) 41;53° opposite

(argument) 24° and you subtract one third of the difference from it, namely 0;34°: the remainder is 41;19°, which is the azimuth. If (the difference) had been marked positive we would have added them. Know this!"

The simplest formula for compiling such a table of a(H,h) is:

$$a(H,h) = \text{arc Sin } \{ [\text{Sin } h - \text{Sin } h_0(H)] \cdot \text{Tan } \phi / \text{Cos } h \} ,$$

where:

$$\text{Sin } h_0(H) = R \text{ Sin } (H - \bar{\phi}) / \text{Sin } \phi ,$$

$h_0(H)$ being the altitude of the sun in the prime vertical on a day when it has meridian altitude H.

Although several other solar azimuth tables are known from the Islamic sources (**I-5**), mostly with arguments h and λ; al-Khalīlī's is the only one with arguments H and h. Tables of the time since sunrise were often computed for arguments H and h, the advantage being that only about 3,000 entries have to be computed, as compared with about 10,000 entries in tables with arguments h and λ. al-Khalīlī's azimuth table was soon replaced by Shihāb al-Dīn al-Ḥalabī's solar azimuth table (**11.2**) which had the same format as al-Khalīlī's hour-angle tables and which was occasionally included in copies of the main Damascus corpus.

10.3* The universal auxiliary tables for finding the solar azimuth

The following tables are mentioned by al-Khalīlī in the introduction to his main universal auxiliary tables (**10.4**) but the unique surviving copy come to my attention only in January, 2001.[15] I have mentioned them briefly in **I-8.5.1a** and **I-9.4a**, and show here how they fit into al-Khalīlī's development.

MS Bursa Haraççioğlu 1177,4 (fols. 72r-90r), copied *ca.* 1450, is a unique copy of a set of tables by al-Khalīlī, not identified as the compiler, and entitled simply *Jadwal al-samt li-kulli 'rtifāʿ fī ʿurūḍ al-aqālīm al-sabʿa*. The set contains sub-tables for the following functions:

$$f'_\phi(\lambda) = R \text{ Sin } \delta(\lambda) / \text{Cos } \phi \quad \text{and} \quad g_\phi(h) = \text{Sin } h \text{ Tan } \phi / R$$

for the domains:

$$\lambda \text{ and } h = 1°, 2°, \dots , 90° ,$$
$$\phi = 1°, 2°, \dots , 48°, \text{ as well as } 33;30° \text{ (Damascus)} ,$$

and:

$$K(x,h) = \text{arc Sin } \{ R \cdot x / \text{Cos } h \}$$

for the domains:

$$x = 1, 2, \dots , 59 \text{ and } h = 0°, 1°, \dots , n(x) ,$$

where n(x) is the largest integer such that $x \leq \text{Cos } n(x)$. The functions f'_ϕ and g_ϕ are labelled *jayb saʿat al-mashriq* and *ḥiṣṣat al-samt* by al-Khalīlī, and it is clear that they represent Sin ψ(λ) and k(h) – see **5.0** and **8.5**. The value of ε underlying the first table is about 23;30°, possibly 23;31°.

The instructions (see **Fig. 10.3***) describe how to determine the *taʿdīl al-samt* using the first two tables, and then how to find the azimuth from the third table. These universal auxiliary tables represent the ultimate solution to the problem of determining the solar azimuth a(h,λ,φ).

[15] The manuscript is mentioned in İhsanoğlu *et al.*, *Ottoman Astronomical Literature*, II, pp. 805.

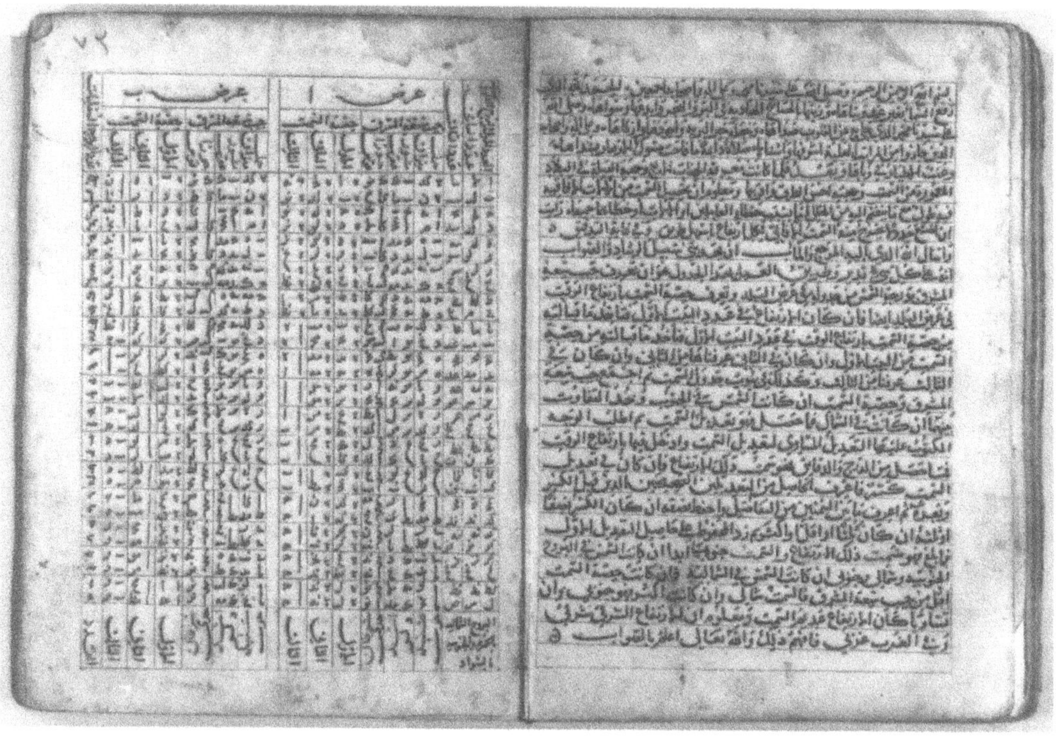

Fig. 10.3*: The introduction to al-Khalīlī's tables and the first two sub-tables of f_ϕ and g_ϕ for latitudes 1° and 2°. See also **Figs. I-9.4*a-b**. [From MS Bursa Haraççioğlu 1177,4, fols. 72v-73r, courtesy of the Genel Kütüphanesi.]

It is clear that all al-Khalīlī had to do to compile his splendid tables described in **10.7** was to replace the table of $f'_\phi(\lambda)$ with a new one of $f_\phi(\theta)$ and rewrite the instructions!

There is a note on fol. 90v of the Bursa manuscript in the hand of (the Damascus astronomer) Muḥammad ibn Muḥammad al-Tīzīnī stating that he had looked at this table in the year 867 H [= 1462/63].[16] al-Tīzīnī himself compiled a set of tables for sundial construction for different latitudes which is extant in MS Vatican ar. 105,3.[17]

10.4 The main Damascus corpus: the introduction and solar tables

The full title of the Damascus corpus of tables for timekeeping is given on MS Paris BNF ar. 2558, fol. 1r. It translates: "Tables of the hour-angle and the operations of night and day for latitude 33;30° computed by *al-Shaykh* Shams al-Dīn Muḥammad ibn Muḥammad ibn Muḥammad al-Khalīlī, muezzin at the Umayyad Mosque, followed by a table for finding the azimuth of the qibla and a universal table which can be used in all localities."

Fol. 1v of MS Paris BNF ar. 2558 contains virtually all that remains of an introduction to

[16] On al-Tīzīnī see Suter, *MAA*, no. 450; Brockelmann, *GAS*, II, p. 160, and SII, p. 484; *Cairo ENL Survey*, no. C95; and also **10.4**.

[17] These are not listed in any reference work from Suter, *MAA*, to Rosenfeld & İhsanoğlu, *MAIS* (no. 903).

some tables for calendar conversion and for finding the solar longitude.[18] The bulk of the introduction and the tables themselves have apparently been removed and replaced by a later set (fols. 2v-9r) in a different hand. At this time fol. 1v was apparently glued to an empty fol. 2r. More recently this introductory fragment was separated from fol. 2r with the result that a mirror image of part of the text may be read. Although the introduction and the missing tables are anonymous they are most probably the work of al-Khalīlī himself; they were already outdated in 1408, when the manuscript was copied. It is clear from this fragment that al-Khalīlī had prepared a table for converting Hijra lunar dates into dates in the Syrian solar calendar and another table displaying the solar longitude for each day of an artificial Syrian year which needed to be corrected from an auxiliary table. This correction provided the necessary adjustment for leap years in the Syrian calendar and was tabulated as a function of the excess Hijra years over al-Khalīlī's epoch (not stated) reckoned modulo 60.

The new introduction and tables in MS Paris BNF ar. 2558, fols. 2v-9r, are the work of al-Tīzīnī (**10.3***). They are similar in conception to those of al-Khalīlī, and tabulate the Syrian dates corresponding to the beginnings of each lunar month from 851 to 1000 Hijra. al-Tīzīnī then presents a table of solar longitude of the *shabaka* variety for an artificial Syrian year to which a correction must be applied. The correction is to be taken from a sub-table copied around the main table; values are given for each year after 850 Hijra from 1 to 150 (corresponding, of course, to years 851 to 1000 Hijra). al-Tīzīnī then goes on to discuss various timekeeping operations which need not concern us here.

Solar tables arranged according to the date in a solar calendar are attested in other Islamic sources both earlier and later, but none of these has been investigated for structure or accuracy.[19] It is likely that al-Khalīlī's solar longitude table would have been computed using the solar tables in the *Zīj* of Ibn al-Shāṭir, and that the entries would have been accurate to within a few minutes.

MS Paris BNF ar. 2558, fol. 50v, contains a simple table for correcting the solar longitude at midday to yield the solar longitude for any other time of day. Corrections $\Delta\lambda$ are given as a double-argument function of the daily solar motion λ (Arabic, *buht*) and the time t before or after midday. Values to one digit (minutes) for each 3° of t from 3° to 180°, and each unit of the daily solar motion from 57 min/day to 62 min/day are given. The underlying formula[20] is simply:

$$\Delta\lambda \;=\; \lambda \times t \,/\, 360.$$

MS Paris BNF ar. 2558, fol. 64r, contains a table of solar declination $\delta(\lambda)$ with values to degrees and minutes for each degree of solar longitude. The underlying value of the obliquity is 23;31°.

Finally, in MS Paris BNF ar. 2558, fols. 18v-20r, there is a set of equatorial coordinates (α',Δ) for 81 stars. This table is not the work of al-Khalīlī: the entries are, as stated in the heading, those of Ibn al-Shāṭir's star catalogue, with 0;15° added to the normed right ascensions.[21] Of course the correction for precession should be added to the stellar longitudes, so we have here a very crude and rather unsuccessful attempt to update the tables. A precessional

[18] On calendars in Islamic astronomy see n. **I**-1:3.
[19] See nn. 1:43.
[20] See Kennedy, "*Zīj* Survey", p. 162b
[21] On Ibn al-Shāṭir's star catalogue see Kennedy, "*Zīj* Survey", p. 164a.

correction of 0;15° corresponds to about 15-20 years using the standard medieval values (1° in 66 or 70 solar years) for this motion. Ibn al-Shāṭir's catalogue is for 765 H [=1363/64].

For our present purposes it is important to realize that al-Khalīlī assumed that the user of his tables for timekeeping knew either the longitude or declination of the sun or the stellar declination.

10.5 The hour-angle tables

al-Khalīlī's corpus contains a complete set of tables of the hour-angle $t(h,\lambda)$ for Damascus based upon the parameters:

$$\phi = 33;30° \text{ and } \varepsilon = 23;31°.$$

There is no mention of these tables, or the prayer-tables (**10.6**), in the introduction. Values of the hour-angle are given for arguments:

$$h = 1°, 2°, \ldots , 80° \quad \text{and} \quad \lambda = 271°, 272, \ldots , 359, 0, 1, 2, \ldots , 90°.$$

Note that because of the format of the tables no entry is given for the winter solstice, that is, $\lambda = 270°$. There are about 3,000 entries. Extracts are shown in **Figs. I-2.1.4a-b**.

The standard of al-Khalīlī's computational accuracy is remarkably high. He may have used his first set of auxiliary tables (**10.3**) to compile these hour-angle tables. The procedure would be equivalent to:

$$t(h,\lambda) = \text{arc Vers } \{ [\text{Sin } h(\lambda) - \text{Sin } h] \,/\, B'(\lambda) \} \,.$$

Alternatively he may have tabulated the functions $B(\lambda)$ and $C(\lambda)$ for this latitude and then used the simple relationship:

$$t(h,\lambda) = \text{arc Cos } \{ R \cdot [\text{ Sin } h - C(\lambda)] \,/\, B(\lambda) \} \,.$$

The hour-angle tables are contained in MS Paris BNF ar. 2558, fols. 20v-50r, and at least seven later copies (see **10.2**). In MS Cairo MM 71,1, copied *ca.* 1600, triplets of entries (t,T,a) for Damascus are given for each pair of arguments (h,λ). The values of both t and T are attributed to al-Khalīlī, and the values of a are attributed to Shihāb al-Dīn al-Ḥalabī (**11.2**). Some tables for the time remaining until moonset based on al-Khalīlī's tables of $T(h,\lambda)$ for the sun are described below (**10.11**).

MSS Paris BNF ar. 2558, fol. 33r, and Oxford Seld. Supp. 100, fol. 16v, of the corpus contain two tables by al-Khalīlī without instructions for their use. The two tabulated functions are labelled *qaws al-aṣl bi-'l-murī* and *qaws al-aṣl bi-ghayr murī*, "arc of the base with / without a movable marker". These functions are denoted by b_1 and b_2, and inspection reveals that for al-Khalīlī's parameters:

$$b_1(\lambda) = \text{arc Sin } \{ B(\lambda) \} \quad \text{and} \quad b_2(\lambda) = \text{arc Tan } \{ B(\lambda) \}.$$

where:

$$B(\lambda) = \text{Cos } \delta(\lambda) \text{ Cos } \phi \,/\, R \,.$$

These tables, discussed in more detail in **I-6.5.1**, were designed to facilitate the marking of $B(\lambda)$ on an instrument called the sine quadrant (Arabic: *al-rubʿ al-mujayyab*), a computational device which in different forms was popular among Muslim astronomers from the 9th to the 19th century. Complicated trigonometric computations can be made with the sine quadrant without any calculations whatsoever.

10.6 Tables for regulating the times of prayer

The main set of prayer-tables compiled by al-Khalīlī, for which there is likewise no introduction (**10.5**), consists of tables displaying some twelve functions for each degree of solar longitude, beginning with Capricorn 1°. The tables are entitled *jadwal aʿmāl al-layl wa-'l-nahār*, that is, "tables for the operations of night and day", and are based on the same parameters: an extract is shown in **Fig. 10.6**. The functions tabulated are:

H	the solar meridian altitude (*ghāyat al-irtifāʿ*);
D	half the diurnal arc (*niṣf qaws al-nahār*);
$2D^h$	the number of hours of daylight (*sāʿāt al-nahār*);
h_a	the solar altitude at the beginning of the *ʿaṣr* (*irtifāʿ al-ʿaṣr*);
t_a	the hour-angle at the beginning of the *ʿaṣr* (*dā'ir al-ʿaṣr*);
T_a	the time between the beginning of the *ʿaṣr* and sunset (*mā bayn al-ʿaṣr wa-'l-ghurūb*);
t_b	the time between midday and the end of the *ʿaṣr* (*mā bayn al-ẓuhr wa-ākhir waqt al-ʿaṣr*);
2N	the duration of night (*qaws al-layl bi-kamālihi*);
s	the duration of evening twilight (*ḥiṣṣat al-shafaq*);

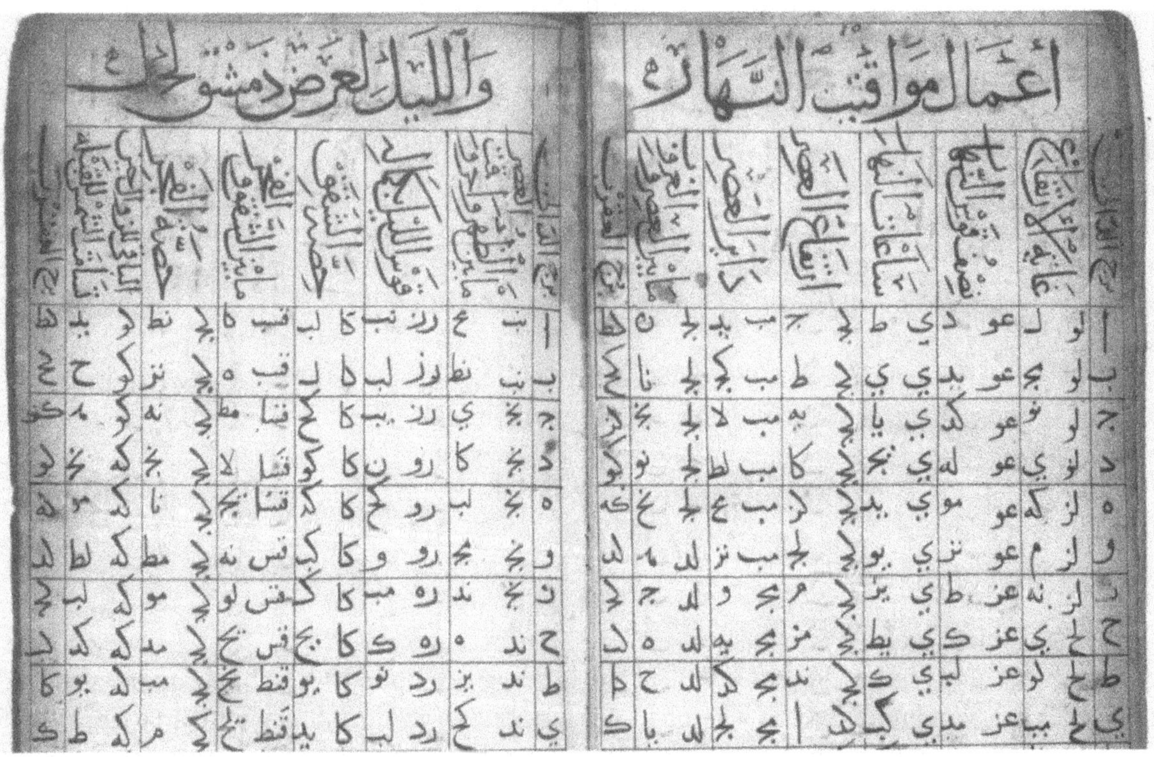
أعمال مواقيت النهار والليل لعرض دمشق

Fig. 10.6: The sub-table for solar longitudes in Aquarius in al-Khalīlī's corpus of prayer-tables. [From MS Paris BNF ar. 2558, courtesy of the Bibliothèque Nationale de France.]

n the duration of darkness (*mā bayn al-shafaq wa-'l-fajr*);
r the duration of morning twilight (*ḥiṣṣat al-fajr*);
t_q the time remaining until midday when the sun is in the azimuth of the qibla (*al-bāqī li-'l-zawāl ḥīna tusāmit al-shams al-qibla* (text: *li-'l-qibla*)

Entries are to two sexagesimal digits and are arranged side by side in columns for each zodiacal sign. Since the vertical arguments run downwards from 1° to 30° for each sign from Capricorn to Gemini and upwards from 0° to 29° in the same tables from Sagittarius to Cancer, there is no entry for the winter solstice. (The different format of the Cairo corpus means that there is no entry for the equinoxes.) The total number of entries in al-Khalīlī's prayer-tables is 2,160.

These tables appear in MS Paris BNF ar. 2558, fols. 9v-14r, and at least 15 later copies (see **10.2**). In MS Princeton Yahuda 861,1 they have the format of the Cairo corpus in a collection of prayer-tables for all latitudes. In MSS Berlin Ahlwardt 5758, 5759/5771 and 5760/5772, they are preceded by the solar longitude tables of ʿAbd al-Laṭīf called Ibn al-Kayyāl (**11.4**). In MS Damascus Ẓāhiriyya 9233 they form part of a set of tables attributed to Muḥammad ibn Muṣṭafā al-Ṭanṭāwī (**11.13**).

al-Khalīlī's tables defining the interval for the afternoon prayer are based on the following formulae for the solar altitudes at the beginning and end of the interval:

$$h_a = \text{arc Cot} \{ \text{Cot } H + R \} \quad \text{and} \quad h_b = \text{arc Cot} \{ \text{Cot } H + 2R \} .$$

The values of $t_a(\lambda)$ are more accurate than those that can be derived by linear interpolation using the argument $h = h_a$ in the tables of $t(h,\lambda)$. Nevertheless, the values of t_a, and also T_a and T_b, are often incorrect by several digits in the second sexagesimal place. Possibly al-Khalīlī used his table of $h_a(H)$ to generate the values of $h_a(\lambda)$ and the resulting values suffered from the interpolation scheme used. (See **11.3** below for some additional, rather extensive tables based on al-Khalīlī's tables for the *ʿaṣr*.)

al-Khalīlī defined the *ʿaṣr* as beginning when the shadow of a vertical object had increased by the length of the gnomon and ending when the shadow increase was double this amount. This is the definition usually associated with the Ḥanafī legal school, to which al-Khalīlī adhered, and presumably it corresponded to the currently-accepted practice at the Umayyad Mosque in his time.

The tables defining the length of twilight and darkness are based on the parameters 19° and 17° for morning and evening twilight. The values of $r(\lambda)$ and $s(\lambda)$ can be derived from the tables of $T(h,\lambda)$ for 19° and 17°, using:

$$r(\lambda) = T(19°, \lambda^*) \quad \text{and} \quad s(\lambda) = T(17°, \lambda^*), \qquad \text{where } \lambda^* = \lambda + 180°.$$

The table defining the hour-angle when the sun is in the azimuth of the qibla is, of course, based on a value for the azimuth of Mecca at Damascus. This value is not stated in any of the manuscripts in the corpus, but al-Khalīlī's values of $t_q(\lambda)$ correspond most closely to the parameter $q = 28;57°$ (**10.9**).

In some of the manuscripts there are separate tables of certain additional functions:

(a) Tables of $\delta(\lambda)$ (*al-mayl*) (*e.g.*, MS Paris BNF ar. 2558, fol. 64r), $\psi(\lambda)$ (*saʿat al-mashriq*) and $h_0(\lambda)$ (*al-irtifāʿ al-ʿadīm al-samt*) (*e.g.*, fol. 51v of the Paris copy) to two digits for each degree of λ.

(b) Tables of $\text{Cot}_R\, h$ (*al-ẓill al-mabsūṭ*) to two sexagesimal digits for each degree of h, for the bases R = 20 (only in the Paris copy, fol. 51v) and R = 10 (only in MS Berlin

Wetzstein 1138 (Ahlwardt 5754-6), fol. 59r). The use of these bases is remarkable. No explanation is provided in the manuscript and none is forthcoming from this author.[22]

(c) Tables of the function $h_a(H)$, which is universal as it is independent of terrestrial latitude (*al-ʿaṣr al-āfāqī*) (*e.g.*, fol. 64r of the Paris copy). This is computed to two digits for each degree of H from 1° to 90°, and is based on the standard formula. al-Khalīlī's values of $h_a(\lambda)$ (see above) may have been computed using this table, although they would have been more accurate had they been computed directly.

(d) Tables of two functions which we label $b_1(\lambda)$ and $b_2(\lambda)$ for use in timekeeping calculations with quadrants (only in the Paris copy of the corpus, fol. 33r): see already **10.5**.

(e) Most of the sources (*e.g.*, fols. 104v-105r, 17v-18r, 15v-16r, 16v-17r, of the Paris copy) contain four tables of ascensions, with values in degrees and minutes for each degree of ecliptic longitude. The functions tabulated are:

$$\alpha'(\lambda),\quad \alpha_\phi(\lambda),\quad \alpha_s(\lambda) \text{ and } \alpha_r(\lambda).$$

Each table is arranged in twelve columns of thirty entries, with the name of the zodiacal sign at the head of each column and a vertical argument running from 1 to 30. The actual arrangement of the columns differs for each table, beginning with the columns for the quadrant of the ecliptic at the beginning of which the function tabulated is between 0° and 90°. Thus the first columns in the table are for Capricorn, Aries, Libra, and Cancer, respectively.

The tables of α_r and α_s are based on the relations:

$$\alpha_r(\lambda) = \alpha_\phi(\lambda_H) = \alpha_\phi(\lambda) - r(\lambda) \quad \text{and}$$

$$\alpha_s(\lambda) = \alpha_\phi(\lambda_H) = 360° - [\, \alpha_\phi(180° - \lambda) - s(\lambda)\,] \ .$$

In the first case H is the ascendant at daybreak, and in the second it is the ascendant at nightfall.

(f) The Paris copy (fols. 18v-20r) contains a star catalogue attributed to Ibn al-Shāṭir listing the equatorial coordinates of 81 stars.

Each of the sub-tables in this set is rather carefully computed, with the exception of the tables for the *ʿaṣr* (noted above).

10.7 Universal auxiliary tables

al-Khalīlī's most original contribution was his second set of auxiliary tables for solving all of the standard problems of spherical astronomy for all latitudes:[22a] see **I-9.5** and **VIa-9.1** and the extracts illustrated there. Over a dozen copies are now known, of which the best is on fols. 61v-104r of the superior Paris manuscript. al-Khalīlī tabulated three auxiliary functions, each to two sexagesimal digits. For the first two, the horizontal argument is terrestrial latitude, and the functions are denoted by f_f and g_f, where f is the local latitude. The vertical argument is

[22] The same holds for the use of 10 by Ibn al-Ādamī in the 10th century: see **I-4.2.3**. On the use of base 20 see n. **I-1**:35, and the text to n. **II-10**:39a.

[22a] The first study of these was in King, "al-Khalīlī's Universal Tables" (1973).

defined in the instructions, and the functions are called *al-maḥfūẓ al-awwal* and *al-maḥfūẓ al-thānī*, "first and second functions", respectively. The third function, which is denoted by K, has a quantity called *jayb al-tartīb*, perhaps best rendered as "the auxiliary Sine", as horizontal argument. The vertical argument is again defined in the instructions. The *jayb al-tartīb* used here is not the same as the one defined in the minor auxiliary tables above (**10.3**).

The functions are not explained in the instructions, but are in fact:

$$f_\phi(\theta) = R \text{ Sin } \theta \ / \text{ Cos } \phi \ , \quad g_\phi(\theta) = \text{Sin } \theta \text{ Tan } \phi \ / \ R \ , \quad \text{and}$$

$$K(x,y) = \text{arc Cos } \{ \ R \cdot x \ / \text{ Cos } y \ \},$$

for appropriate domains (see below), and the trigonometric functions are to base R = 60. The functions are defined in a marginal note in MS Cairo MM 43, fol. 5r:

> "The first function is the quotient of the Sine of an arc divided by the Cosine of the latitude. The second (function) is the product (*musaṭṭaḥ*, lit., rectangle) of the Sine of the arc and the Tangent of the latitude, (both expressed) to base sixty (*sittīniyy^{an}*) The function tabulated in the table labelled *jayb al-tartīb* (*ḥāṣil jayb al-tartīb*) is the complement of the arc (*i.e.* inverse Cosine) of the quotient [multiplied by a factor of sixty] of (the *jayb al-tartīb*) and the Cosine of the arc."

The first two functions are tabulated for the domains θ = 1°, 2°, ..., 90°, and ϕ = 1°, 2°, ..., 55°, as well as 21;30° (Mecca) and 33;30° (Damascus). In some sources (*e.g.* MS Paris ar. 2558) there is no table for Mecca, and in others (*e.g.* MSS Escorial ár. 931 and London BL Add. 9599,31) the argument ϕ runs only to 50°. The arguments for the third functions are x = 1, 2, 3, ... , 59 and y = 0°, 1°, ... , n(x), where n(x) is the largest integer such that $x \leq$ Cos n(x). The values for each function are arranged in columns each of thirty entries. On a given double opening in the first set of tables one finds values of f and g for two consecutive values of ϕ. In the second set one generally finds values of K for four consecutive values of x. However, when the values of K are, in modern terms, non-real, no entry is given in the table. Columns which would be empty are omitted; thus, for example, the values of K for x = 52, 53, ... , 59 fit on a double opening in MS Paris ar. 2558, fols. 103v-104r. If the values of K were displayed in a single table it would be trapezoidal in shape.

The first two tables each contain 5,040 entries and the third about 3,500, all computed with remarkable precision. Most of the entries are accurate, and the error in the remainder is usually only ±1 in the second digit, and occasionally ±2. Errors in the K table are slightly more frequent than in the f_ϕ and g_ϕ tables, and larger isolated errors do occur. Some entries in the table of K(x,y) for small values of x and large values of y are rather inaccurately computed. This author is unable to explain how these errors might have arisen.

However, Glen Van Brummelen, armed with an impressive combination of mathematical and statistical skills and not lacking in historical sensitivity, has investigated these error patterns in an attempt to establish precisely how al-Khalīlī computed his tables; this was a topic beyond the scope of the present study, and I cannot claim to understand all that Van Brummelen has published on this subject.[23] He would be the first to admit that al-Khalīlī's tables still actually defy explanation. He has, however, applied various statistical procedures in his investigation

[23] Van Brummelen, "al-Khalīlī's Auxiliary Tables", and *idem* & Butler, "Interdependence of Astronomical Tables", pp. 46-48.

of the structure of the auxiliary tables and – to our joint disappointment, for I do not think it was too much to hope for – was also unable to come up with precisely the procedures that were used by al-Khalīlī. The interested reader should consult Van Brummelen's paper, of which I here cite the conclusion (using the mathematical notation of this study):[24]

> "We have found in al-Khalīlī's auxiliary tables an interesting mix of ingenious insights and rudimentary errors. The $g_\phi(\theta)$ table was computed from two-place sine and tangent values on an interpolation grid of 5° of θ, and some of the remaining entries were filled in using a well chosen variant of linear interpolation. Many of the entries, however, seem to be unexplainable other than by simple guesswork. The $f_\phi(\theta)$ table was computed from the $g_\phi(\theta)$ table by a clever application of prosthaphairesis. The K(x,y) table was calculated by using only two places for the value of Cos y, followed by an extremely careful and precise calculation of the quotient Rx / Cos y and the arc Cosine."

Van Brummelen has shown that the errors in the entries for f_ϕ and g_ϕ in a large part of the tables, if not throughout, are linked by the following prosthaphaeretical relation:

$$f_\phi(\bar{\phi} \pm n) = \text{Cos } n \pm g(\phi,n) ,$$

and he suggests that al-Khalīlī, having computed the table of $g_\phi(\theta)$ and noted that $f_\phi(\bar{\phi}) = R = 60$, used this relation to generate a major part of the table of $f_\phi(\theta)$. When he published this in 1991, I found the implications surprising. However, we now have independent confirmation of the fact that al-Khalīlī did tabulate $g_\phi(\theta)$ first, because he simply took over the table from his universal auxiliary tables for finding the solar azimuth.

We are on safer ground regarding what al-Khalīlī intended with his tables. The rules for their use which he outlines in words will now be compared with the medieval formulae listed in **I-1.2**. Most of the rules can be explained by transforming standard medieval expressions into a form which is equivalent to al-Khalīlī's rule, given the definitions of the functions f, g, and K. The arguments to be used in the tables are considered positive, and although al-Khalīlī's introduction (MS Paris ar. 2558, fols. 61v-64r) gives detailed instructions to cover all possible cases, modern notation renders these superfluous in the commentary below.

> "In the name of God, the Merciful and Compassionate: Praise be to God, by whose grace good things are achieved, and to whose might the earth and the heavens are obedient, and by whose wisdom the months and the times (of prayer) are known, and to whose order(s) the sun, moon and stars are subservient. May God bless our Lord Muḥammad, master of brilliant miracles, and his family and his companions, (all of whom) possess great virtue and noble qualities, with a blessing such that he who pronounces it be saved from misfortunes in life and in death.
>
> The aspect of astronomy most deserving of investigation is that by which the times of the five prayers and the qibla can be found and which deals with other information useful for guiding oneself on land and sea. The instruments which achieve this are very numerous, and the best of them are those which are easy to use and which are not specific to one latitude, but there are only a few people who can use these properly. Finding (the times of prayer and the qibla) by means of calculation is more precise

[24] Van Brummelen, "al-Khalīlī's Auxiliary Tables", p. 689.

but it involves more work, and only someone skilled in this discipline can do it properly. Thus if a table is prepared by calculation it is better to use such a table than to use instruments, although (??) the table is easier to comprehend than (the instruments).

Previously I had compiled two universal tables, from one of which one could determine the times, and from the other, the azimuth. Now God has enabled me to compile a table for finding all that could be found from those two tables and more besides. In addition, it is simpler than them. Praise and thanks be to God.

Section on the half diurnal or nocturnal arc for the sun or a star. Find the longitude of the sun from an ephemeris or some other tables and use it to find the declination from the declination table, or find the declination of the star corrected for the epoch in which you are working from the table of stellar coordinates. Then look for the page on which is written the local latitude and find the value of the second function corresponding to the solar or stellar declination. If the declination is in the first, second or third column, take the value of the function from the first, second or third column, respectively. Enter the result in the table of the third function (as the horizontal argument) and the declination as the (vertical) argument. What you find in the first, second or third box will be the half diurnal arc if the declination is southerly and the half nocturnal arc if it is northerly. If one of them is subtracted from 180°, the result will be the other one.

Section on the hour-angle. Find the value of the first function on the page corresponding to the terrestrial latitude with the altitude as (vertical) argument and the value of the second function as before in the section on the half arc. If the declination is southerly add the two values, otherwise take the difference. The result is the auxiliary sine. Enter it in the table of the third function with the solar or stellar declination as (vertical) argument, and you will find the hour-angle. If the declination is northerly and the first function is less than the second, subtract the result from 180° and the remainder will be the hour-angle. If the two functions are equal then the hour-angle will be 90°.

Section on the maximum altitude, the local latitude, the altitude of (the sun) and the hour-angle at the *ʿaṣr*, the time remaining from the *ʿaṣr* until sunset, and the duration of evening and morning twilight. Add the northern declination to 90° and subtract the southern declination: the result will be the sum of the maximum altitude and latitude. Then find the local latitude by observation or otherwise and subtract it from this sum: the result will be the maximum altitude. If it is greater than 90° subtract it from 180°: the remainder will be the maximum altitude to the north of the zenith (i.e. measured from the north point). If the maximum altitude has been found by observation and it is subtracted from the sum, the remainder will be the local latitude. If the latitude is added to the maximum altitude and the difference between the sum and 90° is taken, the result will be the solar or stellar declination. If the maximum altitude is northerly with respect to the zenith, subtract it from 180° and subtract the remainder from the sum: the result will be the local latitude, or add the latitude to the remainder, subtract 90° from the sum, and the remainder will be the declination.

Next enter with the maximum altitude of the sun in the universal *ʿaṣr* table and you will find the altitude of the *ʿaṣr*. Find the hour-angle for this altitude: it will be the time from midday to the *ʿaṣr*. Subtract this from the half diurnal arc: the remainder will be the time from the *ʿaṣr* to sunset. Next you find the value of the first function on the page for the latitude corresponding to (vertical argument) 17° for nightfall or 19° for daybreak. Then add to it the previous value of the second function for the half diurnal arc if the declination is northerly, otherwise subtract it from the arc: the result will be the auxiliary sine. Enter this in the table of the third function and the declination as (vertical) argument. Subtract what you find from the half nocturnal arc: the result will be the duration of twilight.

Section on the rising amplitude, the altitude in the prime vertical, and the azimuth corresponding to a particular altitude. Find the value of the second function for latitude 45° with (vertical argument equal to) the declination and enter the result as argument in the table of the third function with the local latitude or its complement as vertical argument, and you will find the complement of the rising amplitude or [complement of] the altitude in the prime vertical. Another method: Find the value of the first function for the solar or stellar declination on the page corresponding to the latitude and enter whatever it is as (vertical) argument in the table of the second function for latitude 45° or in the page corresponding to the latitude. The corresponding argument will be the rising amplitude or the altitude with no azimuth.

Then find the value of the first function for the solar or stellar declination on the page corresponding to the latitude, and find the value of the second function for the altitude. Add the two values (if the declination is) southerly and take the difference between them (if it is) northerly. The result is the auxiliary sine: enter it in the table of the third function with the altitude as the (vertical) argument and you will find the azimuth measured from the meridian. If the declination is southerly the azimuth is southerly. If it is northerly and the value of the first function is greater than that of the second, the azimuth will be northerly; if it is less, then (the azimuth) will be southerly. If the two values are equal then there is no azimuth.

Section on finding the declination by observation given that the latitude is known and finding both the latitude and the declination. Determine the meridian or the prime vertical either by means of an azimuth found from an altitude observation (*samt al-irtifāʿ*) or by using the Indian circle or a qibla-box (*ḥuqq al-qibla*) or by some other well-known means.[25] When one of these two lines is known use it to find the azimuth of a particular altitude by measuring the altitude at that time to obtain (a pair of) altitude and azimuth (values). Then look in the table of the third function for a box such that if the azimuth is entered in it (as one argument) the altitude would be the corresponding (vertical) argument: the auxiliary sine is (the argument) above the box. Then find the value of the second function with the altitude (as argument) on the page for the latitude. Subtract the value of the second function from the auxiliary sine for southern

[25] On the Indian circle see n. 2:23. On qibla-boxes in Egypt *ca.* 1300 see Schmidl, "Earliest Arabic Sources on the Compass", and King, *Mecca-Centred World-Maps*, pp. 113-114.

declinations, or add the two for northern declinations if the azimuth is northerly, otherwise take the difference between them. The result is a value of the first function, so find the corresponding argument: it will be the declination.

If you do not know either the latitude or the declination, make an observation of the azimuth and altitude and find the auxiliary sine as before, then observe another azimuth and altitude and find the auxiliary sine (for this pair) also. If the azimuths are in the same direction, take the difference between the two, otherwise add them: the result will be the *ḥiṣṣa*. Next find the values of the second function for latitude 45° and for each of the altitudes and take the difference between them. Enter with the result in the (table of the) second function for latitude 45° and find the corresponding argument: the result will be the *taʿdīl*. Then look for a page of (the tables of) the second function so that if you enter with the *taʿdīl* as argument, the value is the *ḥiṣṣa*: (you do this) to obtain the latitude at the top of the page. Then find the value of the second function for the larger altitude on that page and add to it the auxiliary sine for the larger altitude if the azimuth is northerly, otherwise take the difference between them: enter the result in (the table of) the first function on the (same) page, and find the corresponding argument to obtain the declination.

Conclusion on operations with stars. Enter the longitude of the sun in the table of (normed) right ascensions and you will find the (normed) right ascensions for that day. Subtract from these the half diurnal arc if possible, otherwise add a complete revolution (of 360° and then subtract): the result will be the oblique ascensions. If you add the half diurnal arc to the (normed) right ascensions, the result will be the (oblique) ascensions of the opposite point of the ecliptic. If the sum is greater than one revolution then the excess will be the ascensions of the opposite point. Subtract the half diurnal arc of a star or its eastern hour-angle from its ascensions – if this is not possible add a revolution to the ascensions: the result will be the ascensions of midheaven when the star is rising or when its altitude was measured. If the half diurnal arc or the hour-angle is added to the ascensions (of the star) the result will be the ascensions of midheaven when (the star) is setting or when its altitude was measured.

If the ascensions of midheaven are found by these operations or by means of a star which is culminating, subtract from them the ascensions of the opposite point of the ecliptic and the remainder will be how much of the night has passed. If the ascensions of the opposite point are added to the ascensions of midheaven, the culmination will be before sunset by the amount of the sum (lit. increase). If you subtract the ascensions of midheaven from the oblique ascensions, the remainder will be the time remaining of the night. If you add the ascensions of midheaven to the oblique ascensions, the culmination is after sunrise by the amount of the sum (lit. increase).

Section on the declination of a star. Look in the table of the third function for a box such that if you enter in it with the longitude of the star measured from the beginning of Capricorn or Cancer the corresponding (vertical) argument would be the latitude (of the star): you obtain the auxiliary sine above that box. Then enter in (the table of) the second function for latitude 49° the altitude of the star, and double the result and add it to the auxiliary sine if the longitude of the star and its latitude are in the same direction, otherwise take the difference. Enter one half of the result as a value of the first function for latitude 37° and find the corresponding argument: it will be the declination of that star. The direction of the declination is the same as the direction of the longitude and the latitude (when these are) in the same direction. If they are in different directions and if the auxiliary sine is less than the value of the second function, I mean the doubled value, then the direction of the declination is the same as that of the longitude. If they are equal then there is no declination and no direction.

Section on the ascensions of a star. Enter with the latitude of the star in (the table of) the first function for latitude 37°, double what you find and keep it in mind. Next enter the declination of the star in (the table of) the second function, double what you find and keep it in mind. Then add the two quantities which you kept in mind if the latitude of the star and its declination are in different directions and take the difference between them if they are in the same direction: the result will be the auxiliary sine. Enter this in the table of the third function with the declination of the star as the (vertical) argument, and see what you find in the table. If the latitude and the declination are both northerly and the value of the first function is more than that of the second, and the longitude is between Capricorn and Cancer, then the result is the ascensions of the star. If the longitude is between 3 (signs) and 9 subtract the result from 360°: the remainder will be the ascensions. Likewise for (both latitude and declination) southerly. If the value of the first function is less than that of the second and the longitude is between 3 and 9, then subtract the result from 180°, but add it to 180° if the longitude is between 3 and 9 to obtain the ascensions. Likewise for both (latitude and declination) in the south if the value of the first quantity is greater than that of the second. If the latitude is northerly and the declination is southerly and the longitude is greater than Capricorn, then the result is the ascensions of the star, and if (the longitude) is less then subtract the result from 360° to obtain the ascensions. If the latitude is southerly and the declination is northerly and the longitude is less than Cancer, subtract the result from 180°, and if (the longitude) is more, add the result to 180° to obtain the ascensions of the star. God knows best."

Commentary:

(i) To find the half arc of daylight, D, or night, N, of a celestial body with declination δ, given the terrestrial latitude ϕ, use:

$$\text{D or N} = K\{\ g_{\phi}(\delta),\ \delta\ \}$$

for $\delta \gtrless$% 0, respectively. From **F7** in **I-1.2** for all δ:

$$D = 180° - N = 90° + \text{arc Sin}\ \{\ \text{Tan}\ \delta\ \text{Tan}\ \phi\ /\ R\ \}$$

so that:

$$\text{D or N} = \text{arc Cos}\ \{\ \text{Sin}\ \delta\ \text{Tan}\ \phi\ /\ \text{Cos}\ \delta\ \}\ \text{for}\ \delta < \text{or} > 0,\ \text{respectively.}$$

(ii) To find the hour-angle, t, given the solar altitude, h, for any declination and terrestrial latitude, use:

$$t = K\ \{\ [\ f_{\phi}(h) - g_{\phi}(\delta)\ ],\ \delta\ \}.$$

To illustrate the correctness of this procedure, we transform **F12** in **I-1.2** into:

$$t = \text{arc Cos}\ \{\ R\ [\ R\ \text{Sin}\ h\ /\ \text{Cos}\ \phi - \text{Sin}\ \delta\ \text{Tan}\ \phi\ /\ R\ ]\ /\ \text{Cos}\ \delta\ \}$$

(iii) To find the lengths of evening and morning twilight, use:

$$r \text{ or } s = N - K\{ [f_\phi(x) + g_\phi(\delta)], \delta \}$$

for $x = 19°$ and $17°$, respectively. These are the parameters for the angle of depression at twilight used in al-Khalīlī's prayer-tables.

(iv) To find the solar rising amplitude, ψ, use:

$$\psi = 90° - K\{ g_{45}(\delta), \phi \} ,$$

or solve:

$$g_{45}(\psi) = f_\phi(\delta) .$$

Now **F14** in **I-1.2** can be expressed in the form:

$$\psi = 90° - \text{arc Cos } \{ \text{Sin } \delta \text{ Tan } 45° / \text{Cos } \phi \}$$

and also:

$$\text{Sin } \psi \text{ Tan } 45° / R = R \text{ Sin } \delta / \text{Cos } \phi ,$$

so the procedure is valid. Note that $g_{45}(x)$ is the sine function to base 60.

(v) To find the altitude of a celestial body in the prime vertical, h_0, use:

$$h_0 = 90° - K \{ g_{45}(\delta), \bar{\phi} \}$$

(the Paris copy incorrectly states that $h_0 = K \{ g_{45}(\delta), \bar{\phi} \}$, where $\bar{\phi}$ is the complement of ϕ) ,

or solve:

$$g_{45}(h_0) = f_\phi(\delta) .$$

Now **F13** in **I-1.2** yields:

$$h_0 = \text{arc Sin } \{ \text{Sin } \delta \text{ Tan } 45° / \text{Cos } \bar{\phi} \} ,$$

and also:

$$\text{Sin } h_0 \text{ Tan } \phi / R = R \text{ Sin } \delta / \text{Cos } \phi ,$$

so that again al-Khalīlī's procedure is correct.

(vi) To find the solar azimuth, a, measured from the meridian, use:

$$a = K \{ [g_\phi(h) - f_\phi(\delta)] , h \}.$$

The standard medieval formula **F15** in **I-1.2** can be rendered:

$$a = \text{arc Cos } \{ R [\text{Sin } h \text{ Tan } \phi / R - R \text{ Sin } \delta / \text{Cos } \phi] / \text{Cos } h \}$$

which is mathematically equivalent.

(vii) To find the solar declination from an observed pair of values (h,a), when the local latitude is known, first solve for x in:

$$a = K(x,h)$$

and then solve for δ in:

$$f_\phi(\delta) = g_\phi(h) - x .$$

This procedure is the reverse of that in (vi) above.

(viii) To find both the solar declination and the local latitude from two such pairs, subscripted 1 and 2, observed on the same day, first solve $a = K(x,h)$ for x_1 and x_2. Then define the quantity:

$$\Delta x = x_2 - x_1$$

called the argument (*ḥiṣṣa*), and the quantity:

$$\Delta g = g_{45}(h_2) - g_{45}(h_1) .$$

Next find the quantity ξ, called the correction (*taʿdīl*), defined by:

$$g_{45}(\xi) = \Delta g .$$

Then solve for ϕ in:

$$g_\phi(\xi) = \Delta x .$$

The declination may then be found from $f_\phi(\delta) = g_\phi(h) - x$, as in (vii) above. These procedures are easily verified by projecting the celestial sphere onto the meridian plane.

(ix) To find the declination of a star, Δ, from its ecliptic coordinates (λ,β), first find λ'', the longitude difference between the star and the nearest solstice, and then solve for x in:

$$\lambda'' = K(x,\beta) .$$

The declination can then be found by solving:

$$f_{37}(\Delta) = {}^1/_2 \{ 2g_{49}(\beta) + x \} .$$

To explain this procedure, let ε denote the obliquity of the ecliptic. The modern formula[26] can be transformed thus:

$$\sin \Delta / \sin \varepsilon = \sin \beta \cos \varepsilon / \sin \varepsilon + \cos \beta \sin \lambda .$$

In order to adapt this formula to serve al-Khalīlī's tables, write:

$$R \text{ Sin } \Delta / \text{Cos } \phi_1 = {}^1/_2 \{ 2 \text{ Sin } \beta \text{ Tan } \phi_2 / R + \text{Cos } \beta \text{ Sin } \lambda / R \}.$$

It follows that the values ϕ_1 and ϕ_2 are defined by:

$${}^1/_2 \cos \phi_1 = \sin \varepsilon \quad \text{and} \quad 2 \tan \phi_2 = \cot \varepsilon .$$

For al-Khalīlī's parameter $\varepsilon = 23;31°$, the accurate values of ϕ_1 and ϕ_2 are 37;4° and 48;58°: he used the approximations 37° and 49°.

(x) To find the normed right ascension α' of a star, use:

$$\alpha'' = K \{ [2f_{37}(\beta) - 2g_{49}(\Delta)], \Delta \}$$

where α'' is the normed right ascension measured from $\alpha' = 0°$ or $\alpha' = 180°$. To justify this, we write the modern formula for the right ascension as:

$$\alpha = \text{arc Sin} \{ R [\text{Sin } \Delta \text{ Cot } \varepsilon / R - R \text{ Sin } \beta / \text{Sin } \varepsilon] / \text{Cos } \Delta \},$$

so that, with ϕ_1 and ϕ_2 as defined in (ix) above:

$$\alpha' = \text{arc Cos} \{ - R [2 \text{ Sin } \Delta \text{ Tan } \phi_2 / R - 2 R \text{ Sin } \beta / \text{Cos } \phi_1] / \text{Cos } \Delta \}$$
$$= \text{arc Cos} \{ R [2 R \text{ Sin } \beta / \text{Cos } \phi_1 - 2 \text{ Sin } \Delta \text{ Tan } \phi_2 / R] / \text{Cos } \Delta \},$$

al-Khalīlī's handling of these problems can, by any criteria, only be described as brilliant!

10.8 The universal qibla table

Three copies of al-Khalīlī's corpus contain a qibla table, namely, MSS Paris ar. 2558, fols. 53v-61r, copied 811 H [= 1411]; Berlin Ahlwardt 5754 (Wetzstein 1138, fols. 47v-55r), *ca.* 1700; and Paris BNF ar. 2560,11, fols. 148v-159v, *ca.* 1750.[27] The tables in the first two of these sources are preceded by a short introduction but in the third source there is no introduction and indeed no indication of the nature of the table.

The function tabulated by al-Khalīlī is here denoted by q, although in fact it defines the *acute* angle which the arc joining a given locality to Mecca makes with the local meridian. (See **Fig. VIa-3.1.**) Values of the function $q(\phi,L)$ are given in two sexagesimal digits for the domains:

$$\phi = 10°, 11°, \ldots , 56° \text{ and } 33;30°, \text{ and } \Delta L = 1°, 2°, \ldots, 60°.$$

[26] Smart, *Spherical Astronomy*, p. 40.

[27] My detailed study listed as King, "al-Khalīlī's Qibla Table" (1975) contains a critical edition of the table based on the first Paris manuscript and the Berlin one (pp. 87-94). The second Paris manuscript came to my attention too late to be used for the edition (see *ibid.*, p. 84, n. 10).

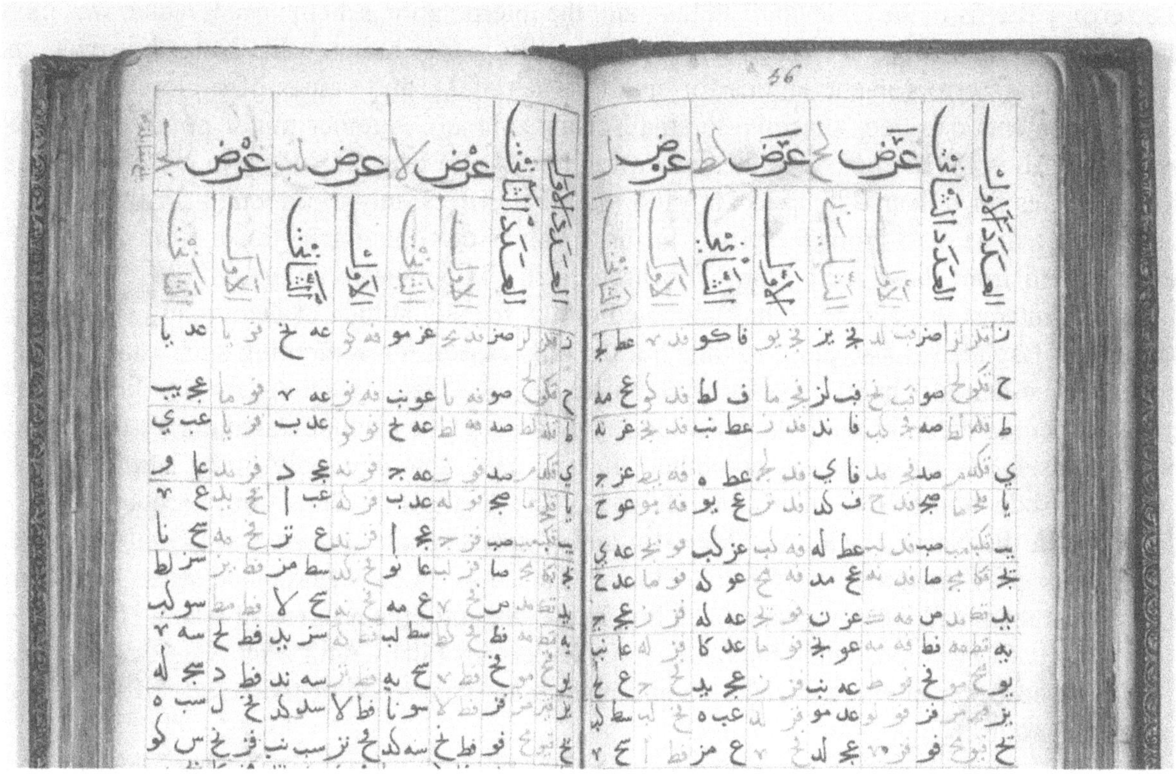

Fig. 10.8: An extract from al-Khalīlī's universal qibla table serving latitudes 28°-33°. The entries in light (actually red) ink indicate that the qibla is northerly. [From MS Paris BNF ar. 2558, courtesy of the Bibliothèque Nationale de France.]

For each value of ϕ there is a sub-table in which L is entered vertically – see **Fig. 10.8** for an extract. The double vertical argument runs from 7° to 66° (longitudes west of Mecca, for which the qibla is easterly) and from 127° to 68° (longitudes east of Mecca, for which the qibla is westerly). These two sets of longitudes correspond to a change in ΔL from 60° to 1°, since the longitude of Mecca is taken as 67;0°, and the underlying latitude of Mecca is found by inspection (see below and also **10.9**).

The entries for each latitude are arranged in two columns of thirty entries, and two facing pages of the manuscripts serve six consecutive latitudes. There are 2,880 entries in the entire table.

In the introduction (see below) he mentions that the qibla is southerly if the entry in the table is in red ink and northerly if it is in black ink. In the first sub-tables all entries are northerly, and in the remaining five sub-tables all entries above the broken line are also northerly; all other entries are southerly.

al-Khalīlī's qibla values are remarkably accurate: the errors in most of the table are less than ±0;2°, although larger errors do occur in certain sections. It is reasonable to assume that al-Khalīlī would have computed a matrix of values for, say, each degree of ϕ and each 5° of ΔL, and that he would have then used an interpolation scheme to compute intermediate values.

The error patterns in the table confirm this, and the interpolation scheme used, which was more sophisticated than a second order scheme, generally works rather well. Note, however, the relatively large errors in the entries for small values of ΔL and values of ϕ close to ϕ_M. For these arguments q is very sensitive to small changes in ΔL. Another minor point to be noted is that for $\Delta L = 1°$ (L = 66°, 68°) the entries for values of ϕ from 21° down to 17° are identical to the entries for ϕ from 22° up to 26°, and the qibla differs only in direction for these values close to ϕ_M. It is clear from the errors in these values that the entries for ϕ from 21° to 17° were copied from those for ϕ from 22° through 26°, which were computed.

It is of interest that al-Khalīlī has tabulated q(ϕ,L) for the latitude of Damascus, 33;30°. The latitude of Baghdad is also close to this, but the only reason for computing such a table would appear to be that al-Khalīlī had already tabulated his auxiliary functions for this latitude and he used them to compile his qibla table. On the other hand, he also tabulated q(ϕ,L) for latitude 56°, but as far as we know, he computed the auxiliary functions only up to latitude 55°.

The introduction to the qibla table (fols. 52v-53r of the first Paris copy) reads:

> "In the Name of God the Merciful and Compassionate. Praise be to God who made us pray towards His Sacred House (*i.e.*, the Kaʿba). May God bless our Lord Muhammad, who was sent (as a Prophet to institute) the religion of Islam, His Family, His Companions, and His Wives – noble persons (all) – with a blessing that will heal whosoever utters it of sickness and make him want to be in the Abode of Peace (*i.e.* Paradise).
>
> I have not seen a better method for (determining) the azimuth of the qibla than the one mentioned by Abū ʿAlī al-Marrākushī in the sixty-seventh chapter of his "Treatise on the Sine (Quadrant)" (*Risālat al-Jayb*). If the moon is at some instant (directly) above the Kaʿba beginning to be eclipsed then people looking at it from (different) parts of the earth will see it in different parts of the sky relative (to the meridian of) the (particular) location. One person will see before it culminates, another will see it culminating in the south or the north, another will see it after it culminates. Anyone who faces the moon at that time will be facing the qibla. If the declination of the moon at that time is known, and it is equal to the latitude of Mecca, and the hour-angle corresponding to its position (relative to the meridian) in any locality is known, and it is equal the difference in longitude (between the locality and Mecca), the altitude of the moon will be known, and similarly its azimuth, that is, the azimuth of the qibla. If the azimuth of the qibla is subtracted from 90°, the remainder will be the azimuth of the qibla measured from the local meridian (*al-inḥirāf*).
>
> Almighty God has enabled me to compile a table for finding the meridian azimuth of the qibla from the longitude and latitude of the locality on the basis that the longitude of Mecca – may God honour it – is 67° and the latitude is 21;30°. The way to use this table is to look for the page on which is written the latitude of the locality and to enter in the first or second argument with its longitude. Then move vertically and horizontally until the column and linc meet and you will find the azimuth of the qibla for that locality. If the longitude is in the first longitude argument column then take the first entry for the azimuth, and similarly for longitudes in the second column. The

azimuth is southerly if the entry is written in black and northerly if it is written in red. It is easterly if the longitude of the locality is less than 67°, otherwise it is westerly.

Example for a locality with longitude 25° and latitude 35°. We find its qibla azimuth from the table to be 81;12° east of south. I performed (the calculations for) this example with a sine quadrant using the method of al-Marrākushī and I found the altitude (of the zenith of Mecca above the local horizon) to be 51° and the qibla to be 81;10°. I performed the operation using another method involving the differences in longitude and latitude and I found (the qibla) to be 68;50°, which is an atrocious error.

I ask God for success (in achieving a goal) which He will like and with which He will be pleased. Verily, He is the Excellent and the Noble. Praise be to God alone. May God bless and save our Lord Muḥammad, his Family and his Companions."

To illustrate how the arguments are to be entered in the table, al-Khalīlī considers the case $\phi = 35°$ and $L = 25°$. The value given in the table for these coordinates (which do not correspond to those of any specific locality) is 81;12° east of south, and is, in fact, accurate. al-Khalīlī also mentions that he used a sine quadrant (*rubʿ mujayyab*), to perform the calculation and that he found the results h = 51;0° and q = 81;10°. Actually this value of h is in error by only -0;2°, as is, fortuitously, the final value of q. al-Khalīlī does not describe the procedure he used to arrive at these results, nor does he tell us the size of the instrument he used, although it is known that the quadrants of his colleague al-Mizzī are between 15 and 20 cm. in radius.[28]

al-Khalīlī goes on to state that when he used the "well-known" approximate method involving latitude and longitude differences he found the "terribly inaccurate" value 68;50° for q. Now there are five known approximate methods involving ΔL and $\Delta\phi$,[29] four of which can be represented in modern notation by:

$$q = \arctan(x / y),$$

where:

(1) $x = \Delta L$ $\quad y = \Delta\phi$

(2) $x = \Delta L \cos \phi_M$ $\quad y = \Delta\phi$

(3) $x = \sin \Delta L$ $\quad y = \sin \Delta\phi$

(4) $x = \sin \Delta L \cos \phi_M$ $\quad y = \sin \Delta\phi$

The fifth method involves a more complicated formula, nowhere explicitly stated. Recomputation yields:

72;11° 70;56° 70;46° 69;27° 70;1°

instead of al-Khalīlī's 68;50°. However, if we assume $\phi_M = 21;0°$ instead of 21;30° the corresponding values are:

71;34° 70;21° 70;7° *68;50°* 69;24°

Thus al-Khalīlī's result was probably computed with formula (4), but it seems that he either made the computation using this other well-attested value for the latitude of Mecca, or he simply overlooked the fraction in the value of ϕ which he used to find $\Delta\phi$.

---

[28] On al-Mizzī see n. 9:21 above. On his quadrants see n. 9:22 and **X-6**.

[29] See King, "Earliest Qibla Methods", p. 90, and *idem*, *Mecca-Centred World-Maps*, pp. 57-59 and 335-336, on these procedures.

Unfortunately al-Khalīlī does not describe the method by which he compiled his table. However, he does remark that he knows of no better method for finding the qibla than the one advocated by Abū ʿAlī al-Marrākushī (**6.7**).[30] This involves first finding the altitude h of the zenith of Mecca above the local horizon using:

$$\text{Sin } h(\phi,L) = \text{Sin } (\bar{\phi} + \phi_M) - \text{Vers } \Delta L \text{ Cos } \phi_M \text{ Cos } \phi \ / \ R^2 \ ,$$

and then finding the corresponding azimuth q using:

$$q(\phi,L) = \text{arc Cos } \{ \ R \ [\ \text{Sin } h \text{ Tan } \phi \ / \ R \ - \ R \text{ Sin } \phi_M \ / \ \text{Cos } \phi \ ] \ / \ \text{Cos } h \ \}.$$

al-Khalīlī's universal auxiliary tables can also readily generate qibla values for a particular value of ϕ. First calculate $h(\phi,L)$, an operation simplified by the fact that:

$$\text{Sin } h = c_1 + c_2 \text{ Cos } \Delta L,$$

where c_1 and c_2 are constants. Then compute the quantity:

$$x(\phi,L) = g_\phi(h) - f_\phi(\phi_M)$$

using the first two auxiliary tables and enter this as horizontal argument in the third table since:

$$q(\phi,L) = K \ \{ \ x(\phi,L) \ , \ h \ \}.$$

Although it cannot be established that this procedure was actually followed by al-Khalīlī, by using his auxiliary tables it is possible to generate qibla values which are less accurate than, the same as, or more accurate than his own.[31] Van Brummelen has noted that the errors in al-Khalīlī's table of $f_\phi(\theta)$ for $\phi = \phi_M$ are not reflected in his qibla table, and he agrees that al-Khalīlī did not use his auxiliary tables to compute the qibla table.[32]

10.9 List of qibla values

Some copies of the Damascus corpus, for example, MS Paris ar. 2558, fol. 51v, contain a table of geographical coordinates "copied from the handwriting of ... al-Mizzī" with associated qibla values "computed by Shams al-Dīn al-Khalīlī": see **Fig. 10.9**.[33] In this copy the first four entries (Mecca, Medina, Cairo and Alexandria) are doubtless later additions, firstly because the title in the Egyptian copy MS Cairo MM 167, fol. 204r (dated 989 H [= 1581/82]), indicates that the localities belong to "the climate of Syria to the (further) edge of al-ʿIrāq", but secondly because the qiblas are carelessly computed. The entries in the table as found in the Paris and Cairo copies are presented below. Clearly the final qibla values offer some measure of control over copyists' mistakes.

The final column in the edited table displays values computed using linear interpolation in al-Khalīlī's qibla table. It is clear that these are slightly more accurate than those in the list of qiblas, which indicates that al-Khalīlī did not use his qibla table to compile them.

---

30 For details of al-Marrākushī's procedure and an attempt to recompute al-Khalīlī's qibla values see King, "al-Khalīlī's Qibla Table", pp. 99-108.

31 See the examples listed *ibid*., pp. 106-108.

32 Van Brummelen, "al-Khalīlī's Auxiliary Tables", pp. 688-689, and *idem* & Butler, "Interdependence of Astronomical Tables", pp. 46-48.

33 Kennedy & Kennedy, *Islamic Geographical Coordinates* (pp. xxii *sub* KHL and xxv *sub* MIZ) have determined that the list is related to an early set of geographical tables recorded by Abu 'l-Fidā' (*fl.* Hama, *ca*. 1310 – see n. 2:33). This table is now published in King, *Mecca-Centred World-Maps*, App. F7.

Fig. 10.9: The geographical tables in al-Khalīlī's Damascus corpus. [From MS Paris BNF ar. 2558, fols. 51v-52r, courtesy of the Bibliothèque Nationale de France.]

The geographical table of al-Mizzī
with qibla-values computed by al-Khalīlī
A: MS Paris BNF ar. 2558; B: MS Cairo MM 167

Notes: The entries Δq are the errors compared with computation using the accurate formula. The entries q* are those derived by linear interpolation in al-Khalīlī's qibla-table and Δq* the corresponding errors.

	Locality	L	φ	q	Δq	q*	Δq*
1	Shawbak	56; 0	31; 0	48;42	[+3]	48;42	[+3]
2	Gaza	57; 0	32; 0	42;46	[+1]	42;46	[+1]
3	Ramla	56;50	32;10	42;56[a]	[+8]	42;50	[+2]
4	Hebron	56;30	31;35	45;21	[-6]	45;29	[+2]
5	Jerusalem	56;30	32; 0	44;14	[-2]	44;16	[0]
6	Nablus	57;30	32;10	40;38	[-6]	40;44	[0]
7	Baysan	58; 0	32;50	37;[2]5[b]	[0]	37;26	[+1]
8	Tiberias	58;15	32; 5	38;34	[+4]	38;30	[0]
9	Ajlun	58;10	32;10	38;35	[+1]	38;34	[0]
10	Salt	58;10	32; 0	39; 2	[+2]	39; 0	[0]
11	Adhraat	60; 0	31;55	32;40	[+3]	32;39	[+2]
12	Bosra	60; 0	32;15[c]	31;52	[+4]	31;51	[+3]
13	Sarkhad	60;20	32;15[d]	30;31	[-1]	30;33	[+1]
14	Damascus	60; 0	33;30[e]	29; 4	[+1]	29; 7	[+4]
15	Safad	57;35	32;30[f]	39;35	[0]	39;36	[+1]
16	Shaqif Tayrun	57;40	33; 5	37;53	[+1]	37;51	[-1]
17	Baniyas	59; 0	33; 0	33;40	[0]	33;40	[0]
18	Sidon	58;15	33; 0	36;13	[+2]	36;10	[-1]
19	Beirut	59;15	33;20	31;59	[-4]	32; 4	[+1]
20	Baalbek	60; 0	33;50	28;24[g]	[0]	28;27	[+3]
21	Tripoli Sh	59;40	34; 0	29;13[h]	[-1]	29;14	[0]
22	Buluniyas	60; 0	34;45	26;47	[+4]	26;46	[+3]
23	Homs	61; 0	34;20	23;53[i]	[-2]	23;59	[+4]
24	Salamiyya	61;20	34; 0	23;14	[-1]	23;15	[0]
25	Barin	60;45	34;20	24;49	[0]	24;52	[+3]
26	Ḥisn al-Akrād	60;30	34;10	25;58	[-2]	26; 2	[+2]
27	Hama	61;45	34;45	20;32	[-2]	20;35	[+1]
28	Maarra	61;45	35; 0	20;12	[-1]	20;13	[0]
29	Sarmin	61;50	35;45	18;50	[-7]	19; 0	[+3]
30	Lattakia	60;40	35;25	23;21	[-3]	23;26	[+2]
31	Sahyun	60;10	35;25	25; 3	[-1]	25; 6	[+2]
32	Antioch	60;35	35;50	23; 4	[0]	23; 6	[+2]
33	Aleppo	62;10[j]	35;50	17;42	[0]	17;44	[+2]
34	al-Bāb / Buzaa	62;15	36;10	17; 1	[-2]	17; 4	[+1]
35	Bira	62;30	36;50	15;31	[-1]	15;33	[+1]
36	al-Rahba	64;30	36; 0	9;14	[0]	9;13	[-1]
37	Mardin	64; 0	37;55	9;48	[0]	9;48	[0]
38	Sinjar	66; 0	36;20	3;38	[0]	3;41	[+3]
39	Baghdad	70; 0	33;25	13;19	[0]	13;19	[0]
40	Anbar	69;30	33;15	11;19	[+1]	11;18	[0]
41	Kufa	69;30	31;30	13;12	[+1]	13;12	[+1]

42	Basra	74; 0	30; 0	38;11	[+2]	38; 7	[-2]
43	Tustar	74;30	31;30	35;41	[+4]	35;37	[0]
44	Shiraz	78; 0	29;36	53; 8	[-8]	53;20	[+4]
Additional entries in A only:							
45	Mecca	67; 0	21;30	-			
46	Medina	65;20	24;40	??;40	[?]	26;23	[+13]
47	Cairo	54;40	30; 0	55;58[k]	[+34]	55;24	[0]
48	Alexandria	51;54	31; 0	58;35	[+5]	58;30	[0]

Notes: [a] B: 42;36°; [b] A/B: 37;35°; [c] B: 32;55°; [d] B: 32;55°; [e] B: 32;55° (!), a value not attested in other known source; [f] B: 33;30°; [g] B: 28;30°; [h] A: 29;53°; [i] B: 26;47°; [j] B: 60;10°; [k] The standard approximate formula yields q*: 55;19°, and the 55;58° given here may be a copyist's error for 55;18°.

Notice that the qibla value 29;4° for Damascus (in error by only +1′) is at variance with the value 28;57° which underlies the table of solar altitude in the azimuth of the qibla in al-Khalīlī's prayer-tables (**10.6**). Different approximate formulae as well as the exact formula have been applied to a number of sets of medieval geographical coordinates for Damascus and Mecca to no avail. Two pairs of values, however, which yield precisely 28;57° using the exact formula are:

$$L = 60;\ 0^\circ \text{ and } \phi = 33;33^\circ \qquad L_M = 67;\ 0^\circ \text{ and } \phi_M = 21;30^\circ$$

$$L = 60;\ 0^\circ \text{ and } \phi = 33;24^\circ \qquad L_M = 67;\ 0^\circ \text{ and } \phi_M = 21;20^\circ$$

The value 33;33° is not attested in any known geographical tables, but the value 33;24° is found in an early anonymous *zīj* which unfortunately has no coordinates for Mecca,[34] and is also used by Ibn al-Mushrif (**I-7.3.1** and **9.8** and **II-9.10**). The value 21;20° was used by Ibn al-Shāṭir for Mecca. Perhaps al-Khalīlī used the second set of parameters to derive 28;57° for the qibla at Damascus; perhaps he did not.

Opposite this geographical table in the first Paris manuscript (and also in MS Paris BNF ar. 2560,13, fol. 164v, *ca.* 1750) is a list of stations on the pilgrim road from Damascus to Mecca (*manāzil al-Ḥijāz al-Sharīf ʿala 'l-darb al-Shāmī*) with their respective latitudes: see also **Fig. 10.9**.[35] It is unlikely that al-Khalīlī made these measurements himself, and the list should be compared with similar ones in other sources (*e.g.* MS Cairo MM 167, fol. 203v). In some (*e.g.* Cairo MM 167, fol. 203r), we find also a list of pilgrim stations on the road from Cairo to Mecca. See also **8.4**.

10.10 The arithmetical and trigonometric tables available to al-Khalīlī

There can be no doubt that for his calculations al-Khalīlī used a sexagesimal multiplication table displaying products:

$$m \times n \quad \text{for} \quad m, n = 1, 2, 3, \ldots, 59 .$$

Such tables had been in use amongst Muslim astronomers since at least the 10th century. He may also have made use of a table of sexagesimal quotients m/n for the same arguments. The

[34] Kennedy & Kennedy, *op. cit.*, p. 473 (*sub* UTT). See also n. **I**-7:28.

[35] On the pilgrim routes see EI_2, III, p. 34 (in the article "Ḥadjdj" by Bernard Lewis), and Brice *et al.*, eds., *Atlas*, p. 22.

more extensive Islamic tables for arithmetical operations with sexagesimals were probably compiled after his time. See **I-1.2**.

The contemporary *Zīj* of Ibn al-Shāṭir contains tables of the Sine function (base 60) to four sexagesimal places for each degree of argument, and the Tangent function (base 60) to three places for each half degree. Both tables also display first differences. In addition Ibn al-Shāṭir tabulated the Cotangent to bases 7 and 12 to three places for each degree of argument and the inverse Cotangent function.[36]

Tables of the Sine and Cotangent function with values for each minute of arc computed to four or five sexagesimal places had been available since the 10th century. The *muwaqqit*s of Cairo also used tables of the Sine and Cotangent for each minute of argument to three places.[37] Such tables would have been available to al-Khalīlī. Yet the only table of the Sine function in the available manuscripts gives values to only two sexagesimal digits for each degree of argument (MS Dublin CB 4091, fol. 1v), and, rather surprisingly, is not without mistakes. This may be compared with the values of al-Khalīlī's function $g_\phi(\theta)$ (**10.7**) for $\phi = 45°$, which is identical to the Sine function. Van Brummelen has determined that al-Khalīlī used Sine and Cotangent tables with values to two sexagesimal places in the computation of his auxiliary functions.[38]

It is also possible that al-Khalīlī used tables of the Secant function and the inverse Sine / Cosine function since these functions are fundamental to his tables. But while such tables are attested in the Islamic sources,[39] they do not occur in any known copies of the Damascus corpus.

Only two other trigonometric tables are found in the various manuscripts of al-Khalīlī's tables, both relating to the Cotangent function. In the first (MS Paris ar. 2558, fol. 51r) the base used is 20. Values are given to two digits for each 1° of altitude from 1° to 80° (this being the maximum integral solar altitude at Damascus). It is likely that al-Khalīlī had a gnomon divided into 20 units for which such a table could have been useful. I know of only one other attestation of the use of base 20 in the medieval sources.[39a]

In the second trigonometric table (MS Berlin Wetzstein 1138 (Ahlwardt 5754-6), fol. 59r) the base used is 10. Values are given to two digits for each 1° of altitude from 1° to 90°. No reason is given for the choice of 10 as base, not even in the instructions (fols. 60r-61r) on the use of the shadow tables. The base 10 was also used in the trigonometric tables of the 16th-century Istanbul astronomer Taqi 'l-Dīn ibn Maʿrūf.[40] In the tables in the Damascus corpus the entries are expressed sexagesimally, whereas in Taqi 'l-Dīn's tables they are expressed decimally.

Possibly al-Khalīlī used some kind of interpolation scheme to compile his tables, which contain some 40,000 entries. Several descriptions of second-order interpolation schemes are

---

36 Kennedy, "*Zīj* Survey", pp. 162b-163a.

37 For copies see *Cairo ENL Survey*, no. C137.

38 See the text to n. 10:24.

39 For examples of Islamic tables of the Secant function see **I-6.9** and **8.4**.

39a The 15th-century Egyptian astronomer al-Wafāʾī, in his treatise on the universal astrolabe of Ibn al-Sarrāj (see n. 9:12), proposes 20 for the length of the gnomon: see Charette & King, *Universal Astrolabe of Ibn al-Sarrāj* (forthcoming).

40 On these tables by Taqi 'l-Dīn (**I-2.3.6** and **II-14.9**) see King, "Islamic Multiplication Tables", B, pp. 413-414.

known from the Islamic sources,[41] including one advocated by the 10th-century Cairo astronomer Ibn Yūnus and ideally suited for generating tables of trigonometric functions.[42] Nevertheless Glen Van Brummelen's detailed analysis of the error patterns in al-Khalīlī's tables has not cast any light on the procedures that he used.[43]

10.11 The influence of al-Khalīlī's tables

Since al-Khalīlī's auxiliary tables for solar timekeeping survive in their original form only in a single manuscript (MS Dublin CB 4091), it is probable that they were not widely used in Syria. However it is known that the tables travelled as far as Tunis, Tlemcen, and Cairo, where they were either extended or modified. In each case the attribution to al-Khalīlī was suppressed (or at least is not found in the surviving copies).

Tunisian redaction

MS Cairo DM 689 contains an extensive set of auxiliary tables copied in an elegant Maghribi hand *ca.* 1600.[44] They conclude with a star catalogue dated 801 H [= 1398], and appear to have been compiled in Tunis. The title folio, instructions, and first few tables are missing from the manuscript, which begins with the last page of a set of tables displaying the solar longitude for each day of a period of four Syrian years. The tables for solar timekeeping follow.

The main functions tabulated are al-Khalīlī's $B'(\lambda,\phi)$ and $V'(x,y)$. In fact these Tunisian tables are merely an extension of al-Khalīlī's tables, and the corresponding entries in both sets are the same. The Tunisian tables of B' are computed for each degree of ϕ from 1° to 48° and also 21;40° (Mecca) and various other non-integral latitudes between 30° and 40° intended to serve localities in Ifdrīqiyya and the Maghrib and perhaps also Sicily. al-Khalīlī had a separate table for latitude 33;30° (Damascus), and the Tunisian set has separate tables for latitudes:

30;30°, 31;30, 32;30, 33;30, 33;40, 34;30, 35;30,
36;30, 36;40, 37;10, 37;30, 38;30 and 39;30.

The Tunisian tables of $V'(x,y)$ are simply those of al-Khalīlī rearranged so that the horizontal argument is increasing.

The tables of B' are preceded by tables of the function Sin $H(\lambda)$ (called *jayb al-ghāya*) computed to two digits for each degree of λ and the latitudes between 30° and 38° for which B' is tabulated. These were probably used by the anonymous Tunisian astronomer to compile his tables of $B'(\lambda)$, since:

$$B'(\lambda,\phi) \;=\; \tfrac{1}{2}\,[\ \sin H(\lambda,\phi) + \sin H(\lambda^*,\phi)\ ],$$

where $\lambda^* = 180° + \lambda$. The underlying value of ε is 23;35°, whereas al-Khalīlī used 23;31°.

However, the change in ε hardly affects the values of $B'(\lambda)$ given to two digits. The Tunisian tables also contain a table of Sin $H(\lambda)$ for $\phi = 0$, which is simply Cos $\delta(\lambda)$. See further **13.4**.

---

[41] See n. **I**-1:21. MS Cairo Zakiyya 917,4 (pp. 50-54, copied 1163 H) contains some notes on interpolation attributed to al-Khalīlī but I have been unable to obtain a microfilm of it.

[42] King, "Astronomical Timekeeping in Medieval Cairo", pp. 354-357, and Hamadanizadeh, "Islamic Interpolation Schemes", p. 146.

[43] See the text to nn. 10:23-24.

[44] Listed in *Cairo ENL Survey*, no. F30.

Tlemcen extract

MS London BL Or. 411,2 contains the only known complete copy of a commentary on the astronomical poem of the late-14th-century Maghribi writer ʿAbd al-Raḥmān ibn Muḥammad al-Jādarī.[45] This anonymous commentary was written in Tlemcen and contains several spherical astronomical tables computed for parameters:

$$\phi = 35;0° \text{ (Tlemcen) and } \varepsilon = 23;35° .$$

The same manuscript also contains interesting historical accounts of trepidation, twilight, and the obliquity of the ecliptic. Among the spherical astronomical tables two display the functions:

$$\text{Sin } H(\lambda) \text{ and } {}^1/_2\, B(\lambda),$$

and another displays the function:

$$t\ \{\ {}^1/_2\, B, (\text{Sin } H - \text{Sin } h)\ \}$$

for the same latitude. Unfortunately it has not been possible to compare these tables with the Tunisian corpus of auxiliary tables (see above), from which they may have been lifted. See further **13.6**.

Egyptian version

Two Egyptian manuscripts (Istanbul S. Esad Efendi Medresesi 119,2 and Cairo DM 644,1) contain an anonymous set of auxiliary tables entitled *Fatḥ al-Karīm al-Bāqī fī maʿrifat al-dāʾir wa-faḍlihi āfāqī* ("The Victory of God the Noble and Eternal for Finding the Time since Sunrise and Hour-Angle for all Latitudes").[46] The main functions tabulated are:

$$B'(\lambda,\phi) = {}^1/_2\, B(\lambda,\phi) = {}^1/_2 \text{ Cos } \delta(\lambda) \text{ Cos } \phi \ / \ R$$

for the domains:

$$\lambda = 1°, 2°, \dots, 90°;\ \phi = 3°, 6, \dots, 27, 28, 30, 32, 33;30, 36, 38, \dots, 48°$$

and:

$$V'(x,y) = \text{arc Vers } \{\ R \cdot y \ / \ 2x\ \}$$

for the domains:

$$x = 19, 20, \dots, 30 \text{ and } y = 0;5, 0;10, \dots, 60;0.$$

The first function is labelled *al-aṣl*, "(half) the base" and the arguments in the table of V′ are called *al-aṣl al-muʿaddal*, "the modified base", and *faḍl al-jaybayn*, "the difference between the two Sines". The entries in both tables were simply coped from al-Khalīlī, in whose set B′ is tabulated for some 50 values of ϕ and the increment in the argument y for the tables of V′ is 0;10 rather than 0;5. Also, this later Egyptian set gives no horizontal differences in the tables of V′, but the simplified format makes them slightly easier to use. See further **I-9.11**.

al-Khaṭāʾī's auxiliary tables

The unique source MS Vatican Borg. ar. 217,2 (fols. 6r-7r) contains part of a set of auxiliary tables intended for use in Cairo, attributed to Muḥammad ibn al-Amīr Fakhr al-Dīn ʿUthmān al-Khaṭāʾī (*ca.* 1475). Some minor tables display:

$$\text{Sin } \theta, \text{ Sin } H(\lambda), \text{ and } B(\lambda),$$

---

[45] On al-Jādarī see **I-6.15.2** and **II-13.6**.

[46] Listed in *Cairo ENL Survey*, no. C144.

each to three digits for each degree of argument. The second and third functions are computed for parameters:

$$\phi = 30;0° \text{ (Cairo) and } \varepsilon = 23;35° .$$

The main set, incomplete in the Vatican manuscript, displays the function:

$$V(x,y) = \text{arc Vers } \{ R \cdot y / x \} ,$$

with values to two digits for the domains:

$$x = 47;37, 47;57, \ldots , 51;57 \text{ and } y = 1, 2, \ldots , Y(x),$$

where Y(x) is a certain maximum defined below (see an extract in **Fig. I-9.9b**). The instructions for finding t(h,λ) indicate that one should first find B(λ) and z(h,λ) (*i.e.*, Sin H(λ) - Sin h) and enter these arguments in the main table. This is valid since:

$$t(h,\lambda) = V \{ B(\lambda), z(h,\lambda) \} .$$

The argument x runs in intervals of 0;20 between the limits of B(λ) for Cairo, and the argument y runs in unit intervals up to Y(x), the greatest integer less than the value of Sin H(x). The table originally contained about 825 entries, but in the Vatican manuscript only the page for y > 31 remains. The entries are fairly accurately computed.

Being devised for latitude 30° al-Khaṭā'ī's tables are not as useful as al-Khalīlī's minor auxiliary tables for solar timekeeping. It is probable that al-Khaṭā'ī knew of al-Khalīlī's tables, but the former's tables were redundant anyway because by the 13th century there existed a complete set of tables of t(h,λ) for Cairo. See further **1-6.15.1** and **II-6.8**.

Prayer-tables

With the passage of time only al-Khalīlī's calendrical tables and the solar longitude table needed to be updated. The decreasing obliquity (23;30° observed in *ca.* 1440 by Ulugh Beg[47] and 23;29° used by later Ottoman astronomers[48]) was not considered sufficient cause to recompute the corpus, or even the prayer-tables.

After the time of al-Tīzīnī, the Damascus *muwaqqit* Ibn al-Kayyāl provided a new set of calendrical and solar tables to precede al-Khalīlī's tables for timekeeping.[49] These are extant in numerous copies including MSS Berlin Ahlwardt 5758 (Sprenger 1858), 5759/5771 (Wetzstein 1146), and 5760/5772 (Wetzstein 1148), none of which credit al-Khalīlī as the author of the main tables. The same is true of the 19th-century Damascus *muwaqqit* Muḥammad ibn Muṣṭafā al-Ṭanṭāwī,[50] whose recension is contained in several copies including MSS Damascus Ẓāhiriyya 9233, Cairo ṬM 173, Cairo TR 129, and Cairo DM 1007. al-Ṭanṭāwī, however, modified the tables so that the entries displayed actual times in hours and minutes according to the Ottoman convention that sunset is 12 o'clock. Thus al-Khalīlī's tables survive in three versions: the original, and the two recensions of Ibn al-Kayyāl and al-Ṭanṭāwī. See further **11.4** and **11.13**.

Besides these modifications to al-Khalīlī's corpus, various new sets of prayer-tables were compiled for a variety of localities after the 14th century (see further **11**). Mention may be made

---

47 Kennedy, "*Zīj* Survey", p. 166b.
48 See, for example, King, "Astronomical Timekeeping in Ottoman Turkey", p. 251.
49 On Ibn al-Kayyāl see n. 11:8.
50 On al-Ṭanṭāwī see n. **I**-2:27.

of the anonymous prayer-tables for latitude 34;20° (Tripoli ?) in MS Cairo ṬM 228,3; for latitudes 41;15° (Istanbul) and 36° (Aleppo) in MS Cairo ṬM 255,6; for latitude 33;45° (Tripoli) in MS Damascus Ẓāhiriyya 4893; and for latitude 41;0° (Istanbul) and 21;30° (Mecca) in MS Cairo DM 184. All of these are based on obliquity 23;31°, which betrays the influence of Ibn al-Shāṭir and al-Khalīlī. On the other hand, ʿAbd al-Fattāḥ al-Dīsṭī's tables for latitude 34;30° (Lattakia) in MSS Aleppo Awqāf 911 and Leiden Or. 2808,2 are based on obliquity 23;35°, which betrays Egyptian influence, and have the format of the Cairo corpus. The later tables in MS Aleppo Awqāf 943 for latitude 35;50° (Aleppo) are based on obliquity 23;30°, which is the parameter of Ulugh Beg (rounded to minutes).

In MSS Oxford Seld. Supp. 100 and Paris BNF ar. 2521 there are separate tables of various supplementary functions in the same format as the Cairo corpus. These functions are:

$$(r+D),\ (D+2N),\ (2N-s),\ (2N-s+D),\ (D+s), \text{ and } (D+2n-r)\ .$$

Although not attributed to al-Khalīlī, the entries are based on those in his main set of prayer-tables.

Similarly, MS Leipzig 814, fols. 31r-31v, is one remaining folio of a set of prayer-tables for Damascus which was drawn up using al-Khalīlī's tables. The functions displayed are:

$$Z_{(12)},\ H,\ D,\ z_{a(7)},\ \tilde{h},\ \delta,\ d,\ h_a,\ t_a,\ T_a,\ h_b,\ t_b,\ 2N,\ N,\ \psi,\ s,\ r,\ n,\ t_q, \text{ and } 2N^h\ .$$

No other examples of these prayer-tables are known. The function t_b is labelled *ākhir waqt al-ikhtiyār*, "the last time for choosing (to begin the afternoon prayer)" and h_b is called *irtifāʿ al-ʿaṣr al-thānī*, "the altitude at the second afternoon prayer". In al-Khalīlī's main set t_b is called *mā bayn al-ẓuhr wa-ākhir waqt al-ʿaṣr*, "the interval between midday and the last time for the afternoon prayer".

Hour-angle tables

al-Khalīlī's tables of t(h,λ) are supplemented in various copies by a set of tables for T(h,λ). Although it is not certain that these are the work of al-Khalīlī himself (they are not contained in MS Paris ar. 2558), evidence that they might be is contained in MS Cairo MM 71, where both the tables of t(h,λ) and T(h,λ) are attributed to al-Khalīlī while the table of a(h,λ) is attributed to Shihāb al-Dīn al-Ḥalabī. al-Ṭanṭāwī, besides converting the prayer-tables as mentioned above, also converted al-Khalīlī's hour-angle tables so that the entries (given for solar altitudes in both the east and west) were expressed in hours and minutes according to the Ottoman convention, as, for example, in MSS Cairo ṬM 173 and DM 1007. See further **11.13**.

Finally, in MS Damascus Ẓāhiriyya 7387, fols. 57v-60r, of an abridgement of the *Zīj* of Ibn al-Shāṭir by ʿAbd al-Raḥīm al-Qazwīnī of Damascus (*fl. ca.* 1610),[51] there is a set of tables entitled *jadwal al-dā'ir li-rtifāʿ al-shams wa-yuʿraf minhu qaws al-ru'ya min al-makth idhā dakhalta bi-naẓīr al-burj* ..., displaying the time remaining till moonset for lunar altitudes:

$$h = 7°,\ 8°,\ \ldots\ ,\ 16°,$$

and each degree of ecliptic longitude (symmetrically arranged). The entries are from al-Khalīlī's tables of T(h,λ), and, as stated in the title, are to facilitate calculations relating to lunar

[51] On al-Qazwīnī see n. 11:5.

crescent visibility. The tables can be used to find the arc of visibility (*qaws al-ru'ya*), that is, the altitude of the moon above the horizon at sunset, from the difference in setting times of the sun and the moon (*qaws al-makth*). The longitude of the point of the ecliptic which sets with the moon, λ', is computed using the longitude and latitude of the moon and the latitude of the locality, and then the difference in setting times can be determined using tables of oblique ascensions.[52] The tables can then be used to find the lunar altitude h corresponding to this time s, with the modified longitude λ' entered as argument.

Miscellaneous related tables

Shihāb al-Dīn al-Ḥalabī's solar azimuth tables

al-Khalīlī's tables of $t(h,\lambda)$ and $T(h,\lambda)$ for Damascus were supplemented in *ca.* 1425 by Shihāb al-Dīn Aḥmad ibn Ibrāhīm al-Ḥalabī[53] with a complete set of tables of the function:

$$a(h,\lambda)$$

based on al-Khalīlī's parameters. These are contained in MS Cairo MM 71, where values are tabulated side by side with al-Khalīlī's functions, and in MSS Damascus Ẓāhiriyya 9227 and Cairo K 8525 where they are tabulated separately. al-Ḥalabī clearly felt the need to compile a table of $a(h,\lambda)$ to supplement the Damascus corpus even though al-Khalīlī had already tabulated $a(H,h)$ (**10.3c**). Possibly al-Ḥalabī was aware that the tables of $a(h,\lambda)$ for Cairo computed by Ibn Yūnus had been incorporated into some copies of the Cairo corpus, with values of the functions t/T/a tabulated together.[54] See further **11.2**.

Ibn Barakāt's tables for the afternoon prayer

MS Damascus Ẓāhiriyya 9227 of al-Ḥalabī's solar azimuth tables also contains a set of tables attributed to Ibn Barakāt, who is otherwise unknown to us. These tables display the time τ_a remaining to the beginning of the interval for the afternoon prayer as a function of solar longitude and solar altitude in the east, and are computed for the latitude of Damascus. The function tabulated is simply:

$$\tau_\alpha(\lambda,h) = t_a(\lambda) - t(h,\lambda) .$$

For each integral degree of solar longitude the arguments for solar altitude are the integral values such that:

$$h_a < h < H.$$

Similar tables were computed for Cairo.

In MS Berlin Wetzstein 1138 (Ahlwardt 5754-6), fols. 62v-94r, appended to al-Khalīlī's tables, there are also some anonymous tables of a function described as *al-maktūb bi-'l-sawād al-bāqī li-'l-ghurūb wa-bi-'l-ḥamra li-'l-ʿaṣr*, "time remaining until sunset written in black, or time remaining until the beginning of the afternoon prayer, written in red". These tables have the same format and argument domains as al-Khalīlī's hour-angle tables, and it is understood that the altitudes are eastern. The tabulated function is simply:

$$f(h,\lambda) = T(h,\lambda) \quad \text{for} \quad h < h_a(\lambda) \quad \text{or} \quad \tau_\alpha(h,\lambda) \quad \text{for} \quad h \geqslant h_a(\lambda).$$

---

52 For the procedures see King, "Tables for Lunar Crescent Visibility", pp. 186-189.

53 On al-Ḥalabī see n. **I**-2:42.

54 On this see **5.6** (illustrated).

(Note the different format of these tables in MSS Damascus Ẓāhiriyya 9227 and Berlin Wetzstein 1138 (Ahlwardt 5754-6).) See further **11.3**.

Universal auxiliary tables
al-Khalīlī's main auxiliary tables were used for several centuries not only in Syria but also in Egypt, the Maghrib, and Turkey.

Egyptian copies
MSS Cairo MM 43, MM 98, DM 758, and Princeton Yahuda 861,2 are Egyptian copies of these tables. Evidence exists that the leading Egyptian astronomers in the 15th century, men such as ʿIzz al-Dīn ʿAbd al-ʿAzīz ibn Muḥammad al-Wafāʾī[55] and Shams al-Dīn Muḥammad called Ibn Abi ʾl-Fatḥ al-Ṣūfī,[56] had seen the tables. From a passing remark by al-Wafāʾī in his treatise *Khulāṣat al-durar* (MS Manchester Rylands 361, fol. 45r) it can be inferred that the expressions "first and second functions" (*al-maḥfūẓ al-awwal* and *al-thānī*) were well known to contemporary Egyptian astronomers. al-Wafāʾī advocates the use of these functions without even defining them. See further **6.14**.

The Egyptian astronomer Muḥammad ibn Aḥmad al-Bilbaysī wrote a commentary on the auxiliary tables,[57] stating that he found al-Khalīlī's instructions to be long-winded. This commentary, extant in the unique MS Cairo DM 442, copied *ca.* 1700, includes Sharaf al-Dīn al-Khalīlī's instructions about the use of his uncle's tables to compute functions for predicting lunar crescent visibility (also found in MS Berlin Wetzstein 1138 (Ahlwardt 5754-6).

MS Cairo MM 167,8, copied in 989 H [= 1581/82], contains a fragment of an anonymous Egyptian treatise on the sine quadrant with many numerical examples, mostly for latitude 30° (*i.e.* Cairo) and obliquity 23;35°. The author deals at length with the functions f and g and the determination of the qibla (ϕ_M = 21°). He mentions various earlier Mamluk astronomers such as ʿAlāʾ al-Dīn Ṭībughā al-Baklamīshī, (Taqi ʾl-Dīn) Abū Ṭāhir, and Jamāl al-Dīn ʿAbdallāh ibn Khalīl al-Māridīnī,[58] but not al-Khalīlī. MSS Cairo MM 177,2, and Cairo DM 624,2, both copied *ca.* 1700-1750, are two other treatises dealing with al-Khalīlī's functions f and g, but neither of them mentions his name.[59]

Maghribi copy
MS London BL Add. 9599,31 is a late Maghribi copy of al-Khalīlī's universal auxiliary tables. This could be taken as evidence that it also travelled to the Islamic West, but the possibility of a Maghribi copyist working in, say, Cairo, should not be excluded. See further **13.4**.

Turkish copies
MS Istanbul Hamidiye 1453/3 is a copy of the tables prepared about 1465 in Edirne, while MS Istanbul Ayasofya 2590 contains another copy of the tables compiled in 1491 by the

55 On al-Wafāʾī (Suter, no. 437, and *Cairo ENL Survey*, no. C61) see n. **I**-9:29.
56 On al-Ṣūfī (Suter, nos. 447/460, and *Cairo ENL Survey*, no. C98) see n. **I**-9:30.
57 On al-Bilbaysī see *Cairo ENL Survey*, no. C80.
58 On these see *Cairo ENL Survey*, nos. C53, C56, and C47 (also Suter, *MAA*, no. 421).
59 *Ibid.*, nos. C145 and D232.

Istanbul *muwaqqit* Muḥammad ibn Kātib Sinān,[60] preceded by a Turkish translation of al-Khalīlī's instructions. An unnumbered manuscript formerly (*ca.* 1970) in the private collection of the late Professor Buhairi, American University of Beirut contains tables of f and g for $\phi = 41°$ (Istanbul) and also of K for x = 41. The hapless individual who put these three tables together was unaware that the argument x is in no way related to the latitude. See further **14.6**.

Qibla table and list of qibla values

al-Khalīlī's qibla table was not widely used after his time, and no references to it have been found in treatises by any contemporary or later Muslim astronomers. On the other hand, his list of qiblas appears in several later Syrian and Egyptian manuscripts. Even so, the mathematically computed qibla values for various Syrian cities were not always taken seriously by those who built mosques.

In Syria early mosques faced the traditional qibla direction of due south, while later mosques are often oriented at variance to the directions which can be computed from medieval geographical coordinates. This is particularly evident in the new Mamluk city of Tripoli, where the religious architecture faces several different directions.[61] The modern qibla value for Damascus (based on the correct longitude difference from Mecca) is some 15° closer to south than al-Khalīlī's value, but this is irrelevant to any discussion of the orientation of medieval mosques.

No new tables of any real consequence were compiled in Syria after the time of al-Khalīlī (see **11**), just as no new Syrian *zīj*es of consequence were compiled after the time of Ibn al-Shāṭir.

---

[60] On Muḥammad ibn Kātib Sinān see nn. **I**-2:43 and **II**-14:28.
[61] On the whole problem of mosque orientation see now **VIIa**.

CHAPTER 11

LATER SYRIAN TABLES FOR TIMEKEEPING

11.0 Introductory remarks

After the destruction of Damascus in 1401 there was little original in Syrian astronomy, but the activity certainly continued. In the early 15th century Shihāb al-Dīn al-Ḥalabī prepared a redaction for Damascus of the *Īlkhānī Zīj* (Maragha, mid 13th century);[1] at the end of that century ʿAbd al-Raḥmān al-Ṣāliḥī prepared a redaction for Damascus of the *Zīj* of Ulugh Beg (Samarqand, *ca.* 1430);[2] and in the mid 16th century Ibn al-Kayyāl prepared a redaction for Damascus of al-Kāshī's *Khāqānī Zīj* (Samarqand, *ca.* 1438).[3] Recensions of the *Zīj* of Ibn al-Shāṭir (Damascus, *ca.* 1350) were prepared in the mid 16thcentury by Ibn Zurayq,[4] around 1600 by al-Qazwīnī, and even in the 18th century by al-Mukhallālātī.[5] There was also some activity in astronomical timekeeping. Shihāb al-Dīn al-Ḥalabī contributed some solar azimuth tables to the main Damascus corpus (**11.2**), and Ibn al-Kayyāl compiled some calendrical and solar longitude tables which were used with al-Khalīlī's prayer-tables until the 19th century (**11.4**). Various sets of prayer-tables prepared in order to serve other cities in Syria such as Aleppo, Tripoli, Nablus and Lattakia (**11.5-9**). In the 19th century al-Ṭanṭāwī made an attempt to revive traditional astronomy in Damascus, and brought up to date the calendrical and solar tables, which were used alongside al-Khalīlī's prayer-tables (**11.13**).

11.1 Anonymous prayer-tables for Damascus

In MS Paris BNF ar. 2521, fols. 69v-75r, penned in the 16th century, there is an anonymous set of prayer-tables which the title states are for latitude 33;6°. Elsewhere in the same manuscript (fols. 14v-16r), there are prayer-tables based on those of al-Khalīlī (discussed in **10.11**), for parameters ϕ = 33;30° and ε = 23;31°. The anonymous set gives values of the following functions for each degree of λ, a page of tables serving each zodiacal sign:

$2D^h$, δ, H, D, t_a, h_a, T_a, n, $2N$, s and r .

---

[1] On al-Ḥalabī (**5.2**) see n. 11:6. His *Zīj* is preserved in MS Cairo M 226,1 (see *Cairo ENL Survey*, no. C69/ 2.1.12).

[2] On al-Ṣāliḥī see Suter, *MAA*, no. 454; *Cairo ENL Survey*, no. C87; and İhsanoğlu *et al.*, *Ottoman Astronomical Literature*, I, pp. 204, 205, and 225.

[3] On Ibn al-Kayyāl (**11.4**) and n. 11:8. This work of his is extant in MS Dublin CB 4677.

[4] On Ibn Zurayq see Suter, *MAA*, no. 426; Mayer, *Islamic Astrolabists*, suppl., p. 296; *Cairo ENL Survey*, no. C116; and İhsanoğlu *et al.*, *op. cit.*, I, pp. 155-158, no. 79. On various recensions of Ibn al-Shāṭir's *Zīj* see my article "Ibn al-Shāṭir" in *DSB*.

[5] On al-Qazwīnī see *Cairo ENL Survey*, no. D38; and İhsanoğlu *et al.*, *op. cit.*, I, pp. 301-302, no. 160. On al-Mukhallalātī see *Cairo ENL Survey*, no. D90, and İhsanoğlu *et al.*, *op. cit.*, II, pp. 544 545.

It is immediately apparent that the tables are very corrupt, and analyzing them is made difficult by the fact that the individual who compiled them did not understand the argument system used in the tables from which he was copying. Thus the entries in his tables do not display the symmetry which the tabulated functions naturally enjoy.

The table of δ is based on ε = 23;35°, but the values of H are not consistent with the corresponding values of δ or any particular value of ϕ. It seems, however, that the compiler wished his tables to work for ϕ = 33;30°, because most of his values are lifted from al-Khalīlī's prayer-tables. It is also curious that the second digit in these entries for D assumes values only between 0 and 5, which suggests that the units used are sixths of a degree.

An analysis of the entries for a particular longitude will suffice to illustrate their deficiency. We consider λ = 45°, and first note that the corresponding entries for λ = 135°, which should be the same, differ in several instances. The entries 13;23° (λ = 45°) and 13;25° (λ = 135°) for $2D^h$ differ from al-Khalīlī's value 13;30°. The entry 16;26° for δ is accurate for ε = 23;35°, and the entry 72;56° for H corresponds to ϕ = 33;30°. For D our table has 100;2° (λ = 45°), which seems to mean 100;20°, and 100;3° (λ = 135°), that is 100,30°: the accurate value is about 101;15°. The entries 55;38° and 45;48° for t_a and T_a are al-Khalīlī's values for λ = *46°*, but the entry 37;25° for h_a is his for λ = 45°. The entries 108;59° and 157;34° for n and 2N are likewise al-Khalīlī's for λ = 45°. The 23;4° (λ = 45°) and 22;56° (λ = 135°) for s differ from al-Khalīlī's 22;50° (λ = 45°) and 22;54° (λ = 46°), but the 25;50° (λ = 45° and 135°) for r is al-Khalīlī's for λ = *46°*.

The tables are thus full of inconsistencies, and it was doubtless from tables of this kind originally computed for Cairo that those described in **7.6** were copied out in words. Two sets of these corrupt prayer-tables now preserved in Syria, MSS Aleppo Aḥmadiyya 1310 and Damascus 9227, appear to be based on tables computed for Cairo.

11.2 Shihāb al-Dīn al-Ḥalabī's solar azimuth and twilight tables

The Damascus astronomer Shihāb al-Dīn Aḥmad ibn Ibrāhīm al-Ḥalabī (d. 1455)[6] computed a set of tables of the function a(h,λ), measured from the prime vertical, for each integral degree of h and each degree of λ. The underlying parameters, that is, the values used for ϕ and ε, are the same as those used by al-Khalīlī (see **10.5-6**). These tables are preserved in MSS Cairo K 8525, copied in al-Ḥalabī's own hand and thus datable *ca.* 1425, Cairo MM 71,1, copied *ca.* 1600, and Damascus 9227. The first manuscript contains al-Khalīlī's prayer-tables, and al-Khalīlī's values of t(h,λ) are tabulated side by side with al-Ḥalabī's value of a(h,λ). In the second the function a(h,λ) is tabulated separately, and in the third, triplets of values (t,T,a) are given, the first pair of functions being attributed to al-Khalīlī.

al-Ḥalabī's azimuth tables, which are fairly accurately computed, are similar in conception to those of Ibn Yūnus for Cairo, computed over four centuries previously (**4.4** and **5.2**), and to the later azimuth tables computed for Alexandria (**8.5**), but they have al-Khalīlī's format. al-Khalīlī had also compiled azimuth tables for the latitude of Damascus, but having arguments H and h rather than h and λ (**10.3**).

---

[6] On al-Ḥalabī see Suter, *MAA*, no. 434; Brockelmann, *GAL*, II, p. 159; and *Cairo ENL Survey*, no. C69.

In MS Cairo K 8525, copied by al-Ḥalabī himself, he also reproduces al-Khalīlī's tables of α', and α_ϕ and α_s for Damascus. His table of $\alpha_r(\lambda)$ is presented as his own calculation, and differs from that of al-Khalīlī in that it is based on the parameter 20°. al-Ḥalabī further includes (fols. 14r-20r) a set of tables displaying the time remaining until twilight in equatorial degrees and minutes as a function of solar longitude when any of seven prominent stars (which vary from sign to sign) are culminating (*cf.* **I-2.7.1**). In his brief introduction to these tables he states that daybreak (*al-fajr*) is the time of the prayer-call (*al-ādhān*) in Ramaḍān and of the *salām* (written *s-l-m*) in the other months of the year (see further **4.10** on the institutions associated with twilight in medieval Cairo).

11.3 Ibn Barakāt's tables for the afternoon prayer

MS Damascus Ẓāhiriyya 9227 of Shihāb al-Dīn al-Ḥalabī's solar azimuth tables (**11.2**) also contains a set of tables attributed to Muḥammad ibn Barakāt ibn al-Buṣrawī, on whom I have no further information.[7] These display $\tau_a(\lambda,h)$, the time to the beginning of the interval for the afternoon prayer as a function of solar longitude and solar altitude in the eastern sky over Damascus. The entries are based on the simple relation:

$$\tau_a(\lambda,h) \quad = \quad t_a(\lambda) - t(h,\lambda) \ ,$$

and for each degree of λ' the arguments for h are the integral values such that $h_a(\lambda) \leq h \leq H(\lambda)$. A marginal note by al-Ṭanṭāwī (**11.13**) explains the method of computing the entries in the table from those of $t_a(\lambda)$ and $t(h,\lambda)$. See also **7.3** on some similar tables for Cairo.

In MS Berlin Ahlwardt 5754/5/6, fols. 62v-94r, there are some anonymous tables appended to others by al-Khalīlī (**10.5**) of a function which is described as *al-maktūb bi-'l-sawād al-bāqī li-'l-ghurūb wa-bi-'l-ḥamra li-'l-ʿaṣr*, "time remaining until sunset if written in black ink or time remaining until the beginning of the afternoon prayer if written in red ink". The tables have the same format and the same argument domains as al-Khalīlī's hour-angle tables, and it is understood that the altitudes are eastern. The tabulated function is simply:

$$f(h,\lambda) \quad = \quad T(h,\lambda) \text{ for } h < h_a(\lambda) \quad \text{and} \quad \tau_a(h,\lambda) \text{ for } h \geq h_a(\lambda) \ .$$

Note that the tables of τ_a in the Damascus and Berlin manuscipts have different formats.

11.4 Ibn al-Kayyāl's prayer-tables

MSS Damascus Ẓāhiriyya 4893, Istanbul Esat Efendi 1990, Berlin Ahlwardt 5759/5771 (Wetzstein 1146), *ca.* 1700, and 5760/5772 (Wetzstein 1148), *ca.* 1800, are copies of al-Khalīlī's prayer-tables (**10.6**) in which the main tables are preceded by an introduction and some calendrical tables by the mid-16th-century Damascus astronomer Ibn al-Kayyāl.[8] There is no mention of al-Khalīlī.

---

[7] Ibn Barakāt, whose full name is given in MS Damascus Ẓāhiriyya 9227, is not listed in the modern sources.

[8] On Ibn al-Kayyāl see Suter, *MAA*, no. 474; Brockelmann, *GAL*, II, p. 469; *Cairo ENL Survey*, no. D60; and İhsanoğlu *et al.*, *Ottoman Astronomical Literature*, I, p. 127, no. 59.

11.5 Anonymous prayer-tables for Aleppo and Istanbul

MS Cairo ṬM 255,6, copied 1060 H [= *ca.* 1650], contains al-Khalīlī's introduction to his prayer-tables followed by some anonymous prayer-tables for latitudes:

ϕ = 36° (Aleppo) and 41;15° (Istanbul).[9]

Mention of these was omitted from the first version of this text, and all that is available to me now is a photo of an extract from the tables for Aleppo (**Fig. 11.5**). This shows tables of the same kind as those of al-Khalīlī but for ε = 23;35° rather than his more up-to-date value 23;31°. The functions tabulated are:

H, Z, D, h_a, t_a, T_a, N, n, r, s, $\tilde{h}$, d and t_q .

The quantities t_a and T_a are labelled *ḥiṣṣat al-ʿaṣr* and *ḥiṣṣat al-ghurūb*, expressions which I do not recall having seen elsewhere (though see below). I have not investigated the parameters used for r, s and q. See also **14.8**.

Fig. 11.5: An extract from an anonymous set of prayer-tables for the latitudes of Aleppo and Istanbul. [From MS Cairo ṬM 255,6, fols. 78v-79r, courtesy of the Egyptian National Library.

MS Aleppo ʿUthmāniyya 14, copied *ca.* 1750, contains another anonymous set of prayer-tables for Aleppo. Values of the following functions are displayed in equatorial degrees and minutes for each degree of λ from Capricorn 1°:

D, t_a, T_a, s, n and r

and I have not investigated the underlying parameters. The entries are written in Arabic num-

[9] *Cairo ENL Survey*, no. C142.

erical notation. The five functions are labelled *ḥiṣṣat al-zuhr*, *ḥiṣṣat al-ʿaṣr*, *ḥiṣṣat al-maghrib*, *ḥiṣṣat al-ʿishāʾ*, and *ḥiṣṣat al-fajr*.

11.6 Anonymous prayer-tables for Tripoli

MS Damascus Ẓāhiriyya 4893 contains calendrical and solar tables due to Ibn al-Kayyāl (**11.4**), followed by some anonymous prayer-tables for the latitude of Tripoli, stated to be 34°. (See also **9.3** on two earlier sets of prayer-tables for the latitude of Tripoli, whioch may ber related to each other and to some of these.) The functions tabulated are:

$$D,\ t_a,\ T_a,\ s,\ n,\ r \text{ and } z_{a(7)}\ .$$

The entries are arranged in six columns of thirty, beginning with the column for Capricorn/Aquarius, although the entries for D(λ) begin with the column for Aries/Virgo. Analysis reveals firstly that the table of D(λ) is based on the parameters:

$$\phi = 33;45° \text{ and } \varepsilon = 23;31°.$$

Some of the entries are garbled in such a way as to give the impression that they were copied from another set with a different format: the original table appears to have been rather accurately computed. Secondly, certain of the other tables, including those of t_a, r and s, are the same as those for latitude 34° in MS Princeton Yahuda 861,1 of the *Natīja* attributed to al-Wafāʾī (**8.1**). Thus they are supposedly based on a value of 23;35° for ε, but the entries were so carelessly computed that the value of ε is of little significance. Finally, the table of n(λ) is based on the above-mentioned table of D(λ) (computed for ϕ = 33;45° and ε = 23;31°) and the anonymous values for r(λ) and s(λ) in the Princeton manuscript (computed for ϕ = 34;0° and ε = 23;35°)!

11.7 Anonymous prayer-tables for Nablus

MS Berlin Ahlwardt 5765 [= Wetzstein 1149, fols. 85v-88v], copied *ca.* 1850, consists of anonymous prayer-tables compiled for the latitude of Nablus, taken as 32°.[10] The tables are preceded by a short introduction in which the title *Natījat al-afkār fī aʿmāl mawāqīt al-layl wa-'l-nahār* (compare the title of al-Lādhiqī's tables for Cairo in **7.8**) is mentioned. The functions tabulated for each degree of solar longitude are:

$$D,\ t_a,\ T_a,\ s,\ n \text{ and } r\ ,$$

expressed in equinoctial hours and minutes. Values are displayed side by side in columns, and are written in modern Arabic numerals in an untidy hand – see **Fig. 11.7** for an extract. They are generally inconsistent for a given solar longitude. Consider for example, the entries for Gemini 1° (λ = 61°), which are:

$$D:\ 7;4^h \quad t_a:\ 3;56^h \quad T_a:\ 3;3^h \quad s:\ 1;49^h \quad n:\ 8;40^h \quad r:\ 1;33^h\ .$$

---

[10] The only medieval source which gives 32;0° for Nablus (32;10° is a common value) is labelled "*qiyās*" ("measurement") by Abu 'l-Fidāʾ: see Kennedy & Kennedy, *Islamic Geographical Coordinates*, pp. xxx and 235. Given the crude nature of the tables, this may not be significant: possibly 32° was simply rounded from 32;10°.

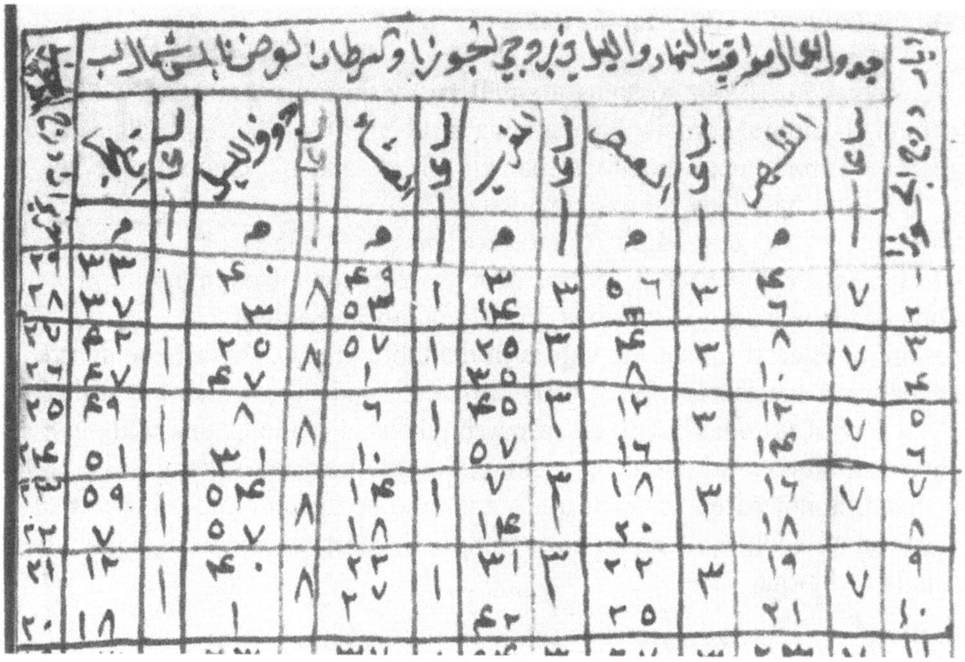

Fig. 11.7: An extract from some messy prayer-tables for Nablus. [From MS Berlin Ahlwardt 5765 [< Wetzstein 1149], fols. 87v-88r, courtesy of the Deutsche Staatsbibliothek (Preußischer Kulturbesitz).]

Notice that the sum of these values, which should correspond to a complete day, is over 26 hours.

11.8 al-Dīsṭī's prayer-tables for Lattakia

MSS Aleppo Awqāf 911 and Leiden Or. 2808(2) are copies of a work entitled *Asna 'l-ghāyāt fī ʿilm al-mīqāt,* "The Best Ends in Timekeeping", compiled by ʿAbd al-Fattāḥ ibn Ibrāhīm al-Mālikī al-Azharī al-Dīsṭī,[11] an Egyptian astronomer who was a student of Riḍwān Efendī (**7.10**). After a brief introduction in 3 *bāb*s al-Dīsṭī presents a set of prayer-tables for parameters:

$$\phi = 34;30° \text{ (Lattakia) and } \varepsilon = 23;35°.$$

The format of the tables is the same as that of the main Cairo corpus. The functions tabulated are:

$$\delta,\ d,\ H,\ D,\ h_a,\ t_a,\ T_a,\ s,\ r,\ \Delta D,\ B,\ C,\ h_0,\ 2N,\ \alpha \text{ and } \alpha_\phi\ .$$

The values of ΔD are those for Cairo found, for example, in MS Cairo DM 108 of the main Cairo corpus (**4.11**). The work concludes with a table of the solar motion between the prayer-times as in the tables of Shams al-Dīn al-Lādhiqī (**7.8**), and a star catalogue for 1160 H [= 1747].

[11] On al-Dīsṭī (**I-4.8.7**, *etc.*) see Brockelmann, *GAL*, SII, pp. 1017, where the manuscript of his prayer-tables which is now in Leiden is listed; *Cairo ENL Survey*, no. D62 (there incorrectly as al-Daystī without a dot); and İhsanoğlu *et al.*, *Ottoman Astronomical Literature*, II, p. 514, no. 352 (now as al-Daysaṭī).

11.9 al-Ḥakīm al-Lādhiqī's tables for Lattakia

Cairo ṬM 92, copied *ca.* 1750, contains a small *zīj* by an individual named Muḥammad ibn ʿAbd al-Maḥmūd al-Lādhiqī compiled in Lattakia in 1170 H [= 1756/57].[12] This contains tables of three spherical astronomical functions, namely, δ, α', and α_ϕ, computed to two digits for each degree of λ and based on the parameters:

$$\phi = 34;30° \text{ (Lattakia) and } \varepsilon = 23;35°.$$

MS Istanbul Bağdatli Vehbi Efendi 887 contains a selection of anonymous planetary tables as well as others for α' and α_ϕ attributed to "al-Ḥakīm al-Lādhiqī".

The tables are of interest in that the values in the table of $\alpha_\phi(\lambda)$ have been derived by linear interpolation between the entries for λ = 0°, 30°, 60°, and 90°, which are rather accurately computed.[13] I know of no other Islamic ascension tables as crude as this, although see **8.6** on some late Egyptian tables also based on a simplistic interpolation scheme. The table of d(λ), on the other hand, is not based on such a scheme, and so al-Lādhiqī had tables of $\alpha'(\lambda)$ and d(λ) at his disposal with which he could have compiled a proper table of $\alpha_\phi(\lambda)$, but he preferred his own curious technique.

11.10 al-Manāshīrī's tables for Damascus, Cairo, Mecca, and Istanbul

MS Cairo DM 184, penned in 1171 H [= 1757/58], is the only copy known to me of a treatise on astronomy compiled in the early 17th century by Muḥammad ibn Maḥmūd al-Manāshīrī, otherwise known for his work in the Qurʾānic sciences.[14] al-Manāshīrī presents various calendrical and planetary tables and also four sets of prayer-tables, for Damascus, Cairo, Mecca, and Istanbul. The tables for Damascus are those of al-Khalīlī (**10.6**), based on parameters:

$$\phi = 33;30° \text{ and } \varepsilon = 23;31° ,$$

and are attributed to him. The functions tabulated include various ones not generally displayed in the main Damascus corpus (**10.6**):

$$\text{Sin } H, \text{ Sin } \delta, \text{ and Sin } \psi .$$

al-Manāshīrī tabulates some twenty six functions side by side for each degree of λ starting at Capricorn 1°, namely:

$$H, \text{ Sin } H, Z_{(12)}, Z_{(7)}, D, 2D^h, \bar{h}, d, \delta, \text{ Sin } \delta, h_a, t_a, T_a, h_b,$$
$$2N, N, 2N^h, \bar{h}', s, r, \psi, \text{ Sin } \psi, h_0, n \text{ and } t_q .$$

In the column for Sin ψ ($\delta > 0$) al-Manāshīrī tabulates al-Khalīlī's values. However in the column for Sin ψ ($\delta < 0$) a completely different function appears, which I am unable to identify. Sample values from this garbled table are:

1°	89;3	31°	76;31	61°	165;25°
15	89;1	45	69;43	75	60;29°
30	88;56	60	62;31	91	57;36°

---

[12] On this al-Lādhiqī see *Cairo ENL Survey*, no. D66.
[13] See n. **I**-1:21.
[14] On al-Manāshīrī (**I-6.1.2**) see Brockelmann, *GAL*, I, p 427; *Cairo ENL Survey*, no. D8; and İhsanoğlu *et al.*, *Ottoman Astronomical Literature*, I, pp. pp. 274-275, no. 138.

The prayer-tables for Cairo are taken from the main corpus, although this is not stated, and are thus based on the parameters:

$$\phi = 30;0° \text{ and } \varepsilon = 23;35° .$$

The columns are ruled for functions:

$$H, D, h_a, t_a, T_a, s, r, 2N \text{ and } n ,$$

but only the values for D, t_a, T_a, s and r have been filled in.

The prayer-tables for Mecca and Istanbul are not attested in any other source known to me. The functions tabulated are:

$$H, D, h_a, t_a, T_a, s, r, \text{ and } 2N .$$

It is stated that the underlying latitudes are 21;30° and 41;15° but in fact the tables for Istanbul are based on latitude 41;0°. The obliquity is 23;31° in both cases, and the parameters for morning and evening twilight are 19° and 17°. It seems unlikely that al-Manāshīrī compiled the prayer-tables for Mecca and Istanbul himself. However, in view of the fact that both sets are based on obliquity 23;31°, a value used in *mīqāt* tables only – as far as we know – by al-Khalīlī and Ibn al-Shāṭir, it seems probable that these tables were compiled in Damascus.

al-Manāshīrī states in his introduction that his calendrical tables are based on the techniques of ʿAbd al-Laṭīf ibn Ibrāhīm al-Kayyāl (**11.4**), which were in turn taken over from "al-Suyūṭī al-Jalāl ibn al-Kamāl" (?).[15] In his discussion of the determination of the prayer-times he refers to the statement of Sharaf al-Dīn al-Khalīlī concerning the observations made on Jabal Qāsiyūn with his uncle Shams al-Dīn al-Khalīlī.

11.11 Ottoman-type prayer-tables for Aleppo

MS Damascus Ẓāhiriyya 7564, fols. 53r-55v, contains an anonymous set of prayer-tables of the same type as those of Darendelī (**14.4**), computed for latitude 36°. These are followed (fol. 56r) by a set of geographical coordinates in which no locality is listed with latitude 36°. I presume the tables were intended for use in Aleppo, for which the geographical tables in the Damascus manuscript give the widely-accepted latitude 35;50°. The functions tabulated are:

$$2D, m', a', s', r' \text{ and } R' ,$$

with values in equinoctial hours and minutes, rather accurately computed. The entries are written in standard Arabic numerals, with some unusual abbreviations. The parameters for morning and evening twilight are 19° and 18°.

In the top margin at the beginning of these tables it is stated that:

(a) the interval between the appearance of the zodiacal light[16] and the time to extinguish the candles (*bayn ṭulūʿ al-raʾs wa-ṭafy al-qanādīl*) is 0 45 minutes;

[15] This can hardly be a reference to the famous 16th-century polymath Jalāl al-Dīn al-Suyūṭī (Suter, *MAA*, no. 449; *Cairo ENL Survey*, no. 103). His treatise on the references to astronomy in the *Qurʾān* and *ḥadīth* has been published in Heinen, *Islamic Cosmology*. However, al-Suyutī is not known to have written anything of consequence on the calendar. On his poem on the midday shadows for each month of the Coptic year see **III-9.7a**.

[16] See the article "Shafaḳ" on twilight in EI_2.

(b) the interval between the time to extinguish the candles and the time to start fasting in Ramaḍān (*bayn al-ṭafy wa-'l-imsāk*) is 0 25 minutes; and

(c) the interval between the time to start fasting and daybreak (*bayn al-imsāk wa-'l-fajr*) is 0 45 minutes.

I am not sure what these intervals mean, as the time between the appearance of the zodiacal light and the true dawn can hardly be 0;45 + 0;25 + 0;45 = 1;55 hours. See **4.10** on these institutions as cultivated in Cairo.

11.12 Anonymous timekeeping tables for Jerusalem and Nablus

MS Cairo ṬM 81,1 (fols. 1r-84v), copied *ca.* 1900, contains a set of tables of the function $T(\lambda,h)$ for altitudes in the east and west, and with entries in hours and minutes for each degree of both arguments. The underlying latitude is found by inspection to be $\phi = 32°$ (Jerusalem), and values of $D(\lambda)$ are given on each sub-table of entries: see **Fig. I-2.2.6**. These tables are followed by a set of tables of the function $t(h,\lambda)$ with values in degrees and minutes, for parameters:

$$\phi = 32;0° \text{ (Jerusalem) and } \varepsilon = 23;35° .$$

These tables are copied without the solar longitude arguments, and without the degrees of the entries, except at the head of each column. These are probably the tables computed by Ibn al-Rashīdī (**I-2.1.5** and **II-6.12**). The manuscript concludes with tables of the three functions:

$$N, d \text{ and } h_0$$

for latitude $\phi = 32;10°$ (Nablus).

I suspect that all of these tables in the Cairo manuscript were copied from an original in the Khālidiyya Library in Jerusalem.

11.13 al-Ṭanṭāwī's prayer-tables for Damascus

Muḥammad ibn Muṣṭafā ibn Yūsuf al-Ṭantāwī al-Ṭanda'tā'ī al-Azharī al-Dimashqī was a *muwaqqit* in the Umayyad Mosque at Damascus who died in 1889, and thus represents the last period of "medieval" Islamic astronomy in Syria.[17] Numerous of the astronomical manuscripts in the Ẓāhiriyya Library in Damascus are written in his hand: in particular, two manuscripts are actually attributed to him. The first, MS 9353, is a collection of tables for constructing sundials, in which he used on the parameters (ϕ and ε) of Ibn al-Shāṭir. It was al-Ṭanṭāwī who by accident broke the sundial of his illustrious predecessor which embellished the main minaret (*al-ʿarūs*) of the Umayyad Mosque in Damascus: whilst trying to realign it: he constructed a new one with identical markings which is still in position on a balcony of the minaret.[18] The second, MS 9233, contains a brief introduction explaining the calendrical and solar tables which form a small part of the manuscript (pp. 1-21). These are followed by a complete set of al-Khalīlī's hour-angle tables, prayer-tables and tables of ascensions (pp. 22-156), but the only reference to al-Khalīlī is a statement that it was he who computed the tables

[17] He is not mentioned in the modern sources.
[18] See n. 9:29.

of Cot_n H(λ) and Cot_n h_a(λ) (n = 7 and 12). These are not contained in MS Paris ar. 2558 of al-Khalīlī's prayer-tables (**10.2**). MS Damascus 9233 concludes with some tables (pp. 157-161) for the lunar mansions, computed for the years 1292 and 1294 H [= 1875/77]. Finally, MS 7387 contains a set of tables attributed to al-Ṭanṭāwī displaying the function T(λ,h) expressed in equinoctial hours and minutes for each integral degree of both arguments. I have little doubt that these were computed from the corresponding values of T(h,λ) in equatorial degrees and minutes in the tables of al-Khalīlī (*cf.* **10.5**) by a simple division by 15. Note that the format of the two sets of tables is different.

MS Cairo ṬM 173, copied *ca.* 1900, is the only one of the manuscripts associated with al-Ṭanṭāwī that I have located which contains a set of prayer-tables for Damascus with times expressed according to the Ottoman convention (**14.0**). Here:

s′, i′, r′, R′, m′ and a′

are displayed side by side for each degree of λ starting with Capricorn 1°. The time of the *imsāk*, i′, is always 15^m before daybreak, r′.

Several of the manuscripts containing tables associated with al-Ṭanṭāwī, such as MSS Damascus Ẓāhiriyya 3116, fols. 1v-2v, 9233, p. 37, 10076, fol. 16r, and 7388, fol. 16v, contain a table computed by him entitled *daqā'iq al-ikhtilāf wa-niṣf quṭr al-shams li-ʿarḍ lām-jīm lām*, "the difference minutes and the radius of the sun for latitude 33;30°", which purports to represent the interval in equatorial minutes between the time when the centre of the sun is on the true astronomical horizon, and the time when the solar disc has disappeared below the visible horizon at Damascus. Values are given for each degree of λ to one or two digits and are derived by linear interpolation from three significant entries, namely:

EQ: 67 SS: 84 WS: 51 .

In two of the sources, MSS Damascus Ẓāhiriyya 9233, p. 9, and 7388, fol. 16v, it is stated that the value used for the radius of the sun is 0;15°, and if this is subtracted from the entries in the table the resulting values are the time taken for the centre of the sun to move between the two horizons, that is, the ΔD(λ) of earlier Egyptian sources (**4.11**, **5.7**, **6.10**, **7.1**, **8.2**, and **8.8**). Indeed, if we subtract 15 minutes from the above values we obtain:

EQ: 52 SS: 69 WS: 36 .

If we multiply the corresponding values for Cairo, namely, 47, 62, and 32 minutes, by the ratio of the latitude of Damascus to that of Cairo, namely, 33;30°/30°, according to the method of ʿAbd al-Qādir al-Minūfī (**8.2**), we obtain:

EQ: 52;29 SS: 69: 14 WS: 35;44 .

which, when rounded, are the same as those underlying al-Ṭanṭāwī's tables. (If we assume that the distance between the true horizon and the circle 0;15° below the visible horizon is 0;40° + 0;15° = 0;55°, then the time in equatorial minutes taken by the sun to move between these two positions is actually 66 at the equinoxes and 74 at both solstices.)

In MS Damascus 9233, p. 9, there are some notes in the hand of al-Ṭanṭāwī which explain the use of this table to modify the various functions of timekeeping which relate to horizon phenomena. The procedures he outlines are along the same lines as those of Muḥammad al-Minūfī (**7.1**). al-Ṭanṭāwī concludes with the statement that all the *muwaqqits* use a fixed amount to adjust for horizon phenomena. They call this quantity, which does not vary throughout the year, *tamkīn*. al-Ṭanṭāwī complains that the *muwaqqit*s do not know why they do this and they fall into error. What surprises him more, he says, is that they do not know the truth and will not accept it from one who does. al-Ṭanṭāwī was very much in control of his subject, but his hapless colleagues represent the ultimate decline of Syrian astronomy, and the end of the brilliant tradition associated with al-Mizzī, Ibn al-Shāṭir, and al-Khalīlī.

In 1971 the Director of the Umayyad Mosque in Damascus told me that he had never heard of al-Ṭanṭāwī, the last traditional *muwaqqit* at the Mosque not dead a century, let alone al-Khalīlī. The manuscripts in the nearby Ẓāhiriyya Library had just been catalogued for the first time, but the name of al-Khalīlī does not appear there: the tables attributed to Ibn al-Kayyāl and al-Ṭanṭāwī were listed but had not inspired any interest locally.

11.14 ʿAbdallāh al-Ḥalabī's prayer-tables for Aleppo

MS Aleppo Awqāf 943, copied *ca.* 1800, contains an anonymous set of anonymous auxiliary tables for computing the hour-angle (**I-6.14.4**), and a set of prayer-tables for parameters:

$$\phi = 35;50° \text{ (Aleppo) and } \varepsilon = 23;30°$$

attributed to ʿAbdallāh al-Ḥalabī, *muwaqqit* at the Umayyad Mosque in Aleppo *ca.* 1750.[19] The same prayer-tables occur in MS Aleppo Awqāf 962. The functions tabulated are the following:

N	*niṣf qaws al-layl al-ḥaqīqī*
D	*niṣf qaws al-nahār al-marʾī*
m′	*waqt al-ẓuhr al-sharʿī*
R′	*ṭulūʿ al-shams bi-ʾl-ufuq al-marʾī*
H	*ghāyat irtifāʿ al-shams*
Sin H	*jayb al-ghāya*
Z = Cot H	*ẓill al-ghāya*
h_a	*irtifāʿ al-ʿaṣr al-awwal*
Sin h_a	*jayb al-ʿaṣr al-awwal*
Sin H - Sin h_a	*al-faḍl bayn al-jaybayn*
Vers t_a	*sahm dāʾir al-ʿaṣr*
t_a	*faḍl dāʾir al-ʿaṣr*
a′	*waqt al-ʿaṣr al-awwal*
h_b, ... , b′	ditto (6 functions) for *al-ʿaṣr al-thānī*
Vers (D + r)	*sahm faḍl dāʾir al-fajr*
D + r	*faḍl dāʾir al-fajr*
r′	*waqt al-fajr*

Values are given for each degree of λ beginning with Cancer 0°. The times are given according to the Ottoman convention, and are expressed in hours, minutes and seconds, and values of other functions are given to two digits. Separate tables display values of:

---

[19] On ʿAbdallāh al-Ḥalabī (**I-6.2.7**) see Ihsanoğlu *et al.*, *Ottoman Astronomical Literature*, II, pp. 560-561, no. 398.

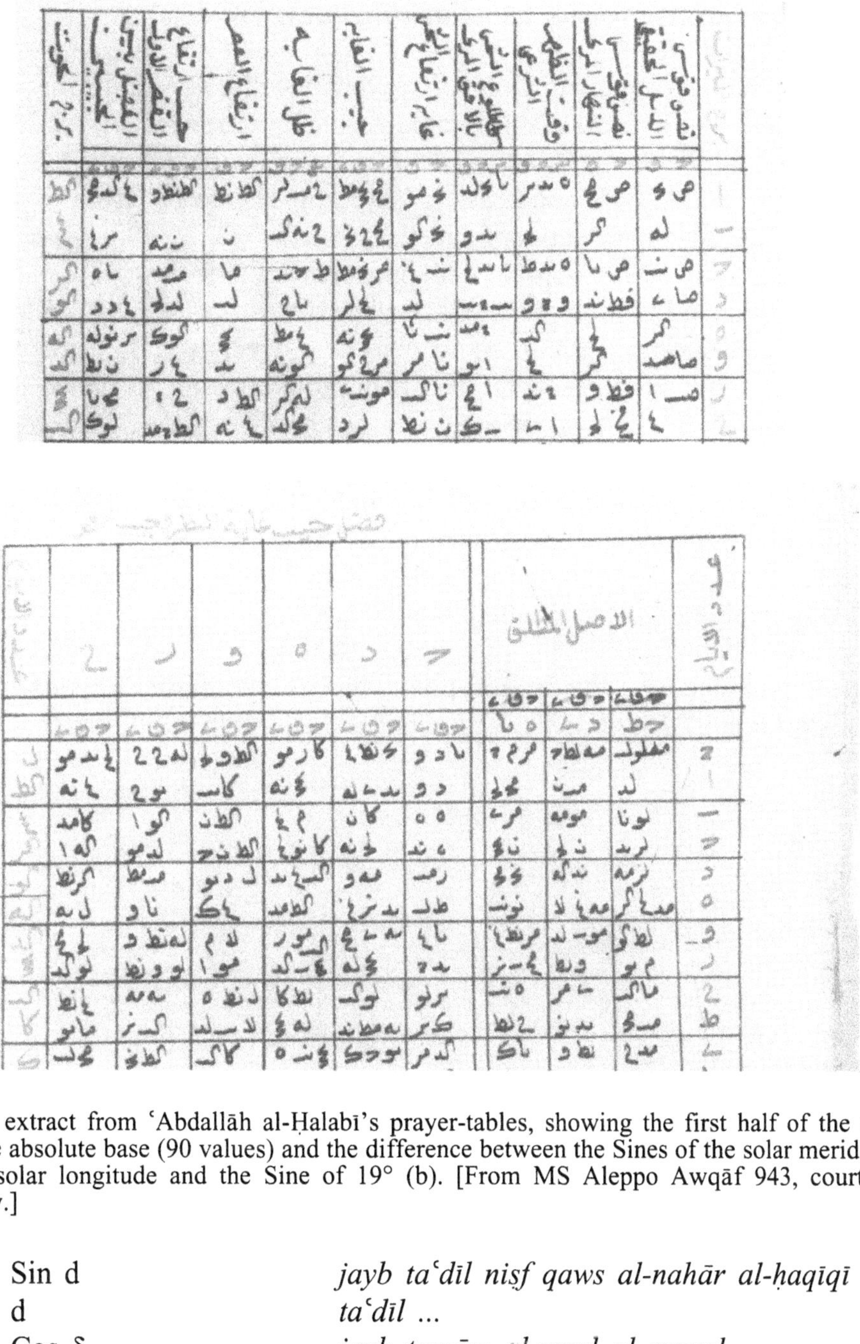

Figs. 11.14a-b: An extract from ʿAbdallāh al-Ḥalabī's prayer-tables, showing the first half of the entries (a), and his tables of the absolute base (90 values) and the difference between the Sines of the solar meridian altitude for 180° plus the solar longitude and the Sine of 19° (b). [From MS Aleppo Awqāf 943, courtesy of the Aḥmadiyya Library.]

Sin d	*jayb taʿdīl niṣf qaws al-nahār al-ḥaqīqī*
d	*taʿdīl ...*
Cos δ	*jayb tamām al-mayl al-awwal*
B	*al-aṣl al-muṭlaq*
Sin H(λ*) - Sin 19°	*faḍl jayb ghāyat al-naẓīr wa-jayb al-fajr*

Extracts are shown in **Figs. 11.14a-b**. ʿAbdallāh al-Ḥalabī's tables are of interest because they display the various stages of his calculations of the prayer-times. See **8.3** on the related tables for twilight of Ibn Abi 'l-Khayr al-Ḥusnī.

11.15 Anonymous auxiliary tables for computing prayer-times

The late Syrian treatise on the 15 times of day with religious significance preserved in MS Aleppo Awqāf 970 (see already **2.11**) concludes with a set of auxiliary tables for timekeeping. The functions are labelled *mayl*, *sahm al-mayl* (= *sahm al-aṣl al-muṭlaq*), and *nisba sahmiyya*, and I have not investigated their character. Values appear to be given to *six decimal places* (!), and the underlying parameters are stated to be:

$$\phi = 35;45° \text{ (Aleppo) and } \varepsilon = 23;28°.$$

Some numerical examples follow, as well as a computation of the qibla at Jerusalem as 45;27°.

The anonymous author states that the scholars of astronomy and geometry have established that the visible horizon is half a degree below the true horizon. At the solstices the legal day is 2° longer than the day derived mathematically, the difference being twice the sum of the solar radius and the time taken by the sun to move between the two horizons, both $^1/_2$°. For the summer and winter solstices he advocates respectively differences of 3° and 4°. This means that he is assuming the following corrections for the "difference minutes":

EQ: $^1/_2$° SS: 1° WS: $1^1/_2$°

For $\phi = 36°$ and $\Delta h = {}^1/_2°$, the accurate values are:

EQ: 0;37,5° SS: 0;42,41° WS: 0;42,31° .

Thus the underlying theory is the (inappropriate) one used by earlier Egyptian astronomers (see, for example, **4.11** and **7.1**) and some earlier, but still late, Syrian astronomers (**11.8** and **11.13**).

CHAPTER 12

YEMENI AND HEJAZI TABLES FOR TIMEKEEPING

12.0 Introductory remarks

In the Yemen there was considerable activity in mathematical astronomy between 900 and 1700.[1] More than 15 *zīj*es were compiled and several sets of tables for timekeeping were prepared during this period. This colourful Yemeni tradition is of interest to the history of science not least because it was based mainly on Egyptian, ʿIrāqī, and Tunisian sources, some of which are no longer extant.in their original form. The earliest Yemeni tables which survive are those on the *Muẓaffarī Zīj* of Muḥammad ibn Abī Bakr al-Fārisī, based on Iranian sources.[2] About 700 H [≈ 1300] the Yemeni astronomer Abu ʾl-ʿUqūl compiled an enormous corpus of tables for timekeeping for the latitude of Taiz (**12.1**); he was obviously more influenced by the Mamluk Eyptian tradition. There are tables for instrument construction and timekeeping in the two treatises on instruments and mathematical astrology by the late-13th-century Rasulid Sultan al-Ashraf (**12.2**). The late-14th-century Rasulid Sultan al-Afḍal has left us a disordered pile of tables in his astronomical miscellany (**12.4**). Some simpler tables are found in a remarkable Yemeni ephemeris from 1405 (**12.5**). An early-14th-century Yemeni astronomer apparently named Ibn al-Mushrif, who cannot be identical with the early-15th-century Egyptian astronomer of the same name (**I-9.8,** *etc.*), may have compiled a complete set of tables for the latitude of Zabid (**12.1**), incorporating some of Abu ʾl-ʿUqūl's tables for Taiz. Another Yemeni astronomer named ʿAbdallāh ibn Ṣalāḥ Dāʿir (**12.6**) compiled a set of tables for timekeeping for Sanaa, of which only a fragment survives. The prayer-tables of al-Thābitī (**12.7**), compiled for the Yemen (latitude of Sanaa), are for timekeeping by lunar mansions. Those of al-Wāsiʿī *ca.* 1940 (**12.12**) give no hint of any awareness of the achievements of his predecessors in Rasulid times.

As far as I know, no astronomical works of scientific consequence were compiled in Arabia outside the Yemen. (See, however, the highly-significant treatise on folk astronomy by Ibn Raḥīq (**2.3**).) Some prayer-tables for Mecca (**6.10** and **11.10**) were probably compiled in Cairo and Damascus, respectively. Some late prayer-tables for Mecca and Medina have been located (**12.8-10**).

---

[1] A survey of Yemeni astronomers and their surviving works, based on about 100 manuscripts, is in King, *Astronomy in Yemen*. New information is contained in my review of the facsimile edition of al-Afḍal's miscellany cited in n. 12:9.

[2] On al-Fārisī (**I-7.1.7** and **II-2.2**) see King, *Astronomy in the Yemen,* no. 9.

12.1 Abu 'l-ʿUqūl's corpus of tables for Taiz

The extensive corpus of tables for timekeeping entitled *Mirʾāt al-zamān*, "The Mirror of Time", was compiled for the latitude of Taiz by the Yemeni astronomer Muḥammad ibn Aḥmad al-Ṭabarī, known as Abu 'l-ʿUqūl, *ca.* 1300.[3] Most but not all of the tables in this corpus are extant in MSS Berlin Ahlwardt 5720 (Mq. 733,3), copied in Mukhkha in 1209 H [= 1795]. MS Milan Ambrosiana C84 is a late anonymous abridgement of the tables, and a manuscript in a private collection in Sanaa of an astronomical miscellany by the Yemeni Sulṭān al-Afḍal compiled *ca.* 1375 (**12.4**) contains a few isolated tables from the corpus. Abu 'l-ʿUqūl also compiled two *zīj*es. The first, called the *Mukhtār Zīj*, is extant in MS London BL Or. 3624, copied in 1008 H [= 1599], and is based mainly on an early *zīj* of Ibn Yūnus. The second is no longer extant, but a few tables "from the *Zīj* of Abu 'l-ʿUqūl" which are not from the *Mukhtār Zīj* are found in the Sanaa manuscript.

The title folio and beginning of the introduction are missing from the Berlin manuscript, and the tables in it are in fact anonymous. The attribution to Abu 'l-ʿUqūl is secured from references in the Sanaa manuscript. The introduction (fols. 16r-23v) is in two *maqāla*s of nine and seven *bāb*s dealing with the various operations of timekeeping by day and night, respectively. All the tables are based on the parameters:

$$\phi = 13;40° \text{ (Taiz) and } \varepsilon = 23;35° \ .$$

The tables in the Berlin manuscript are bound in disorder, but we can distinguish the following main sets:

(a) Tables of $t(h,\lambda)$ and $T(h,\lambda)$, the hour-angle and time since sunrise as a function of solar altitude and solar longitude (fols. 128v-166r). See further **I-2.1.2**, with extract.

(b) Tables of $\lambda_H(h,\lambda)$, the longitude of the horoscopus as a function of solar altitude (fols. 26r-31v, 33r-10v, 122r-123v). See further **I-3.1.1**and the extract illustrated there.

(c) Tables of coordinates of fixed stars and lunar mansions, and tables of $h(\lambda_H)$, the altitudes of various fixed stars at daybreak as a function of the longitude of the horoscopus (fols. 32r-32v, 24r-25v, 108v-120v). See further below and also **I-4.5.1** (illustrated).

(d) Some spurious tables for Sanaa, computed by Ibn Dāʿir (**12.6**).

Various minor tables by Abu 'l-ʿUqūl relating to timekeeping are contained in the various manuscripts mentioned above. Firstly, the Sanaa manuscript contains an anonymous set of spherical astronomical tables displaying the functions:

$$\delta,\ 2D,\ 2D^h,\ H,\ 2N \text{ and } \tilde{h}' \ ,$$

side by side for each degree of λ. Values are given to two digits and the underlying parameters are:

$$\phi = 13;40° \text{ (Taiz) and } \varepsilon = 23;35° \ .$$

Details of the rising, culmination and setting of certain fixed stars, as well as agricultural information, are also given for each 1° of λ, roughly corresponding to each day of the year. It is stated that this information is based on the positions of the stars in 631 H [= 1233/34]. The tables are in fact based on those of the *Mukhtār Zīj* (MS London BL Or. 3624, fols. 176v-184r), in which each of the functions:

[3] On Abu 'l-ʿUqūl (**I-2.1.2**, *etc.*) see King, *Astronomy in Yemen*, no. 9.

Fig. 12.1: A table displaying the oblique ascension of the horoscopus at daybreak, most probably by Abu 'l-ʿUqūl. [From a manuscript in a private collection in Sanaa; taken from Varisco & Smith, eds., *al-Afḍal's Anthology*, p. 317, with permission of Professor Daniel Varisco.]

$$\tilde{h},\ 2D^h,\ H,\ d \text{ and } \alpha_\phi$$

is tabulated to two digits for each degree of λ for latitudes 13;0° (Aden), 13;40° (Taiz), 14;0° (Zabid), and 14;30° (Sanaa). Tables of the functions:

$$H,\ D^h \text{ and } d$$

computed for ϕ = 13;40° and ε = 23;35° also occur on fol. 107r of the Berlin copy of the *Mirʾāt al-zamān*. I conclude that the tables in the Sanaa manuscript were also compiled by Abu 'l-ʿUqūl. The additional information on the stars and on cultivation of crops formed the basis of an almanac, extant in several copies, of which I have examined MS Milan Ambrosiana C46 (fols. 52-58).[4]

Elsewhere in the Sanaa volume there is a table of the functions:

$$Z_{(7)} \text{ and } h_a(H)$$

with entries to two digits for each degree of H from 1° to 90°. It is stated that this table is taken from the *Mirʾāt al-zamān* by Abu 'l-ʿUqūl, and the same table occurs on fol. 108r of the Berlin copy.

A table for twilight which is probably by Abu 'l-ʿUqūl is extant only in the Sanaa manuscript (see **Fig. 12.1**), where it is not actually attributed to him. It is, however, rather accurately computed for his parameters and is based on other tables compiled by him. The function tabulated is simply the oblique ascension of the horoscopus at daybreak, $\alpha_r(\lambda)$, as in the Cairo and Damascus corpuses (**4.10** and **10.6**), and the underlying parameter for morning twilight

[4] See now Varisco, *Yemeni Almanac*.

is 20°. The function is called *al-dāʾir min al-falak ʿinda ṭulūʿ al-fajr*, literally, "the amount the celestial sphere has revolved at daybreak", and is defined by:

$$\alpha_r(\lambda) = \alpha_\phi(\lambda) - r(\lambda) = \alpha_\phi(\lambda) - T(20°,\lambda^*) .$$

Values of $\alpha_\phi(\lambda)$ for the above parameters are given in the second twilight tables described below, and a separate table of $\alpha_\phi(\lambda)$ with the same entries occurs in MS Paris BNF ar. 2523, fol. 95v, of an anonymous late-14th-century Yemeni *Zīj* (**12.3**). The table of $\alpha_\phi(\lambda)$ for Taiz in MS London BL Or. 3624 of Abu ʾl-ʿUqūl's *Mukhtār Zīj* is based on $\phi = 13;40°$. Abu ʾl-ʿUqūl tabulated $t(h,\lambda)$ and $T(h,\lambda)$ for each degree of both arguments (see above), and his table of $T(h,\lambda)$ for $h = 20°$ is found on fol. 138r of the Berlin manuscript. The values of $\alpha_r(\lambda)$ in the Sanaa manuscript can be derived from the appropriate values of $\alpha_\phi(\lambda)$ in the same Berlin manuscript and Paris BNF ar. 2523, and of $T(h,\lambda)$ for $h = 20°$ again in the Berlin manuscript.

The other twilight tables of Abu ʾl-ʿUqūl are found on fols. 24v-25v, 32r-32v, *etc.*, of the Berlin copy. They show the altitudes at daybreak of about eight prominent stars in the eastern and western sky as a function of the longitude of the horoscopus. For each degree of λ_H, values to two digits are given for the stellar altitudes, and also for $\alpha_\phi(\lambda_H)$ and the duration of morning twilight $r(\lambda_H)$, and for the corresponding solar longitude. I have not investigated the parameters used for the various stars, but there is a table of right ascensions of thirty stars on fol. 24r of the Berlin manuscript which Abu ʾl-ʿUqūl states he used for the twilight tables. On fol. 108v there is another table giving the ecliptic coordinates and declination, and, for the latitude of Taiz, the meridian altitude, half arc of visibility, and the co-ascendants and co-descendants of the stars. No doubt these are the coordinates which underlie all of Abu ʾl-ʿUqūl's tables for timekeeping by the stars. The star catalogue was originally dated, but the year has been deliberately erased. However, the stellar positions correspond to about 1300. In the Sanaa volume there is a similar star catalogue compiled for a later date, with an additional set of entries displaying the meridian altitudes at Zabid: this table is stated to have been taken from a *Zīj* by Ibn al-Mushrif, otherwise unknown to the literature for he cannot be identical with the early-15th-century Egyptian astronomer with this name (**I-9.8**). In the Sanaa manuscript there are also various astrological tables by Abu ʾl-ʿUqūl, some of which originally belonged to the *Mirʾāt al-zamān*, such as tables of the longitudes of the astrological houses.[5]

In all, the Taiz corpus contains over 100,000 entries, and is the largest single set of tables for a specific locality known to have been compiled by a medieval astronomer. Abu ʾl-ʿUqūl was well named as *al-ḥāsib*, "the calculator", although this actually meant simply "the astronomer" in some circles.

12.2 The Sulṭān al-Ashraf's tables for timekeeping

The major astronomical works of the Rasulid Sultan al-Ashraf (*reg.* 1295-96),[6] compiled whilst he was still a prince, were his treatise on the construction of astrolabe and sundials, as well as the qibla-compass,[7] extant in the autograph MSS Cairo TR 105 (149 fols.) and a late copy

[5] On the houses see n. I-3:2.
[6] On al-Ashraf (**I-3.3.1**, *etc.*) see Suter, *MAA*, no. 394, and King, *Astronomy in Yemen*, no. 8.
[7] See now Schmidl, "Early Arabic Sources on the Compass".

thereof in a library in Tehran (unidentified), and his astrological work extant in MS Oxford Bodleian Hunt. 233, copied in the 14th century (?). See **I-3.3.1**, **4.1.4**, **4.2.5**, and **5.4.6**, on various rather unusual tables contained in these.

In the former treatise the qibla for Taiz, Zabid, and central Yemen is said to be 27°-29° W. of N. (p. 145 of the Cairo copy). The modern value for Taiz is 26;47°! This good news is somewhat vitiated by the fact that the diagram of the qibla-compass in the same manuscript shows the qibla of Taiz at 20° West of North and the qibla at Aden at 20° East of North: see **Fig. X-9.2.2**.

12.3 A *ṭaylasān* table for all latitudes in a Yemeni *zīj*

The anonymous Yemeni *Zīj* preserved in MS Paris BNF ar. 2523 (late 14th century), which was compiled in Taiz in 775 H [= 1374], contains a table of T(H,h) (fols. 97v-100v). Entries are computed to two digits for the domains:

$$H = 46°, 47°, \ldots , 90°, \text{ and } h = 1°, 2°, \ldots , H .$$

In the same work there is a table (fol. 86r) of the function h(T,H) computed for the domains T = 1, 2, ... , 11sdh and H = 51°, 52°, ... , 90°. Entries are given to two digits. Note that both tables are adequate for use in the Yemen, where the minimum value of the solar altitude is about 50°. On these tables see also **I-2.5.3** and **I-4.3.3**.

12.4 The astronomical tables in Sulṭān al-Afḍal's astronomical miscellany

A precious manuscript in a private collection in Sanaa contains an enormous collection of treatises and notes on a whole range of subjects ranging from lexicography, grammar, poetry, Rasulid history, genealogy, biography, medicine and pharmaceutics, and far more besides, to astronomy and astrology, copied in the hand of the Rasulid Sulṭān al-Afḍal *ca.* 1375.[8] This precious manuscript is now available in a facsimile edition, and a survey of the astronomical contents would be a valuable contribution but would merit a doctoral dissertation. See **I-2.1.2**, **3.3.2**, **4.2.6**, **4.5.1**, **6.16**, **7.1.8**, **II-9.1**, and **III-4.7** on various tables relating to timekeeping in this work.[9] There is also an agricultural almanac by al-Afḍal for the latitude of the Yemen (see **Fig. 12.4**),[10] containing values of the following functions to sexagesimal minutes for each degree of solar longitude:

$$\delta, 2D, 2D/15, H, 2N \text{ and } 2N/15 .$$

12.5 An anonymous almanac for Taiz, 808 H

MS Cairo TR 274 (152 pp.) is an anonymous Yemeni almanac for the year 808 H [= 1405/

[8] On al-Afḍal and his astronomical miscellany (**I-2.1.2**) see King, *Astronomy in Yemen*, no. 18.

[9] See Varisco & Smith, eds., *al-Afḍal's Anthology*, and also my review in *Yemeni Update* (www.aiys.org/webdate/kngrev.html), which includes of a survey of research over the past 20 years on Yemeni astronomy.

[10] This has been studied in Varisco, "Yemeni Almanac II". The entire text of the almanac is reproduced in Varisco & Smith, eds., *al-Afḍal's Anthology*, pp. 97-109.

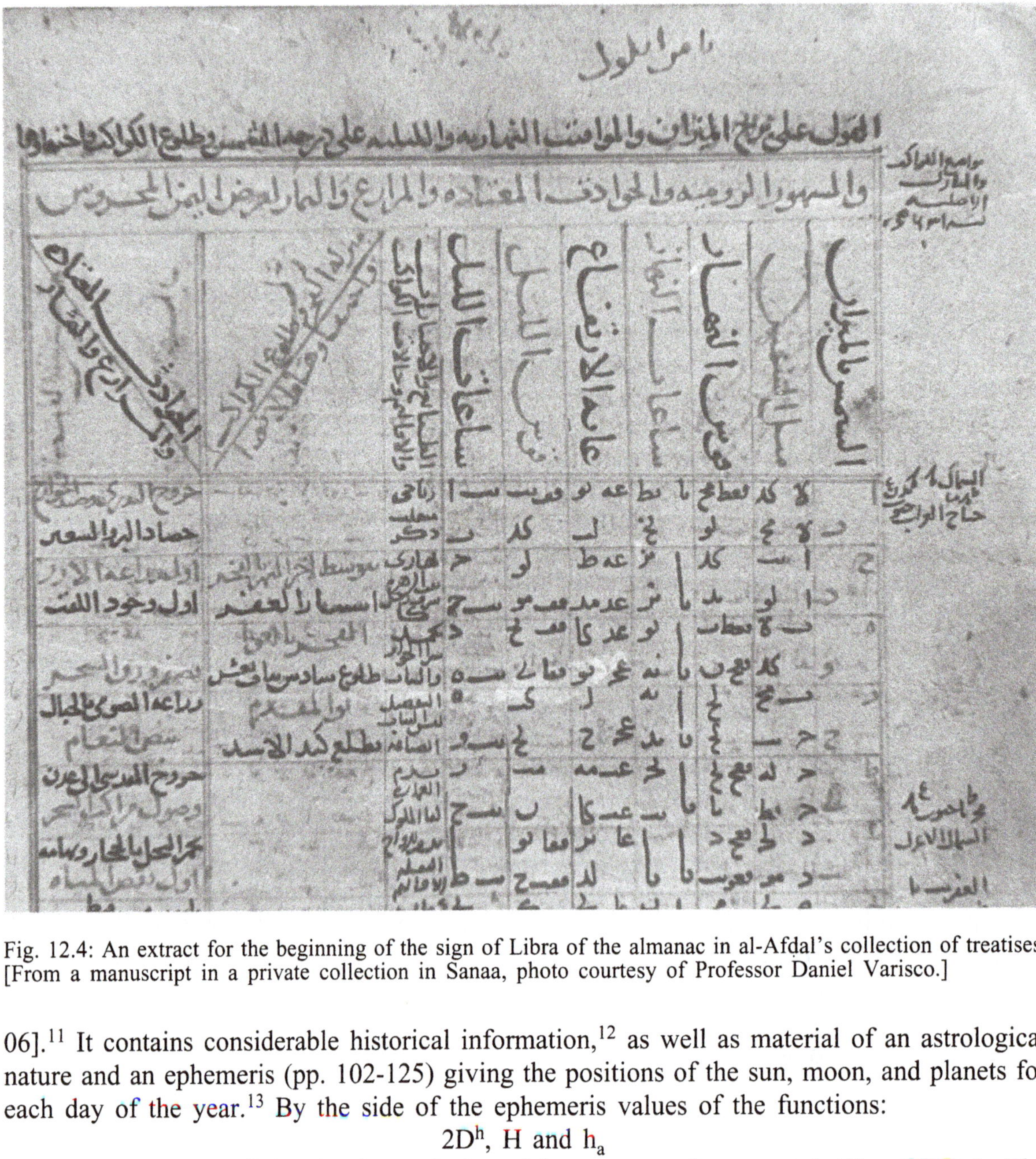

Fig. 12.4: An extract for the beginning of the sign of Libra of the almanac in al-Afḍal's collection of treatises. [From a manuscript in a private collection in Sanaa, photo courtesy of Professor Daniel Varisco.]

06].[11] It contains considerable historical information,[12] as well as material of an astrological nature and an ephemeris (pp. 102-125) giving the positions of the sun, moon, and planets for each day of the year.[13] By the side of the ephemeris values of the functions:

$$2D^h, \ H \text{ and } h_a$$

are displayed to two digits also for each day of the year: see the extract in **Figs. 12.5a-b**. The underlying parameters are:

$$\phi = 13;40° \text{ (Taiz) and } \varepsilon = 23;35°.$$

---

[11] On this almanac see further *Cairo ENL Survey*, no. E11, and King, *Astronomy in Yemen*, no. 22.

[12] See Ayman Fuad al-Sayyid, *Sources de l'histoire du Yémen à l'époque musulmane*, Cairo: Institut Français d'Archéologie Orientale du Caire, 1974, pp. 159-160.

[13] This almanac, together with another Yemeni work of the same genre for the year 727 H [= 1326/27] (King, *Astronomy in Yemen*, no. 11), is currently being investigated by Michael Hofelich of Frankfurt. See his article "Taḳwīm. i" in EI_2.

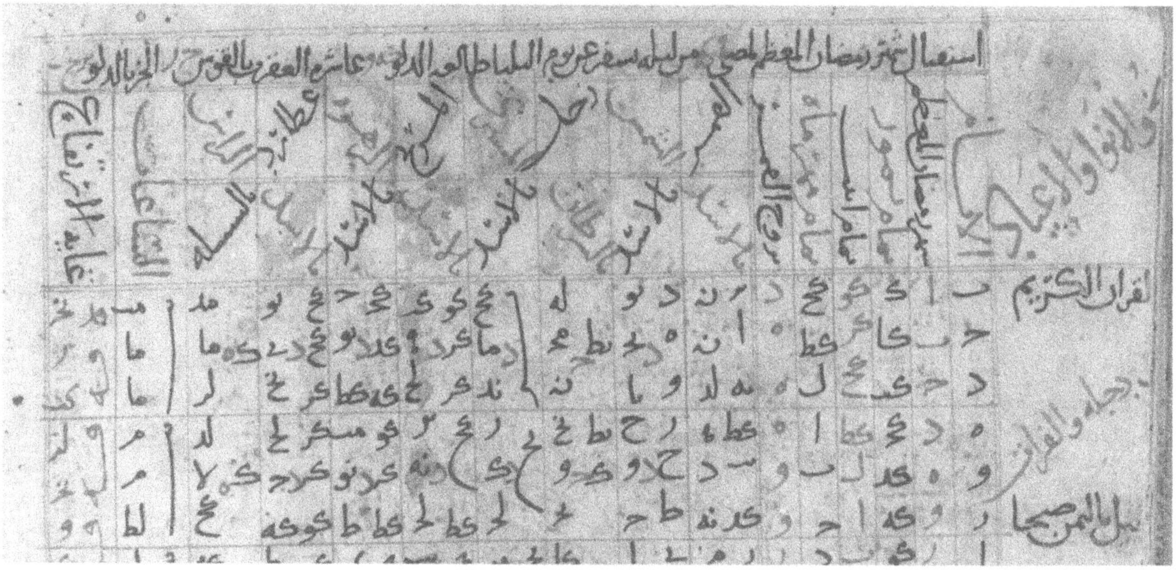

Figs. 12.5a-b: An extract from the ephemeris (a) and the associated astrological consequences (b) for the first few days of Ramaḍān, 808 H. [From MS Cairo TR 274, courtesy of the Egyptian National Library.]

I know no other tables of $h_a(\lambda)$ for Taiz.

In the only other table specifically for timekeeping (p. 96), the following linar mansions are shown for each 13 days of the Coptic year starting with Barmūda 23:

manzilat al-fajr, the lunar mansion rising at dawn

mutawassiṭ rubʿ al-layl, etc., the lunar mansions culminating at one-quarter, one-third, one-half, two-thirds, and three-quarters of the night

mutawassiṭ al-fajr, the lunar mansion culminating at dawn

ghārib al-fajr, the lunar mansion setting at dawn

manzilat al-shams, the lunar mansion of the sun.
Only the first half of this simple table, containing 14 rows of entries, is found in the manuscript now.

12.6 Ibn Dāʿir's timekeeping tables for Sanaa

MS Berlin Ahlwardt 5720 of the corpus of tables for Taiz (**12.1**) contains some additional tables of a less sophisticated type for the latitude of Sanaa (fols. 121v, 124r-127r). These are all for timekeeping by the stars, and I suspect that they were originally part of a larger corpus of such tables for Sanaa. The functions tabulated are: (a) the longitude of the horoscopus as a function of the altitude of 16 stars; (b) the longitude of the horoscopus as a function of each hour of visibility of 18 stars; and (c) the altitudes of 18 stars at each hour of visibility. The hours of visibility represent twelfths of the arc of visibility for each star and hence differ for each star. In the heading of the first table it is stated that the stellar coordinates were based on the observations of Ibn Yūnus, but the name of the compiler and the date for which he computed his stellar coordinates have been deliberately erased. However, it is just possible to read the name ʿAbdallāh ibn Ṣalāḥ Dāʿir,[14] which also appears on fol. 121r in reference to some instructions on the use of the third table. On these tables and their author see further **I-3.2.3**, and also **I-3.4.1** and **I-4.6.1**.

12.7 al-Thābitī's prayer-tables for the Yemen

MSS Berlin Ahlwardt 5769 (Mq. 733,2), fols. 8v-15v, copied in Mukhkha in 1209 [= 1795], Vatican ar. 962, fols. 13r-19r, and Sanaa GML *majāmī*ʿ 27,1, contain a set of prayer-tables compiled for the latitude of the Yemen by Muḥammad ibn ʿAbd al-Laṭīf al-Thābitī. MS Algiers Fagnan 1485,3, which I have not examined, contains the same tables.[15] al-Thābitī was a Syrian who lived in Zabid, and he compiled these tables in 1047 H [= 1637/38].[16]

For each day of the Syrian year al-Thābitī tabulates the solar longitude and which 13th part of the appropriate lunar mansions is:

(a) culminating at sunset,
(b) rising at daybreak,
(c) culminating at midnight, and
(d) culminating at daybreak.

Entries are given as M_1, M_2, ... , M_{13}, where M is the name of a given mansion. A rearrangement of al-Thābitī's column of entries for the lunar mansions rising at daybreak is found in an almanac called *Ḥisāb al-Shibāmī*, which was still in use in the Ḥadramawt in the mid 20th century.[17]

---

[14] On Ibn Dāʿir (**I-3.2.3**, *etc.*) see King, *Astronomy in Yemen*, no. 31, and now Ihsanoğlu *et al.*, *Ottoman Geographical Literature*, I, pp. 76-78, no. 39.

[15] *Algiers BN Catalogue*, p. 409.

[16] On al-Thābitī see King, *Astronomy in Yemen*, no. 33.

[17] This table is published on p. 435 of Serjeant, "S.W. Arabian Almanac", where it is implied that it displays the lunar mansions corresponding to the position of the sun. See King, *Astronomy in Yemen*, no. 44.

al-Thābitī further tabulates the hours of day and night, giving entries to two digits, where the second represents *thirtieths of an hour*, as well as the shadows to base $6^{1}/_{2}$ corresponding to the solar altitude at midday and at the beginning of the afternoon prayer, giving entries to two digits, where the second now represents *twelfths of a shadow digit.* The midday shadows are computed according to a linear zigzag scheme,[18] with a maximum of $5^{5}/_{12}$ at the winter solstice and a minimum of $2^{2}/_{12}$ (south) at the summer solstice. Note that these extremal values are equivalent to 10 and 4 digits for a gnomon of 12 digits, which suggests that al-Thābitī converted these from base 12 to base $6^{1}/_{2}$. In fact, for $\phi = 15°$ the extremal values are closer to 10 and 2 digits, al-Thābitī's afternoon shadow lengths are always $2^{2}/_{12}$ feet more than (the absolute value of) his midday shadow lengths. See further **III-4.7**.

12.8 al-Hattārī's prayer-tables for Medina

MS Cairo Sh 89,11-15 (fols. 157r-322v), copied 1070 H [= 1659/60], is the only known copy of some scientific treatises by Ḥusayn ibn Shāmī al-Hattārī, otherwise known only for his writings on religious topics.[19] Some of the former, which deal with the standard topics of *zīj*es, the astrolabe, and trigonometric and astrolabic quadrants, as well as arithmetic, are here copied in al-Hattārī's own hand. Scattered throughout all of his treatises on astronomy are some spherical astronomical tables computed for $\phi = 25;0°$ (Medina).

Some of the tables are based on $\varepsilon = 23;35°$ and others are based on $\varepsilon = 23;30(,17?)°$, and I have not systematically investigated their accuracy. The functions tabulated include:

$$\delta,\ \delta_2,\ \alpha,\ \alpha',\ \alpha_\phi,\ \psi,\ H,\ d,\ D,\ N,\ t_a,\ T_a,\ s,\ r,\ n \text{ and } \alpha_\sigma\ ,$$

and al-Hattārī also presents some simple trigonometric tables, and some tables for trepidation (fols. 203r and 218v-220r).[20] An extract from the prayer-tables is shown in **Fig. 12.8**. MS

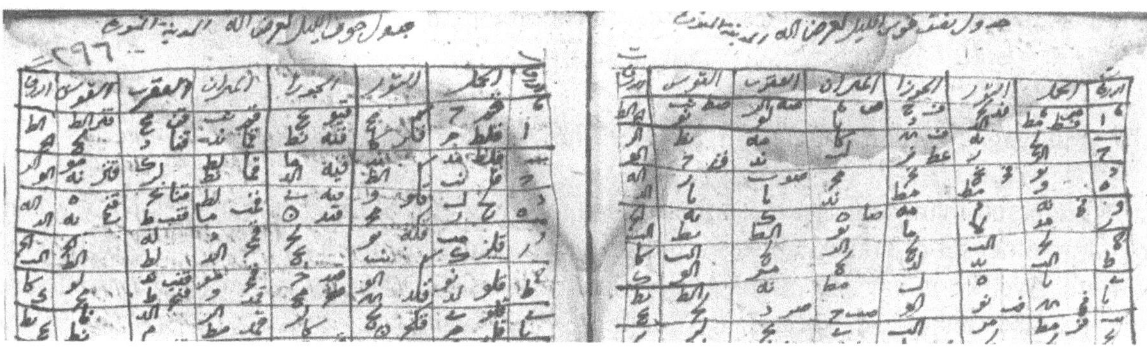

Fig. 12.8: This extract from al-Hattārī's prayer-tables for Medina shows the sub-tables for the functions $N(\lambda)$ and $n(\lambda)$, that is, the length of night from sunset to sunrise and the length of darkness of night from nightfall to daybreak. [From MS Cairo Sh 89, fols. 295v-296r, courtesy of the Egyptian National Library.]

---

[18] Such a crude procedure is virtually unknown in medieval Islamic astronomy. For one example see the study by Kennedy and Ukashah cited in n. 1:9.

[19] On al-Hattārī see Brockelmann, *GAL*, II, p. 523, and SII, pp. 543 and 1039; *Cairo ENL Survey*, no. D42; and İhsanoğlu *et al.*, *Ottoman Astronomical Literature*, I, pp. 388-389, no. 252.

[20] See n. **I**-6:22.

Leiden Or. 2538 is another copy of one of al-Hattārī's treatises entitled *al-Sirāj al-wahhāj* which also contains some of the tables listed above.

12.9 Ḥusayn Ḥusnī's timekeeping tables for Mecca

MS Cairo K 4002, 40 fols., penned in 1249 H [= 1833/34], is the only known copy of a set of tables for timekeeping compiled for the latitude of Mecca, taken as 21;45°, by the Ottoman astronomer Ḥusayn Ḥusnī ibn Aḥmad Ṣabīḥ.[21] The main set of tables displays the time of day according to the Ottoman convention as a function of solar longitude, with entries for solar altitudes in the east and west given in hours, minutes and seconds. See further **I-2.2.4** (illustrated). These are followed (fols. 35v-38r) by a set of prayer-tables for the same latitude. For each degree of solar longitude the following functions are displayed with entries to minutes:

$$2D,\ 2N,\ m',\ a',\ b',\ s' \text{ and } r'$$

I have not investigated the parameters used for twilight.

12.10 Tāj al-Dīn's Ottoman-type prayer-tables for Mecca

MS Damascus Ẓāhiriyya 27 contains a set of prayer-tables for latitude 25° (Medina) similar to those for Aleppo in MS Damascus 7564 (**11.11**). The entries are expressed in hours and minutes and are written in standard Arabic numerals, without the unusual numerical abbreviations found in the Aleppo tables. Values of α' and α_r are also given, expressed in equinoctial hours and minutes. The parameters used for morning and evening twilight are 19° and 17°.

These tables are attributed to Tāj al-Dīn, a *muwaqqit* at the Prophet's Mosque in Medina,[22] and were apparently compiled in 1256 H [= 1840]. A star catalogue giving equatorial coordinates for the year 1100 H [= 1688/89] is also contained in the manuscript and perhaps the prayer-tables were adapted from an earlier source. Tāj al-Dīn also wrote a commentary on the *Ruznāme* of Shaykh Vefa, extant in Cairo.

12.11 Anonymous Ottoman-type prayer-tables for Sanaa

An unnumbered manuscript in the Grand Mosque Library in Sanaa is a comparatively recent copy of an older set of prayer-tables for the latitude of Sanaa, stated as $\phi = 15;0°$. The following times are given in hours and minutes according to the Ottoman convention (**14.0**), written in modern Arabic numerals, for each day of the Syrian year:

$$R',\ z',\ a',\ b',\ s',\ r',\ m', \text{ and also } 2N \text{ and } 2D\ .$$

There are separate tables of the functions:

$$z_{z(7)}\ (h_z = H),\ z_{a(7)},\ z_{b(7)},\ H \text{ and } \delta$$

---

[21] On Ḥusayn Ḥusnī see n. I-2:25.

[22] Tāj al-Dīn is mentioned in Brockelmann, *GAL*, SII, p. 538, but the work attributed to him there is apparently by al-Hattārī, on whom see **12.8**. He wrote a commentary to the prayer-tables of Shaykh Vefā (**14.3**): see *Cairo ENL Survey*, nos. D239 and H2; and İhsanoğlu *et al.*, *Ottoman Astronomical Literature*, 1, p. 53. See also **III-3.4** on his shadow-scheme for timereckoning.

for each degree of solar longitude. The first three of these five tables are computed to two digits, but the remaining two are computed to four digits (ε = 23;35°) and were clearly lifted from some earlier source.

I have not investigated parameters used for morning and evening twilight. The entries for z′ (*ẓuhr*) are in fact simply (12-D^h). The entries for m′ (*niṣf al-nahār*, midday) have been made earlier than z′ by a few minutes (about 8^m at the equinoxes), and the entries for R′, sunrise, have been made earlier by double this correction (about 18^m at the equinoxes). These corrections represent a generous allowance for horizon phenomena (see also **14.15**). The entries for all functions but R′ and m′ correspond quite closely to recomputed values.

12.12 al-Wāsiʿī's prayer-tables for the Yemen

The work *Kanz al-thiqāt fī ʿilm al-mīqāt*, "The Treasure of Reliable Scholars for the Science of Timekeeping", prepared by Shaykh ʿAbd al-Wāsiʿ ibn Yaḥyā al-Wāsiʿī[23] for the Imām Yaḥyā was published in Cairo (Maṭbaʿat Ḥijāzī) in 1358 H [= 1939/40]. It contains two sets of tables, the first for finding the *feria* of a date in the Muslim calendar (from 1358 H to 1596 H) as well as the corresponding day number of the solar year, and the second for reckoning the prayer times, apparently based on ϕ = 15° (Sanaa) although the actual latitude of Sanaa is 15;23°.

In this second set of tables the following information is given for each day of the solar year:

manzilat al-shams, the 13th division of the lunar mansion of the sun
ṭāliʿ awwal al-layl, the ascendant at the beginning of night (in signs and degrees)
maʿālim al-zirāʿa, information about the rising or setting of a prominent star or constellation
daraj al-shams fī 'l-burūj, the solar longitude (to the nearest degree)
al-māḍī min al-shuhūr al-rūmiyya, the date in the Syrian calendar
aqdām al-ẓuhr / al-ʿaṣr wa-banānuhu wa-sāʿatuhu wa-daqāʾiquhu, the shadow length in feet and fingers at the midday / afternoon prayer and the corresponding times in hours and minutes
ṭulūʿ al-fajr, daybreak
shurūq al-shams, sunrise.

The times of daybreak, sunrise, midday and the afternoon prayer are given in hours and minutes according to the Ottoman convention. Daybreak is made to precede sunrise by 1^h10^m throughout the year, which corresponds to a parameter of about 17° at the equinox. The shadow lengths are to base 7 and the fractional twelfths are called *banān*, fingers, as in al-Thābitī's prayer-tables (**12.7**). The shadow lengths are computed according to a very primitive scheme. al-Wāsiʿī's shadow has the constant value $^4/_{12}$ feet from Taurus 8° to Cancer 18°. Since the midday shadow in the Yemen actually becomes southerly in the summer, al-Wāsiʿī's values represent a gross distortion. The equinoctial and solstitial values are given as:

VE: 2 SS: $^4/_{12}$ AE: $1^4/_{12}$ WS: 5 ,

[23] On al-Wāsiʿī see King, *Astronomy in Yemen*, no. 47. A more detailed investigation of his shadow-schemes is in **III-4.10**.

whereas for $\phi = 15°$ they should be:

VE: 1;53 SS: 1;03S AE: 1;53 WS: 5;35 .

Note that al-Wāsiʿī even gives different values for the two equinoxes. The afternoon shadow lengths are always 7 feet more than the midday shadows, so that the summer shadows are not considered negative.

al-Wāsiʿī's times for sunrise, midday and the afternoon prayer were adopted in the prayer-tables currently in use in the Yemen when I visited there in 1974. These also showed nightfall at 1 o'clock (Ottoman convention!) throughout the year.

In the introduction to the tables al-Wāsiʿī mentions a method for timekeeping by night (p. 16):

> "To find the hours by the stars. If you want to know how many hours of the night have passed, first find the ascendant at the beginning of the night (from the tables), and then the ascendant at the time in question (by observation), and then see how many lunar mansions there are between the two ascendants. Multiply this by six and then divide the product into sevenths, taking each seven sevenths as one hour and the remainder less than seven as a fraction of an hour."

This crude method ignores the fact that the mansions are measured along the ecliptic and time is measured along the celestial equator. The factor 6/7 results from the fact that 14 mansions rise in the 12 seasonal night-hours between sunset and sunrise.

al-Wāsiʿī concludes his work (pp. 69-70) with a brief discussion of the qibla, which for the Yemen is, he says, due north. Thus one can use either the pole star or a compass to find it. Elsewhere, he says, the compass is useless for finding the qibla. Compare **12.2**!

On the cover of the work it is stated that the author worked for years to prepare the two tables, and consulted several *zīj*es. Now that we know that the Yemen was an important centre of astronomy from the 10th to the 17th century we can judge al-Wāsiʿī in the light of this early tradition and can only regard his scientific output as rather pathetic.

CHAPTER 13

MAGHRIBI TABLES FOR TIMEKEEPING

13.0 Introductory remarks

I know of no tables relating to astronomical timekeeping from al-Andalus, other than the simple shadow-schemes of the kind found in the 10th-century *Calendar of Cordova*, which I have discussed elsewhere (**III-5.1**). We can, however, be sure that some tables for *mīqāt* were compiled there, not least because the earliest European tables of this kind, some from medieval Chrisitan Spain, show some Islamic influence (**I-10.1**). It is therefore to the Maghrib that we turn next, a region whose distinctive astronomical tradition was influenced successively over the centuries by Baghdad, then by al-Andalus and then by Egypt and Syria, whence came also the *Zīj* of Ulugh Beg.

Only recently has the history of astronomy in the Maghrib been surveyed.[1] According to the late 14th-century Tunisian historian Ibn Khaldūn,[2] the *zīj* that was commonly used in his time (in Tunis or in Morocco or both?) was that of Ibn Isḥāq, who worked in Tunis in the early 13th century. A manuscript of this *zīj* was discovered only in 1978 and has been subjected to detailed study in Barcelona.[3] An abridgement of it was made by Ibn al-Bannā᾿ of Marrakesh *ca.* 1300, and this survives in a number of copies.[4] Four 14th-century Maghribi *zīj*es have survived, namely three *zīj*es by Ibn al-Raqqām, compiled in Tunis,[5] and a small *zīj* by al-Qusanṭīnī, compiled in Fez.[6] This last is of particular interest because it contains planetary equation tables based on Indian rather than Ptolemaic theory. The later Tunisian *zīj*es of Ḥusayn Quṣʿa and Muḥammad Sanjaqdār are based on the 15th-century *Zīj* of Ulugh Beg of Samarqand.[7] Also, a recension of the 14th-century *zīj* of Ibn al-Shāṭir was prepared for Algiers.[8] The three Tunisian sets of tables for timekeeping described here (**13.2-4**) were probably also compiled in the 14th century, and reflect the influence of the Syrian tradition in astronomical timekeeping. Certain later Maghribi sources are full of surprises (**13.5-8**).

In the past hundred or so years Gäetan Delphin, Henri-Paul-Joseph Renaud, and Georges Colin conducted a series of studies on astronomy in medieval times in what is now Morocco.[9]

---

[1] See King, "Astronomy in the Maghrib", already to some extent superseded by various works of the Barcelona school, such as Samsó, "Maghribi *zīj*es", and *idem*, "Astronomical Observations in the Maghrib". See also King & Samsó, "Islamic Astronomical Handbooks and Tables", pp. 60-64.

[2] Franz Rosenthal, *Ibn Khaldūn: The Muqaddimah – An Introduction to History*, 2nd edn., 3 vols., Princeton, N.J.: Princeton University Press, 1967, III, pp. 133-137.

[3] On Ibn Isḥāq see n. **I**-6.17, and especially the new studies of Àngel Mestres there cited.

[4] On Ibn al-Bannā᾿ see n. **I**-6:18.

[5] On Ibn Raqqām see n. **I**-6:19.

[6] On al-Qusanṭīnī (Suter, *MAA*, no. 371) and his tables see Kennedy & King, "Indian Astronomy in Fourteenth-Century Fez".

[7] See n. **I**-4:20.

[8] *Cairo ENL Survey*, nos C30/2.1.19 and F66.

[9] *Cf.* Delphin, "L'astronomie au Maroc", and Colin & Renaud, "Abū Miqraʿ".

Their investigations suggested that the most celebrated *muwaqqits* there such as the 13th-century scholar Abū Miqraʿ (**2.0**) were innocent of the developments in astronomical timekeeping further East (which the French scholars did not know about anyway). However, the numerous sets of prayer-tables mentioned below, mostly unknown to modern scholarship before the 1970s, show that there was more activity in *mīqāt* in Morocco than was previously thought and that some of it was rather sophisticated.

The *mīqāt* specialists of the Maghrib were involved with three times of day, two of which were not generally used by their colleagues further East. These three times are graphically displayed on an elegant 14th-century sundial from Tunis.[10] The first of these was the *ḍuḥā* prayer, beginning around mid-morning, or, more precisely, at a time before midday equal to the time after midday of the beginning of the time for the *ʿaṣr*.[11] The *ḍuḥā* was also considered important in Ottoman Turkey (**Ch. 14**). The second was the *taʾhīb*, a "preparation" for the communal prayers on Friday, which took place one equinoctial hour or 15 equinoctial degrees before midday.[12] The third time is the beginning of the *ẓuhr*, which was taken not as immediately after midday as it was in the East but rather as the moment when the shadow of any object z has increased by one-quarter of its length over its midday minimum Z. This definition of the *ẓuhr* is attested in some early Eastern Islamic sources (see, *e.g.*, **2.4**), but remained standard in Maghribi and Andalusī practice.[13] The corresponding solar altitude h_z is defined by

$$h_z = \text{arc Cot}_n \ \{ Z_{(n)} + {}^1\!/_4\, n \} \ .$$

In this chapter I use the subscripts ∂, t and z to refer to the *ḍuḥā*, *taʾhīb* and *ẓuhr*.

13.1 The spherical astronomical tables for Tunis in the *Zīj* of Ibn Isḥāq

MS Hyderabad Āṣafiyya 298, copied *ca.* 1400, is a unique copy of a voluminous (*ca.* 400 pp.) Syrian recension of the *Zīj* of Ibn Isḥāq.[14] During my brief encounter with this manuscript in Hyderabad in 1978, I was able to record the existence of some auxiliary tables for calculations in timekeeping and computing ascensions also found in earlier works of Ibn al-Zarqālluh and the later works of Ibn Isḥāq, Ibn al-Bannāʾ and Ibn al-Raqqām: see further **I-6.9.1*** and **7.1.5***. The only other tables specifically relating to timekeeping in this manuscript are tables of the functions:

$$2D^h,\ \tilde{h},\ H \text{ and } \delta^*,$$

with entries in degrees and minutes for each degree of λ' based on parameters:

$$\phi = 36;40° \text{ (Tunis) and } \varepsilon = 23;33° .$$

Two tables whose titles indicate that they were to display r(λ) and s(λ) in hours and minutes have no entries (see further **XI-5.3**). Another table, unique of its kind, displays the half-excess

---

[10] See King, "Tunisian Sundial", for an analysis of this instrument and (pp. 193-196) a first interpretation of the institutions represented by the markings (including the *ḍuḥā*, *taʾhīb*, *ẓuhr* and *ʿaṣr*), accomplished with the help of contemporaneous Tunis tables for timekeeping. More information is in **IV-4-6**.

[11] On the *ḍuḥā* see *ibid.*, pp. 189 and 193-194, and **IV-5.**

[12] On the *taʾhīb* see *ibid.*, p. 190, and **IV-6.2**.

[13] *Cf.* Wiedemann & Frank, "Gebetszeiten". On the origin of this definition see **IV-4.4**.

[14] On Ibn Isḥāq see n. **I**-6:17.

of the arc of visibility of the moon for each of the seven climates and for each degree of its declination from 1° to 30° (see further **Fig. VIa-7.2**).

Whilst this manuscript may be somewhat disappointing for the history of astronomical timekeeping it provides exciting new material for the history of most other aspects of mathematical astronomy and astrology.

13.2 Anonymous corpus of timekeeping tables for Tunis

MS Berlin Ahlwardt 5724 (Wetzstein 1150), fols. 1r-41r, copied *ca.* 1700, contains an introduction and a set of tables for timekeeping computed for the latitude of Tunis. The work is anonymous but is entitled *ʿUmdat al-nuẓẓār fī mawāqīt al-layl wa-'l-nahār*, roughly "The Support of those who look closely of the Times of Night and Day". The author dedicated his work to the Ḥafṣid ruler Abū Fāris (al-Mutawakkil) ʿAbd al-ʿAzīz ibn Abi 'l-ʿAbbās Aḥmad (*reg.* 1394-1434).

The first tables (fols. 11r-13r) are for calendar conversion and for finding the solar mean longitudes (from 600 H to 900 H) and equations. These are followed (fols. 13v-14r) by Sine and Cotangent (base 12) tables, and (fols. 14v-15v) by tables of:

$$\delta, \text{ H and D}$$

for parameters:

$$\phi = 37° \text{ (Tunis) and } \varepsilon = 23;35° ,$$

which also underlie the remaining tables. There follow (fols. 16r-28v) tables displaying the hour-angle t(H,h) for each degree of both arguments (h ≤ H) (**I-2.3.5**, illustrated). A full set of prayer-tables follows (fols. 29r-41r), in which the following functions are displayed:

$$h_0,\ \psi,\ \text{Sin}\ \psi,\ \tilde{h},\ D^h,\ 2D^h,\ h_t,\ h_z,\ h_a,\ (D+s),\ (D+2N-r),\ \alpha' \text{ and } \alpha ,$$

as well as k(h) and sec h for azimuth calculations (**I-8.1.3** and **8.4.1**). There is also a separate table (fols. 3v-38r – see **Fig. 13.2**) with entries to one digit for the functions:

$$D,\ (D+s),\ (D+2N-r) \text{ and } (D-r) .$$

Those tables with entries for $\delta \geq 0$ begin with Capricorn 1°. The tables for the solar altitude at the time of the *ẓuhr* and *taʾhīb* are based on the definitions noted in **13.0**. The tables for twilight are based on parameters 20° and 18° for morning and evening, respectively. The entries in all of the tables are rather accurately computed to two digits. The remainder of the tables in the Berlin manuscript are based on latitude 36;40° for Tunis and are the same as those in the *Zīj* of Ḥusayn Quṣʿa (see the next section).

13.3 Anonymous prayer-tables for Tunis

A second set of prayer-tables for the latitude of Tunis, based on parameter 36;40°, is attested in a number of sources. The prayer-tables for latitude 37;0° in MS Berlin Ahlwardt 5724 (**13.2**) are followed (fols. 42r-53r) by this second set, as well as a star catalogue for the year 773 H [= 1371/72] attributed to Ibn al-Shaṭīr (fols. 53v-54v), and a table for reckoning time from stellar altitudes (fols. 55r-56r), on which see **I-2.4.1**. The prayer-tables for latitude 36;40° also occur in MS Princeton Yahuda 147c, pp. 113-132, of the later *Zīj* of Ḥusayn Quṣʿa, based

Fig. 13.2: An extract from the prayer-tables in the main Tunis corpus. [From MS Berlin Ahlwardt 5724, fol. 36r, courtesy of the Deutsche Staatsbibliothek (Preußischer Kulturbesitz).]

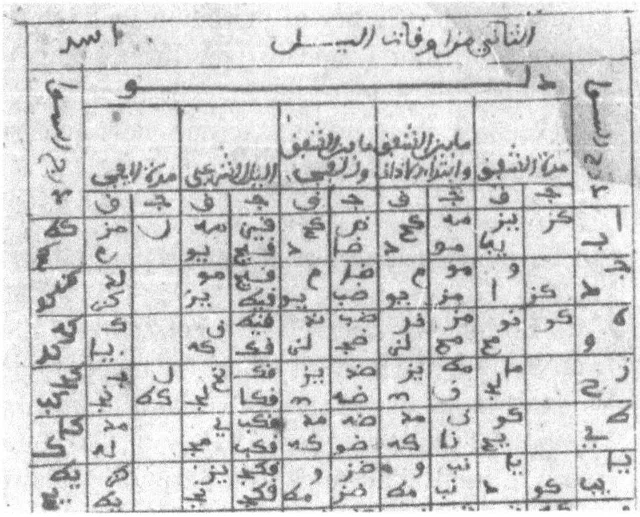

Fig. 13.3: An extract from anonymous prayer-tables for Tunis serving Aquarius / Leo. [From MS Berlin Ahlwardt 5724, fol. 56v. [Courtesy of the Deutsche Staatsbibliothek (Preußischer Kulturbesitz).]

mainly on the 15th-century *Zīj* of Ulugh Beg, and the prayer-tables are likewise not original.

The tables for latitude 36;40° include some for the *ta'hīb* and *ẓuhr* like those for latitude 37;0°. The functions t_z and τ_{za} denote the time from midday to the *ẓuhr* and from the *ẓuhr* to the *ʿaṣr*. Altogether these prayer-tables display the following functions, reasonably accurately computed to two digits for each degree of solar longitude:

(Berlin and Princeton): h_t, H, Sin H, $Z_{(12)}$, $z_{z(12)}$, $z_{a(12)}$, h_z, t_z, τ_{za}, h_a, t_a, T_a, s, (n-45), n, (2N-r), r

(Princeton only): D, $\tilde{h}$, D^h, (D+2n-r), and (D+s) .

The entries begin with those for either Capricorn 1° or Cancer 1°. An extract from part of the tables in the Berlin manuscript is shown in **Fig. 13.3**. The function (2N-r), representing the time from sunset to daybreak, is referred to as *qaws al-layl al-sharʿī*, "length of night as defined by religious law", an expression also used in the tables for latitude 37;0°. The function (n-45) is referred to as *mā bayn al-shafaq wa-'btidā' al-ādhān*, "the time between nightfall and the beginning of the prayer-call". This means that the muezzin would perform a prayer-call three hours before daybreak. Probably this particular table was intended for use only in the festive nights of Ramaḍān, to announce the time for the *suḥūr*. A similar table for Damascus, allowing more time for sleep, is contained in MS Berlin Ahlwardt 5758, fol. 49v, of the Damascus corpus (**10.2**).

A manuscript in the Topkapı Library, whose number is not available to me, is another anonymous copy of some of these tables without introduction or title. The following functions are represented:

h_a, t_a, T_a, $Z_{(12)}$, $z_{z(12)}$, $z_{a(12)}$, ψ, Sin ψ, s, (n-45°), n, (2N-r) and r .

It is noted that the tables of ψ and Sin ψ are based on ϕ = 37° rather than 36;40°.

13.4 Anonymous auxiliary tables for timekeeping by the sun

MS Cairo DM 689 contains an extensive set of auxiliary tables copied in an elegant Maghribi hand *ca.* 1600.[15] The tables conclude with a star catalogue dated 801 H [= 1398], and they appear to have been compiled in Tunis. The title folio and instructions are missing from the manuscript, which begins with the last page of a set of tables displaying the solar longitude for each day of a period of four Syrian years (**Fig. I-1.2d**). The auxiliary tables are simply those of al-Khalīlī for timekeeping by the sun (**10.3**), with a few additional tables of B′ for the latitude of Mecca, taken as 21;40°, and various unspecified localities in the Maghrib, al-Andalus and Sicily (?) with latitudes:

30;30°, 31;30°, 32;30°, 33;40°, 34;30°, 36;30°, 36;40°, 37;10°, 37;30°, 38;30° and 39;30°

and also tables of Sin H(λ) for these latitudes, from which those of B′(λ) are easily derived. See further **I-6.2.5**, **I-6.3.3**, and **I-9.7** (illustrated).

13.5 al-Jannād's prayer-tables for the Western Maghrib

MS Cairo TR 338,2 (pp. 204-287), penned *ca.* 1850, is the only known copy of a set of prayer-tables for various latitudes in what is now Morocco.[16] The tables are preceded by a short introduction in which the author is named as Abū ʿAbdallāh Muḥammad ibn Muḥammad ... al-Jannād al-Anṣārī al-Andalusī, and the work is dedicated to the ruler (called *imām* and *sulṭān*) Nāṣir al-Dīn Abū ʿAbdallāh Muḥammad ibn ʿAbdallāh ibn Ismāʿīl, surely the ʿAlawī *sharīf* Muḥammad III ibn ʿAbdallāh ibn Ismāʿīl who ruled Morocco during 1171-1204 H = 1757-1790].[17]

al-Jannād's tables begin with a table of solar longitude for each day of the Christian year. According to this table the sun enters Aries 0° on March 8-9, and in the title of this table and of all those which follow it is stated that the operations are all *li-ḥarakat yḥ*, which shows that al-Jannād accepted that the equinox had moved 18° from the sidereally-fixed Aries 0°, as a result of the (false) notion known as the trepidation of the equinoxes.[18] A table of geographical coordinates promised in the introduction is no longer contained in the manuscript. The remaining tables consist of three sets for latitudes:

34° (Meknes, Zarhun, Fez, and Sakka), 31° (Marrakesh, Kairouan (!), Alexandria, *etc.*), and 30° (Sijilmasa, Cairo) ,

with underlying obliquity ε = 23;30°. An extract is shown in **Fig. 13.5**. The functions displayed are in each case:

δ, H, h_a, Sin h_a, C, B, d, $t_a{}^h$, $2D^h$, $2N^h$, D^h, $(D{+}s)^h$, $(N{-}r)^h$ and N^h .

Values of the first seven functions are given to two sexagesimal digits, and all times are given in hours, degrees and sexagesimal minutes thereof, except the lengths of day and night, which are tabulated to one extra digit.

---

[15] *Cairo ENL Survey*, no. F30.
[16] *Ibid.*, F35.
[17] Boswell, *Islamic Dynasties*, p. 53. See also the article "ʿAlawīs" in EI_2 by the Editors, especially p. 356.
[18] See n. **I**-6:22, and also n. 13:20 below.

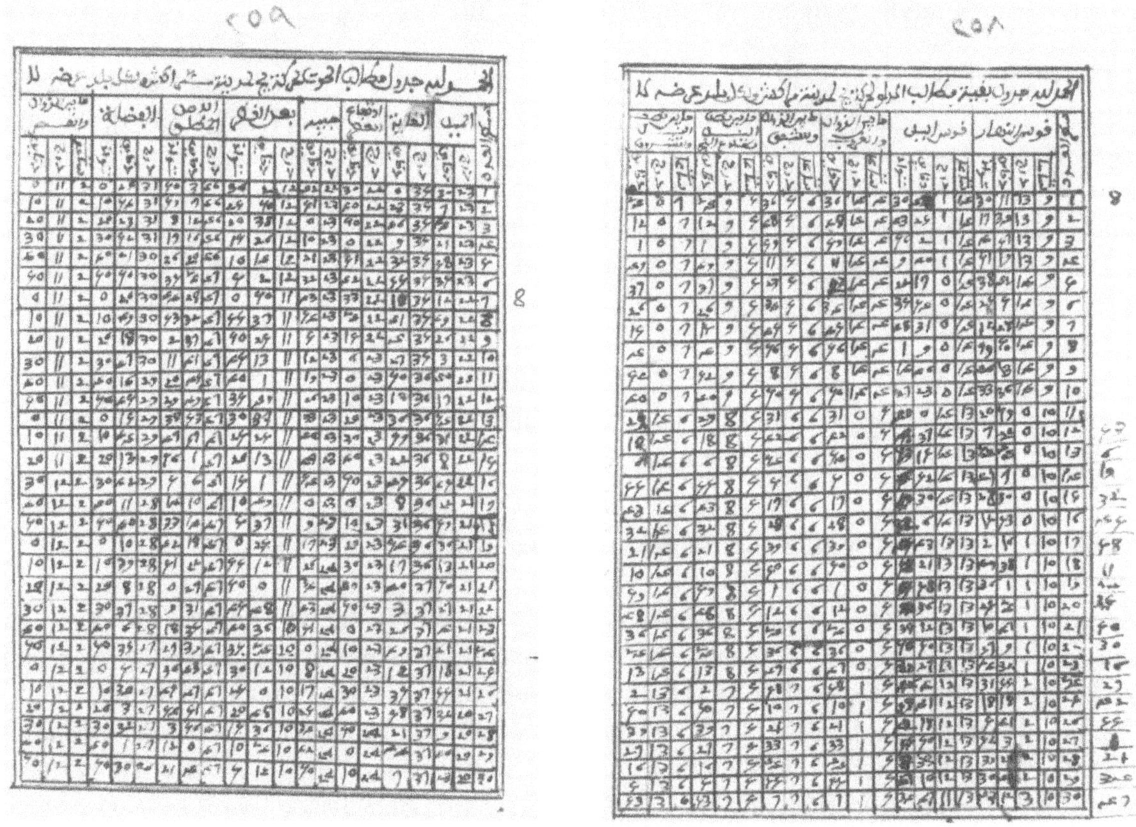

Fig. 13.5: An extract from al-Jannād's prayer-tables for the Maghrib. [From MS Cairo TR 338,2, pp. 258-259, courtesy of the Egyptian National Library.]

13.6 Anonymous prayer-tables for Tlemcen

One of the numerous commentaries on al-Jādarī's poem on timekeeping *Rawḍat al-azhār* (**2.0**, also **I-6.15.2**)[19] is an anonymous work entitled *Natā'ij al-afkār*. This commentary is of interest for the historical information it contains about the medieval Islamic tradition of obliquity measurements, twilight parameters, and theories of trepidation.[20] It is extant in two copies, MSS London BL Or. 411,2, and Cairo K 4311, dated 1183 H [= 1769/70]. The first of these contains a set of tables for timekeeping scattered in the text of the commentary; the second contains no tables. All of the tables based on local latitude are computed for:

$$\phi = 35;0° \text{ (Tlemcen)} .$$

The tables include some for the solar mean motion and the solar equation, simple trigonometric functions, and a star catalogue. The tables relating specifically to timekeeping display the functions:

$$H,\ \alpha_{\phi},\ \alpha,\ D,\ D^{h},\ h_{z},\ h_{a}\ t_{a} \text{ and } T_{a}\ ,$$

[19] On al-Jādarī see n. **I**-6:21.

with values given in degrees and minutes for each degree of solar longitude. There are also tables of various auxiliary functions for use in determinations of the solar azimuth and hour-angle – see **I-6.15.2** and **I-8.1.4**.

13.7 Anonymous prayer-tables for Algiers

MS Algiers Fagnan 1472 contains six pages of prayer-tables for the latitude of Algiers. I have not been able to inspect the manuscript, but according to the catalogue,[21] the prayer-times are arranged in eight columns, values are given for each day of the year, and the manuscript itself dates from the end of the 18th century.

MS Cairo ṬFT 9,1, penned *ca.* 1900, is the only copy known to me of an extensive anonymous almanac and set of prayer-tables for the latitude of Algiers, stated to be 36;40°.[22] The introduction to the work is in Turkish, and the author pays his respects to Shaykh Vefā (**14.3**) and ʿAlī Efendī (the latter presumably made some alterations to Shaykh Vefā's almanac),[23] and then proceeds to mention Ulugh Beg, ʿAlī al-Maghribī (perhaps Abū ʿAlī al-Marrākushī of **6.7** is intended), Abū Miqrāʿ (**2.0**) and Ibn al-Bannāʾ (**13.0**). His almanac contains extensive calendrical tables, and a table for the solar declination in a four-year cycle, as well as three sets of tables for timekeeping. The first of these displays for each day of each month in the Syrian calendar the duration of day (*al-nahār*) and night (*al-layl*). The second displays likewise – see **Fig. 13.7** – the following ten functions:

$$2D,\ 2N,\ (2D+r),\ (2N-r),\ r,\ D,\ t_z,\ t_a,\ T_a,\ s\ .$$

The third and fourth functions are labelled *al-nahār al-sharʿī* and *al-layl al-sharʿī*, as opposed to the first and second, labelled ... *al-ʿurfī*. In both of these sets values are given in hours and minutes. The third set of tables – see **Fig. IV-5.5** – displays the time in hours and minutes, according to the Turkish convention, of the three institutions, the *ṣanjāq*, the *ʿaṣr*, and the *imsāk*. These tables are not based directly on those in the second set. The term *ṣanjāq*, which here refers to a time shortly before midday, is new to me. These tables merit more detailed investigation: see **IV-5.4**.

13.8 Anonymous tables for timekeeping for Sfax

MS Cairo K 7584,1, copied in 1257 H [= 1841/42] in an elegant Maghribi hand, contains a substantial fragment of an anonymous Tunisian treatise on sexagesimal arithmetic and time-keeping compiled in Sfax.[24] The treatise, which probably dates from the 17th or 18th centuries, contains various auxiliary functions for timekeeping (**I-6.4.16, I-6.10.13, I-7.2.3** and **I-8.4.2**), as well as trigonometric tables, and tables of the functions:

$$H \text{ and } 2D^h$$

---

[20] See nn. 2:12 and 13:18.
[21] *Algiers BN Catalogue*, p. 406.
[22] *Cairo ENL Survey*, no. F68.
[23] He is apparently too early to be identified with ʿAlī Efendī, the *muwaqqit* at the Laleli Mosque in Istanbul in the 19th century (İhsanoğlu *et al.*, *Ottoman Astronomical Literature*, II, p. 632).
[24] *Cairo ENL Survey*, no. F69.

Fig. 13.7: An extract from the anonymous prayer-tables for Algiers. See also **Fig. IV-5.5**. [From MS Cairo ṬFT 9,1, fols. 34v-35r, courtesy of the Egyptian National Library.]

with entries to four digits for each degree of λ, and based on the parameters:

$$\phi = 34;48° \text{ (Sfax) and } \varepsilon = 23;30,17° .$$

13.8* Miscellaneous tables for timekeeping from the Maghrib

As this book was going to press I came across some photocopies of photos from a Maghribi manuscript labelled "Ms. 7, v. 3. Zijes. Manuscript", with the pages numbered 446 to 474. The original document dates from about 1800. I have no recollection of the source of these documents. There are some interesting tables for timekeeping, which I can only briefly list here: see **Figs. 13.8*a-d**.

These tables may have been taken from a larger set; certainly, there is no indication of a compiler or a date. The tables are stated to serve latitude 34;10°, which is not an attested value for any location in the Maghrib, and may well be the result of a new set of observations for Fez, whose latitude is accurately 34°05′, but was traditionally taken as 33° or 33;40°, so that was certainly need for a new measurement.[25] The underlying obliquity is 23;29°, which may also have been derived by observation, although it is an attested Ottoman (and European) value. See **I-6.4.20** and **6.10.17** for related tables.

The first group of tables displays the following functions side-by-side for each degree of λ' (1° to 90°):

$$\delta, \text{ Sin } \delta, \bar{\delta}, \text{ Sin } \bar{\delta}, d, \text{ Sin } d,$$
$$\text{arc Sin } C, C, \text{ arc Sin } B, \text{ and } B,$$

[25] Kennedy & Kennedy, *Islamic Geographical Coordinates*, pp. 117-118.

Figs. 13.8*a-d: Four extracts from the remarkable anonymous prayer-tables for Fez that survive in an apparently unique copy, whose location is a mystery to me. Possibly the compiler is al-Warjānī – see **I-6.4.20**. [From a set of photocopies of a manuscript labelled "Ms. 7, v. 3, Zijes, Manuscript", provenance unknown.]

with values to two digits. The functions arc Sin C and arc Sin B are called *qaws buʿd al-quṭr* and *qaws al-aṣl*, respectively; on their purpose see **I-6.5**.

The second group displays in a similar fashion, but now for λ from 270° to 90°, the functions:

$$H,\ Cot_{12}\ H,\ Sin\ H,\ h_a,\ Cot_{12}\ h_a,\ \text{and}\ Sin\ h_a\ .$$

The third set displays two sets of functions labelled *ḥiṣaṣ* – and *sāʿāt* – (sometimes *sawāyiʿ*), also for λ from 270° to 90°. The *ḥiṣaṣ* are for daybreak (*al-fajr*) and sunrise (*al-shurūq*), both measured from midnight, and for the afternoon prayer (*al-ʿaṣr*), sunset (*al-ghurūb*), and nightfall (*al-ʿishā*), all three measured from midnight, followed by the hour-angle at nightfall (*faḍl dāʾir al-ʿishā*). The *sāʿāt* are for the same five times, now expressed in hours and minutes, followed by the time between nightfall and midnight (*mā bayna 'l-ʿishā wa-niṣf al-layl*). The *sāʿāt* are written in Maghribi "Hindu-Arabic" notation, whereas the *ḥiṣaṣ* are in *abjad* notation.

Apart from a single table showing the solar longitude to the nearest degree from the date in the Western calendar, the remaining tables are of the same format, and are of a kind not known from any other source. A series of functions:

h_a (*irtifāʿ al-aṣr*), $t_a{}^h$ (*ḥiṣṣat al-ʿaṣr*), $(D+s)^h$ (*ḥiṣṣat al-ʿishā*),
$(N+r)^h$ (*ḥiṣṣat al-fajr*), and d^h (*niṣf al-faḍla*)

are tabulated for each half degree of solar declination (entered vertically) for each half degree of latitude (entered horizontally) from 30° to 35;30°, then 35;45°. The parameters underlying the entries for evening and morning twilight are stated to be 18° and 19°, respectively. Again the entries are in Maghribi "Hindu-Arabic" numerals. Since only the first of three pages for the last function is present in the photocopy at my disposal it may be that the set originally served yet other functions.

In the light of the existence of this late set of sophisticated tables for timekeeping from the Maghrib, and no less the set described in **13.5**, it is perhaps all the more remarkable that H.-J.-P. Renaud, who wrote the first serious historical essays on Maghribi astronomy (see **13.0**), was not aware of the existence of any tables of this kind.

13.9 Miscellaneous notes

A motley collection of corrupt approximate rules for the prayer-times is contained in a late Maghribi source that I have consulted, but forgot where. The rules outlined are as follows:

$$t_a = 50° + {}^1/_4\ d \quad \text{or} \quad (45° + {}^1/_4\ \varepsilon - 1°) + {}^1/_4\ d \ (\text{where } \varepsilon = 24°!)$$
$$\text{or} \quad 51° + {}^1/_4\ d \quad \text{or} \quad {}^4/_7\ [90° - {}^1/_2\ d]$$
$$r = s + 3°\ .$$

The anonymous author also explains how to find the time of the *ḍuḥā* using the fact that $t_ə = t_a$.

CHAPTER 14

TURKISH TABLES FOR TIMEKEEPING

14.0 Introductory remarks

The only astronomical activity amongst the Muslim community in Anatolia before the rise of the Ottomans is restricted to some legendary accounts of observations in specific *madrasas* of Kirshehir and Kütayha documented by Aydın Sayılı,[1] and the sojourn of the well-known astronomer Quṭb al-Dīn al-Shīrāzī, who compiled some of his astronomical works in Konya, Sivas and Malatya in the middle of the second half of the 13th century.[2] If my assumption that the two sets of tables discussed below (**14.1-2**) are of Seljuk Anatolian provenance is correct, then clearly further research in this field would be worthwhile.

The later Ottoman tradition in astronomy has until recently been generally neglected by historians, despite the vast amount of source material. The popular *Ruznāme* of Shaykh Vefā (**14.3**) has never been properly studied, in contrast with the less widely-used *Rūznāme* (almanac) of Darendelī (**14.4**), on which three descriptive studies were published in Europe between 1676 and 1804.[3] Some brief notes on Ottoman astronomical works were published by J.-B. Toderini in 1789,[4] and an eloquent study by Adnan Adıvar, *La science chez les Turcs Ottomans*, alas devoid of any serious scientific content, was published in Paris in 1939. Otherwise, apart from a series of articles on specific topics that were published over many decades by Aydın Sayılı and his former student, Sevim Tekeli,[5] the notes which follow represented in the 1970s the first serious attempt to investigate the activities of the astronomers of Ottoman Turkey.[6] Now, however, we have a monumental survey of the available literature prepared by Ekmeleddin İhsanoğlu and his colleagues in Istanbul, which will serve as a starting-point for all future research.[7]

In the sequel (**14.3-14**), I describe several Ottoman sets of tables for timekeeping in general and prayer-tables in particular. These are of two kinds: in the first, time intervals are expressed in equatorial degrees and/or equinoctial hours, and in the second the times of day are expressed in equinoctial hours according to the convention that sunset is 12 o'clock. This latter convention, which in the 1970s was still used in the Levant, and more especially in Arabia,

---

[1] Sayılı, *The Observatory in Islam*, pp. 253-255.

[2] *Ibid.*, pp. 215-217, *etc.*, on his association with the Observatory at Maragha. See also the article by Seyyed Hossein Nasr in *DSB*. MS Paris BNF ar. 2516 of his *Tuḥfa shāhiyya* on mathematical cosmology is in his own hand and dated Sivas, 684 H [= 1285/86]: see *Cairo ENL Survey*, no. G27.

[3] See nn. 14:20-21.

[4] Jean-Baptiste Toderini, *De la littérature des Turcs, traduit de l'Italien en Français par Cournant*, 3 vols., Paris, 1789, I, pp. 147-148 and 406.

[5] A list of Sayılı's publications is in *Sayılı Memorial Volumes*, I, pp. 22-29. Many of Tekeli's publications are listed in İhsanoğlu *et al.*, *Ottoman Astronomical Literature*, II, pp. 956-957.

[6] See King, "Astronomical Timekeeping in Ottoman Turkey".

[7] İhsanoğlu *et al.*, *Ottoman Astronomical Literature*, and also *Ottoman Mathematical Literature*.

including the Yemen, seems to be of Ottoman Turkish origin, although, of course, the Islamic day has always been considered to begin at sunset.[8] It is the reckoning of time from sunset in equinoctial hours rather than seasonal hours that characterizes Ottoman practice. An obvious disadvantage of this convention in modern usage is that clocks must (in theory) be reset every evening at sunset.[9] In the sequel I use prime notation to indicate those times which are expressed in this convention (see also **1.3** above).

The most popular prayer-tables prepared for Istanbul were the *rūznāmes* of Shaykh Vefā and Darendelī, already mentioned above. I show (**14.3** and **14.4**) that they are based on different values for the latitude of Istanbul, and that the entries are expressed in different units. Both of these almanacs also contain calendrical tables, information on religious festivals, the seasons, and astrology, none of which con-cerns the present study. More sophisticated prayer-tables were prepared for Istanbul and Edirne, probably in the 16th and 17th centuries (**14.7**, **14.8**, and **14.11**). Particularly impressive are the contributions to our subject by Muḥammad ibn Kātib Sinān and Taqi 'l-Dīn (**14.5** and **14.9**). The extensive corpus of tables for timekeeping by the sun compiled by Ṣāliḥ Efendī (**14.12**) in the late 18th century, by which time Ottoman astronomers had come into contact with European astrology and had "translated" the *Zījes* of Cassini and Lalande into Turkish and Arabic and adapted the tables to their own meridians. Various Ottoman-type prayer-tables for other localities in the Ottoman Empire, such as Cairo, Aleppo, Sanaa, Algiers, Mecca, Crete and Yarqand, have been located in the manuscript sources (see **7.11**, **11.11**, **12.11**, **13.7**, **12.10**, **14.14** and **3.17**, respectively). Doubtless other Ottoman sets of tables of this type for yet other localities are available amongst the several thousand mainly-uncatalogued astronomical manuscripts preserved in Turkey.

14.1 Anonymous Seljuk prayer-tables

MS Istanbul Süleymaniye 1037/32 (fols. 282v-285v) is an apparently unique copy of an anonymous set of prayer-tables, probably of Seljuk origin, computed for some unspecified latitude in Anatolia. The tables are bound amidst a collection of later astronomical works for Istanbul. The following functions are tabulated side by side for each degree of λ from 271° to 90°:

$$H, D, \tilde{h}, h_a, t_a, T_a, h_b, t_b, T_b, s, 2N, n \text{ and } r .$$

Values are given to two digits and the underlying parameters are:

$$\phi = 38;30° \text{ and } \varepsilon = 23;35°.$$

This main set of prayer-tables is followed by a smaller single table displaying values of the functions:

$$\delta, C, B \text{ and } d$$

to two digits for each degree of λ and the same parameters.

The twilight tables are of particular interest. Whilst the values of s correspond quite closely to recomputed values using the accurate formula and parameter 17°, the values of r are based

[8] See Charles Pellat's *EI*$_2$ article "Layl and Nahār", esp. p. 708b.

[9] See Kurz, *European Clocks*, pp. 83-84.

on an approximate formula different from the standard approximate one (see **1.4** and, for example, **3.3**), and they correspond most closely to values recomputed with parameter 20°. I have not been able to determine the underlying formula. Whereas the table has values:

26;30° 25;39° 32;56°

for arguments λ = 271°, 1°, and 90°, recomputation with the accurate formula and parameter 20° gives:

27;46° 25;57° 35;11°

and recomputation with the standard approximate formula and parameter 20° gives:

25;24° 25;59° 36;30° .

This anomalous set of entries of course affects the values of n (=2n-r-s). Otherwise the tables are rather accurately computed.

14.2 Anonymous Seljuk prayer-tables (II)

MS Istanbul Nuruosmaniye 2782, copied in Sivas in 773 H [= 1371/72] by Zayn al-Munajjim "the astronomer" ibn Sulaymān al-Qūnawī, consists mainly of calendrical and astronomical tables of a non-numerical character. The accompanying text, which is in Persian, makes mention of al-Khwārizmī, Kūshyār and al-Bīrūnī. Amongst the tables is one displaying the solar longitude for each day of the Syrian year (fol. 39v, the entry for Ādhār I is 11^s 20°), followed by a set of tables for timekeeping (fols. 53v-54r), here labelled in Persian *jadwal-i ṭāliʿ-i awqāt*, which means "table showing the horoscopus at different times of day". It is stated that the entries were derived using a sextile astrolabe (*asṭurlāb sudsī*).[10] Another copy of this table came to light after this study had been completed (the first time round). It is contained in the unique MS Cambridge Browne O.1, fol. 179r, of the Persian *Zīj-i Mufrad* of the 11th(?)-century astronomer Muḥammad ibn Ayyūb al-Ṭabarī.[11] However, this table – see **Fig. IV-5.3** – is not necessarily original to the *Zīj,* since it is appended to the main tables in a different hand. It is not certain whether the Persian or the Arabic terminology is original, and the precise significance of the *chāsht* or *ḍaḥwa* is uncertain. The names of the functions tabulated, that is, the ten times of day, are given in the two sources as follows:

1	*ṣubḥ-i kādhib*	*al-fajr al-kādhib*	false dawn
2	*ṣubḥ-i-ṣādiq*	*al-fajr al-ṣādiq*	true dawn
3	*chāsht-i kūchak*	*al-ḍaḥwa 'l-ṣughrā*	the small (= first) *chāsht* or *ḍaḥwa*
4	*chāsht-i-miyān*	*al-ḍaḥwa 'l-wusṭā*	the middle *chāsht* or *ḍaḥwa*
5	*chāsht-i-buzurg*	*al-ḍaḥwa 'l-kubrā*	the large (= last) *chāsht* or *ḍaḥwa*
6	*nīm-i-rūz*	*anṣāf al-nahār*	midday
7	*miyān-i-dū-namāz*	*mā bayna 'l-ṣalātayn*	(the time) between two prayers
8	*awwal-i-namāz-i-dīgar*	*awwal al-ʿaṣr*	beginning of the other (= afternoon) prayer

[10] See Hartner, "Astrolabe", and also **X-4.1**.
[11] On al-Ṭabarī see Kennedy, "*Zīj* Survey", no. 65; and Storey, *PL*, II:1, pp. 3-4 and 43-44.

9	*ākhir-i-namāz-i-dīgar*	*ākhir al-ʿaṣr*	end of the afternoon prayer
10	*mughīb-i-shafaq*	*maghīb al-shafaq*	end of evening twilight

The underlying latitude is found by inspection to be about $\phi = 38°$, and it is not possible to determine the value used for ε. This latitude corresponds to Konya, although the possibility that the table was lifted from an earlier Iranian source cannot be excluded (see below). The three sub-tables for twilight (1, 2, and 10) are based on parameters 23°, 18°, and 18°, respectively. The remaining seven sub-tables correspond to the division of the interval from sunrise to sunset into eight equal parts, *i.e.*:

$^1/_8$, $^1/_4$, $^3/_8$, $^1/_2$, $^5/_8$, $^3/_4$ and $^7/_8$.

This is the only source in which such a division of the day is attested; on the significance of this division see **IV-5.3**.

14.3 The *Rūznāme* of Shaykh Vefā

The *Rūznāme-yi Vefāʾī* exists in dozens of manuscript sources, of which I have examined MSS Istanbul Hamidiye 842, Istanbul Nuruosmaniye 2914,1, Paris BNF turque 186, 187, 188, 194, turque supp. 537, Vienna 1426 and 1427, and several copies in Cairo, including Cairo KhMT 3: see **Fig. 14.3a-b**. See The identity of the compiler of this almanac has not previously been established with any certainty: the work has previously been ascribed to Shaykh Vefā, a celebrated saint who lived in the time of Muḥammad II and Bāyezīd II and died in the year 1491,[12] and to Shaykh Vefāʾī Muḥammad, who wrote a history of Murād III *ca.* 1585. A new source, MS Cairo K 4037, copied *ca.* 1700,[13] contains an anonymous commentary on the *Rūznāme* and confirms the former attribution, providing the following biographical information (fols. 1v-2r):

> "Actually his name is Muṣṭafā ibn Aḥmad ibn al-Ṣīrawi (?) al-ʿĪsawī, known as Vefā. This is as found in his handwriting on some of his copies (of the *Ruznāme*), according to the statement of the author of the (book entitled) *al-Sh[aq]āʾiq*.[14] [Note: the rest of these notes are probably taken from that work.] Vefā was the name of his father, but (the son) was also known by (this name). He studied mysticism under Shaykh ʿAbd al-Laṭīf al-Qudsī and with him completed the stages of the order.[15] (al-Qudsī) authorized him (to give) instruction (himself). (Shaykh Vefā) – may God Almighty have mercy on him – was a Ḥanafī, versed in all of the phenomenological and esoteric sciences and also in the science of timekeeping. He had noble behaviour because of his blessing. He also had a complete knowledge of music and rhetoric in poetry and prose. He was contemporary with the Sulṭān Muḥammad al-Fātiḥ [reg. 1444-1446] and he died in the time of the Sulṭān Bāyezīd (II) [reg. 1481-1512] in Constantinople.

[12] See *Cairo ENL Survey*, no. H2; and İhsanoğlu *et al.*, *Ottoman Astronomical Literature*, I, pp. 51-54, no. 19 (with references to many more manuscripts).

[13] For the Arabic text see *Cairo ENL Catalogue*, II, pp. 234-235.

[14] This is the biographical encyclopedia *al-Shaqāʾiq al-nuʿmāniyya fī ʿulamāʾ al-dawlat al-ʿUthmāniyya* by Ṭashköprüzāde (d. 1561 – see the EI_2 article by F. Babinger updated by Christine Woodhead).

[15] See the article "Taṣawwuf" in EI_2.

Fig. 14.3a-b: More extracts from another copy of the *Rūznāme* of Shaykh Vefā: calendrical tables and mystical interpretations of the Kaʿba, mentioning the *rijāl al-ghayb*, intermediaries between God and Man (b), and the first two of twelve pages of tables for the times of prayer (c). [From MS Cairo KhMT 3, fols. 1v-2r and 2v-3r, courtesy of the Egyptian National Library.]

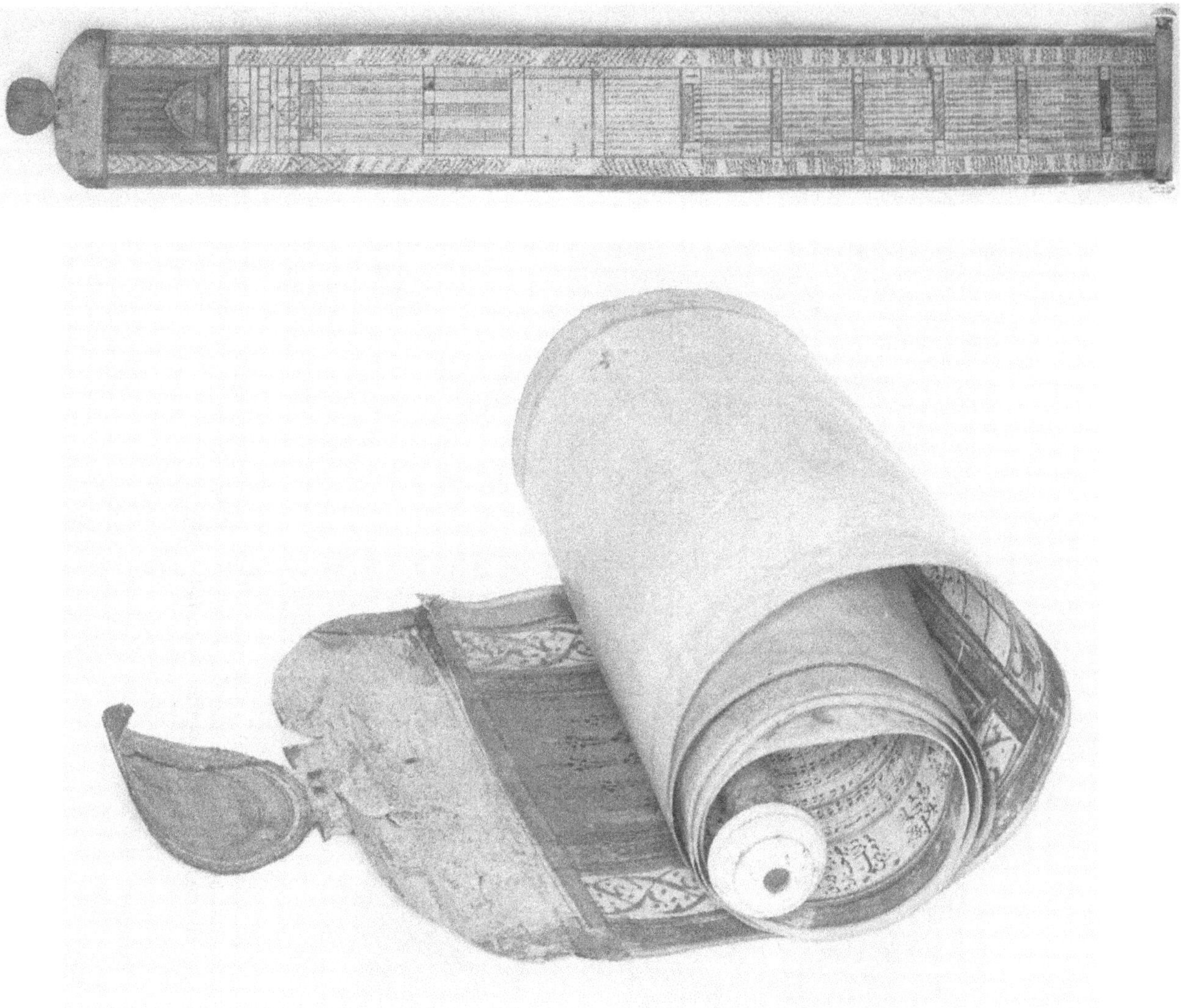

Fig. 14.4b-c: The Oxford copy of Darendelī's prayer-tables, apparently dated 1795. [From MS Oxford MHS 84-33, courtesy of the Museum of the History of Science.]

(This was) after he had journeyed on the pilgrimage by sea. The Christians took him prisoner and incarcerated him in the Citadel of Rhodes.[16] He compiled these tables in prison. The *amīr* Ibrāhīm Bey ibn Qarāmān[17] bought him from (the Christians), then he went home to Constantinople and died there. He has a prayer room and mosque which are known, and his tomb is in front of the mosque. It is famous and is visited (by the people) to obtain blessing from him.[18] The date of his death is the year eight hundred and ninety six of the Hijra of the Prophet [= 1490/91] ... ."

[16] On the history of Rhodes under the Ottomans see the EI_2 article "Rodos" by S. Soucek.

[17] Perhaps the grandson of the better known Ibrāhīm Beg (d. 1464): see the EI_2 article "Ḳarāmān-Og͟hullari (Ḳarāmānids)" by F. Sümer, esp. pp. 619 and 622.

[18] İhsanoğlu, *Ottoman Astronomical Literature*, I, p. 51.

All copies of the *Rūznāme* contain calendrical tables. However, MSS Istanbul Hamidiye 842 and Nuruosmaniye 2914,1, Paris BNF turque supp. 537, and Vienna 1426 contain in addition a set of prayer-tables. In MS Vienna 1426 there are twelve sub-tables, one for each month of the Syrian year. In this particular Paris manuscript the tables for each month have been copied on both sides of a single folio. The solar longitude corresponding to each day of the year is given in the argument column in MS Vienna 1426, and in a separate table in the Paris manuscript, which is described below. The functions tabulated are the following, and the values are expressed in *equinoctial hours* (*sāʿāt*) *with fractions thereof in equatorial degrees* (*daraja*):

2D	*nahār*, length of daylight
2N	*layl*, length of night
D	*ẓuhr*, time from sunrise to midday or midday to sunset
t_a	*ʿaṣr*, time from midday to the afternoon prayer
T_a	*maghrib*, time from the afternoon prayer to sunset
s	*ʿishāʾ*, duration of evening twilight.

The entries are written in Eastern Arabic numerical characters. Pretty designs (MS Vienna 1426) or the Arabic word *khālī* meaning "empty" (MS Paris BNF turque supp. 537) are used for units which are zero. Half degrees are indicated by the symbol ι. The parameters underlying these prayer-tables are:

$$\phi = 41;30° \text{ and } \varepsilon \approx 23;29° \ ,$$

with 19° and 17° for morning and evening twilight. In other Ottoman tables the latitude 41;30° is used for Edirne (**14.7**), but it seems certain that the *Rūznāme-ye Vefāʾī* was intended for use in Istanbul, where the latitude is in fact 41°.[19]

In MS Paris BNF turque supp. 537 there is a separate table of two functions for regulating the *ifṭār* and *imsāk* in Ramaḍān, that is, the time of the evening meal after the day's fast and the beginning of fasting the next day. The functions are labelled *ḥiṣṣat-i ghurūb* and *ḥiṣṣat-i ṣuḥūr* (*sic* for *suḥūr*) and values are given to one digit for each day of the Syrian year. The solar longitude is also given for each day. In fact the main functions are simply T_a and r, expressed in equatorial degrees.

MS Istanbul Esat Efendi 1973, fol. 148v, contains a table entitled *jadwal mīqāt imsākiyya,* which means "table for the time of fasting in Ramaḍān". It is stated that the *tamkīn*, the time before daybreak when one should start fasting, is 12^m. The entries in the table are given in hours and minutes according to the Ottoman convention, for each degree of λ from Capricorn 1° to Gemini 30°. The entries are symmetrically arranged, but both vertical arguments run from 1° to 30°, whereas the upward arguments should run from 0° to 29°. The entries are 10^m (not 12^m) less than the times for daybreak given in the *Rūznāme-yi Vefāʾī*. The remaining tables in this manuscript consist of planetary tables for Istanbul based on the parameters of Ulugh Beg.

---

[19] See n. 14:25.

14.4 The *Rūznāme* of Darendelī

The *Rūznāme* of Darendelī has been published three times, by Velschius (1676), d'Ohsson (1787), and Navoni (1804).[20] From an examination of the material published by d'Ohsson and Navoni, as well as of MSS Istanbul Aşir Efendi 470, Istanbul Kandilli 440, Oxford MHS 84-33, and an unnumbered manuscript in the Yale Medical School Historical Library, New Haven, Conn., it is clear that at least these copies are recensions of the same work. The relatively small number of surviving manuscripts suggests that this *rūznāme* was less popular than that of Shaykh Vefa (**14.3**). Navoni and d'Ohsson attributed the almanac to an individual named Darendelī (Arabic, al-Darandawī), as did Toderini, who stated in 1789 that the almanac had been in use in Turkey for a century.[21] Muḥammad ibn ʿUmar ibn ʿUthmān Darendelī[22] is also known as the author of a treatise on the astrolabic quadrant,[23] and he died in 1739: his name indicates that he or his family was associated with the town of Darende in central Anatolia.[24] Judging by the number of extant copies his almanac was less popular than that of Shaykh Vefā (**14.3**).

The lithograph copy of Darendelī's almanac published by d'Ohsson and also the Yale and Oxford manuscripts are in the form of scrolls, about 10 cm wide and 1 m long: see **Fig. 14.4a-b**. The non-numerical material is written around the tables in Osmanli Turkish, and the entries in the tables are written in Eastern Arabic numerals as in the *Rūznāme-ye Vefāʾī*. An entry such as 2512 (in Arabic numerals) indicates 25 minutes past 12 o'clock. Zero is represented by a dot in the number 10, but by three dots ∴ when it is one of the units of an entry. One-half is represented by a tick or by /. Time is reckoned in equinoctial hours and minutes, according to the Ottoman convention.

The prayer-tables of Darendelī consist of six sub-tables for each pair of zodiacal signs: see **Fig. IV-5.4**. The vertical argument in each generally runs from 1° to 30°, but extra entries enable the user to find the prayer times for each day of the solar year using a second argument column. Nine functions are tabulated side by side in each sub-table. These are:

2D	*nahār*, the length of daylight
2N	*layl*, the length of night
M′	*ẓuhr*, the time of midday
a′	*ʿaṣr-i awwal*, the earliest time for the first afternoon prayer
b′	*ʿaṣr-i thānī*, the earliest time for the second afternoon prayer
s′	*ʿishāʾ*, the time of nightfall
r′	*imsāk*, the time of daybreak
t_q'	*qibla*, the time when the sun is in the azimuth of Mecca
∂'	*ḍaḥwa*, mid-morning

[20] G. H. Velschius, *Commentarius in ruzname naurus*, Augsburg, 1676; M. d'Ohsson, *Tableau général de l'Empire Othoman*, vol. I, Paris, 1787; and J. B. Navoni, "Rouz-name ou calendrier perpetuel des Turcs", *Fundgruben des Orients (Mines de l'Orient)* 4 (Vienna, 1814), pp. 38-67, 127-153, 253-277, and appendix.

[21] See n. 14:4.

[22] İhsanoğlu *et al.*, *Ottoman Astronomical Literature*, II, pp. 406-410, no. 271, also p. 464.

[23] In *Cairo ENL Survey*, no. D176, this treatise is incorrectly attributed to "Muḥammad Efendī Darandar", listed under Egyptian (and Syrian) authors.

[24] See the EI_2 article "Lāranda", and also King, *Mecca-Centred World-Maps*, p. 75 and n. 65.

The entries in the tables are based on the parameters:

$$\phi = 41;0°, \ \varepsilon \approx 23;29°, \ q = 42;0°$$

with 19° and 17° for morning and evening twilight. The value for the latitude of Istanbul is very accurate, and much better than that used by Shaykh Vefā (**14.3**) or the first value derived by Taqi 'l-Dīn (**14.9**), let alone the earlier (and absurd) Islamic and Byzantine value 45°.[25] The value of q indicates that the longitude difference between Istanbul and Mecca was taken as about 17°: in fact, it is only about 11°, and the modern value of q is approximately $28^1/_2$°.

The time called *imsāk* is of particular significance during Ramaḍān. The word means "abstinence" and denotes the beginning of the fast which lasts until sunset. In these tables the *imsāk* is identified with daybreak (*fajr*). However, at the foot of each of the three copies of Darendelī's almanac which I have consulted, it is stated that the *imsāk* should be 15^m before the time given in the table. In al-Gedūsī's treatise on the sine quadrant, a late Ottoman source (**14.15**), it is stated that the *imsāk* should be 16^m before dawn.[26] On the time called *ḍaḥwa*, Navoni (1814) remarked:[27]

> "[Le] Zahve(?), vulgairement dit *Kouschlouk,* marque un autre temps entre le lever du soleil et le midi, mais sans une prière particulière et d'obligation."

In fact, the *ḍaḥwa* divides the interval between daybreak and sunset in half. I do not know whether any religious activity was associated with the *ḍaḥwa* in Ottoman Turkey. On the *ḍuḥā* or *ḍaḥwa* in late Islamic astronomical sources see further **IV-5.4**.

14.5 Muḥammad ibn Kātib Sinān's tables for timekeeping by the stars

Muḥammad ibn Kātib Sinān was a *muwaqqit* in Istanbul in the late 15th century.[28] One of his works is an enormous table for timekeeping by the stars, preserved in MSS Istanbul Ayasofya 2710 and Istanbul Topkapı T 3046 (Ahmet III 3515). The table occupies some 500 pages and contains about 240,000 entries, and is arranged so that one can feed in both the normed right ascension of a star which is culminating and the solar longitude, and read off:

(1) the time since sunset,
(2) the time remaining until sunrise,
(3) the time remaining until daybreak, and
(4) the time remaining until midday .

The tables are computed for latitude $\phi = 41;0°$ (Istanbul): see further **I-2.7.2**. On a much smaller Turkish table for timekeeping by the stars, or rather, one particular star, see **I-2.8.1**.

---

[25] On this see my notes in "Byzantine Astronomy", pp. 117-118; and King, "Geography of Astrolabes", p. 12.

[26] See n. 14:40.

[27] Navoni, "Rouz-name des Turcs" (cited in n. 14:20), p. 58.

[28] On Muḥammad ibn Kātib Sinān (**I-2.7.2**) see Suter, *MAA*, no. 455; *Cairo ENL Survey*, no. H8; and İhsanoğlu *et al.*, *Ottoman Astronomical Literature*, II, pp. 84-90, no. 46.

14.6 Three Ottoman copies of al-Khalīlī's universal auxiliary tables

MS Istanbul Hamidiye 1453,3 (fols. 232v-266v) contains al-Khalīlī's universal auxiliary tables (**10.7**) and was copied in Edirne in 869 H [= 1464/65] by ʿUmar ibn ʿUthmān al-Ḥusaynī. The tables follow a number of treatises on timekeeping and quadrants (see, for example, **2.5** and **2.6**) and are preceded by al-Khalīlī's instructions written in Arabic (see below). The functions f_ϕ and g_ϕ are tabulated only for arguments:

$$\phi = 20°, 21°, \ldots, 49°, \text{ as well as } 21;30° \text{ (Mecca) and } 33;30° \text{ (Damascus)} .$$

The main tables are preceded by a small table (fols. 233v-234r) displaying the qibla as a function of local latitude and longitude, $q(\Delta\phi,\Delta L)$, which, as I have shown elsewhere, is ultimately of Abbasid origin;[29] a table of $\delta(\lambda)$ to two digits, which, although for $\lambda = 90°$ it has entry 23;32°, a value of ε nowhere else attested in the known Islamic sources, is apparently based on $\varepsilon = 23;31°$; and tables for finding the solar longitude from the date in the Muslim and Syrian calendars (the solar longitude for Ādhār 1 is given as Pisces 20°;23°). The main tables are followed by some simple trigonometric tables of the functions:

$$\text{arc Sin}(x),\ \text{arc Vers}(x),\ \text{Cot}_{12}\ h \text{ and } \text{Tan}_{60}\ h ,$$

as well as a table of max $d(\phi)$ based on $\varepsilon = 23;35°$.

MS Istanbul Ayasofya 2590 is a second Ottoman copy of al-Khalīlī's universal auxiliary tables in a recension prepared by the *muwaqqit* Muḥammad ibn Kātib Sinān (**14.5**) for the Sulṭān Bāyazīd, and dated 897 H [= 1491]. Ibn Kātib Sinān translated al-Khalīlī's introduction into Turkish and then copied the tables in their entirety. He also copied al-Khalīlī's parameters 19° and 17° for twilight in the introduction, although elsewhere (as in his treatise on the astrolabe preserved in MS Istanbul Ayasofya 2708) he advocated 20° and 16°. The main set is preceded by others of $\delta(\lambda)$ and $\delta_2(\lambda)$ computed to three digits for $\varepsilon = 23;35°$ and followed by tables of $h_a(H)$ and $h_b(H)$.

In a late Ottoman miscellany of astronomical treatises preserved in an unnumbered manuscript formerly (*ca*. 1970) in the private collection of the late Professor Buhairi of the American University of Beirut, there are tables of $f_\phi(\theta)$ and $g_\phi(\theta)$ for $\phi = 41°$ (Istanbul) and a single table of $K(x,y)$ for $x = 41$. There are no accompanying instructions, and the hapless individual who thought fit to copy these three tables together was unaware that the argument x is unrelated to the local latitude.

14.7 Anonymous timekeeping tables for Edirne

MS Oxford arab. e. 93, fols. 3v-25r, contains an undated set of anonymous tables of the functions $t(\lambda,h)$ and $T(\lambda,h)$ expressed in equatorial degrees: see **I-2.2.2** (illustrated). Values are given for each degree of λ starting at the winter solstice and each degree of h up to $[H(\lambda)]$, and the city of Edirne is specifically mentioned at the head of some of the tables. The underlying parameters are:

$$\phi = 41;30° \text{ and } \varepsilon \approx 23;30° .$$

[29] See King, "Earliest Qibla Methods", p. 118.

There are no prayer-tables accompanying this set, but notice that the *Rūznāme-ye Vefā'ī* for Istanbul (**14.3**) is based on the same parameters. See also **14.11** on some other tables for Edirne.

14.8 Anonymous prayer-tables for Istanbul

MS Cairo ṬM 255,6, copied 1060 H [= *ca.* 1650], contains al-Khalīlī's introduction to his prayer-tables and some anonymous Syrian prayer-tables for latitudes:

ϕ = 36° (Aleppo) and 41;15° (Istanbul).[30]

Mention of these was omitted from the first version of this text, and all that is available to me now is a photo of an extract from the tables for Aleppo (**Fig. 11.5**). This shows tables of the same kind as those of al-Khalīlī but for ε = 23;35° rather than his more up-to-date value 23;31°. The functions tabulated for Aleppo are:

$H, Z, D, h_a, t_a, T_a, N, n, r, s, \tilde{h}, d$ and t_q ,

MS Istanbul UL T1824,1 (fols. 3r-9r) contains another set of prayer-tables for Istanbul, based on the parameters:

ϕ = 41;15° and ε = 23;30° .

No compiler is mentioned. Notice that the value of the latitude is the same as that used by Taqi 'l-Dīn in his early work (**14.9**), and that the value of the obliquity is the one usually associated with astronomers who relied on Ulugh Beg (see, for example, **7.1**). The set begins with a table of $\delta(\lambda)$ computed to two digits for each degree of λ. The main prayer-tables display some 22 functions tabulated side by side for each degree of λ from Capricorn 1° to Gemini 30°, as in the tables of al-Khalīlī (**10.6**). Values are given to two digits. The functions tabulated are the following:

$D^h, C, H, D, 2D^h, h_a, h_b, t_a, t_b, T_a, T_b, 2N^h,$
$s, (2N\text{-}r), r, (r\text{+}D), (D\text{+}t_a), h_q, t_q, \tilde{h}, B$ and b_1.

MSS Paris BNF ar. 2544,14 and Cairo DM 36 contain an extract from these tables displaying only the functions:

$D, 2D^h, t_a, T_a, s, (2N\text{-}r), r$ and $\tilde{h}$.

No compiler is mentioned in either source.

These tables are rather accurately computed. The parameters used for twilight are 19° and 17°, and I have not investigated the value of the qibla underlying the tables of h_q and t_q. Note that in the geographical tables of Taqi 'l-Dīn (**14.9**), we find the entries:

Istanbul	ϕ: 41;15°	L: 60;0°
Mecca	21;30	77;0°

so that the qibla at Istanbul, computed according to the accurate formula, would be 41;10° E of S. On the tables of B, b_1 and C see **I-6.4.9**, **I-6.5.2** and **I-6.10.7**. The fact that a table of b_1 was included in the set I take to be further evidence that its compiler was inspired by the tables of al-Khalīlī.

MS Istanbul Hamidiye 842 contains another set of less extensive prayer-tables based on the same parameters, but with slightly different entries. Values of the following functions are given

[30] *Cairo ENL Survey*, no. C142.

to two digits in modern Arabic numerical notation for each degree of λ from Capricorn 0° to Gemini 29°:

$$H,\ Z_{(12)},\ D,\ h_a,\ t_a,\ T_a,\ s,\ r,\ \tilde{h} \text{ and } 2N^h\ .$$

MS Cairo MM 100 (13 fols., *ca.* 1750) contains some anonymous prayer-tables for Istanbul arranged according to the Byzantine months, which I have not investigated.

14.9 Taqi 'l-Dīn on some aspects of timekeeping

Taqi 'l-Dīn (*ca.* 1525-1585)[31] was the principal astronomer associated with the founding of the short-lived observatory in Istanbul.[32] He wrote a number of works of considerable interest, which have not yet received the attention which they deserve. MS Istanbul Kandilli 208 is a copy of some of Taqi 'l-Dīn's works in his own hand, and contains a *ṭaylasān* table displaying the hour-angle and time since sunrise as functions of the meridian altitude and the instantaneous altitude for parameters:

$$\phi = 41° \text{ (Istanbul) and } \varepsilon = 23;30°\ .$$

I see no reason yet to doubt that this was computed by Taqi 'l-Dīn himself (see further **I-2.3.6**). It may be that Taqi 'l-Dīn also compiled the prayer-tables discussed in **14.8** since he used the latitude 41;15° for Istanbul in his earlier works, such as his treatise on sundial construction. In his later work, however, he used latitude 40;58° for Istanbul and obliquity 23;28,54°: these are the results of his "new Murād Khān observations" (see, for example, **I-6.4.8**).

In his *Zīj* entitled *Jarīdat al-durar wa-kharīdat al-fikar*, of which I have examined MS Istanbul Esat Efendi 1976, Taqi 'l-Dīn advocates the parameters 19° and 17° for twilight and then mentions certain corrections (*taʿdīl*) which should be applied to the length of twilight or the diurnal arc. The following is a translation of the relevant section in Ch. 98 of the *Zīj*:

> "For a day of fasting it is necessary to modify the duration of twilight and the length of daylight by the quantities which I determined by observation and calculation. This is explained in detail in my treatise called *al-Qawl al-marʾī fī 'l-nahār al-sharʿī*. In brief, I say that if Ramaḍān occurs when the sun is at the beginning of Aries then the amount which must be added to the semi diurnal arc in the west is 1;3°, likewise, Taurus 1;20°, Gemini 1;43°, Cancer 1;39°, Leo 1;23°, Virgo 0;59°, Libra 0;54°, Scorpio 0;54°, Sagittarius 0;39°, Capricorn 0;29°, Aquarius 0;59°, and Pisces 1;3°. To the semi diurnal arc in the east should be likewise added, in the order of signs: 0;56°, 0;58°, 1,8°, 1;15°, 1;16°, 1;8°, 1;6°, 1;7°, 1;6°, 1;5°, 0;59°, and 0;58°. In the middle of the signs one should use linear interpolation. Modifying the length of daylight obviates the need to modify the duration of morning twilight, because the latter is defined in terms of the apparent horizon and the light of the upper edge of the disk of the sun (*ḥājib al-shams*). The size (of the sun) never exceeds the sum of the two corrections, that is, a maximum 2;51° in Gemini, which is about 3°, and a minimum 1;34° in Capricorn, which is approximately $1^1/_2$°, an amount that is not at all small (*i.e.*, negligible)."

[31] On Taqi 'l-Dīn see n. **I**-2:35. See also the next note.

[32] See Sayılı, *The Observatory in Islam*, pp. 289-305, and also Ünver, *Istanbul Observatory* (in Turkish).

Unfortunately I do not know of any copies of Taqi 'l-Dīn's treatise *al-Qawl al-marʾī* ... . The title means something like "A Statement as Clear and Visible as Light of Day concerning the Length of Daylight as Defined by the Religious Law".[33] It would be interesting to know how he derived these two sets of values. Suffice it to say for the time being that the mean of the two sets is suspiciously close to the values given by ʿAbd al-Qādir al-Minūfī (**8.2**) for $\phi = 41°$. These mean values are:

1;0 1;19 1;26 1;27 1;20 1;4 1;0 1;1 0;53 0;47 0;59 1;1 ,

and al-Minūfī has:

1;4 1;11 1;18 1;25 1;18 1;11 1;4 0;57 0;51 0;44 0;51 0;57 .

Note that al-Minūfī Jr.'s tables for refraction were compiled in Cairo in 975 H [= 1567] and Taqi 'l-Dīn's *Zīj* was compiled in Istanbul about 990 H [= 1582].

In a short treatise preserved in MSS Istanbul Kandilli 208 (fols. 89v-90r) and Istanbul Kandilli 176, of which the former is in his own hand, Taqi 'l-Dīn claims to discuss the difference of opinion between the Cairo *muwaqqit*s concerning the precise length of daylight, nightime, and twilight. He makes no mention of the nature of the difference of opinion, but does state that the visible horizon is 0;2,13° below the true horizon. This number is related to the numerical tradition attributed to Ibn al-Haytham (see **7.1**, also **9.3**). Otherwise the treatise contains no information of historical or scientific consequence.

MS Istanbul Kandilli 441 of the shorter version of Ṣāliḥ Efendī's tables for timekeeping (**14.13**), copied in 1266 H [= 1849/50], contains a simple table for evening twilight (p. 62) and a set of geographical coordinates (p. 73), both of which are attributed to Taqi 'l-Dīn. The following values are given for the duration of twilight in equatorial degrees for each zodiacal sign:

23 25 29 *32* 29 25 23 23 24 25 24 23

If we round the values given in MS Istanbul UL T1824,1 (**14.8**), which are based on Taqi 'l-Dīn's first value 41;15° for the latitude of Istanbul and parameter 17°, we obtain:

23 25 29 *31* 29 25 23 23 24 25 24 23 .

All but the fourth value correspond precisely to Taqi 'l-Dīn's values.

Since there is a dearth of Ottoman geographical tables that have been published,[34] and since the Kennedys did not use any in their survey of Islamic geographical tables, I present the table attributed to Taqi 'l-Dīn in full:

Locality		ϕ	L
1	Istanbul	41;15°	60; 0°
2	Gallipoli	41;15	58; 0
3	Mitilene	39; 0	55; 0
4	Salonica	39;30	50; 0
5	Kavalla	40;30	46;30
6	Cos	38; 0	55; 0
7	Izmir	38; 0	56;20

[33] No copies of this work are mentioned in İhsanoğlu *et al.*, *Ottoman Astronomical Literature*, I, pp. 199-217, but Taqi 'l-Dīn's treatises listed as nos. 3, 12 and 16 (pp. 204, 209-210 and 211) relate to the same subject and would repay investigation.

[34] This table is now featured in Sezgin, *GAS*, X, pp. 182-184. See also King, *Mecca-Centred World-Maps*, p. 86-87, on another Otoman table giving only the qiblas of 90 localities.

8	Rhodes	36;50	54;30
9	Crete	35; 0	54; 5
10	Cyprus	35;30	68;30
11	Alexandria	31; 0	61;55
12	Cairo	30;10	64;55
13	Jerusalem	32;10	66;30
14	Damascus	33;35	70; 0
15	Aleppo	36; 0	[3]6;10 [a]
16	Medina	24;30	75;20
17	Mecca	21;30	77; 0
18	Kayseri	39; 0	69; 0
19	Kastamonu	41;30	65; 0
20	Eskişehir	39;30	62;50
21	Konya	38; 0	65;10
22	Bursa	40; 0	58;10
23	? (illegible)	41; 0	60;15
24	Dile	41;30	61; 0
25	Eregli	41;30	67; 0
26	Sinop	40;30	69;30
27	*B-ʾ-r-ṭ-y-z* (?)	41;30	64; 0
28	Trabzon	40;50	71;50
29	Edirne	41;30	58;30
30	*Q-r-q k-l-y-ʾ* (?)	41;30	58;30

[a] text: 26;10°

14.10 Anonymous prayer-tables for Anatolia

MS Cairo KhMT 2, copied about 1750, consists mainly of a set of planetary tables based on the *Zīj* of Ulugh Beg and a few spherical astronomical tables specifically for $\phi = 41°$, that is, Istanbul. Included amongst these (fols. 55v-58r) is a set of prayer-tables for latitude 38°, locality unspecified. The functions tabulated are:

$$H,\ 2D,\ D,\ t_a,\ t_{a/b},\ t_b,\ T_b,\ s,\ 2N,\ n \text{ and } r\ ,$$

and the values are given in Eastern Arabic numerical notation to the nearest half degree, one-half being indicated by the symbol ι. I have not investigated the parameters underlying the values of s and r. The function $t_{a/b}$ is refered to as *bayn al-ʿaṣrayn*, and displays the hour-angle at a certain time about halfway between the two *ʿaṣr* prayers. The tables display some discrepancies: for example, the values of t_b and T_b do not total D, and the values of s and n and r do not total 2N.

MS Istanbul Hüsnü Paşa 1286, copied around 1700, contains an anonymous set of tables (fols. 66r-74r) of the functions:

$$2N,\ N,\ B,\ C,\ H,\ \delta,\ d,\ \psi,\ h_0,\ D^h,\ D,\ 2D,\ \alpha_\phi \text{ and } \alpha'\ ,$$

computed for parameters:

$$\phi = 42° \text{ (Edirne) and } \varepsilon = 23;31° \text{ (?)}.$$

On other tables for Edirne see **14.7**.

14.11 An anonymous *ṭaylasān* table for Edirne (?)

MS Cairo KhMT 2,1, copied about 1750 (**14.10**), contains a flyleaf (fol. 1r) on which part of a *ṭaylasān* table has been copied. The table is entitled *jadwal faḍl al-dāʾir* and displays the hour-angle t(H,h) for the argument domains:

$$H = 71;29°,\ 71°,70°,\ \dots\ ,\ 48° \text{ and } h = 0°,\ 1°,\ \dots\ ,\ H\ .$$

Values are given to two sexagesimal digits. Since the entry for H = 48° and h = 0° is 90;0°, the underlying parameter is:

$$\phi = 42;0° \quad (\text{with } \varepsilon = 23;29°)\ .$$

Not all of the entries have been filled in, and the page of tables for $\delta < 0$ is not contained in this manuscript. There is no mention of the locality for which the table was prepared; this may have been Edirne. See also **I-2.3.7**.

14.12 Aḥmad Efendī's timekeeping tables for Istanbul

MS Istanbul Kandilli 196 is the only copy known to me of a set of hour-angle tables for Istanbul, computed in 1095 H [= 1684] by "Aḥmad Efendī, known as Miṣrī al-Islāmbūlī", "the Egyptian in Istanbul".[35] The function t(h,λ) is tabulated for the domains:

$$h = 1°,\ 2°,\ \dots\ ,\ 72° \quad \text{and} \quad \lambda = 1°,\ 2°,\ \dots\ ,\ 90°\ (\delta \gtrsim 0)$$

and the underlying parameters are:

$$\phi = 41;0° \text{ (Istanbul) and } \varepsilon = 23;30°\ .$$

The tables have the same format as those in the main Cairo corpus (**4.0**) and are quite accurately computed: see already **I-2.1.7**. Aḥmad Efendī prefered to use obliquity 23;30°, as in the late Egyptian tradition (**7.1**), rather than Taqi 'l-Dīn's value 23;28,54° (**14.9**). A century later Ṣāliḥ Efendī recomputed t(λ,h) for latitude 41° using Taqi 'l-Dīn's parameter (**14.13**).

14.12* Prayer-tables for Tillo by Ismāʿīl Fahīm

In February, 2003, as I was preparing the final illustrations for this book, I came across a negative of some prayer-tables in a Cairo manuscript that were not yet described in the text. Checking my 1986 *Survey* of the Cairo manuscripts, I found that I had promised a description in *SATMI*. The following must suffice.

MS Cairo DMT 2,1 (fols. 1r-26v, 1202 H [= 1787/88]) is a copy of a set of prayer-tables by Ismāʿīl Fahīm, son of the better-known Ibrāhīm Ḥaqqī who was the author of a well-known encyclopaedia entitle *Maʿrifat-nāme*.[36] Five other manuscripts are preserved in Turkey. The tables are preceded by an introduction in Turkish and the work is dated 1193 H [= 1779] and dedicated to the father. The underlying latitude is stated to be 38;30°, and apparently no locality is specified. However, in the colophon of some of the copies mention is made of the village

[35] I did not find mention of Aḥmad Efendī or these tables in İhsanoğlu *et al.*, *Ottoman Astronomical Literature*.

[36] On the author see *Cairo ENL Survey*, no. H39; and İhsanoğlu *et al.*, *Ottoman Astronomical Literature*, II, 627-628 (no. 462), where the other manuscripts of these tables are listed. On his father see *Cairo ENL Survey*, no. H38.

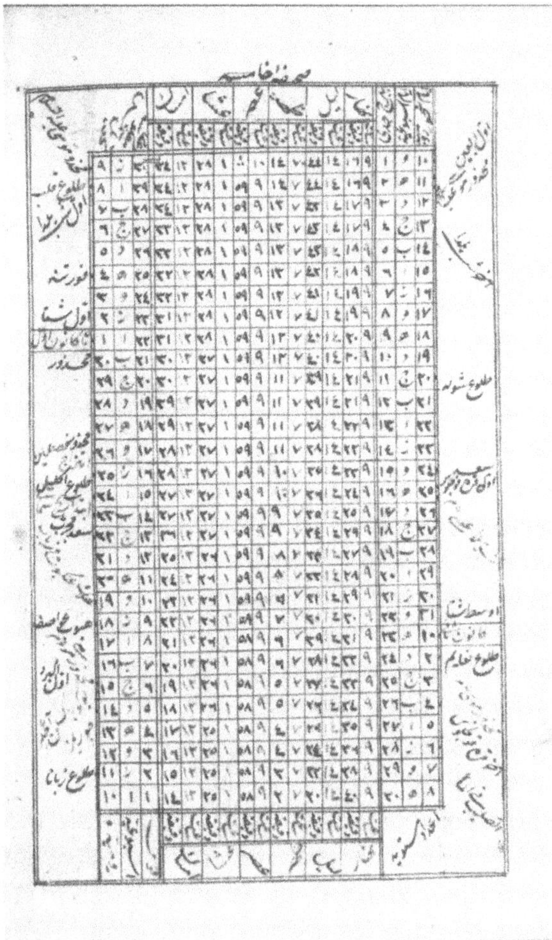

Fig. 14.12*: An extract from the prayer-tables of Ismāʿīl Fahīm. [From MS Cairo DMT 2,1, courtesy of the Egyptian National Library.

of Tillo, where the father, originally from Erzerum, was living at the time. Ekmeleddin İhsanoğlu kindly informs me that Tillo is in the district of Aydınlar in S. E. Turkey, directing me to www.tillo.net. The tables display for each degree of each sign the functions:

$$2D^h,\ 2N^h,\ z',\ a',\ s' \text{ and } i' .$$

I have not investigated the underlying twilight parameters.

14.13 Ṣāliḥ Efendī's corpus of timekeeping tables for Istanbul

Ṣāliḥ Efendī Miʿmārī "the architect" (*fl. ca.* 1700) compiled an extensive corpus of tables for timekeeping by the sun, extant in numerous copies.[37] MSS Princeton Yahuda 353, Istanbul Aşir Efendi 224, Istanbul Kandilli 219, Vienna 2379 (Mixt. 989), Cairo ṬM 151 and 215, and

[37] On Ṣāliḥ Efendī see n. I-2:23.

Cairo K 18199, are copies of this corpus; all but the last are complete. The title is simply *al-jadwal al-kabīr*, "the large table", which is appropriate since the corpus contains over 80,000 entries. Ṣāliḥ Efendī's tables, and simpler versions thereof, were widely used in Istanbul in the 19th century.

The main tables are for timekeeping by the sun; the remainder is for regulating the times of prayer. All are based on the parameters:

$$\phi = 41;0° \text{ (Istanbul) and } \varepsilon = 23;28,54° .$$

This distinctive value of the obliquity is the value derived by Taqi 'l-Dīn (**14.9**), who worked in Istanbul two centuries prior to Ṣāliḥ Efendī.

In the main tables of the corpus Ṣāliḥ Efendī tabulated the hour-angle and time since sunrise T(λ,h) and t(λ,h) for each degree of 1° starting at the winter solstice and for each degree of h up to [H(λ)]. A given sub-table displays entries for one particular solar longitude, with the times for solar altitudes in the east and west expressed in equatorial degrees (*al-dāʾir*, *faḍl al-dāʾir*), equinoctial hours (*al-mustawiya*), seasonal hours (*al-zamāniyya*), and in equinoctial hours according to the Ottoman convention (*al-muwāfiqa*). Entries are given to three sexagesimal digits in each case and are very carefully computed. The way in which the tables were compiled is clear from the nature of certain auxiliary functions also tabulated by Ṣāliḥ Efendī. See further **I-2.2.3**. Various functions are displayed around each of these sub-tables, often in different coloured inks. These are the following:

$$H, 2D, 2N, a', b', r' \text{ and } s' .$$

The parameters used for twilight are 19° and 17°, and the entries, which but for the first function H are in hours, are expressed to three sexagesimal digits.

At the end of the main set of tables the functions:

2D	*al-nahār*	length of daylight
r′	*al-fajr*	time of daybreak
$a' = T_a'$	*al-ʿaṣr al-awwal*	the first afternoon prayer
$b' = T_b'$	*al-ʿaṣr al-thānī*	the second afternoon prayer
2N	*al-layl*	length of daylight
s′	*al-shafaq*	time of nightfall

are tabulated separately (the functions T being labelled *al-bāqī ila 'l-ghurūb*, "time remaining until sunset", and the times *al-muwāfiqa*, meaning "time according to the Ottoman convention"), with entries expressed in equinoctial hours to three digits. A given page of tables serves a pair of zodiacal signs symmetrical with respect to the solstices.

MS Istanbul Husrev Paşa 232 contains a set of tables attributed to Muḥammad Ṣādiq Jihāngīrī, *muwaqqit* in the Mosque called Wālidat al-Sulṭān (Valide Sultan Camii) in Istanbul.[38] This displays the two times when the sun has a particular altitude on the assumption that *midday is 12 o'clock.* Here we see European influence at work. Values are given in equinoctial hours to three digits and are derived from those of Ṣāliḥ Efendī. Anonymous copies of these same tables are contained in MSS Istanbul Bağdatli Vehbi Efendi 990, Istanbul Lala Ismail 287, Istanbul Kandilli 220, 440 and 441, Cairo ṬM 120, and anonymous copies of similar tables

[38] *Ibid.*, II, pp. 561-562, no. 399, and also p. 456 and 684. See the illustration in King, "Astronomical Timekeeping in Ottoman Turkey", pl. 9.

with entries to two digits rather than three are contained in MSS Istanbul Esat Efendi 1979, Istanbul UL T1963 and T1964.

MS Baghdad Awqāf 325/12248 contains some 50-odd pages of tables attributed to Jihāngīrī. I have not been able to inspect this manuscript, but it is catalogued under the title *ikhtilāf mā bayna 'l-ufuq al-ḥaqīqī wa-'l-mar'ī*, which refers to corrections for refraction at the horizon. This title probably relates to something written on the first page of the manuscript rather than the main tables.

On the flyleaf of MS Istanbul Lala Ismail 287, copied in 1195 H (= 1781), is a set of numbers for each of the zodiacal signs labelled *daqā'iq ḥiṣṣat al-qurṣ li-ʿarḍ mīm-alif,* "minutes of the correction for the solar disc at latitude 41°". The numbers are as follows, in the order of signs:

3 2 2 2 2 2 3 4 5 5 4 3 .

14.14 Anonymous Ottoman-type prayer-tables for Crete

MS Istanbul Kandilli 522 contains a set of prayer-tables for the year 1290 H [= 1873/74] computed for the latitude of Khania in Crete (see already **8.8** on some other tables for Crete).[39] The following functions are tabulated for each day of the year, expressed in equinoctial hours and minutes:

$$R', m', a', s' \text{ and } r' .$$

The underlying parameters are found by inspection to be:

$$\phi = 35;20° \text{ and } \varepsilon \approx 23;29° ,$$

and the parameters used for morning and evening twilight are 19° and 17°. The entries for m′, a′ and s′ correspond very closely to recomputation without taking horizon phenomena into consideration. However both the times r′ and R′, daybreak and sunset, are earlier than the recomputed times by an amount which varies between 16^m at the equinoxes and 20^m at the solstices. This appears to be a generous correction for refraction and the size of the solar disc (see also **12.11**), half of which the anonymous calculator should also have applied to the times m′ and a′.

14.15 Gedūsī's twilight tables for different latitudes

At the end of the treatise on the trigonometric quadrant by the 19th-century Ottoman astronomer al-Gedusī,[40] printed in Istanbul in 1311 H [= 1893/94], there is a set of *imsāk* tables for different latitudes. The time of the *imsāk* in hours and minutes, Turkish time, is displayed for each 2° of solar longitude and for latitudes:

21° (Mecca), 24° (Medina) and 35°, 36°, ... , 45°.

I have not investigated the underlying value of the solar depression and the definition used

[39] This source is mentioned in İhsanoğlu *et al.*, *Ottoman Astronomical Literature*, II, p. 807.

[40] On Sulaymān Murād ibn ʿUmar ibn Aḥmad Saʿdī al-Gedūsī see İhsanoğlu *et al.*, *Ottoman Astronomical Literature*, II, pp. 601-602, no. 436. A translation of this treatise and a commentary were published in Würschmidt, "Gedosi über den Quadranten". Würschmidt mentions the tables (p. 154) and apologizes for not reproducing them. Copies of the tables were kindly prepared for me by the late M. Alain Brieux of Paris.

for the *imsāk*. In Gedūsī's treatise the parameters used for morning and evening twilight are 19° and 17°, and the *imsāk* is taken as being 16^m before daybreak.[41]

14.16 Saʿīd Beg Zāde's auxiliary tables for computing the prayer-times for all latitudes

It is appropriate that we should conclude the present study with a few remarks on a set of tables compiled in Istanbul in the 19th century for computing the prayer-times for all latitudes. The compiler of these tables, which are extant in MS Istanbul Kandilli 226, was one Saʿīd (or possibly Sayyid) Beg Zāde Mīr Ibrāhīm, on whom I have no further information.[42] The tables are preceded by an introduction in Turkish and followed by some worked examples. The notation used by Saʿīd Beg Zāde is rather unfortunate: most of his functions are called *taʿdīl*, "equation" or "correction". Also *he uses logarithms to base sixty expressed sexagesimally.* Nevertheless his tables are of interest because they are really still part of the medieval tradition, and indeed bear some resemblance to the auxiliary tables of al-Wafāʾī compiled several centuries previously (**6.15**). One of the two known copies of al-Wafāʾī's tables (MS Nuruosmaniye 2921,2) is preserved in Istanbul, and I consider it likely that Saʿīd Beg Zāde had seen such a copy. He advocates the use of parameters 20° and 17°. He also gives tables for finding the time of prayers at the two major religious holidays in Islam (*ṣalāt-i ʿīdayn*), based on the assumption that the prayer should take place when the solar altitude is 6° in the east. I know of no other tables for this prayer-time, or reference to this particular definition in the Islamic sources. I have outlined the principle underlying these tables in **I-9.13**.

The first set of tables displays three functions of λ, with values for each degree of argument expressed to two sexagesimal digits in the case of the first and to three digits in the case of the second and third. These functions are:

$\delta(\lambda)$ ($\varepsilon = 23;28°$)	*mayl awwal*	first declination
$L(H) = \log\{\mathrm{Sin}\ H\}$	*taʿdīl-i ẓuhr*	equation of midday
??	*taʿdīl-i mayl*	equation of the declination

The first main table displays values of the following functions for each degree of argument (*ʿadad*) H, or H* in the case of the table for twilight, from 1° to 90°. Values are given to three sexagesimal digits, except for the last function whose values are given to two digits.

$L_a(H) = \log\{\mathrm{Sin}\ H - \mathrm{Sin}\ h_a(H)\}$	*taʿdīl-i ʿaṣr-i awwal*	equation of the first afternoon prayer
$L_b(H) = \log\{\mathrm{Sin}\ H - \mathrm{Sin}\ h_b(H)\}$	*taʿdīl-i ʿaṣr-i thānī*	equation of the second afternoon prayer
$L_r(H) = \log\{\mathrm{Sin}\ H^* - \mathrm{Sin}\ 16°\}$	*taʿdīl-i shafaq*	equation of nightfall
$L_s(H) = \log\{\mathrm{Sin}\ H^* - \mathrm{Sin}\ 20°\}$	*taʿdīl-i fajr*	equation of daybreak
$L_\iota(H) = \log\{\mathrm{Sin}\ H - \mathrm{Sin}\ 6°\}$	*taʿdīl-i ṣalāt al-ʿīdayn*	equation of the prayers at the two religious festivals
$S(h) = \mathrm{Sin}\ h$	*taʿdīl-i irtifāʿ*	equation of the altitude

(No values are given for $L_r(H)$ for $H^* < 16°$, *etc.*)

---

41 *Ibid.*, pp. 152-153.

42 I did not find him in İhsanoğlu *et al.*, *Ottoman Astronomical Literature*.

The second main table displays the following functions for each 0;15 of argument *x* (*taʿdīl-i awwal*) up to 120 in the case of the second and third functions, and up to 60 in the case of the remainder. Values are given to two sexagesimal digits except for the second function whose values are given to three:

$S^{-1}(s)$ = arc Sin (s)	*taʿdīl-i thānī*	second equation
L(x) = log x	*taʿdīl-i muʿaddal*	modified equation
V(x) = $^1/_{15}$ arc Vers (x)	*sāʿāt-i dāʾir*	hours of time
C′(x) = $^1/_{15}$ arc Cos (x)	*ẓuhr-i shamālī*	midday for northern latitudes
12^h - C′(x)	*ẓuhr-i janūbī*	midday for southern latitudes

A major disadvantage of these tables, as we shall see, is that one must find V from L in a table where both functions are tabulated with x as argument. I introduce the function

$$V*(x) = {}^1/_{15} \text{ arc Vers (alog x)} ,$$

which Saʿīd Beg Zāde would have done well to tabulate. The third main table displays the function:

$$G*(\lambda,\phi) = \log G(\lambda,\phi) ,$$

called *taʿdīl-i ʿarḍ*, "the equation of the latitude". Values are given to three significant sexagesimal digits for each 5° of λ and each 1° of φ up to 90°.

The prayer-times expressed according to the Ottoman convention can be found using these functions in the following way:

ẓuhr:	$m' = V* \{ L(H) + G*(\lambda,\phi) \}$
ʿaṣr 1 and 2:	$t_{a/b}' = m' + V* \{ L_{a/b}(H) + G*(\lambda,\phi) \}$
shafaq:	$s' = m' - V* \{ L_s(H*) + G*(\lambda, \phi) \}$
fajr:	$r' = m' + V* \{ L_r(H*) + G*(\lambda,\phi) \}$
ṣalāt-i ʿīdayn:	$\iota' = m' + 12^h - V* \{ L_\iota(H) + G*(\lambda,\phi) \}$

To find the hour-angle corresponding to a given solar altitude h, form:

$$L \{ H'(H,h) \} = L \{ S(H) - S(h) \}$$

and then:

$$t(h,H,\lambda,\phi) = V* \{ L \{ H'(H,h) \} + G(\lambda,\phi) \} .$$

Saʿīd Beg Zāde gives numerical examples for φ = 41° and λ = 60°. The prayer-times agree with the corresponding entries in the prayer-tables of Ṣāliḥ Efendī (**14.13**) to within a minute of time. He also demonstrates how to compute the qibla, and using coordinates:

Istanbul	L: 56;33°	φ: 41;0°
Mecca	68;30	21;15

he somehow derives the value 66;31° (measured from the prime vertical). Accurately computed for these coordinates the qibla is 59;16°. As noted above (**14.4**) the qibla at Istanbul is in fact closer to $61^1/_2$° S of E.

Postscript: When I first started writing **Part II** of this book some 30 years ago, tables were available from the local authorities in each region of the Islamic world displaying the prayer times. Some examples are presented in **Fig. V-13.1** and in the article "Mīḳāt ii. Astronomical Aspects" in EI_2. Nowadays one can download tables for any locality from the Internet: see, for example, http://prayer/al-islam.com (2003).

Part III

A survey of arithmetical shadow-schemes for time-reckoning

Dedicated to Otto Neugebauer in 1990
on the occasion of his 90th birthday

ACKNOWLEDGEMENTS AND NOTES ON THIS VERSION

In numerous medieval Islamic treatises on traditional folk astronomy and legal texts various simple arithmetical shadow-schemes are found for regulating the times of the *ẓuhr* and *ʿaṣr* prayers. Less frequently, there occur schemes for the shadows at each of the seasonal hours of daylight. This material represents a development virtually independent of the highly sophisticated astronomical timekeeping (*ʿilm al-mīqāt*) practiced by the astronomers of Islamic world, and it is here investigated for the first time.

As we shall see, some of this material derives from Hellenistic folk astronomy, notably, from the shadow schemes which have been studied by Otto Neugebauer. Also an approximate Indian rule for reckoning time of day from shadow lengths had considerable influence in Islamic folk astronomy, to the extent that it even underlies the definitions for the *ẓuhr* and the *ʿaṣr* which became standard in later practice. The Yemeni sources provide the most original materials, but all of the material discussed here has a distinct Islamic flavour. In passing I mention the shadow-schemes in medieval European folk astronomy, which were also studied by Neugebauer.

Otto Neugebauer (1899-1990) was the leading historian of the exact sciences in the 20th century. His monumental contributions to ancient Babylonian and Greek astronomy and mathematics are well known, his contributions to Islamic, Byzantine and medieval European astronomy perhaps less so. The interested reader should consult Noel Swerdlow, "Otto E. Neugebauer", *Proceedings of the American Philosophical Society* 137:1 (March 1993), pp. 137-165, or a summary available on the Internet under www.nap.edu/html/biomems/oneugebauer.html, as well as the bibliography compiled by Janet Sachs and Gerald Toomer in *Centaurus* 22 (1979), pp. 257-280.

But Neugebauer should also be remembered for his contributions to folk astronomy, Hellenistic, Ethiopic and medieval European. Most historians of science have no conception of folk science, but the fact that it flourished alongside the exact sciences wherever these were cultivated, not least in two of the most significant civilisations in human history, that is, the Hellenistic and the Islamic, suggests that perhaps we historians of science should take it seriously. For an introduction, the interested reader may consult any of Neugebauer's publications listed below:

- "Astronomische Papyri aus Wiener Sammlungen: II. Über griechische Wetterzeichen und Schattentafeln", *Sitzungsberichte der philosophisch-historischen Klasse der Österreichischen Akademie der Wissenschaften* 240:2 (1962), pp. 29-44.
- "On Some Aspects of Early Greek Astronomy", *Journal of the American Philosophical Society* 116 (1972), pp. 243-251, repr. in *idem*, *Essays*, pp. 361-369.
- *Ethiopic Astronomy and Computus*, Vienna: Österreichische Akademie der Wissenschaften, 1979.
- *Abu Shaker's "Chronography" – A Treatise of the 13th Century on Chronological, Calendrical, and Astronomical Matters, written by a Christian Arab, preserved in Ethiopic*, Vienna: Österreichische Akademie der Wissenschaften, 1988.

- "Astronomical and Calendrical Data in the Très Riches Heures," (originally published 1974), repr. in *idem*, *Astronomy and History – Selected Essays*, New York, *etc.*: Springer, 1983, pp. 507-520.

(For writings by other scholars on Islamic folk astronomy see the notes to **1.1** below.)

Most people could only stand in awe of Neugebauer's academic achievements, but those who knew the man will remember him most for his warmth and his inimitable humour. He for one knew that eagles fly along the Bosphorus, could identify those who liked a good "Spiel und Tanz", and recognized that Kindi played a "Doppelhorn". I was happy that he could see the first published version of this study, presented to him on the occasion of his 90th birthday. This appeared as "A Survey of Medieval Islamic Shadow Schemes for Simple Timereckoning", *Oriens* 32 (1990), pp. 191-249.

Most of the sources for this study were gathered during my survey of the scientific manuscripts in the Egyptian National Library conducted during the years 1974-76. My activities at the American Research Center in Egypt during 1972-79 were made possible mainly by the Smithsonian Institution and (U.S.) National Science Foundation, and my research on Islamic folk astronomy during 1983-85 by the (U.S.) National Endowment for the Humanities. This support is gratefully acknowledged. It is also a pleasure to record my appreciation to the Egyptian National Library for the privilege of working with their vast manuscript holdings, and my thanks are also due to the other libraries where the sources are preserved.

To my friend Daniel Varisco I owe much of my enthusiasm for folk astronomy and most of the references to modern practices relating to irrigation. Dr. Abdallah Nasif (University of Riyadh) kindly provided me with copies of relevant sections of his thesis. Dr. Tom Zuidema (University of Illinois) drew my attention to the basic documentation on the Gresik sundial. For remarks on the penultimate draft of this study – and for much more – I am indebted to Ted Kennedy.

Since the original study was published, two new studies have appeared that deal with timekeeping by night using the lunar mansions. These are Miquel Forcada, "*Mīqāt* en los calendarios andalusíes", *al-Qanṭara* 11 (1990), pp. 59-69, and Petra Schmidl, *Volksastronomische Abhandlungen aus dem mittelalterlichen arabisch-islamischen Kulturraum*, doctoral dissertation, Institute for History of Science, Johann Wolfgang Goethe University, Frankfurt am Main, 2003.

In this new version I have simplified the footnotes by using the bibliographical abbreviations listed at the beginning of this book and inserted cross-references to **I-II** and **IV**.

TABLE OF CONTENTS

CHAPTER 1

INTRODUCTION

1.1 Islamic folk astronomy and mathematical astronomy

In the medieval Islamic world there were two separate traditions of astronomical knowledge. The first was a simple, non-technical, essentially practical tradition, devoid of theories. The second was a mathematical tradition in which theories, tables and computation featured prominently. Since the distinction between the traditions is not generally recognized we shall begin by considering each of the traditions separately.

First, the folk tradition.[1] The Arabs of the Peninsula in the time before Islam possessed an intimate knowledge of the apparent motions of the sun, moon, and stars across the heavens, the months and seasons, the changing night sky throughout the year, and associated meteorological phenomena. A distinctive feature of this tradition was the division of the year into 13-day periods (*anwāʾ*) defined by the (acronychal) settings and (heliacal) risings of the 28 lunar mansions (*manāzil*).[2] Another was the custom of describing the time of day with reference to the length of a person's shadow.[3] The Arabs also had names for various divisions of the day and night, including names for the seasonal hours, that is, the twelve hours of daylight and the twelve hours of night.[4]

With the advent of Islam in the early 7th century, the Muslims expanded out of the Peninsula to establish a commonwealth from Spain to Central Asia. At various stages they incorporated some of the simple and non-technical procedures that they found in use in these regions. The resulting Islamic tradition of folk astronomy was widely practiced until the 19th century, and vestiges of it can be detected to this day in rural areas from Mauritania to Indonesia.

Besides the obvious fact that any man in the street could master the fundamentals of folk astronomy, there were two main reasons for its appeal. First, the sun, moon and stars were mentioned in the *Qurʾān*, and a knowledge of the basics of their apparent courses constituted

---

[1] All Arabic sources dealing with folk astronomy *per se* and compiled prior to *ca*. 1050, as well as all previous studies thereon, are surveyed in Sezgin, *GAS*, VII, pp. 336-370. Some of the sources were seriously investigated for the first time in Nallino, *Scritti*, V, pp. 152-197. A contribution of a more philological kind on the subject is the article "Layl and nahār" by Charles Pellat in EI_2. A new overview is Varisco, "Islamic Folk Astronomy". Other recent contributions, particularly by Paul Kunitzsch and Miquel Forcada, are mentioned below. On certain distinctive features see also nn. 1:2-4 and on various practical applications see n. 1:9.

[2] On the lunar mansions in Islamic astronomy see the introductory articles "Anwāʾ" by Charles Pellat and "Manāzil" by Paul Kunitzsch in EI_2, repr. in *idem*, *Studies*, XX, and the overview in Varisco, "Islamic Folk Astronomy". On the Andalusī tradition in particular see now Forcada, "*Anwāʾ* Books in al-Andalus".

[3] Arent J. Wensinck *et al.*, *Concordance et indices de la tradition musulmane*, 8 vols., Leiden: E. J. Brill, 1936-1988, (repr. in 4 vols., Leiden: E. J. Brill, 1992), IV, pp. 78-79, and V, pp. 211-213, lists references to the terms *ẓill* and *fayʾ* for "shadow" in the *ḥadīth* literature. See also **2.1** below.

[4] On the seasonal hours in general see, for example, Neugebauer, *HAMA*, III, p. 1069; *idem*, *Ethiopic Astronomy and Computus*, pp. 167-170; Kennedy, "*Zīj* Survey", p. 141; and **IV-2.1**. On these names (which are not mentioned in Pellat's article cited in n. 1:1) see **IV-8**.

part of a standard religious training. Even a distinctively Islamic cosmology developed over the centuries.[5] Second, the *Qur'ān* and Islamic tradition advocate three duties incumbent on every Muslim which are connected with astronomy, at least with the moon, the sun, and the earth, namely:

- Observance of a sacred month of fasting and of other religious festivals regulated by a strictly lunar calendar in which the beginnings of the months are determined by the first visibility of the lunar crescent;[6]
- Performance of five ritual prayers at times determined by the sun's position relative to the local horizon, defined in terms of shadows during the day, by sunset and sunrise, and by twilight phenomena;[7] and
- Performance of various ritual obligations, including prayer, but also recitation of the *Qur'ān*, the call to prayer, slaughter of animals, and burial of the dead, in a sacred direction towards the Kaʿba in Mecca (called in Arabic, *qibla*).[8]

The basic stipulations of each of these duties – correct timing and correct orientation – could be achieved by applications of the techniques of folk astronomy, with virtually no computation beyond simple arithmetic.[9] These procedures were advocated by the scholars of the sacred law and are to be found recorded not only in legal texts, but also in works on folk astronomy and popular works such as lexicons and encyclopaedias. They involved respectively:

- Actual sighting of the lunar crescent and, in the case of bad weather, counting months of roughly $29^1/_2$ days each in pairs;[10]
- Timekeeping by day using rough estimates of shadow lengths (as described in the present study) and by night using the risings and settings of the lunar mansions; direct observation of sunset, nightfall, daybreak, and sunrise; and
- Adopting as indicators of the local sacred direction the north or south points or the risings and settings of various prominent stars or of the sun at the equinoxes or solstices.[11]

The early Muslims must have been particularly happy to come across arithmetical rules for determining the shadows at midday and at the seasonal hours: these, as we shall see, they used not only to formulate precise definitions for the times of the daylight prayers prescribed in Islam, but also – in certain circles – for timekeeping in general.

Second, mathematical astronomy. In the 8th and 9th centuries the Muslims enjoyed an encounter with a very different approach to astronomy.[12] This they found documented mainly

---

[5] On this Islamic tradition see Heinen, *Islamic Cosmology*. On the corresponding "scientific" tradition see Jachimowicz, "Islamic Cosmology".

[6] See, for example, the articles "Ramaḍān" by M. Plessner and "Ṣawm" by C. C. Berg in EI_1, updated by the editors in EI_2.

[7] See the article "Mīḳāt, i: ritual and legal aspects" by A. J. Wensinck in EI_2.

[8] See the article "Ḳibla, i: ritual and legal aspects" by A. J. Wensinck in EI_2.

[9] See King, "Islamic Folk Astronomy", A-B, for overviews of the folk astronomical tradition and the ways in which it was applied to regulating the calendar, timekeeping, and determining the sacred direction towards Mecca.

[10] The best source of information on this subject is still Renaud, "Sur les lunes du Ramadan" (mainly on the practice in the Maghrib).

[11] See my forthcoming monograph *Sacred Geography of Islam*, summarized in the article "Makka. As Centre of the World" in EI_2, repr. in King, *Studies*, C-X, for a survey of these procedures, and **VIIa** on the consequences for the orientation of Islamic religious architecture.

[12] For an overview of the mathematical tradition in Islamic astronomy see King, "Islamic Astronomy". On

in texts of Indian and Greek provenance. Here the emphasis was on careful observation, preferably with instruments; geometrical models for the sun, moon and planets; and on computation. Greek and Indian astronomers, like their Babylonian predecessors, had a passion for tables and no fear of extensive calculations. Muslim astronomers adopted this earlier Greek and Indian material and modified it for their own use. They conducted new observations, compiled new tables, devised new instruments, and made progress in all aspects of mathematical astronomy within its ancient and medieval framework. Only since the middle of this century has it become clear to us to what extent they too had a passion for tables and instruments, and the basic documentation of their achievements is still in progress.

Muslim astronomers working in the mathematical tradition also treated the three problems mentioned above. Their solutions included:

- Developing limiting conditions on such quantities as the apparent elongation of the moon from the sun, the difference in setting times of the moon and sun, or the height of the moon above the local horizon at sunset, and preparing tables based on such conditions for predicting lunar crescent visibility in specific latitudes;[13]
- Applying trigonometric formulae, either complicated accurate ones or handy approximations, for reckoning time from solar and stellar altitudes and compiling tables for timekeeping by the sun and stars for different latitudes, sometimes containing tens of thousands of entries, based upon such formulae (**I-II**); also developing instruments such as the astrolabe, quadrant and sundial for the same purpose (**X**).
- Developing cartographic, geometric and trigonometric solutions to the problem of determining the direction of one point on a sphere to another; tabulating the direction of Mecca as a function of terrestrial longitude and latitude; listing the qiblas of cities throughout the Islamic world; and devising cartographic techniques for reading the qibla directly from maps.[14]

These approaches are a far cry from the methods advocated by the legal scholars. Only when we realize how different they were do we find ourselves in a position to assess Islamic science on its own terms.

After the formative period of Islamic astronomy in the 8th to 10th centuries there developed regional schools with their own specific interests and authorities.[15] It was in the Yemen in particular that the two traditions of astronomy flourished side by side. Likewise in the Maghrib there were several widely-respected authorities in folk astronomy, this in spite of the

---

the beginnings of this tradition see Pingree, "Indian Influence" and "Greek Influence", and more recently Saliba, "Arabic Science and the Greek Legacy". Pingree's article "ʿIlm al-hayʾa" in EI_2 provides an overview of the activities of the astronomers in all fields except those related to Islamic ritual (on which see the overview in King, "Islamic Folk Astronomy", A-B). For the manuscript sources from the early period see Sezgin, *GAS*, VI.

[13] See King, "Tables for Lunar Crescent Visibility", on the earliest tables, and the article "Ruʾyat al-hilāl" in EI_2 for an overview of Muslim activity in this field.

[14] An overview is in the article "Ḳibla (astronomical aspects)" in EI_2, repr. in King, *Studies*, C-IX. More information is in *idem*, *Mecca-Centred World-Maps*.

[15] On the schools in the Islamic East see Kennedy, "Exact Sciences in Iran", A-C; on those in Egypt and Syria, the Yemen, and the Maghrib see King, "Astronomy of the Mamluks"; *idem*, *Astronomy in Yemen*, and *idem*, "Astronomy in the Maghrib", respectively. On astronomy in al-Andalus see Samsó, *Ciencias en al-Andalus*, and *idem*, *Studies*, and many other publications of the Barcelona school.

availability there of sophisticated tables relating to mathematical astronomy. In Egypt and Syria there were apparently no such authorities.

Now what concerns us here is the second obligation to pray at the right times, particularly in the case of the two prayers performed during the day. It seems that in the first centuries of Islam the times of prayer were regulated by the muezzins who performed the call to prayer five times a day. They were chosen for the excellence of their voices and their good character; they then had to learn the basics of folk astronomy, which was all that they needed in addition to perform their task adequately. However, the 13th century witnessed the rise of the institution of the *muwaqqit*, that is, the astronomer associated with a mosque for the purposes of regulating the times of prayer, as well as the prediction of lunar crescent visibility and the determination of the qibla (**V**). Some impressive scientific achievements were made by certain *muwaqqit*s, particularly in Egypt and Syria. Tables displaying the times of prayer for each day of the year for specific localities became commonplace. Instruments – mainly sundials, astrolabes and quadrants – also bore markings for the times of prayer. We do not know how the *muwaqqit*s functioned in such proximity to legal scholars who were advocating simple techniques of folk astronomy; our sources are silent on this point. Certainly I am not aware of any legal text in which it is suggested that one should consult an astronomer on the prayer-times or use any of the astronomical tables or instruments that were available for this purpose. It would be naïve to suppose that there was any reason why a legal scholar should have consulted an astronomer.

In this study I present a survey of all known Islamic shadow-schemes for regulating the time of day in general and the times of the midday and mid-afternoon prayers in particular. This material belongs almost exclusively to the folk tradition in Islamic science, and most of it has never been studied previously. It has been found mainly in books dealing with the regulation of the prayer-times by non-technical means (*kutub al-mawāqīt*), in popular almanacs, and in books on the sacred law of Islam. Not a few of the schemes are recorded as curiosities in works belonging to the mathematical tradition, or simply as marginalia or notes on flyleaves.

I make no claim to have exhausted the available material (especially legal texts), but I have included all the relevant material currently known to me. I have not come across any Turkish or Persian sources, so the Ottoman heartland and the Islamic East are not well represented in my survey.

1.2 The seasonal hours and the times of the *ẓuhr* and *ʿaṣr* prayers

A minority of our sources displays shadow lengths for each seasonal hour of the day during each month of the year, but most of the Islamic shadow-schemes that I shall discuss were intended to be used primarily for regulating the times of the midday (*ẓuhr*) and afternoon (*ʿaṣr*) prayers. It is not clear how these were regulated during the first few decades of Islam: the *Qurʾān* and the *ḥadīth* do not between them provide precise definitions (**IV-1-3**). But from the 8th century onwards, the times of these two prayers were defined in terms of shadows, or – more specifically – shadow increases: see **Fig. 1.2a**.[16] The *ẓuhr* began shortly after midday

[16] On the definitions of the prayers in the astronomical sources see also Wiedemann & Frank, "Gebetszeiten"; Kennedy, "al-Bīrūnī on Prayer-Times"; and now **IV-4** and **7**.

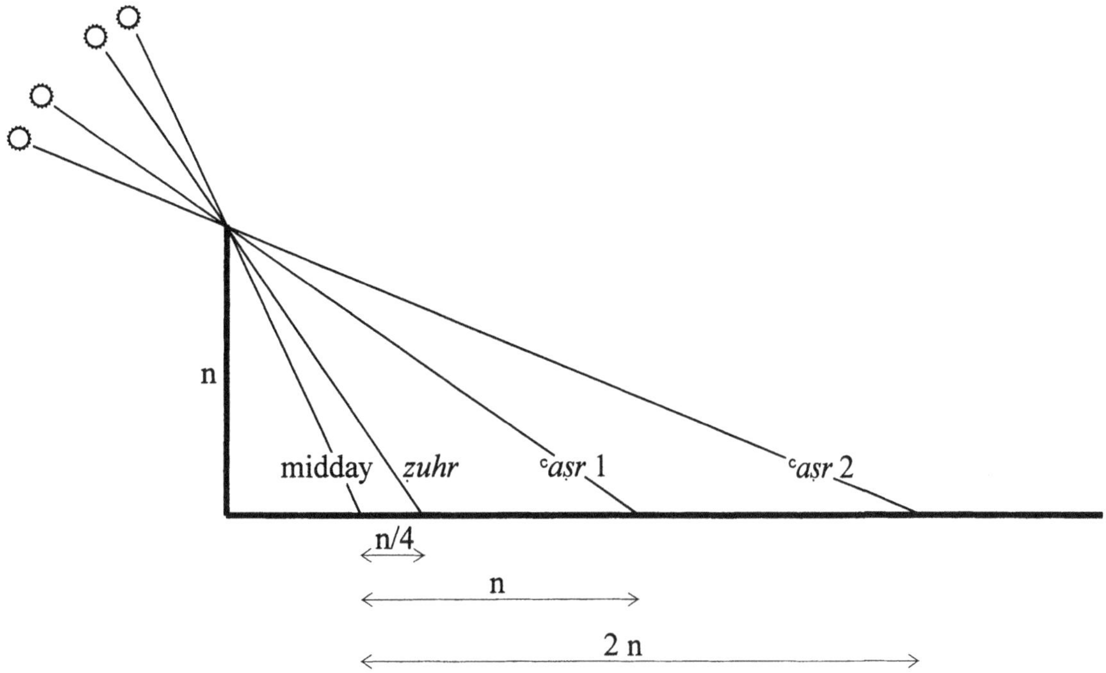

Fig. 1.2a: The standard definitions of the times of the beginning of the *ẓuhr* and the beginning and end of the *ʿaṣr*.

when the shadow was observed to increase, and the *ʿaṣr* began when the shadow had increased over its midday minimum by the length of the gnomon. Variant definitions for the *ẓuhr* in terms of shadow increases are also attested. Likewise, the end of the interval permitted for the *ʿaṣr* is often defined as being when the shadow has increased by twice the length of the gnomon. As I have shown in detail elsewhere (**IV**), these definitions represent practical means of linking the prayers to the seasonal hours, the link being provided by an approximate Indian formula for timekeeping first attested in the Islamic sources of the 8th century – see below and also **2.1**, **2.3** and **3.2**.

Underlying most of the schemes is the implicit notion that one should measure one's own shadow in terms of the length of one's feet. The most common Islamic base for the gnomon length was 7 (*qadam*s or feet) although sometimes 6, $6^1/_2$ and $6^2/_3$ feet were used. These values between 6 and 7 were thought to correspond to the ratio of the height of a man (*qāma*) to the length of his feet.[17]

A gnomon the height of a man is still *in situ* in the courtyard of the 7th-century Mosque of Janad in the Yemen – see **Fig. 1.2b**. It is known locally as the stick (*ʿaṣā*) of Muʿādh ibn Jabal, who was appointed by the Prophet Muḥammad as the first governor of the Yemen.[18]

[17] Kennedy, *al-Bīrūnī's* Shadows, I, pp. 68-79, and II, pp. 25-31, *ad* al-Bīrūnī, *Shadows*, Ch. 7. In the mathematical tradition two other bases were used, namely 12 (*iṣbaʿ*s or digits) and – less frequently – 60: see Schoy, *Schattentafeln*, and Kennedy, "*Zīj* Survey", p. 140; and also the last paragraphs of **I-1.1** and **II.1.3**.

[18] See also King, *Astronomy in Yemen*, p. 8.

Fig. 1.2b: The gnomon in the courtyard of the Mosque of Janad. [Photo courtesy Professor Daniel M. Varisco.]

The 13th-century legal scholar al-Aṣbaḥī (see **4.1**) states explicitly that he measured the midday shadows in Janad for every 13 days of the Syrian year, that is, for each of the *anwāʾ*. I do not know the age of the present gnomon, but there has clearly been a gnomon there for centuries.

Usually the schemes are presented in words or mnemonics in standard medieval Arabic alphanumerical (*abjad*) notation.[19] The schemes for the hours were sometimes represented in tabular form, but underlying them we also find trivial arithmetical schemes. Indeed, virtually all of the material is arithmetical, devoid of any influence of trigonometry. I have included a few shadow-lists that were observed or derived by calculation; these stem from the murky and unclearly-defined border between the two traditions.

Only rarely do competent astronomers mention these primitive schemes. The celebrated scientist al-Bīrūnī (see **2.4**) devoted a whole book to the topic of shadows,[20] and in this he discussed two Indian schemes, but only because he found them interesting, not because he

[19] On this convention see Irani, "Arabic Numeral Notation", and also Berggren, *Islamic Mathematics*, pp. 41-42.

[20] An uncritical edition of the *Shadows* is listed under al-Bīrūnī, *Shadows*. For translation and commentary see Kennedy, *al-Bīrūnī's* Shadows.

would have approved of their use. On the other hand, two masters of spherical astronomy, Najm al-Dīn al-Miṣrī and Ibn al-Shāṭir (see **9.2** and **8.2**), do advocate using the simple formula of Indian provenance mentioned above as a handy approximation, and they can be forgiven for doing this because in spite of its simplicity the formula is remarkably accurate for most practical purposes (see **1.3**). Before we consider the Islamic material in detail we should briefly mention the earlier traditions (**1.4**).

1.3 A simple arithmetical rule for timekeeping

The approximate method for timekeeping advocated by al-Fazārī (see **2.3**) and the equivalent shadow-schemes for the hours in several later sources investigated below (see **2.1**, **3.2**, *etc.*) are actually of Indian origin[21] although this is not explicitly stated (see, however, **2.4**). It is equivalent to the formula:

$$T \approx 6\ n\ /\ (\Delta z + n)\ .$$

Note that, by definition, for $T = 0$, $\Delta z \to \infty$, and for $T = 6$, $\Delta z = 0$. The formula is the simplest means of assuring that these two boundary conditions are satisfied. For the general situation $0 < T < 6$ the formula provides a reasonable approximation.[22] This was the formula mentioned in **1.1** and **1.2** that was used to derive the standard definitions of the times of the daylight prayers in Islam.

1.4 Shadow-schemes from Antiquity

Simple arithmetical schemes representing the shadow lengths at midday for each month of the year, and tables displaying the shadows at each seasonal hour for each month of the year or each zodiacal sign also based on a simple arithmetical scheme, are attested already in early Greek astronomy, and their use continued in the Byzantine, Coptic, and Ethiopic traditions, as well as in the medieval Islamic world and Europe. It is to Otto Neugebauer that we owe the first detailed investigation of this material.[23]

---

[21] On this formula in the already known Islamic sources, see Pingree, "al-Fazārī", pp. 121-122 (al-Fazārī), Kennedy, *al-Bīrūnī's* Shadows, II, pp. 116-121 (al-Bīrūnī), and **IV-2.4**. In this study I refer to this formula as the "standard" Indian formula although the Muslim astronomers also used an accurate formula for timekeeping also of Indian provenance – see Davidian, "al-Bīrūnī on the Length of Day", for details. There is a substantial secondary literature on timekeeping by shadows in Babylonian and Indian astronomy, much of which seems to confuse shadows with shadow increases.

In **XI**, "the standard approximate formula" is the trigonometric formula, also apparently of Indian origin, referred to in n. 4:5 below.

[22] An investigation of its accuracy is in Davidian, *op. cit.*, p. 334. See also **XI-A2**.

[23] See Neugebauer, "Astronomische Papyri"; *idem*, "Early Greek Astronomy", pp. 243-246; and esp. *idem*, *HAMA*, II, pp. 737-746.

For the Ethiopic tradition, based on Hellenistic schemes, with numerous absurdities (*e.g.*, 70 instead of 0 for the twelfth hour, resulting from the misreading of the Greek zero-sign meaning "no shadow" as an *omicron*, that is, 70), see *idem*, *Ethiopic Astronomy and Computus*, pp. 209-215.

See also Kennedy, "Overview of the History of Trigonometry", esp. pp. 5-6, for the first comparison of some of the approximate rules discussed in the present study.

The absurdity of attempting to analyze these purely arithmetical schemes from a modern astronomical point of view is well illustrated in Bremner, "The Shadow Table in Mul.Apin". In this 7th-century B.C.E. source,

The midday shadows are usually symmetrically arranged so that seven values suffice to define them for the whole year, thus for the months of the Julian calendar (type M):

XII	XI/I	X/II	IX/III	VIII/IV	VII/V	VI

or for the zodiacal signs between the winter and summer solstices (type Z):

♑	♒/♐	♓/♏	♈/♎	♉/♍	♊/♌	♋

The midday shadow-scheme:

2	3	4	5	6	7	8

appears to be of early (5th or 4th century B.C.) Greek provenance. Later modifications include *inter alia*:

2 3 4 6 8 10 13 and 1 2 3 4 5 7 8 ,

as well as:

0 $1^1/_2$ 3 $4^1/_2$ 6 $7^1/_2$ 9 .

Also attested are pairings of all the months or signs to yield six rather than seven significant values. The sources usually do not specify the length of the gnomon casting the shadows or the locality for which the values are intended. For the hours it is assumed that the shadows increase at the same rate throughout the year, thus:

hours of daylight:	5/7	4/8	3/9	2/10	1/11
increase each hour:	1	2	3	4	10 (=1+2+3+4)
cumulative increase:	1	3	6	10	20

Each of the classical and medieval traditions was remarkably prone to errors by those who conceived of the tables and yet more by those who copied them, but this feature by no means lessens the interest of the material for the history of science. In his monumental *History of Ancient Mathematical Astronomy* Neugebauer was able to cite just one example of a shadow-scheme from a published Islamic source,[24] and, as might be expected, the unpublished sources offer many more.

1.5 A classification of the Islamic material

I have arranged the sources chronologically according to geographical provenance, as far as this is possible. The monthly schemes are for the Syrian calendar, the Julian (al-Andalus and the Maghrib), or the Coptic (Egypt); I have not found any schemes based on the Persian calendar.[25] We can distinguish seven different kinds of schemes and tables:

a) midday shadow-schemes for the hours based on the approximate Indian formula relating the seasonal hours to the increase of the shadow over its midday minimum (**2.3**, **2.4**, **3.2**, *etc.*);

---

a time function T in degrees (or units of 4 minutes) is tabulated for each cubit of the shadow z from 1 to 10. The values of T(z) are simply T_1/z where T_1 are the times for z=1 at the equinoxes (75) and the solstices (60 in summer and 90 in winter).

[24] Neugebauer, *HAMA*, II, p. 743 (on the scheme listed in **6.3** below).

[25] The various calendars in use in the medieval Islamic world are discussed, for example, in Adolf G. Grohmann, *Arabische Chronologie*, Leiden: E. J. Brill, 1966; the article "Zamān" by Willy Hartner in EI_1; Kennedy, "*Zīj* Survey", p. 139; and the article "Ta'rikh iv" by Benno van Dalen in EI_2.

b) midday shadow-schemes for the months based on Greek schemes, also attested in Coptic sources, and modifications thereof (**7.2**, *etc.*);
c) shadow tables for the hours and for each month based on Greek models (**6.2**, **6.7**, **9.7a**, *etc.* – see also **4.9a** on a possible Himyaritic scheme);
d) midday shadow-schemes for each 13-day period of the year (*anwāʾ*) (**4.1**, *etc.*);
e) simple lists of midday shadows for a given locality, derived by observation (**3.2**, **4.1**, *etc.*);
f) simple lists of midday shadows derived by calculation (for given latitude and obliquity) (**5.1**, *etc.*); and
g) simple shadow tables for the hours and for each zodiacal sign or for each few days of the year based on calculation with an approximate trigonometric formula (only in **4.2**).[26]

1.6 Some technicalities

In the analysis, I use the following notation freely:

AE	autumnal equinox
EQ	equinoxes
M	lunar mansion (numbered with subscripts 1-28)
N	north
S	south
SS	summer solstice
VE	vernal equinox
WS	winter solstice
– – – – – – – – – –	
h	solar altitude
H	solar meridian altitude
n	length of the gnomon
sdh	as a superscript, seasonal day hours
T	seasonal hours since sunrise or remaining until sunset
z	shadow length
Z	midday shadow length
Δz	increase in shadow length over midday minimum
ε	obliquity of the ecliptic
ϕ	local latitude
+	"n+" indicates a number "a little more than" n (Arabic *rājiḥ*)
-	"n-" indicates a number "a little less than" n (Arabic *shāḥḥ*)
m/n	for seasonal hours the slash indicates two hours for which the sun has the same altitude (n = 12-m), and for zodiacal signs two signs for which the sun has the same declination

The astronomical theory underlying the schemes is simple indeed. The altitude of the celestial pole is a measure of the local latitude ϕ. The meridian altitude of the sun at the

[26] See **I-2.5** and **4.3**, for examples of more extensive tables based on this formula, and also **XI-3** and **4**.

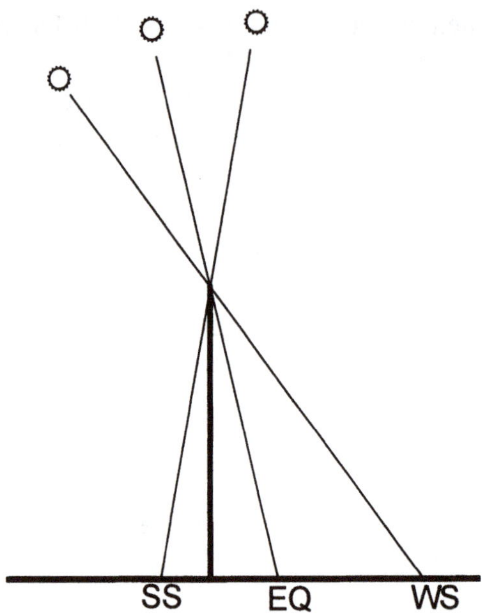

Fig. 1.4a: Solstitial and equinoctial shadows for a tropical locality such as the Yemen (here, specifically Taiz).

equinoxes is $\bar{\phi} = 90° - \phi$. Thus the shadow lengths at the winter solstice, equinoxes, and summer solstice are respectively:

$$n \cot(\bar{\phi} - \varepsilon), \quad n \cot \bar{\phi}, \quad \text{and} \quad n \cot(\bar{\phi} + \varepsilon).$$

Fig. 1.4a shows the solstitial shadows for a tropical locality ($\phi < \varepsilon$) such as the Yemen or Mecca: these vary between almost a full gnomon length north at midwinter to a small fraction of the gnomon length south at midsummer. For the convenience of the reader, I present in **Table 1.5** the approximate shadow lengths at the solstices and equinoxes for the standard gnomon lengths 7 and 12 for latitudes corresponding to the Yemen, Mecca, Cairo, Aleppo, Cordova and Istanbul (assuming $\varepsilon = 23^{1}/_{2}°$). Values are rounded to the nearest digit, as in most of our sources. It is not always sensible to try to establish the latitude for which a given scheme might have been computed.[27]

Table 1.5

Latitude	Midday altitudes			Midday shadows					
				n = 7			n = 12		
	WS	EQ	SS	WS	EQ	SS	WS	EQ	SS
15°	$51^{1}/_{2}°$	75°	$98^{1}/_{2}°$	6	2	1(S)	10	3	2(S)
$21^{1}/_{2}$	45	$68^{1}/_{2}$	92	7	3	0(S)	12	5	0(S)
30	$36^{1}/_{2}$	60	$83^{1}/_{2}$	9	4	1	16	7	1
36	$30^{1}/_{2}$	54	$77^{1}/_{2}$	12	5	2	20	9	3
38	$28^{1}/_{2}$	52	$75^{1}/_{2}$	13	5	2	22	9	3
41	$25^{1}/_{2}$	49	$72^{1}/_{2}$	15	6	2	25	10	4

[27] See Neugebauer, *HAMA*, II, p. 739, on the problems.

In our sources fractions are invariably given as combinations of unit fractions of the form $^1/_n$ ($n < 9$) so that, for example (see **Tables 4.1** and **4.3**), $^1/_{12}$ is expressed as *nisf suds*, that is, $^1/_2$ of $^1/_6$, and $^5/_6$ is expressed as *nisf wa-thulth*, that is, $^1/_2$ plus $^1/_3$. The one exception to the unit fractions is $^2/_3$ (Arabic has a dual form). This is a feature of simple arithmetic in medieval Islam, reminiscent of the practice in ancient Egypt, and it persisted not least because of the fact that Arabic has no names for unit fractions with denominator greater than ten.[28] Occasionally numbers with fractions are written sexagesimally in the standard manner of the astronomers; a number rendered in the form "a b" in the texts is rendered "a;b" in the commentary, and this means "$a + {}^b/_{60}$".

---

[28] See Neugebauer, *ESA*, pp. 74-77, and the article "ʿIlm al-ḥisāb" by Abdelhamid I. Sabra in EI_2, esp. III, pp. 1140-1141.

CHAPTER 2

EARLY SOURCES

2.1 The Prophetic *ḥadīth*

There are various references to shadow lengths in the statements attributed to the Prophet Muḥammad (known as *ḥadīth*). I shall not attempt to give a survey of these; rather, I shall restrict my attention to just two examples.[1]

First, the Prophet is reported to have said that the angel Gabriel led him in prayer on two consecutive days. On the first day they prayed the *ẓuhr* when the shadow was equal to the width of a thong of a sandal and the *ʿaṣr* when the shadow of every object was the same as its length. These situations correspond to $z = \delta$ ($\delta << n$) and $z = n$. On the second day they prayed the *ẓuhr* when the shadow of every object was the same as its length and the *ʿaṣr* when the shadow of every object was twice its length. These correspond to $z = n$ and $z = 2n$.

As pointed out by al-Bīrūnī, these definitions in terms of shadow lengths could not be applied universally.[2] Only in tropical latitudes can the midday shadow be relatively small, and in non-tropical latitudes the solar altitude does not always attain the 45° necessary to ensure that sometime in the afternoon the condition $z = n$ will be satisfied. The definitions for the *ẓuhr* in terms of shadow increases that were adopted in later practice, namely:

$$\Delta z = {}^{1}/_{12}\, n\ ,\ {}^{1}/_{5}\, n \quad \text{or} \quad {}^{1}/_{4}\, n\ ,$$

ensure that the shadow has increased by an observable amount. The definition for the *ʿaṣr* in terms of shadow increases which was adopted by three of the four major legal schools, namely, $\Delta z = n$, and that adopted by the fourth (the Ḥanafīs), namely, $\Delta z = 2n$, can be applied in any habitable latitude. The *ḥadīth* in a sense justifies both of these amounts – n and 2n – for the *ʿaṣr*. Yet other evidence indicates that the reason for adopting these definitions was to regulate the prayers by the seasonal hours. The formula:

$$T \;=\; 6\, n\, /\, (\Delta z + n)$$

yields the following shadow increases at the seasonal hours:

T	0	1	2	3	4	5	6
Δz	∞	5n	2n	n	$^{1}/_{2}$ n	$^{1}/_{5}$ n	0

so that the definitions $\Delta z = n$ for the *ʿaṣr* (and occasionally for a prayer at mid-morning called the *ḍuḥā*) represent practical means of linking the *ʿaṣr* to the beginning of the tenth hour (and the *ḍuḥā* to the beginning of the fourth). Likewise, the definitions:

$$\Delta z \;=\; {}^{1}/_{12}\, n\ ,\ {}^{1}/_{5}\, n \quad \text{or} \quad {}^{1}/_{4}\, n$$

are practical means of relating the *ẓuhr* to one-half or one hour after midday.

---

[1] See **IV-1** and **4.1-2**, for more information on these two *ḥadīth*. For references to shadows in the *ḥadīth* literature see n. 1:3.

[2] al-Bīrūnī, *Shadows*, p. 168, and Kennedy, *al-Bīrūnī's* Shadows, I, p. 218, and II, pp. 137-138.

The second report about the Prophet states that he prayed the *ẓuhr* when $5 < z < 7$ in winter and when $3 < z < 5$ in summer. Since the winter midday shadow in Mecca and Medina is about $7^1/_2$ ($n = 7$), this cannot be a description of a daily routine.

2.2 ʿAlī ibn Abī Ṭālib

al-Aṣbaḥī (see **4.1**) relates a tradition (MS Cairo DM 948, fol. 41r) on the authority of Yaḥyā ibn Saʿīd[3] about the fourth Caliph ʿAlī ibn Abī Ṭālib.[4] The latter – to whom numerous achievements of a mildly scientific nature are attributed in Islamic legend – is reported to have said that the day on which the midday shadow at Basra was $6^1/_2$ feet (the shadow of any object being then equal to its length) marked the beginning of winter. I do not know of any other sources in which winter is thus defined[5] (see also **4.1** on a specific date for the winter solstice – as opposed to the middle of winter – attributed to ʿAlī). The report continues to list the shadow lengths at each eighth of the year; these are shown in the following table:

Table 2.2

Season	Days elapsed	Midday shadow
Beginning of winter	0	$6^1/_2$
midwinter	45	$9 - {}^1/_8$
end of winter	46	$6^1/_2$
middle of spring	45	$3^1/_2$
beginning of summer	46	$1^1/_2+$ *
midsummer	45	$^1/_2+$
end of summer	47 (leap year: 48)	$1^1/_2+$
midautumn	45	$3^1/_2+$
end of autumn	46	$6^1/_2$

Since for Basra, $\phi \approx 30°$, the shadow (base $6^1/_2$) actually varies between

$$8^3/_4 \text{ (WS)}, \quad 3^3/_4 \text{ (EQ)}, \quad \text{and} \quad ^3/_4 \text{ (SS)},$$

and so the values given in the text are reasonable. Note that at the latitude of Mecca the midday shadow at the winter solstice – rather than the middle of winter – is equal to the length of the gnomon (see **Table 1.5** and **Section 3.3**).

Ibn Raḥīq reports that, according to ʿAlī, (the end of) the ninth hour of daylight marks the beginning of the *ʿaṣr* prayer – see **3.2**.

[3] On possible candidates with this name, both having a connection with al-ʿIrāq, see Sezgin, *GAS*, I, p. 407 (on Abū Saʾīd Y. b. S. al-Anṣārī, a *tābiʿī* from Medina, d. 760 in Basra (!)), and p. 293 (on Abū Ayyūb Y. b. S. al-Umawī, b. *ca.* 732 in Kufa, d. 809 in Baghdad, author of a work on the campaigns of the Prophet).

[4] On ʿAlī ibn Abī Ṭālib see the article by L. Vecchia Vaglieri in EI_2. On various achievements in the sciences attributed to him see Sezgin, *GAS*, VII, pp. 8, 10, 11, 26, and 184; *Cairo ENL Survey*, no. B1; King, "A Medieval Arabic Report on Algebra before al-Khwārizmī" and n. **IV**-2:8.

[5] See, however, Lane, *Lexicon*, V, p. 1916a, where the word *ẓill*, "shadow", is associated with the beginning of winter.

2.3 al-Fazārī

The extant fragments of the writings of the mid-8th-century astronomer al-Fazārī, now gathered and interpreted by David Pingree, display the eclectic nature of early Islamic astronomy.[6] For timekeeping al-Fazārī advocated:

$$T \approx 6\,n / (\Delta z + n) \qquad (n = 12),$$

a formula which was to be extremely influential in later Islamic folk astronomy and in the formulation of the practical definitions for the times of the daylight prayers which are still in use to this day (see **2.1**, **3.2**, **4.1**, **6.4**, **6.7**, **8.2**, **9.2**, *etc.*).

2.4 al-Bīrūnī

Two numerical shadow-schemes are presented by al-Bīrūnī in his treatise *On Shadows*.[7] The account of the first is garbled, and it was interpreted by E. S. Kennedy thus:

T	1	2	3	4	5	6
[Δz-12]	60	24	12	2;50	0	?
[Δz	72	36	24	14;50	12	0]

However, it seems more likely that the text should be restored to yield a scheme as follows:

T	1	2	3	4	5	6
[Δz]	60	24	12	[6]	[$2^2/_5$]	0

This emendation requires only inserting a value 6 for the fourth hour and reading *wa-khumsayn* (= $^2/_5$) rather than *wa-khamsūn* (= 50) after the 2 for the fifth hour. Now the scheme corresponds precisely to the Indian formula for n = 12.

The second shadow-scheme mentioned by al-Bīrūnī is also Indian. It relates the shadow increase to the passage of time in *muhurtas*, one-fifteenth divisions of the time between sunrise and sunset. The values given are as follows:

T	1/14	2/13	3/12	4/11	5/10	6/9	7/8	$7^1/_2$
Δz	90	60	12	6	5	3	2	0

The same scheme is described in al-Bīrūnī's *India*, with 96 for the first entry.[8]

---

[6] On al-Fazārī see the article by David Pingree in *DSB*. For his use of this formula see Pingree, "al-Fazārī", pp. 121-122.

[7] On al-Bīrūnī see the article by E. S. Kennedy in *DSB*. On this passage see al-Bīrūnī, *Shadows*, pp. 158-159, and E. S. Kennedy's translation and commentary, I, pp. 195-196, and II, pp. 121-122, as well as *idem*, "Overview of the History of Trigonometry", esp. pp. 5-6.

[8] Sachau, *Alberuni's India*, pp. 338-339.

CHAPTER 3

HEJAZI SOURCES

3.1 Aḥmad al-Rāzī

See **Section 3.2**.

3.2 Ibn Raḥīq

Abū ʿAbdallāh Muḥammad ibn Raḥīq ibn ʿAbd al-Karīm wrote a treatise on the calendar, lunar mansions, and simple time-reckoning, which is extant in the unique MS Berlin Ahlwardt 5664 (Landberg 108), 71 fols., copied *ca.* 1350. The work appears to have been compiled in Mecca in the 11th century.[1] Of particular relevance to the present study are, first, a set of midday shadows at Mecca for each of the Coptic months and, second, a scheme for reckoning the seasonal hours by shadow increases.

Ibn Raḥīq quotes Abū ʿAlī ʿAraqa, a muezzin at the Mosque of ʿAmr in Fustat, as having stated that Aḥmad al-Rāzī, a muezzin at the Sacred Mosque in Mecca, had observed the sun over a period of seven years and had determined the shadow lengths for the Coptic months starting with Bāʾūna (X), expressed in terms of the height of a man (*qāma*). His results (fols. 12r-12v) are as follows:

Table 3.2a

Month	Shadow
X	$^1/_3 \bullet {^1/_6}$ (S)
XI	0
XII	$^1/_6$
I	$^1/_3 + (^1/_2 \bullet {^1/_6})$ [a]
II	$^1/_2$
III	$^2/_3 + {^1/_4}$
IV	$1 + (^1/_6 \bullet {^1/_8})$
V	$1 - (^1/_2 \bullet {^1/_8})$
VI	$^1/_3 + {^1/_4}$
VII	$^1/_4 + {^1/_8}$
VIII	$^1/_6$
IX	0

[a] text has $^1/_3 + {^1/_2} + {^1/_6}$

Since the latitude of Mecca is about $21^1/_2$°, the meridian altitudes at the winter and summer

[1] On Ibn Raḥīq see King, *Astronomy in Yemen*, no. 2, also **II-2.3** and **IV-1.5**. His treatise, as well as those of al-Aṣbaḥī and al-Fārisī (nn. 4:1 and 4:4), are currently being investigated in Schmidl, *Islamische volksastronomische Abhandlungen*.

solstices, measured from the south point, are 45° and 92°, and one would expect maximum shadows in the north of about 1 *qāma* and in the south of about $^1/_{30}$ *qāma*. I have no information on the date of al-Rāzī.

Elsewhere in his treatise (fols. 28v-29r – see **Fig. IV-2.1**) Ibn Raḥīq presents the scheme:

T	1	2	3	4	5
Δz	39	$19^1/_2$	$6^1/_2$	$3^1/_4$	1+ ,

and he uses gnomon length $6^1/_2$. (The value for T = 5 is given as *qadam rājiḥ*.) If we use the standard Indian formula for T = 5, 4, and 3 seasonal hours, we obtain for this value of n:

$$\Delta z = 1^3/_{10}\,,\ 3^1/_4\,,\ \text{and}\ 6^1/_2\,,$$

as in the text. When T = 2 and 1, Δz is large compared with n, and if we apply the very crude approximation:

$$T \approx 6n / \Delta z\,,$$

we obtain for these values of T and n:

$$\Delta z = 19^1/_2 \quad \text{and} \quad 39\,.$$

Each of Ibn Raḥīq's values is thus explained. The above approximation is nowhere attested, explicitly or implicitly, in any of the other Islamic sources known to me.

Ibn Raḥīq adds a remark that (the end of) the ninth hour marks the beginning of the *ʿaṣr* prayer according to ʿAlī ibn Abī Ṭālib (see **2.1**). It was this remark that first led me to investigate the standard definitions of the times of the daylight prayers in Islam and find that they were practical means of regulating the prayer-times in terms of the seasonal hours.

The last part of Ibn Raḥīq's treatise (fols. 40r-71v) consists of an almanac arranged according to the lunar mansions rising at daybreak. For each 13-day period, various kinds of information are given, including the midday shadow lengths at Mecca. Part of the almanac is missing (some 20 fols. between fols. 43 and 44), and the shadow lengths that I have been able to extract are shown in **Table 3.2b**.

3.3 al-Shillī

MS Cairo DM 130,2 (fols. 14v-15v, copied *ca.* 1800 – see **Fig. 3.3**) contains a short treatise by Muḥammad ibn Abī Bakr ibn Aḥmad al-Shillī Bā ʿAlawī, a scholar from the Hadramawt who worked in Mecca in the 17th century.[2] al-Shillī discusses the disagreements between scholars about the midday shadow lengths throughout the year at Mecca and Medina and then presents his own opinion on the shadows at Mecca, whose latitude he takes as 21°.

In brief, al-Shillī states that the sun is overhead at Mecca on "the *fifth* day of the sign of Gemini" and "the twenty-*sixth* day of the sign of Cancer".[3] The sun is overhead (in tropical localities, *i.e.* where $\phi < \varepsilon$) when its declination equals the local latitude. For solar longitude 65° and ε = 23;35° (the standard value, which I would have expected al-Shillī to have used) I compute the declination as 21;16°; for longitude 64° the corresponding value is 21;4°, which

[2] On al-Shillī see Brockelmann, *GAL*, II, p. 502, and SII, p. 516; *Cairo ENL Survey*, no. D47; and King, King, *Astronomy in Yemen*, p. 59.

[3] *Cf.* the discussions in Lee, "Astronomical Tables of Al Farsi" (1822), p. 265 (Sanaa); and King, "al-Bazdawī on the Qibla in Transoxania", p. 19 (Samarqand).

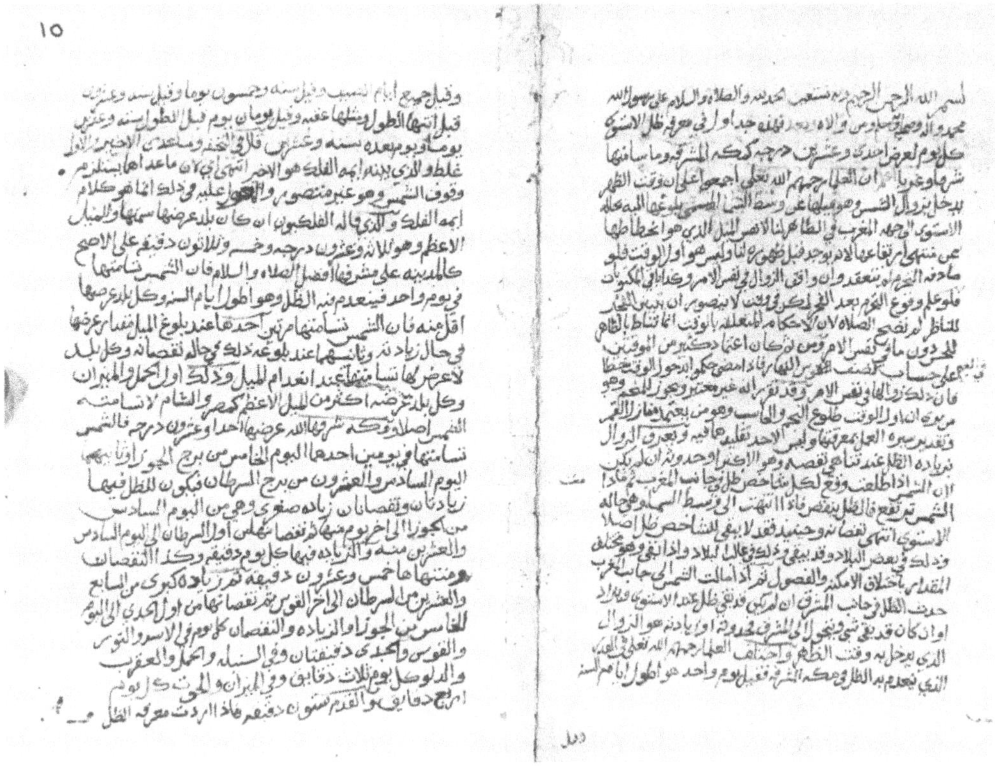

Fig. 3.3: The full text of al-Shillī dealing with shadow-lengths. [From MS Cairo DM 130,2, fols. 14v-15v, courtesy of the Egyptian National Library.]

is close to his value for the latitude of Mecca. However if we compute with longitude 64° and obliquity 23;30° (rounded from the value of Ulugh Beg) the declination is exactly 21°, which is very nice. (Had al-Shillī taken the trouble to measure the latitude he would have no doubt been surprised to find that it is closer to 21;30°.) He goes on to say that the "small increase (in midday shadow)" (*al-ziyāda al-ṣughrā*) occurs between these two days and reaches a maximum of 25 minutes (*i.e.* sixtieths of a foot, where the gnomon is 7 feet long). He states that the daily increase in the shadow is 1 minute. The "large increase" (*al-ziyāda al-kubrā*) occurs at the winter solstice, and no value is given. al-Shillī states that the daily shadow changes in minutes in the signs are as follows:

♉/♌/♐/♑ : 2 ♍/♈/♏/♒ : 3 ♎/♓ : 4 ,

and from this one could infer that the shadow at the winter solstice was about:

$$4 \times 1 + 30 \times 2 + 30 \times 3 + 30 \times 4 + 30 \times 3 + 30 \times 2 = 424 \text{ minutes,}$$

that is, about 7 feet (see **2.2**). This is reasonable because the solar meridian altitude is close to 45° at this time. The author promises a table displaying the midday shadow lengths at Mecca for (solar longitudes in) each degree of each zodiacal sign, but there is no table appended to the treatise in the Cairo manuscript.

3.4 Tāj al-Dīn

MS Damascus Ẓāhiriyya 5588 (16 fols., copied *ca.* 1900) contains some prayer-tables for Medina by Tāj al-Dīn, *muwaqqit* at the Sacred Mosque there. This individual is otherwise known only by a commentary on the almanac of Shaykh Vefā for Istanbul (*ca.* 1475), of which the unique Cairo copy is dateable to *ca.* 1700.[4]

One of Tāj al-Dīn's tables (fol. 14v) displays midday shadows in feet (*aqdām*) and tenths thereof (*aʿshār*) for "some days of the solar year for the latitude of Medina ...". The entries beginning with Kānūn II (= January) are shown in the following table.

Table 3.4

Month / Day	I	II	III	IV	V	VI
1	7 0	5 3	3 5	1 9	0 8	0 2
10	6 6	4 6	3 0	1 5	0 5	0 1
20	6 0	4 1	2 5	1 2	0 3	0 2

For $\phi = 25°$ the maximum shadows for bases 7 and $6^{1}/_{2}$ are 7.9 and 7.3. For $\phi = 24°$, a more common medieval value for Medina, the shadows are 7.6 and 7.1. The use of decimal fractions is not attested elsewhere in Islamic tables except in the trigonometric tables of the 16th-century Istanbul astronomer Taqi 'l-Dīn.[5]

[4] *Cairo ENL Survey*, nos. D239 and H2.
[5] King, "Islamic Multiplication Tables", B, pp. 413-415.

3.5 al-Amrānī / al-Imrānī (?)

In the unique copy of the prayer-tables for Medina by Tāj al-Dīn (see **3.4**) there is an introduction by an individual named al-Amrānī or al-Imrānī, who is new to me. In this (fol. 8r, also 2r) we find the following shadow-scheme for the *ẓuhr*:

طزه جبا أبجه حي

9	7	5	3	2	1	1	2	3	5	8	10

advocated for the (Syrian ?) months, with 7 to be added for the *ʿaṣr*. Elsewhere (fol. 4r), the following scheme is given for the hours:

1	2	3	4	5	6	7	8	9	10	11
28	18	9	6	3	0	3	6	9	18	28

as in the treatise of Ibn Tūmart (see **6.1**). The units are *aqdām*, but no indication is given whether these are shadow lengths or shadow increases.

CHAPTER 4

YEMENI SOURCES

4.1 al-Aṣbaḥī

Ibrāhīm ibn ʿAlī al-Janadī al-Aṣbaḥī worked in Janad in the Yemen in the mid 13th century and compiled an extensive treatise on folk astronomy.[1] This is extant in several copies, notably MS Baghdad Awqāf 2982/6276, apparently executed in Taiz in 680 H [= 1281/82]. I have used MS Cairo DM 948,1, copied in 1320 H [= 1902/03] but nevertheless reliable.

al-Aṣbaḥī's treatise is a valuable source of information on Islamic folk astronomy. Of particular interest to the present study is the fact that he advocates (fols. 38r-38v) the shadow-scheme:

T	1	2	3	4	5
Δz (in multiples of n)	5	2	1	$^1/_2$	$^1/_5$

which corresponds precisely to the Indian rule. See also **2.1**, **2.2**, **3.2**, and **6.4**.

al-Aṣbaḥī concludes his treatise with the statement (fol. 51v) that it is the custom of those astronomers who author works (on folk astronomy) to conclude their treatises with a statement on the (midday) shadow at the location where they work. He therefore adds to his treatise a list of shadow lengths observed by him at 13-day intervals throughout the Syrian year and he specifically states that he made the observations in the Mosque of Janad in the year 654 H [= 1256]. The values are displayed in **Table 4.1**. al-Aṣbaḥī neglects to mention the length of the gnomon, but for the latitude of Janad (about 13;40°) his solstitial values correspond best to $6^1/_2$. His values are expressed in *qadam*s and occasionally also a smaller unit *qīrāṭ*, not usually used as a unit of length (see also **Table 4.5**).[2] Some of the values (indicated by an asterisk in **Table 4.1**) are labelled *shāḥḥ*, a term I have not encountered elsewhere but which seems to mean "just less than".[3] The first value is for XII 15, which al-Aṣbaḥī says is the date of the winter solstice according to ʿAlī ibn Abī Ṭālib (see **2.2**).

[1] On al-Aṣbaḥī see King, *Astronomy in Yemen*, no. 5, and **IV-1.5**, *etc.* His treatise is currently being investigated by Petra Schmidl: see n. 3:1.

[2] On the use of *qīrāṭ* as a unit of weight see Hinz, *Islamische Masse und Gewichte*, p. 27, and the same author's article "*Ḳīrāṭ*" in *EI*$_2$. For another attestation of the term as a unit of length, this time in a 16th-century treatise on timekeeping, see Wiedemann, *Schriften*, II, p. 1039. Wiedemann, on what authority it is not clear, states that it is equal to a finger's breadth.

[3] Thus, for example, the value for XII 15 is *khamsat aqdām illā thumn shāḥḥ*, and the value for X 11 is *thalāthat aqdām shāḥḥu*. The verb *shaḥḥa* means, amongst other things, "to become short, to run out, to decrease, to dwindle". The term *shāḥḥ* thus has the opposite meaning of the more common term *rājiḥ*, "just more than" (see, for example, **III-3.2** and **IV-7.5**). Since sometimes the adjective *shāḥḥ* is applied only to a negative fractional part of the entry I have preferred not to use the notation "n-" in **Table 4.1**.

Table 4.1

Date	Mansion	Midday Shadow
XII 15 (WS)	21	$5 - {}^1/_8$*
XII 28	22	$5 - {}^1/_3$*
I 10	23	$4^1/_3$
I 23	24	$3 + {}^1/_2 + {}^1/_4 + q$
II 5	25	$3^1/_4$
II 18	26	$2 + {}^1/_2 q$
III 3	27	$2 - ({}^1/_2 \cdot {}^1/_6) +$
III 16 (VE)	28	$1^1/_3 +$
III 29	1	$1 - {}^1/_6 +$
IV 11	2	0
IV 24	[3]	*deest*
V 7	4	$1 - {}^1/_6 + q$ [S]
V 20	5	$1 - ({}^1/_2 \cdot {}^1/_6)$* [S]
VI 2	6	$1^1/_2$* [S]
VI 15 (SS)	7	$1 + ({}^1/_2 \cdot {}^1/_6) +$ [S]
VI 30	8	$1 + ({}^1/_2 \cdot {}^1/_6)$* [S]
VII 12	9	$1 - {}^1/_6 +$ [S]
VII 25	10	${}^1/_3 + ({}^1/_2 \cdot {}^1/_6)$* [S]
VIII 7	11	0
VIII 20	12	${}^1/_2$
IX 2	13	$1 + ({}^1/_2 \cdot {}^1/_6)$
IX 15 (AE)	14	$1 + {}^1/_3 - q$
IX 28	15	$2 + {}^1/_6$
X 11	16	3*
X 24	17	$3^1/_2$
XI 6	18	$4 + ({}^1/_2 \cdot {}^1/_6)$
XI 19	19	$4 + {}^2/_3 - ({}^1/_2 \cdot {}^1/_8)$
XII 2	20	$5 - {}^1/_8 +$

4.2 al-Fārisī

The Yemeni astronomer Muḥammad ibn Abī Bakr al-Fārisī,[4] author of a *zīj* compiled for the Rasulid Sultan al-Muẓaffar *ca.* 1295, also wrote a treatise on folk astronomy entitled *Tuḥfat al-rāghib.* This is extant in the unique complete copy MS Milan Ambrosiana Griffini 37 (unfoliated, copied *ca.* 1500).

In this work, which was compiled in Aden, al-Fārisī presents the following shadow-scheme for the hours, written out in words (see further below):

Table 4.2a

T Sign	1	2	3	4	5	6
♈ / ♎	26	12	7	4	$2^1/_2$	1
♏ / ♓	28	13	8	5	3	$2^1/_2$
♐ / ♒	31	15	9	6	5	4
♑	32	$15^1/_2$	10	7	$5^1/_2$	5

[4] On al-Fārisī (Suter, *MAA*, no. 349 and 349N and Kennedy, "*Zīj* Survey", no. 54) and his works see King, *Astronomy in Yemen*, no. 6. Another anonymous, incomplete copy of the *Tuḥfat al-rāghib* is MS Berlin Ahlwardt 5731. This treatise is currently being investigated by Petra Schmidl: see n. 3:1.

He then states that an approximate scheme for the shadows at the hours throughout the year is:

28 17 15 7 3 0 .

As can be seen from comparing these values with those in **Table 4.2a**, this scheme is approximate indeed.

Elsewhere his work al-Fārisī presents the following values of the midday shadows for various solar longitudes, also written out in words; he specifically mentions the town of Aden, for which he uses latitude 13°:

Table 4.2b

Sign	Shadow	Direction
♈ / ♎	$1 + {}^1/_4 + {}^1/_5 + {}^1/_6$	N
♉ / ♍	${}^1/_6 + {}^1/_6 \bullet {}^1/_{10}$	N
♉ 10° / ♌ 25°0	-	
♊ / ♌	${}^4/_5 + {}^1/_2 \bullet {}^1/_6$	S
♋	$1 + {}^1/_4 + {}^1/_3 \bullet {}^1/_5$	S
♎ / ♓	$3 + {}^1/_6 + {}^1/_6 \bullet {}^1/_{10}$	N
♏ / ♒	$4 + {}^1/_2 + {}^1/_{10}$	N
♑ 0°	$5 + {}^1/_5$	N

In a section on the prayer times al-Fārisī uses $\Delta z = 7$ and 14 to define the beginning and end of the *ʿaṣr*, so that he is using $n = 7$. Now if we take al-Fārisī's parameters:

$$\phi = 13° \quad \text{and} \quad \varepsilon = 23;35°,$$

the meridian altitudes at the summer solstice, equinox, and winter solstice measured from the south point are approximately

$100^1/_2$° 77° $53^1/_2$°

and the corresponding shadows to base $n = 7$ are:

1;18 S 1;37 N 5;11 N .

These correspond quite closely with al-Fārisī's values in **Table 4.2b**, these being equivalent to:

1;19 S 1;37 N 5;12 N .

To compute the shadows at the seasonal hours s_i al-Fārisī has used a simple approximation to find the corresponding altitudes h_i equivalent to:

$$\sin h_i = \sin H \sin (i \bullet 15°) ,$$

a procedure derived originally from Indian sources and used by certain Muslim astronomers.[5] For the solstices, using $\phi = 13°$ and $n = 7$, I compute the values:

26.6	12.4	7.2	4.3	2.3	1.3
32.9	15.9	10.1	7.2	5.7	5.2

so that al-Fārisī has in general truncated the results of his calculations.

[5] See n. 1:21. The history of this formula, as well as various tables and instruments based on it, in both the Near Eastern and the European traditions, is traced in **XI**. See already Goldstein, *Ibn al-Muthannā on al-Khwārizmī*, pp. 82-83 and 207-209 (on al-Khwārizmī's use of the formula), and King, "al-Khwārizmī", pp. 7, 11.

4.3 al-Sulṭān al-Ashraf

The Rasulid Sultan al-Ashraf (*reg.* 1296-97) compiled two major works on astronomy. The first was a lengthy treatise on the construction of astrolabes and sundials and also on the use of the magnetic compass. The second was an equally lengthy volume on astrology, which also contains tables for timekeeping for Sanaa and an agricultural almanac.[6]

This second work, entitled *Kitāb al-Tabṣira fī ʿilm al-nujūm*, survives in the unique MS Oxford Hunt. 233, copied *ca.* 1400, and in the almanac on fols. 97r-108v, shadow lengths for the *ẓuhr* and the beginning and end of the *ʿaṣr* are given for each half-month of the Syrian year. The values are written out in words at the side of the almanac; they are reproduced in the following:

Table 4.3

Notes: All fractions are additive unless otherwise stated.
An asterisc denotes that an approximation has been made in a particular set of entries.
A double asterisc denotes that there is an error in the calculation of either of the entries for the *ʿaṣr*.

Month	*ẓuhr*	*ʿaṣr* 1	*ʿaṣr* 2	
I	5	$11\ {}^{2}/_{3}$	$18\ {}^{2}/_{3}$	
	$4\ {}^{1}/_{2}$	$11\ {}^{1}/_{6}$	$18\ {}^{1}/_{6}$	
II	$3\ {}^{1}/_{2}\ {}^{1}/_{4}$	$10\ {}^{1}/_{4}\ {}^{1}/_{6}$	$17\ {}^{1}/_{4}\ {}^{1}/_{6}$	
	3	$9\ {}^{2}/_{3}$	$16\ {}^{2}/_{3}$	
III	$2\ {}^{1}/_{4}$	$8\ {}^{2}/_{3}\ {}^{1}/_{4}$ [a]	$16\ {}^{2}/_{3}\ {}^{1}/_{4}$	
	$1\ {}^{1}/_{2}$	8	$15\ {}^{1}/_{5}$	**
IV	${}^{1}/_{2}\ {}^{1}/_{4}$	$7\ {}^{1}/_{3}\ {}^{1}/_{10}$	$14\ {}^{1}/_{3} \bullet {}^{1}/_{10}$	**
	${}^{1}/_{4}$	$6\ {}^{2}/_{3}\ {}^{1}/_{4}$	$13\ {}^{2}/_{3}\ {}^{1}/_{4}$	
V	${}^{1}/_{3}$	7	14	
	${}^{1}/_{2}$	$7\ {}^{1}/_{4}\ {}^{1}/_{6}$	$14\ {}^{1}/_{4}\ {}^{1}/_{6}$	**
VI	$1\ {}^{1}/_{2} \bullet {}^{1}/_{10}$	$7\ {}^{2}/_{3}$	$14\ {}^{2}/_{3}$	*
	$1\ {}^{1}/_{10}$	$7\ {}^{1}/_{2}\ {}^{1}/_{4}$	$14\ {}^{1}/_{2}\ {}^{1}/_{4}$	*
VII	1	$7\ {}^{2}/_{3}$	$14\ {}^{2}/_{3}$	
	${}^{1}/_{2}\ {}^{1}/_{4}$	$7\ {}^{1}/_{3}$	$14\ {}^{1}/_{3}$	**
VIII	${}^{1}/_{5}$	$6\ {}^{1}/_{2}\ {}^{1}/_{3}$	$13\ {}^{1}/_{2}\ {}^{1}/_{3}$	*
	1	7	14	**
IX	$1\ {}^{1}/_{2} \bullet {}^{1}/_{10}$	7	$14\ {}^{2}/_{3}$ [b]	**
	$1\ {}^{1}/_{2}\ {}^{1}/_{4}$	$8\ {}^{2}/_{5}$	$15\ {}^{2}/_{5}$	**
X	$2\ {}^{1}/_{3}\ {}^{1}/_{4}$	$9\ {}^{1}/_{4}$ [c]	$16\ {}^{1}/_{4}$	
	$3\ {}^{1}/_{3}$	10	17	
XI	$4\ {}^{1}/_{6}$	$10\ {}^{1}/_{2}$	$17\ {}^{1}/_{2}\ {}^{1}/_{3}$	**
	$4\ {}^{1}/_{2}\ {}^{1}/_{4}$	$11\ {}^{1}/_{4}\ {}^{1}/_{6}$	$18\ {}^{1}/_{4}\ {}^{1}/_{6}$	
XII	$5\ {}^{1}/_{6}$	$11\ {}^{1}/_{2}\ {}^{1}/_{3}$	$18\ {}^{1}/_{2}\ {}^{1}/_{3}$	
	$5\ {}^{1}/_{4}$	$11\ {}^{2}/_{3}\ {}^{1}/_{4}$	$18\ {}^{2}/_{3}\ {}^{1}/_{4}$	

[a] text: $3\ {}^{2}/_{3}\ {}^{1}/_{4}$; [b] text: fraction partially illegible (2 for numerator is clear); [c] text: $6\ {}^{1}/_{4}$

We notice immediately that, apart from a few calculating errors, the entries for the end of the *ʿaṣr* are larger than those for the beginning of the *ʿaṣr* by 7, and that the latter are larger than those for the *ẓuhr* by $6^{2}/_{3}$. This is curious indeed, and I have no explanation to offer. Elsewhere in his treatise al-Ashraf presents a table of cotangents to base $6^{2}/_{3}$, and in his timekeeping tables he uses obliquity 24° (14;30° for the latitude of Sanaa). For these parameters the maximum

[6] On al-Ashraf (Suter, *MAA*, no. 394) and his astronomical works see King, *Astronomy in Yemen*, no. 8, and King, "Yemeni Astrolabe". His shadow-scheme was brought to my attention by Daniel Varisco, whose detailed study of al-Ashraf's almanac is now published as Varisco, *Yemeni Almanac*. See also n. 4:12.

midday shadow is 5;18, which corresponds quite well to the maximum value in the table, *viz.* $5{}^1/_4$. Note that the fact that the midday shadows around the summer solstice are southerly is overlooked in the computation of the shadows at the *ʿaṣr*.

4.4 al-Sulṭān al-Afḍal

A precious manuscript preserved in a private library in Sanaa contains an encyclopaedic work compiled *ca.* 1375 by the Yemeni Sultan al-Afḍal.[7] In some marginalia to the astronomical sections of this work there are two sets of notes, both in an untidy hand, apparently due to al-Afḍal himself. The two notes give the shadows (*ẓill*) for each hour as follows, the values being stated to be in cubits (*dhirāʿ*):

T	1	2	3	4	5	6
z	60	30	15	$7{}^1/_2$ (?)	$3{}^1/_4$	${}^2/_3$
z	60	30	19	$7{}^1/_2$ (?)	$3{}^1/_4$	1(?)

These values are clearly garbled. One would expect that the values would be for Δz and that they should be five in number. The entries for T = 4 (read 3?) should probably be $6{}^1/_2$.

4.5 Bā Makhrama

A late Ottoman navigational manual (see **10.2**) contains a table of shadow lengths, said to be *ʿalā manāzil al-Shibāmī al-shaykh ʿAbdallāh ibn ʿUmar Bā Makhrama*, that is, "in accordance with the lunar mansions as described by ... Bā Makhrama". This individual was a Yemeni scholar active *ca.* 1500, and is otherwise known by a treatise on folk astronomy entitled *al-Shāmil*, which survives in a unique manuscript.[8] I have not noticed the shadow table in this treatise. The epithet al-Shibāmī indicates his family's provenance from Shibām in the Ḥaḍramawt.[9]

In the table the 28 mansions are divided in four columns between the seasons (*al-Hanʿa* = M_5 = first mansion of spring), and the shadow lengths in *qadam*s at midday and the *ʿaṣr* are written out in words. The difference is invariably 7, which is clearly the value used for n. The entries are displayed in the following table:

Table 4.5

M	Z	M	Z
6	$5 + {}^1/_8$	20	$1 - {}^1/_{20}$
7	$5 - {}^1/_6$	21	$1 - {}^1/_5$
8	$4 + {}^1/_3$	22	${}^1/_2$
9	$4 - {}^1/_5$	23	${}^1/_8$ [e]
10	$3 + {}^1/_8$	24	${}^1/_3$
11	$2 + {}^1/_2 - {}^1/_{30}$ [a]	25	$1 - {}^1/_6$
12	$2 - {}^1/_6$	26	$1 + {}^2/_3$

[7] On the astronomical activity of al-Afḍal see King, *Astronomy in Yemen*, no. 18, and **II-12.4**.
[8] King, *Astronomy in Yemen*, no. 26.
[9] Another Shibāmī of renown in Yemeni folk astronomy is listed *ibid.*, no. 43.

13	$1 + {}^{1}/_{5}$	27	$2 + {}^{1}/_{30}$
14	${}^{2}/_{3}$	28	$2 + {}^{2}/_{3}$
15	${}^{1}/_{8}$ [b]	1	$3 + {}^{1}/_{3}$
16	${}^{1}/_{4} + q - {}^{1}/_{4}q$ [c]	2	4
17	${}^{1}/_{3} - {}^{1}/_{20}$ [d]	3	$4 + {}^{1}/_{2} + {}^{1}/_{30}$
18	$1 - {}^{1}/_{8}$	4	$5 - {}^{1}/_{20}$
19	$1 - {}^{1}/_{30}$	5	5 [f]

[a] text: *qadamayn wa-niṣf illā thulth ʿushr*; [b] text adds remark: "on the sixth day of this mansion the midday shadow is zero"; [c] here q = *qīrāṭ* (text: *rubʿ qadam wa-qīrāṭ illā rubʿ*) – see **4.1**; [d] for the *ʿaṣr*: $8 - {}^{1}/_{3}$ of ${}^{1}/_{2}$ of ${}^{1}/_{10}$; [e] for the *ʿaṣr*: $7 + {}^{1}/_{3}$; [f] for the *ʿaṣr*: $11 + {}^{1}/_{6}$

Notice that the extremal values are $5 + {}^{1}/_{8}$ and $1 - {}^{1}/_{30}$ (S), which are mutually inconsistent. They correspond most closely to latitude 13°, commonly used for Aden: for this latitude the approximate values are $5 + {}^{1}/_{6}$ and $1 + {}^{3}/_{20}$ (S).

In some notes appended to MS Cairo DM 948 (fols. 101v-102v) there are some values of the midday shadow (base not stated) at Aden, given for each ten days of the Syrian year. These are specifically attributed to Bā Makhrama. They are followed by an incomplete, anonymous set for Taiz (with $n = 6^{1}/_{2}$), also for each ten days of the year (fols. 102v-103r).

4.6 al-Ṭawāshī

A Yemeni treatise on folk astronomy entitled *Miftāḥ al-asrār fī ʿilm al-falak al-dawwār* is attributed in both of the extant manuscripts, MSS Cairo DJ 709,6, (fols. 29r-53v, copied *ca.* 1750) and Princeton Garrett (Hitti) 1016 (61 B) (copied *ca.* 1875), to Nūr al-Dīn ʿAlī ibn ʿAbdallāh al-Ṭawāshī, an individual otherwise known to the modern literature.[10] The treatise contains a poem on the midday shadows for the lunar mansions (fol. 35r of the Cairo manuscript): the values are a simplified version of those given by Ibn Tūmart (see **6.1**), varying linearly between a maximum of $5^{1}/_{2}$ feet for *al-Balda* and a minimum of zero between *al-Haqʿa* and *al-Jabha*.

4.7 al-Thābitī

Muḥammad ibn ʿAbd al-Laṭīf al-Thābitī was a Syrian who lived in Zabid.[11] In 1047 H [= 1637/38] he compiled a set of tables for timekeeping by the lunar mansions and by shadow lengths, of which I have examined MS Vatican ar. 962 (fols. 13r-19r). The shadow-tables display the shadow at midday and the *ʿaṣr*: values to base $6^{1}/_{2}$ are given to two digits, the second called *banān*, fingers, representing *twelfths of a shadow digit*. The midday shadows are computed according to a linear zigzag-scheme, with a maximum of $5^{5}/_{12}$ at the winter solstice and a maximum of $2^{2}/_{12}$ (south) at the summer solstice. These extremal values are equivalent to 10 and 4 digits for a gnomon length of 12 digits, which suggests that al-Thābitī converted a set of values from base 12 to base $6^{1}/_{2}$. In fact, for $\phi = 15°$ and base 12 the extremal values are

[10] *Ibid.*, no. 27.
[11] *ibid.*, no. 33, and **II-12.7**.

closer to 10 and 2 digits. al-Thābitī's afternoon shadow lengths are always $6^1/_2$ feet more than the absolute value of his midday shadow lengths.

In MS Cairo DM 899,2 (fols. 9r-9v, copied *ca.* 1700), appended to an abridged version of the treatise of al-Aṣbaḥī (**4.1**) by Abū Ḥamad Maḥfūẓ ibn ʿAbd al-Raḥmān al-Ḥaḍramī, there is a list of midday shadows to base $6^1/_2$ written out in words for days in Syrian calendar, specifically intended for Zabid. The values are the same as those given by al-Thābitī. Values are also given for the *ʿaṣr* and are all correctly $6^1/_2$ feet more than the midday values.

4.8 Abū Shukayl (?)

On fol. 97v of MS Cairo DM 948 (see **4.1**), amidst some notes on timekeeping, there is a reference to a commentary (*sharḥ*) by one Abū Shukayl (?), not otherwise known to me. In this work the midday shadow is given as zero on Ayyār 1; it has a maximum (southerly) value $2\ ^1/_6$ on Ḥazīrān 22, and a maximum (northerly) value $5\ ^1/_4 + ^1/_6$ on Kānūn I 22. No locality is specified, but the extremal values are the same as those of al-Thābitī (see **4.7**).

4.9 Miscellaneous anonymi

(a) In MS Oxford Marsh 134, copied 855 H [= 1451/1452], of the treatise of folk astronomy by al-Aṣbaḥī (**4.1**), there is a shadow-scheme amidst some notes at the end of the treatise (fol. 52v). The scheme is presented in the form of a circle and values of the midday shadow lengths are given for each of the Syrian months; the corresponding names of the Himyaritic (ancient South Arabian) months are also given.[12] The scheme is simply:

5 4 3 2 1 0 1 2 3 4 5 6 ,

and is probably very old. One may wonder whether the scheme was known in the Yemen before the advent of Islam. In the view of what has come to light of Graeco-Roman influence in South Arabia,[13] I suspect that it was indeed known there. The only other Islamic source in which I have come across such a scheme is from Ḥiṣn Kayfā (now in Southern Turkey), where one could presume Byzantine influence – see **7.2**.

(b) MS Aleppo Awqāf 968 (16 pp., copied *ca.* 1600) contains an anonymous treatise entitled *Kitāb al-Īḍāḥ al-shāfī bi-'l-itqān fī maʿrifat al-manāzil wa-'l-azmān*. The author mentions the *Qāmūs* of Majd al-Dīn (al-Fīrūzābādī) so that he must postdate *ca.* 1400. He also mentions the *Kitāb al-Yawāqīt* by al-Janadī, otherwise known as al-Aṣbaḥī (**4.1**). Two shadow-schemes are presented for the seasonal hours. The shadows (in feet), not shadow increases, are as follows:

---

[12] A. F. L. Beeston, "New Light on the Himyaritic Calendar", *Arabian Studies* 1 (1974), pp. 1-6; George Saliba, "A Medieval Note on the Himyarite Calendar", *JAOS* 105 (1985), pp. 715-717, based on MS Oxford Hunt. 233 of al-Ashraf's *Tabṣira* (**4.3**); and Varisco, *Yemeni Almanac*, pp. 64-71, *etc.*

[13] See, for example, A. H. Masry, *An Introduction to Saudi Arabian Antiquities*, Riyadh, 1975; A.R. al-Ansary, *Qaryat al-Fau. A Portrait of Pre-Islamic Civilisation in Saudi Arabia*, Riyadh, 1982, and Annie Caubet, *Aux sources du monde Arabe: L'Arabie avant l'Islam – Collections du Musée du Louvre*, Paris: Institut du Monde Arabe & Réunion des musées nationaux, 1990.

38	18	9	6	3	0
35	15	9	5	3	0

The numbers are displayed in the text in Arabic numerical notation.

(c) In an anonymous Egyptian source MS Cairo Dār al-Kutub K 3769, (fols. 10v-11r, copied 1910), appended to a late Yemeni treatise on timekeeping, the scheme is:

1	2	3	4	5	6	7	8	9	10	11
26(!)	13	7	5	3	-	3	5	7	13	24(!)

4.10 al-Wāsiʿī

Shaykh ʿAbd al-Wāsiʿ ibn Yaḥyā al-Wāsiʿī published a set of prayer-tables for the Yemen in the year 1939/40.[14] Amongst the functions tabulated for each day of the year are the shadow lengths at midday and the *ʿaṣr* given to base 7 in feet and fingers (as with al-Thābitī). The shadow lengths are computed according to a very primitive scheme. From Taurus 8° to Cancer 18° the midday shadows have a constant value $^4/_{12}$ feet north instead of south, and different values are given for the vernal equinox and autumnal equinox, as follows:

VE: 2 SS: $^4/_{12}$ AE: 1 $^4/_{12}$ WS: 5 .

For $\phi = 15°$ and $n = 7$ these are actually closer to:

$1\,^{10.5}/_{12}$ $1\,^5/_{12}$ (north!) $1\,^{10.5}/_{12}$ $5\,^7/_{12}$.

---

14 On al-Wāsiʿī see King, *Astronomy in Yemen*, no. 47. His tables were published in Cairo, Maṭbaʿat Ḥijāzī, 1358 H [= 1939], under the title *Kanz al-thiqāt fī ʿilm al-mīqāt*. For an analysis see **II-12.12**.

CHAPTER 5

ANDALUSĪ SOURCES

5.1 ʿArīb ibn Saʿd

The earliest known almanac for al-Andalus is the *Kitāb al-Anwāʾ* of Abu ʾl-Ḥasan ʿArīb ibn Saʿd al-Kātib compiled in the late 10th century. This work, known as the *Calendar of Cordova*, was translated into Latin and was transmitted to medieval Europe. The Judaeo-Arabic and Latin texts have been published by Reinhart Dozy and translated into French by Charles Pellat.[1]

ʿArīb ibn Saʿd gives information on the lunar mansions, the duration of day and night and twilight, as well as midday solar altitudes and shadows (expressed in words in terms of the gnomon length), for each 13, 14, 15 or 16 days. This information has recently been investigated by Julio Samsó.[2] The following table displays the values given for the last two quantities (the equinoxes are at III 16 and IX 18):

Table 5.1

Date	H	z ÷ n	Date	H	z ÷ n
I 2	$29^{2}/_{3}{}^{\circ}$	$1^{3}/_{4}$	VII 16	$73^{2}/_{5}$	$^{11}/_{36}$
I 16	$32^{1}/_{2}$	$1^{3}/_{5}$	VIII 1	$69^{2}/_{3}$	$^{3}/_{8}$
II 2	$36^{5}/_{6}$	$1^{1}/_{3}$	VIII 16	$65^{1}/_{4}$	$^{1}/_{2}$
II 15	$40^{1}/_{2}$	$1^{1}/_{6}$	IX 1	$55^{1}/_{4}$ (*sic*)	$^{7}/_{12}$
III 2	$46^{5}/_{6}$	$^{15}/_{16}$	IX 15	$54^{1}/_{2}$	$^{2}/_{3}$
III 18	$52^{1}/_{2}$	$^{3}/_{4}$	X 1	$47^{2}/_{3}$	$^{11}/_{12}$
IV 2	$58^{5}/_{6}$	$^{7}/_{12}$	X 8	*deest*	1
IV 17	$63^{4}/_{5}$	$^{1}/_{2}$	X 16	$41^{11}/_{12}$	$1^{1}/_{8}$
V 2	*deest*	*deest*	XI 1	$36^{1}/_{4}$	$1^{1}/_{3}$
V 14	$72^{4}/_{5}$	$^{1}/_{3}$	XI 15	$32^{2}/_{5}$	$1^{7}/_{12}$
VI 2	$75^{1}/_{3}$	$^{11}/_{40}$	XII 2	$29^{2}/_{3}$	$1^{3}/_{4}$
VI 16	$76^{1}/_{3}$	$^{1}/_{4}$	XII 16	$28^{2}/_{3}$	$1^{5}/_{6}$
VII 1	$75^{3}/_{4}$	$^{1}/_{4}$			

The midday solar altitudes are clearly computed, and underlying them are to be detected the parameters:

$$\phi = 37;30^{\circ} \quad \text{and} \quad \varepsilon = 23;50^{\circ} .$$

The first was no doubt intended to serve Cordova, although it is not attested in any other known medieval Islamic sources;[3] the accurate value is 37;53°. The second is an approximation for

---

[1] On the context of all of this Andalusī material see Forcada, "*Anwāʾ* Books in al-Andalus". On the *Calendar of Cordova* see Pellat, *Le Calendrier de Cordoue*, pp. xiii-xiv, and, more recently, J. Martinez Garquez and J. Samsó, "Una nueva traducción latina del calendrio de Córdoba (siglo XIII)", in Vernet, ed., *Textos y estudios*, A, pp. 9-78, Samsó, "Sobre los materiales astronómicos en el 'Calendrio de Córdoba' y en su versión latina del siglo XIII", in Vernet, ed., *Textos y estudios*, B, pp. 125-138; and Forcada, *op. cit.* See also the article on the author by Charles Pellat in EI_2, and Sezgin, *GAS*, VII, pp. 355-356.

[2] See Samsó, *op. cit.*, esp. pp. 126 ff. for details.

[3] Kennedy & Kennedy, *Islamic Geographical Coordinates*, p. 95.

Ptolemy's 23;51,20°. Likewise computed are the ratios of the shadows to the gnomon. The tabulation of such quantities derived by computation borders on the mathematical tradition of Islamic astronomy.

5.2 al-Umawī

MS Escorial ár. 941 (38 fols., copied *ca.* 1300) is a unique but incomplete copy of a treatise on folk astronomy by al-Ḥasan ibn ʿAlī al-Umawī al-Qurṭubī, who was born in Cordova in 514 H [= 1120/21] and died in Seville in 602 H [= 1205/06].[4] The work is divided into two main parts, both incomplete, the first discussing *anwā*ʾ for each of the *ʿajamī* (Christian) months, and the second giving information on a variety of topics including the times of prayer, the qibla, the winds and the seasons.

Our present concern is with the shadow lengths for the hours given at the end of the sections for each month. Unfortunately only three sets of values are preserved in the Escorial manuscript – see the following table – but these are adequate to reconstruct the entire set for the year.

Table 5.2

Month \ T	1/11	2/10	3/9	4/8	5/7	6	*ʿaṣr*
January	27	17	13	10	9	7	14
February	26	16	12	9	8	6	[13] [a]
March	25	15	11	8	7	5	12

[a] text: 14

The pattern is similar to, but not identical with, that of the Greek tables of type M – see **1.4** and also **7.2**. We may suppose, for example, that the midday shadows vary linearly between 8 (Dec.) and 2 (June). The increases for each hour are:

2 1 3 4 10

instead of the usual

1 2 3 4 10 ;

Neugebauer has noted similar absurdities in Latin tables.[5]

5.3 Ibn Sabʿīn (?)

In MS Cairo MJ 202,2 (fol. 33v, copied 1681), immediately following an astrological treatise *Kitāb al-Daraj* by the 13th-century Andulusian mystic Ibn Sabʿīn,[6] there is a rule for finding the time of day by shadows. Values for the shadow lengths are given in Eastern Arabic numerical notation, and the case-ending of the word *qadam*, "feet", indicates the the values given in the text are probably as originally intended. These are:

---

[4] On al-Umawī see Suter, *MAA*, no. 323.

[5] Neugebauer, *HAMA*, II, p. 745.

[6] On Ibn Sabʿīn see the article by A. Faure in EI_2, and also Brockelmann, *GAL*, I, p. 611, SI, p. 844, and SII, p. 1017, and *Cairo ENL Survey*, no. F15. His various astrological works deserve detailed investigation.

1	2	3	4	5	6	7	8	9	10	11	12
22	18	7	6	3	*	3	6	7	8	18	28

No base is stated. For the sixth hour the sun is said to be in "the pole of the sphere" (*quṭb al-falak*), perhaps here meaning the zenith (the scheme is not necessarily of Andalusī origin).[6] Notice the lack of symmetry and the fact that the compiler could not resist including a value for the twelfth hour. A similar set of values, albeit symmetrical, is presented by Ibn Tūmart – see **6.1**.

---

[6] Even before the advent of Islam the Arabs were aware of the celestial pole, which they called *quṭb* – see the article "al-Ḳuṭb" by Paul Kunitzsch in EI_2, repr. in *idem*, *Studies*, XVIII.

CHAPTER 6

MAGHRIBI SOURCES

6.1 Ibn Tūmart

Muḥammad ibn ʿAlī ibn Tūmart al-Andalusī was the author of a treatise entitled *Kitāb Kanz al-ʿulūm wa-'l-durr al-manẓūm fī ḥaqāʾiq ʿilm al-sharīʿa* and extant in several manuscripts.[1] The work deals with religion, law and the sciences, and is of considerable interest. It has been studied by Georges Vajda, who established that the author can not predate the beginning of the 13th century. I have used MS Cairo ḤJ 47,1 (fols. 1v-85r, copied *ca.* 1750), in which there is a section on simple timekeeping by shadows on fols. 80r-81v: see **Fig. 6.1**.

Although Ibn Tūmart was writing in al-Andalus or the Maghrib, the first shadow-scheme that he presents is for the Yemen and is introduced as such. It gives the midday shadows as the sun progresses through the 28 lunar mansions. For *al-Balda* (M_{21}) the shadow is $5^1/_2$ feet, decreasing $^1/_2$ foot for each mansion until *al-Dabarān* (M_4) when there is no shadow (*yanʿadim al-ẓill*). From *al-Haqʿa* (no. I) till *al-Jabha* (M_{10}) there is a very small midday shadow (*adnā ziyāda*), and then in *al-Zubra* (M_{11}, the text has incorrectly *al-Dabarān*) the shadow is $^1/_2$ foot, increasing $^1/_2$ foot each mansion until *al-Balda* (M_{21}). For each mansion the shadow at the *ʿaṣr* is $6^1/_2$ feet more. The maximum midday shadow length of $5^1/_2$ corresponds to $\phi = 15°$ for n $= 6^1/_2$, but then the shadow should increase to 1 foot south in the mansions close to the summer solstice.

The above information is contained in the text of Ibn Tūmart's treatise and is summarized in a poem. Our author then presents a shadow-scheme in feet for the hours, which is as follows:

28 18 9 6 3 0 (*istiwāʾ*)

This scheme was also used by al-Amrānī / al-Imrānī (**3.5**). Ibn Tūmart concludes this section with a discussion of time-reckoning by night by means of the lunar mansions.

6.2 Abū Miqraʿ

Abū Miqraʿ was an authority on folk astronomy in the Maghrib in the 13th century.[2] His teachings are known to us through a poem called *al-Muqniʿ fī 'khtiṣār ʿilm Abī Miqraʿ* written by the 17th-century Maghribi astronomer Muḥammad ibn Saʿīd al-Sūsī al-Marghīthī, and they

---

[1] On Ibn Tūmart see the article in EI_2 by J. F. P. Hopkins, and Brockelmann, *GAL*, I, pp. 506, and SI, p. 697, and I, p. 274 and SI, p. 424 (are these references to the same individual?), as well as *Cairo ENL Survey*, no. F14. His treatise is investigated in Georges Vajda, "Une synthèse peu connue de la révélation et de la philosophie: le 'Kanz al-ʿUlūm' de Muḥammad ibn ʿAlī Ibn Tūmart al-Andalusī", in *Mélanges Louis Massignon*, Damas: Institut Français de Damas, 1957, vol. III, pp. 359-374.

[2] Colin & Renaud, "Abū Miqraʿ"; Renaud, "Additions et corrections à Suter", no. 531, and *Cairo ENL Survey*, no. F17.

Fig. 6.1: The section of the treatise of Ibn Tūmart dealing with time-keeping by shadows. [From MS Cairo ḤJ 47,1, fols. 80r-81v, courtesy of the Egyptian National Library.]

are outlined in greater detail in various commentaries, of which I have consulted one called *al-Mumtiʿ fī sharḥ al-Muqniʿ* by al-Sūsī.[3]

In MS Cairo ṬJ 201,2 (fols. 7r-12v, copied *ca.* 1600) of the *Mumtiʿ*, al-Sūsī quotes Abū Miqraʿ (fol. 11r) as having advocated the standard (Indian) formula for timekeeping (with n = $6^1/_2$), and the scheme:

9 7 5 3 2 1 1 2 4 ? 8 10

for the midday shadows for the months of the (Christian) calendar. The values are given in *abjad* notation, and the value for month X looks like a European 6, but this could be read as either a 5 or a 6. In MS Cairo DM 415 (55 fols., copied *ca.* 1700) of the *Mumtiʿ* the same values are given (fol. 41r) but with a *yāʾ* = 10 for month X; elsewhere (fol. 41v) the sequence 1 2 4 5 is mentioned for months VII to X, and it seems that a 5 was indeed intended for month X. al-Sūsī adds that the above values were specifically for latitude 32°, Marrakesh (the actual value is 31;49°), and that al-Jādarī had proposed the scheme:

10 8 5 3 2 1 1 2 4 6 8 10

for Fez. See further **6.4**.

6.3 Pseudo-Ibn al-Bannāʾ

Ibn al-Bannāʾ was a Morroccan astronomer and mathematician of some renown who worked in Marrakesh about 1300 and whose major work was a *zīj*, extant in several copies.[4] An almanac entitled *Risāla fī ʾl-Anwāʾ* is also attributed to him and has been published by Henri-Paul-Joseph Renaud.[5] Along with diverse information on agriculture, meteorology, and astronomical folklore given for each day of the year, this almanac displays the lengths of day and night for each 15 days of the solar year based on a latitude of about 38°, that is, Cordova, as well as the lengths of the gnomon shadow in feet (*aqdām*) at the *ẓuhr* and *ʿaṣr* for each month. Since Ibn al-Bannāʾ worked in Marrakesh ($\phi \approx 32°$) there are problems with the attribution of this work to him, but I shall not pursue these here.

Renaud used five manuscripts (S, P_1, P_2, A_1 and A_2[6]) for his translation of the almanac and noted the variants in the shadow lengths; there are clearly three main traditions – see **Table 6.3a**. Neugebauer, perhaps overeagerly, restored some of the values in the scheme represented by MS P_2 to produce the primitive scheme of Greek origin shown in **Table 6.3b**.[7]

---

[3] Renaud, *op. cit.*, no. 540.

[4] On Ibn al-Bannāʾ see Suter, *MAA*, no. 399; Renaud, *op. cit.*, no. 399; *idem*, "Ibn al-Bannāʾ"; *Cairo ENL Survey*, no. F23; and the article by J. Vernet in *DSB*.

[5] Renaud, *Calendrier d'Ibn al-Bannāʾ*, esp. pp. 22-23. See now Forcada, "Les sources andalouses du Calendrier d'Ibn al-Bannāʾ".

[6] These stand for: S – Salé (private owner); P_1 – Paris BNF ar. 4764; P_2 – Paris BNF ar. 6020; A_1 – Algiers 941; A_2 – Algiers 1468 (see Renaud, *op. cit.*, p. 26).

[7] Neugebauer, *HAMA*, II, pp. 743-744.

Table 6.3a

	I	II	III	IV	V	VI	VII	VIII	IX	X	XI	XII
(i) MS S:												
ẓuhr	7	6	4	3	2	1	1	4	5	7	$9^1/_5$	11
ʿaṣr	14	13	11	10	9	8	8	10	10	11	15	17
(ii) MSS P_1, A_1 and A_2:												
ẓuhr	7	6	6	4	3	2	4	4[a]	5	7	$9^1/_5$	11
ʿaṣr	14	13	11	10	9	8	8	10[b]	10	11	15	15
(iii) MS P_2:												
ẓuhr	7	6	5	4	3	3	3	4	5	7	$9^1/_5$	11
ʿaṣr	14	13	11	10	9	8	8	10	10	11	15	15

[a] P_1 has $2^1/_3$ [b] P_1 has 8

Table 6.3b

Modifications to the scheme in MS P_2 made by Neugebauer are underlined.

ẓuhr	7	6	5	4	3	2	3	4	5	<u>6</u>	<u>7</u>	<u>8</u>
ʿaṣr	14	13	11	10	9	8	<u>9</u>	10	1<u>2</u>	1<u>3</u>	1<u>4</u>	15
Δz	7	7	6	6	6	6	<u>6</u>	6	<u>7</u>	<u>7</u>	<u>7</u>	<u>7</u>

The *ẓuhr* entries for months XI and XII, and perhaps also month X, are garbled. Virtually all of the *ʿaṣr* shadows are garbled since, at least according to the standard definition of the *ʿaṣr*, they should each be in excess of the corresponding midday entries by 7. (Even in Neugebauer's reconstruction the increases are either 6 or 7.) The *ẓuhr* shadows are clearly all midday shadows, with perhaps one exception (see below), rather than midday shadows augmented by one-quarter of the length of the gnomon, which was the standard definition of the time of the *ẓuhr* in al-Andalus and the Maghrib (see also **6.4**).[8]

The anomalous but well-attested value $9^1/_5$ for the *ẓuhr* in month XI appears to represent an attempt to add one-fifth of the gnomon length to the gnomon shadow. This corresponds to a time one seasonal hour after midday using the standard Indian formula (see **2.1** and **3.2**).[9] The value of the gnomon length varies between 6 and 7 feet in the table, but the value $9^1/_5$ could be derived from a midday shadow of 8 augmented by one-fifth of a gnomon length of six digits.

6.4 Abū ʿAbdallāh al-Muʾaddib

MS Oxford Pococke 249 (63 fols., copied *ca.* 1400) is a unique copy of a treatise on folk astronomy by a Maghribī astronomer named Abū ʿAbdallāh Muḥammad ibn ʿAbdallāh ibn ʿAmr al-Muʾaddib ("the teacher"), not otherwise known to the modern literature. On fols. 46v-47r there is a table displaying the shadow lengths in digits (n = 12) at each seasonal hour and at the *ẓuhr* and *ʿaṣr* for each date of the solar year when the shadow has an integral value – see **Fig. 6.4** and **Table 6.4** for extracts.

[8] On the motivation for the value $\Delta z = {}^1/_4$ n see **IV-4.4**.

[9] See further King, "Tunisian Sundial", p. 195, and now **IV-2**.

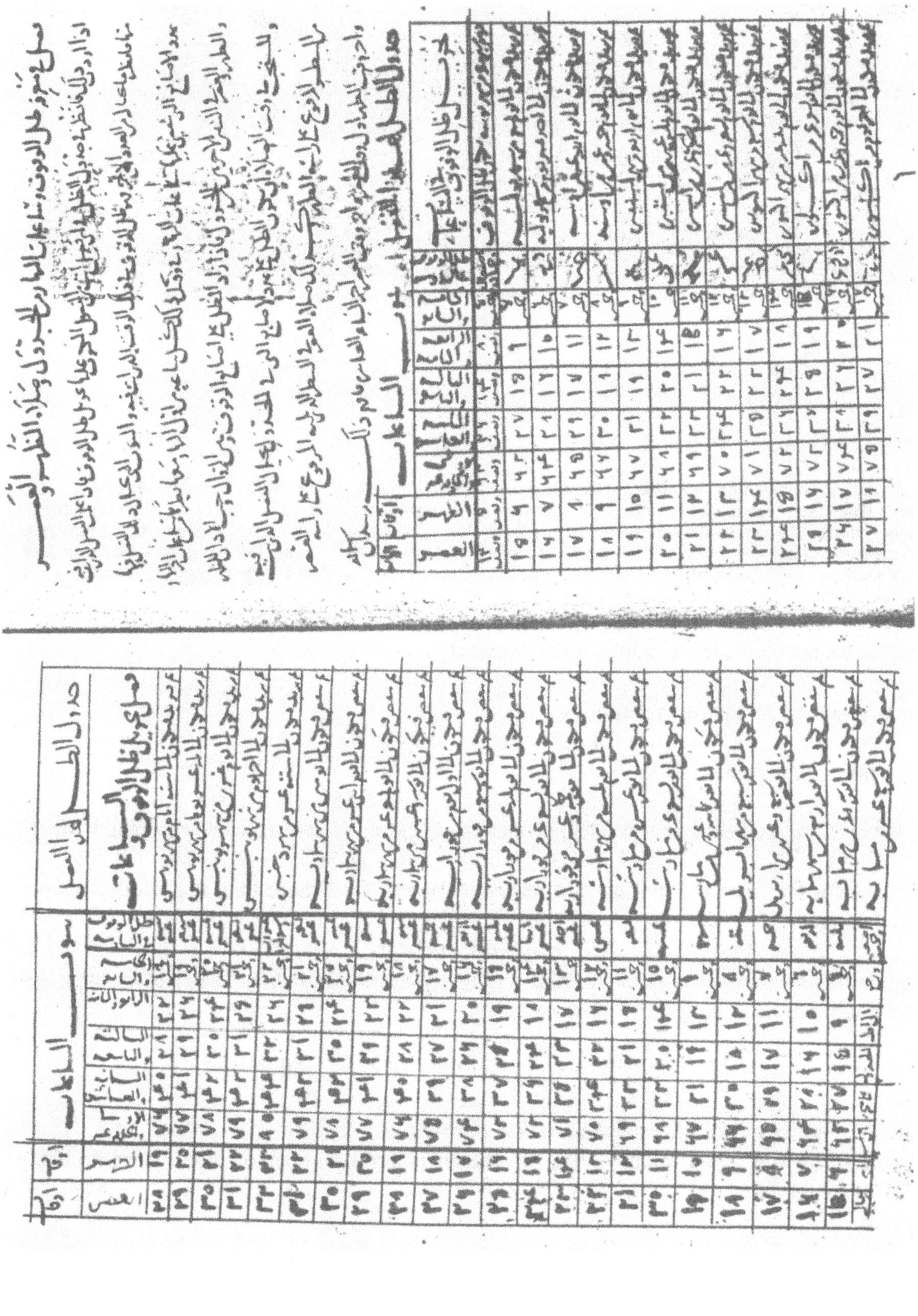

Fig. 6.4: A table for timekeeping by Abū ʿAbdallāh al-Muʾaddib. [From MS Oxford Pococke 249, fols. 46v-47r, courtesy of the Bodleian Library.]

Table 6.4

Date[a] | Hours	6	5	4	3	2	1	*ẓuhr*	*ʿaṣr*
VI 17	$2^{2}/_{5}$ [b]	5- [c]	$8^{1}/_{2}$	$14^{1}/_{2}$	$26^{1}/_{2}$	$62^{1}/_{2}$	$5^{1}/_{2}$	$14^{1}/_{2}$
VII 9	3	$5^{2}/_{5}$	9	15	27	63	6	15
VII 11 [d]	4	$6^{2}/_{5}$	10	16	28	64	7	16
VIII 14 - 25		.....						
IX 4 - 13 - 22 - 29		.....						
X 7 - 13 - 19 - 25 - 31		.....						
XI 6 - 13 - 20		.....						
XI 30	19	$21^{2}/_{5}$	25	31	43	79	22	31
XII 16	20;7 [e]	$22^{2}/_{5}$	26	32	44	80	23	32
I 2 19	$21^{2}/_{5}$	25	31	43	79	22	31	
I 12 - 19 - 26		.....						
II 1 - 7 - 13 - 19 - 25		.....						
III 3 - 10 - 19 - 28		.....						
IV 7 - 27 [f]		.....						
V 4		.....						
V 16(?)	3	$5^{2}/_{5}$	9	15	27	63	6	15
V 27(?)	$2^{2}/_{5}$	* [g]						

[a] the first date is followed by the phrase: *yakūn ẓill al-wuqūf* ... , meaning: the midday shadow will be ... . The remaining dates are preceded by the phrase: *thumma yazīd / yanquṣ fa-yakūn ilā* ... , meaning: then it increases / decreases until it becomes (amount) until (date). [b] *iṣbaʿān wa-khumsān.* [c] fraction illegible; one could read a *khumsayn* (= $^{2}/_{5}$) but there is more. [d] read 21? [e] *ʿishrūn wa-sabʿat ajzāʾ.* [f] read 17? [g] *wa-rajaʿ* (read: *wa-rjaʿ* ?) *ilā awwal al-jadwal ʿalā* ... (??) *in shāʾa llāh taʿālā*, meaning: go back to the beginning of the table ... ... (??) if God wills.

For the solstices non-integral values occur, namely:

$$2^{2}/_{5} \quad \text{and} \quad 20;7 \ .$$

These correspond to meridian altitudes:

$$78;42^\circ \quad \text{and} \quad 30;49^\circ,$$

whence one can derive:

$$\phi \approx 35;15^\circ \quad \text{and} \quad \varepsilon \approx 23;55^\circ \ .$$

It is, however, more realistic to conclude that these extremal values were computed for:

$$\phi \approx 35^\circ \sim 36^\circ \quad \text{and} \quad \varepsilon \approx 23;30^\circ \ .$$

The values for the *ẓuhr* and the *ʿaṣr* are respectively 3 and 12 more than the midday values, corresponding to:

$$\Delta z = {}^{1}/_{4}n \quad \text{and} \quad n \ ,$$

the standard definitions used in al-Andalus and the Maghrib – see **6.3**. The shadow increases for the hours are according the following scheme:

T	5	4	3	2	1
Δz	$2^{2}/_{5}$	6	12	24	60
(=	$^{1}/_{5}n$	$^{1}/_{2}n$	n	2n	5n) ,

so that they are based on the standard Indian formula.

The dates on which the midday shadows are an integral number of digits were probably derived by observation.

6.5 al-Jādarī

ʿAbd al-Raḥmān ibn Abī Ghālib Muḥammad al-Jādarī was famous as the author of a short poem on folk astronomy compiled in 794 H [= 1391/92] and entitled *Rawḍat al-azhār*, which enjoyed several commentaries at the hands of later Maghribi astronomers.[10] In this poem there is no shadow-scheme, but one specifically for Fez is attributed to al-Jādarī by the 17th-century Maghribi astronomer al-Sūsī. This scheme is:

10 8 5 3 2 1 1 2 4 6 8 10

and is presented, for example, in MS Cairo DM 415, fol. 41v, of al-Sūsī's treatise entitled *al-Mumtiʿ*. See further **6.2**.

The only other known work of al-Jādarī is an almanac for Fez, extant in MS Istanbul Lalelı 3748,10. In this work values of the solar altitude at midday and the corresponding shadows to base $6^2/_3$ are given for each few days of the year, and I have not investigated these. However, the scheme

9 7 5 3 2 1 1 2 4 5 8 10

for the shadows at the *ẓuhr* beginning with January with corresponding values increased by 7 for the *ʿaṣr* is found in a note on fols. 131v, immediately following al-Jādarī's almanac.

6.6 al-Ṣafāqusī

ʿAlī al-Sharafī of Sfax compiled a navigational atlas *ca.* 1579.[11] I have examined two of the four available copies, namely, MSS Paris BNF ar. 2273 and Oxford Marsh 294 (both completed *ca.* 1600); I label these A and B, respectively.

Both copies contain a table displaying the shadow lengths at significant times of day – see **Table 6.6**. The table in A displays shadow lengths for the *ẓuhr* and the *ʿaṣr*, with difference six units: clearly the length of the gnomon is supposed to be 6. The scheme is not attested in any other known source. The table in B displays values for midday, the *ẓuhr* and the *ʿaṣr*. The difference between the first and second is 2 units. The difference between the first and the third is 7 units: clearly this is the length of the gnomon. The same scheme is found in other Maghribi sources (see **6.4** and **6.6**). A marginal note states that the shadow increase for the *ẓuhr* could be two units but that the value of three digits is precautionary (*iḥtiyāṭ*). These increases of 2 and 3 units probably represent approximations for $1^3/_4$ and $3^1/_2$ units, that is, one-quarter and one-half the length of the gnomon 7. In A, but not in B, the lengths of day and night for each month are also tabulated but they do not concern us here. The entries in both sources are displayed in the following table:

[10] On al-Jādarī see Suter, *MAA*, no. 424a; Renaud, "Additions et corrections à Suter", no. 424a; Brockelmann, *GAL*, SII, pp. 217-218; and *Cairo ENL Survey*, no. F26.

[11] On al-Ṣafāqusī and his work see Nallino, *Scritti*, V, pp. 533-548; and Brockelmann, *GAL*, SI, p. 710. A new study of these manuscripts is currently in preparation by Mónica Herrera Casais (La Laguna, Tenerife, and Frankfurt).

Table 6.6

Month	MS A day	MS A night	MS A *ẓuhr*	MS A *ʿaṣr*	MS B midday	MS B *ẓuhr*	MS B *ʿaṣr*
III	$10\frac{1}{3}$	$13\frac{2}{3}$	7	13	5	8	12
IV	$12\frac{2}{3}$	$11\frac{1}{3}$	4	9	3	6	10
V	$13\frac{4}{5}$	$10\frac{1}{5}$	2	8	2	5	9
VI	$14\frac{2}{3}$	$9\frac{1}{3}$	1	7	1	4	8
VII	$14\frac{1}{2}$	$9\frac{1}{2}$	$1\frac{1}{3}$	$7\frac{1}{3}$	1	4	8
VIII	$13\frac{4}{5}$	$9\frac{1}{5}$	$2\frac{1}{3}$	$8\frac{1}{3}$	2	5	9
IX	$13\frac{2}{3}$	$11\frac{1}{3}$	4	10	4	7	11
X	$11\frac{1}{3}$	$12\frac{2}{3}$	5	11	5	8	12
XI	$10\frac{1}{3}$	$13\frac{2}{3}$	8	14	8	11	15
XII	$9\frac{2}{3}$	$14\frac{1}{3}$	10	16	10	13	17
I	$10\frac{1}{4}$	$13\frac{3}{4}$	9	15	9	11	16
II	$10\frac{1}{3}$	$13\frac{2}{3}$	7	13	7	10	14

6.7 Miscellaneous anonymi

(1) The standard approximate rule (with n = 7 and n = 12) is mentioned in an anonymous Maghribi treatise on the astrolabe extant in the unique MS Cairo DM 1169,6 (fols. 45v-47v, copied 1158 H [= 1745/46])[12] (see esp. fols. 47r-47v).
(2) Hoest (1781), Delphin (1891), Renaud (1948), and Roche (1965) have noted the use of the mnemonic

طزه جبا أبده حي

9 7 5 3 2 1 1 2 4 5 8 10

in the Maghrib.[13] This differs from the Egyptian version (see **9.7a**) only in that the shadow for month X is 5, not 6. Hoest records the text of an almanac in which these lengths occur. Also given in this source, but not the others, is the length of day and night for each month:

9;30 10+ 12 13 14 15;40 14;30 13;20 12 10;30 9;20 8;55
14;30 13+ 12 11 10 8;20 9;30 10;40 12 13;30 14;40 15; 5

Hoest reported that his informants had told him that they thought that the almanac had come to the Maghrib from Istanbul. The winter cold in the Maghrib was not so severe as implied by the almanac, and the shadows there were a little shorter.

[12] See **IV**, Arabic text 2.8.
[13] G. Hoest, *Nachrichten von Marokos und Fes im Lande selbst gesammelt, in den Jahren 1760 bis 1768*, Copenhagen: Christian Gottlob Prost, 1781, pp. 252-257; Delphin, "L'astronomie au Maroc", pp. 199-200; Renaud, *Calendrier d'Ibn al-Bannā'*, p. 23; and P. Roche, "L'irrigation et le statut juridique des eaux au Maroc", *Revue juridique et politique d'indépendance et coopération* 19 (1965), pp. 55-120 and 537-561, esp. p. 85, n. 68.

CHAPTER 7

ʿIRĀQĪ SOURCES

7.1 ʿAbd al-Qādir al-Kīlānī

ʿAbd al-Qādir ibn ʿAbdallāh al-Kīlānī was a celebrated *ṣūfī*, and founder of the Qādiriyya order.[1] He was born in 1078 in Gilan, studied in Baghdad, and died in 1167. His major work *al-Ghunya* contains some passages on reckoning time by shadow lengths, which I came across in MS Cairo MJ 180,17 (fols. 146v + 148r-149r, copied *ca.* 1750).[2]

al-Kīlānī first presents the midday shadow lengths for the Syrian months, beginning with Aylūl (= September), as follows:

IX	X	XI	XII	I	II	III	IV	V	VI	VII	VIII
5	6	7	8	7	6	5	4	3	2	3	4 .

He then presents "another opinion", which is that the midday "shadow of a person" is 3 feet on Ādhār (= III) 19, decreasing to $^1/_2$ by Ḥazīrān (= VI) 19, then increasing at a rate of 1 foot each 36 days to 3 feet on Aylūl (= IX) 19 and then at 1 foot each 14 days to $7^1/_2$ on Kānūn I (= XII) 19.

Yet another scheme is presented next. The shadows at midday and the *ʿaṣr* for the months beginning with Ḥazīrān are:

VI	VII	VIII	IX	X	XI	XII	I	II	III	IV	V
3	4	5	6	7	8	$10^1/_2$	9	$7^1/_2$	6	$4^1/_2$	$3^1/_2$
$9^1/_2$	$10^1/_2$	$11^1/_2$	$12^1/_2$	$13^1/_2$	$14^1/_2$	17	15	$14^1/_2$	$12^1/_2$	11	10

This scheme is clearly a combination of two linear schemes. The midday shadow increases from 3 in June to 8 in November at a rate of 1 foot each month and decreases from $10^1/_2$ in December to $4^1/_2$ in May at a rate of $1^1/_2$ feet each month. No tables displaying this feature are recorded by Neugebauer. The values for the *ʿaṣr* are in general, but not always, $6^1/_2$ units more than those for the *ẓuhr*. Since for n = 6;30 the minimum shadow 3 corresponds to a solar altitude of 65° and the maximum shadow $10^1/_2$ corresponds to a solar altitude of 32°, even the limiting values of the shadow are inconsistent with reality for any latitude. The same scheme is recorded by Ibn Qudāma – see **8.1**.

7.2 al-Ḥiṣkafī

MS Paris BNF ar. 2540,2 (fols. 16r-27v, copied *ca.* 1450) contains a set of planetary tables compiled about the year 1400 by Nāṣir al-Dīn ibn ʿĪsā al-Ṭabīb al-Ḥiṣkafī (from Ḥiṣn Kayfā

[1] Article "ʿAbd al-Ḳādir al-Djīlānī" by W. Braune in EI_2, and Brockelmann, *GAL*, I, pp. 560-563, and SI, pp. 777-779.

[2] The passages extracted in this manuscript correspond to ʿAbd al-Qādir al-Kīlānī, *al-Ghunya*, 2 vols., Cairo, 1304 H [= 1886/87], II, pp. 89-90.

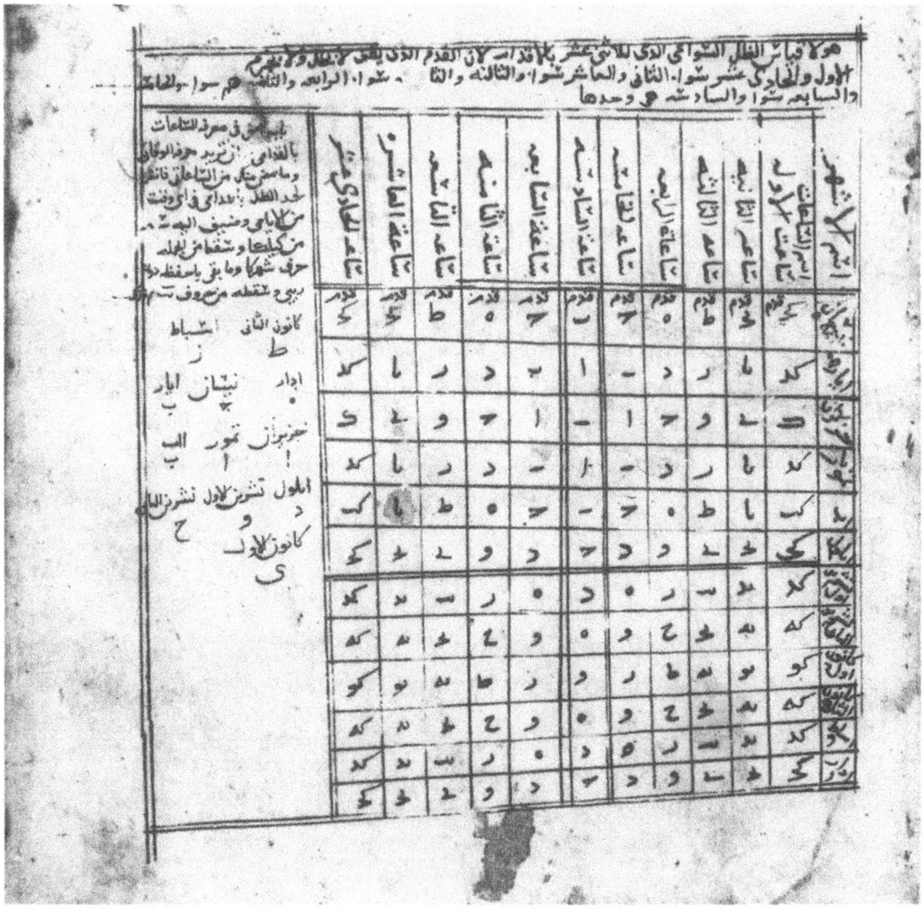

Fig. 7.2: A table for the seasonal hours based on a Hellenistic model and preserved in the planetary tables of al-Ḥiṣkafī. [From MS Paris BNF ar. 2540, fol. 27v, courtesy of the Bibliothèque Nationale de France.]

on the Tigris).[3] Another copy of this work is in MS Vatican ar. 1499 (copied 1576/77). I label these two sources A and B.

Both manuscripts contain (A: fol. 27v, B: fol. 13r) a table of the shadows at each seasonal hour from 1 to 11 for each month of the Syrian year starting with Nīsān (= April). See **Fig. 7.2** for the table as it appears in A. The values given in the table for hours 1 to 6 are displayed in **Table 7.2a**; those for 1 to 5 are confirmed by the identical entries for 11 to 7. The underlying scheme for midday is simply:

5 4 3 2 1 0 1 2 3 4 5 6 .

but the entries for the hours are in some cases, particularly for the first and third, corrupt. **Table 7.2b** shows a matrix of values based on the above midday scheme and increases at the hours of:

1 2 3 4 10 (individual) or 1 3 6 10 20 (cumulative).

[3] al-Ḥiṣkafī and the Paris copy of his tables are listed in Brockelmann, *GAL*, SI, p. 869.

The entries in **Table 7.2a** different from these are italicized. The entries in **Table 7.2b** are less by 2 than the corresponding ones in the standard classical tables of type M (see **1.4**) reproduced in **Table 7.2c**.[4]

Table 7.2a

Month \ T	1	2	3	4	5	6
IV	*23*	13	*9*	5	3	2
V	*24*	11	7	4	2	1
VI	*20* [a]	10	6	3	1	0 [b]
VII	*24*	11	7	4	2	1
VIII	22	11	*9*	5	3	2
IX	23	13	*10*	6	4	3
X	24	14	*12*	7	5	4
XI	25	15	*13*	8	6	5
XII	26	16	*15*	9	7	6
I	25	15	*13*	8	6	5
II	24	14	*12*	7	5	4
III	23	13	*10*	6	4	3

[a] thus in both A and B; [b] A: 2 (*sic*)

Table 7.2b

22	12	8	5	3	2
21	11	7	4	2	1
20	10	6	3	1	0
21	11	7	4	2	1
22	12	8	5	3	2
23	13	9	6	4	3
24	14	10	7	5	4
25	15	11	8	6	5
26	16	12	9	7	6
25	15	11	8	6	5
24	14	10	7	5	4
23	13	9	6	4	3

Table 7.2c

Month | T	1	2	3	4	5	6
I	24	14	10	7	5	4
II	23	13	9	6	4	3
III	22	12	8	5	3	2
IV	23	13	9	6	4	3
V	24	14	10	7	5	4
VI	25	15	11	8	6	5
VII	26	16	12	9	7	6
VIII	27	17	13	10	8	7
IX	28	18	14	11	9	8
X	27	17	13	10	8	7
XI	26	16	12	9	7	6
XII	25	15	11	8	6	5

Note that the midday June shadow of zero serves the tropic of Cancer, not the upper reaches of the Tigris. Since it is unlikely that the table was intended to display increases of the shadows

[4] Neugebauer, *HAMA*, II, p. 738.

over the summer midday minimum, we are left with the conclusion that it is simply the work of an incompetent.[5]

By the side of the main table in both copies is another in which the midday shadows for the Syrian months beginning with Kānūn II (= January) are given as:

9 7 5 3 2 1 1 2 4 6 8 10

This scheme is identical to the one attested in the Coptic and Arabic sources described in **9.7a**.

---

[5] Compare the Ethiopic tables (based on limits 0 and 9 for Z) in *idem*, *Ethiopic Astronomy and Computus*, pp. 212-213.

CHAPTER 8

SYRIAN SOURCES

8.1 Ibn Qudāma

MS Damascus Ẓāhiriyya 10732 (fols. 2r-4v, copied *ca.* 1600) contains a short treatise on the zodiacal signs, lunar mansions and simple timekeeping by Muwaffaq al-Dīn Abū Muḥammad ʿAbdallāh ibn Aḥmad ibn Muḥammad ibn Qudāma al-Maqdisī (1147-1223), a scholar of law and theology who worked mainly in Jerusalem and Damascus.[1]

The tract includes some prayer-tables (the underlying latitude is 30°, which corresponds more happily to Cairo),[2] and also a section on the shadows at the *ẓuhr* and the *ʿaṣr* for each month of the Syrian year beginning with Ḥazīrān (VI). The values given are the same as those in the third scheme of al-Kīlānī (see **7.1**) with midday values missing for months X and V and an *ʿaṣr* value $15^1/_2$ for month I.

In his major legal treatise, *al-Mughnī*, Ibn Qudāma presented a different scheme.[3] This is described as an approximation (*taqrīb*) and is attributed to Abu ʾl-ʿAbbās al-*Ṣ-n-j-y* (?), an individual otherwise unknown to me. For the middle of each of the Syrian months, the shadow lengths in *qadam*s are as follows:

VI	VII/V	VIII/IV	IX/III	X/II	XI/I	XII
$1\,^1/_3$	$1 + ^1/_2 + ^1/_3$	3	$4^1/_2$	$6^1/_2$	9	$10\,^1/_6$

These are said to be valid for the provinces of Iraq and Syria and others with similar latitudes. The extremal values are roughly consistent with a latitude of *ca.* 34° for $n = 6^1/_2$.

8.2 Ibn al-Shāṭir

Ibn al-Shāṭir was the head of the astronomers at the Umayyad Mosque in Damascus in the mid-14th century, and he is well known to the history of science for his work in planetary astronomy, spherical astronomy and timekeeping.[4]

The introduction to a set of prayer-tables for latitude 34° by Ibn al-Shāṭir survives in the unique MS Leiden Or. 1111 (fols. 108r-113r, copied 1601), and the prayer-tables themselves survive in MS Cairo DM 1170,2 (fols. 11r-22v, copied *ca.* 1650). In the introduction to the

---

[1] On Ibn Qudāma see the article "Ibn Ḳudāma" by George Makdisi in EI_2. Only one scientific work is (incorrectly) attributed to him by Brockelmann (SI, p. 689, no. 20), namely, a work on comets which is actually by a later Egyptian writer also called ʿAbdallāh ibn Aḥmad al-Maqdisī (on whom see Brockelmann, *GAL*, II, p. 470, and SII, p. 486, and *Cairo ENL Survey*, no. D45).

[2] These are analyzed in **II-6.2**.

[3] Ibn Qudāma, *al-Mughnī*, 12 vols., Cairo, 1341-48 H [= 1922-30], I, p. 270.

[4] On Ibn al-Shāṭir (**II-9.3**) see my article in the *DSB* and the bibliography there cited (to which add Kennedy & Ghanem, eds., *Ibn al-Shāṭir*), and also King, "Astronomy of the Mamluks", pp. 538-539 and 547-548.

tables, which themselves are quite sophisticated and based on accurate formulae (**II-9.3**), Ibn al-Shāṭir mentions some approximate methods for timekeeping, including (fol. 111v) the standard approximate formula for n = 7.[5]

8.3 Anonymous

MS Princeton Yahuda 1168 (= Mach 4872), copied *ca.* 1800 (?)), contains a short text on the shadows at midday for the Syrian months, accompanied by a table: see **Fig. 8.3**. The text states that shadows in feet (*qadam*s, base 7) can be measured by one's feet lengthwise, and shadows in digits (*iṣba*ʿs, base 12) can be measured by one's feet widthwise (it is not completely clear what is intended here).

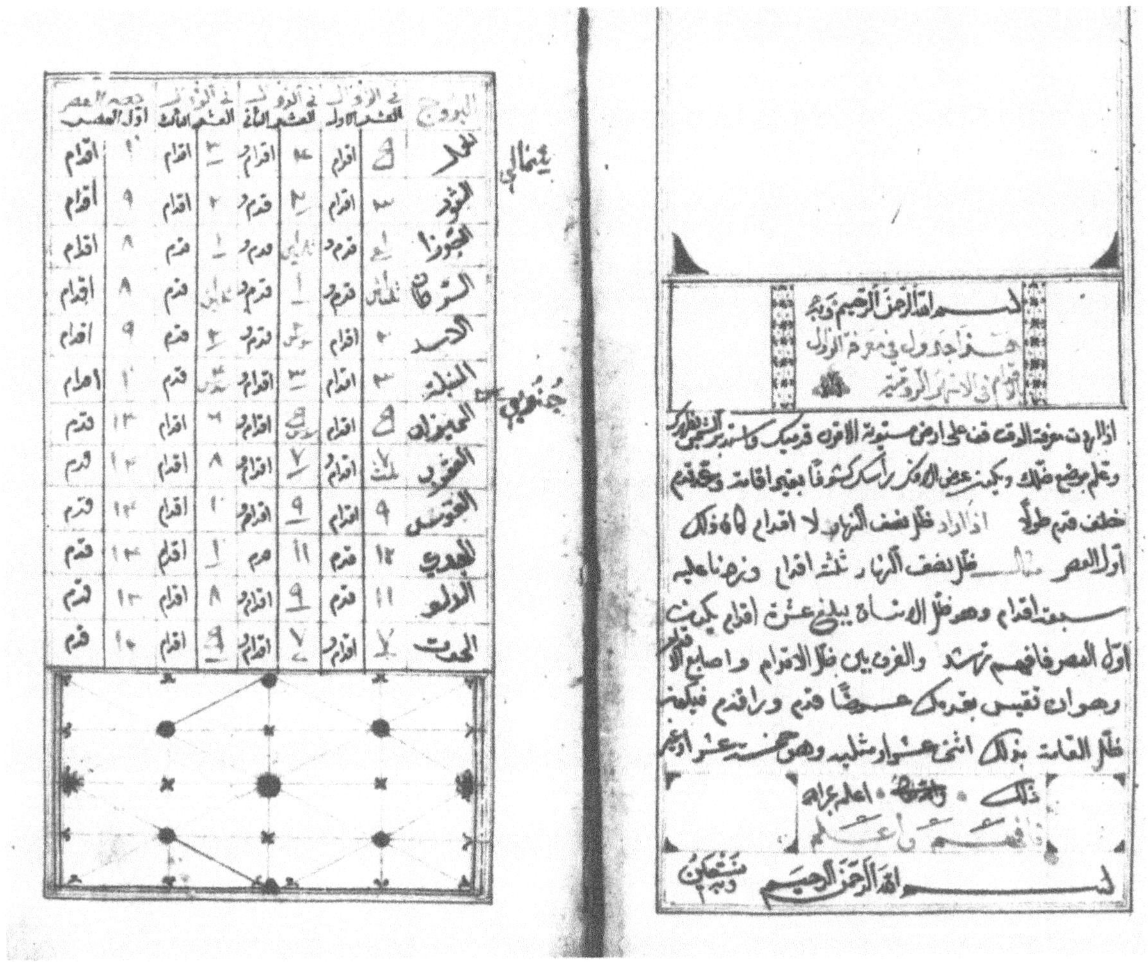

Fig. 8.3: The entire text of a short anonymous Syrian treatise on timekeeping by shadow-lengths. [From MS Princeton Yahuda 1168, courtesy of the Special Collections, Princeton University Library.]

[5] **IV**, Arabic text 2.9.

For each month in the Syrian calendar values are given in *qadam*s for the shadows at midday and the *ʿaṣr*. Those values of the midday shadow serve the first, middle and last ten days of each month.

Table 8.3

Months / Days	1-10	10-20 Midday	20-30	*ʿaṣr*	[*ʿaṣr* - 7]
I	5	4	$3^1/_2$	10	3
II	3	$2^1/_2$	2	9	2
III	$1^3/_4$	$1^2/_3$	$1^1/_2$	8	1
IV	$1^2/_3$	$1^1/_2$	$1^2/_3$	8	1
V	2	$2^1/_6$	$2^1/_2$	9	2
VI	3	$3^1/_2$	$4^1/_6$	10	3
VII	5	$5^1/_6$	6	13	6
VIII	$7^1/_3$	$7^1/_2$	8	14	7
IX	9	$9^1/_2$	10	14	7
X	12	11	$10^1/_2$	14	7
XI	11	$9^1/_2$	8	13	6
XII	$7^1/_2$	$7^1/_4$	$5^1/_2$	12	5

These midday shadows range between $1^1/_2$ and 12, which correspond quite well to the latitude of Northern Syria ($\phi \approx 36°$). The values for the *ʿaṣr* are unrelated to these midday values. If we subtract 7 from each value we obtain a series of values for midday ranging from 1 to 7; these correspond to no terrestrial latitude.

CHAPTER 9

EGYPTIAN SOURCES

9.1 Ibn al-Ḥammāmī

A treatise on timekeeping entitled *Kitāb al-Fuṣūl al-muʿlama fī mawāqīt al-ṣalāt wa-'l-qibla* by an unidentified Egyptian astronomer named Ibn al-Ḥammāmī (**II-2.4**) is extant in the unique MS Istanbul Haci Mahmoud Ef. 5713,1 (fols. 1r-10v, copied *ca.* 1100). The author presents a series of shadow lengths for $\phi = 30°$ for each pair of zodiacal signs between the solstices as follows (fol. 3r):

♋	♌/♊	♍/♉	♎/♈	♏/♓	♐/♒	♑
1;21	2	4	7	10	14	16;16

The values 1;21 and 16;16 are accurately computed for $\phi = 30°$ and $n = 12$ (with $\varepsilon = 23;35°$). The others are, as Ibn al-Ḥammāmī says, rounded (*majbūr*). Our author also prescribes the standard approximate formula with $n = 12$.

9.2 Najm al-Dīn al-Miṣrī

The early-14th-century Egyptian astronomer Najm al-Dīn al-Miṣrī compiled a treatise on aspects of timekeeping (**II-2.5**), which is extant in the unique copy MS Istanbul Hamidiye 1453 (fols. 219r-228v, copied 869 H [= 1464/65]). In this he gives a similar set of values for Cairo, with limiting values 1;20 and 16;15. He also presents the standard Indian formulae for timekeeping with bases $6^2/_3$ and 12.

Najm al-Dīn was not limited to such simple procedures: his major work was a universal table for timekeeping by day and night serving all latitudes and containing close to half a million entries (**I-2.6.1** and **9-3a**).

9.3 Anonymous (*Kitāb al-Durar wa-'l-yawāqīt*)

In a 13th-century Egyptian treatise on simple time-reckoning entitled *al-Durar wa-'l-yawāqīt* (**II-6.1**), extant in the unique copy MS Oxford Bodley 133,2 (fols. 94v-130r, copied 734 H [= 1333/34]), the author, who has not yet been properly identified, gives a set of shadow lengths for Egypt as follows:

1;21	2	4	7	10	13	16 .

See **9.1** above.

9.4 al-Dīrīnī

The early-13th-century Egyptian mystic al-Dīrīnī wrote a treatise on folk astronomy, partly in verse, which is extant in several manuscripts (**II-2.6**), including MS Cairo DM 651,5 (fols. 41r-58r, copied 828 H [= 1424/25]). al-Dīrīnī presents the following shadow-scheme in feet (*aqdām)* for the zodiacal signs:

♋	♌/♊	♍/♉	♎/♈	♏/♓	♐/♒	♑
$^2/_3$	1	2	$3^1/_2$	$5\ ^1/_4 + {}^1/_8$	$7\ ^1/_4$	$8\ ^1/_8$

These values correspond roughly to latitude 30° with base 6. al-Dīrīnī also presents the daily shadow increases for each sign in minutes (*daqā'iq)* thus:

$1^1/_3$	4	6	$7^1/_2$	$7^1/_2$	$3^1/_2$

but these values correspond to base 12!

The same set of values together with daily rates of increase occurs on the inside cover of MS Cairo MM 168, which contains some miscellaneous tables from the 13th-century Egyptian *Muṣṭalaḥ Zīj* and some anonymous material on folk astronomy, copied *ca*. 1400-1500 probably in Egypt but possibly in the Yemen.[1]

9.5 Sirāj al-Dīn

MS Cairo ṬM 127 (59 fols., copied 1255 H [= 1839/40]) is a unique copy of a treatise entitled *Maṣābīḥ al-anwār wa-mafātīḥ al-asrār fī aʿmāl al-layl wa-'l-nahār* on folk astronomy and simple timekeeping. The work is attributed to an individual named simply Sirāj al-Dīn wa-'l-Dunyā.[2] It is indirectly related to a treatise on the same subject preserved in MS Princeton Yahuda 4657 (Mach 4983, fols. 40v-80r, copied *ca.* 1350), which – according to a spurious note at the beginning of the treatise – purports to be an abridgement of the *Maṣābīḥ al-anwār* by Sirāj.

The Cairo manuscript contains (fols. 50r-50v) a section on reckoning the time of day by shadows in which the schemes:

9	7	5	3	2	1	1	2	4	6	8	10

and

40	20	10	6	3	2

are advocated for the midday shadows for each month of the Syrian year and shadow increases for the seasonal hours, respectively. I suspect that this treatise is of Egyptian provenance, perhaps from the early 13th century; the same is the case for the treatise in the Princeton manuscript.

Of particular interest is a remark by the author of the treatise in the Cairo manuscript that the author of a book entitled *Kitāb Ḥall al-mushkilāt fī maʿrifat al-daqā'iq wa-'l-daraj wa-'l-sāʿāt* – this title is new to the modern literature – devised a sundial called *al-basīṭa al-*

---

1 *Cairo ENL Survey*, no. C12.

2 The author – on whom see *Cairo ENL Survey*, no. C4 – is the celebrated Ḥanafī legal scholar Sirāj al-Dīn al-Sajāwandī (*fl. ca.* 1200 – see Brockelmann, *GAL*, I, pp. 470-471, and SI, pp. 650-651, and the article "al-Sadjāwandī" by Rudolf Sellheim in EI_2. See further **IV-3.3**.

sab'īniyya because it bore seventy (*sab'īn*) lines for regulating time of day according to these two rules. I do not understand the text of this passage, which merits further investigation. (See **10.3** on a sundial based on another simple shadow-scheme.) Sirāj al-Dīn points out that the author of this treatise, whom he calls *al-ra'īs al-fāḍil*,[3] used n = 7 and mentions that the longest shadow represented on the sundial was at one hour before sunset in the 4th Coptic month of Kayhak, namely, 10 + 60 = 85 (??). This suggests that the author was using a scheme for the hours other than the

40 20 10 6 3 2

scheme advocated by Sirāj al-Dīn. There is some confusion here, not least because $\Delta z = 60$ at $T = 1$ corresponds to a gnomon length of $n = 12$.

9.6 al-Ṣiqillī

MS Dublin CB 4538 (*ca.* 30 fols., copied *ca.* 1300) is a unique copy of a work on timekeeping by the lunar mansions. The title page is barely legible. I read *Anwā' ... ibn Hārūn* (? written *Hrwn*) *al-Ṣiqillī*, that is, "(Book on) the periods of the year arranged according to the lunar mansions (*naw'*) by ... ibn Hārūn al-Ṣiqillī". The name of this author is new to the modern literature. Although he or his family hailed from Sicily, his work was clearly compiled for use in Egypt. Perhaps he is identical with one Abu 'l-Qāsim ibn 'Abdallāh ibn 'Abd al-Raḥmān ibn Ḥasan al-Qurashī al-Ṣiqillī, recently identified by Julio Samsó.[4] This man's date is uncertain, but he is clearly early enough to be a candidate. He is known to have made serious twilight measurements in Cairo and Alexandria and to have authored a book entitled *al-Intiqā' min al-yawāqīt fī 'ilm al-yawāqīt*, otherwise unknown to us but obviously dealing with more than timekeeping by night with the lunar mansions.

Each pair of pages is devoted to one lunar mansion: see **Fig. 9.6** for an example. Tables display the configuration of the lunar mansions with respect to the horizon and meridian at each of thirteen times during the night. Dates corresponding to each of the days of the *naw'* are given in the Coptic, Syrian and Western (Andalusī) calendars. Solar meridian altitudes are also given for each day. This information is accompanied by a diagram of the stars constituting the mansion and various additional items of information, such as the shadow lengths for the *naw'* in question. Values are given in digits (*iṣba'*) and sixtieth divisions (*daqīqa*) thereof.

From the extremal values of the solar altitude, which are given to the nearest degree, namely, 83° and 36°, I conclude that these are truncated from values based on

$$\phi = 30;0° \text{ (Cairo)} \quad \text{and} \quad \varepsilon = 23;35° ,$$

although they might also be computed with $\varepsilon = 24°$. The values of the shadows are as follows:

[3] This is a common title given to the early-11th-century scholar Ibn Sīnā, on whom see the article by A. M. Goichon in EI_2. However, it is somewhat unlikely that Ibn Sīnā would have used the Coptic calendar.

[4] Samsó, "Astronomical Observations in the Maghrib", p. 174.

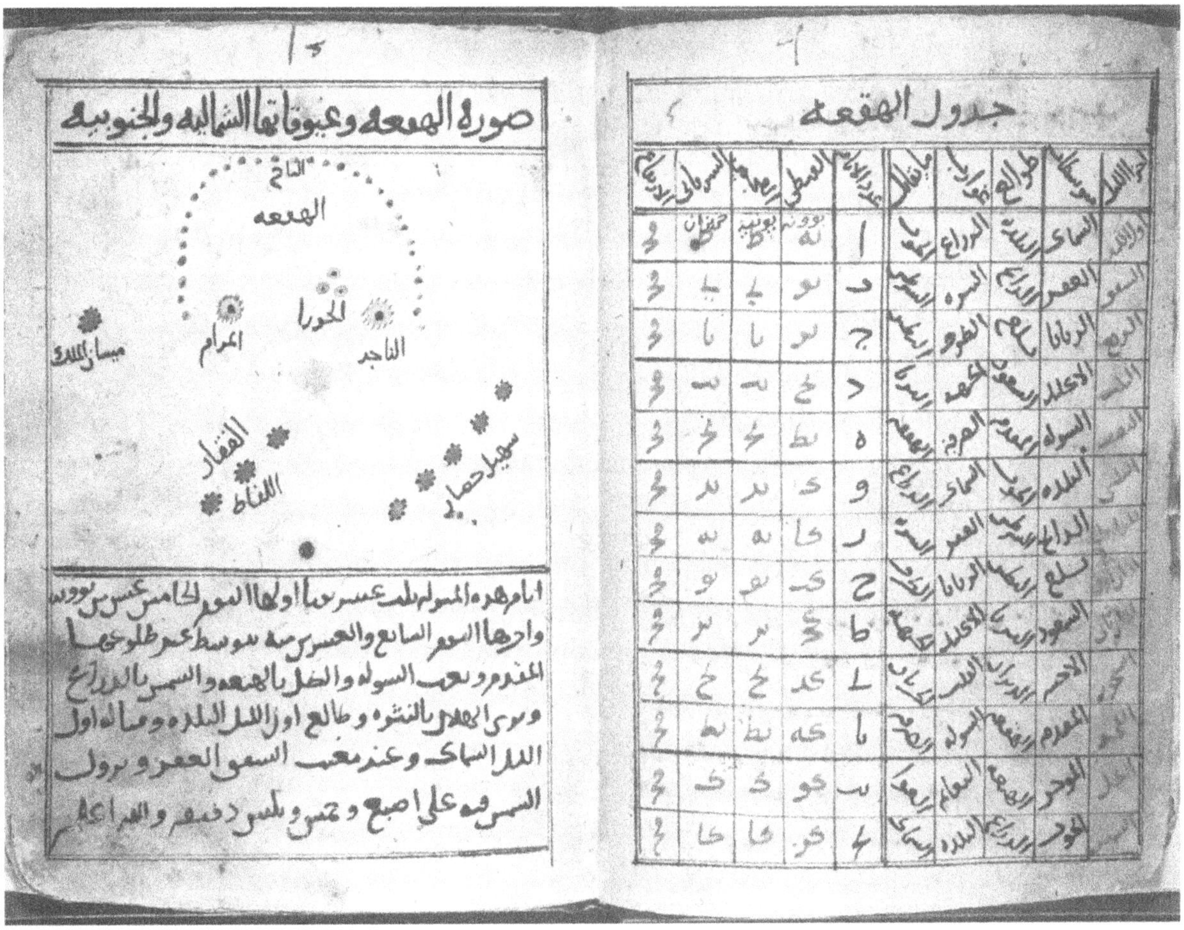

Fig. 9.6: An extract, one double page out of 28, from al-Ṣiqillī's tables for timekeeping by the lunar mansions with additional information about midday shadow lengths throughout the year. [From MS Dublin CB 4538, courtesy of the Chester Beatty Library.]

Table 9.6

M	Z	M	Z	M	Z	M	Z
1	3;?5	8	2;13	15	10;7	22	14;36
2	3;1	9	3;1	16	12;12	23	12;53
3	2;17	10	3;53	17	*deest*	24	11;4
4	1;15	11	5; 5	18	*deest*	25	9;20
5	1;35	12	6;20	19	16;50	26	7;43
6	1;33	13	7;45	20	16;49	27	6;16
7	1;49	14	9;23	21	16;1	28	5; 2

They were apparently computed (rather than observed), but extremal values 1;15 and 16;50 do not correspond properly to those for the above parameters, namely, 1;21 and 16;16. The corresponding values for $\varepsilon = 24°$ are 1;16 and 16;31. Thus al-Ṣiqillī's values are not fully explained.

9.7a Miscellaneous

Neugebauer has already drawn attention to a midday shadow-scheme for the months occurring in an Arabic text written for the use of the Copts and ascribed to the 12th patriarch of Alexandria, Demetrius (d. A.D. 230) but copied only in 1768.[5] The shadows for the twelve months beginning with Ṭūba (V) are:

9 7 5 3 2 1 1 2 4 6 8 10

As Neugebauer has pointed out, there are two arithmetical progressions underlying this primitive scheme. For latitude $\phi = 30°$ and $\varepsilon = 24°$ and base n = 7 we should expect maximum and minimum shadows of:

0;44 and 9;38 .

Neugebauer labelled the scheme as "badly distorted", but its simplistic nature did not detract from its popularity in medieval Egypt, as the following new sources show. See also **6.7** on a related scheme in various Maghribi sources and **7.2** on a Syrian version.

First I should mention that I have not found any such scheme in the one Coptic astronomical handbook from the medieval period which I have inspected, namely the work of the 13th-century Copt al-Asʿad ibn al-ʿAssāl,[6] extant in MS Cairo DM 910 (86 fols., copied *ca.* 1805). This work, which is of considerable interest and remains to be studied, contains solar and lunar tables and tables for calendar conversion and determining the feasts. However, it does not contain any material relating to timekeeping.

The same scheme for the Coptic months starting with Ṭūba that was noted by Neugebauer, now usually formulated as a mnemonic:

طزه جبا أبد وحي

is attested in the following sources:

(1) a poem on the midday shadows for the Coptic months attributed to the late-15th-/early-16th-century Egyptian polymath Jalāl al-Dīn al-Suyūṭī,[7] extant in MS Cairo MJ 39,3 (fol. 58r, copied 1069 H [= 1658/59]);
(2) a poem attributed to one Aḥmad al-Jālī in MS Cairo MM 171,6 (fols. 59v-60r, copied 1145 H [= 1732/33]);
(3) a treatise on timekeeping by the 17th-century Egyptian savant Shihāb al-Dīn al-Qalyūbī[8] in MS Cairo DM 203 (fol. 3r, copied 1094 [= 1683]);
(4) at the end of an anonymous almanac in MS Cairo MM 33,1 (fol. 5v, copied ca. 1600) – see **Fig. 9.7a**;
(5) amongst some anonymous notes in MS Cairo MM 181,1 (fol. 4r, copied *ca.* 1620); and

[5] G. Sobhy (Bey), "The Coptic Calendrical Computation and the System of Epacts known as "The Epact Computation" Ascribed to Abba Demetrius the XIIth Patriarch of Alexandria", *Bulletin de la Société d'Archéologie Copte* 8 (1942), pp. 169-199, and 9 (1943), pp. 237-252, p. 187 (French), p. 250 (Arabic), cited in Neugebauer, *HAMA*, II, p. 743.

[6] On Ibn al-ʿAssāl see the article by A. Atiya in EI_2, and *Cairo ENL Survey*, no. C10.

[7] On al-Suyūṭī see Suter, *MAA*, no. 449; Brockelmann, *GAL*, II, pp. 180-204, and SII, pp. 178-198; *Cairo ENL Survey*, no. D103; the article by E. Geoffroy in EI_2, and also n. **IV**-8:26.

[8] Brockelmann, *GAL*, II, pp. 478-479, and SII, pp. 492-493, and *Cairo ENL Survey*, no. D30.

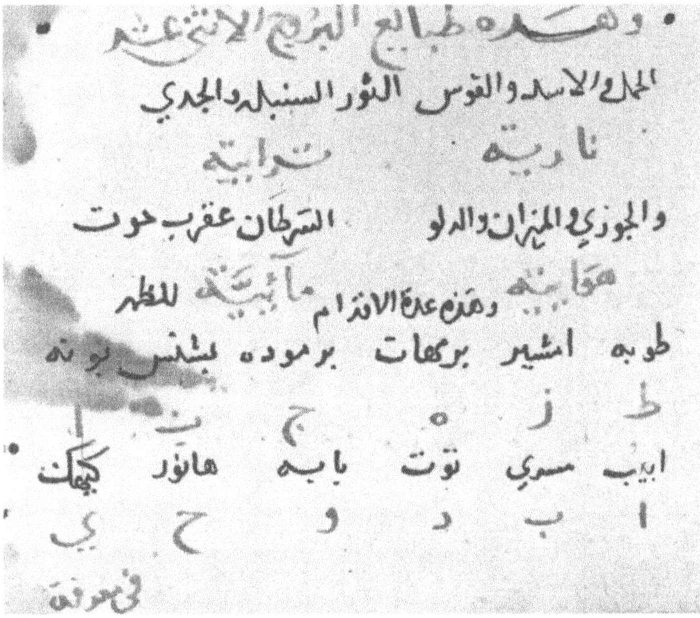
وهذه طبايع البروج الاثني عشر
الحمل والاسد والقوس الثور السنبله والجدي
نارية ترابية
والجوزي الميزان والدلو السرطان عقرب حوت
هوايه وهذه عدة الاقدام مايية للظهر
طوبه امشير برمهات برموده بشنس بونه
ط ز ه ج ب
ابيب مسري توت بابه هاتور كيهك
ا ب د و ح ي

Fig. 9.7a: Midday shadow lengths for the Coptic months in a late Egyptian source. [From MS Cairo MM 33,1, courtesy of the Egyptian National Library.]

(6) a legal treatise by al-Bājūrī, who was *Shaykh al-Azhar* in Cairo *ca.* 1850. This source was noted by Eilhard Wiedemann, but the mnemonic as interpreted by him is garbled.[9]

A variation of this scheme is given in MS Cairo DJ 323,9 (fol. 39r, copied *ca.* 1900) in a poem attributed to one Abū Bakr al-Mīqātī, on whom I have no further information. The midday shadows for the Coptic months beginning with Tūt (I) are stated to be:

3 5 8 10 9 8 5 3 2 1 1 2 ,

these numbers being given in *abjad* notation thus:

جه حي طح هج با اب

In MSS Cairo DM 119,4 (fol. 9v, copied 1109 H [= 1697/98]) – see **Fig. 9.7b** – and 1112 (fol. 10v, copied 1310 H [= 1892/93]), both manuscripts of Egyptian provenance, there are notes (*fā'ida*) on the shadows increases for each hour to be added to the midday shadow, as follows:

T	1	2	3	4	5
Δz	30	12	6	3	1

For n = 6 the Indian formula yields the values:

30 12 6 3 $1^1/_5$,

so the rule is explained. The use of the base 6 is uncommon in post-Abbasid Islamic astronomy. The scheme

40 20 10 6 3 2

is presented in an anonymous Egyptian almanac containing mainly material relating to folk astronomy, preserved in MS Cairo DM 187 (recto of penultimate folio, copied *ca.* 1850).

[9] Wiedemann, *Aufsätze*, II, p. 781. On al-Bājūrī see the article "al-Bā<u>dj</u>ūrī" by Th. W. Juynboll in EI_2.

Fig. 9.7b: In these notes appended to a copy of a treatise on a special type of astrolabic quadrant by the late-15th-century Cairo astronomer Sibṭ al-Māridīnī, we find a reference to the sundial of the Caliph ʿUmar ibn ʿAbd al-ʿAzīz (**IV-7.1**), a poem on the prayer-times attributed to the Imām al-Shāfiʿī, elsewhere attributed to his descendant, the Fatimid astronomer Ibn Yūnus (**II-2.1**), and a simple shadow-scheme for timereckoning. [From MS Cairo DM 119,2-4, fols. 8v-9v, courtesy of the Egyptian National Library.]

Another mnemonic is attested in various Egyptian sources for the shadow lengths at the seasonal hours. This is:

مكي وجب

which represents the sequence

40	20	10	6	3	2

for the hours 1 to 6. In notes found in MSS Cairo MM 169,2 (fol. 39v, copied *ca.* 1750) and Cairo DM 830,2 (fol. 18r, copied 1257 H [= 1841/42]), the scheme is incorrectly represented as follows:

1	2	3	4	5	6	7	8	9	10	11	12
40	20	10	6	3	2	2	3	6	10	20	40

That this nonsense was actually intended is confirmed by the accompanying Arabic text in the second source; here the mnemonic:

مكي وجب بجو يكم

occurs. The sequence:

20	10	6	3

is reminiscent of the shadow increases:

20	10	6	3	1

for the seasonal hours in Greek shadow tables, in which the minimum midday shadow is also 2.[10] Notice also that, whereas in the Greek tables the hourly increases are:

1	2	3	4	10 (= 1 + 2 + 3 + 4) ,

in this scheme they are:

1	3	4	10	20 (= 1 + 2 + 3 + 4 + 10) .

9.7b Miscellaneous (time-reckoning by fingers)

The following procedure for reckoning time of day with the hand, using the thumb as gnomon, is outlined in MS Cairo DJ 323,7 (fol. 34r, copied *ca.* 1900). The text occurs amidst various astronomical treatise of Egyptian provenance – see **Fig. 9.7c.**[11]

> Translation: "A note on finding the hours with the shadow of the thumb (*ibhām*). You raise the thumb straight up with the shadow (falling) on the index finger (*sabbāba*). As long as the shadow has not reached the tip (*unmula*) of the index finger you are in the first hour. If the shadow falls on the tip of the index finger then you are in the second hour. If it has reached the first joint (*ʿuqda*) you are in the third hour, and if it has reached the second joint you are in the fourth hour. If it has reached the third mark (*ḥazza*) you are in the fifth hour, and if it has reached the fourth mark you are in the sixth hour. Then (use) the opposite (procedure) until the end of the day."

On an adult hand this corresponds approximately to a shadow-scheme

T	1	2	3	4	5	
z	5	4	3	2	1	(n = $2^1/_2$ inches)

I know of no other attestation of such a procedure in the Islamic sources. al-Bīrūnī, it is true,

[10] Neugebauer, *HAMA*, II, p. 738.
[11] *Cairo ENL Survey*, no. D228.

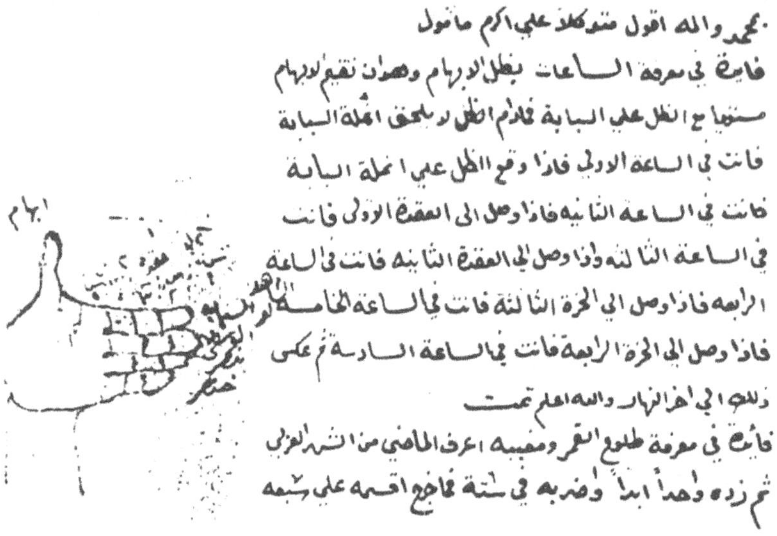

Fig. 9.7c: A scheme for reckoning the hours with the thumb and fingers. [From MS Cairo DJ 323,7, courtesy of the Egyptian National Library.]

Fig. 9.7d: The scheme for timekeeping using the hand in a treatise by Jacob Koebel entitled *Eyn künstliche sonn Uhr inn eynes jeden menschen Lincken handt*, printed in Mainz in 1532. [From Gockerell, "Telling Time without a Clock", p. 133.]

mentions in the *Shadows* that he had observed the Indians using their middle finger extended vertically as a gnomon to cast shadows along their forearms held horizontally, measuring the shadows in finger lengths with the other hand.[12] But, as he says regretfully, he neglected to ask them the details. A method similar, but not identical, to that described in the Cairo manuscript is attested in a 16th-century German text (see **Fig. 9.7d**),[13] and no doubt such procedures were more widespread both in the Near East and in Europe than we could ever know from surviving textual evidence.

9.7c Miscellaneous (lost)

In the course of preparing this study I came across the following shadow-schemes for the monthly midday shadows in my notes on Ibn al-Bannāʾ's almanac, the first marked "Egypt". I have no recollection in which Cairo manuscript I originally found these schemes, which are as follows:

9	7	5	3	2	1	1	4	8	8	10	10
7	6	4	3	2	1	1	3	3	4	8	8

[12] al-Bīrūnī, *Shadows*, pp. 33-34, and Kennedy, *al-Bīrūnī's* Shadows, I, pp. 73-74, and II, p. 28.
[13] Gockerell, "Telling Time without a Clock", p. 133.

CHAPTER 10

MISCELLANEOUS

10.1 Sundry medieval European schemes

Neugebauer has described various wretched European shadow-schemes, concluding: "As usual the western tradition represents the lowest level in our material."[1] I include a brief summary:

(a) Two Latin tables:
midday shadows: 7 6 5 4 3 2 , hour differences: 1 3 7(!) 10 20 ;

(b) Three Spanish Mozarab (Latin) texts (11th century):
8 7 5 4 3 2 / 1 3 6 10 20 ;

(c) Palladius (4th-5th century):
9 7 5 4 3 2 / 1 3 6 10 20 ;

(d) Various missals (10th and 11th centuries):
11 9 7 5 3 1 with $\Delta z = 6$ for $T = 3^{sdh}$;

(e) Pseudo-Bede (date ?):
11 9 7 5 3 1 / 2 4 6 8 18 ;

(f) The German monk Wandalbert (9th century):
13 11 9 7 5 3 / 2 4 6 8 11 .

Did monks actually use these schemes to regulate their prayers at (the ends of) the third, sixth and ninth seasonal hours? For their sake, one could hope that they did not. In any case, not one of these schemes bears any trace of Islamic influence.

It would be amiss not to mention here the possibility of the transmission of some of the other material surveyed in this study to Europe. Only the scheme of the *Calendar of Cordova* is known to have been rendered into Latin, and the shadow tables compiled by Nicholas of Lynn in Oxford in the 14th century (strictly in the mathematical tradition) are a development of tables of this kind (**I-10.1**). The early European tables of solar altitudes or shadows at the hours specifically for marking cylindrical sundials also have their counterparts in yet earlier Islamic sources (**XI-4.4**). At least now that the ancient and medieval Islamic material is identified and classified we stand a better chance of understanding the medieval European material.

10.2 A late Ottoman navigational manual

MS Cairo DM 570 (182 fols., *ca.* 1875?) is a unique copy of a collection of texts, tables and diagrams relating to navigation. The work was apparently compiled *ca.* 1860, and it awaits

---

[1] Neugebauer, *HAMA*, II, pp. 745-746. For the table of Pseudo-Bede see also Ginzel, *Chronologie*, III, p. 88. On yet simpler forms of reckoning the hours in medieval Europe see Gockerell, "Telling Time without a Clock" (n. 9:13).

detailed study.[2] Of concern to the present investigation is a table (fol. 17v) of shadow lengths attributed to *al-shaykh* ʿAbdallāh ibn ʿUmar Bā Makhrama, a Yemeni scholar who flourished *ca.* 1500: see further **4.5**.

10.3 A sundial from Java

A horizontal sundial constructed in 1722 in the town of Gresik in Java (see **Fig. 10.3**) was published by the Batavian savant Tjondro Negoro in 1882.[3] The instrument measures *ca.* 60 × 60 cms., and its surface is covered with texts in Javanese, which were reproduced and rendered into Dutch by Negoro. It has since been discussed by various scholars of ethnography and ethnoastronomy interested mainly in the associated calendar, which is defined by means of midday shadows.[4]

The sundial, called a *bencet*, consists of a pair of perpendicular scales shaped like a "T". The shorter scale measures shadows at midday. It bears markings for each unit between 2 units north and 4 units south: the underlying gnomon length must be 7, as we see from the longer perpendicular scale. This measures shadows at the *ʿaṣr* (Javanese: *asar*) based on the standard definition $\Delta z = n = 7$, and it is marked for each unit from 1 up to 11, although it is only the markings from 7 to 11 which are relevant to the function of the scale. In order to use the scale for the *ʿaṣr*, the sundial should ideally be rotated away from the meridian so that the afternoon shadow can be compared with the markings on the scale, but we are not dealing with a precision instrument.

The year is divided in twelve periods called *mangsa*s, beginning with the summer solstice. These are defined as the intervals during which the shadow changes by 1 unit. The *mangsa*s vary in length between 23 and 43 days, an unusual feature for any calendar, and this has disturbed modern investigators. The inspiration for the calendar is clearly the fact that (for this latitude and gnomon length) the midsummer shadow is 4 units south and the midwinter one 2 units north, the total change in shadow length thus being 6 units. The midday shadows at the beginning of the *mangsa*s (m_i) are thus:

m_1	m_2/m_{12}	m_3/m_{11}	m_4/m_{10}	m_5/m_9	m_6/m_8	m_7
4 S	3 S	2 S	1 S	0	1 N	2 N

The limits 4 S and 2 N correspond reasonably well to the latitude of Gresik, namely *ca.* -7° (actually 4.12 S and 2.07 N for n = 7). The Dutch savant van der Stok, unaware of standard

[2] *Cairo ENL Survey*, no. Z41.

[3] See R. M. A. A. Tjondro Negoro, "De koperen zonnewijzer van Gresik", *Tijdschrift van het Bataviaasch Genootschap van Kunsten en Wetenschappen* (Batavia/Jakarta) 27 (1882), pp. 47-68; and also F. van den Bosch, "Der javanische Mangsa-Kalendar", *Bijdragen tot de Taal, Land en Volkenkunde* 126 (1980), pp. 248-282 and figs. 13-17, espeically pp. 251-253.

[4] See, for example, N. Daldjoeni, "Pranatamangsa, the Javanese Agricultural Calendar", *The Environmentalist* 4 (1984), supp. no. 7, pp. 15-17; *idem* and B. Hidayat, "Astronomical Aspects of "Pranotomongso" of the 19th-Century Central Java", in *New Delhi 1985 Colloquium Proceedings*, pp. 249-252; G. Ammarell, "Sky Calendars of the Indo-Malay Archipelago", *ibid.*, pp. 241-248; as well as A. Maas, "Sternkunde und Sterndeuterei im malaiischen Archipel", *Tijdschrift voor Taal, Land en Volkenkunde* (Koninklijke Bataviaasch Genootschaap van Kunsten en Wetenschappen, Batavia / Jakarta), 64 (1924).

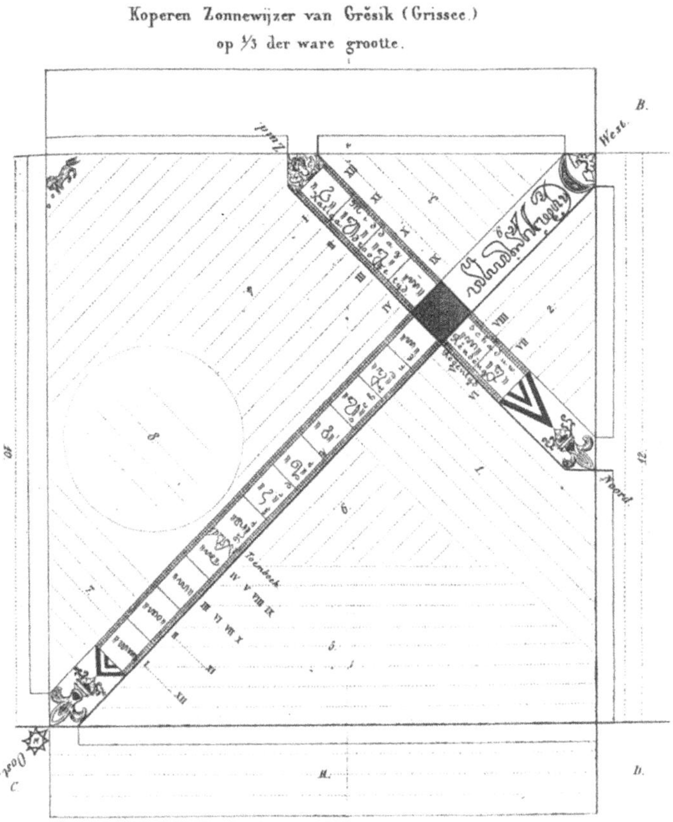

Fig. 10.3: The Gresik sundial. [From Negoro, "De koperen zonnewijzer van Gresik" (1882).]

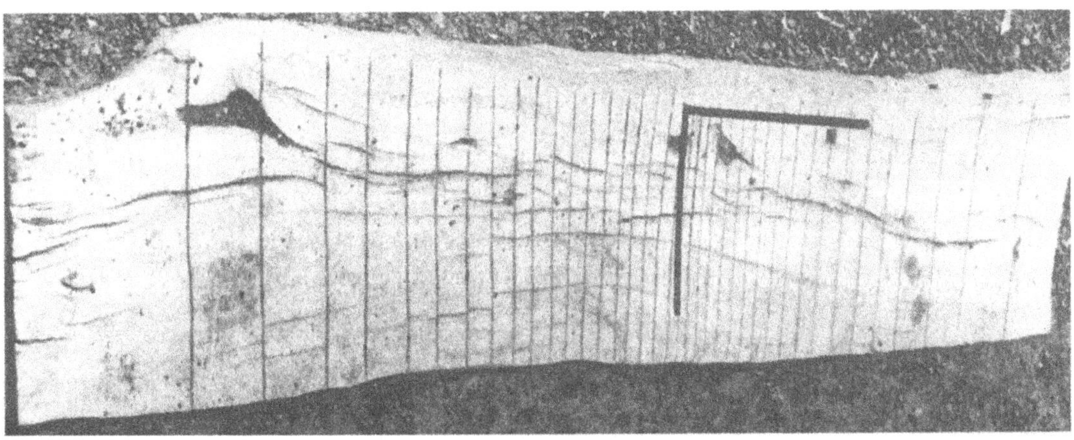

Fig. 10.4: A sundial for regulating water distribution from Muscat and Oman. I have no information on its precise location or the way in which it was used. The divisions are neither equal nor uniform, which may have made some *fallāḥīn* rave wildly if, as a result of this, they were deprived of their fair share of water. [From the archives of Alain Brieux, Paris.]

medieval Islamic practice, determined the most appropriate gnomon length as 7.11 units (but then the limiting shadows are 4.19 S and 2.1 N).[5]

The development of a calendar out of a shadow-scheme is not, as far as I know, attested in other societies. See, however, **6.4** for a first step in this direction.

10.4 20th-century Arab shadow-schemes for water regulation

From both Northern and Southern Arabia, as well as Egypt, we have accounts of the regulation of water for irrigation by means of folk astronomical techniques.[6] I do not doubt that similar practices were, and indeed in some cases still are, in vogue elsewhere in the rural areas of the Islamic world.

In the Oasis of al-ʿUlāʾ in North-Western Arabia, for example, the length of daylight is divided into 24 *asdās* (plural of *suds*, lit. sixth), and the passage of the *asdās* is regulated by observing the shadow of a 2-metre high wall pass over sets of stone markers for each *suds*.[7] There are three sets, two for the solstices and one for the equinoxes. The positions of the stones for the solstices, as recorded recently, approximate rather well to the hyperbolae of a horizontal sundial.

In the Buraimi Oasis of Northern Oman, the length of daylight is likewise divided into 24 *asdās*, but their passage is regulated by observing changes in the length of a man's shadow.[8] This is said to shorten by 40 *qadam*s during the first *suds* (that is, after the shadow becomes clearly identifiable), 20 in the second, 5 in the third, 3 in each of the fourth and fifth, 2 in the sixth and seventh, and 1 in each of the remaining *asdās* until midday. (Note that the clear shadow is thus 79, or around 80 *qadam*s long.)

In 1917 W. J. H. King published an account of a most curious procedure for regulating the distribution of water in the Dakhla Oasis to the west of the Nile.[9] The water supply for an entire day from sunrise to sunset is divided into 31 *qadam*s. Two men are involved in reckoning the passage of the *qadam*s. The meridian is marked and one man stands on it with his back to the sun and "bottom joint of his forefinger against the tip of his nose". The second man measures with his feet the distance from the shadow of the finger to the meridian. The *qadam*s are then regulated according to the information recorded in the following table.

---

[5] J. P. van der Stok, "Opmerkingen omtrent bovengemelden zonnewijzer", *Tijdschrift van het Bataviaasch Genootschap van Kunsten en Wetenschappen* (Batavia / Jakarta) 27 (1882), pp. 62-68, esp. p. 66.

[6] On irrigation in the Islamic world see the articles "Māʾ" in EI_2. A new study of links between irrigation and simple timekeeping is Thomas F. Glick and Simonne Teixeira, "Azaira, alhetma: Two medieval Arabisms Reflecting the Allocation of Irrigation Water", *Suhayl* 3 (2002-03), pp. 213-219. The duration of the *ẓuhr* in n. 3 on p. 214 is incorrectly cited.

[7] See A. A. Nasif, "Qanats at al-ʿUlāʾ", *Proceedings of the Seminar for Arabian Studies* 10 (1980), pp. 75-80, esp. pp. 76-77, and *idem*, *A Historical and Archaeological Survey of al-ʿUlāʾ with Special Reference to its Irrigation System*, 2 vols., unpublished Ph.D. thesis, Victoria University of Manchester, 1981, I, pp. 246-252, and II, pl. CCLIII.

[8] J. C. Wilkinson, *Water and Tribal Settlement in South-East Arabia: A Study of the* Aflāj *of Oman*, Oxford: Clarendon Press, 1977, pp. 106-110.

[9] See W. J. Harding King, "Irrigation in the Dakhla Oasis", *Geographical Journal* 50 (1917), pp. 358-364, for details.

Table 10.4

Morning (shadow west)	Perpendicular in *qadams*	Afternoon (shadow east)
1	∞ ← 26-26 → ∞	31
2	26-22	30
3	22-18	29
4	18-14	28
5	14-12	27
6	12-10	26
7	10-9	25
8	9-8	24
9	8-7	23
10	7-6	22
11	6-5	21
12	5-4	20
13	4-3	19
14	3-2	18
15	2-1	17
16	1W-1E	16

As King noted, the 16th *qadam* provides a rather generous allowance, but he obtained his information from an old *fellāḥ* who "raved wildly on the beauties of the system in such a vile patois that even his village headman had some difficulty in following". Even in the 1910s the headman possessed a watch, and the old methods are most probably no longer used nowadays. King also described some methods of regulating water distribution by night using the moon. **Fig. 10.4** shows a sundial related to this kind of procedure from Muscat and Oman.

E. Stack, in his account of a journey through Persia in the late 19th century, presented a detailed account of contemporary irrigation practices,[10] and then – alas, without further information – added:

> "In some villages of Faraidan I found the water-distribution by day regulated by the length of a man's shadow, measured in feet; and at night by the stars. Elsewhere I believe a dial is used by day."

10.5 A late-20th-century almanac from Qaṭar

The almanac entitled *al-Taqwīm al-Qaṭarī* which is published annually in the Arabian Gulf state of Qaṭar (latitude *ca.* 25°) contains a set of prayer tables and agricultural information for each zodiacal sign and lunar mansion rising at dawn. I have examined a copy published in the 1980s. Occasional midday shadow lengths in *qadam*s are given and they were clearly lifted from some earlier work in a rather haphazard fashion. The following values are given for various signs (S) and mansions (M):

M_{10}: $7^1/_2$	M_2: $^1/_2$	S_{11}(♒): 6
M_{10}: 3	S_{12}(♓): 5	M_{12}: $5^1/_2$
M_{26}: $2^1/_2$	M_{14}: 6	M_{28}: 1

This information is, for all practical purposes, quite useless.

---

[10] E. Stack, *Six Months in Persia*, 2 vols., London, 1882, II, p. 269 (cited in EI_2, V, p. 870a).

10.6 A shadow-table from 18th-century Sind

For the sake of future researchers I include a reference to a shadow scheme for the solar months in Sindi verse that was in circulation in Sind in the 18th century.[11] It came to the attention of the jurist Makhdūm Muḥammad Hāshim of Tatta (d. 1760), who pointed out that the shadows had been calculated for some region other than Sind. He conducted his own observations in 1719-20 on a weekly / bi-weekly basis to produce to a table of the midday shadows in feet (*per*) for each two weeks of the solar year. Values are said to be valid for Lower Sind and are given to the nearest half a foot.

[11] See Baloch, "Measurement of Space and Time in Lower Sind", pp. 185-186, where the table is reproduced.

Index of Monthly Shadow-schemes

Scheme (beginning with longest shadows)												Section
13	11	9	7	5	3	3	5	7	9	11	13	10.1
13	10	8	6	4	3	2	3	4	6	8	10	1
11	9	7	5	3	1	1	3	5	7	9	11	10.1
11	7	6	6	4	3	2	4	4 / $2\frac{1}{3}$	5	7	$9\frac{1}{5}$	6.3
11	7	6	5	4	3	3	3	4	5	7	$9\frac{1}{5}$	6.3
11	7	6	4	3	2	1	1	4	5	7	$9\frac{1}{5}$	6.3
$10\frac{1}{2}$	9	$7\frac{1}{2}$	6	$4\frac{1}{2}$	$3\frac{1}{2}$	3	4	5	6	7	8	7.1, 8.1
$10\frac{1}{6}$	9	$6\frac{1}{2}$	$4\frac{1}{2}$	3	$1\frac{5}{6}$	$1\frac{1}{3}$	$1\frac{5}{6}$	3	$4\frac{1}{2}$	$6\frac{1}{2}$	9	8.1
10	10	9	7	5	3	2	1	1	4	8	8	9.7c
10	10	8	5	3	2	1	1	2	4	6	8	6.2, 6.4
10	9	8	5	3	2	1	1	2	3	5	8	9.7a
10	9	7	7	4	2	1	$1\frac{1}{3}$	$2\frac{1}{3}$	4	5	8	6.5
10	9	7	5	3	2	1	1	2	4	5	8	6.4, 6.5, 6.6
10	9	7	5	3	2	1	1	2	4	6	8	7.2, 9.5, 9.7a
10	9	7	5	3	2	1	1	2	4	?	8	6.2
10	9	7	5	3	2	1	1	2	3	5	8	3.5
9	7	5	4	3	2	2	3	4	5	7	9	10.1
8	8	7	6	4	3	2	1	1	3	3	4	9.7c
8	7	6	5	4	3	2	3	4	5	6	7	1, 5.2, 7.1
8	7	5	4	3	2	1	2	3	4	5	7	1
8	7	5	4	3	2	2	3	4	5	7	8	10.1
7	6	5	4	3	2	2	3	4	5	6	7	10.1
6	5	4	3	2	1	0	1	2	3	4	5	4.9, 7.2

Index of Hour Schemes

Type	Values						Section
T	6	5/7	4/8	3/9	2/10	1/11	
z	0	3	7	15	17	28	4.2
Δz	0	1+	$3\frac{1}{4}$	$6\frac{1}{2}$	$19\frac{1}{2}$	39	3.2
Δz / z (?)	0	3	6	7	18	22	5.3
"	3	6	7	8	18	28	"
z (?)	0	3	6	9	18	28	3.5, 6.1
z	0	3	6	9	18	38	4.9
"	0	3	5	9	15	35	"
Δz	0	1	3	6	10	20	1, 7.2, 10.1
Δz	0	1	3	6	12	30	9.7a
Δz	0	2	3	6	10	20	5.2
Δz / z (?)	0	3	5	7	13	26	4.9
"	0	3	5	7	13	24	"
z (?)	2	3	6	10	20	40	9.5, 9.7a
Δz	0	$\frac{1}{5}$n	$\frac{1}{2}$n	n	2n	5n	2.1, 4.1
Δz	0	$2\frac{2}{5}$	6	12	24	60	2.4, 6.4
z (?)	$\frac{2}{3}$	$3\frac{1}{4}$	$7\frac{1}{2}$	15	30	60	4.4
"	1	$3\frac{1}{4}$	$7\frac{1}{2}$	19	30	60	"
Δz	0	1	2	3	4	5	9.7b
Δz	?	?	?	?	?	60	9.5
Δz		1	3	7	10	20	10.1
"		2	4	6	8	18	"
"		2	4	6	8	11	"

See also the scheme for *muhurtas* in **2.4**.

Index of Solstitial and Equinoctial Shadow Lengths

Scheme			Section
WS	EQ	SS	
$8\,{}^{7}/_{8}$	$3\,{}^{1}/_{2}$	${}^{1}/_{2}$	2.2
$(1\,{}^{1}/_{48})$n	-	${}^{1}/_{18}$n (S)	3.2
ca. 7	-	0;25 (S)	3.3
7	-	${}^{1}/_{10}$	3.4
$5\,{}^{1}/_{5}$	-	1+ ${}^{1}/_{4}$ + ${}^{1}/_{3}$ • ${}^{1}/_{5}$ (S)	4.2
5	-	1 (S)	4.2
5 - ${}^{1}/_{8}$	$1\,{}^{1}/_{3}$	1 + ${}^{1}/_{12}$ + (S)	4.1
$5\,{}^{1}/_{4}$	$1\,{}^{1}/_{2}$	$1\,{}^{1}/_{10}$ (S)	4.3
$5\,{}^{1}/_{8}$	-	1 - ${}^{1}/_{30}$ (S)	4.5
$5\,{}^{1}/_{2}$	-	0	4.6
5 + ${}^{1}/_{4}$ + ${}^{1}/_{6}$	-	$2\,{}^{1}/_{6}$ (S)	4.6, 4,7
10	-	4 (S)	4.7
5	2 or $1\,{}^{4}/_{12}$ (!)	${}^{4}/_{12}$ (N!)	4.10
$1\,{}^{5}/_{6}$	${}^{3}/_{4}$	${}^{1}/_{4}$	5.1
$5\,{}^{1}/_{2}$	-	${}^{1}/_{2}$ (N!)	6.1
$7\,{}^{1}/_{2}$	3	${}^{1}/_{2}$	7.1
$10\,{}^{1}/_{6}$	-	$1\,{}^{1}/_{3}$	8.1
12	-	$1\,{}^{1}/_{2}$	8.3
16;16	7	1;21	9.1
16;15	7	1;20	9.2
16	7	1;21	9.3
$8\,{}^{1}/_{8}$	$3\,{}^{1}/_{2}$	${}^{2}/_{3}$	9.4
16;50	-	1;15	9.6
4 (S)	-	2 (N)	10.3
20;7	-	$2\,{}^{2}/_{5}$	6.4

See also the schemes for Aden and Taiz cited in **4.5**.

Manuscripts consulted

N.b.: An asterisk denotes that I have not been able to consult the manuscript or obtain a copy thereof. On the manuscripts of the almanac of Ibn al-Bannāʾ which I have not consulted, see n. 6:6.

Manuscript	Section
Aleppo Awqāf 968	4.9
Baghdad Awqāf 2982/6276*	4.1
Berlin Staatsbibliothek Ahlwardt:	
5664	3.2
5731	4.2
Cairo Egyptian National Library:	
DM = Dār al-Kutub *mīqāt*	
119	9.7
130	3.3
187	9.7
203	9.7
415	6.2, 6.5
570	4.5, 10.2
651	9.4
830	9.7
899	4.7
910	9.7
948	2.2, 4.1, 4.8
1112	9.7
1169	6.7
1170	8.2
DJ = Dār al-Kutub *majāmīʿ*	
323	9.7
709	4.6
K = astronomy	
3769	4.9
Ḥ J = Ḥalīm *majāmīʿ*	
47	6.1
MJ = Muṣṭafā Fāḍil *majāmīʿ*	
39	9.7
180	7.1
202	5.3
MM = Muṣṭafā Fāḍil *mīqāt*	
33	9.7
168	9.4
169	9.7
171	9.7
181	9.7
ṬJ = Ṭalʿat *majāmīʿ*	
201	6.2
ṬM = Ṭalʿat *mīqāt*	
127	9.5
Damascus Ẓāhiriyya	
5588	3.4, 3.5
10732	8.1
Dublin Chester Beatty 4538	9.6
Escorial ár. 941	5.2
Istanbul (Süleymaniye):	
Haci Mahmoud Ef. 5713	9.1
Hamidiye 1453	9.2
Lalelı 3748	6.5
Leiden Universiteitsbibliotheek 1111	8.2
Milan Ambrosiana Griffini 37	4.2
Oxford Bodleian Library:	
Bodley 133	9.3
Hunt. 233	4.3
Marsh 134	4.9
" 294	6.6
Pococke 249	6.4
Paris BNF	
ar. 2273	6.6
ar. 2540	7.2
Princeton University Library:	
Garrett 1016	4.6
Yahuda 1168	8.3
" 4657	9.5
Sanaa, private collection	4.4
Vatican	
ar. 962	4.7
ar. 1499	7.2

Part IV

On the times of Muslim prayer

To the memory of
Peter Kilburn, Hakim Mohammed Said
and Annemarie Schimmel

ACKNOWLEDGEMENTS AND NOTES ON THIS VERSION

This study, written by a Westerner on an aspect of Islamic ritual, is dedicated to the memory of two friends – one a Westerner and the other a Muslim – both of whom were murdered by ignorant religious or racist fanatics of one sort or another. With these acts, these murderers wasted on the one side an American who was a true friend of the Arabs, and on the other, one of the leading Muslim historians of Islamic science. It is also dedicated to the memory of a German friend who died as the last few touches were being made to the text of this book, a scholar who knew the Islamic world, medieval and modern, so well that she could even think like a Muslim and talk about many aspects of Islamic civilization in all of the major languages of the Muslim world.

The American, Peter Kilburn, was acquisition librarian at the American University of Beirut from 1963 until his tragic death in 1986.[1] He had an M.A. in Anthropology from the University of Washington, Seattle, and before coming to Lebanon had lived in Morocco and Turkey. He personally built up a rich collection of books on the Near East at A.U.B.'s Jafet Library, much appreciated by students, faculty and visiting scholars, and he was himself extremely well-read in many areas of Near Eastern Studies. Peter was a devotee of the Arab world in general and of Lebanon in particular. In 1984 he wrote to his friend Richard Dewey, librarian at the American University in Cairo: "The situation in Afghanistan is dreadful: there isn't a village left intact." Peter was always broke, mainly because of the expenses he incurred financing the education of refugee children. At the time of his murder, he was sheltering in his apartment several homeless Lebanese children, who again lost their home. My wife and I first met Peter when we went to Beirut for the academic year 1970-71. His friendship and inimitable humour greatly enriched our stay there, and his professional services greatly facilitated my research. He was one of several innocent victims killed in retaliation for the U.S. bombing of Libya in April 1986. But he considered Ronald Reagan "a cretin", whose policies made it "embarrassing to be an American in Lebanon". In 2003, the situation is far worse.

Hakim Mohammed Said was born in Delhi in 1920 and moved to Karachi in 1948. A scholar and a benefactor, he was the founder of the Hamdard Foundation in Karachi. Hakim Mohammed Said was a serious Muslim scholar who neither disdained nor underestimated Western scholarship on Islam. He hosted the International Congress of the History and Philosophy of Science in Islamabad in 1979.[2] He had a profound respect for the way in which a small group of modern scholars of widely different backgrounds were using the primary

---

[1] A brief tribute by Richard Dewey and this author appeared in the *Bulletin of the Middle East Studies Association of North America* 9:3 (Fall, 1987). See also, amongst numerous press reports, *Eagle Times*, (Claremont NH), April 22, 1986, pp. 1 and 8, and, on the context of Kilburn's death, Sandra Mackey, *Lebanon: Death of a Nation*, New York & Chicago: Congdon & Weed, 1989, pp. 214-216.

[2] See the conference proceedings in Hakim Mohammed Said, ed., *History and Philosophy of Science – Proceedings of the International Congress of the History and Philosophy of Science, Islamabad, 8-13 December, 1979*, Nazimabad, Pakistan: Hamdard Foundation Press, n.d. (*ca.* 1980).

sources to the unveiling of Muslim achievements in the sciences. I first met Hakim Mohammed Said in Aleppo in 1975, and our paths crossed several times thereafter. He had charisma, and, behind his soft-spokenness, authority. He was always impeccably dressed in traditional white *reshti*, and all eyes would turn towards him whether in Aleppo or Istanbul or in Frankfurt. He was assassinated in Karachi in 1998.[3]

The German, Annemarie Schimmel, started studying Arabic at the age of 15, and eventually mastered a dozen languages, including besides Arabic, Persian, Turkish and Urdu. She authored over 50 books on Islamio calligraphy, art, literature, and mysticism, even to the role of cats in Islamic literature. She was as beloved in the Muslim world as she was respected by the Western community of Islamicists.[4] It was my pleasure to meet her at the Columbia Arabic Seminars in New York in the early 1980s. She was heartbroken when the German press misinterpreted her stance with respect to the fate of the British novelist, Salman Rushdie, and, in her words, "the sinister holy war against (him)". I met her again in Bonn in 2001, where she delivered a speech at the occasion of the award of the International Book Prize for 2000 by the Islamic Republic of Iran to a British orientalist. Her life's work was devoted to better understanding between East and West, and this she achieved amongst those fortunate to have known her or heard her or read her. Since she died at age 80 in January, 2003, she fortunately did not have to witness the latest crusade.

The materials for this study were gathered at the American Research Center in Egypt during the period 1972-79, when my research was supported mainly by the Smithsonian Institution and the (U.S.) National Science Foundation, Washington, D.C. My further research on Islamic folk science, conducted at New York University, was supported by the (U.S.) National Endowment for the Humanities (1980-1983). This support, provided by two U.S. Government agencies with a sincere belief in the promotion of knowledge for its own sake, is gratefully acknowledged.

A summary of Chapters 1-3 of the present article was read at the 187th Annual Meeting of the American Oriental Society in Ithaca, New York, April, 1977, and a summary of the entire paper was read at the 192nd Annual Meeting of the Society in Austin, Texas, April, 1982. A preliminary account of some of my findings was published in 1977 in my paper "A Fourteenth-Century Tunisian Sundial for Regulating the Times of Muslim Prayer", in *Hartner Festschrift*, pp. 187-202 (reprinted in King, *Studies*, B-XVIII), especially pp. 193-196. Preliminary drafts of the first full version were read by my father, Dr. Henry C. King; my former teacher, Professor Franz Rosenthal of Yale University; my former student, Dr. John O'Malley of New York, and my colleagues Professor Paul Kunitzsch of Munich and Professor David Powers of Cornell University. For their comments and corrections I am profoundly grateful but I alone am responsible for the content.

---

[3] A tribute by Asghar Ali is in *Studies in History of Medicine and Science* (Hamdard Nagar, New Delhi) 15 (1997/98), pp. 193-196.

[4] One of many tributes was published in *The Guardian* on February 5, 2003. A more detailed *vita* by Ronny Baier, with bibliography, is in *Biographisch-bibliographisches Kirchenlexikon*, Band XXIII (2004), available at http://www.bautz.de/bbkl/s/s1/schimmel_a.shtml (21.09.2003). I owe these references to the kindness of Dr. Martina Müller-Wiener, Bonn.

The photograph of the Tunisian sundial (**Fig. 5.2**) was provided by the late Alain Brieux of Paris. Various museums and libraries kindly provided other photographs and credit is given *ad loc*. To Kurt Maier of Frankfurt my sincere thanks for his help in proof-reading the English and Arabic texts the first time round. The penultimate version was checked for consistency by François Charette of Frankfurt, to whom I am indebted for more and more each day. In early 2002 Gerald Hawting of London read the penultimate version and saved me from a few pitfalls. There are some dangers associated with sitting on a paper for 20-odd years and not publishing it. It was only in February, 2002, that I learned from Gerald Hawting of an article by Uri Rubin of Tel Aviv University dealing with the *ḍuḥā*: "Morning and Evening Prayers in Early Islam", *Jerusalem Studies in Arabic and Islam* 10 (1987), pp. 40-64. Almost with disbelief, I realized that Uri Rubin and myself had been working on the *ḍuḥā* (and also on the Kaʿba) at the same time without either knowing. Interested readers should consult that important article, whose contents I have surveyed very briefly here (**Section 1.1***). Any interpretation of the voluminous source material that Uri Rubin presents is entirely beyond my competence.

Readers of my study should keep in mind that I am not trying to write about the early history of the institution of prayer in Islam, as Uri Rubin has done, but rather to explain the definitions of the *times* of the prayers that became standard in the 9th century and to address some of the questions raised by these explanations.

TABLE OF CONTENTS

INTRODUCTION

"... ʿAbdallāh ibn Masʿūd said: I asked the Messenger of God – may God bless Him and grant Him salvation: "Which works are most favoured by God?" He replied: "Prayer at the appropriate time." *Ḥadīth* recorded by al-Bukhārī, Aḥmad ibn Ḥanbal and al-Tirmidhī.

"But considre wel that I ne usurpe not to have founden this werk of my labour or of myn engyn. I n'am but a lewd compilator of the labour of olde astrologiens, and have translatid in myn Englissh oonly for thy doctrine. And with this swerd shal I sleen envie." Geoffrey Chaucer (d. 1400), in the introduction to his treatise on the astrolabe, quoted from John North, *Chaucer's Universe*, p. 38

In Islam there are five daily prayers, and their times are defined in terms of astronomical phenomena.[1] Why are there five prayers? Why are their times astronomically defined? Muslim tradition has a ready answer to the first question in the legend of the Prophet's night-journey (*al-miʿrāj*). But the second question is only partly answered by the traditional sources. It is the daylight prayers that merit particular attention because the definitions of their times in terms of shadow increases, found neither in the *Qurʾān* nor in the *ḥadīth*, are clearly secondary. What was the motivation behind these definitions? What can we learn from them?

Previous studies of the origin and development of the institution of prayer in Islam have sifted the evidence from the *Qurʾān*, the *ḥadīth* literature and texts on the sacred law of Islam (**Ch. 9**), but have overlooked the significance of the times prescribed for the performance of the prayers.[2] Likewise, previous studies of the definitions of the times of prayer based on scientific sources[3] have not questioned the original purpose of these definitions. The aim of this study is to present evidence from a collection of medieval sources most of which have never been investigated before. These deal mainly with the *times* of the prayers, and they provide the explanations of the definitions. Also they treat of six prayers rather than five. This new material thus casts light on the early development of one of the most important aspects of Islamic religious practice.

My study begins with an account of the standard definitions of the prayer-times (**Ch. 1**). The explanation that our sources provide for the motivation behind the standard definitions is in full accord with the astronomical knowledge of the Muslims in the 8th century when these

---

[1] For an introduction see the article "Mīḳāt" in EI_2 by A. J. Wensinck (legal aspects) and D. A. King (astronomical aspects), the latter reprinted in King, *Studies*, C-V. On traditional notions of time in general see the article "Layl and Nahār" by Charles Pellat in EI_2.

[2] In the article "Ṣalāt" in EI_2 by G. Monnot the definition of the time of the *ʿaṣr* is incorrect and the *ḍuḥā* is mentioned only in passing! I would be the first to acknowledge that there is more to Muslim prayer than the times for the individual prayers, but I would maintain that the number and times of the prayers are highly significant in determining the early development of the prayer ritual.

In several cases the articles in EI_1 are often more reliable in places (or altogether) than those in EI_2. And in other articles in EI_1, through no fault of the authors, the aspects of the subject which interest us here are overlooked altogether: thus, for example, the article "Sāʿa" does not deal with hours at all. See n. 9:28.

definitions were first formulated. The first encounter of the Muslims with mathematical astronomy was from Indian sources. We shall see that the definitions are practical means of regulating the prayer-times at the seasonal hours, and there is evidence that they were sometimes actually defined in terms of the seasonal hours. The connection between shadow increases and seasonal hours is provided by a simple Indian formula for timekeeping (**Ch. 2**). The occasional alternative definitions of the times of the midday and the afternoon prayers in selected early sources confirm the original purpose of the standard definitions (**Chs. 3 and 4**). But there was also another prayer performed at mid-morning in some circles, and there were special prayers on Fridays as well as during the two major religious festivals; in each case the times were astronomically defined (**Chs. 5 and 6**). Some additional evidence for the regulation of the prayer-times by means of the seasonal hours in early practice is available in the form of astronomical instruments (**Ch. 7**). Also there is a connection between the names of the prayers and the names of the corresponding seasonal hours (**Ch. 8**). Once these aspects of the prayer-times are recognized, it is interesting to look at the scheme of prayers in Islam in the light of the orders of early Syrian Christianity; the relationship is seen to be somewhat closer than has been noted previously (**Ch. 9**).

I am well aware that some of the material presented in this study raises as many problems as it solves. The major problem is one that besets any investigation of the development of religious and legal notions in Islam (or any other religion), namely, the historicity of the sources.[4] The problems associated with material attributed to the Prophet Muḥammad, Companions of his like Abū Hurayra, the "pious" Caliph ʿUmar ibn ʿAbd al-ʿAzīz, and the main authorities of the Shīʿites, namely, ʿAlī ibn Abī Ṭālib and the Imām Jaʿfar al-Ṣādiq, are well known.[5] The authenticity of such material even when it occurs in early sources is often problematic enough, let alone when it occurs in relatively late ones, such as 13th-century Yemeni texts on folk astronomy. Then there is the problem of the authenticity of the materials associated with "the Arabs" of pre-Islamic Arabia as they are presented by scholars active in an urban environment already long exposed to sophisticated new scholarly influences and techniques. How much of what they report is authentic? What has been added by the scholars who recorded and transmitted this information? As we shall see, when scholars in the 9th and 10th century list the names of the 24 hours, we cannot be sure (a) that *sāʿāt* really meant hours, (b) that the order in which the *sāʿāt* are presented was original, and least of all (c) that the organization of this lexicographical information into 24 hours is not due to those who recorded and transmitted it. Similar problems are associated with the transmission of star-names and the names of the lunar mansions as recorded and interpreted by medieval Muslim astronomers, notably the celebrated ʿAbd al-Raḥmān al-Ṣūfī.[6] We simply cannot always take the early

---

[3] See Wiedemann & Frank, "Gebetszeiten" (general), a groundbreaking study which also took account of information in legal sources; Kennedy, "al-Bīrūnī on the Prayer-times", summarized from the author's translation and commentary on al-Bīrūnī's *Shadows* (see n. 1:25); and King, "Tunisian Sundial", pp. 193-196.

[4] On the authenticity of such sources for the purpose of writing history (rather than the development of religious institutions) see R. Stephen Humphreys, *Islamic History – A Framework for Inquiry*, Minneapolis, Minn.: Bibliotheca Islamica, 1988, pp. 68-98. See also n. 0:9.

[5] See the references cited in nn. 1:9 (Abū Hurayra), 7:2 (ʿUmar), 2:8 (ʿAlī), and 2:10 (Jaʿfar).

[6] See Kunitzsch, *Sternnomenklatur der Araber*, p. 14 (on problems with star-names in al-Ṣūfī's book on the constellations) and p. 25:

sources at face value, let alone the late sources such as the popular *Lisān al-ʿArab* by Ibn Manẓūr and the dictionary of Classical Arabic by Edward W. Lane based on medieval Arabic dictionaries.[7] On the other hand, it is naïve to expect that any single early authority would have recorded the entire corpus of pre-Islamic lore on any given subject. [8]

Records of the pronouncements of early Muslim authorities on matters relating to religion, language, law and history consitute an imposing corpus of historical sources, unique of its kind. As Bernard Lewis has written:[9]

> "The careful scrutiny (by scholars of Islamic tradition) of the chains of transmission, and their meticulous collection and preservation of variants in the transmitted narratives, give to medieval Arabic historiography a professionalism and a sophistication without precedent in antiquity and without parallel in the contemporary medieval West."

Nevertheless when we stumble across "new" material of this kind, new in the sense that it has not previously caught the attention of modern historians, we must beware of taking it at its face value. I would, however, contend that in some cases it barely matters whether a particular story is historical or not; the fact that it was circulated at all proves that one particular interest group considered its propagation worthwhile. This is particularly the case when Sunnī and Shīʿī *ḥadīth* are mutually contradictory, which situation, as we shall see, prevails for the prayer at mid-morning. Since I am not a *ḥadīth* specialist I have not combed the literature for other attestations of the various stories that I cite: when I say, for example, that a story is related by al-Bīrūnī or al-Aṣbaḥī, this simply means that this is where I found it; it does not mean that it is not to be found elsewhere. Likewise I have not investigated *isnād*s when these are given. My hope in presenting this material together with some comments about its possible significance is that this will inspire an exhaustive investigation of the relevant sources. Whoever

---

> "Es ist also zu schließen, daß die Namen der Mondstationen ursprünglich selbständig und beziehungslos dastanden und erst später von den *šuʿarā'* in dichterischer Freiheit oder gar erst von einigen Philologen in bewußter Spekulation auf bestimmte Tierkreisbilder, in deren Bereich sie sich befanden, bezogen wurden."

Arab interest in the mansions was inspired not least because the *manāzil* are mentioned in the *Qur'ān*: see Paul Kunitzsch's article "Manāzil" in EI_2, reprinted in *idem*, *Studies*, XX. On al-Ṣūfī see the same author's article in *DSB*. See also n. 8:19.

[7] The situation has been eloquently summarized by Manfred Ullmann:

> "Es dürfte klar geworden sein, daß die Worterklärung des ibn Manẓūr und all seiner Kollegen oft nicht das Papier wert sind, auf das sie gedruckt sind. Daß Lane all diese chaotischen Bedeutungsangaben mit umständlicher Genauigkeit wiedergegeben hat, entsprach seinem Konzept, und man würde die Situation, in der sich die Forschung im vorigen Jahrhundert befand, falsch einschätzen, wenn man es Lane zum Vorwurf machen würde, daß er die arabischen Lexika nachgeahmt hat. Aber wenn von Lane's Werk heute noch gesagt wird: "It is difficult to conceive a better dictionary in the accuracy of its definitions, and the fullness of its examples" (Haywood, *Arabic Lexicography*, p. 125), so spiegelt dieses Urteil dasselbe Problembewußtsein, das die Herausgeber arabischer Texte besitzen, die ihre Editionen mit Fußnoten verzieren, in denen die Worterklärungen des *Lisān al-ʿarab* reproduziert sind."

This is quoted from Ullmann's article "Der Werwolf – Ein griechisches Sagenmotiv in arabischer Verkleidung", *Wiener Zeitschrift für die Kunde des Morgenlandes* 68 (1976), pp. 171-184, esp. p. 182; on the theme see the same author's EI_2 article "Ḳuṭrub".

[8] Compare the situation with the names and limits of the winds amongst early Muslim authorities. Thus, for example, none of the folk-scientific material surveyed in the EI_2 articles "Maṭlaʿ" by D. A. King and "Rīḥ" by Miquel Forcada, or in Schmidl, "Qibla und Winde", is contained in the *Kitāb fī Asmā' al-rīḥ / Kitāb al-Rīḥ* of Ibn Khālawayh (d. 980 – see Sezgin, *GAS*, VII, p. 357, and VIII, p. 180, and the EI_2 article by A. Spitaler).

[9] See his penetrating essay review of Stephen Humphrey's book cited in n. 1:4, entitled "In Search of Islam's Past" and published in *The New York Review*, Dec. 5, 1991, 37-40, esp. p. 37b-c.

undertakes such a study should not lose sight of the importance of the times advocated for the prayers: see further **Ch. 10**.

The materials used for this study are taken mainly from medieval Arabic treatises, some published but most still in manuscript form, and of these the majority had never been investigated previously in modern times. Others are published, notably al-Bīrūnī's *On Shadows* (see **1.5** below), but have not attracted the attention that they merit. It should not surprise the reader if some of these materials challenge our previous conceptions. He or she should keep in mind that the history of Islamic civilization is constantly being rewritten and revised through research on the vast manuscript resources available for the further study of Islamic civilization.[10] I have taken care to provide the Arabic texts, which the reader will do well to ponder before evaluating my presentation and analysis of these. Also the subject is not new: the fact that the standard definitions for the day-time prayers which have been in use since the 8th century are not identical with the definitions outlined in the *ḥadīth* was discussed already by al-Bīrūnī in the early 11th century.

In certain earlier studies I have pointed to the fact that science in medieval Islam functioned on two levels: a mathematical tradition cultivated by the scientists of Islam and familiar only to the select few, and a popular "folk" tradition cultivated by the legal scholars of Islam and documented in more "popular" works. In particular, the ways in which science was applied to service religious needs – the regulation of the lunar calendar, the organization of the times of prayer, and the determination of the sacred direction – were twofold. (See further **1.4**.)

In various other studies I have attempted to show how the same kind of non-technical scientific sources, this time treatises on the sacred geography of Islam and treatises on the qibla from the point of view of the legal scholars (a genre called *kutub dalāʾil al-qibla*), can explain how the early Muslims determined the qibla in ways different from the Muslim scientists and hence account for the diverse orientations of medieval mosques and religious architecture.[11] In that case it was the astronomical orientation of the Kaʿba, discussed in the medieval texts but long forgotten in Islamic tradition thereafter, but now a cornerstone in the relatively new discipline of archaeoastronomy, which was most significant. In other words, one cannot understand the orientation of medieval Islamic religious architecture unless one knows about the orientation of the Kaʿba. The Muslims who laid out mosques in medieval times knew about the orientation of the Kaʿba, and this is why they laid out the mosques in the ways they did.[12] There are many analogies in the present study.

In this study I shall not expect much technical knowledge of astronomy on the part of the reader. We shall be preoccupied with the shadows cast by a vertical object on a horizontal surface. When the sun rises the shadow is indeterminate; during the morning the shadow decreases, reaching its minimum value at midday, when the sun is on the meridian; the shadow then increases during the afternoon becoming indeterminate again at sunset. The situations

---

[10] See the various contributions to *The Significance of Islamic Manuscripts*, John Cooper, ed., London: Al-Furqan Islamic Heritage Foundation, 1992, also available in Arabic.

[11] On the qibla problem see the articles "Ḳibla" and "Makka. iv. As Centre of the World" in EI_2, reprinted in King, *Studies*, C-IX-X, and, most recently, *idem*, *Mecca-Centred World-Maps*, pp. 47-127, and **VIIa**.

[12] See Hawkins & King, "Orientation of the Kaʿba", and King, "Astronomical Alignments in Medieval Islamic Religious Architecture"; as well as **VIIb**.

before and after midday are symmetrical with respect to the meridian, and it will be convenient for us to think of them together: when the sun is just above the horizon the shadow is long, and it decreases as the sun gets further from the horizon, reaching a minimum at midday. The length of daylight is divided into 12 equal "seasonal" hours, whose length varies throughout the year. If we denote the shadow length by z (for *ẓill*), its minimum by Z (with the inevitable disadvantage that z is always larger than Z), and the time in seasonal hours since sunrise or time before sunset by T ($0 \leq T \leq 6$), then:

$$\text{when } T \to 0 \text{ then } z \to \infty \quad \text{and} \quad \text{when } T = 6 \text{ then } z = Z \,.$$

If we denote the increase of the shadow over its midday minimum by $\Delta z = z - Z$, then:

$$\Delta z \to \infty \text{ as } T \to 0 \quad \text{and} \quad \Delta z = 0 \text{ when } T = 6 \,.$$

The determination of z and Δz in terms of T by means of accurate procedures is a complicated process that does not concern us here.[13] Rather we shall encounter a simple approximate rule relating time of day and shadows, more specifically, the time in seasonal day hours and the shadow increases, that is, T and Δz.

[13] See **I-4** on the determination of the solar altitude at the hours throughout the year.

CHAPTER 1

THE STANDARD DEFINITIONS OF THE TIMES OF PRAYER IN ISLAM

> "Signalons, enfin, que le Coran ne décrit en détail ni la Prière elle-même ni tout ce qui touche de près ou de loin". Michel Choumet, "Interrogeons le Coran à propos de la prière rituelle" (cited n. 1:1), p. 84.

> "(The) time (of the *ʿaṣr*) begins when the shadow of objects is equal to their height." G. Monnot, article "Ṣalāt" in EI_2 (VIII, p. 928b).

1.1 The times of the prayers in the Qurʾān

In the *Qurʾān* there are no specific instructions for the timing of the prayers, and not all of the five daily prayers of later practice are specifically mentioned. The Qurʾānic verses that mention prayer at certain times are the following (my translations):

> " ... Celebrate the praises of your Lord before the rising (*ṭulūʿ*) of the sun and before its setting (*ghurūb*). Celebrate (them) for some parts (*ānāʾ*) of the night, and at the ends (*aṭrāf*) of the day ... ." (20.130)
>
> " ... Celebrate the praises of your Lord when you arise (from sleep) and during the night (*mina 'l-layl*) praise him, and when the stars are disappearing." (52.48)
>
> "Give glory to God when you reach evening (*tumsūna*) and when you rise in the morning (*tuṣbiḥūna*) – to Him is praise in the Heavens and on earth – and in the late afternoon (*ʿashiyy^{an}*) and when you are at midday (*tuẓhirūna*)." (30.17-18)
>
> "Perform the prayer at the two ends of the day (*ṭarafayi 'l-nahār*) and at the approaches (*zulaf*) of the night ... ." (11.114)
>
> "Perform the prayer from the sun's decline (from the meridian) (*dulūk al-shams*) until the darkness of night (*ghasaq al-layl*), as well as the recitation at daybreak (*qurʾān al-fajr*) ... and perform the night vigil for some part of the night (*wa mina 'l-layl fa-tahajjad bihi*) as an optional duty for you (*nāfila^{tan} laka*)." (17.78-79)
>
> "Take care to perform the prayers (*al-ṣalawāt*), and the middle prayer (*al-ṣalāt al-wusṭā*) ... ." (2.238)

In the last of these verses "prayers" (*ṣalawāt*) are mentioned in the plural, and the Muslim commentators took this as referring to the five canonical prayers. In *Qurʾān* 103.1, the *ʿaṣr* is mentioned (or rather, invoked in an oath) but here it is Time in general rather than a time of day or a specific prayer-time that was intended. The Qurʾānic injunctions concerning prayer listed above clearly reflect an early stage of development of the prayer ritual.[1] Scholars have

[1] M. Th. Houtsma, "Iets over den dagelijkschen Çalat der Mohammedanen", *Theologisch Tijdschrift* 24 (1890), pp. 127-134, summarized in Rudi Paret, *Der Koran*, I: Übersetzung, and II: Kommentar und Konkordanz, 2. Auflage, Stuttgart, *etc.*: W. Kohlhammer, 1980, II, pp. 50-51. See also, for example, Michel Choumet ("Interrogeons le Coran à propos de la prière rituelle "aṣ-Ṣalāt"", in *Mélanges offerts au professeur Dominique Sourdel*, in *Revue des Études islamiques* (Paris), 54 (1986), pp. 77-84.

debated the meaning of the mysterious phrase *al-ṣalāt al-wusṭā* for over thirteen centuries, relating it to the morning and midday prayer and even to the night vigil,[2] but fortunately the matter is of little concern here.

1.1* The times of prayer amongst the first Muslims

Note: This brief section, added in May, 2002, is condensed from Uri Rubin's article, "Morning and Evening Prayers in Early Islam", mentioned in the preface. That article, to which I have barely done justice here, contains a wealth of references and further information and details, and the interested reader is strongly recommended to consult it.

The early Islamic historical sources reveal that the first times of prayer observed by the Muslims before the introduction of the five prayers were the *ḍuḥā* and the *ʿaṣr*. The time of the *ḍuḥā*, here that part of the day immediately following sunrise, functioned as a time of ritual practices amongst the Meccans, some of which practices were associated with the Kaʿba. The *ṣalāt al-ḍuḥā* as a thanksgiving ritual was particularly in vogue amongst the Bedouins. The *ʿaṣr* is here the end of the day immediately before sunset, apparently the same as the *ʿashī*, mentioned several times in the *Qurʾān*. According to some reports, Muḥammad was not able to pray the *ʿaṣr* openly on account of the objections of the Quraysh, who had no problem with praying at the *ḍuḥā*. In contrast to the *ḍuḥā*, the *ʿaṣr* did not have any roots in pre-Islamic practice (compare, however, the Jewish *minḥah*). Only later did the *ʿaṣr* become the most sacred time of the Islamic day. Meanwhile the prayer at the *fajr* (compare the *shaḥarith*) replaced the prayer at the *ḍuḥā*. And even after the five prayers had been introduced, the most important ones remained the *fajr* and the *ʿaṣr* (jointly called *al-ʿaṣrāni*). Some followers liked to delay the *ʿaṣr*, and this accounts for the special definition that we shall find associated with Abū Ḥanīfa.

1.2 The *ḥadīth* about the Archangel Gabriel

According to Islamic tradition, as recorded in the biography of the Prophet by Ibn Isḥāq, the five prayers were established during Muḥammad's night journey to Heaven.[3] Muḥammad is reported to have told his followers upon his return that God had made obligatory upon them five prayers every day and night.

The authority for the definitions of the *times* for these prayers in later practice is usually given as a *ḥadīth*[4] attributed to the Prophet (**Text 1.1**) describing an encounter with the

---

[2] Paret, *Koran* (n. 1:1), II, p. 51. On the night vigil see also the EI_1 articles "Ṣalāt" (esp. IV, p. 100b) and "Tahadjdjud" by A. J. Wensinck (the latter taken over in EI_2). See also n. 2:15.

[3] Alfred Guillaume, *The Life of Muhammad: A Translation of Ibn Isḥāq's Sīrat Rasūl Allāh with Introduction and Notes*, Karachi: Oxford University Press (Pakistan Branch), 2nd edn., 1968, pp. 184-187. On the night journey see also the articles "Miʿrādj" by Joseph Horovitz in EI_1 and by various authors in EI_2.

[4] On the *ḥadīth* literature in general see the article "Ḥadīth" by James Robson in EI_2. This particular *ḥadīth* is indexed in Arent J. Wensinck *et al.*, *Concordance et indices de la tradition Musulmane*, 7 vols., Leiden: E. J. Brill, 1936-1969, paper-back reprint in 4 vols., n. d. (*ca.* 1992), I, p. 85b, and III, p. 116b, and is here quoted from al-Aṣbaḥī, *al-Yawāqīt* (see n. 1:22), fols. 4v-5r. It is even recorded as one of the 40 favourite *ḥadīth* in the popular 14th-century compilation *Mishkāt al-maṣābīḥ* (see Brockelmann, *GAL*, I, p. 448, *etc.*) which has

Archangel Gabriel,[5] the bearer of the Revelation to Muḥammad:

> "(The Prophet) said: Gabriel came to me and led me in prayer (on two successive days. On the first day) the *ẓuhr* was when the sun had declined (from the meridian) (*zālati 'l-shams*) and the shadow (*fay'*) was equal to (the width of) a thong of a sandal, the *ʿaṣr* when the shadow of every object was the same as its (length), the *maghrib* when the sun had set and when a person fasting would have started to break the fast, the last *ʿishā'* when evening twilight had disappeared, and the *ṣubḥ* when morning twilight had appeared.
>
> The next day he lead me in prayer (again): the *ẓuhr* when the shadow of every object was the same as (its length), the *ʿaṣr* when the shadow of every object was the same as twice (its length), the *maghrib* when the sun had set and a person fasting would have started to break the fast – and he did not delay (this prayer), the last *ʿishā'* when one-half (*shaṭr*) of the night had passed – or one-third of the night in another version, and the *fajr* when the sky had begun to glow."

Another version (**Text 1.2**) reads:[6]

> " ... ... Gabriel came to me and led me in prayer: the *ẓuhr* when the sun had declined (from the meridian so that the shadow was) the amount of a thong of a sandal (*zālati 'l-shams qadra shirāki 'l-naʿl*), the *ʿaṣr* when the shadow (*ẓill*) of every object was the same as its length ... *etc.*"

In the first version of this *ḥadīth*, the term *fay'* in the first definition of the *ẓuhr* refers to the shadow-length rather than to the increase over an unspecified minimum.[7] As we shall see in **Ch. 3**, the early legal scholars seemed to think that the midday shadow must have been zero on this day and so the width of the thong – surely intended to represent the smallest observable increase – was the actual length of the shadow. Elsewhere in this first version, however, *fay'* is used in the same way as is *ẓill*, *i.e.*, simply "shadow cast by an object", in the second version.

If this *ḥadīth* genuinely reflected a pronouncement of the Prophet about inspiration he received, one would expect more unity in early Muslim practice than the historical evidence suggests. A similar story about ʿAlī ibn Abī Ṭālib was also in circulation. It is recorded in the *Nahj al-balāgha*, the most authoritative Shīʿite collection of stories about ʿAlī (**Text 1.3**):[8]

> "From a letter of (ʿAlī) – peace be upon him – to the governors of the (various) regions (*umarā' al-bilād*) concerning the practice (*maʿnā*) of prayer. "Lead the people in prayer (at the following times): the *ẓuhr* when the sun casts a shadow inside the goat pen (!), the *ʿaṣr* when the sun is (still) white and live at a time of day when it is still very

---

been translated by James Robson (*Mishkāt al-Maṣābīḥ*: *English Translation with Explanatory Notes*, 2 vols., Lahore: Sh. Muhammad Ashraf, 1970-81).

[5] See the article "Djabrā'īl" by J. Pedersen in EI_2.

[6] Recorded, for example, in the *Muqaddima* of the 10th-century Ḥanafī scholar Abu 'l-Layth al-Samarqandī: see A. Zajaczkowski, ed., *Le traité arabe* Muḳaddima *d'Abou-'l-Lait as-Samarḳandī en version mamelouk-kiptchak (Ms. Istanbul, Aya Sofya 1451)*, Warsaw: Panstwowe Wydawnictwo Naukowe, 1962, pp. 79-83. On the author see the article by Joseph Schacht in EI_2 and Sezgin, *GAS*, I, pp. 445-450.

[7] See Lane, *Lexicon*, VI, col. 2468b, and V, cols. 1915a-1916a, on *fay'* and *ẓill*.

[8] Abu 'l-Ḥasan Muḥammad al-Raḍī ibn al-Ḥasan al-Mūsawī, *Nahj al-balāgha*, ed. Muḥammad Muḥyi 'l-Dīn ibn ʿAbd al-Ḥamīd, 3 vols., Cairo: Istiqāma Press, n. d., II, pp. 91-92. On the problematic authorship see the three EI_2 articles: "al-Murtaḍā" by C. Brockelmann; "ʿAlī ibn Abī Ṭālib" by L. Veccia Vaglieri, esp. p. 385b; and "Nahdj al-balāgha" by M. Djebli.

bright (*fī ghuḍuww mina 'l-nahār*, printed text has: ʿ-*ḍ-w*) (and) when one can still ride two *farsakh*s, the *maghrib* when a person who is fasting breaks his fast and a pilgrimsets out for (*yadfaʿ ilā*) Minā, the *ʿishā*ʾ from the time when evening twilight disappears from sight until one-third of the night has passed, and the *ghadāt* when a man can recognize the face of his companion ... .""

It would be of interest to know which of these stories was formulated first. Also the Caliph ʿUmar ibn al-Khaṭṭāb is reported to have written to his governors on the times of the five prayers; there are several versions of this story; in one of these he is said to advocate a period for the *ẓuhr* from when the increase of a man's shadow is one cubit (**4.4**) until when the shadow is equal to his height (the *ʿaṣr* is here not defined in terms of shadows). Last but not least, the Companion Abū Hurayra[9] is reported to have said that the *ẓuhr* should be prayed from the time when the shadow was the same as a man's height and the *ʿaṣr* when it was twice this (**3.1**).

The *ḥadīth* about the Archangel Gabriel was most likely coined in order to impress the number five upon the community, to provide a clear textual basis for the timing of the prayers, and to give sanction to various diverse practices within certain limits.

1.3 The standard definitions

The definitions that I now describe appear to have been adopted by the major legal schools in the 8th century.[10] Hence they were used throughout the medieval period, by legal scholars and astronomers alike; indeed, they remain in use to this day.

The Islamic day begins at sunset since the Islamic calendar, being lunar, is regulated by the sighting of the lunar crescent shortly after the sun has set. Each of the prayers is to be performed within a specific interval of time, the endpoints of which are astronomically defined – see **Fig. 1.3a**; it is considered preferable to perform the prayers as early as possible within the appointed intervals.[11]

The permitted interval for the first prayer (*al-maghrib*) begins at sunset and ends at nightfall. The time allowed for the second prayer (*al-ʿishā*ʾ) begins at nightfall and ends when one-third or one-half of the night is passed. The third prayer (*al-fajr*) begins at daybreak and must be completed by sunrise. The fourth prayer (*al-ẓuhr*) begins either when the sun has been observed to start its decline from the meridian or when the shadow of any vertical object has increased beyond its midday minimum by a specific amount. The permitted interval ends when that for the fifth prayer (*al-ʿaṣr*) begins. This moment is also defined by means of shadow increases. The end of the fifth prayer is either when the shadow has increased again by a specific amount or at sunset.

Thus the prayers at sunset, nightfall and daybreak are defined in terms of horizon or twilight phenomena. Of particular interest and concern to the present study are the definitions for the

---

[9] On Abū Hurayra see the article by James Robson in EI_2, and also G. H. A. Juynboll, *The Authenticity of the Tradition Literature: Discussions in Modern Egypt*, Leiden: E. J. Brill, 1969.

[10] For an unhappy projection of the later definition of the *ʿaṣr* back into early 8th-century Mecca see Maranyi, "Zwischen *ʿaṣr* and *mağrib* in Mekka", p. 115.

[11] See the article "Mīḳāt. i. Legal Aspects" by A. J. Wensinck in EI_2, especially p. 27a.

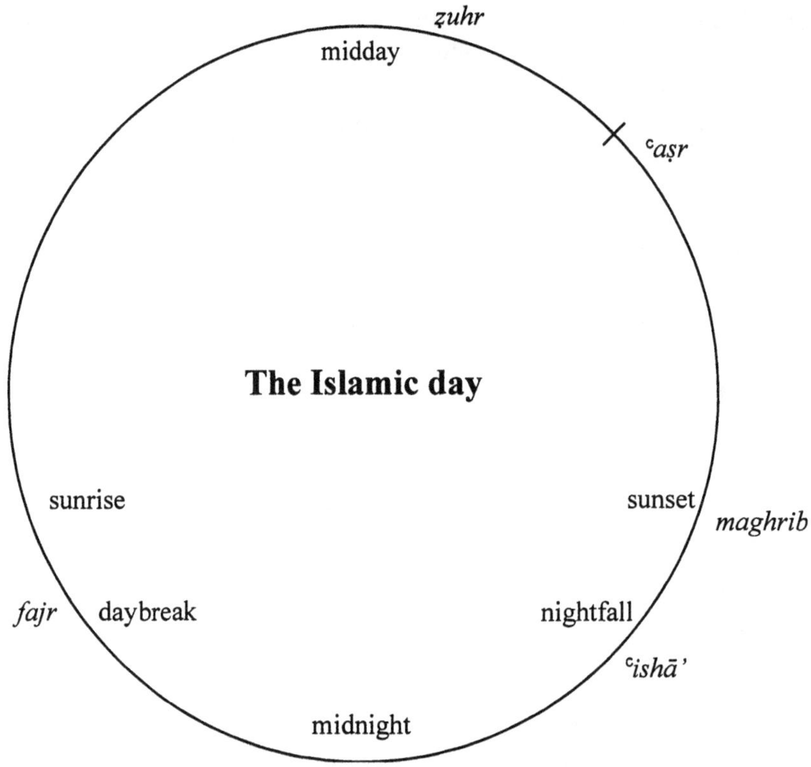

Fig. 1.3a: The Islamic day.

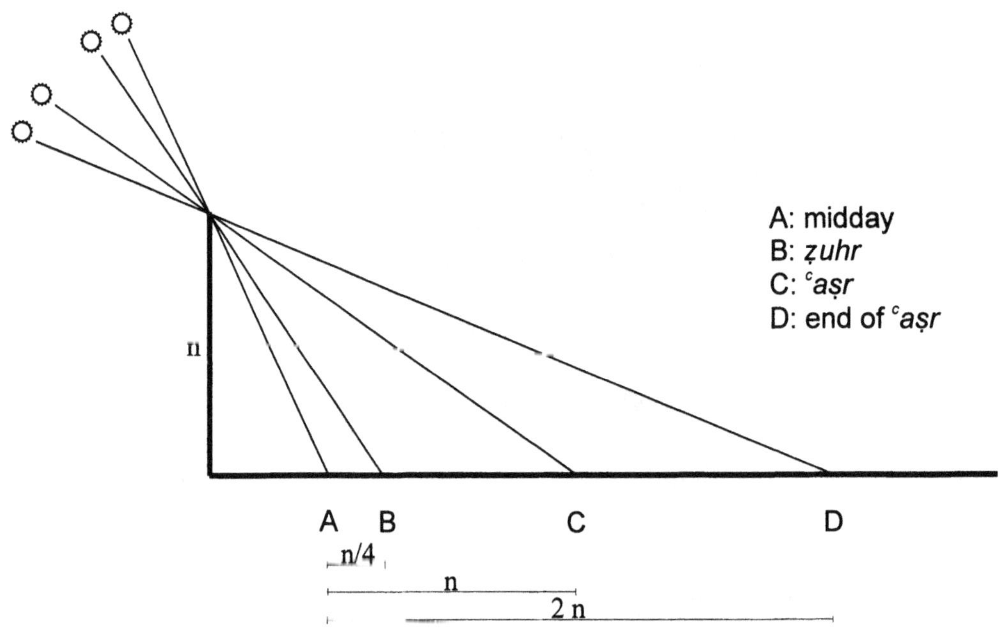

Fig. 1.3b: The definitions of the beginning of the *ẓuhr* (Andalusī and Maghribī practice) and the beginning and end of the *ʿaṣr* in terms of shadow-increases over the midday minimum.

ends of the intervals during which the two day-time prayers are to be performed. These are expressed in terms of shadow lengths, or, more precisely, in terms of *the increase of the length of the shadow of a vertical object over its midday minimum*. The gnomon, or object casting the shadow, may be a man, in which case he can measure his shadow in terms of his own feet, his height (*qāma*) being considered as about seven "feet". Or it may be a wooden or metal object erected vertically on a horizontal surface.[12]

The beginning of the interval for the *ẓuhr* prayer in medieval Maghribi and Andalusī, and also early Egyptian, practice, in other words, in Mālikī practice, is the moment when the shadow has increased beyond its midday minimum by one-quarter of the length of the gnomon ($\Delta z = {}^{1}/_{4}n$). The beginning of the interval for the *ʿaṣr* prayer, according to all but the Ḥanafī school, is the moment when the shadow has increased by the length of the gnomon ($\Delta z = n$). For the Ḥanafīs, the *ʿaṣr* begins when the shadow has increased by twice the length of the object ($\Delta z = 2n$) – see **Fig. 1.3b**. The Shīʿites in modern-day Iran perform the *ẓuhr* and the *ʿaṣr* prayers (and also the *maghrib* and *ʿishāʾ* prayers) consecutively,[13] so that there are only three periods for prayer (see **Fig. V-13.1e**).

1.4 The regulation of the prayer-times in later practice

In the Islamic Middle Ages there were two virtually independent traditions of astronomy, each of which was applied to the practical problem of regulating the times of the prayers according to the above definitions.[14]

The first, folk astronomy, was by far the most widely practiced. It was based solely upon phenomena that could actually be observed in the sky, and it was approved by the religious authorities.[15] Folk astronomy was practiced all over the Islamic commonwealth but it was in the Yemen that it was cultivated most avidly. In this tradition, simple methods were used for timekeeping and regulating the times of prayer, namely, arithmetical shadow schemes for daytime and observation of the lunar mansions by night (**III**). The schemes give for each month of the year the shadows at the *ẓuhr* and the *ʿaṣr* to the nearest "foot" for a gnomon of length about 7 "feet", *i.e.* the height of a man. The information is often given in the form of mnemonics and is sometimes rather inaccurate. These techniques are recorded in the so-called *kutub al-mawāqīt*, non-technical books on astronomical timekeeping, as well as in legal texts. All that one needed to regulate the daylight prayers was to observe one's own shadow. Tables and

---

[12] On the divisions of the gnomon see al-Bīrūnī, *Shadows*, pp. 29-41, Kennedy, *Shadows*, I, pp. 68-80, and II, pp. 25-31, as well as the last paragraphs of **I-1.1** and **II-1.3**.

[13] S. H. Nasr, trans., *Shiʿite Islam by ʿAllamah Sayyid Muhammad Husayn Tabatabaʾi*, Albany, N.Y.: State University of New York Press, 1975, p. 231, and M. Momen, *An Introduction to Shīʿī Islam*, New Haven, Conn. & London: Yale University Press, 1985, pp. 178 and 237.

[14] See King, "Science in the Service of Religion: The Case of Islam", *impact of science on society* (UNESCO), no. 159 (1991), pp. 245-262, reprinted in *idem*, *Studies*, C-I, for an overview of astronomy in the service of Islam; *idem*, "Islamic Folk Astronomy", A-C, for the applications of the folk tradition to the same ends; also **III-1.1**.

[15] A list of "acceptable" uses of (folk) astronomy is recorded in Heinen, *Islamic Cosmology*, p. 25 (quoting the 11th-century authority al-Kha«ºb al-Baghd,dº): see **V-5.3**.

instruments were eschewed. In the courtyard of the the restored 7th-century mosque of Janad in the Yemen there is a gnomon about six feet high, used for regulating the times of the *ẓuhr* and *ʿaṣr*; this provided a standard *qāma*, and there is no trace of a meridian line drawn through its base, if ever there was one (see **Fig. III-1.2b**).[16] The legal scholars considered the techniques of folk astronomy to be adequate for practical purposes.

The second tradition, mathematical astronomy, was at first based mainly on the Indian and Hellenistic astronomical traditions and then developed specifically Islamic features.[17] It was practiced by a select few in various locations over the centuries. From the 9th century onwards, mathematical tables were used in conjunction with instruments such as the astrolabe, quadrant and sundial for reckoning time by day and night from observed solar and stellar altitudes as well as for regulating the times of prayer (**I-II** and **VIII**). The 13th century saw the rise of the institution of the mosque astronomer and professional timekeeper known as the *muwaqqit* (**V**). These individuals, working in the major centres of Islamic civilization, sometimes used sophisticated tables for timekeeping and for regulating the times of prayer often containing tens of thousands of entries, but it is not clear to what extent their activities were approved by the religious authorities. Certainly the other activities of the astronomers in medieval Islam did not meet with their approval, not least because of the close association of mathematical astronomy and astrology.

The legal scholars Ibn Raḥīq and al-Aṣbaḥī (see below) state (**Texts 1.4** and **1.5**)[18] that one should determine the prayer-times "by observation with one's own eyes, not using an astrolabe or any of this astronomy nonsense (*tanjīm*)", and they warn their readers against the techniques of the astronomers, who "take their knowledge from Euclid and (the Indian tradition of) the *Sindhind*, and from Aristotle and other philosophers, all of whom were infidels."

An analogous situation prevailed for finding the *qibla*, the sacred direction in Islam. The astronomers formulated sophisticated mathematical procedures and compiled extensive tables for determining the qibla, and the legal scholars preferred to use astronomical horizon phenomena. This accounts for the wide variation in orientation of medieval mosques in each region of the Islamic world (**VIIa**).

1.5 New sources

In the *kutub al-mawāqīt* timekeeping and the regulation of the prayer times are discussed in terms of simple folk astronomy rather than sophisticated mathematical astronomy.[19] The earliest works of this kind, dating perhaps from the 9th and 10th century, are no longer extant, but the several later works that do survive record and elaborate the same tradition. Most of the extant

---

[16] The gnomon is popularly called *ʿaṣā Muʿādh ibn Jabal*, that is, the stick of the celebrated Companion of the Prophet despatched by Him to the Yemen. It is not clear how long there has been a gnomon of this kind, but al-Aṣbaḥī stated in the 13th century that he made regular measurements of the midday shadows (**III-4.1**).

[17] For various introductions see the article "ʿIlm al-hay'a" by David Pingree in *EI*$_2$; Saliba, "Islamic Astronomy and Astrology"; and King, "Islamic Astronomy". On the earliest Muslim encounters with Indian and Hellenistic astronomy see David Pingree, "Indian Influence", and "Greek Influence".

[18] Ibn Raḥīq, *al-Mawāqīt*, fol. 8r, and al-Aṣbaḥī, *al-Yawāqīt*, fols. 6r-6v.

[19] Various works of this genre are listed in *Cairo ENL Catalogue*, II, 1.4 and 3.2.

works of this genre are from the Yemen or the Hejaz; others are of Egyptian and Maghribi provenance.[20]

The main representative works of this genre that I have used for the present study are two in number. Firstly, a treatise on different aspects of folk astronomy by an 11th-century writer in Mecca named Abū ʿAbdallāh Muḥammad ibn Raḥīq ibn ʿAbd al-Karīm. This work, whose precise title is unknown, is extant in a unique manuscript in the Deutsche Staatsbibliothek in Berlin.[21] Secondly, the *Kitāb al-Yawāqīt fī ʿilm al-mawāqīt* by the 13th-century Yemeni scholar Ibrāhīm ibn ʿAlī ibn Muḥammad al-Aṣbaḥī, extant in at least four manuscripts, of which I have used the copy in the Egyptian National Library in Cairo.[22] A further source from the same milieu, the *Tuḥfat al-rāghib wa-turfat al-ṭālib* ... of the late-13th-century Adeni scholar Muḥammad ibn Abī Bakr al-Fārisī, also provides useful information.[23] Each of these three works is currently under investigation in a doctoral thesis by Petra Schmidl, with the purpose of establishing what they contribute to our knowledge of folk astronomy as applied to the determination of the prayer-times and the sacred direction.[24]

Another important source of information on the prayer-times is the book on timekeeping entitled *Ifrād al-maqāl fī amr al-ẓilāl* by the great early-11th-century scientist of Khwarazm and Ghazna, Abu 'l-Rayḥān al-Bīrūnī. In this treatise al-Bīrūnī discussed the definitions of the prayer-times at length. This important work is now published, with English translation and a valuable commentary by Ted Kennedy.[25]

---

[20] On the astronomical traditions in medieval Egypt, Yemen, and the Maghrib, see King, "Astronomy in Fatimid Egypt", *idem*, "Astronomy of the Mamluks"; *idem*, *Astronomy in Yemen*; and *idem*, "Astronomy in the Maghrib", respectively.

[21] All that is legible of the title is ... *ʿalā madhhab al-ʿArab*, which is probably to be rendered "according the the Arabs before Islam". On the manuscript (Berlin Ahlwardt 5664), hereafter for convenience referred to as Ibn Raḥīq, *al-Mawāqīt*, see *Berlin Catalogue*, pp. 149-150. It was copied *ca.* 1350, probably in the Hejaz or the Yemen. For a brief account of the contents see King, *Astronomy in Yemen*, no. 2.

[22] On al-Aṣbaḥī (**III-4.1**) see King, *Astronomy in Yemen*, no. 5, and *Cairo ENL Survey*, no. E5. Four manuscripts of his treatise are:

(1) MS Baghdad Awqāf 2982/6276, copied in 1281/82 in Taiz (I have not been able to inspect this);
(2) MS Oxford Marsh 134, fols 9r-52r, copied in 1451/52;
(3) MS Cambridge Arberry Supp. 110(a) = Or. 1236(11), copied *ca.* 1650; and
(4) MS Cairo DM 948,1, fols. 1v-53r, copied in 1902/03.

The Cairo manuscript, hereafter referred to as al-Aṣbaḥī, *al-Yawāqīt*, is reliable in spite of its late date of copying – see further *Cairo ENL Catalogue*, I, *sub* DM 948, and II, 1.4.8+, as well as King, *Astronomy in Yemen*, p. 68, *sub* TH.

[23] On al-Fārisī (**III-4.2**) see Suter, *MAA*, no. 349; Kennedy, "*Zīj* Survey", no. 54; and King, *Astronomy in Yemen*, no. 6. He was the author of a *Zīj* prepared for the Yemeni Sultan al-Muẓaffar, and his treatise on folk astronomy is extant in MS Milan Ambrosiana X73 = Griffini 37 (O. Löfgren and R. Traini, *Catalogue of the Arabic Manuscripts in the Biblioteca Ambrosiana*, vol. I, Vicenza: Neri Pozza Editore, 1975, no. CCLXX).

[24] See Schmidl, *Islamische volksastronomische Abhandlungen*. On aspects of folk astromomy applied to the determination of the qibla as treated in these texts see already *eadem*, "Zur Bestimmung der Qibla mittels der Winde" (n. 0:8).

[25] On al-Bīrūnī see the splendid article by E. S. Kennedy in *DSB*. For his work on shadows see al-Bīrūnī, *Shadows* (text) and Kennedy, *al-Bīrūnī's* Shadows (translation and commentary). A preliminary account of al-Bīrūnī's writings on the prayer-times was given in Kennedy, "Prayer-times".

For the extracts I have relied more on the uncritical published text than on a poor-quality copy of a microfilm of the unique manuscript (MS Bankipore 2468, copied 1234/35) generously provided by Dr. Jan Pieter Hogendijk of Utrecht (see his "Rearranging the Arabic Mathematical and Astronomical Manuscript Bankipore 2468", *JHAS* 6 (1982), pp. 133-159, esp. p. 145). Some years ago I recommended that this manuscript be published in a

My other sources are sundry astronomical, folk scientific, legal, historical and lexicographical texts, as well as a few astronomical instruments.

Added in proof: Jean-Charles Ducène of the Université de Bruxelles kindly informed me in April, 2001, that the 10th-century Shāfiʿī legal scholar Ibn al-Qāṣṣ[26] in his treatise on the determination of the qibla by non-mathematical means describes two gnomons for determining the *ẓuhr* and the *ʿaṣr* (MS London BL Or. 13315, fols. 31v and 33r) and presents the Indian formula for timekeeping and part of al-Fazārī's poem (fols. 33v-34r) as well as a list of names for the 24 seasonal hours (fol. 30r). Some of this material is also found in the work *al-Muʿrib ʿan baʿḍ ʿajāʾib al-Maghrib* by the 12th-century scholar Abū Ḥāmid al-Gharnāṭī, or at least in the Madrid manuscript used by the editor of the text, Ingrid Bejarano (Madrid 1991, pp. 37, 39-40, and 41). More information can be found in Jean-Charles Ducène's doctoral dissertation. I have refrained from inserting any of this material in the present study.

---

facsimile edition – see King, "Early Islamic Astronomy" [an essay review of Sezgin, *GAS*, VI], *JHA* 12 (1981), pp. 55-59, esp. p. 56, and *idem*, "Islamic Scientific Manuscripts", p. 134 – but alas no institution has undertaken the preparation of facsimile editions that would be of use to scholars.

[26] See now Ducène, "*Kitāb Dalāʾil al-qibla* d'Ibn al-Qāṣṣ", and a 2002 Brussels doctoral dissertation on the same subject.

CHAPTER 2

THE SEASONAL HOURS AND THE TIMES OF THE *ẒUHR* AND *ʿAṢR* PRAYERS

2.1 The seasonal hours

In what follows I shall demonstrate that the standard definitions of the times of the *ẓuhr* and *ʿaṣr* prayers are practical means of regulating the prayer-times defined in terms of the seasonal hours (Arabic, *al-sāʿāt al-muʿwajja*, later also *al-zamāniyya*).

The seasonal day and night hours are twelfth divisions of the day and night, respectively, and the lengths of the hours vary throughout the year and also depend on the local latitude.[1] Seasonal hours were used throughout the lands bordering the Eastern Mediterranean in antiquity. Graeco-Roman sundials marked with the seasonal hours were in use in Syria, Palestine and Northern Arabia in late Antiquity.[2] Both through their contacts with Jewish communities in Arabia[3] and their trading ventures to the north[4] the Arabs of the Hejaz could have known of the seasonal hours long before the advent of Islam. The quotes in **2.2** below, together with the lexicographical evidence presented in **Ch. 8**, raise the question whether the seasonal hours were also known in some circles in Arabia in the early 7th century. It is clear that they were *not* widely used in practice.[5]

The seasonal hours represent a less precise practical means of measuring time than the equinoctial hours, which are twenty-fourth divisions of the complete day and night and hence constant in length. The seasonal hours were maintained along with the equinoctial hours in Islamic astronomical sources, although the latter were generally preferred in astronomical timekeeping, if mainly implicitly since the passage of time was generally reckoned in equatorial degrees (15 of which correspond to one hour).

---

[1] On the origins of the seasonal hours see Neugebauer, *ESA*, pp. 81 and 86, and *idem*, *HAMA*, II, p. 561, and the references there cited. On the Jewish usage see Myron Pleasure, "Time in Jewish Tradition", *Antiquarian Horology* 8 (1974), pp. 619-624. On the Islamic usage see, for example, the articles "Zamān" by Willy Hartner in EI_1 (reprinted in his *Studies*, A, pp. 260-263) and "Layl and Nahār" by Charles Pellat in EI_2. See also n. 2:5 below.

[2] Sharon L. Gibbs, *Greek and Roman Sundials*, New Haven, Conn., and London: Yale University Press, 1976, pp. 308-312.

[3] On the Jews in Medina see EI_2, V, pp. 994b-995a.

[4] See the article "Makka, i: The Pre-Islamic and Early Islamic Periods" by W. Montgomery Watt in EI_2, and also Patricia Crone, *Meccan Trade and the Rise of Islam*, Princeton, N.J.: Princeton University Press, 1987 (to be used with some caution).

[5] Both Willy Hartner and Charles Pellat (see the references cited in n. 2:1 above) have stated categorically that that the Arabs before Islam were not familiar with the seasonal hours.

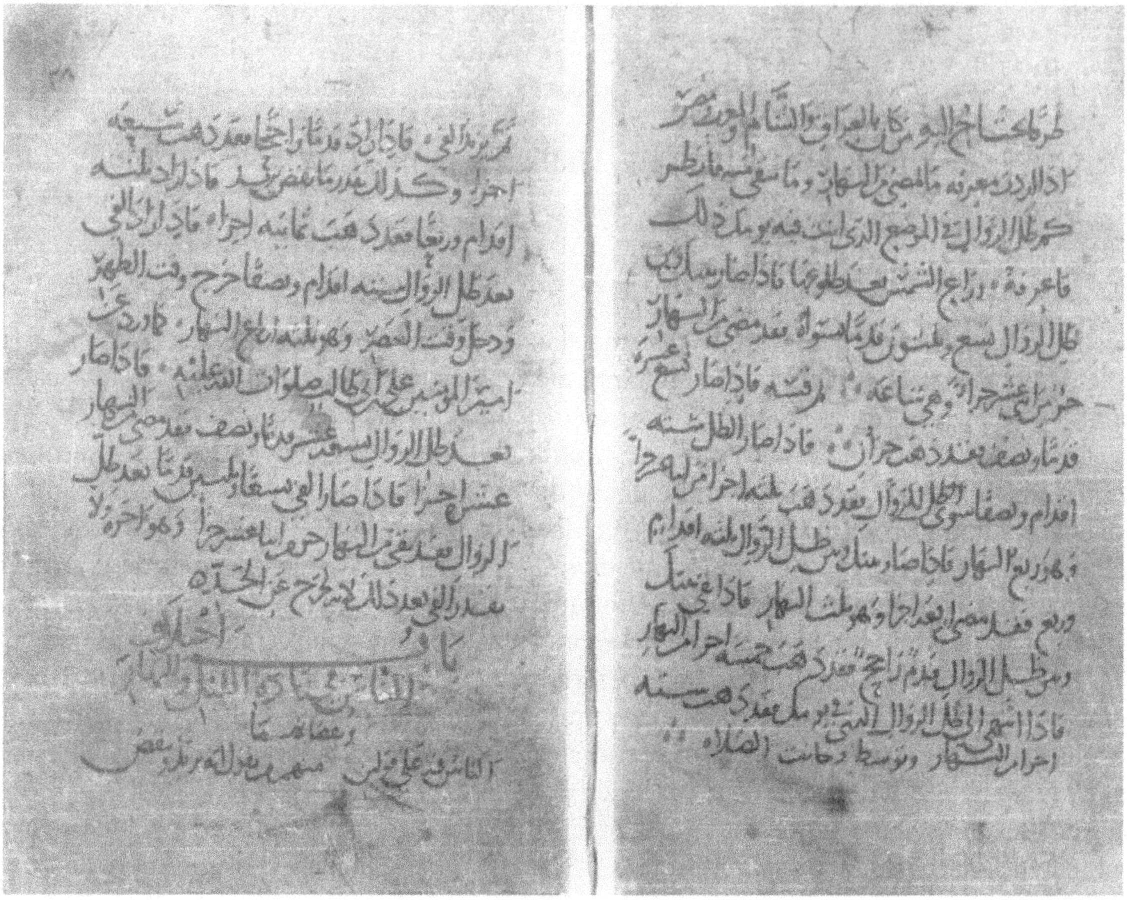

Fig. 2.1: The text of Ibn Raḥīq on a simple shadow-scheme and the remark attributed to ʿAlī ibn Abī Ṭālib. [From MS Berlin Ahlwardt 5664, fols. 28v-29r, courtesy of the Deutsche Staatsbibliothek (Preußischer Kulturbesitz).]

2.2 The prayers and the seasonal hours

al-Aṣbaḥī records two stories, which, if taken at face value, would constitute direct evidence that in the first few decades in Islam the prayer-times were sometimes related to the seasonal hours without reference to shadow lengths.

A statement attributed by al-Aṣbaḥī to the Prophet Muḥammad (**Text 2.1**)[6] translates:

> "Daylight is twelve hours (*sāʿa*) ... . Prepare for each hour (*sāʿa*) one bowing and two prostrations, thereby vowing yourself away from all sins. Also, further we say that the hours of the day in winter are light and the night hours are heavy, and that the hours of the day in summer are heavy and the night hours are light ... ."

Another significant statement, attributed to the Companion Ibn Sīrīn by al-Aṣbaḥī (**Text 2.2**), translates:[7]

[6] al-Aṣbaḥī, *al-Yawāqīt*, fols. 18v-19r. This is not found in the canonical Sunnī collections, and I have not pursued it elsewhere.

[7] al-Aṣbaḥī, *al-Yawāqīt*, fol. 4r. On Ibn Sīrīn see the article by Toufic Fahd in EI_2.

> "This hour is for finding the time of this prayer (*inna hādhihi 'l-sāʾa li-mīqāt hādhi 'l-ṣalāt*): there is no other way for finding its time (*mā lahā mīqāt ghayruhu*)."

Both Ibn Raḥīq and al-Aṣbaḥī (**Fig. 2.1** and **Texts 2.3** and **2.4**) record that in the opinion of ʿAlī ibn Abī Ṭālib the *ʿaṣr* should be performed at the beginning of the tenth hour, when three-quarters of the day have passed.[8] (This report about ʿAlī is not recorded in the *Nahj al-balāgha*.) An Andalusī mystic of the early 13th century named Ibn Abī Wāṣil predicted that Jesus will come again at the time of the *ʿaṣr*, defined in the same way.[9]

al-Bīrūnī attributes to the Shīʿite *Imām* Jaʿfar al-Ṣādiq (d. 765)[10] a statement relating the times of prayer to the odd-numbered hours of day and night (**Text 2.5b**),[11] thus:

> "The *ẓuhr* is at the beginning of the seventh hour (that is, midday), the *ʿaṣr* is at the beginning of the ninth hour (that is, two hours later), the *maghrib* is at the beginning of the first hour (that is, at sunset), the *ʿatama* is at the beginning of the third hour, and the *ghadāt* is at the beginning of the eleventh hour (that is, one hour before sunrise)."

However, immediately before this, al-Bīrūnī had attributed to Jaʿfar a statement (**Text 2.5a**) that:

> " ... the time for the *ʿaṣr* is long since there is no fixed sign in the heavens similar to those four signs (for the other prayers defined in terms of clearly visible celestial phenomena), so He (= God) made a limit for it that would not be too restricting (?). He (= Jaʿfar?) said that (it extends) from the (time when the) shadow of anything becomes equal to itself up to (the time that the shadow becomes) twice itself ... ."

Whilst we have no assurance that these statements were actually made by Jaʿfar, they are of interest in that they refer to regulating the prayers both by seasonal hours and by shadow lengths. See **3.1** on more material attributed to Jaʿfar al-Ṣādiq in specifically Shīʿite sources.

al-Aṣbaḥī provides similar information (**Text 2.6**):[12]

> "Daylight is twelve hours always ... (These hours) are called "crooked" (*muʿwajja*) because they are short in the daylight of winter and long in the daylight of summer. If six hours have passed, the time of the *ẓuhr* starts with the beginning of the seventh and lasts until the end of the ninth. Then starts the preferred time for the *ʿaṣr* at the beginning of the tenth, lasting until the end of (that hour). Then starts the (last)

---

[8] Ibn Raḥīq, *al-Mawāqit*, fol. 29r, and al-Aṣbaḥī, *al-Yawāqīt*, fols. 38r-38v. On ʿAlī ibn Abī Ṭālib see the article by L. Veccia Vaglieri in EI_2. On various achievements in the sciences attributed to him see Sezgin, *GAS*, VII, pp. 8, 10, 11, 26, and 184; *Cairo ENL Survey*, no. B1; King, "A Medieval Arabic Report on Algebra before al-Khwārizmī"; and **III-2.2**.

[9] The tradition is recorded by the 14th-century historian Ibn Khaldūn: see Franz Rosenthal, *Ibn Khaldûn: The Muqaddimah, an Introduction to History*, 3 vols., 2nd edn., Princeton, N.J.: Princeton University Press, 1967, II, p. 193. On Ibn Abī Wāṣil see *ibid*., pp. 187-188. His prediction is in the same spirit as one by the 11th-century *Qurʾān* commentator al-Zamakhsharī (see n. 8:15): he anticipated that when Jesus comes again he will perform his prayers in the direction of Mecca, as if he were a member of the Muslim community (Helmut Gätje, *The Qurʾān and its Exegesis: Selected Texts with Classical and Modern Muslim Interpretations*, trans. by Anthony T. Welch, Berkeley and Los Angeles: University of California Press, 1976, p. 83).

[10] See the article "Dja‘far al-Ṣādiḳ" by Marshall G. S. Hodgson in EI_2. As Hodgson points out, "so many sayings on all sides of all controversial issues have been ascribed to him that such reports are almost useless for determining his actual opinions in a given case". See also n. 3:9.

[11] al-Bīrūnī, *Shadows*, p. 163, and Kennedy, *Shadows*, I, p. 213, and II, p. 134.

[12] al-Aṣbaḥī, *al-Yawāqīt*, fol. 18r.

permitted time (for the ʿaṣr), according to the more correct of the two viewpoints (?), with the beginning of the eleventh and it lasts until sunset ... ."

There is other evidence that the daylight prayers were regulated using the seasonal hours – see below and also **Chs. 5-8**. Also we have vestiges of an association of the night prayers with fractions of the night, which would in practice be most easily regulated by the seasonal hours.[13] It is perhaps of some relevance that the *ʿishāʾ* prayer can be performed until a certain fraction, sometimes one-third, sometimes one-half, of the night has passed.[14] Likewise the night vigil, the *tahajjud*, said to have been practiced by the Prophet and some of His Companions for some ten years, but later made voluntary, is associated already in Qurʾān 73.20 with fractions of the night: "just less than two-thirds", "one-half" and "one-third".[15] We should recall that the *ʿaṣr* was to begin when one-half of the afternoon had passed. The seasonal hours are particularly suited for dividing the day and night into watches of one sort or another.

2.3 A shadow scheme for the prayers at the seasonal hours

al-Aṣbaḥī describes the following procedure for regulating the daylight prayers. He is clearly quoting an earlier source, perhaps the *Kitāb al-Mawāqīt* of Abū Jaʿfar al-Baṣrī, compiled in Basra *ca.* 885, which is mentioned in the text.[16] He first presents a scheme for the morning that is symmetrical to the following one for the afternoon (**Text 2.4**):[17]

> "Finding the hours. ... ... When your shadow decreases to the extent that there remains only the midday shadow of the previous day, six hours have passed, and the middle of the day (has come). The beginning of the time for the *ẓuhr* (prayer) is the beginning of the seventh (hour). When the afternoon shadow has increased one fifth of a length seven hours have passed. When it has increased by half a length, eight hours have passed. When it has increased by one length, nine hours have passed which is three-quarters of the day. Thus it was related of ʿAlī – may God be pleased with him – at that time is the beginning of the *ʿaṣr* (prayer) at the beginning of the tenth hour and the end of the *ẓuhr*, and there is no period of time between them. When (the afternoon shadow) increases by two lengths, ten hours have passed and the optimum time for the *ʿaṣr* is ended. The permitted time according to the more correct of the two viewpoints (??) continues until sunset. When it increases by five lengths, eleven hours have passed. The afternoon shadow is not determined thereafter until sunset at the twelfth (hour)."

---

[13] On timekeeping by night using simple folk astronomical techniques, see Forcada, "*Mīqāt* en los calendarios andalusíes". On timekeeping by night using highly sophisticated tables see **I-II**. On twilight in particular see also the EI_2 article "Shafaḳ".

On a candle-clock described in a 12th-century Arabic source, so designed that one candle goes out at each seasonal night hour throughout the year, see the references cited in n. **II**-1:9.

[14] See further nn. 2:23 and 8:36.

[15] See the EI_2 article "Tahadjdjud" (n. 1:2), and also n. 8:4.

[16] al-Aṣbaḥī, *al-Yawāqīt*, fol. 38r. Neither author nor title is listed in Sezgin, *GAS*, I-IX.

[17] al-Aṣbaḥī, *al-Yawāqīt*, fols. 38r-38v.

On the reference to ʿAlī ibn Abī Ṭālib see **2.2**. Notice that the *ẓuhr* begins at the beginning of the seventh hour and the *ʿaṣr* begins when nine hours have passed and lasts one hour. We represent the other information in al-Aṣbaḥī's scheme in tabular form as follows:

Time (since sunrise)	6	5/7	4/8	3/9	2/10	1/11	0/12
T	6	5	4	3	2	1	0
Δz	0	$^1/_5$n	$^1/_2$n	n	2n	5n	∞

and we do not have to look far for the underlying principle.

2.4 The Indian rule for timekeeping

A simple approximate Indian rule for timekeeping was known to the first generation of Muslim astronomers in the 8th century. It is recorded, for example, in the earliest Arabic astronomical writings, namely, those of the mid-8th-century Baghdad astronomer al-Fazārī (**Text 2.7**).[18] It is also found occasionally in later technical astronomical literature (**Texts 2.8** and **2.9**),[19] and more frequently in the literature on folk astronomy. The most detailed discussion is in al-Bīrūnī's *Shadows*.[20] (Both the early Muslim astronomers and their Indian predecessors were of course aware of more complicated exact formulae for timekeeping.)

The Indian rule relates the increase of shadow to the time of day in seasonal hours and may be rendered algebraically as follows. If T is the seasonal hours from sunrise in the morning or before sunset in the afternoon and Δz is the increase of the shadow over its midday minumum for a gnomon length n, then T is given by:

$$T = 6\,n \,/\, (\Delta z + n) .$$

This formula is, of course, approximate,[21] but it is the simplest possible rule for ensuring that T varies inversely with z subject to the end conditions:

$$\Delta z \rightarrow \infty \text{ when } T = 0 \quad \text{and} \quad \Delta z = 0 \text{ when } T = 6 .$$

The approximate shadow increases for each hour given by this simple rule are precisely as recorded by al-Aṣbaḥī.

Here then we have an explanation of the two different definitions for the *ʿaṣr* in the main legal schools as well as of an unusual definition for the *ẓuhr*. The shadow increase of one length for the *ʿaṣr* in the Mālikī, Shāfiʿī, and Ḥanbalī schools corresponds (using this approximate rule) to a time three seasonal hours before sunset, that is, exactly halfway between midday

---

[18] On al-Fazārī see the article by David Pingree in *DSB*. He is here quoted from al-Bīrūnī, *Shadows*, pp. 143-144. See further Pingree, "al-Fazārī", esp. pp. 121-122, and Kennedy, *al-Bīrūnī's* Shadows, II, pp. 116-117 and 118-121 on this passage.

[19] The first text is taken from MS Cairo DM 1169,6, fols. 57r-57v (on which see *Cairo ENL Catalogue*, I, *sub* DM 1169,6; and II, 4.3.30) of an anonymous Maghribi or Andalusī treatise on the use of the astrolabe, in which the rule is given for both n = 12 and 7 (see also n. 4:18).

The second is from MS Leiden Or. 1111, fol. 111v, of a treatise on timekeeping by the 14th-century Syrian astronomer Ibn al-Shāṭir (on the author see the article in *DSB*, and on this text see further **II-9.3** and **III-8.2**).

[20] See Kennedy, *al-Bīrūnī's* Shadows, II, pp. 116-122. On the accuracy of the formula see Marie-Louise Davidian, "al-Bīrūnī on the Time of Day", esp. p. 334.

[21] The "shadow increases" associated with the actual seasonal hours through this formula are approximate, and the "hours" that are defined by the actual shadow increases are not of the same length.

and sunset.[22] The shadow increase of two lengths in the Ḥanafī school corresponds (again using the Indian rule) to a time one seasonal hour later, that is, two seasonal hours before sunset. Furthermore, in the Shāfiʿī school the *ʿaṣr* begins when the shadow increase is one gnomon length and ends when it is two gnomon lengths. This corresponds precisely (using the Indian rule) to the duration of the tenth seasonal hour of daylight.

Finally, we should mention that the sources often advocate performing the *ʿishāʾ* prayer until one-third or one-half of the night is passed (see **Text 1.1**). Implicit here is the notion that the prayer should last until the end of the fourth or sixth hour of night. [23]

We now turn to the earliest legal texts to see how the definitions are developed in these.

---

[22] Occasional accounts by 19th-century Western travellers in the Near East have implied that the *ʿaṣr* prayer takes place in the middle of the afternoon. Thus, for example, Edward Lane stated that the *ʿaṣr* is "about mid-time between noon and nightfall", adding in a footnote that the time is actually defined by shadow-lengths ($\Delta z = 2n$ for the Ḥanafī rite, and $\Delta z = n$ for the others): see *An Account of the Manners and Customs of the Modern Egyptians*, first published in 1836, reprint of the 1895 edn., The Hague and London: East-West Publications, and Cairo: Livres de France, 1978 (also available in paper-back from Dover Pubications, Inc., New York), p. 76. C. Snouck Hurgronje (*Mekka*, 2 vols., The Hague: Martinus Nijhoff, 1888-89) stated: " ... ungefähr drei Stunden nach Mittag verkünden die sieben Mu'eddin's den Anfang ... des ... Nachmittags-çalat ... ." Again, T. P. Hughes in *A Dictionary of Islam*, London: W. H. Allen, 1935 (reprint of 1885 edn.), p. 469, stated that the *ʿaṣr* was midway between the *ẓuhr* and the *maghrib*. Perron, in his translation of the account of a visit to Dārfūr by the early-19th-century Muslim traveller Muḥammad ibn ʿUmar al-Tūnisī stated that the *ʿaṣr* was "à trois ou quatre heures après midi" and "environ quatre heures après midi" (Perron, *Mohammed ibn Omar el-Tounsy: Voyage au Ouaday*, Paris, 1851 (not consulted), pp. 53 and 107, cited in Dozy, *Supplément aux dictionnaires*, II, p. 134a). Also L. A. Sédillot (*fils*), in his summary of the treatise on astronomical instruments of al-Marrākushī (*fl* . Cairo *ca.* 1280 – see EI_2, *s.v.* and also **I-4.2.4** and **II-2.7**), stated without citing any authority (Sédillot-fils, "Mémoire", p. 170, n. 6): "le dohre [*al-ẓuhr*] est le moment le plus chaud de la journée et à une égale distance du zaoual [*al-zawāl*] ou vrai midi et de l'ashre [*al-ʿaṣr*]."

[23] See nn. 2:13 and 8:36.

CHAPTER 3

THE TIMES OF THE *ẒUHR* AND *ʿAṢR* PRAYERS IN THE EARLIEST LEGAL TEXTS

3.1 The opinions of Abū Ḥanīfa, Mālik and al-Shāfiʿī

The legal sources that I have consulted on the prayer times clearly merit the attention of a specialist, but they do not seem to contribute in a significant way to our understanding of the development of the standard definitions. I have not ventured into the early legal sources beyond consulting the major works of Mālik (*ca.* 710-795), al-Shāfiʿī (767-820) and Aḥmad ibn Ḥanbal (780-855).[1] There are problems about the attribution of works usually associated with Abū Ḥanīfa (699-767),[2] but his opinions are quoted elsewhere. At some risk I ignore the writings of early legal scholars such as al-Awzāʿī and al-Ṭabarī whose schools had only limited influence (though see **3.3** below).

The very first topic in the legal *summa* entitled *al-Muwaṭṭaʾ*, which is based on the doctrines of Mālik, is the times of the prayers.[3] Yet the treatment is devoid of personal opinion save what can be inferred from the selection of *ḥadīth*s and *khabar*s that are listed. The story about the Archangel Gabriel is not included, but a story about the Companion Abū Hurayra that the *ẓuhr* and the *ʿaṣr* should begin when the shadow (of a man) (*ẓill*) equals his height and twice that (*i.e.* z = n and 2n) is recorded. Three versions of a story about the Caliph ʿUmar ibn al-Khaṭṭāb[3a] writing to his governors about the times of prayer are also featured. The times are vaguely defined in terms of the appearance of the sun and the distances covered by a rider, but there is no mention of shadows (**Text 3.1**).

The recension of *al-Muwaṭṭaʾ* by Mālik's student Muḥammad ibn al-Ḥasan al-Shaybānī (749-805)[4] is of more historical interest for his remarks on the *ḥadīth* about Abū Hurayra (**Text 3.2**). Here the time for the *ẓuhr* was when z = n and the time for the *ʿaṣr* was when z = 2n. al-Shaybānī states that this was also Abū Ḥanīfa's definition for the *ʿaṣr*, but his own opinion was that *the shadow should have increased by twice the gnomon length*, *i.e.* $\Delta z = 2n$. Here we see the development of the definition for the *ʿaṣr* between Abū Ḥanīfa and al-Shaybānī.

---

[1] See the article "Fiḳh" by Joseph Schacht in EI_2 for a general introduction to Islamic law. The sources and bibliography are documented in Sezgin, *GAS*, I. Wiedemann & Frank, "Gebetszeiten", pp. 7-32, contains an overview of the opinions of the *imāms* on the times of prayer culled from *later* legal texts and various astronomical sources.

[2] See the article on Abū Ḥanīfa by Joseph Schacht in EI_2, and Sezgin, *GAS*, I, pp. 409-410.

[3] Mālik ibn Anas, *al-Muwaṭṭaʾ*, ed. Muḥammad Fūʾād ʿAbd al-Bāqī, 2 pts. in 1 vol., Istanbul: Çağrı Yayınları, 1401 H [= 1981], pp. 3-17, esp. pp. 3-7. On Mālik see the article by Joseph Schacht in EI_2, and Sezgin, *GAS*, I, pp. 457-464.

[3a] On ʿUmar ibn al-Khaṭṭāb see the in EI_2 article by G. Levi Della Vida / M. Bonner.

[4] Muḥammad ibn al-Ḥasan al-Shaybānī, *Muwaṭṭaʾ al-Imām Mālik*, ed. ʿAbd al-Wahhāb ʿAbd al-Laṭīf, vol. 1, Cairo: al-Majlis al-aʿlā li-'l-shuʾūn al-Islāmiyya, 1381 H [= 1967], I, pp. 31-33. On al-Shaybānī see the article in EI_2 by E. Chaumont, and Sezgin, *GAS*, I, pp. 421-433. In his *Aṣl* (published in Hyderabad, 1969-1973) there is no material on the times of prayer.

Unfortunately al-Shaybānī gives no reason for this modification. The 12th-century astronomer al-Kharaqī of Merw[5] stated that $\Delta z = n$ was preferred by the *imāms* of the Ḥijāz and $\Delta z = 2n$ by the *imāms* of al-ʿIrāq (**Text 3.3**): he is referring to the Mālikīs and the Ḥanafīs respectively. We may infer that the Ḥanafīs just wanted to be different from the Mālikīs of the Ḥijāz. The *ḥadīth* about the Archangel Gabriel justifies both opinions.

al-Shāfiʿī in his major work, the *Kitāb al-Umm*, is more original and informative (**Text 3.4**).[6] According to him, the *ẓuhr* begins as soon as the midday shadow is seen to increase, and he gets involved in a consideration of the midday shadow for different seasons. No locality is mentioned and although he lived more than half of his life in Mecca, Medina, and the Yemen, and later worked in Baghdad and Egypt, he is clearly considering the case of a locality in the tropics. He states that the midday shadow in summer can "fade away" (*yataqallaṣ*), by which he means that the shadow can be of zero length, but in other seasons it behaves differently; it is clear that he has only a vague idea about the behaviour of shadows. According to him, the *ẓuhr* begins when the sun is observed to have declined from the meridian and ends when the shadow is the same as the length of the object ($z = n$).

al-Shāfiʿī appears to wish to differentiate between the cases at mid-summer when $z = 0$ at midday and the other seasons when at midday $z > 0$ (the meaning is not completely clear to me and I offer two alternative translations, each underlined). He describes the end of the *ẓuhr* at seasons other than mid-summer as being "when it is known that *the shadow although it is different from the summer shadow* OR *the shadow less the summer shadow* has reached an amount such that the shadow of any object is the same as the object in the summer" (*idhā ʿulima an qad balagha 'l-ẓill maʿa khilāfihi ẓilla 'l-ṣayf qadra mā yakūnu ẓillu kulli shayʾ mithlahu fī 'l-ṣayf*). He is obviously concerned that both the *ẓuhr* and the *ʿaṣr* should begin shortly after the situations when $z = Z$ and $z = n$, but he ignores completely the problem that these definitions cannot be applied universally.

Ibn Ḥanbal makes no mention of the times of prayer in his treatise *al-Risāla al-Sunniyya fī 'l-ṣalāt wa-mā yalzamuhā*,[7] and I have not pursued his opinions elsewhere. Some later Ḥanbalī legal scholars, such as Ibn Qayyim al-Jawziyya (1292-1350), the author of a *Kitāb al-Ṣalāt*,[8] were apparently not concerned with the times of prayer *per se*, only with the consequences for people who delay prayer.

Thus Abū Ḥanīfa, Mālik, al-Shāfiʿī and Ibn Ḥanbal do not themselves, at least in these works consulted, specifically advocate times for the prayers defined in terms of shadow *increases*. But al-Shaybānī does. I suspect that he was advised by people who realized that the definitions mentioned in the *ḥadīth* could indeed not be applied universally.

---

[5] On al-Kharaqī see Suter, *MAA*, nos. 276 and 401, and *Cairo ENL Survey*, no. G6. This opinion of his is cited in Wiedemann & Frank, "Gebetszeiten", p. 8. I have used MS Escorial ár. 955 (128 fols., copied 1320-21) of his *Kitāb al-Tabṣira fī ʿilm al-hayʾa*, in which this passage occurs on fols. 99v-100r.

[6] al-Shāfiʿī, *Kitāb al-Umm*, 7 vols., Cairo, 1321-25 H [= 1903-07], I, pp. 62-63. On al-Shāfiʿī see the article in EI_2 by E. Chaumont, and Sezgin, *GAS*, I, pp. 484-490.

[7] Ibn Ḥanbal, *al-Risāla al-Sunniyya fī 'l-ṣalāt wa-mā yalzamuhā*, no. 7 in *Majmūʿat al-ḥadīth*, Riyadh: Maktabat al-Riyāḍ al ḥaditha, n. d. On the author see the article "Aḥmad ibn Ḥanbal" by Henri Laoust in EI_2, and Sezgin, *GAS*, I, pp. 502-509.

[8] Ibn Qayyim al-Jawziyya, *Kitāb al-Ṣalāt*, no. 8 in *Majmūʿat al-ḥadīth*, Riyadh: Maktabat al-Riyāḍ al-ḥaditha, n. d. On the author see the article "Ibn Ḳayyim" by Henri Laoust in EI_2.

I have not investigated Shīʿite legal texts beyond the influential work entitled *al-Kāfī* by the Baghdad scholar al-Kulaynī (d. 939).[9] (Here it should be noted that various opinions attributed to ʿAlī ibn Abī Ṭālib and Jaʿfar al-Ṣādiq occur in the astronomical sources – see especially **2.2**.) al-Kulaynī (**Text 3.5**) records that some followers of ʿAlī reported that they used to measure shadow lengths in Medina in cubits (*kunnā naqīs al-shams bi-'l-Madīna bi-'l-dhirāʿ*) until Abū ʿAbdallāh, that is, Jaʿfar al-Ṣādiq, told them he had a better way, namely, after the sun started to decline, the time for the *ẓuhr* could begin when one had performed a superogatory prayer (*subḥa*), that could be as long or as short as one wished. (al-Aṣbaḥī, on the other hand, records some observations of the midday shadow at Basra for each month of the year, which he says were made by ʿAlī: see **III-2.2**.) In another story related by al-Kulaynī a follower asks Abū ʿAbdallāh, again Jaʿfar al-Ṣādiq, about reports that one should perform the *ẓuhr* when the shadow was a man's height or twice that, one cubit or two, one foot or two. How could this be (*fa-matā hādhā wa-kayfa hādhā*), when the shadow might only be half a foot? It is clear from this and other stories related by him that al-Kulaynī knew of the story involving the Archangel Gabriel, although he did not include it. I am not aware of any other definitions involving two cubits, or one or two feet.

3.2 al-Bīrūnī on the opinions of the *Imāms*

In addition to what the *Imām*s and their students wrote themselves, it is useful to consider what later authorities thought they advocated. Unfortunately the unique manuscript of al-Bīrūnī's *Shadows* contains a lacuna in his discussion of the times of the *ẓuhr* and the *ʿaṣr* (**Text 3.6**).[10] A translation of the surviving part of his text follows:

> "[The first daylight prayer is the *ẓuhr* ... ... (*lacuna*!)]. (The authorities) have differed on what was preferable. al-Shāfiʿī was of the opinion that one should prostrate oneself (*al-taqlīs*) for a little while (*naẓr*[an]) (whilst the shadow was ??) shorter than one cubit. *But (I say that) the cubit must be the increase over (the shadow at) midday, otherwise the whole business is messed up because of the (different situations prevailing at) different seasons and localities.*
>
> The first point of view corresponds better to the above-mentioned report (about Gabriel) and is more pertinent to it in the sense that no midday shadow is mentioned for the *ʿaṣr* along with the one or two gnomon lengths (*mithl*). It is thus clear that (the definition of) the time requires the moving of the shadow (from one side of the meridian to the other) and its fading away (to zero) and that when it had become equal to the amount of (the width of) the thong of a sandal this was an indication of the decline (of the sun from the meridian). If (the midday shadow) had been of any (substantial) magnitude, this would have been mentioned along with the one or two

---

[9] Abū Jaʿfar Muḥammad ibn Yaʿqūb ibn Isḥāq al-Kulaynī, *al-Furūʿ min al-Kāfī*, ed. ʿAlī Akbar al-Ghaffārī, 8 vols., Tehran, 1381 Hijrī Shamsī (= 1961), III, esp. pp. 276 and 277. On the author see p. 24b of the article "Ḥadīth" by James Robson in EI_2 and article "al-Kulaynī" by Wilferd Madelung, as well as Sezgin, *GAS*, I, pp. 540-542. On Jaʿfar al-Ṣādiq see n. 2:10.

[10] al-Bīrūnī, *Shadows*, pp. 168-169; Kennedy, *al-Bīrūnī's* Shadows, I, pp. 218-220, and II, pp. 137-139.

> gnomon lengths. *In the majority of localities the midday (shadow) is more than one gnomon length on most days of the year.* If it is not mentioned then midday would be the beginning of the time for the *ʿaṣr*; indeed, the time for the *ẓuhr* would become invalid since there would be no end-point for it, as there should be in the best-known (*mashhūr*[11]) versions on the authority of Abū Ḥanīfa. (These state that the end of the *ẓuhr* is) when the shadow of any object is twice its length plus (*baʿd*) its midday shadow. There are other reports (that he advocated) one length instead of two (for the end of the *ẓuhr*); this is also the opinion of Abū Yūsuf, Muḥammad al-Shaybānī, and al-Shāfiʿī.
>
> The second (daylight) prayer is the *ʿaṣr*. It begins at the end of the time for the *ẓuhr*. Hence it begins according to the best-known versions on the authority of Abū Ḥanīfa when the shadow is twice the gnomon length plus the midday shadow, or, according to the other version and in the opinion of Abū Yūsuf, Muḥammad (al-Shaybānī), and al-Shāfiʿī, when it is equal to (the gnomon) (plus the midday shadow ??). Some versions on the authority of Abū Ḥanīfa state that if the shadow is equal to the gnomon plus the midday shadow then the time for the *ẓuhr* is past, and that the time for the *ʿaṣr* does not begin until the shadow reaches twice the gnomon length plus the midday shadow. But this is not an accepted tradition.
>
> The time for the *ʿaṣr* according to al-Shāfiʿī lasts until the shadow (reaches) twice the length of the gnomon. Whoever delays (praying) beyond this time (may do so): (there enters) the optional time (for the *ʿaṣr*) (*waqt al-ikhtiyār*). It is reported on the authority of ʿAṭāʾ, Ṭāwūs and Mālik that the end of the *ʿaṣr* is clearly revealed in the Qurʾānic verse (XVII.78): "Perform the prayer from the sun's decline until the darkness of night (*ghasaq al-layl*)." (They assert that it is acceptable) to delay the (*ʿaṣr*) prayer until dark (*ẓalām al-layl*), so that the interval from the moment when the shadow has increased by one gnomon length until sunset belongs to both the *ẓuhr* and the *ʿaṣr*, and the time before that is exclusively for the *ẓuhr*."

We learn from this that Abū Ḥanīfa was also reported to have advocated $\Delta z = 2n$ for the time of the *ẓuhr* and $\Delta z = n$ for the time of the *ʿaṣr*. In view of the inconsistency of these reports it is comforting that the *Qurʾān* itself allows for prayer the entire afternoon.

3.3 The testimony of Sirāj al-Dīn

Another commentator, Sirāj al-Dunyā wa-'l-Dīn, who wrote a treatise on folk astronomy in 1210 (**Texts 3.7** and **3.8**),[12] argued that the phrase "when the shadow became like the thong

[11] The term *mashhūr* also has a technical meaning in the *ḥadīth* literature (EI_2, III, p. 25b) but that does not seem to apply here. It should be borne in mind that al-Bīrūnī is quoting different versions (*riwāyāt*) of the basic legal works; on that concept in literature see the article "Riwāya" by Stefan Leder in EI_2.

[12] On this author see already **III-9.5**. A unique copy, alas defective, of his treatise is contained in MS Princeton Yahuda 4657 (fols. 40v-80r, copied *ca.* 1350) – see *Princeton Catalogue*, no. 4983. The authorship is indicated only in a spurious title *Mukhtaṣar al-hayʾa wa-'l-mafātīḥ fī ʿilm al-awqāt wa-'l-maṣābīḥ li-'l-Sirāj*. This work is related to another entitled *Maṣābīḥ al-anwār wa-mafātīḥ al-asrār fī aʿmāl al-layl wa-'l-nahār* attributed to Sirāj al-Dunyā wa-'l-Dīn, extant in MS Cairo ṬM 127 (59 fols., copied 1839/40): see *Cairo ENL Catalogue*,

of a sandal" in the *ḥadīth* implied that the midday shadow was of zero length, which, as he says, is possible at certain times of the year in localities in the tropics. In order to make the definitions valid for other localities the midday shadow was incorporated into the definitions. The author's discussion is somewhat unconvincing, but he continues to describe how to set up a vertical gnomon and regulate the *ẓuhr* and *ʿaṣr* prayers, the first beginning when the smallest observable increase (*adnā ziyāda*) in the shadow is observed after midday and ending when $\Delta z = n$, and the latter beginning when the smallest observable increase occurs thereafter. He attributes these definitions to various early legal authorities as follows (**Text 3.7**):

> "This is the opinion of al-Shāfiʿī ... , Mālik, al-Awzāʿī, Sufyān (al-Thawrī), Aḥmad (ibn Ḥanbal), Abū Yūsuf (Yaʿqūb ibn Ibrāhīm), and Muḥammad (al-Shaybānī) ... . (ʿAbdallāh) Ibn Mubārak and Isḥāq (= Ibn Rāhawayh ?) said "after the shadow of any object becomes the same as itself by the amount of four genuflections at the beginning of the times of both the *ẓuhr* and *ʿaṣr* prayers." (Abū Jaʿfar) Muḥammad ibn Jarīr al-Ṭabarī and Mālik said "after the shadow of any object becomes the same as itself until it becomes twice itself ... ." Abū Ḥanīfa ... said "the time of the *ẓuhr* (lasts) until the shadow of any object becomes twice itself apart from the shadow at midday, then the time of the *ʿaṣr* begins." According to al-Shāfiʿī the time of the *ʿaṣr* does not last beyond (the time) when the shadow of the wood(en gnomon) is twice itself apart from the amount when the sun declines, and if (the shadow) passes that and increases by the least amount, the preferred (*mukhtār*) time of the *ʿaṣr* has finished according to al-Shāfiʿī ... ."

According to this late source, several early authorities, including al-Shāfiʿī and al-Shaybānī, advocated the definitions:

$$\Delta z = \delta \quad (\delta << n) \quad \text{and} \quad \Delta z = n$$

for the *ẓuhr* and the *ʿaṣr*. Also Abū Ḥanīfa *himself* (rather than al-Shaybānī) is supposed to have advocated the definitions:

$$\Delta z = n \quad \text{and} \quad \Delta z = 2n$$

for the *ẓuhr* and the *ʿaṣr*. The information about al-Shāfiʿī's opinion concerning the time when $\Delta z > 2n$ contradicts the report of al-Bīrūnī cited in **3.2**.

3.4 An anecdote about the duration of the *ʿaṣr*

We conclude this discussion with a story recorded in several of the canonical Sunnī *ḥadīth* collections (**Text 3.9a**):[13]

I, *sub* ṬM 127, and II, 3.2.4; as well as *Cairo ENL Survey*, no. C4. However, the passages on the prayer times in these two works are different: *cf.* **Texts 3.7** and **3.8**. The author is probably the celebrated Ḥanafī legal scholar Sirāj al-Dīn Muḥammad ibn Muḥammad al-Sajāwandī (on whom see the article "al-Sadjāwandī" by Rudolf Sellheim in EI_2) who lived around 1200 and who authored a treatise on algebra and arithmetic: see *Cairo ENL Survey*, no. G11.

The Princeton manuscript contains a remarkable world-map with a grid – see King, *Mecca-Centred World-Maps*, pp. 90-91 – and a record of the first determination of the length of 1° on the meridian compiled by the *qāḍī* Yaḥyā ibn Aktham commissioned by the Caliph al-Maʾmūn to oversee the project – see *idem*, "Earliest Muslim Geodetic Measurements".

[13] Wensinck *et al.*, *Concordance* (n. 1:4), V, p. 365. The story appears to be an Islamic echo of the parable of the labourers in the vineyard recorded in the Gospel of Matthew (20.1-16). It first appears in the *Muwaṭṭaʾ*

"The Prophet said to his followers: With you and the people of the two Books it is like a man who was hiring some day-labourers. He asked: "Who will work for me from morning (*ghadwa*) until midday (*niṣf al-nahār*) for one *qīrāṭ* (each)?" and the Jews worked for him. Then he asked "Who will work for me from midday until the *ʿaṣr* prayer for a *qīrāṭ* (each)?" and the Christians worked for him. Then he asked "Who will work for me from the *ʿaṣr* until sunset for two *qīrāṭ*s (each)?" and (addressing the Muslims,) he said "You are they." The Jews and the Christians got angry and said: "We worked longer and received less." He said: "Did I treat you unjustly with regard to the pay?" They said: "No." He said: "That is a benefit I bestow on whom I wish."

The 16th-century Ḥanafī legal scholar Ibn Nujaym wrote a treatise on the times of the *ʿaṣr* and the *ʿishāʾ*.[14] Commenting on this story (**Text 3.9b**), he remarked that the Christians would have worked longer only if the *ʿaṣr* was defined according to the definition associated with Abū Ḥanīfa ($\Delta z = 2n$). He also asserted that anyone who claims that the interval from midday until when $\Delta z = 2n$ is longer than the interval from when $\Delta z =$ [2n, apparently missing in the text] until sunset in order to prove that the Christians worked longer, should bear in mind that the difference between these intervals is small and only known to those versed in computation (*al-ḥussāb*). Ibn Nujaym is thus using his own ignorance of astronomy to defend Abū Ḥanīfa's definition. In the *ḥadīth* as recorded in the manuscript I have consulted the payment for the third interval is apparently given as two *qīrāṭs* rather than one so that the Jews and the Christians are made to appear to have had a particularly raw deal.

We have seen that the material in the legal texts is somewhat far from presenting a unified picture. Nowhere are the seasonal hours mentioned, and for this reason at least I have considered the possibility that the standard definitions were introduced without any thought for relating the prayers to the seasonal hours. In my opinion, however, the evidence presented in **Chs. 2** and **4-8** speaks against this. The early legal scholars were not sufficiently knowledgeable of mathematical astronomy[15] to realize that definitions of the form $z = x$ could be made universal by modifying them to $\Delta z = x$. Their own utterances cited above prove this. The definitions of the form $\Delta z = x$ *must* have been suggested by someone or some group of people who realized their utility. In Baghdad both al-Shaybānī and al-Shāfiʿī could have come into contact with people familiar with the Indian rule expounded several decades previously by al-Fazārī.

---

of Mālik in the *riwāya* of al-Shaybānī (*bāb al-tafsīr*) – see the EI_2 article "Indjīl" by Bernard Carra de Vaux updated by Georges C. Anawati, esp. p. 1206a-b. There is an allusion to the story in al-Bīrūnī, *Chronology* (see p. 9, and Sachau's notes on p. 369). I here quote Ibn Nujaym – see the next note.

[14] On the author see Brockelmann, *GAL*, II, p. 401, and SII, p. 426 (esp. no. 5). I have used notes made in the 1970s from MS Cairo ṬJ 200,5 (fols. 27r-30r, copied 1733/34), listed in *Cairo ENL Survey*, no. C104.

[15] This is not to say they were not familiar with the kind of non-numerical folk astronomy represented by, say, the *Kitāb al-Anwāʾ* of Ibn Qutayba (see nn. 1:15 and 8:11).

CHAPTER 4

OTHER EARLY DEFINITIONS OF THE TIME OF THE *ẒUHR* PRAYER

4.1 The thong of a sandal

In the time of the Prophet Muḥammad it was apparently already established that the *ẓuhr* prayer should not begin when the sun was actually on the meridian. Indeed Islamic tradition appears to be unanimous that there are three times of day, sometimes called *sāʿāt*, when prayers should not be performed, namely sunrise and sunset, and midday when the sun is culminating on the meridian (**9.3**). The prayer of the *ẓuhr* was therefore made to begin when the sun had started to decline, and the smallest perceivable increase in the shadow would establish that the sun had indeed crossed the meridian. In the *ḥadīth* about the Archangel Gabriel (**1.2**) a shadow of the width of a thong of a sandal is mentioned. An attempt to be more specific is recorded by al-Aṣbaḥī in a statement (**Text 4.1**) traced to the Companion ʿAbdallāh ibn Masʿūd:[1]

> "Some have estimated (the width of) the thong of the sandal to be about two digits, which is a sixth of a foot."

4.2 The Prophet and the *ẓuhr*

Whilst we are discussing the *ẓuhr* it is appropriate to consider another story associating the regulation of this prayer by means of shadow lengths with the Prophet himself. This *ḥadīth* is recorded by al-Aṣbaḥī and is traced back to the same Companion Ibn Masʿūd (**Text 4.2**):[2]

> "The length of the shadow when the Prophet – may God bless Him and grant Him salvation – prayed the *ẓuhr* was from five feet to seven in the winter and from three feet to five in the summer."

Let us investigate these shadows. It makes little difference whether we take the length of the gnomon (the height of a man) as 6 or 7 feet, or whether we take the latitude of Mecca (≈ $21^1/_2$°) or Medina (≈ $24^1/_2$°). For convenience we take the latitude of the tropic (≈ $23^1/_2$°) and base n = 7. Then the midday altitudes at midwinter, midday and midsummer are respectively:

$$90° - \phi - \varepsilon = 43° , \quad 90° - \phi = 66° , \quad \text{and} \quad 90° - \phi + \varepsilon = 90° ,$$

and the corresponding shadows are:

$$7^1/_2 , \quad 3 , \quad \text{and} \quad 0 .$$

Thus it is not possible that the Prophet could have prayed the *ẓuhr* throughout the winter at

---

[1] al-Aṣbaḥī, *al-Yawāqīt*, fol. 8r.

[2] *Ibid.*, fols. 2v-3r, repeated on fol. 13v (**Text 7.6**), also Robson, trans., *Mishkāt al-maṣābīḥ* (see n. 1:4), I, p. 120. On ʿAbdallāh ibn Masʿūd see the article "Ibn Masʿūd" by J.-C. Vadet in EI_2.

the time when the shadow was increasing from 5 to 7 feet. (It would be possible for, say, Kufa.) It may be that we should not consider this report as anything more than an attempt to justify the postponement of the *ẓuhr* in the summer months. However, if the winter midday shadow were 5 feet, then the 2-feet ($\approx {}^2/_7$n) increase would correspond to about one hour of time. Likewise, since the summer midday shadow is zero, the 3-feet shadow ($\approx {}^1/_2$n) would correspond roughly to a time two hours after midday, and the 5-feet shadow ($= {}^5/_7$n) would correspond roughly to a time half an hour thereafter. I do not think it is profitable to speculate further on the possible significance of the shadows cited in this *ḥadīth*.

4.3 Specific small shadow increases for the *ẓuhr*

Instances in which astronomers went so far as to mention a small minimum value of Δz ($< {}^1/_4$ n) for the *ẓuhr* are rare, and I can cite only one example. al-Fārisī stated (**Text 4.3**) that the *ẓuhr* begins when the shadow increase is ${}^1/_2 \times {}^1/_6$ ($= {}^1/_{12}$) of a foot (*niṣf suds qadam*) and ends when it is 7 feet, and that this is also the time of the beginning of the *ʿaṣr*, which ends when the shadow is 14 feet.[3] These three values of Δz can be expressed as:

$${}^1/_{12}\ {}^n/_7 \quad , \quad n \ , \quad \text{and} \quad 2n \quad (n = 7)\ .$$

The first definition is of particular interest, and is not explicitly attested in any other non-Yemeni sources known to me, although it is found in various late Yemeni astronomical sources.[4] It probably results from an attempt to divide the length of a foot by the standard division for the span of the hand into 12 digits, and then taking one of these "digits" to represent a very small amount.[5]

4.4 Larger shadow increases for the *ẓuhr*

In one source it seems that we can associate the call to prayer for the *ẓuhr* with the definition $\Delta z = {}^1/_5$n and the *ẓuhr* itself with the definition $\Delta z = {}^1/_2$n. In the *Shadows* al-Bīrūnī relates (**Text 4.4**) that some people performed the prayer-call for the *ẓuhr* when the shadow had increased by one digit (*iṣbaʿ*), the *ẓuhr* prayer when the increase was 3 digits, and the *ʿaṣr* when the increase was 13 digits.[6] He states that they are ignorant of the nature of digits, and offers an unsatisfactory explanation of their practice. However, the shadow scheme makes some sense if one substitutes the unit *qadam*, base $6^1/_2$, for *iṣbaʿ*, base 12. Indeed the values:

$$\Delta z \quad = \quad 1, \quad 3, \quad \text{and} \quad 13$$

are the integral values corresponding to 1, 2, and 4 seasonal day hours after midday using the Indian formula and base $6^1/_2$. For the *ẓuhr* with $\Delta z = {}^1/_4$n (see below) rather than $\Delta z = {}^1/_5$n we should have $1^1/_2$ or 2 *qadam*s.

---

[3] On al-Fārisī see n. 1:23. This passage occurs in MS Milan Ambrosiana X73 = Griffini 37, fols. 22r-22v.

[4] For example, in the *zīj* entitled *Zād al-musāfir* by al-Daylamī (King, *Astronomy in Yemen*, no. 28), and the *zīj* entitled *Ghāyat itqān al-ḥarakāt* by ʿAbdallāh al-Sarḥī (*ibid*., no. 37).

[5] Note that if the value of Δz were ${}^1/_{12}$n rather than ${}^1/_{12}$ of a foot, then the time would correspond roughly to a time one-half of a seasonal hour after midday using the Indian formula. In fact for $T = 5^1/_2$, $\Delta z = {}^1/_{11}$n using the formula, so that ${}^1/_{12}$n would be a nice approximation. But these considerations appear to be irrelevant.

[6] al-Bīrūnī, *Shadows*, p. 185, and Kennedy, al-Bīrūnī's *Shadows*, I, pp. 235-236, and II, p. 147.

We now investigate the factor $^1/_4$ in the definition of the time of the *ẓuhr* that became standard in Andalusī and Maghribi practice.[7] The Indian formula gives a factor $^1/_5$ for the first hour after midday, but I think that the factor is not adequately explained as an approximation to this, even though it is more convenient for use with a gnomon length of 12 units since 4 divides evenly into 12 and 5 does not.[8]

Rather, perhaps, the explanation is as found in a late Fatimid or early Mamluk Egyptian (12th or 13th century?) work on folk astronomy: the anonymous author (**Text 4.5**) states that $\Delta z = 3$ digits ($= {}^1/_4 n$ for $n = 12$) results from the general agreement (*ijtimāʿ al-nās*) that the shadow should increase by a cubit (*dhirāʿ*).[9] A cubit or the length of a forearm is indeed one-quarter of the height of a man.

The association of the time of the *ẓuhr* with a shadow increase of one cubit seems to be a very early development. al-Bīrūnī, and also al-Aṣbaḥī, relates that the Caliph ʿUmar ibn al-Khaṭṭāb wrote to his governors in the 17th year of the Hijra (**Text 4.6**):[10]

> "Pray the noon prayer from the time when the increase of the shadow (*fayʾ*) is a cubit up to the time when the shadow (*ẓill*) of anyone of you is equal to (his height), and the afternoon prayer when the sun is still high and a pure white (and there remains a time till sunset) such that a rider can still travel two or three *farsakh*s ... ."

However, al-Bīrūnī notes that the Caliph also wrote to the Companion Abū Mūsā al-Ashʿarī:

> "Pray the noon prayer when the sun has declined, and the afternoon prayer when the sun is still a pure white, before any yellowness begins to appear on it ... ."

By citing an Egyptian work containing the definition $\Delta z = {}^1/_4 n$ for the *ẓuhr*, I have shown that this definition was not entirely restricted to the Islamic West. However, I have found it explicitly in only one other Eastern Islamic astronomical source, namely, another 12th(?)-century Egyptian treatise on timekeeping by an astronomer named Ibn al-Ḥammāmī (**Text 4.7**).[11] The definition is not mentioned at all by al-Bīrūnī in his chapter on the prayer-times in the *Shadows*.

A remark by al-Bazdawī, a prominent 11th-century Ḥanafī legal scholar of Samarqand, stating (**Text 4.8**) that his community there performed the *ẓuhr* in the summer when the shadow of an object was one-half the length of the object,[12] could be interpreted as corresponding to

---

[7] Wiedemann & Frank, "Gebetszeiten", pp. 761-762 (citing the 13th-century Andalusī *Libros del saber*); King, "Three Sundials from Islamic Andalusia" (on markings for the *ẓuhr* based on this definition); *idem*, "Astronomy in the Maghrib", pp. 38-39; and now **II-13.2-3** (on various Maghribi tables, notably those in MS Berlin Ahlwardt 5724, based on this definition). See also Carandell, *Ibn al-Raqqām sobre los cuadrantes solares*, pp. 163-169.

[8] Thus the remarks in King, "Tunisian Sundial" (n. 0:3), p. 194, are to be modified.

[9] MS Oxford Bodley 133 (copied 1333/34), fols. 124v and 125v. On this work see **II-6.1**, **III-9.3** and **V-5**.

[10] On the Caliph see n. 3:3a above. al-Bīrūnī, *Shadows*, p. 162, and al-Aṣbaḥī, *al-Yawāqīt*, fol. 3v. See also Robson, trans., *Mishkāt al-maṣābīḥ* (n. 1:4), I, p. 120.

[11] MS Istanbul Haci Mahmud Ef. 5713 (copied in the 12th century), fols. 3v and 19r-19v. The importance of the manuscript depends very much on the date of copying, since it contains the earliest known treatise on the use of the astrolabic quadrant, yet the author makes no claim to have invented the instrument (see **XIIa**).

[12] MS Cairo B 19385 (12 pp., copied in 1936 from a manuscript then in Sohag but now lost), esp. p. 11 – see King, "al-Bazdawī on the Qibla in Transoxania", pp. 11 and 20. On al-Bazdawī see Brockelmann, *GAL*, I, p. 460 and S I, pp. 637-638 (where this treatise is not mentioned), and *Cairo ENL Survey*, no. B88.

the definition $\Delta z = {}^1/_4 n$, since the midday shadow in Samarqand at midsummer is itself about ${}^1/_4 n$. However, this remark should not be taken as evidence that the definition $\Delta z = {}^1/_4 n$ was actually used in Samarqand since al-Bazdawī does not mention the practice at any other time of the year.

4.5 Other technical definitions

When discussing the time of the *ẓuhr*, most Muslim astronomers outside al-Andalus and the Maghrib state simply that it begins when the sun begins to decline from the meridian (*al-zawāl*). Rarely do they apply strictly astronomical conditions to the beginning of the *ẓuhr*. Here we note four exceptions.

We first consider the definition of the time of the *ẓuhr* in a treatise on the use of the astrolabe by the early-9th-century astronomer al-Khwārizmī.[13] The author states (**Text 4.9**) that the *ẓuhr* begins when the altitude of the sun has decreased by 7°, giving no explanation whatsoever for this definition.[14] I have not found such a definition for the *ẓuhr* anywhere else in the medieval Arabic literature. However, if we take the solar midday altitude at Baghdad at the equinoxes and two solstices and compute the gnomon shadows,[15] add one-fifth of the length of the gnomon

[13] On Muḥammad ibn Mūsā al-Khwārizmī (**I-7.1.1**) see the article by Gerald Toomer in *DSB*. MS Berlin Ahlwardt 5793 (fols. 77v-97r, copied *ca.* 1500) contains two treatises by him, one on the construction of the instrument which has been investigated in King, "al-Khwārizmī", pp. 23-27, and the other on the use of the instrument which has been translated in Frank, "al-Khwārizmī über das Astrolab" (Although in addition to an explicit reference to Baghdad (p. 7) the latitude 37° is implicitly mentioned with that of Kufa in an example (p. 14), there can be little doubt that the entire treatise is due to al-Khwārizmī; in other words, the reservations on the authenticity of these passages expressed by Carl Schoy in a review published after the reprint (pp. 160-161) are unfounded. Likewise Schoy's contention that "... vor 1300 n. Chr. kaum Schriften über Gebetszeiten verfaßt sind ..." is incorrect. al-Khwārizmī's various treatises on instruments are edited with translation and commentary in a forthcoming study by François Charette and Petra Schmidl.

[14] *Ibid.*, p. 14, translated from MS Berlin Ahlwardt 5793, fol. 89v. For a discussion of the curious definition of the *ʿaṣr* in the same text see **II-3.1**. See also Wiedemann & Frank, "Zirkel zur Bestimmung des Gebetszeiten", on the instrument described by al-Khwārizmī for determining the *ʿaṣr*, using the definition $\Delta z = n$.

[15] For latitude $\phi = 33°$, the value used by al-Khwārizmī for Baghdad, and obliquity $\varepsilon = 23;51° \approx 24°$ also used by him (see King, "al-Khwārizmī", p. 2), we have solar meridian altitudes as follows:

SS: 81° EQ: 57° WS: 33°

with corresponding shadows to base 12:

1;54 7;47 18;29

We now add 1/5 n or 2;24 and obtain shadows:

4;18 10;11 20;53

for which the corresponding altitudes are:

70° 50° 30°

so that the solar altitude has decreased by amounts:

11° 7° 3°

If we had added 1/4 n rather than 1/5 n we should have obtained the following decreases in solar altitude:

13° 9° 4°.

On the next folio of the Berlin manuscript (**Text 4.9b**) a different definition of the end (?) of the time of the *ẓuhr* is given: see Frank, "al-Khwārizmī über das Astrolab" (n. 4:13), p. 15. The method is to find the altitude of the sun when its azimuth is 14° or 54° from the west-point (the Arabic is unpointed and the value can be either 14° or 54°). I derive an azimuth of 52° from west for the sun with altitude 50° at latitude 33° at the equinox, but it does not seem fruitful to pursue the investigation further since the text is obviously corrupt.

– not one-quarter – and then convert back to solar altitude, we find that the solar altitude has decreased by the following amounts at the first seasonal hour:

SS: 11° EQ: 7° WS: 3° .

Notice that not only is the value for the equinox 7°, but also the mean of solstitial values is 7°, although this is probably less significant. These calculations demonstrate that al-Khwārizmī's curious definition for the *ẓuhr* is intended to define a time one seasonal hour after midday using the Indian formula. His table for the shadow length at the *ẓuhr* is not based on this approximation (but rather on another that has yet to be explained): see **II-3.1**.

al-Khwārizmī's other writings on the *ẓuhr* and the *ʿaṣr* include a description of a simple scale for measuring shadows together with a table displaying the shadow lengths at the *ẓuhr* and at the beginning and end of the *ʿaṣr*.[16] Values in this table are given for each 6° of solar longitude (corresponding roughly to each six days of the year), and are computed specifically for the latitude of Baghdad (taken as 33°). They correspond roughly, but not with any convincing consistence, to the definitions:

$$\Delta z = 0, \quad n, \quad \text{and} \quad 2n.$$

The table remains to be properly explained.

The late-10th-century astronomer al-Khujandī of Rayy advocated a curious definition for the *ẓuhr* in his treatise on the universal dial (*al-āla al-shāmila*).[17] He implies (**Text 4.10**) that it begins after midday by 3°, that is, $^1/_5$ of an equinoctial hour. The motivation for this particular definition eludes me; for $T = 6 - {}^1/_5 = 5^4/_5$, the Indian formula yields $\Delta z = {}^1/_{30}n$, which does not seem relevant, and the interval of one-fifth of an hour is more than adequate to ensure that the sun has moved past the meridian – see below. The text continues with a description of how to find the altitude of the sun at the end of the *ẓuhr* and the beginning of the *ʿaṣr* ($\Delta z = n$).

As a third example of an attempt to be specific about the small interval of time after midday for the performance of the *ẓuhr* I quote the author of an anonymous Maghribi or Andalusī treatise on the use of the astrolabe (**Text 2.8**).[18] The author prescribes the use of the approximate Indian formula (for both n = 12 and 7), and then discusses the prayer-times and states that "the time of the *ẓuhr* is when the sun has declined by the amount of one degree", which implies that the solar altitude should decrease by one degree. However, the author continues "and six hours have passed by the falling of the point opposite the solar longitude (on the ecliptic of the rete) on the first degree of the seventh (hour)", which implies that one equatorial degree (4 minutes) of time should have elapsed since midday. Here again we have a practical definition intended to ensure that the sun has indeed crossed the meridian.

Finally, we note the definition for the *ẓuhr* given by a Syrian astronomer of the 14th century,

[16] See Wiedemann & Frank, "Zirkel" (n. 4:14); King, "al-Khwārizmī", pp. 7 and 9; and **II-3.1**.

[17] On al-Khujandī see Sezgin, *GAS*, V, pp. 307-308, and VI, pp. 220-222, and *Cairo ENL Survey*, no. B50. Five copies of this work are known, of which I have examined MSS Oxford Hunt. 566 = Uri no. 970 (128 fols., two hands, original *ca.* 1200, additions *ca.* 1500), Birmingham Mingana 560 (35 fols., 13th century), and Cairo DM 970 (42 fols., copied *ca.* 1750), in which the text occurs on fols. 38v, 16v and 14r, respectively. This treatise is one of the most important works on instrumentation from the entire Middle Ages and awaits detailed investigation. The author was the leading instrument-maker of his age; for a spectacular astrolabe made by him in 374 H [= 984/85] see **X-4**.

[18] MS Cairo DM 1169,6, esp. fols. 57r-57v – see n. 2:19.

Sharaf al-Dīn al-Khalīlī.[19] In his treatise on the use of the astrolabe (**Text 4.11**), al-Khalīlī observed:

> "If the interval for half the disk of the sun to cross (the meridian), namely, one-third of a degree, is added to the time of astronomical midday, the resulting time will be the time of midday according to the religious law."

This means that al-Khalīlī was considering the maximum apparent diameter of the sun to be $^2/_3$°. But he clearly intended to ensure that the entire disk of the sun had crossed the meridian, and since the apparent diameter of the sun is about 1°, it would be tempting to restore *thulth* (= $^1/_3$) to *thulthā* (= $^2/_3$). Against this emendation, however, we can cite two anonymous, roughly contemporary, treatises on quadrants: in these the *ẓuhr* is said to begin when the sun has moved past the meridian by "at least $^1/_4$° but no more than $^1/_3$°" and "about $^1/_3$°", respectively.[20]

---

[19] On Sharaf al-Dīn al-Khalīlī see Suter, *MAA*, no. 427, and *Cairo ENL Survey*, no. C38. I have used MS Princeton Yahuda 1168 (= Mach 4913) of his treatise on the use of the astrolabe.

[20] Wiedemann & Frank, "Gebetszeiten", p. 5.

CHAPTER 5

THE TIME OF THE *ḌUḤĀ* PRAYER

5.1 The *ḍuḥā* in the *ḥadīth*

A third day-time prayer is mentioned in the *ḥadīth* literature and is referred to in isolated historical and astronomical texts from the medieval period. This is the prayer called *al-ḍuḥā* and it occurs at mid-morning. (The oath on the *ḍuḥā* in *Qur'ān* 93.1, refers to the light of mid-morning.) The *ḍuḥā* is in fact a pre-Islamic institution which was adopted by the first Muslims, but was later replaced by the *fajr* prayer: see **1.1***. Some of the canonical Sunnī *ḥadīth* make Muḥammad recommend the *ḍuḥā* and perform it regularly, and even state that it was a *farīḍa* or religious obligation for him, and *sunna* or usage sanctioned by tradition for the Muslim community.[1] Others say that Muḥammad performed this prayer only once or that the authority in question saw him do it once, or that he performed it only upon returning from a journey. Some of the early authorities such as the Caliphs Abū Bakr and ʿUmar did not perform the *ṣalāt al-ḍuḥā* and the latter's son, a *ḥadīth* specialist, went so far as to call it *bidʿa*, heretical innovation. In one Shīʿī *ḥadīth* recorded by al-Kulaynī, the Prophet himself declares the prayer *bidʿa* (**Text 5.1**).[2]

Ibn Ḥanbal, who is one of the main authorities asserting that Muḥammad performed the *ḍuḥā* regularly, records a *ḥadīth* that the Prophet used to perform the *ṣalāt al-ḍuḥā* when the sun had risen the same distance from its rising point as it is distant from its place of setting at the *ṣalāt al-ʿaṣr*. This is hardly a working definition, but is precisely the one found in some later astronomical sources.

5.2 The *ḍuḥā* in early astronomical sources

This definition of the *ḍuḥā* is most clearly displayed in a 14th-century Tunisian sundial shown in **Fig. 5.2**. This instrument, made by al-Qāsim ibn Ḥasan ibn al-Shaddād in 746 H [= 1345/46], is unique amongst the known surviving sundials from medieval Islam in that it shows only the times of prayer, not the hours of daytime.[3] The markings on the sundial include a curve for each of the *ʿaṣr* and the *ẓuhr* (based on the definitions $\Delta z = n$ and $\Delta z = {}^{1}/_{4}n$) and a curve

[1] The information on the *ḍuḥā* recorded here is taken from A. J. Wensinck's article "Ṣalāt" in EI_1 (IV, pp. 100b-101a), suppressed in the new article in EI_2. In 2002 Gerald Hawting kindly alerted me to the important article by Uri Rubin, "Morning and Evening Prayers in Early Islam", *Jerusalem Studies in Arabic and Islam* 10 (1987), pp. 40-64. Here the interested reader will find much more material on the *ḍuḥā* in early Islamic practice, including references to the Prophet performing it in the early morning (*awwal-al-nahār*) and to its being later declared unlawful because of its closeness to sunrise. See n. 10:2 below.

[2] al-Kulaynī, *al-Kāfī* (n. 3:9), III, p. 453.

[3] See King, "Tunisian Sundial" (n. 0:3), for a detailed description of this sundial.

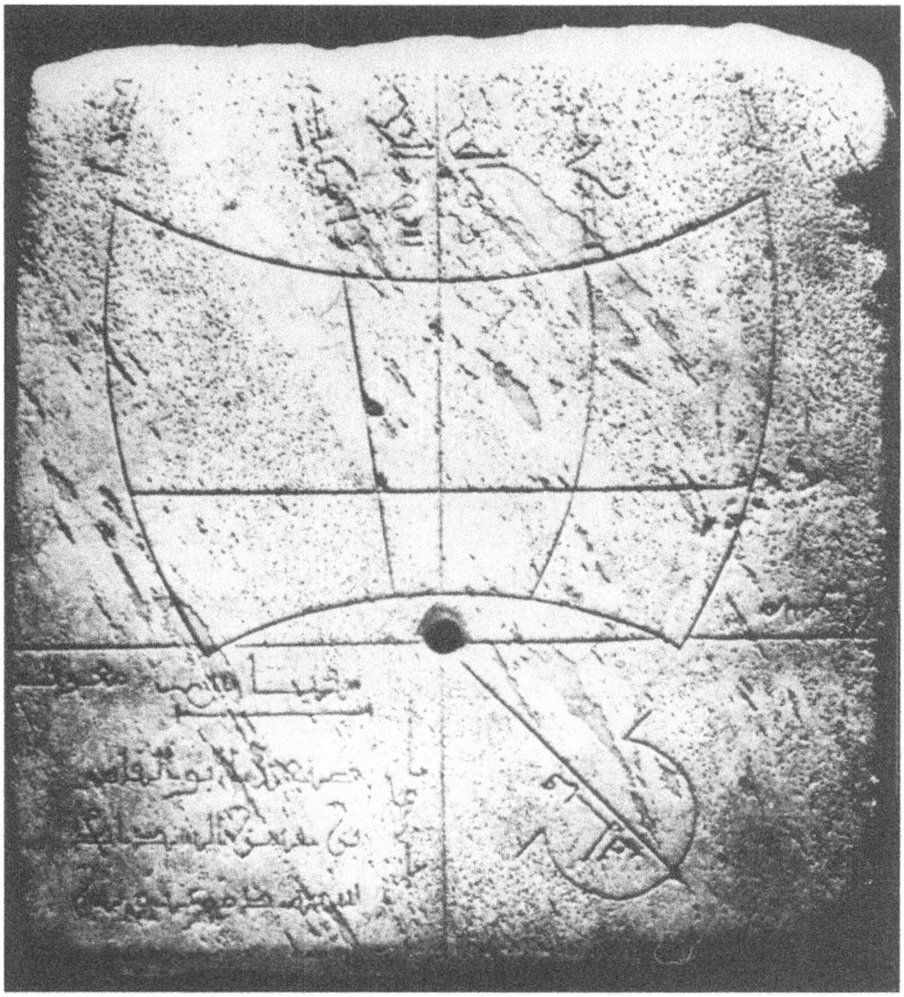

Fig. 5.2: A 14th-century Tunisian sundial showing markings for four times of day with religious significance (# 7341). These are (from left to right) the *ḍuḥā* (when $\Delta z = n$), the *taʾhīb* (one equinoctial hour before midday), the *ẓuhr* (when $\Delta z = {}^1/_4 n$), and the *ʿaṣr* (when $\Delta z = n$). [Carthage, Musée, photograph courtesy of the late Alain Brieux, Paris.]

for the *ḍuḥā* that is the mirror image of the *ʿaṣr* curve with respect to the meridian. Thus the time of the *ḍuḥā* as defined in the *ḥadīth* of Ibn Ḥanbal and as represented, for example, on the Tunisian sundial, corresponds to the definition $\Delta z = n$ or, using the Indian formula, to the beginning of the fourth seasonal hour of daylight or midway between sunrise and midday. al-Bīrūnī mentions a group of people who used the curve for the (end of) the third seasonal hour on astrolabe plates for determining the *ḍuḥā*: see further **7.2**.

The only astronomical text in which I have found this definition is an anonymous Algerian commentary on a poem on timekeeping by the late-14th-century Maghribi astronomer al-Jādarī.[4] The author states (**Text 5.2**) that:

[4] The treatise is extant in MS London B.L. Or. 411,2 (see also n. 6:12), and another copy, which lacks the various

"the solar altitude at the preferred time of prostration for the *ḍuḥā* is the same as the solar altitude at the beginning of the time of the *ʿaṣr* ... ."

As we shall see (**7.2**), al-Bīrūnī in the early 11th century refers to some people at an unspecified place and time who used the curve for the end of third hour on astrolabe plates to determine the *ḍuḥā*. Furthermore, the *ḍuḥā* prayer was practised in Egypt in the 10th century, as we know from a reference to the Fatimid Caliph al-Ḥākim by the 15th-century historian Ibn Taghrībirdī (**Text 5.3**).[5]

From the sources mentioned above we know that it was practised in the Maghrib in the 14th century. From an early astronomical source, which we discuss next, we know that there were three times called *ḍaḥwa* (= *ḍuḥā*) with religious significance in Anatolia or Iran some time between the 10th and the 14th century; here times of the *ḍaḥwa*s were determined not by shadow lengths but by equal divisions of the time between sunrise and midday.

5.3 An anomalous table for the prayers

The only astronomical context in which this simple division of the length of day-light without reference to shadows occurs is in a table preserved in two different works. The first source is the unique manuscript of the *Mufrad Zīj* of the 11th(?)-century astronomer Muḥammad ibn Ayyūb al-Ṭabarī, and the second is a unique manuscript of a 14th-century Anatolian astrological work.[6] Only further investigation might establish whether this table, which is in a different hand from the main part of this manuscript, dates back to al-Ṭabarī himself; in any case the date of copying of the table in the first manuscript precedes the compilation of the Anatolian work preserved in the second.

The table (see **Fig. 5.3**) displays for an unspecified latitude (about 38°) the longitude of the horoscopus to the nearest degree at ten different times of day and night for each six degrees of solar longitude. In the first manuscript the names are given in Arabic, while in the second they are given in Persian. Three of the times are related to twilight phenomena and are named as follows:

al-fajr al-kādhib	= *ṣubḥ-i kādhib*	false dawn (zodiacal light?)
al-fajr al-ṣādiq	= *ṣubḥ-i ṣādiq*	true dawn
maghīb al-shafaq	= *maghīb-i shafaq*	end of evening twilight.

tables for Tlemcen contained in the London manuscript, is MS Cairo K 4311 (*cf. Cairo ENL Catalogue*, I, *sub* K 4311, and II, 3.2.7). On al-Jādarī (**III-6.5**) see Suter, *MAA*, no. 424a; Renaud, "Additions et corrections à Suter", no. 424a; and *Cairo ENL Survey*, no. F26. The commentary on his poem is anonymous in both manuscripts, and so the attribution to Ibn al-Ḥabbāk (*cf.* Suter, *MMA*, nos. 424a and 435) is not certain.

[5] Jamāl al-Dīn Abu 'l-Maḥāsin ... ibn Taghrībirdī, *al-Nujūm al-zāhira fī mulūk Miṣr wa-'l-Qāhira*, 12 vols., Cairo: Ministry of Culture, 1348-75 H [= 1928-56], IV, p. 222.

[6] On al-Ṭabarī see Suter, *MAA*, no. 360; Krause, "Stambuler Handschriften", no. 360; Storey, *PL*, nos. 5 (pp. 3-4) and 79 (pp. 43-44); Sezgin, *GAS*, V, pp. 385-386 (inadvertently omitted from VI); and *Cairo ENL Survey*, no. B92. His *Zīj* (Kennedy, "*Zīj* Survey", no. 65) is preserved in MS Cambridge Browne Or. O.1 (180 fols., copied before 1300), where the table illustrated in **Fig. 5.2** occurs on fol. 179r. The anonymous Anatolian astronomical work (not mentioned in Krause, "Stambuler Handschriften") is contained in MS Istanbul Nuruosmaniye 2782, copied in Sivas in 1371/72, where the table occurs on fols. 53v-54r. See already King, "Tunisian Sundial" (n. 0:3), pp. 195-196, and **I-3.3.3** and **II-14.2**, on this table.

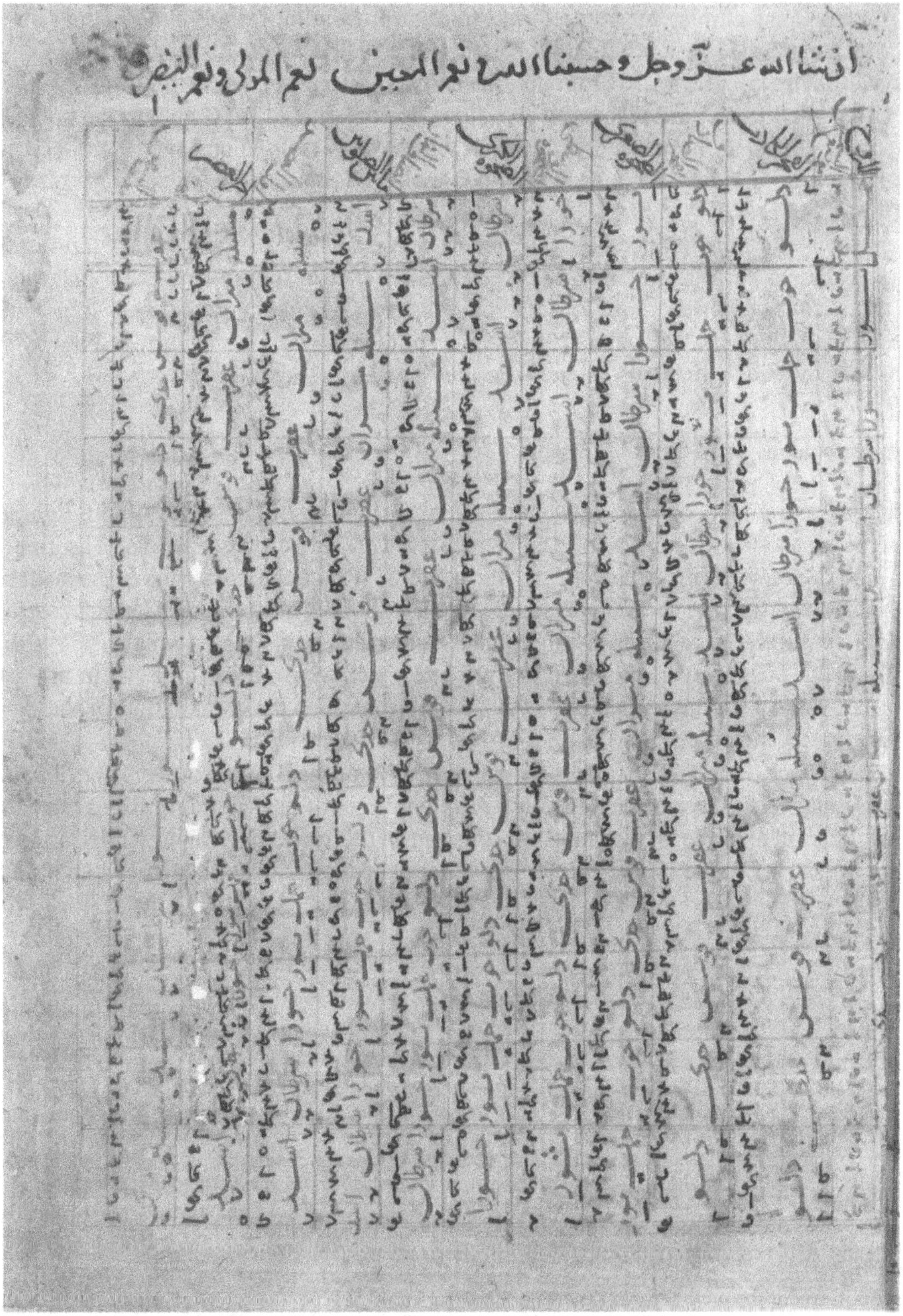
انشا الله عز وجل وحسبنا الله ونعم المعين نعم المولى ونعم النصير

Fig. 5.3: A table displaying the ascendant at nine times of day with religious significance appended to a copy of the *Mufrad Zīj*. [From MS Cambridge Browne O.1, fol. 179r, courtesy of Cambridge University Library.]

The other seven times divide the length of daylight into eight equal parts (of $1^1/_2$ seasonal hours each). These seven times are labelled:

al-ḍaḥwa al-ṣughrā	= *chāsht-i kūchek*	the small (= first) *ḍaḥwa*
al-ḍaḥwa al-wusṭā	= *chāsht-i miyān*	the middle *ḍaḥwa*
al-ḍaḥwa al-kubrā	= *chāsht-i buzurg*	the large (= last) *ḍaḥwa*
anṣāf al-nahār	= *nīmrūz*	midday
mā bayn al-ṣalātayn	= *miyān-i dō namāz*	(the time) between the two prayers
awwal al-ʿaṣr	= *awwal-i namāz-i dīgar*	beginning of the other (afternoon) prayer
ākhir al-ʿaṣr	= *ākhir-i namāz-i dīgar*	end of the afternoon prayer.

In this table there are three times of religious significance in the morning: at $1^1/_2$, 3, and $4^1/_2$ hours after sunrise. The association of a prayer with midday is established by the name of the fifth time, specifically referred to as "(the time) between the two prayers", *i.e.*, midway between the *ẓuhr* and the *ʿaṣr*. The latter prayer begins at nine hours after sunrise and ends at ten hours. Thus the definitions of the prayer-times underlying this table are quite different from the standard definitions in terms of shadow increases but, as I have shown, ultimately related, insofar as they are based on a simple division of the length of daylight.

5.4 The *ḍuḥā* in later astronomical sources

The institution called *ḍuḥā* in Arabic was practiced in Ottoman Turkey as the *żaḥwe*. Tables for the prayer-times in a popular almanac by Darendelī, compiled probably in the 16th century, display the time (according to the Turkish convention that sunset is 12 o'clock) for the *żaḥwe* along with the other times of prayer, based on the assumption that the *żaḥwe* begins half-way between daybreak and sunset: see **Fig. 5.4**.[7] Navoni, who published Darendelī's almanac in 1814, remarked:[8]

> "(le) *Zahve*, vulgairement dit *Kouschlouk*, marque un autre temps entre le lever du soleil et le midi, mais sans une prière particulière et d'obligation."

Precisely what took place at the *żaḥwe* is therefore not certain.

In an early-19th-century Syrian treatise on the 15 times of the Muslim day with religious significance the author distinguishes between *al-ḍaḥwa al-ṣughrā*, which is when the solar altitude is $4^1/_2°$, and *al-ḍaḥwa al-kubrā*, which is halfway between daybreak and sunset (**Text 5.4b**).[9] The lesser *ḍaḥwa* is stated to be the time of the prayers at the two religious festivals (*ṣalāt al-ʿīd*) and of the *ḍuḥā*. Neither of the two times corresponds to the original *ḍuḥā*, and

---

[7] On the Ottoman convention for measuring time with respect to sunset, see Würschmidt, "Osmanische Zeitrechnung". On Ottoman tables for timekeeping see King, "Astronomical Timekeeping in Ottoman Turkey", and now **II-14**. Darendelī's tables are published in M. D'Ohsson, *Tableau général de l'Empire Othoman*, 2 vols., Paris, 1787, I, facing p. 192, and J. B. Navoni, "Rouz-name ou calendrier perpetuel des Turcs", *Fundgruben des Orients* (*Mines de l'Orient*) 4 (Vienna, 1814), pp. 38-67, 127-153, 253-277, and appendix. An analysis is in **II-14.4**. The extract shown in **Fig. 5.4** is taken from MS Cairo ṬFT 29,2 (copied *ca.* 1800).

[8] Navoni, *op. cit.*, p. 58. In *İslam Ansiklopedisi*, 11 vols., Istanbul: Milli Eğitim Basimevi, 1940-1986, III, p. 654, the *ḍuḥā* is interpreted simply as "the time of the *kuşluk*".

[9] *Cf.* MS Aleppo Awqāf 970, esp. fols. 5r and 6r. A more detailed account of the contents of this manuscript is given in **II-2.11**. See also n. 6:4.

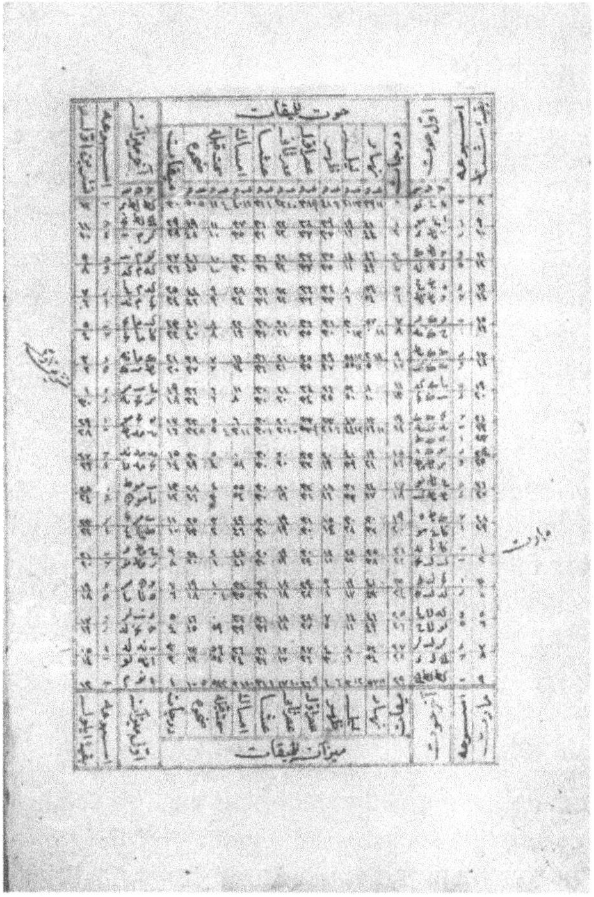

Fig. 5.4: An extract from the prayer-tables for Istanbul by the 16th(?)-century Ottoman astronomer Darandelī. This table serves the two zodiacal signs Pisces (reading down the table) and Libra (reading up the table), and displays the length of day and night and the times of the *ẓuhr* ($\Delta z = 0$), *ʿaṣr* 1 ($\Delta z = n$), *ʿaṣr* 2 ($\Delta z = 2n$), *ʿishā* and *imsāk*, as well as the time when the sun is in the direction of Mecca and the time of the *żaḥwe*. Here the *imsāk* is at daybreak, and the *żaḥwe* is taken as halfway between daybreak and sunset. [From MS Cairo ṬMT 29,2, courtesy of the Egyptian National Library.]

Fig. 5.5: An extract from the anonymous Ottoman almanac for the latitude of Algiers. It shows for each day of January the times of the *ṣanjāq* (?), the afternoon prayer, and the beginning of fasting in Ramadan (*imsāk*). The term *ṣanjāq* is puzzling, being neither Arabic not Turkish, and may be the result of a copy-ist's error. [From MS Cairo ṬFT 9, fol. 40v, courtesy of the Egyptian National Library.]

we shall return to the curious value given for the solar altitude in **Ch. 6**. Ottoman quadrants often bear markings for finding the time of the *ḍaḥwa*,[10] and a 19th-century Ottoman treatise on the use of the quadrant describes how to use such *ḍaḥwa* curves, and all the other curves on the quadrant that relate to the times of prayer.[11] In the prayer-tables in use in Jerusalem during the 1970s it was stated that the *ḍuḥā* (and the *ṣalāt al-ʿīdayn*) are at 20 minutes after sunrise.

[10] See, for example, Janin & Rohr, "Deux astrolabes-quadrants turcs", fig. 5 facing p. 116, and p. 120.
[11] See Würschmidt, "Gedosi über den Quadranten", esp. pp. 142-143.

A late Algerian almanac written in Turkish contains tables for the *imsāk* in Ramaḍān, the *ʿaṣr*, and a time of day labelled *ṣanjāq*.[12] The word *ṣanjāq* may be a scribal error for *ḍuḥā*, the additional *q* being a standard medieval abbreviation for *daqāʾiq*, minutes. Inspection of the table reveals that the time for the institution, whatever it may have been called, was about 15 minutes before midday.

In Indonesia in the 19th century a prayer corresponding to the *ḍuḥā* (Malay: *loeha*, Javanese: *loeka*) was apparently performed some 3 to 4 hours before midday.[13] In modern Pakistan the middle of the "Islamic day", which begins at daybreak and ends at sunset, is called *żaḥwat ul-kubrā*, but whilst this is no longer the *ḍuḥā* of early Islamic practice, the preferred time for the *chāsht* prayers in Pakistan is after one-quarter of the Islamic day has passed, so that it is the *chāsht* that corresponds roughly to the original *ḍuḥā*.[14]

There were other times of day with especial religious significance whose times were also astronomically defined. To these we now turn.

---

[12] MS Cairo ṬMT 9,1 (fols. 1v-48v, copied *ca.* 1900) – see *Cairo ENL Survey*, no. F68, and the illustration in pl. XLIb on p. 261. A preliminary analysis is in **II-13.7**. See also n. 5:7.

[13] Th. W. Juynboll, *Handleiding tot de kennis van de Mohammedaansche wet volgens de leer der Sjafiʿitische school*, 4th edn., Leiden: E. J. Brill, 1930, p. 75.

[14] On modern practice in Pakistan see Bagvi, *Qibla and Timings*, esp. pp. 26-28 and 54-55.

CHAPTER 6

THE TIMES OF THE *ṢALĀT AL-ʿĪDAYN* AND THE FRIDAY PRAYER

6.1 The *ṣalāt al-ʿīdayn*

We now consider the times of the communal prayers at the two main religious festivals (*ʿīds*)[1] and the time of preparation for the Friday prayers occurring in certain medieval sources. A note on the time of the prayer at the festivals, which can be dated to about the year 1500, states that (**Text 6.1**):[2]

> "The best (procedure) is to delay the *ʿīd* prayer until the sun has risen (above the eastern horizon) by the amount of a lance (*qadr rumḥ*), as was related in the two (books called) *al-Ṣaḥīḥ* (by al-Bukhārī and Muslim) of what (the Prophet) – may God bless Him and grant Him salvation – did. But the custom when it is cloudy is that the prayer of the *ʿīd al-fir* is when one-quarter of an hour has passed and the prayer of the *ʿīd al-aḍḥā* is when one-fifth of an hour has passed. God knows best."

The meaning of the description of this practice attributed to the Prophet is by no means clear from the context. Likewise mysterious are the times specified for the two *ʿīd* prayers.

In later Ottoman practice, a specific solar altitude was adopted. In one source, a set of auxiliary tables for computing prayer-times for any latitude compiled by Saʿīd Beg Zāde, the solar altitude at the *ʿīd* prayers is taken as 6°.[3] In another Ottoman source, namely, the early-19th-century Syrian treatise on the times of prayer mentioned in **5.4**,[4] the author states (**Text 5.4a**) that the lesser *ḍaḥwa*, the time before which prayer is forbidden and also the time for the *ʿīd* prayer and the *ḍuḥā*, is when the solar altitude is $4^1/_2°$ in the east. Likewise the latest time for the *ʿaṣr* prayer is when the solar altitude is $4^1/_2°$ in the west. No explanation is offered for this value of the solar altitude. A curve for determining the time of the *ṣalāt al-ʿīdayn* often occurs on Ottoman quadrants.[5]

The connection between these various traditions is provided by Edward Lane's *Arabic-English Lexicon*.[6] In the entry for *rumḥ* Lane states that "as a measure in astronomy, according to modern Arabian astronomers, it is four degrees and a half, (or) the eightieth part of a great-circle", adding that "according to various works on practical law, (the *rumḥ*) consists of twelve *shibr*s or spans". The *rumḥ* is indeed used, if rarely, as a measure of distance in the sky: the 10th-century astronomer ʿAbd al-Raḥmān al-Ṣūfī states that the distance between the two stars

---

[1] See the three EI_2 articles "ʿĪd ... " by Eugen Mittwoch (taken over from EI_1).

[2] MS Cairo DM 534,2, fol. 19v – see *Cairo ENL Survey*, no. C146.

[3] On these tables, extant in MS Istanbul Kandilli 226, see further **II-14.16**.

[4] MS Aleppo Awqaf 970, fol. 6r. See also n. 5:9.

[5] See, for example. *Paris IMA 1993-94 Catalogue*, p. 438, no. 333, and pp. 442-443, no. 337.

[6] Lane, *Lexicon*, III, cols. 1153a-b, also recorded in Joseph G. Hava, *al-Farāʾid*: *Arabic-English Dictionary*, Beirut: Catholic Press, 1964, p. 269a. See, however, the sentiments expressed in n. 0:7.

surrat al-faras or *raʾs al-musalsala* (α Andromedae) and *janāḥ al-faras* (γ Pegasi) is one *rumḥ*: his translator Hans Schjellerup measured this, however, as 14°, which is too large to be relevant here.[7] Perhaps the required time is when an object of length one span casts a shadow equal to the length of a lance. It may not be a coincidence that a man can throw a lance of length equal to his own height about twelve times the length of the lance.[8] In other words, the sun would be at the same height as the top of a vertical lance viewed from the ground at the point from which it was thrown. In any case, the corresponding solar altitude is close to $4^1/_2°$ (≈ arc tan $[^1/_{12}]$). The time since sunrise, calculated using the Indian formula, would be about half an hour (when $\Delta z = 11n$ then $T = {}^1/_2$).

The other definitions for the *ʿīd* prayers at one-fifth or one-quarter of an hour after sunrise are less easily explained. If we revert to the Indian formula we find that the shadow increases are respectively 23n or 29n, that is, about 24n and 30n, so that the shadows are approximately twice and two and one-half times the shadow that yields the solar altitude of $4^1/_2°$. As mentioned in **5.4**, that the time currently used for the *ṣalāt al-ʿīdayn* in Jerusalem is 20 minutes, that is, one-third of an equinoctial hour, after sunrise.

6.2 Preparations for Friday worship

In medieval Maghribi practice we find an institution called *taʾhīb* or *taʾahhub*, literally "preparation", and relating to the communal prayers on Fridays. In the corpus of tables for timekeeping compiled for the latitude of Tunis in the late 14th century, there are tables displaying the solar altitude at the *taʾhīb* in which the entries correspond to a time one equinoctial hour before midday: see **Fig. 6.2**.[9] On the contemporary Tunisian sundial illustrated in **Fig. 5.2** the straight line to the left of the meridian displays the time of the *taʾhīb* defined in the same way.[10] In an anonymous Moroccan treatise on sundial construction the author refers to the first and second *taʾahhub* on Friday (**Text 6.2**) but does not specify the times at which these take place.[11]

In the anonymous Algerian commentary on the poem on timekeeping by al-Jādarī referred to in **5.2**, the author mentions a first and second *indhār*, "warning", on Friday mornings, the first at two and one-half seasonal hours and the second at four seasonal hours after sunrise (**Text 5.2**).[12] No reason is given in any of these sources for choosing these times, and none occurs to me. On the term *ṣanjāq* referring to a time shortly before midday (but perhaps a corruption of *ḍuḥā*) see **5.4**.

---

[7] Schjellerup, *Description des étoiles fixes par al-Sûfi*, pp. 26-27, esp. p. 27, and p. 112. On al-Ṣūfī see the *DSB* article by Paul Kunitzsch. On the *rumḥ* see also Kunitzsch's notes appended to Rada, "Comets in Arabic Literature", p. 91.

[8] I tried this myself on the beach at ʿAgamī near Alexandria in the mid 1970s, albeit with a harmless *rumḥ* and without witnesses.

[9] On these tables, extant in MS Berlin Ahlwardt 5724 (83 fols., copied *ca.* 1700), see already King, "Tunisian Sundial" (n. 0:3), pp. 192-193, and *idem*, "Astronomy in the Maghrib", pp. 38-39. A full analysis is in **II-13.2-3**.

[10] King, "Tunisian Sundial" (n. 0:3), p. 190.

[11] MS Cairo Ḥalīm *mīqāt* 19, esp. fol. 47v – see *Cairo ENL Survey*, no. F4.

[12] MS London B.L. Or. 411,2, fol. 45v. See also n. 5:4.

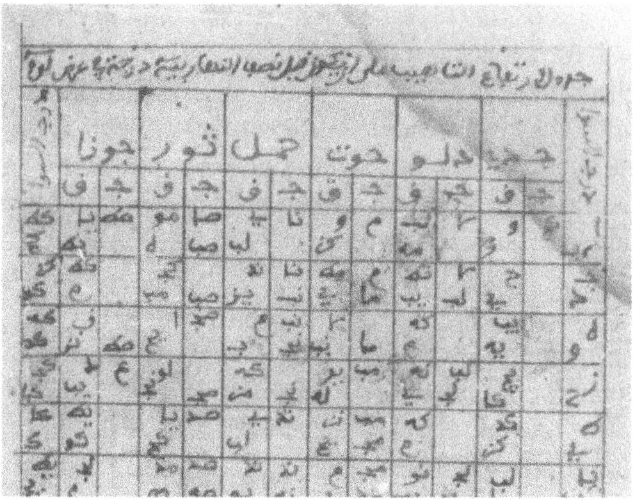

Fig. 6.2: A table displaying the altitude of the sun at the time of the *ta'hīb*, one equinoctial hour before midday. This table occurs in the corpus of tables for timekeeping that was used in medieval Tunis. [From MS Berlin Ahlwardt 5724, fol. 43v, courtesy of the Deutsche Staatsbibliothek (Preußischer Kulturbesitz).]

Finally we note that Ibn al-Ajdābī, a philologist and specialist in folk astronomy of the 13th century (**Text 6.3**),[13] states that the beginning and end of the *ẓuhr* and the *ʿaṣr* are defined by:

$$\Delta z = 0+ \text{ and } n \quad \text{and} \quad \Delta z = n \text{ and } 2n.$$

He adds that the communal prayer on Friday (the text apparently has *ṣalāt al-jamāʿa* – perhaps, but not necessarily, one should read *ṣalāt al-jumʿa*) should take place half-way through the *ẓuhr*, when the shadow increase is ... . Either there was a lacuna in the unique manuscript used for the published text, or the editor was doubtful about the curious value, which must have been $^1/_3$n (corresponding to T = $4^1/_2$).

There is an obvious need for further investigation of these institutions in medieval Islamic practice, and contemporaneous local legal texts might be a good place to start.

[13] Ibn al-Ajdābī, *Kitāb al-Azmina wa-'l-anwā'*, ed. ʿIzzat Ḥasan, Damascus: Wizārat al-thaqāfa wa-'l-irshād al-qawmī, 1964, pp. 117-118. On the author see the article by Charles Pellat in *EI*$_2$ (under "Ibn al-Adjdābī").

CHAPTER 7

SOME INSTRUMENTS FOR REGULATING THE TIMES OF PRAYER IN SEASONAL HOURS

During the period from the 9th to the 15th centuries Muslim astronomers devised numerous instruments for observation and calculation.[1] Several varieties, notably sundials and compendia, bore markings for regulating the times of prayer. Our present concern is only with a small subset of such instruments that cast light on our subject.

7.1 The sundial of ʿUmar II

al-Aṣbaḥī relates three stories about the Caliph ʿUmar ibn ʿAbd al-ʿAzīz (born Medina in 682/83, *reg. ca.* 717-720, d. 720 in Aleppo), each of which is traced back to al-Awzāʿī, a contemporary of the Caliph (**Text 7.1**):[2]

> " ... ʿUmar ibn ʿAbd al-ʿAzīz set up sundials (*naṣaba sāʿāt*) in order to know the times of prayer and the remainder of the day. His muezzin used to make the call for the *ẓuhr* prayer if six hours had passed and the beginning of the seventh had been reached, which is when the sun starts to decline, and (the muezzin) used to make the call for the *ʿaṣr* prayer when nine hours had passed and the beginning of the tenth hour had been reached. Then he would give the order for the prayer to be performed and it would be performed, and he would pray (*wa-yaʾmur bi-'l-ṣalāt fa-tuqām wa-yuṣallī*)."
>
> " ... ʿUmar ibn ʿAbd al-ʿAzīz set up sundials in order to know how much of his day had passed ... ."
>
> " ... ʿUmar ibn ʿAbd al-ʿAzīz set up sundials (*sāʿāt*) in order to know from them how much of his day had passed on him and how much of it remained and these hours (*sāʿāt*) are light in winter and heavy in summer ... ."
>
> " ... ʿUmar ibn ʿAbd al-ʿAzīz never stopped learning the quantities of the sun (*maqādīr al-shams*) whilst he was caliph until he died – may God have mercy upon him ... ."

The term *sāʿa* is unusual as a name for an instrument, and we may assume that the author(s) of these reports could not think of a better name. By the 10th century various Arabic or Arabicized terms were available for various devices for timekeeping (see **Text 7.2**).[3] But the only such instrument for timekeeping by the sun that would have been available to the Caliph

[1] On Islamic instrumentation see **X**. See also n. 7:15 on a simple gnomon for determining the *ẓuhr* and the *ʿaṣr*, and Wiedemann & Frank, "Gebetszeiten", p. 4, on a water-clock marked for the prayer-times (apparently not defined in terms of the seasonal hours).

[2] al-Aṣbaḥī, *al-Yawāqīt*, fols. 3v, 13v, 38v and 3r. On the Caliph see the article "ʿUmar (II) b. ʿAbd al-ʿAzīz" by P. M. Cobb in EI_2, and on al-Awzāʿī see the article by Joseph Schacht in EI_2.

[3] Quoted from the encyclopaedia of the 10th-cen-tury scholar Abū ʿAbdallāh al-Khwārizmī; see al-Khwā-

in Syria or in the Hejaz in the early 8th century would have been a sundial. Sundials of different kinds were known throughout the Graeco-Roman world in Antiquity; indeed, conical sundials displaying the lines for seasonal hours on the inside had been used in Northern Arabia at least as far back as the 1st century A.D. by the Nabateans. There is no reason to doubt that Graeco-Roman sundials were still in use in Damascus at the time of the Muslim conquest.[4]

Slightly different versions of the same story are recorded in yet other sources. The 11th-century Andalusī legal scholar and histo-rian Ibn ʿAbd al-Barr, in his commentary *al-Tamhīd* on the *Muwaṭṭa*ʾ of Mālik,[5] confirms that the Caliph had his own opinions about the way in which the times of prayer should be determined (see **Text 7.3** and also **Fig. III-9.7b**):

> "One of the ancestors went in to ʿUmar ibn ʿAbd al-ʿAzīz – may God be pleased with him – and found that he had not yet prayed the *ʿaṣr* even though its time had begun and was firmly established. He said to (the Caliph): "O Prince of the Believers, the Caliphs – may God be pleased with them – did not delay the *ʿaṣr* prayer in this way. And (the Caliph) replied: "(Not) until I know (the right time)." He (?) said: ʿUmar ibn ʿAbd al-ʿAzīz kept on (the practice of) finding the time of the prayer by a mark (*ʿalāma*) and determine it with an instrument (*āla*) until he left the lower world."

An anonymous commentator on this passage points out that "some of the ancestors" did not approve of using instruments for regulating the times of prayer until they saw this passage in the *Tamhīd*.

7.2 The hour curves on astrolabes

al-Bīrūnī in the *Shadows* (**Text 7.4**)[6] records the same kind of information about ʿUmar II and shows how his practice was adopted by others – presumably well over a century after the time of ʿUmar II[7] – to justify regulating the prayer-times using the hour-curves on astrolabes:[8]

> "When (the story) was related concerning ʿUmar ibn ʿAbd al-ʿAzīz that he used to make the prayer-call for the *ẓuhr* at the (beginning of the) seventh hour and that he

rizmī, *Mafātīḥ al-ʿulūm*, p. 235 (reprint p. 136), for the text, and the article "al-Khuwārizmī" by Juan Vernet in *DSB* on the author. See also Wiedemann, *Aufsätze*, I, pp. 136-138 and 201-202 and *idem*, *Schriften*, III, p. 1217 for a discussion. In this text *tarjahāra* and *dabba* refer to water-clocks, *rukhāma* and *mukḥula* to horizontal and conical sundials, and the precise meaning of *ṣandūq al-sāʿāt* and *lawḥ* is not clear, although I strongly suspect that the latter is some kind of universal horary dial of the kind found, for example, on the later medieval European *navicula de Venetiis*, a brilliant computer device whose use was not fully understood by Europeans in the 13th and 14th century (and is also misinterpreted in the modern literature) – see King, *Mecca-Centred World-Maps*, pp. 351-359, **X-9.1**, and **XIIb**.

[4] On extant Greek and Roman sundials see Gibbs, *Greek and Roman Sundials* (n. 2:2), esp. p. 311 on some Nabatean dials. (The common hemispherical type from Antiquity, was not, as far as we know, made by Muslim astronomers.)

[5] This passage is contained in MS Cairo DM 119,2, fol. 8v (illustrated in **Fig. III-9.7b**). See further *Cairo ENL Catalogue*, I, *sub* DM 119, and II, 3.5.2. On Ibn ʿAbd al-Barr see the article by Charles Pellat in EI_2.

[6] *Cf.* al-Bīrūnī, *Shadows*, p. 185, and I, p. 235 of Kennedy's translation. Kennedy wrote in his commentary (II, p. 147): "this passage simply points out that since the definitions of the times of prayer and the hours are essentially different, it is erroneous to identify any particular prayer-time limit with any particular hour".

[7] Only in the mid 8th century did the Muslims begin making astrolabes (see next note). al-Bīrūnī records this passage in a section on markings for the prayer-times on astrolabes.

[8] On the astrolabe, and the standard markings for the seasonal hours and prayer-times, see the article "Asurlāb" by Willy Hartner in EI_2, reprinted in *idem*, *Studies*, A, pp. 312-318 (see also pp. 287-311); North, "Astrolabe"; *Washington NMAH Catalogue*, esp. pp. 52-54; and now **X-4-5**.

> used to perform the prayer at the (beginning of the) eighth hour, and (that he used to perform) the ʿaṣr at the (beginning of the) tenth hour, these hours being undoubtedly seasonal hours, *some of the people changed over from (using) shadows to (using) the curves (of the seasonal hours on astrolabes)* (*ʿaddala baʿḍuhum ʿan al-ẓilāl ilā khuṭūṭihā*). Thus the curve of the tenth seasonal hour was used for the end of the time of the *ʿaṣr* and the curve of the ninth (hour) for the beginning (of the time of the *ʿaṣr*). *Likewise the third of (the curves) was used for the time of the ḍuḥā prayer.* This is contrary to what has been laid down in the *sunna*, and it should not be followed."

al-Bīrūnī overlooked the opportunity to comment on the relationship between the seasonal hours and the definitions in terms of shadow increases in standard practice.

A problem in the interpretation of this story related by al-Bīrūnī, and other Arabic texts that I have consulted, is the ambiguity of the terminology. Thus, for example, expressions such as *ʿalā sabʿ sāʿāt* or *fī 'l-sāʿati 'l-sābiʿa* can mean at the beginning of the seventh hour, or during that hour, or at the end of that hour. However, it seems that the time used by the Caliph for the performance of the *ẓuhr* prayer was at the beginning of the eighth hour, and that the call to prayer was one hour earlier, which seems surprising. Also the Caliph began the *ʿaṣr* at the same time favoured by ʿAlī ibn Abī Ṭālib (**2.2**). The remaining information apparently refers to those who took their initiative from the Caliph, rather than to the Caliph himself. The time used by these people for the performance of the *ʿaṣr* prayer corresponds precisely to the tenth hour, and the time given for the *ḍuḥā* prayer corresponds precisely to the beginning of the fourth hour.

No indication is given in this account of where or when the incident took place. And in any evaluation of this report, one should keep in mind that there is a plethora of stories which use ʿUmar II as a source of *sunna*, especially in the Mālikī tradition, so that the authenticity of many of them is questionable. The people who used the hour-curves on astrolabes did so when they learned of the practice of the Caliph ʿUmar II, so the *terminus post quem* is the time when the Muslims began making astrolabes[9] while the *terminus ante quem* is the time of al-Bīrūnī. One could take this story as indicating official sanction for regulating the prayers by the seasonal hours in early-8th-century Damascus (though not yet, of course, using astrolabes). It furthermore concerns a group of people somewhere at some later time who observed the *ḍuḥā* prayer.

Perhaps it was precisely such sanction that led to the formulation of the definitions in terms of shadow increases prescribed for the *ʿaṣr* by the four main legal schools in the late 8th and early 9th century. As I have shown, these definitions presume a familiarity with the Indian formula, and it is surely significant that the earliest occurrence of this formula in an Arabic source is in the writings of the mid-8th-century Baghdad astronomer al-Fazārī.

7.3 Prayer curves on astrolabes

The difference between the times of the prayers defined in terms of shadow increases and the seasonal hours with which these prayers were originally associated is, of course, a result of

[9] See n. 7:7.

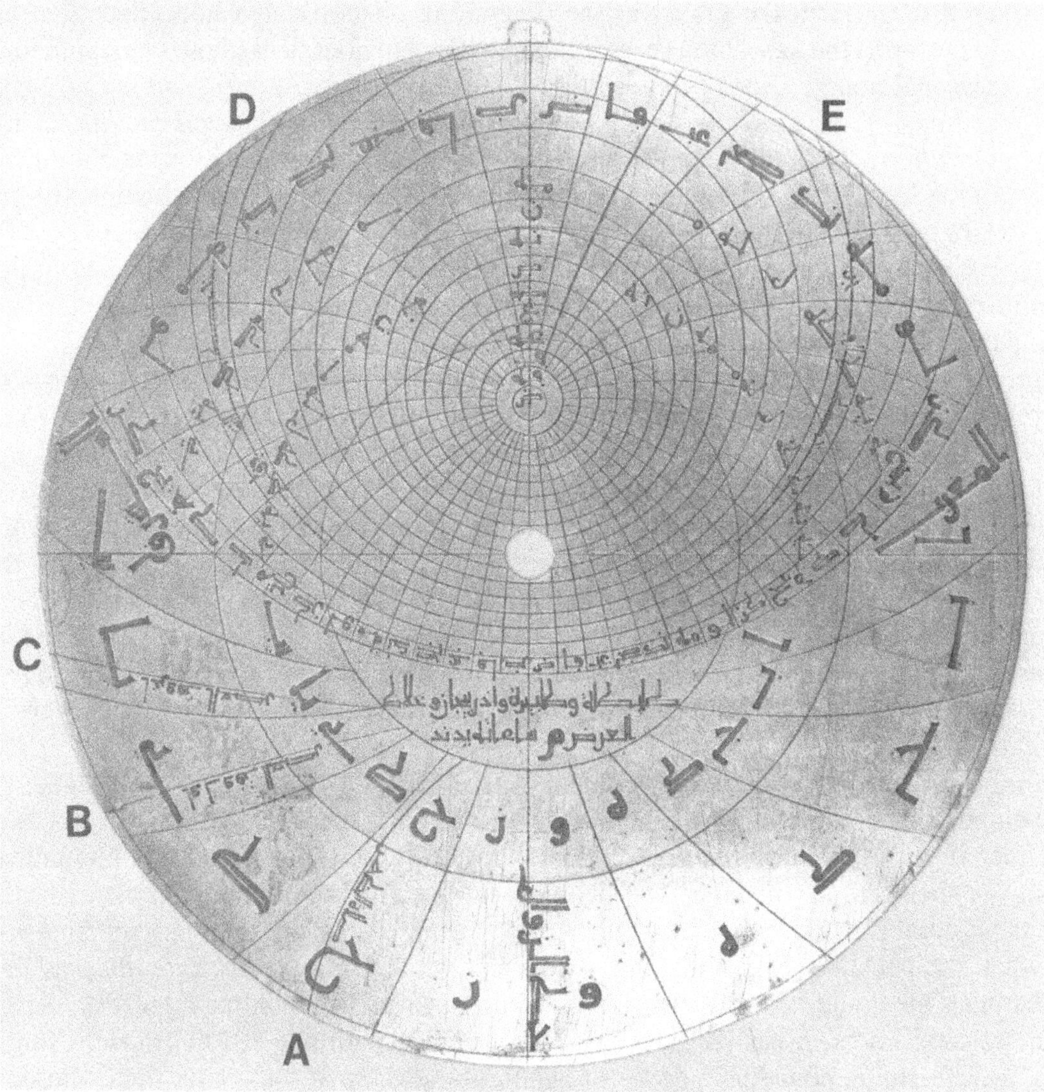

Fig. 7.3: A plate from an 11th-century Andalusi astrolabe (#118) displaying curves for the *ẓuhr* (A: $\Delta z = {}^1/_4 n$), the beginning of the *ʿaṣr* (B: $\Delta z = n$), and the end of the *ʿaṣr* (C: $\Delta z = 2n$), amidst curves for each seasonal hour of day-light from 1 to 12. The times of the prayers are here defined by means of shadow increases, and they therefore no longer correspond to the definitions in terms of the seasonal hours. There are also markings above the horizon for the night-time prayers (D: nightfall, for the *ʿishāʾ*; and E: daybreak, for the *fajr*), it being understood that with these one should use the point opposite the sun. The plate serves latitude 40° and Toledo, Talavera, Azerbeijan and Akhlat are specifically mentioned. [Courtesy of the Museum of the History of Science, Oxford.]

the inaccuracy of the Indian formula. Consider, for example, the standard astrolabe plate illustrated in **Fig. 7.3**. This bears curves for the seasonal hours, marked 1, 2, ... , 11, and also three curves (here marked A, B, and C) for the *ẓuhr*, the beginning of the *ʿaṣr*, and the end of the *ʿaṣr*, based respectively on the definitions:

$$\Delta z \quad = \quad {}^1/_4 n, \quad n, \quad \text{and} \quad 2n.$$

The plate shown was constructed in Toledo by Ibrāhīm ibn Saʿīd al-Sahlī in 460 H [= 1068] specifically for the latitude of Toledo, taken as 40°.[10] Observe first that the curves for the prayers do not coincide with the curves for the corresponding hours. In fact, the *ẓuhr* curve defines a time that varies from about $7^1/_2$ hours (= seasonal day hours) from sunrise at the winter solstice (represented by the circumference of the plate marked WS) to close to 8 hours at the equinoxes (represented by the concentric circle marked EQ), and to about $7^1/_2$ hours at the summer solstice (represented by the concentric circle marked SS). The times displayed by the other two curves are closer to the ninth and tenth hours with which the definitions of the times of prayer were originally associated. The same point could be made with a typical Andalusī sundial bearing markings for the seasonal hours and for the *ẓuhr* and the *ʿaṣr*.[11]

7.4 Instruments for reckoning seasonal hours and prayer times

There are two implicit references in the Islamic astronomical sources to the association of the prayer-times with the seasonal hours. The first occurs in a treatise on the construction of the conical sundial (*mukḥula*) written by an individual named ʿAbdallāh ibn Qāsim al-Ṣiqillī, and known from a unique 13th-century manuscript.[12] Nothing is known of the author, save that his name indicates that he or his family hailed from Sicily. al-Ṣiqillī appears to suggest (**Text 7.5**) that the *ẓuhr* begins at the end of the seventh or beginning of the eighth hour of daylight while the *ʿaṣr* begins at the end of the ninth or beginning of the tenth hour; however, it may well be that he considered these only as convenient approximations for the times of prayer.

Another case in which the prayer-times are associated with the seasonal hours occurs on a portable sundial made by Abu ʾl-Faraj ʿĪsā in Syria in 554 H [= 1159/60]: see **Figs. 7.4a-b**.[13] The maker wrote in his inscription that the two sides of the sundial were for "finding the seasonal hours and the times of prayers" for the latitudes of 33° and 36°, serving Damascus and Aleppo, respectively. It is of interest to the present discussion that there are markings on the sundial only for the seasonal hours, and that not even the words *ẓuhr* and *ʿaṣr* are inscribed on the instrument. I take this to mean that in Ayyubid Syria these two prayer-times were associated, at least in some circles, with the seasonal hours, as in the case of the text of al-Ṣiqillī.

---

[10] On al-Sahlī see Mayer, *Islamic Astrolabists*, pp. 50-52. This particular instrument (no. II in Mayer's list) is now in the Lewis Evans Collection in the Museum of the History of Science, Oxford.

[11] On such sundials see King, "Three Sundials from Islamic Andalusia" (n. 4:7); *Madrid MAN 1992 Exhibition Catalogue*, pp. 89-102; and Labarta & Barceló, "Ocho relojes de sol hispanomusulmanes". Most of these Andalusī sundials bear markings for the prayers which are either careless or incompetent. For a treatise on gnomonics by a 14th-century Tunisian astronomer see Carandell, *Ibn al-Raqqām sobre los cuadrantes solares*.

[12] Published from an unnumbered manuscript in the Université de St Joseph, Beirut, in Cheikho, "al-Mukḥula" (in Arabic, see especially p. 77), translated in Wiedemann & Würschmidt, "Arabische kegelförmige Sonnenuhr", and further studied in Livingston, "Islamic Conical Sundial". The St. Joseph manuscript was already several decades ago in a state of disintegration – I owe my microfilm of the remains to the generosity of Pro-fessor E. S. Kennedy.

[13] For descriptions of this instrument see Casanova, "Cadran solaire syrien"; and *Paris IMA 1993 Exhibition Catalogue*, pp. 436-437 (no. 332).

Figs. 7.4a-b: A portable vertical sundial from Ayyubid Syria (#7315) which purports to serve for "finding the seasonal hours and the times of the prayers", but there are no separate markings for the prayers. The implication is that the prayers take place at the seasonal hours. The markings on either side of the instrument serve latitudes 36° (a) and 33° (b), intended for Aleppo and Damascus, respectively. [Paris, BNF, Département des Cartes et Plans, photographs courtesy of the Bibliothèque Nationale de France.]

7.5 A special gnomon for the *ẓuhr* and the *ʿaṣr*

Simple gnomons erected on horizontal bases for determining the *ẓuhr* and *ʿaṣr* are described by al-Khwārizmī[14] and by al-Bīrūnī.[15] The latter labels his device *mīzān*, literally "balance". Both authors use the definitions:

[14] Described in Wiedemann & Frank, "Zirkel zur Bestimmung der Gebetszeiten" (n. 4:14). See also King, "al-Khwārizmī", p. 7.

[15] al-Bīrūnī, *Shadows*, p. 110, and Kennedy, *al-Bīrūnī's* Shadows, I, p. 155, and II, pp. 82-83.

$$\Delta z \quad = \quad 0 + \quad \text{and} \quad n$$

for the two prayer-times.

Now al-Aṣbaḥī describes an instrument also called *mīzān* for regulating the prayer-times, which he says was first made by Yaḥyā ibn Saʿīd al-Zubayrī (**Text 7.6**).[16] I have no further information on al-Zubayrī, but it would be of great interest if he had lived as early as the 8th century. His *mīzān* consists of a vertical gnomon on a horizontal wooden base on which the shadow can be measured, and al-Aṣbaḥī states that to use the *mīzān al-aqdām*, in which the gnomon is divided into $6^1/_2$ units, one simply waits until the shadow increase is:

$$0\ +,^* \quad 2^1/_6, \quad 6^1/_2, \quad 9^1/_{12}\ +, \quad \text{and} \quad 13 \text{ units}$$

(* text: *ḥīna tazūlu 'l-shams*, literally "when the sun starts to decline")

to determine the beginning and midpoint of the *ẓuhr*, the end of the *ẓuhr* which is the same as the beginning of the *ʿaṣr*, and the midpoint and end of the *ʿaṣr*. The value $2^1/_6$ corresponds to $^1/_3$n (n = $6^1/_2$), as noted by al-Aṣbaḥī himself, and is the value of z midway between midday and the time to Δz = n according to the Indian formula (T = $4^1/_2$). The value "a little over $9^1/_{12}$" (*tisʿat aqdām wa-niṣf suds rājiḥ*) is an approximation for $9^1/_{10}$ feet, which, as $^7/_5$n, is the shadow increase at the time midway between the times corresponding to Δz = n and Δz = 2n according to the Indian formula (T = $1^1/_2$). For the *mīzān al-aṣābiʿ*, in which the gnomon is divided into 12 units, al-Aṣbaḥī gives no such values; he simply states the equivalent of the Indian formula for base 12.

At this point it is appropriate to cite an anecdote recorded by al-Bīrūnī in the *Shadows* (**Text 7.7**).[17] He mentions that some of the Karrāmiyya, a sect associated with Abū ʿAbdallāh ibn Karrām,[18] preferred to take a "middle path" with regard to the definition of the *ʿaṣr*. However, they could not decide whether they should begin the prayer-time when the shadow increase was the mean of the values n and 2n, or at the time halfway between the two times defined by Δz = n and Δz = 2n, al-Bīrūnī pointed out that if the two times defined in these two ways were ever the same, it would be by chance.

With this account of nitpicking over the prayer-times we conclude this chapter. I shall now pursue the relationship between the daylight prayers and the seasonal hours further, firstly to explain the origin of the names of these prayers (**Ch. 8**), and secondly to investigate the connection between Islamic and earlier Syrian Christian practice (**Ch. 9**).

---

16 al-Aṣbaḥī, *al-Yawāqīt*, fols. 10v-11r, 13v-14r, and 16v-17r.
17 al-Bīrūnī, *Shadows*, p. 171, and Kennedy, *al-Bīrūnī's* Shadows, I, p. 222, and II, p. 140.
18 See the article "Karrāmiyya" by C. Edmund Bosworth in EI_2.

CHAPTER 8

ON THE ORIGIN OF THE NAMES OF THE PRAYERS

8.1 On the meaning of the names of the prayers

Where do the names of the prayers come from? The names associated with such astronomical phenomena as *al-maghrib*, sunset, and *al-fajr*, daybreak, are of course self-explanatory. But it is the names *ḍuḥā*, *ẓuhr* and *ʿaṣr* that merit some investigation.

From Lane's *Arabic-English Lexicon*, based entirely on the medieval lexicographers, it is already obvious that these terms have other meanings.[1] For *ḍuḥā* Lane gives: the early part of the forenoon, after sunrise; when the sun is somewhat high; when the sun shines brightly; *etc.* For *aḍḥa 'l-ḍuḥā*, he has: to perform the supererogatory act of prayer in the time called the *ḍuḥā*, *etc.* For *ẓuhr*: midday, or noon; the time when the sun declines from the meridian; and for *ṣalāt al-ẓuhr*: the prayer of midday or noon, or of the time after the declining of the sun. For *ʿaṣr*: an hour, or a time (*sāʿa*) of the day; the morning, before or after sunrise; the afternoon; the evening; the last part of the day; until the sun becomes red, *etc.*

al-Bīrūnī quoted the opinions of various earlier authorities on the etymologies of the names of the prayers, which, as he said, are connected with (*muḍāfa ilā*) the times at which they are to be performed.[2] The word *ẓuhr*, he said, was from *ẓahīra*, which means "violent heat", taken most probably from *ẓahīr*, "strong-backed camel". *ʿAṣr*, according to one opinion, means *ʿashī*, "late afternoon", and according to another "killing in order to extract something". These are typical of etymologies found in other medieval sources: see, for example, **Text 6.3**.[3]

8.2 The names of the divisions of day and night

The Arabs of the peninsula before the advent of Islam had names for various divisions of the day and night. There were names for each third, fourth and fifth part of the night,[4] some of which were later used – as we shall see – to designate more specific periods. The term *sāʿa* was used loosely for such divisions.[5] For information on these names we turn to the extensive

---

[1] See Lane, *Lexicon*, V, cols. 1773a-b, 1929b, and 2062a-b, with the reservations expressed in n. 0:7.

[2] al-Bīrūnī, *Shadows*, pp. 164-166, and Kennedy, *al-Bīrūnī's* Shadows, I, pp. 215-217, and II, pp. 134-136.

[3] Taken from Ibn al-Ajdābī, *al-Anwāʾ* (n. 6:13), pp. 117-118.

[4] See, for example, Lane, *Lexicon*, II, col. 708a (*khaḍar*), and IV, col. 1467a (*sawʿ*). A brief overview is in the article "Layl and nahār" by Charles Pellat in EI_2, V, pp. 708b-709a. Some of the vague expressions are still in use today in traditional regions of the Near East: see, for example, Robert B. Sergeant and Ronald Lewcock, eds., *Ṣanʿāʾ: An Arabian Islamic City*, London: The World of Islam Festival Trust, 1983, p. 53, on those used in the Yemen.

[5] See Lane, *Lexicon*, IV, cols. 1467c-1468a. On the origin of the word *sāʿa* in Arabic see also Arthur Jeffery, *The Foreign Vocabulary of the Qurʾān*, Baroda: Oriental Institute, 1938, pp. 157-158.

corpus of Arabic works on philology, particularly works dealing specifically with names for intervals of time.[6]

One of the earliest surviving writings of this genre, the *Kitāb al-Ayyām wa-'l-layālī* of al-Farrā' (761-822),[7] does not actually explore in any detail the names of the different times of daylight but does devote a chapter to the names of the times (*sā'āt*) of night. Here some nineteen different names are introduced, without any indication of precisely to which intervals of the night they refer. For al-Farrā', *'aṣr* is simply a period of time during the day, and the variant *ṣalāt al-shāhid* is stated to be a dialectal variant (Asad and Kalb) for *ṣalāt al-maghrib*.[8] The same information, and more of the same kind, is recorded by the philologists Quṭrub (d. after 825) in his *Kitāb al-Azmina*,[9] and by Abū 'Umar al-Zāhid (874-957) in his *Kitāb Yawm wa-layla*.[10] The celebrated scholar Ibn Qutayba (828-889) does not mention the divisions of day and night in his treatise on folk astronomy, the *Kitāb al-Anwā'*, but does give a disordered list of divisions in his *Kitāb Adab al-kātib*.[11] The so-called *Kitāb al-Jarāthīm*, of uncertain authorship but compiled after *ca.* 900, lists nine different parts of the night.[12]

al-Hamadhānī (d. 932), author of a lexicographical work entitled *Kitāb al-Alfāẓ al-kitābiyya* gives lists of sixteen times of the day and seven or twenty times of the night (**Text 8.1** and **Table 8a**), apparently, but not necessarily, in order (the conjunction is *thumma*, "then" or "next", or simply "also").[13]

The *Kitāb al-Azmina* of the philologist al-Marzūqī (d. 1030) contains a lengthy discussion of the names of different parts of the day and night without arranging them,[14] and the exegetist and linguist al-Zamakhsharī (d. 1143/44), in his *Muqaddimat al-adab*, gives a long disordered list of the names of the *sā'āt*.[15] The major lexicographical works such as the *Mukhaṣṣaṣ* of

---

[6] Fuat Sezgin (*GAS*, VII and VIII) has documented the available sources on folk astronomy and lexicography up to *ca.* 1050.

[7] On the author see the article by Régis Blachère in EI_2. For this discussion see Abū Zakariyā Yaḥyā ibn Ziyād al-Farrā', *al-Ayyām wa-'l-layālī wa-'l-shuhūr*, ed. I. al-Ibyārī, Cairo: al-Maba'a al-Amīriyya, 1956, pp. 47-49.

[8] *Ibid.*, pp. 50 and 63.

[9] Edited by Ḥātim Ṣāliḥ al-Ḍamīn in *al-Mawrid* (Baghdad) 13 (1984), pp. 109-172, esp. pp. 151-158 (parts of the night), 158-160 (parts of the day), also printed Beirut: Mu'assasat al-Risāla, 1985, esp. pp. 58ff. On the author see the EI_2 article by G. Troupeau.

[10] Abū 'Umar al-Zāhid, *Kitāb Yawm wa-layla fī 'l-lugha wa-'l-gharīb*, ed. M. J. Mu'aybid, in *Revue de l'Institut des Manuscrits Arabes (Cairo)* 24:2 (1978), pp. 229-338, esp. pp. 276ff. (pp. 281-297 are missing from the copy which I used). On the author see Sezgin, *GAS*, VII, pp. 354-355, and VIII, pp. 154-158. Unfortunately his *Kitāb al-Sā'āt* (Sezgin, *GAS*, VIII, p. 157) is lost.

[11] On Ibn Qutayba see the articles by G. Lecomte in EI_2 ("Ibn Ḳutayba") and by Paul Kunitzsch in *DSB*, as well as Sezgin, *GAS*, VII, pp. 350-351, and VIII, pp. 161-165. For his discussion of the hours see Max Grünert, ed., *Ibn Kutaiba's Adab-al-Katib nach mehreren Handschriften herausgegeben*, Leiden: E. J. Brill, 1900, reprinted photo-mechanically Beirut: Dār Ṣādir, 1967, pp. 97-98.

[12] Louis Cheikho, "*Nukhba min Kitāb al-Jarāthīm li-'Abdallāh ibn Muslim*", appended to the Beirut, 1885 edition of al-Tha'ālibī, *Kitāb Fiqh al-lugha wa-sirr al-'arabiyya*, p. 354. On this work see Sezgin, *GAS*, VIII, pp. 258-260.

[13] On al-Hamadhānī see the article by Régis Blachère in EI_2, and Sezgin, *GAS*, VIII, p. 193. The text was published by Cheikho: see al-Hamadhānī, *Kitāb al-Alfāẓ al-kitābiyya*, ed. Louis Cheikho, Beirut, 1913, pp. 287-289.

[14] al-Marzūqī, *Kitāb al-Azmina wa-'l-amkina*, 2 vols., Hyderabad: Maba'at Majlis Dā'irat al-Ma'ārif, 1332 H [= 1913/14], I, pp. 321-341. On the author and his work see Sezgin, *GAS*, VII, pp. 361-363, and VIII, pp. 230-231.

[15] On al-Zamakhsharī see the article by Carl Brockelmann in EI_2. I have consulted MS Cairo Dār al-Kutub *adab* 636 (271 fols., copied *ca.* 1200) of his *Muqaddima*, in which the list occurs on fols. 4v-5r.

Ibn Sīda (Murcia, d. 1066),[16] and the *Lisān al-ʿArab* of Ibn Manẓūr (Tripoli, Libya, d. 1311),[17] contain much information on the names of these divisions of day and night. But lexicographical information of a different kind and more relevant to our topic is contained in other sources.

These imply that "the Arabs" had names for the seasonal hours of day and night.[18] Indeed there were several such sets of names for the hours; as we shall see, the situation is so complicated that most of the names used for the hours have several meanings. The question is only: how old is the association of these names with the seasonal hours? Was it simply the case that someone like Quṭrub in the early 9th century cited the *sāʿāt* one after the other, with a *thumma* and various literary citations in between, and then someone else a few decades later simply arranged them in a list, making sure that there were twelve for each the day and night? Yet another example of the systematization of knowledge by early Muslim scholars? Another endeavour by later authors to interpret old materials collected by the early philologians in a rational way?[19] Whatever may be the case, the names adopted for the prayers in Islam have their counterparts in the names for the corresponding "hours" in these schemes. Indeed the different traditions for the definitions of the prayer-times, such as whether the *ʿaṣr* should be performed at the end of the ninth hour or the tenth are reflected in the different schemes. Also the fact that sometimes precisely one hour is advocated for the length of the interval permitted for the *ʿaṣr* prayer seems to result from the fact that the *ʿaṣr* is in some traditions precisely the tenth hour.

Which came first, the names of the prayers or the names of the hours? There is no indication of the origin of these schemes of twenty-four names. More than one scheme that I have found dates from the early 10th century, and it seems to me probable that the schemes are much earlier and that the divergences might be dialectal. Seasonal hours were used throughout the Hellenistic world, including Palestine and Northern Arabia (**2.1**). The Jews and Christians of the Ḥijāz would have been familiar with the seasonal hours (**Ch. 9**), and with them the pagan

[16] Ibn Sīda, *al-Mukhaṣṣaṣ*, 1-17 parts in 7 vols., Būlāq: al-Mabaʿa al-Kubrā al-Amīriyya, 1316-21 H [= 1898-1903], also Beirut: al-Maktab al-tijārī li-'l-ṭibāʿa wa-'l-tawzīʿ wa-'l-nashr, n. d., IX, pp. 37-59. On the author see the article by M. Talbi in EI_2.

[17] Ibn Manṣūr, *Lisān al-ʿArab*, 15 vols., Beirut: Dār Sādir, n. d., VIII, pp. 169-170 (*sub s-w-ʿ*). On the author see the article by Johannes Fück in EI_2. His main source was the *Ṣiḥāḥ* of al-Jawharī (d. *ca.* 1010), on whom see the article "al-Djawharī" by L. Kopf in EI_2.

[18] These names were overlooked by Charles Pellat in his article "Layl and nahār" in EI_2. The phenomenon is not restricted to the Islamic world: see Francis R. Maddison and Anthony J. Turner, "The Names and Faces of the Hours", in *North Festschrift*, pp. 125-156 (to the bibliography add François Masai, "Les noms des heures et les textes de Cassien intéressant l'histoire de prime", *Archivum latinitatis medii aevi* (Brussels) 19 (1945), pp. 23-37).

I was surprised to find my tables of hour-names published in Maddison & Turner, *op. cit.*, pp. 136-139. I had sent the tables to my friend, Anthony Turner, without asking what he proposed to do with them. The presentation of the tables is sloppy (my six tables are carelessly suppressed into five!) and the transliteration is imprecise. In addition, "translations" of the names by Emilie Savage-Smith are given for the Arabic terms, some of which are incorrect and others absurd: from them the innocent reader might get the impression that dusk immediately precedes daybreak, that stars twinkle during daylight, or that people chew *qāt* in the middle of the night. I take this opportunity to point out that I had nothing to do with these "translations".

[19] Paul Kunitzsch has drawn my attention to al-Ṣūfī and his attempt at exact identification of Arabic names for stars and asterisms with precise stars in Ptolemy's star catalogue. The results are partly acceptable, but there are many instances where al-Ṣūfī misunderstood the old traditions or where he was technically unable to find the correct interpretation (since he lived too far to the north, he was unable to observe and hence to identify celestial objects with considerable southern declinations). See already n. 0:6.

Arabs of such urban centres as Yathrib and Mecca, if not also some of the Bedouin, more perhaps for philological interest rather than practical purposes.[20]

8.3 The names of the seasonal hours according to Ibn Raḥīq

Let us first consider the names given by Ibn Raḥīq (**Text 8.2**), displayed in **Table 8b**.[21] In some cases the name that he gives refers to an astronomical phenomenon occuring at the beginning of the hour, and in others to a phenomenon occuring at the end of the hour. Thus, for example, the first hour is called *al-shurūq*, which refers to sunrise at the *beginning* of the first hour, and the sixth hour is called *al-zawāl*, which refers to the sun declining from the meridian after the *end* of the sixth hour. It is tempting to think that the names given for the sixth and seventh hours, *zawāl* and *ẓuhr*, have been switched. In a later source, *ẓuhr* is the name of the sixth hour, and *zawāl* is the name of the seventh (see **8.6**). Of particular relevance to the present discussion is the fact that the seventh hour, which begins at midday, is called *al-ẓuhr*, and that the tenth hour, which begins when three-quarters of the day have passed, is called *al-ʿaṣr*. Thus the tenth hour called *al-ʿaṣr* begins when $\Delta z = n$ and ends when $\Delta z = 2n$ if we use the Indian formula. Elsewhere Ibn Raḥīq relates the performance of the *ʿaṣr* at this time to ʿAlī ibn Abī Ṭālib – see **2.2**.

In a treatise on folk astronomy attributed to Aḥmad ibn Fāris, a lexicographer of the late 10th century, similar names of the hours are given, but some are illegible in the available copy.[22] The lists presented by these two authors, Aḥmad ibn Fāris in al-Andalus in the 10th century and Ibn Raḥīq in the Ḥijāz in the 11th or 12th, go back to a common, as yet unidentified, source.

8.4 The names according to Ibn al-Naḥḥās

Two lists of names for the day-hours and one for the night-hours are presented by the Egyptian litterateur al-Nuwayrī (d. 1332) in his encyclopaedic work *Nihāyat al-arab*.[23] These, presented in **Table 8c**, are attributed to the Egyptian grammarian Ibn al-Naḥḥās (d. 950).[24] Here the *ḍuḥā* is the third or fifth hour and the *ʿaṣr* is apparently the eighth hour. The same names are also recorded by the Egyptian encyclopaedist al-Qalqashandī (*fl. ca.* 1400) in his *Ṣubḥ al-*

---

[20] See n. 2:1 on the seasonal hours in Jewish and Islamic practice. In passing we note the existence of a set of mystical names for each of the 24 hours from a Christian Arabic source recorded in Renan, "Testament d'Adam" (cited in n. 9:27 below), esp. p. 461.

[21] Ibn Raḥīq, *al-Mawāqīt*, fols. 3v-4r.

[22] See Kunitzsch, "*Anwāʾ*-Tradition", p. 15. On Aḥmad ibn Fāris see Sezgin, *GAS*, VII, p. 360-361; and on his treatise see now Forcada, "The *Mukhtaṣar min al-Anwāʾ* of Aḥmad ibn Fāris"; and *idem*, "Folk Astronomy of Aḥmad b. Fāris", pp. 116-120, where Forcada compares the names presented by Aḥmad ibn Fāris and Ibn Raḥīq.

[23] al-Nuwayrī, *Nihāyat al-arab fī funūn al-adab*, *ca.* 30 vols. to date, Cairo: Mabaʿat Dār al-Kutub, 1923 to present, I, pp. 131-132 and 147. On the author see the article by M. Chapoutot-Remadi in EI_2.

[24] See the article "Ibn al-Naḥḥās" by J.-C. Vadet in EI_2; Brockelmann, *GAL*, I, p. 138, and S I, p. 201; and Sezgin, *GAS*, II, p. 125. al-Qalqashandī (see n. 8:25), but not al-Nuwayrī, states that this information derives from his *Ṣināʿat al-kuttāb*, which is no longer extant but which was also used by other Mamluk scholars (see Brockelmann, *GAL*, SI, p. 201).

aʿshā[25] (**Text 8.3**) and by the Egyptian polymath Jalāl al-Dīn al-Suyūṭī (d. 1511).[26]

al-Qalqashandī adds a remark from the *Manāhij al-fikr* (of Jamāl al-Dīn al-Waṭwāṭ (d. 1318))[27] that the first person to divide daylight into twelve hours was Adam, and that he had advised his son Seth which acts of worship should be performed at each hour: see further **Ch. 9**.

8.5 The names according to Ḥamza al-Iṣfahānī and al-Thaʿālibī

A different list of names is given by the philologist Ḥamza al-Iṣfahānī (*ca.* 893-970) in his *Kitāb al-Khaṣāʾiṣ wa-'l-muwāzana*.[28] Ḥamza promises 24 names but the published text (**Text 8.4**) is defective in that it contains only 23, as shown in **Table 8d**: one of the names of the day-hours is missing. However, the explanation for this is found in the philological treatise *Kitāb Fiqh al-lugha* (**Text 8.5**) by the litterateur al-Thaʿālibī (*ca.* 960-1038).[29] His list, for which he says he has relied on Ḥamza, contains 24 names. The one missing from Ḥamza's list in the Cairo manuscript is the ninth day-hour *al-qaṣr*, which follows the eighth hour *al-ʿaṣr*. The similarity of the appearance of these two names in unpointed Arabic clearly led to some confusion. al-Nuwayrī (**8.4**) also records these names, citing both Ḥamza and al-Thaʿālibī; he has the twelve day-hours correctly but has shuffled the first few night-hours.

All three of these sources give the name of the fourth day-hour as *al-ḍuḥā*, which corresponds to the definition in terms of shadows for the *ḍuḥā* prayer, namely $\Delta z = n$. According to both of them – given the restoration of *al-qaṣr* – the eighth hour is *al-ʿaṣr* and the ninth hour is *al-qaṣr*, but the standard definition of the time for the beginning of the *ʿaṣr* prayer in terms of shadows, namely $\Delta z = n$, is difficult to relate to the beginning of the eighth hour. Even supposing that *al-qaṣr* and *al-ʿaṣr* should be reversed, the definition $\Delta z = n$ applies only to the end of the ninth hour. Note that the Andalusī and Maghribi practice, attested already in the practice of the Caliph ʿUmar II (**7.1**), was to perform the *ẓuhr* prayer in this eighth hour.

In the statement attributed to Jaʿfar al-Ṣādiq by al-Bīrūnī (**2.2**), the times for the *ẓuhr* and *ʿatama* prayers correspond precisely to the beginnings of the hours with the same names in the list of Ḥamza and al-Thaʿālibī. The *maghrib* prayer begins with the first hour of sunset, called *al-shafaq* by them, but the name of the *ghadāt* prayer beginning two hours before sunrise has no parallel in their names for the night-hours before sunrise. Likewise, only if we assume

---

[25] al-Qalqashandī, *Ṣubḥ al-aʿshā*, II, pp. 347-349. On the author see the article ("al-Ḳalḳashandī") by C. Edmund Bosworth in EI_2.

[26] On al-Suyūṭī (n. **III**-9:7) see the article by E. Geoffroy in EI_2. I have not been able to consult his recension of the *Kitāb al-Kanz al-madfūn wa-'l-fulk al-mashḥūn* (by the 14th-century scholar Sharaf al-Dīn Yūnus al-Mālikī), printed in Būlāq 1288 H and 1321 H [= 1871/72 and 1903/04] (see Brockelmann, *GAL*, II, p. 90, and SII, p. 81). I owe the reference to the editor of al-Marzūqī's *Kitāb al-Azmina wa-'l-amkina* (n. 8:14) at pp. 335-336, n. 1.

[27] On al-Waṭwāṭ see Brockelmann, *GAL*, SII, pp. 53-54, and EI_2, VIII, 158b, *etc.*

[28] MS Cairo Dār al-Kutub *adab* 90 (49 fols., copied *ca.* 1300), fol. 5r. On Ḥamza al-Iṣfahānī see Sezgin, *GAS*, I, pp. 336-337, VI, pp. 210-211, and VIII, pp. 200-201, and the article by Franz Rosenthal in EI_2.

[29] al-Thaʿālibī, *Fiqh al-lugha*, Cairo n. d. edn., p. 211, Beirut 1885 edn. (n. 8:12), pp. 328-329. On the author (Abū Manṣūr al-Thāʿālibī) see the article by Everett K. Rowson in EI_2.

that the order for the eighth and ninth hours was *al-qaṣr* followed by *al-ʿaṣr* can we relate the *ʿaṣr* prayer to the ninth hour.

al-Thaʿālibi's names for the hours are recorded by ʿAbd al-Raḥmān ibn ʿĪsā al-Murshidī, Ḥanafī *muftī* in Mecca, in his treatise on folk astronomy compiled in the year 1597 and entitled *Barāʿat al-istihlāl.*[30] In both manuscripts of this work that I have consulted there are only eleven names for the day-hours (see **Text 8.6**): *al-qaṣr* has again been omitted! A similar list is given in a modern Lebanese primary-school textbook.[31] The list is given in the first lesson, to be learned by heart by the student (although the names have apparently not been used for a millennium). It contains only 23 names, and differs only in two instances from the list of Ḥamza and the list attributed to al-Thaʿālibī by al-Murshidī (see **Table 8d**). I wonder how many students notice year-by-year that there are only 23 names, and how this is explained by their teachers!

8.6 Other sets of names

al-Murshidī also quotes two short poems attributed to the father of the celebrated Mamluk legal scholar and administrator ʿIzz al-Dīn ibn Jamāʿa (d. 767 H [= 1365/66]), namely Badr al-Dīn Muḥammad ibn Jamāʿa (d. 733 H [= 1332/33]),[32] in which the names of the day and night-hours are recorded. The names for the day-hours are closely related to those given by Ibn Raḥīq with one particularly interesting difference: Ibn Jamāʿa has *ẓuhr* for the sixth hour and *zawāl* for the seventh, whereas Ibn Raḥīq has *zawāl* for the sixth (which is certainly curious, given the meaning of *zawāl*) and *ẓuhr* for the seventh. Ibn Jamāʿa's names for the night-hours are, in the main, a combination and permutation of those of Ibn Raḥīq and Ḥamza / al-Thaʿālibi – see **Table 8e**. al-Murshidī goes on to quote the opinions of Ibn Qutayba expressed in his *Adab al-kātib* (see **8.2**).

Another set of 24 names for the hours, some of which differ from the main sets of Ibn Raḥīq and Ḥamza / al-Thaʿālibi, is found in a late Medinan astronomical work by Muḥammad ibn ʿAbd al-Raḥmān al-Ḥusaynī al-Āmidānī (**Text 8.7**).[33] The readings are corrupt but the names of the day-hours, when restored, are identical to those given by Ibn Raḥīq, and the names of the night-hours are a mixture of those given by Ibn Raḥīq and Ḥamza / al-Thaʿālibi in a new arrangement, as is evident from **Table 8f**.

If these lists of 24 names had been found only in *late* medieval texts I should not dare to overestimate their importance. But the fact that three, if not four, different lists were known already to scholars in the first half of the 10th century leads me to suspect a still earlier provenance for the idea of naming the hours. Also, it may not be insignificant that the two main authorities of the Shīʿites are mentioned in this regard. I should not be surprised to find

---

[30] MSS Cairo KhM 4, fols. 47r-47v and DM 15, fols. 92v-93r. On al-Murshidī see Brockelmann, *GAL*, II, p. 499, and SII, p. 513, Azzawi, *Astronomy in Iraq*, p. 339, and *Cairo ENL Survey*, no. D5.

[31] Ḥannā al-Fākhūrī, *al-Jadīd fī 'l-inshāʾ al-ʿarabī (al-sana al-rābiʿa min marḥalat al-taʿlīm al-ibtidāʾī)*, Beirut: Makta-bat al-madrasa wa-dār al-kitāb al-Lubnānī, 1964 edn., p. 6. This reference was kindly shown to me by Mary-Helen Kennedy.

[32] On the Ibn Jamāʿa family see the article "Ibn Djamāʿa" by Kamal S. Salibi in EI_2.

[33] MS Damascus Ẓāhiriyya 5588, fol. 4r. See also **III-3.4-5** on this source.

these names attested in any of the many other relatively early lexicographical sources that have not been available to me. But we should not expect that any single early authority should record the entire corpus of pre-Islamic knowledge on a particular topic.[34] Likewise, any specialist in Arabic philology could surely add new sources to the ones cited above.[35]

As noted above (**2.4**), the *ʿishāʾ* prayer may be performed at least until the end of the fourth seasonal hour of night. The Qurʾānic injunction is to perform this prayer until "darkness of night" (*ghasaq al-layl*) – see **1.1**, and as can be seen from **Tables 8a-f** the term *ghasaq* was used by different authorities for the first, second or third seasonal hour of night. The first third of the night embraces all of these, but I see no other motivation for the choice of this limit.[36]

We now investigate the possible connection of the *ḍuḥā*, *ẓuhr* and *ʿaṣr* at mid-morning, midday and mid-afternoon with the corresponding times of worship in the early Christian church.

Table 8a
The names of the seasonal hours according to al-Hamadhānī

Day-hours	Night-hours (I)	Night-hours (II)
ṣabāḥ (before sunrise)	*shafaq*	*ghasaq*
bukūr (before sunrise)	*ʿishāʾ*	*faḥma*
ghadā	*ʿatama*	*ʿashwa*
ḍuḥā, raʾd al-ḍuḥā	*suḥra*	*jahma*
ishrāq	*ghalas*	*ghabash*
ḍaḥāʾ	*balja*	*ghaṭash*
shurūq (!)	*tanwīr baʿda ʾl-ṣalāt*	*ẓulam al-layl wa-hanādisuhu wa-khtilāṭuhu*
zawāl, junūḥ		*hadʾa*
hājira, hajīra		*jinḥ*
ẓahīra		*qiʿṭa*
rawāḥ		*suwāʿ*
aṣīl		*hazīʿ*
masāʾ		*buhra*
ʿaṣr, qaṣr		*sāʿ*
ṭufūl, ṭafal		*siʿw*
ʿashiyya		*wahn*
		mawhin
		zulfa
		ruʾba
		suḥra

[34] See n. 0:8.

[35] I have deliberately not included here the names presented by Aḥmad ibn Fāris which came to my attention after the first version of this study was completed. The interested reader may consult the writings of Miquel Forcada cited in n. 8:22. In addition see the new studies of the treatise of Ibn al-Qāṣṣ by Jean-Charles Ducène noted in n. 1:26.

[36] See nn. 2:13 and 2:23.

Table 8b

The names of the seasonal hours according to Ibn Raḥīq

	Day-hours		Night-hours
1	*shurūq*	1	*ghasaq*
2	*raʾd*	2	*faḥma*
3	*mutūʿ*	3	*ʿashwa* or *hadʾ*
4	*tarḥīl*	4	*hadʾa* or *hazīʿ*
5	*hajīr*	5	*surāʿ*
6	*zawāl*	6	*junḥ* or *jawz*
7	*ẓuhr*	7	*hazīʿ*
8	*junūḥ*	8	*juhma* or *ʿaqʿas*
9	*ibrād*	9	*buhra*
10	*ʿaṣr*	10	*hadīʿ* (?) or *suḥra*
11	*ṭaṣīl*	11	*zulfa* or *sudfa*

Table 8c

Names of the hours attributed to Ibn-Naḥḥās by al-Nuwayrī

	Day-hours		Night-hours
1	*bukūr, dhurūr*	1	*shāhid*
2	*shurūq, buzūgh*	2	*ghasaq*
3	*ishrāq, ḍuḥā*	3	*ʿatama*
4	*raʾd, ghazāla*	4	*faḥma*
5	*ḍuḥā, hājira*	5	*mawhin*
6	*mutūʿ, zawāl*	6	*qiṭʿ*
7	*hājira, dulūk*	7	*jawshan*
8	*aṣīl,* ʿaṣr*	8	*hutka*
9	*ʿaṣr,* aṣīl*	9	*tabāshīr*
10	*ṭafal, ṣabūb*	10	*al-fajr al-awwal ****
11	*ʿashī,** ḥadūr*	11	*al-fajr al-thānī*
12	*ghurūb, ghurūb*	12	*al-fajr al-muʿtariḍ*

* Switched by al-Suyūṭī (according to the editor of al-Marzūqī (I, pp. 335-336, n. 1)).
** Omitted by al-Suyūī, according to *idem*.
*** Omitted in al-Qalqashandī, *Ṣubḥ al-aʿshā*, restored by the editor from a parallel list of al-Nuwayrī.

Table 8d

The names of the seasonal hours according to Ḥamza al-Iṣfahānī (H) and al-Thaʿālibī (T) / al-Nuwayrī (N)

	Day-hours H / T - N		Night-hours H / N - T
1	*shurūq*	1	*shafaq - jahma*
2	*bukūr*	2	*ghasaq - shafaq*
3	*ghadūw*	3	*ʿatama - ghasaq*
4	*ḍuḥā*	4	*sudfa - ʿatama*
5	*hājira*	5	*jahma sudfa*
6	*ẓahīra*	6	*zalla*
7	*rawāḥ*	7	*zulfa*
8	*ʿaṣr*	8	*buhra*
9	*aṣīl - qaṣr*	9	*saḥar*
10	*ʿashī - aṣīl*	10	*fajr*
11	*ghurūb - ʿashī*	11	*ṣubḥ*
12	*deest - ghurūb*	12	*ṣabāḥ*

Variants to "Ḥamza" in Lebanese schoolbook (al-Fākhūrī, *Inshāʾ* (see n. 8:31)): Day-hour 3: *ghudwa*; day-hour 9 omitted; night-hour 5: *faḥma*

Table 8e
The names of the seasonal hours
according to Ibn Jamāʿa's father

	Day-hours		Night-hours
1	[*shurūq*]	1	*shafaq*
2	*raʾd*	2	*ʿashwa*
3	*matāʿ*	3	*ghasaq*
4	*tarḥāl* → [*tarḥīl*]	4	*hadʾa*
5	*hajīr*	5	*surāʿ*
6	*ẓuhr*	6	*junḥ*
7	*zawāl*	7	*zulfa*
8	*rawāḥ*	8	*hazīʿ*
9	*junūḥ*	9	*ghalas*
10	*ʿaṣr*	10	*saḥar*
11	*aṣīl*	11	*fajr*
12	*tafāl* → [*tafīl*]	12	*ṣubḥ*

Table 8f
The names of the seasonal hours according to al-Āmidānī
(reconstructed from the corrupt Arabic text)

	Day-hours		Night-hours
1	*shurūq*	1	*maghrib* or *ḥumra*
2	*raʾd*	2	*shafaq*
3	*mutūʿ*	3	*ghasaq*
4	*tarḥīl*	4	*faḥma*
5	*hājira*	5	*sudfa*
6	*zawāl*	6	*juhma*
7	*ẓāhira* (for *ẓahīra*?)	7	*ḥudwa* (?)
8	*junūḥ*	8	*zulfa*
9	*ibrād*	9	*buhra*
10	*ʿaṣr*	10	*suḥūr*
11	*aṣīl*	11	*fajr*
12	*ṭafal*	12	*ṣabāḥ*

CHAPTER 9

SOME REFLECTIONS ON THE ISLAMIC PRAYER RITUAL

> "Five daily prayers: this is the rhythm of *ṣalāt*. It is known that Ignaz Goldziher saw here, not a simple coincidence but the result of a "Persian" influence ... . It seems, indeed, inappropriate to attach too much importance to the number of prayers, the result of contingent evolution rather than of deliberate organisation. A comparison between religions would more usefully examine the integration of the cosmos in the prayer, or the converse ... ." G. Monnot, article "Ṣalāt" in EI_2 (VIII, p. 932a).

9.1 Jewish and Christian prayer

We now seek other reasons why the prayer-times were linked with the seasonal hours, a pursuit that brings us to the problems surrounding the very origin of the institution of prayer in Islam. It is clear from the outset that the Islamic prayer ritual was in some way related to the earlier Jewish and Christian practices but was distinguished by various features resulting from Muslim initiative.

Jewish practice owes its inspiration to a verse from the *Psalms* (LV.17): "Evening and morning, and at noon, will I pray, and cry aloud ...", and a verse in the *Book of Daniel* (VI.10): " ... (Daniel) went into his house, and his windows being open to Jerusalem, he kneeled upon his knees three times a day, and prayed ... ." The three daily prayers of the Jews were called *shaḥarīth*, *minḥa*, and *ʿarbīth*.[1]

Early Christian practice was a combination of these Jewish prayers, performed three times a day at morning, midday, and evening, and the so-called apostolic hours, that is, the third, the sixth and the ninth seasonal hours of daylight.[2] The apostolic hours are traceable to two verses in the *Acts of the Apostles* (III.1 and X.9): "Now Peter and John went up together into the temple at the hour of prayer, being the ninth hour ...", and " ... Peter went up upon the housetop to pray about the sixth hour."

The clearest evidence for external influence in the earliest system of prayers is of course the fact that both the Qurʾānic word *ṣalāt* and expression *iqāmat al-ṣalāt* for prayer and its

---

[1] On the times of Jewish prayer see, for example, *Encylopaedia of Judaism*, 12 vols., New York & London: Funk & Wagnalls, 1901-06, X, pp. 169-170, and Eugen Mittwoch, "Zur Entstehungsgeschichte des islamischen Gebets und Kultus", *Abhandlungen der Preussischen Akademie der Wissenschaften* 2 (1913), pp. 7-9, as well as Kennedy, *al-Bīrūnī's* Shadows, II, p. 141.

[2] For an introduction to early Christian practice see the article "Worship" in J. Hastings, ed., *Encyclopaedia of Religion and Ethics*, 12 vols., Edinburgh: T. & T. Clark, 1921, and the following articles in F. L. Cross, ed., *The Oxford Dictionary of the Christian Church*, Oxford: Oxford University Press, 1957: "Breviary", "Compline", "Lauds", "Mattins", "None", "Office, Divine", "Prime", "Sext", "Terce-Sext-None", "Vespers", and "Vigil". A useful survey of the various practices of the early Church in the East and the West is R. Taft, S.J., *The Liturgy of the Hours in East and West – The Origins of the Divine Office and Its Meaning for Today*, St. John's Abbey, Collegeville, Minn.: The Liturgical Press, 1986.

performance are Aramaic. The borrowing in this case was apparently from Christian, rather than Jewish, sources.[3]

Various Western scholars have investigated the origin and early development of the institution of the five prayers of Islam. Ignaz Goldziher identified the two prayers of the first Muslims in the morning and afternoon, but his assertion that "only later were the other three added" cannot be taken at face value because the morning prayer (*ḍuḥā*) did not become one of the canonical five.[4] He later suggested that the five prayers were instituted under the influence of the five prayer times of the Persians, the first three having been appointed by Muḥammad after the Jewish pattern.[5] Some 75 years later the Zoroastrian connection is taken for granted by Mary Boyce.[6] Eugen Mittwoch pointed only to the Jewish precedents for Islamic practice. The closest that he came to our subject was in noting the correspondence between the *ẓuhr* and the *ʿaṣr* and the two Jewish *minḥa*s, a minor one at about half an hour after midday and a major one three or four hours later.[7] Charles Torrey was of the opinion that "the (Islamic) regulations concerning prayer are very obviously derived in the main from Jewish usage".[8] L. A. Ginzberg claimed that the Arabian Jews prayed five times a day and that the Islamic practice was "undoubtedly ordained by Muḥammad as a result of this early Jewish practice".[9] In this connection, it should be remembered that in Islamic tradition the number five was negociated down from 50 by Muḥammad on the advice of Moses, and Muḥammad settled for five even though Moses thought it was too many.[10]

Tor Andrae, Richard Bell and Alois Mingana, on the other hand, favoured the possibility of Christian influence in the formation of the Islamic prayer ritual,[11] although John Bowman,

---

[3] Jeffery, *Foreign Vocabulary* (n. 8:5), pp. 198-199. See also Carl Brockelmann, "*Iqāmat as-ṣalāt*", in G. Weil, ed., *Festschrift Eduard Sachau ...* , Berlin: Georg Reinmer, 1915, pp. 314-320; Goitein, *Studies* (cited in n. 9:14 below), pp. 75-76; and, more recently, I. K. A. Howard, "The Development of the *Adhān* and *Iqāma* of the *Ṣalāt* in Early Islam", *Journal of Semitic Studies* 26:2 (1981), pp. 219-228.

[4] Ignaz Goldziher, *Muhammedanische Studien*, I, Halle: Max Niemeyer, 1889, p. 37; English translation edited by Samuel M. Stern in Goldziher, *Muslim Studies*, I, London: George Allen & Unwin, 1971, p. 43.

[5] *Idem*, "Islamisme et parsisme", *Revue de l'histoire des religions* 43 (1901), pp. 1-29 (repr. in *idem*, *Gesammelte Schriften*, Hildesheim, 1970, IV), esp. p. 15 (repr. p. 246). See also his article "Islam" in the *Jewish Encyclopaedia*, vol. 6, New York: Funk & Wagnalls, Co., 1925, pp. 651-659, esp. p. 653, cited in A. I. Katsh, *Judaism in Islam: Biblical and Talmudic Backgrounds of the Koran and its Commentaries – Suras II and III*, New York, N.Y.: Bloch Publishing Co. for New York University Press, 1954, introduction (*sub* Prayer). Goldziher's view was repeated by Carl Brockelmann (Katsh, *ibid.*). See also the quote at the beginning of this chapter, in which his view is treated with appropriate caution.

[6] See her passing remark in *A History of Zoroastrianism*, vol. I: *The early period*, Leiden and Cologne: E. J. Brill, 1975, p. 259, n. 36:

> "The Muslims followed the same practice (as the Zoroastrian laity who performed prayers before sleeping and others before dawn so that they could have an unbroken night's rest) when they adopted the five daily prayers from Zoroastrianism."

[7] Mittwoch, "Gebet" (n. 9:1), esp. p. 8.

[8] Charles C. Torrey, *The Jewish Foundation of Islam*, 2nd edn., New York, N.Y.: Ktav Publishing House, 1967, pp. 135-136.

[9] L. A. Ginzberg, *A Commentary on the Palestinian Talmud*, New York, N.Y., 1941, p. 73 (in Hebrew, not consulted), quoted in Katsh, *op. cit.* I have not consulted a recent article by Meir and Menahem Kister (again in Hebrew) in which they also speculated that the Jews of Arabia may have prayed five times a day.

[10] See the EI_2 article "Miʿrādj. 1. In Islamic exegesis ... " by Joseph Horovitz, updated by B. Schrieke., esp. p. 99b.

[11] Tor Andrae, *Der Ursprung des Islams und das Christentum*, Uppsala & Stockholm: Almqvist & Wiksells, 1926, esp. p. 194; Richard Bell, *The Origin of Islam in its Christian Environment*, London: Macmillan and Co.,

in his article on the debt of Islam to Monophysite Syrian Christianity, did not include the institution of prayer at all.[12] Joseph Chelhod mentioned the times of prayer only in passing in his survey of ritual prayer in Islam, even though he delved into some aspects of the origin of the prayers.[13] More recently S. D. Goitein, in a penetrating essay on prayer in Islam, attempted to account for the five prayers of Islam. He suggested that it might have been in their role as a "community in the middle" (*umma*^tan^ *wasaṭ*^an^, as in *Qur'ān* II.143) that the Muslims adopted "a reasonable median between the seven 'hours' of the Christian monks and the three services of the Synagogue".[14] But the answer to the question why there came to be five prayers in Islam is not to be found in such deliberations.

Arent J. Wensinck surveyed the relevant *ḥadīth* dealing with times of prayer[15] but none of the other scholars mentioned above has seriously considered the definitions of the *times* of the prayers that became standard.[16] It was M. Th. Houtsma who already in 1890 noticed that if one drops the optional *ḍuḥā* and *tahajjud* from the "full" set of seven prayers one obtains the number five.[17] Having now established that the "standard" definitions of the daylight prayers were originally intended to link the *ḍuḥā*, *ẓuhr* and *ʿaṣr* to the third, sixth and ninth hours, I would simply point out that we have bridged a gap to the practice of the monastic orders of the Syrian Christian Church in the centuries before Islam.[18] The bridge connects the *times* of Syrian Christian prayer and the *times* of Muslim prayer.

9.2 The orders of the Syrian Christians

It may be useful to briefly outline here the development of the Christian institution of prayer in the Eastern Mediterranean region.[19]

1926, p. 143; and Alois Mingana, "Syriac Influence on the Style of the Ḳur'ān", *Bulletin of the John Rylands Library, Manchester* 2 (1927) (reprinted Nendeln, Liechtenstein: Kraus Reprint Limited, 1967), pp. 77-98, esp. pp. 85-87 (this last source deals only with Qur'ānic religious vocabulary – I remember reading as a student something by Mingana on the Islamic prayers but cannot relocate it).

[12] See John Bowman, "The Debt of Islam to Monophysite Syrian Christianity", in *Essays in Honour of Griffithes Wheeler Thatcher 1863-1950*, E. C. B. MacLaurin, ed., Sydney: Sydney University Press, 1967.

[13] Joseph Chelhod, "La prière rituelle dans l'Islam", *Revue de l'histoire des religions* 156 (1959), pp. 161-187, esp. p. 167.

[14] S. D. Goitein, chapter "Prayer in Islam", in *Studies in Islamic History and Institutions*, Leiden: E. J. Brill, 1966, pp. 74-89, esp. p. 85.

[15] Articles "Mīḳāt" in EI_1, repeated in EI_2, and "Ṣalāt" in EI_1. (See n. 0:1 above.)

[16] This also holds for G. Monnet, author of the extensive article "Ṣalāt" in EI_2.

[17] Houtsma, "Çalat" (n. 1:1), p. 133:

> "Hoe is men nu tot een aantal van juist vijf dagelijksche *çalâts* gekomen? Wel man liet allererst vallen het voor-middag gebed (*ṣalāt al-ḍuḥā*) en de nachtelijke vigilien (*tahajjud*), waarover boven reeds het een en ander is opgemerkt. Nog steeds zijn beide *çalâts* gebruikelijk, dat is *sonnat*-voorschriften, maar verplicht zijn zij niet. Op die wijze blijven van zelf vijf *çalats* over."

Houtsma did not venture to seek the origin of the seven prayers.

[18] Andrae, *Ursprung* (n. 9:11), pp. 195-196, drew attention to the ritual of the Syrian Church but did not pursue the relation of the individual prayers to the corresponding Muslim prayers. See also Hughes, *Dictionary of Islam* (n. 2:22), pp. 177-178.

[19] For this section I have relied mainly on G. E. Gingras, *Egeria: Diary of a Pilgrimage*, New York: Newman Press, 1970, pp. 27-28 and notes. See also C. W. Dugmore, *The Influence of the Synagogue upon the Divine Office*, (first published 1944), 2nd edn., Leighton Buzzard, Beds. (U.K.): The Faith Press, 1964, pp. 59-70, and P. E. Bradshaw, *Daily Prayer in the Early Church*, New York: Oxford University Press, 1982, as well as the more recent study Taft, *Liturgy of the Hours* (n. 9:2)

From the 2nd century onwards, in the lists of times to be especially observed by the Christians, the third hour was linked to the sixth and ninth hours as one of the traditional times of prayer. The *Didache*, apparently compiled between 50 and 70 A.D., calls for prayer at these three times. Tertullian (160?-230?) in the *De oratione* speaks of the third, sixth and ninth hours as times of prayer, and in the *De ieiunio* he calls them "Apostolic Hours" because certain events in the life of Christ were associated with them: the descent of the Holy Spirit, Peter's prayer, and the entry of the two Apostles, Peter and John, into the Temple.

Basil of Cappadocia (*ca.* 330-379) wrote of five hours as suitable for monks to offer prayer: morning, the third, sixth and ninth hours, and evening. From the writings of the pilgrim Egeria, who visited Jerusalem *ca.* 385, we learn that there were five services in the weekday liturgy in Jerusalem: nocturns, the morning office, sexts, nones and vespers. An office at the third was performed only in Lent. *The Apostolic Constitutions* (*ca.* 380) speak of six hours of prayer – in the morning, at the third, sixth and ninth hours, in the evening, and at cock-crow. The monks in the Monastery of St. Paula (d. 404) in Bethlehem observed these same hours, as we learn from the monk Jerome: *mane, hora tertia, sexta, nona, vespera, noctis medio per ordinem psalterium canebant.* John Cassian (360?-435?), who sojourned in the monasteries of Syria, and Egypt, and founded a monastery and convent in Marseilles, also referred to six regular hours of daily prayer.

The Syrian Christian tradition known as the "Order of the Apostles", advocated seven times of prayer. It is described, for example, in the *Canons* of Maruta, a bishop of Mayyāfāriqīn in northern Syria *ca.* 410 A.D.[20] The scheme includes prayers in the morning, at the third, sixth and ninth hours, at the time of "the table", in the evening, and in the night. The inspiration for this was the remark of David (*Psalms* CXIX.164): "Seven times a day do I praise Thee because of Thy righteous judgements." I have no information on the way in which the Christian communities of the Near East regulated their prayers at the seasonal hours.

Table 9a

The Prayers of the Syrian Christians

Prayer	Maruta (5th c.)	al-Bīrūnī (11th c.)
daybreak	*ṣafrā*	*iṣbāḥ*
mid-morning	*tlot*	*ḍuḥā*
midday	*ġet*	*niṣf al-nahār*
mid-afternoon	*tġaʿ*	*ʿaṣr*
sunset	*futūrā*	*ghurūb*
nightfall	*ramġa*	*ʿatama*
midnight	*lilyā*	*niṣf al-layl*

al-Bīrūnī, when describing "the prayers of the Christians" (**Text 9.1**),[21] is clearly referring to the "Order of the Apostles". He identifies seven times of prayer:

---

[20] A. Vööbus, *Syriac and Arabic Documents regarding Legislation relative to Syrian Asceticism*, Stockholm: Etse, 1960, p. 142.

[21] al-Bīrūnī, *Shadows*, p. 175; Kennedy, *al-Bīrūnī's* Shadows, I, p. 225 and II, p. 141. (al-Bīrūnī's remarks on the Jewish prayers are garbled – see *ibid.*)

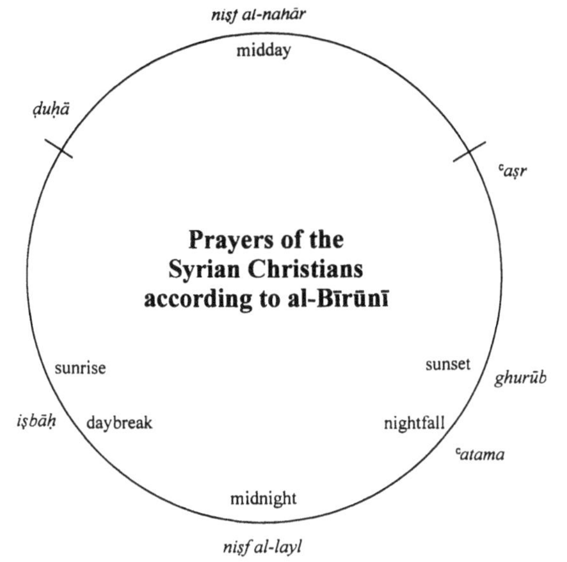

Fig. 9a.

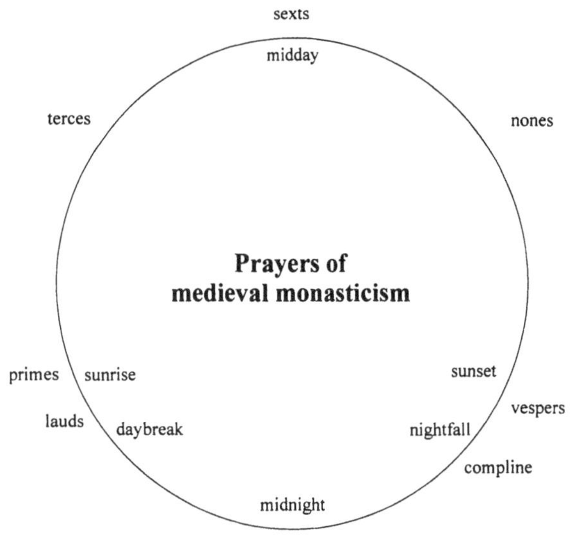

Fig. 9b.

> " … … at midnight (*niṣf al-layl*), early morning (*iṣbāḥ*), mid-morning (*ḍuḥā*), midday (*niṣf al-nahār*), mid-afternoon (*ʿaṣr*), sunset (*ghurūb*), and nightfall (*ʿatama*)."

See **Table 9a** for a comparison of these names with those used by Maruta.

I think that it is useful to compare the seven prayers of the "Order of the Apostles" with the prayers in Islam at the same times, namely, the optional night vigil known as *tahajjud*, and the six prayers called *ṣubḥ*, *ḍuḥā*, *ẓuhr*, *ʿaṣr*, *maghrib* and *ʿishāʾ*. In particular, the *ḍuḥā*, *ẓuhr* and *ʿaṣr* are to be compared with the prayers performed at the apostolic hours. At the risk of gross over-simplification of the highly complicated details of each tradition, I display the canonical prayer-times in Islam together with the orders of the early Syrian Church and the *horae canonicae* common to most orders in medieval Europe schematically in **Figs. 1a**, **9a** and **9b**. In all three traditions the prayer at the third hour was the one most likely to be abandoned.

al-Bīrūnī also discussed the (three) prayers of the "Magians", without dwelling on their times; he further discussed the seven times of prayer of the Manichaeans, which – at noon (*al-ẓuhr*), afternoon (*al-ʿaṣr*), nightfall (*al-ʿatama*), half an hour later, midnight, dawn and sunrise – may be more relevant to our subject, namely, prayer at 3, 6 and 9 seasonal hours of daylight.[22] However, other evidence of any connection appears to be lacking.

In view of the Christian polemical treatises that appeared in the early Abbasid period I can only imagine that the conscious or unconscious imitation of the apostolic hours in Muslim practice occurred somewhat earlier, that is, already in the 7th and early 8th centuries.

---

[22] al-Bīrūnī, *Shadows*, p. 175; and Kennedy, *op. cit.*, I, p. 226 and II, pp. 141-142.

9.3 Times at which prayer is forbidden

This comparison with Christian practice leaves us with more prayers than are found in the standard Islamic ritual. As noted in **Ch. 5**, some early authorities were against performing the *ḍuḥā*, and it survived only in isolated instances. There are also three times during the day when it is forbidden in Islamic law to perform the ritual prayers: at sunrise (when the sun is rising over the horizon), at midday (when the sun is at its highest point in the sky), and at sunset (when the sun is setting over the horizon).

A. J. Wensinck regarded the three times as the moments when the early Muslims attributed to the sun a particularly great magical influence, hence the prohibition of prayer at these times.[23] However, if the sanctity of the locality or of the time is especially significant, the prohibition is suspended; thus there are traditions authorizing prayer at Mecca at all three times. Islamic practice maintained the midday (*ẓuhr*) and sunset (*maghrib*) prayers, but it was generally stipulated that the *ẓuhr* should begin a short time *after* astronomical midday. Thus even in the *ḥadīth* about the Archangel Gabriel (**1.2**), we see the shadow had been observed to have increased by a noticeable amount. Likewise the *maghrib* prayer should begin only when the sun has actually disappeared over the horizon.

Prayer at sunrise was specifically forbidden. The Prophet is reported to have pronounced: "The sun rises between the two horns of the Devil."[24] As I have stated elsewhere, I suspect that the qibla of the pagan Arabs was towards the east, that is, towards the rising sun.[25] If this was indeed the case, one would expect that the time of prayer would be at sunrise, and I am inclined to see the Prophetic prohibition of prayer at sunrise as directed against such an earlier practice. In the period immediately preceding the advent of Islam, the eclectic nature of the worship and pilgrimage rites conducted at the Kaʿba may have led to the adoption of a qibla towards Jerusalem, but the texts are silent on such earlier practice, and speculation without new evidence is futile.

The night vigil, for which Wensinck, Torrey and Goitein see mainly Jewish precedents, was made optional already in the lifetime of the Prophet. The Qur'ānic verse XVII.79 cited above (**1.1**) already distinguishes the *tahajjud* from the other prayers mentioned in the previous verse as a *nāfila* or optional duty. It is reported that ʿĀ'isha, the Prophet's wife, used to get out a mat for his prayers, but the Prophet became angry when his followers found out that she did this, lest they think that the nightprayer was obligatory and not optional.

9.4 The religious significance of the hours

We may also enquire what was so special about certain seasonal hours that they should be selected for prayer. Was the Islamic scheme of prayers merely adopted in principle from early

[23] A. J. Wensinck, "Quelques remarques sur le soleil dans le folk-lore des Semites", in *Mémorial Henri Basset*, 2 vols., Paris: Paul Geuthner, 1928, II, pp. 267-277, esp. p. 273.

[24] *Ibid*., pp. 273-274. See also Rubin, "Morning and Evening Prayers" (n. 5:1), p. 45.

[25] King, "Astronomical Alignments in Medieval Islamic Religious Architecture" (see n. 0:12), pp. 308-309. See also Barthold, "Orientierung der ersten Moscheen".

Christian practice, with a very definite Islamic flavour provided by the *Qur'ān* and *ḥadīth*, let alone by the definitions of the times of the day-time prayers in terms of shadow lengths? Or was there in early Islam a belief, also inevitably borrowed from earlier sources, in the religious significance of certain seasonal hours? We recall that two Mamluk scholars recorded the legend that Adam told his son Seth which acts of worship should be performed at each hour (**8.4**).

In his study of the significance of the afternoon prayer in Islam, Goldziher refrained from any mention of the standard definition of the time of the *ʿaṣr* and (wrongly) associated the beginning of the *ʿaṣr* with the seventh hour.[26] However, he did notice the special significance of the seventh hour in the so-called *Cave of Treasures* or *Testament of Adam* of the Syrian Christians.[27] Even though his conclusions were vitiated by his association of the *ʿaṣr* with the wrong seasonal hour, they may serve as a guide to anyone who would investigate the relationship further.

We have already noted a medieval tradition that it was Adam who first divided the day and night into twelve hours each (**8.4**). It is also of interest that in the *Testament of Adam* (**Text 9.2**), the seventh hour, the beginning of which sometimes marks the beginning of the *ẓuhr*, is stated as being a time when the prayers of every living thing rise up to God, and that also the fourth hour of daylight, corresponding to the beginning of the *ḍuḥā*, is associated with the worship of God by people (as opposed to worship by other living things in the fifth hour). Finally, the ninth hour, to which the *ʿaṣr* corresponds in some Islamic traditions, is a time when the angels worship around the throne of God.[28]

Since it has been shown that some of the non-Biblical elements in the stories about Adam in the *Qur'ān* correspond to those in the *Testament of Adam*,[29] it is possible that the attributes of the hours presented in it may have been familiar to Muḥammad, even though the definitions for the times in terms of shadow increases probably do not pre-date the 8th century. With such reflections, however, I stray from my original purpose, and perhaps also from what may be considered firm ground.

---

[26] *Cf.* Goldziher, "Nachmittagzeit im Islam", pp. 301-302, also Rubin, "Morning and Evening Prayers" (n. 5: 1), p. 55.

[27] The Syriac version is published in Ernst Renan, "Fragments du livre gnostique intitulé Apocalypse d'Adam, ou Penitence d'Adam ou Testa-ment d'Adam, publiés d'après deux versions syriaques", *Journal Asiatique*, 5e série, II (1853), pp. 427-471, and the Ethiopic and Arabic versions are published in Carl Bezold, "Das arabisch-äthiopische Testamentum Adami", in *idem*, ed., *Orientalische Studien Theodor Nöldeke ... gewidmet ...* , Giessen: A. Topelmann, 1906, II, pp. 893-912 (see esp. p. 898).

[28] Some *ḥadīth* attributed to the Prophet opinions on the suitability of different times of the day, such as his answer to the question *hal min sāʿa aqrab ila 'llāh ... min ukhrā* ? (Wensinck *et al.*, *Concordance* (n. 1:4), III, p. 286), but these *sāʿāt* do not correspond to the seasonal hours. On *sāʿa* in classical Arabic see Lane, *Lexicon*, IV, cols. 1467c-1468a, and on its eschatological meaning see Louis Massignon, "Time in Islamic Thought", in J. Campbell, ed., *Man and Time*: *Papers from the Eranos Yearbooks*, (London, 1958), III, pp. 108-11, esp. p. 109.

[29] Article "Ādam" by J. Pedersen in EI_2.

CHAPTER 10

CONCLUDING REMARKS

In any investigation of early Islamic institutions, we should realize that there are aspects that were specifically Arabian or that were initiated by Muḥammad or his immediate followers. This is clear from my investigations of the original function of the Kaʿba, and of the ways in which Muslims have determined the sacred direction towards the Kaʿba in popular practice (*i.e.*, without computation).[1] But there were outside influences as well. I like to think that we have progressed far beyond the stage when Islamicists sought only to identify Jewish or Christian or other foreign antecedents for Muslim practice. However, my investigations of the times of the daylight prayers have unearthed evidence that some aspects of the early Islamic ritual *ṣalāt*, namely, prayers associated with the third, sixth and ninth hours, owed at least some of their inspiration to *Christian* practice, and that some 8th-century refinements of the definitions of these daylight prayers, which resulted in the definitions that became standard, owed their ultimate inspiration to an *Indian* formula for timekeeping. The result was just what was wanted: a highly distinctive and clearly-defined ritual for prayer which has served the community for well over a millennium. And over the ensuing centuries Muslims have paid far more attention to praying at the prescribed times (and in the prescribed direction) than have their Christian neighbours. Nowadays, while Muslims still pray five times a day, to most Christians the hours of the early Christian communities and medieval monasticism are of little more than historical interest (**VIII**).

The institution of Muslim prayer had a long and complex development, only part of which is hinted at in the *Qurʾān* and the canonical *ḥadīth*, and only part of which – the times of the prayers – has been treated here. Before we can write further regarding the development of the physical trappings of the institution of prayer in Islam, we need to consider the various phases of development,[2] namely:

(1) the situation prevailing in Mecca prior to the advent of Islam;
(2) the institutions introduced and practiced during the lifetime of the Prophet;
(3) the practice of the early Muslim community in Mecca and Medina and thereafter in the new Islamic commonwealth of the late 7th and the 8th centuries;
(4) the practice of the community as advocated in the law-books of the 8th and early 9th century, which defined the standard practice thereafter; and
(5) the numerous local deviations from standard practice attested from the 9th century to the present day.

---

[1] See nn. 0:11-12 above.

[2] On the prayers of the early Muslim community see already Rubin, "Morning and Evening Prayers" (n. 5:1). In 2001 I became aware of another study by I. K. A. Howard, "Morning and Evening Prayers in Early Islam", to appear in a new volume on early Islamic institutions edited by G. R. Hawting and published by Variorum-Ashgate in 2002. As this book of mine goes to press, I have not had access to this new study.

In such an undertaking it should be borne in mind that the relevant sources include the kind of traditional, legal, lexicographical and, last but not least, folk scientific material whose surface I have but scratched. Indeed, I can claim only to have worked seriously on the last two items of the above list. And it should not be forgotten that the nature of the sources precludes exact results.[3]

[3] *Cf.* Goitein, *Studies* (n. 9:14), p. 79.

APPENDIX: ARABIC TEXTS

Nota bene: For the convenience of the reader I present the most important texts, both published and unpublished. Single asterisks are used to separate different parts of the same text, double ones to indicate texts taken from different parts of the same work. Medieval Arabic does not always obey the formidable rules of Arabic grammar, and whilst trying to remain faithful to the original sources I have provided *hamza*s throughout, not least to distinguish the preposition *fī* from *fay*ʾ, "shadow"! There are several cases where the texts make little sense.

List of Arabic texts:

Note: The symbol # indicates that I have not been able to consult the original or a copy thereof in the final proof-reading.

1.1 al-Aṣbaḥī, *al-Yawāqīt*, fols. 4v-5r
1.2 Abu 'l-Layth al-Samarqandī, *al-Muqaddima*, pp. 79-83
1.3# al-Mūsawī, *Nahj al-balāgha*, II, pp. 91-92
1.4 Ibn Raḥīq, *al-Mawāqīt*, fol. 8r
1.5 al-Aṣbaḥī, *al-Yawāqīt*, fols. 6r-6v
2.1 al-Aṣbaḥī, *al-Yawāqīt*, fols. 18v-19r
2.2 al-Aṣbaḥī, *al-Yawāqīt*, fol. 4r
2.3 Ibn Raḥīq, *al-Mawāqīt*, fol. 29r
2.4 al-Aṣbaḥī, *al-Yawāqīt*, fols. 38r-38v
2.5 al-Bīrūnī, *Shadows*, p. 163 (MS Bankipore 2468, fol. 227r)
2.6 al-Aṣbaḥī, *al-Yawāqīt*, fol. 18r
2.7 al-Bīrūnī, *Shadows*, p. 143-144 (MS Bankipore 2468, fol. 223r)
2.8# MS Cairo DM 1169,2, fols. 57r-57v
2.9# MS Leiden Or. 1111, fol. 111v
3.1 Mālik, *al-Muwaṭṭa*ʾ, pp. 6-7
3.2 al-Shaybānī, *Muwaṭṭa*ʾ *Mālik*, I, pp. 31-33
3.3 MS Escorial ár. 955, fols. 99v-100r
3.4 al-Shāfiʿī, *al-Umm*, I, pp. 62-63
3.5 al-Kulaynī, *al-Kāfī*, III, pp. 276 and 277
3.6 al-Bīrūnī, *Shadows*, pp. 168-169 (MS Bankipore 2468, fol. 228r)
3.7 MS Princeton Yahuda 4657 = Mach 4983, fols. 90r-91r
3.8 MS Cairo ṬM 127, fol. 52r
3.9 MS Cairo TJ 200,5, *ca*. fol. 29
4.1 al-Aṣbaḥī, *al-Yawāqīt*, fol. 8r
4.2 al-Aṣbaḥī, *al-Yawāqīt*, fols. 2v-3r
4.3# MS Milan Ambrosiana X73 (= Griffini 37), fols. 22r-22v
4.4 al-Bīrūnī, *Shadows*, p. 185 (MS Bankipore 2468, fol. 231v)

4.5# MS Oxford Bodley 133, fols. 124v and 125v
4.6 al-Bīrūnī, *Shadows*, p. 162 (MS Bankipore 2468, fol. 226v)
4.7 MS Istanbul Haci Mahmud Efendi 5713, fol. 3v
4.8 MS Cairo B 19385, p. 11
4.9 MS Berlin Ahlwardt 5793, fols. 89v and 90r
4.10 MSS Oxford Hunt 566, fol. 38v, Birmingham Mingana 560, fol. 16v, and Cairo DM 970, fol. 14r
4.11 MS Princeton Yahuda 1168 (= Mach 4913)
5.1 al-Kulaynī, *al-Kāfī*, III, p. 453
5.2# MS London BL 411,2, fol. 45v
5.3 Ibn Taghrībirdī, *al-Nujūm al-Ẓāhira*, IV, p. 222
5.4 MS Aleppo Awqāf 970, fols 5r and 6r
6.1# MS Cairo DM 534,2, fol. 19v
6.2 MS Cairo ḤM 19, fol. 47v
6.3 Ibn al-Ajdābī, *al-Anwā'*, pp. 117-118
7.1 al-Aṣbaḥī, *al-Yawāqīt*, fols. 3v, 13v, 38v and 3r
7.2 al-Khwārizmī, *Mafātīḥ al-ʿulūm*, p. 235
7.3 MS Cairo DM119,2, fol. 8v
7.4 al-Bīrūnī, *Shadows*, p. 185 (MS Bankipore 2468, fol. 231v)
7.5 Cheikho, "Mukhula", p. 77, checked using a defective microfilm
7.6 al-Aṣbaḥī, *al-Yawāqīt*, fols. 10v-11r, 13v-14r, and 16v-17r
7.7 al-Bīrūnī, *Shadows*, p. 171 (MS Bankipore 2468, fol. 228v)
8.1# al-Hamadhānī, *al-Alfā'*, pp. 287-289
8.2 Ibn Raḥīq, *al-Mawāqīt*, fols. 3v-4r
8.3 al-Qalqashandī, *Ṣubḥ al-aʿshā*, II, pp. 347-349
8.4 MS Cairo Dār al-Kutub *adab* 90, fol. 5r
8.5 al-Thaʿālibī, *Fiqh al-lugha*, Cairo n.d. edn., p. 211, and Beirut 1885 edn., pp. 328-329
8.6# MSS Cairo KhM 4, fols. 47r-47v, and Cairo DM 15, fols. 92v-93r
8.7# MS Damascus Ẓāhiriyya 5588, fol. 4r
9.1 al-Bīrūnī, *Shadows*, p. 175 (MS Bankipore 2468, fol. 229v)
9.2 Bezold, "Testament of Adam", p. 898

1.1 al-Aṣbaḥī, *al-Yawāqīt*, fols. 4v-5r

... قال أتاني جبريل فصلى بي الظهر حين زالت الشمس والفئ مثل الشراك وصلى بي العصر حين صار فئ كل شئ مثله وصلى بي المغرب حين غابت الشمس وحل فطور الصائم وصلى بي العشاء الأخيرة حين غاب الشفق وصلى بي الصبح حين طلع الفجر المعترض ثم صلى بي الظهر في الغد حين صار فئ كل شئ مثله وصلى بي العصر حين صار فئ كل شئ مثليه وصلى بي المغرب حين غابت الشمس وحل فطور الصائم ولم يؤخرها وصلى العشاء الأخيرة حين مضى شطر الليل أو قال ثلث الليل وصلى بي الفجر حين أسفر ...

1.2 Abu 'l-Layth al-Samarqandī, *al-Muqaddima*, pp. 79-83

فصل وإنما قلنا بان الوقت شرط بالكتاب والسنة أما الكتاب فقوله تعالى فسبحان الله حين تمسون وحين تصبحون وله الحمد في السموات والأرض وعشيا وحين تظهرون والمراد به أوقات الصلوات هكذا ورد في التفسير وأما السنة فما روي عن رسول الله صلى الله عليه وسلم أنه قال أمني جبرئل عليه السلام بإزاء باب الكعبة يومين صلى الفجر في اليوم الأول حين طلع الفجر الثاني وصلى الظهر حين زالت الشمس مقدار شراك النعل وصلى العصر حين صار ظل كل شئ مثله وصلى المغرب حين غربت الشمس وصلى العشاء حين غابت الشفق والشفق هو البياض الذي في الأفق بعد الحمرة عند أبي حنيفة رضي الله عنه وقال أبو يوسف ومحمد والشافعي رحمهم الله هو الحمرة وصلى الفجر في اليوم الثاني حين أسفر جدا وصلى الظهر حين صار ظل كل شئ مثله وصلى العصر حين صار ظل كل شئ مثليه وصلى المغرب حين يفطر الصائم وصلى العشاء حين مضى ثلث الليل ثم التفت إلي فقال لي يا محمد هذا وقتك ووقت الأنبياء من قبلك ووقت أمتك ما بين هذين الوقتين

1.3 al-Mūsawī, *Nahj al-balāgha*, II, pp. 91-92

... ومن كتاب له عليه السلام إلى أمراء البلاد في معنى الصلاة أما بعد فصلوا بالناس الظهر حتى تفئ الشمس من مربض العنز وصلوا بهم العصر والشمس بيضا حية في عضو من النهار حين يسار فيها فرسخان وصلوا بهم المغرب حين يفطر الصائم ويدفع الحاج إلى منى وصلوا بهم العشاء حين يتوارى الشفق إلى ثلث الليل وصلوا بهم الغداة والرجل يعرف وجه صاحبه وصلوا بهم صلاة أضعفهم ولا تكونوا فتانين ...

1.4 Ibn Raḥīq, *al-Mawāqīt*, fol. 8r

باب مواقيت الصلاة إعلم وفقنا الله وإياك أن الأوقات لا تعرف إلا بالعيان والرصد لا بالأسطرلاب[1] والتجيم كما ذكر بعض المتأخرين وإنما تحققناها بما ورد به الشرع عن النبي صلى الله عليه وعلى آله حين جاءه جبريل عليه السلم فعلمه أوقات الصلاة في اليومين الذين ذكرهما البخاري ومسلم رحمهما الله فأما من أخذ علمه من الكفرة والسندهند ومن يقول بقولهم فلا يلتفت إليه ولا يقبل قوله ...

1.5 al-Aṣbaḥī, *al-Yawāqīt*, fols. 6r-6v

... إن أوقات الصلوة لا تعرف بدرج الأسطرلاب ولا هي من علم المنجمين تعرف بالحساب وإنما تعرف بالعيان كما جاء عنه صلى الله عليه وسلم أن جبريل أتاهي من عند الله تعالى فعلمه أوقات الصلوة في يومين متواليين والمنجمون إنما أخذوا علمهم عن أقليدس والسندهند وأرسطاطاليس وغيرهم من الفلاسفة وهم كلهم كفرة يكذبون بما جاءت به الأخبار عنه صلى الله عليه وسلم فيكذبون العيان الذي نصّ عليه الرسول صلى الله عليه وسلم وقالوا حسابنا أصدق منه وهذا مذهب قوم يقال لهم السوفسطانيون ...

2.1 al-Aṣbaḥī, *al-Yawāqīt*, fols. 18v-19r

... والإجماع أن النهار اثنتا[2] عشرة ساعة وكذلك الليل بدليل ما رواه يحيى بن الحكم الكسائي عن عبادة بن صهيب عن عثمان بن عطاء الخراساني عن أبيه عن أبي ذر عن النبي صلى الله عليه وسلم أنه قال يا با ذر النهار اثنتا[3] عشرة[4] ساعة فأعد لكل ساعة ركعة وسجدتين نذرا بها عنك كل ذنب غير أنا نقول ساعات نهار الشتاء خفاف وساعات ليله ثقال وساعات نهار الصيف ثقال وساعات ليله خفاف ...

2.2 al-Aṣbaḥī, *al-Yawāqīt*, fol. 4r

... قال ابن سيرين ... والذي لا إله إلا هو إن هذه الساعة لميقات هذه الصلوة ما لها ميقات غيره ...

2.3. Ibn Raḥīq, *al-Mawāqīt*, fol. 29r

... فاذا زاد الفئ بعد ظل الزوال ستة أقدام ونصفا خرج وقت الظهر ودخل وقت العصر وهو ثلثة أرباع النهار كما ورد عن أمير المؤمنين علي بن أبي طالب صلوات الله عليه ...

2.4 al-Aṣbaḥī, *al-Yawāqīt*, fols. 38r-38v

... معرفة الساعات فإذا أردت عرفت ظل زوال يومك فإذا كان من الغد رصدت ظل الشمس في أول النهار فإذا صار ظلك خمس قامات غير ظل الزوال فقد مضت ساعة وهي جزء من النهار من اثني عشر جزءا فإذا صار ظلك قامتين غير ظل الزوال فقد مضت ساعتان فإذا صار ظلك قامة غير ظل الزوال فقد مضت ثلاث ساعات فإذا صار ظلك نصف قامة غير ظل الزوال فقد مضت أربع ساعات فإذا صار ظلك خمس قامة وهي قدم وثلث إلا أربعة أخماس قيراط أو أصبع وخمسا أصبع غير ظل الزوال فقد مضت خمس ساعات وإذا فقد ظلك ولم يبق إلا ظل الزوال بالأمس فقد مضت ست ساعات وانتصف النهار ودخل أول وقت[5] الظهر في أول السابعة فإذا صار الفئ خمس قامة من موضع الزيادة فقد مضت سبع ساعات فإذا صار نصف قامة من موضع الزيادة فقد مضى ثماني ساعات فإذا صار قامة من موضع الزيادة فقد مضت تسع ساعات وهو ثلاثة أرباع النهار هكذا جاء عن علي رضي الله عنه وحينئذ دخل أول وقت العصر في أول العاشرة وخرج وقت الظهر ولا فاضل بينهما من زمان فإذا صار قامتين من موضع الزيادة فقد مضت عشر ساعات وخرج وقت الاختيار من العصر وبقي وقت الجواز على أصح الوجهين مستديما إلى المغرب فإذا صار خمس قامات من موضع الزيادة فقد مضت إحدى عشرة ساعة ثم لا ينضبط الفئ إلى الغروب في الثانية عشر

2.5 al-Bīrūnī, *Shadows* (A: MS Bankipore 2468, fol. 227r; B: published text, p. 163)

... ووسع فقت العصر إذ ليس له في السماء علامة كظاهر هذه الأربع العلامات فحد لها حدا غير مضيق فيه وقال بين أن يصير طل كل شئ مثله إلى مثليه[6] ... ... وروي عنه (اي عن جعفر الصادق) أيضا أنه قال فروض الصلوات في الساعات الأفراد الظهر في أول السابعة والعصر في أول التاسعة والمغرب في أول الأولى والعتمة في أول الثالثة والغداة في أول الحادية عشرة ...

2.6 al-Aṣbaḥī, *al-Yawāqīt*, fol. 18r

... والنهار اثنتا عشرة ساعة أبد الدهر ... وقد يسمونها المعوجة إذ تقصر في النهار الشتاء وتطول في نهار الصيف فإذا مضت ست ساعات دخل وقت الظهر مع أول السابعة مستديما إلى آخر التاسعة ثم يدخل وقت الاختيار من العصر في أول العاشرة مستديما إلى آخرها ثم يدخل وقت الجواز على أصح الوجهين مع أول الحادية عشرة مستديما إلى الغروب ...

2.7 al-Bīrūnī, *Shadows* (A: MS Bankipore 2468. fol. 223r; B: published text, pp. 143-144)

... وكذلك نظمه محمد بن إبراهيم الفزاري قصيدته النجومية فقال فيها في الماضي من النهار ... ... وجل هذه المنظومات هكذا يزاد على الظل الموجود اثني عشر أبدا ويلقى من المبلغ ظل نصف النهار ويقسم على الباقي اثنان وسبعون أصل لا يتغير فتخرج الساعات الماضية قبل الزوال من أول النهار أو الباقية بعده إلى آخر النهار ونسبته ...

2.8 Anonymous Maghribi or Andalusi treatise on the astrolabe: MS Cairo Dār al-Kutub *mīqāt* 1169,6, fols. 57r-57v

معرفة الساعات من الظلال بعد أن تحصل ظل الزوال المبسوط من غاية الارتفاع سواء كان الظل أصابع أو أقدام يعلم ذلك بالضرب أو غيره كما سبق ولا يكون العمل إلا بالمبسوط فإن كان عملك بالأصابع فاعرف ظلها المبسوط للوقت وزد عليه اثني عشر أبدا واعرف الفضل بينه وبين ظل الزوال مبسوطا واقسم اثنين وسبعين على ذلك الفضل يخرج لك ما مضى من الساعات الزمانية إن كنت قبل الزوال وما بقي إن كنت بعده وإن كان عملك بالأقدام فاعمل بها كالأصابع بأن تزيد على ما معك سبعة بتقريب أبدا وتعرف الفضل بينه وبين ظل الزوال أبدا وتقسم على الفضل اثنين وأربعين تخرج لك الماضية قبل الزوال والباقية بعده والله أعلم معرفة أوقات الصلوات وقت الظهر إذا زالت الشمس بمقدار درجة ومضت ست ساعات بوقوع النظير على أول درجة السابعة وإذا وضعت النظير على خط الظهر في الثامنة أو على خط العصر في العاشرة خرج الارتفاع لذلك فتعرف منه ظله معرفة الحصة الشفقية بوضع النظير على مقنطرة ثمانية عشر في جهة المغرب ويوضع جزء الشمس على خط الشفق في الساعة الثانية فتعلم على المري علامة وترد النظير إلى الأفق فما بين المري والعلامة فهي الحصة وكذلك تفعل بالنظير في جهة المشرق وتخرج الحصة الفجرية والله أعلم ...

2.9 Ibn al-Shāṭir on timekeeping by approximate rule: MS Leiden Universiteitsbibliotheek 1111, fol. 111v

... وأما الماضي والباقي من الساعات بالأقدام فقس ظلك وزد عليه ٧ أبدا فما بلغ فأسقط منه ظل زوال يومك وما بقي اقسم عليه مب فما خرج فساعات إن كان صحيحا وإن كان فيه كسر فجزء من ساعة فإن كان قبل الزوال فهو الماضي من النهار وإن كان بعده فهو الباقي للغروب ...

3.1 Mālik ibn Anas, *al-Muwaṭṭa'*, pp. 7-8

... حدثني عن مالك عن نافع مولى عبد الله بن عمر أن عمر بن الخطاب كتب إلى عماله إن أهم أمركم عندي الصلاة فمن حفظها وحافظ عليها حفظ دينه ومن ضيعها فهو لما سواها أضيع ثم كتب أن صلوا الظهر إذا كان الفئ ذراعا إلى أن يكون ظل أحدكم مثله والعصر والشمس مرتفعة بيضاء نقية قدر ما يسير الراكب فرسخين أو ثلاثة قبل غروب الشمس والمغرب إذا غربت الشمس والعشاء إذا غاب الشفق إلى ثلث الليل فمن نام فلا عينه والصبح والنجوم بادية مشتبكة * حدثني عن مالك عن عمه أبي سهيل عن أبيه أن عمر بن الخطاب كتب إلى أبي موسى أن صل الظهر إذا زاغت الشمس والعصر والشمس بيضاء نقية قبل أن تدخلها صفرة والمغرب إذا غربت الشمس وأخر العشاء ما لم تنم (؟) وصل الصبح والنجوم بادية مشتبكة واقرأ فيها بسورتين طويلتين من المفصل * حدثني عن مالك عن هشام بن عروة عن أبيه أن عمر بن الخطاب كتب إلى أبي موسى الأشعري أن صل العصر والشمس بيضاء نقية قدر ما يسير الراكب ثلاث فراسخ وأن صل العشاء ما بينك وبين ثلث الليل فإن أخرت فإلى شطر الليل ولا تكن من الغافلين ...

3.2 al-Shaybānī, *Muwaṭṭa' Mālik*, pp. 31-33

... باب وقوت الصلاة قال محمد بن الحسن أخبرنا مالك عن يزيد بن زياد مولى لبني هاشم عن عبد الله ابن رافع مولى أم سلمة زوج النبي صلى الله عليه وسلم عن أبي هريرة أنه سأله عن وقت الصلاة قال أبو هريرة أنا أخبرك صل الظهر إذا كان ظلك مثلك والعصر إذا كان ظلك مثليك والمغرب إذا غربت الشمس والعشاء ما بينك وبين ثلث الليل فإن نمت إلى نصف الليل فلا نامت عينك وصل الصبح بغلس قال محمد وهذا قول أبي حنيفة في وقت العصر وكان يرى الإسفار بالفجر وأما في قولنا فإنا نقول إذا زاد الظل على المثل فصار مثل الشئ وزيادة من حين زالت الشمس فقد دخل وقت العصر وأما أبو حنيفة فقال لا يدخل وقت العصر حتى يصير الظل مثليه أخبرنا مالك أخبرني ابن شهاب الزهري عن عروة قال حدثتني عائشة أن رسول الله صلى الله عليه وسلم كان يصلي العصر والشمس في حجرتها قبل أن تظهر * أخبرنا مالك أخبرني ابن شهاب الزهري عن أنس بن مالك أنه قال كنا نصلي العصر ثم يذهب الذاهب إلى قباء فيأتيهم والشمس مرتفعة أخبرنا مالك أخبرنا اسحاق بن عبد الله بن أبي طلحة عن أنس بن مالك قال كنا نصلي العصر ثم يخرج الإنسان إلى بني عمرو بن عوف فيجدهم يصلون العصر * قال محمد تأخير العصر أفضل عندنا من تعجيلها إذا صليتها

والشمس بيضاء نقية لم تدخلها صفرة وبذلك جاءت عامةطرثار وهو قول أبي حنيفة وقال بعض الفقهاء إنما سميت العصر لأنها تعصر وتؤخر ...

3.3 al-Kharaqī, *al-Tabṣira fī ʿilm al-hayʾa*: MS Escorial ar. 955, fols. 99v-100r

... وإذا انتهى الظل بنهايته كل يوم عند غاية ارتفاع الشمس فهو [7] أول وقت صلوة الظهر وإذا ازداد ... [8] بمثل المقياس فهو أول وقت صلاة العصر عند أئمة الحجاز وإذا ازداد مثلي المقياس فهو أول وقت صلوة العصر عند أئمة العراق ...

3.4 al-Shāfiʿī, *al-Umm*, I, pp. 62-63

... قال الشافعي رحمه الله تعالى وأول وقت الظهر إذا استيقن الرجل بزوال الشمس عن وسط الفلك وظل الشمس في الصيف يتقلص حتى لا يكون لشئ قائم معتدل نصف النهار ظل بحال وإذا كان ذلك فسقط للقائم ظل ما كان الظل فقد زالت الشمس وآخر وقتها في هذا الحين إذا صار ظل كل شئ مثله فإذا جاوز ظل كل شئ مثله بشئ ما كان فقد خرج وقتها ودخل وقت العصر لا فصل بينهما إلا ما وصفت والظل في الشتاء والربيع والخريف مخالف له فيما وصفت من الصيف وإنما يعلم الزوال في هذه الأوقات بأن ينظر إلى الظل ويتفقد نقصانه فإنه إذا تناهى نقصانه زاد فإذا زاد بعد تناهي نقصانه فذلك الزوال وهو أول وقت الظهر ثم آخر وقتها إذا علم أن قد بلغ الظل مع خلافه ظل الصيف قدر ما يكون ظل كل شئ مثله في الصيف وذلك أن تعلم ما بين زوال الشمس وأول وقت الظهر أقل مما بين أول وقت العصر والليل فإن برز له منها ما يدله وإلا توخى حتى يري أنه صلاها بعد الوقت واحتاط ** ... وقت العصر قال الشافعي رحمه الله تعالى ووقت العصر في الصيف إذا جاوز ظل كل شئ مثله بشئ ما كان وذلك حين ينفصل من آخر وقت الظهر وبلغني عن بعض أصحاب ابن عباس أنه قال معنى ما وصفت وأحسبه ذكره عن ابن عباس وأن ابن عباس أراد به صلاة العصر في آخر وقت الظهر على هذه المعنى أنه صلاها حين كان ظل كل شئ مثله يعني حين تم ظل كل شئ مثله ثم جاوز ذلك بأقل ما يجاوزه وحديث ابن عباس محتمل له وهو قول عامة من حفظت عنه وإذا كان الزمان الذي لا يكون الظل فيه هكذا قدر الظل ما كان ينقص فإذا زاد بعد نقصانه فذلك زواله ثم قدر ما لو كان الصيف بلغ الظل أن يكون مثل القائم فإذا جاوز ذلك قليلا فقد دخل أول وقت العصر ويصلى العصر في كل بلد وكل زمان وإمام جماعة ينتاب من بعد وغير بعد ومنفرد في أول وقتها لا أحب أن يؤخرها عنه ...

3.5 al-Kulaynī, *al-Kāfī*, III, pp. 276 and 277

... الحسين بن محمد الأشعري عن عبد الله بن عامر عن علي بن مهزيار عن فضالة بن أيوب عن الحسين بن عثمان عن ابن مسكان عن الحارث بن المغيرة وعمر بن حنظلة ومنصور بن حازم قالوا كنا نقيس الشمس بالمدينة بالذراع فقال أبو عبد الله عليه السلام ألا أنبئكم بأبين من هذا إذا زالت الشمس فقد دخل وقت الظهر إلا أن بين يديها سبحة وذلك إليك إن شئت طولت وإن شئت قصرت ... ** ... علي بن إبراهيم عن أبيه عن صالح بن سعيد عن يونس عن بعض رجاله عن أبي عبد الله عليه السلام قال سألته عما جاء في الحديث أن صل الظهر إذا كانت الشمس قامة وقامتين وذراعا وذراعين وقدما وقدمين من هذا ومن هذا فمتى هذا وكيف هذا وقد يكون الظل في بعض الأوقات نصف قدم قال إنما قال ظل القامة ولم يقل قامة الظل وذلك أن ظل القامة يختلف مرة يكثر ومرة يقل والقامة قامة أبدا لا يختلف ثم قال ذراع وذراعان وقدم وقدمان فصار ذراع وذراعان تفسير القامة والقامتين في الزمان الذي يكون فيه ظل القامة ذراعا وظل القامتين ذراعين فيكون ظل القامة والقامتين والذراع والذراعين متفقين في كل زمان معروفين مفسرا أحدهما بالآخر مسددا به فإذا كان الزمان يكون فيه ظل القامة ذراعا كان الوقت ذراعا من ظل القامة وكانت القامة ذراعا من الظل فإذا كان ظل القامة أقل أو أكثر كان الوقت محصورا بالذراع والذراعين فهذا تفسير القامة والقامتين والذراع والذراعين ...

3.6 al-Bīrūnī, *Shadows* (A: MS Bankipore 2468, fol. 228r; B: published text, pp. 168-169)

... (نقصان) ... وإنما اختلفوا في الأفضل فذهب الشافعي فيه إلى التقليس نزرا يقصر [9] عن ذراع وإما أن يكون الذراع للزيادة دون نصف النهار نفسه وإلا فسد الأمر باختلاف الزمان والمكان * والوجه الأول أولى بالخبر المتقدم وأليق به من

جهة أنه لم يذكر للعصر فئ الزوال مع المثل والمثلين فاتضح أن الوقت كان يقتضي انتقال الظل واضمحلاله وأنه لما ظهر منه قدر الشراك كان دليلا على الزوال ولو كان له حينئذ قدر لذكر مع المثل والمثلين فهي في أكثر البلاد يزداد في الزوال على المثل في أكثر أيام السنة فإذا لم يلحق به ذكره كان الزوال أول وقت العصر بل ربما بطل وقت الظهر بعدم آخره الذي هو في مشهور الروايات عن أبي حنيفة حين يصير ظل كل شئ مثليه بعد فئ الزوال وروي عنه أيضا بدل المثلين مثلا واحدا وهو قول أبي يوسف ومحمد والشافعي أيضا * والصلوة الثانية صلوة العصر وأول وقتها آخر وقت الظهر ولذلك يتعلق بكون ظل كل شئ مثليه بعد فئ الزوال عند أبي حنيفة في الرواية المشهورة أو مثله في الرواية الأخرى وعند أبي يوسف ومحمد والشافعي وروي عن أبي حنيفة في بعض الروايات أن الظل إذا صار مثل الشئ بعد فئ الزوال فقد خرج وقت الظهر ولا يدخل وقت العصر ما لم يصر الظل مثلي الشئ بعد فئ الزوال ولكن الرواية ليست بمشهورة ووقت العصر عند الشافعي قائم إلى ظل المثلين فمن تجاوزه فإنه وقت الاختيار وحكي عن عطاء وطاؤس ومالك في آخر وقت العصر وهو ظاهر التنزيل في قوله تعالى أقم الصلوة لدلوك الشمس إلى غسق الليل بل تجاوزها أيضا إلى ظلام الليل وبهذه صار من لدن زيادة الفئ مثلا واحدا إلى الغروب مشتركا للظهر والعصر وما قبله محضا للظهر ...

3.7 Sirāj al-Dunyā wa-ʾl-Dīn on the times of prayer: MS Princeton Yahuda 4657 (= Mach 4983), fols. 90r-91r

... صفة زوال الشمس زوال الشمس هو انحطاطها من كبد السماء إلى جانب الغرب فإن الشمس إذا طلعت يقع لكل شئ ظل طويل من جانب الغرب فكلما تزاد الشمس ارتفاعا يزداد الظل نقصانا فإذا استوت الشمس في كبد السماء في أطول يوم من السنة حين تكون في الذراع لا يبقى لكل شئ ظل في البلاد التي هي على خط الاستواء[10] إلى جناب الشمال بل يرجع الظل يرى إلى جانب الجنوب على قدر ارتفاعات البلدان التي هي على خط الاستواء حتى لا يرى للكعبة ظل إلى جانب الشمال إلا ربع قدم وإذا كان في تلك البلاد في ذلك الفصل من السنة كان الخطب يسيرا إذا زالت الشمس يظهر الفئ مثل الشراك وهذا هو معنى قوله عليه السلام فصل بي الظهر حين صار الفئ مثل الشراك وأما في البلاد التي ليست على خط الاستواء في غير ذلك الفصل إذا استوت الشمس يبقى لكل شئ ظل في جانب الغرب يختلف ذلك باختلاف البلدان كل بلد هو أقرب إلى وسط الأرض يكون الظل فيه أقصر ثم يقف ذلك الظل ساعة لطيفة فإذا انحطت الشمس يتحرك ذلك الظل من جانب الغرب إلى جانب الشرق وهو الفئ بعد ذلك ثم يشرع في الزيادة فإذا تحول وأخذ في الزيادة زالت الشمس ودخل وقت الظهر * فصل وإن أردت معرفة الزوال فاعمد إلى أرض مستوية وانصب فيها عودا نصبا معتدلا ولا يميل إلى جانب بعد أن تقسمه سبعة أقسام متساوية وانصب فيها عودا نصبا معتدلا وارصد ظله واعمل على انتهاء ظله علامة فما دام الظل ينقص إلى الخشبة فالنهار في الزيادة والشمس لم تزل بعد فإذا تناهى نقصان الظل وازداد بعد تناهي نقصانه فقد وقع النهار في النقصان وزالت الشمس وهو أول وقت الظهر ثم لا يزال وقت الظهر قائما إلى أن يصير ظل الخشبة المنصوبة مثلها سوى المقدار الذي زالت عليه الشمس فإذا جاوز ذلك وزاد أدنى زيادة فقد خرج وقت الظهر ودخل أول وقت العصر هذا قول الشافعي رحة الله عليه ومالك والأوزاعي وسفيان وأحمد وأبي يوسف ومحمد رضي الله عنهم وقال ابن المبارك وأسحق بعد ما صار ظل كل شئ مثله بقدر أربع ركعات وقت الصلاتين جميعا للظهر والعصر وقال محمد بن جرير الطبري ومالك بعد ما صار ظل كل شئ مثله إلى أن يصير ظل كل شئ مثليه لم يدخل وقت الصلاتين جميعها وقال أبو حنيفة رضي الله عنه وقت الظهر إلى أن يصير ظل كل شئ مثليه سوى في الزوال ثم يدخل وقت العصر ثم عند الشافعي لا يزال وقت العصر قائما إلى أن يصير ظل الخشبة مثليها سوى المقدار الذي زالت عليه الشمس فإذا جاوز ذلك وزاد أدنى زيادة فقد خرج وقت العصر المختار عند الشافعي رضي الله عنه والعمل على ميزان الشمس هكذا يكون والله أعلم بالصواب

3.8 Sirāj al-Dunyā wa-ʾl-Dīn on the times of prayer: MS Cairo Dār al-Kutub Ṭalʿat *mīqāt* 127, fol. 52r

... والآن نشرع في المقصود فنقول أول الأوقات وقت الظهر بدأت بها كغيري لأنها نهارية اتفاقا وأول صلاة ظهرت في الإسلام وأول صلاة وقع الإجماع على وجوبها وسميت بذلك لفعلها وقت الظهيرة وتسمى الأولى لما مر وتسمى الهاجرة لتسمية وقتها بذلك والهاجرة شدة الحر وفي تسميتها حكمة هي ظهور الإسلام وشدته وقوته فلها ثلاثة أسماء ويدخل وقتها

عقب زوال الشمس بالإجماع ويعرف بوجود الظل بعد عدمه وعليه حمل حديث أمنى جبريل عند البيت فصلى بي الظهر حين كان الظل قدر الشراك أي سير النعل والعصر حين كان ظل الشئ مثله أو بزيادة الظل على الأقدام المحسوبة له في محله أو بزيادته بعد تناهي قصره أو بميل الظل المبسوط لرأس الشاخص القائم على الحائط القائم على خط نصف نهار البلد إلى جهة المشرق أو بميل الظل المنكوس لرأس الشاخص القائم على الحائط القائم على خط نصف نهار البلد إلى جهة المشرق بشرطه ويخرج وقته عند غير الإمام أبي حنيفة في إحدى الروايتين عنه وغير الإمام مالك في نحو أرباب الأعذار بتمام ظل المثل بعد ظل الزوال إن كان وظل المثل يسمى ظل القامة وهو سبعة أقدام تقريبا بقياس قدم الشخص لظله على الهيئة السابعة * ثم وقت العصر سميت به لمعاصرتها وقت الغروب أو لعصرها النهار بضيقه أو لأن فضيلة المداومة عليها كفضيلة العصر أي الدهر أو لغير ذلك ويسمى بالوسطى وبالبرد بسكون الراء المهملة لبرد النهار عندها فلها ثلاثة أسماء ويدخل وقتها على ما مر بالزيادة على ظل المثل أو بما هو علامة عليه كالخطوط الموضوعة لذلك ويخرج وقتها على الأصح بغروب جميع قرص الشمس أي بستره عن النظر بنواحي الأفق المرئ وإن بقي شعاعه المتصل به المسمى بضوء حاجب الشمس الأعلا ...

3.9 MS Cairo Dār al-Kutub Ṭalʿat *majāmīʿ* 200,5, ca. fol. 29

... وهو ما ثبت في الصحيحين أنه عليه الصلاة والسلام قال مثلكم ومثل أهل الكتابين كمثل رجل استأجر أجيرا فقال من يعمل لي من غدوة إلى نصف النهار على قيراط فعملت اليهود ثم قال من يعمل لي من نصف النهار إلى صلوة العصر على قيراط فعملت النصارى ثم قال من يعمل لي من العصر إلى غروب الشمس على قيراطين فأنتم هم فغضبت اليهود والنصارة وقالوا كنا أكثر عملا وأقل عطاء قال هل ظلمتكم من أجركم شيئا قالوا لا قال فذلك فضل أعطيه من أشاء * ووجه الاستدلال أنه ضرب قصر المدة لقلة العمل مثلا فكانت مدة العصر أقصر يجب أن تكون النصارى أكثر عملا إلا على قول أبي حنيفة فإن قيل من الزوال إلى صيرورة ظل كل شئ مثليه أكثر من وقت صيرورة ظل كل شئ مثليه[11] إلى آخر النهار فيتحقق كون النصارى أكثر عملا على هذا التقدير أجيب بأن التفاوت بين هذين الوقتين يسير لا يعرفه إلا الحساب والمراد من الحديث تفاوت يظهر لكل واحد من الأمة فثبت بهذه الأحاديث مذهب أبي حنيفة رحمه الله[12]

4.1 al-Aṣbaḥī, *al-Yawāqīt*, fol. 8r

... وقد قدر بعضهم شراك النعل بأصبعين على التقريب وهو سدس قدم ...

4.2 al-Aṣbaḥī, *al-Yawāqīt*, fols. 2v-3r

... روى عن ... عبد الله بن مسعود قال كان قدر صلوة النبي صلى الله عليه وسلم الظهر في الشتاء من خمسة أقدام إلى سبعة وفي الصيف من ثلاثة أقدام إلى خمسة ...

4.3 al-Fārisī on the times of prayer: MS Milan Ambrosiana X73 (= Griffini 37), fols. 22r-22v

... وأما معرفة أوقات الصلوة فأولها الظهر وأول وقته إذا زنت على ظل زوال ذلك اليوم نصف سدس قدم وآخر وقته أن تزيد على ظل الزوال سبعة أقدام وأما العصر فأول وقته آخر الظهر في كل يوم وآخر وقته إذا زدت على ظل الزوال أربعة عشر قدما وهو ضعف القامة[13] ...

4.4 al-Bīrūnī, *Shadows* (A: MS Bankipore 2468, fol. 231v; B: published text, p. 185)

... وذهب بعضهم في وقت الأذان للظهر إلى ازدياد الفئ أصبعا واحدا وفي وقت الإقامة لها إلى زيادة ثلاثة[14] أصابع وفي وقت[15] العصر إلى ازدياد ثلاثة عشر أصبعا فإن سلمنا له وقت أذان الظهر لم تكن زيادة المثل على ظله وإنما هو على فئ الزوال نفسه على أن الأمر غير مسلم له فإنه صادر عن غير معرفة بالأصابع التي هي أنصاف أسداس الشخص سواء كان شبرا أو كان عمود جبل دنباوند (أي: جبل دماوند) مثلا فالأصابع[16] المذكورة منها في أذان الظهر أولى أن

تكون هي[17] أصابع اليد فبالأقل منها يدخل وقت الظهر والأصابع المذكورة في وقت العصر ممتزجة من هذه الواحدة ومن أجزاء المقياس الاثني عشر ...

4.5 Anonymous Egyptian manual for muezzins: MS Oxford Bodleian Bodley 133, fols. 124v and 125v

باب ٩ في معرفة الظهر والعصر إذا أردت ذلك فاعلم أن وقت الظهر زوال الشمس عن وسط السماء لاكن الوقت يؤخذ في المساجد لاجتماع الناس إلى أن يزيد الفئ ذراعا وهو أن يزيد الظل على ظل نصف نهار يومك ثلاث أصابع والثلاثة من اثني عشر ربع وهو وقت صلاة الظهر في المساجد واعلم أن آخر وقت [الظهر هو أو]ل وقت العصر وهو أن يزيد الظل على ظل نصف ال[نهار اثـ]ني عشر أصبعا وهي القامة وآخر وقت العصر إذا زاد الظل على [ظ]ل نصف النهار ليومك أربعة وعشرين أصبعا وهي قامتان فهذا آخر وقت العصر فافهم ترشد ...

4.6 al-Bīrūnī, Shadows (A: MS Bankipore 2468, fol. 226r; B: published text, p. 162)

... وكتب عمر بن الخطاب إلى عماله في سنة سبع عشرة للهجرة أن صل الظهر إذا كان الفئ ذراعا إلى أن يكون ظل أحدكم مثله والعصر والشمس مرتفعة بيضاء نقية قدر ما يسير الراكب فرسخين أو ثلاثة والمغرب إذا غربت الشمس والعشاء إذا غاب الشفق إلى[18] ثلث الليل والصبح والنجوم بادية مشتبكة * وكتب إلى أبي موسى الأشعري أن صل الظهر إذا زاغت الشمس والعصر والشمس بيضاء نقية قبل أن تدخلها صفرة والمغرب إذا غربت الشمس وآخر العشاء ما لم تنم والصبح والنجوم بادية واقرأ فيها سورتين طويلتين من الفصل ...

4.7 Ibn al-Ḥammāmī on the times of prayer: MS Istanbul Haci Mahmud Efendi 5713, fol. 3v

... وأما ظل اختيار الظهر فهو أن تزيد على ظل الزوال ربع قامة وهي ج أصابع على المشهور فما بلغ فهو المطلوب وأما ظل آخره فهو إن تزيد على ظل الزوال قامة فما بلغ فهو المطلوب وأما ظل أول وقت العصر فهو إن تزيد على ذلك أدنى زيادة وظل آخره بزيادة قامتين على ظل الزوال وأما الماضي والباقي من الساعات فهو أن تعلم عدد أصابع الظل للوقت الذي تريد وتزيد عليه قامة ثم أسقط من المجتمع ظل زوال يومك فما بقي اقسم عليه أو افن به (؟؟) عب فما خرج فهو الماضي إن كنت قبل الزوال والباقي إن كنت بعد الزوال والله تعالى أعلم

4.8 al-Bazdawī on the times of prayer: MS Cairo Dār al-Kutub B 19385, p. 11

... فإذا صار ظل كل شئ بعد الزوال مثل نصفه وهو وقت يؤدي صلاة الظهر أصحاب أبي حنيفة رحمهم الله في ذلك الزمان فإنه زمان الصيف فمن استقبل قرص الشمس في ذلك الوقت في بلدة من هذه البلاد فهو قبلة تلك البلدة ...

4.9 al-Khwārizmī on the times of prayer: MS Berlin Ahlwardt 5793, fols. 89v and 90r

... معرفة وقت الظهر والعصر إذا أردت ذلك بالارتفاع فارصد الشمس حتى تقف على خط وسط السماء ثم انقص من الارتفاع سبع درج فهو الظهر وأما وقت العصر خذ ارتفاع نصف نهار يومك فانقصه من تسعين فما بقي خذ عشره وزد العشر على نصف ارتفاع نصف نهار يومك فما خرج فهو ارتفاع وقت العصر إن شاء الله ... * ... وقت الظهر إذا أردت أن تعرف آخر وقت الظهر فضع درج الشمس على ند[19] من السمت من ناحية المغرب ثم انظر على كم وقعت من ارتفاع المقنطرات فما كان فهو ارتفاع آخر وقت الظهر من ناحية المغرب فإذا أردت آخر وقت العصر فانقض ارتفاع نصف النهار من أبدا فما بقي فخذ عشره فما كان فزده على نصف ارتفاع نصف النهار فما بلغ فهو ارتفاع آخر وقت العصر فاعلم ذلك ...

4.10 al-Khujandī on the times of the *ẓuhr*: A: MS Oxford Bodleian Hunt. 566, fol 38v; B: MS Birmingham Mingana 560, fol. 16r; C: MS Cairo Dār al-Kutub *mīqāt* 970, fol. 14r

الباب الثاني والعشرون في معرفة أول وقت الظهر[20] محققا[21] إذا اردنا أن نعلم ذلك رصدنا الشمس[22] عند نصف النهار فإذا وافى نظير درجة الشمس[23] على خط نصف الليل نظرنا إلى راس المري[24] أين وقع من أجزاء معدل النهار وادرنا[25] الصفيحة ثلاثة[26] أجزاء ونرصد الشمس حتى يدخل ضوؤها في ثقبتي الهدفتين جميعا فإذا اتفق ذلك فهو أول وقت الظهر محققا[27]

4.11 Sharaf al-Dīn al-Khalīlī on midday: MS Princeton Yahuda 1168 (= Mach 4913), fol. 10r

... وإذا زيد على الزوال[28] الاصطلاحي مقدار مرور نصف جرم الشمس ومقداره ثلث درجة حصل وقت الزوال الشرعي ...

5.1 al-Kulaynī, *al-Kāfī*, III, p. 453

... علي بن إبراهيم عن أبيه عن حماد بن عيسى عن حريز عن زرارة والفضل عن أبي جعفر وأبي عبد الله صلوات الله عليهما أن رسول الله صلى الله عليه وآله قال صلاة الضحى بدعة ...

5.2 al-Jādarī on the *ḍuḥā*: MS London, British Library 411,2, fol. 45v

... وأما معرفة ارتفاع وقت ركوع الضحى المختار فارتفاعه مثل ارتفاع أول وقت العصر في كل يوم أبدا إلا أن جهة ارتفاعه شرقية أبدا فاعلمه وأما معرفة الإنذار الأول من يوم الجمعة فهو على ساعتين ونصف زمانية والإنذار الثاني على أربع ساعات زمانية ...

5.3 Ibn Taghrībirdī, *al-Nujūm al-zāhira*, IV, p. 222

... وأطلق (الحاكم) صلاة التواريح والضحى وغير الأذان وجعل مكان حي على خير العمل الصلاة خير من النوم وركب بنفسه إلى جامع عمرو بن العاص وصلى فيه الضحى ...

5.4 Anonymous late Syrian treatise on timekeeping: MS Aleppo Awqāf 970, fols. 5r and 6r

... وأما وقت الكراهة بعده (أي: بعد العصر) فهو أن تعرف فضل الدائر[29] لارتفاع أربع درجات ونصف وتزيد ساعاته على ساعات الزوال تحصل ساعات الوقت المطلوب ... ** ... وأما الضحوة الصغرى وهي التي تحرم الصلاة قبلها فحصل الدائر لارتفاع أربع درجات ونصف كما تقدم في الكراهة بعد العصر واطرحه من ساعات الظهر يحصل الوقت المطلوب وهو وقت صلاة العيد والضحاء وأما الضحوة الكبرى وهي التي تصح نية الصوم ولا صلاة الضحاء بعدها فهي نصف ساعات قوس النهار الشرعي الذي ابتداؤه طلوع الفجر الصادق إلى غروب الشمس فإذا أردت وقتها فنصف ذلك القوس وإن شئت فخذ نصف ساعات وقت الفجر الصادق يحصل المطلوب ...

6.1 Anonymous note on the times of the prayers or the two *ʿīd*s: MS Cairo Dār al-Kutub *mīqāt* 534,2, fol. 19v

فائدة الأفضل تأخير صلاة العيد إلى[30] أن ترتفع الشمس قدر رمح كما روي في الصحيحين من فعله صلى الله عليه وسلم لكن جرت العادة إذا كان غيما أن تصلى صلاة عيد الفطر لمضي ربع ساعة وفي صلاة عيد الأضحى لمضي خمس ساعة والله أعلم

6.2 Anonymous Maghribi treatise on sundial construction: MS Cairo Dār al-Kutub Ḥalīm *mīqāt* 19, fol. 47v

... وقد ذكرنا ذلك في كتاب تخطيط البلاطة التي احتوى على ما في هذه البلاطة وزاد عليها بالساعات الزمانية والمعدلات وبخط الاعتدال وبخط االتأهب الأول والثاني في يوم الجمعة ...

6.3 Ibn al-Ajdābī, *al-Anwā*', pp. 117-118

... ثم تزول الشمس عن خط وسط السماء وتنحط إلى جهة المغرب وذلك عند انقضاء الساعة السادسة من النهار ودخول الساعة السابعة فيزيد الظل حينئذ وينتقل إلى جهة المشرق ويسمى فيئا لأنه فاء أي رجع من جانب إلى جانب وذلك أول وقت الظهر وسميت ظهرا لأن وقتها يدخل في الظهيرة والظهيرة أشد أوقات النهار حرا وهي وقت القائلة ثم تزداد الشمس انحطاطا والظل زيادة فإذا صارت زيادة الظل على ما كان عليه في نصف النهار[31] فتلك وسط الظهر وهو الوقت المستحب لصلاة الجماعة وإذا صارت الزيادة مثل طول الشخص فذلك آخر وقت الظهر وهو نفسه أول وقت العصر يكون مشتركا بينهما فإذا زاد الفئ على طول الشخص شيئا خرج وقت الظهر واختص الوقت بالعصر ثم إذا بلغت زيادة الفئ مثلي[32] طول الشخص كان ذلك آخر وقت العصر وسميت العصر عصرا باسم الوقت الذي يصلى فيه والعصر عند العرب العشاء وابتداؤه من وقت امتداد الظلال وانكسار حر النهار يقال جئتك عصرا أي عشاء حين راح النهار ...

7.1 al-Aṣbaḥī, *al-Yawāqīt*, fols. 3v, 13v, 38v and 3r

... عن الأوزاعي أيضا أن عمر بن عبد العزيز نصب ساعات ليعرف بها أوقات الصلوة وما بقي عليه من نهاره وكان مؤذنه يؤذن للظهر إذ مضت ست ساعات ودخل في أول السابعة وذلك حين تزول الشمس ويؤذن للعصر إذ مضت تسع ساعات ودخل أول العاشرة ويأمر بالصلوة فتقام ويصلى ... ** ... عن الأوزاعي أن عمر بن عبد العزيز نصب ساعات ليعرف بها ما مضى عليه من نهاره ... ** ... عن الأوزاعي أن عمر بن عبد العزيز نصب ساعات ليعرف بها ما مضى عليه من نهاره وما بقي منه وهذه الساعات تكون في الشتاء خفافا وفي الصيف ثقالا ... ** ... عن الأوزاعي قال ما زال عمر بن عبد العزيز يتعلم مقادير الشمس وهو خليفة حتى مات رحمه الله ...

7.2 Abū ʿAbdallāh al-Khwārizmī, *Mafātīḥ al-ulūm*, p. 235

... الآت الساعات كثيرة فمنها الطرجهارة ومنها صندوق الساعات ومنها دبة الساعات ومنها الرخامة ومنها المكحلة ومنها اللوح ...

7.3 Quote from the *Tamhīd* of Ibn ʿAbd al-Barr: MS Cairo Dār al-Kutub *mīqāt* 119,2, fol. 8v

وفي التمهيد عن ابن عبد البر أن بعض السلف دخل على عمر بن عبد العزيز رضي الله عنه فوجده لم يصل العصر وقد دخل وقته وتمكن فقال له يا أمير المؤمنين ما كانت الخلفاء رضي الله تعالى عنهم يؤخرون صلاة العصر كهذا فقال له حتى اعرفه قال فما زال عمر بن عبد العزيز يعلم وقت الصلاة بعلامة ويحرره بآلة حتى فارق الدنيا رحمه الله وقد كان بعض السلف ينكر على من يستدل على الأوقات[33] بهذه الآلات لكونه لم يقف على دليل لهم من أثر السلف ولا من غيره حتى اطلع على ما في التمهيد من قصة عمر بن عبد العزيز فزال إنكاره واعتراضه لأن هذه الآلات يعرف بها الأوقات ...

7.4 al-Bīrūnī, *Shadows* (A: MS Bankipore 2468, fol. 231v; B: published, p. 185)

... ولما حكي عن عمر بن عبد العزيز أنه كان يؤذن للظهر على سبع ساعات ويصليها على ثماني ساعات والعصر على عشر ساعات وكانت تلك الساعات[34] معوجة لا محالة عدل بعضهم عن الأظلال إلى خطوطها فجعل خط الساعة العاشرة من خطوط الساعات المعوجة لآخر وقت العصر وخط التاسعة لأول وقته كما جعل خط الثالثة منها لوقت صلوة الضحى وذلك مخالفة للمسنون واجب أن لا يعمل به ...

7.5 Cheikho, "Mukhula", p. 77 (checked against microfilm of the manuscript)

... ثم يلتقي في ٦ خطوط لست ساعات نازلة ولست ساعات طالعة تمام اثني عشر ساعة والخط السادس من هذه الخطوط يسمى خط الاستواء وخط نصف النهار وهو الخط الذي ينتهي إلى الدارة[35] التي في أسفل المكحلة وعلى الخط الخامس من هذه الخطوط نقط وعليها مكتوب ظهر وعلى الخط الثالث منها نقط أيضا مكتوب عصر ...

7.6 al-Aṣbaḥī, *al-Yawāqīt*, fols. 10v-11r, 13v-14r and 16v-17r

... فإذا قدرت الظل فرأيته قد ابتدأ في الزيادة فأول ما يبتدئ وتراه بقينا فاعلم أن الشمس قد زالت ودخل أول الوقت فاعرف الظل كم هو من قدم حينئذ وصل الأولى وما كان من الظل في الميزان فهو مثل ظل الرجل والميزان أصح وأسرع فإذا[36] زاد الفئ ستة أقدام ونصفا وهو فئ كل شئ مثله أعني من موضع الزيادة فقد خرج وقت الظهر ودخل أول وقت العصر فحينئذ صل العصر فإذا زاد الظل ستة أقدام ونصفا وأخرى فصار ثلاثة عشر قدما من موضع الزيادة فهو آخر وقت العصر في الشتاء والصيف والخريف والربيع لا يختلف والميزان أصح ما قيس به لأنه ثاني وقت في السنة إذا ارتفعت لم يمكنك أن تقدر ظلك ظل بدنك برجليك لأنه قصير لا يقدر أن ينظر اليه أو تنكس رأسك فإذا نكست فسد عليك ما تقدره وأول من عمل الميزان يحيى بن سعيد الزبيري وربما كان الرجل في سفينة أو على ظهر دابة لا يمكنه أن يقدر ظل بدنه برجليه فالميزان أصح وأسرع ... ** ... فصل واعلم أن الوقت وقتان وقت رضوان ووقت عفو هكذا جاء في الخبر عن سيد البشر إذ قال صلى الله عليه وسلم أول الوقت رضوان الله وآخره عفو الله قال الشافعي رضي الله عنه إنما يكون الرضوان للمحسنين ويشبه أن يكون العفو للمقصرين وقد أردت بتوفيق الله أن أشرح لك الوقتين بالفئ والأقدام ... وأشرحها بما رواه الوليد بن مسلم عن الأوزاعي أن عمر بن عبد العزيز نصب ساعات ليعرف بها ما مضى عليه من نهاره فوقت الرضوان من الظهر هو من حين تزول الشمس إلى أن يصير الفئ ثلث قامة وهو قدمان وسدس من موضع الزيادة وهذا القدر يستغرق نصف الوقت وهو وقت الرضوان الذي كان يصلي فيه النبي صلى الله عليه وسلم كما روى أحمد بن محمد بن غالب الباهلي رفعه إلى ابن مسعود قال كان قدر صلاة النبي صلى الله عليه وسلم الظهر في الشتاء من خمسة أقدام إلى سبعة وفي الصيف من ثلاثة أقدام إلي خمسة فمعنى الظهر ليحصل على فضل النصف الأول من الوقت إذ هو وقت الرضوان هو أن يدخل في الصلوة في أول جزء من الوقت لا فاصل بينهما ... فإذا صار الفئ قدمين وسدسا من موضع الزيادة كما عرفتك فقد خرج وقت الرضوان من الظهر ودخل وقت العفو مستديما إلى أن يصير فئ كل شئ مثله وهو ستة أقدام ونصف من موضع الزيادة ثم يخرج الوقتان من الظهر ويدخل وقت الرضوان من العصر مستديما إلى أن يصير الفئ تسعة أقدام ونصف سدس راجح من موضع الزيادة أيضا فحينئذ انقضاء نصف وقت الاختيار من العصر وهو وقت الرضوان منه ثم يدخل وقت العفو مستديما إلى أن يصير الفئ ثلاثة عشر قدما من موضع الزيادة وهو فئ كل شئ مثلاه ثم يخرج الوقتان من العصر وقت الرضوان ووقت العفو ويدخل وقت الجواز على أصح الوجهين مستديما إلى الغروب ... ** ... فصل نذكر فيه صورة الميزانين اللذين تتصل بهما إلى معرفة الزوال بالسهولة[37] اعلم ثبتك الله أن الميزان هو أن يأخذ قطعة من خشب طولها نصف شبر وعرضها أصبعان من كل جهة يسمونها ميلا في ميل وتأخذ ميلين مربعين طول كل واحد منهما ثلثا أصبع فتنزلهما في طريق الخشبة متوسطا بين القرنتين تنزيلا صحيحا حتى ينفذ من أسفل الخشبة والباقى منها تقسمه ستة أقسام ونصفا وتسميها أقداما وتاخذ قياس قدم منها بمخدر النقطة فتقسم تلك الخشبة من أصل الميل إلى أصل الثاني قسما صحيحا حتى يأتي آخر نقطة ملصقة مع أصل الميل ولو قسمته قبل أن تنزل الميلين لسهل عليك ولا تلتفت إلى عدد الأقدام وتخدر فيه أربعة أخدار على القرن من الخشبة وتولج خيوطا متساوية وتجمع أطرافها بعقد فحينئذ كمل الميزان...

7.7 al-Bīrūnī, *Shadows* (A: MS Bankipore 2468, fol. 228v; B: published text, p. 171)

... وأما أصحاب أبي عبد الله بن كرام فقد شاهدت منهم من يريد أن يتخذ في وقت العصر بين ذلك سبيلا متوسطا فيأخذ[38] بأوساط الأمور ولم يكد يقطع على هذه الواسطة أيجعلها للظل فتكون زيادة على فئ الزوال مثلا واحدا ونصف مثل أم يجعلها للوقت المتوسط بين وقتي الرأيين فيه فكل واحدة من واسطتي الظل والوقت تخالف الأخرى بالزمان وإن اتفقتا مرة فباتفاق ...

8.1 al-Hamadhānī, *al-Alfāẓ*, pp. 287-289

باب ساعات النهار يقال لأول ساعة من النهار الصباح ثم البكور قبل طلوع الشمس ثم الغداة بعد طلوعها ثم الضحى ورأد الضحى (أصل اللحى والضحى ممدود أي ارتفاع الشمس) ثم الإشراق ثم الضحاء ثم الشروق ثم الزوال والجنوح ثم الهاجرة والهجيرة (وذلك إذا استوت الشمس في كبد السماء) ثم الظهيرة (إذا زالت ساعة) ثم الرواح بعد ذلك (إذا برد النهار وراح) ثم الأصيل ثم المساء بعد ذلك ثم العصر والقصر ثم الطفول والطفل ثم العشية (وهو آخر ساعة من النهار) * (ويقال) لأول ساعة من الليل الشفق وهو وقت صلاة المغرب ثم العشاء بعدما يغيب الشفق ثم العتمة بعد ذلك (إذا اشتدت ظلمة الليل وهدأت العيون) ثم السحرة بعد ذلك ثم الغلس ثم البلجة ثم التنوير بعد الصلاة ... ** باب الظلمة والليل والغسق والفحمة والعشوة والجهمة والغبش والغطش وظلم الليل وحنداسه واختلاطه والهذأة والجنح والقطع والسواع والهزيع والبهرة والساع والسعو والوهن والموهن والزلفة والرؤبة والسحرة (قطعة من الليل) ...

8.2 Ibn Raḥīq, *al-Mawāqīt*, fols. 3v-4r

أسماء الساعات الأولى الشروق والثانية الرأد والثالثة المتوع والرابعة الترحيل والخامسة الهجير والسادسة الزوال والسابعة الظهر والثامنة الجنوح والتاسعة الإبراد والعاشرة العصر (و)الحادية عشر[39] الأصيل (و)الثانية عشر الطفل[40] * ساعات الليل الأولى الغسق والثانية الفحمة والثالثة العشوة ويقال الهدؤ والرابعة الهدأة ويقال الهزيع والخامسة السراع والسادسة الجنح ويقال الجوز[41] وجوز الليل وسطه والسابعة الهزيع والثامنة الجهمة ويقال العقعس والتاسعة البهرة والعاشرة الهديع وتسمى السحرة (و)الحادية عشر الزلفة ويقال السدفة (و)الثانية عشر البلجة والتنوير عند الصلاة

8.3 al-Qalqashandī, *Ṣubḥ al-aʿshā*, II, pp. 347-349

الجملة الخامسة في ساعات الليل والنهار قال أصحاب الهيئة لما كان الفلك متحركا حركات متعددة يتلو بعضها بعضا جعل مقدار كل حركة منها يوما ولما كانت الشمس في حركة من هذه الحركات تارة تكون ظاهرة لأهل الربع المعمور وتارة مستترة عنهم بحدبة الأرض انقسم لذلك مقدار تلك الحركة إلى الليل والنهار فالنهار عبارة عن الوقت الذي تظهر فيه الشمس على ساكني ذلك الموضع من العمور والليل عبارة عن الوقت الذي تخفى عنهم فيه فإنه يوجد وقت الصبح في موضع وقت طلوع الشمس في موضع آخر وفي موضع آخر وقت الظهر وفي موضع آخر وقت المغرب وفي موضع آخر وقت نصف الليل * ولما كانت منطقة البروج مقسومة إلى اثني عشر برجا وكل برج إلى ثلاثين درجة وكانت الشمس تقطع هذه المنطقة بحركة فلك الكل لها في زمان اليوم الجامع لليل والنهار قسم كل واحد منهما إلى اثني عشر جزءا وجعل قسط كل جزء منها خمس عشرة درجة وسمي ساعة ثم لما كان الليل والنهار يزيد أحدهما على الآخر ويتساويان في الاعتدالين على ما مر اضطر إلى أن تكون الساعات نوعين متسوية وتسمى المعتدلة وزمانية وتسمى المعوجة فالمستوية تختلف أعدادها في الليل والنهار وتتفق مقاديرها بحسب طول النهار وقصره فإنه إن طال كانت ساعاته أكثر وإن قصر كانت ساعاته أقل مقدار كل ساعة منه خمس عشرة درجة لا تزيد ولا تنقص والمعوجة تتفق أعدادها وتختلف مقاديرها فإن زمان النهار طال أو قصر ينقسم أبدا إلى اثنتي عشرة ساعة مقدار كل واحدة منها نصف سدس الليل والنهار وهي في النهار الطويل أطول منها في القصير والذي كانت العرب تعرفه من ذلك الزمانية دون المتسوية فكانوا يقسمون كلا من الليل والنهار إلى اثنتي عشرة ساعة ووضعوا لكل ساعة من ساعات الليل والنهار أسماء تخصها * فأما ساعات الليل فسموا الأولى منها الشاهد والثانية الغسق والثالثة العتمة والرابعة الفحمة والخامسة الموهن والسادسة القطع والسابعة الجوشن والثامنة الهتكة والتاسعة التباشير[42] والحادية عشرة الفجر الأول والثانية عشرة الفجر المعترض * وأما النهار فسموا الساعة الأولى منه الذرور والثانية البزوغ والثالثة الضحى والرابعة الغزالة والخامسة الهاجرة والسادسة الزوال والسابعة الدلوك والثامنة العصر والتاسعة الأصيل والعاشرة الصبوب والحادية عشرة الحدود[43] والثانية عشرة الغروب * ويروى عنهم على وجه آخر فيقال فيها البكور ثم الشروق ثم الإشراق ثم الرأد ثم الضحى ثم المتوع ثم الهاجرة ثم الأصيل ثم العصر ثم الطفل بتحرك الفاء ثم العشي ثم الغروب ذكرهما ابن النحاس في صناعة الكتاب * قال في مناهج الفكر ويقال إن أول من قسم النهار إلى اثنتي عشرة ساعة آدم عليه السلام ضمن ذلك وصية لابنه شيث عليه السلام وعرفه ما وظف عليه كل ساعة من عمل وعبادة والله أعلم

8.4 Ḥamza al-Iṣfahānī, *al-Khaṣāʾiṣ wa-ʾl-muwāzana*: MS Cairo Dār al-Kutub *adab* 90, fol. 5r

... ساعات النهار والليل على أربع وعشرين لفظة الشروق ثم البكور ثم الغدو ثم الضحى ثم الهاجرة ثم الظهيرة ثم الرواح ثم العصر ثم الأصيل ثم العشي ثم الغروب وساعات الليل الشفق ثم الغسق ثم العتمة ثم السدفة ثم الجهمة ثم الزلة[44] ثم الزلفة ثم البهرة ثم السحر ثم الفجر ثم الصبح ثم الصباح وباقي سائر الأوقات يجئ بتكرير[45] الألفاظ المختلفة التي معانيها متفقة ...

8.5 al-Thaʿālibī, *Fiqh al-lugha* (A: undated Cairo edition, p. 211; B: Beirut 1885 edition, pp. 328-329)

فصل[46] في تعديد ساعات النهار والليل على أربع وعشرين لفظة عن حمزة بن الحسن وعليه عهدتها ساعات النهار الشروق ثم البكور ثم الغدوة ثم الضحى ثم الهاجرة ثم الظهيرة ثم الرواح ثم العصر ثم القصر ثم الأصيل ثم العشى ثم الغروب ساعات الليل الشفق ثم الغسق ثم العتمة ثم السدفة ثم الفحمة[47] ثم الزلة ثم الزلفة ثم البهرة ثم السحر ثم الفجر ثم الصبح ثم الصباح وباقي أسماء الأوقات يجئ بتكرير الألفاظ التي معانيها متفقة

8.6 al-Murshidī, *Barāʿat al-istihlāl* (A: MS Cairo Dār al-Kutub Khalīl Āghā *mīqāt* 4, fols. 47r-47v; B: Dār al-Kutub *mīqāt* 15, fols. 92v-93r)

... وقد اعتبرت العرب النهار اثنتي عشرة ساعة والليل كذلك وسمت كل ساعة من النهار وكذلك من الليل باسم خاص سواء وافق ذلك الساعات الفلكية كما في وقتي الاعتدال أم خالفها كما في غير ذلك فساعات النهار الشروق ثم البكور ثم الغدو ثم الضحى ثم الهاجرة ثم الظهيرة ثم الرواح ثم العصر ثم الأصيل ثم العشى ثم الغروب كذا في فقه اللغة للثعالبي وقيل فيها غير ذلك قال العز بن جماعة[48] أنشدني والدي لنفسه في ساعات النهار قوله

وأسماء ساعات[49] النهار فهاكها شروق وراد والمتاع[50] وترحال
هجير وظهر والزوال وراحهم[51] جنوح وعصر والأصيل وتطفال

* وساعات الليل الشفق ثم الغسق[52] ثم العتمة ثم السدفة[53] ثم الجهمة[54] ثم الزلة ثم الزلفة ثم البهرة[55] ثم السحر[56] ثم الفجر ثم الصبح ثم الصباح كذلك قاله الثعالبي وقيل فيها غير ذلك قال العز بن جماعة[57] أنشدني والدي لنفسه في أسماء ساعات[58] الليل شعر

هاك ساعات ليلنا باسامي ها[59] لمن ليس بالعلوم يشح
شفق عشوة وقل غسق ثم هدأة شراع تتلوه جنح[60]
زلفة والهزيع مع[61] غلس قل سحر بعدهن فجر وصبح

انتهى * قال ابن قتيبة في أدب الكاتب يقال مضى هزيع من الليل وهدء من الليل وذلك من أوله إلى ثلثه[62] وجوز الليل وسطه وجهمة[63] الليل أول ما أخبره[64] والبلجة آخره وهو مع السحور والسدفة مع الفجر والسحرة والسحر الأعلى والتنوير عند الصلاة[65] والخيط الأبيض بياض النهار والخيط الأسود سواد الليل والهاجرة من الزوال إلى قرب العصر وما بعد ذلك الأصيل والعصر والقصر[66] إلى تطفيل الشمس ثم الطفل والجنوح إذا اجنحت الشمس للمغيب وهما شفقان الأحمر والأبيض فالأحمر[67] من لدن غروب الشمس إلى وقت صلاة العشاء ثم يغيب ويبقي الأبيض إلى نصف الليل انتهى

8.7 al-Āmidānī on the names of the hours: MS Damascus Ẓāhiriyya 5588, fol. 4r

ساعات النهار الشروق البرد (!) المتوح (!) الرجل (!) الهاجرة الزوال الظاهرة الجنوح الأبرص (!) العاحر (!) الأصيل الغفل (!) * ساعات الليل المغرب أو الحمرة الشفق الغسق الفحة (؟؟) الصرفة (؟) الجهمة الحدوة الزلفة النهرة (!) السحور الفجر الصباح

9.1 al-Bīrūnī, Shadows (A: MS Bankipore 2468, fol. 229v; B: published text, p. 175)

... وصلوات النصارى سبع هي لنصف الليل والإصباح والضحى ونصف النهار والعصر والغروب والعتمة...

9.2 Bezold, "Testament of Adam", p. 898

... إعلم يا بني أن الساعة الأولى من النهار ارتفاع تسبحة أولادي إلى الله وفي الساعة الثانية تكون صلوات الملائكة ودعاؤهم وفي الساعة الثالثة تمجيد الطير وفي الساعة الرابعة عبادة الروحانيين[68] وفي الساعة الخامسة عبادة سائر الحيوان وفي الساعة السادسة طلبة الكاروبين وتضرعهم وفي الساعة السابعة الدخول إلى الله والخروج من عنده لأن فيها ترتفع إلى الرب صلوات كل حي وفي الساعة الثامنة تكون عبادة السماويين والنورانيين وفي الساعة التاسعة تكون تشمسة ملائكة الله الذين يقومون بين يدي كرسي وقاره والساعة العاشرة للماء فإن فيها ترفرف روح القدس ويطلع على سائر المياه ... ... وفي الساعة الحادية عشرة تكون بهجة وفرح للصديقين وفي الساعة الثانية عشرة تضرع البشر ودعاؤهم مقبول بين يدي الله

---

[1] MS: بالسطلاب
[2] MS: اثـنـي
[3] MS: اثـنـي
[4] MS: عشـر
[5] MS: لوقـت وقـت
[6] so in A; in the printed text: مـثلها
[7] MS: وهو
[8] MS: غـابه ; what is meant is clearly الظل
[9] last two words A: نزر تقصر, printed text: نـزر القصـــر – the sense is unclear
[10] MS: الاســـتوى
[11] om. original
[12] om. original
[13] last two words MS: ضـعف ظل القامة
[14] MS and printed text: ثلاث
[15] om. printed text
[16] MS: فـاصـــبع
[17] MS: illegible letter followed by ى ; in the printed text: مع
[18] last two words MS: الشـــمس على ; but corrected in the printed text
[19] MS: بد
[20] A: الظاهر
[21] om. A and C
[22] om. C
[23] om. C
[24] A: لموري
[25] A: فـادرنـا
[26] C: ٣
[27] om. C; A: ان شـــا الله
[28] the last two words are repeated in the MS
[29] MS: الدائـرة
[30] MS: om. original
[31] so in the printed text; there seems to be a lacuna here
[32] printed text: مثل
[33] MS: الاقـوات
[34] MS: لسـاعة

[35] so in the original
[36] MS: فلاذ
[37] MS: بالسهولية
[38] so in the MS; in the printed text: لياخذ
[39] MS, here and hereafter: عشرة
[40] MS: الطفا
[41] MS: الحود
[42] lacuna; the editor remarks that in the *Nihāyat al-arib* the term التباشير is followed by the words الفجر الأول ثم الفجر الثاني ثم المعترض ما هنا وبه تعلم
[43] Editor's comment: perhaps this should be الحدور
[44] MS: الدلة
[45] MS: بتبكير. See Text 8.5
[46] B: الفصل السابع عشر
[47] B: الجهمة ; in the notes: وفي بعض الرويات الهجمة والفحمة وكلاهما غلط
[48] the last three words in A: العزيز جماعة
[49] the last two words in A: واسماعات
[50] the last three words in A: وقل دون رادوا المتاع ; in B: وقل دون راد والمتاع
[51] A: روحهم
[52] A: الفسق
[53] A: الشرقة
[54] A: الهجمة
[55] A: السهرة
[56] B: السحرة
[57] See note 48
[58] the last two words in A: اسماعات
[59] the last two words in A: وباسا مها ; and in B: باسام يها
[60] so the whole half line in A and B
[61] the last two words in A: الهرمع ; and in B: مع الهزيع
[62] A: ثلاثة
[63] A: وجمة
[64] A: اخيره
[65] B: الصلوة
[66] A: والعصر
[67] the last two words are missing in A
[68] in the printed text: الروحانين

MANUSCRIPTS CONSULTED

Notes: The references are to the footnotes in **IV** in which the manuscripts are mentioned.

Aleppo Aḥmadiyya Awqāf 970 5:9, 6:4
Baghdad 2982/6276 1:22
Bankipore 2468 1:25, *etc.*
Beirut, Université de St Joseph (unnumbered, defective microfilm available) 7:12
Berlin, Deutsche Staatsbibliothek
  Ahlwardt 5664 1:21, *etc.*
  5724 4:7, 6:9
  5793 4:13, 4:14
Birmingham, Selly Oak Colleges,
  Mingana Coll. 560 4:17
Cairo, Dār al-Kutub al-Miṣriyya (Egyptian National Library)
  adab 90 8:28
  adab 636 8:15
  mīqāt (DM) 15 8:30
  mīqāt 119,2 7:5
  mīqāt 534,2 6:2
  mīqāt 948,1 1:22, *etc.*
  mīqāt 970 4:17
  mīqāt 1169,6 2:19, 4:18
  B 19385 4:12
  K 4311 5:4
  Ḥalīm *mīqāt* (ḤM) 19 6:11
  Khalīl Āghā *mīqāt* (KhM) 4 8:30
  Ṭalʿat *falak turkī* (ṬFT) 29 5:7
  Ṭalʿat *majāmīʿ* (ṬJ) 200,5 3:14
  Ṭalʿat *mīqāt* (ṬM) 127 3:12
  Ṭalʿat *mīqāt turkī* (ṬMT) 9 5:12
Cambridge University Library
  Arberry Supp. 110(a) 1:22
  Browne Or. O.1 5:6
Damascus Ẓāhiriyya 5588 8:33
Escorial ar. 955 3:5
Istanbul (Süleymaniye)
  Haci Mahmud Ef. 5713 4:11
Istanbul Kandilli Observatory 226 6:3
Istanbul Nuruosmaniye 2782 5:6
Leiden Universiteitsbibliotheek
  1111 2:19
London British Library Or.
  411,2 5:4, 6:12
Milan Ambrosiana X73
  (Griffini 37) 4:3
Oxford Bodleian
  Bodley 133 4:9
  Hunt. 566 4:17
  Marsh 134 1:22
Princeton University Library
  Yahuda 1168 (Mach 4913) 4:19
  Yahuda 4657 (Mach 4983) 3:12

Part V

On the role of the muezzin and muwaqqit in medieval Islamic societies

Dedicated to Bashi Sabra
in 1995

ACKNOWLEDGEMENTS AND NOTES TO THIS VERSION

This study is dedicated to Professor Abdelhamid I. Sabra of Harvard University, known to some of his colleagues and friends as "Bashi". I have long respected him both as a man and as a scholar, but I am glad that I did not follow his advice and get involved with philosophical aspects of Islamic science; there are plenty of others who can do that much better. I first encountered Bashi on paper in 1970 when I was writing a seminar paper at Yale on the theory of parallels in Islamic mathematics, and shortly after when he came to Yale to give a lecture I distinguished myself by cleaning the blackboard for him without realizing that he himself had laboriously filled the blackboard with optical extravaganza for his lecture. Our paths have crossed many times since. I can claim to have taught Bashi one very useful thing, namely, that if you go round the back of the breakfast buffet at our favourite hotel in Kuwait, behind the cornflakes and the *ersatz* bacon and all that, you can find a serious Arab breakfast with *fūl*, *ṭaʿmiyya*, *zaytūn*, real *khubz* and many other delights.

Bashi and I discussed the muwaqqits and their activities during his visit to Cairo around 1975, and I decided on the spot to put together all of the information I had gathered. It was long overdue to dedicate this study to Bashi at the conference held in his honour at the University of Oklahoma at Norman in 1995. It was published in F. Jamil Ragep and Sally P. Ragep, with Steven J. Livesey, eds., *Tradition, Transmission, Transformation: Proceedings of Two Conferences on Premodern Science Held at the University of Oklahoma*, Leiden, New York & Cologne: E. J. Brill, 1996, pp. 285-346. It is also a pleasure to thank Sally Ragep for typing the Arabic text into a computer.

In this new version, many of the footnotes of the original have been suppressed, and a few cross-references to **I-II** on tables and **X** on instruments have been inserted. Several of the illustrations in the original article are now found elsewhere in this book and are cross-referenced appropriately.

TABLE OF CONTENTS

CHAPTER 1

INTRODUCTION

For well nigh 1400 years Muslims have been performing their prayers at appointed times; these times are astronomically determined and vary throughout the year as well as from one locality to another. Timekeeping by the sun and stars, particularly the regulation of the times of prayer, was called *ʿilm al-mīqāt* in medieval Arabic. The word *mīqāt*, with plural *mawāqīt*, means "time (of a prayer)", and is attested already in the *Qurʾān* and the *ḥadīth* literature.

Similarly, the office and duties of the muezzin (*muʾadhdhin*), whose responsibility it was to announce the call to prayer from the minaret, can be traced back to the time of the Prophet Muḥammad. Indeed, the history of the institution is fairly well documented in the primary and secondary literature.[1] On the other hand, the role of the muwaqqit, or mosque official responsible for regulating the times at which the muezzin should perform, is considerably more obscure: the primary sources are disappointing for the information they provide, and the secondary sources reflect this. In the first edition of the *Encyclopaedia of Islam* (1913-34), repeated in the second (1989), close to three pages are devoted to the duties of the muezzin, but the muwaqqit enjoys only one sentence:

> "Closely associated with the muezzin is the muwaqqit, the astronomer, whose task it was to ascertain the qibla and the times of prayer; sometimes the chief muezzin did this."

A related appellation dating from a later period of Islamic astronomy is *mīqātī*, meaning a specialist in *ʿilm al-mīqāt*, but not necessarily a member of the staff of a mosque.

It has long been known that a series of astronomers who functioned as muwaqqits in various mosques (mainly in Cairo and Damascus) in the 14th and 15th centuries produced a large number of treatises on simple timekeeping and on the use of instruments, mainly the sine and astrolabic quadrants. One need only sift through the relevant sections of the bio-bibliographical reference works of Suter and Brockelmann to find their names and institutions and lists of their known works. The only historian of science to have written on the office of the muwaqqit, however, was the Turkish scholar Aydın Sayılı. In his valuable history of the observatory in Islam, published in 1960,[2] Sayılı pointed out that simple rules for regulating the prayer-times were laid out in the *ḥadīth*, and suggested that:

> "community prayers created a tendency and tradition of defining these times with greater precision, and the Moslem community did not take the problem lightly. Such activities were entrusted to special astronomers called muwaqqit, and the larger mosques had each such a specialist attached to its staff."

There is in fact no direct evidence that this was the case before the Mamluk period. Similarly,

---

[1] See the following articles in EI_2: "Masdjid" by J. Pedersen and others, esp. VI, pp. 675-677, and "Mīḳāt (legal aspects)" by A. J. Wensinck.

[2] See Sayılı, *Observatory in Islam*, esp. pp. 24, 338, and 315. Interested readers should be aware of the *Sayılı Memorial Volumes*, published recently in honour of Aydın Sayılı.

Sayılı's assertion that because of their value to the community "the offices of muwaqqit ... enjoyed continuous existence" – as opposed to the observatories in Islam, most of which were short-lived – is not based on direct evidence.

Another Turkish scholar, A. Süheyl Ünver, made a survey of the buildings labelled muvakkithanes that were attached to various mosques in Istanbul.[3] Here the muwaqqits kept their books and instruments (see **10**). Ünver's study illustrates the physical background to the activities of the muwaqqits of Istanbul, but, as an art historian, he was unfamiliar with the technical side of their work, to a large extent documented in the manuscripts that abound in the libraries of Istanbul.

Since the religious side of the activities of the muezzins is already quite well documented, the present study will concentrate on the activities of the muwaqqits. It should be borne in mind here that I am not writing about astronomical timekeeping *per se*; indeed I shall mention only briefly the theoretical treatment of spherical astronomy and isolated tables for timekeeping that are found in early Islamic astronomical handbooks, and the detailed account of timekeeping in al-Bīrūnī's monumental book *Ifrād al-maqāl fī amr al-ẓilāl* (= *A Unique Statement on Shadows*).[4] Nor do I intend to present lists of known muwaqqits since these could be easily compiled from the bibliographical sources.[5] Rather, I propose to present the few relevant primary sources that have come to my attention, and to summarize the results of my investigations of the most significant works of the major muwaqqits. Ideally one would like to start with the knowledge of when and where the office of muwaqqit originated, but evidence for this is altogether lacking. The earliest evidence found indicates that the office was already functioning in Egypt in the 13th century. However, in that same century there were mosque officials performing the same function in al-Andalus and the Maghrib but bearing a different title. No materials are available from al-ʿIrāq and points eastward. Are we to conclude that the office of muwaqqit did not extend eastwards of what was later to become the Ottoman Empire?

There were two main traditions in Islamic astronomy: "folk" or "ethno-" astronomy and "mathematical" astronomy.[6] The first was widely known; the second was practised only by a select minority. As a result, there were two main streams in astronomical timekeeping in Islam: one operating at an extremely simple – and sometimes even primitive – level, the other involving complicated – and sometimes highly sophisticated – procedures. The first of these traditions was practised throughout the Islamic world;[7] the second seems to have been restricted mainly to Egypt (13th century onwards), Syria, the Yemen, and also Tunis (14th century onwards), and finally Istanbul (15th century onwards).[8] There is little evidence on the practice in the rest of the Islamic world, notably the Maghrib, al-ʿIraq, Iran, Central Asia and India.

---

[3] See Ünver, "Muwaqqits' Quarters" (in Turkish).

[4] See al-Bīrūnī, *Shadows*, with E. S. Kennedy's translation and commentary.

[5] See in particular Brockelmann, *GAL*; Suter, *MAA*; and *Cairo ENL Survey*.

[6] For surveys of "Islamic astronomy" and "astronomy in the service of Islam" written for a general audience, see Saliba, "Islamic Astronomy and Astrology"; and King, "Islamic Astronomy", A-B, and *idem*, "Science in the Service of ... Islam" (first published 1991), repr. in *idem*, *Studies*, C-I.

[7] See now the survey in Varisco, "Islamic Folk Astronomy". A survey of simple Islamic techniques for timekeeping is **III**.

[8] All known Islamic tables for timekeeping are described in **I-II**. On *mīqāt* in Egypt, Syria, and Turkey see already King, *Studies*, A-IX, X, and XII, and on the Maghrib see King, "Maghrib". Further information on all of the Egyptian and Syrian astronomers mentioned in this study is contained in King, "Astronomy of the Mamluks"; on the Yemeni astronomers see King, *Astronomy in Yemen*.

CHAPTER 2

ON THE TIMES OF PRAYER IN ISLAM

In standard Islamic practice, each of the five daily prayers must be performed within a certain interval of time, the endpoints of which are astronomically determined.[1] The times of prayer in a specific locality vary throughout the year. When expressed in local time they vary with terrestrial latitude, and when expressed in mean time they also vary with terrestrial longitude.

The Islamic day begins at sunset: this is because the calendar is lunar and is regulated by the sighting of the lunar crescent shortly after sunset (see **11**). The interval for the first prayer (*maghrib*) begins at sunset and ends at nightfall. The interval for the second prayer (*ʿishāʾ*) begins at nightfall and may last until one-third or one-half of the night has passed. The interval for the third prayer (*fajr*) begins at daybreak (not to be confused with the appearance of the zodiacal light) and the prayer must be completed by sunrise. According to some authorities the permitted interval for the fourth prayer (*ẓuhr*) begins when the sun starts to decline from the meridian, that is, when the shadow of any object used as a gnomon has been observed to have increased by the slightest amount over its midday minimum. According to other authorities (mainly in al-Andalus and the Maghrib), the interval for the fourth prayer begins when the shadow of the gnomon has increased over its midday minimum by one-quarter of the length of the gnomon. In both cases, the end of the interval is when the time for the fifth prayer (*ʿaṣr*) begins. This is defined as the moment when the shadow has increased by the full length of the gnomon. The end of the interval for the *ʿaṣr* is either when the shadow has increased by an amount equal to twice the length of the gnomon or at sunset.

In some quarters in early Islam (and in isolated localities in later centuries) there was an additional prayer called the *ḍuḥā* which took place during the morning. The *ḍuḥā* began before midday by a period of time equal to the interval between midday and the *ʿaṣr* prayer.[2]

The times of the five prayers that became obligatory are specified neither in the *Qurʾān* nor in the *ḥadīth* literature. In the latter there are conflicting definitions for the times of the daylight prayers in terms of shadow lengths as opposed to shadow increases. The definitions outlined above appear for the first time in 8th-century books on the sacred law. There is evidence indicating that in early Islam the times of the daylight prayers were regulated by specific seasonal hours of the day.[3] (Seasonal hours are one-twelfth divisions of the length of daylight and hence vary throughout the year and with terrestrial latitude.) Ibn Sīrīn, the well-known Companion of the Prophet, is reported to have stated:[4]

---

[1] For further information see Wiedemann & Frank, "Die Gebetszeiten im Islam"; the article "Mīḳāt" in EI_2; E. S. Kennedy's translation of and commentary on al-Bīrūnī's *Shadows*, I, pp. 210-230, and II, pp. 132-143; and *idem*,"Al-Bīrūnī on the Muslim Times of Prayer".

[2] On the *ḍuḥā* in the astronomical sources see **IV-5**.

[3] **IV-2, 7** and **8**.

[4] Recorded by al-Aṣbaḥī (see n. 3:22 below) in MS Cairo DM 948, fol. 4r.

> "This (particular) hour is for regulating (*li-mīqāt*) this (particular) prayer; there is no other (way of) regulating (it)."

In fact, the times of the *ḍuḥā*, *ẓuhr* and *ʿaṣr* are closely related to the apostolic hours of early Syrian Christianity, which were performed at the third, sixth, and ninth seasonal hours of the day. The later definitions of these prayers in terms of shadow increases represent a practical means of regulating the prayers in terms of the seasonal hours. They result from the application of a simple, approximate Indian formula for timekeeping, known to the Muslims in the 8th century, which relates the time of day in seasonal hours to shadow increases.[5]

It is related that the Caliph ʿUmar II used a sundial for regulating his prayers: there can be little doubt that it must have been a Graeco-Roman sundial marked with the seasonal hours.[6] Even after the 8th century there were some who regulated the prayer-times in terms of the seasonal hours. al-Bīrūnī mentions people who used the seasonal hour curves on astrolabes to regulate the daylight prayers;[7] an 11th(?)-century set of prayer tables from Northern Iran indicates times of day with religious significance by dividing the period from sunrise to sunset into eight equal intervals;[8] and an Ayyubid vertical sundial purports to regulate "the hours and the times of prayer" but it bears no special markings for the prayers.[9]

For fixing the time of the sunset prayer, the muezzin in the minaret had an optimum vantage point for ascertaining that the solar disc had disappeared over the local horizon. (A problem arises if the local horizon is not flat. The largest corpus of timekeeping tables for any specific latitude is for Taiz in the Yemen, a city surrounded by mountains on all sides save the northern one; however, no instructions are given on how to observe celestial altitudes with respect to the horizon.) Again, for the two night prayers, which are regulated by twilight phenomena described in Islamic legal works, observation by a muezzin without tables or instruments would suffice.

---

[5] **III-1.3**, *etc.*, and **IV-2.4**.

[6] **IV-7.1**, quoting al-Aṣbaḥī (see n. 3:22 below).

[7] **IV-7.3**, quoting al-Bīrūnī's *Shadows*, for which see E. S. Kennedy's translation and commentary, I, p. 235, and II, p. 147.

[8] **IV-5.3**.

[9] **IV-7.4**.

CHAPTER 3

FOLK ASTRONOMY *VERSUS* MATHEMATICAL ASTRONOMY

Among Muslim traditionalists there has always been an interest in the heavens, which are seen as reflecting the glory of God. The sun, moon and stars are mentioned in the *Qur'ān* in this respect, and the opinions of the earliest religious authorities constituted a sacred cosmology that was documented in a series of texts compiled over the ensuing centuries.[1]

All treatises that are concerned with the sacred law discuss the duty of praying at the right time and in the right direction, and always do so in terms of folk astronomy, never with reference to mathematical methods or tables. Also, two genres of Arabic legal literature called *kutub al-mawāqīt* and *kutub dalā'il al-qibla* arose in the first few centuries of Islam and flourished particularly in Yemen and the Hejaz in later centuries.[2]. These works are based upon verses from the *Qur'ān*, the sayings and practices of the Prophet recorded in the *ḥadīth* literature, and the pronouncements of early authorities concerning the times and direction of prayer. They are characterised by their non-mathematical approach to the subject. Occasionally, simple schemes are presented for time-reckoning, using shadows by day and the lunar mansions by night (see **5**),[3] and rather inaccurate procedures are prescribed for finding the qibla by the sun, the stars and the winds (see **12**). No doubt those who used such books were invariably at odds with the astronomers and "real" muwaqqits.

In his treatise against astrology, the 12th-century ʿIrāqī religious scholar al-Khaṭīb al-Baghdādī discussed the aspects of astronomy which were acceptable, listing them as follows:[4]

> "The science of star nomenclature, the appearances of the stars, their risings and settings, their courses across the sky, using them for finding direction, the movements of the Bedouin from their watercourses or wells according to the times indicated by the stars, their choice of times for their young animals and for fertilisation by their male animals, their knowledge of the rains according to the changing night sky, their methods for telling the good from the bad, fixing the direction of the qibla by means of the stars, and finding the times of prayer and the hours of the night by the appearances and settings of the stars."

These are all aspects of folk astronomy, and there is no mention of systematic observation, theory or calculation, on which mathematical astronomy is based.

The 13th-century Yemeni scholar al-Aṣbaḥī was associated with the Mosque of al-Janad, a few miles north of Taiz. Although he does not call himself a muwaqqit, he compiled a work

---

[1] See Heinen, *Islamic Cosmology*, on this tradition.

[2] For some examples of these genres of literature see King, *Astronomy in Yemen*, pp. 11-12. See also the detailed study in Schmidl, *Islamische volksastronomische Abhandlungen*.

[3] See n. 1:7.

[4] Quoted in Heinen, *Islamic Cosmology*, p. 25.

entitled *Kitāb al-Yawāqīt fī maʿrifat al-mawāqīt* (= *Sapphires Concerning the Times of Prayer*),[5] in which he wrote:

> "The times of prayer are not to be found by the degrees of the astrolabe and not by calculation using the science of the astronomers; they are only to be found by observation ... . The astronomers took their knowledge from Euclid and (the authors of) the Sindhind, and from Aristotle and other philosophers: all of them were infidels."

In the early 11th century, the greatest scientist of Islam, al-Bīrūnī, had little time or patience for those who consulted books of law on what he considered to be scientific matters. In his *Ifrād al-maqāl fī amr al-ẓilāl* he stated that there were two kinds of muezzins.[6] The first concerned themselves with shadows, measuring time by using their own bodies as gnomons and measuring the shadows with their own feet. This confirms our suspicion that simple shadow-schemes for time-reckoning were rather popular in medieval Islam. The second kind of muezzin is from the common people:

> "their hearts are disgusted by the mention of shadows, or altitude, or sines, and they get goosepimples at the mere sight of computation or scientific instruments. With them it reaches such an extent that one cannot trust them with anything of the sort, much less with the times of prayer, not because of their unfaithfulness or treason, but because of excessive ignorance."

al-Bīrūnī records that he once encountered a muezzin who was upset because the astrolabe with which he wanted to determine the times of the two day-time prayers was not marked with the Muslim months, and consequently he did not want to take it into a mosque. al-Bīrūnī said to him: "The Christians also eat food and walk around in the market: do not imitate them in these two things (either)", and when the hapless muezzin still did not see the point al-Bīrūnī told him he had a disease for which there was no cure, whereupon the man solved his own problem by breaking the astrolabe.

Elsewhere in his treatise al-Bīrūnī discussed the kind of astronomical knowledge he thought the muezzin needed, emphasising spherical astronomy and careful observations, but without reference to either timekeeping by the lunar mansions or to the use of tables. He concluded on a more positive, if unrealistic, note by recommending prospective muezzins to read the works of Euclid and Ptolemy.

---

[5] On this author and his work see **III-4.1** and **IV:1.5**. The relevant passage from MS Cairo DM 948, fols. 6r-6v, is presented as **Text 1** in the Appendix.

[6] al-Bīrūnī, *Shadows*, pp. 35-37, and 175-180; and E. S. Kennedy's translation and commentary, I, pp. 75-76 and 226-230, and II, pp. 28-29 and 142-143.

CHAPTER 4

THE DUTIES OF THE MUEZZIN

In a guide written *ca.* 1300 by the Egyptian Ibn al-Ukhuwwa to aid the *muḥtasib*, or official responsible in medieval Islam for law enforcement (*ḥisba*), there is a section on mosque-attendants and muezzins.[1] Here it is stated that the *muḥtasib* should test the muezzin on his knowledge of timekeeping, which, it is claimed, can be learned from texts on the sacred law. The guide goes on to say that the muezzin should know the lunar mansions and the formations of the stars described therein in order to determine the time of night and the time of daybreak. Ibn al-Ukhuwwa says this involves knowledge (*ʿilm*) and calculation that requires much explanation. An earlier treatise on the same subject by the late-12th-century Syrian scholar ʿAbd al-Raḥmān al-Shayzarī[2] adds that whoever is interested should consult the *Kitāb al-Anwāʾ* by Ibn Qutayba.[3] This early-9th-century compilation indeed contains all that one needs to know about the lunar mansions, but it is hardly a good source for the regulation of the times of prayer since these are not even mentioned.

The following is a translation of the relevant section in Ibn al-Ukhuwwa's treatise:

> "Only an honorable, reliable, and trustworthy man who is acquainted with the times of prayer may pronounce the call to prayer from the minaret, because the Prophet – may God bless Him and grant Him salvation – said: "The muezzins who call the faithful to prayer (*al-muʾadhdhinūn*) are (to be) reliable, and the *imāms* who lead the faithful in prayer (*al-aʾimma*) are (to be) responsible. So may God guide the *imāms* aright and grant forgiveness to the muezzins."[4]
>
> The *muḥtasib* must test (the muezzins) in their knowledge of the times of prayer. Anyone who does not know (the times properly) should be prevented from performing the call to prayer until he does know them because he may do it at the wrong time and so (cause) a person who is fasting to break his fast, or (cause someone else) to do what he has sworn not to do, or (cause other people) to pray before the proper time. Their prayer would then be incorrect and (the muezzin) would be responsible for ruining their status (before God). So he must know the times of prayer and should read the chapters on the prayer-call and on performing the prayer (*al-iqāma*) in (the books on) the sacred law.

---

[1] On Ibn al-Ukhuwwa see the article by Claude Cahen in EI_2. The relevant passage from his *Maʿālim al-qurba*, pp. 176-178, is presented as **Text 2**.

[2] On al-Shayzarī see King, *Astronomy in Yemen*, no. 4, and the references there cited. For this remark see the editorial note at the end of **Text 2**.

[3] On Ibn Qutayba see the articles in *DSB* by Paul Kunitzsch and EI_2 (*sub* Ibn ƒutayba) by G. Lecomte.

[4] This *ḥadīth* is listed in Wensinck *et al.*, *Concordance* (n. **IV**-1:4), I, p. 44a.

It is preferable that the muezzin have a good voice, and the *muḥtasib* must forbid him from pronouncing the prayer-call improperly, either indistinctly or with conceit or by singing it. He must order (the muezzin) when he ascends the minaret to keep his gaze lowered so as not to look at the women-folk or into people's houses; a promise to this effect should be exacted. No one should ascend the minaret except the muezzin (and that only) at prayer-times.

The muezzin must know the lunar mansions and the shapes of the star-groups in them, so that he may be able to tell the time at night. There are twenty-eight mansions. [A list follows.] Daybreak occurs in each one of these mansions for thirteen days, then it moves to the following one. If the muezzin knows in which mansion dawn is breaking and looks at the intervening one (?? *al-muʿtariḍa*) on the meridian, he will be able to tell (which lunar mansions are) rising and setting (over the horizon), and know how long it is until daybreak. This involves knowledge (*ʿilm*) and calculation (*ḥisāb*), which it would take too long to explain (here). [Clearly the author knew very little about the lunar mansions and was wise to change the subject at this point!]

The muezzin is required to be a male Muslim of sound mind. A prayer-call made by an unbeliever, a woman, a lunatic or a drunkard, is not valid, but that from a minor with power to discriminate is valid. Ritual purity is recommended for (those) performing the prayer-call, but (it) will be valid without. Disapprobation in the case of someone ritually unclean (*junub*) (to act as a muezzin) is stronger.

The muezzin must know the times of prayer, as we have already said. The prayer-call is in two parts (*muthannā*), whereas the performance (*iqāma*) of the prayer itself should be uninterrupted (*maʿa 'l-idrāj*). (The muezzin) should stand facing the qibla, and should glance to the north and south respectively for the two calls of (the formula) *Ḥayya ʿala 'l-ṣalāt*, "come to prayer", (keeping) his body in the direction of the qibla.[5] It is essential when performing the call to prayer to raise the voice. It is also a requirement that the various formulae (*al-kalimāt*) shall be in their correct order; but if the order is reversed the formulae should not be repeated. But if the error was committed in jest or through (deliberate) carelessness, the muezzin must be severely punished.

The call to prayer should be made only for the obligatory prayers; for other occasions, such as prayers on the occasion of an eclipse, prayers at the two major festivals (*ʿīd*s), or prayers for rain, the call shall be included with that for the ordinary prayers.

The muezzin must prolong his utterance of the *tadhkār* and the *tasbīḥ* during the nights of Ramadan.[6] He must first announce the dawn, then drink water, then announce the approach of daybreak and then the approach of the call to prayer. Then he must extinguish the lamp (lit on the minaret during the nights of Ramadan) and finally

---

[5] The implication is that the qibla is due east. Yet the qibla in Egypt was generally taken as 27° S. of E. (winter sunrise) or 37° S. of E. (computed using available geographical coordinates and an accurate trigonometric formula): see **VIIb**. I suspect that the author is simply thinking of the qibla as the direction of sunrise: see **12** below.

perform the call to prayer. If he is unable to extinguish the lamp he must break it, because if a person is too distant to hear the call to prayer, all that he has to rely on (to regulate the time of) his eating and drinking will be the sight of the lamp and whether it is (still) burning or (already) extinguished.

The muezzin may receive payment for his duties. People who lead the faithful in prayer may not (receive payment) for their services, but if anything is paid to them without being legally due it is permissible for them to accept it as a gift. Even if they are maintained by the public treasury, the correct opinion is that (their receiving such a gift) is (still) permissible."

CHAPTER 5

SIMPLE TECHNIQUES FOR TIMEKEEPING BY DAY AND NIGHT

Simple arithmetical schemes representing the shadow lengths at midday for each month, and tables displaying the shadows at each seasonal hour for each month (or zodiacal sign) likewise based on a simple arithmetical scheme, are attested in early Greek astronomy, and their use continued in the Byzantine, Coptic, and Ethiopic traditions, as well as in Islamic folk astronomy. The Islamic material is found in the texts of almanacs; in the *kutub al-mawāqīt*, that is, books dealing with the regulation of the prayer-times by non-technical means; and in books on the sacred law of Islam.[1] Sometimes the schemes are presented in tabular form, more often in words or mnemonics in standard medieval Arabic alphanumerical notation. Only rarely do competent astronomers mention these schemes.

Most Islamic shadow-schemes were intended to be used primarily for regulating the times of the midday (*ẓuhr*) and afternoon (*ʿaṣr*) prayers, both defined in terms of shadows. Only rarely do the tables give values for the seasonal hours. The associated texts recommend observing the length of one's own shadow to regulate the daylight prayer-times; they seem never to suggest that one should consult an astronomer or use any of the astronomical tables or instruments, such as sundials, that were available for this purpose. The schemes were particularly vulnerable to copyists' mistakes and often the length of the gnomon is not mentioned. The most common Islamic bases for the gnomon length were 7 feet (*qadam*s) and 12 digits (*iṣbaʿ*s), although sometimes 6, $6^1/_2$ and $6^2/_3$ feet were used, and occasionally also 60 units. Values between 6 and 7 feet were thought to represent the ratio of the height of a man (*qāma*) to the length of his feet.[2] A gnomon the height of a man is still in situ in the courtyard of the 7th-century Mosque of Janad in the Yemen. This gnomon is known locally as the stick (*ʿaṣā*) of Muʿādh ibn Jabal, the first governor of the Yemen in the 7th century. The 13th-century legal scholar al-Aṣbaḥī stated that he measured the midday shadows in the Mosque every 13 days of the Syrian year, but it is not known whether the present gnomon is the one he used.

We can distinguish seven different kinds of schemes and tables for timekeeping by day:

(1) midday shadow-schemes for the hours based on an approximate Indian formula relating the seasonal hours to the increase of the shadow over its midday minimum;
(2) midday shadow-schemes for the months based on Hellenistic schemes, also attested in Coptic sources, and modifications thereof;
(3) shadow tables for the hours and for each month based on Hellenistic models;
(4) midday shadow-schemes for the 13-day periods of the year (*anwāʾ*) as regulated by the risings and settings of the 28 lunar mansions;

---

[1] See n. 1:7.

[2] On these values see Kennedy on al-Bīrūnī's *Shadows*, I, pp. 68-79, and II, pp. 25-31.

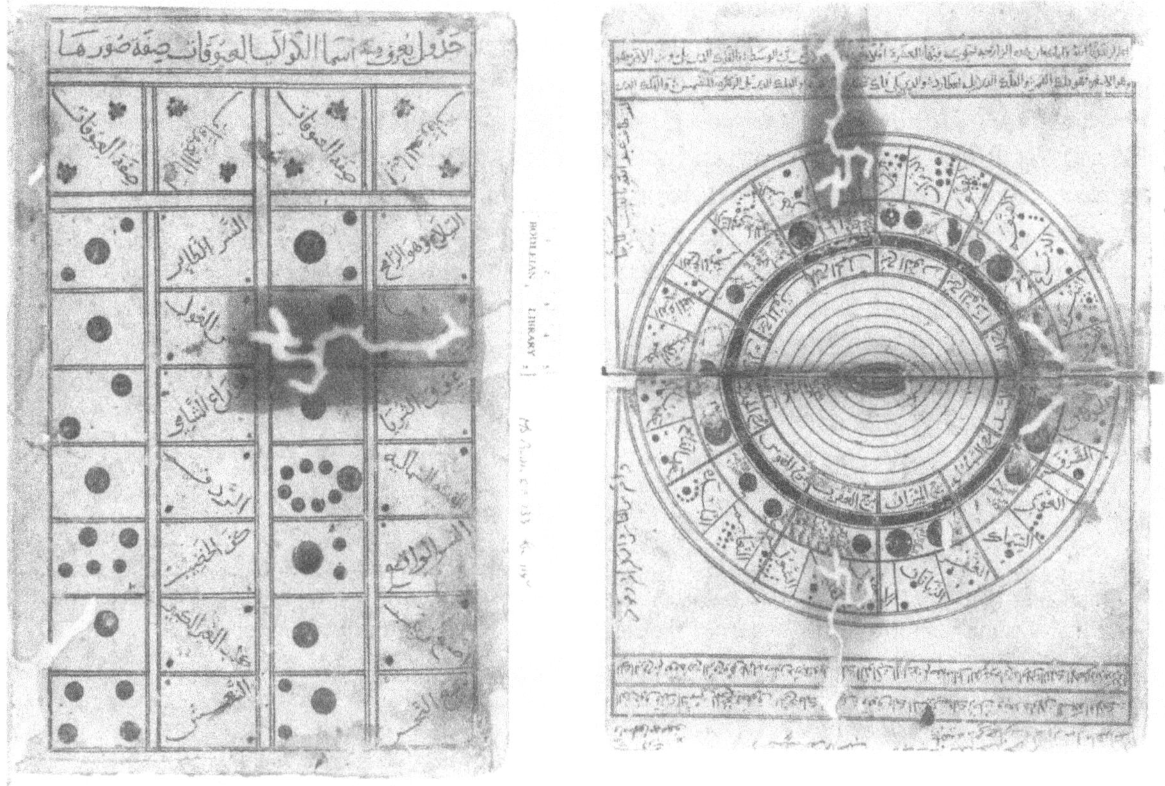

Figs. 5.1a-b: Diagrams showing the appearances of various star-groups (a) and a simplistic representation of the seven spheres, together with the signs of the zodiac and the lunar mansions (b) from an early-13th-century muezzin's manual. See also **Figs. II-6.1** and **XI-9.4c**. [From MS Oxford Bodley Or. 133, fols. 114v and 117v-118r, courtesy of the Bodleian Library.]

(5) midday shadow-schemes derived by observation (for a given locality);
(6) midday shadow-schemes derived by calculation (for given latitude and obliquity of the ecliptic); and
(7) shadow tables for the hours and for each zodiacal sign or for each few days of the year based on calculation with an approximate formula.

For timekeeping by night some of the sources advocate using the lunar mansions. One keeps track of the lunar mansion that is rising at sunset, and at some time during the night observes that a particular mansion is rising. One can then estimate the time since sunset on the basis that 14 mansions rise in 12 seasonal night hours.[3]

Unfortunately no early practical handbooks specifically intended for muezzins survive. The title of such a work from 12th-century al-ʿIrāq or Iran is known however, namely *al-Manār (?) fī ʿilm mawāqīt al-layl wa-'l-nahār*, (= *The Guiding Light (?) on the Science of Timekeeping by Night and Day*), by an individual known as Ibn al-Munajjim (and possibly also as Ibn al-

[3] A valuable new study of some of this material is Forcada, "*Mīqāt* en los calendarios andalusíes".

Muqābala) who specialized in inheritance laws and the regulation of the prayer-times.[4]

On the other hand, an early-13th-century Egyptian manual for timekeeping entitled *Kitāb al-Durar wa-'l-yawāqīt fī maʿrifat al-raṣad wa-'l-mawāqīt* (= *The Book of Pearls and Jewels on Astronomical Observation and the Times of Prayer*) has been preserved in a unique manuscript: see **Fig. 5.1**.[5] It gives us an indication of the kind of information that a high-class muezzin or low-class muwaqqit would have known, although such material was soon superseded in Egypt (if not in such places as al-Andalus, al-ʿIrāq and Iran, and Central Asia) by the more sophisticated tables and texts prepared by the muwaqqits. The work contains information on the lunar mansions and simple non-numerical tables for reckoning the time of night by the mansions as well as some simple and rather inaccurate tables displaying the shadow length at midday and the *ʿaṣr* prayer for each day of the Coptic year. The tables (which are not based on accurate mathematical formulae) also display the duration of daylight and the solar altitude at the beginning of the *ʿaṣr* prayer and at the time when the sun is in the direction of Mecca.

---

[4] See Ibn Ṭāwūs: *Faraj al-mahmūm*, p. 206; Zeina Matar, *The* Faraj al-mahmūm *of Ibn Ṭāwūs: A Thirteenth-Century Work on Astrology and Astrologers*, Ph.D. dissertation, New York University, 1987, pp. 241, 321 and 480; and Etan Kohlberg, *A Medieval Scholar at Work: Ibn Ṭāwūs and his Library*, Leiden, *etc*.: E. J. Brill, 1992, pp. 252-253.

[5] MS Oxford Bodleian Hunt. 133, fols. 94v-130r, copied 734 H [= 1333/34]. A more detailed discussion of this work is in **II-6.1**.

CHAPTER 6

FROM MUEZZIN TO MUWAQQIT AND *MĪQĀTĪ*

The appellation muwaqqit is not attested in the Arabic astronomical sources, as far as can be ascertained, before the late 13th century. The earliest use of the term that has come to my attention is in a passage describing some astronomers of the 13th and early 14th centuries by a contemporaneous Egyptian astronomer, in a manuscript preserved at Leiden.[1] In this treatise, entitled *Kanz al-yawāqīt fī 'stīʿāb al-mawāqīt* (= *The Treasure of Sapphires Concerning the Comprehension of the Times of Prayer*), the anonymous author mentions some of his ancestors, including one Abu 'l-Ḥasan ʿAlī ibn ʿAbd al-Malik ibn Samʿūn (d. 685 H = 1286/87), his son Muḥammad known as al-Wajīh (d. 701 H = 1301/02), and his grandson Muḥammad known as al-Majd (< Majd al-Dīn), all of whom were muwaqqits at the Mosque of ʿAmr in Fustat. The first-mentioned had held the position for thirty years, the second for sixty. Probably from the same family came the astronomer Nāṣir al-Dīn Muḥammad ibn Samʿūn (d. 737 H [= 1337]),[2] called *al-shaykh al-imām* but not *al-muwaqqit*, who wrote a quiz on *ʿilm al-mīqāt* and several treatises on unusual instruments; indeed, he may well be the author of the *Kanz al-yawāqīt*. If the institution of the muwaqqit did in fact arise in Egypt, it was not long in spreading to Palestine and Syria. A manuscript of some treatises on unusual instruments by Nāṣir al-Dīn ibn Samʿūn preserved in Cairo was copied in 706 H [= 1306/07] by Ibrāhīm ibn Aḥmad, who refers to himself as muwaqqit at the Sacred Mosque in al-Ḥaram al-Khalīlī (Hebron) – see **Fig. 6.1**.[3]

On the other hand, there is evidence of a similar office in al-Andalus at the end of the 13th century, albeit with a different title.[4] The functions of the father and son team Aḥmad and Ḥusayn ibn Bāṣo are variously labelled in the available manuscripts. In one (MS Escorial ár. 961, fol. 69v), the father is called *al-mu'adhdhin al-mubārak* and the son *al-imām al-[mu]ʿaddil al-mubārak*, the term *muʿaddil* apparently meaning "astronomer", that is, one who is *au fait* with the operations of computing solar, and lunar and planetary positions (*ʿaddala*) by means of the mean motions and their equations (*taʿdīl*, plural *taʿādīl*). In another (MS Cairo TR 159,2,

---

[1] MS Leiden Or. 468, on which see P. Voorhoeve, *Handlist of Arabic Manuscripts in the Library of the University of Leiden ...* , Leiden: University Press, 1957, p. 153. The treatise is important not least because the author quotes from the lost *Kitāb Ṭabaqāt al-ḥukamā'* of Ṣāʿid al-Andalusī – see King, "Universal Astrolabe", pp. 250 and 252, and Samsó, *Ciencias en al-Andalus*, p. 412. A detailed study of this work is being prepared by Margarita Castells of the Universitat de Barcelona. The relevant passage (fol. 91r) is extracted below as **Text 3**.

[2] On this author see Suter, *MAA*, no. 398, and *Cairo ENL Survey*, no. C24.

[3] On this manuscript see *Cairo ENL Catalogue*, I, *sub* QM 2. The copyist's signature is reproduced in *Cairo ENL Survey*, p. 323 (pl. CIIIb), with caption on p. 215.

[4] The available evidence is collected in Calvo, *Ibn Bāṣo and his Universal Plate*. See Mayer, *Islamic Astrolabists*, pp. 34-36, and Samsó, *Ciencias en al-Andalus*, pp. 412-413. The original sources are extracted below as **Text 4**.

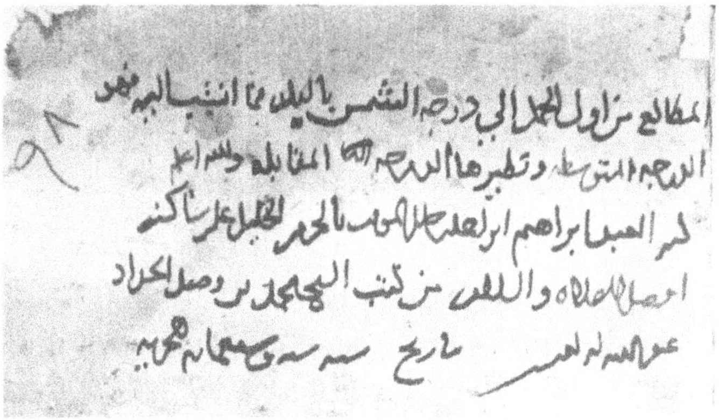

Fig. 6.1: The colophon of a manuscript containing treatises on unusual instruments, mentioning the name of the copyist, Ibrāhīm ibn Aḥmad ibn Khalīl, the muwaqqit at the Sacred Mosque in Hebron (*al-Ḥaram al-Khalīlī*), and the date of copying, 706 H [= 1306/07], as well as the owner, Muḥammad ibn Wāṣil al-Jarrād. [From MS Cairo QM 2, fol. 98r, courtesy of the Egyptian National Library.]

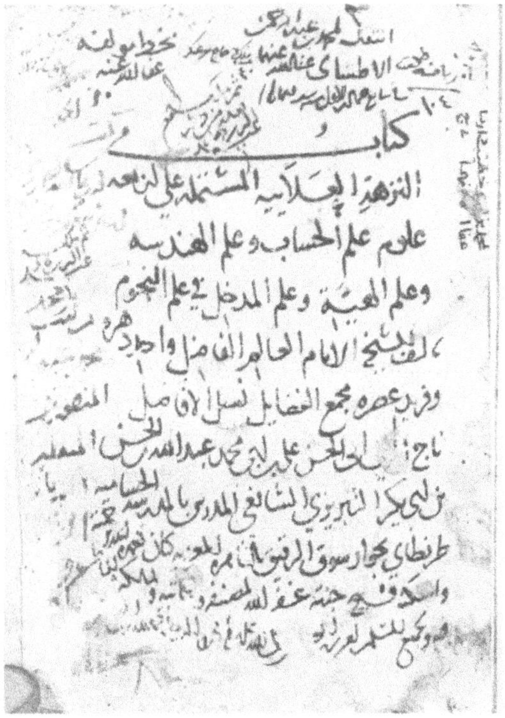

Fig. 6.2: The title-folio of the unique copy of the treatise *al-Nuzha al-ʿAlāʾiyya*, on dealing with arithmetic, geometry, theorietical astronomy and astrology, by Tāj al-Dīn al-Tabrīzī, teacher in the madrasa of Ḥusām Ṭaranṭāy near the slave market in Cairo. The manuscript was copied *ca.* 1400 and the notes on the title-folio identify a series of owners. [From MS Cairo ṬR 104, courtesy of the Egyptian National Library.]

p. 63), the son's name is confused (Abu 'l-Ḥasan ʿAlī instead of Abū ʿAlī Ḥusayn) and he is referred to as *al-shaykh al-muʿaddil bi-Gharnāṭa wa-ṣāḥib al-awqāt bihā*, that is, "the astronomer in Granada and the official responsible for regulating the times of prayer there". Yet another source (MS Rabat 451,1, not consulted) refers to Aḥmad ibn Yūsuf ibn Bāṣo as *amīn al-awqāt*, that is, "the official responsible for the times of prayer", in Cordova. In the historical work *al-Iḥāṭa* by the 14th-century Andalusī scholar Ibn al-Khaṭīb the father Aḥmad is said to be the muwaqqit in the Grand Mosque at Granada, but this may be a later rendering of the expression *ṣāḥib al-awqāt*. In this connection it is worth noting the appellation *ṣāḥib al-qibla* applied to an astronomer and mathematician Muslim ibn Aḥmad ibn Abī ʿUbayda of Cordova, who died in the year 908.[5] He was so-called because of his intensive preoccupation with the qibla, and there was indeed a problem with the orientation of mosques in al-Andalus, which he was apparently unable to influence.[6] The man who in 1317 constructed the second water-clock in the Qarawiyyīn Mosque in Fez, namely, Muḥammad al-Ṣanhājī, was also known as *al-muʿaddil*, probably meaning simply "astronomer".[7]

[5] Sezgin, *GAS*, VI, p. 171.

[6] On the problem of the qibla in al-Andalus see King, *Studies*, B-XV, pp. 370-387 (pages not numbered consecutively); Samsó, *Ciencias en al-Andalus*, pp. 60-67; and the definitive study Rius, *La Alquibla en al-Andalus y al-Magrib*.

[7] See Mayer, *Islamic Astrolabists*, p. 77 (*sub* Muḥammad al-Ṣinhājī) and p. 73 (*sub* Muḥammad ibn

Two other instances of the term muwaqqit being applied by later writers to astronomers because they worked on timekeeping can be cited. Firstly, Colin and Renaud (1933) have written on the identity of a 13th-century "muwaqqit" from Morocco called Abū Miqraʿ, but I suspect that the terminology may be their own.[8] This Abū Miqraʿ wrote a poem on simple timekeeping, but one should not conclude from this that he was called muwaqqit and he certainly did not merit the title *mīqātī*. Secondly, in a report by the late-15th-/early-16th-century Egyptian historian Ibn Iyās about the Ayyubid al-Malik al-Kāmil (*ca.* 1225), the Sultan is said to have received a mechanical device (a brass candelabra housing a little man who popped out at daybreak with a greeting) from the *mīqātiyya* of the Yemen.[9] The word *mīqātiyya* is here used as a plural of *mīqātī*, but we cannot be sure that the term goes back to the donors. Indeed, in the various Yemeni astronomical sources examined by the present author (well over one hundred) not a single astronomer called *mīqātī* or muwaqqit has been identified. Even Abu 'l-ʿUqūl, who in about 1300 compiled a corpus of tables for Taiz containing over 100,000 entries, was not called al-muwaqqit, but simply (and, in light of his achievements, appropriately) *al-ḥāsib*, literally "the computer", although perhaps this appellation is better rendered "the astronomer".[10] As we shall see in **8** and **9** some muezzins and muwaqqits put their names to astronomical instruments or authored treatises on astronomy.

In passing, it should be noted that not only the professional astronomers were involved in serious scientific activity at the mosques: see, for example, **Fig. 6.2**. Furthermore, there were also specialists in the laws of inheritance present in some mosques and madrasas: the most distinguished scholar of this genre was probably Ibn al-Hāʾim (d. 1412) of the Ṣāliḥiyya Madrasa in Jerusalem.[11]

Muḥmmad ibn al-ʿArabī, who repaired the clock *ca.* 1347). Mayer cites the 14th(?)-century historian of Fez, al-Jaznāʾī, *Zahrat al-âs*, (ed. Alfred Bel, Algiers, 1923), pp. 38 (text) and 92-93 (transl.). On the present form of the clock see Price, "Mechanical Water Clocks in Fez".

[8] Colin & Renaud, "Abū Miqraʿ".

[9] Ibn Iyās, *Taʾrīkh Miṣr*, p. 78, cited in Mayer, *Islamic Astrolabists*, p. 21. The relevant passage is presented as **Text 5**.

[10] On Abu 'l-ʿUqūl see King, *Astronomy in Yemen*, no. 10, and Varisco, *Yemeni Almanac*, p. 13. On his tables see now **I-2.1.2**, *etc.*, and **II-12.1**. See also n. 8:63.

[11] On Ibn al-Hāʾim see Suter, *MAA*, no. 423; Brockelmann, *GAL*, II, pp. 153-155, and SII, pp. 154-155; and *Cairo ENL Survey*, no. C58. Curiously, he is not listed amongst the most important *farāʾiḍ* authors in Rebstock, *Rechnen im islamischen Orient*, p. 276.

CHAPTER 7

ON THE SALARIES OF MUEZZINS AND MUWAQQITS AND *MĪQĀTĪ*S

Of considerable interest to the present study is the information that can be derived from historical endowment documents known as *waqfiyya*s. As an example we can cite the *waqfiyya* of the Madrasa of Sultan Barqūq in Cairo, established as a Ṣūfī *khānqāh* or "convent" in 1386. This document informs us about the officials associated with the *khānqāh* and also about their salaries as provided by the *waqf*.[1] The Ḥanafī *shaykh* received 500 *dirham*s a month, his counterpart from one of the other schools: 300; the *shaykh* of *ḥadīth*: 150; the *shaykh* versed in the variant readings of the *Qur'ān*: 100; *etc.*, *etc.*, ... *etc.*, down to the cleaners: 30; the cook: 20, the students: 20 each, the man in charge of carpets: 10; the doorman: 10, *etc.*, *etc.* Amongst the officials who received low salaries are six muezzins and two scholars of *ʿilm al-mīqāt* with 15 *dirham*s a month each. (Another two muezzins are mentioned as an afterthought toward the end of the *waqfiyya*.) Their salaries seem to be in addition to a basic 10 *dirham*s a month for all Ṣūfīs associated with the *khānqāh*.

The muezzin's duties as outlined in the *waqfiyya* were to announce the prayers, to pronounce the formulae *Subḥāna 'llāh* and *Allāhu akbar*, to recite praises upon the Prophet on the nights before Fridays and to announce the times for beginning the fast of Ramadan.[2] The duties of the resident astronomers were to inform (the muezzins?) of the times of prayer and the times for pronouncing the formulae specified above, as well as the times for reciting the creed, and for the congregation to go before the preacher on Fridays. Unfortunately no more details are given.

E. Strauss, in a valuable study of prices and salaries in Mamluk Egypt published in 1949, has extracted lists of salaries taken originally from *waqfiyya*s of various religious institutions.[3] A summary of his findings on the salaries (in *dirham*s per month) of muezzins and muwaqqits in Cairo appears below. Unfortunately Strauss often omitted the actual number of people provided for in the *waqf*. Unless otherwise indicated, I am not aware of the names of any individuals who were attached to these institutions as muezzins or muwaqqits or *mīqātī*s.

(1) Mausoleum of Qalā'ūn (1284):- *imām*: 80; chief muezzin: 40; muezzin: 30, *etc.*

---

[1] See Felicitas Jaritz, "Auszüge aus der Stiftungsurkunde des Sultans Barqūq", Part IV of S. Lamei Mostafa, *Madrasa, Ḫānqāh und Mausoleum des Barqūq in Kairo*, *Abhandlungen des Deutschen Archäologischen Instituts Kairo*, *Islamische Reihe*, Band 4, Glückstadt: J. J. Augustin, 1982, for further information. The sections on muezzins and *muwaqqit*s (for which see **Text 6** below) occur on pp. 129 and 137 (text), and pp. 153 and 163-164.

[2] See **I-4.10** and **V-4**.

[3] Eduard Strauss, "Prix et salaires à l'époque mamlouke: une étude sur l'état économique de l'Egypte et de la Syrie à la fin du moyen âge", *Revue des Études Islamiques* 17 (1949), pp. 49-94, esp. pp. 77-83. On the salaries at the Mosques of al-Mu'ayyad and Qāyitbāy see also William Popper, *Egypt and Syria under the Circassian Sultans 1382-1468 A.D.: Systematic Notes to Ibn Taghrī Birdī's Chronicles of Egypt*, Berkeley and Los Angeles: University of California Press, 1955, pp. 120-121.

(2) Madrasa of Qalā'ūn (1284):- *shaykh*: 200; *imām*: 80; chief muezzin: 40; muezzin: 30; *etc.*
(3) Mausoleum of Muḥammad ibn Qalā'ūn:- *imām*: 80; chief muezzin: 40; muezzin: 25; *etc.*
(4) Madrasa of Muḥammad ibn Qalā'ūn (1303):- *shaykh*: 200; *imām*: 80; librarian: 30; *etc.*; no muezzins listed.
(5) Mosque of al-Mu'ayyad (*waqfiyya* dated 1420):- Ḥanafī *shaykh*: 650; Shāfiʿī *shaykh*: 150; *shaykh* of *ḥadīth*: 150;. ... ; muezzin: 15, *etc.* (It is curious that no *mīqātī* is mentioned here because the astronomer ʿIzz al-Dīn al-Wafā'ī (died *ca.* 1470) was a muwaqqit at this mosque and also signed himself al-*mīqātī*.[4])
(6) Mosque and Khānqāh of Sultan Barsbāy (established 1429):- *imām*: 1000; preacher: 700; muezzin: 300; *etc.*
(7) Mosque of the Amīr Qānem (document dated 1446):- *imām*: 900; preacher: 500: muezzin: 200; muwaqqit: 200; servant: 300; *etc.* (Here for the first time in the documents considered here, a distinction is made between the muezzin and the muwaqqit.)
(8) Mosque of Sultan Qāyitbāy (document dated 1474, published by L. A. Mayer):- chief ṣūfī: 3000; *imām*: 500; chief muezzin: 250; nine muezzins (!): 200 each; librarian: 200, *etc.*
(9) Mosque of the Amīr Qurqumās (1513-14):- Chief ṣūfī: 900; *imām*: 600; preacher: 600; muwaqqit: 600; muezzin: 200; servant: 150; *etc.* If it was only one muwaqqit who received these 600 *dirham*s per month, we would have here the first instance of the qualifications of a muwaqqit being recognized as worth more than those of a muezzin!
(10) Mosque of al-Ghawrī (1516):- *imām*: 1200; muezzins and muwaqqits between them (16? in number): 1400; *etc.* (The texts used by Strauss are somewhat confused here.)

Similar information on the salaries of muezzins and muwaqqits in Syrian institutions does not seem to be available in the secondary literature.[5] Indeed, neither the primary nor secondary literature refers to the intense activity in *mīqāt* at the Umayyad Mosque in Damascus in the 14th century. A 17th-century Syrian historian states that in his time there were 75 muezzins attached to the mosque.[6]

---

[4] On al-Wafā'ī (**I-9.9** and **VIb-14**) see *Suter*, MAA, no. 442, and *Cairo ENL Survey*, no. C61, and also nn. 8:74 and 9:84.

[5] Unfortunately no information on the salaries of *muwaqqit*s is contained in Eliyahu Ashtor, *Histoire des prix et des salaires dans l'Orient médiéval*, Paris: S.E.V.P.E.N., 1969.

[6] Quoted in Morley, "Arabic Quadrant", p. 328, n. 1. Randi Deguilhem-Schoem (*Winder Festschrift*, p. 72) draws attention to a 19th-century *waqfiyya* which mentions muezzins at the mosque but no *muwaqqit*s.

CHAPTER 8

ASTRONOMICAL TIMEKEEPING AS A SCIENCE

Perhaps the most surprising thing about the statements of Ibn al-Ukhuwwa on the duties of the muezzin (see **4**) is that they were recorded not long after the compilation in Cairo of the major Islamic work on astronomical timekeeping and astronomical instruments, the *Kitāb al-Mabādiʾ wa-ʾl-ghāyāt fī ʿilm al-mīqāt* (= *An A to Z of Astronomical Timekeeping*) of Abū ʿAlī al-Marrākushī.[1] This treatise is in general highly eclectic, but is nevertheless the work of a competent astronomer. It consists of about 400 folios of text dealing with theoretical spherical astronomy and including numerous tables, some of them specifically computed for Cairo, followed by an extensive treatment of the construction and use of astronomical instruments, and concluding with a quiz. al-Marrākushī is referred to in later astronomical sources as *al-shaykh al-imām*, not as al-*mīqātī*. Significant as this work is, Ibn al-Ukhuwwa ignored it altogether.

The 14th-century Egyptian scholar Ibn al-Akfānī, author of an encyclopaedia and several treatises on medicine, wrote in the former on *ʿilm al-mawāqīt* as follows:[2]

> "The science of astronomical timekeeping is a branch of knowledge for finding the hours of the day and night and their lengths and the way in which they vary. Its use is in finding the times of prayer and in determining the direction in which one should pray, as well as in finding the ascendant and the right and oblique ascensions from the fixed stars and the lunar mansions. This science is also concerned with shadow lengths and the altitudes of celestial bodies, and with the orientation of one city from another. A short work on this science is the *Nafāʾis al-yawāqīt*, and a more extensive one is *Jāmiʿ al-mabādiʾ wa-ʾl-ghāyāt* by Abū ʿAlī al-Marrākushī."

This is an intelligent summary of the subject. Unfortunately the treatise *Nafāʾis al-yawāqīt* ("The Most Precious Sapphires") is no longer extant: one can safely assume that it was a work similar to the *Kitāb al-Durar wa-ʾl-yawāqīt* that is briefly described in **5** above.[3]

Even Ibn al-Akfānī neglects to mention the existence of a corpus of astronomical tables in

---

[1] On al-Marrākushī (**I-4.2.4** and **II-2.7** and **6.7**) see the article in EI_2, and King, "Astronomy of the Mamluks", pp. 539-540. An early Istanbul manuscript of his treatise was published in facsimile by the Institut für Geschichte der Arabisch-Islamischen Wissenschaften, Frankfurt in 1984. The part of his treatise dealing with spherical astronomy and sundials has been published in Sédillot-père, *Traité*, and the part dealing with other instruments summarized in Sédillot-fils, *Mémoire*. On the sundial theory in this work see Schoy, *Gnomonik der Araber*.

[2] On Ibn al-Akfānī see Brockelmann, *GAL*, II, p. 171, and SII, pp. 169-170. This passage, taken from Jan Just Witkam, *De egyptische arts Ibn al-Akfānī (gest. 749/1348) en zijn indeling van de wetenschappen ...*, Leiden: Ter Lugt Pers, 1989, p. 59 (Arabic text) is reproduced below as **Text 7**.

[3] Another work on timekeeping which could be important is by the 13th-century Cairene legal scholar Shihāb al-Dīn al-Qarāfī, who also wrote on the qibla in his book entitled *al-Dhakhīra* (ed. ʿAbd al-Wahhab ʿAbd al-Laṭīf and ʿAbd al-Samīʿ Aḥmad Imām, vol. I, Cairo: Maṭbaʿat Kulliyyat al-Sharīʿa, 1961), but no manuscripts of this are known to me. The work is mentioned in M. ibn ʿAbd al-Wahhāb, *Ruʾyat al-hilāl* (on which see n. 11:6). On the author see Brockelmann, *GAL*, I, pp. 481-482, and SI, pp. 665-666.

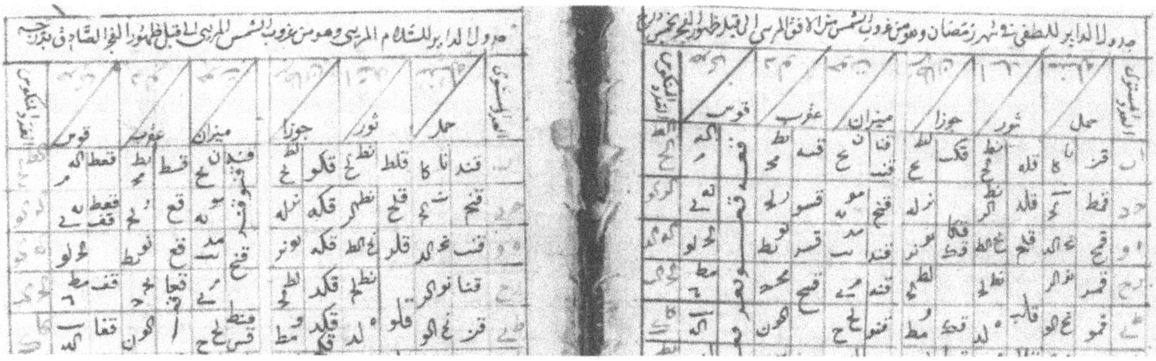

Fig. 8.1: An extract from the tables in the main Cairo corpus for timekeeping. These two tables display the length of time from sunset to the time when the candles on the minarets in Ramadan must be extinguished and the time from sunset until the muezzin should pronounce the benediction on the Prophet. [From an unidentified Cairo manuscript, courtesy of the Egyptian National Library.]

ca. 100 folios containing some 30,000 entries for timekeeping by the sun and regulating the times of prayer specifically computed for the latitude of Cairo.[4] This corpus was made up of tables compiled by Ibn Yūnus in the 10th century and supplemented by others compiled *ca.* 1280 by Shihāb al-Dīn al-Maqsī, who also prepared extensive tables for sundial construction (see **9**), and other Mamluk astronomers: see **Fig. 8.1** for an example. al-Maqsī was not, however, called *al-mīqātī,* but rather *al-shaykh al-imām.*[5] Also available in Cairo in the time of Ibn al-Akfānī was a set of astronomical tables containing over 440,000 entries for timekeeping by the sun and stars for all latitudes: this was compiled by Najm al-Dīn al-Miṣrī, who was referred to simply as *al-shaykh al-imām al-ʿālim* rather than *al-mīqātī.*[6]

Thirty years ago only one or two of these tables were known to exist, but we can now trace their origin to 9th-century al-ʿIrāq, and follow their development down to the 19th century. The earliest known Islamic tables for regulating the *ẓuhr* and the *ʿaṣr* for the latitude of Baghdad probably date back to al-Khwārizmī.[7] The mid-10th-century Baghdad astronomer ʿAlī ibn Amājūr compiled two trapezoidal (*ṭaylasān*) tables displaying the time since sunrise as a function of meridian altitude and instantaneous solar altitude: one of these was based upon the accurate formula and was computed especially for the latitude of Baghdad, the other upon an approximate formula serving all latitudes. In general, extensive tables for timekeeping are not found in the category of Islamic astronomical literature known as *zīj*es, these being handbooks with tables for planetary and spherical astronomy together with explanatory text.[8] Exceptions are the 13th-century *Zīj* of Abu 'l-Qāsim ibn Maḥfūẓ al-Baghdādī, which preserves

[4] On the Cairo corpus see **II-4-5**.

[5] On al-Maqsī (Suter, *MAA*, no. 383, and *Cairo ENL Survey*, no. C15) see **I-4.1.3** and **II-4.1.2**, and on his sundial tables also King, "Astronomy of the Mamluks", pp. 540, 541, 547, and **X-7.1**.

[6] On Najm al-Dīn al-Miṣrī (Suter, *MAA*, no. 460 (confused), and *Cairo ENL Survey*, no. C16) see **I-2.1.6**, **9.3.a** and **II-2.5**; Charette, "Najm al-Dīn's Monumental Table", and also n. 8:12 below.

[7] See King, "al-Khwārizmī", pp. 7-11, and **II-3.1**.

[8] Kennedy, "*Zīj* Survey", contains information on some 125 Islamic *zījes*, some of which contain hundreds of pages of tables and explanatory text. The number of *zīj*es now known exceeds 200. See, most recently, King & Samsó, "Islamic Astronomical Handbooks and Tables".

for us the earlier tables of ʿAlī ibn Amājūr, and the *Zīj-i Īlkhānī* of Naṣīr al-Dīn al-Ṭūsī, which contains a *ṭaylasān* table for the latitude of Maragha.[9]

Considerable impetus to the compilation of timekeeping tables was provided in the late 10th century by Ibn Yūnus. As noted above, he contributed some of the tables that formed the corpus used in Cairo until the 19th century. Major developments in *mīqāt* were made in the late 13th century, by Abū ʿAlī al-Marrākushī and al-Maqsī in Cairo,[10] and also by Abu 'l-ʿUqūl in Taiz.[11] The early-14th-century Egyptian astronomer Najm al-Dīn al-Miṣrī has left us a remarkable treatise on all of the instruments known to him and all those devised by himself, many of which have universal applications.[12] In the early 14th century we find a lone scholar in Aleppo, namely, Ibn al-Sarrāj, who made impressive contributions in the form of instruments specifically designed to solve the problems of spherical astronomy and timekeeping for all latitudes.[13] About the same time, the Syrians al-Mizzī and Ibn al-Shāṭir came to Egypt to study astronomy and then returned to Damascus, where they both worked as muwaqqits at the Umayyad Mosque. It was there in the mid-14th century that *ʿilm al-mīqāt*, of which the basics were documented in the writings of al-Marrākushī, with some "luxury items" in the works of Ibn al-Sarrāj and Najm al-Dîn al-Miṣrī, reached its zenith with the works of al-Khalīlī and Ibn al-Shāṭir.[14]

The achievements of the Damascus school are impressive by any standards. al-Khalīlī compiled an extensive corpus of tables for timekeeping and regulating the prayer-times for Damascus as well as auxiliary tables for solving problems of spherical astronomy for any latitude.[15] His colleague Ibn al-Shāṭir constructed the magnificent horizontal sundial that until the last century adorned the main minaret of the Umayyad Mosque, and he also prepared a set of tables for regulating the times of prayer for an unspecified locality in Syria, probably for the new Mamluk city of Tripoli.[16] Ibn al-Shāṭir's main contribution, however, was in the field of planetary astronomy. Continuing the tradition of the Maragha astronomers he developed new models for representing the motions of the sun, moon and planets, basing these as well as his new parameters on his own observations. How an astronomer occupied mainly in timekeeping, with his education in astronomy gained in Egypt, could achieve so much in this virtually unrelated field has yet to be explained.[17]

In view of these achievements it is somewhat ironic that a contemporary Syrian legal scholar

---

[9] See **I-2.3.1-2**.

[10] See nn. 8:1 and 8:5.

[11] See n. 6:10.

[12] On Najm al-Dīn al-Miṣrī (n. 8:6) see now Charette, *Mamluk Instrumentation*.

[13] On Ibn al-Sarrāj see Suter, *MAA*, no. 508 (confused), and *Cairo ENL Survey*, no. C26. On his instruments see King, "Astronomy of the Mamluks", pp. 544-545, and *idem*, "The Astronomical Instruments of Ibn al-Sarrāj", published for the first time in *idem*, *Studies*, B-IX. See also n. 9:5 below.

[14] On *mīqāt* in 14th-century Syria see now **I:9-10**.

[15] On al-Khalīlī see the article in *DSB*, supp. I, and on his various tables see **II:10**.

[16] On Ibn al-Shāṭir see the article in *DSB*, and also Kennedy & Ghanem, eds., *Ibn al-Shāṭir*, which includes (pp. 107-121) a reprint of Janin, "Cadran solaire de Damas".

[17] On Ibn al-Shāṭir's planetary theory see the various articles from the 1950s by E. S. Kennedy and colleagues reprinted in Kennedy & Ghanem, eds., *Ibn al-Shaṭir*, and Kennedy *et al.*, *Studies*. More recent studies of the Maragha school and its influence, especially on Ibn al-Shāṭir, are reprinted in Saliba, *Studies*. George Saliba is currently preparing an edition and translation of and commentary on Ibn al-Shāṭir's treatise on theoretical astronomy. See also the remarks in n. **I**-1:15.

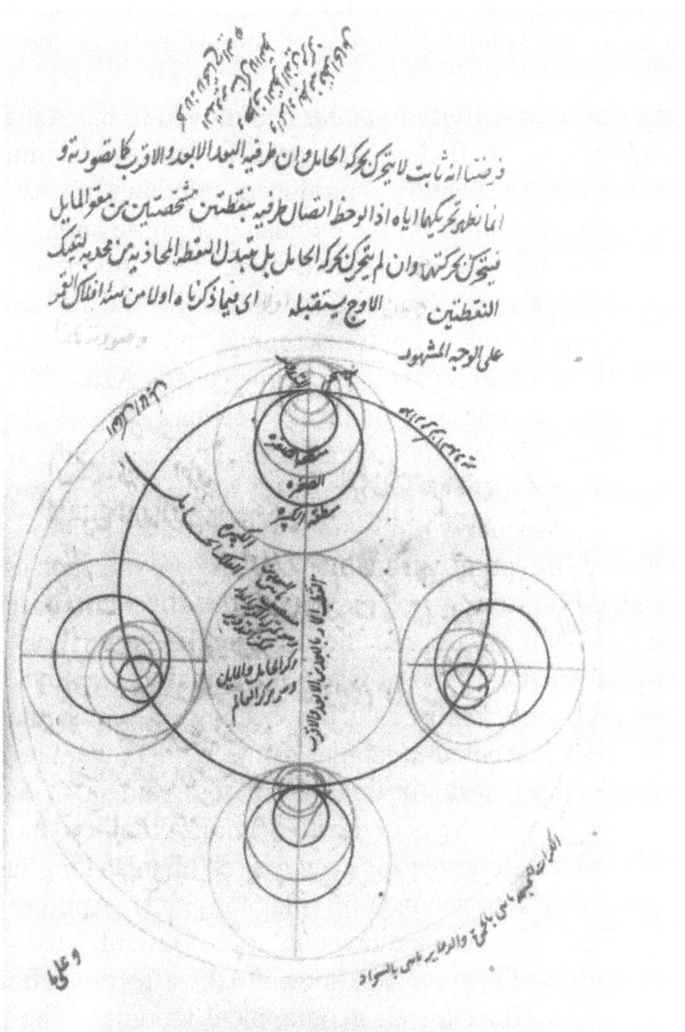

Fig. 8.2: An example of a sophisticated model for the moon in a 13th-century Iranian treatise on theoretical astronomy. [From an unidentified Cairo manuscript, courtesy of the Egyptian National Library.]

should pour scorn on the activities of the Damascus muwaqqits. The *qāḍī* Tāj al-Dīn al-Subkī stated that theoretical astronomy (*fann al-hay'a*) was important for muwaqqits; this either reflects some confusion in his mind or shows that he was familiar with the ground-breaking research in theoretical astronomy being conducted at that time by Ibn al-Shāṭir at the Umayyad Mosque. He complained that astrologers and magicians proliferated amongst the muwaqqits.[18] We may presume that al-Subkī performed his prayers according to the times announced from

[18] On al-Subkī see Brockelmann, *GAL*, II, pp. 108-110, and SII, pp. 105-107. His remarks about muezzins and *muwaqqit*s, from D. M. Myhrman, ed., *Kitāb Muʿīd al-niʿam wa-mubīd al-niqam ... by Tāj al-Dīn ... al-Subkī*, London: Luzac & Co., 1908, p. 165, are extracted as **Text 8** below. See now Ragep, "Freeing Astronomy from Philosophy", p. 53, n. 14, where it is noted that al-Subkī was also hostile to philosophy, with the exception of logic, and, for a general overview, Saliba, "Astronomy and Astrology in Medieval Arabic Thought".

the Umayyad Mosque, if not in the Mosque itself. These times were regulated by highly competent astronomers using the splendid sundial of Ibn al-Shāṭir or the sophisticated tables of al-Khalīlī.

By the 15th century the centre of activity in *mīqāt* had moved to the Azhar Mosque in Cairo, but the candle had already started to flicker. The prolific Sibṭ al-Māridīnī, muwaqqit at the Azhar Mosque in the latter half of the century,[19] produced a plethora of treatises on timekeeping that were widely used in Cairo thereafter, to the exclusion of more sophisticated works on the subject. Thus, no doubt unwittingly, he contributed more than anyone else to the decline of serious astronomy in Egypt, and also in Syria and the Hejaz. By the 18th century it was mainly commentaries upon his various treatises on timekeeping, quadrants, and sexagesimal arithmetic, that formed the basis of astronomical instruction at the Azhar.[20]

Two other leading Egyptian astronomers of the 15th century, Ibn al-Majdī and Ibn Abi 'l-Fatḥ al-Ṣūfī, both wrote extensively on *mīqāt*, at a higher level than Sibṭ al-Māridīnī, and also dealt with planetary astronomy, which he avoided. But neither was a muwaqqit, and neither – so far as is known – was associated professionally with any religious institution.[21] Less productive, but still competent, Egyptian astronomers of this century, such as al-Kawm al-Rīshī, ʿIzz al-Dīn al-Wafāʾī, and al-Karādīsī, were muwaqqits at different Cairene institutions, the first two at the Mosque of al-Muʾayyad, and the third at the Ashrafiyya Madrasa.[22]

Some Cairene muwaqqits, such as ʿAbd al-Qādir al-ʿAjmāwī, employed at the Madrasa of Sultan Ḥasan, spent time preparing treatises on timekeeping in which the prayer-times were written out in words rather than presented in tabular form.[23] Others went beyond realistic limits in the compilation of "new" tables. Ibrāhīm ibn Qāyitbāy, a student of the muwaqqit Shams al-Dīn al-Minūfī at the Ghawriyya Madrasa in the 16th century, took the most significant tables of the Cairo corpus, where values are given for each degree of solar longitude, and "stretched" them to display values for every few seconds of solar longitude. Another 16th-century astronomer, ʿAbd al-Qādir al-Minūfī, who was a muwaqqit at the Qawṣūn Madrasa, appears to have lost his sanity, perhaps as a result of his investigations into the effect of refraction at the horizon and his attempts to tabulate this effect for all geographical latitudes. The reader will by now be aware that the Muslim astronomers had a passion for compiling tables, but we can only speculate on the extent to which certain of the more exotic tables were ever used.

Interest in traditional *mīqāt* continued in Egypt until the 19th century, but no innovations of consequence were made after the 15th. Likewise the Umayyad Mosque in Damascus continued to be the centre of *mīqāt* in Syria, but no muwaqqit came up with any tables or instruments to surpass those of the Damascus school of the 14th century.

---

[19] On Sibṭ al-Māridīnī see Suter, *MAA*, no. 445; *Cairo ENL Survey*, no. C97; and also **II-2.10** and **5.8**.

[20] J. Heyworth-Dunne, *An Introduction to the History of Education in Modern Egypt*, London: Luzac & Co., 1939, pp. 63-64, quoting ʿAbd al-Raḥmān al-Jabartī, whose *History of Egypt* is now available in English translation (Thomas Phillipp, Moshe Perlmann and Guido Schwald, eds. and comm., 3 vols., Stuttgart: Franz Steiner, 1994).

[21] On these two individuals see Suter, *MAA*, nos. 432 and 447, and *Cairo ENL Survey*, nos. C62 and C98.

[22] On al-Wafāʾī see nn. 7:4 and 9:7. On al-Kawm al-Rīshī and al-Karādīsī see Suter, *MAA*, nos. 437 and 428, and *Cairo ENL Survey*, nos. C41 and C90, respectively.

[23] See **II-7-8** for details of the tables mentioned in this paragraph.

Muslim activity in the preparation of tables for timekeeping did not stop in the 14th and 15th centuries. On the contrary, it continued all over the Ottoman Empire. The Ottomans cultivated *mīqāt* with considerable zeal, and Ottoman astronomers either converted existing tables so that the times were displayed according to the Turkish convention that sunset is 12 o'clock,[24] or they compiled new sets for localities where no tables were available.[25]

---

[24] On this convention see Würschmidt, "Osmanische Zeitrechnung".

[25] On astronomical timekeeping in Ottoman Turkey see now **II-14**.

CHAPTER 9

INSTRUMENTS FOR TIMEKEEPING

In the days before the introduction of the mechanical clock, using tables displaying either the times of the prayers or their duration required in addition an instrument with which to measure the passage of time and hence to ascertain that the moment for a particular prayer had arrived. Islamic instruments are currently being catalogued, many for the first time. We now have a measure of control over these instruments – globes, astrolabes, quadrants, sundials, and others – such as never before.[1]

For the daytime prayers, a sundial marked with curves for the prayer-times obviated the use of any tables; indeed, a single sundial in a mosque could actually replace the services of a muwaqqit during the daytime provided the sun was shining. Muslim contributions to the theory and construction of sundials are impressive. Already in the early 9th century Ḥabash compiled a set of tables for drawing the hour lines on horizontal sundials for a number of different latitudes. Thābit ibn Qurra and his grandson, Sinān ibn Ibrāhīm, wrote on the theory of sundials oriented in any plane, and a 10th-century Baghdad astronomer Ibn al-Ādamī compiled a set of auxiliary tables for facilitating the construction of the hour lines on vertical sundials inclined at any angle to the meridian.

The treatise of al-Marrākushī displays the sophistication of 13th-century Islamic gnomonics,[2] but it also marked the beginning of a new period of interest in the subject, particularly by a series of *mīqātī*s and muwaqqits in Cairo and Damascus. al-Marrākushī's contemporary in Cairo was al-Maqsī, who compiled a set of tables for constructing horizontal and vertical sundials, and sundials inclined to the horizon or to the prime-vertical or skew to both.[3] Several later muwaqqits in Cairo and Damascus compiled tables for sundials displaying the hour-angle in equatorial degrees (rather than time since sunrise in seasonal hours) as well as the time remaining until the various prayers. A treatise on sundial theory by Ibn al-Raqqām indicates the sophistication of Maghribi gnomonics in the 14th century.[4]

We can assume that from about the 13th century onwards most mosques had a sundial, although few medieval sundials survive to this day. Vertical sundials on mosque walls, out of reach of passers-by, became more popular than horizontal sundials, which people could lean or sit on, or from which they could pilfer the gnomon. Yet most surviving vertical sundials have lost their gnomons as well. Others, observed in Istanbul, have the gnomons bent to support electric cables and telephone wires; one in Damascus even has a drainpipe hiding its meridian line! It was only in 1958 that three fragments of Ibn al-Shāṭir's sundial were unearthed in the drainage system of the Umayyad Mosque.

---

[1] See **X** for references to some of the instrument-types and specific instruments mentioned in this chapter.
[2] See n. 8:1.
[3] See n. 8:5.
[4] Carandell, *Ibn al-Raqqām sobre los cuadrantes solares.*

Within decades of their first encounter with the astrolabe in the 8th century, Muslim astronomers had made considerable advances in its theory and construction. In the early 9th century al-Khwārizmī tabulated a function for generating coordinates to draw the altitude and azimuth circles on an astrolabe plate, and al-Farghānī actually compiled a set of tables of such coordinates for marking astrolabe plates for each degree of terrestrial latitude. Ḥabash al-Ḥāsib developed the plate of horizons with which problems of spherical astronomy involving the horizon can be solved with a rete for any latitude. Also in the 9th century Muslim astronomers developed the trigonometric quadrant and the horary quadrant with fixed and movable cursor.

In the 10th century a succession of Muslim astronomers considered solutions to the problem of the rete of the rectangular planispheric astrolabe not being symmetrical with respect to the east-west line. In Toledo during the 11th century ʿAlī ibn Khalaf and Ibn al-Zarqālluh developed a universal astrolabe that does not need any plates and a universal plate in which the rete is replaced by a special alidade. Before the 13th century – it is not clear precisely when or where – Muslim astronomers developed a series of quadrants specifically for time-reckoning. All of these instruments, with the notable exception of the universal astrolabe of ʿAlī ibn Khalaf, are described by Abū ʿAlī al-Marrākushī in his monumental A-Z on *ʿilm al-mīqāt*. But this work was by no means the culmination of Islamic instrumentation. Rather, for those astronomers who specialized in *mīqāt* it was only a beginning and challenged them to come up with more useful instruments.

One of the most successful Muslim inventors of astronomical instruments was the early-14th-century Aleppo astronomer Ibn al-Sarrāj.[5] Although some of his works deal with mathematics, most deal with *mīqāt*, yet he did not call himself *mīqātī* and he was apparently not a muwaqqit. He reinvented the universal astrolabe of ʿAlī ibn Khalaf and devised a number of semicircular and quadrant grids for solving problems of trigonometry and hence spherical astronomy. Ibn al-Shāṭir and al-Khalīlī took up the same challenge, but the best solutions had already been obtained by Ibn al-Sarrāj. Ibn al-Shāṭir devised a universal astrolabe in which a grid of horizons rotates over a fixed stellar and ecliptic projection, and an astronomical "compendium" (a multi-functional instrument) called *ṣundūq al-yawāqīt* for timekeeping by day and night for all latitudes and for determining the qibla for all locations.[6] al-Khalīlī achieved rather less with instruments than with tables (see **8**).

At some point (the evidence presently available suggests Ayyubid Cairo) a Muslim astronomer realised that the problems that can be solved with an astrolabe can also be solved by using one half of the markings on an astrolabe plate for a fixed latitude and a thread with a moveable bead rather than an astrolabe rete. The earliest treatises on the astrolabic quadrant date from 14th-century Syria, but the instrument was clearly invented earlier. The Cairo astronomers of the 14th and 15th centuries inherited the instruments described by al-Marrākushī as well as the instruments invented by Ibn al-Sarrāj and the Damascus school; however, their only contribution of consequence was an equatorial armilla called *dāʾirat al-muʿaddil*, which was invented by the Cairo muwaqqit al-Wafāʾī.[7]

---

[5] See n. 8:13.
[6] This instrument is described in Janin & King, "Ibn al-Shāṭir's Compendium".
[7] See nn. 7:4 and 8:22.

Some muezzins and muwaqqits put their names to astronomical instruments or authored treatises on astronomy. The following are some examples. al-Sarrāj, a muezzin of Damascus *ca.* 1250, is known by three astrolabes that do credit to their maker.[8] Again, Saʿdū ibn ʿAlī al-Muʾadhdhin made an horary quadrant for latitude 30° (Cairo);[9] the instrument is undated but could well be from the 13th or 14th century. ʿAlī ibn Ibrāhīm al-Jazzār, a muezzin of Taza, was responsible for a highly sophisticated universal astrolabe dated 728 H [= 1327/28].[10] al-Mizzī and Ibn al-Shāṭir, both muwaqqits at the Umayyad Mosque in Damascus in the first half of the 14th century, made several instruments each: from the hand of the former we have five quadrants as well as a copy of one of his astrolabes,[11] from Ibn al-Shāṭir we have an astrolabe, two universal plates and one compendium.[12] Abu ʾl-Fatḥ ʿAlī, muwaqqit at the Sulṭān Mehmet Khān Mosque in Istanbul *ca.* 1750, made at least two equatorial semi-circles.[13] The vast

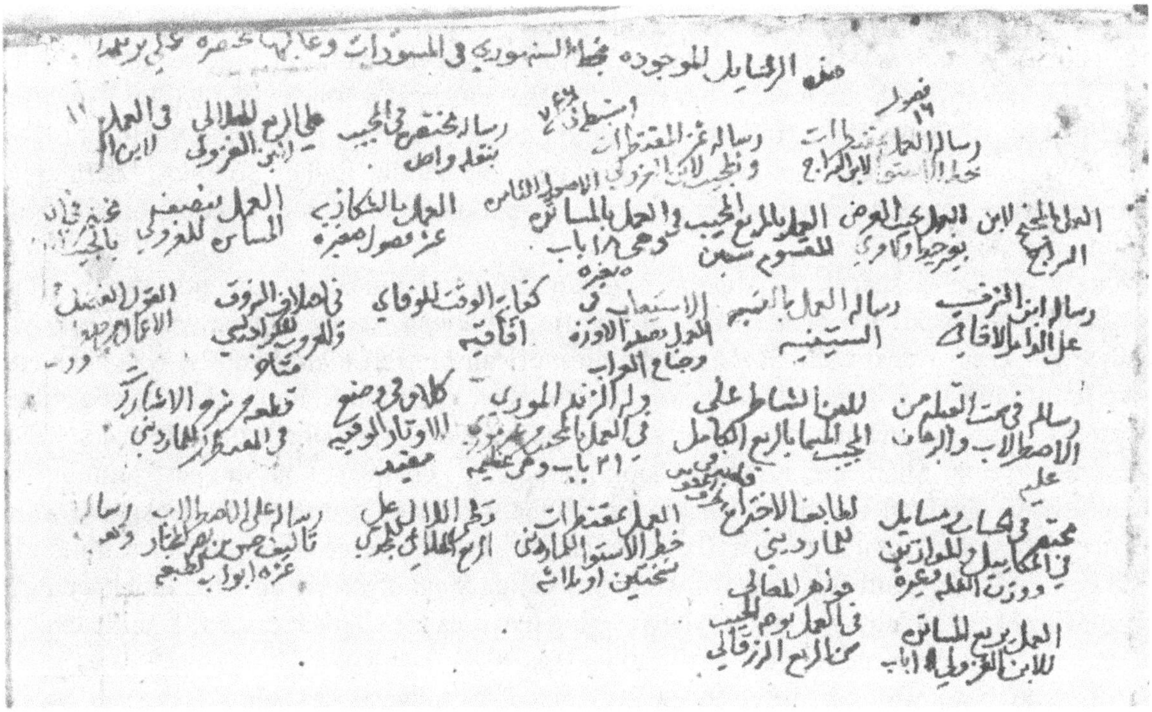

Fig. 9.1: A list of treatises on the use of various kinds of quadrants. It is stated that the treatises were all copied in the hand of one al-Sanhūrī, and the list is datable *ca.* 1550. Several of the treatises are new to us. [From MS Cairo DM 504,2, courtesy of the Egyptian National Library.]

---

[8] King, *Studies*, B-XII, p. 83, mentions two of these.

[9] The maker is not listed in Mayer, *Islamic Astrolabists*. His instrument is described in King, "Kuwait Astrolabes", pp. 91-93.

[10] Mayer, *Islamic Astrolabists*, suppl., p. 294. The instrument is illustrated in Vol. 2.

[11] Mayer, *op. cit.*, pp. 61-62, lists four of these. One one see Morley, "Arabic Quadrant".

[12] See Mayer, *op. cit.*, pp. 40-41 and 42. The astrolabe now in the Observatoire de Paris (listed on p. 42) is also by Ibn al-Shāṭir: see *Paris IMA 1993-94 Exhibition Catalogue*, p. 435, no. 331.

[13] One is described in Dizer, "Equatorial Semicircle"; another in *Christie's London 26.9.1991 Catalogue*, p. 38 (lot 80). For the latter see **Fig. VIb-14.1**.

Fig. 9.2: A double astrolabic quadrant serving Cairo and Damascus fashioned in ivory in 741 H [= 1348/49] by Abū Ṭāhir (#5009), also known as the author if some treatises on unusual instruments. [Courtesy of the Benaki Museum, Athens.]

Fig. 9.3: Another astrolabic quadrant by Abū Ṭāhir (#5010) on which the markings have not yet been interpreted. The same holds for the trigonometric quadrant on the back (the markings are very difficult to see and the available photo is too dark to include here). [Courtesy of the Chester Beatty Library, Dublin.]

majority of Islamic instruments are signed by people who have no professional relation to any mosque.

Although al-Mizzī and his colleague Ibn al-Shāṭir still constructed astrolabes, among later muwaqqits the astrolabic and trigonometric quadrant took over from it as the most popular instrument for solving problems in spherical astronomy. A quadrant with astrolabic markings

Fig. 9.4: This unsigned 14th-century astrolabic place from Egypt or Syria (#4038) bears on its two sides a total of eight sets of quadrant markings. Part of each set is ingeniously folded over so that the whole fits within a quarter circle, and several different arrangements of the markings are used. The latitudes served here are 33° (Damascus), 36° (Aleppo), 39° and 41;30° (Istanbul), and on the other side 22° (Mecca), 24° (Medina), 30° (Cairo), and 45°. The form of the now missing part of the instrument that would have rotated over the markings is a matter of speculation. For operations with the sun, there is already an ecliptic scale on each set of markings and a (graduated?) diametrical sighting alidade would have sufficed. For operations with the stars, a full standard astrolabic rete may have been provided. [Photo courtesy of the Deutsche Staatsbibliothek, Preußischer Kulturbesitz (Sprenger 2049), Berlin.]

on one side and a trigonometric grid on the other is by no means inferior to an astrolabe as an instrument for timekeeping, but the plethora of late treatises on the use of these two instruments and on sundial construction gives the impression that their authors were uninterested in any other instruments.[14]

Fig. 9.1 shows an index of close to 30 treatises dealing mainly with quadrants of one sort or another copied *ca.* 1550: alas these treatises themselves are no longer contained in the manuscript in which the index is preserved.[15] The remarkable variety of Islamic quadrants is a testimonial to both the ingenuity of the Muslim astronomers (see **Figs. 9.2-4**), mainly, in this case, Egyptian and Syrian muwaqqits, in mathematical methods and to their preoccupation with aids to timekeeping.

---

[14] See Janin & Rohr, "Deux quadrants", for two late quadrants, and Würschmidt, "Schriften Gedosis", on their use. See also *Paris IMA 1993-94 Exhibition Catalogue*, pp. 442-443, no. 337, and *Christie's London 13.12.1996 Catalogue*, pp. 40-41, lot 599, for descriptions of a quadrant from Damascus *ca.* 1895, recently purchased by the British Museum.

[15] See *Cairo ENL Catalogue*, II, pp. 369-370 (4.1.5), and *Cairo ENL Survey*, no. C147 and pl. LXVI.

CHAPTER 10

THE REMAINS OF A MUWAQQIT'S LIBRARY

No systematic study has been made of the provenance of the surviving manuscripts of the works of the leading Cairo and Damascus muwaqqits of the 13th and 14th centuries (or indeed of any other group of Islamic scientific manuscripts). However, we can obtain a glimpse of one Cairo muwaqqit's library by surveying those of his manuscripts that survive in the Egyptian National Library (numbers in the notes preceded by # are references to my survey of those manuscripts). The man in question is ʿAbd al-Raḥmān ibn ʿAbdallāh al-Ishkīrī al-Ṭūlūnī, muwaqqit and *imām* at the Mosque of Ibn Ṭūlūn in the middle of the 17th century.[1] The room in which he perhaps kept those books was situated on the roof of the domed fountain in the centre of the courtyard of that mosque – see **Fig. 10.1**.

First we list al-Ṭūlūnī's known works:

- A star catalogue appended to the small set of tables for regulating the prayer-times by al-Lādhiqī (#C132) – MM 15.
- A treatise in about 70 folios entitled *Taḥṣīl al-intifāʿ* ... on different kinds of instruments, mainly quadrants and sundials, compiled in 1034 H – MM 21, DM 445 and DM 21.
- A short treatise on the construction of a special kind of sundial – DM 982 (also MS Princeton Mach 5012).
- A short treatise in about 10 folios on sundial theory with tables for the latitude of Cairo – DM 421.

These are unimpressive when viewed in the light of the works of his illustrious predecessors.

The Cairo manuscripts bearing his mark of ownership are the following:

Treatises on theoretical astronomy (*ʿilm al-hayʾa*)

- A unique copy of a treatise on theoretical astronomy by ʿAbd al-Raḥmān al-Bakrī (unidentified) – MS Cairo University Library 18918.

Tables for planetary astronomy (*ʿilm al-zīj*)

- A fine copy of the auxiliary tables entitled *al-Durr al-yatīm* ... for computing planetary ephemerides,[2] compiled by Ibn al-Majdī (#C62) – DM 391.

Tables for timekeeping for the latitude of Cairo

- A fine copy of the tables of time of day as a function of solar longitude and solar altitude,

---

[1] On ʿAbd al-Raḥmān al-Ṭūlūnī see Brockelmann, *GAL*, II, p. 480, and SII, p. 493, and King, *Cairo ENL Survey*, no. D16. On his son see *ibid.*, no. D17. The information on the Cairo manuscripts surviving from their libraries is taken from King, *Cairo ENL Catalogue*, I, pp. 745 and 746. Numbers preceded by the symbol # refer to my *Cairo ENL Survey*.

[2] On these tables see King, *Studies*, A-VI (with E. S. Kennedy).

Fig. 10.1: The small room at the side of the dome over the fountain of the Mosque of Ibn Ṭūlūn, which would have housed the books and instruments of the muezzins and muwaqqits, possibly including those of ʿAbd al-Raḥmān al-Ṭūlūnī. [From Georg Ebers, *Aegypten in Bild und Wort*, Stuttgart and Leipzig, 1879, pl. 251, courtesy of the late Professor Michael Meinecke, Berlin.]

probably compiled by Shihāb al-Dīn al-Maqsī (#C15) in the 13th century and forming part of the Cairo corpus of tables for timekeeping – DM 444.

- A table displaying the time taken by the sun to move between the visible and true horizons, with values for each minute of solar longitude and a table of the half length of daylight adjusted for the effect of refraction at the horizon, each with values for each minute of solar longitude, both computed and copied by Ibrāhīm ibn Qāyitbāy al-Ashrafī (#C122) *ca.* 1020 H – DM 682 and 33. Tables of the ascensions at the beginning of the *ʿaṣr* prayer (*maṭāliʿ awwal al-ʿaṣr*), compiler unknown (#C123) – DM 45,1. Tables of the ascensions at the time of the benediction on the Prophet (*maṭāliʿ al-salām*), with values for each 3 minutes of solar longitude, compiler unknown (#C123) – DM 158. On these manuscripts and tables see **II-7.2** (illustrated).
- Tables of the corrections to the descensions of the moon and planets as a function of their latitude, compiler unknown – DM 57.

Treatises on instruments (*ʿilm al-ālāt*)

- A treatise on the use of an unusual variety of trigonometric grid known as *ṣadr al-wizza wa-janāḥ al-ghurāb*, compiled by Sibṭ al-Māridīnī (#C97) – DM 135,3.
- A short treatise on the "perfect" (*tāmm*) quadrant by Ibn al-Shāṭir (#C30) – QM 2,2.

- A short treatise on an unusual instrument called *rubʿ al-musātira* by the late-13th-century Cairo astronomer (and muwaqqit?) Nāṣir al-Dīn Muḥammad ibn Samʿūn (#C24) – QM 2,3.
- A treatise on the trigonometric quadrant entitled *Qaṭf al-zuhūr* ... , by Sibṭ al-Māridīnī (#C97) – K 4047.
- An anonymous treatise in Turkish on the trigonometric quadrant (#H89) – DMT 19.
- A marginal note in al-Ṭūlūnī's distinctive hand is found in a copy of treatise and tables on vertical sundials by the late-13th-century Cairo astronomer Shihāb al-Dīn al-Maqsī (#C15) – DM 597,4.
- A defective copy (lacking illustrations) of a treatise on mechanical clocks entitled *al-Ṭuruq al-saniyya fī 'l-ālāt al-rūḥāniyya* by the 16th-century Istanbul astronomer Taqi 'l-Dīn (#H12) – DM 557,4.

Miscellaneous

- Notes on the Jewish calendar and the computation of planetary longitudes – MM 183,2.

Given his interest in sundials it may be that al-Ṭūlūnī was responsible for demolishing the sundial in the Mosque of Ibn Ṭūlūn that had been constructed in the late 13th century but which had an incorrectly-drawn curve for the beginning of the *ʿaṣr* prayer, the most important time of the day.[3] al-Ṭūlūnī's son ʿAbdallāh, also an astronomer of sorts, compiled a treatise in about 10 folios on an instrument comprising four horizontal sundials for different latitudes[4] and a circle for determining the location of the mysterious *rijāl al-ghayb*, intermediaries between God and man who were popular amongst some Ṣūfīs (extant in MS Cairo ṬJ 811,6). Furthermore he left his mark of ownership on three manuscripts,[5] of which the first was previously owned by his father:

- The short treatise on the "perfect" (*tāmm*) by Ibn al-Shāṭir (#C30) – QM 2,2).
- A fine copy of an anonymous treatise on the universal (*jāmiʿ*) quadrant invented by Ibn al-Shāṭir (#C30) – DM 17.
- The main treatise on the trigonometric quadrant by Sibṭ al-Māridīnī (#C97) – DJ 273,2.

If these might seem lean pickings it should be borne in mind that the more substantial manuscripts from ʿAbd al-Raḥmān al-Ṭūlūnī's library are now in European libraries. The following are a few examples from my own collection of microfilms bearing the mark of his ownership:

- A unique copy of a treatise on folk astronomy by the 11th-century Hejazi or Yemeni author Muḥammad ibn Raḥīq, one of the most important works ever composed on the subject and a gold-mine of historical information – MS Berlin Ahlwardt 5664 (see **Fig. 10.2**).[6]

[3] This instrument is described in Janin & King, "Cadran d'Ibn Ṭūlūn".
[4] Published in Janin & King, "Ibn al-Shāṭir's Compendium", pp. 251-255.
[5] *Cairo ENL Catalogue*, I, p. 746.
[6] On Ibn Raḥīq (Brockelmann, *GAL*, I, p. 257, and King, *Astronomy in the Yemen*, no. 2) see now **III-3.2** and **IV-1.5**.

Fig. 10.2: The title-folio of the sole surviving copy of the treatise on folk astronomy by Ibn Raḥīq, with al-Ṭūlūnī's mark of ownership on the lower right corner. This folio is in very bad shape, and parts of it have been taped over, surely in Egypt before it was brought to Berlin. On the contents see **II-2.3** and **III-3.2**. [From MS Berlin Ahlwardt 5664, fol. 1r, courtesy of the Deutsche Staatsbibliothek (Preußischer Kulturbesitz).]

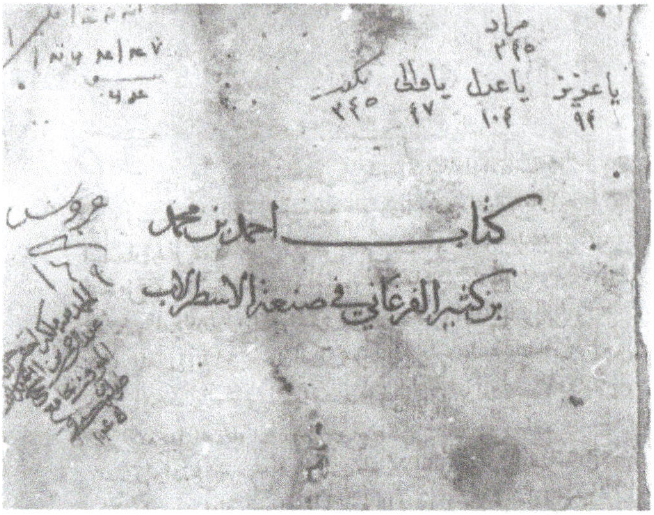

Fig. 10.3: Part of the title-folio of a fine copy of the treatise on astrolabe construction by the 9th-century Baghdad scholar al-Farghānī, showing al-Ṭūlūnī's mark of ownership dated 1045 H [= 1635/36]. [From MS Berlin Ahlwardt 5791, fol. 1r, courtesy of the Deutsche Staatsbibliothek (Preußischer Kulturbesitz).]

- A fine copy of various treatises on instruments, including Sharaf al-Dīn al-Ṭūsī on the linear astrolabe (rare),[7] Naṣr ibn Zurayr on unusual astrolabes (unique),[8] and al-Bīrūnī's treatise *al-Istīʿāb* on the astrolabe[9] (originally this manuscript also contained Theon's treatise on the armillary sphere) – MS Leiden Or. 561.
- A fine copy of the treatise on astrolabe construction by the 9th-century Baghdad astronomer al-Farghānī[10] – MS Berlin Ahlwardt 5791: see **Fig. 10.3**.
- A copy of the treatise on the hemispherical dial by the 10th-century astronomer al-Khujandī,[11] the leading instrument maker of his time – MS Manchester Mingana 560.
- A copy of the treatise by the 12th-century scholar al-Samaw'al al-Maghribī on the mistakes made by his predecessors, dealing *inter alia* with observational problems, notably those caused by refraction at the horizon, interpolation methods, and calendrical conversion.[12] This, MS Oxford Bodleian Hunt. 538,2, was copied in 988 H [= 1580] by Muḥammad ibn Yūsuf ibn ʿAbd al-Qādir ibn Muḥammad al-Dimyāṭī al-Ḥanafī, muwaqqit at the *madrasa* of al-Sulṭān al-Malik al-Ashraf Barsbāy, otherwise unknown to us.
- al-Ṭūlūnī also copied the treatise *al-Manhal al-sākib* on timekeeping by the stars, compiled by the 16th-century Cairo astronomer Muḥammad ibn Abi 'l-Khayr al-Ḥusnī (#C124) – MS Princeton Garrett 1017.

---

[7] See Brockelmann, *GAL*, I, p. 622, and II, pp. 858-859, and the article "al-Ṭūsī, Sharaf al-Dīn" by Adel Anbouba in *DSB*.

[8] This treatise and its author are apparently not documented in the secondary literature.

[9] Sezgin, *GAS*, VI, p. 268.

[10] *Ibid.*, VI, p. 151.

[11] *Ibid.*, VI, p. 221

[12] On al-Samaw'al see Suter, *MAA*, no. 302, and Sezgin, *GAS*, VI, pp. 65-66, and also nn. **II**-8:5 and **V**-10:12.

François Charette has kindly brought to my attention two further examples:

- MS Birmingham 1893 of some anonymous tables of the Versed Sine;
- MS St. Petersburg 83 (Rosen, no. 189) of the *Zīj* of Ibn al-Shāṭir, copied by ʿAlī ibn Aḥmad *al-muwaqqit* al-Ṣafadī in 854 H [= 1450].

These, then, are some of the manuscripts from the Ṭūlūnīs' libraries. There are probably dozens more on the shelves of other modern libraries. Alas no instruments from his collection are known. There are two points to be made here: firstly, that cataloguing manuscripts and looking out for such details as owners' marks brings rewards; and secondly, that the Cairo muwaqqit ʿAbd al-Raḥmān al-Ṭūlūnī owned a singularly impressive library.[13] That even with such a library at his disposal he was unable to compile anything substantial of his own is an indication of the fact that all of the major problems of traditional astronomy had been solved long before his time.

---

[13] See Kohlberg, *Ibn Ṭāwūs* (n. 5:4), for a detailed study of the library of a medieval Muslim scholar, alas not particularly rich in scientific works. François Charette has surveyed treaties compiled by, manuscripts owned by, and others copied by the Egyptian astronomer Muṣṭafā Ṣidqī (d. 1769) who was resident in Istanbul (*Cairo ENL Catalogue*, no. D81, and Ihsanoğlu *et al.*, *Ottoman Astronomical Literature*, II, pp. 466-467, and *Ottoman Mathematical Literature*, I, pp. 248-249) – see Charette, "The Library of Muṣṭafā Ṣidqī". The results are astounding: new editions of Greek works, copies of early Abbasid treatises not otherwise known to us, illustrations of early Islamic and Renaissance European astronomical instruments, *etc.*, *etc.*

CHAPTER 11

ON THE REGULATION OF THE LUNAR CALENDAR

There is no textual evidence that the muwaqqits were consulted on the expected visibility of the lunar crescent at the beginning of the lunar months. Clearly, however, the average muwaqqit could have predicted first visibility with reasonable accuracy, for the subject was discussed in every *zīj* and most treatises on *mīqāt*.[1] Unfortunately we have very little information on the astronomers who compiled annual ephemerides giving positions of the sun, moon, and planets for each day of the year and also information on the lunar crescent at the beginning of each month.[2] Did they have official positions? Were they muwaqqits?

The basic mathematical procedure for determining lunar visibility with reasonable hope of success is quite simple. See **Fig. 11.1**. Many Muslim astronomers used an Indian condition, namely, that the difference in setting times between the sun and the moon over the local horizon should be at least 12 equatorial degrees (= 48 minutes of time). Tables displaying the minimum solar-lunar longitude difference satisfying this condition were compiled for different latitudes from the 9th century onwards. From daily ephemerides, which were prepared by Muslim astronomers from the 9th to the 19th century, one could read off the solar and lunar longitudes for midday, correct these to sunset, and then use such a visibility table to determine whether the longitude difference was adequate to ensure visibility. Other more complicated conditions were formulated and no doubt applied in practice: those advocated by Ibn al-Shāṭir, al-Khalīlī, and various Cairo muwaqqits are in the tradition of the leading astronomers of earlier generations such as Kūshyār and Ibn Yūnus. Although most surviving Islamic ephemerides give various items of information on the crescent at the beginning of each month, only two manuscripts containing a set of calculations for each month of an individual year or for a series of years are known to us: see **Fig. 11.2** for one of these.[3] Other sources give simply the results of the calculations: see **Fig. 11.3**.

There is a wealth of material available on the regulation of the calendar that needs to be

---

[1] Various theories of lunar crescent visibility advocated by Muslim astronomers are analyzed in Kennedy *et al.*, *Studies*, pp. 140-163; King, "Lunar Crescent Visibility Tables", and "Ibn Yūnus' Lunar Crescent Visibility Theory", repr. in *idem*, *Studies*, C-II-III (also n. 11:3 below); Hogendijk, "Three Islamic Lunar Crescent Visibility Tables"; and Morelon, *Thābit ibn Qurra*, pp. xciii-cxviii, 93-117, and 230-259. See also my survey article "Ru'yat al-hilāl" in EI_2.

[2] These ephemerides were usually called called *taqwīm*, pl. *taqāwīm*, in Arabic: see the EI_2 article "Taḳwīm" by Michael Hofelich. In Turkish the term *ruznāme* was used for an almanac, ephemeris, or set of calendrical tables together with prayer-tables, but the Ottoman title *ruznāmejī* does not refer to an astronomer (see Stanford J. Shaw, *Ottoman Egypt in the Eighteenth Century*, Cambridge, Mass.: Harvard University Press & Harvard Center for Middle Eastern Studies, 1962, pp. 35-36). Dozens of examples of Ottoman ephemerides from the 17th to the 19th century, all anonymous, are preserved in the Egyptian National Library and are listed in *Cairo ENL Catalogue*, II, 2.3.12, and *Cairo ENL Survey*, no. D209; others, mainly from libraries in Istanbul, are listed in İhsanoğlu *et al.*, *Ottoman Astronomical Literature*, II, pp. 885-937.

[3] See King, *Studies*, C-IV, esp. pp. 235-239.

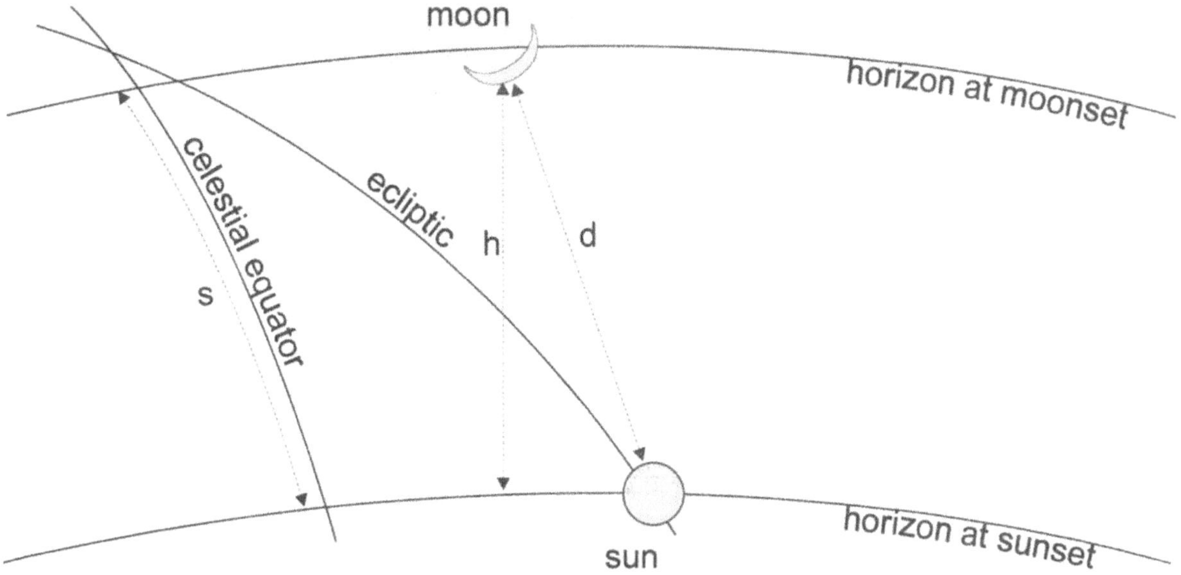

Fig. 11.1: The positions of the sun and moon relative to the western horizon.

investigated. The recent study of Klaus Lech on the early development of the Ramadan fast[4] invites a sequel to illustrate the way in which the religious calendar was regulated in practice. The one work of this kind, by Henri-Paul-Joseph Renaud,[5] was based only on Maghribi sources. More can be said on the subject, but not until the available sources have been tapped. A remarkable publication by a modern Moroccan *faqīh* contains a wide variety of original sources that could serve as a starting point.[6]

In the Islamic world today there is considerable confusion over the regulation of the calendar: witness the heated discussions in the local newspapers at the beginning of Ramadan each year. Modern astronomers can compute the time of the first appearance of the crescent with reasonable precision,[7] but generally their predictions are ignored. Different groups of Islamic countries rely on different authorities, and politics often plays a rôle as well. The result is that, for example, the beginning of Ramadan may be announced a day or even two days early in certain countries. For example, the beginning of Ramadan in 1973 was announced in some countries on an evening when the moon had actually set before the sun, a situation inconceivable in medieval times and hardly in keeping with the spirit of Islam.[8] (The reader should keep in mind that the Islamic day begins at sunset.) This is a sensitive subject, so

[4] Klaus Lech, *Geschichte des islamischen Kultus*, Band I: *Das ramadan-Fasten*, 1. Teil, Wiesbaden: Otto Harrassowitz, 1979.

[5] See Renaud, "Sur les lunes du Ramadan".

[6] See Muḥammad ibn ʿAbd al-Wahhāb al-Marrākushī, *al-ʿAdhb al-zulāl fī mabāḥith ruʾyat al-hilāl*, Qatar: Maṭbaʿat Idārat al-Shuʾūn al-Dīniyya, 1977.

[7] Mohammad Ilyas, *A Modern Guide to the Astronomical Calculations of Islamic Calendar, Times & Qibla*, Kuala Lumpur: Berita, 1984, provides an overview of the topic by a modern Muslim astronomer.

[8] See *al-Ahrām*, Sept. 26 and 27, 1973.

sensitive that even Muslim astronomers have to be careful what they write on the fact that Ramadan is often announced on the wrong day as a result of the incompetence of the responsible authorities.[9]

Fig. 11.2: Extracts from an anonymous set of tables displaying calculations for predicting the first visibility of the lunar crescent during the years 1125-1130 H [=1713-18]. For each month the ecliptic coordinates of the moon at sunset, the altitude of the moon above the horizon at Cairo, and the time between sunset and moonset are calculated for the first day of the civil calendar. Then a prediction is made: the crescent will be seen, it will be seen faintly, or it will probably be seen, or it will not be seen at all. In the last case the month begins the next day. [From MS Cairo DṢ 166,2, courtesy of the Egyptian National Library.]

Fig. 11.3: The appearance of the lunar crescent for each month of the solar year 1614/15 C.E. with respect to the horizon at Cairo. [From MS Cairo DM 141,3, courtesy of the Egyptian National Library.]

[9] See, for example, Imad-ad-Dean Ahmad, *Signs in the Heavens: A Muslim Astronomer's Perspective on Religion and Science*, Beltsville, Md.: Writers' Inc.-International, 1992, pp. 133-134.

CHAPTER 12

ON THE DETERMINATION OF THE SACRED DIRECTION

Another topic discussed in most works on *mīqāt* as well as in the *zīj*es was the determination of the qibla, or local direction of Mecca. Muslim astronomers from the 9th century onwards had at their disposal both approximate methods by which the qibla could be found with great facility and accurate methods involving complicated trigonometric formulae.[1] The most informative source on the subject is al-Bīrūnī's *Taḥdīd nihāyāt al-amākin*, in which he sets out to calculate the qibla at Ghazna (in what is now Afghanistan) by first establishing the necessary geographical data and then calculating the qibla in accordance with several different procedures.[2]

Fig. 12.1: An extract from a table displaying the qibla as a function of the latitude difference (16°-44°) and longitude difference (1°-12°) from Mecca. Values are in degrees and minutes and are based on on accurate procedure. This table is accompanied by another with the same format displaying the distance (in degrees) to Mecca. Both are contained in a treatise on the use of the sine quadrant compiled by ʿAlāʾ al-Dīn ʿAlī ibn Ṭaybughā, muwaqqit at the Umayyad Mosque in Aleppo *ca.* 1400. [From MS Cairo DM 832, fol. 90v, courtesy of the Egyptian National Library.]

[1] For a brief introduction to the subject see the article "Ḳibla (astronomical aspects)" in EI_2, repr. with corrections in King, *Studies*, C-IX, and the references there cited. More details are in King, *Mecca-Centred World-Maps*, pp. 47-193, and **VIIa**.

[2] See al-Bīrūnī, *Taḥdīd*, and the translation by J. Ali and commentary by E. S. Kennedy.

All *zīj*es contained geographical tables listing the longitudes and latitudes of major localities; often the qibla would be tabulated as well. Also, tables were available from the 9th century onwards displaying the qibla for each degree of latitude and longitude, based on either approximate or exact formulae: see **Fig. 12.1**. From the 14th century onwards magnetic compasses with qibla-indicators for various localities were in use, and these proliferated in Ottoman Turkey and particularly in Safavid Iran: see **Fig. 12.2**.[3] Particularly impressive are three world-maps centred on Mecca engraved on brass and made in Isfahan *ca*. 1675; they each bear a cartographic grid that preserves direction and distance to Mecca.[4]

Yet it is clear that in the medieval period astronomers were seldom consulted on the matter of the qibla. Firstly, when dealing with the obligation to face the Kaʿba in prayer, the legal texts do not suggest that one should consult an astronomer or a muwaqqit. Rather, they advocate entirely different non-mathematical methods for finding the qibla. Secondly, it is clear from the orientations of many medieval mosques that various non-mathematical methods were used in each of the major regions of the Islamic world. Only a minority of such mosques face the directions computed by astronomers.

The procedures advocated by legal scholars and outlined in detail in books of the genre *kutub dalāʾil al-qibla* (see **3**) are based either on tradition or on folk astronomy. By "traditional" qiblas I refer, for example, to the use of south in places as far apart as al-Andalus, Syria and Palestine, and Central Asia, following the first generation of Muslims in imitating the practice of the Prophet, who had faced south when he was praying in Medina. The other qiblas advocated in the non-scientific literature are directions of astronomical horizon phenomena such as winter sunrise (Egypt and al-Andalus), winter sunset (al-ʿIrāq, Central Asia) or the risings and settings of various fixed stars.

Muslim scholars from the 9th to the 16th century produced a variety of schemes of sacred geography, in which the world was divided into sectors about the Kaʿba.[5] Visually these schemes represent the supreme significance of the Kaʿba as a pointer to the presence of God. Each sector of the world was associated with a particular segment of the perimeter of the Kaʿba, and the qibla in each sector was defined in terms of an astronomical horizon phenomenon – see **Fig. 12.3**. The crudity of the geographical information in these schemes contributed to the wide variety of qiblas used in each region of the Islamic world.

Medieval mosques bear witness to this profusion of directions used for the qibla.[6] Although various texts on the divergence of mosque orientation in different regions survive and have been studied, not one of these was written by a muwaqqit, and indeed only one instance of muwaqqits being consulted on the matter of the qibla can be cited. This was in a *fatwa* drawn up in Cairo in the 16th century by a group of legal scholars and *mīqātī*s: the theme was the particularly confused orientation of the mosques in the Maghrib.[7] The importance of all this

---

[3] For some examples of such devices see King, *Mecca-Centred World-Maps*, pp. 107-124, and **X-9.2-4**.

[4] A detailed study is *ibid*., pp. 195-364. See **VIIc** on the third world-map, which came to my attention in 2001.

[5] See my forthcoming monograph *The Sacred Geography of Islam* and the summary in the article "Makka. iv. As centre of the world)" in EI_2, repr. in King, *Studies*, C-X.

[6] See **VIIa** for an overview, and especially Rius, *La Alquibla en al-Andalus y al-Magrib*.

[7] On the manuscript see *Cairo ENL Catalogue*, I, *sub* DM 540, and II, 3.3.11, and also *Cairo ENL Survey*, no. C110.

Fig. 12.2: An Egyptian qibla-compass described and depicted in a treatise on spherical astronomy by Ibn Samʿūn of Cairo *ca.* 1300. There follows a list of qibla-values for 160-odd localities. [From MS Leiden UB Or. 468, fol. 190r, courtesy of the Universiteitsbibliotheek.]

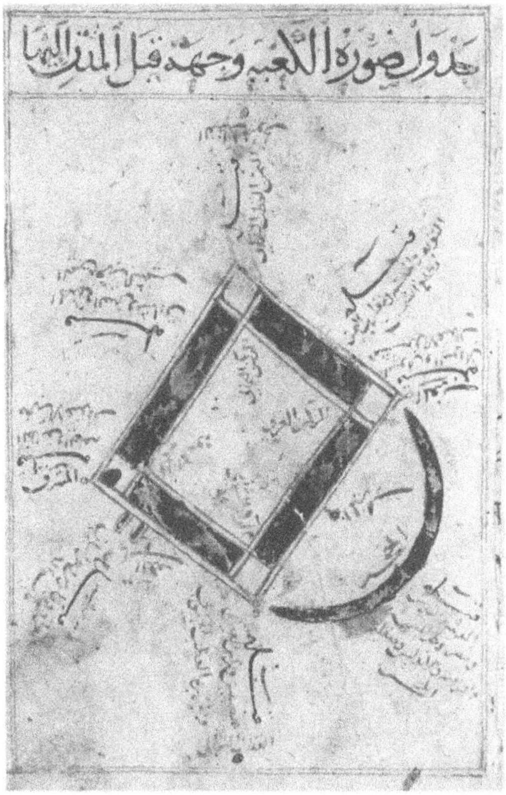

Fig. 12.3: A scheme of sacred geography as found in the cosmography of Pseudo-Ibn al-Wardī (Aleppo, *ca.* 1420). Here the walls of the Kaʿba are divided into eight regions about the Kaʿba, and the associated qiblas are the cardinal directions and solstitial limits for sunrise. In other copies of this work schemes with 12 sectors of the world about the Kaʿba are presented. [From MS Paris BNF ar. 2186, fol. 44r, courtesy of the Bibliothèque Nationale de France.]

material for the study of the orientation of Islamic religious architecture and even city-planning should be obvious, but historians of Islamic architecture have yet to react to it.

Since medieval values of terrestrial longitudes were generally rather inaccurate, the values of the qibla computed using a mathematically accurate formula or taken from a table based on such a formula, would also be "inaccurate". Only in the 18th century, when terrestrial longitudes could be measured fairly accurately, did it become apparent that even medieval mosques oriented in the directions computed by the astronomers were in general not properly oriented towards Mecca. Nowadays many medieval mosques that are still in use have secondary *miḥrāb*s to indicate the "true" qibla. Most new mosques are, I assume, erected in the qibla computed by some responsible authority such as the local survey department. Yet in rural areas people still use astronomical horizon directions: Egyptian peasants I have consulted told me they prayed towards the direction of sunrise, regardless of the season.

In the 1970s an Egyptian engineer in Saudi Arabia published a universal qibla-table generated by electronic computer,[8] but this is far more than the average Muslim needs. Lately there have appeared on the market magnetic compasses with booklets containing information about the qiblas of major cities all over the world. I understand that in the 1990s Saudia Airlines purchased a million such qibla-indicators from a company in Switzerland for free distribution to its passengers.

[8] See the three volumes of mathematical tables and graphs listed as Kamal al-Din, *Tables*, published *ca.* 1970, esp. vol. I dealing with the qibla, and also Ilyas, *Modern Guide ... , etc.* (n. 11:7).

CHAPTER 13

MODERN TABLES FOR REGULATING THE TIMES OF PRAYER

The times of the five daily prayers in the modern Islamic world are displayed on wall-calendars, in pocket diaries, or in daily newspapers: see **Fig. 13.1.**[1] In Egypt in the 1970s the Survey Department would compute sets of prayer-tables for any locality in which these are requested by a Muslim community. By that time in Turkey, however, the electronic computer was being used to generate prayer-tables for any location on earth, and several other computer centres around the world can offer such services to Muslim communities. Nowadays prayer-times for any locality are available on the Internet.

Thus one might think there is no more need for the muwaqqit: the muezzin can take his cue from tables and read the time from his watch or, in wealthy environments, from his lap-top computer. With tape-recorded prayer-calls that a custodian can simply switch on, it might be thought that there is no need for a muezzin either. Ingenious electronic clocks are now available that blurt out prayer-calls at the appropriate times. I do not know how they work or whether they work properly. But I do know that it is far more pleasant to be alerted to the times of prayer by a human voice, and I think most Muslims would agree.

→

Fig. 13.1a-e: Extracts from various 20th-century tables displaying the times of the prayers for different localities: Istanbul, 1910, with times according to the "Turkish" convention on the right, and according to the Western convention on the left (a); Brunei, 1997 (b); Tashkent, 1984 (c); and various cities in the former Soviet Union, photo taken by the author from the tables in the main mosque in Moscow in 1983 (d). The last set (e), for the month of *Rabīʿ al-thānī* of the year 1379 *Hijrī shamsī* / 1431 Hijra / June 2000, serves six cities in Iran. Here the times reflect the Shīʿī practice of combining the two daylight prayers into one (*ẓuhr*) and the two evening prayers into one (*maghrib*).

[1] Other examples are illustrated in the article "Mīḳāt" in EI_2.

a

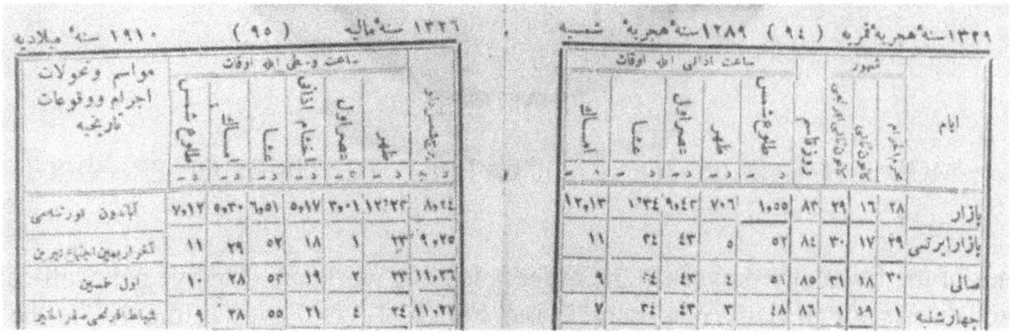

b

Prayer Times in Negara Brunei Darussalam MAY / JUNE 1997

Tarikh	Imsak	Subuh	Terbit Matahari	Zuhur	Asar	Maghrib	Isyak
1	4.39 / 4.33	4.49 / 4.43	6.08 / 6.06	12.19 / 12.20	3.36 / 3.44	6.27 / 6.32	7.38 / 7.45
2	4.39 / 4.33	4.49 / 4.43	6.08 / 6.06	12.19 / 12.20	3.36 / 3.45	6.27 / 6.32	7.39 / 7.45
3	4.38 / 4.33	4.48 / 4.43	6.08 / 6.06	12.19 / 12.20	3.37 / 3.45	6.27 / 6.32	7.39 / 7.46
4	4.38 / 4.33	4.48 / 4.43	6.07 / 6.06	12.18 / 12.20	3.37 / 3.45	6.27 / 6.33	7.39 / 7.46
5	4.38 / 4.33	4.48 / 4.43	6.07 / 6.06	12.18 / 12.20	3.37 / 3.45	6.27 / 6.33	7.39 / 7.46
6	4.37 / 4.33	4.47 / 4.43	6.07 / 6.07	12.18 / 12.20	3.37 / 3.46	6.27 / 6.33	7.39 / 7.47
7	4.37 / 4.33	4.47 / 4.43	6.07 / 6.07	12.18 / 12.21	3.38 / 3.46	6.27 / 6.33	7.39 / 7.47
8	4.37 / 4.33	4.47 / 4.43	6.07 / 6.07	12.18 / 12.21	3.38 / 3.46	6.27 / 6.34	7.39 / 7.47
9	4.36 / 4.33	4.46 / 4.43	6.07 / 6.07	12.18 / 12.21	3.38 / 3.47	6.27 / 6.34	7.40 / 7.47
10	4.36 / 4.33	4.46 / 4.43	6.06 / 6.07	12.18 / 12.21	3.38 / 3.47	6.27 / 6.34	7.40 / 7.48
11	4.36 / 4.34	4.46 / 4.44	6.06 / 6.07	12.18 / 12.21	3.39 / 3.47	6.27 / 6.34	7.40 / 7.48
12	4.35 / 4.34	4.45 / 4.44	6.06 / 6.07	12.18 / 12.21	3.39 / 3.47	6.27 / 6.34	7.40 / 7.48
13	4.35 / 4.34	4.45 / 4.44	6.06 / 6.08	12.18 / 12.22	3.39 / 3.48	6.27 / 6.35	7.40 / 7.49
14	4.35 / 4.34	4.45 / 4.44	6.06 / 6.08	12.18 / 12.22	3.40 / 3.48	6.27 / 6.35	7.41 / 7.49
15	4.35 / 4.34	4.45 / 4.44	6.06 / 6.08	12.18 / 12.22	3.40 / 3.48	6.27 / 6.35	7.41 / 7.49
16	4.35 / 4.34	4.45 / 4.44	6.06 / 6.08	12.18 / 12.22	3.40 / 3.48	6.27 / 6.35	7.41 / 7.49
17	4.34 / 4.34	4.44 / 4.44	6.06 / 6.08	12.18 / 12.23	3.40 / 3.49	6.28 / 6.36	7.41 / 7.50
18	4.34 / 4.35	4.44 / 4.45	6.06 / 6.08	12.18 / 12.23	3.41 / 3.49	6.28 / 6.36	7.42 / 7.50
19	4.34 / 4.35	4.44 / 4.45	6.06 / 6.09	12.18 / 12.23	3.41 / 3.49	6.28 / 6.36	7.42 / 7.50
20	4.34 / 4.35	4.44 / 4.45	6.06 / 6.09	12.18 / 12.23	3.41 / 3.49	6.28 / 6.36	7.42 / 7.50
21	4.34 / 4.35	4.44 / 4.45	6.06 / 6.09	12.18 / 12.23	3.41 / 3.50	6.28 / 6.36	7.40 / 7.51
22	4.34 / 4.35	4.44 / 4.45	6.05 / 6.09	12.18 / 12.24	3.42 / 3.50	6.28 / 6.37	7.43 / 7.51
23	4.33 / 4.36	4.43 / 4.46	6.05 / 6.09	12.19 / 12.24	3.42 / 3.50	6.28 / 6.37	7.43 / 7.51
24	4.33 / 4.36	4.43 / 4.46	6.05 / 6.10	12.19 / 12.24	3.42 / 3.50	6.28 / 6.37	7.43 / 7.51
25	4.33 / 4.36	4.43 / 4.46	6.06 / 6.10	12.19 / 12.24	3.43 / 3.50	6.28 / 6.37	7.43 / 7.51
26	4.33 / 4.36	4.43 / 4.46	6.06 / 6.10	12.19 / 12.24	3.43 / 3.51	6.28 / 6.37	7.44 / 7.52
27	4.33 / 4.36	4.43 / 4.46	6.06 / 6.10	12.19 / 12.25	3.43 / 3.51	6.28 / 6.38	7.44 / 7.52
28	4.33 / 4.37	4.43 / 4.47	6.06 / 6.11	12.19 / 12.25	3.43 / 3.51	6.29 / 6.38	7.44 / 7.52
29	4.33 / 4.37	4.43 / 4.47	6.06 / 6.11	12.19 / 12.25	3.44 / 3.51	6.30 / 6.38	7.44 / 7.52
30	4.33 / 4.37	4.43 / 4.47	6.06 / 6.11	12.19 / 12.25	3.44 / 3.51	6.30 / 6.38	7.45 / 7.52
31	4.33	4.43	6.06	12.19	3.44	6.30	7.45

NOTE: For the Belait District, 3 minutes should be added. One minute should be added in Tutong.

c

d

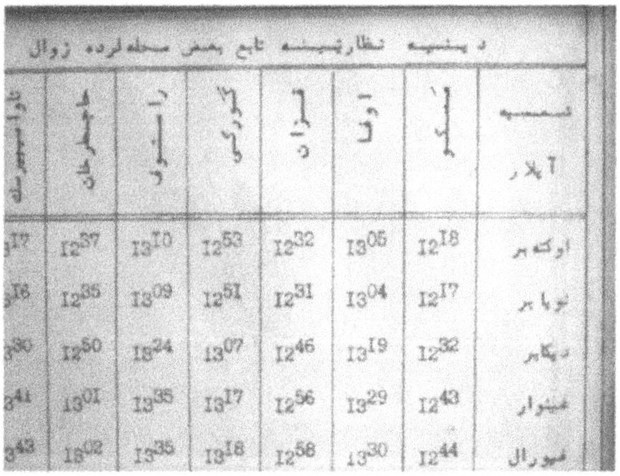

e

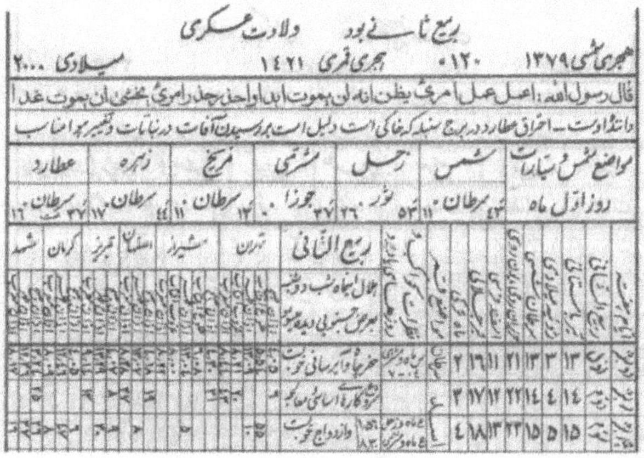

CHAPTER 14

CONCLUSION

The importance of the material discussed in this study for our understanding of medieval religious ritual and of the reasons why that ritual developed in the way it did will be obvious to any student of Islamic civilization. For anyone who might have doubted that astronomers working primarily towards determining the times and direction of prayer could contribute much to the development of hard-core science, let us mention some topics that such astronomers were the first to handle. It was primarily those working on *mīqāt* who established the different values for the solar depression at daybreak and nightfall and who compiled tables displaying the duration of morning and evening twilight. They developed a mathematical theory – albeit not yet fully understood – for the effect of refraction at the horizon. They devised tables and instruments for solving all of the standard problems of spherical astronomy for any latitude. They prepared tables of coordinates for marking curves on astrolabes and sundials and made substantial progress in instrumentation. They developed the astrolabic quadrant from the astrolabe, but this was only one of a couple of dozen varieties of quadrant that they conceived as computational devices. They devised graphical representations of the relationship between the altitude of the sun and the hours of daylight. Finally it was also they who developed the magnetic compass as a qibla-indicator and discovered magnetic variation, and instrument-makers with a passion for qibla-indicators who produced even a cartographic grid on which both distance and direction to Mecca are correctly represented.

For the history of science as practiced from the traditional Western perspective, Islamic timekeeping is indeed not so important: medieval Europe knew virtually nothing of contemporary Islamic timekeeping and did not profit by it. But for the history of science on the global scale *ʿilm al-mīqāt* in Islam was as important as *ʿilm al-hayʾa* (theoretical astronomy), *ʿilm al-zījāt* (mathematical astronomy), or *ʿilm aḥkām al-nujūm* (astrology). A *ḥadīth* attributed to the Prophet Muḥammad asserts that: "the necks of the muezzins will be stretched on the Day of Judgement" (**Text 9**).[1] Thus even those who merely announced the time of prayer to the faithful now hold their heads high, and surely the muwaqqits who ensured that this took place at the right time should enjoy similar rewards in Paradise.

[1] This *ḥadīth* is listed in Wensinck *et al.*, *Concordance*, I, p. 44a.

APPENDIX

ARABIC TEXTS

(١) الأصبحي: مخطوطة دار الكتب المصرية ميقات ٩٤٨، ق ٦و - ٦ظ

... إن أوقات الصلوة لا تعرف بدرج الأسطرلاب ولا هي من علم المنجمين تعرف بالحساب وإنما تعرف بالعيان ... والمنجمون إنما أخذوا علمهم عن أقليدس والسندهند وأرسطاطاليس وغيرهم من الفلاسفة وهم كلهم كفرة ...

(٢) ابن الأخوة ، معالم القربة ، ص ١٧٦- ١٧٨

... فصل ولا يؤذن في المنارة إلا عدل ثقة أمين عارف بأوقات الصلوات لأن النبي صلعم قال المؤذنون أمناء والأئمة ضمناء فأرشد ا الأئمة وغفر للمؤذنين وينبغي للمحتسب أن يمتحنهم بمعرفة الأوقات فمن لم يعرف ذلك منعه من الأذان حتى يعرفها لأنه ربما أذن في غير الوقت فأفطر الصائم أو فعل المحلوف عليه أو يصلون قبل الوقت فلا تصح صلاتهم فيكون هو السبب في إفساد أحوالهم فيجب عليه معرفة الوقت ويقرأ باب الأذان والإقامة في الفقه ويستحب أن يكون المؤذن حسن الصوت وينهاه المحتسب عن النغي في الأذان وهو التمطيط الفاحش والتطريب ويأمره إذا صعد المنارة أن يغض بصره عن النظر إلى حريم الناس ودورهم ويأخذ عليه العهد في ذلك ولا يصعد إلى المنارة غير المؤذن في أوقات الصلوات وينبغي للمؤذن أن يكون عارفا بمنازل القمر وشكل كواكب المنازل ليعلم أوقات الليل ومضي ساعاته وهي ثمانية وعشرون منزلة الشرطين والبطين ... وبطن الحوط وهو الرشاء فهذه جملة عدد المنازل والصبح يدوم يطلع في كل منزلة من هذه ثلاثة عشر يوما ثم ينتقل إلى المنزلة المعترضة في وسط السماء فيعرف حينئذ الطالع والساقط وكم بينه وبين الصبح وهذا فيه علم وحساب يطول شرحه * ومن شرط المؤذن أن يكون مسلما عاقلا ذكرا فلا يصح أذان كافر أو امرأة أو مجنون أو سكران ويصح أذان الصبي المميز ويستحب الطهارة في الأذان ويصح بدونها والكراهية في الجنب أشد وليكون المؤذن عارفا بالأوقات كما تقدم والأذان مثنى والإقامة فرادي مع الإدراج وأن يكون قائما وهو مستقبل القبلة ويلتفت في الحيعلتين يمينا وشمالا وصدره إلى جهة القبلة ورفع الصوت في الأذان ركن من أركانه والترتيب في كلمات الأذان شرط من شروطه فلو عكس لم يعتد به وإن فعله استهزاء أو استهتارا عزر تعزيرا بليغا ولا يؤذن الا لفريضة وما سوى ذلك من صلاة الكسوف والعيدين والاستسقاء ينادي لها الصلاة جامعة وينبغي للمؤذن أن يزيد في التذكار والتسبيح في ليالي رمضان وينادي بالسحور أولا ثم يشرب الماء ثم ينادي قرب الصباح ثم قرب الأذان ثم بعد ذلك يطفئ الفانوس ثم يؤذن ومتى تعذر عليه طفئ الفانوس كسره فإن من لا يسمع الأذان لبعده اعتماده في أكله وشربه على رؤية الفانوس وإيقاده وطفئه ويجوز للمؤذن أخذ الأجرة على الأذان وأما الأئمة فلا يجوز لهم أخذ الأجرة على الصلوات والإمامة فإن دفع للإمام شيء من غير شرط جاز له أخذه على سبيل الهبة وإن رزقوا من بيت المال جاز على الصحيح ...

تعليق للمحرر : * زيادة في كتاب نهاية الرتبة لعبد الرحمن الشيزري : فمن أراد ذلك فعليه بكتاب الانواء لابن قتيبة

(٣) ابن سمعون: مخطوطة ليدن ٤٦٨ ، ق ٩١و

... ومن فضلاء المائة السابعة من الهجرة في هذا الفن أبو الحسن علي بن عبد الملك بن سمعون بالجامع العتيق بمصر نحو ثلاثين سنة وولي بعده مكانه بالجامع المذكور ولده محمد عرف بالوجيه وهو جدنا لأبينا توفي سنة إحدى وسبع مائة مكث موقتا بالجامع المذكور ستين سنة وولي بعده مكانه ولده محمد عرف بالمجد وهو عمنا وكان جد والدي أبو الحسن علي رحمه(؟) توفي قبل ابنه الوجيه بنحو سبع عشر سنة سنة خمس وثمانين وستمائة ...

(٤) نصوص عن ابن باص

(ا) مخطوطة اسكوريال ٩٦١

ق ٦٩و : كتاب فيه رسالة الصفيحة الجامعة ... تصنيف الشيخ الإمام الأجل الفاضل المبارك أبي علي الحسين ابن أحمد ابن ماص (كذا) الأسلمي المغربي ...

حاشية : وجدت في نسخ أخرى أبي علي الحسين بن أبي جعفر أحمد بن يوسف بن باص

ق ٦٩ظ : بسم الله الرحمن الرحيم ... قال الشيخ الإمام العدل (كذا) المبارك أبو علي حسين ابن الشيخ المؤذن المبارك أبي جعفر

أحمد بن يوسف ابن ماص (كذا) الأسلمي أما بعد ... فإنه سنح بخاطري على منذ اعوام سالفة قبل تاريخنا هذا وهو عام ٦٧٣ أن أضع في معرفة الأوقات وما يتعلق بها صفيحة للأسطرلاب المسطح يكون جامعة لجميع العروض عاملة في جميع البلدان ...

(ب) مخطوطة دار الكتب المصرية تيمور رياضة ١٥٩،٢ (١٣٢٠ هجرية عن نسخة ١١٢٩ هجرية محفوظة في المكتبة الخالدية بالقدس)

ص ٦٢: رسالة في الصفيحة الجامعة مختصرة من رسالة ابن باص

ص ٢٣: ... هذه رسالة مختصرة من رسالة الشيخ المعدل بغرناطة صاحب الأوقات بها أبو (كذا) الحسن علي ابن أبي جعفر ابن أحمد ابن يوسف بن ناصر الاسلي (كذا) رحمه تعالى وصنعها على صفيحة ذكر أنه استنبطها لجميع العروض ...

(ج) ابن الخطيب ، الإحاطة ، ج ١ ، ص ٨٥

أحمد بن حسن (كذا) بن باضة (كذا) السلمي (كذا) الموقت بالمسجد الأعظم بغرناطة أصله من شرق الأندلس وانتقل إليها والده يكنى أبا جعفر كان نسيج وحده وقريع دهره معرفة بالهيئة وإحكاما للالة الفلكية ... أخذ ذلك عن والده الشيخ المتفنن شيخ الجماعة في هذا الفن

(٥) ابن إياس ، تأريخ مصر ، ص ٧٨

... وقال بعض المؤرخين إن ملوك اليمن أهدت إلى الملك الكامل محمد شمعدانا من نحاس يخرج منه عند طلوع الفجر شخص من نحاس لطيف الخلقة يخاطب الملك قائلا صبحك ا بالخير قد طلع الفجر أو صفير هذا معناه وكان هذا الشمعدان من صنعة الميقاتية فأقام في حواصل الملوك إلى أيام الملك الناصر محمد بن قلاوون ثم فقد ...

(٦) من وقفية مدرسة السلطان برقوق: جاريتز ، ص ١٢٩ و ١٣٧

... المؤذنون ستة في شهر تسعون لكل منهم خ ع (= خمسة عشر درهما) ويرتب الناظر ستة من الصوفية المذكورين حسني الأصوات يعلنون بالأذان والتسبيح والإذكار في أوقات الأسحار وإقامة الصلوات وتبليغ التكبير خلف الإمام والتسليم على النبي صلى ا لله عليه وسلم في ليالي الجمع وأوقات الأسحار يفعلون ذلك بالتناوب على حسب ما يرتبهم الناظر في هذا الوقت ويصرف لكل واحد منهم في كل شهر خمسة عشر درهما نقرة مضافا لما هو مقرر لهم عن وظيفة التصوف ...

... وما يصرفه لرجلين فاضلين عالمين بعلم الميقات من جملة الصوفية المنزلين بالخانقاه المدرسة(؟) المشار اليها اعلاه يرتبهما الناظر رئيسين بالنوبة كل نوبة يومين وليلتين على أن يفعل كل منهما ما جرت العادة به من الإعلام بدخول أوقات الصلوات والتسبيح والإذكار وقراءة العقيدة والخروج أمام الخطيب في يوم الجمعة ثلثون درهما نقرة بالسوية بينهما لكل واحد منهما خمسة عشر درهما نقرة زيادة على ما هو مرتب له من معلوم التصوف في كتاب الوقف المسطر بأعاليه وما يصرف لرجلين من الصوفية المشار اليهم في كتاب الوقف المسطر بأعاليه يكونان حسني الأصوات يرتبهما الناظر مؤذنين بالمدرسة المذكورة في كتاب الوقف المسطر بأعاليه على أن يفعل كل منهما ما شرط في حق المؤذنين المذكورين أعلاه ثلثون درهما نقرة لكل منهما خمسة عشر درهما نقرة زيادة على ما هو مرتب له عن معلوم التصوف المشار إليه أعلاه ...

(٧) ابن الأكفاني ، إرشاد القاصد ، ص ٥٩

... علم المواقيت علم يتعرف منه أزمنة الأيام والليالي وأحوالها وكيفية التوصل إليها ومنفعته معرفة أوقات العبادات وتوخي جهتها والطوالع والمطالع من أجزاء البروج ومن الكواكب الثابتة التي منها منازل القمر ومقادير الظلال والارتفاعات وانحراف البلدان بعضها عن بعض وسموتها ومن الكتب المختصرة فيه نفائس اليواقيت ومن المبسوطة جامع المبادئ والغايات لأبي علي المراكشي ...

(٨) السبكي ، معيد النعم ، ص ١٦٥

... المثال الخامس والستون المؤذن عليه معرفة الوقت وإبلاغ الصوت يؤذن للصبح من نصف الليل وعند دخول الوقت ولذلك يسن للصبح مؤذنان المثال السادس والستون الموقت ولا بد من معرفته علم الميقات فليحقق فن الهيئة وجهة القبلة على الخصوص وقد كثر في هذه الطائفة المنجمون والكهان نعوذ بالله منهم ...

(٩) حديث نبوي مسجل في ونسينك ، ج ١ ، ص ٤٤

... المؤذنون أطول الناس أعناقا يوم القيامة ...

Part VIa

Universal solutions in Islamic astronomy

Dedicated to Asger Aaboe in 1987

ACKNOWLEDGEMENTS AND NOTES ON THIS VERSION

This study is dedicated to a man whose generosity, humour, scholarship and impeccable taste, not only in things mathematical, have always impressed me. I first met Asger Aaboe when I drifted into the Department of the History of Science at Yale University one day in 1968. I was seeking a kind of *maison secondaire*, an alternative from my home base at the Department of Near Eastern Languages and Literatures. With great pleasure I remember Asger at the enormous table in the so-called Bernie Library, surrounded by many papers and books and pipes and cups in some disorder, but not lacking a steady supply of pipe tobacco and coffee. The conversations ranged far beyond Babylonian columns for this or that function, planetary models with or without equants, and the respective merits of Levi ben Gerson and Ibn Yūnus. One of my favorite topics of conversation was the date of the next trip to the History of Mathematics Department at Brown, to which we drove once a month in cars with or without roofs. And it was there that, thanks to Asger and Bernard Goldstein, I met Otto Neugebauer, David Pingree, and Gerald Toomer, and the person who was to influence my academic life more than any other, namely, Ted Kennedy.

This study originated a few years later. One winter evening in our Cairo apartment, as Ted was engrossed in his notes on Persian manuscripts gathered that morning in the freezing, window-pane-less wastes of the *Dār al-Kutub*, I had nothing else to do but reflect. My thoughts of that evening, for what they were worth, were eventually presented as a paper delivered at the XVth International Congress of the History of Science in Bucharest in 1981. It is a pleasure to acknowledge my debt to Jan Hogendijk (Utrecht) as well as Ted Kennedy for their comments on the penultimate version of the first public manifestation of the paper, which was published as "Universal Solutions in Islamic Astronomy", in J. Lennart Berggren & Bernard R. Goldstein, eds., *From Ancient Omens to Statistical Mechanics: Essays on the Exact Sciences Presented to Asger Aaboe, Acta Historica Scientiarum Naturalium et Medicinalium* (Copenhagen) 39 (1987), pp. 121-132, reprinted in 1993 in my *Studies*, C-VI.

In this new version I have added some new material, notably two instruments for determining time of day and night, the first corresponding to the universal horary dial on the medieval European *navicula*, which I am fairly confident was devised by Ḥabash, and the second which was certainly devised by him, as well as the grids for determining the direction and distance to Mecca, which I strongly suspect was also originally due to our friend Habash. I have also updated the references, although those to materials from Mamluk Egypt and Syria are now in the form of cross-references to **VIb**. To François Charette my gratitude for new materials on Najm al-Dīn al-Miṣrī and for checking the new text for consistency.

TABLE OF CONTENTS

INTRODUCTION

> "Cette science dite arabe, qu'a-t-elle d'arabe en réalité? La langue, rien que la langue. ... Cette science n'est pas arabe. Est-elle du moins musulmane? L'islamisme a-t-il offert à ces recherches rationelles quelque secours tutélaire? Oh! En aucune façon! Ce beau mouvement est tout entier l'oeuvre de parsis, de chrétiens, de juifs, de harraniens, d'ismaéliens, de musulmans intérieurement révoltés contre leur propre religion. ... L'islamisme, en réalité, a donc toujours persécuté la science et la philosophie. Il a fini par les étouffer." Ernest Renan, "L'Islamisme et la science" (1883), in *idem*, *Œuvres complètes*, 10 vols., Paris, 1947-61, I, pp. 953-955.

Medieval Islamic mathematical astronomy was divided into three main branches: planetary astronomy, spherical astronomy and mathematical astrology. Spherical astronomy, with which we are concerned here, is the study of problems relating to the apparent daily rotation of the celestial sphere. The main problems of this discipline that confronted the astronomer were the determination of the time and the azimuth of a celestial body from an observed altitude, given its declination (or distance from the celestial equator), and the local terrestrial latitude (measured by the altitude of the celestial pole above the local horizon).

Timekeeping was important to the Muslim astronomers not only because of their need to accurately record and predict celestial phenomena but also because the times of the five daily prayers in Islam are astronomically defined. Another problem of major concern was the prediction of the appearance of the lunar crescent, upon which depends the regulation of the lunar calendar. Yet another was the determination of the qibla, or direction of Mecca from any locality; the qibla is the direction of prayer and of various other ritual acts. The problem of finding the qibla from the appropriate geographical coordinates belongs properly to mathematical geography, but Muslim scientists often treated it as a problem of spherical astronomy: it is then required to find the azimuth of the zenith of Mecca on the local horizon.

The Muslim astronomers had at their disposal five main practical means of obtaining the solutions to problems of spherical astronomy:

1. Observational instruments (such as an equatorial armilla), which could be aligned so that the time could be read from the instrument without any calculation whatsoever.
2. Analogue computer devices (such as an astrolabe), which could be made to simulate an instantaneous celestial configuration relative to the local horizon and provide an instant solution.
3. Geometric procedures (such as the analemma), by which solutions to specific numerical examples could be found by two-dimensional constructions.
4. Trigonometric procedures, derived either from geometric constructions or by the application of spherical trigonometry. Tables of the sine/cosine and tangent/cotangent functions were available for calculating the solutions to specific numerical examples.
5. Trigonometric computational devices (such as a trigonometric quadrant), for deriving without calculation the solution to a specific numerical example, given the appropriate formula.

Tables or instruments providing solutions to problems in spherical astronomy for all latitudes rather than one specific latitude will hereafter be labelled universal. What I mean by universal solutions are instruments, mathematical procedures and tables which aim at providing solutions for all terrestrial latitudes to those problems of spherical astronomy in which latitude is involved.

I am fully aware that I am imposing a modern categorisation on the source material, but it seems to me useful to consider these solutions together and thereby to try to add a new dimension to our understanding of the achievements of the Muslim scientists. The expression 'universal solution' is not a particularly happy one, and I use it for lack of anything better. In our sources, the notion is usually but not always restricted to the *oekoumene*; I have resisted the temptation to use the term oecumenical in my title even though Asger Aaboe might have enjoyed it. To illustrate the problem of terminology, we note that, for example, all geometric and trigonometric solutions are general and thus universal, but these I shall not treat. Given the formula, calculations still have to be performed for different latitudes. Instruments designed to facilitate such computations I shall discuss. Tables of various spherical astronomical functions computed for each of the climates of classical geography (see **Fig. 0.1**) are in a sense not universal, but these will be considered. Tables which serve each degree of latitude between the Yemen and the Bosphorus I regard as universal in that, as remarked by Abū Naṣr (Central Asia, *ca.* 1000), there was no-one beyond these bounds who was interested in spherical astronomy anyway.[1] I hope that the reader will gain a clearer notion of precisely what I mean by universal solutions as he or she reads on.

The problems of spherical astronomy essentially involve converting co-ordinates from one orthogonal co-ordinate system to another. Of particular interest to Muslim astronomers was the determination of time from an observed solar or stellar altitude, that is, converting from a horizon- to an equator-based co-ordinate system. This and other problems can be solved using either an analogue computer or trigonometry. With the former, the instantaneous celestial configuration relative to the horizon is represented in three of two dimensions; then the time is read from a scale. Otherwise the three-dimensional problem is solved by spherical trigonometry or by reducing it to two dimensions and solving by plane trigonometry; then the resulting formula is applied to the numerical data, either to calculate the solution by means of trigonometric tables or to derive it without computation by means of a trigonometric grid.

This one aspect of Muslim scientific creativity that I shall describe was a direct continuation of various Greek developments along the same lines. One example of a 'universal solution' inherited by the Muslims from the Greeks is the planisphaeric astrolabe with plates for each of the seven climates or for each few degrees of latitude.[2] The astrolabe can be made to simulate the apparent daily rotation of the heavens over the horizon of any locality. The underlying theory of stereographic projection, together with a numerical investigation of the size and position of the various markings on such an instrument, is presented in Ptolemy's *Plani-*

[1] Jensen, "Abu Naṣr's *Table of Minutes*", p. 5.

[2] On the climates see Neugebauer, *HAMA*, II, pp. 725-733, and on their use in Islamic astronomical instrumentation see King, "Geography of Astrolabes", pp. 6-9, and *idem*, *Mecca-Centred World-Maps*, pp. 27-28. See also n. **I**-1:11.

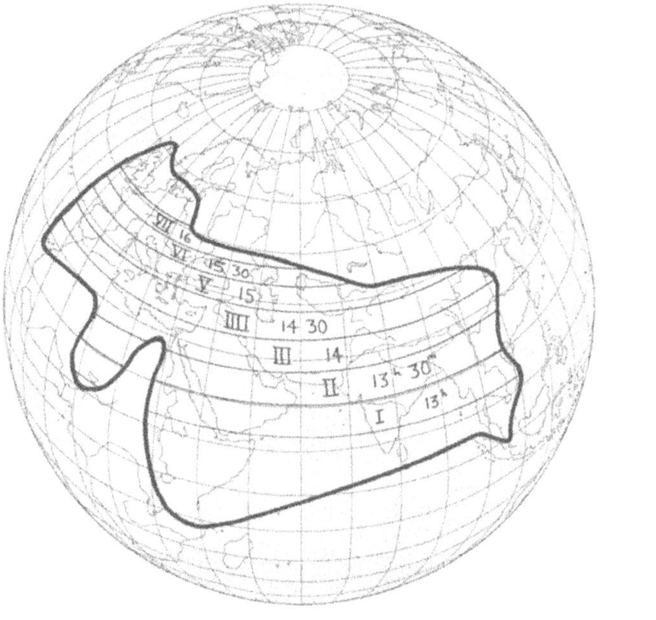

Fig. 1: The seven climates of classical geography, whose limits and middles were defined by the maximum length of daylight. The climates were of prime importance in medieval Islamic astronomy and instrumentation. [Prepared by a student at Frankfurt University, who alas did not sign this production.]

sphaerium.[3] Other examples are Ptolemy's tables in *Almagest* II.8 and II.13 of oblique ascensions and the angles between the ecliptic and the horizon and meridian for each of the climates;[4] his universal dial for coordinate conversion and tables for this purpose (and hence also for constructing sundials) described in the *Analemma*;[5] and the tables for planetary visibility and parallax in the *Handy Tables*.[6] Of these works, only the *Analemma* was not translated into Arabic (contrary to what is often maintained in the secondary literature).[7]

The main centres of the activity I shall describe were Baghdad in the 9th century, al-Andalus in the 11th, and from the 13th to the 15th, Egypt and Syria.[8] Some of the tables occur in the astronomical handbooks called *zīj*es,[9] but many are found in other compilations.[10] Most of the material I survey here is described in greater detail in other papers, some not yet published.

---

[3] On the astrolabe in the Islamic context see the article "Asṭurlāb" by Willy Hartner in EI_2, and also **X-4**. On the theory of stereographic projection see Neugebauer, *HAMA*, II, pp. 857-879.

[4] *Ibid.*, I, pp. 34-37 and 45-52.

[5] *Ibid.*, II, pp. 839-856.

[6] *Ibid.*, II, pp. 1017-1025 and 990-999.

[7] The procedure known as the analemma was, however, to become a powerful tool in the hands of the most capable of the Muslim scientists. For a survey of the published literature see King, *Mecca-Centred World-Maps*, p. 15.

[8] The more detailed survey of Mamluk (13th to 15th century Egyptian and Syrian) universal solutions published in 1988 in the *Festschrift* for R. Bayly Winder has been updated in **VIb**.

[9] Some 125 *zīj*es are surveyed in Kennedy, "*Zīj* Survey". See now King & Samsó, "Islamic Astronomical Handbooks and Tables", and the shorter version in the EI_2 article "Zīdj".

[10] The categories of tables not contained in *zīj*es are briefly described in King, *Studies*, A-II. See now King & Samsó, *op. cit.*, pp. 82-96.

UNIVERSAL TABLES

Chapter 1 Tables of spherical astronomical functions

In the Islamic astronomical handbooks known as *zīj*es, spherical astronomical functions were generally tabulated only for the latitude of the locality where the *zīj* was compiled. Occasionally certain functions were tabulated "for all latitudes" – see **Fig. 1.1**. For example, several Muslim

Fig. 1.1: A simple anonymous Abbasid table showing solar altitudes for each pair of zodiacal signs and for each of the climates. [From MS Istanbul Aya Sofya 4830, courtesy of the Süleymaniye Library, Istanbul.]

astronomers from the 10th century onwards presented in their *zīj*es extensive sets of oblique ascension tables for each degree of terrestrial latitude (up to about 40°).[11] Sometimes auxiliary

[11] Kennedy, "*Zīj* Survey", pp. 140-141 and 170.

tables were presented with which tables of such functions could be conveniently generated for all latitudes.[12]

Chapter 2 Tables for timekeeping

Muslim astronomers compiled numerous tables for timekeeping, mostly for specific latitudes.[13] However, already al-Khwārizmī (Baghdad, *ca.* 825) tabulated a function of the meridian altitude and the time since sunrise in seasonal hours.[14] Ibn Amājūr (Baghdad, *ca.* 900) compiled a table intended to serve all latitudes displaying the time in seasonal hours as a function of the observed altitude and the meridian altitude.[15] The formula underlying these two tables was necessarily approximate.[16] However, Ibn Amājūr also compiled a table of time since sunrise for the latitude of Baghdad as a function of the same arguments based on an exact formula.[17] Both of his tables are extant in a unique manuscript preserved in Paris.

Various Muslim astronomers compiled similar and still more sophisticated tables for timekeeping by the sun and stars for specific latitudes, often containing tens of thousands of entries. One, compiled by Najm al-Dīn al-Miṣrī (Cairo, *ca.* 1300), was different from all the others in that it was not only universal but could be used for timekeeping by the sun or by any non-circumpolar star.[18] The two halves of the unique manuscript are preserved in Cairo and Oxford. In this table, one enters with the meridian altitude, the observed altitude, and the half-arc of visibility for the celestial body in question, and reads off the time since rising. There are over 440,000 entries, a feature of the table that would not attract many copyists. See further **3** on a further use of this remarkable table.

Chapter 3 Auxiliary tables for solving problems of spherical astronomy

Muslim astronomers over several centuries devised and tabulated a set of trigonometric functions with no direct astronomical significance, but such that successive applications thereof would yield the solutions of some of the standard problems of spherical astronomy.[19] The first of these were Ḥabash, al-Nayrīzī and Abū Naṣr, but although their tables were highly ingenous, they were not idealy suited for solving problems of timekeeping.[20] Several later astronomers tried their hand at compiling such auxiliary tables. Najm al-Dīn al-Miṣrī provided a full account of the way in which his three-argument table (see **2**) could be used in this way.[21] However, the ultimate solution was achieved by al-Khalīlī (Damascus, *ca.* 1360).[22] al-Khalīlī tabulated

---

[12] All such auxiliary tables are surveyed in **I-6-8**.
[13] See **I** for references to specific tables.
[14] **I-4.2.1**, **4.3.2**, and **VII-9.3**, and Hogendijk, "al-Khwārizmī's Sine of the Hours".
[15] **I-2.5.1**.
[16] On the formula see now **VII**.
[17] **I-2.3.1**.
[18] **I-2.6.1** and **VIb-3**.
[19] All known Islamic auxiliary tables are surveyed in **I-9**.
[20] **I-9.1-3**.
[21] See n. 18.
[22] **I-9.5** and **VIb-9**.

three double-argument functions with over 13,000 entries. With these tables, the most complicated problems of spherical astronomy for both the sun and stars reduce to simple arithmetical operations. al-Khalīlī's tables were known in Egypt and Turkey as well as in Syria, and they survive in several copies. A few later astronomers compiled auxiliary tables, but none was as sophisticated or as useful as those of al-Khalīlī. Indeed his tables represent the ultimate numerical solution to problems of spherical astronomy in the medieval context. With singular ingenuity, he described even the way to use the tables for converting ecliptic and equatorial coordinates.

Chapter 4 Tables for determining the times of prayer

The times of the five daily prayers in Islam are astronomically defined, the daytime prayers in terms of shadow lengths and the night-time ones in terms of twilight phenomena. From the 9th century onwards, Muslim astronomers produced tables for regulating the times of prayer acccording to these definitions, mostly for specific latitudes.[23]

A few of these tables were universal. Ibn Isḥāq (Tunis, *ca.* 1225) compiled a table displaying the duration of twilight for all latitudes, alas not contained in the sole surviving manuscript of his *zīj*.[24] al-Wafā'ī (Cairo *ca.* 1450) is supposedly the compiler of an extensive set or prayer-tables, extant in a unique manuscript in Princeton, with tables for each degree of latitude from 15° to 40°, another universal solution.[25] Also Ibn Abi 'l-Khayr al-Ḥusnī (Cairo, *ca.* 1550) compiled a table displaying the duration of twilight at the equinoxes and solstices for each half climate, extant in a unique manuscript in Cairo.[26] Certain Egyptian astronomers attributed to their illustrious predecessor Ibn Yūnus (*ca.* 990) – correctly or not, it is still not clear – a theory for the effect of refraction on the arc of daylight, with numerical values for the equinoxes and solstices at Cairo. ʿAbd al-Qādir al-Minūfī (Cairo, *ca.* 1550) compiled a table purporting to display the correction for refraction at the horizon for each degree of terrestrial latitude, extant in a unique manuscript in Istanbul.[27] Various tables for the prayer-times serving each degree of latitude within the Maghrib, and thus of a limited universal type, are known from a Maghribi manuscript copied *ca.* 1800.[28] Finally, the 19th-century Ottoman astronomer Gedūsī tabulated the time of the *imsāk* in Ramaḍān for Mecca and Medina and for each degree of latitude between 35° and 45°.[29]

Chapter 5 Tables for determining the qibla

The first mathematical solutions to the problem of finding the qibla or local direction of Mecca, apparently dating from the early 9th century, were derived by cartographic considerations, and

[23] All known Islamic prayer-tables are surveyed in **II**.
[24] **II-13.1** and **XI-5.4**.
[25] **II-8.1** and **VIb-14**.
[26] **II-8.3** and **VIb-19**.
[27] **II-8.2** and **VIb-20**.
[28] **II-13.8***.
[29] **II-14.15**.

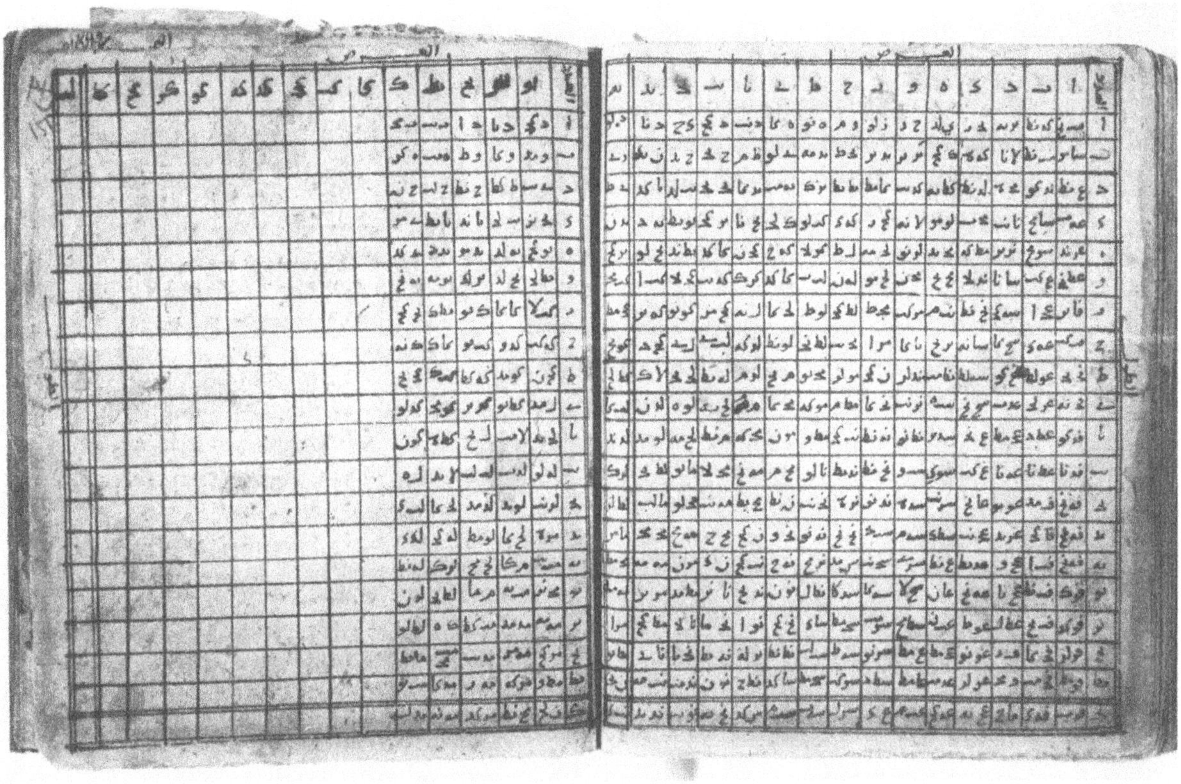

Fig. 5.1: A copy of the anonymous Abbasid qibla-table The latitude-difference argument has been extended beyond 20° by mistake to 30°. This table is not to be confused with a contemporaneous table with the same format but based on the standard approximate formula (see **Fig. VIIa-3.7**). [From MS Tashkent 177,3, fols. 150v-151r, courtesy of the Oriental Institute.]

were approximate. Exact solutions derived by means of an analemma or by spherical trigonometry were formulated in Baghdad within a few decades.[30] An unidentified 9th-century or early-10th-century Baghdad astronomer prepared a table displaying the qibla as an angle counted from the meridian for each degree (up to 20°) of two arguments that are stated to be longitude and latitude difference from Mecca.[31] The entries in the table were based on an accurate procedure, but for the stated arguments the entries are approximate: see **Fig. 10.5**. (The second argument must be slightly modified in order that the table yield exact qibla values.)

Ibn al-Haytham (Cairo, *ca.* 1025) authored three treatises on the qibla.[32] In one he solved the problem geometrically by means of an analemma, in another by spherical trigonometry considering 16 possible cases. In the third, known only from a reference in his autobiography,

[30] See the articles "Kibla (astronomical aspects)" and "Makka (centre of the world)" in EI_2, and now also King, *Mecca-Centred World-Maps*, pp. 46-193, and **VIIa**, for an overview of the qibla problem from the points of view of the astronomers on the one hand and the legal scholars on the other.

[31] See King, "Earliest Qibla Methods", esp. pp. 118-129.

[32] On these see Schoy, "Ibn al-Haytham über die Qibla"; Dallal, "Ibn al-Haytham's Universal Solution for the Qibla"; and King, *Mecca-Centred World-Maps*, p. 61, respectively.

he tabulated the qibla as a function of latitude and longitude differences. Various other qibla tables, based on either approximate or exact formulae, were compiled over the centuries (including one for longitudes up to 60° on either side of the meridian of Mecca). But it was al-Khalīlī again who produced the ultimate qibla table. Based on an exact formula, and with arguments simply latitude and longitude, his table contains over 3,000 entries, each accurately computed or in error by at most 1′ or 2′. It is available to us in three manuscripts and was apparently not widely known in later astronomical circles.[33]

Chapter 6 Tables displaying the qibla for many localities

Over the centuries Muslim astronomers compiled over 100 sets of geographical tables giving the longitudes and latitudes for dozens, but usually hundreds of localities.[34] In a minority of these the qibla for each locality is also tabulated. The compilers of some of these qibla-lists

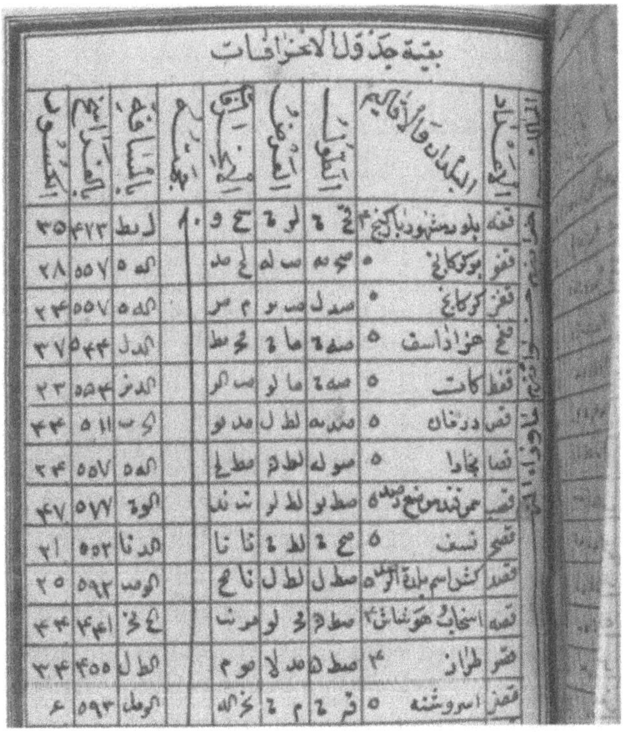
بقية جدول الانحرافات

Fig. 6.1: An extract from a modified version of the anonymous Timurid geographical table displaying, in addition to the longitudes and latitudes, the qiblas and Mecca-distances of 274 localities. The last two quantities are mainly accurate to the nearest minute, but the original table had entries to minutes, and the evidence points to these having been correct to within a few minutes. Both Samarqand and nearby Kish are noted as being the localities of "the observations", and this may mean that these tables were compiled in Kish about the time of the Samarqand observations of the team of Ulugh Beg *ca.* 1430. [From MS London Add. 7489, fol. 57r, courtesy of The British Library.]

---

[33] See King, "al-Khalīlī's Qibla Table", and **II-10.8**.

[34] See Kennedy & Kennedy, *Islamic Geographical Coordinates.*

were not particularly successful. One such, surprisingly enough, was ʿAbd al-Raḥmān al-Khāzinī (Marv, *ca.* 1125), whose geographical table appears to be derived from a world-map by al-Bīrūnī (Ghazna, *ca.* 1025): for about 250 localities he gives in addition qibla-values in degrees and minutes, which are sometimes in error by a few degrees. Apparently he computed these by himself using a defective qibla-table and a heavy-handed interpolation technique.[35] Several 13th- and 14th-century Egyptian tables share this feature.[36] However, in the mid 14th century the Syrian astronomer al-Khalīlī computed the qiblas of some 44 localities in Syria and Palestine to within a minute or two.[37] But in the early 15th century an unidentified Timurid astronomer, apparently active in Kish near Samarqand, computed the qiblas and distances to Mecca (in equatorial degrees) for some 274 localities, giving values to minutes, mainly accurately to within a few minutes. Only a simplified version of his table, with these values rounded to minutes, has been preserved for us: see **Fig. 6.1**.[38] It was this table, or another corrupt version of it, that was used by Safavid craftsmen when they engraved geographical gazetteers on astrolabes (see **16**).

Chapter 7 Tables for determining lunar and planetary visibility

The Muslim calendar is lunar and is regulated by the sighting of the lunar crescent. Numerous Muslim astronomers adopted the simple Indian condition that visibility occurs if the difference in setting times between the sun and moon is 12 equatorial degrees. Already in the 9th century a table displaying the minimum ecliptic distance between the two luminaries to ensure visibility according to this Indian condition was computed with values for each zodiacal sign and specifically for the latitude of Baghdad. Later, more complicated theories were formulated and other lunar visibility tables were compiled, mostly for specific latitudes.[39]

One astronomer, however, probably in al-Andalus during the 11th or 12th century but possibly earlier, compiled a table seemingly based on Indian theory for each of the climates, another universal solution. This is available in several extremely corrupt copies mainly of Maghribī and Egyptian provenance – see **Fig. 7.1**; a reconstruction of the original table has not yet been successful.[40] Another universal table for each of the climates is contained in the Hyderabad manuscript of the recension of the *Zīj* of Ibn Isḥāq (*fl.* Tunis *ca.* 1225): it displays the half-excess of the arc of visibility of the moon for each degree of its declination – see **Fig. 7.2**.[41]

The *Zīj*es of Ḥabash and al-Battānī (Raqqa, *ca.* 910) contain planetary visibility tables for each of the climates borrowed from the Handy Tables. The *Zīj* of al-Khāzinī (Merv, *ca.* 1125), however, has a different set based on an Indian theory and computed by one of his predecessors.[42]

---

[35] King, *Mecca-Centred World-Maps*, pp. 71-75 and App. D.
[36] *Ibid.*, pp. 76-84, and also **II-6.6**.
[37] *Ibid.*, pp. 84-86, and App. F7, and also **II-10.9**.
[38] *Ibid.*, pp. 149-168, and App. A.
[39] See the survey article "Ruʾyat al-hilāl" in EI_2.
[40] King, "Tables for Lunar Crescent Visibility", pp. 197-207, contains an edition of the wretched extant versions of this table.
[41] On Ibn Isḥāq see **II-13.1**.
[42] Kennedy *et al.*, *Studies*, pp. 137-140.

Fig. 7.1: This anonymous table for predicting lunar crescent visibility for each of the climates exists in several manuscripts. It seems to be early Andalusī in origin, and it was known in Tunis in the 13th century, and in Cairo in Ottoman times. The fact that a table was known in several places over the centuries is no proof that it was ever used. In this case, all of the available copies are so corrupt that the table is, in effect, useless. [From MS Cairo MM 23, fol. 9r, courtesy of the Egyptian National Library.]

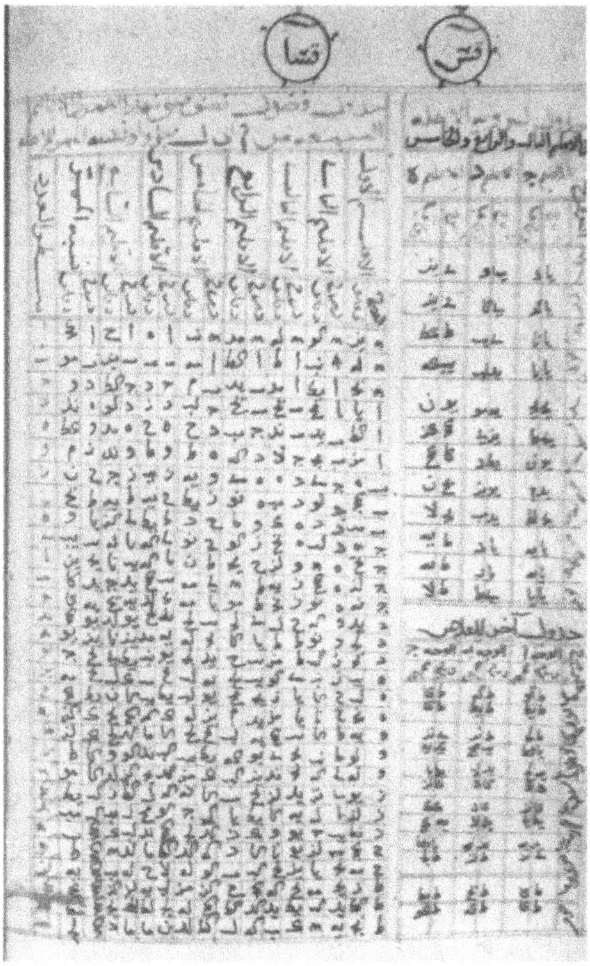

Fig. 7.2: A table for the half-excess of the arc of visibility of the moon in each of the climates with values for each degree of lunar declination. Presumably this was compiled by Ibn Isḥāq. On the same page we find an extract from the table in **Fig. 7.1** serving the 2nd (not 3rd as stated), 4th and 5th climates, as well as a table for an unspecified latitude in al-Andalus attributed to one al-Qallās, which is none other than the one which ended up in the Latin version of the Andalusī recension of the *Zīj* of al-Khwārizmī. [From MS Hyderabad 298 (unfoliated), tables 160-161, courtesy of the Andra Pradesh State Library.]

Chapter 8 Tables for parallax

Theon's tables of parallax components for each of the climates were incorporated into several Islamic *zīj*es. The corresponding tables in the *Zīj* of Ulugh Beg (Samarqand, *ca.* 1425), however, are computed for his newly-found value of the obliquity and for each 5° of latitude from 25° to 50°.[43]

---

[43] *Ibid.*, p. 41.

Chapter 9 Tables for astrolabe construction

The well-known iterative geometrical constructions for marking the various circles on astrolabe plates was not used as often as the modern literature would have us believe.[44] An alternative means of constructing the markings on astrolabes was to compute the radii of the various circles and the positions of their centres relative to the centre of the instrument. al-Khwārizmī (Baghdad, *ca.* 825) devised an auxiliary function for generating values of such radii and centre-distances. His contemporary al-Farghānī went so far as actually to tabulate both of these functions for each degree of latitude from 15° to 50°, with over 8,000 entries for each degree of altitude for the altitude circles and each degree of azimuth for the azimuth circles – see **Figs. 9.1** and **9.2**. He thus achieved a universal solution to the problem of astrolabe construction.[45]

Later Muslim astronomers, such as the Yemeni Sultan al-Ashraf (*ca.* 1295), compiled astrolabe tables for a set of latitudes of specific localities. The Egyptian astronomer al-Bakhāniqī (*ca.* 1325) completed al-Farghānī's tables so that they served each degree of terrestrial latitude from 1° to 90°.[46] Astrolabe tables were apparently widely used in medieval Egypt and Syria, if not elsewhere. In Iran, a grid called *dastūr* which displayed graphically values of the auxiliary function tabulated by al-Farghānī seems to have been more popular.[47]

Chapter 10 Tables for sundial construction

The tables for coordinate conversion in Ptolemy's *Analemma* could be used to construct markings for the seasonal hours on different kinds of planar sundials for all climates. Although the Muslims, as far as we know, were not familiar with this work of Ptolemy, they did develop

→

Fig. 9.1: An extract from al-Farghānī's tables for astrolabe construction. These sub-tables show the centre-distance and radii of the altitude circles for arguments 1°-90° for latitude 21°. [From MS Berlin Ahlwardt 5791, courtesy of the Deutsche Staatsbibliothek, Preußischer Kulturbesitz.]

Figs. 9.2a-c: To plot the altitude circles on an astrolabe plate it suffices to know their radii and distance from the centre of the plate (a). al-Farghānī used an auxiliary function f(θ) measuring the segment on the celestial equator corresponding to the projection of an arc θ from the pole (b). We then view in cross-section the stereographic projections of the horizon of a locality with latitude ϕ and an altitude circle with altitude h (c). The projection of the altitude circle PQ in the plane of the celestial equator is UV. The functions tabulated by al-Farghānī are precisely r(h,ϕ) and d(h,ϕ) for each degree of h from 0° to 90° (that is, from the horizon to the zenith) and each degree of ϕ from 15° to 50°, and these can be generated by the simple arithmetical relations:

$$r(h,\phi) = {}^1/_2 UV = {}^1/_2(UO+OV) = {}^1/_2\{f(180°-\phi-h) + f(\phi-h)\}$$

$$d(h,\phi) = OU - r = f(180°-\phi-h) - {}^1/_2\{f(180°-\phi-h) + f(\phi-h)\} = {}^1/_2\{f(180°-\phi-h) - f(\phi-h)\} .$$

The same trigonometric procedure was proposed in 1947 by Henri Michel in his *Traité de l'astrolabe* (1947), without realizing that al-Farghānī had done the same, and more, over a millennium previously.

---

[44] Michel, *Traité de l'astrolabe*, pp. 62-63; and Hartner, "Astrolabe".

[45] On the tables of al-Khwārizmī see King, "al-Khwārizmī", pp. 22-27, superseded by Charette & Schmidl, "al-Khwārizmī on the Astrolabe" (forthcoming). On those of al-Farghānī see King, "Islamic Astronomical Tables", pp. 53-55. An edition of al-Farghānī's treatise on astrolabe construction has been prepared by Richard Lorch of Munich.

[46] On the tables of al-Ashraf see King, "Yemeni Astrolabe", pp. 108-111. On those of al-Bakhāniqī see **VIb-7**.

[47] See King & Samsó, "Islamic Astronomical Handbooks and Tables", pp. 91-92, and also **X-4.2**, as well as Charette & King, "Survey of Islamic Tables for Astrolabe Construction" (forthcoming). On the *dastūr* see, for example, Michel, *Traité de l'astrolabe*, pp. 47-48.

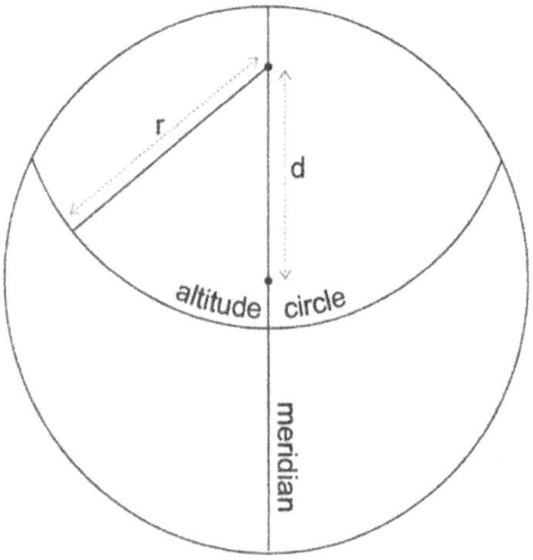

9.2a

9.1

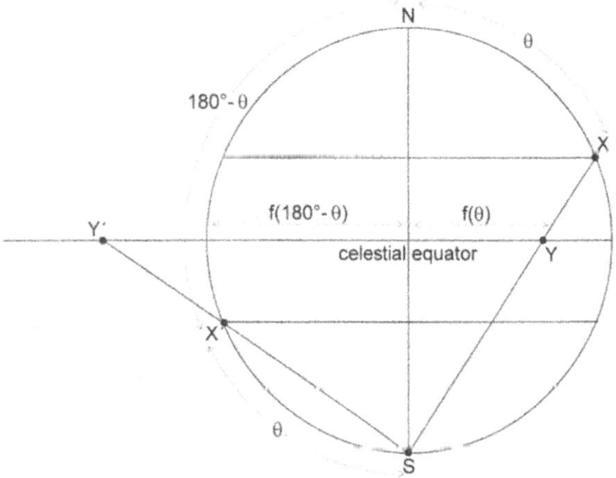

9.2b

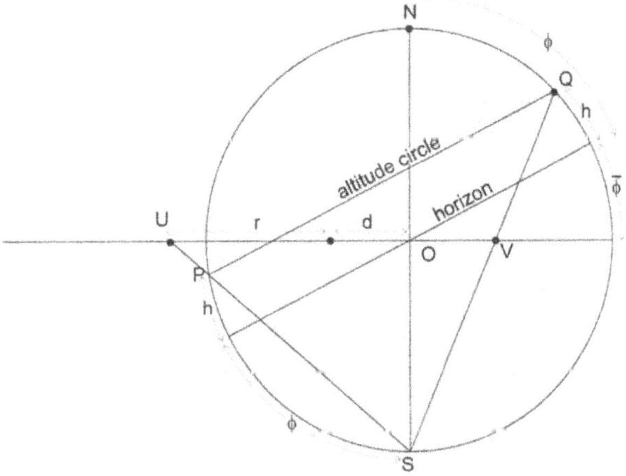

9.2c

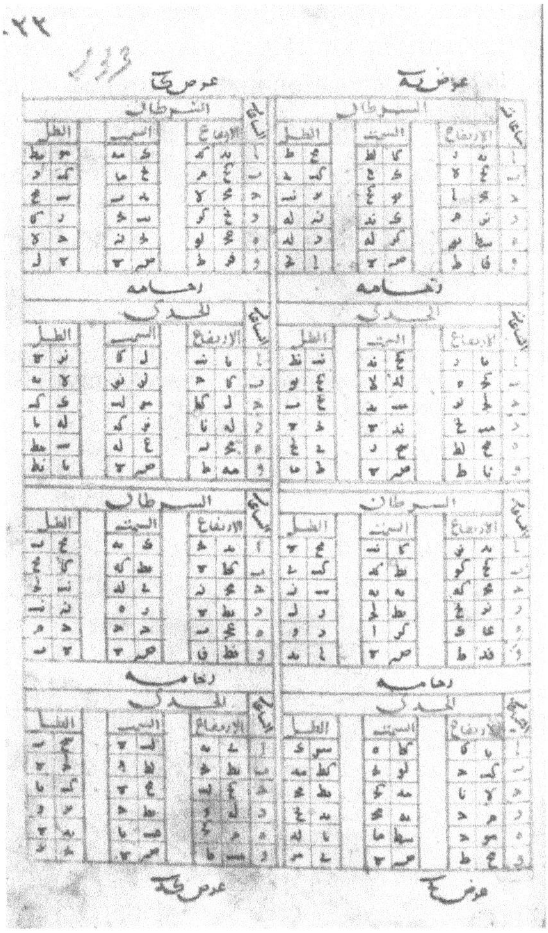

Fig. 10.1: An extract from the tables for sundial construction of Ḥabash al-Ḥāsib, which started a tradition that lasted a millennium. The extract shows the sub-tables for latitudes 15°, 18°, 21° and 24°, with values of the solar altitude, the azimuth and the shadow at each hour for both solstices. The word *rukhāma* (lit., "marble") in the middle of each sub-table simply means "horizontal sundial". See also **Fig. I-4.1.1a**. [From MS Istanbul Ayasofya 4830, fol. 233r, courtesy of the Süleymaniye Library, Istanbul.]

more practical tables for sundial construction. Ḥabash al-Ḥāsib compiled a set of tables for making the curves on horizontal sundials for a series of latitudes – see **Fig. 10.1**.[48] His tables display radial coordinates for the intersections of the hour lines with the solstitial curves for each of these latitudes. He also presented a table for making a "polar" sundial (see **15**).

al-Farghānī wrote on sundial theory, but unfortunately his treatise is lost. Thābit ibn Qurra (Baghdad, *ca.* 875) produced a treatise on sundial theory (without tables) along the same lines

[48] See King, "al-Khwārizmī", pp. 17-22, and now **I-4.1.1**. In my earlier writings I did not question the attribution of these tables to al-Khwārizmī in the unique Istanbul manuscript. However, it now seems clear that the author is Ḥabash; for details see now King, *Mecca-Centred World-Maps*, pp. 349-350. As this book goes to press, I find that François Charette has adduced evidence that these tables may indeed by al-Khwārizmī after all: see his *Mamluk Instrumentation*, p. 182.

as Ptolemy's *Analemma*, apparently independently.[49] His contemporary Ibn al-Ādamī compiled a set of tables to facilitate the marking of curves on vertical sundials inclined at any angle to the meridian for any latitude.[50] Later astronomers compiled sundial tables for latitudes of specific localities such as Cairo, Damascus and Jerusalem; such tables remained in use for centuries in Egypt and Syria.[51]

---

[49] Thābit's treatise was first published with German translation by Karl Garbers and commentary by Paul Luckey (see Sezgin, *GAS*, VI, 168, no. 9). See now Morelon, *Thābit ibn Qurra*, pp. cxix-cxxv, 118-129 and 260-264, and cxxvi-cxl, 130-169 and 265-295.

[50] See **I-4.2.3**.

[51] A survey of Islamic sundial tables would be a useful contribution; see already King, "Islamic Astronomical Tables", pp. 51-53 (and the works of Carl Schoy there cited), *idem* & Samsó, "Islamic Astronomical Handbooks and Tables", pp. 92-94, and further **VIb-13** and **X-7.1**.

Chapter 11 Universal aspects of the astrolabe

The inclusion inside the standard astrolabe of a series of plates for different latitudes represents an attempt to render the instrument universal. The idea stems originally from the Greek tradition of having seven surfaces engraved for each of the seven climates; both the earliest Islamic astrolabe and some of the earliest medieval European astrolabes have plates for the seven climates. Later the plates were constructed to serve either a whole range of latitudes or a narrower range serving localities of immediate concern to the maker or his client.[52]

Another universal solution that was first developed in 9th-century Baghdad was the universal plate (*ṣafīḥa āfāqiyya*), a single astrolabe plate displaying stereographic projections of horizons for each few degrees of latitude: see **Fig. 11.1**. Ḥabash devised such a plate for use with an astrolabe rete, and various later astronomers in the Islamic East wrote on its use.[53]

In the early 14th century the scholar and instrument-maker Ibn Bāṣo of Granada devised a modification of the *shakkāziyya* grid for use with a standard astrolabic rete.[54] On this device, brilliant in its conception, each one of three sets of markings is associated with a specific but variable astronomical function (declination, altitude, *etc.*) depending on the problem to be solved. Plates bearing these markings became a standard feature in later Maghribi astrolabes, and similar markings are found on a 15th-century astrolabe from Vienna.

The horary quadrant that became a common feature of the backs of Islamic astrolabes and a standard feature of European astrolabes before the Renaissance is in fact a universal instrument. It has been somewhat maligned in the modern literature, but is indeed a most elegant graphical solution to the standard approximate formula for timekeeping.[55] One needs only to insert the solar meridian altitude and the instantaneous solar altitude and the device yields at a glance the time since sunrise or remaining till sunset for any latitude. The device was first invented in 9th-century Baghdad, both with a declination scale for inserting the meridian altitude as a function of the date, and without (the instrument functions just as well without the scale), and is, of course, approximate.

Chapter 12 The universal astrolabe and quadrant

More familiar to historians of science is the universal astrolabe devised in 11th-century Toledo by ʿAlī ibn Khalaf al-Shajjār. In this, known only from the 13th-century *Libros del saber*, the

[52] See King, "Geography of Astrolabes", esp. pp. 8-11 and 14-16.

[53] This item of information on Ḥabash is taken from Morley, "Astrolabe of Shāh Ḥusayn", p. 7. A treatise on the universal plate by al-Sijzī (*ca.* 975) is listed in Sezgin, *GAS*, VI, p. 226 (no. 6).

[54] Comprehensive studies of the texts and numerous instruments are in Calvo, *Ibn Bāṣo and his Universal Plate*, "Ibn Bāṣo's Universal Plate", and "The Use of Ibn Bāṣo's Universal Plate". See also **X-5.2**.

[55] See now **XIIa**.

Fig. 11.1: Markings for the horizons on the solitary mater of an unsigned undated astrolabe which can be perhaps be associated with the early-14th-century Aleppo astronomer and instrument-maker Ibn al-Sarrāj (**VIb-5**). Each of eight "bundles" of half-horizons serves a different set of latitudes, starting with 13°, 14°, ... , 20°, and increasing in 10°-intervals (*e.g.*, 13°, 23°, ... , 53°). Together the markings serve each degree of latitude from 13° to 60°, except for those latitudes with 1 and 2 in the second digit: this not a particularly satisfactory arrangement. [Object now in the Khalili Collection, London; photo from the archives of the late Alain Brieux, Paris.]

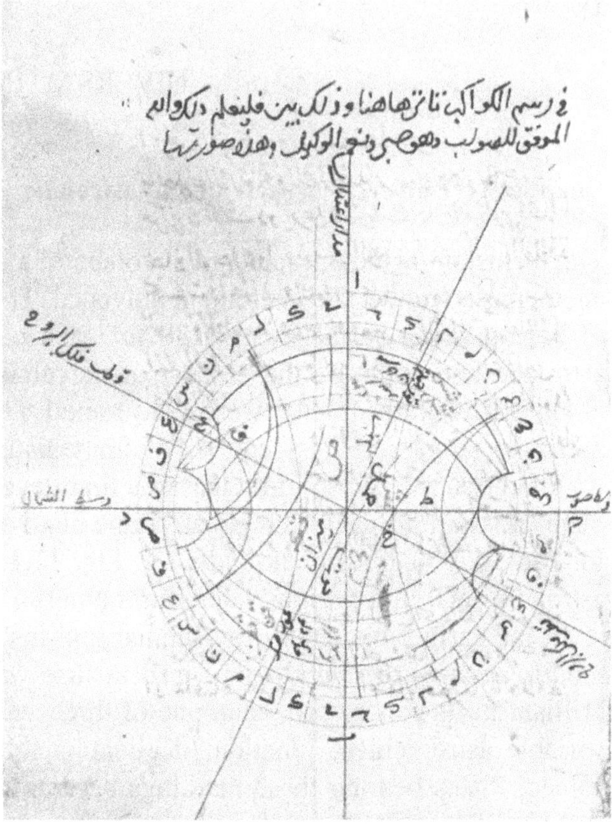

Fig. 12.1: A device for converting celestial coordinates from the ecliptic system to the equatorial system. The grid is called *zarqālliyya* after its inventor Ibn al-Zarqālluh. [Taken from MS Cairo ṬM 155,1, courtesy of the Egyptian National Library.]

plate bears so-called *shakkāziyya* markings based on a stereographic projection for latitude zero and hence independent of terrestrial latitude. The rete consists of a semi-circle of *shakkāziyya* markings and a semi-circle fitted with star pointers (if not an ecliptic!). Clearly, ʿAlī ibn Khalaf's astrolabe is extremely useful for solving problems of spherical astronomy, but the instrument no longer simulates the apparent daily rotation of the heavens over any horizon. His better-known contemporary Ibn al-Zarqālluh replaced the rete by an alidade fitted with a movable perpendicular cursor to produce a universal plate, with which, however, problems of timekeeping can be solved universally, but only approximately. On another plate called

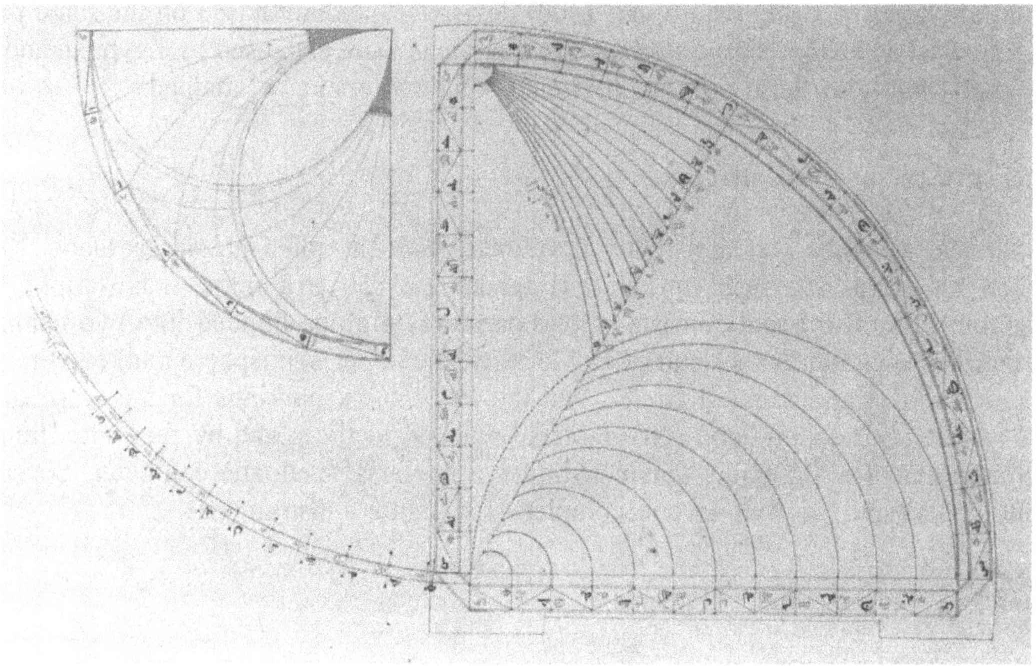

Fig. 12.2: Illustrations in an 18th-century manuscript of two universal instruments devised by Ibn al-Sarrāj, modified from the basic markings of the universal horary quadrant. [From MS Cairo MR 40,2 (*Cairo ENL Survey*, no. D81/4.1.8), courtesy of the Egyptian National Library.]

zarqālliyya for converting ecliptic and equatorial coordinates he introduced a second set of *shakkāziyya* markings inclined to the first at an angle equal to the obliquity.[56]

A universal astrolabe similar to that of ʿAlī ibn Khalaf (but now with an ecliptic) was reinvented by Ibn al-Sarrāj (Aleppo, *ca.* 1325), and we have a treatise by him on its use, extant in a unique manuscript in Cairo. (The relationship between this instrument and the one described by Najm al-Dīn al-Miṣrī is still unclear.[57]) A second, more sophisticated version, designed and constructed by Ibn al-Sarrāj himself, is now housed in the Benaki Museum in Athens: see **Fig. 12.2**. A treatise by al-Wafā'ī (Cairo, *ca.* 1450), who owned the very instrument now preserved in Athens, enables us to understand what Ibn al-Sarrāj intended to achieve with it. This splendid astrolabe could be used universally in five different ways. First and second, it could be used in the same way as the astrolabe of al-Shajjār exploiting the features of either of the two halves of the rete; third, the rete could be used with a series of quarter plates inscribed with astrolabic markings for each 3° of latitude; fourth, the rete could be used with a set of horizons for each 3° of latitude; and fifth, there is on the back a highly ingenious trigonometric grid and alidade (see below).[58]

---

[56] For more details see my article "Shakkāziyya" in EI_2 and **X-5.2**, and the various studies of José María Millás Vallicrosa and Roser Puig there cited.

[57] Charette, *Mamluk Instrumentation*, II-2.6.

[58] See King, *Studies*, B-IX; *Paris IMA 1993-94 Exhibition Catalogue*, pp. 434-435, no. 330; and also **VIb-5** and **14** on Ibn al-Sarrāj and al-Wafā'ī. François Charette and I are preparing for publication a detailed study of this astrolabe.

Jamāl al-Dīn al-Māridīnī (Cairo, *ca.* 1400) devised a quadrant based on the same principle as the universal astrolabe. Single *shakkāziyya* quadrants were also used by Egyptian and Syrian astronomers, but with these only approximate solutions could be attained.[59]

Chapter 13 The spherical astrolabe

The spherical astrolabe was apparently developed from the planispheric astrolabe. It is first described in a treatise attributed to Ḥabash (Baghdad, *ca.* 850) in a unique manuscript preserved in Istanbul.[60] The instrument consists of two parts: (1) a globe divided into two hemispheres by a circle representing the horizon of any locality, the upper hemisphere marked with altitude circles, a meridian and azimuth circles, and the lower with markings for the seasonal hours; and (2) a rete with ecliptic and star pointers, attached to the globe by pins fitted into holes on the meridian. The ensemble thus constitutes a universal analogue computer. Several later descriptions survive, as well as one complete 15th-century instrument.[61]

Chapter 14 Quadrants and trigonometric grids

al-Khwārizmī or one of his contemporaries developed a trigonometric quadrant with a set of lines drawn parallel to one axis. This could be used with a thread attached at the centre and a movable marker to perform trigonometric calculations graphically. Already in the 10th-century, Abū Jaʿfar al-Khāzin in Baghdad devised a grid with two orthogonal sets of parallel lines (corresponding to the 60 divisions of each radial scale), which provided greater flexibility in trigonometric "calculations".[62] The horary quadrant with a movable cursor for timekeeping by the sun for any latitude is likewise first attested in a treatise from 9th-century Baghdad, a copy of which is now preserved in Cairo. Some later horary grids were designed for specific latitudes, but others were suitable for use in any latitude, the procedure being approximate. Such sine and universal horary quadrants are often found on the backs of Islamic astrolabes.[63]

Several Egyptian and Syrian astronomers devised different kinds of trigonometric grids, of which the most ingenious (if least practical) is the one on the back of the astrolabe of Ibn al-Sarrāj (**Fig. VIb-5.1**). Such grids could be used to yield numerical solutions to all of the standard problems of spherical astronomy for any latitude, although only the simpler

[59] The single and double *shakkāziyya* quadrants are discussed in Samsó & Catalá, "Cuadrante *šakkāzī*", and Samsó, "Cuadrante *šakkāzī*"; and King, "al-Māridīnī's Universal Quadrant", respectively. On al-Māridīnī see further **VIb-12**.

[60] Sezgin, *GAS*, VI, p. 175.

[61] Seemann, "Das kugelförmige Astrolab", is a survey of Islamic writings on the spherical astrolabe. Ḥabash's treatise is listed in Sezgin, *GAS*, VI, p. 175 (no. 4); Qusṭā ibn Lūqā's treatise – if such it was – is confused with his writings on the celestial globe *ibid.*, p. 181. On the surviving instrument see Maddison, "15th-Century Spherical Astrolabe".

[62] On the sine quadrant see Schmalzl, *Geschichte des Quadranten*, pp. 71-99; King, "al-Khwārizmī", pp. 28-31; *idem*, "al-Khalīlī's Qibla Table", pp. 111-115; and now **X-6.1** and **XI-8.1**. On Abū Jaʿfar al-Khāzin see **X-5.3**.

[63] On the universal horary quadrant see Lorch, "Universal Horary Quadrant"; King, "*Quadrans vetus*", and now **XIIb**. The early treatise is listed in *Cairo ENL Survey*, no. B105 (see also no. F6).

trigonometric quadrant was widely used for such purposes.[64] One grid has not been properly understood: a combination of standard sexagesimal trigonometric markings and *shakkāziyya* markings is found on the back of an astrolabic quadrant made by Abū Ṭāhir *ca.* 1350; the astrolabic markings on this piece have not yet been explained either (see **Fig. V-9.3**).

Chapter 15 Universal sundials

Ḥabash included in his tables for constructing sundials (see **10**) one for a sundial for latitude 0° – see **Fig. 15.1**. His treatise does not mention its potential application as a universal polar sundial.[65] However, Ibn al-Shāṭir had such a device on his "compendium" (see **17**) and a substantial sundial of this kind from the Ottoman period survives in the mosque of Acre – see **Fig. 15.2**.[66] al-Maqsī (Cairo, *ca.* 1280) included in his treatise on sundial construction a table for a sundial set up in the plane of the celestial equator.[67] Najm al-Dīn al-Miṣrī (Cairo, *ca.*

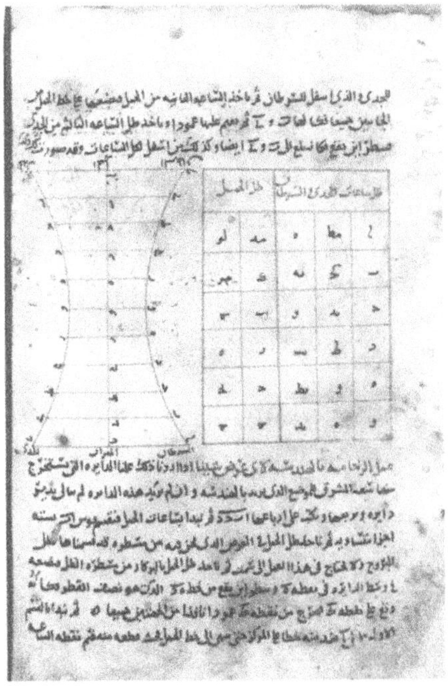

Fig. 15.2: A universal polar sundial in the courtyard of the mosque in Acre, erected in 1201 H [1786-87]. A similar, albeit much smaller, sundial is found in the compendium of the 14th-century Syrian astronomer Ibn al-Shāṭir – see **Fig. X-9.3.1**. [From King & Janin, "Ibn Ṭūlūn Sundial", pl. 6.]

↑
Fig. 15.1: A table for the construction of a sundial for latitude 0° and an illustration thereof in the treatise on sundial construction by Ḥabash. The sundial can be used universally if it is aligned in the plane perpendicular to the celestial equator. [From MS Istanbul Aya Sofya 4830, fol. 234v, courtesy of the Süleymaniye Library, Istanbul.]

---

[64] On some by Ibn al-Sarrāj see King, *Studies*, B-IX, and **VIb-5**. On two by Ibn al-Shāṭir see Schmalzl, *Geschichte des Quadranten*, pp. 100-108, repr. in Kennedy & Ghanem, eds., *Ibn al-Shāṭir*, pp. 27-35, and **VIb-8** (illustrated). In Charette, *Mamluk Instrumentation*, numerous other varieties are presented.

[65] King, "al-Khwārizmī", pp. 17, 19. See also Sédillot-père, *Traité*, II, p. 488, for a similar table by al-Marrākushī.

[66] Illustrated in Michel & Ben-Eli, "Cadran solaire remarquable".

[67] See **VIb-2**.

1300) proposed several universal sundials based on an approximate formula for timekeeping,[68] but it seems doubtful that these were ever constructed. Likewise al-Khalīlī (Damascus, *ca.* 1360) seems to have been the first to conceive of a universal inclined sundial, also known to al-Wafā'ī and al-Ṣūfī in Cairo in the 15th century.[69] al-Wafā'ī preferred to put a universal equatorial dial on his compendium (see again **17**).

Chapter 16 Universal instruments for finding the time of day and night

In a unique manuscript in Oxford there is a treatise by Ḥabash on an instrument for finding the time of night by the stars.[70] In the form of an astrolabe with throne, the instrument bears mathematical data relating to stellar coordinates, inserted graphically – see **Figs. XIIb-12a-d**. With the aid of trigonometric grids and various non-uniform scales on the alidade, this data can be manipulated to yield the time of night and the azimuth of the star from its altitude for any latitude, without any calculation. The operations are alas not described in the treatise, which deals solely with the construction and is anyway not complete, and have so far defied satisfactory explanation. Two things are clear: first, that the instrument was not widely used, if at all, and second, that one does not devise such a complicated, indeed brilliant, device for timekeeping by the stars for any latitude if one has not already devised a simpler device for timekeeping by the sun.

No such device with which one can find the time of day from the solar altitude for any latitude without any calculation is known from the Muslim world. However, the European universal horary dial for timekeeping known as the *navicula de Venetiis*, "little ship of the Venetians", of which a few survive, mainly from 14th-century England,[71] has long been thought to be an Islamic invention.[72] I have recently assessed the scant evidence that the mathematical components of the *navicula*, not necessarily organized in the form of a ship, have their origin in the mind of Ḥabash in 9th-century Baghdad.[73] The most convincing evidence is still hypothetical, namely, (a) that one would only devise an instrument for the stars if one had already devised one for the sun, and (b) that the universal horary dial for the sun as found on the *navicula* has a set of components which are the same as those on Ḥabash's universal instrument for timekeeping by the stars.

---

[68] See **VIb-5** and **XI-7**, and, for more details, Charette, *Mamluk Instrumentation*, pp. 113-208

[69] On al-Khalīlī see n. 22 above and on al-Wafā'ī and al-Ṣūfī see **VIb-14** and **17**. On their sundials see the two manuscripts mentioned in King, *Mecca-Centred World-Maps*, p. 210.

[70] See King, *Mecca-Centred World-Maps*, pp. 354-358; Charette & Schmidl, "Ḥabash's Universal Plate for the Stars"; and now **XIIb-12**.

[71] Several examples are described in Brusa, "Navicelli di Venezia".

[72] In 1960 the late Derek de Solla Price thought that he had found the origin of the *navicula* in an Islamic instrument also shaped like a ship (see Price, "Little Ship of Venice", esp. pp. 399-400). The latter device, actually part of an astrolabe rete serving as a horizon for a specific latitude, he found described in the treatise of al-Marrakushī (Sédillot-*fils*, *Mémoire*, p. 182 and pl. 89, cited in Price, *op. cit.*, p. 399, n. 2), but it is merely one of the many kinds of non-standard astrolabe rete and plate combinations devised by Muslim astronomers mainly in the 10th century.

[73] King, *Mecca-Centred World-Maps*, pp. 351-359, and now King, "*Navicula*", with a longer version in **XIIb**.

Chapter 17 Compendia and gazetteers

Ibn Samʿūn (Cairo, *ca.* 1300)[74] devised a magnetic compass fitted with a sundial for a fixed latitude which was a predecessor of certain later Egyptian, Syrian and Turkish instruments. In his description of this instrument, extant in a unique manuscript in Leiden, he also presents a list of qiblas of 160 localities: see **Fig. V-12.2.**[75]

Ibn al-Shāṭir (Damascus, *ca.* 1350) devised a compendium, that is, a multi-functional instrument, in the form of a box with a hinged lid and various movable parts. Some parts of this "jewel-box" made by Ibn al-Shāṭir himself survive in Aleppo and a single plate fronm another oner made by al-Wafāʾī is preserved in Istanbul.[76] A treatise preserved in Berlin explains the way in which the ensemble is to be used. The box contained a magnetic compass for aligning it in the cardinal directions and a universal polar sundial on a plate that could be attached to the lid and erected in a plane perpendicular to the celestial equator for any of a series of latitudes. Also, sights could be attached to the lid to measure the hour-angle at any of these latitudes, and a set of horizons engraved on the lid enabled the user to compute the instantaneous configuration of the ecliptic relative to the appropriate horizon. In addition, the plate bore markings indicating the qiblas or local directions of Mecca for a series of localities.

al-Wafāʾī (Cairo, *ca.* 1450) devised a universal equatorial semi-circle with a sighting device, inclined at an adjustable angle to a base that could be oriented in the cardinal directions by means of a compass. Several examples of this instrument and copies of al-Wafāʾī's treatise on its use are available.[77] The geographical information recorded on these instruments is invariably taken from the folk tradition of Islamic sacred geography (in other words, the qiblas shown are not derived by computation).

These instruments were the precursors of the qibla-indicators that were used in Ottoman times.[78] Any influence on the development of compendia in Europe, if such there was, has yet to be established.

A different tradition is attested in Iranian instruments. There is, for example, a unique ceramic compass bowl made in Damascus in the early 16th century with the qiblas of 40 localities marked on it. The geographical data are extremely corrupt and is clearly of Iranian provenance, and I have argued that they go back to the 13th century.[79] The extensive gazetteers on 17th and early-18th-century Safavid astronomical instruments are derived from the 15th-century Timurid table mentioned above (see **6**), or a corrupt version thereof.[80]

---

[74] On Ibn Samʿūn see Suter, *MAA*, no. 398, and *Cairo ENL Survey*, no. C24.

[75] Schmidl, "Early Arabic Sources on the Compass"; and King, *Mecca-Centred World-Maps*, pp. 80-83 and 113-114.

[76] See Janin & King, "Ibn al-Shāṭir's Compendium", and **VIb-8**.

[77] See Tekeli, "Equatorial Semicircle", for al-Wafāʾī's treatise, and also **VIb-14**; and Brice & Imber & Lorch, "Equatorial Semicircle", Dizer, "Equatorial Semicircle", and *London Khalili Catalogue*, I, pp. 277-279, for some examples.

[78] On two of these see King, *Mecca-Centred World-Maps*, p. 117.

[79] *Ibid.*, pp. 110-114 and 168-170, and App. B1.

[80] *Ibid.*, pp. 170-186.

Chapter 18 World-maps for finding the qibla and distance to Mecca

It is appropriate that we conclude this survey with the most-recently identified "universal solutions", namely, three late-17th-century Safavid Mecca-centred world-maps fitted with complex grids which serve to enable the user to read off the qibla of any locality on the circumferential scale and the distance to Mecca (in *farsakh*s) on the diametrical rule. These three instruments came to light only in 1989, 1995 and 2001, and at least the first two have been studied in detail, an operation which involved a reassessment of Muslim achievements in mathematical cartography.[81] The maps are clearly in a tradition that goes back several centuries: the coordinates of the close to 200 localities on each map are taken from the 15th-century Timurid table mentioned in above (see **6**), and the mathematics underlying the grids was first derived by Ḥabash in the 9th century. Indeed, I strongly suspect that it was he who conceived the grid associated with maps, although neither in his known writings or those of al-Bīrūnī, who was well versed in his achievements, is there any mention of grids or world-maps of this kind. However, only two out of a dozen works on mathematical geography by al-Bīrūnī survive. On the other hand, it was Ḥabash who first devised an astrolabe based on a mapping preserving distance and direction to a central (polar) point.[82] In my study of the first two of these maps I have assessed the various possibilities regarding the inspiration of the grids on them. See now **VIIc**.

The instrument discovered in 1989 is clearly missing an attachment, which fortunately survives on the two instruments that were discovered thereafter. This is a universal inclined dial of European inspiration for measuring time of day: the annulus with its hour-scale can be elevated so that it functions for any latitude – see **Fig. 18.1**. Whilst this kind of dial serves to recognise midday, and hence the beginning of the *ẓuhr*, it is useless for determining the time of the *ʿaṣr*, unless used in conjunction with prayer-tables. What is ironic is that this kind of sundial, a European invention which was very popular in Europe in the 17th century, was actually first conceived in 14th-century Syria and known in Egypt in the 15th century (see **15**) but never became popular in the Muslim world, except in its European manifestation.

Concluding remarks

One of the remarkable features of the materials that we have described is that the universal tables invariably survive in unique manuscripts and the universal instruments are known to us either because unique examples have survived or descriptions of them have been found in unique manuscripts. Our knowledge of the achievement of Muslim astronomers in this particular field of scientific endeavour would be the poorer but for the chance survival of this or that manuscript or this or that instrument. And our knowledge can progress in leaps and bounds when a new manuscript or instrument relevant to our subject shows up "out of nowhere". We should also bear in mind that Ḥabash was redicovered only in the 1950s, and al-Khalīlī, Ibn al-Sarrāj and Najm al-Dīn al-Miṣrī only in the 1970s-90s. The most important

---

[81] *Ibid.*, pp. 197-371. On the third see **VIIc-2**.

[82] See now Kennedy & Kunitzsch & Lorch, *The Melon Astrolabe of Ḥabash al-Ḥāsib*.

Fig. 18.1: The second Safavid world-map, complete with all of its original components. [Private collection; photo courtesy of Christie's of London.]

works of all of these except for al-Khalīlī survive in unique manuscripts. We should further keep in mind that Muslim knowledge of science was not cumulative but very much dependent upon which works were available at the time. And a chance encounter of a medieval scholar with an old manuscript might enable him to revive ideas that had long lain dormant.[83]

The question remains: are there any more relevant materials "out there" waiting to be discovered. My answer is an unreserved "Yes!", but I am at a loss to predict what they might be.[84] After all, if anyone had been asked in the 1960s whether any Muslim scientist could have produced auxiliary trigonometric tables more sophisticated than those of Regiomontanus and Magini, or if anyone had asked me before 1989 whether any Muslim scientist could have produced a world-map preserving direction and distance to Mecca at the centre, the answer would have been, in both cases, an unconditional "No!".

[83] On this theme see King, *Mecca-Centred World-Maps*, pp. 358-359.

[84] At least in the case of al-Khalīlī's universal auxiliary tables for finding the solar azimuth (see **VIb-9**), discovered only in 2001 in the manuscript from Bursa, the existence of the tables was already known thanks to a passing reference by the compiler in a later work.

Part VIb

Universal solutions to problems of spherical astronomy from Mamluk Syria and Egypt

To the memory of R. Bayly Winder

ACKNOWLEDGEMENTS AND NOTES ON THIS VERSION

This study is dedicated to the memory of a man who was highly influential in my life without even trying and without hardly knowing it. I first met Bayly Winder in the summer of 1968 when I took a graduate course on the *Qurʾān* with him at the University of Pennsylvania in Philadelphia. In the summer of 1979, when I was in Alexandria confronted with the bleak reality that my seven-year research grant was on the point of expiring, I received the offer of an appointment at New York University. I tried to reach Bayly in Princeton to seek information and advice, and eventually tracked him down at breakfast in a hotel at London, somewhat surprised to hear from his former student. In no small measure thanks to his encouragement, I was within weeks on Washington Square. A year later I was informed, without prior warning, by the department chairman, Frank Peters, that he had put me down to conduct a graduate seminar on the *Qurʾān* to begin the next week. (Frank was like that, but I never regretted that decision of his, and in the seminar we read what al-Zamakhsharī and al-Suyūṭī had to say on certain *sūra*s, inspired by Bayly, who himself was hooked on *Sūrat Yūsuf*.) It was a pleasure working alongside Bayly at NYU during the period 1979-1985, and I shall never forget his generosity and his contagious love for the Arab world. I think that he recognized in me someone already afflicted.

The materials for this study were mainly gathered during my research on the history of Islamic science conducted at the American Research Center in Egypt during the period 1972-79. This activity was sponsored mainly by the Smithsonian Institution and National Science Foundation, Washington, D.C. (1972-79), the American Philosophical Society (1972-74), and the Ford Foundation (1976-79). This support is gratefully acknowledged. An abstract of the first version of this study was presented at the 188th Annual Meeting of the American Oriental Society, held at Toronto in April, 1978, and the full text appeared as "Universal Solutions to Problems of Spherical Astronomy from Mamluk Egypt and Syria", in Farhad Kazemi & Robert B. McChesney, eds., *A Way Prepared: Essays on Islamic Culture in Honor of Richard Bayly Winder*, New York: New York University Press, 1988, pp. 153-184, reprinted in 1993 in my *Studies*, C-VII. My thanks are also due to Jan Hogendijk (Utrecht) and Richard Lorch (Munich) for their assistance in reaching that stage.

In this new version I have rearranged the old material, dealing first with tables and then with instruments, and have added some new material. I have replaced many of the bibliographical references relating to tables with cross-references to the sections of **I-II**, and for references to instruments I now refer the reader to **X**. For various non-Mamluk universal solutions noted in passing, I have given cross-references to **VIa**. My thanks go once again to François Charette, who provided the information on Najm al-Dīn's use of his universal auxiliary table and controlled the final version for consistency. Many more references could have inserted to his new book *Mamluk Instrumentation* (2003). The choice of illustrations, which is not particularly happy, was partly dictated by the fact that many relevant illustrations are scattered elsewhere in this book.

TABLE OF CONTENTS

INTRODUCTION

> " ... it is true to say that *the Muslims excelled in every branch of scientific knowledge* ... . On the other hand, it is also true [*sic*] that the main concern of the Arabs was astrological, and here we come to another obstacle in the way of real progress." Patrick Moore, in the foreword to C. Walker, ed., *Astronomy before the Telescope*, p. 12. [The words italicized are taken from my article "Islamic Astronomy" in that volume, which Patrick Moore appears not to have read before writing his foreword.]

There was considerable interest in astronomy in Egypt and Syria during two centuries of Mamluk rule, especially from the middle of the 13th century to the middle of the 15th. This Egyptian-Syrian tradition began in late-13th-century Cairo, reached its culmination in 14th-century Aleppo and Damascus, and continued with occasional innovations in 15th-century Cairo. My 1983 essay on Mamluk astronomy illustrates new Mamluk material that attests to the sophistication of their activities and provides part of the background for the 2003 publication of François Charette.[1]

Some of the activity of the Egyptian and Syrian astronomers was in the field of planetary astronomy, both theoretical, in the form of modifications to the planetary models of Ptolemy, and practical, in the form of new *zīj*es or astronomical handbooks compiled for Cairo and Damascus. These achievements do not concern us here. However, Egyptian and Syrian astronomers also made notable advances in the field of spherical astronomy, particularly astronomical timekeeping. It has long been clear that these astronomers were mostly *muwaqqit*s – astronomers associated with mosques and responsible for regulating the astronomically defined times of Muslim prayer (**V**).

My purpose here is to draw attention to one particular aspect of the Mamluk scientific tradition – the interest in instruments, computational devices, and mathematical tables, which facilitate the universal solution of problems in spherical astronomy. In another study of a more general nature (**VIa**), I have noted Greek achievements in the field of universal solutions and

---

[1] King, "Astronomy of the Mamluks", and Charette, *Mamluk Instrumentation*. On the earlier background see also King, "Astronomy in Fatimid Egypt". All of the tables for spherical astronomy mentioned in this study are analyzed in **I-II**. On the different categories of instruments see **X**.

I purposely refrain from referring to the earlier tradition in Egyptian astronomy exemplified by the works of the late-10th-century Cairo astronomer Ibn Yūnus (on whom see the article in *DSB*), or the astronomers of the 12th-century Afḍal / Baṭā'iḥī observatory (on which see Sayılı, *Observatory in Islam*, pp. 167-175), or the astronomers in 13th-century Cairo who compiled the astronomical handbook known as the *Muṣṭalaḥ Zīj* (on this see Kennedy, "*Zīj* Survey", no. 47, and King, "Astronomy of the Mamluks", pp. 535-536), which remained popular in Egypt until it was eventually replaced by Egyptian recensions of the *Zīj*es of Ibn al-Shāṭir and Ulugh Beg (see Kennedy, "*Zīj* Survey", nos. 11 and 16, and King, "Astronomy of the Mamluks", p. 536).

Likewise I refrain from any discussion of the earlier tradition of astronomy in Syria. After the activity of Ḥabash (*ca.* 830; see the article by Willy Hartner in *DSB*), there is a large gap in our knowledge of mathematical astronomy in Syria until the mid-13th century when there appeared Ibn al-Lubūdī (Suter, *MAA*, no. 365 and Kennedy, "*Zīj* Survey", nos. 86 and 87), who prepared two *zīj*es, neither of which is extant, and Muḥyi 'l-Dīn al-Maghribī (Suter, *MAA*, no. 376 and Kennedy, "*Zīj* Survey", nos. 41 and 108), who compiled a *zīj* for Damascus before taking off to work at the observatory in Maragha (see King, "Astronomy of the Mamluks", pp. 533-534). Thereafter there is another smaller gap before the appearance of al-Marrākushī on the Cairo scene.

surveyed the Muslim activity in this area (notably in 9th-century Baghdad, 11th-century al-Andalus, and Mamluk Egypt and Syria).

The material presented here shows that a succession of highly competent astronomers in Mamluk Egypt and Syria turned their attention to universal solutions and also came up with some impressive instruments and tables. I shall discuss each of these Mamluk astronomers in chronological order, indicating their contribution to universal solutions. The reader should bear in mind that only one aspect of Muslim astronomy is being considered; many important Mamluk astronomers are not mentioned because they did not, as far as we know, write on this one aspect of spherical astronomy. Also, even for those who did, I have refrained from mentioning their productions on universal instruments that are well known, such as the planispheric astrolabe with a complete set of plates, the spherical astrolabe, and the trigonometric quadrant.

In what follows I shall assume that the reader is familiar with the basic problems of medieval spherical astronomy and their solution by means of an astrolabe or a trigonometric quadrant. I shall use the following notation freely:

a	azimuth (measured from the prime vertical)
h	instantaneous altitude
H	meridian altitude
D	half arc of visibility
L	local longitude
q	qibla (measured from the meridian)
t	hour-angle
T	time since rising
δ	celestial declination
ϕ	local latitude
λ	solar longitude

I shall also refer to the medieval trigonometric functions using the standard capital notation. As far as trigonometric tables are concerned, it should be borne in mind that tables of the sine and cotangent functions with values to three sexagesimal digits for each minute of argument were standard equipment of the Mamluk astronomers.[2]

[2] See further *Cairo ENL Survey*, nos. C72 and C137.

Chapter 1 al-Marrākushī

Abū ʿAlī al-Marrākushī was an astronomer of Moroccan origin who worked in Cairo about the year 1280.[3] He compiled a *summa* on spherical astronomy and astronomical instruments, which is one of our most valuable sources for the history of Islamic astronomical instrumentation. In this work, which exists in several manuscripts copied in Egypt, Syria and Turkey, al-Marrākushī presents a detailed discussion of spherical astronomy and then deals first with different kinds of plane sundials, and then other instruments such as the armillary sphere, the planisphaeric astrolabe, the universal plate known as the *shakkāziyya*, the trigonometric grid called *mujayyab*, and a variety of quadrants for determining time from solar altitude. The first half of al-Marrākushī's treatise, dealing with spherical astronomy and sundials, was translated by Jean-Jacques Sédillot, and the second half, dealing with instruments, was summarized in a rather chaotic fashion by Louis-Amélie Sédillot. al-Marrākushī's sundial theory was studied by Carl Schoy.

al-Marrākushī was not concerned with the kind of solutions I label universal, but those Mamluk astronomers who were knew of his work. For the modern reader who wishes to obtain an overview of spherical astronomy in 13th-century Egypt the treatise of al-Marrākushī, as presented by the Sédillots and Schoy, is the best introduction. A task for the future would be the preparation of a detailed commentary on this work and an investigation of its sources (particularly the sections lifted wholsale from other works), not all of which are stated.

The activities of al-Marrākushī in Cairo seem to have set the scene for the flowering of the science of astronomical timekeeping in both Egypt and Syria. There was continuous exchange between the two main regions of the Mamluk domains. We shall encounter a contemporary of al-Marrākushī, the Egyptian al-Maqsī (**2**), and a few decades later we find the Egyptian Najm al-Dīn al-Miṣrī (**3**) and the Syrian Ibn al-Sarrāj (**5**), apparently both working in Aleppo.[4] Not long thereafter, we find the Syrian astronomers Ibn al-Shāṭir (**8**) and al-Mizzī both going to Cairo and Alexandria to study astronomy and then returning to Damascus to practice their profession.

Chapter 2 al-Maqsī

In the late 13th-century, the Egyptian astronomer Shihāb al-Dīn al-Maqsī[5] compiled a set of tables of the time since sunrise for each degree of solar altitude and solar longitude, with values computed for the latitude of Cairo, taken correctly as 30°. There are indications that al-Maqsī based these tables of $T(h,\lambda)$ on some earlier tables by the 10th-century Egyptian astronomer

[3] On al-Marrākushī (**I-4.2.4** and **II-2.7** and **6.7**) see Suter, *MAA*, no. 363, *Cairo ENL Survey*, no. C17, and the article in EI_2. On his treatise see Sédillot-*père*, *Traité*, and Sédillot-*fils*, *Mémoire*, and Schoy, *Gnomonik der Araber*.

[4] The relationship between al-Marrākushī and al-Maqsī is by no means clear. Neither mentions the other. Najm al-Dīn al-Miṣrī mentions al-Maqsī in the introduction to his universal tables, and al-Marrākushī in his treatise on astronomical instrumentation (as someone who has already died). The relationship between the two contemporaries Najm al-Dīn and Ibn al-Sarrāj is even more obscure.

[5] On al-Maqsī see Suter, *MAA*, no. 383, and *Cairo ENL Survey*, no. C15. See also King, "Astronomy of the Mamluks", p. 540.

Ibn Yūnus, and the precise nature of al-Maqsī's own contribution is obscure.[6] Certainly the format of these tables is that of the solar azimuth tables of Ibn Yūnus, which display the function $a(h,\lambda)$ for each degree of both arguments, likewise for the latitude of Cairo.

al-Maqsī's tables of $T(h,\lambda)$, together with a related set displaying the hour angle $t(h,\lambda)$ and Ibn Yūnus' tables of $a(h,\lambda)$, formed a major part of the corpus of tables for timekeeping that was used in Cairo from the 13th century onward. The other tables in the corpus display the astronomically-defined intervals during which the five daily prayers may be performed. The functions tabulated include the duration of morning and evening twilight for regulating the prayers at daybreak and nightfall, and the times between sunrise and midday, midday and mid-afternoon (defined in terms of shadow lengths), and mid-afternoon and sunset, for regulating the prayers at midday, mid-afternoon, and sunset. Each of these functions is displayed for each degree of solar longitude.

The corpus contains about 200 pages of tables with more than 30,000 entries,[7] each computed to degrees and minutes. Although the corpus survives in over two dozen manuscripts, no two contain exactly the same tables, and not one contains a reliable statement of the authorship of each of the tables. These tables and others based on them were used in Cairo until the 19th century.

al-Maqsī's tables of the time since sunrise are the earliest known tables for timekeeping computed for arguments h and λ. All earlier Islamic tables displaying the time of day (or night) as a function of solar (or stellar) altitude have the meridian altitude as the second argument.[8] No such tables of $T(h,H)$ are known to have been compiled for Cairo. However, it was surely al-Maqsī's tables of $T(h,\lambda)$ for timekeeping by the sun and the Islamic tradition of compiling tables of $T(h,H)$ for specific latitudes which inspired Najm al-Dīn al-Miṣrī (**3**) to compile a universal table displaying the time since rising of the sun or any star for all latitudes.

al-Maqsī also wrote a treatise on sundials, including an introduction on sundial theory and tables for constructing horizontal sundials for Mecca, Cairo, Jerusalem, and Damascus, as well as more than 100 tables specifically for Cairo for marking sundials in various planes inclined to the meridian and to the prime vertical and skew to both.[9] These tables are considerably more extensive than those of al-Marrākushī.

al-Maqsī included tables for constructing sundials in the plane of the celestial equator and the perpendicular plane, and pointed to their universal application. The former have the disadvantage that they serve only northern or southern declinations. The latter, known as polar sundials, were discussed by later astronomers such as Ibn al-Sarrāj, Ibn al-Shāṭir, and al-Tīzīnī (**5**, **8**, and **13**), and a unique Islamic example of such a sundial, made in the year 1786/87, is preserved in Acre.[10]

---

[6] These tables, and indeed the entire Cairo corpus of tables for timekeeping, are in several of the available manuscripts attributed to Ibn Yūnus (on whom see the article in *DSB*). The corpus is analyzed in King, "Astronomical Timekeeping in Medieval Cairo", and in **II-4**, and the attribution of each of the various tables is discussed in **II-5**.

[7] Tables of the functions $T(h,\lambda)$, $t(h,\lambda)$, or $a(h,\lambda)$ require about 10,000 entries to display values for each degree of both arguments. Tables of $T(h,H)$ with values for each degree of both arguments ($h \leq H$) for a given latitude require about 3,000 entries to display values for each degree of both arguments.

[8] On these tables see **I-2**.

[9] See further King, "Islamic Astronomical Tables", pp. 51-53, and *idem* & Samsó, "Islamic Astronomical Handbooks and Tables", p. 93.

[10] Illustrated in Michel & Ben-Eli, "Cadran solaire remarquable", and **Fig. VIa-15.2**.

Chapter 3 Najm al-Dīn al-Miṣrī

The Egyptian astronomer Najm al-Dīn al-Miṣrī,[11] working in Cairo a few decades after al-Marrākushī and al-Maqsī (**1-2**), decided to make a table for timekeeping which would serve all latitudes and could be used for timekeeping by both the sun and any non-circumpolar star.[12] He tabulated the time since rising as a function of the observed altitude and two variables that depend on the local latitude and celestial declination – the meridian altitude and half arc of visibility. Thus the function tabulated is T(H,h,D) – see **Fig. I-2.6.1**. Values are given for each degree of both arguments, and number over 440,000. The computation of the entries is not difficult by virtue of the choice of argument, and the errors that I have investigated in spot-checks amount to no more than ±0;15°, that is, one minute of time. The underlying formula for the hour-angle t (= D - T) is:

$$\text{Vers } t = \{ 1 - \text{Sin } h / \text{Sin } H \} \cdot \text{Vers } D .$$

The table is a highly ingenious extension of the Islamic tables for timekeeping for specific latitudes. It is extant in a unique contemporaneous manuscript, the two halves of which are now preserved in Cairo and Oxford. This manuscript may be in the hand of Najm al-Dīn himself. I find it somewhat surprising that Najm al-Dīn's enormous table is not mentioned in later astronomical sources, and think that it is safe to assume that the table was scarcely used. With its enormous size, the table would be unlikely to attract many copyists, and besides, tables for timekeeping by the sun were already available for Cairo (see **2**).

However, Najm al-Dīn conceived of his table in more general terms than simply for timekeeping, and he was not at a loss to describe the ways in which it could be used to solve all of the problems of spherical astronomy for any latitude, not only determining time and azimuth, but also solving complicated operations as the determination of the qibla using the exact formula and conversion between ecliptic and equatorial coordinates.[13] In addition, Najm al-Dīn actually used his own table to prepare other tables for instrument construction, and it has been shown that some of the numerous tables in his remarkable treatise describing over 100 different kinds known to him from earlier sources and invented by himself are actually based on his enormous auxiliary table.[14] See **Fig. 3.1** for just one of the instruments described by Najm al-Dīn.

Chapter 4 al-Abharī

In the year 1320 the Syrian astronomer Amīn al-Dīn al-Abharī compiled a treatise on instruments, discussing four instruments in particular – the *shakkāziyya* plate, the armillary

[11] On Najm al-Dīn al-Miṣrī see Suter, *MAA*, no 460 (confused); *Cairo ENL Survey*, no. C16; and King, "Astronomy of the Mamluks", pp. 540-541; but more especially Charette, "Najm al-Dīn's Monumental Table", and *idem*, *Mamluk Instrumentation*.

[12] A summary of Charette, "Najm al-Dīn's Monumental Table", is given in **I-2.6.1**.

[13] A summary of his instructions based on Charette, "Najm al-Dīn's Monumental Table", is in **I-9.3***.

[14] *Idem*, *Mamluk Instrumentation*, pp. 53-56. In several early publications I attributed this treatise to Ibn al-Sarrāj – see, for example, King, *Mecca-Centred World-Maps*, pp. xxix and 23, n. 43, *etc*. However, François Charette has established that the author is without doubt Najm al-Dīn.

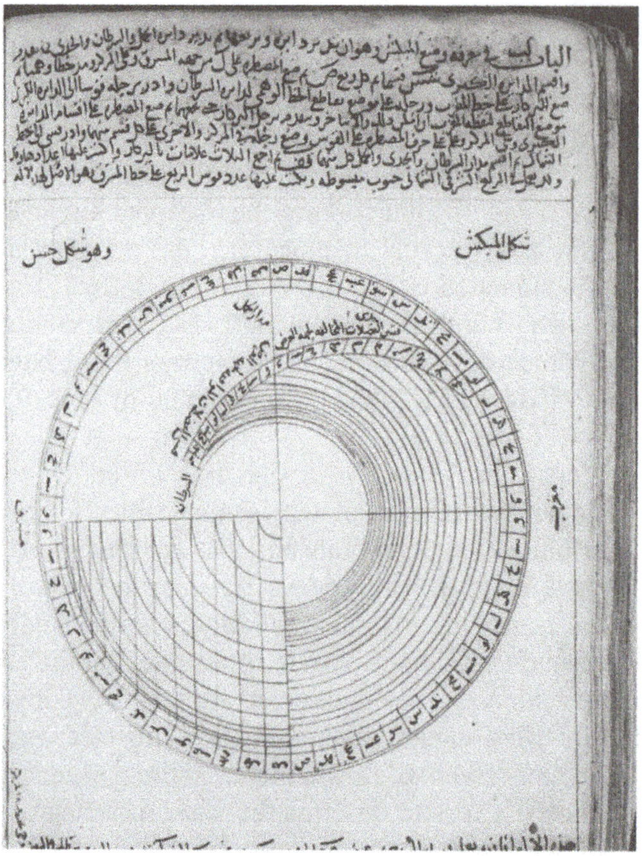

Fig. 3.1: An illustration of the universal instrument called *al-mubakka/ish* (?) by Najm al-Dīn al-Miṣrī. The curious name is unexplained. The device serves to find the time since sunrise in *seasonal* hours from the time in *equatorial* hours using the half-excess of daylight as an additional argument, entered on the scale across the upper half of the markings. In the lower left are markings for one of several varieties of trigonometric quadrants. All of the over 100 instruments featured by Najm al-Dīn have now been subjected to a detailed analysis by François Charette. [From MS Dublin CB Persian 102,2, fol. 60v, courtesy of the Chester Beatty Library.]

sphere, the circular *mujayyab* trigonometric grid, and the *āfāqiyya* plate of horizons.[15] These instruments are all universal – they are not restricted to a specific latitude – and it seems likely that it was this feature which led al-Abharī to treat them together. His treatise was not widely known, and survives in only two manuscripts, now preserved in El Escorial and Gotha.

Chapter 5 Ibn al-Sarrāj

Ibn al-Sarrāj worked in Aleppo in the first half of the 14th century.[16] He wrote on mathematics and astronomy, but his main interest was in astronomical instruments. He was a first-class

[15] On Amīn al-Dīn al-Abharī see Suter, *MAA*, no. 393, and **II-2.8**. He is not to be confused with Athir al-Dīn al-Abharī (*ibid.*, no. 364).

[16] On Ibn al-Sarrāj see Suter, *MAA*, no. 508, Mayer, *Islamic Astrolabists*, p. 34, and *Cairo ENL Survey*, no. C26. See also King, "Astronomy of the Mamluks", pp. 544-545, and **X-5.2**.

instrument-maker, and displayed considerable ingenuity in the design of new varieties of instruments.

Ibn al-Sarrāj re-invented the universal astrolabe of the 11th-century Andalusī astronomer ʿAlī ibn Khalaf.[17] There is no evidence that ʿAlī ibn Khalaf's instrument or the treatise he wrote on its use was known outside al-Andalus except in Europe. Ibn al-Sarrāj certainly knew of the single *shakkāziyya* plate, and may have known of the universal astrolabe of the Andalusī astronomer Ibn Bāṣo,[18] who was a contemporary of al-Marrākushī. However, Ibn al-Sarrāj probably hit upon the idea independently, and wrote a treatise on the use of the instrument which he called *Sarrājiyya*, after himself. In this astrolabe, described in a unique manuscript preserved in Cairo, a half *shakkāziyya* grid could rotate over a whole *shakkāziyya* plate, and the other half of the rete consisted of a mixed north-south astrolabe projection of the ecliptic and position of selected stars, with which one could solve problems of spherical astronomy for all latitudes. Another anonymous Syrian or Egyptian wrote a treatise on the same instrument, calling it *al-asṭurlāb al-mughnī*, "the astrolabe which does not need [any plates]".

A universal astrolabe made by Ibn al-Sarrāj himself in 1329 survives in the Benaki Museum in Athens. However, only the front of this astrolabe and the lower part of the back correspond to the instrument that Ibn al-Sarrāj described in this treatise. The surviving astrolabe of Ibn al-Sarrāj contains a set of plates for each three degrees of latitude which could be used with the half rete bearing the ecliptic and star pointers. The altitude circles for each latitude are folded over so that each set takes up a quarter of each plate. Ibn al-Sarrāj's instrument is a universal astrolabe which can be used universally in five ways; as such it ranks as one of the most interesting, if not *the* most interesting, of medieval astronomical instruments – see **Fig. X-5.2**. Ibn al-Sarrāj did not write on the use of this particular instrument, but the Egyptian astronomer al-Wafāʾī (**14**), who owned the astrolabe about a century after Ibn al-Sarrāj made it, did write a treatise on its use, extant in two copies in Princeton and Dublin.[19]

Fortunately, al-Wafāʾī included an account of the operations that can be performed with the alidade and the mysterious grid on the upper back of Ibn al-Sarrāj's astrolabe (**Fig. 5.1**). Ibn al-Sarrāj devised a number of other grids in the shape of semicircles and quadrants, the purpose of which was to solve the standard problems of spherical astronomy for all latitudes. With the grid on the back of his astrolabe and the special alidade one can very simply effect a transformation of the form (a,b) → (c,d), where

$$c = \text{arc sin (sin a sin b)} \quad \text{and} \quad d = \text{arc cos (sin a cos b)}\ ,$$

an operation which has several applications in spherical astronomy.

Another of Ibn al-Sarrāj's inventions consists of a semicircular plate with scales along the diameter and around the circumference and two threads with movable markers attached at both ends of the diameter. The semicircle needs no other markings: Ibn al-Sarrāj called it *al-jayb al-ghāʾib*, "the trigonometric grid without any grid". A related instrument in the form of a quadrant and again without any markings was called by Ibn al-Sarrāj *al-rubʿ al-jāmiʿ*, "the

[17] On this instrument see the article "*Shakkāziyya*" in EI_2, and **VIa-11**.

[18] On this see further **VIa-11**.

[19] See the text to n. 50 below. All of Ibn al-Sarrāj's instruments are discussed in detail in an unpublished study, which includes texts and translations of all the related treatises.

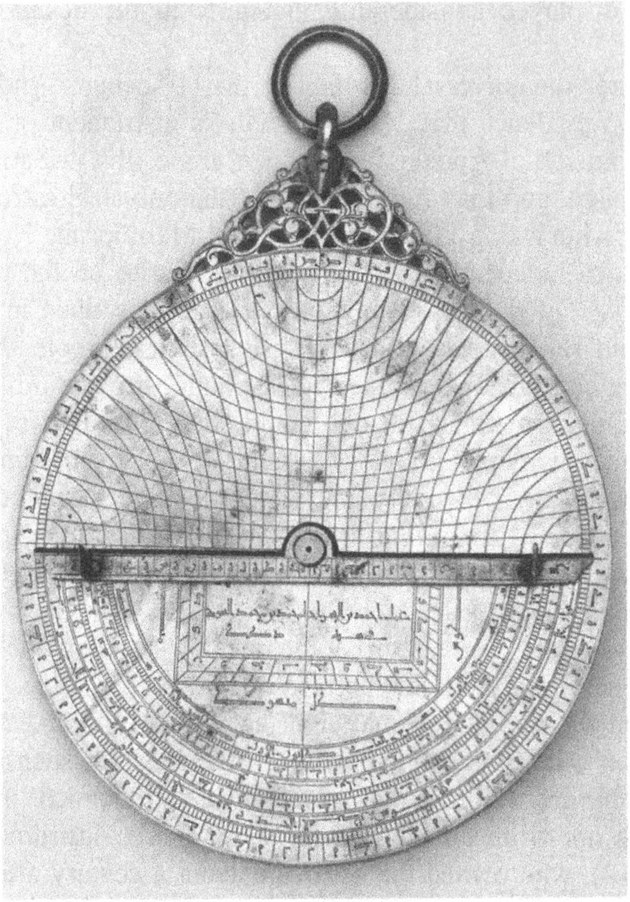

Fig. 5.1: The back of the universal astrolabe of Ibn al-Sarrāj (#140), displaying a remarkable semi-circular grid based on, but different from, standard markings of the *shakkāziyya* type. Together with the non-uniform scales on the diametrical rule, the grid can be used for all manner of calculations in spherical astronomy, or in general spherical trigonometry. In addition, there is a solar / calendrical scale of a kind that is rare on Eastern instruments, but is indeed ultimately of Eastern Islamic, rather than Andalusī, origin. [Courtesy of the Benaki Museum, Athens.]

universal quadrant". Yet another quadrant called *al-mujannaḥ,* "the winged quadrant", bore two families of arcs of circles.[20] All of these instruments can be used, with the kind of ingenuity which Ibn al-Sarrāj possessed and which is in evidence in all his treatises on their use, to solve all of the standard problems of spherical astronomy for any latitude.

Chapter 6 Ṭaybughā al-Baklamshī

A contemporaneous Syrian scholar, Ṭaybughā al-Baklamshī,[21] appears to have been the first to note that the problems of spherical astronomy can be solved using just one quadrant of

[20] See now Charette, "Der geflügelte Quadrant".
[21] On Ṭaybughā see Brockelmann, *GAL*, II, pp. 168-169, and SII, p. 167, and *Cairo ENL Survey*, no. C53.

shakkāziyya markings with a thread attached at the centre. Ṭaybughā wrote a treatise on the use of such an instrument in which he claimed to be its inventor. One disadvantage of this instrument is that the determination of time from celestial altitude is approximate.[22]

Various Syrian and Egyptian astronomers in the next century, such as Ibn Ṭaybughā (**11**), al-Shādhilī and Abū Ṭāhir, wrote treatises on the use of the single *shakkāzī* quadrants.[23]

Chapter 7 al-Bakhāniqī

The 14th-century Egyptian astronomer al-Bakhāniqī[24] took the tables of al-Farghānī (*fl. ca.* 825) displaying coordinates for marking astrolabe plates for each degree of latitude from 15° to 50°[25] and extended them to cover the complete range 0° to 90°. He did this for a Yemeni ruler who had probably noticed that al-Farghānī's tables did not serve latitudes in the Yemen. His tables survive in a unique manuscript now in Dublin.

Another set of tables compiled by al-Bakhāniqī, which survives in a unique manuscript in Cairo, displays the hour-angle at the beginning of the mid-afternoon prayer for each 10° of solar longitude at each degree of terrestrial latitude from 10° to 50°: see **Fig. 7.1**.[26]

Chapter 8 Ibn al-Shāṭir

Ibn al-Shāṭir was the head of the muwaqqits associated with the Umayyad Mosque in Damascus in the mid-14th century and worked mainly in planetary astronomy.[27] His contribution to planetary theory is now well known[28] although none of his treatises or his new *zīj* for Damascus has been published. Ibn al-Shāṭir was inevitably also concerned with spherical astronomy and timekeeping; indeed, he constructed the magnificent sundial on the minaret of the Umayyad Mosque in Damascus.[29] He devised various instruments for solving problems in spherical astronomy for all latitudes; however, he faced the problem that Najm al-Dīn al-Miṣrī and Ibn al-Sarrāj had already almost exhausted the possibilities.

One of Ibn al-Shāṭir's inventions was a compendium consisting essentially of two universal sundials, the first a horizontal sundial for the equator which could be inclined to serve as a sundial in any latitude, and the second an equatorial plate which could be inclined in the plane of the celestial equator in any latitude and used to measure the hour-angle of the sun by day or the stars by night. Ibn al-Shāṭir called this instrument *ṣandūq al-yawāqīt*, "box of sapphires",

---

22 This problem is investigated in Samsó & Catalá, "Cuadrante *šakkāzī*". See further **XI-8.2**.

23 On these astronomers see *Cairo ENL Survey*, nos. C54, C55 and C56.

24 On al-Bakhāniqī see King, *Astronomy in Yemen*, pp. 34-35; *Cairo ENL Survey*, no. C28; and **II-4.1.5**.

25 On al-Farghānī see the article in *DSB* by A. I. Sabra, and Sezgin, *GAS*, V, pp. 259-260 and VI, pp. 149-151, where these tables are not mentioned. Various Mamluk tables for making astrolabe plates are analyzed in Charette & King, "Survey of Islamic Tables for Astrolabe Construction" (forthcoming).

26 Analyzed in **II-6.13**.

27 On Ibn al-Shāṭir see the article in *DSB* and the references there cited, to which add Kennedy & Ghanem, eds., *Ibn al-Shāṭir*, and *Cairo ENL Survey*, no. C30. See also the remarks in n. **I**-2:15.

28 See various articles by Kennedy, Roberts and Abboud reprinted in both Kennedy & Ghanem, eds., *Ibn al-Shāṭir*, and Kennedy *et al.*, *Studies*, pp. 50-107.

29 On this sundial see Janin, "Cadran solaire de Damas".

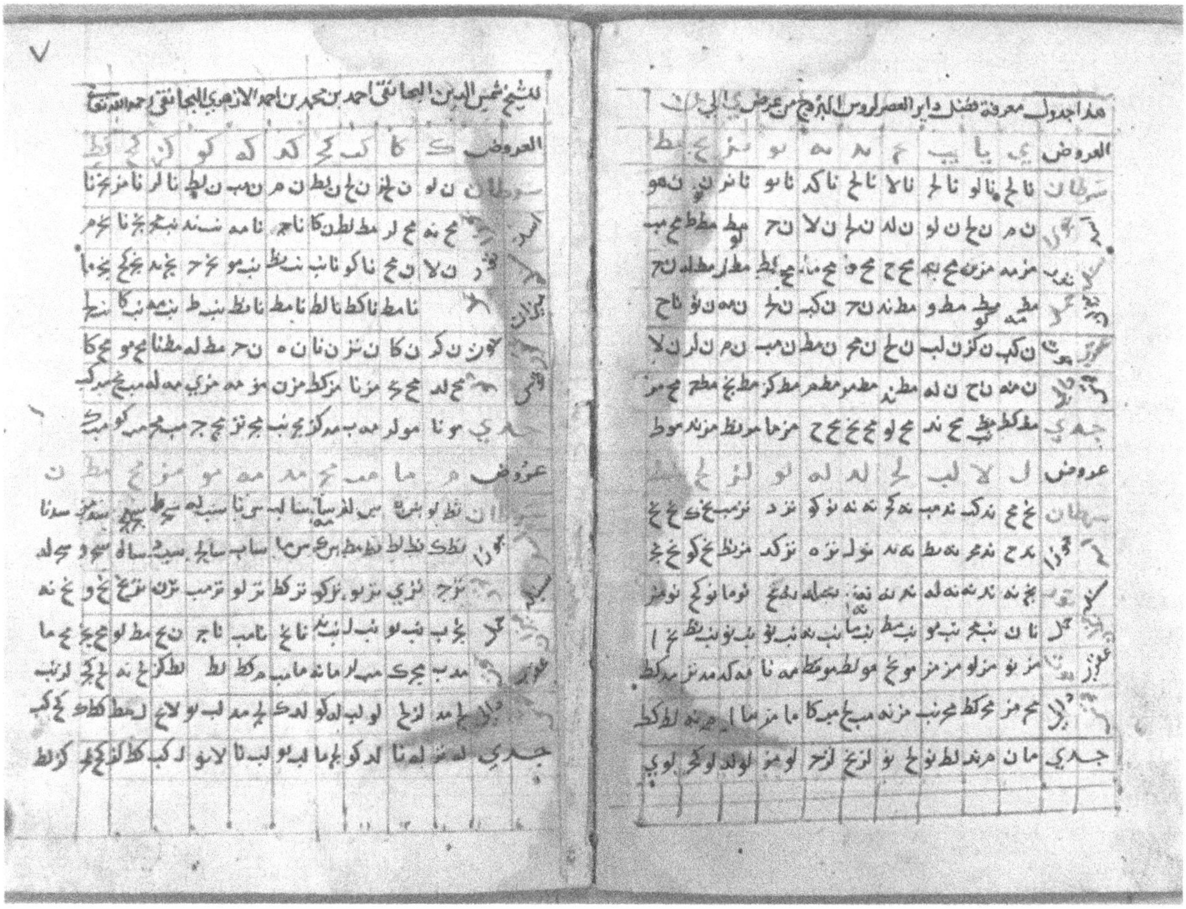

Fig. 7.1: The tables of al-Bakhāniqī displaying the hour-angle at the beginning of the *ʿaṣr* prayer for all latitudes. [From MS Cairo MM 33,2, fols. 6v-7r, courtesy of the Egyptian National Library.]

and wrote a treatise on its use, as did the later Egyptian astronomer Ibn Abi ’l-Fatḥ al-Ṣūfī (**17**). Both of these treatises are preserved in a unique manuscript now in Berlin. Ibn al-Shāṭir devised other different trigonometric grids,[30] including one in the form of an equilateral triangle which he labelled *al-rubʿ al-tāmm*, “the complete quadrant” (see **Fig. 8.1**). Another labelled *al-rubʿ al-ʿAlāʾī* in the form of a quadrant with an orthogonal grid of markings and one family of parallel diagonals, and yet another in the form of a square.[31] For each he prepared a treatise showing how it could be used to solve problems of trigonometry and spherical astronomy.

Ibn al-Shāṭir also designed a circular plate, one side of which displayed two orthogonal grids superimposed at an angle equal to the obliquity of the ecliptic. The second side consisted of a fixed stereographic projection of the celestial equator and solstitial day-circles and ecliptic over which could rotate a grid of horizons for all latitudes: an astrolabe in reverse. An example of this instrument, made by Ibn al-Shāṭir himself and unfortunately now missing the rete and

30 This instrument is analyzed in detail in Janin & King, “Ibn al-Shāṭir’s Compendium”.

31 On the “complete” and ʿAlāʾī quadrants see Schmalzl, *Geschichte des Quadranten*, pp. 100-108, repr. in Kennedy & Ghanem, eds., *Ibn al-Shāṭir*, pp. 27-35.

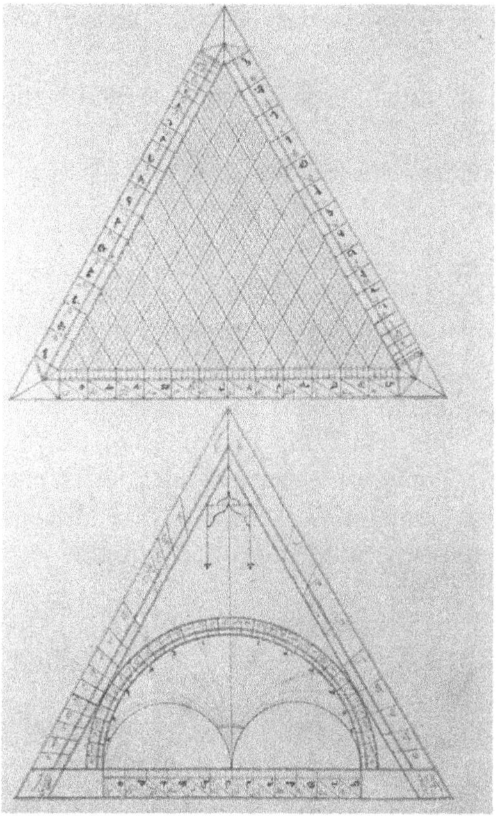

Fig. 8.1: The complete quadrant invented as a universal trigonometric grid by Ibn al-Shāṭir. [Photo by Owen Gingerich, from MS Cairo MR 40,2, courtesy of the Egyptian National Library.]

Fig. 8.2: The trigonometric grid on the back of Ibn al-Shāṭir's astrolabic plate. [Photo courtesy of the late Alain Brieux, Paris.]

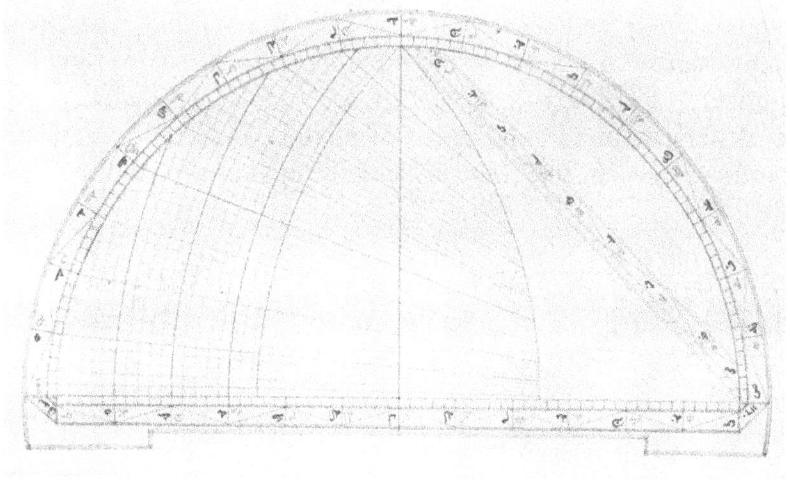

Fig. 8.3: Another calculating device developed by Ibn al-Shāṭir, again for solving all of the problems of shpeorical astronomy for any latitude. [Photo by Prof. Owen Gingerich, from MS Cairo MR 40,2, courtesy of the Egyptian National Library.]

alidade, is preserved in Cairo (see **Fig. 8.2**). His treatise on its use survives in manuscript form in St. Petersburg, Aleppo, Cairo, and Princeton.

Ibn al-Shāṭir, like Ibn al-Sarrāj, did not, as far as is known, prepare any universal auxiliary tables for solving the problems of spherical astronomy.

Chapter 9 al-Khalīlī

Shams al-Dīn al-Khalīlī[32] was a muwaqqit at the Umayyad Mosque and a contemporary of Ibn al-Shāṭir. He made singularly impressive contributions to spherical astronomy, mainly in the form of astronomical tables.

al-Khalīlī compiled a corpus of tables for timekeeping for Damascus,[33] including tables of $t(h,\lambda)$ for each degree of both arguments and tables for regulating the times of prayer, all computed specifically for Damascus. A few years previously some prayer-tables for Damascus had been compiled by al-Mizzī,[34] but al-Khalīlī used the parameters of Ibn al-Shāṭir. al-Khalīlī's tables were used in Damascus until the 19th century and exist in about two dozen manuscripts, the best of which is preserved in Paris.

al-Khalīlī also devised a set of tables for solving all the standard problems of spherical astronomy for any latitude.[35] The functions tabulated are:

$$f_\phi(\theta) = R \text{ Sin } \theta \ / \text{ Cos } \phi \ ,$$
$$g_\phi(\theta) = \text{Sin } \theta \text{ Tan } \phi \ / \ R \ ,$$
$$K(x,y) = \text{arc Cos } \{ \ R \bullet x \ / \text{ Cos } y \ \} \ .$$

and values are given for each degree or unit of argument. There are more than 13,000 entries in the complete set (see **Fig. 9.1**). With these functions the determination of the hour-angle t reduces to an application of the formula:

$$t(h,\Delta,\phi) = K \ \{ \ [f_\phi(h) - g_\phi(\Delta)], \ \Delta \ \} \ ,$$

which is easily shown to be equivalent to the modern formula

$$t = \text{arc cos } \{(\sin h - \sin \Delta \ \sin \phi) \ / \ \cos \Delta \ \cos \phi \ \} \ .$$

al-Khalīlī's *Universal Table*, called in Arabic *al-jadwal al-āfāqī*, exists in several copies of the Damascus corpus and was known to later astronomers in Syria, Egypt, and Turkey, and even in the Maghrib.

al-Khalīlī had earlier compiled another set of auxiliary tables for timekeeping by the sun which was less well known. It survives in a unique manuscript preserved in Dublin.[36] The functions tabulated are:

$$B'(\phi,\lambda) = \text{Cos } \delta(\lambda) \text{ Cos } \phi \ / \ 2R = {}^1/_2 \ B(\phi,\lambda) \qquad (\varepsilon = 23;31°)$$
$$V'(x,y) = \text{arc Vers } \{ \ R \bullet y \ / \ 2x \ \}$$

One should find Sin H and Sin h in order to establish their difference:

$$H'(h,\lambda,\phi) = \text{Sin } H(\lambda,\phi) - \text{Sin } h \ ,$$

---

[32] On al-Khalīlī (Suter, *MAA*, no. 418) see the article in *DSB* (supplement), and *Cairo ENL Survey*, no. C32. See also King, "Astronomy of the Mamluks", pp. 542-543.

[33] The Damascus corpus is analyzed in **II-10**.

[34] On al-Mizzī see Suter, *MAA*, no. 406; Mayer, *Islamic Astrolabists*, pp. 61-62; and *Cairo ENL Survey*, no. C34. On his tables see **II-9.2**.

[35] Analyzed in King, "al-Khalīlī's Auxiliary Tables" *Studies*, A-XI. See also **II-10.4**.

[36] Analyzed in **II-10.3**.

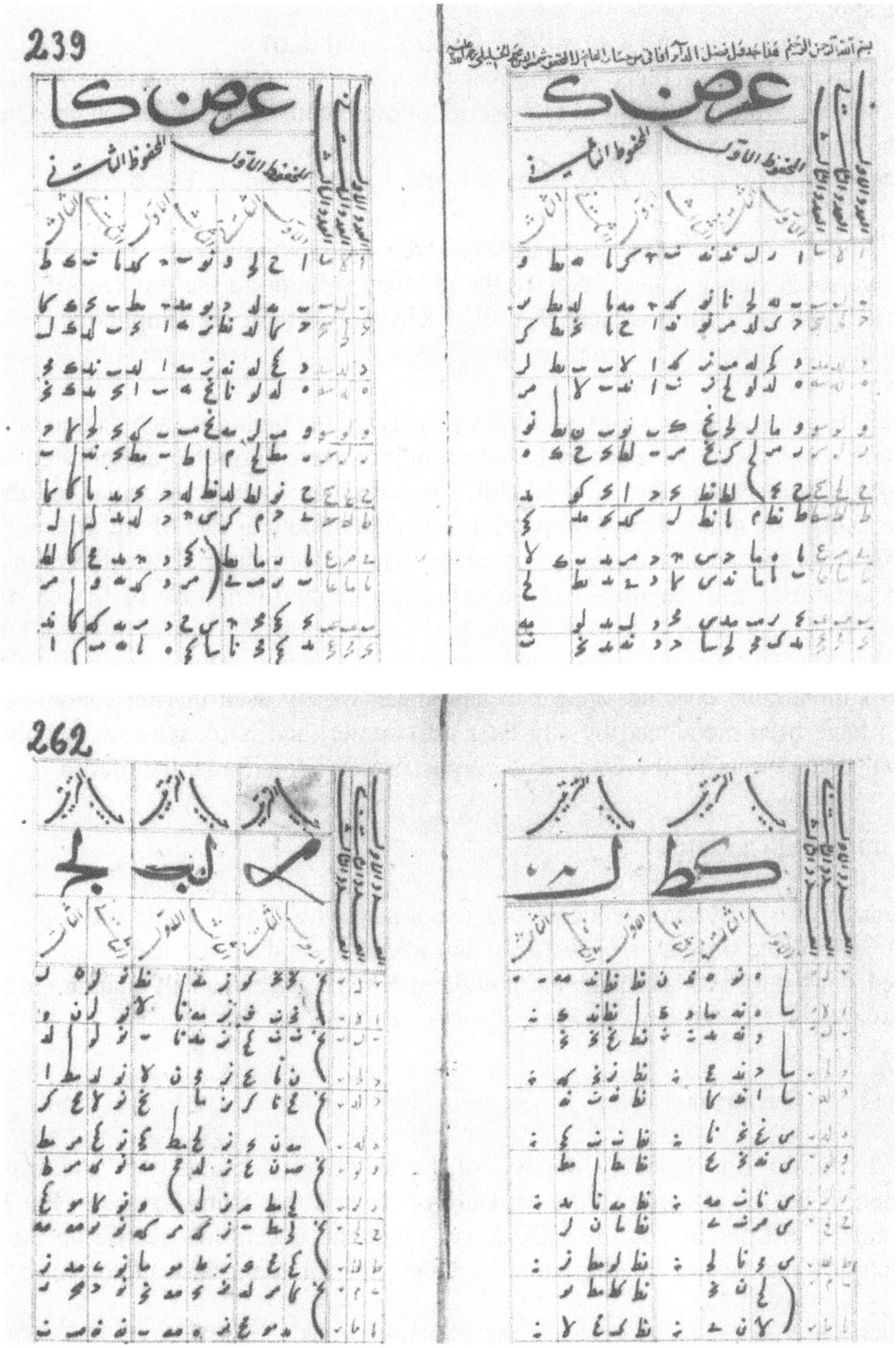

Fig. 9.1: Some extracts from the *Universal Table* of al-Khalīlī, showing the first and second functions for latitudes 20°-21°, and the third function for arguments 29-33. Notice that the format of the table for the third function adjusts itself to avoid columns of zero entries. [From MS Istanbul Hamidiye 1453,3, fols. 261v-262r and 261v-262r, courtesy of the Süleymaniye Library, Istanbul.]

and then apply:

$$t(h,\lambda,\phi) = V' \{ B'(\phi,\lambda) , H'(h,\lambda,\phi) \} .$$

al-Khalīlī also mentioned another set of universal tables in the introduction to his main universal auxiliary tables but the unique Bursa manuscript of these come to my attention only in January, 2001. The functions tabulated are:

$$f'_\phi(\lambda) = R \text{ Sin } \delta(\lambda) / \text{Cos } \phi \quad \text{and} \quad g_\phi(h) = \text{Sin } h \text{ Tan } \phi / R$$

and:

$$K(x,h) = \text{arc Sin } \{ R \cdot x / \text{Cos } h \}$$

These universal auxiliary tables represent the ultimate solution to the problem of determining the solar azimuth $a(h,\lambda,\phi)$. It is clear that all al-Khalīlī had to do to compile his better-known corpus of universal tables was to replace the table of $f'_\phi(\lambda)$ with a new one of $f_\phi(\theta)$ and rewrite the instructions!

al-Khalīlī's *pièce de résistance* was a table displaying the qibla for each degree of terrestrial latitude and longitude.[37] Qibla tables based on approximate formulae are attested in several earlier Islamic sources, but that of al-Khalīlī is based on an accurate formula, and the entries, more than 3,000, are all correctly computed to within a minute or two of arc. It is not yet clear how al-Khalīlī achieved this degree of accuracy. The modern formula for the qibla in terms of the local latitude and longitude (ϕ,L) and the latitude and longitude of Mecca (ϕ_M,L_M) is:

$$q = \text{arc cot } \{ [\sin \phi \cos \Delta L - \cos \phi \tan \phi_M] / \sin \Delta L \}$$

(where $\Delta L = L \sim L_M$), so that the computation of a single qibla value is not a trivial exercise. al-Khalīlī's qibla table does not appear to have been widely used in later centuries. It is not known to have been mentioned by any later astronomer, and is extant now in only three of the known manuscripts of the Damascus corpus preserved in Paris and Berlin.

Chapter 10 Ibn al-Ghazūlī

Ibn al-Ghazūlī was a Syrian astronomer contemporary with Ibn al-Shāṭir and al-Khalīlī (nos. 8 and 9).[38] He wrote on the *jayb ghāʾib* of Ibn al-Sarrāj, and on the use of an octant which he devised, consisting of a half quadrant of *mujayyab* markings and a half quadrant of astrolabic markings.

Chapter 11 Ibn Ṭaybughā

ʿAlī ibn Ṭaybughā was probably the son of Ṭaybughā al-Baklamshī (**6**) and was also an astronomer.[39] Like Ibn al-Sarrāj (**5**) he worked in Aleppo rather than Damascus. Ibn Ṭaybughā was the author of a treatise on the *shakkāziyya* quadrant, and in his treatise on the sine and astrolabic quadrants he presented a table displaying the qibla and also the distance to Mecca

---

[37] Analyzed in King, "al-Khalīlī's Qibla Table". See also **II-10.8**.

[38] On Ibn al-Ghazūlī see Suter, *MAA*, no. 412; Mayer, *Islamic Astrolabists*, pp. 66-67; and *Cairo ENL Survey*, no. C33.

[39] On Ibn Ṭaybughā see *Cairo ENL Survey*, no. C54. The Cairo manuscript of the treatise used in Samsó & Catalá, "Cuadrante *šakkāzī*", is the only evidence for the existence of a treatise on the *shakkāziyya* quadrant by Ibn Ṭaybughā.

for each degree of latitude and longitude. This table is based on an accurate formula but is independent of that of al-Khalīlī (**9**). The values serve a smaller range of arguments, but are likewise very carefully computed.[40]

Chapter 12 al-Māridīnī

The late-14th-century astronomer Jamāl al-Dīn al-Māridīnī was born in Syria but later moved to Egypt.[41] He was the author of several treatises on instruments, one of which described the use of a double *shakkāzī* quadrant in which a *shakkāziyya* grid rotates over a solid *shakkāziyya* plate. This can be used to solve all the problems of spherical astronomy for any latitude (see **Fig. 12.1**).[42] The inspiration for such a quadrant was surely the universal astrolabe of Ibn al-Sarrāj (**5**). al-Māridīnī's treatise is extant in three manuscripts in Damascus and Istanbul. He also devised a set of very simple and somewhat impractical auxiliary tables for solving problems of spherical astronomy, which are extant in two manuscripts in Paris and Cairo.[43]

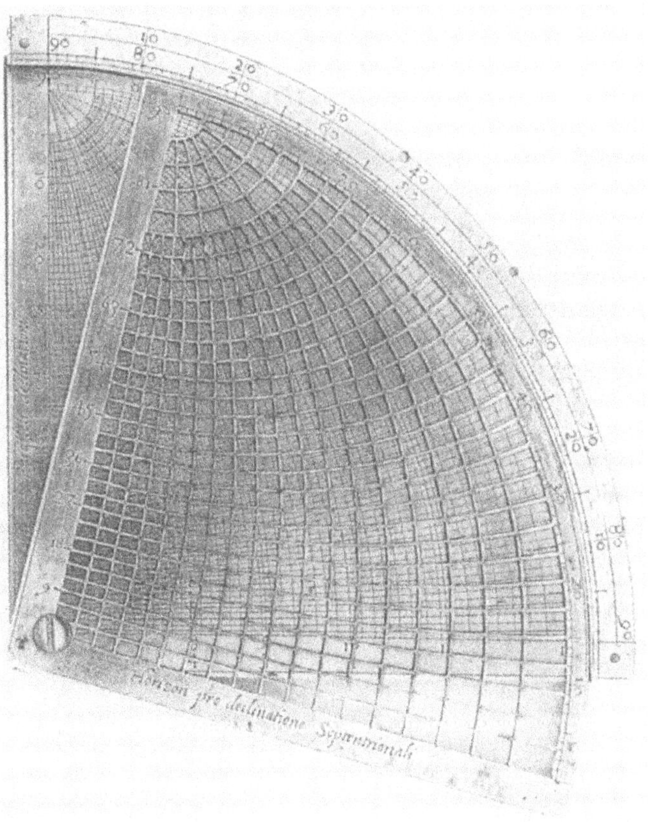

Fig. 12.1: A European version of the double *shakkāziyya* quadrant first described by al-Māridīnī. The instrument is from late-16th-century Spain and demonstrates the influence of the contemporaneous Louvain school. The question of the inspiration for such a device amongst European instrument-makers remains open. (Courtesy of the late Roderick Webster, Adler Planetarium, Chicago, Ill.)

---

[40] On this table see now King, *Mecca-Centred World-Maps*, pp. 66-67.

[41] On al-Māridīnī see Suter, *MAA*, no. 421, and *Cairo ENL Survey*, no. C47.

[42] This treatise is published with translation and commentary in King, "al-Māridīnī's Universal Quadrant".

[43] Analyzed *ibid.*, pp. 231-240. See also **I-9.6**.

Chapter 13 Miscellaneous (sundial tables)

Since universal tables were compiled by Mamluk astronomers for astrolabe construction, regulation of the prayer times, timekeeping by the sun and stars, and solving problems of spherical astronomy, it is somewhat surprising that there is no set of tables of coordinates for marking sundials for all latitudes. However, one such set had been compiled by Ḥabash in the early 9th century and this was known in early Mamluk Syria.[44] Also Ibn al-Sarrāj (**5**) is known to have compiled an extensive work on sundial theory with tables, but this is unfortunately no longer extant. As we have noted, the sundial tables compiled by al-Maqsī (**2**) include isolated tables for Mecca, Jerusalem, and Damascus. All later Mamluk Egyptian sundial tables were computed specifically for Cairo, by Ibn Sūdūn, Ibn al-Muhallabī and Sibṭ al-Māridīnī,[45] and for other localities in Egypt such as Alexandria, Damietta, and Assiut. The only known set from Mamluk Syria was compiled by the Damascus astronomer al-Tīzīnī[46] in the late 15th century: in the unique Vatican copy we find extensive tables for marking inclined vertical hour-angle dials for the latitudes of Damascus, Aleppo, and Cairo, the last-mentioned being lifted from the tables of Ibn al-Muhallabī. As far as I know, the only attempt at universality in Mamluk sundial tables is a small table prepared by Ibn Abi 'l-Fatḥ al-Ṣūfī (see **17**).

Chapter 14 al-Wafā'ī

The Egyptian astronomer ʿIzz al-Dīn al-Wafā'ī lived in the early 15th century and was a muwaqqit at the Mosque of al-Mu'ayyad in Cairo.[47] He was the author of several treatises, mostly dealing with instruments. He wrote on universal instruments such as the octant of Ibn al-Ghazūlī (**10**) and the complete quadrant of Ibn al-Shāṭir (**8**). Another of his treatises deals with an instrument of his own invention, an equatorial armilla in which a graduated semicircle affixed to a horizontal plate could be oriented in the plane of the celestial equator, and used to measure the hour-angle of the sun or any star in any latitude (see **Fig. 14.1**);[48] this instrument was called *dā'irat al-muʿaddil*, "the equatorial (semi-)circle", and became rather popular in later Egypt, Syria, and Turkey. Several examples survive, as do numerous copies of al-Wafā'ī's treatise and various later treatises on the instrument. In an earlier version of this instrument, called *al-muqawwar*, al-Wafā'ī put all of the component parts in a box built on the base: this confirms my impression that he was influenced by Ibn al-Shāṭir's "box of sapphires".[49]

---

[44] On Ḥabash's tables, previously attributed to al-Khwārizmī, see n. 48 to **VIa**. The unique Istanbul manuscript was copied in Damascus in 1223/24 – see King, "al-Khwārizmī", p. 1.

[45] On Ibn Sūdūn, Ibn al-Muhallabī, and Sibṭ al-Māridīnī (Suter, *MAA*, no. 445) see *Cairo ENL Survey*, nos. C82, C67, and C97. The sample extracts from the tables of Sibṭ al-Māridīnī published in Schoy, "Spätarabische Sonnenuhren", pp. 352-360, are in fact by Ibn al-Muhallabī and are independent of Sibṭ al-Māridīnī's introduction translated by Schoy on pp. 344-351. Schoy was aware of a problem between text and tables (*cf.* his pp. 342-347), but had only one defective manuscript at his disposal. On Ibn al-Muhallabī's treatise on sundial theory see Janin & King, "Cadran solaire d'Ibn Ṭūlūn", pp. 351-352.

[46] On al-Tīzīnī see Suter, *MAA*, no. 450, and *Cairo ENL Survey*, no. C95.

[47] On al-Wafā'ī see Suter, *MAA*, no. 437, and *Cairo ENL Survey*, no. C61.

[48] On the *dā'irat al muʿaddil* see Tekeli, "Equatorial Armilla", where al-Wafā'ī's text is published with translation, and also Brice & Imber & Lorch, "Equatorial Semicircle", and Dizer, "Equatorial Semicircle". See also Janin & King, "Ibn al-Shāṭir's Compendium", pp. 216-217, where this instrument is compared with the earlier instrument of Ibn al-Shāṭir.

[49] *Cf.* Janin & King, "Cadran solaire d'Ibn Ṭūlūn", App., n. 3.

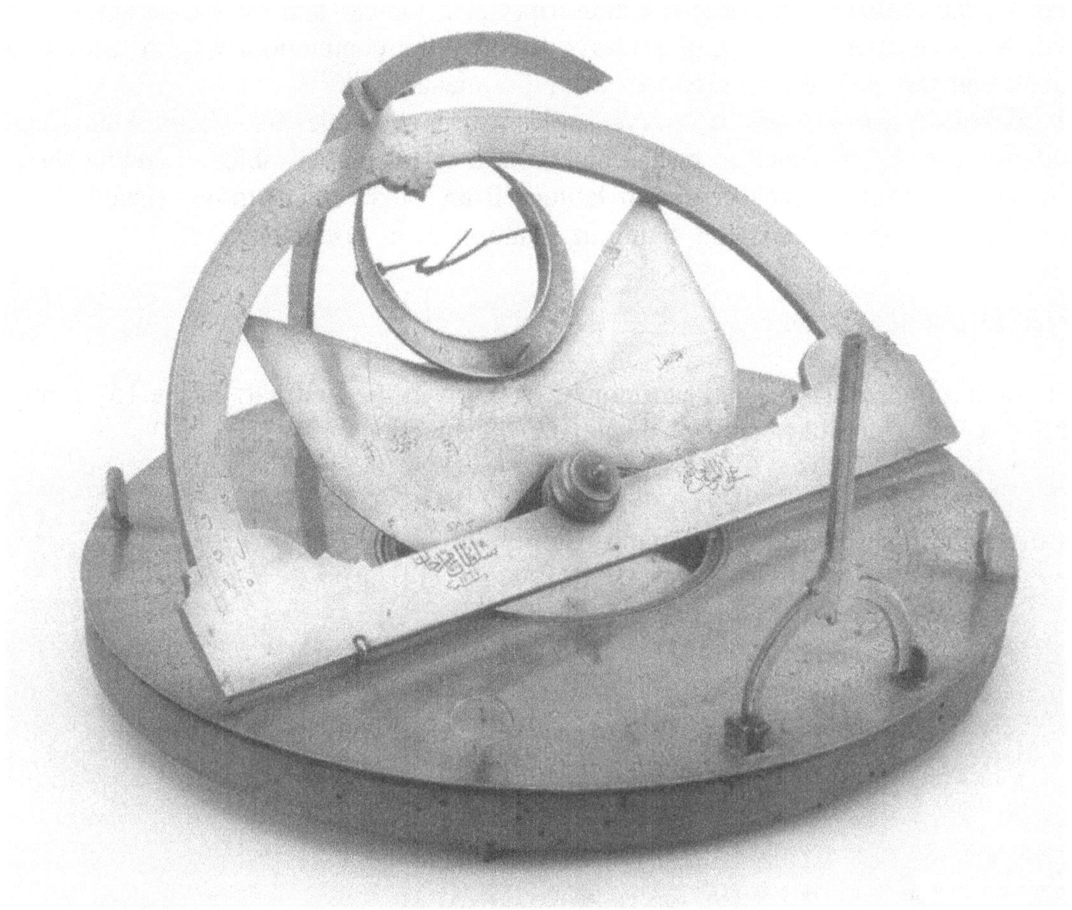

Fig. 14.1: A late Ottoman version of the *dāʾirat al-muʿaddil* invented by al-Wafāʾī. This one is signed by one Abu ʿl-Fatḥ, *muwaqqit* at the Sulṭān Mehmet Khān Mosque in Istanbul and dated 1161 H [= 1748]. The main semi-circular equatorial scale is cluttered with two other sundials, both superfluous. Around the base of the instrument is a scheme of sacred geography based on the folk tradition. See n. **V**-9:13. [Photo courtesy of Christie's of London.]

al-Wafāʾī became the owner of the universal astrolabe of Ibn al-Sarrāj (**5**) and wrote a treatise on the use of all its component parts, extant in two manuscripts in Dublin and Princeton. His mastery of the various functions of the instrument is impressive.[50]

al-Wafāʾī also compiled a set of auxiliary tables for solving the problems of spherical astronomy for a range of latitudes.[51] The functions tabulated are:

$$G(\phi,\theta) = R^3 / [\ \text{Cos}\ \phi\ \text{Cos}\ \theta\ ] \quad \text{and} \quad V(x) = \text{arc Vers}(x)\ ,$$

To find $t(h,\Delta,\phi)$ one first computes the quantity:

$$H'(h,\Delta,\phi) = \text{Sin}\ H(\Delta,\phi) - \text{Sin}\ h = \text{Sin}\ (\ \bar{\phi} + \Delta\ ) - \text{Sin}\ h$$

and then uses the tables to find:

$$t(h,\Delta,\phi) = V\ \{\ H'(h,\Delta,\phi) \bullet G(\phi,\Delta)\ \}\ .$$

---

50 See the text to n. 19 above.

51 **I-9.9**.

al-Wafāʾī's tables survive in a unique manuscript in the Vatican and were apparently not widely known. A commentary on the tables was written by the contemporary Cairo astronomer al-Karādīsī, and this is likewise extant in a unique manuscript.[52]

By al-Wafāʾī's time, sets of prayer-tables were available for Cairo, Damascus, and Jerusalem. al-Wafāʾī is stated to be the author of a set of prayer-tables including these three sets as well as those for each degree of latitude from 15° (Yemen) to 41° (Istanbul).[53] These universal prayer-tables survive in a unique manuscript in Princeton.

Chapter 15 Ibn al-Mushrif

Abu Bakr Ibn al-Mushrif was an astronomer in Cairo in the first half of the 15th century. He compiled a set of auxiliary tables[54] displaying the functions:

$$B(\phi,\Delta) = \mathrm{Cos}\ \phi\ \mathrm{Cos}\ \Delta\ /\ R, \quad H'(H,h) = \mathrm{Sin}\ H - \mathrm{Sin}\ h$$

$$V'(x,y) = \mathrm{arc\ Vers}\ \{\ R \cdot y\ /\ x\ \}$$

With these the hour-angle can be found using:

$$t(h,\Delta,\phi) = V'\ \{\ B(\phi,\Delta)\ ,\ H'(H,h)\ \}\ .$$

Fig. 16.1: An extract from the tables for astrolabe construction by al-Aqfahsī, serving latitudes 28°-29° south. No astrolabes with mixed north-south projections (**X-5.1**) are known from Mamluk Egypt. [From MS Berlin 5692, fol. 12r, courtesy of the Deutsche Staatsbibliothek, Preußischer Kulturbesitz.]

---

[52] On al-Karādīsī see Suter, *MAA*, no. 442, and *Cairo ENL Survey*, no. C90.
[53] **II-8.1**.
[54] **I-9.8**.

Recensions of these tables were prepared by Sūdūn al-Bashtakī and Ibn Abi 'l-Fatḥ al-Ṣūfī (**17**). The original tables and each of the recensions survive in unique manuscripts, all preserved in Cairo.

Chapter 16 al-Aqfahsī

The mid-15th-century Cairo astronomer al-Aqfahsī[55] compiled a set of tables for constructing astrolabe plates for an extensive range of southern latitudes. These tables are extant in a unique manuscript in Berlin: see **Fig. 16.1**.

Chapter 17 Ibn Abi 'l-Fatḥ al-Ṣūfī

Ibn Abi 'l-Fatḥ al-Ṣūfī was one of the leading astronomers in Cairo in the second half of the 15th century.[56] His major work was a *zīj* for Cairo based mainly on the *zīj* of Ulugh Beg for Samarqand.[57] His contributions to spherical astronomy include an extensive set of auxiliary tables for solving problems of spherical astronomy, extant in a unique manuscript preserved in Oxford.[58] These tables display the functions:

$$C(\xi,\eta) = \text{Sin}\ \xi\ \text{Sin}\ \eta\ /\ R \quad \text{and} \quad B(\xi,\eta) = \text{Cos}\ \xi\ \text{Cos}\ \eta\ /\ R\ ,$$

$$S(x,y) = \text{arc Cos}\ \{\ R \cdot x\ /\ y\ \}\ ,$$

with which, for example,

$$t(h,\Delta,\phi) = S\ \{\ [\text{Sin}\ h - C(\phi,\Delta)] \cdot B(\phi,\Delta)\ \}\ .$$

They represent no improvement upon the more versatile tables of al-Khalīlī and Ibn Mushrif (nos. 9 and 15).

al-Ṣūfī wrote several short treatises on sundial theory, one of which contains a table for facilitating the construction of hour-angle lines on any vertical dial.[59] The table displays the solar azimuth at the equinoxes as a function of time, for all latitudes. Using these azimuth values, one can determine the orientation of the hour-angle lines on a vertical dial inclined at any angle to the meridian for any latitude. The function tabulated by al-Ṣūfī is:

$$a(\text{Sin}\ \phi,\ T) = \text{arc Tan}\ (\text{Sin}\ \phi \cdot \text{Tan}\ T\ /\ R)$$

Tables of a(T) for $\phi = 30°$ (Cairo) exist in earlier Egyptian sources.

Amongst al-Ṣūfī's numerous works on instruments are treatises on the *mujannaḥ* quadrant of Ibn al-Sarrāj (**5**), and universal ones such as the "box of sapphires" of Ibn al-Shāṭir (**8**)[60] and each of the two equatorial armillas of al-Wafā'ī (**14**).

---

55 On al-Aqfahsī see Suter, *MAA*, no. 440, and *Cairo ENL Survey*, no. C68.
56 On al-Ṣūfī see Suter, *MAA*, no. 447, and *Cairo ENL Survey*, no. C98.
57 *Cf.* Kennedy, "*Zīj* Survey", nos. 12 and 37.
58 **I-9.10**.
59 **I-5.5.2**.
60 His treatise is published with translation and commentary in Janin & King, "Ibn al-Shāṭir's Compendium".

Chapter 18 Anonymous

In 1966 Emmanuel Poulle cited a 15th-century European report that the instrument known as the sexagenarium was of contemporaneous Egyptian provenance.[61] The sexagenarium is a quadrant with a trigonometric grid on one side, and a series of circular scales for computing mean longitudes and anomalies for the sun, moon, and planets on the other. The trigonometric quadrant serves not only to provide numerical solutions to problems of spherical astronomy but also to compute equations of the sun, moon, and planets. (The latter function serves only to challenge the wits of the user, because auxiliary tables for computing the equations obviate the need for such tedious calculations.) While the trigonometric quadrant and associated instructions for performing computations in spherical astronomy were well-known in 15th-century Cairo,[62] no such planetary dials are known from there. However, such a dial was described by al-Kāshī (Samarqand, *ca.* 1425)[63] although, as far as I know, he did not write on the trigonometric quadrant and his planetary dial was not used by later Muslim astronomers.

Chapter 19 Ibn Abi 'l-Khayr al-Ḥusnī

Muhammad ibn Abi 'l-Khayr al-Ḥusnī worked in Cairo in the 16th century.[64] He was the author of several astronomical works, including a set of tables displaying the duration of morning and evening twilight at the equinoxes and solstices for each half climate. These tables are extant in a unique manuscript in Cairo.[65]

Chapter 20 ʿAbd al-Qādir al-Minūfī

One last example of the Egyptian astronomers' penchant for universal solutions is a table prepared in the year 1567 by the Cairo astronomer ʿAbd al-Qādir al-Minūfī.[66] This table displays the amount to be added to the solar day-circle to account for the effect of refraction at the horizon. Values of this function ΔD are displayed for each zodiacal sign and each degree of terrestrial latitude from 1° to 60°, and are based on the notion that ΔD increases linearly with latitude and assumes the values for latitude 30° (Cairo) attributed to Ibn Yūnus.[67] al-Minūfī's table survives in a unique manuscript in Istanbul but the underlying theory is attested in several later writings of Syrian and Turkish provenance. Some of al-Minūfī's writings indicate that he was slightly deranged.

---

61 Poulle, "Sexagenarium", p. 130.
62 Schmalzl, *Geschichte des Quadranten*, pp. 71-99, and King, "al-Khalīlī's Qibla Table", pp. 111-115.
63 Kennedy, *Planetary Equatorium of al-Kāshī*, pp. 189-190.
64 On al-Ḥusnī see Suter, *MAA*, no. 511, and *Cairo ENL Survey*, no. C124.
65 **II-8.3** (illustrated).
66 On ʿAbd al-Qādir al-Minūfī see Suter, *MAA*, no. 479, and *Cairo ENL Survey*, no. C121.
67 **II-8.2**.

Concluding remarks

The astronomers of Egypt and Syria during the Ottoman period, that is from the 16th century onward, did not share the interests of their Mamluk predecessors in universal solutions. To be sure, the manuscripts of the Mamluk works were copied during this period and presumably used, but I know of no serious activity of any originality in this area. Rather, the astronomers in Ottoman Egypt and Syria, and also Istanbul, which became the centre of astronomical timekeeping in the 16th century, preferred to compile tables for specific latitudes and use instruments with the same limitations. Interest in astronomy did not subside in Ottoman times, but creative activity did. The candle was burning low, and another one was burning more brightly in the West.

Part VIIa

The orientation of medieval Islamic religious architecture and cities

Dedicated to Eleanor Sims and Ernst Grube

ACKNOWLEDGEMENTS AND NOTES ON THIS VERSION

This study is dedicated to two long-time friends, both art historians. I first met Eleanor Sims in Philadelphia in the (memorable) summer of 1969, and Ernst Grube a few years later. Our paths have crossed many times since, mainly in New York and London but also in Frankfurt and at conferences all over the place. I have long admired their indefatigable interest in Islamic Studies generally and Islamic Art in particular. And it was always a pleasure staying with them in London, enjoying their invigorating company, browsing in their extensive library, and all so close to points of exit from London. They have done much to encourage me in my investigations of orientations.

The first version of this study was published in *Journal for the History of Astronomy* 26 (1995), pp. 253-274. In the present version I have expanded the introductory sections on the determination of the qibla by folk astronomy and mathematical astronomy, not least to better serve the reader of **I-II** and **V-VI**. The interested reader will find more information on the qibla in my *World-Maps for Finding the Direction and Distance to Mecca: Innovation and Tradition in Islamic Science*, Leiden: E. J. Brill, and London: Al-Furqan Islamic Heritage Foundation, 1999.

TABLE OF CONTENTS

CHAPTER 1

INTRODUCTORY REMARKS

> "We [God] see thee [Man] turning thy face towards the heavens. Now we shall make thee face a qibla that will please thee. Turn then thy face towards the Sacred Mosque: wherever ye [all Mankind] are, turn your faces towards it ... ." *Qur'ān* II.144.

> "The Kaʿba is the qibla for the Sacred Mosque, the Sacred Mosque is the qibla for the sacred precincts (of Mecca and its environs), and the sacred precincts are the qibla for the inhabitants of the whole world from where the sun rises to where it sets." Ibn al-Qāṣṣ (*ca.* 975), *Kitāb Dalā'il al-qibla.*

The sacred direction in Islam, called the qibla in all languages of the Islamic commonwealth, has been of fundamental importance in the life of every Muslim for close to one and a half millennia.[1] The *Qur'ān*, regarded by Muslims as the Word of God, and Islamic tradition between them prescribe the performance of both prayer and a variety of other ritual acts towards the sacred Kaʿba in Mecca.[2] Since Muslims pray towards the qibla, and since the *miḥrāb* or prayer-niche in mosques symbolizes the direction of the qibla, mosques are qibla-oriented. However, the actual orientation of medieval Islamic religious architecture constitutes a problem that has long concerned Muslim scholars and for a much shorter period intrigued Western historians.[3]

Muslim scientists were able – within the limitations of medieval geography – to determine the direction to Mecca of any locality correctly to within a few minutes. Thus one might be forgiven for anticipating that most of this architecture would face Mecca more or less in the way we moderns think it should. Yet a glance at a plan of any medieval Islamic city gives the impression that some of the religious architecture does not face Mecca; often, in fact, one can identify in one and the same city a variety of directions used for mosque orientations. Just how far colleagues in the history of Islamic architecture misunderstand this situation and its implications is revealed by a remark by one of the most distinguished: "The orientation of mihrabs may often be faulty". They are not faulty: they simply do not always correspond to what we would expect. Indeed, a misunderstanding of certain orientations has misled two modern scholars in the enthusiasm of their youth to conclusions that are absurd. For they took the fact that the earliest mosques in al-ʿIrāq and Egypt faced a point in Northern Arabia as evidence confirming the existence of a "Hagaritic shrine" towards which these mosques must have been oriented.[4] In fact the mosques in question were oriented towards summer sunset and

---

[1] See the article "Ḳibla, i: Ritual and legal aspects" by A. J. Wensinck in EI_2.

[2] On the Kaʿba see the article "Kaʿba" by A. J. Wensinck in EI_1 as well as the revised version by *idem* and J. Jomier in EI_2 (in which some material from the former has been omitted). A useful recent study, which does not mention the orientation of the edifice, is Rubin, "The Kaʿba".

[3] Thus, for example, Annemarie Schimmel, in her paper "The Sacred Geography of Islam" (1991), mentions the qibla only in passing (p. 164), commenting that although "every mosque has to face exactly the qibla, ... it seems that the direction was not everywhere perfectly correct".

[4] See Patricia Crone and Michael Cook, *Hagarism – The Making of the Islamic World*, Cambridge, *etc.*:

Fig. 1.1: Muslims facing the Kaʿba during the pilgrimage. [Source unknown.]

sunrise, respectively: these, as we now know, were the directions used for the qibla. In other words, the mosques in both localities face the Kaʿba.

One principal mistake in modern scholarship has been to think of the qibla as the local direction of Mecca. It is the direction of a specific edifice, the Kaʿba, in the heart of that city. We need, then, to understand how Muslims other than the scientists thought that they could face a distant building that they could not see. Another mistake has been to overlook the fact that medieval geographical coordinates do not correspond to modern coordinates. Even the most capable medieval scientists could not have derived the modern, "accurate" qiblas, for these are computed using modern geographical coordinates. In fact, modern qibla-values are irrelevant to any discussion of the orientation of medieval architecture. And yet another mistake has been to ignore the existence of a second tradition in medieval Islamic astronomy, that is, folk astronomy.[5] In this tradition, which flourished alongside the mathematical one, not least because it was favoured by the scholars of the religious law, there is no advanced mathematics, and the qibla was determined by completely different techniques of a non-mathematical nature.

The state of documentation of orientations of medieval Islamic architecture is in general abysmal: most authors on Islamic architecture to this day do not take the trouble to measure the orientations, and if they do mark their plans of mosques with a direction, they usually do not state how the orientation was measured.[6] And, alas, to measure the orientation of a mosque

---

Cambridge University Press, 1977, pp. 23-24, and *eadem*, *Meccan Trade and the Rise of Islam*, Princeton, N.J.: Princeton University Press, 1987, p. 198. The authors have since recognized their error in interpreting the two qiblas in this way.

[5] See King, "Islamic Folk Astronomy", and the articles "Maṭlaʿ" and "Rīḥ" in EI_2, which supplement material surveyed in the articles "Anwāʾ" and "Layl and nahār" by Charles Pellat.

[6] In the early 1980s I conducted an armchair survey of over 1000 published plans of mosques and other religious architecture. Other than being a useful educational experience, this proved to be mainly a waste of time as far as my immediate purpose was concerned, the information on orientations which I gathered being so unreliable that nothing useful could be done with it. See below for some of the few exceptions.

Fig. 1.2: The Ka'ba and the Sacred Mosque around it. This view from the south was taken in 1889. [Courtesy of the Bildarchiv Preußischer Kulturbesitz, Berlin.]

properly one cannot simply stand in front of the qibla-wall with a compass and take a reading. Of concern here are not only mosques and madrasas, but also cemeteries (the dead are buried so as to rise facing the qibla on the Day of Resurrection), occasional private houses and even facilities for performing bodily functions (necessarily perpendicular to the qibla). In addition, as we shall see, the qibla, appropriately defined according to some medieval criterion, can dominate the plan of parts of a city or even of an entire city. Since we do not yet have a reliable corpus of basic data on the orientation of the monuments themselves, we have been fortunate to find some relevant textual sources. These will eventually enable us to better understand the orientiations when they are measured properly. (It is but a small consolation that the documentation of orientations of Christian and Jewish architecture in the Near East and also in Europe is no better, and our understanding of them at a yet more elementary level.)

These new sources that have been studied reveal why different directions were adopted for the qibla in one and the same place, and why particular interest groups in these places favoured one or the other qibla at different times. The purpose of this study is to summarize the results of some recent research and to point to the vast amount of work waiting to be done. It was originally inspired by a paper published in 1990 by Michael Bonine,[7] in which he considered

[7] Bonine, "Sacred Direction and City Structure in Morocco", and *idem*, "The Morphogenesis of Iranian Cities", for an analogous study of another region of the Islamic world.

the orientations of religious architecture and of entire cities in the medieval Maghrib (N. W. Africa). Bonine was not only familiar with the recent research but added his own insights as a geographer, taking the trouble to measure the orientations of numerous edifices in different urban centres, and also invoking the layout of the land in question, that is, its slope. His results were, as he stated, preliminary, and more recently much more has been discovered about orientations in the Maghrib, as well as in al-Andalus (see below).

CHAPTER 2

THE DETERMINATION OF THE QIBLA BY FOLK ASTRONOMY

Already before the advent of Islam the four corners of the Kaʿba were associated with four main regions of the surrounding world: Syria, al-ʿIrāq, the Yemen and "the West". How does one face a corner of an edifice or a segment of a wall when one is so far from that edifice that one can no longer see it? The answer, according to the legal scholars of Islam if not the Muslim astronomers (see below), is straightforward: one stands in the same direction one would stand if one were in front of the Kaʿba itself, at the part of the edifice associated with one's own region. Now the Arabs of Mecca in the 7th century knew that the Kaʿba was astronomically aligned, that is, they knew that if one stood in front of any of its four walls one would be facing significant astronomical horizon phenomena – see **Figs. 2.1** and **2.2.**[1] The major axis of the rectangular base is aligned towards the rising of Canopus, and the minor axis towards summer sunrise and winter sunset, these directions being roughly perpendicular at the latitude of Mecca. (More precisely the minor axis is oriented towards the furthest limits of the moon at the solstices, but lunar alignments are not known from Islamic architecture.) The first generation of Muslims also knew that the Prophet Muhammad, when he was in Medina, north of Mecca, had prayed due south.

The first mosques, built from al-Andalus to Central Asia within a few decades of the death of the Prophet, were oriented according to these two main principles. Either their orientations were in some way related to the orientation of the Kaʿba itself or they were oriented due south, in imitation of the practice of the Prophet when he was in Medina. In other words, we should not be surprised to find early mosques in the Islamic West facing the same direction as the main axis of the Kaʿba, that is, towards the rising of Canopus (or towards winter sunrise), or mosques in the Islamic East facing the direction of the minor axis of the Kaʿba, that is, towards winter sunset, or mosques from al-Andalus to Central Asia facing due south – see **Fig. 2.3**. Certain mosques in such locations as Jerusalem and Damascus were aligned with pre-existent

---

[1] Hawkins & King, "Orientation of the Kaʿba", and King, "Astronomical Alignments in Medieval Islamic Religious Architecture" (both repr. in King, *Studies*, C-XII and XIII).

Robert G. Hoyland (*Seeing Islam as Others Saw it*, pp. 572-573) cannot accept the relevance of the kind of medieval Arabic textual materials I have been using for the past 30 years to the early period of Islam. He asserts, for example, that the earliest Muslims would not have known which main regions of the world (meaning Syria, al-ʿIrāq, Yemen and "the West") were to be associated with the walls of the Kaʿba, and implies that the orientation of the Kaʿba might have been changed during one of the several demolitions and rebuildings of the edifice. Further, he sees the associations of the winds with the Kaʿba as later devlopments. But the sacred cosmology attested in a series of works over the centuries, preserved for us only by al-Suyūṭī in the 16th century (edited with an imperfect translation – *ḥajar* (stone) for *ḥijr* ("apse" wall), to cite just one example – in Heinen, *Islamic Cosmology*, which Hoyland cites), is based mainly on 7th-century sources, whose authenticity is not in question (because they have no religio-political implications and because nobody claimed the opposite).

Fig. 2.1: The orientation of the Kaʿba is described in this late-13th-century Yemeni treatise on folk astronomy by Muḥammad ibn Abī Bakr al-Fārisī. The 11th chapter deals with the determination of the qibla by the methods of folk astronomy and includes a statement about the winds and their astronomically-defined limits, which correspond to the alignments of the walls of the Kaʿba. There is also a note about the luni-solar alignment of the Kaʿba. The remainder of this chapter deals with the 12 regions of the world about the Kaʿba and their astronomically-defined qiblas. For more details see Hawkins & King, "Orientation of the Kaʿba". [From MS Milan Ambrosiana 73 Sup., unfoliated, courtesy of the Biblioteca Ambrosiana.]

architecture; this was viewed without criticism because the resultant qibla-directions (which chanced to be in both cases due south) were deemed legally acceptable.[2]

In the 10th century, if not earlier, a sacred geography was developed in which every region of the world was associated with a segment of the perimeter of the Kaʿba and the qibla was more precisely defined, albeit still in terms of the cardinal directions or astronomical horizon phenomena – see **Figs. 2.4-6** (also **Figs. V-12.3** and **VIIc-9.3-4**). Early schemes of a very simple variety are attested, but the first detailed schemes are to be associated with the legal scholar Ibn Surāqa, who flourished about the year 1000. Several different schemes in this tradition were proposed over later centuries, with the result that between the various sources different directions were proposed for each locality. Islamic sacred geography constitutes a fascinating chapter in the history of Islamic folk- or ethno-astronomy, a practical, popular astronomy

[2] Certain mosques in al-Andalus and India, which do not appear to face the qibla, have been declared to have been built on pre-existent architecture. Most of these pronouncements, which are not worth citing, are rendered void by the findings presented here.

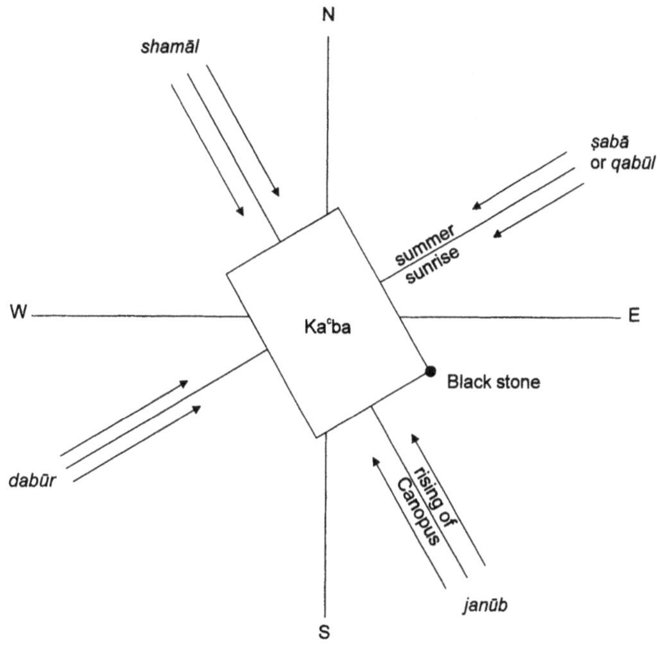

Fig. 2.2: The orientation of the rectangular base of the Kaʿba with its major axis towards the rising of Canopus and its minor axis towards the rising sun at the summer solstice, as recorded in various medieval sources. The "cardinal" winds are also shown, each one striking a wall of the Kaʿba head-on.

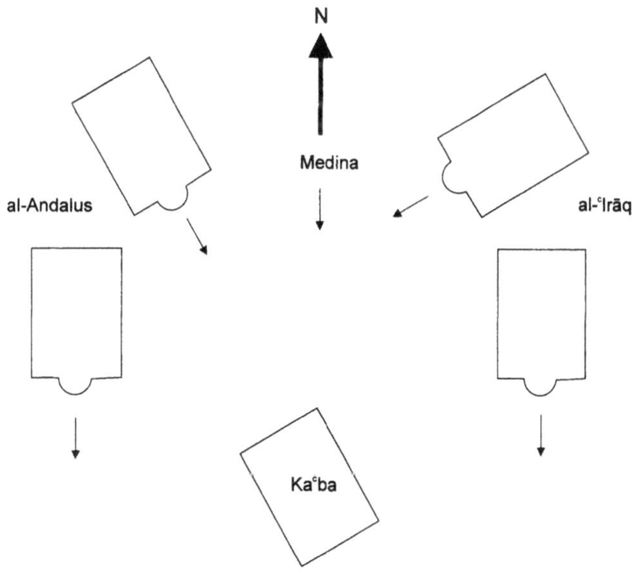

Fig. 2.3: Early mosques in al-Andalus and the province of al-ʿIrāq were oriented so that their qibla-walls were "parallel" to the appropriate wall of the Kaʿba, or so that they were facing the same direction adopted by the Prophet in Medina, namely, due south.

Fig. 2.4: A scheme of sacred geography with eight sectors of the world around the Kaʿba. The scheme is found only in a 15th-century manuscript of a geographical treatise by the 10th-century Abū ʿAbdallāh al-Muqaddasī (but not in earlier copies) and is clearly even earlier in origin. The information is extremely corrupt, in some sectors the place-names have been omitted and one sector has been left blank. The qiblas are defined in terms of the Pole Star, and the risings and settings of Canopus, Altair, Vega, the Pleiades, and the stars of the Great Bear. In two of the sectors there are remnants of the information on the size of the regions, as presented in the geographical sections of the work of the 8th-century astronomer al-Fazārī. For more information see King, *Sacred Geography of Islam*. [From MS Berlin Ahlwardt 6034 (= Sprenger 5), fol. 34r, courtesy of the Deutsche Staatsbibliothek (Preussischer Kulturbesitz).]

devoid of any underlying theory and unencumbered by any calculation beyond the most trivial arithmetical procedures. All available texts relating to sacred geography, some 30 in number and presenting some 20 different schemes, have been edited and prepared for publication; a summary is already available.[3] The different qiblas proposed by the legal scholars for each of the major regions of the Islamic world have been identified; they all involve the cardinal directions or the risings and settings of the sun or various fixed stars. Sometimes the winds are invoked for identifying directions, and here it should be borne in mind that the limits of the winds in Arab folklore were defined by astronomical risings and settings.[4]

The legal scholars were equal to the task of confronting these palettes of different directions for each region: they pronounced that whilst facing the actual direction (*ʿayn*) of the Kaʿba was optimal, it was legally acceptable to use as qibla any direction (*jiha*) within the limits of

[3] King, *The Sacred Geography of Islam*, to appear, summarized in the article "Makka. iv. As the centre of the world" in EI_2, repr. in King, *Studies*, C-X. See also Schmidl, "Qibla Schemes in a Medieval Yemeni Text".

[4] See my EI_2 article "Maṭlaʿ" (repr. in King, *Studies*, C-XI) and the article "Rīḥ" by Miquel Forcada, and also Schmidl, "Qibla und Winde".

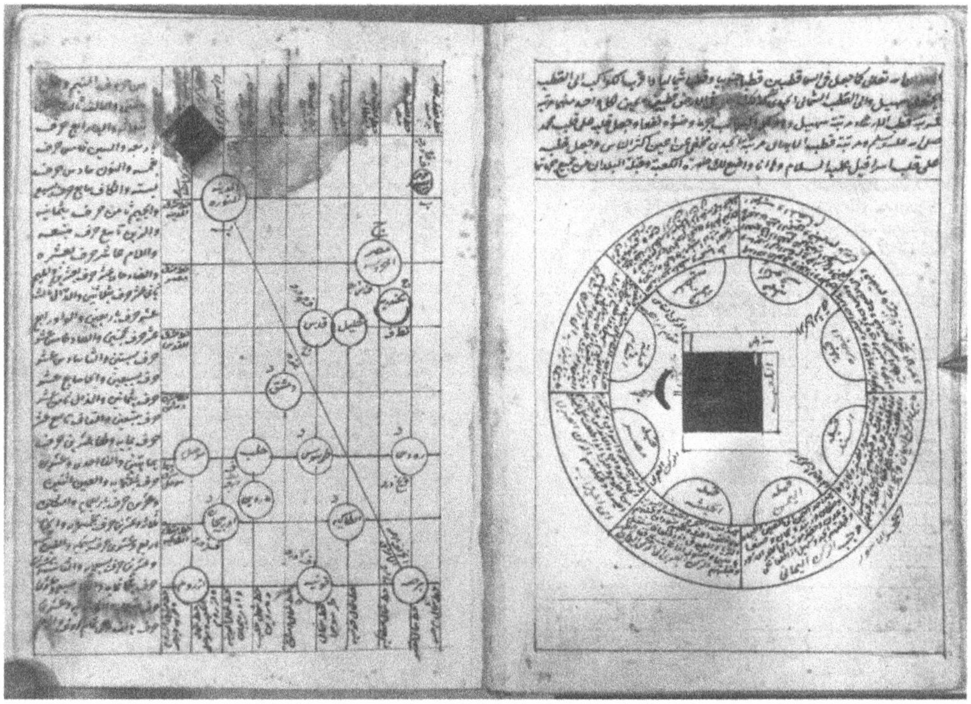

Fig. 2.5: Another early scheme of sacred geography in eight sectors found only in an 18th-century Ottoman Egyptian manuscript. I would date it to the 9th or 10th century, for it is independent of the main tradition associated with Ibn Surāqa and found elsewhere only in a Persian astrological work. On the left is a simplistic cartographic grid with various localities indicated, as well as the Kaʿba in the upper left corner, appropriately shown with its rectangular base inclined to the local meridian. [From MS Cairo ṬJ 811,7, fols. 60v-61r, courtesy of the Egyptian National Library.]

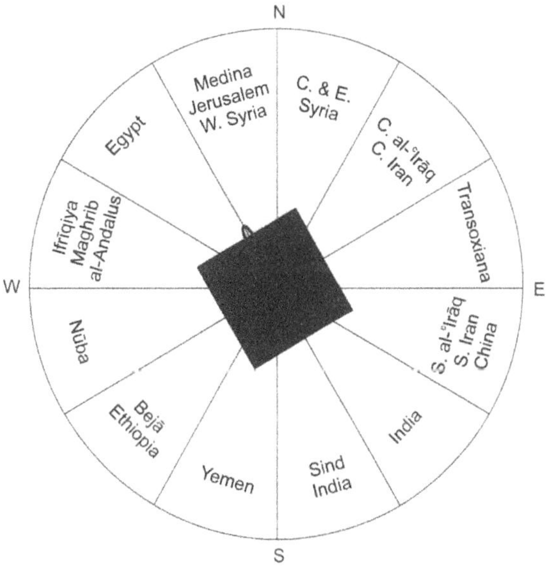

Fig. 2.6: A simplified version of the 12-sector scheme of sacred geography found in some manuscripts of the popular geographical work *Āthār al-bilād* by al-Qazwīnī (d. 1283). See also **Fig. VIIb-9.4**.

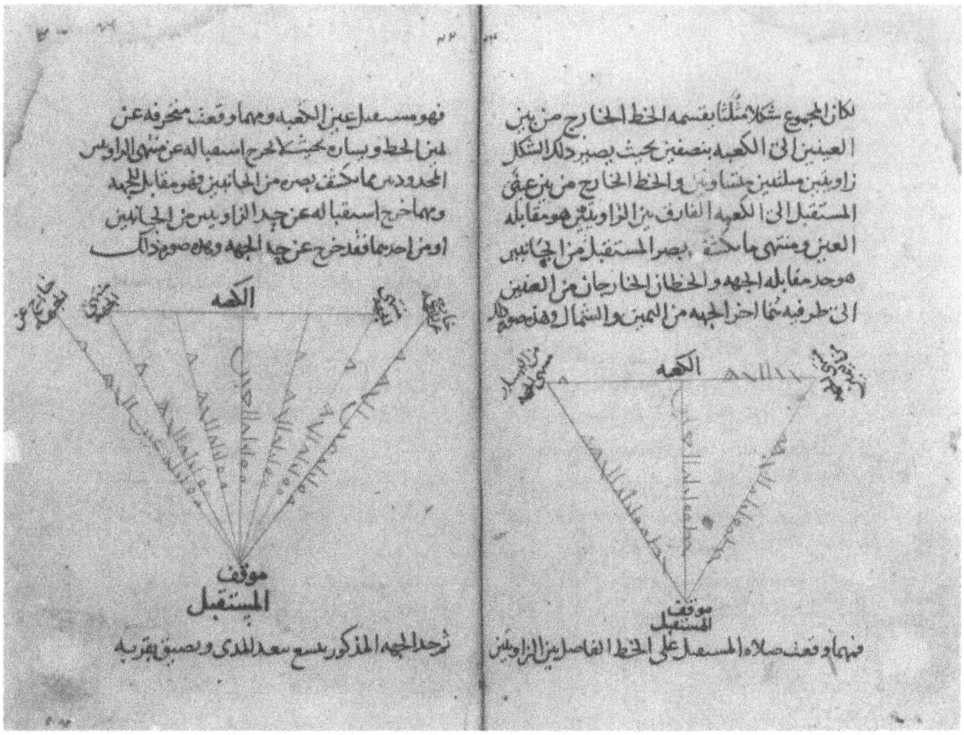

Fig. 2.7: Diagrams explaining the notions of "actual" qibla (on the right) and "legally permissible" qiblas (on the left) in the treatise on the qibla by the 12th-century Egyptian legal scholar al-Dimyāṭī. The person praying is at the bottom apex of the triangles, the Kaʿba at the foot of the perpendicular to the opposite side. [From MS Oxford Bodleian Marsh 592, fols. 23v-24r, courtesy of the Bodleian Library.]

one's vision, that is, about a quadrant of the horizon centred on the "actual" direction – see **Fig. 2.7**. In other words, for example, if the qibla is somewhere in the south-east, it is permissible to pray due east or due south.

CHAPTER 3

THE DETERMINATION OF THE QIBLA AS A PROBLEM OF MATHEMATICAL GEOGRAPHY

> "{The} table {for finding the qibla} prepared by the fourteenth-century Syrian astronomer al-Khalīlī ... is the most sophisticated trigonometric table known to me from the entire medieval period." D. A. King, "al-Khalīlī's Qibla Table" (1975), p. 82.
>
> "Now I can claim only that {al-Khalīlī's qibla-table} is one of several highly sophisticated trigonometric tables compiled by Muslim scientists." *Idem*, "Earliest Qibla Methods and Tables" (1986), p. 133.

Already in the 8th century that the Muslims had come in contact with the very different notions of mathematical geography, to which we now turn. Muslim astronomers defined the qibla as the direction of the great circle joining a specific locality to Mecca, measured as an angle to the local meridian. There is already a substantial modern literature on various medieval mathematical methods, tables and instruments for finding the qibla.[1]

From the 9th century onwards, they calculated the direction of Mecca for various localities. These calculations required knowledge of latitudes and longitudes, originally adopted from Ptolemy's *Geography*, and also involved the application of complicated trigonometric formulae or geometrical constructions. Most Islamic *zīj*es contain a chapter on the determination of the qibla by such procedures. Independent treatises dealing only with qibla determination were also compiled. The earliest determinations of the qibla by mathematical means, dating from the 9th – if not the 8th – century, were approximate, but were nevertheless adequate to within a degree or two for localities as far from the meridian of Mecca as Egypt and Iran.

Fig. 3.1 illustrates the problem of determining the qibla on the terrestrial sphere. AB is the equator and P is the north pole. X represents a given locality and M represents Mecca. The latitudes of the two localities are measured by XA (ϕ) and MB (ϕ_M), and their longitude difference by AB (ΔL). The arc XM defines the qibla, its direction being measured by the angle AXM (q). To determine q from the two variables ϕ and ΔL given ϕ_M is clearly a problem of mathematical geography which can most conveniently be solved by spherical trigonometry, but it is easily transformed into a problem of spherical astronomy: see **Fig. 3.2**. The problem is by no means simple, and the solutions proposed by Muslim astronomers between the 8th and the 14th century constitute an important chapter in the history of applied mathematics.

These early procedures were cartographic in the sense that they were based on notions – I hesitate to use the term projections – of representing of the terrestrial sphere in two dimensions. One procedure involved simply joining the positions of any locality and Mecca

[1] See various articles in Schoy, *Beiträge*; Kennedy *et al.*, *Studies*; and King, *Studies*, A-XIII and C-XIV; as well as al-Bīrūnī's *Taḥdīd* (the most important surviving medieval work on mathematical geography), and the translation by Jamil Ali and commentary by E. S. Kennedy. A brief summary is in the article "Ḳibla. ii. Astronomical Aspects" in EI_2, reprinted with corrections of many printing errors in King, *Studies*, C-IX, and a more detailed overview is in King, *Mecca-Centred World-Maps*, pp. 47-193.

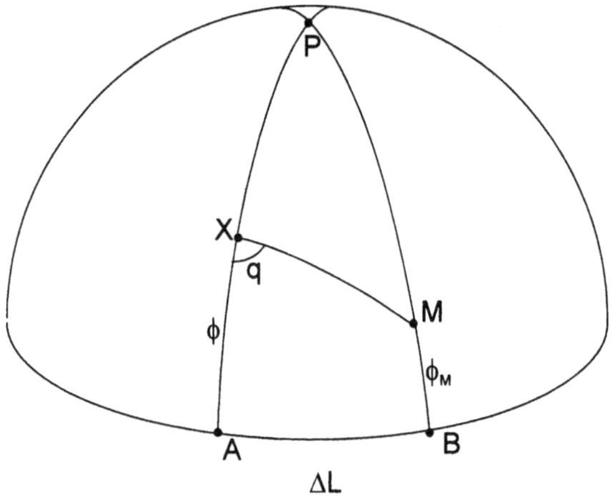

Fig. 3.1: In this diagram AB represents the equator and P the North Pole. It is required to find the qibla or direction of Mecca M from any locality X. We denote the qibla measured from the local meridian by q = ∠ AXM and the distance by d = XM. The latitudes of both localities are represented by ϕ_M = MB and ϕ = XA and their longitude difference by ΔL = AB.

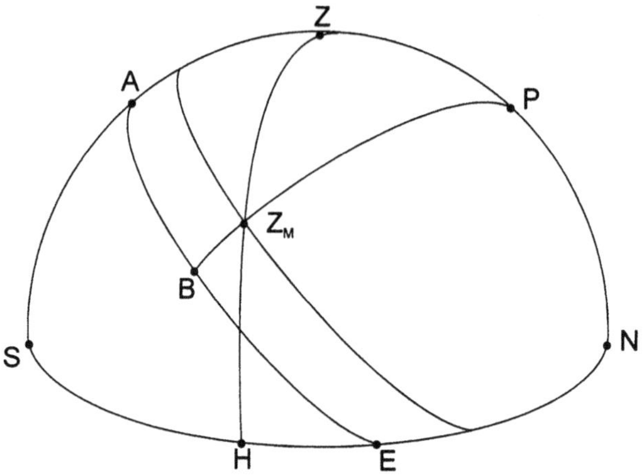

Fig. 3.2: The qibla-problem transferred to the celestial sphere. It is required to find the azimuth of the zenith of Mecca Z_M on the horizon of a locality with latitude ϕ. Z_M is a point having declination ϕ_M and hour-angle ΔL.

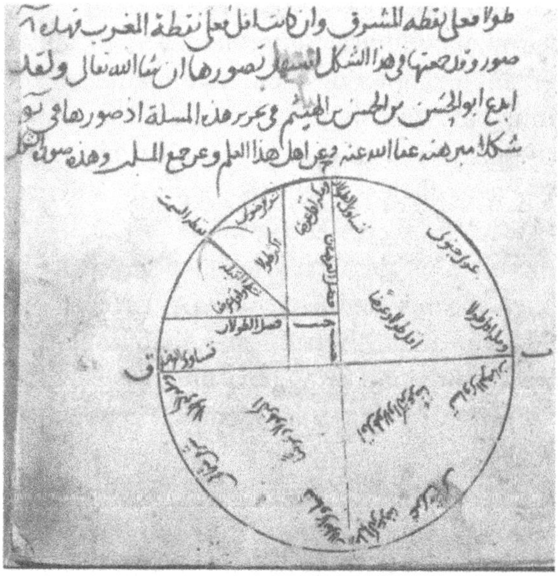

Fig. 3.3: A diagram for finding the qibla by this standard approximate method in a treatise on the use of the sine quadrant by Abū Bakr ibn Shihāb al-Dawla al-Miṣrī al-Karakī, *muwaqqit* at the al-Aqṣā Mosque in Jerusalem in the late 13th or early 14th century. al-Karakī does not specifically state that the method is approximate, but he does mention that the early-11th-century scholar Ibn al-Haytham had presented 16 different diagrams to display the various cases (for an exact solution). [From MS Dublin CB 4833, fol. 8v, courtesy of the Chester Beatty Library.]

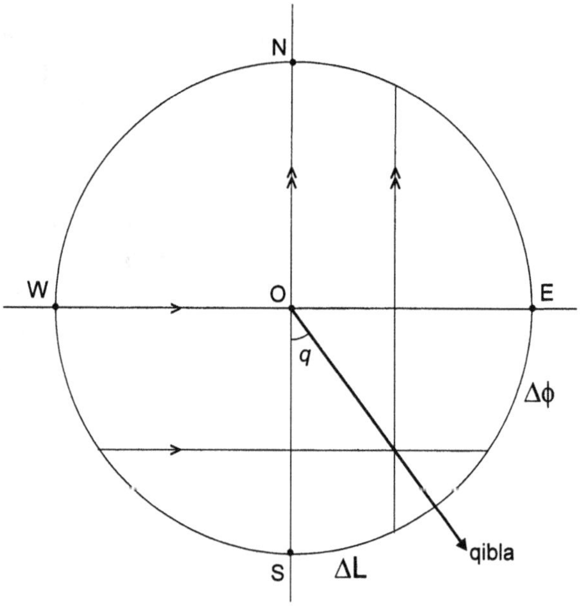

Fig. 3.4: The standard approximate method for finding the qibla, used for over a millennium.

on a map with a square or rectangular grid (see the example in **Fig. 2.5**). Another involved marking the latitudes and longitudes of a locality and of Mecca on a circular diagram. Latitudes were marked vertically and the 180° longitude span of the inhabited world was marked horizontally. The qibla could then be found by investigating where the line joining the two points cuts the circumference of the circle: see **Figs. 3.3-4**. This geometrical procedure, which I label "the standard approximate procedure", is equivalent to the formula:

$$q(\Delta L,\Delta\phi) \quad \approx \quad \text{arc Tan } \{ \text{ R Sin } \Delta L \ / \text{ Sin } \Delta\phi \ ) \ .$$

Other approximate cartographic procedures were also devised but were not widely used.

In the 9th and 10th centuries, more sophisticated, accurate methods were derived by means of plane or solid geometry (**Fig. 3.5**) or by the techniques of spherical trigonometry. Usually the problem was transferred to the celestial sphere and solved as one of spherical astronomy. Thus it was required to find the azimuth a of the zenith of Mecca with respect to the local horizon, this zenith being defined by $\delta = \phi_M$ and $t = \Delta L$. The solutions found by the medieval astronomers are ultimately equivalent to an application of the modern formula:

$$q = \text{arc cot } \{ \ [\ \sin\phi \cos\Delta L - \cos\phi \tan\phi_M \ ] \ / \ \sin\Delta L \ \} \ .$$

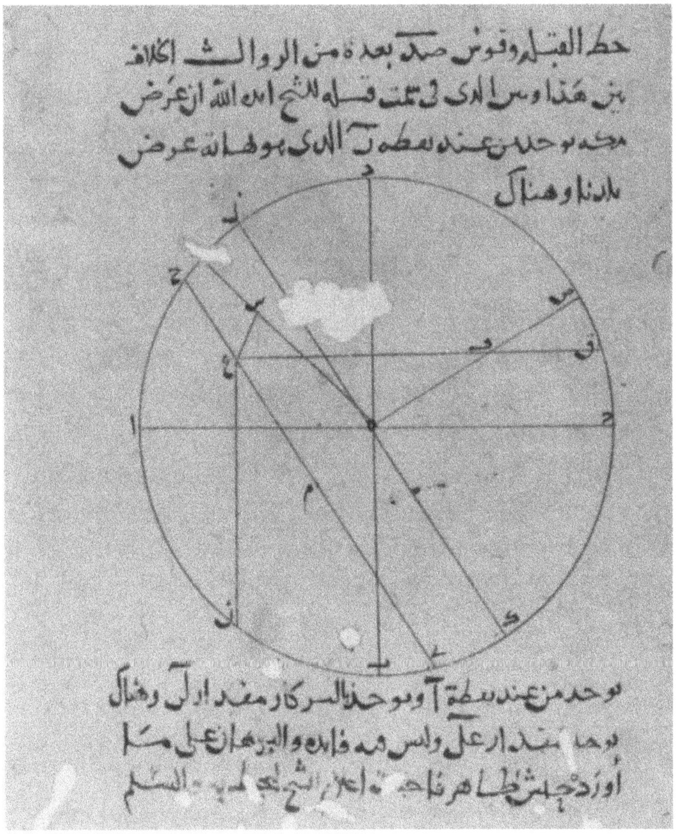

Fig. 3.5: One of the solutions to the qibla problem devised by the 9th-century astronomer Ḥabash al-Ḥāsib involved this two-dimensional geometrical construction based on an analemma. For more details see King, *Mecca-Centred World-Maps*, pp. 62-63. [From MS Leiden UB Or. 168,16, fol. 137r, courtesy of the Universiteits-bibliothek.]

The qibla problem was first solved by spherical trigonometry in the late 9th century. For example, al-Faḍl ibn Ḥātim al-Nayrīzī proposed using four consecutive applications of the cumbersome Theorem of Menelaos.[2] Later Muslim astronomers, such as Abu 'l-Wafā' al-Būzajānī, also used the Sine Theorem and Tangent Theorem to solve the problem in essentially the same way. Others, such as Ibn Yūnus, preferred the use of projection methods, while others, preferred to use spherical trigonometry. Ibn al-Haytham and Abu 'l-Rayḥān al-Bīrūnī presented both kinds of solutions.

The most popular procedure involving spherical trigonometry was known as the "method of the *zīj*es", and is recorded in several works from the 9th to the 15th century. This solution was developed in the mid-9th century, possibly by Ḥabash al-Ḥāsib, who is the first known author to have proposed it explicitly. It was also favoured by various later authors such as Ibn Yūnus, and a series of authors in the Islamic East from ʿAbd al-Raḥmān al-Ṣūfī to Abu 'l-Wafā' and Abū Sahl al-Qūhī to al-Kāshī (*ca.* 1425), as well as at least two in the Islamic West, namely, Ibn Muʿādh (al-Andalus, *ca.* 1075) and Ibn Isḥāq (Tunis, *ca.* 1225).[3] The procedure was first derived by spherical trigonometry (anonymous 9th-century author, maybe Ḥabash), then by projection methods (Ḥabash), then by spherical trigonometry (various 10th-century scholars). It involves finding three auxiliary quantities:

1) the 'modified longitude difference', $\Delta L'$, defined by:
$$\sin \Delta L' = \sin \Delta L \cos \phi_M \,;$$
2) the 'modified latitude (of Mecca)', ϕ_M', found from:
$$\sin \phi_M' = \sin \phi_M \,/\, \cos \Delta L' \text{ ; and}$$
3) the 'modified latitude difference', $\Delta\phi'$, given by:
$$\Delta\phi' = \phi - \phi_M' \,.$$

With these d and q are determined by:
$$\cos d = \cos \Delta L' \cos \Delta\phi' \quad \text{and} \quad \sin q = \sin \Delta L' \,/\, \sin d \,.$$
What is involved here is essentially a conversion of equatorial/polar coordinates (longitude and latitude) to coordinates centred on a specific point on the sphere (qibla and distance). The last formula underlies the grids on three 17th-century Mecca-centred world-maps which preserve direction and distance to the centre (see below).

Muslim astronomers from the 9th century onwards also computed tables displaying the qibla as a function of terrestrial latitude and longitude. These were based on either approximate formulae, or – in a few cases – on the accurate formula: see **Figs. 3.6-7** and also **II-10.8**. Some eight different tables are known from the manuscript sources, and that of the 14th-century

[2] See now Hogendijk, "al-Nayrīzī on the Qibla at Baghdad", which contains an explanation of the way in which poor al-Nayrīzī derived a qibla-value for Baghdad that was in error by over 15° inspite of the fact that he used correct formulae. (The error results from minor errors of calculation at singularly sensitive places.)

[3] The "method of the *zīj*es" is discussed in Debarnot, "*Zīj* of Ḥabash", p. 49; Berggren, "Four Analemmas for the Qibla", and *idem*, "Origins of the 'Method of the *Zīj*es'"; as well as E. S. Kennedy, commentary to al-Bīrūnī, *Taḥdīd*, pp. 211-214, and Debarnot, *al-Bīrūnī*'s Maqālīd, pp. 260-264, and the summary in King, "Earliest Qibla Methods and Tables", pp. 115-117. The Western Islamic sources are studied in Julio Samsó and Honorino Mielgo, "Ibn Isḥāq al-Tūnisī and Ibn Muʿādh al-Jayyānī on the Qibla" (not published previously), in Samsó, *Studies*, VI. Berggren, in his second article cited above, demonstrates how the method involves essentially the transformation of coordinates from one set of system to another, and shows how it relates to procedures of Greek sundial theory.

Damascus astronomer al-Khalīlī (**II-10.8**) is the most successful of these.[4] It is highly probable that al-Khalīlī was familiar with at least some of these earlier qibla tables. Also, the tables of geographical coordinates that were a feature of every Islamic astronomical handbook occasionally included qibla values for each locality (see **Fig. VIa-6.1**).

The use of the magnetic compass for finding the qibla is outlined already in a 13th-century Yemeni astronomical treatise, and one 14th-century astronomical instrument originally fitted with a compass-needle survives.[5] We also have a Syrian ceramic bowl from *ca.* 1520 for a compass-needle floating on water, but the corrupt geographical data on it indicate that it is a copy of a much earlier instrument of the same kind from Iran (**Fig. X-9.2.1**).[6]

Treatises on the use of instruments such as the astrolabe and the different varieties of quadrants usually also included a chapter on finding the qibla by means of the instrument. From the 14th century onwards, compass boxes were available bearing lists of localities with their qibla directions (**Fig. X-9.3-4**).[7]

Three remarkable world-maps fitted with grids preserving both the direction and the sine of the distance to Mecca at the centre survive from 17th-century Iran (**VIIc** and **X-9.5**).[8] The qibla of any locality is correctly indicated on a circular scale, and the corresponding distance on a non-uniform diametral rule (which actually measures the sine of the angular distance linearly). These two maps are clearly representative of an earlier tradition. The question is: how early? The geographical data on the maps is from 15th-century Central Asia; the basic mathematics underlying the grid was derived in the 9th-century Baghdad. Since 2001 we have two texts – one from 10th-century Baghdad and the other from 11th-century – that mention the solution of the qibla problem by means of conic sections. However, we still do not know who proposed the ingenious use of circular arcs on the grids to approximate the latitude curves, which are in reality ellipses.

Of course, the accuracy (judged by modern criteria) of a qibla direction computed by correct mathematical procedures for a particular locality depended on the accuracy of the available geographical data. Although medieval latitude determinations were usually accurate to within a few minutes, estimates of the difference in longitude between Mecca and various localities might be in error by several degrees, and hence the qibla might also be off the mark by a few degrees.

In short, any Muslim astronomer worth his salt could have determined the qibla of any locality in the Islamic world as accurately as the available geographical data would allow. The qiblas accepted by the medieval astronomers are either known or can be found very easily by using the appropriate medieval geographical coordinates and the appropriate trigonometric

---

[4] On these see King, "Earliest Qibla Methods", pp. 107-126 and 133-141, and also *idem*, *Mecca-Centred World-Maps*, pp. 64-70 and 74.

[5] King, *Astronomy in the Yemen,* p. 29 and pl. 7; Janin & King, "Ibn al-Shāṭir's Compendium", pp. 195 and 216; King, *Mecca-Centred World-Maps*, pp. 109-115; and now Schmidl, "Early Arabic Treatises on the Compass".

[6] The description in *Paris IMA 1993-94 Exhibition Catalogue*, pp. 440-441, no. 336, was written before the geographical information had been analyzed. See now King, *Mecca-Centred World-Maps*, pp. 110-114, 168-170 and 478-480.

[7] *Ibid.*, pp. 107-124.

[8] See King, *Mecca-Centred World-Maps*, pp. 195-371, on the first two and **VIIc-2**, on the third.

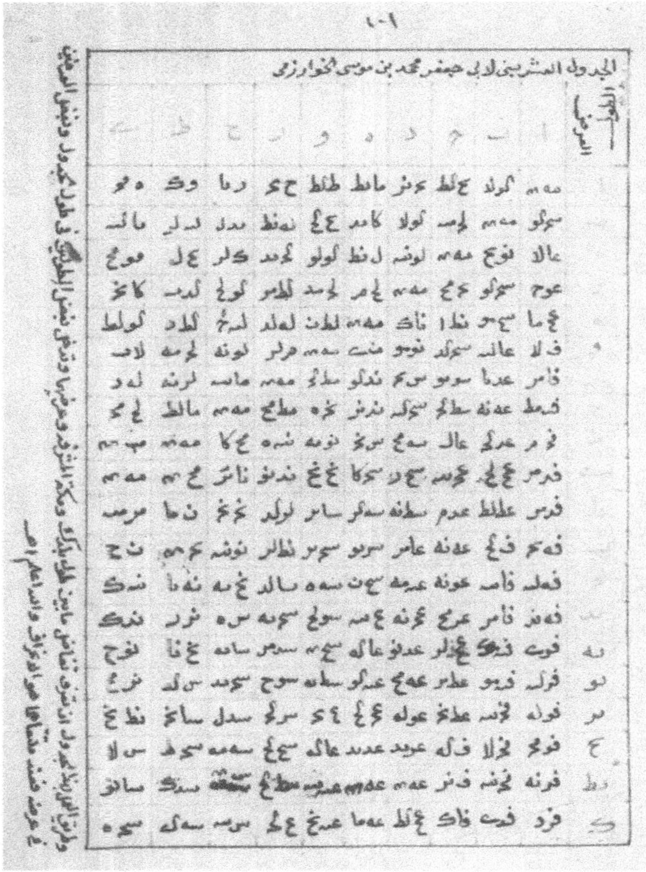

Fig. 3.6: The first half of a qibla-table based on the standard approximate formula, so that the entries for equal latitude and longitude diffferences from Mecca are all 45;0°. The table exists in a dozen-or-so copies, and in this one, but in no other, it is attributed to the 9th-century astronomer al-Khwārizmī. Nevertheless the first version of this table is probably indeed from the 9th century, although it is not to be confused with another contemporaneous table with similar format (see **Figs. 3.7** and **VIa-5.1**). [From MS Cairo TR 103,2, p. 101, courtesy of the Egyptian National Library.]

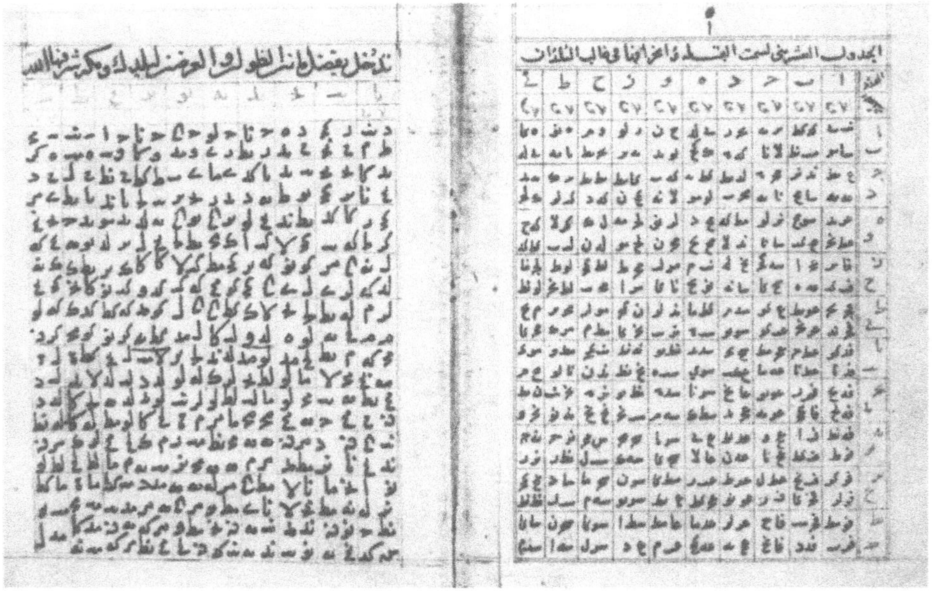

Fig. 3.7: An Abbasid table of the same format as the one shown in **Fig. 3.6** is at first sight based on another approximate formula, but, in fact, it is based on an accurate formula for a slightly different latitude argument than the one indicated. This table was widely known, and all of the nine known copies (from Iran, Central Asia, Yemen, Syria, Egypt and Turkey) have the same problem with the unmodified latitude argument. The qibla values are nevertheless adequately accurate for practical purposes. See **Fig. VIa-5.1** for another copy of this table, and also **Fig. II-10.8** for the splendid table of al-Khalīlī, which is based on the acccurate formula. In fact, the table shown here is appended to a copy of al-Khalīlī's universal auxiliary tables, penned in Edirne in 869 H [= 1464/65]. [From MS Istanbul Hamidiye 1453, fols. 233v-234r, courtesy of the Süleymaniye Library.]

formula. Rarely, though, were astronomers consulted on the qibla when it came to planning the layout of religious architecture; it was generally the legal scholars who decided on the qibla. In a few cases, however, we do find medieval architecture aligned in the qibla-directions computed by the astronomers. Since medieval longitudes were in general not particularly accurate even these qibla-values may differ by a few degrees from the modern qibla. Thus modern religious architecture, oriented in the modern qibla, will in general be slightly skew to even those edifices laid out in the mathematically-computed qiblas of yesteryear. But, as noted above, the modern qiblas are otherwise irrelevant to any investigation of the orientation of medieval architecture.

CHAPTER 4

REGIONAL SURVEYS OF ORIENTATIONS

The qiblas proposed in treatises on sacred geography are for various regions of the Islamic world. Also the medieval discussions of the appropriateness of particular qiblas are mainly regional. In the following survey, which is essentially only bibliographical, I have followed such a regional organization.

Arabia

The orientation and astronomical alignment of the Kaʿba is mentioned in medieval texts and the information in these has been shown to be accurate.[1] An understanding of the orientation of the Kaʿba is essential to an understanding of orientations generally. The orientation of the earliest mosques in Arabia has been investigated using the sparse written sources.[2]

Yemen

Information on qiblas is available from the astronomical and folk-astronomical sources,[3] and I believe that Professor Ronald Lewcock of Harvard University has prepared a list of actual orientations. A comparison has yet to be conducted.

Egypt

Most of the available information in astronomical and legal and historical works (notably al-Maqrīzī) has been gathered and interpreted.[4] Several different qibla-directions are attested in these sources for medieval Cairo – see **Fig. 4.1**. A text of major importance from the 12th century by a legal scholar named al-Dimyāṭī was discovered more recently and has yet to be published.[5] A corpus of reliable plans of Mamluk mosques in Cairo is now available but has not yet been exploited for the information it can cast on orientations.[6]

---

[1] See n. 2:1 above.

[2] See Barthold, "Orientierung der ersten Moscheen", and also King, "Astronomical Alignments in Medieval Islamic Religious Architecture", pp. 308-309.

[3] King, *Astronomy in Yemen*, esp. pp. 92 (index *sub* qibla) and 97 (caption to Pl. 7).

[4] See **VIIb-6** and **7**.

[5] Extant in MS Oxford Bodleian Marsh 592. See already **VIIb-6**, nn. 17-18. This was a source for al-Maqrīzī's information.

[6] Michael Meinecke, *Die mamlukische Architektur in Ägypten und Syrien*, 2 pts., Glückstadt: J. J. Augustin, 1992.

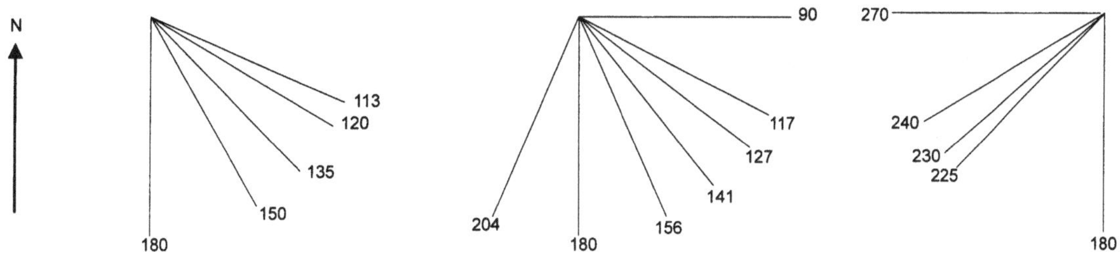

Fig. 4.1: The Egyptian historian al-Maqrīzī (d. 1442) mentioned these qiblas used for mosque orientation in Cairo: 90°, due east (not explicitly mentioned); 117°, winter sunrise (the qibla of the Companions of the Prophet); 127°, the qibla computed by the 10th-century astronomer Ibn Yūnus according to the accurate formula; 150°, the qibla of the Mosque of Ibn Ṭūlūn (variously explained); and *ca.* 156° - *ca.* 204°, any direction between the rising and setting of Canopus in Cairo. See further **VIIb-6**.

Fig. 4.2: According to a 12th(?)-century Andalusī treatise on the astrolabe, mosques in Cordova were oriented in these different directions: 113°, the qibla computed by the standard approximate formula (which works well for localities in the central regions of the Islamic world, but not for al-Andalus); 120°, winter sunrise; 135°, south-east (a compromise direction between due east and due south); 150°, the direction of the Great Mosque,which is "parallel" to the major axis of the Kaʿba; and 180°, due south (not specifically mentioned in this text).

Fig. 4.3: The legal scholar al-Bazdawī (d. 1089) reported these qiblas used for mosque orientation in Sa-marqand: 270°, due west, used by the Ḥanafī legal school and corresponding to the direction in which the road to Mecca left the city; 240°, winter sunset, as used for the Great Mosque; 230°, the qibla computed using the accurate formula (not specifically mentioned); 225° south-west (underlies a table presented by al-Bazdawī, lifted from some earlier source); and 180°, due south, used by the Shāfiʿī legal school, corresponding to the qibla of the Prophet in Medina.

Cairo provides a particularly interesting example of a city facing Mecca: the three main parts of the medieval city are oriented in three different qiblas – see **Fig. VIIb-7.4**, and some medieval architectural complexes are aligned in more than one qibla.[7]

In brief, the city founded in the year 969 alongside the canal (first laid out in Pharaonic times) has an orthogonal street-pattern aligned towards winter sunrise (*ca.* 117°), which was the qibla favoured by the Companions of the Prophet who first settled in Egypt. The "City of the Dead" to the east of the old city is aligned along a wadi and has an orthogonal street pattern facing the qibla determined by the astronomers (*ca.* 127°). Here, then, in both cases slope has played a role in the layout of the urban plan, if not in the very choice of the sites in question. The southern part of the city has a basic orientation of due south, and it is not a coincidence that the tomb of the legal scholar al-Shāfiʿī (d. 820) is there; his school favoured a qibla of due south, after the practice of the Prophet in Medina.

In some of the Mamluk (13th-15th-century) architecture in the old city of Cairo the outer wall is aligned with the street-plan even though the qibla of the astronomers was favoured by that time. A simple architectural expedient was adopted: although the outer qibla-wall faces the qibla of the Companions, that is, winter sunrise at *ca.* 117°, the inner wall with the *miḥrāb*

[7] This phenomenon is noted in the three papers by Christel Kessler listed as "Mecca-Oriented Architecture in Cairo", A-C, and to some extent explained in **VIIb-7**.

was built to face the qibla of the astronomers at *ca.* 127°. One can see the 10° difference in orientations at the places where the walls are pierced by windows. The modern qibla of Cairo is *ca.* 135°, but this is irrelevant to any discussion of medieval architecture in the city.

Syria and Palestine

No lists of orientations are available. The qibla towards the south was favoured in both Jerusalem and Damascus, although the mathematically-computed qiblas in these two cities are closer to 135° and 150°, respectively.[8] The religious architecture of the new Mamluk city of Tripoli shows the complexity of the problem of orientations in this region: here mosques and madrasas were built apparently after the model of buildings elsewhere in Syria, their orientations being adopted too, and that in spite of the fact that any Mamluk astronomer would have known the mathematically-computed qibla of Tripoli.

al-Andalus

All of the currently available information in various textual sources has been studied and reliable data is available on mosque orientations: I refer the reader to the new book of Mònica Rius.[9] The Great Mosque in Cordova faces the direction perpendicular to summer sunrise, its main axis being "parallel" to that of the Kaʿba. In other words, it is aligned towards the rising of Canopus, or rather, "parallel" to the rising of Canopus in Mecca – see **Fig. 2.2**. From textual sources we know that some early mosques faced winter sunrise and others faced the qibla computed by the approximate procedure advocated by al-Battānī (Raqqa, Syria, *ca.* 910), which works well for the central lands of Islam but for Cordova yields a qibla differing by 10° from that which could have been derived by exact procedures. Yet other mosques faced south, following the Prophetic practice in Medina, and others south-east, as a kind of compromise between due south and due east – see **Fig. 4.2**.

The Maghrib

Numerous orientations have been measured and various texts are available: again Mònica Rius is the best source.[10] A predominant direction is that of *ca.* 150°, as in the case of the Great

---

[8] On mosque orientations in Jerusalem see, for a modest start, the information gathered in King & Walls, "Jerusalem Sundial", p. 21. Plans are now available in Michael H. Burgoyne, *Mamluk Jerusalem: An Architectural Study*, London: World of Islam Festival Trust, 1987.

[9] Some initial studies are King, "Some Medieval Values of the Qibla at Cordova", pp. 370-387 (improperly paginated), an appendix to *idem*, "Andalusī Sundials", repr. in *Studies*, B-XV (with corrections in the addenda), based mainly on information in an early Andalusī treatise on the use of the astrolabe, and Samsó, *Ciencias en al-Andalus*, pp. 60-67. See more recently Jiminez, "La qibla extraviada"; and especially Rius, *La Alquibla en al-Andalus y al-Magrib*.

[10] See Bonine, "Maghrib" (cited in n. 1:7), as well as King, "Astronomy in the Maghrib", pp. 145-146. The problems of Maghribi orientations were first discussed in George Sarton, "Orientation of the mihrab in mosques", *Isis* 20 (1933), pp. 262-264, and in a reply by Henri-Paul-Joseph Renaud in *Isis* 34 (1942), p. 24, later also in M. Philibert, *Qibla des mosquées maghrébines*. See now ʿAmmār Saʿīd al-Shabīb, *Kitāb Dalāʾil al-Qibla*

Mosque in Cordova. Fez constitutes a particularly interesting example of orientations, with some mosques oriented due south and others due east. A qibla of south-east was also adopted as a kind of compromise.

al-ʿIrāq and Iran

With few exceptions, notably data compiled several decades ago by Myron B. Smith,[11] little reliable information on mosque orientations has been collected. Some early mosques in the province of al-ʿIrāq are oriented towards winter sunset (at *ca.* 240°). See also below on the newly-excavated mosque at Ayla near Aqaba (which has an orientation more appropriate for a mosque in al-ʿIrāq). An important 10th-century text on the problems of early mosque orientations in Iran has yet to be studied.[12]

Central Asia

No reliable information on mosque orientations is available. A most informative 11th-century treatise on the problems of the qibla in Transoxania has been published and evaluated.[13] From this we learn that whilst the earliest mosque in Samarqand was oriented towards winter sunset, later adherents of the Shāfiʿī and Ḥanafī legal schools favoured due south and due west, respectively (following the Prophetic practice in Medina on the one hand and the direction of the road to Mecca on the other), whereas others favoured south-west and the astronomers computed *ca.* 230° – see **Fig. 4.3**.

Turkey, Greece and the Balkans

Orientations in Turkey and Greece have been measured,[14] but no relevant texts have been identified. The Balkans are not promising for field research these days.

– *Abū ʿAlī al-Matījī*, unpublished doctoral dissertation, Universitat de Barcelona, 1992, uncritical edition with some useful appendixes, and especially Rius, *La Alquibla en al-Andalus y al-Maghrib* (n. 4:9).

[11] See, for a useful treatment, the kind of information recorded in Myron Bement Smith, "Material for a Corpus of Early Iranian Islamic Architecture", a series of papers published in *Ars Islamica*, starting in vol. 2 (1935). Even so, Smith was under the impression that the modern qibla, with which he compared the actual orientations, was somehow relevant. But the point is: he realized that the orientations are important and he took the trouble to measure them properly. Numerous orientations of individual edifices are given in Heinz Gaube and Eugen Wirth, *Der Bazar von Isfahan*, Wiesbaden: Ludwig Reichert, 1978, but it is not clear how they were measured.

[12] MS Tashkent Oriental Institute 177, fols. 97r-144v, by ʿAbd al-Qāhir al-Baghdādī – see *Cairo ENL Survey*, p. 46 (no. B46) and King, "Earliest Qibla Methods", p. 84, n. 6.

[13] King, "al-Bazdawī on the Qibla in Transoxania", on a treatise by the legal scholar Abu 'l-Yusr al-Bazdawī, extant in Cairo (see *Cairo ENL Survey*, p. 50, no. B88).

[14] Barmore, "Turkish Mosque Orientation", and MacKay, "The Qibla of Ottoman Mosques in Greece" (unpublished).

India and the Far East

Most plans of mosques show the qibla-wall facing due west; few are to be trusted. Relevant texts have yet to be identified.

East Africa

Mosque orientations have been measured, and display a range from *ca.* 300° to *ca.* 60°.[15] No relevant texts have been identified, but this wide variation can be explained in the light of the literature on sacred geography: clearly two different walls of the Kaʿba, namely, those to the south-east and the south-west, as well as the southern corner, were associated with the region by different groups.

West Africa

No information is available, though it seems that most historical mosques face roughly east.[16]

Orientation of cities

In my earlier work I pointed out that some cities may be said to be qibla-oriented in the sense that their city-plans (defined by a main axis, which may be the major axis or the minor axis of a rectangular orthogonal street-pattern) are influenced by the qibla.[17] My study of orientations in medieval Cairo (from texts, confirmed by maps) is now supplemented by Bonine's study of Maghribi cities (from on-site measurements, confirmed by texts). Were the sites of some cities, such as Taza in Morocco and Cairo, actually chosen because by virtue of the hydrographical layout they were by chance qibla-oriented? I suspect that the answer is a most definite yes, but I have yet to find evidence in historical texts, and I wonder if one should expect to find it anyway.

---

[15] Private communication from Dr. Thomas H. Wilson, Lamu Museum, Kenya, in 1980.

[16] See, for example, Dorothee Gruner, "Der traditionelle Moscheebau am mittleren Niger: Bemerkungen zur historischen und aktuellen Entwicklung", *Paideuma* 23 (1977), pp. 101-140, esp. p. 117, where the author maintains that all the mosques she had investigated were oriented within 10° of either side of due east.

[17] King, "Astronomical Alignments in Medieval Islamic Religious Architecture", p. 304.

CHAPTER 5

CONCLUDING REMARKS

What remains to be done is clear. Various available texts as yet unstudied must be published and evaluated.[1] We must continue the search for other relevant textual sources, the vast corpora of texts on sacred law and folk astronomy being the most obvious place to look. At the same time it would be worthwhile to measure and document the orientations of Islamic religious architecture. Individual orientations must be compared with others in the same city and region, as Michael Bonine has done for Morocco, and they can then be interpreted in the light of the textual sources.

There is material available for several doctoral dissertations, involving library- or field-work or both. But where are the candidates? The importance of the subject has already been made known to the history of science and the flourishing new field of archaeoastronomy. In the latter discipline the Islamic material is unique in being both archaeological and textual, to the envy of those who work on megalithic and other astronomically-aligned architecture in other cultural contexts.[2] But now it is time that the subject of orientations be recognized as an important chapter in the history of Islamic architecture; it certainly has much to contribute to that discipline.

---

[1] See the texts of al-Dimyāṭī, al-Matījī and al-Baghdādī, mentioned in nn. 4:5, 4:10 and 4:12, as well as some pronouncements on orientations in the medieval Maghrib by the *ʿulamāʾ* of Cairo in the 16th century, extant in MS Cairo DM 540 (see *Cairo ENL Survey*, no. C110).

[2] The papers listed as King, "Astronomical Alignments in Medieval Islamic Religious Architecture", repr. in *Studies*, C-XIII, and "Islamic Folk Astronomy", were presented at conferences on tropical archaeoastronomy (1982) and ethnoastronomy worldwide (1990), respectively.

Part VIIb

Architecture and astronomy: the ventilators of medieval Cairo and their secrets

To the memory of Franz Rosenthal

ACKNOWLEDGEMENTS AND NOTES TO THIS VERSION

I have an enormous debt to my former teacher Professor Franz Rosenthal of Yale University, who adopted me in 1968 when I knew only newspaper Arabic, and raised me to face the rigours of medieval manuscripts and to savour the delicacies of classical Arabic. For his friendship, guidance, and inspiration over the years, I am profoundly grateful. With his passing on April 8th, 2003, the field of Islamic Studies lost one of the major scholars of the 20th century. A bibliography of his numerous writings was published in *Oriens* 36 (2001), pp. xiii-xxxiv. David C. Reisman has published a tribute in *Aleph – Historical Studies in Science and Judaism* 3 (2003), pp. 329-342, and Dimitri Gutas is publishing "Knowledge Triumphant: The Life of Franz Rosenthal, 1914-2003", in both *Proceedings of the American Philosophical Society* and *Bulletin of Middle East Medievalists*.

My research on orientations of medieval Islamic religious architecture was supported by grants SES 8007145 and 8204520 from the National Science Foundation (1980-83); this support is gratefully acknowledged. It is also a pleasure to express my gratitude to the Egyptian National Library in Cairo and the Chester Beatty Library in Dublin and the Bodleian Library in Oxford for unlimited access to their rich collections of medieval scientific manuscripts and for permission to publish the photographs of the manuscripts used in this study.

This study is a revised and up-dated version of a paper with the same title published in the *Festschrift* for Franz Rosenthal (*Studies in Islam and the Ancient Near East Dedicated to Franz Rosenthal*, Jeannette Wakin, ed., a special issue of the *Journal of the American Oriental Society* 104 (1984), pp. 97-133). All opinions expressed herein are my own responsibility, although I have profited greatly from discussions with colleagues in other fields. It is a pleasure to record my particular gratitude to various friends who have guided me to material related to my investigations, most especially the late Felicitas Jaritz of the Swiss Institute in Cairo, and also James Allen (the Egyptologist), Carol Bier, William Chittick, Michael Dols, Horst Jaritz, the late Michael Meinecke, and Viktoria Meinecke-Berg. The Arabic text was prepared by Mónica Herrera in Frankfurt.

It is with particular appreciation that I acknowledge my surprise and pleasure at receiving a letter from my friend John Rodenbeck of the American University in Cairo (dated 14th October, 1984) about this study of mine on the ventilators of medieval Cairo. Four single-spaced pages of well-informed commentary, some critical, some complimentary, and often posing questions I had never considered. I have corrected the mistakes that John pointed out, but I have not addressed many of the new questions he raised. Some of these have also been posed, and answered, by Christel Kessler.

My purpose in writing this paper was to show how much astronomical sources can contribute to our understanding of far more mundane topics. No historian of Islamic architecture had ever addressed the problem of the orientation of a single mosque or any Islamic city with any understanding of medieval notions about the qibla. The astronomical and legal sources dealing with the qibla help us understand the often curious orientations of medieval Islamic religious architecture (and sometimes of entire cities), and historians of Islamic architecture will do well not to ignore these sources.

In April, 2003, a *nadīm* in Frankfurt, who alas cannot be identified, drew my attention to a study by Olivier Jaubert: "Capteurs de vents d'Égypte – Essai de typologie", *Annales islamologiques* (Institut Français d'Archéologie Orientale du Caire) 25 (1995), pp. 169-231. This is an investigation of the Cairo ventilators from the point of view of a historian of Islamic architecture. The author was blissfully unaware of the previous studies of Franz Rosenthal (1978) and myself (1984) on the Cairo ventilators, contending that "aucune étude spécifique ne leur a été consacrée". He provides an impressive list of some 70 ventilators from Islamic Cairo, some known only from uncapped vertical shafts, others only from textual references. The list is based mainly on the writings of Creswell and a series of publications on palaces and private houses in the *Mémoires* of the Institut Français d'Archéologie Orientale in Cairo between the 1930s (Pauty), early 70s (Lezine), and thereafter into the early 1980s (mainly Maury), continued during the 1980s at C.N.R.S. in Paris. Alas, only one Cairo ventilator, a 19th-century example, is illustrated! But neither is there any serious discussion of the orientation of the ventilators, any investigation of the fact that the ventilators had a Persian name, any awareness of medieval texts dealing with the different kinds of ventilators, or any recognition of the fact that the best-preserved example, that of the *Musāfirkhāne*, deliberately has one side open to the winds as well as the front. Nevertheless, there are numerous references that should be consulted by any interested reader of my text that follows, and by any future researcher. Certainly, Jaubert's article goes beyond the present (1984!) study in providing more evidence on the earlier Ancient Egyptian and Coptic connection, updated references to documentation of such edifices as the *Musāfirkhāne* (Jaubert's no. 54), and more references to early photographs of Cairo showing ventilators now lost for all time. The recent historiography of the Cairo ventilators shows clearly that there is a need for better access to scholarly writings on individual subjects in Islamic Studies: an author in the mid 1990s had no idea that two others had written on "his" chosen subject in the late 70s and early 80s, and the present author had no idea before April, 2003, that anyone had written a major research paper on "his" subject in the mid 1990s.

TABLE OF CONTENTS

CHAPTER 1

THE VENTILATORS OF MEDIEVAL CAIRO

In the early medieval period many houses in Fustat and Cairo had several storeys, and a large wind-catcher on the roof over a vertical shaft conveying air to each storey was a common feature of such houses.[1] The ʿIrāqī scholar ʿAbd al-Laṭīf al-Baghdādī, who visited Cairo about the year 1200, described the ventilators as follows:[2]

> "(The Egyptians) make the opening of their houses exposed to the agreeable winds from the north, and one sees hardly any houses without ventilators. These ventilators are tall and wide, and open to every action of the wind; they are erected carefully and with much skill. One can pay between one hundred and five hundred dinars for a single ventilator, but small ones for ordinary houses cost no more than one dinar each."

Ventilators were also a feature of the architecture of religious institutions in medieval Cairo. Several of the Cairene buildings discussed by K. A. C. Creswell in his monumental survey of medieval Egyptian architecture have air-shafts from which the wind-catchers at the upper ends have now disappeared.[3]

A European traveller to Cairo in the 16th century, Prosper Alpin, also observed the ventilators there. He wrote as follows (de Fenoyl's translation of the original Latin):[4]

> "Cette ville est faite de très hautes maisons dont les toits s'étendent si largement sur les rues qu'ils cachent presque entièrement le ciel au-dessus d'elles et empêchent le soleil d'incommoder les passants (car, à toutes les heures du jour, l'ombre recouvre et protège les rues). De plus, on utilise de vastes conduits, semblables à de grandes trompes, et placés à l'intérieur de toutes les maisons pour recevoir l'air froid. Ils s'élèvent au milieu des maisons, avec une ouverture d'environ six coudées; ils montent droit en l'air et atteignent le sommet, où ils se terminent en s'élargissant beaucoup

---

[1] For an introduction to ventilation in the medieval Islamic world see Richard Ettinghausen, "The Man-made Setting", in Bernard Lewis, ed., *Islam and the Arab World*, New York, N.Y.: Alfred A. Knopf, in association with American Heritage Publishing Co., Inc., 1976, pp. 57-116, esp. p. 71, and Guy T. Petherbridge, "Vernacular Architecture: The House and Society", in George Michell, ed., *Architecture of the Islamic World: Its History and Social Meaning*, New York, N.Y.: William Morrow and Company, Inc., 1978, pp. 193-208, pp. 201-204. See also n. 1:17 below.

[2] Kamal Hafuth Zand *et al.*, trans., *The Eastern Key: Kitāb al-Ifādah wa-'l-Iʿtibār of ʿAbd al-Laṭīf al-Baghdādī*, London: George Allen & Unwin, 1965, pp. 178-179.

[3] K. A. C. Creswell, *The Muslim Architecture in Egypt*, 2 vols., Oxford: Clarendon Press, 1952-59, I, pp. 45, 226, and 284-285, and II, pp. 244-245, and index, *sub malqaf* (II, p. 291). See also **Fig. 7.3** below.

[4] Prosper Alpin, *La médecine des Égyptiens*, trans. by R. de Fenoyl, Cairo: Institut Français d'Archéologie Orientale, 1980, vol. 1, pp. 35-36. According to de Fenoyl (*ibid.*, p. 36, n. 21), there is an illustration of a ventilator in Alpin's *De rerum aegyptiarum* (vol. IV, pl. 1 on p. 35). I owe this reference to the kindness of the late Professor Michael Dols. It is rather curious that 16th-century European drawings of Cairo do not feature ventilators at all: see Viktoria Meinecke-Berg, "Eine Stadtansicht des mamlukischen Kairo aus dem 16. Jahrhundert", *Mitteilungen des Deutschen Archäologischen Instituts, Abteilung Kairo*, Band 32 (1976), pp. 113-132 and pls. 33-39, on several such drawings. See also the discussion of the lack of representation of ventilators in Persian miniatures in Rosenthal, "Bādhanj" (cited in n. 1:18 below), p. 5.

plus, à la manière d'une panse de cloche. Par cette ouverture, tournée vers le nord, ils reçoivent la brise fraîche, qu'ils conduisent dans les parties basses de la maison: ainsi se trouvent refroidis le sous-sol et le rez-de-chausée."

The 17th-century Turkish scholar Evliya Chelebi,[5] in his monumental 10-volume description of the Ottoman Empire entitled *Seyāḥatnāme*, made the following observations about the ventilators of Cairo:[6]

"The lower city (*i.e.*, below the citadel) has buildings pressed up against one another, so the roads are narrow and dark. For this reason, all the houses overlook one another and every room has a ventilator on the roof to catch the wind, like a sail. These ventilators are said to go back to Ebu Ali Sina. They catch all the winds of Cairo through the science of alchemy (!). Everyone according to his means buys these wind-presses (?), which were invented by Ebu Ali Sina, for one gold piece each. They put the wind-presses on their roofs, the wind enters them, and the people of Cairo can breathe. These ventilators are said to go back to Ebu Ali Sina, and because of them the air in the lower city is not uniform, but changes ten times a day."

Whatever the problems of this reference, not least the association of them with the great philosopher Ibn Sīnā, it is clear that Evliya was impressed by the many ventilators of Cairo and their efficacy. They were still common in Cairo just a century or two ago. In the early 19th century, Edward W. Lane stated in his celebrated work, *Manners and Customs of the Modern Egyptians*:[7]

"Many houses (in Cairo) have, at the top, a sloping shed of boards, called a "malkaf", directed towards the north or northwest, to convey to a "fes-hah" or "fesahah" (an open apartment) below the cool breezes which generally blow from these quarters."

A photograph of Cairo taken in the 19th century published by F. Frith in 1860,[8] as well as a contemporary sketch (**Fig. 1.1**), a painting by Jean Léon Gérôme (**Fig. VIII-2.1**), and a late-

[5] See the article "Ewliyā Čelebi" by I. Mélikoff in EI_2.

[6] This reference was kindly brought to my attention by Professor Robert Dankoff of the University of Chicago, who provided the translation of the relevant passage from MS Istanbul Üniversitesi türkçe 5973, fols. 241a-241b.

The association with the eminent scholar Ibn Sīnā is completely spurious. My friend Professor Dimitri Gutas of Yale University assures me that Ibn Sīnā wrote nothing on the winds or on ventilators, or on alchemy, but his reputation in subsequent centuries, mainly in Iran and also in Turkey, was such that he was attributed with universal knowledge, not least of things that were useful. I have not encountered the association of ventilators with Ibn Sīnā in any of the secondary literature on ventilators in Iran (see n. 1:17), although, as we shall see, the ventilators in Cairo had a Persian name, rather than an Arabic one. The phrase here translated "through the science of alchemy" (*ʿilm-i sīmyā ile*) is rendered as *ʿilm ile*, "scientifically", in the printed text: this is incorrect, even though it makes more sense, for, as we shall see, the Cairo ventilators were indeed oriented according to notions of folk science.

Prof. Dankoff points out that there is a problem with the expression rendered here as "wind-presses". The manuscript has *maṣra* (carefully vowelled), which is not in the classical dictionaries, but might be Egyptian colloquial influenced by Turkish; perhaps it is colloquial for *maʿṣara*, so that Evliya has coined the curious phrase *maṣra-i rūzgār* to mean "wind-press" or something of that sort. Elsewhere, however, he gives that word properly with an *ʿayn* (*maʿṣara*) and defines it as an oil-press.

[7] Edward W. Lane, *An Account of the Manners and Customs of the Modern Egyptians*, repr. of the 1895 ed., The Hague & London: East-West Publications, and Cairo: Livres de France, 1978, p. 29.

[8] Francis Frith, *Egypt, Sinai, Jerusalem and the Pyramids of Egypt*, London: J. S. Virtue, 1860, pl. 11, already cited in Rosenthal, "Bādhanj" (see n. 1:18 below), p. 1. I have not seen this photograph.

Fig. 1.1: The skyline of Cairo in the mid-19th (?) century viewed across the canal known as al-Khalīj as illustrated in an unidentified work. Note that the ventilators are open to a direction parallel to that of the Khalīj; if they are correctly displayed, then the view is across the Khalīj away from the old city. [From Michell, *Architecture of the Islamic World* (cited in n. 1:1), p. 179.]

19th-century postcard (**Fig 1.2**) show row upon row of air-shaft covers, comparable with a forest of television antennas on the skyline of a modern city.[9]

Since none of the houses from Fatimid, Ayyubid or Mamluk Cairo has survived to this day, it is not surprising that no early ventilators survive either. In Cairo, new houses have been built on the ruins of old ones; in Fustat, destroyed in the 12th century, little remains of the houses besides the foundations and the drainage systems.[10] Most of the ventilators that do survive in Cairo today are comparatively late Ottoman constructions, and a dozen or so of these have been discussed by A. Lézine.[11] **Figs. 1.3a-b** show one of the Ottoman ventilators in Cairo, a splendid specimen on the roof of the two-storey 18th-century private house now known as the *Musāfirkhāne*.[12] Ventilators do not appear on 20th-century Cairene buildings, and the only use

[9] Houses with ventilators are inadvertently omitted in the illustrations in Ettinghausen, "The Man-made Setting" (cited in n. 1:1), p. 92, of house-types in early-19th-century Egypt. The original illustrations of house-types in Cairo in the *Description de l'Égypte* show ventilators on each house.

[10] See Paul Casanova, *Essai de reconstruction topographique de la ville d'al-Foustat ou Misr*, 2 vols., Cairo: Institut Français d'Archéologie Orientale, 1913-19, and George T. Scanlon, "Preliminary Report: Excavations at Fusṭāt", *Journal of the American Research Center in Egypt* 4 (1965), pp. 7-30, continued in later issues of the same journal. See also A. R. Guest, "The Foundation of Fustat and the Khittahs of that Town", *Journal of the Royal Asiatic Society*, Jan., 1907, pp. 49-83, and the article "Fusṭāṭ" by J. Jomier in EI_2 on Fustat and its early history.

[11] See A. Lézine, "La Protection contre la chaleur dans l'architecture musulmane d'Égypte", *Bulletin d'Études Orientales* (Damascus) 24 (1971), pp. 122-128.

[12] The *Musāfirkhāne* (on which see A. Patricolo, "Service de conservation des monuments: exposé des travaux

Fig. 1.2: A profusion of ventilators on Mamluk religious architecture, still visible on this late-19th-century postcard, but long demolished. [Original source unknown.]

→
Figs. 1.3a-b: The ventilator on the roof of the *Musāfirkhāne*, the most impressive of the very few historical ventilators of Cairo which survive to this day. Notice that the western side of the wind-catcher is open to the wind, whereas the eastern side is blocked. [Fotodienst Abū Max al-Zamalkāwī.]

---

exécutés, avec 67 notices historiques", *(Bulletin du) Comité de Conservation des monuments de l'art Arabe*, XXXII, Cairo, 1915-29 (1922), pp. 187-190, pls. CCXXIV-CCXXX; and E. Pauty, "Les palais et les maisons d'époque musulmane au Caire", (*Mémoires publiés par les membres de l'Institut Français d'Archéologie Orientale du Caire*, LXII), Cairo, 1932, pp. 59, 77, fig. 42, pls. 5/B and 8-XI) is distinguished from the majority of historical buildings in Cairo by the fact that it has been restored and that until recently it was used as a centre for artists. However, during a visit to the building in May, 1979, to take the photographs presented here as **Figs. 1.3a-b**, I was distressed to find the building no longer occupied and already displaying signs of neglect. On this problem of Cairo's architectural heritage see Paul Bergne, "Cairo: can the medieval city be saved?", *The*

of them in modern Egypt known to me is in the outstanding work of the Egyptian architect, Hassan Fathy.[13]

The early history of the ventilators of medieval Cairo is still a matter of some speculation. It is well known that ventilators were featured in domestic architecture in ancient Egypt. However, the notion that the ventilators of Cairo represent a purely Egyptian development of the ancient Egyptian ventilators[14] is probably to be abandoned because of philological considerations. It is already well-established that the medieval Egyptian name for ventilators was a Persian word *bādahanj*, and more than one scholar has suggested that the Egyptian ventilators were inspired by those used in Abbasid al-ʿIrāq.[15] However, the Persian scholar Nāṣir-i Khosraw, who visited Egypt in the middle of the 11th century and observed that most houses in Cairo had five or six storeys, did not mention the ventilators on these houses.[16] This is perhaps all the more surprising because ventilators were, and indeed still are, a prominent feature of Iranian architecture.[17]

Architectural Review, 114, no. 978, August 1978, and the various articles in Michael Meinecke, ed., "Islamic Cairo: Architectural Conservation and Urban Development of the Historic Centre", a special issue of *art and archaeology research papers* (June, 1980). My photos reappear on p. 13 of Elfried R. Knauer, "Wind Towers in Roman Wall Paintings?" *Metropolitan Museum Journal* (The Metropolitan Museum of Art, New York) 25 (1990), pp. 5-20.

[13] See, for example, Hassan Fathy, *Gourna: A Tale of Two Villages*, Cairo: Ministry of Culture, 1969, pp. 68-69. Fathy remarks (p. 68) that such wind-catchers as he incorporated in his architectural designs "may be set at precisely the right angle to catch the wind, irrespective of the orientation of the house".

[14] Such a view is expressed in A. Badawy, "Architectural Provision against Heat in the Orient", *JNES* 17 (1958), pp. 122-128, and also Hassan Fathy, "The Qaʿa of the Cairene Arab House ... ", *Colloque International sur l'Histoire du Caire, 27 mars - 5 avril 1969*, Cairo: General Egyptian Book Organisation, 1969, pp. 135-152, pp. 143-144. It is maintained, in spite of the linguistic evidence to the contrary, in Knauer, "Wind Towers in Roman Wall Paintings?" (cited in n. 1:12), esp. p. 14. On the ancient Egyptian connection see also **Ch. 7** below.

[15] Rosenthal, "Bādhanj" (see n. 1:18) and C. Edmund Bosworth in the article "Bādgīr" in EI_2, Supplement 1.

[16] Charles Schefer, *Sefer Nameh: Relation du Voyage de Nassiri Khosrau*, Paris, 1881 (repr. in *Islamic Geography*, vol. 170 (1994)), p. 132. See also the article "Fusṭāṭ" by J. Jomier in EI_2, esp. p. 958, and further Janet L. Abu-Lughod, *Cairo: 1001 Years of the City Victorious*, Princeton, N.J.: Princeton University Press, 1971, p. 19.

[17] On ventilators in Iran see Bosworth, "Bādgīr" (cited in n. 1:15) and the literature there cited, to which add, for example, E. Beazley, "Some Vernacular Buildings of the Iranian Plateau", *Iran* 15 (1977), pp. 89-102; R. Rainer, *Traditional Building in Iran*, Graz: ADVA, 1977; M. Siroux, "Caravanserails d'Iran et petites constructions routières", (*Mémoires publiés par les membres de l'Institut Français d'Archéologie Orientale du Caire*, LXXXI), Cairo, 1949; Michael E. Bonine, "Aridity and Structure: Adaptations of Indigenous Housing in Central Iran", in *Desert Housing: Balancing Experience and Technology for Dwelling in Hot Arid Zones*, Kenneth N. Clark and Patricia Paylore, eds., Tucson, Az.: Office of Arid Lands Studies, 1980, pp. 193-219. Nasr, *Islamic Science*, pp. 220-231; and also Mehdi N. Bahadori, "Passive Cooling Systems in Iranian Architecture", *Scientific American* 238:2 (Feb., 1978), pp. 144-154, for a scientific investigation of their function. The film entitled "Man and Nature" prepared on the occasion of the "Festival of the World of Islam" (London, 1976) provides a useful visual account of these Iranian ventilators and their function. Susan Roaf, "The Windcatchers of the Middle East", in *Islamic Architecture and Urbanism,* Aydin Germen, ed., Dammam, KSA, 1983, pp. 257-268, deals mainly with Iran, but also Dubai. On the latter see also, for example, A. Coles and P. Jackson, "A Wind Tower House in Dubai", *art and architecture research papers* (June 1975), pp. 1-28. On ventilators in Sind see Jean-Louis Bourgeois, "Welcoming the Wind", *Natural History* 89:11 (Nov., 1980), pp. 70-75, where the orientation and shape of the wind-catchers are mentioned. Daniel Dunham, "The Courtyard House as a Temperature Regulator", *The New Scientist* (September 8, 1960), pp. 663-666, deals with houses in the Maghrib.

In a study published in 1978 Professor Franz Rosenthal presented a series of translations of references to ventilators in medieval Egyptian poetry, where their aesthetic and erotic features are extolled. This valuable study provides a wealth of new material on the ventilator and new insight into its origin. The references gathered by Franz Rosenthal are mainly from Ayyubid and Mamluk times, but some occur already in Fatimid sources, and they all illustrate that the ventilator achieved a "modest measure of literary celebrity" in medieval Egypt.[18]

New information on the ventilators of medieval Cairo is available, and this constitutes the subject matter of the present study. This new material is contained in what may at first sight seem a rather unlikely source, namely, medieval Egyptian astronomical treatises and tables. There was a vigorous tradition of mathematical astronomy in Egypt in Fatimid, Ayyubid, and Mamluk times, as well as in Ottoman times, and most of the many available sources for the history of this tradition were investigated only during the 1970s and early '80s.[19]

In the present study I have assembled all the material on the ventilators of Cairo that has come to my attention, and used it to establish some rather astounding connections between this prominent architectural feature of medieval Cairo, the street plan of the Fatimid-Mamluk city, and medieval traditions of folk meteorology and astronomy. As we shall see, the ventilators of Cairo lead us to an exciting new perspective in Islamic urban and architectural planning.

First I shall discuss the modern term *malqaf* for ventilator and the medieval term *bādahanj* (**2**). Then I shall show that the ventilators were not aligned towards the north but were astronomically aligned, open to the direction perpendicular to midwinter sunrise (**3**). The astronomical sources that provide us with this information also lead us simultaneously to the first known quantitative measurement of the refraction at the horizon, a topic of some interest to the history of science (**4**). Other astronomical sources contain information on the shape of the ventilators and the names of the different varieties of Cairene ventilators (**5**). The direction adopted for the ventilators was, in fact, perpendicular to one of the two main directions that were accepted for the qibla or local direction of Mecca in medieval Cairo (**6**). This ventilator orientation was also used for aesthetic reasons; the entire Fatimid city of al-Qāhira, built alongside the Pharaonic Red Sea canal, was fortuitously aligned in the astronomically-defined qibla direction. The ventilators were thus aligned on the roof tops in accordance with the more or less orthogonal street plan (**7**). However, the ventilators were not aligned in this fashion solely for aesthetic reasons: the direction of the winds in medieval Egyptian folklore also played a role in the orientation and design of the ventilators (**8**). Finally, I shall show that the first Muslims in Egypt used an astronomical direction for the qibla because they wanted to face one particular side of the Kaʿba in Mecca, which is itself astronomically aligned (**9**).

Within the broader context of medieval Islamic architecture and city-planning in general, the medieval city of Cairo, with its qibla-oriented religious architecture and astronomically-aligned ventilators, is just one example, albeit probably the most interesting one, of an Islamic city oriented about the Kaʿba.

[18] See Franz Rosenthal, "Poetry and Architecture: the Bādhanj", *Journal of Arabic Literature* 8 (1978), pp. 1-19, esp. pp. 4-5.

[19] For an overview of this activity see King, "Astronomy of the Mamluks", with new information in Charette, *Mamluk Instrumentation*, pp. 5-39, *et passim*.

CHAPTER 2

THE TERMS *MALQAF* AND *BĀDAHANJ*

Modern historians of Islamic architecture, following Lane, usually refer to the Cairo ventilators by their modern Arabic name, *malqaf*,[1] from an uncommon root *l-q-f* meaning "to gather". However, the medieval name was not *malqaf* at all, but an Arabicized Persian word *bādahanj*, with variants *bādhahanj*, *bādhāhanj*, *bādāhanj*, and *bādhanj*,[2] and in one text (no. 6), a plural bawādahanj. The Persian word, most correctly spelled *bādāhanj*, was apparently less common in Iran itself than bādgīr for ventilator, and is derived from bād, wind, and the verb *āhanjīdan*, to pull out or extract.[3] The fact that a long a vowel occurs in some of the sources indicates that the word was borrowed in its correct form *bādāhanj* or yet more correct *bādhāhanj*. In fact, however, the relevant manuscript sources I published in 1983 use *bādahanj* and *bādāhanj* with equal frequency and sometimes use both forms. They never use a *dhāl* rather than a *dāl*, and the *d* or *dh* of the Arabicized form would both be pronounced as a *d* in Middle Arabic anyway. The poet Burhān al-Dīn al-Qīrāṭī (*fl. ca.* 1350) specifically implies the use of a short second a rather than a long one and of the Arabic letter *dāl* rather than *dhāl* when he renders the letters of the word (*b-ʾ-d-h-n-j*) in numerical notation as 2-1-4-5-50-3.[4] For the purposes of this study I shall use the form *bādahanj*.

Where *malqaf* occurs as an architectural feature in a medieval Arabic text, it apparently does not mean *bādahanj*, although precisely what it does mean in this particular context is not yet clear. It may refer only to the sloping part of the cover at the top of the *bādahanj*, and relate to the function of this particular feature in gathering the air, but even this association is not certain.[5] The term *malqaf* is not listed at all in Lane's *Arabic-English Lexicon*, although it should be remembered that in this work words whose roots begin with *lām* are only sketchily treated. In Dozy's *Supplément* to Lane's *Lexicon* and other dictionaries, *malqaf* is listed with the meaning "shield" with four references to the *Thousand and One Nights* (one would expect milqaf rather than *malqaf* for such a meaning), and *milqaf* as a kind of tool used by a burglar in the same source.[6] The terms *malqaf* / *milqaf* are not listed in the major Arabic dictionaries

---

[1] See, for example, K. A. C. Creswell (references cited in n. 1:3); the article "al-Ḳāhira" by J. Michael Rogers in EI_2, esp. P. 434b; and A. Lézine, "La Protection contre la chaleur" (n. 1:11), *passim*. Lézine mentions the term *bādhanj* in passing on p. 13, n. 1.

[2] On the term *bādhanj* see Rosenthal, "Bādhanj" (cited in n. 1:18), esp. pp. 2-5. Franz Rosenthal settled for *bādhanj* in his 1978 study; in my 1983 article I settled for *bādahanj*. Blachère *et al.*, *Dictionnaire Arabe-Français-Anglais*, lists *bādinj* and *bādinjān* (I, p. 344b), quoting Ibn Baṭṭūṭa and R. Dozy, respectively.

[3] I owe this information, which is based on the articles *bādgīr* and *bādahanj* in the *Lughat-nāme* (Tehran, 1964 onwards), to the kindness of Dr. William Chittick.

[4] See Rosenthal, "Bādhanj" (n. 1:18), p. 7. On the Arabic alphanumerical system see Irani, "Arabic Numeral Forms".

[5] The term *malqaf* occurs in the *waqfiyya* of the madrasa of the Sultan Barqūq, now published in facsimile, and edited and translated by Felicitas Jaritz in S. Lamei Mustafa, *Madrasa, Hanqah und Mausoleum des Sultans Barqūq in Kairo* (see n. V-7:1).

[6] Dozy, *Supplément aux dictionnaires arabes*, II, p. 553b.

such as the *Lisān al-ʿArab* and the *Tāj al-ʿarūs*. The term *bādahanj* for ventilator is used by ʿAbd al-Laṭīf al-Baghdādī in the passage cited at the beginning of this paper, in all of the references gathered by Professor Rosenthal, and also in several other medieval Egyptian sources, including the *Thousand and One Nights*.[7] Again, however, it is listed by Dozy but not by Lane.[8] The same term *bādahanj* is also attested in one medieval Egyptian source with the meaning of "a fashionable garment with wide openings in the sleeves".[9] The various references to *malqaf* / *milqaf* and *bādahanj* in the historical sources that have come to my attention are collected in **Text 1** at the end of this paper.

---

[7] Some reference to *bādahanj*s in the historical sources are the following:

(1) Ibn Taghrī Birdī's description of the Azhar Mosque (*al-Nujūm al-zāhira fī mulūk Miṣr wa-'l-Qāhira*, 16 vols., Cairo: Dār al-Kutub al-Miṣriyya, 1929-76, IV, p. 102);

(2) the *waqfiyya* of the *madrasa* of Qāyitbāy (Leo A. Mayer, *The Buildings of Qāytbāy as described in his Endowment Deed*, London: Arthur Probsthain, 1938, pp. 40 and 43);

(3) the history of Ibn Ḥabīb (*Tadhkirat al-nabīh fī ayyām al-Manṣūr wa-banīh*, 3 vols., Cairo: Dār al-Kutub al-Miṣriyya, 1976-86, II, p. 345); and

(4) al-Maqrīzī's *Kitāb al-Khiṭaṭ* (2 vols., Cairo, 1853 edn., repr. 1970, II:1, p. 222), and his *Kitāb al-Sulūk li-maʿrifat duwal al-mulūk* (4 vols., Cairo, 1970-73, III:1, p. 281).

On the story of the corpse being lowered down the shaft of the *bādahanj* in the *Thousand and One Nights* see Rosenthal, "Bādhanj" (n. 1:18), p.6

[8] Dozy, *Supplément aux dictionnaires arabes*, II, p. 47b.

[9] al-Qalqashandī, *Ṣubḥ al-aʿshā fī ṣināʿat al-inshāʾ*, 14 vols., Cairo: al-Maṭbaʿa al-Amīrujya, 1913, IV, p. 43. This term is not listed in Reinhard Dozy, *Dictionnaire détaillé des noms des vêtements chez les Arabes*, Amsterdam: Jean Müller, 1845, repr. Beirut, n.d.

CHAPTER 3

A MEDIEVAL ASTRONOMICAL TABLE RELATING TO VENTILATORS

The professional astronomical timekeepers (*muwaqqits*) who were associated with mosques and madrasas in Cairo from early Mamluk times onward used a corpus of astronomical tables for determining time by the sun and for regulating the astronomically-defined times of the five daily prayers.[1] Some of these tables were compiled by the celebrated Fatimid astronomer Ibn Yūnus; others were added by various Mamluk astronomers.[2] Among these tables, which were rediscovered only about 30 years ago, was a table displaying the altitude of the sun in degrees and minutes for each degree of solar longitude (corresponding roughly to each day of the solar year) when the sun was in the direction of the *bādahanj* (see **Fig. 3.1**).[3] The entries, which are expressed in standard Arabic alphanumerical *abjad* notation, are given for each degree (read vertically) of each zodiacal sign (read horizontally),[4] this format being standard in the entire corpus of tables. Analysis of the table reveals that the back of the *bādahanj* was intended to be aligned with the direction of the rising sun at midwinter, namely, *ca.* 27°30′ S. of E. for Cairo. This is curious indeed. Why should one align a ventilator in an astronomically-defined direction, which one would think has nothing to do with wind directions? Now this direction of the winter sunrise was known to be the qibla of the first Muslims in Egypt (see **Ch. 5**). But this raises another question, namely, why was an astronomically-defined direction chosen for the qibla? One would also think that the direction of winter sunrise has nothing to do with the direction of Mecca. Furthermore, the value given in the table for the solar altitude at the winter solstice is not zero, as one might have expected – since the Muslims were not known to have been familiar with the quantitative effect of refraction at the horizon – but rather a fraction of a degree. These features of the table were of obvious interest, and it took several years for me to fully appreciate their significance.

Some of the surviving ventilators in Cairo, notably the one on the roof of the *Musāfirkhāne*, appeared to be aligned in the direction underlying the table.[5] If the orientation of ventilators was so important, why then is the west side (but not the east side) of the *Musāfirkhāne* wind-

---

[1] On the tables of the Cairo corpus see now **II-4-5** and the next note. On the role of the muwaqqit in Islamic societies see **V**.

[2] On Ibn Yūnus see my article in *DSB* and **II-4.1.1** and **5.1**. On the various later Mamluk astronomers who contributed to the corpus see **II-4.1** and **5**.

[3] On this table see King, "Astronomical Timekeeping in Medieval Cairo", esp. pp. 371-373, where, however, several errors of interpretation occur, and now **II-4.8**.

[4] On the notation see Irani, "Arabic Numeral Forms", and for more details on the format see King, "Astronomical Timekeeping in Medieval Cairo", pp. 351-353.

[5] According to Lézine, "La Protection contre la chaleur" (n. 1:11), p. 14, the ventilators examined by him face forth, north-east, north-west, east, and even south-west, and south (?). From these he reasonably concludes: "Il n'y a pas de règle pour la place du malqaf." On the other hand, both Edward Lane (1834) (see n. 1:7) and Prisse d'Avennes (*L'art arabe après les monuments du Kaire*, Paris: Veuve A. Morel, 1877, p. 155) noted the northern and north-western orientation of the *malqafs* of Cairo (Rosenthal, "Bādhanj" (n. 1:18), p. 16, line 75).

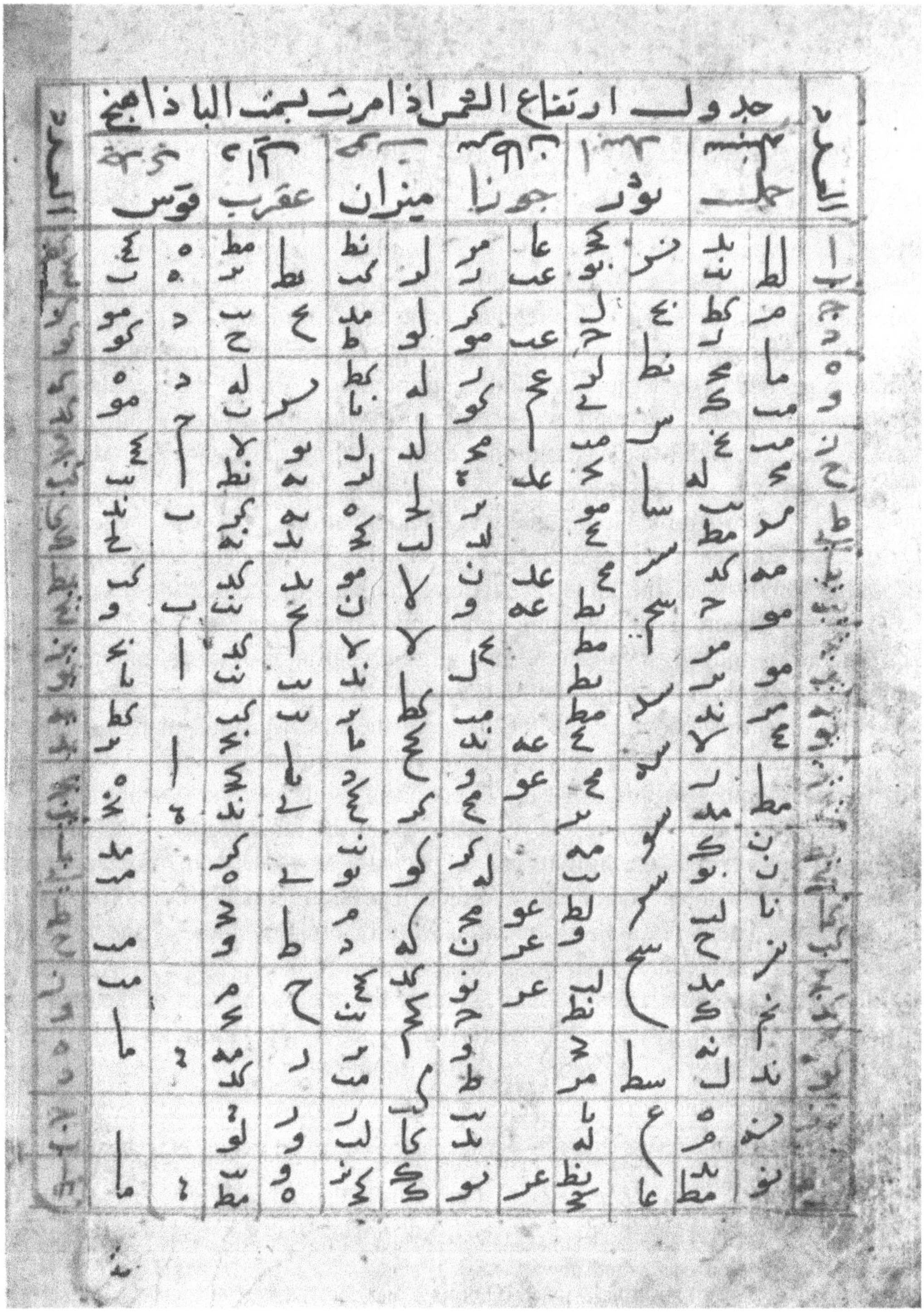

Fig. 3.1: The table for the altitude of the sun in the direction of the ventilator found in the corpus of tables for timekeeping that was used in Cairo throughout the medieval period. Notice that the entry for the winter solstice (at the lower left-hand corner of the table) is 0;41°. This represents a correction to an altitude of 0° above the true astronomical horizon to take into consideration the effect of refraction at the horizon. [From MS Cairo DM 444, courtesy of the Egyptian National Library.]

catcher also open to the winds? A slightly more northerly orientation could perhaps have achieved a more symmetrical and more durable structure. However, even this feature of the Cairo ventilators can now be explained.

The direction of the winter sunrise was used for the qibla of the mosque of ʿAmr in Fusṭāṭ, the first mosque to be built in Egypt (see **6**). It was the qibla of this mosque, built in the winter of the year 641-642, which was known as the *qiblat al-ṣaḥāba*, that is, the qibla of the Companions of the Prophet. This qibla was toward winter sunrise, that is, *ca.* 27° S. of E.; it was occasionally favoured in later times over the qibla that was computed in the 10th century according to a correct mathematical procedure and based on the available geographical data, viz. *ca.* 37° S. of E. Most medieval mosques in Cairo are oriented in one direction or the other (or, as we shall see, in both!). Only one of the references to the *bādahanj* in the medieval Egyptian poetry cited by Professor Rosenthal contains any mention of the direction in which they are aligned. Burhān al-Dīn al-Qīrāṭī wrote a line of poetry which translates:[6]

"I see that the love of air has turned (the *bādahanj*) away from the qibla of Islam."

I take this as a confirmation that the *bādahanj* was intended to face the direction perpendicular to the *qiblat al-ṣaḥāba*.

---

[6] Rosenthal, "Bādhanj", p. 16, line 11.

CHAPTER 4

A DIGRESSION: ISLAMIC QUANTITATIVE ESTIMATES OF THE EFFECT OF REFRACTION AT THE HORIZON

The table displaying the solar altitude in the azimuth of the *bādahanj*, that is, at $27^{1}/_{2}°$ S. of E., exists in a corpus of tables attributed in most of the available copies to the Fatimid astronomer Ibn Yūnus.[1] However, it was not Ibn Yūnus who computed all of these 200 pages of tables.[2] Whether Ibn Yūnus himself actually computed this particular table for the *bādahanj* is not certain and seems to me unlikely; in any case, there are actually two versions of the table in the various manuscripts of the corpus.

The first, which appears to be the original version and is the more common of the two, is distinguished by the fact that at the winter solstice (when the sun is at Capricorn 0°) the solar altitude is 0;41° above the horizon, rather than 0° as one might expect, given that the azimuth of the *bādahanj* is the azimuth of the rising sun. This value 0;41° represents an attempt to incorporate the effect of atmospheric refraction at the horizon, since the visible horizon is indeed about $^{2}/_{3}°$ above the true horizon.[3] The value 0;41° is thus intended to be the altitude of the sun above the visible horizon and it accords with the theory of refraction at the horizon attributed elsewhere in the medieval sources to Ibn Yūnus. In some copies (see **Fig. 4.1**) this first table is attributed specifically to an early-14th-century Egyptian astronomer named Ibn al-Rashīdī.[4] He is known to have corrected certain tables of Ibn Yūnus where they had become corrupt after much copying. However, it is precisely this first version of the table that occurs in certain copies of Ibn Yūnus' tables as they were before Ibn al-Rashīdī modified them.

In the second version of the table, the solar altitude decreases to zero at the winter solstice. In the 16th century the Cairo corpus of tables was modified to include the effect of refraction at the horizon, but Muḥammad al-Minūfī, the muwaqqit at the Ghawrī Madrasa who was responsible for this, attributes the underlying theory to Ibn Yūnus.[5] Furthermore, al-Minūfī did not adjust any of the tables involving solar altitudes; rather he adjusted only those tables of functions involving times of horizon phenomena. He did, however, present a table of the solar altitude in the azimuth of the *bādahanj*, but it is simply a table with the solar altitude zero at the winter solstice (see **Fig. 4.2**). al-Minūfī claims to have recomputed his tables for the new obliquity found by Ulugh Beg in the Samarqand observations *ca.* 1430, but, in fact, for the *bādahanj* table he merely "fixed up" the last column of entries, which serves solar

---

[1] See n. 3:3 above.
[2] See n. 3:2 above.
[3] See already King, "Astronomical Timekeeping in Medieval Cairo", pp. 373-376, on this, and now **II-4.11**.
[4] On Ibn al-Rashīdī (*Cairo ENL Survey*, no. C39) and his tables see **II-4.1.3**.
[5] See King, "Astronomical Timekeeping in Medieval Cairo", p. 373 (somewhat confused). The various Cairo manuscripts mentioned there are now analyzed in **II-4**. On al-Minūfī see *Cairo ENL Survey*, no. C120, and on his tables see **II-7.1**.

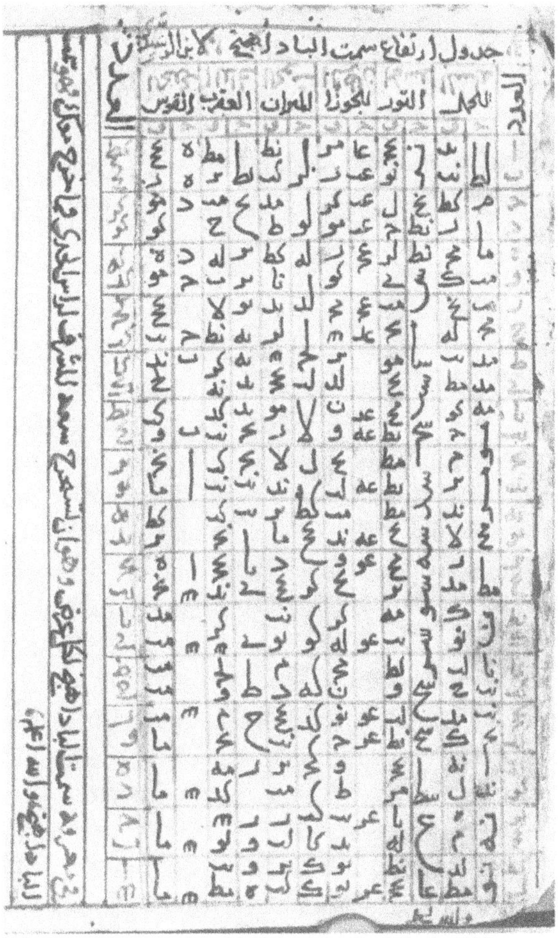

Fig. 4.1: Not a few copies of the Cairo corpus of tables for timekeeping are attributed to Ibn Yūnus. But this table for orienting ventilators in one particular copy of the corpus is attributed to a later Egyptian astronomer, Ibn al-Rashīdī. It bears a marginal note of some interest, which implies different orientations for ventilators in Alexandria, Cairo and Upper Egypt. [From MS Cairo DM 758, copied *ca.* 1650, fol. 4r, courtesy of the Egyptian National Library.]

longitudes up to 30° on either side of the winter solstice. Associated with the *bādahanj* table is another that is standard in the various copies of the Cairo corpus, and which displays the altitude of the sun in the azimuth of the qibla at Cairo, namely, 37° S. of E.[6]

Before the rediscovery of the Cairo corpus it was not known that the Muslims concerned themselves at all with quantitative estimates of refraction at the horizon. More recent investigations of other collections of tables for timekeeping that were used in other Muslim cities have revealed that some of these also incorporate such modifications for refraction.[7] I have been able to trace quantitative estimates of the effect of refraction to the celebrated 11th-century scholar Ibn al-Haytham and the less well-known 12th-century scholar Samaw'al al-Maghribī.[8] This feature of the table for the *bādahanj*, while of considerable interest to the history of science, is, however, of little relevance to the present study.

---

[6] On this and related tables see King, "Astronomical Timekeeping in Medieval Cairo", p. 368, and now **II-4.7**.

[7] For details of this tradition see **II-4.11**, **5.7**, **7.1** and **7.2**.

[8] Samaw'al's work on astronomy and mathematical methods, extant in MSS Leiden Or. 68,1, and Oxford Bodleian Hunt. 538,2, merits detailed study: see nn. **II**-8:5 and **V**-10:12.

CHAPTER 5

MORE ON THE *BĀDAHANJ* IN THE MEDIEVAL EGYPTIAN ASTRONOMICAL LITERATURE

The table for the *bādahanj* illustrated in **Fig. 4.1** is of particular interest. It is found in a 17th-century Egyptian copy of the auxiliary tables for solving problems of spherical astronomy for all latitudes by the 14th-century Syrian astronomer Shams al-Dīn al-Khalīlī.[1] On the left-hand side of the table there is a note (**Text 2**), which reads:

> "To find the azimuth of the *bādahanj* for every latitude, you determine the rising amplitude (of the sun) at the first point of Capricorn, and the result is the azimuth of the *bādahanj*, (but) God knows better."

By this remark it is implied that the alignment of ventilators with winter sunrise was not practised in Cairo alone. For different latitudes the direction of winter sunrise is different; from this text we might anticipate that the azimuth of the *bādahanj* is *ca.* $27^3/_4°$ S. of E. for Alexandria, $27^1/_4°$ for Cairo, and *ca.* 25° for Aswan. See further **7** and **8**.

Another reference of the same kind is found in a treatise on the use of the trigonometric quadrant by an unidentified author named Muḥammad Ḥattāta al-Fāriskūrī. The unique copy of this treatise was copied *ca.* 1700 and is preserved in Cairo. The section relating to our subject (**Text 3**) translates as follows:[2]

> "The twelfth chapter: on finding the azimuth of the *bādahanj*. Find the rising amplitude of the day-circle of Capricorn for the azimuth of Mecca, and then move south by this amount (from the east point) and this will be the *bādahanj*. God knows better."

Unless the text is garbled, the author either was no astronomer or was simply careless. The expression "for the azimuth of Mecca" makes no sense whatsoever in this context.

In a collection of short treatises attributed to Ibn Yūnus and copied *ca.* 1300, which is preserved in the Biblioteca Ambrosiana in Milan, we find instructions for finding the direction of the *bādahanj* (**Text 4**):[3]

> "To mark the direction of the *bādahanj* first establish the four cardinal directions, and then count from the east point southwards by the amount of the rising amplitude of (the sun at the first point of) Capricorn. Next extend a line (from the centre) in this direction, and this will be the direction for the ventilator. Form a rectangle with another line, and set up the *maḥilla* on this rectangle (*rabbiʿhu bi-khaṭṭ ākhar wa-aqimi l-maḥilla ʿalā dhālika 'l-tarbīʿ*). A good procedure (for laying out the direction of the *bādahanj*) is to divide the front in ten parts and make the side five and one-half, according to the technical convention of the craftsmen. God Almighty grants success."

---

[1] On al-Khalīlī see my article in *DSB*, Supplement vol. 2, and also now **II-10**.

[2] On al-Fāriskūrī see *Cairo ENL Survey*, no. D181. His treatise is contained in MS Cairo DM 639,5, fol. 46v-51r, copied *ca.* 1700, and the remarks on the qibla are found on fol. 50r.

[3] On this manuscript, numbered 281e, see King, "Astronomical Timekeeping in Medieval Cairo", p. 372. The passage on the *bādahanj* occurs on fol. 110v.

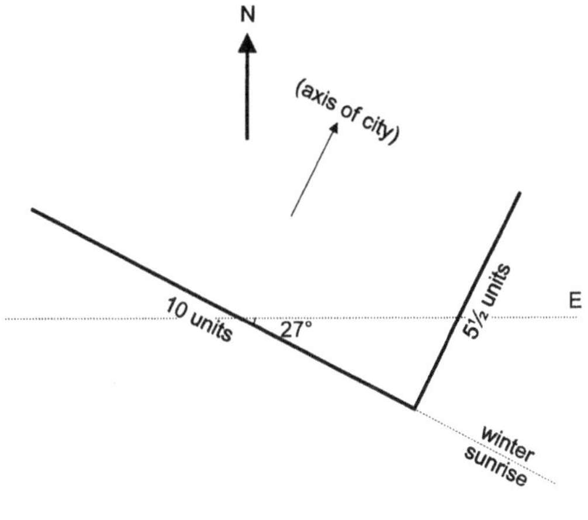

Fig. 5.1: The rule for constructing ventilators outlined in the Ambrosiana text.

From this text we might conclude that the term *maḥilla*, not attested in any medieval or modern Arabic dictionary known to me,[4] refers to the entire wind-catching device corresponding to the modern word *malqaf*. Other sources, however, do not confirm this (see below and also **Fig. 5.2**). The ratio for the lengths of the front and side of the *bādahanj*, namely, 10 : $5^1/_2$, ensures that the diagonal is precisely east-west. Observe also that the western side of this *bādahanj* is open to the winds; it appears that this design was in accordance with a wind-scheme (see further **8**). The plan and orientation of the *bādahanj* as outlined in this text is displayed in **Fig. 5.3**.[5] Notice that the *bādahanj* on the *Musāfirkhāne* is about twice as wide as it is deep, as one can see by counting the wooden arches around the base. However, two other sources point to a different arrangement for the base of the *bādahanj*.

A second method of constructing the base of the *bādahanj* is provided by the early-14th-century Egyptian astronomer Najm al-Dīn al-Miṣrī.[6] His treatise on instruments was discovered in 1982 in the Chester Beatty Library in Dublin. The manuscript, numbered 102 amongst the Persian manuscripts (!) in the collection, is copied in a 14th-century hand, and the discussion of the *bādahanj* (**Text 5**) occurs on fol. 52v (see **Fig. 5.2**). (I have previously attributed this anonymous treatise to the Aleppo astronomer Ibn al-Sarrāj (*ca.* 1325), who is now well established in the history of science as the inventor of several varieties of astrolabes and quadrants, to which I have drawn attention elsewhere.[7] I was misled by the fact that it contains

[4] The form is *maf'ila* for the geminated root *ḥ-l-l*, signifying a *nomen loci* (see W. Wright, *A Grammar of the Arabic Language*, (first published 1859-62), 3rd edn., 2 vols. in 1, Cambridge: Cambridge University Press, 1967, II, pp. 124-125 and 128-129), but the precise meaning is obscure to me.

[5] This procedure is misinterpreted in King, "Astronomical Timekeeping in Medieval Cairo", p. 372, n. 66. It is reminiscent of various architects' rules for determining the qibla: see al-Bīrūnī, *Taḥdīd*, comm. by E. S. Kennedy, pp. 213-214; and Lorch, "Al-Khāzinī's 'Sphere That Rotates by Itself'", esp. pp. 314-317, on some rules for laying out the qibla at Ghazna and Marw, respectively.

[6] On Najm al-Dīn al-Miṣrī and his treatise see now Charette, *Mamluk Instrumentation*.

[7] On Ibn al-Sarrāj (*Cairo ENL Survey*, no. C26) see already King, *Studies*, B-IX: "The Astronomical Instruments of Ibn al-Sarrāj: A Brief Survey". On the confusion with Najm al-Dīn al-Miṣrī see King, *Mecca-Centred World-Maps*, p. xxix. On his universal astrolabe see King & Charette, *The Universal Astrolabe of Ibn al-Sarrāj* (forthcoming). See also **X-5.2**, *etc.*

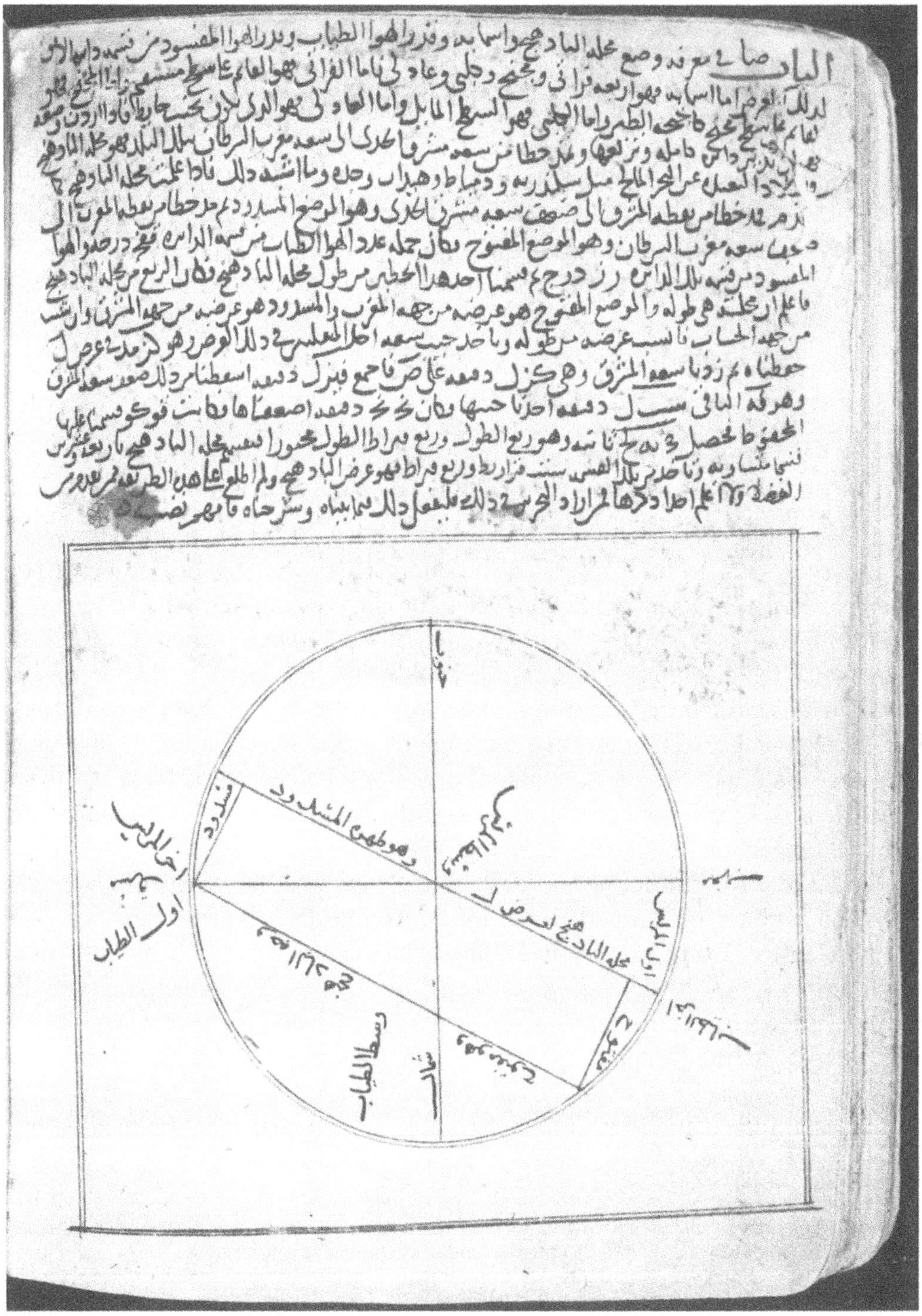

Fig. 5.2: A diagram for laying out the *bādahanj* in Cairo in the treatise on astronomical instrumentation by Najm al-Dīn al-Miṣrī (Cairo, *ca.* 1325). [Taken from MS Dublin CB Persian 102, fol. 52v, courtesy of the Chester Beatty Library.]

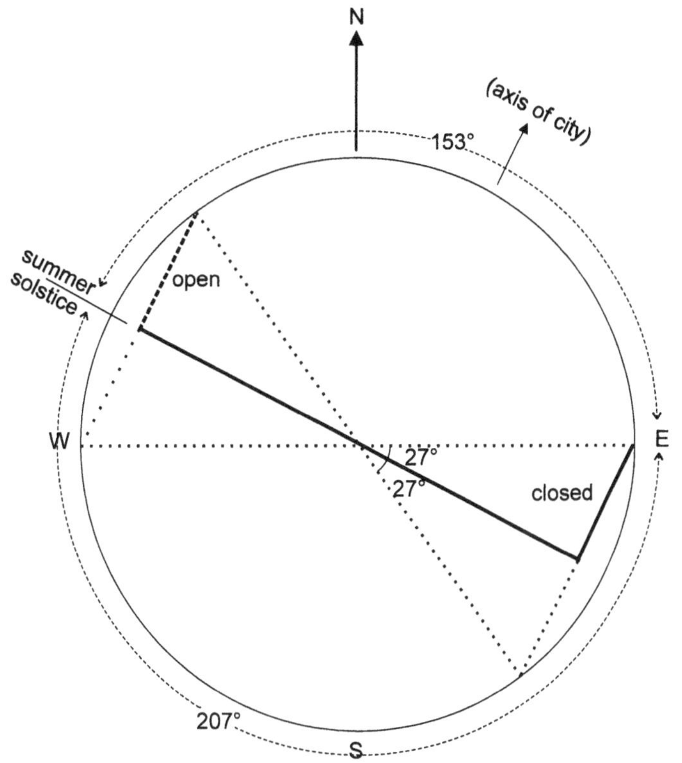

Fig. 5.3: The rule for constructing ventilators outlined by Najm al-Dīn al-Miṣrī and Ibn al-Qāṣiḥ.

tables for latitude 36°, serving, in the Mamluk context, Aleppo.[8]) The text translates as follows:

"The 91st chapter on knowing how to set up the *maḥilla* of the *bādahanj* and the names of the (various kinds) and the amounts (measured) on the horizon circle for that latitude which are (open) to favourable winds and (closed) to unfavourable winds. The names of the four (kinds of *bādahanj*) are *furātī*, *mujannaḥ*, *killī* and *ʿādilī*. The *furātī* is the one which stands on a flat surface (*al-qāʾim ʿalā saṭḥ mustaqīm*). The *mujannaḥ* (literally, winged) is the one which stands on a surface "winged" like the wings of a bird (*qāʾim ʿalā saṭḥ mujannaḥ ka-ajniḥat al-ṭayr*). The *killī* (that is, like a veil) is (read: has?) the sloping surface (*al-saṭḥ al-māʾil*). The *ʿādilī* is the one which is by the side of a wall (*yakūn bi-janb ḥāʾiṭ*).

If you want to set up (the *bādahanj*), draw a full circle and divide it into four parts (by marking the cardinal directions). Then draw a line from (the point corresponding to) the rising amplitude of (the sun at the beginning of) Capricorn to (the point corresponding to) the setting amplitude of (the sun at the beginning of) Cancer {that is, from winter sunrise to summer sunset} in that locality. This will be the *maḥilla* of the *bādahanj* [in] localities which are far from the sea. [For localities which are on

[8] It is now certain that Najm al-Dīn simply belonged to a tradition, also represented by al-Bīrūnī, in which the middle of the fourth climate, representing the middle of the classical world, was favoured for didactic reasons.

the sea] such as Alexandria, Damietta, ʿAydhāb, Jedda and similar places [the *maḥilla* of the *bādahanj* is to be facing the sea]. {*N.b.*: the text is corrupt here.}

When you have made the *maḥilla* of the *bādahanj* as I have described, draw a line from the east point to twice the rising amplitude of (the sun at the beginning of) Capricorn and this will be the closed part (*al-mawḍiʿ al-masdūd*); then draw a line from the west point to twice the setting amplitude of Cancer and this will be the open part (*al-mawḍiʿ al-maftūḥ*). The total number of degrees on the (horizon) circle corresponding to the favourable winds is 183° {*sic*, read 153°} and the number of degrees on the circle corresponding to the unfavourable winds is 207°.

Then we divide one of these two lines [which are perpendicular to] (text has: *min*) the length of the *maḥilla* of the *bādahanj* [into two parts] and it will be one quarter of the *maḥilla* of the *bādahanj*. You should know that the *maḥilla* is the length (of the *bādahanj*) and the open part is its width on the western side and the closed (part) is its width on the eastern side.

If you want to do this by calculation, take the ratio of its width to its length. (Then) find the Sine of the rising/setting amplitude at either of the solstices at that latitude, which is 27;44 {*sic*, the correct value is 27;42} at latitude 30°, which we keep in mind. Then we add the rising amplitude, which is 27;30°, to 90° to obtain 117;30° and subtract from that twice the rising amplitude which is 55° and the remainder is 62;30°. We take the Sine of this, namely, 53;13 {the text has 53;53, a scribal error}, double it and obtain 106;26. We divide this by the quantity which we kept in mind and obtain 0;15,38, which is a quarter of the length and, when rounded (to 0;15), equals a quarter of a *qīrāṭ* if the length is taken as one *qīrāṭ* (?). Then we divide the *maḥilla* of the *bādahanj* in 24 equal parts and take from this division six and a quarter *qīrāṭ*s which will be the width of the *bādahanj*.

I have not seen this method in (the writings of) any of the virtuous scholars who preceded me and I know of nobody who mentioned it. Whoever wants to do this properly let him do it as I have explained. Understand this and you will get it right."

Najm al-Dīn does not deserve top marks for his mathematical presentation. The essence of his calculation is to determine the breadth B in terms of the length L using $B = \frac{1}{2} L \tan \psi$, where ψ is the solstitial rising amplitude, correctly stated as being equal to 27;30° for the latitude of Cairo.[9] In accordance with medieval convention he uses Sines to base 60 (indicated by capital notation) rather than to base 1 as we use nowadays, and also he does not avail himself of the tangent function. Values of the Sines are expressed in the same way as values of angles, that is, sexagesimally (to base 60) in the standard alphanumerical abjad notation.[10]

The base advocated by Najm al-Dīn is only half the width of that described in the Ambrosiana text. The two angles to which he ascribes values of 153° (= 180° - 27°) and 207° (= 180° + 27°) relate to the open and closed parts of the *bādahanj*, and imply that the western side of

[9] See King, "Astronomical Timekeeping in Medieval Cairo", pp. 359 and 371-372, and now **II-4.3(l)**. (This value is based on Ibn Yūnus' (remarkably accurate) value for the latitude of Cairo-Fustat, namely, 30;0°, and his value for the obliquity of the ecliptic, namely, 23;35°: see **I-5.1.1** and **5.6.2**, and **II-5.1-2**.)

[10] See n. 2:4 above.

the erection be open to the winds, which is also indicated in his diagram. His directions appear to accord with a different wind-scheme from that underlying the Ambrosiana text (**8**). In view of the fact that the part of his text dealing with different kinds of ventilators is garbled, I assume that for this he was merely quoting an earlier source. The same source is quoted by a later writer (see below). The information provided by Najm al-Dīn on the four different varieties of *bādahanj* is not known to me from other sources besides the related one mentioned below,[11] and will be of interest to scholars concerned with vernacular architecture and its transmission. The term *furātī* applied to one variety suggests a connection with the Euphrates.

Another source relating to the *bādahanj* is a treatise on the use of the astrolabic quadrant by a scholar named Ibn al-Qāṣiḥ (1316-1399), otherwise known for his writings on the *Qurʾān*ic sciences.[12] This treatise exists in a unique manuscript copied about the time of the death of the author and now preserved in the Egyptian National Library in Cairo. Chapter 63 deals with the setting up of *miḥrāb*s and *bādahanj*s (in the manuscript the singular is given as both *bādahanj* and *bādāhanj* and the plural as *al-bawādahanj*), and gives information on the different kinds of *bādahanj*s that were used, as well as their shape. Part of the text bears marked resemblance to that of Najm al-Dīn, and it seems that both were quoting a common source (see **Fig. 5.3**). The relevant section of Ibn Qāṣiḥ's treatise (**Text 6**) translates as follows:

> "The way to set up the *bādahanj* is to draw the four cardinal directions on a flat slab (*balāṭa*) as described above, and move from the east point towards the south by the amount of the azimuth of the *bādahanj* in that locality by (using) the degrees of the altitude arc. You make a mark on the altitude arc (of the quadrant), then you place the ruler on the centre of the quadrant and on the mark and draw a line on the slab through the end of the ruler: the line will be the azimuth of the *bādahanj*, which is the position of the *maḥilla* [vowelled thus in the text]. There are (different) kinds of *bādahanj*s:
> (1) *furātī*, which stands on a straight surface (*qāʾim ʿalā saṭḥ mustaqīm*);
> (2) *mujannaḥ*, which stands on a surface like the wings of a bird (*qāʾim ʿalā saṭḥ ka-ajniḥat al-ṭayr*);
> (3) *ʿādilī*, which is by the side of a wall (*yakūn bi-janb ḥāʾiṭ*); and
> (4) *killī*, which has a sloping surface (*wa-huwa māʾil al-saṭḥ).*
> If the place where you put the *bādahanj* is on the shores of the sea, such as Alexandria or Damietta or somewhere similar, the *maḥilla* of the *bādahanj* is to be facing the sea. If you have made the *maḥilla* of the *bādahanj*, extend a line from the rising point of the equinoxes to twice the rising amplitude of (the sun at the beginning of) Capricorn which is in the direction of the closed part (of the *bādahanj*) and then extend a line

---

[11] No such information is contained in the other literary sources; see Rosenthal, "Bādhanj" (n. 1:18), p. 6.

[12] On this author, whose full name was Abu 'l-Baqāʾ ʿAlī ibn ʿUthmān ibn Muḥammad ibn al-Qāṣiḥ, see Suter, *MAA*, no. 419 and Brockelmann, *GAL*, II, p. 521, and SI, pp. 725-726, and, more especially, II, p. 214, and SII, p. 212, and the biographical sources there cited. His treatise on the astrolabic quadrant is preserved in MS Cairo DM 26, copied *ca.* 1400 – see *Cairo ENL Survey*, no. C51 – and the passage on the *bādahanj* occurs on fols. 22v-23r. Brockelmann (*GAL*, II, p. 214) lists another manuscript in Princeton which bears a similar title, but this is in fact not a copy of Ibn al-Qāṣiḥ's treatise. He also (SII, p. 212) lists another astronomical treatise by Ibn al-Qāṣiḥ which I have not consulted, namely, a treatise on the use of the sine quadrant extant in the Vatican Library.

> from the west point to twice the setting amplitude of Cancer, which is (the direction of) the open part. The total required angle on the horizon circle is 153 degrees, and the blocked air is 207 degrees. If we divide one of the sides by the length of the *maḥilla*, it will be one quarter of it. You know that the *maḥilla* of the *bādahanj* is its length, and its width is a quarter of its length. Know also that the azimuth of the *bādahanj* in Cairo is 27 degrees and 45 minutes. The azimuth of the qibla at Cairo is 37 degrees, and it is also said to be 37 degrees and 30 minutes because there is a difference of opinion concerning the latitude of Mecca: some say it is 21 degrees and others say 21 degrees and 30 minutes or yet other values.[13] There is also a difference of opinion concerning the latitude of Cairo: some say it is 30 degrees and others 29 degrees and 55 minutes. The longitude of Cairo is 55 degrees and the longitude of Mecca is 67 degrees, but some say other values. Cairo is in the north-west quarter and Mecca is in the south-east quarter. The inhabitants of Cairo and Alexandria up to Kairouan and Tahert and al-Sūs al-Aqṣā and the Atlantic Ocean, and all places in that azimuth, all face in their prayers the section of the Kaʿba from the western corner to the waterpipe (at the middle of the top of the north-west wall of the Kaʿba). I have mentioned in a short compilation the ways to find the direction of the Kaʿba by the blowing of the four winds."

Unfortunately, Ibn al-Qāṣiḥ's treatise on the qibla is not preserved in the known manuscript sources, although various other treatises dealing with the qibla from a non-scientific point of view do survive. Such treatises, which belong to a category called *kutub dalāʾil al-qibla*, date from the 9th century onwards. In some of these treatises and others relating to geography and cosmography, different geographical areas are associated with specific sections of the perimeter of the Kaʿba (see further **9**) and Ibn al-Qāṣiḥ follows this tradition.[14] None of the treatises on the *dalāʾil al-qibla* currently known to me makes mention of the orientation of the *bādahanj*.

Ibn al-Qāṣiḥ states that the azimuth of the *bādahanj* is 27;45° S. of E. This value is either based on the solar rising amplitude at the solstice, namely, 27;30°, with 0;15° added to take care of the apparent solar radius and thus to represent the azimuth of the right-hand (southern) edge of the solar disc as it rises at midwinter, *un véritable excès de délicatesse*, or it is an error for the 27;44 given as the Sine of the solar rising amplitude by Najm al-Dīn. Ibn al-Qāṣiḥ's remarks about setting up the *bādahanj* on the *maḥilla* are as obscure as those of Najm al-Dīn.

Another source for us to consider is the short treatise entitled *Tuḥfat al-aḥbāb fī naṣb al-bādāhanj wa-'l-miḥrāb* ("The gift of the loved ones on setting up *bādāhanj*s and *miḥrāb*s") by the 15th-century Egyptian astronomer Ibn al-Majdī.[15] However, this is devoted only to methods for marking the azimuth of the *bādahanj* at $27^1/_2$° S. of E. and the qibla at 37° S.

---

[13] On Islamic values for the latitude of Mecca see King, "Earliest Muslim Geodetic Measurements", pp. 225-226.

[14] For a discussion of his remarks in the context of other Islamic notions of sacred geography see my forthcoming monograph *The Sacred Geography of Islam*, summarised in my article "Makka. iv. As centre of the world" in *EI*$_2$, repr. in King, *Studies*, C-X.

[15] On Ibn al-Majdī (Suter, *MAA*, no. 432, and *Cairo ENL Survey*, no. C62), see Kennedy & King, "Ibn al-Majdī's Tables for Calculating Ephemerides". On this treatise see King, "Astronomical Timekeeping in Medieval Cairo", p. 372. The treatise survives in several copies, of which I have used MS Cairo DM 183,2, fols. 7r-11r, copied *ca.* 1450.

of E. on a plane horizontal surface (**Text 7**). The only astronomical source currently known to me which actually identifies the qibla of the *ṣaḥāba* with the orientation of the *bādahanj*, both at 27° S. of E., is a treatise on the use of the astrolabic quadrant by Zakariyyāʾ ibn Yaḥyā al-Bilbaysī, an Egyptian astronomer of the 14th or 15th century (**Text 8**).[16]

The 15th-century Egyptian historian al-Maqrīzī discussed the problems of mosque orientation in Cairo (**6**), but as far as I know, did not mention the orientation of the *bādahanj*. After his time, a new corpus of tables for timekeeping was compiled by al-Minūfī,[17] which still contained a table for orienting the *bādahanj* in the direction of winter sunrise (**4**). The original corpus of tables for timekeeping attributed to Ibn Yūnus, and to a lesser extent the new tables of al-Minūfī, were used in Cairo until the 19th century.

---

[16] See, for example, MS Cairo ṬJ 377,3, fols. 20v-21v, copied 1586-87. On al-Bilbaysī see Suter, *MAA*, no. 522, and *Cairo ENL Survey*, no. C64.

[17] On al-Minūfī see n. 4:5 above.

CHAPTER 6

ON THE QIBLA IN EARLY MUSLIM EGYPT

There were several directions accepted for the qibla in medieval Cairo-Fustat. The most popular directions were the qibla of the *ṣaḥāba*, facing winter sunrise at 27° S. of E., and the qibla of the astronomers, facing 37° S. of E. Both of these are mentioned by a 16th-century Egyptian (?) astronomer Ghars al-Dīn al-Ḥalabī (**Text 9**), in the following rather simplistic terms:[1]

> "Know that in the year eighteen of the Hijra [= 639 A.D.], in the time of the Caliphate of ʿUmar ibn al-Khaṭṭāb – may God be pleased with him –, ʿAmr ibn al-ʿĀṣ conquered Fustat (*Miṣr*), and then he built his mosque (which was then) known as the New Mosque. The Companions of the Prophet – may God bless him and grant him salvation – who were in Egypt when he built (the mosque) were more than thirty (in number), and there were about sixteen thousand Muslim fighters of the holy war with them. The azimuth of the *miḥrāb* of this mosque was about 27 degrees (south of east), and no one amongst those mentioned contested this, rather (the *miḥrāb* was placed in this direction) with their unanimous agreement. The astronomers, when they determined the azimuth (of the qibla) by proven mathematical means, decided that the azimuth in Fustat is 37 degrees (south of east). They (further) decided that the *miḥrāb* (which we) mentioned was wrong. Now there is great danger in this, since it (amounts to) saying that the Companions have erred. (The Prophet) – may God bless him and grant him salvation – said of them: "My companions are like stars; whichever of them you follow, you will be (rightly) guided."
>
> This story has many parallels in respect to the shrines and graves and mosques of the Companions. Those who hold the opinion that one should face directly towards the Kaʿba (*iṣābat al-ʿayn*) think that Kufa, Baghdad, Hamadan, Qazwin and Rayy have one and the same azimuth. The (real) situation is that there is a difference of about 4 (*sic*, read 14 ??) degrees between them. As a result of this and similar (problems), many people have become confused and mixed up (?). As a result of what we have presented (in this treatise), we hope by the grace of God that this obscure problem may be solved.

We shall have occasion to return in **Ch. 9** to the notion of *iṣābat al-ʿayn* to which al-Ḥalabī refers. The story of the foundation of the mosque of ʿAmr and its orientation is more complicated than al-Ḥalabī knew. The historical sources have already been investigated by K. A. C. Creswell, who failed to understand them.[2] The mosque was built in Fustat by the military

[1] On al-Ḥalabī see Suter, *MAA*, no. 465 and *Cairo ENL Survey*, no. C88. His treatise is contained in MS Cairo MM 114, copied *ca.* 1600, in which this passage occurs on fols. 5r-5v.

[2] Creswell, *Muslim Architecture in Egypt* (cited in n. 1:3), II:1, pp. 149-151. See also A. J. Butler, *The Arab Conquest of Egypt and the Last Thirty Years of the Roman Dominion*, (first published 1902), 2nd edn., Oxford: Clarendon Press, 1978, p. 343.

commander ʿAmr ibn al-ʿĀṣ in the winter of 641/42. The 15th-century historian al-Maqrīzī relates two stories about the way in which ʿAmr had the qibla laid out.[3] The first account translates (**Text 10**):

> "ʿAmr sent the two men (named in the text) to set up the qibla and said to them: "Stand when the sun is beginning to decline" – or, in another version, "when the sun is on the meridian" – "and have (the sun) at your two eyebrows", and they did (this)."

The text as published here and translated makes no sense at all unless we assume that *ḥājibaykumā*, "your two eyebrows", is a corruption of *jānibaykumā*, "your two sides." If this is the case, then the men were to be facing due east, perpendicular to the direction of the sun when it was on the meridian.

The second version translates:

> "ʿAmr was laying out ropes so that the qibla of the mosque could be set up, and he said: "Make the qibla towards the east (*sharriqu 'l-qibla*) and you will face the Kaʿba." ... ... (The qibla) was too far to the east."

Of course, *sharriqu 'l-qibla* can also mean "make the qibla towards sunrise". al-Maqrīzī relates that when ʿAmr prayed in the mosque, he prayed almost towards east (*nāḥiyat al-sharq illa 'l-shayʾ al-yasīr*), and that when he prayed in a church he would pray almost in the qibla of the Christians, *i.e.*, due east (*lam yanṣarif ʿan qiblatihim illā qalīl*[an]). Ibn Taghrībirdī and also al-Qalqashandī, but not al-Maqrīzī (although the published text seems to be defective), report that when the mosque was enlarged by Qurra ibn Sharīk in 673 it was actually pulled down and rebuilt on a larger scale in a different direction, more to the south.[4] Ibn Taghrībirdī's remarks (**Text 10**) translate as follows:

> "The qibla was much too far to the east. When Qurra ibn Sharīk pulled down his mosque and (re)built it in the time of al-Walīd ... he turned (the qibla) a little to the south."

Now al-Maqrīzī seems to be saying that the mosque of ʿAmr was built facing due east. Ibn Taghrībirdī and also al-Qalqashandī (and probably al-Maqrīzī as well, if indeed the published text is defective) seem to be saying that Qurra rebuilt the mosque in a more southerly direction, but they do not mention that this new direction was towards winter sunrise. A change of qibla from due east to the direction of winter sunrise can hardly be described as a change "a little to the south".

In an earlier publication I have suggested that the qibla of the polytheist Arabs of the Hejaz in the time before the introduction of the qibla towards Jerusalem, which occurred some time before the advent of Islam, was towards the east.[5] Also, Wilhelm Barthold has argued that the qibla of the earliest mosques of the Hejaz (Medina and Qubāʾ) was originally towards the east.[6]

---

[3] al-Maqrīzī, *al-Khiṭaṭ* (cited in n. 2:7), II, p. 247. The same stories are recorded by the comtemporaneous (?) Egyptian writer Murtaḍā ibn Khafīf, whose treatise is available only in an unreliable French translation prepared in 1666: see Gaston Wiet, ed., *L'Égypte de Murtadi fils du Gaphiphe – introduction, traduction et notes*, Paris: Imprimerie Nationale & Paul Geuthner, 1953, pp. 252ff. of Pierre Vattier's translation.

[4] Ibn Taghrībirdī, *al-Nujūm al-zāhira* (cited in n. 2:7), II, p. 67, ll. 1-16. The earlier historian Ibn ʿAbd al-Ḥakam states that Qurra pulled the mosque down (*hadama*) but does not mention its orientation either before this event or after his reconstruction; see Charles C. Torrey, ed., *The History of the Conquest of Egypt, North Africa and Spain known as the Futūḥ Miṣr of Ibn ʿAbd al-Ḥakam*, (Yale Oriental Series, Researches, 3), New Haven, Conn.: Yale University Press, 1922, p. 131, l. 16.

[5] See King, "Astronomical Alignments in Medieval Islamic Religious Architecture", repr. in *idem*, *Studies*, C-XIII, pp. 308-309, and also **VIIa**.

[6] Barthold, "Orientierung der ersten Moscheen".

Apparently the earliest qibla used in Egypt was also due east, shortly thereafter changed to winter sunrise. Soliman Bashear has collected considerable evidence from early Islamic sources for Muslim prayer facing east, sometimes in churches.[7] The direction of due west was used for the qibla in early Islamic al-ʿIrāq, Iran and Transoxania, but the direction of winter sunset was also favoured.[8]

I have not yet been able to check the precise orientation of the present mosque of ʿAmr, and none of the modern plans of the mosque, all of which go back to Corbet and Creswell,[9] is to be trusted. The present mosque is not rectangular in shape, which adds to the difficulty of identifying the precise qibla used in its orientation. From the map of Cairo published by the "Survey of Egypt 1950" I measure the north-east wall to be at *ca.* 33° S. of E.[10] However, I have not investigated the altitude of the hills to the south east. The mosque should be surveyed afresh, together with the appropriate horizon conditions; for the time being I shall assume that it is, as the medieval authors say, aligned toward winter sunrise.

No Egyptian treatises on the qibla are known from the 9th or early 10th century. In the late 10th century, the astronomer Ibn Yūnus, who worked for both Caliphs al-ʿAzīz and al-Ḥākim, wrote on the determination of the qibla by purely scientific means.[11] He made no mention of the *qiblat al-ṣaḥāba*. Rather, using a series of correct trigonometric formulae and the geographical coordinates of Cairo and Mecca, he derived a value of about 37° S. of E. for the qibla at Cairo. (This differs from the modern qibla in Cairo, which is about 45° S. of E., because the medieval value for the longitude difference between Cairo and Mecca was inaccurate.) Ibn Yūnus also compiled tables displaying for each day of the year the altitude of the sun when it was in the direction of Mecca,[12] and possibly he was the compiler of the similar table for the *bādahanj* which we discussed in **Chs. 3**, **4** and **5**. The qibla direction found by Ibn Yūnus is advocated by most later Egyptian astronomers who wrote on the subject.[13]

The religious scholars, on the other hand, appear to have favoured the *qiblat al-ṣaḥāba*. When the Fatimid Caliph al-Muʿizz tried to change the qibla of the mosque of ʿAmr in 956-957 he encountered considerable opposition, and was made to see the error of his ways by various supposedly miraculous happenings.[14] The 14th-century Egyptian legal scholar al-Zarkashī, in his book on mosques, states that it is permissible to pray to the right or left of the *miḥrāb* in a mosque, except in the main mosques in Medina, Kufa, Basra, Damascus, and also Fustat, because the Prophet or his Companions prayed in these.[15]

---

[7] Bashear, "Qibla Musharriqa".

[8] See King, "al-Bazdāwī on the Qibla in Transoxania", for details.

[9] See Creswell, *Muslim Architecture in Egypt* (cited in n. 1:3), I:1, p. 150 (Corbet), and II, opposite p. 188 and p. 192 (Creswell ?).

[10] This map is included inside the cover of Creswell, *op. cit.*

[11] On Ibn Yūnus see n. 3:2. For details of this calculation see King, *Ibn Yūnus*, III-3.28.

[12] See n. 4:6 above.

[13] There is evidence that some astronomers preferred 34° S. of E., a value derived from different geographical coordinates for Cairo and Mecca. The value 34;2° occurs on MS Paris BNF ar. 2558, fol. 51r, of al-Khalīlī's list of qibla-values (see n. 5:1 above concerning the author) and underlies the qibla-indicator on an early-14th-century Egyptian sundial preserved in the Victoria and Albert Museum in London (illustrated in King, *Studies*, B-XVI, pl. 7 of the reprint, but not in the original article).

[14] Creswell, *Early Muslim Architecture*, 2 vols., II (Oxford, 1940), p. 213, quoting the 17th-century historian al-Bakrī.

[15] Muḥammad ibn ʿAbd Allāh al-Zarkashī, *Iʿlām al-sājid bi-aḥkām al-masājid*, ed. by Abu l-Wafa Mustafa

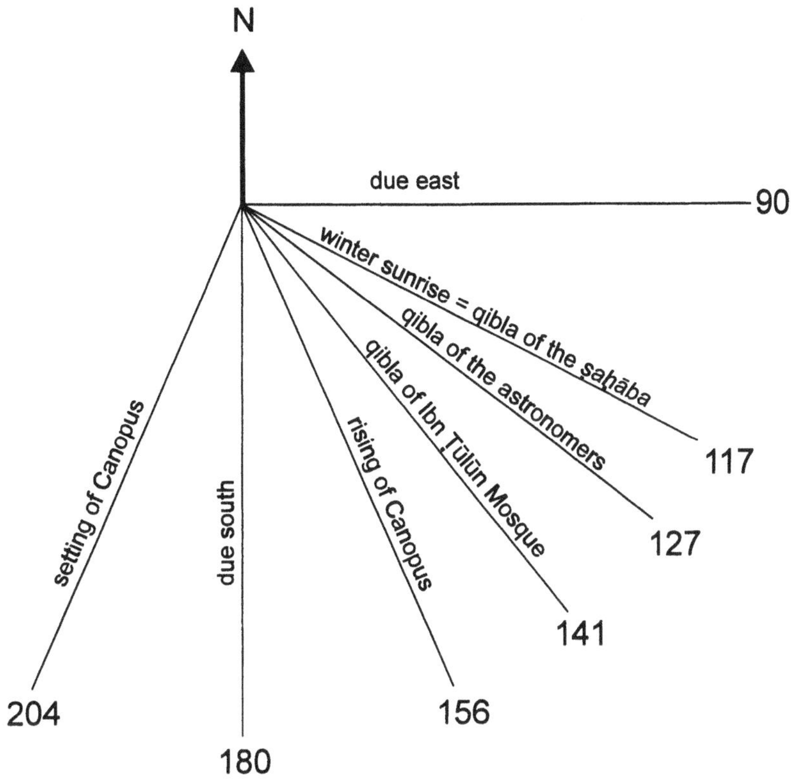

Fig. 6.1: The various qibla directions used in Egypt according to al-Maqrīzī.

Now there were in fact more than two qibla directions used in medieval Egypt. al-Maqrīzī discussed the divergences of the *miḥrāb*s in Egyptian mosques and identified four different *miḥrāb* orientations (see **Fig. 6.1**).[16] The first was the qibla of the *ṣaḥāba*, namely, the direction of the rising sun of midwinter, which he stated was used for the mosque of ʿAmr in Fustat and the *jāmi*ʿ mosques in Giza, Bilbays, Alexandria, Qus and Aswan. The second (al-Maqrīzī's third) was the mathematically-computed qibla, which he stated was used for the Azhar mosque. The third *miḥrāb* orientation (al-Maqrīzī's second) was that of the mosque of Ibn Ṭūlūn, which al-Maqrīzī points out is 14° S. of E. of that mathematically-computed qibla, namely, 127° + 14° = 141°. Two somewhat fanciful explanations of this are given by al-Maqrīzī. One involves a man sent by Ibn Ṭūlūn to Medina to measure the orientation of the Prophet's Mosque there: he found that the mosque was "about ten degrees to the south" of the qibla "which can be determined by technical means". As a result Ibn Ṭūlūn made a similar adjustment for his mosque in Cairo. The other story is that the Prophet Muḥammad traced out the *miḥrāb* for Ibn Ṭūlūn whilst the latter was asleep.

Maraghi, Cairo: al-Majlis al-aʿlā li-'l-shu'ūn al-Islāmiyya, Lajnat iḥyā' al-turāth al-Islāmī, 1384 H [= 1964], p. 363.

[16] See al-Maqrīzī, *al-Khiṭaṭ* (n. 2:7), IV, pp. 21-23.

The fourth *miḥrāb* orientation recorded by al-Maqrīzī was was due south or even a few degrees west of south. This orientation, he says, was attested in Qarāfa and in "the villages". The modern maps of Cairo confirm this southerly orientation of religious edifices in the Qarāfa area. The reasons which al-Maqrīzī gives for such orientations are, first, that various mosques were converted from churches that faced east and a *miḥrāb* was installed on the south wall; and, second, that the Egyptians were using the same qibla as their Syrian neighbours, namely, due south. He also mentions that mosques were built "toward Canopus", stating that this star rises "a little to the south of the rising point of the sun at mid-winter", culminates due south, and sets "at a small inclination to due south". He adds that: "the rising point of Canopus is approximately in the azimuth of the qibla for Egypt". Here al-Maqrīzī combines several errors: first, Canopus rises in Cairo some 40° south of the rising point of the sun at mid-winter; second, it rises and sets at *ca.* 23° east and west of due south; third, its rising point is some 30° south of the mathematically-computed qibla for Cairo. However, as we shall see in **Ch. 9**, the Egyptians had good reason to use either the winter sunrise or the rising point of Canopus for orienting their mosques. At the end of his discussion al-Maqrīzī displays at once both his piety and his lack of understanding of the qibla problem by referring to "the correct *miḥrāb*s of Egypt which were laid out by the Companions".

Another Egyptian, the 12th-century legal scholar Zayn (?) al-Dīn al-Dimyāṭī, wrote a short treatise on the qibla in Egypt.[17] In this work, which is essentially an open letter to his colleagues in Damietta, al-Dimyāṭī also discussed the different *miḥrāb* orientations in the mosques of Egypt, but this treatise contains less useful details than the discussion of al-Maqrīzī.[18] al-Dimyāṭī confessed that he knew that the qibla of the Companions was wrong when compared with the qibla that could be derived by computation. However, he had been to Mecca and had conferred with scholars there on the part of the Kaʿba that was facing Egypt. They told him that it was the part between the western corner and the water-pipe, and so he stayed up with some other Egyptians at the Kaʿba by night and stood facing that section of the Kaʿba, noting various features of the night sky. They found that the Pole was at the side of the left shoulder and the Pleiades set behind the back, that Vega set at the right side, and that Scorpio (that is, the star α Scorpionis) rose in front of the face (**9**). When they returned to Egypt they found that the same phenomena were observable when they faced the qibla in the mosque of ʿAmr at Fustat. So, al-Dimyāṭī concludes, the Companions were right after all.

His astronomical knowledge was such that he apparently did not realize that the night sky at Mecca looks somewhat different from the night sky at Cairo, and that the directions of the risings and settings of the sun and stars vary with latitude. Elsewhere in his treatise, he

---

[17] al-Dimyāṭī and his writings on the qibla are not mentioned in the modern bio-bibliographical soures known to me. His shorter treatise on the qibla is extant in the unique copy MS Damascus Ẓāhiriyya 5579 (17 fols., copied *ca.* 1350), and has been translated by my former graduate student Bill Smyth. For a discussion of the qibla by a 13th-century Egyptian legal scholar see Shihāb al-Dīn Aḥmad ibn Idrīs al-Qarāfī, *al-Dhakhīra*, vol. I, Cairo, 1961, pp. 489-508.

[18] In the summer of 1982 I came across a copy of al-Dimyāṭī's longer treatise on the qibla, extant in MS Oxford Bodleian Marsh 592. This is the most important single source on the qibla ever compiled by a Muslim legal scholar. A preliminary translation of the entire manuscript was prepared with my former students Gregory Grant, Margaret Nito, Nadia Rawjee and Christopher Noey during a graduate seminar at New York University in the spring of 1983. It is now clear that al-Maqrīzī's account is to some extent based on al-Dimyāṭī.

advocated a qibla for Egypt that is inclined a little away from the rising point of the sun at mid-winter toward the rising point of Canopus. This is indeed a happy compromise, being halfway between the qibla of the *ṣaḥāba* and the qibla used in early mosques in Qarāfa, and also being closer to the qibla of the astronomers. There is a lot more that could be said about the qibla in Egypt, but we should return to the problem of the ventilators.

CHAPTER 7

ON THE ORIENTATION OF MEDIEVAL CAIRO

Although the ventilators of medieval Cairo cannot represent a purely Egyptian development of the ventilators of ancient Egypt, not least because they had a Persian name, the connection between ancient Egypt and the ventilators of medieval Cairo is, I believe, stronger than one might suspect. Orientation of religious architecture in ancient Egypt was generally toward the River Nile, although there do exist temples aligned toward the rising sun at a particular day of the year. There are also examples where, as a result of the local river orientation, the temples facing the river also roughly face the direction of winter sunrise; such is the case for the temple complexes at Thebes, Karnak and Luxor.[1] Yet it is not with ancient Egyptian architectural orientations that the ventilators of medieval Cairo have a connection.

The site selected for the new city of al-Qāhira in the latter half of the 10th century was alongside the ancient Egyptian canal joining the Nile to the Red Sea. This canal, re-opened once already by the Romans, was reactivated for a second time by ʿAmr after the Muslim conquest. The section of the canal alongside which al-Qāhira was built was straight, and its direction was of course necessitated by the local topography.[2] However, its direction is roughly 27° E. of N. The Canal was thus fortuitously perpendicular to the *qiblat al-ṣaḥāba* (see **Fig. 7.1**). Since the Fatimid city was built with a roughly orthogonal street-plan, it is aligned with the *qiblat al-ṣaḥāba*, the main axis, now roughly defined by the thoroughfare known as *al-Qaṣaba*, being perpendicular to the *qiblat al-ṣaḥāba*. The Fatimids, however, apparently preferred the qibla computed by the astronomers: the Azhar mosque (970) and the mosque of al-Ḥākim (1003) face the qibla derived by Ibn Yūnus, and are some 10° skew to the city plan (see **Fig. 7.2**). Some Fatimid religious architecture, on the other hand, faces the qibla of the *ṣaḥāba*.[3]

Here it should be pointed out that because of the hills to the south-east of Cairo, the direction of actual sunrise over the horizon will vary from one part of the city to another. This fact is never referrred to in the medieval texts, and the expressions *mashriq al-shitāʾ* or *mashriq al-*

---

[1] The astronomical orientation of ancient Egyptian temples and the significance of such orientations is still a matter of debate. According to a survey of temple orientations conducted by A. Badawy (*A History of Egyptian Architecture*, Berkeley & Los Angeles, Ca.: University of California Press, 1968, pp. 184-185), the general impression of predominantly south-eastern orientations in the examples chosen is superficial since most of the temples studied are in Eastern Thebes (which face roughly south-east, the River Nile flowing roughly north-east). See, more recently, Gerald S. Hawkins, "Astronomical Alinements in Britain, Egypt and Peru", in F. R. Hodson, ed., *The Place of Astronomy in the Ancient World*, London: Oxford University Press (for The British Academy), 1974, pp. 157-167; E. C. Krupp, "Astronomers, Pyramids, and Priests", in *idem*, ed., *In Search of Ancient Astronomies*, Garden City, N.Y.: Doubleday, 1977, and also the article "Orientierung" by G. Vitmann in *Lexikon der Ägyptologie*, 7 vols., Wiesbaden: Otto Harrassowitz, 1975-92, IV, cols. 607-609.

[2] On the topography of the area see Abu-Lughod, *Cairo* (n. 1:16), p. 9, fig. 2.

[3] See, for example, Kessler, "Mecca-Oriented Architecture in Cairo", B, p. 258, fig. 1.

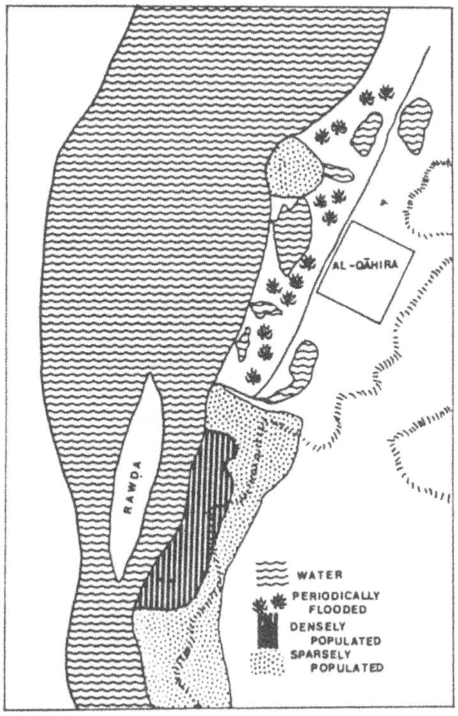

Fig. 7.1: The site of Cairo and its development at the time of the Fatimid invasion and the founding of al-Qāhira. Notice the orientation of the Red Sea Canal at *ca.* 27° E. of N. [After Abu-Lughod, *Cairo* (cited in n 1:16), p. 7).]

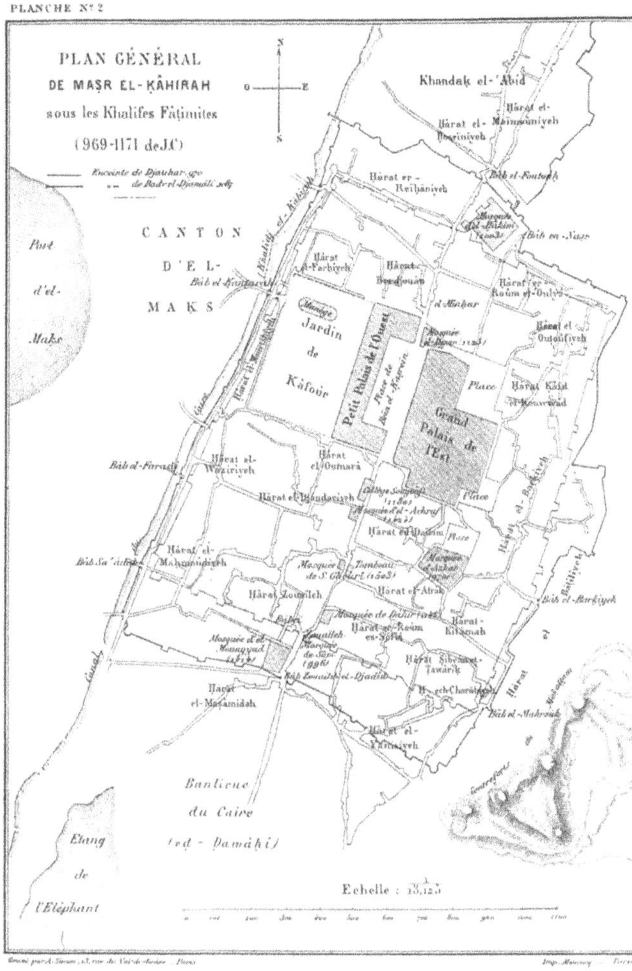

Fig. 7.2: A plan of the Fatimid city of Cairo, showing clearly the orthogonal street plan, with the major axis roughly parallel to the canal and the minor axis in the *qiblat al-ṣaḥāba*. Notice how the Azhar and al-Ḥākim Mosques are oriented in the qibla of the astronomers, rather than in accord with the street pattern. [Reproduced from Pauty, "Palais et maisons au Caire" (cited in n. 1:12).]

jady (winter sunrise, that is, the rising point of the sun when it is at the beginning of Capricorn) were taken by the astronomers to be the rising point of the sun at mid-winter over the true horizon rather than the actual horizon. In other words, it is the computed value of 27° S. of E. which was used for the *qiblat al-ṣaḥāba*.

In some of the Mamluk architecture flanking the main thoroughfare of the Fatimid city, such as the early-14th-century madrasa of al-Sulṭān al-Nāṣir Muḥammad (see **Fig. 7.3**), the facade of the building is aligned according to the street, and the inside of the building is skew to this orientation, being in the astronomically-defined qibla. The thoroughfare known as *al-Qaṣaba* is of course not exactly parallel to the canal. However, some of the kinks in it result from a deliberate attempt to align the outsides of the buildings on it with the *qiblat al-ṣaḥāba*.[4] We

[4] See N. al-Sayyad, *Streets of Islamic Cairo: A Configuration of Urban Themes and Patterns*, (*Studies in*

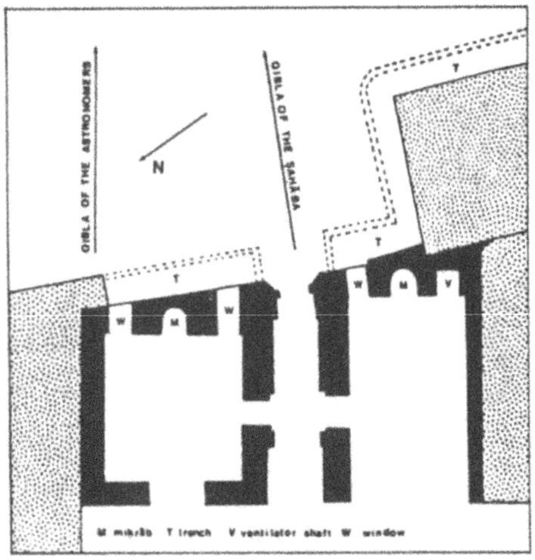

Fig. 7.3: Sketch of part of the madrasa of al-Ṣāliḥ Muḥammad on the main thoroughfare of medieval Cairo (based on Creswell). The outside wall of the madrasa is aligned with the main street plan, and is thus perpendicular to the *qiblat al-ṣaḥāba*. The facade is not aligned with the street *per se*, but rather with the axis of the city. The internal orientation of the various parts of the complex is in the mathematically-determined qibla. Thus, for example, the outside and inside wall of the facade are inclined to each other at 10°, which is the difference between the two qiblas. Notice also the ventilator-shaft next to the *miḥrāb* on the right; the shaft is oriented in the qibla of the astronomers, and presumably the wind-catcher that once adorned the top of the shaft would have been 10° skew to the shaft and aligned with the outside wall of the madrasa.

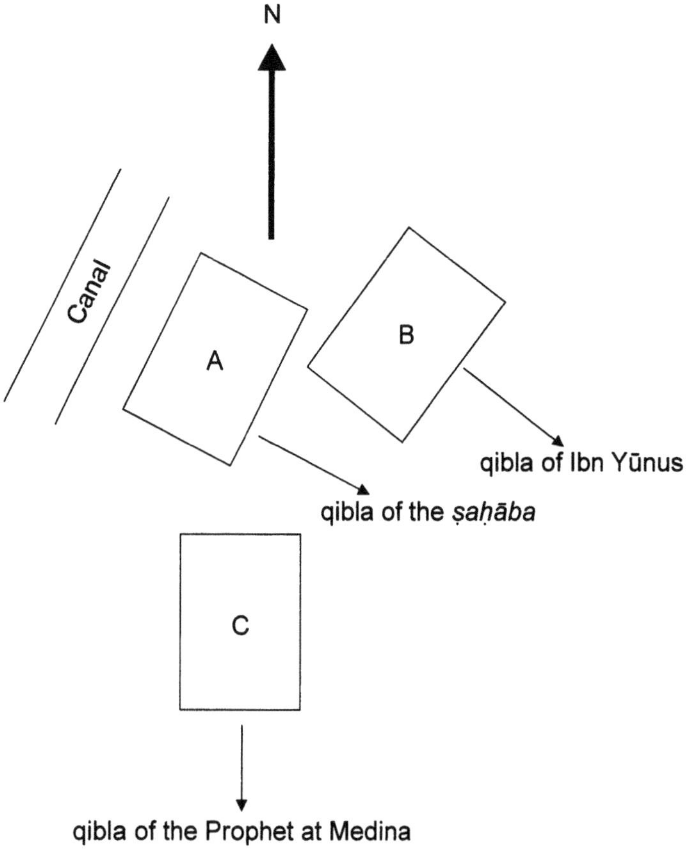

Fig. 7.4: Sketch-map of Mamluk Cairo showing the basic orientations of the three major areas. The Fatimid city (A) is aligned in the *qiblat al-ṣaḥāba*; the Mamluk "City of the Dead" (B) is aligned in the qibla of the astronomers; and the district of al-Qarāfa (C) is aligned towards the south.

notice also that the street-pattern of the Mamluk "City of the Dead" is aligned with the qibla of the astronomers (see **Fig. 7.4**), so that the mausolea flanking the streets are oriented, both internally and externally, in this qibla.

From the astronomical sources one would expect that *bādahanj*s erected on buildings on either side of streets parallel to the major axis of the Fatimid city would have had their openings parallel to the street, perpendicular to the *qiblat al-ṣaḥāba*, and that, likewise, *bādahanj*s erected on buildings on streets perpendicular to the main axis of the city would have had their openings facing the direction perpendicular to the street and hence also perpendicular to the *qiblat al-*

Islamic Architecture, no. 2), Cambridge, Mass.: Aga Khan Program for Islamic Architecture at Harvard University and Massachusetts Institute of Technology, 1981, for a penetrating study of the main thoroughfare of medieval Cairo, in which the qibla is not completely overlooked as it is in most modern writings on Islamic city-planning.

ṣaḥāba. **Fig. 1.1** shows how ventilators were aligned parallel to the canal. However, as we shall see in **Ch. 8**, it was not only aesthetic considerations that inspired this practice.

These features of the orientations of medieval Cairo and its architecture have not been noted before. al-Maqrīzī's remarks on orientations were completely ignored by K. A. C. Creswell in his monumental survey of early Egyptian architecture, but Creswell generally ignored orientations anyway. More recently, Christel Kessler has written on Mecca-oriented architecture in Cairo, at first without reference to what the medieval Egyptians thought was the direction of Mecca,[5] but later distinguishing between the *qiblat al-ṣaḥāba* and the qibla of the astronomers.[6] Clearly a historian of Cairene architecture has other interests and preoccupations than my own, but Kessler's writings on this subject raise all manner of important questions about the role and purpose of the architecture that she considers and no less about its surroundings (that is, the milieu in which the architect had to make it fit). And Kessler is the only colleague amongst historians of Islamic architecture ever to have addressed the issue of architectural orientations.[7] The recent monumental publication on Mamluk architecture in Cairo by the late Michael Meinecke contains numerous reliable plans but, unfortunately, no discussion of orientations or of the qibla.[8]

---

[5] See Kessler, "Mecca-Oriented Architecture in Cairo" A-B.

[6] *Eadem*, "Mecca-Oriented Architecture in Cairo", C.

[7] I have not been able to consult Injy Roy Shediac (Shami), *Qibla Orientation versus Street Alignment in the Mosques and Madrasas from Qaybay to the End of the Mamluk Period*, unpublished Master's thesis, American University in Cairo, 1984, because the A.U.C. would not provide me with a copy without the author's permission, which was not forthcoming.

[8] Michael Meinecke, *Die mamlukische Architektur in Ägypten und Syrien*, 2 pts., Glückstadt: J. J. Agustin, 1992.

CHAPTER 8

ON THE FOLKLORE OF THE WINDS IN MEDIEVAL EGYPT

We have seen that the ventilators of medieval Cairo were aligned in accordance with the street-pattern, which was fortuitously aligned with the *qiblat al-ṣaḥāba*, that is, with winter sunrise. But we have also seen a medieval text which implies that the *bādahanj*s should be aligned with winter sunrise anywhere in Egypt (**Ch. 4**). This we can now explain.

In a substantial number of Arabic texts relating to folk astronomy rather than to mathematical astronomy, we find statements about the winds: their number, their names, and the directions from which they blow.[1] Similar details are recorded by the Arab lexicographers.[2] The limits for the winds, that is, the boundaries of the sectors of the horizon from which they are thought to blow, are always astronomically defined in terms of the cardinal directions, or sunrise and sunset, or the risings and settings of particular stars. The schemes are not generally associated with any particular locality, but most of them seem to be of Hejazi provenance. These Arab wind-schemes constitute a meteorological tradition quite independent of the classical traditions represented in the writings of Aristotle and Theophrastos, which were translated into Arabic in the 8th and 10th centuries, respectively.[3]

I am not aware of the existence of any treatise in which the limits of the winds in Egypt are discussed in detail. The 10th-century ʿIrāqī scholar al-Masʿūdī discusses the winds in Egypt, and implies that they blow from the cardinal directions when he observes that each one corresponds to a side (the term used is *rukn*, actually "corner") of the pyramids.[4] Likewise, al-Maqrīzī implies that the four winds blow from the cardinal directions.[5] However, the late-14th-century Cairene scholar al-Qalqashandī in his encyclopaedic work *Ṣubḥ al-aʿshā* presents a different scheme for the limits of the winds.[6] In this scheme the winds blow from the quadrants defined by the cardinal directions. The relevant passage (**Text 11**) is as follows:

> "The main winds are four. The first is the *ṣabā* which is the one that comes from the east {*sic*, read: from between the east and the north}. The author of the book *Ṣināʿat al-kuttāb* said that the Egyptians call it the *sharqiyya* because it comes from the rising point (*mashriq*) of the sun. ... The second is the *dabūr* and it blows from between the setting point of the sun and the direction of the south celestial pole ... . The third is the *shamāl* ... and it blows from between the direction of the north celestial pole and

[1] See the *EI*$_2$ articles "Maṭlaʿ" (repr. in King, *Studies*, C-XI), and "Rīḥ" by Miquel Forcada.

[2] See, for example, Lane, *Lexicon*, *sub shamāl*, *qabūl*, *janūb* and *dabūr*.

[3] On Arab meteorology see Sezgin, *GAS*, VII, pp. 203-370.

[4] See al-Masʿūdī, *al-Tanbīh wa-'l-ishrāf*, edited by M. J. De Goeje, Leiden: E. J. Brill, 1894, translated by B. Carra de Vaux as *Maçoudi: Le livre de l'avertissement et de la revision*, Paris: Imprimerie Nationale, 1896, both repr. in *Islamic Geography*, vol. 41 (1992), text, pp. 16-21, transl., pp. 25-36.

[5] See al-Maqrīzī, *al-Khiṭaṭ* (n. 2:7), II, p. 263.

[6] al-Qalqashandī, *Ṣubḥ al-aʿshā* (n. 2:9), II, pp. 175-177. The author of the work entitled *Ṣināʿat al-kuttāb* appears to be the early-10th-century Egyptian grammarian Ibn al-Naḥḥas (n. **IV**-8:24).

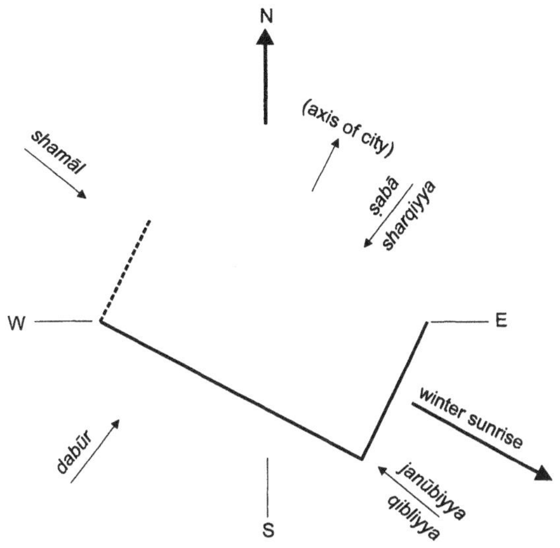

Fig. 8.1: The limits of the winds in medieval Egyptian folklore, as reported by al-Qalqashandī, and their relation to the orientation and shape of ventilators.

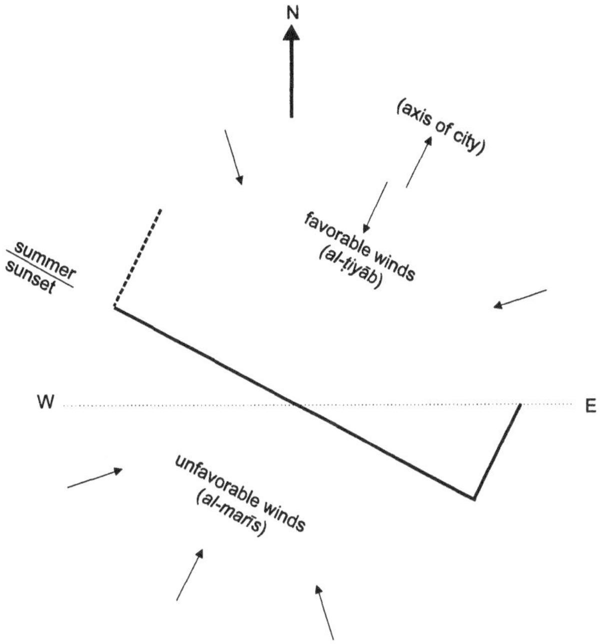

Fig. 8.2: The limits of the favourable and unfavourable winds in Egyptian folklore, as described by Najm al-Dīn al-Miṣrī (see **Fig. 5.2**).

> the setting point of the sun. ... The fourth is the *janūbiyya* and it blows from between the direction of the south celestial pole and the rising point of the sun. It is called *qibliyya* in Egypt because it comes from the qibla there ... and it is the worst of the winds according to the Egyptians."

Fig. 8.1 shows al-Qalqashandī's wind-scheme and its relation to the kind of *bādahanj* described in the Ambrosiana manuscript. Notice that both the open and the closed parts correspond to 180° on the horizontal scale. It appears that some *bādahanj*s in Cairo such as that on the *Musāfirkhāne*, having the width equal to one half of the length, were constructed to catch both the *shamāl* and the *ṣabā*, in conformity with this tradition of local folklore.

As we have seen, Najm al-Dīn and Ibn al-Qāṣiḥ describe a different kind of ventilator, with the width equal to one quarter of the length. In this variety, angles of 153° and 207° for the open and closed parts of the *bādahanj* are mentioned. Najm al-Dīn's text reveals that the western side of the wind-catcher is open to the wind as well as the northern side, as in the case of the *bādahanj* in the *Musāfirkhāne* described above. His diagram indicates the reason: the favourable northerly winds were thought to blow from points between due east and summer sunset, and the unfavourable southerly *marīs* winds from points between summer sunset and due east. **Fig. 8.2** shows this second wind-scheme and its relation to the kind of *bādahanj* described by Najm al-Dīn and Ibn al-Qāṣiḥ.

We now see why the sources cited in **Ch. 5** state that the ventilators in Egypt as a whole, rather than just Cairo, were to be erected in line with winter sunrise. Only in Cairo did the street plan also lie in this direction.

CHAPTER 9

CONCLUSION: CAIRO, A CITY FACING THE KAʿBA

The first Muslims in Egypt used winter sunrise for the qibla because they wanted to be facing one particular side of the Kaʿba, which itself is astronomically aligned.[1] The base of the Kaʿba is rectangular; its minor axis is roughtly solstitially aligned towards summer sunrise in the east and winter sunset in the west, and its major axis points roughly toward the local rising point of the star Canopus. (The minor axis is actually aligned exactly to the farthest southerly setting point of the new moon at the winter solstice, and, for the latitude of Mecca, these lunar and Canopic alignments are roughly perpendicular.) Also, one diagonal of the base of the Kaʿba is almost north-south.

The reason for these approximate solar and Canopic alignments appears to be that the underlying directions were the limits for the winds in a prevailing pagan Arab tradition. Hints of this association are contained in various Arabic folk astronomical and lexicographical texts relating to the winds and their names.[2] My own findings on the Kaʿba reinforce those of Joseph Chelhod who, albeit from different premises, also observed that it was "a microcosm of the universe".[3]

The Kaʿba, originally just a walled enclosure, was a pagan Arab shrine, perhaps associated with worship of the sun and moon, which was built in a valley surrounded by sacred mountains near the site of a sacred spring, with sacred stones embedded in its walls. It was oriented astronomically to face the four winds whose directions were defined according to local tradition. The east wind was favoured by the Arabs, and the south wind yet more so, because it was the wind which brought rain. One of the sacred stones, a black stone probably of meteoric origin, was at some time set in the most sacred corner of the enclosure, namely, the south-east corner, enjoying exposure to both the east and the south winds. These pre-Islamic meteorological associations of the Kaʿba were not mentioned in the *Qurʾān* and were perhaps deemed of little consequence to the new Muslim community, for they are barely mentioned in the oldest Muslim traditions.

---

[1] On the Kaʿba see the articles "Kaʿba" in EI_1 by A. J. Wensinck and in EI_2 by A. J. Wensinck and Jacques Jomier. The latter omits the legends on the origin of the edifice documented in the former; for a popular overview of those legends see Grant Alexander, "The Story of the Kaʿba", *Muslim World* 28 (1938), pp. 43-53. Also of interest is Elsebeth Thomsen, "New Light on the Origin of the Holy Black Stone of the Kaʿba", *Meteoritics* 15:1 (1980), pp. 87-91. An important investigation of the ritual functions in Pre-Islamic and early Islamic times is Rubin, "The Kaʿba".

The alignment of the Kaʿba is discussed in Hawkins & King, "Orientation of the Kaʿba".

[2] See, for example, the definitions of the winds given in Lane's *Arabic-English Lexicon*, which is based on the medieval Arab lexicographers. For the specific association of the winds with the Kaʿba see, for example, al-Bīrūnī, *Astrology*, transl. R. R. Wright, p. 19.

[3] See Joseph Chelhod, "A Contribution to the Problem of the Pre-eminence of the Right, Based upon Arabic Evidence", in Rodney Needham, ed., *Right & Left*, Chicago, Ill. & London: University of Chicago Press, 1973, pp. 239-262, esp. pp. 248-253.

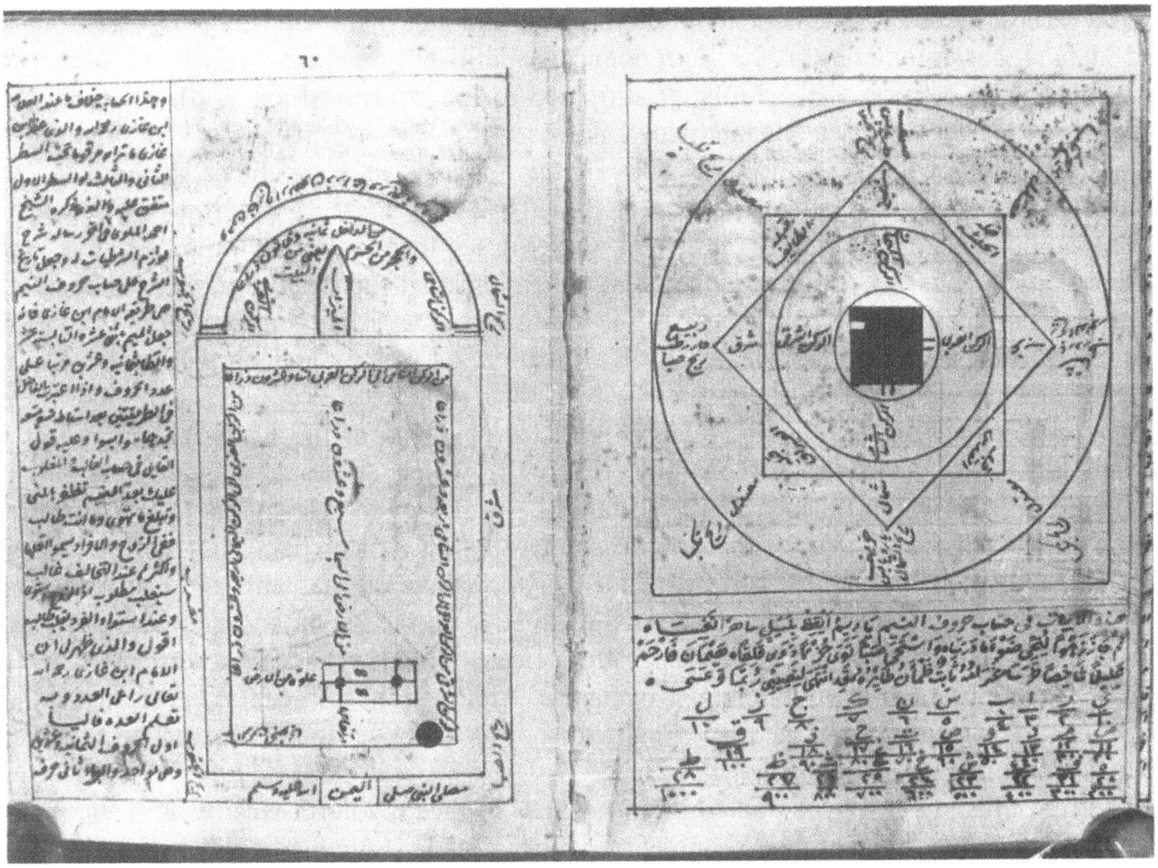

↑
Fig. 9.1: Two diagrams in a treatise of uncertain origin preserved in an 18th-century Egyptian manuscript. The diagram on the right shows the Kaʿba at the centre of two diagonally-superimposed squares usually associated with the representation of the four elements and their qualities. The seasons and their weather conditions, the winds, and the qiblas of four main geographical areas feature on the diagram. The diagram on the right shows in an exaggerated fashion the rectangular base of the Kaʿba, with the Black Stone set in its south-east corner. [Reproduced from MS Cairo ṬJ 811, fols. 59v-60r, courtesy of the Egyptian National Library.]

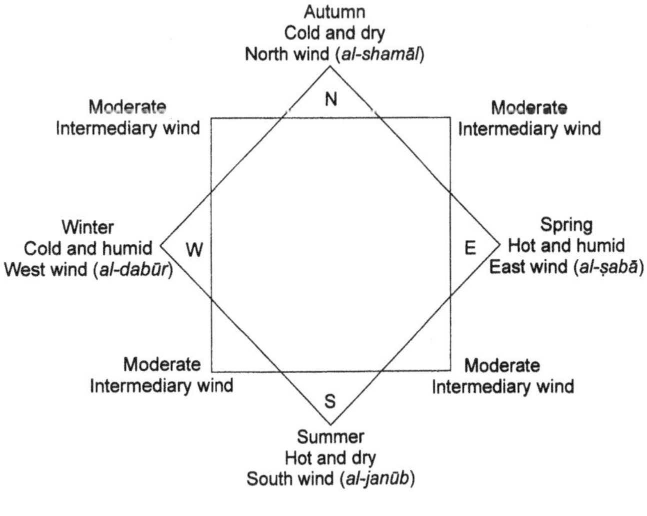

→
Fig. 9.2: The diagram on the right of Fig. 9.1 interpreted.

The earliest recorded statement about the limits of the four main winds (shamāl, ṣabā, janūb and dabūr) corresponding to the astronomically-defined directions which are elsewhere associated with the axes of the Kaʿba is attributed to Ibn ʿAbbās (619-687/8), the celebrated Companion of the Prophet Muḥammad. Likewise, the earliest mention of an association between the four winds and the four sides of the Kaʿba is in a statement attributed to the celebrated religious personality, al-Ḥasan al-Baṣrī (642-728).[4] I see no reason to suspect that these traditions were deliberately fabricated; rather, I see them as reflecting pre-Islamic Arab meteorological beliefs. Certainly, there are no indications that such traditions were deliberately suppressed in later times. They appear in the genre of books called *kutub al-ʿaẓama*, dealing with the greatness of God as revealed by His creation, as well as lexicographical works and treatises dealing with folk astronomy.

In a different guise these astro-meteorological associations of the Kaʿba also reappear in certain Ṣūfī texts. Consider, for example, **Fig. 9.1**, which shows a diagram relating the orientation of the Kaʿba (falsely displayed) and the qiblas of four locations to the directions of the winds and the weather of the four seasons, using an Aristotelian-type configuration of two diagonally-superimposed squares which usually represent the elements and their qualities.[5] This illustration is found in an anonymous Ṣūfī treatise on the Kaʿba and the qibla preserved in a late-18th-century Egyptian manuscript, but is clearly of earlier origin.[6] The treatise, which is in Arabic, appears to be of Ottoman provenance. Although the 13th-century Andalusī mystic Ibn al-ʿArabī is mentioned in the preceding treatise, I have not yet found any earlier references to this kind of representation of the Kaʿba.

The first Muslims in Egypt wanted their mosques to face the north-west wall of the Kaʿba, and to do this, they thought they should face winter sunrise. Some later Egyptians thought it more reasonable to face the rising point of Canopus. When one stands in front of the north-east wall of the Kaʿba in Mecca one is facing the rising point of Canopus, and al-Maqrīzī informs us that this latter direction was also adopted for the qibla in Egypt. Indeed, as we have seen in **Ch. 6**, al-Maqrīzī himself actually confuses mid-winter sunrise and the rising-point of Canopus. It is also worth mentioning that the 14th-century Moroccan astronomer Ibn al-Bannāʾ stated that the mosque of ʿAmr was erected towards the rising point of the star α Scorpionis,[7] which is roughly the same direction as winter sunrise, and recalling that the 9th-century Andalusī jurist Ibn Ḥabīb stated that the qibla in Cordova was towards the rising point of this star "because it rises at the corner of the Black Stone."[8] The precise meaning of this remark is not clear to me, because if one stands with one's back to this corner one faces *ca.* 15° S. of E.,

[4] These remarks are recorded in a treatise by the 15th/16th-century Egyptian polymath Jalāl al-Dīn al-Suyūṭī – see Heinen, *Islamic Cosmology*, pp. 157 and 158. al-Suyūṭī was quoting the *Kitāb al-ʿAẓama* by the 10th-century scholar Abu 'l-Shaykh.

[5] On this symbolism see Price, *Science since Babylon*, pp. 76-78.

[6] The diagrams illustrated in **Fig. 9.1** are found in MS Cairo ṬJ 811,7, fols. 59v-60r, copied 1783/84.

[7] On Ibn al-Bannāʾ see the article by Juan Vernet in *DSB* and the references there cited. This quotation from his lost *Kitāb Dalāʾil al-qibla* is contained in MS Cairo TR 132, copied *ca.* 1850, p. 132, on which see *Cairo ENL Survey*, nos. F23 and F62.

[8] On Ibn Ḥabīb see the article in *EI*$_2$ by A. Huici Miranda. This quotation of his is recorded in Renaud, "Astronomie et astrologie marocaines", p. 58, but is not found in the text edited in Kunitzsch, "Ibn Ḥabīb's *Book on the Stars*".

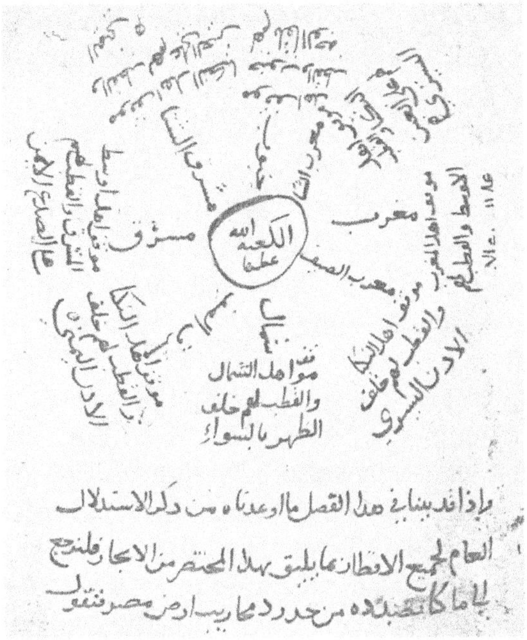

9.3

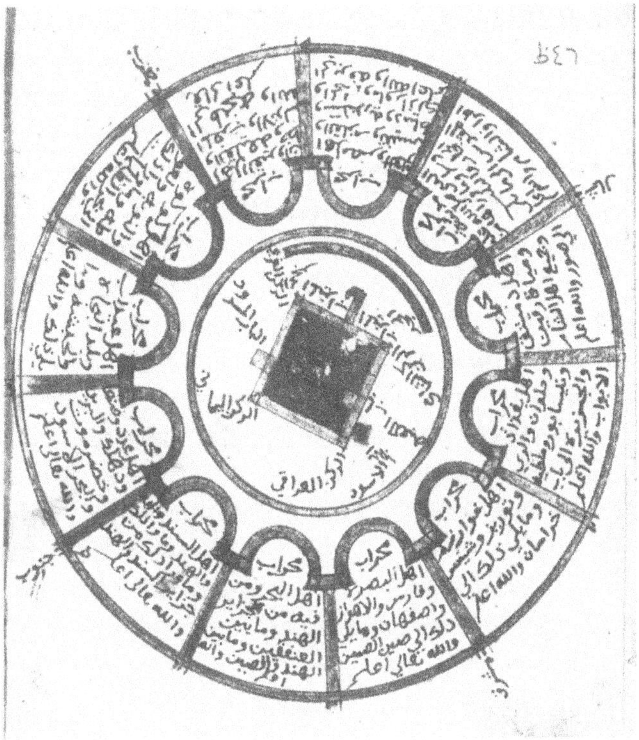

9.4

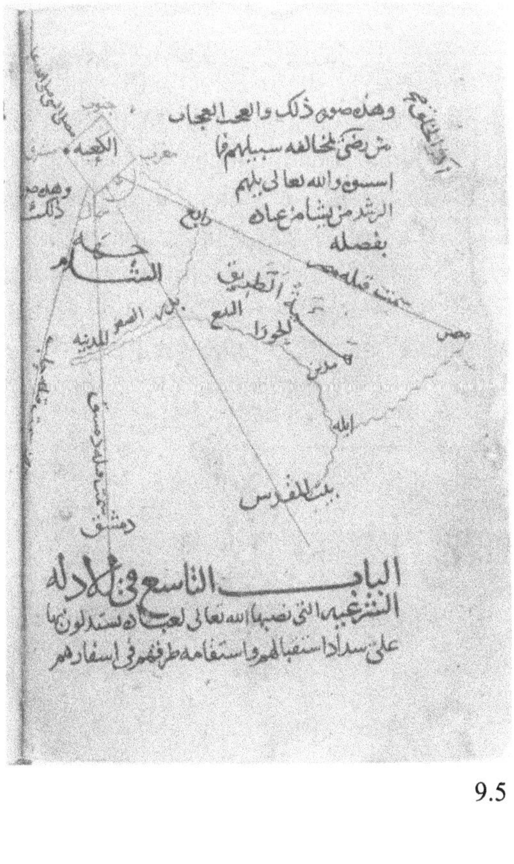

9.5

Fig. 9.3: A diagram in the shorter treatise on the determination of the qibla by non-mathematical means by the 12th-century legal scholar al-Dimyāṭī. Eight regions are shown about the Kaʿba, each one associated with a wind. The qiblas in each region are defined in terms of the cardinal directions and the directions of sunrise and sunset at the solstices, as well as in terms of the way in which one should stand with respect to the Pole Star. Underlying this scheme is the notion that all of the sides of the Kaʿba are solstitially aligned. See further Hawkins & King, "Orientation of the Kaʿba" (cited in n. 9:1), pp. 104-105. [From MS Damascus 5579, fol. 14r, courtesy of the Ẓāhiriyya Library.]

Fig. 9.4: A diagram relating the qiblas of 12 regions of the world about the Kaʿba to the perimeter of the astronomically-oriented Kaʿba. This is found in an Egyptian manuscript datable *ca.* 1450 and is without attribution, but is taken from the geographical works of Yāqūt (*ca.* 1200) or al-Qazwīnī (*ca.* 1250). Originally the information in the 12 sectors included statements about which astronomically-defined directions one should face in each sector. This kind of information was first documented by the Yemeni scholar Ibn Surāqa *ca.* 1000. [From MS Cairo DM 634,4, fol. 46v, courtesy of the Egyptian National Library.]

Fig. 9.5: A diagram in al-Dimyāṭī's longer treatise on the qibla showing the relative positions in Egypt, Palestine and Syria with respect to the Kaʿba. [From MS Oxford Marsh 592, fol. 88v, courtesy of the Bodleian Library.]

whereas the star rises *ca.* 25° S. of E. in Mecca (epoch 900 A.D.). However, al-Dimyāṭī (**6**) states that if one faces the section of the north-west wall of the Kaʿba between the western corner and the water-pipe then the star a Scorpionis rises directly in front of one.

On the other hand, the first Muslims in Iran, al-ʿIrāq and Transoxania faced the direction of winter sunset in order to face the north-east wall of the Kaʿba, and when standing in front of that wall, one is indeed facing winter sunset. The first Muslims in Egypt and al-Andalus thus assumed a kind of symmetry in the orientation of the Kaʿba that in fact does not exist. Such a symmetry is, however, explicitly described by the 13th-century scholar Ibn al-Ajdābī of Tripoli (Libya)[9] who, in his treatise on folk astronomy, presents a description of eight regions of the world about the Kaʿba, and writes of their qiblas as if the corners of the Kaʿba were cardinally aligned and the sides were all solstitially aligned.[10] A cruder scheme based on the same underlying notion is actually illustrated in the shorter treatise of al-Dimyāṭī (see **Fig. 9.3**). This arrangement of the world in eight sectors about a cardinally/solstitially-aligned Kaʿba is less common in the medieval sources currently known to me than the arrangement with twelve sectors about a Kaʿba with its minor axis aligned towards summer sunrise, as shown in **Fig. 9.3**. Such schemes for what I call "the sacred geography of Islam" are attested in more than thirty different sources, and I surveyed them in a study prepared in the 1980s but not yet published.[11]

As we have seen, different directions were used for the qibla in Egypt between the 7th and 15th centuries. al-Maqrīzī was upset by their divergence, but the legal scholars were in general less concerned. Several medieval legal treatises, as well as al-Maqrīzī in his discussion of the qiblas in Egypt, deal with the determination of the qibla by two non-mathematical methods, referred to as *iṣābat al-ʿayn* and *iṣābat al-jiha*, or simply as *ʿayn* and *jiha*.[12] The essence of the first method is that one should stand in such a way that, if one could see the Kaʿba, one would be looking along a side of the Kaʿba. The essence of the second is that when one is standing in that direction, then the sector of the horizon that one can see, namely, roughly a quadrant, serves as the qibla. The qibla directions prescribed for the first method are always in terms of astronomical risings and settings, equated with those that one would be facing if one were standing directly in front of the appropriate section of the perimeter of the Kaʿba. However the *ʿayn* might be determined, the *jiha* would be the entire quadrant that was most appropriate. Thus in Egypt, as well as the Maghrib and al-Andalus,[13] the limits of the *jiha* were east and south, so that any qibla in the south-eastern quadrant would be legally acceptable

---

[9] On Ibn al-Adjābī see the article by Charles Pellat in EI_2 (under "Ibn al-Adjdābī"). For his discussion of the sectors of the world around the Kaʿba see Ibn al-Ajdābī, *al-Azmina wa-'l-anwā'*, ed. Izzat Hassan, Damascus: Wizārat al-Thaqāfa wa-'l-Irshād, 1964, pp. 119-125.

[10] See also Hawkins & King, "Orientation of the Kaʿba", p. 105.

[11] Texts, translations and analysis are in my unpublished monograph *The Sacred Geography of Islam* (cited in n. 5:14). Several such schemes, mostly corrupt, were presented in Konrad Miller, *Mappae Arabicae: Arabische Welt- und Länderkarten*, 5. Band: Weltkarten, Stuttgart: Selbstverlag (?), 1931, (repr. in *Islamic Geography*, vols. 240-241 (1994)), pp. 149-155.

[12] See, for example, the discussion in al-Ghazzālī, *Iḥyā' ʿulūm al-dīn*, Cairo, 1967, section on the qibla; al-Qarāfī, *al-Dhakhīra* (cited in n. 6:17), pp. 489-508, and al-Maqrīzī, *al-Khiṭaṭ* (n. 2:7), II, pp. 256-264.

[13] See King, "Some Medieval Values of the Qibla at Cordova", an appendix to *idem*, "Andalusī Sundials", pp. 370-387, for a discussion of this situation in al-Andalus, as recorded by a 12th-century astronomer.

according to those who favoured the *jiha* over the *ʿayn*. These notions are clearly illustrated in **Fig. VIIa-2.7**. A modest knowledge of geography would inevitably have obviated the need for such deliberations – see **Fig. 9.5**.

To sum up, we have shown first that the various orientations of mosques in medieval Cairo were all expressions of a desire to have the mosques (or at least their *miḥrāb*s) facing the Kaʿba. Secondly, we have shown that the orientation of the Fatimid city of al-Qāhira, determined by the orientation of the Pharaonic Red Sea canal, which in turn was determined by topography, was fortuitously perpendicular to the astronomically-determined *qiblat al-ṣaḥāba*, which deliberately faced the north-western wall of the Kaʿba. It was the orientation of the lost *bādahanj*s of medieval Cairo as described in the medieval astronomical sources that provided the clue to these discoveries. The *bādahanj*s, veritable status symbols of medieval Cairo, were set up, as the poet said, "turned away from the qibla", but not only "for the love of of air".

(١) ملتقطات من بعض النصوص التاريخية التي فيها ذكر الملقف والبادهنج

ألف ليلة وليلة ، ج ٢ ، ص ١٠٣ - ١٠٤

... ثم سحب سيفه في يمينه وأخذ ملقفه في يساره وأقبل على قاعة الجلوس التي للخليفة ونصب سلم التسليك ورمى ملقفه على قاعة الجلوس فتعلق بها وطلع على السلم إلى السطوح ورفع طابق القاعة ونزل فيها فوجد الطواشية نائمين فبنجهم وأخذ بدلة الخليفة والسبحة والنمشة والمنديل والخاتم والمصباح الذي بالجواهر ثم نزل من الموضع الذي طلع منه ...

وقفية مدرسة السلطان برقوق (محكمة الأحوال الشخصية بشبرا القاهرة ، وثيقة رقم ٥١ ، سطر ٩٦ - ٩٨)

... إيوان مسقف نقيا مدهون كافوريا بصدره شباك حديد كبير مطل على الدركاة المقدم ذكرها أمامه دور قاعة منصوبة بالغرد [؟] بملقف بها أربعة أبواب كل منها مربع ...

ابن تغري بردي ، النجوم الزاهرة ، ج ٤ ، ص ١٠٢

... وأما خطبة الخليفة (المعزّ) في شهر رمضان ... وكان إذا أراد أن يخطب يتقدم متولي خزانة ألفرش إلى الجامع ويغلق المقصورة التي برسم الخليفة والمنظرة وأبواب مقاصيرها وباذهنج المنبر ثم يركب متولي بيت المال آلخ آلخ ... ... وكان ذلك بجامع الأزهر قبل أن يبني الحاكم جامعه ثم صار بعد ذلك بجامع الحاكم ...

ماير ، وقفية قايتباي ، ص ٤٠

... بصدر الإيوان مرتبة بواجهتها قنطرة معقودة يعلوها بيت باذاهنج ...

ماير ، ص ٤٣

... فأما الإيوان الكبير فإنه على يسرة الداخل به خرستانان يعلوها باذاهنج ...

ابن حبيب ، تذكرة النبيه ، ص ٣٤٥

... والباذاهنجات علو القيسارية المذكورة ...

المقريزي ، السلوك ، ج ٣:١ ، ص ٢٨١

... فأشارت إلى بادهنج البيت فوجدوا السلطان قد لبس ثياب النساء واختفى فيه ...

المقريزي ، الخطط ، ج ٢:١ ، ص ٢٢٢

... وإذا بالنار ابتدأت من أعلا البادهنج وكان ارتفاعه من الأرض زيادة على مائة ذراع بذراع العمل وراوا فيه نفطا قد عمل فيه فتيلة كبيرة ...

القلقشندي ، صبح الأعشى ، ج ٤ ، ٤٣

... وقد ذكر في مسالك الأبصار أن أكابرهم كانوا يجعلون في أكمامهم بادهنجات مفتوحة وقد صار ذلك الآن قاصرا على ما يلبسونه من التشاريف ...

(٢) تعليق في هامش جدول ارتفاع الشمس في سمت البادهنج يرد في مخطوطة دار الكتب المصرية ميقات ٧٥٨، ق ٤ظ

في معرفة سمت البادهنج لكل عرض وهو أن تستخرج سعة المشرق لرأس الجدي فما خرج معك فهو سمت البادهنج والله اعلم

(٣) قطعة من رسالة محمد حتاتة الفارسكوري منقولة من مخطوطة دار الكتب المصرية ميقات ٦٣٩، ٥، ق ٥٠و

... الباب الثاني عشر في معرفة سمت البادهنج اعرف سعة المشرق على مدار الجدي لسمت مكة ثم ابعد بقدره في الجنوب فما كان فهو سمت البادهنج والله أعلم ...

(٤) قطعة من مجموعة فوائد منسوبة إلى ابن يونس منقولة من مخطوطة ميلانو امبروزيانا ٤٩ج، ق ١١٠ ظ

... معرفة إخراج البادهنج استخرج الجهات الأربع ثم عد من نقطة المشرق طالبا الجنوب بمقدار سعة مشرق الجدي ثم مد عليه خطا هو خط البادهنج ربعه بخط آخر واقم المحلة على ذلك التربيع وأحسن ما يكون أن تقسم وجهه بعدة أجزاء مثلا وتجعل جنبه منها خمسة ونصف هذا الذي وقع عليه الاصطلاح من أهل الصناعة والله تعالى الموفق ...

(٥) قطعة من رسالة نجم الدين المصري في الآلات الفلكية كما وردت في مخطوطة دبلن تشستر بيتي ١٠٢ فارسي، ق ٥٢ ظ

الباب صا في معرفة وضع محلة البادهنج وأسمائه وقدر الهواء الطياب وقدر الهواء المفسود من قسمة دائرة الأفق لذلك العرض أما أسمائه (!) فهو (!) أربعة فراتي ومجنح وكلي[1] وعادلي فأما الفراتي فهو القائم على سطح مستقيم وأما المجنح فهو القائم على سطح مجنح كأجنحة الطير وأما الكلي فهو السطح المائل وأما العادلي فهو الذي يكون بجنب حائط فإذا أردت وضعه فهو أن تدير دائرة كاملة وتربعها وتمد خطا من سعة مشرق الجدي إلى سعة مغرب السرطان بتلك البلد فهو محلة البادهنج في البلاد البعيدة عن البحر المالح [ملاحظة: يوجد نقص في الأصل] مثل سكندرية ودمياط وعيذاب[2] وجدة وما أشبه ذلك فإذا عملت محلة البادهنج كما تقدم فمد خطا من نقطة المشرق إلى ضعف سعة مشرق الجدي وهو الموضع المسدود ثم مد خطا من نقطة المغرب إلى ضعف سعة مغرب السرطان وهو الموضع المفتوح فكان جملة عدد الهواء الطياب من قسمة الدائرة **قفج** درجة والهواء المفسود من قسمة تلك الدائرة رز درج ثم قسمنا أحد هذا (!) الخطين من طول محلة البادهنج فكان الربع من محلة البادهنج فاعلم أن محلته هي طوله والموضع المفتوح **هو** عرضه من جهة المغرب والمسدود **هو** عرضه من جهة المشرق وإن شئت من جهة الحساب فانسب عرضه من طوله وتأخذ جيب سعة أحد المنقلبين في ذلك العرض وهو **كز مد في** عرض **ل** حفظناه ثم زدنا سعة المشرق **وهي كز ل** دقيقة على **ص** فما جمع **فيز ل** دقيقة أسقطنا من ذلك ضعف سعة المشرق **وهو نه** الباقي **سب ل** دقيقة أخذنا جيبها فكان **نج نج** دقيقة أضعفناها فكان **قو كو** قسمنا عليها المحفوظ فحصل m **يه لح** ثانية وهو ربع الطول وربع قيراط الطول محبورا فتقسم محلة البادهنج بأربعة وعشرين قسما متساوية وتأخذ من تلك القسمة ست قراريط وربع قيراط فهو عرض البادهنج ولم اطلع على هذه الطريقة ممن تقدم من الفضلاء ولا أعلم أحدا ذكرها فمن أراد التحرير في ذلك فليفعل ذلك كما[3] بيناه وشرحناه فافهم تصب

التحقيق: [1] في الأصل: وجلى (؟)؛ [2] في الأصل: هبداب؛ [3] في الأصل: فيما

(٦) قطعة في نصب البادهنج ماخوذة عن تحفة الطلاب في العمل بربع الأصطرلاب لأبي البقاء على بن عثمان ... بن القاصح كما ترد في مخطوطة دار الكتب المصرية ميقات ٢٦، ق ٢٢ظ - ٢٣و

الباب الثالث والستون في معرفة نصب المحاريب والبوادهنج ... ... نصب البادهنج فطريقة أن تخط الجهات الأربع في بلاطة كما تقدم وتبعد عن نقطة المشرق إلى جهة الجنوب بقدر سمت البادهنج في ذلك البلد من أجزاء قوس الارتفاع وتعلم في قوس

الارتفاع علامة ثم تضع المسطرة على مركز الربع وعلى تلك العلامة ثم تخط خطا في البلاطة على حرف المسطرة فيكون ذلك الخط هو سمت البادهنج وهو موضع المَحِلّة والبادهنج على أنواع فراتي وهو الذي قائم على سطح مستقيم ومجنح وهو الذي قائم على سطح كأجنحة الطير وعادلي وهو الذي يكون بجنب حائط وكلي وهو الذي هو مائل السطح وإن كانت البلدة التي تضع فيها البادهنج على سواحل البحر المالح[1] كالإسكندرية ودمياط ونحو ذلك فإن محلة البادهنج مستقبلة البحر فإذا عملت محلة البادهنج فمد خطا من نقطة مشرق الاعتدالين إلى ضعف سعة مشرق الجدي وهو الموضع المسدود ثم تمد خطا من نقطة المغرب إلى ضعف سعة مغرب السرطان وهو الموضع المفتوح فكان جملة المطلوب من قسمة[2] دائرة الأفق **قنج** درجة والهواء المفسود رز درجات فإذا قسمنا أحد[3] هذه الجوانب من طول محلة البادهنج فكان الربع منها فتعلم أن محلة البادهنج هي طوله وعرضه هو ربع طوله واعلم أن سمت البادهنج بمصر **كز** درجة **مه** دقيقة وسمت القبلة بمصر **لز** وقيل **لز ل** لأن عرض مكة اختلف فيه فقيل **كا** وقيل **كا ل** وقيل غير ذلك **ه**[4] واختلف في عرض مصر فقيل **ل** وقيل **كط نه** وطوله **نه** وطول مكة **سز** وقيل غير ذلك وإن مصر في الربع الغربي الشمالي ومكة في الربع الشرقي الجنوبي **ه**[4] وإن أهل مصر وسكندرية إلى القيروان إلى تاهرت والسوس الأقصى إلى البحر الأسود وما كان على سمت ذلك من البلاد يستقبلون في صلاتهم من ركن الكعبة الغربي إلى ميزابها وقد ذكرت الاستدلالات على جهات الكعبة بمهبات الرياح الأربع في تصنيف لطيف

التحقيق : [1] في الأصل : كاالمالح ؛ [2] في الأصل : تسمية قسمة ؛ [3] في الأصل : ابعد ؛ [4-4] أي : انتهى

(٧) قطعة من رسالة تحفة الأحباب في نصب البادهنج والمحراب لابن المجدي منقولة من مخطوطة دار الكتب المصرية ميقات ١٨٣ ، ٢ ، ق ٧و

... وبعد فهذه رسالة لطيفة سميتها بتحفة الأحباب في نصب الباداهنج والمحراب نذكر ذلك في عرض مخصوص ويقاس غيره عليه فنقول اعلم أن سمت القبلة بمدينة مصر حماها الله على سبعة وثلاثين درجة وسمت الباداهنج سبعة وعشرين ونصف وكلاهما في الربع الشرقي الجنوبي ...

(٨) قطعة من رسالة زكرياء البلبيسي في العمل بربع المقنطرات منقولة من مخطوطة دار الكتب المصرية طلعت مجاميع ٣،٣٦٧, ق ٢٠ظ

... وقد تحددت لك الجهات الأربع فإذا خرجت لك الجهات الأربع فعد من قوس الارتفاع بقدر سمت القبلة وهو سبعة وعشرون درجة في مصر[1] أو بقدر سمت الباداهنج وهو مثل سعة المشرق لبلدك طالبا[2] الجنوب وضع الخيط عليه يكون سواء كان جهة القبلة أو جهة الباداهنج[3] ثمة الجهة المطلوبة[4]

التحقيق : [1] في الأصل : المصر [!] ؛ [2] في الأصل : الباداهيخ [!] ؛ [3] في الأصل : كالبا ؛ [4] في الأصل : المطلوب

(٩) قطعة من رسالة في معرفة القبلة لغرس الدين خليل بن أحمد الحلبي منقولة من مخطوطة دار الكتب المصرية ميقات م ١١٤ ، ق ٥و - ٥ظ

... المقصد الثاني اعلم أنه في سنة ثمانية عشر من الهجرة في زمن خلافة عمر بن الخطاب رضي الله عنه فتح عمرو بن العاص مصر ثم بنى مسجده المشهور بالجامع الجديد وكانت أصحاب رسول الله صلى الله عليه وسلم يومئذ في مصر حال بناه نيفا وثلثين وكان معهم[1] من المسلمين المجاهدين نحو ستة عشر ألفا وقد كان سمت المحراب للمسجد المذكور نحو سبعة[2] وعشرين درجة ولم ينكر أحد من المذكورين على ذلك بل كان بإجماعهم وأهل الحساب لما أخرجوا سمته بالطرق الحسابية المبرهنة حكموا بأن السمت بمصر سبعة وثلاثون وحكموا بتخطية المحراب المذكور وفي ذلك خطر عظيم إذ في ذلك تخطية الصحابة الذين قال صلى الله عليه وسلم في حقهم أصحابي كالنخوم بأيهم اقتديتم[3] اهتديتم[4] ولهذه القصة نظائر كثيرة من مشاهد الصحابة وقبورهم ومساجدهم وأيضا ذهب القائلون بإصابة العين إلى أن الكوفة وبغداد وهمذان وقزوين[4] والري سمتهم واحد والحال أن بينهم تفاوت نحو أربع درج ومن هذا أو أمثاله وقع كثير من الناس في الحيرة والخبط ومما قدمناه نرجو من فضل الله أن ينحل هذا الإشكال

التحقيق : [1] في الأصل : منعهم ؛ [2] في الأصل : تسعة ؛ [3] في الأصل : اقتدتم ؛ [4-4] مكرر في الأصل ومشطوب

(١٠) المقريزي ، السلوك ، ج ٢ ، ص ٢٤٧

... وقال دأود بن عقبة إن عمرو بن العاص بعث ربيعة بن شرحبيل بن حسنة وعمرو بن علقمة القرشي ثم العدوي يقيمان القبلة وقال لهما قوما إذا زالت الشمس أو قال انتصفت الشمس فأجعلاها عل حاجبيكما ففعلا وقال الليث إن عمرو بن العاص كان يمتد الحبال حتى أقيمت قبلة المسجد وقال عمرو بن العاص شرقوا القبلة تصيبوا الحرم قال فشرقت جدا فلما كان قرة بن شريك [ملاحظة: يوجد هنا نقص في النص المطبوع] تيامن بها قليلا وكان عمرو بن العاص إذا صلى في مسجد الجامع يصلى ناحية الشرق إلا الشيء اليسير وقال رجل من تجيب (؟) رأيت عمرو بن العاص دخل كنسية فصلى فيها ولم ينصرف عن قبلتهم إلا قليلا ... ...

ابن تغري بردي ، النجوم الزاهرة ، ج ٢ ، ص ٦٧

... وكانت القبلة مشرقة جدا وإن قرة بن شريك لما هدم المسجد المذكور وبناه في زمان الوليد بن عبد الملك بن مروان تيامن بها قليلا ...

(١١) القلقشندي ، صبح الأعشى ، ج ٢ ، ص ١٧٥ - ١٧٧

... وأصول الرياح أربعة الأولى الصبا وهي التي تأتي من المشرق ... ... قال في صناعة الكتاب وأهل مصر يسمونها الشرقية لأنها تأتي من مشرق الشمس ... ... والثانية الدبور ومهبها من مغرب الشمس إلى حد القطب الجنوبي ... ... والثالثة الشمال ... ... ومهبها من حد القطب الشمالي إلى مغرب الشمس ... ... الرابعة الجنوبية ومهبها من حد القطب الأسفل إلى مطلع الشمس وتسمى بالديار المصرية القبلية لأنها تأتي من القبلة فيها ... ... وهي أردأ الرياح عند أهل مصر ...

Part VIIc

Safavid world-maps centred on Mecca

A third example and some new insights on their original inspiration

To Jan Hogendijk

NOTES ON THIS VERSION:

Most of this study was prepared in the last months of 2001, when there was considerable action relating to the Safavid world-maps. First, a third example was located. Second, two texts mentioning the use of conic sections, *i.e.*, ellipses, for the determination of the qibla were discovered: one from 10th-century Baghdad and the other from 11th-century Isfahan.

This paper is dedicated to the colleague who found those two texts and will probably soon find a few more of the same. I first met Jan Hogendijk in the early 1980s when he came down to New York from Brown University, where he was doing a doctorate on Ibn al-Haytham's completion of the *Conics* of Apollonius and where he later taught. In 1985 we both came to Frankfurt, where without much delay he found himself in a research institute where he was free to do his own work. He volunteered a course such as had never been taught in our Institute before and has not been since: a one-semester course on the History of Mathematics from Antiquity to the Renaissance. He started with 20 students and at the end of the semester could still count the same 20 souls. I have followed his research with interest ever since, as that of those colleagues in the same tradition. Hogendijk is one of those historians of what has been called the "hidden" Islamic mathematics,[1] who publish texts and interpret them sensibly and build onto or in places restructure the edifice constructed in the 19th century by the likes of Franz Woepcke and Heinrich Suter. Now, as we shall see, it is Jan Hogendijk who has rediscovered the primary sources that confirm my hypothesis of the early Islamic origin of the mathematics underlying cartographic grids preserving direction and distance to the centre.

---

[1] This concept was first formulated by Len Berggren in *JAOS* 120 (2000), pp. 282-283.

TABLE OF CONTENTS

CHAPTER 1

INTRODUCTION

> "In the tenth century A.D., a number of Arabic geometers believed that only straight lines, circles, and conic sections should be allowed in geometry. Other curves were objectionable, because they were produced "by instrument and motion," and motion was not allowed in geometry (which supposedly dealt with constant truths). Curves other than circles and conic sections are extremely rare in the Arabic geometrical tradition. This situation would have been different if Archimedes' *On Spirals* had been transmitted into Arabic." J. Hogendijk, "Arabic and Greek Geometry", p. 50, miscited in King, *World-Maps*, p. 240 (where we unfortunately read " ... **not** produced "by instrument and motion" ... "), corrected in Berggren (n. 3:1), p. 513b.

In 1999 I published a detailed description of two Safavid – that is, 17th-century Iranian – world-maps centred on Mecca with grids so devised that direction and the sine of the distance to the centre were preserved: see **Fig. 1a** for the first of these and **Fig. VIa-18.1** for the sec-

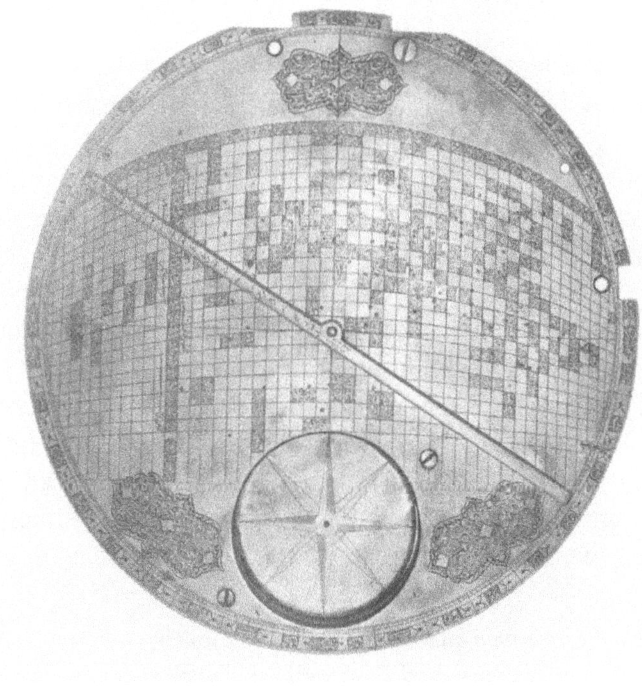

Fig. 1a: The first Safavid world-map (**A**), which came to light in 1989. Its rediscovery necessitated a re-evaluation of Muslim achievements in mathematical geography and cartography. Some 150 localities are marked and named on a highly sophisticated grid, with Mecca at the centre. One simply lays the diametrical rule at the marker for any locality, and can then read off the qibla on the circumferential scale and the distance to Mecca on the non-uniform scale on the rule. This piece lacks the universal sundial found on the other two examples discovered more recently (see **Fig. VIa-18.1**). [Property of the Dār al-Āthār al-Islāmiyya, Kuwait; photo courtesy of Christie's, South Kensington.]

ond.[1] The maps are engraved on metal. Maps of this kind are not mentioned in any of the known Islamic sources, and the first European maps preserving direction and distance to a cen tral non-polar locality date from the early 20th century.[2] In my book I showed that:

(1) the two maps are copies of a prototype (date uncertain) which is lost;
(2) the geographical data on the maps was taken from a Timurid – 15th-century Central Asian – source;
(3) Ḥabash al-Ḥāsib in 9th-century Baghdad and al-Bīrūnī in 11th-century Central Asia had written on projections preserving direction and distance to a central point; and
(4) the mathematical formula underlying the conception of the grids was first derived by Ḥabash, although no direct connection between him and the grids could be established.

Although no cartographical application of that particular accurate formula was known from any century, I also showed that:

(5) various cartographical procedures based on different approximate formulae are attested from the 9th century onwards.

The original inspiration for the grids on the Safavid world-maps remained an enigma for a year or so. Personally I was convinced that it pre-dates the construction of the world-maps by several centuries. In any case I presented all of the evidence I could adduce at the time, even for a European inspiration behind the grids. Some deficiencies in the book may be pointed out here, the first two serious:

(a) First, not enough attention was paid to the nature of the accurate latitude curves for a map preserving both direction to and the sine of the distance to the centre: these are, in fact, segments of ellipses, and the arcs of circles on the Safavid grids are approximations to these.[3] And as we shall see in **4a**, it is in the literature on conics that we find references to finding the qibla by means of conics. Nevertheless, it was standard practice in Islamic instrumentation to approximate curves of a non-trivial nature by arcs of circles,[4] or to construct curves such as hyperbolae through points whose radial or orthogonal coordinates had been computed in advance.[5]
(b) Second, the basic data – longitudes and latitudes – in the 15th-century Timurid geographical table I showed were ultimately derived with modifications from a mysterious *Kitāb al-Aṭwāl wa-'l-ʿurūḍ*. This I dated to the 12th or, at the latest, to the early 13th century, though not without enquiring if it might possibly derive from the activities in Isfahan around 1080 associated with ʿUmar al-Khayyām, and I concluded that there was no evidence either way.[6] As we shall see in **4c** below, there

[1] *World-Maps for Finding the Direction and Distance to Mecca: Innovation and Tradition in Islamic Science*, Leiden: E. J. Brill, and London: Al-Furqan Islamic Heritage Foundation, 1999, hereafter *W-M*.
[2] *W-M*, pp. 325-328. See now the reference cited in n. 3:4 below.
[3] See further Hogendijk and Dekker (n. 3:1 below) *ad W-M*, Ch. 6.
[4] For one example see *W-M*, pp. 186-191.
[5] See, for example, King, "al-Khwārizmī", pp. 17-22, and Janin & King, "Cadran solaire d'Ibn Ṭūlūn", pp. 335-337.
[6] *W-M*, pp. 42-43 and 156-159, especially p. 157, n. 79. On the recognition of this work as a source for the 14th-century geographer-prince Abu 'l-Fidā' by Robert Greaves in early-17th-century Oxford see Mercier, "English Orientalists and Islamic Astronomy", pp. 171-173.

were at least two versions of the *Kitāb al-Aṭwāl wa-'l-ʿurūḍ* in circulation in Isfahan in the mid 11th century. I still have no idea about the original compiler.

Some other errors are less critical.

(c) The radial lines on the grids – I surmised that the one surviving line on map **B** might indicate that such lines were originally drawn for, say, each 30° or even each 10° – may be likened to "loxodromes" only, of course, if the earth is flat. My use of this term in the context of the Safavid world-maps was inappropriate.[7]

(d) I mistakenly stated that no treatise on sundial construction is attributed to the 9th-century astronomer al-Khwārizmī by the 10th-century bibliographer Ibn al-Nadīm.[8] I still think that the treatise on sundial construction in MS Istanbul Ayasofya 4830 attributed to al-Khwārizmī, which consists mainly of tables of radial coordinates for drawing hyperbolae, is by Ḥabash.[9]

(e) A map of greater Iran bears an incorrect caption attributing it to Adam Olearius, 1656. It is, as stated on the map itself, by Adrianus Relandus, and it dates from 1695.[10]

(f) The expression *navicula de Venetiis* was unfortunately rendered as "little ship of the Venetians (?actually Venices?)".[11] Also the illustration of the *navicula* shows the instrument now in Geneva, not, as stated, the one in Greenwich.[12]

This short list, of course, no means exhausts the minor errors. On the "errors" which others have found see **3** below. In the sequel I shall assume that the reader is familiar with my earlier study, and I shall use the same notation (L for longitude, ϕ for latitude, q for qibla and d for distance to Mecca (in degrees)).

---

[7] See Dekker (n. 3:1 below), p. 115a-b, *ad W-M*, p. 311. On the diameter on map **B** see *ibid.*, pp. 220-221.

[8] See *W-M*, pp. 349-350, overlooking the reference in Sezgin, *GAS*, VI, p. 143, to Ibn al-Nadīm. (The reference to Sezgin is found in King, "al-Khwārizmī", p. 33, n. 2, written at a time when I believed the attribution in the Ayasofya manuscript.) This omission is pointed out in Charette, *Mamluk Mathematical Instrumentation*, p. 181.

[9] And not only because the latitude 34° for Samarra is favoured, as it is in several other works by Ḥabash (see *W-M*, p. 350, n. 85). But rather because the tables which form the bulk of this work are much more in the spirit of Ḥabash's treatises on his universal auxiliary tables, his melon astrolabe, and his universal plate for timekeeping by the stars, than in the spirit of anything we know by al-Khwārizmī. François Charette (*op. cit.*) still prefers to accept al-Khwārizmī as author.

[10] See Fig. 8.3.4 on *W-M*, p. 310. This was kindly pointed out to me by my colleague Professor Fuat Sezgin of Frankfurt.

[11] *W-M*, p. 351. The absurdity of this, which resulted from a dictation over the telephone (something like: "yes, Venetians, actually from Venice, plural. Put in a question mark in brackets."), was appropriately questioned in North, "Review" (n. 3:1), col. 749. The *navicula* is addressed again in **XIIb**.

[12] *W-M*, fig. 9.5.2 on p. 354. The error was pointed out to me by Sven Ruhberg. The photos were provided by Sotheby's of London.

CHAPTER 2

A THIRD SAFAVID WORLD-MAP

The two world-maps that came to light in 1989 and 1995 I labelled, for convenience, **A** and **B**. Now I briefly discuss a third world-map of precisely the same kind which first came to my attention in August, 2001: I shall label this **C** and shall assume in the following remarks that the reader has access to my description of **A** and **B**. The new instrument **C** is complete with a compass and sundial (as on **B**, both missing on **A**), and has an original leather case. It is in a private American collection and is currently on temporary loan to the Arthur M. Sackler Museum at Harvard University.[1]

There are four cartouches rather than three as on **A** and **B**. In the fourth the maker is identified as Ḥasan Ḥusayn. I had suspected that a person with such a name – my "pseudo-Ḥasan Ḥusayn" – might have been behind various qibla-indicators *cum* sundials engraved in Iran apparently from components made in England in the early 18th century.[2] Alas, no further information is available on this person. See below on Merv and Isfahan as the provenance of this map. The following minor details are also of interest:

- The diameter is 22.6 cm., precisely the same as **A** and **B**.[3]
- There is a scale for d in equatorial degrees along the other half of the diametrical rule, not found on either **A** or **B**.[4] The arguments 10 - 20 - 30 - 40 are on the top of the rule with 5-unit subdivisions across the scale.
- The 1°-divisions on the rim of **C** are more carefully engraved than those on **A** and **B**.[5]
- The longitude scale is now "below" the grid (between ϕ = 8° and 10°), probably to accommodate the fourth inscription.[6]
- As on **B**, there is no meridian at longitude 77;10°, so that the grid is centred on L = 77°. (On **A** there is such a meridian just to the right of that for 77°[7])
- There is limited information on the climates along the latitude scale (as on **A**, information on **B** mainly superfluous).[8]
- The latitude curves are not always smooth at L = 77°, suggesting that the two halves were constructed separately (see, for example, the curve for ϕ = 46°).[9]

---

[1] I am grateful to Dr. Mary McWilliams, Norma Jean Calderwood Curator of Islamic and Later Indian Art, Harvard University Art Museums, for drawing my attention to this piece and providing photos for study purposes. Also Melanie Michailidis kindly answered various questions about the third map.

[2] *W-M*, pp. 260-262 and 295-296.

[3] *W-M*, p. 205.

[4] On each of **A**, **B** and **C** one-half of the rule bears a non-uniform distance scale in *farsakh*s. On **A** and **B** the other half bears no scale. See *W-M*, pp. 237-239.

[5] *W-M*, pp. 235-236.

[6] *W-M*, p. 236.

[7] *Ibid.*

[8] *W-M*, pp. 230-234.

[9] *W-M*, pp. 249-252.

- The meridians are extended below the longitude scale to a straight line drawn for $\phi = 0°$, as on **A** (but not on **B**).[10]
- There are no localities in the far south-east, such as the Sarandīb in **A**.[11]
- The name of the city Balanjar is written *xxḥrd* (where *x* represents the carrier of a medial *bā'*, *nūn*, *etc.*) as on **A** and **B** and on the sundials of "pseudo-Ḥasan Ḥusayn", and its coordinates (84;20°/46;40°) are essentially the same as on **A** and **B**, *i.e.*, the longitude is 1° less than the value in the Timurid table (TMR).[12]
- The names for Cordova and Constantinople are the "older" forms as on **A**, rather than the "new" forms as on **B**.[13]
- The name of the city of Meshed (*Mashhad-i muqaddas*) seems to have been given extra prominence.[14] If this is the case, then the scene of the production of these instruments is perhaps to be moved from Isfahan to Meshed, a possibility I had suspected.[15]
- The compass on **C** is similar to that on **B**,[16] with two main differences. First, the four pointers for the intermediary directions are decorated with a fir-tree type pattern. Second, the needle has an elongated heart-shaped pointer at one end.
- The sundial on **C** is similar to that on with **B**,[17] with two main differences. First, the hours are numbered to 8 rather than 7. Second, the bar supporting the collapsible gnomon and resting underneath the diametrical base of the gnomon is more substantial than that on **B**, being shaped like a spatula, attached across the entire rim (as opposed to at the rim), and stretching just beyond the centre.

---

[10] *W-M*, p. 236.
[11] *W-M*, pp. 217, 218, and 220.
[12] *W-M*, pp. 217, 224, 225, and 262.
[13] *W-M*, pp. 228-229.
[14] This is not a reference to the size of the inscription (here two 2°×2° cells), but to the intensity of the engraving. (The inscriptions for Dongola and Constantinople, for example, fill three cells.)
[15] *W-M*, pp. 272-274.
[16] *W-M*, pp. 205-207 and 282-284.
[17] *W-M*, pp. 202, 207-212 and 284-289.

CHAPTER 3

SOME REFLECTIONS

Since 1999 various reviews of my book on **A** and **B** have appeared: I here address only those of Elly Dekker, Emilie Savage-Smith, John North, and Owen Gingerich, and turn later (in **4a**) to that of Jan Hogendijk.[1]

Elly Dekker, in a serious attempt to come to terms with the world-map grids, was unable to live with my proposal that they were simply a graphical representation of a formula relating q and sin d which was first derived by Ḥabash in 9th-century Baghdad (although not necessarily proposed by him for a map-grid).[2] She also rejected my claim that the latitude curves on the grids were arcs of circles, even though I had established that such arcs provided the best fit with the markings on the instruments and in spite of the fact that the centre of the arc with the smallest radius is still visible on **A**.[3] She argued that on instruments of this modest size (diameter 22.5 cm) it is impossible to distinguish between ellipse segments and arcs of circles. Indeed, she prefers to consider the grids as resulting from a retro-azimuthal projection derived by some clever European or Safavid Iranian from purely cartographic considerations.[4] Anybody who has the slightest suspicion that Dekker has cracked the problem should read on till the end of this study.

I further supposedly underestimated the innovative potential of the Safavid astronomers. However, neither of the reviewers who felt I was somewhat condescending towards the Safavids, namely, Savage-Smith and Gingerich, mentioned that I had taken the trouble to investigate dozens of Safavid treatises without finding anything original in any of them.[5] I also stand accused, this time by John North, of underestimating the astronomers of medieval England by suggesting that they were not smart enough to have invented the universal horary dial on the medieval instrument known as the *navicula de Venetiis*.[6] Yet in both medieval

---

[1] The main reviews are: **J. Lennart Berggren** in *Journal of the American Oriental Society* 121:3 (2001), pp. 512-514; **Charles Burnett** in *Annals of Science* 59 (2002), pp. 328-329; **Emilia Calvo** in *Suhayl* 1 (2000), pp. 363-366; **Benno van Dalen** in *Journal of the Royal Asiatic Society* (London), Series 3, 12:3 (2002), pp. 371-373; **Elly Dekker**, "Cartographic Grids from Iran: An Early Version of the Retro-Azimuthal Orthographic Projection?", *The Cartographical Journal* 37 (2000), pp. 109-116; **Owen J. Gingerich** in *History of Science* 38:2 (2000), pp. 245-247; **Jan Hogendijk** in *Historia Mathematica* 30 (2003), pp. 85-87; **Muzaffar Iqbal** in *Islam & Science* 1:1 (2003), pp. 135-142 (a Muslim perspective on a Western "positivist" operation); **John D. North** in *Bibliotheca Orientalis* (Leiden), 57 (2000), cols. 747-750; **Roser Puig** in *Isis* 92 (2001), pp. 360-362; **F. Jamil Ragep** in *Journal for the History of Astronomy* 32:2 (2001), pp. 171-172; **George Saliba** in *Mathematical Reviews* (2001), accessible on the Internet under MathSciNet, No. 2001h:01008, pp. 1-5; **Emilie Savage-Smith** in *Bulletin of the Scientific Instrument Society* 66 (2000), pp. 32-35; and **Roberto Tottoli** in *Quaderni di Studi Arabi* 18 (2000), pp. 239-240. These reviews are hereafter referred to only by the name of their author.

[2] See *W-M*, pp. 61-64. Elly Dekker conveniently ignores the fact that Ḥabash's "method of the *zīj*es" provides the relation between sin d and q that underlies the map grids.

[3] *W-M*, p. 240.

[4] See Jackson, "Retro-Azimuthal Projections" (1968) and Hinks, "Retro-Azimuthal Equidistant Projection" (1929), both unknown to me in 1999.

[5] *W-M*, pp. 128-138.

[6] North, cols. 749-750, *ad W-M*, pp. 351-359.

English treatises on the *navicula*, one in Latin and one in Middle English, the procedure for constructing and using the device is over-complicated and somewhat inept (so that it yields an approximate solution). With a simpler construction and some neat fine-tuning in the usage, the device provides an accurate solution. I simply cannot accept that the person who actually *invented* the dial would have proposed such an approximate procedure (not least since an approximate solution is provided by the universal horary quadrant on the back). So we are entitled to look elsewhere for a possible inventor. I suggested Ḥabash al-Ḥāsib because he devised a more complex instrument for timekeeping by the stars which comprises – not surprisingly, given its function – some of the same components as the universal horary dial on the *navicula*. This new treatise is now published,[7] and an overview of all the known material relating to the early history of the universal horary dial is now available in **XIIb**.

Another instrument that has been misunderstood in the modern literature as well as in the review by John North is the universal horary quadrant.[8] In 1981 North pronounced that the markings of this kind on some 40-odd out of the total of 130-odd astrolabes at the Museum of the History of Science in Oxford were incomplete without a solar scale and thus to some extent non-functional, and he apparently still believes this.[9] In fact, all of these markings can be used precisely for the purpose for which they were intended by the person who invented them, with and without a solar scale; this person was a contemporary of Ḥabash in Baghdad: see now **XIIa**. It is still an open question to what extent it was realised in, say, Renaissance England, that these markings gave only an approximate solution to the problem of determining time of day from solar altitude (the maximum error in England is three times what it is in Baghdad), but in England in the 16th century these markings were used only for astrological purposes so that the accuracy achieved by them was of no ultimate importance anyway.[10]

For readers of these notes who are wondering what the universal horary quadrant has to do with world-maps, I bring the discussion back to post-Safavid Iran. In his review Len Berggren asks how an early treatise on a grid preserving direction and distance to a central point, of such potential importance for Islamic civilisation, and the existence of which I had hypothesised (one sentence? one diagram? one set of tables? we have the formula already in the 9th century), could have disappeared to resurface in the 15th or 17th century.[11] I may cite a parallel example. The Arabic treatise on the universal horary quadrant with and without cursor, that is, the *quadrans vetus* of medieval European astronomy, dates from 9th-century Baghdad, but is preserved for us only in a manuscript copied in Meshed *ca.* 1800. However, Iranians in Meshed *ca.* 1800, almost a thousand years after the Baghdad treatise was first penned, did not start producing *quadrantes vetera* because they had absolutely no need for such instruments. Also, addressing Len Berggren's question again, I would repeat that scientific knowledge in medieval Islamic societies was not cumulative.[12]

Emilie Savage-Smith adds yet more confusion to a confused situation relating to geographi-

[7] On Ḥabash's universal plate for the stars see Charette & Schmidl, "Ḥabash's Universal Plate", and for a possible *modus operandi* see **XIIb-12**. The early history of the main markings of the *navicula* is treated *ibid.*

[8] *W-M*, p. 36, n. 68, and p. 351, n. 91; and North, col. 750.

[9] See North, "Hour-Line Ritual", and *idem*, "Review of *W-M*" (n. 3:1), col. 750.

[10] See now Gerard L'E. Turner, *Elizabethan Instrument Makers*, Oxford: Oxford University Press, 2000, pp. 83, 97-98, 100, 130-131, 133, 160 and 163, as well as **XIIb-13**.

[11] Berggren, pp. 513-514.

[12] *W-M*, p. 369.

cal data on astrolabes by confusing Naṣīr al-Dīn al-Shīrāzī with Naṣīr al-Dīn al-Ṭūsī.[13] I had shown that a table in a work by Shīrāzī was probably due to Ṭūsī, because I had found the same table attributed to Ṭūsī in another work. Savage-Smith apparently wants to attribute the table to Shīrāzī. The same reviewer had a problem with my comments on the abilities of ʿAbd al-Raḥmān al-Khāzinī (Merv *ca.* 1125) relating to the qibla. I had found that the errors in the geographical data of al-Khāzinī indicated that he read/misread the positions of the 250-odd localities in his list from a map by or in the tradition of al-Bīrūnī (*ca.* 1025), and furthermore that the additional values which he presented for the qibla of the 250-odd localities were poorly calculated. It seemed to me reasonable to hypothesise that al-Khāzinī had derived his qibla-values from his qibla-table displaying the qibla as a function of longitude and latitude difference from Mecca; this latter table is no longer extant but is mentioned by al-Khāzinī himself.[14] I was able to show that this qibla-table was poorly calculated, and that al-Khāzinī had used some heavy-handed interpolation for qibla-values with non-integral arguments. Savage-Smith was not at all happy about this,[15] but it is in a respectable tradition of modern scholarship on ancient and medieval astronomical tables, initiated by Otto Neugebauer and much practised by Asger Aaboe, E. S. Kennedy and Bernard Goldstein, then by George Saliba, the present author, and Jan Hogendijk, and, more recently, and in yet more spectacular directions, by Benno van Dalen and Glen Van Brummelen.[16]

Two reviewers, namely, Savage-Smith and North, inevitably both based in Oxford, felt that it was not politically correct on my part to make any reference to a local tradition started by Robert Gunther (*Astrolabes of the World*, 1932) and still in vogue. This considered it acceptable to work on Islamic instruments without knowing Arabic properly and without viewing Islamic instruments in the context of the history of Islamic astronomy in general. Yet one particularly regrettable consequence of this unhappy trend is that the best (or only) source for most Islamic instruments in Oxford is still Gunther,[17] and another is that fake Islamic instruments have been confused with genuine ones, and even one of the most important Renaissance European astrolabes has been dubbed a fake.[18]

Finally we consider some remarkable suggestions made in a review by Owen Gingerich.[19] He was upset by my central thesis that the two Safavid maps were based on an earlier Islamic tradition and proposed that the construction of the grids was actually quite simple, indeed trivial. His procedure was to first mark the localities using what are essentially their radial coordinates and construct the grid thereafter. He claimed that by using the Timurid geographical table – which gives longitudes and latitudes, and also the direction and distance to Mecca (but not the sine of the distance!) for some 275 localities – he himself could reconstruct part of the

[13] Savage-Smith, p. 35a, *ad W-M*, pp. 75-76.

[14] *W-M*, pp. 65-66.

[15] Savage-Smith, p. 34b-c, *ad W-M*, pp. 71-75.

[16] For Greek tables see Van Brummelen, "Mathematical Tables in Ptolemy's *Almagest*", and *idem* & Butler, "Interdependence of Astronomical Tables"; and for Islamic tables see van Dalen, *Ancient and Mediaeval Astronomical Tables*, *idem*, "Statistical Method", and van Brummelen, al-Khalīlī's Auxiliary Tables", and other references cited in King & Samsó, "Islamic Astronomical Handbooks and Tables", p. 30. See also **I-1.5**.

[17] Gunther had his own problems at Oxford not least because he was a "Cambridge man".

[18] See my "Review of the *Khalili Collection Catalogue*", and also King & Turner, "Regiomontanus' Astrolabe", a kind of rescue operation.

[19] The nonsense about the construction of the grid was repeated uncritically and in all innocence by Jamil Ragep in his review (p. 172): "Owen Gingerich ... has made the intriguing suggestion that the grids could be produced without any advanced mathematics."

map "without any fancy mathematics whatsoever", that it is then "easy to swing regularly-spaced arcs past the plotted points", and that then "the grid makes itself". His approach shows a total misunderstanding of medieval cartography in general, these maps in particular, and all of the relevant historical evidence I adduced, perhaps somewhere in the "painfully elementary" sections of my book.

In the book, however, I had shown that the localities on the maps were inserted *after* the grids had been prepared:[20] the problem was the conception and construction of the grids. Gingerich wrote further that "some ingenuity is required to think of compressing the distance scale with the sine function", but ignored the fact that the formula relating the sine of the distance to the qibla had been known since the 9th century and was recorded in every serious Iranian astronomical work for the next millennium. There was, therefore, no need for the Safavid craftsmen, who were indeed capable of producing elegant and accurate astrolabes, to arrive, as Gingerich supposes, at the brilliant idea of using the sine of the distance (which alone allowed the construction of the double-purpose grid within the context of medieval Islamic mathematics) "by trial and error".

In short, the problems relating to these world-maps have not been solved by myself, and still less by Owen Gingerich. Alas, not much new light is cast by the "new" world-map **C**. It is not too much to hope that future research will bring to light a few more Safavid world-maps of the same kind as **A**, **B** and **C**. Perhaps one of these might help us to better understand them all. It is unlikely that a (hypothetical) Timurid prototype has survived.[21] But, as I wrote in my book, what would help us more to understand these maps would be something like a (still hypothetical) Arabic or Persian treatise[22] on the construction of the grid or a (hypothetical) set of tables for constructing the grid, or a (hypothetical) entry in a diary of some European resident in or visitor to Iran recording an encounter with such world-maps.[23] The reader may imagine my surprise in learning in 2001 that two early treatises that were relevant to the grids on the world-maps were available, and that a facsimile of one of them was already in press in Frankfurt.

Before discussing these new historical sources, we note that the Safavid world-maps have occasioned a flurry of interest amongst serious specialists in cartographical grids. In a paper on qibla projections, Waldo Tobler, Professor Emeritus of Geography at the University of California, was inspired to present an overview of various useful projections in the tradition of Craig, Hammer, Schoy, and Jackson.[24] Also A. Kimerling has written on the utility of a study of the qibla in geography education.[25]

---

[20] *W-M*, pp. 218-230.

[21] Only a single astrolabe has survived from medieval Central Asia, in fact, from early-15th-century Samarqand: see *W-M*, pp. 106-107 and 108.

[22] I made some suggestions where future researchers might look for relevant Persian manuscripts in *W-M*, pp. 134-137, 218-230, and 256. I would be very, very surprised if a treatise were found in Latin or, say, French.

[23] On the Europeans in Iran in the 17th century see now Sonja Brentjes, "The Interests of the Republic of Letters in the Middle East, 1550-1700", *Science in Context* 12:3 (1999), pp. 435-468.

[24] See Tobler, "Qibla Projections". There are some problems with the bibliographical citations in this article. For example, for an overview of the literature on the mathematical treatment of the qibla problem by Muslim scholars, the reader is referred to Schoy's 1927 article "Kibla" in the 1st edn. of the *Encyclopedia of Islam*, supposedly published in "Leiden, Germany". Also on medieval geographical coordinates the reader is referred to J. K. Wright's 1923 article, and no mention is made of Kennedy & Kennedy, *Islamic Geographical Coordinates*. This is a pity not least because the article is likely to be cited extensively.

[25] Kimerling, "Adding the Qibla to Geographic Education".

CHAPTER 4

SOME NEW INSIGHTS

a The mathematical grids

At the American Mathematical Society meeting in New Orleans in January, 2001, Jan Hogendijk presented a description of a world-map based on the idea of the Safavid world-maps extended to cover "the whole earth" with distances in kilometres. He cautiously observed that the principle was first proposed by "an anonymous Islamic (probably Iranian) mathematician between the 11th and 17th centuries", and had to concede that there are practical problems with finding the qibla in the Americas (and other outlying regions).

In his review Hogendijk went further (mathematical notation suppressed):

> "The mathematical analysis of the instruments can be carried a little further than in King's book. The grid on the instrument can be explained as the orthogonal projection on the horizon plane of two families of circles on the celestial sphere. Any latitude curve is the projection of a circle through the zenith of Mecca parallel to the celestial equator; the position of this circle depends on the latitude and the latitude of Mecca but not on the longitude. Any longitude curve is the projection of a circle parallel to the meridian, and the position of this circle depends on the longitude and on the longitude and latitude of Mecca, but not on the latitude. This projection argument explains why the latitude curves are ellipses in a way which the medieval Islamic mathematicians could have understood. Thus the two instruments can be related to various medieval Islamic texts (e.g. by al-Bīrūnī), and these relations support King's hypothesis of an Islamic origin of the grid."

On December 4th, 2001, Hogendijk informed me in Frankfurt that he had found a reference to the use of conic sections for finding the qibla in a 10th-century Arabic text by a well-known astronomer of Baghdad, an old friend of mine who wrote on mathematics, astronomy and astronomical instruments. This individual knew of Ḥabash's works and was himself quoted by al-Bīrūnī. On December 5th, whilst Hogendijk was still in Frankfurt, he received a present from my colleague Professor Fuat Sezgin, Director of the Institut für Geschichte der Arabisch-Islamischen Wissenschaften, in the form of a facsimile of an important manuscript preserved in Istanbul.[1] This is a late and corrupt copy of several earlier treatises on mathematics, astronomy and astronomical instruments, some of which are unique and all of which are important. (The reader may find it curious that I was not aware of the fact that this facsimile had been published in February, 2001, just 500 metres from my own office.) In the train back to Utrecht on December 5th, Hogendijk skimmed through this important manuscript, and found

[1] Sezgin *et al.*, eds., *Scientific Manuscript*. See also **I-6.4.1**** and **7.2.0*** on some tables in this manuscript.

a second reference to this topic. He kindly drew my attention to this second reference by telephone the next morning, after he had finished with Sint Niklaas. It occurs in a treatise on astronomy by a named author from 11th-century Isfahan (see below), who mentions al-Bīrūnī elsewhere in his text. I write no more on this subject pending an imminent publication by Jan Hogendijk, except to say that even though certain Muslim scientists knew the true nature of the latitude curves on a rectazimuthal grid, they would have, to a man, approximated them as arcs of circles when it came to engraving a family of such curves on brass, rather than playing around with pieces of string to construct ellipse segments.

So it is to Jan Hogendijk that we shall be indebted for the explanation of the origin of the rectazimuthal grids on the Safavid world-maps. I did not find the relevant texts because I was looking in the wrong place, and because I had been looking for some 10 years, in 1999 I more or less gave up. The nature of the information in these newly-discovered texts inevitably takes me completely by surprise: see **Fig. 4a**. But the date and location and cultural context of the

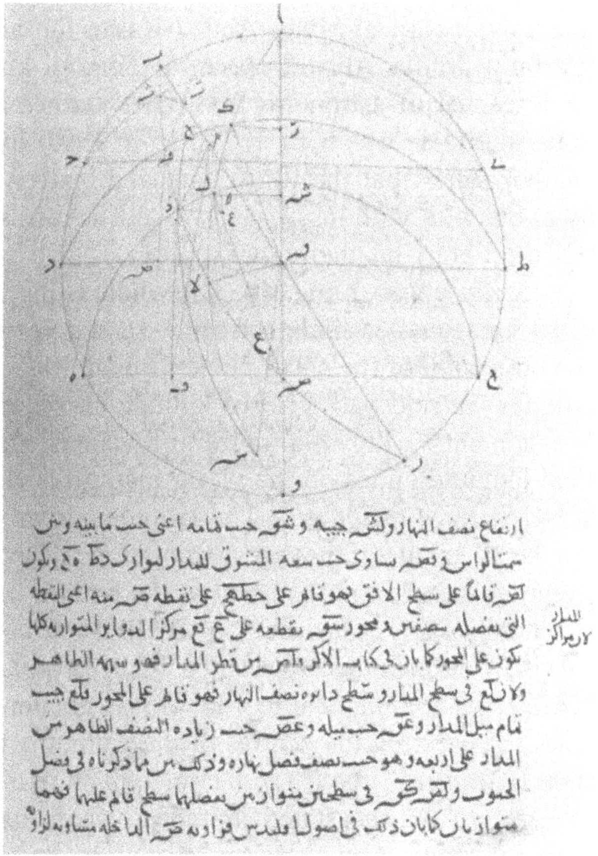

Fig. 4a: Muslim astronomers, especially during the formative period between the 9th and the 14th / 15th centuries, were very much at home with trigonometric solutions to problems deriving from the geometry of the celestial / terrestrial sphere. This particular text dealing with problems of spherical astronomy, copied probably in the 13th century, mentions the works of Theodosios and Euclid. It is in a text such as this, but one mentioning instead the *Conics* of Apollonios, that we can expect to find the original inspiration for the construction of grids preserving direction and distance to the centre. [From an unidentified manuscript, courtesy of the Universiteits-bibliotheek, Leiden.]

compilation of these texts is indeed what I predicted. And now we know what we are looking for, I am confident that we shall find more, or that at least Jan Hogendijk will. This is a particularly pleasing incident in the history of collaboration with Jan Hogendijk that it has been my privilege to enjoy over many years.

b Another early modern grid for finding the qibla for any locality

But I have not finished with Jan Hogendijk yet. For in May, 2002, he sent a copy of a recent Iranian publication to our Institute Library. This was a curious edition of al-Bīrūnī's treatise *al-Istī'āb* on the construction of the astrolabe, published in Meshed in 2001/02.[2] In this there was a foreword by Ayatallāh Ḥasan Ḥasan Zāde al-Āmulī, in which he listed various other instruments known to him from the medieval Islamic sources. One of these, labelled *ṣafīḥat al-Shaʿrānī*, "the plate of al-Shaʿrānī", was a device for finding the qibla at first sight (and with a little imagination) rather similar to what I have termed a Mecca-centred world-map grid. al-Shaʿrānī, whose full name is al-Ḥājj al-Mīrza Abu 'l-Ḥasan ibn al-Mullā Muḥammad ibn al-Mullā Ghulām-Ḥusayn ibn al-Mullā Abi l-'l-Ḥasan al-Ṭihrānī, known as "*al-ʿallāma al-Shaʿrānī*", is identified as a teacher of astronomy and other mathematical sciences, and was apparently the teacher of al-Āmulī. He was born in Tehran in 1902, lived in Tehran and Qum, and died in 1973.[3] Our author states that the illustration and text on it are copied from and instrument constructed by al-Shaʿrānī with inscriptions in al-Shaʿrānī's own hand, and that the instrument is preserved in the library of the author.

The illustration and text are illustrated in **Fig. 4b**. The whole is inscribed on a circular plate. The main markings are fitted in a smaller circle off-centre, and a yet smaller concentric circle that seems to represent the projection of the small-circle through Mecca on the equator. In one quadrant of the main markings we find half a set of latitude curves meeting at $\Delta L = 90°$ but that the orthogonal (longitude?) curves drawn to uniform divisions of the outer qibla scale are not the same as the vertical longitude lines on the world-map grids. (These main markings in fact look like those on a *shakkāziyya* quadrant.[4]) The latitude curves are apparently extended as parallel straight lines on the other side of the meridian of Mecca, labelled "lines of the circles of the azimuths for different latitudes", and the point at the bottom of the central meridian is labelled "degrees of the zenith". The Arabic text on the illustration translates:

> "In the name of (God), may He be exalted. To find the qibla a mark is made on the longitude difference or its complement to 180° if it is greater than 90°. Then look for a point on the lines representing the centres of the azimuths for the latitude required, keeping it as centre (*murakkiz*^an^ *abad*^an^) of a circle which passes through the longitude difference and the zenith for the stated latitude towards its horizon. Made by the (unworthy) servant (of God), Abu 'l-Ḥasan al Shaʿrānī."

---

[2] al-Bīrūnī, *Istīʿāb al-wujūh al-mumkina fī ṣanʿat al-aṣṭurlāb*, al-Sayyid Muḥammad Akbar Jawādī al-Ḥusaynī, ed., with introductory notes by Ayatallāh Ḥasan Ḥasan Zāde al-Āmulī, Mashhad: Islamic Research Foundation, Astan Quds Razavi, 1380 HS [= 2001/02]. The edition is curious because it is illustrated with computer-generated diagrams, many of which bear little relation to the diagrams in the manuscript tradition.

[3] This information was kindly provided by Dr. Mohammad Bagheri of Tehran in May, 2002.

[4] See **X-5.2** and **6.2** and **XI-8.2**.

٢٨ – صفيحة الشعرانى، هى صفيحة لمعرفة سموت قبلة الآفاق، اخترعها
بذهنه الوقّاد مولانا الأعظم، العلاّمة ذو الفنون، وحيد الدهر و
معلّم العصر، جامع العلوم
العقلية والنقلية،
الأستاذ المختص
بالهيئة وسائر
العلوم الرياضيّة
العالية، آية الله
العظمى: الحاج
الميرزه ابو الحسن بن
الملاّ محمد بن الملاّ غلامحسين
ابن الملاّ ابى الحسن الطهرانى، الشهير
بالعلاّمة الشعرانى - رفع الله المتعالى درجاته، وأفاض علينا من بركات أنفاسه النفيسة
القدسية. وتلك الصحيفة المكرّمة المصنوعة بيده المباركة وبخطه الشريف محفوظة فى مكتبتنا
وهذه الصورة تصويرها.

Fig. 4b: Ayatallāh Ḥasan Ḥasan Zāde al-Āmulī's hand-written account of al-Shaʿrānī's plate for finding the qibla. [From the lithograph in the new Meshed edition by al-Sayyid Muḥammad Akbar Jawādī al-Ḥusaynī of al-Bīrūnī's astrolabe treatise, *Istīʿāb al-wujūh al-mumkina fī ṣanʿat al-asṭurlāb*, p. 86.]

So one is to draw a circle whose centre lies on the appropriate latitude line and which passes through the longitude difference marked on the inner circle and the appropriate "zenith". Presumably the point of intersection of this circle with the outer scale of the quadrant with the main markings will indicate the qibla. These instructions were unclear to me until Jan Hogendijk explained them to me. We are, in fact, dealing with a serious attempt to produce a qibla-calculator.[5]

[5] See *W-M*, pp. 359-361, on a not-so-serious attempt, also innocent of the existence of the Safavid world-maps.

c On the ultimate source of Iranian mathematical geography

Our joint debt to our mutual colleague Professor Fuat Sezgin is also great. The treatise in this precious collection now published in a facsimile edition mentioned in **4a** above, whose author is Abū ʿAbdallāh Muḥammad ibn Aḥmad al-Khāzimī (*fl.* Isfahan *ca.* 1050), is important for another reason.[6] The author remarks that he has measured the longitude difference between Alexandria and Isfahan and corrected what was in the book *Kitāb al-Aṭwāl wa-'l-ʿurūḍ* (*ʿalā mā ṣaḥḥah-nāhu fī Kitāb al-Aṭwāl wa-'l-ʿurūḍ*).[7] Here at last we have come a step closer to identifying the author of this monumental work and its historical milieu, which underlies all of later Iranian mathematical geography (and hence much of European geography of Greater Iran during the Renaissance). The value for the longitude difference that al-Khāzimī mentions is 25;10° but the value that can be derived from the entries associated with this title by Abu 'l-Fidā' is 24;46° (Sezgin has incorrectly 21;36°), and that same value underlies all major Iranian tables thereafter (TUS, KAS, ULG). al-Khāzimī does not specifically say that he was the author of this book (*contra* Sezgin), rather he implies that he corrected what he found in that book. Perhaps this correction was made in a recension of the work by himself or at least in his own copy of a manuscript (though this is unlikely); if anyone knew of his correction to 25;10° thereafter, they did not take it seriously. al-Khāzimī's measurement of the obliquity of the ecliptic at Isfahan is mentioned by the late-13th-century scholar Quṭb al-Dīn al-Shīrāzī in his treatise on planetary theory.[8] al-Khāzimī himself was apparently forgotten over the centuries, although perhaps someone was thinking of him when he attributed to al-Khāzi<u>m</u>ī a treatise by the 12th-century astronomer al-Khāzi<u>n</u>ī on a self-rotating sphere.[9]

One thing is clear: there is an urgent need to reassess the astronomical climate in Isfahan in the decades before ʿUmar al-Khayyām.[10] But one should not overestimate al-Khāzimī: his new value of the latitude of Isfahan (32;25°, rounded from 32;26°) was hardly an improvement over the excellent value 32;40° already in the *Kitāb al-Aṭwāl wa-'l-ʿurūḍ* (the accurate value is 32;41°). Nevertheless, the major sources in subsequent centuries (TUS, KAS, ULG) used 32;25°.

In any case, thanks to Professor Sezgin, we now have not only a second reference to the use of ellipses for finding the qibla but also some new information on the mysterious *Kitāb al-Aṭwāl wa-'l-ʿurūḍ*, which confirms that more than one recension of that work was in circulation already in Isfahan in the 11th century. I strongly suspect that the work was compiled in that city, but God alone knows when and by whom.[11]

---

[6] On the author and the available copies (all incomplete?) of his abridgment (in only 25 folios?) of the *Almagest* see Suter, *MAA*, no. 502; Rosenfeld & Matvievskaya, *MAMS*, II, p. 311, no. 259b; and especially Sezgin, GAS, V, p. 92 and also n. 1, where this manuscript is mentioned.

[7] Fuat Sezgin *et al.*, eds., *Scientific Manuscript: Manuscript of Arabic Mathematical and Astronomical Treatises reproduced from MS A.Y. 314 Istanbul Üniversitesi Kütüphanesi, Istanbul*, Frankfurt: IGAIW, 2001, p. 40 of text and pp. v-vi of Sezgin's Introduction.

[8] *Ibid.*, pp. 40, 50 and 52 (all observations at Isfahan, latitude 32;26°); and Sayılı, *The Observatory in Islam*, p. 177. On the problems associated with the name of al-Khāzinī in the medieval sources see the splendid article on him by Robert E. Hall in *DSB*, especially pp. 335a-336b.

[9] Lorch, "Al-Khāzinī's 'Sphere that Rotates by Itself'", pp. 287-288.

[10] See n. 1:6 above.

[11] See also n. 1:6 above. Sezgin, *GAS*, X, p. 174, also has 13th century, because of some internal evidence (clearly from a recension). I have commented on the attribution of this work to *al-Furs*, "the Persians" – see *W-M*, p. 157, n. 80.

Part VIII

Aspects of practical astronomy in mosques and monasteries

To the memory of Harry and Christiane Kühnel

ACKNOWLEDGMENTS AND NOTES ON THIS VERSION

Harry Kühnel and his wife Christiane died tragically as the result of a car crash in the Summer of 1995. Harry Kühnel was the Director of the Institut für Realienkunde des Mittelalters und der Frühen Neuzeit in Krems on the Danube. He was a medievalist who more than any other I have ever met appreciated the value of objects as historical sources.[1] (Innocent readers who know the Middle Ages only from museums probably do not realise that most medievalists work on nothing other than textual sources.) The study of medieval artefacts has made substantial leaps forward in recent years, to a great extent thanks to Harry Kühnel and his associates. There is no single word in any other language for *Realienkunde*, "the study of material objects", which is not even good English and certainly does no credit to one of the most fascinating aspects of Medieval Studies. It is highly appropriate that there should be a first-rate historical research-institute with the term "Realienkunde" in its title. In fact it is the study of "Realien" which reveals what Kühnel, in the spirit of Jacques Le Goff, called "das andere Mittelalter". Inevitably when he invited me to Krems in 1992 it was to deliver a paper on what we can learn from medieval astronomical instruments, which goes far beyond the history of science. The conference was on "Alltag and Sachkulktur", something like "objects in everyday life" in Europe and the Byzantine and Islamic worlds during the Middle Ages.[2] The timing was perfect: there was a splendid exhibition at the magnificent Minoritenkirche in Krems-Stein entitled "Das andere Mittelalter".[3] So now, as I write about a medieval topic, it is thanks to the insights and encouragement of Harry Kühnel that what I have to say is based partly on texts and partly on objects.

This is an updated version of a paper prepared for the Symposium "Science and Theology in Medieval Islam, Judaism, and Christendom", held at the University of Wisconsin at Madison, during April 15-17, 1993. A German translation was published as "Aspekte angewandter Wissenschaften in Moscheen und Klöstern", *Berichte zur Wissenschaftsgeschichte* (*Organ der Gesellschaft für Wissenschaftsgeschichte*) 18 (1995), pp. 85-95 and 137-149. It is, in a sense, a sequel to my contribution "Astronomical Instruments between East and West" to the 1992 conference organized by Harry Kühnel in Krems.

---

[1] Obituaries for Harry Kühnel are: Helmut Hundsbichler, "Univ.Prof. Dr. Harry Kühnel verstorben", *Niederösterreichische Kultur Berichte* (Vienna), 9 (1995), p. 5; Georg Wacha, "Harry Kühnel. Kulturhistoriker und Denkmalpfleger", *Oberösterreichischer Kulturbericht* (Linz), 49:10 (October 1995); and anon., "Würdigungspreis Wissenschaft: posthum an Univ.Prof. Dr. Harry Kühnel. Vernetztes Denken, Interdisziplinarität, Internationalität", in *Kulturpreisträger des Landes Niederösterreich 1995*, Vienna, 1995, pp. 6-7. My thanks to Prof. Dr. Gerhard Jaritz (Krems) for sending copies of these to me.

[2] See the proceedings: *Kommunikation zwischen Orient und Okzident – Alltag und Sachkultur*, Harry Kühnel, ed., (*Sitzungsberichte der Philosophisch-Historischen Klasse der Österreichischen Akademie der Wissenschaften*, Band 619, *Veröffentlichungen des Instituts für Realienkunde des Mittelalters und der Frühen Neuzeit*, Nr. 16), Vienna: Österreichische Akademie der Wissenschaften, 1994.

[3] See Gerhard Jaritz, Harry Kühnel *et al.*, *Das andere Mittelalter – Emotionen, Rituale und Kontraste*, Krems (A): Kunsthalle, 1992.

There are vast areas of early medieval European astronomy with which I have no personal experience, but for which there is now a reliable overview by Stephen McCluskey in his *Astronomies and Cultures in Early Medieval Europe* (1998). The reader should keep in mind that this study was written some years before the publication of that book, and I have made no attempt to expand it. This study is not intended as anything more than a few personal "jottings"; these made some impression on colleagues when illustrated with a hundred folios showing medieval Islamic and European manuscripts and instruments. I certainly do not purport to include an authoritative overview of medieval Christian timekeeping or anything else Christian, but my purpose will be achieved if I can persuade readers that that subject is worthy of further research.

For comments on the penultimate draft of this study I am grateful to both Stephen McCluskey and Julio Samsó, without in any way involving them in the remaining shortcomings.

TABLE OF CONTENTS

CHAPTER 1

INTRODUCTION

Science, mainly astronomy and mathematics, as applied to the needs of daily life in medieval mosques and monasteries, that is, science in the service of Islam and Christendom, is the main subject of this study. But the term "science" is an unhappy one, because, even though the Muslims indeed had a branch of knowledge called "the science of astronomical timekeeping" (*ʿilm al-mīqāt*), "science" really is inappropriate to describe the activities of the Christian monks. Also it is not always easy (but I think that it is necessary) to separate this "science" from that practised outside mosques and monasteries. It seems to me worthwhile, however, to make a first attempt. Even if our subject is not really "science" in the sense most moderns would use the term, which is one reason why it does not feature in works on the "history of science", it certainly was "practical" and it was "applied". We know something about the way this practical astronomy was applied, and, as we shall see, we clearly have a lot more to learn. In this preliminary study, featuring little more than some episodes in two parallel traditions which each lasted for more than a millennium, regional differences within each of the two cultural contexts, and, to a lesser extent, also chronological developments, will be overlooked. And one should not, of course, ignore intellectual developments outside the confines of the religious institutions. So the result is admittedly not satisfactory.

To point out the very different, and yet in some senses remarkably similar, concerns of medieval Muslims and Christians in practical astronomy, I shall use two illustrations. The first is an evocative painting by the 19th-century French painter Jean Léon Gérôme (1824-1904) showing Muslims at prayer on the rooftops of Cairo, with the minarets of numerous mosques in the background: see **Fig. 2.1** below. The other is a charming miniature in a 15th-century manuscript of the French translation of the mystical treatise *Horologium sapientiae* by the Dominican monk Henricus Suso or Heinrich Seuse (*ca.* 1330), showing the author receiving inspiration at the hands of Wisdom personified, this in a room further graced with two weight-driven clocks, an astrolabe, and several other, smaller astronomical instruments: see **Fig. 3.1** below.

Muslims pray at specific times in a special direction, and the most sacred time of the year is Ramadan, the beginning and end of which are astronomically determined by a very natural phenomenon, the appearance of the lunar crescent. A great deal of effort was expended over more than a millennium by Muslim scholars concerned with the organization of the astronomically-defined times of prayer, the determination of the sacred direction (*qibla*) towards the Kaʿba in Mecca, and strict adherence to the lunar calendar. Gérôme's painting, in which all of these aspects of Muslim life are depicted, invites comment in the light of the research conducted over the past 30 years on the way in which this effort was expended. The times of prayer could be determined with an astrolabe or with a sundial, but in general they were not. How, then, were they determined? And are the men at prayer depicted in the painting doing things according to the prescriptions of their faith? As we shall see, Gérôme was not watching his subjects *in situ* when he painted this picture.

Christian monks also prayed at specific times closely related to those of the Muslims, and a special direction was favoured albeit without the consistency of the Muslims with regard to the qibla. The most sacred time of the year is Easter, whose timing is defined by complicated and artificial rules involving the intricacies of the solar and lunar calendars and *salti* of the imagination. Here again much effort was expended over many centuries in an attempt to do things according to the rules. The French miniature invites discussion of some of the many questions that remain regarding astronomy in the monastic context. Why, for example, did the monks apparently not use astrolabes, on which the "hours" of their prayers are actually marked? And how can one regulate their "hours" with a mechanical clock that marks the passage of time in hours of a different kind? As we shall see, the artist had in front of him some real instruments as he painted his picture.

In both cases it is not only illustrations but rather manuscripts and a few published texts, as well as instruments and religious architecture, that can reveal to us the ways in which science – hard-core mathematical science in the tradition of Euclid and Ptolemy, as well as soft-core folk science and simple computus in the traditions exemplified by medieval Islamic handbooks of religious law and certain Christian *Books of Hours* – was used in the service of religion. In the case of the Islamic material we now have some measure of control over the available sources; for the European material we have somewhat further to go before the same can be said. (The parallel situation in Judaism invites similar treatment.)

CHAPTER 2

THE ISLAMIC SCENE

> "There is no Arabian science. The wise men of Mohammedanism were always the more or less faithful disciples of the Greeks, but were themselves destitute of all originality." Pierre Duhem, in an article on the history of physics in *The Catholic Encyclopaedia*, New York: Robert Appleton Company, XI (1911), p. 48.

> "So there was a brilliant rise in Arabian astronomy, but no significant progress. After some centuries it died down. This indeed was the case with Muhammedan culture as a whole. ... The importance of Arabian astronomy lay in the fact that it preserved the science of antiquity in translations, commentaries, interpretations and new observations and handed it down to the Christian world." A. Pannekoek, *A History of Astronomy*, London, 1961, p. 170.

That it is worthwhile to look at the Islamic sources (mainly unpublished) is proven by the fact that we do not have to go far to find material to write a new chapter in the history of science. Some 500 Arabic and Persian manuscripts have now been investigated and between them they contain tables for every conceivable function relating to timekeeping by day and night, including tables for a specific latitude with tens of thousands of entries into which one simply inserts the observed altitude of the sun or a given fixed star and reads the time (**I-II**). We also find tables displaying the qibla for each degree of terrestrial latitude and longitude, as well as tables to facilitate the prediction of the first visibility of the lunar crescent. All manner of instruments were devised to perform the same functions mechanically, that is, without any computation (**X**). All this activity started in late-8th- and 9th-century Baghdad, long before the development of the institution of the muwaqqit, the astronomer associated in an official capacity with the mosque and madrasa. The post is clearly defined only in the central regions of the Islamic world, and there only from the 13th century onwards (**V**). Were these tables and instruments really used?

For anyone who might doubt that astronomers working primarily to determine the times and direction of prayer could contribute much to the development of science, I mention some topics that such astronomers were the first to treat. It was primarily they who established the different values for the solar depression at daybreak and nightfall and who compiled tables displaying the duration of morning and evening twilight; they also developed a mathematical theory for the effect of atmospheric refraction at the horizon. They devised tables and instruments for solving all of the standard problems of spherical astronomy for any latitude. They developed the astrolabic quadrant from the astrolabe, but more impressive is the fact that they erected monumental astrolabic clocks (examples are known from Fez and Damascus, and a description of such a clock is in the 13th-century *Libros del saber*). They prepared tables of coordinates for marking curves on astrolabes and sundials and made substantial progress in all aspects of instrumentation, and devised graphical representations of the relationship between the altitude of the sun and the hours of daylight. They investigated the problem of lunar crescent visibility and produced a series of solutions, some of considerable complexity. Finally it was also they who developed the magnetic compass and first measured magnetic declination.

Some specific examples of achievements by astronomers working in this tradition well illustrate their ingenuity:

In Cordova in the early 13th century a mosque official responsible for regulating the times of prayer devised an astrolabic plate with which one could perform all of the operations of spherical astronomy for any latitude. This was one of many universal instruments devised by Muslim astronomers but was of particular mathematical elegance, in that the various families of curves were associated with different arguments according to the specific problem to be solved. This plate was a common feature of Maghribi astrolabes for many centuries and was also known in Syria and India. A European plate of the same kind is found inside one 15th-century German astrolabe in the tradition of Regiomontanus.

In Cairo in the early 14th century the astronomer Najm al-Dīn al-Miṣrī compiled a table into which one feeds the altitude of the sun or any star and reads the time since rising of the celestial body in question. The table, which works for any latitude, has three arguments and a total of close to half a million entries. The author also describes how to use the table for other purposes, essentially providing a numerical solution to the medieval equivalent of the spherical cosine formula, whereby one enters with the three sides of the triangle and finds the corresponding angles without calculation. Tables similar in conception were prepared in Europe and the US in the 19th and early 20th centuries.

In Aleppo in the early 14th century the astronomer Ibn al-Sarrāj, albeit not a muwaqqit, made an astrolabe that can be used for any latitude in five different ways. The instrument is the most sophisticated astrolabe ever made. Some, but not all, of the components would have been intelligible to the astronomers in 17th-century Flanders who redeveloped some of the ideas relating to astrolabe construction first formulated in Baghdad in the 9th and 10th centuries.

Later in the 14th century the chief muwaqqit at the Umayyad Mosque in Damascus constructed a horizontal sundial of 2 × 1 metres that he claimed would serve each of the five daily prayers. Indeed when one looks carefully at the different families of markings one finds that they serve hours from sunrise and until midday, after midday and before sunset, and before the mid-afternoon prayer. There are also markings indicating times before nightfall and daybreak. And the gnomon is quietly oriented towards the celestial pole. No other sundial as sophisticated as this is known from the time before the late Renaissance. (The maker was the same Ibn al-Shāṭir who developed planetary models mathematically equivalent to those of Copernicus.)

Finally, instrument-makers in Isfahan *ca.* 1675 produced cartographic grids on which both distance and direction to Mecca are correctly represented. I confess that it took me three years to ascertain that the first of the three instruments of this kind now known indeed performed these complicated operations according the exact formulae, if by introducing a simplification in the construction of the grids; and I still do not know how the makers actually constructed the grids. It is, however, quite clear that these instruments represent the culmination of an older cartographical tradition of which we have no other evidence.

These tables and instruments are impressive by any standards. The five examples noted above have all come to light or been studied just recently. The function of the universal plate has been understood only in the past few years (thanks to research by Emilia Calvo). It was only in the 1970s that the universal table was discovered, the quintuply-universal astrolabe studied for the first time, and the sundial published (by Louis Janin). The first of the three universal qibla-indicators came to light at an auction only in 1989. In fact, given the more abundant library and museum resources available to us today, a remark by L. A. Sédillot is still as valid today as when he wrote it in 1845:

> "Chaque jour amène quelque découverte nouvelle et vient démontrer l'extrême importance d'un examen approfondi des manuscrits {add: et des instruments} de l'Orient."

But there is a catch. It was not always the astronomers who had the final say as to what happened in practice. Daily life in medieval Islam was more often regulated by the prescriptions of the scholars of the sacred law, and not one is known to me from over a millennium who said that one should consult an astronomer on any aspect of religious ritual. The textual sources known to me are silent on the way in which the two groups co-existed.

The legal scholars advocated procedures based on folk astronomy, a discipline innocent of theory and unencumbered by computation. So, for example, for the times of the daylight prayer they advocated simple shadow schemes easily mastered in the form of mnemonics: in some of these the midday shadow was given to the nearest foot for each month of the solar year, and the shadow length for the afternoon prayer was simply 7 feet more. These schemes were crude and often incorrect, some adopted from Hellenistic folk astronomy being originally intended for other latitudes; any astronomer worth his salt could have pointed out that they were defective, but I do not know of one who did. For timekeeping by night, arithmetical schemes involving the lunar mansions were used. A particularly interesting phenomenon is the mathematization of folk astronomy that occurred in certain circles, notably amongst the scholars of the Yemen but also elsewhere, for example, in Tunis. Thus we also have tables based on simple approximate arithmetical rules displaying, for example, shadow lengths at the seasonal hours for each few days of the year.

For the qibla the legal scholars advocated the cardinal directions or astronomical horizon phenomena. The idea was not to face Mecca, or the Kaʿba in Mecca, but rather that segment of the perimeter of the Kaʿba associated with one's own region. This was achieved by facing the very same direction that one would be facing if one were standing in front of part of a wall or a corner of the Kaʿba. Since the rectangular base of the Kaʿba itself is astronomically aligned, such alignments were used for the qibla and for the laying out of mosques. (This is the only branch of the new discipline of archaeoastronomy in which we have textual evidence relating to the astronomical orientations of architecture.) Also, since Muslim cities are often built around a mosque, we even have entire cities or sections of cities that are astronomically aligned. Historians of Islamic architecture have wondered for decades why much of their subject material is not properly aligned towards Mecca. In fact it *is* aligned towards Mecca, but to understand this we have to know how Muslims conceptualized the qibla and how they achieved alignments in the qibla. So when the Great Mosque of Cordova faces the deserts of Algeria, it is in fact aligned in the same direction as the rising of Canopus in Mecca, being parallel to the main axis of the Kaʿba. That Canopus is not visible in places as far north as Cordova was not lost on the legal scholars, who developed ingenious practical methods of making sure that their mosques were in precisely the directions they wanted. And they also devised legal strategies to determine which qibla directions were acceptable.

For the regulation of the calendar the legal scholars favoured actual sighting of the crescent and saw no purpose in detailed calculations for predicting it in advance. If it was not possible to see the crescent on the first day of the civil calendar (organised in alternating months of 29 and 30 day) because of meteorological or other problems, then a simple arithmetical rule was applied. We have very little historical information about the way in which such regulation of the calendar was achieved in practice. In the modern world, with its facilities for instant communication, the regulation of the calendar has led to controversies between scientists and religious authorities, and for various reasons (though through no fault of the Muslim astronomers) Ramadan was, at least in the 1970s, in some countries occasionally announced too early.

Islamic sources on folk astronomy likewise contain clues to the original function of the Kaʿba and the positioning of some its features such as the door on its north-eastern wall, the Black Stone in the eastern corner (although they are silent on others, such as the *ḥaṭīm* on the north-western wall). In addition they provide the key to the question why there are five prayers in Islam, not all specifically and individually mentioned in the *Qurʾān*, and why in the 8th century their times were astronomically defined in ways mentioned neither in the *Qurʾān* nor in the Prophetic tradition.

Gérôme's painting

The scene is Cairo. The time is the *maghrib*, the sun has just set, and the muezzins, visible on the two closest minarets, have performed the call to prayer. The setting crescent moon, appropriately inclined to the horizon, is visible behind the men, so that a new month has just begun. It is almost certainly Ramadan, which, when Gérôme was visiting Egypt, began in July. (Egyptians wear heavy woollen garments even when Europeans think that it is still very hot.) Obviously the men at prayer are supposed to be standing in the qibla, facing towards Mecca.

In Ottoman Egypt the beginnings of the months were carefully computed in advance, and in those days it was surely the astronomers who had the say when, for example, Ramadan would begin. To determine the possibility of crescent visibility on the first day of the month in the civil calendar (organized according to an arithmetical scheme) one needs to investigate the positions of the sun and moon on the ecliptic and relative to the horizon and to rely on various conditions. Muslim astronomers had over a millennium of practice at this. The legal scholars, however, demanded attestation of an actual sighting, which the astronomers could usually predict anyway.

The muezzins would have been advised of the time to perform the call to prayer by muwaqqits who had access to astronomical tables for determining the times of prayer, such as had been available in Cairo for several centuries. Although we have no idea how they actually used these tables, extant in numerous manuscripts, and we might think that observing sunset

Fig. 2.1: *Prayer on the Rooftops* (1865) by the French artist Jean Léon Gérôme. This beautiful painting shows men at prayer on the roof of one of the mosques or madrasas of the old Mamluk part of Cairo. The time is the *maghrib*, the sun has just set, and the muezzins, visible on the two closest minarets, have performed the call to prayer. The setting crescent moon, appropriately inclined to the horizon, is just visible behind the men, so that a new month has just begun. Obviously the men at prayer are supposed to be standing in the qibla, facing towards Mecca. But, as we shall see, there are some problems. On other aspects of this painting see Gerald M. Ackerman, *The Life and Work of Jean-Léon Gérôme*, London & New York: Sotheby's, 1986, pp. 67 and 216-217 (no. 152). [Hamburg Kunsthalle, inv. no. 1909, courtesy of the Kunsthalle Hamburg.]

over the hills of Giza might have been a relatively easy affair, the fact is that their tables would have given the exact time of astronomical sunset, perhaps corrected for the effect of refraction at the horizon. Other tables would tell the muezzin when to announce the time of the *ʿishāʾ* prayer at nightfall, defined as the moment when the sun was 17° below the western horizon, when to extinguish the lamps on the minarets during the nights of Ramadan so that fasting could begin, when to pronounce a blessing on the Prophet, and when to announce the *fajr* prayer at daybreak, with the sun now at 19° below the eastern horizon. These tables give the appropriate moments correct to the nearest few seconds of time, for each degree of solar longitude, roughly corresponding to each day of the year. Of course only the most skilled of astronomers, equipped with an astrolabe or quadrant (or better, a watch), would have known precisely when these pre-determined moments had arrived. Similar corpora of tables were compiled for other Islamic cities in the Middle Ages. In this century most of these have been replaced by new tables, now usually generated by computer.

The first Muslims in Egypt adopted an astronomical direction, winter sunrise, for the qibla. By so doing they sought to face the western corner of the Kaʿba. Some time later astronomers computed the direction of Mecca from available geographical coordinates. Inevitably they came up with a different result. This computed direction was known throughout the medieval period, but, because of the strength of religious tradition, it did not completely replace the earlier one. Muslim astronomers had developed accurate solutions to the complicated problem of determining the direction of one locality from another already in the 9th century. The Muslim legal scholars, at least in the Middle Ages, preferred procedures involving astronomical horizon phenomena. Nowadays the local surveying department usually determines the local qibla; usually this is different from the medieval values, being based on modern geographical coordinates.

The scene is set for us to take a closer look at Gérôme's painting. As far as the astronomy is concerned, all is not well. The moon should clearly be somewhere north of west, but it is on the right of the men at prayer, which would require that they be facing south. They are, however, facing roughly the same direction as the openings on the ventilator-shafts that were a common feature of the roof-tops of medieval Cairo, and many of which survived until Gérôme's time if not until our own. In medieval Egyptian scientific folk-lore the favourable north wind was thought to blow mainly from the direction about 27° east of north, perpendicular to the direction of winter sunrise at 27° south of east. So these ventilators faced this first direction, which is actually the same as that of the main axis of the orthogonal street-plan of the old city. Their deliberate alignment in this fashion is confirmed by medieval astronomical texts on, and even astronomical tables for, the construction of ventilators. In the painting the ventilators are correctly shown as being oriented in the same way as the buildings on which they stand. Now the qibla used in medieval Cairo was either 37° south of east, as calculated by the astronomers, or towards winter sunrise at 27°, as established by the first generation of Muslims in Egypt. So the men should actually be facing the direction to their right, but then Gérôme would have lost the lunar crescent.

It was the astronomical tables compiled by the muwaqqits or professional mosque astronomers of medieval Cairo that led to the discovery that the entire city of al-Qāhira, founded in the 10th century, is astronomically aligned. But it was constructed alongside the Pharaonic canal, which at this point fortuitously runs perpendicular to winter sunrise. Was the city built on this site because of this feature of the local topography? I suspect that it was, but the medieval texts are silent on this point. The reason why the city was deliberately astronomically aligned is that the rectangular base of the Kaʿba itself faces, surely not accidentally, significant astronomical directions (including, for example, that of the main axis of Stonehenge). The Mamluks built their imposing religious architecture in Cairo in keeping with the old city-plan, but they favoured the new qibla of the astronomers, at 10° south of the minor axis of the city. This qibla was used *inside* their buildings, so that the walls are of varying thickness. When standing in front of the windows one can see the 10° difference between the outer and inner walls. The "City of the Dead", on the other hand, is laid out so that the main axis lies along a *wādī*, this time conveniently perpendicular to the qibla of the astronomers. A third medieval suburb is laid out due south, corresponding to the qibla favoured by those who followed the practice of the Prophet when he was in Medina, due north of Mecca. Various textual sources discuss the orientations of various mosques in Cairo, perhaps the most interesting example of a city deliberately oriented towards Mecca (although no-one would know it from a map of modern Cairo). Medieval legal texts discuss the acceptability of divergent qiblas, and, in this case, since they are all within a quadrant around the "real" qibla, they are all within bounds.

I shall return later to other aspects of science in the service of Islam not treated here. At least I hope to have given the reader some insights into the wealth and diversity of the available material and its significance, which extends far beyond the history of science.

CHAPTER 3

THE CHRISTIAN MONASTIC SCENE

> " ... that remarkable endeavor of scholars from the Latin West to seek out Greco-Arabic science and assimilate it into their own intellectual tradition ... ." Stephen C. McCluskey, "Monastic Timekeeping", p. 9.

> "Monasteries were ... the principal focus of learning during (the early Middle Ages) and we should look to them if we are to characterize this scientific activity, its context, and its development. The rudimentary techniques of astronomical computus that arose to determine the date of Easter have been a recurring topic of discussion. Less well studied, however, are the observational practices ... that monastic timekeeping employed to regulate the hours of monastic prayer." *Ibid.*

The early European scientific sources (again mainly unpublished) are so inadequately documented that what we know about them is very much a matter of chance. As Neugebauer wrote: "The worst conditions prevail for the mediaeval European scientific literature in Latin". Some of those that are published have still never been approached critically. But the unpublished sources are certainly full of surprises.

Medieval timekeeping, on which I shall concentrate here, is a topic of which only the surface has been scratched by modern scholarship. No overview is possible yet, and I shall simply cite a few examples of procedures and instruments. These supplement those that can be culled from expert writings such as those of Otto Neugebauer, Olaf Pedersen, John North and Stephen McCluskey. None but the last of the examples is new to the literature, but some have been forgotten, and they have certainly never been grouped together. The specialist will notice many omissions in my presentation, but others have treated exhaustively such topics as, say, the astronomy of Gerbert, later Pope Sylvester II.

The earliest known European astrolabe was made in Catalonia in the 10th century. It bears some indications of Islamic influence – notably an alphanumerical number-notation based on the Western Islamic one, azimuth curves on some of the plates, and the choice of Francia (from *bilād al-ifranj*) for one of the plates – but various others – notably, the rete design and choice of latitudes for the plates, especially the choice of Rome – which indicate independent, possibly Roman (??), influence. (See further **X-4.8.**) The milieu in which this astrolabe was made is unknown. Was it perhaps made by a monk?

One of the few astrolabes for which a monastic provenance is fairly certain bears numbers expressed in an ingenious and elegant notation developed by Cistercian monks in Flanders in the 13th century as an aid to scholarship. With this system any number up to 9999 can be uniquely represented by a single cipher. Thus, for example, the year-number 1397 is uniquely defined by the cipher:

 , which is to be interpreted thus: 

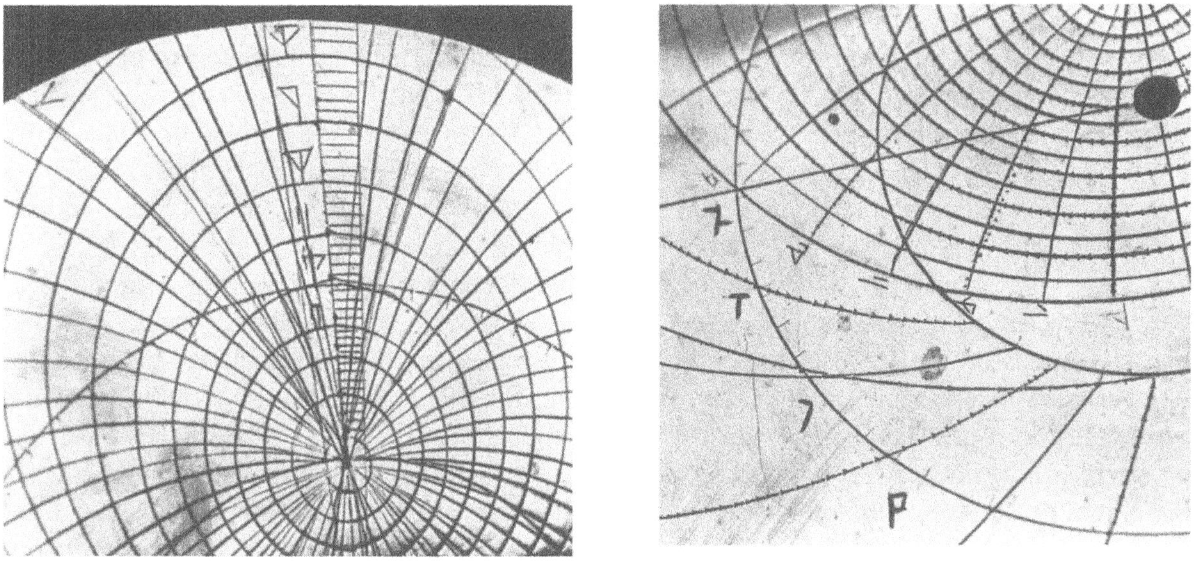

Fig. 3.1: Numerical ciphers on the 14th-century Picard astrolabe. The ciphers lend themselves ideally to engraving on metal because they are composed of straight line segments. Until the 18th century they were used in Bruges to mark the volumes of wine in wooden barrels. [Photos by the author, courtesy of the owner.]

The monks were accustomed to the cumbersome Roman system, and the "new" Arabic numerals were not everywhere received with enthusiasm, tainted as they were with alien associations. The new notation was used in very limited circles in monasteries, not all Cistercian, from France to Sweden and Italy and Spain for foliating manuscripts, for numbering sermons and lines of a staff in musical notation, for representing year-numbers in chronicles, *etc.* It was also used outside the monastic scriptoria for writing year-numbers in astronomical tables and for engraving volumes on wine-barrels, and the like. Inside a monastery or not we may never know, the notation was used for engraving numbers on an astrolabe made *ca.* 1375; some 150 years later this astrolabe was in a Benedictine abbey in Liège. With the introduction of printing the ciphers were reduced to the status of curiosa, and as a result of a mistake in the 16th century they became associated with magic and astrology and came to be labelled Chaldean! In fact they were developed from a simpler set serving numbers up to 99; these simpler ciphers are used for numbering items in lists and as arguments in an Easter table in a single manuscript from the circle of Grosseteste. Now the chronicler Matthew Paris says the English monk John of Basingstoke brought the ciphers from Athens in the early 13th century. The shapes of the Basingstoke ciphers are indeed Greek in origin, for they are described in a 4th century B.C. tablet found on the Acropolis in the 1880s and featuring a system of syllabic shorthand. They reappear in an English *ars notaria* of the late 12th century, in a shorthand that represents an ingenious revival of the Tironean notes. Did Basingstoke himself hit on the clever idea of using them for numbers? Or are the Basingstoke ciphers a Greek numeral notation unknown from the Greek sources? Certainly the Cistercian ciphers are an original invention because the available manuscripts attest to various unsuccessful trial versions.

This remarkable astrolabe, of 14th-century Picard provenance, exhibits minor technical defects, one of which (confusion of some stars on the rete) reflects the degeneracy of the local manuscript tradition, and another (incorrectly-constructed shadow squares) is found elsewhere only on one other instrument, a contemporaneous Northern French astrolabe with similar engraving. This second piece, now in the Science Museum in London, constitutes the missing link in the development of the astronomical clock. It is based on a southern stereographic projection and thus belongs to a tradition not attested – as far as we know at present – in European manuscripts (but started in Baghdad in the 9th century). Now the earliest astronomical clocks – those in Valenciennes, Münster, Prague, Bourges, Doberan, Lübeck, Lund, Stralsund and Ulm – exhibit the same feature. And it is not insignificant that this astrolabe also features a gear mechanism to represent the relative apparent solar and lunar motions. Was this astrolabe too made in a monastery? If so, what are the implications?

There is no evidence that monks used astrolabes on a regular basis for timekeeping. Clearly astrolabes must have been used occasionally, for the instrument is recommended for "celebrating the daily office of prayer" in a 10th-century tract associated with the school of Gerbert. And Walcher, a monk at the Priory of Great Malvern, had an instrument of some kind at hand when a lunar eclipse occurred before dawn on October 18th, 1092. Nevertheless with an astrolabe one can very easily determine the times of the seasonal hours and in particular the times of the Christian prayers. One procedure monks apparently did use for the daylight prayers was observing shadows. If the shadow schemes proposed by the Muslim legal scholars for regulating the times of prayer were crude and often incorrect, the shadow schemes recorded by medieval monks were catastrophic (**III-10.1**). As Neugebauer wrote: "as usual the western tradition represents the lowest level in our material". Thus, for example, the table for the length of daylight presented by Gregory of Tours (late 6th century) serves a latitude corresponding to the Mediterranean rather than Tours. It, like several of the Islamic and European shadow-schemes for timekeeping, derives from classical folk astronomy. As with the Islamic shadow-schemes, so in the European schemes, albeit for quite different historical reasons, one gnomon length is to be added to the midday shadows to determine the shadows at the mid-morning and mid-afternoon prayers.

One such crude shadow scheme is represented graphically on a small vertical sundial from the 10th century excavated at Canterbury Cathedral. The markings for *terce*, *sext* and *none* (not specifically identified as such) are so crude that it has not yet proved possible to identify the underlying scheme. Another, later example with similar markings is known, and I suspect that we are dealing with degenerate examples of an otherwise unknown genre of Roman sundial. In any case, no other instruments are known from the Christian Middle Ages with markings distinguishing the ends of the 3rd, 6th and 9th hours as special times. (This is quite the opposite situation from Islamic instruments, which often indicate the times of Muslim prayer.)

Schemes are attested for timekeeping at night by means of the moon. An 11th-century star-chart has survived in which the direction of a particular star or star-group with respect to certain architectural features of a monastery (windows of the dormitory and refectory, chapels, *etc.*), viewed from a specific location inside the cloister, is defined at the time when the lamps were to be lit and the monks roused. Thus, for example, on the feast of the Purification of the Virgin

Fig. 3.2: The sundial of Acquafredda. [From D. J. L. Benoit, "Un cadran solaire monastique du XIIe siècle".]

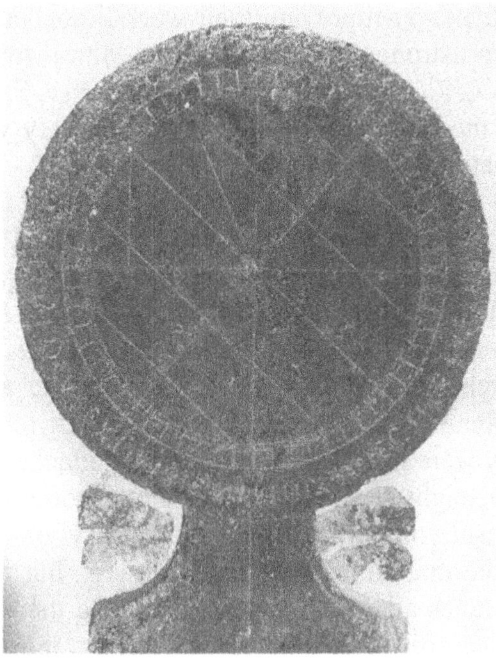

Fig. 3.3: The astronomical markings on the Regensburg dial. [Photo from the Zinner Archives, Institute for the History of Science, Frankfurt University. (Modern colour images do not show the markings as clearly as this.)]

the bright star in the knee of Bootes will be seen at that time between the third and fourth window of the dormitory, closer to the former. The astronomical information given is adequate to reconstruct the plan and orientation of the compound, but this cannot be safely associated with any known monastery.

An equatorial sundial was discovered amongst the remains of the Cistercian Monastery at Acquafredda on Lake Como in 1904, when Benedictines occupied the long-abandoned site. The sundial, 425 mm. in diameter, bears a date 1193 in bold Roman numerals and is hence the oldest dated European sundial. Such an equatorial sundial can only be used in the summer months; two such sundials would have been needed to serve the whole year, the other one erected to face the opposite direction. Such sundials are not known from Antiquity, probably because the Greeks and Romans preferred to measure time with sundials in seasonal hours; they are mentioned in 13th-century Islamic sources, if not in earlier ones, and there is a very fine pair in the early-18th-century observatory in Jaipur in India. However, the Acquafredda sundial was doubtless made equatorial in the late Renaissance or early modern period. It is of the simple vertical scratch-dial variety but when discovered it was mounted so that a replacement gnomon would point towards the celestial pole. Those monks who might have observed the shadow of the horizontal gnomon moving across the twelve equal sectors below the horizontal diameter would probably have been completely confused: the "hours" indicated by the sundial are neither the seasonal hours of daylight by which they would want to regulate their prayer-times and their work-times, nor the equinoctial hours that they might have known from astronomical sources. In addition to the main radial hour-lines there are smaller marks at $4^1/_2$, $7^1/_2$ and $9^1/_2$ "hours" after sunrise. Since there is no plausible explanation for these within the medieval context, I suggest that they were added by the person who rendered the sundial equatorial. If this took place before 1783 – and a new analysis of the ensemble, if it still exists and can be found, might reveal a date for the conversion – that person would have surely been a Cistercian monk. If it took place after 1904 then he would have been a Benedictine. In any case, he had obligations to perform at $1^1/_2$ hours before midday, and at $1^1/_2$ and $3^1/_2$ hours after midday, and he marked these on the sundial. It is easy to imagine that these might relate to a generous lunch break and a modest siesta, a compromise perhaps between the gastronomic delights of the region and the monastic work ethic.

Scratch-dials are an embarrassment to all sundial enthusiasts: nobody really knows what they were used for. A *vertical* dial attached to a south-facing wall, hemispherical in shape with four, eight or twelve sectors and a horizontal gnomon attached at the centre, simply does not measure time properly. On some surviving examples a replacement gnomon has been installed in the direction of the celestial axis, but this will not help much. Scratch-dials are found on churches all over Europe but to date there have only been regional studies. Some in England are carefully constructed and merit a more dignified name. Where do they come from? Now one was also described in a weak moment by the distinguished 10th-century Cairo astronomer Ibn Yūnus, but at least he was aware that the "hours" measured by the sundial were not those in vogue at the time and he investigated their mathematics. Another, this time *horizontal*, is described in a treatise on mechanical devices by the Andalusī Ibn Khalaf al-Murādī and by Maimonides, both in the 12th century, and by the Tunisian astronomer Ibn Isḥāq in the early 13th. What is going on here? I confess that I have no idea.

An important and mysterious historical source of a very different kind is the so-called "astronomisches Lehrgerät" or teaching device from the garden of the Benedictine Monastery of St. Emmeram in Regensburg. Founded in 697 on the site of an earlier Roman building, this monastery was eventually suppressed in 1812. A 2-metre-high column from the monastery with a remarkable astronomical representation on the top is now in the Museum der Stadt Regensburg and has been studied by Ernst Zinner, but apparently by no historian of science before or after him. The column has been dated on art-historical grounds to the 11th century. Before Zinner's time the object was described as an astrolabe and as a sundial; in fact it is neither. On one side of a vertical circular disc is a young man kneeling and presumably looking at the heavens. The Latin inscription on this side mentions Aratus (first half of the 3rd century B.C.), but the markings on the other side bear no relation to the known scientific work of that scholar, to whom, however, much was attributed in the Middle Ages. The astronomical markings consist of an orthogonal projection (from a point at infinity) of the main circles on the celestial sphere in the plane of the local meridian: celestial equator, solstitial and polar circles, traversed by a horizontal diameter representing the horizon and a vertical diameter drawn between the zenith and the nadir. The latter has been estimated as corresponding to latitude 48°, which serves Regensburg (with latitude 49°) quite nicely, but no-one in Regensburg in the early Middle Ages would have known that the latitude there was 48°, and the horizon was surely intended to correspond to the middle of the 7th climate. In addition there is a line between the upper end of the summer solstitial line and the lower end of the winter solstitial line; in the projection this corresponds to the position of the ecliptic at sunrise and sunset at the equinoxes. Also the perpendicular from the upper end of the equator to the horizon is drawn. Not the least remarkable feature is the circular scale divided in 5°-intervals subdivided for each degree. The ensemble is executed with remarkable skill and mathematical accuracy. Zinner noted that the underlying astronomical notions are attested in a diagram appended to the astronomical writings of Wilhelm of Hirsau, and for want of evidence to the contrary assumed that the diagram was due to that author (Zinner, *Die ältesten Räderuhren*, pp. 15-16 and pls. 1-3). Wilhelm was a monk in the Monastery of St. Emmeram before 1071, when he was appointed Abbot of the Benedictine Monastery in Hirsau. It is, however, a complete mystery who erected this remarkable column and to what purpose. One can do nothing with the astronomical markings except talk about them; the instrument is indeed well described by Zinner's term "*Lehrgerät*". But who in Regensburg in the 11th or 12th century knew enough astronomy and mathematics to be able to understand the underlying notions, or to represent them in such a "modern" fashion, and who would be his audience? No such *Lehrgerät* is known from Antiquity (which does not mean that they did not exist). Further the underlying theory was in all probability not known to Aratus, although his astronomical poem, the *Phaenomena*, does include mention of the circles and celestial sphere and the risings and settings of stars. The theory is, however, clearly outlined and further developed in Ptolemy's *Analemma* and in Vitruvius' *De architectura*. Whoever made the *Lehrgerät* clearly felt more secure in citing Aratus.

A small instrument discovered *ca.* 1990 on the estate of a Cistercian monastery in Suffolk is shaped like a ship with a mast. It was called *navicula de Venetiis*, and its origins are obscure. The instrument, which is probably from the 14th century, was purchased by a London museum and the ensuing publicity, partly inspired by the price paid, occasioned someone to come forward with another very similar instrument; this second piece was auctioned at Sotheby's of London in 1993 and is now in Geneva. With another, similar but inferior, piece already in Oxford we now have three English examples of an instrument that is one of the most brilliant computers ever devised before the modern era. (One other contemporaneous example, preserved in Florence, is also apparently of English provenance, but differs from these three.) One feeds in the latitude and the date and holds the thing up towards the sun; then, if one follows an appropriate procedure, one can simply read off the time of day, *correctly given in equinoctial hours*! Who first devised this universal horary dial? (The original instrument need not have looked like a boat, and the Venetian connotation does not necessarily imply an Italian connection.) Richard of Wallingford, an English scholar who did design innovative instruments, is not known to have been associated with this one. No other English astronomer comes to mind, although the texts are associated with a Benedictine monk named Peter of Muchelney and another man named John Slape. Now in the second of these textual traditions relating to the *navicula*, it is described along with the quadrant (with or without cursor) and the cylindrical sundial; each of these other instruments was invented in Baghdad in the 9th century although they are invariably described as European in the modern specialist literature. But what was a *navicula* doing on the grounds of an English monastery? (See now **XIIb**.)

Textual studies have revealed that in France it was only in the middle of the second half of the 14th century that the division of the day into 24 hours became common (although A. J. Turner has pushed this back to the first half of that century). Before this references to time were invariably in terms of the canonical hours. And it was about this time that the first striking clocks appeared on public buildings. Yet in the 14th and 15th century sophisticated tables for timekeeping by means of the solar altitude for specific terrestrial latitudes – like those favoured by Muslim astronomers – were compiled in Europe, but only those of Nicholas of Lynn (Oxford) and pseudo-Enrique de Villena (Jaén) have been studied. Another set was compiled for Paris, probably by Jean Fusoris in the early 15th century, and various later printed tables for localities in Italy are known but have never been studied. So we now have three examples of medieval European tables for timekeeping. (And just 25 years ago only three medieval Islamic examples were known.)

In the 1480s a Benedictine monk from Vienna named Hans Dorn produced some of the most remarkable astronomical instruments that were being made anywhere in the world at that time. We have little information on Dorn's monastic ties, but fortunately several of his instruments survive. His instruments, certainly not made for monastic use, are clearly inspired by those of his Humanist predecessors in Vienna, Peuerbach and Regiomontanus, but they are very much the product of an independent spirit. And whilst they qualify as Renaissance pieces – their elegance, their style, the capital *antiqua* script, the transversal scales – they are still very much medieval – mistakes in the Latin with occasional lapses into vernacular, cumbersome dimensions, impracticality. These remarkable instruments, including two astrolabes, a celestial globe with an astrological plate, a torquetum and several compendia, await detailed study.

In 1522 a Benedictine monk from Liège named Paschasius Berselius presented the Picard astrolabe mentioned above to his friend Amerotius, the second professor of Greek at the newly-founded College of Three Languages at Louvain. How did Berselius come by the astrolabe? Did he know how to use it? Alas the instrument itself can tell us only so much, and the known historical sources are silent on these points. This piece is in fact the second known astrolabe in the history of Liège: the first one was there already five hundred years previously. Ragimbold, the head of the Cathedral School in Cologne, who was a student of Fulbert of Chartres, heard around the year 1025 that there was an astrolabe in Liège, perhaps amongst the legacies of Gerbert and Fulbert, perhaps only on loan. He wrote to his opposite number in Liège and asked to borrow it. Although the monks in Liège at that time were better known for their work in arithmetic and geometry than in astronomy and timekeeping, the reply that Ragimbold received was somewhat supercilious, namely that there was no point in sending him the astrolabe because he would not understand it anyway and if he wanted to see it he should come to Liège.

Note added in 2002: I was surprised to learn that two copies of parts of Euclid's *Elements*, one from the 13th century, the other from the 14th, were originally preserved in the library of the Cistercian abbey of Ter Doest in Flanders (see further *Brussels BR 2001 Exhibition Catalogue*, pp. 156-157).

I conclude this section by mentioning an ivory quadrant made by one Antonius Loffroi in 1556 at the Cistercian Abbey of Ourscamp in Southern Picardy. The piece is an horary quadrant, of a special variety having rectilinear markings for the hours and, necessarily, a non-uniform solar declination scale. Now this kind of quadrant appears to be a European invention (although curiously it does occur on a small number of later Ottoman quadrants that are otherwise completely within in the Islamic tradition). The earliest known description of this variety of quadrant is in a German manuscript from *ca.* 1500, and it was also featured by Stöffler (1452-1530), whose work on instruments was printed in Paris in Latin by 1550 and in French by 1560. In brief, we are dealing with a sophisticated "state of the art" instrument (even though Loffroi used the latitude of Paris rather than that of Ourscamp for the markings). Clearly in this particular monastery in the middle of the 16th century scientific activity was not merely confined to theoretical knowledge obtained from books.

In brief, astronomical timekeeping in medieval Europe, both inside the monastic setting and without, is a subject that cries out for further research. As in the case of Islam, the implications are of interest to more than the history of science.

The French miniature

Heinrich Seuse or Henricus Suso, the Dominican monk, is receiving inspiration from Wisdom (Sapientia or Sagesse) personified. This lady is sometimes identified with the Virgin Mary: the red dress, the blue cloak, and the crown are symbols that prove that the Mother of God was not far from the thoughts of the artist. Of more relevance to our topic are the two weight-driven clocks on either side. The larger bears a dial displaying 24 hours labelled in Roman numerals 1-12 and again 1-12, as on an astrolabe; the smaller lacks a dial. Both incorporate devices for operating bells (the bell for the clock on the left is at the top of the page). There is a good technical reason why the hours are not numbered 1 to 24; in Roman numerals this would look absurd (witness the clock by Paolo Uccello in the Duomo of Florence). The miniature has nothing to do with the contents of the book: this deals with mysticism and it was the division of the book into 24 chapters that seems to have inspired the illustration.

Learned papers and even books have been written about monastic clocks and the regulation of the prayer-times by means of clepsydra and weight-driven clocks, but any that overlook the facts that the monastic orders are defined in term of the seasonal hours and that these cannot be regulated by any kind of mechanical clock alone miss the point. To do things properly with a clock one would also have needed a table of the length of daylight throughout the year, and these do not seem to have been common (only that of Gregory of Tours, an unhappy example, is known to me – see above). And tables displaying the times of monastic prayers for specific latitudes, or non-mechanical clocks that *do* measure time in seasonal hours (one such is attested from 12th-century al-ʿIrāq), are apparently quite unknown. Are we to believe that (in complete contrast to the situation in Islam) the equinoctial hours and the clock came to dominate monastic life, so that monks did indeed perform the *none* at nine o'clock in the morning and *terce* at three o'clock in the afternoon? Most modern works on monasticism imply this, but then they also have sunrise and sunset at 6 o'clock.

There are several manuscript miniatures and works of art in which an astrolabe is depicted in such detail that one could hope to identify a similar instrument in a museum collection. In

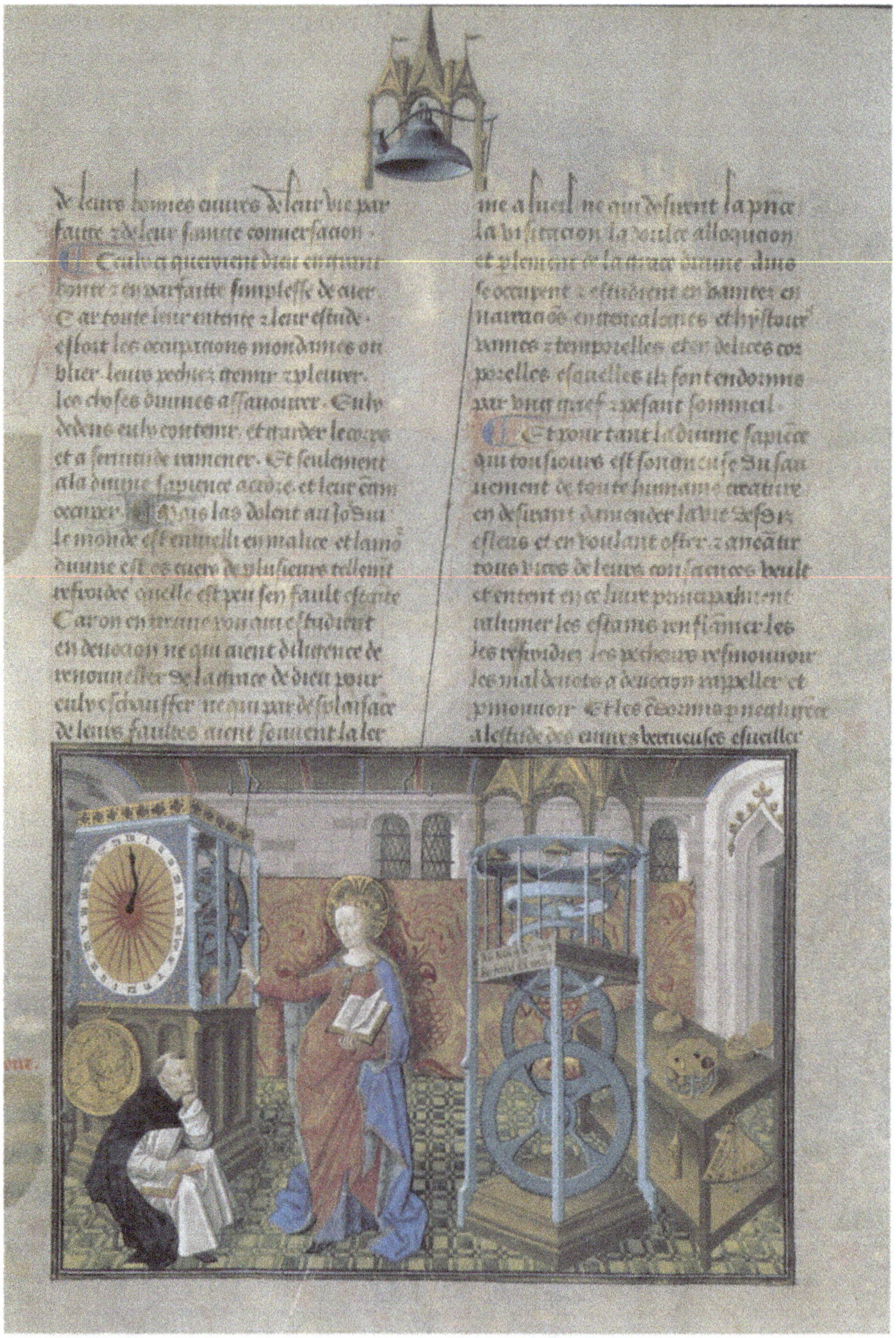

Fig. 3.4: The monk Suso receiving instruction from Sapientia. On this miniature see already Lynn White, chapter "The Iconography of *Temperantia* and the Virtuousness of Technology", first publ. 1969, repr. in *idem*, *Medieval Religion and Technology: Collected Essays*, Berkeley, Los Angeles & London: University of California Press, 1978, pp. 181-204, especially pp. 195-197 and the works there cited, and also Henri Michel, *Images des sciences: les anciens instruments scientifiques vus par les artisites de leur temps*, Rhode-St-Genèse (B): Albert de Visscher, 1977, pp. 124-125, and Anthony J. Turner, ed., *time*, Amsterdam: Tijd voor Tijd Foundation, 1990 (catalogue of an exhibition at the Nieuwe Kirk, Amsterdam, in 1990), p. 105. [From MS Brussels B.R. IV 111, fol. 13v, courtesy of the Bibliothèque Royale Albert 1er.]

the case of the astrolabe depicted in the intarsia of the study of Archduke Frederico of Urbino (middle of 2nd half of the 15th century) a very similar instrument has indeed been located. Here we can identify only the general type of instrument depicted. (It would be of interest to know what kinds of astrolabes, if any, might be illustrated in other French manuscripts of the very popular *Horologium sapientiae*; Lynn White has already pointed out that not even clocks are featured in 14th-century copies of the Latin version.) There is a minuscule throne and a very simple shackle and ring. The rete, which from its style is clearly French, is distinguished firstly by the semi-circular frame on the solstitial bar inside the upper part of the ecliptic and secondly by the flame-shaped star-pointers, two of which are double-pronged. Also the equinoctial bar is counterchanged in the standard way, thus:

The instrument that comes closest to this is in a private collection in England; it bears the distinctive upper frame, the semi-circular frame on the solstitial bar, and several double-pronged pointers, two of which correspond to those in the miniature. The trilobed throne is more substantial than the one in the miniature. Another similar rete is illustrated in a French manuscript on the astrolabe treatise (MS Berlin SB lat. Fol. 610) but this is dated 1276. Several 14th-century French astrolabes of this kind survive in various museum collections (not one properly published), but none is quite so close as the one now in England to that in the illustration – they have either a different shaped upper frame or the semi-circular frame open in the other direction. These astrolabes appear to be the immediate predecessors of those of the Paris workshop of Jean Fusoris (early 15th century), of which 30 survive but only one has been published.

On the right of the miniature there is a table bearing a universal horary quadrant (without a solar scale, but bearing a movable bead on the thread), a cylindrical sundial, a horizontal sundial, and two compendia, that is, multi-functional devices. One of the latter reveals a gear-mechanism, doubtless for reproducing the relative motions of the sun and moon (with a circular aperture on the top to display the age of the moon), and the other is an equatorial dial that can be dismantled to fit into the circular box (which would probably contain a magnetic compass). With but one exception these, like the astrolabe, are instruments that came to Europe originally from the Islamic world. The one piece with gears is "our first indisputable evidence of a spring-driven, fusee-regulated mechanism" (Lynn White, after Henri Michel, confirmed by Anthony Turner), which is indeed a European innovation. But the rest of the instruments came to Europe through either Spain or Sicily, some through the intermediation of Jewish scholars. Each variety of instrument has an individual history that is not yet properly documented. Some of the sources raise new questions rather than solve old ones.

It is still an open question how monks in 15th-century France would have actually regulated their prayer-times. At that time, as in earlier centuries, only one monk with some knowledge of astronomy would be needed in each monastery. He may have been versed solely in the simple techniques of folk astronomy, or have known enough to be able to use an astrolabe, or have been sufficiently in control of the subject to even make his own instruments.

CHAPTER 4

CONCLUDING REMARKS

There are many more examples of science in the service of Islam that could have been cited. I have not dealt at all with the complicated algebra of inheritance, based on *Qur'ānic* prescriptions. For arithmetic as applied to aspects of daily life, including inheritance, we have a valuable study by Ulrich Rebstock. Nor have I dealt with the geometry of Islamic patterns that decorated religious architecture in spite of proscriptions against it. A recent study by Gülrü Necipoglu has dealt with an artisan's handbook of geometric patterns dating from *ca.* 1500. (See also the essay review by George Saliba.) But the patterns themselves are not properly documented, except for an unpublished study of designs in Cairo by Manuel Keene and the splendid book on Andalusī designs by Basilio Pavón Maldonado. Likewise the systematic documentation of the orientation of Islamic religious architecture has so far been conducted only for a few regions of the Islamic world (Morocco, Turkey, Greece, and East Africa). There is plenty of work still to be done, and there is certainly no shortage of available source material, be it scientific, legal or architectural.

Now by the time Gérôme painted his picture, Edward W. Lane had written in his *Manners and Customs of the Modern Egyptians* (1836):

> "Those persons in Egypt who profess to have considerable knowledge of astronomy are generally blind to the true principles of the science: to say that the earth revolves round the sun they consider absolute heresy. Pure astronomy they make chiefly subservient to their computations of the calendar."

It is a fact that science was more lively in Egypt between the 10th and the 15th centuries than it was in the 19th. Many learned writings have addressed the question why "Islamic science", if there ever was such a thing, declined. During the past few decades the date when it "declined" has had a tendency to slip forward through the centuries; early writers liked the 11th century, some now write in terms of the 15th. But none has been aware of any of the material described above, none has addressed the dual nature of science in Islamic civilization or the substantial differences between the regional schools, and few have realized that the legal scholars provided solutions to the fundamental practical problems quite different from those achieved with such ingenuity by Muslim scientists. So there is a lot more that can be said on this question.

Likewise, there are many more examples of "scientific" activity in early medieval Europe inside and outside of the monasteries that one could cite, and I have not dealt at all with related topics such as the determination of Easter, the orientation of churches (is there any better study than that of T. D. Atkinson?), and the geometry of Gothic architecture, all of prime importance for medieval "science" defined in its broadest terms. But of all the examples I have cited the only "real" European invention is the ciphers of the monks (inspired by a Greek model), and this was not intended for any "scientific" purpose whatsoever, and the controlled gear mechanism. The inspiration for the *navicula* may eventually be found in one of the 10,000 available Arabic astronomical manuscripts. But who knows what other surprises await us in the thousands of scientific manuscripts in Latin and the vernaculars, let alone the yet more abundant sources relating to aspects of monastic life?

In the Renaissance European astronomers started to study seriously the original Greek scientific texts, some of which had been recovered in Europe as early as the 12th century. Such was their enthusiasm for these splendid documents that they had neither time nor inclination either to pursue further the medieval Islamic tradition, of which their predecessors had inherited but a modicum, or to build on the rather shaky medieval European tradition based mainly on rudimentary knowledge of parts of the Greek and Islamic traditions. They started afresh, as it were, Regiomontanus in the 15th century essentially redoing in Latin what al-Farghānī and others had done in Arabic with the *Almagest* in the 8th and 9th centuries. Thanks to the invention of printing they were able for the first time to keep the texts under control and to achieve a wide (and no longer sporadic, as was the case in the days of manuscripts) distribution for them. And soon thereafter, with the invention of the telescope, they were able to look for the first time beyond Ptolemy's universe and through it. At the same time muwaqqits in the Islamic world were still using tables for timekeeping already centuries old and yet perfectly adequate for their needs, writing "new" simplified treatises on quadrants and sundials, and making trivial qibla-indicators consisting of compass-boxes with lists of qiblas for important cities. And monks in Europe were trying to adjust to a new world in which the clock and other innovations would eventually dominate daily life.

There has been an inevitable tendency amongst historians of science to concentrate on materials that are important mainly for the transmission of hard-core science to Europe. By so doing we have tended to overlook major works of science that were either not transmitted westwards or were not influential in later circles. And we have tended to treat with disdain or ignore altogether any theory or practice that was crude or "primitive". But the greatest historian of the exact sciences in the 20th century, Otto Neugebauer, who set our understanding of ancient science on a firm basis, was also well aware of the importance of folk science. This explains why he spent many years studying the Greek, Coptic, Ethiopic, and medieval European traditions; others, including myself, have devoted much time to the Islamic tradition. Thus it was Neugebauer who surveyed the astronomy underlying the *Très riches heures* and the shadow-schemes attested in various early European manuscripts. Elsewhere Neugebauer wrote that one only had to compare the *Almagest*, the *Zīj* of al-Battānī and Copernicus' *De revolutionibus* to convince oneself of "the inner coherence of ancient and medieval astronomy", but he had also convinced himself of the unity of ancient and medieval folk science and of the like mentality of monks from Ethiopia to Northumbria whose heads were full of computus. Those three major works of mathematical astronomy are now textually under some measure of control, in that they are published with translations and critical commentaries, but there are many more medieval Arabic, Persian, Hebrew, Latin and vernacular works that are not. I would add to Neugebauer's remark another in the spirit of Sédillot, namely, that one only has to look at a few selected manuscripts and instruments to convince oneself of the diversity and occasional brilliance of medieval science in the context of mosques and monasteries, and to look at a few hundred to realize how much work remains to be done before we can hope to understand it.

PART VIII: SELECTED BIBLIOGRAPHY

The following works arranged alphabetically by author are not intended to be representative of the history of medieval science in general but rather only as a background to the topics of this study. The list is presented in three parts, works relating to the Islamic scene, then those on the medieval Christian scene, and finally works which bridge the gaps between various medieval traditions, some relating these to Ancient science, others to Renaissance science. The list is independent of the main bibliography preceding **Part I** of this book and has not been coordinated with it. Very few of the many writings on medieval European Easter computations are listed; I have preferred to draw attention instead to such studies as those of Neugebauer and Saliba on the Ethiopic, Coptic and Islamic traditions of Easter computation.

Islam

Encyclopaedia of Islam, new edn., 10 vols. to date, Leiden: E. J. Brill, 1960 to present, especially articles "Anwā'" (aspects of folk astronomy); "Asṭurlāb" (astrolabe); "ʿIlm al-hay'a" (astronomy); "Ḳamar" (moon); "Ḳibla" (sacred direction); "Layl and nahār" (aspects of folk astronomy); "Makka. iv. As centre of the world" (sacred geography); "Manzila" (lunar mansions); "Mīḳāt" (time-keeping); "Mizwala" (sundials); "Nudjūm" (stars); "Nudjūm, ʿilm aḥkām al-" (astrology); "Rīḥ" (winds); "Rubʿ" (quadrant); "Ru'yat al-hilāl" (lunar visibility predictions)"; "Samt" (direction); "Shakkāziyya" (universal projections); "Shams" (sun); "Ta'rīkh, iv" (calendars in Islamic astronomy); "Ṭāsa" (magnetic compass); "ʿUṭārid" (Mercury); and "Zīdj" (astronomical handbooks and tables, with an update on *ʿilm al-hay'a*).

Ali, Jamil, *The Determination of the Coordinates of Positions for the Corrections of Distances between Cities – A translation of al-Bīrūnī's Kitāb Taḥdīd nihāyāt al-amākin li-taṣḥīḥ masāfāt al-masākin*, Beirut: American University of Beirut, 1967. See also Kennedy, *al-Bīrūnī's Taḥdīd*.

Baffioni, Carmela, ed., *Religion* versus *Science in Islam – A Medieval and Modern Debate*, in *Oriente moderno* (Rome: Istituto per l'Oriente C. A. Nallino), n.s.,19 (2000).

Berggren, J. Lennart, *Episodes in the Mathematics of Medieval Islam*, New York, *etc.*: Springer, 1986.

Charette, François, "A Monumental Medieval Table for Solving the Problems of Spherical Astronomy for all Latitudes", *AIHS* 48 (1998), pp. 11-64; *Mathematical Instrumentation in 14th-Century Egypt and Syria*, doctoral thesis, Johann Wolfgang Goethe University, Institute for the History of Science, 2001.

Charette & D. A. King, *The Universal Astrolabe of Ibn al-Sarrāj*, to appear.

Dold-Samplonius, Yvonne, "Practical Arabic Mathematics: Measuring the Muqarnas by al-Kāshī", *Centaurus* 35 (1992), pp. 193-242.

Endreß, Gerhard, "Die wissenschaftliche Literatur", Kapitel 8 in *Grundriß der Arabischen Philologie*, Band II: *Litteraturwissenschaft*, Helmut Gätje, ed., Wiesbaden: Dr. Ludwig Reichert, 1987, pp. 400-506.

Miquel Forcada, "*Mīqāt* en los calendarios andalusíes", *al-Qantara* 11 (1990), pp. 59-69.

Edward Stewart Kennedy, *A Commentary upon al-Bīrūnī's Kitāb Taḥdīd* [*nihāyāt*] *al-amākin*, Beirut: American University of Beirut Press, 1973. See also Ali, *al-Bīrūnī's Taḥdīd*; *The Exhaustive Treatise on Shadows by ... al-Bīrūnī*, 2 vols., Aleppo: Institute for the History of Arabic Science, 1976; *Astronomy and Astrology in the Medieval Islamic World*, Aldershot & Brookfield, Vt.: Ashgate-Variorum, 1998.

——, and Imad Ghanem, eds., *The Life and Work of Ibn al-Shāṭir – An Arab Astronomer of the Fourteenth Century*, Aleppo: Institute for the History of Arabic Science, 1976.

——, *et al.*, *Studies in the Exact Sciences in Medieval Islam*, Beirut: American University Press, 1983.

King, David A. *Islamic Mathematical Astronomy*, *Islamic Astronomical Instruments*, and *Astronomy in the Service of Islam*, London, *etc.*: Variorum, 1986 (2nd rev. edn., 1993), 1987 (repr. 1995) and 1993; "Folk Astronomy in the Service of Religion: The Case of Islam", in *Astronomies and Cultures*, Clive L. N. Ruggles and Nicholas J. Saunders, eds., Niwot, Co.: University Press of Colorado, 1993 [published 1994], pp. 124-138; "Islamic Astronomy", in C. B. F. Walker, ed., *Astronomy before the Telescope*, London: British Museum Press, 1996, repr. 1999, pp. 143-174; *World-Maps for Finding the Direction and Distance to Mecca: Innovation and Tradition in Islamic Science*, Leiden: E. J. Brill, and London: Al-Furqan Islamic Heritage Foundation, 1999; *The World around the Kaaba – The Sacred Folk Geography of Medieval Islam*, to appear.

King & Julio Samsó, "Astronomical Handbooks and Tables from the Islamic World (750-1900): An Interim Report", *Suhayl* 2 (2001), pp. 9-105.

Kunitzsch, Paul, *The Arabs and the Stars*, Northampton: Variorum, 1989.

Nallino, Carlo Alfonso, "Sun, Moon, and Stars (Muhammadan)", in J. Hastings, ed., *Encyclopaedia of Religion and Ethics*, 12 vols., Edinburgh: T. & T. Clark, vol. XII (1921), pp. 88-101.

Necipoglu, Gülrü, *The Topkapı Scroll – Geometry and Ornament in Islamic Architecture ...* , Santa Monica, Ca.: The Getty Center for the History of Art and the Humanities, 1995.

Rebstock, Ulrich, *Rechnen im islamischen Orient – Die literarischen Spuren der praktischen Rechenkunst*, Darmstadt: Wissenschaftliche Buchgesellschaft, 1992.

Rius, Mònica, *La Alquibla en al-Andalus y al-Magrib al-Aqṣà*, (*Anuari de Filologia* (Universitat de Barcelona) XXI (1998-99) B-3), Barcelona: Institut "Millás Vallicrosa" de Història de la Ciència Àrab, 2000.

Pavón Maldonado, Basilio, *El arte hispano-musulman en su decoración geometrica,* Madrid: Ministerio de Asunto Exteriores, Instituto Hispano-Árabe de Cultura, 1975.

Saliba, George, "Artisans and Mathematicians in Medieval Islam" [an essay review of Necipoglu, *Topkapı Scroll*], *Journal of the American Oriental Society* 119 (1999), pp. 637-645; and "Astronomy and Astrology in Medieval Arabic Thought", in Roshdi Rashed and Joël Biard, eds., *Les doctrines de la science de l'antiquité à l'âge classique*, Leuwen/Louvain: Peeters, 1999, pp. 131-164.

Samsó, Julio, *Las ciencias de los antiguos en al-Andalus*, Madrid: Mapfre, 1992; *Islamic Astronomy and Medieval Spain*, Aldershot: Variorum, 1994. See also King & Samsó.

Schmidl, Petra, *Volksastronomische Abhandlungen aus dem mittelalterlichen arabisch-islamischen Kulturraum*, doctoral dissertation in preparation (Institute for History of Science, Johann Wolfgang Goethe University, Frankfurt).

Sezgin, Fuat, *et al.*, eds., *Arabische Instrumente in orientalistischen Studien*, 6 vols., Frankfurt am Main: Institut für Geschichte der Arabisch-Islamischen Wissenschaften, 1991.

Varisco, Daniel M., "Islamic Folk Astronomy", in Selin, ed., *Astronomy across Cultures*, pp. 615-650; *Medieval Folk Astronomy and Agriculture in Arabia and the Yemen*, (Variorum Collected Studies Series: CS585), Aldershot & Brookfield, Vt.: Ashgate – Variorum, 1997; *Medieval Agriculture and Islamic Science – The Almanac of a Yemeni Sultan*, Seattle, Wa.: University of Washington Press, 1993.

Vernet Festschrift: *From Baghdad to Barcelona. Studies in the Islamic Exact Sciences in Honour of Prof. Juan Vernet*, Josep Casulleras and Julio Samsó, eds., (*Anuari de Filologia* (Universitat de Barcelona) XX (1996) B-2), 2 vols., Barcelona: Instituto "Millás Vallicrosa" de Historia de la Ciencia Árabe, 1996.

Medieval Christendom

Atkinson, T. D., "Points of the Compass", in James Hastings, ed., *Encyclopaedia of Religion and Ethics*, 13 vols., Edinburgh: T. & T. Clark, 1908-1926, X, pp. 73-88.

Beaujouan, Guy, "Les soi-disant chiffres grecs ou chaldéens (XIIe-XVIe siècles)", *Revue d'histoire des sciences* 3 (1950), pp. 170-174, repr. in *idem*, *Par raison de nombres – L'art du calcul et les savoirs scientifiques médiévaux*, Aldershot (U.K.): Variorum, 1990, XII.

Benoit, J. L., "Un cadran solaire monastique du XII[e#] siècle", *Le Cosmos*, vol. 59, no. 1238 (1908), pp. 435-437.

Bergmann, Werner, *Innovationen im Quadrivium des 10. und 11. Jahrhunderts: Studien zur Einführung von Astrolab und Abakus im lateinischen Mittelalter*, Stuttgart: Franz Steiner, 1985.

Biarne, J., "Le temps du moine d'après les premières règles monastiques d'Occident (IVe-VIe siècles)", in *Le temps chrétien de la fin de l'Antiquité au Moyen-Age, IIIe-XIIIe siècles*, (Colloques Internationaux du Centre National de la Recherche Scientifique, no. 604), Paris, 1984, pp. 99-128. [Not consulted.]

Bilfinger, Gustav, *Die mittelalterlichen Horen und die modernen Stunden – Ein Beitrag zur Kulturgeschichte*, Stuttgart: W. Kohlhammer, 1892.

Bischoff, Bernhard, "Die sogenannten "griechischen" und "chaldäischen" Zahlzeichen des abendländischen Mittelalters", published in 1944, repr. in *idem*, *Mittelalterliche Studien – Ausgewählte Aufsätze zur Schriftkunde und Literaturgeschichte*, 3 vols., Stuttgart: Anton Hiersemann, 1966, I, pp. 67-73.

Borst, Arno, *Astrolab und Klosterreform an der Jahrtausendwende*, Heidelberg: Carl Winter, 1989 [problematic].

Brusa, Giuseppe, "Le navicelle orarie di Venezia", *Annali dell'Istituto e Museo di Storia della Scienza di Firenze* 5 (1980), pp. 51-59.

Cátedra, Pedro M., and Julio Samsó, *Tratado de Astrología atribuido a Enriqve de Villena*, Madrid, Río Tinto Minera, 1980.

Destombes, Marcel, "Un astrolabe carolingien et l'origine de nos chiffres arabes", *Archives Internationales d'Histoire des Sciences* 58-59 (1962), pp. 1-45, repr. in G. Schilder *et al.*, eds., *M. Destombes (1905-1983): Selected Contributions to the History of Cartography and Scientific Instruments*, Utrecht: Hes Publishers, and Paris: A. G. Nizet, 1987, pp. 153-200.

D'Haenens, Albert, "La clepsydre de Villiers (1267) – Comment on mesurait et vivait le temps dans une abbaye cistercienne au XIIIe siècle", in Harry Kühnel, ed., *Klösterliche Sachkultur des Spätmittelalters*, Vienna: Österreichische Akademie der Wissenschaften, 1980.

Dictionary of the Middle Ages, 13 vols., New York: Charles Scribner's Sons, 1982-89, especially articles "Arabic Numerals", "Astrolabe", "Astrology / Astronomy", "Book of Hours", "Calendars", "Computus", "Roman Numerals".

Eisner, S., ed., and *idem* and G. MacEoin, trans., *The Kalendarium of Nicholas of Lynn*, Athens, Ga.: The University of Georgia Press (under the auspices of The Chaucer Library), 1980.

García Avilás, Alejandro, *El tiempo y los astros – Arte, ciencia y religión en la Alta Edad Media*, Murcia: Universidad de Murcia, 2001.

Gockerell, Nina, "Telling Time without a Clock", in Klaus Maurice and Otto Mayr, eds., *The Clockwork Universe: German Clocks and Automata 1550-1650*, New York: Neale Watson, 1980, pp. 131-143.

Gunther, Robert T., *Early Science in Oxford*, vol. II: *Astronomy*, Oxford: Clarendon Press for the Oxford Historical Society, 1923, and vol. V: *Chaucer and Messahalla on the Astrolabe*, Oxford: University Press, 1929.

Hill, G. F., *The Development of Arabic Numerals in Europe Exhibited in Sixty-Four Tables*, Oxford: Clarendon Press, 1915.

King, David A., *The Ciphers of the Monks – A Forgotten Number Notation of the Middle Ages*, (*Boethius: Texte und Abhandlungen zur Geschichte de Mathematik und der Naturwissenschaften*, vol. 44), Stuttgart: Franz Steiner, 2001.

Kren, Claudia, "The Traveller's Dial in the Late Middle Ages: The Chilinder", *Technology and Culture* 18 (1977), pp. 419-435; *Medieval Science and Technology – A Selected, Annotated Bibliography*, New York and London: Garland Publishing, Inc., 1985.

Lacroix, Paul, *Science and Literature in the Middle Ages and Renaissance*, first published in 1878, reprinted New York, N.Y.: Frederick Ungar, 1964.

Lexikon des Mittelalters, 9 vols., Munich & Zurich: Artemis, 1980-99, especially articles "Astrologie", "Astronomie", "Instrumente", "Komputistik", "Mathematik".

Mills, Allan, "The Canterbury Pendant: A Saxon Seasonal-Hour Altitude Dial", *Bulletin of the British Sundial Society* 95.2 (June 1995), pp. 39-44.

Murray, Alexander, *Reason and Society in the Middle Ages*, (first published 1978), Oxford: Clarendon Press, 1991 edn.

McCluskey, Stephen C., "Gregory of Tours, Monastic Timekeeping, and Early Christian Attitudes to Astronomy", *Isis* 81 (1990), pp. 9-22; *Astronomies and Cultures in Early Medieval Europe*, Cambridge, *etc.*: Cambridge University Press, 1998.

Neugebauer, Otto, "Astronomical and Calendrical Data in the Très Riches Heures", (originally published 1974), repr. in *idem*, *Astronomy and History* (cited below), pp. 507-520.

John D. North, *Richard of Wallingford – An Edition of his Writings, with Introductions, English Translation and Commentary*, 3 vols., Oxford: Clarendon Press, 1976; *Chaucer's Universe*, Oxford: Oxford University Press, 1988; *Stars, Minds and Fate – Essays in Ancient and Medieval Cosmology*, London and Ronceverte, W.V., 1989.

Pedersen, Olaf, "Astronomy", in David C. Lindberg, ed., *Science in the Middle Ages*, Chicago, Ill.: University of Chicago Press, 1978, pp. 303-337.

Price, Derek J. de Solla (with R. M. Wilson), *The Equatorie of the Planetis edited from Peterhouse MS. 75.I*, Cambridge: Cambridge University Press, 1955.

Poulle, Emmanuel, *Les instruments astronomiques du Moyen Age*, Paris: Alain Brieux, 1983 (first appeared in *Le Ruban Rouge* no. 32 (March 1967), pp. 18-29; *Un constructeur d'instruments astronomiques au 15e siècle – Jean Fusoris*, Paris: Honoré Champion, 1963.

Samsó: see Cátedra & Samsó.

Schaldach, Karlheinz, and Herbert Rau, "Vertikalsonnenuhren des 6. bis 14. Jahrhunderts", in *Frankurt IGN Festband*, pp. 273-290.

Schechner, "Sundials": Sara Schechner, "The Material Culture of Astronomy in Daily Life: Sundials, Science and Social Change", *JHA* 32 (2001), pp. 189-222.

Sesiano, Jacques, "Un système artificiel de numération au moyen âge", in Menso Folkerts and Uta Lindgren, eds., *Mathemata – Festschrift für Helmuth Gericke* (Reihe "Boethius", Bd. 12), Wiesbaden: Franz Steiner, 1985, pp. 165-196.

Stautz, Burkhard, "Ein Astrolab aus dem Jahr 1420", *NTM – Internationale Zeitschrift für Geschichte und Ethik der Naturwissenschaften, Technik und Medizin* (Basle: Birkhäuser), N.S., 5:3 (1997), pp. 142-159.

Stevens, Wesley M., Guy Beaujouan and Anthony J. Turner, eds., *The Oldest Latin Astrolabe*, (*Physis* 32:2-3), Florence: Leo S. Olschki, 1995.

White, Jr., Lynn, *Medieval Religion and Technology – Collected Essays*, Berkeley, Ca., *etc.*: University of California Press, 1978.

Zenner, Armin, "Der Mond – Volvellen und Aequatorien", *Jahresschrift der Deutschen Gesellschaft für Chronometrie* 40 (2001), pp. 161-178; "Die Sonnenuhr des Gerbert von Aurilac", *ibid.* 38 (1999), pp. 150-158.

Zinner, Ernst, "Das mittelalterliche Lehrgerät für Sternkunde zu Regensburg und seine Beziehung zu Wilhelm von Hirsau", *Zeitschrift für Instrumentenkunde* 43 (1923), pp. 278-282; *Die ältesten Räderuhren und modernen Sonnenuhren*, (28. Bericht der Naturforschenden Gesellschaft, Bamberg), Bamberg, 1939; *Deutsche und niederländische astronomische Instrumente des 11.-18. Jahrhunderts*, Munich: C. H. Beck, 1956, 2nd edn., 1967, repr. 1972; *Alte Sonnenuhren an europäischen Gebäuden*, Wiesbaden: Franz Steiner, 1964.

Bridging the gaps

Bott, Gerhard, ed., *Focus Behaim Globus*, 2 vols., Nuremberg: Germanisches Nationalmuseum, 1992.

Crombie, Alistair C., "Griechisch-arabische Naturwissenschaften und Abendländisches Denken", in Gereon Sievernich and Hendrik Budde, eds., *Europa und der Orient 800-1900*, Gütersloh and Munich: Bertelsmann Lexikon Verlag GmbH, 1989, pp. 102-131.

Ginzel, F. K., *Handbuch der mathematischen und technischen Chronologie*, 3 vols., Leipzig: Hinrichs, 1906-14.

Goldstein, Bernard R., "Descriptions of Astronomical Instruments in Hebrew", in *Kennedy Festschrift*, pp. 105-141; *Theory and Observation in Ancient and Medieval Astronomy*, London: Variorum, 1985.

Gunther, Robert T., *The Astrolabes of the World*, Oxford: Oxford University Press, 1932, repr. in 1 vol., London: Holland, 1976.

Kennedy Festschrift: David A. King and George Saliba, eds., *From Deferent to Equant: A Volume of Studies in the History of Science in the Ancient and Medieval Near East in Honor of E. S. Kennedy*, *Annals of the New York Academy of Sciences* 500 (1987).

King, David A., "Medieval Astronomical Instruments: A Catalogue in Preparation", *Bulletin of the Scientific Instrument Society* no. 31 (Dec., 1991), pp. 3-7, and "1992 – A Good Year for Medieval Astronomical Instruments", *ibid.* no. 36 (March, 1993), pp. 17-18; "Astronomical Instruments between East and West", in Harry Kühnel, ed., *Kommunikation zwischen Orient und Okzident*, Vienna: Österreichische Akademie der Wissenschaften, 1993, pp. 143-198; **XIIa-b** on the universal horary quadrant and universal horary dials; also "An Astrolabe from 14th-Century Christian Spain with Inscriptions in Latin, Hebrew and Arabic – A Unique Testimonial to an Intercultural Encounter", *Suhayl – Journal for the History of the Exact and Natural Sciences in Islamic Civilisation* (Barcelona) 3 (2002/03), pp. 9-156; "The Astrolabe Depicted in the Intarsia of the *Studiolo* of Archduke Federico in Urbino", in *La scienza del Ducato di Urbino – The Science of the Dukedom of Urbino*, Flavio Vetrano, ed., Urbino: Accademia Raffaello, 2001, pp. 101-139; and "A Remarkable Italian Astrolabe from *ca.* 1300 – Witness to an Ingenious Tradition on Non-Standard Astrolabes", in *MUSA MUSAEI: Studies on Scientific Instruments and Collections in Honour of Mara Miniati*, Marco Beretta, Paolo Galluzzi and Carlo Triarico, eds., Florence: Leo S. Olschki, 2003, pp. 29-52.

Paul Kunitzsch, *Arabische Sternnamen in Europa*, Wiesbaden: Otto Harrassowitz, 1959; *Typen von Sternverzeichnissen in astronomischen Handschriften des zehnten bis vierzehnten Jahrhunderts*, Wiesbaden: Otto Harrassowitz, 1966; *Claudius Ptolemäus – Der Sternkatalog des Almagest – Die arabisch-mittelalterliche Tradition*, 3 vols., Wiesbaden: Otto Harrassowitz, 1986-1991.

Landes, David S., *Revolution in Time – Clocks and the Making of the Modern World*, Cambridge, Mass, and London: The Belknap Press of Harvard University Press, 1983

Levin, Daniel Z., "Which Way is Jerusalem? Which Way is Mecca? The Direction-Facing Problem in Religion and Geography", *Journal of Geography* (National Council for Geographic Education, Washington, D.C.) 101 (2002), pp. 27-37.

Lindberg, David C., *The Beginnings of Western Science*, Chicago: University of Chicago Press, 1992.

Michel, Henri, *Images des sciences – Les anciens instruments scientifiques vus par les artistes de leur temps*, Rhode-St-Genèse (B): Albert De Visscher,.1977.

Murdoch, John E., *Album of Science – Antiquity and the Middle Ages*, New York: Charles Scribner's Sons, 1984.

Neugebauer, Otto, *A History of Ancient Mathematical Astronomy*, 3 pts., New York, *etc.*, Springer, 1975; *Ethiopic Astronomy and Computus*, Vienna: Österreichische Akademie der Wissenschaften, 1979; *Astronomy and History – Selected Essays*, New York, *etc.*: Springer, 1983; *Abu Shaker's "Chronography" – A Treatise of the 13th Century on Chronological, Calendrical, and Astronomical Matters, written by a Christian Arab, preserved in Ethiopic*, Vienna: Österreichische Akademie der Wissenschaften, 1988.

Ruggles, Clive L. N., and Nicholas J. Saunders, eds., *Astronomies and Cultures*, Niwot, Colorado: University Press of Colorado, 1993.

Sabra, Abdelhamid I., "The Appropriation and Subsequent Naturalization of Greek Science in Medieval Islam: A Preliminary Statement", *History of Science* 25 (1987), pp. 223-243.

Saliba, George, "Easter Computation in Medieval [Islamic] Astronomical Handbooks", published 1970, repr. in Kennedy *et al.*, *Studies* (cited above), pp. 677-709.

Sarton, George, *Introduction to the History of Science*, 5 vols. in 3, Baltimore: Williams and Wilkins & Co. for the Carnegie Institution, Washington, D.C., 1927-1948, repr. 1962.

Schechner, Sara, "The Material Culture of Astronomy in Daily Life: Sundials, Science, and Social Change", *Journal for the History of Astronomy* 32 (2001), pp. 189-222.

Selin, Helaine, ed., *Astrononomy across Cultures – The* [!] *History of Non-Western Astronomy*, Dordrecht, *etc.*: Kluwer, 2000.

Stautz, Burkhard, "Die früheste bekannte Formgebung der Astrolabien", in *Frankfurt IGN Festband*, pp. 315-328; *Untersuchungen von mathematisch-astronomischen Darstellungen auf mittelalterlichen Instrumenten islamischer und europäischer Herkunft*, Bassum (D): Verlag für Geschichte der Naturwissenschaften und der Technik, 1997; *Die Astrolabiensammlungen des Deutschen Museums und des Bayerischen Nationalmuseums*, Munich: Oldenbourg, 1999.

Swerdlow, Noel, "The Recovery of the Exact Sciences of Antiquity: Mathematics, Astronomy, Geography", in Anthony Grafton, ed., *Rome Reborn – The Vatican Library and Renaissance Culture*, Washington, D.C.: Library of Congress, 1993, pp. 125-168.

Turner, Anthony J., *Astrolabes – Astrolabe Related Instruments*, and *Water-Clocks, Sand-Glasses, Fire-Clocks*, parts 1 and 3 of Bruce Chandler, ed., *The Time Museum – Catalogue of the Collection*: vol. 1, *Time Measuring Instruments*, Rockford, Ill., The Time Museum, 1985; *Early Scientific Instruments – Europe 1400-1800*, London: Sotheby's Publications, 1987; *time*, Amsterdam: Tijd voor Tijd Foundation, 1990 [Catalogue of an exhibition at the Nieuwe Kirk, Amsterdam, in 1990].

Van De Vyver, A., "Les premiers traductions latines (X^e^-XIe siècles) de traités arabes sur l'astrolabe", *Actes du 1er Congrès International de Géographie Historique*, 2 vols., Brussels, 1931, II, pp. 266-290, repr. in *AIOS*, VI, pp. 377-405.

Walker, Chistopher B. F., ed., *Astronomy before the Telescope*, London: British Museum Press, 1996, repr. 1999.

Part IX

When the night sky over Qandahar was lit only by stars ... ...

To Sonja Brentjes,
François Charette and Benno van Dalen,
in gratitude

DEDICATION AND ACKNOWLEDGEMENTS

This study is dedicated to my young colleagues Sonja Brentjes, Benno van Dalen and François Charette, each of whom is doing exciting and ground-breaking research in the history of Islamic science. They are prominent members of the "next generation" of scholars in our field. Sonja is currently working on the two-way transmission of scientific knowledge between turkey and Iran on the one hand and Europe on the other.* François, having recently published a monumental contribution to Islamic astronomical instrumentation, is currently working on the significance of illustrations and tables in Islamic scientific manuscripts. Benno is well advanced on a new survey of the Islamic astronomical handbooks known as *zīj*es, taking the work of Ted Kennedy a monumental step further. In addition, Francois' contribution to the rest of this volume has been substantial, and both he and Benno stepped in at the last minute to provide the final recomputations and derivation of parameters for the remaining problem tables in **I** and **II**.

It was these three colleagues who organized the conference "Certainty, Doubt, Error: The production of knowledge and its impediments in the practice of pre- and early modern science" held in Frankfurt on 17th-18th November, 2001. The proceedings are published as *Certainty, Doubt, Error: Aspects of the Practice of Pre- and Early Modern Science*, in *Early Science and Medicine* (E. J. Brill, Leiden) 7:3 (2002). The conference was remarkable and enjoyable for many reasons, not the least of which was the presence of colleagues and friends from Algeria, Belgium, Canada (B.C. and P.Q.), Czech Republic, Denmark, Egypt, France, Germany, Greece, Iran, Ireland, Israel, Japan, Morocco, the Netherlands, Palestine, Poland, Russia, Spain, Switzerland, the United Kingdom, and the United States. For me the whole business was a very moving experience.

My three colleagues had invited me to say a few words on "science in the service of Islam" but, at the last minute, I changed my mind and talked about some very nice tables instead. I tried to make my contribution have some relevance to the general theme of the conference but my mind was preoccupied with recent events in the U.S. and in Afghanistan, hence the choice of subject. This study is a written version of a short presentation I made at the conference. It could not be included in the proceedings for lack of space so I present it here. I have not adapted the bibliographical citations to the rest of this book.

* See her overview "Between Doubts and Certainties: On the Place of History of Science in Islamic Societies within the Field of the History of Science", *NTM – International Journal of History and Ethics of Natural Sciences, Technology and Medicine*, N.S. 11:2 (2003), pp. 65-79.

WHEN THE NIGHT SKY OVER QANDAHAR
WAS LIT ONLY BY STARS ... ...

The subject of my contribution to the 2001 Frankfurt Conference was supposed to be astronomy in the service of religion in medieval Egypt, Syria and the Yemen, that is, the world of the Mamluks and Rasulids. When I protested that I had already written and said enough on this subject, the organizers, my dear colleagues, kindly told me I might actually talk about anything I liked. So I did indeed end up talking about something a little bit different. And if I allowed myself some license, it was, after all, because I was supposed to be having a good time, and this in spite of what was happening elsewhere in the world in November, 2001.

What is implied by the phrase "astronomy in the service of Islam" that I coined some years ago[1] is the regulation of the lunar calendar, the organization of the times of prayer, and the determination of the direction of Mecca, the three aspects, indeed the only three aspects of Islamic astronomy which are still of concern to Muslims to this day. In fact, it was not astronomy in the service of Islam, but rather Islam in the service of the astronomers.[2] It brought them work and occasional recognition, for which any medieval scientist, like any modern historian of science, would be grateful. Rather, I would like to draw attention to some sources that I have come across which may be of interest to historians of astronomy, historians of mathematics, and, perhaps, particularly at the present time, even to a more general audience. They demonstrate that whatever may be our individual predilections in contributing to the history of Islamic astronomy and mathematics – be it analysing astronomical tables or observation records; establishing the social context of scientific activity; differentiating between the mathematical procedures of the astronomers and the folk scientific procedures of the scholars of the sacred law; investigating the "visual language" of science through illustrations and diagrams in scientific manuscripts; or whatever —, there are still untapped sources out there in the great manuscript collections of the world containing materials which can, for one reason or another, still surprise even seasoned scholars of medieval astronomy. I would maintain that these sources still merit the now old-fashioned "Quellen und Studien" approach to historical scientific materials,[3] and that it will be a sad day for our discipline when such investigations cease.

We turn to a manuscript, now preserved in the Deutsche Staatsbibliothek in Berlin, of the astronomical handbook (*zīj*) of the celebrated astronomer Kūshyār ibn Labbān, who worked in Iran *ca.* 1000. This is an important work that was influential in the Islamic East, but most

---

[1] See my "Science in the Service of Religion: The Case of Islam", *impact of science on society* (U.N.E.S.C.O.), no. 159 (1991), pp. 245-262, repr. in *Astronomy in the Service of Islam*, Aldershot: Variorum, 1993, no. I.

[2] Jamil Ragep, "Freeing Astronomy from Philosophy: An Aspect of Islamic Influence on Science", *Osiris* 16 (2001), pp. 49-71, esp. p. 51.

[3] See Noel Swerdlow, "Otto E. Neugebauer", *Proceedings of the American Philosophical Society* 137:1 (March 1993), pp. 137-165, on the scholar who contributed most to the history of ancient astronomy in the 20th century.

of its contents has not been studied since the 1950s, when Ted Kennedy made a brief survey of the contents of another copy now preserved in Leiden.[4] There is, in fact, not even a detailed overview of the history of astronomy in the Islamic East, and there are already too many sources available for the very few workers in the field.[5]

Already some 50 years ago Ted Kennedy pointed to the existence of the various tables taken from other early Islamic sources that are found at the end of the Berlin copy of Kūshyār's work, which was penned somewhere in Greater Iran *ca.* 1300.[6] All of these tables merit detailed study, and I shall here restrict my attention to three sets for timekeeping by the fixed stars, compiled by three different individuals.[7]

Before looking at these tables we introduce some relevant concepts and notation.[8] The reader familiar with **I-1.1** can skip this paragraph. **Fig. 1** shows the celestial coordinates of a star: the coordinates relative to the ecliptic are the longitude λ and latitude β, and those relative to the celestial equator are the right ascension α and declination Δ. Both α and Δ are measured from the vernal equinox V. The stars move slowly parallel to the ecliptic (at the rate of 1° in about 70 years) so that λ is a function of time (and β is constant); both α and Δ change with time. To convert from (λ,β) to (α,Δ), it is useful to know the "second declination", denoted by $\delta_2(\lambda)$. Tables of (λ,β) and/or (α,Δ) for numerous stars belonged to the standard equipment of medieval astronomers. **Fig. 2** shows a star at A rising over the horizon of a locality with latitude ϕ. The north and south points on the horizon are N and S, P is the celestial pole (at altitude ϕ above N), Z is the zenith, E is the east point, and EQ is the celestial equator. The instantaneous position of the ecliptic is shown: the point R with longitude ρ rises with the star. The oblique ascensions, which are a measure of the time taken by an arc of the ecliptic to rise over the local horizon, are a useful notion. The oblique ascensions of VR are VE, denoted by $\alpha_\phi(\rho)$. Similar diagrams could show the points of the ecliptic culminating over the local meridian or setting over the western horizon: we denote their longitudes by τ and σ. **Fig. 3**

---

[4] E. S. Kennedy, "A Survey of Islamic Astronomical Tables", *Transactions of the American Philosophical Society*, N. S., 46:2 (1956), pp. 123-177, repr. with separate pagination, n.d. [*ca.* 1990], pp. 125 (no. 9) and 156a of the original (hereafter "*Zīj* Survey"). On this genre of literature see further King and Julio Samsó, with a contribution by Bernard R. Goldstein, "Astronomical Handbooks and Tables from the Islamic World (750-1900): An Interim Report", *Suhayl – Journal for the History of the Exact and Natural Sciences in Islamic Civilisation* (Barcelona) 2 (2001), pp. 9-105, summarized in the article "Zīdj" in *The Encyclopaedia of Islam*, new edn., 11 vols. to date, Leiden: E. J. Brill, 1960 to present (hereafter *EI*$_2$).

[5] Brief overviews by Ted Kennedy appeared in the *Cambridge History of Iran* some 30 years ago. For overviews of the regional schools of Islamic astronomy see the references cited in King, *World-Maps for Finding the Direction and Distance to Mecca: Innovation and Tradition in Islamic Science*, Leiden: E. J. Brill, and London: Al-Furqan Islamic Heritage Foundation, 1999, pp. 8-10.

[6] Wilhelm Ahlwardt, *Die Handschriften-Verzeichnisse der Königlichen Bibliothek zu Berlin*, 5. Band: *Verzeichnis der arabischen Handschriften*, Berlin: A. Asher, 1893, pp. 203-206 (*ad* no. 5751), esp. p. 206a. Ahlwardt listed every single table in the Berlin astronomical manuscripts, and these tables are cited again in Kennedy, "*Zīj* Survey" (n. 4), p. 125b.

[7] This information is taken from **Part I**: A Survey of Tables for Timekeeping by the Sun and Stars, **Chapter 3**: Tables of the longitude of the ascendant as a function of solar and stellar altitude, **Section 3.2**: Tables of the longitude of the ascendant as a function of stellar altitude, for a specific latitude; and **Chapter 6**: Tables of auxiliary function for timekeeping, **Section 6.16**: Tables of auxiliary functions for timekeeping by the stars. In the present version I have simplified the text somewhat and have reduced the medieval trigonometric functions from base 60 to base 1.

[8] On stellar coordinates and the concepts of declination, ascensions and ascendant/horoscopus see the articles "Mayl", "Maṭāliʿ" and "Ṭāliʿ" in *EI*$_2$.

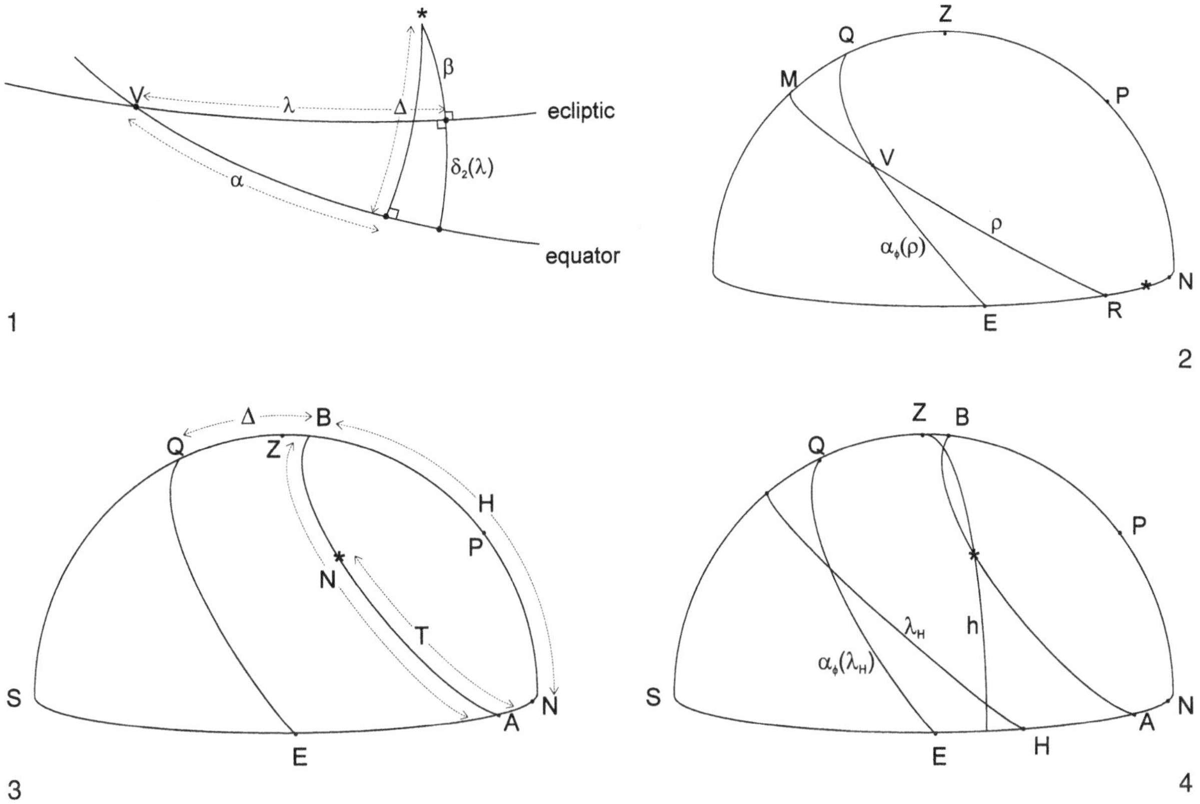

Fig. 1-4: These diagrams show the various concepts necessary to understand the tables. Fig. 1 shows the celestial coordinates of a star, measured with respect to the ecliptic and the celestial equator. In Fig. 2 the star rises over a local horizon and its co-ascendant on the ecliptic is shown. Fig. 3 shows the arc of visibility of the star relative to the celestial equator and the position of a star at a given time. Fig. 4 shows the star with altitude h in the eastern sky at time T; the horoscopus is H with longitude λ_H.

shows the half arc of visibility of the star AB parallel to the celestial equator EQ, with BQ measuring the declination Δ. The star culminates on the meridian NZS at B. Its culminating altitude is NB, denoted by H, and its zenith distance BZ is 90°-H. The half arc of visibility AB is denoted by D, and the time since rising until the star reaches the instantaneous position X, measured by ΛX, is labelled T. **Fig. 4** shows the star at X with observed altitude h, and the point H of the ecliptic rising instantaneously over the horizon, called the horoscopus, with

→

Figs. 5-7: Extracts from the three sets of tables under discussion here. Fig. 5 shows the star catalogue with additional information for timekeeping by the stars, and Fig. 6 shows the sub-tables displaying the instantaneous horoscopus for each degree of altitude of each of Sirius and Capella in the eastern and western sky. The second set was compiled using the first set. Fig. 7 shows another set of auxiliary functions for timekeeping by the stars, with instructions on the mode of using the functions to derive the time of night the instantaneous horoscopus. There are no associated tables for individual stars, and probably none were intended. The tables in Figs. 5-6 are independent of those in Fig. 7, and the two sets are based on different values of the terrestrial latitude. Both latitudes are attested in medieval sources for Qandahar. Illustrations of the full double pages selected here are in **Figs. I-3.2.1a-b** and **I-6.16.1**. [From MS Berlin Ahlwardt 5751, pp. 188-189 and 190-191, and 186-187, courtesy of the Deutsche Staatsbibliothek (Preußischer Kulturbesitz).]

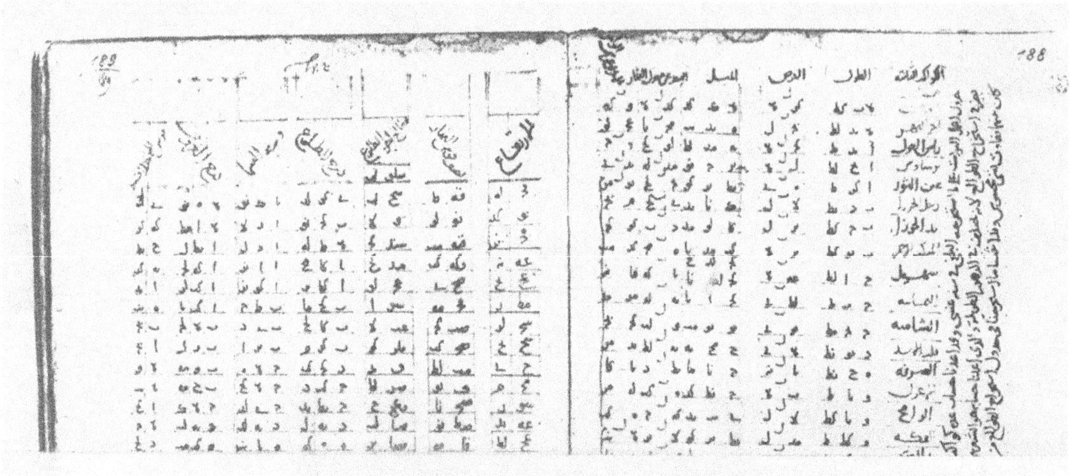

5

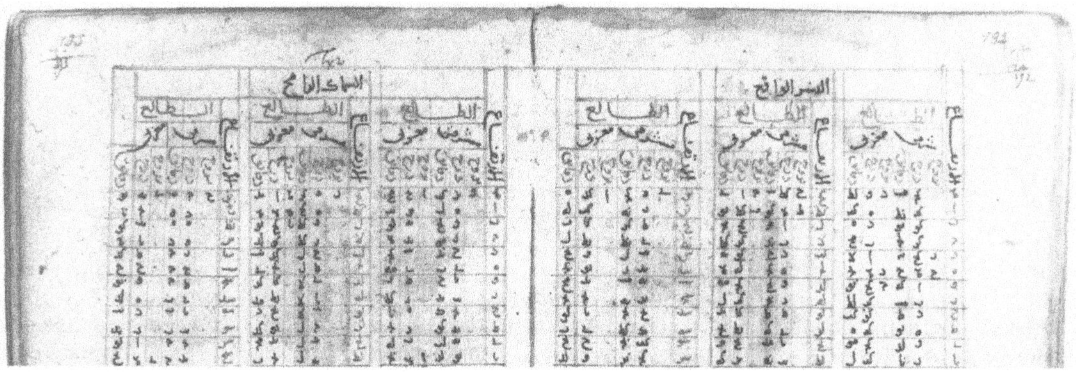

6

7

longitude λ_H and oblique ascensions $\alpha_\phi(\lambda_H)$. The problem that will concern us in the sequel is the determination of the time since rising, T, from the observed altitude, h, using trigonometric formulae, with the ultimate object of determining the longitude of the horoscopus, λ_H. As we shall see, our medieval astronomers were in full control of the formula for determining $T(h,\Delta,\phi)$. A simple consideration of the celestial sphere shows that:

$$\alpha_\phi(\lambda_H) = \alpha_\phi(\rho) + T(h) ,$$

which relationship they could then apply, using standard tables of the function $\alpha_\phi(\lambda)$, to derive λ_H. We return below to the question why one should want to know the horoscopus.

One of the tables that interest us is a star catalogue (pp. 188-189) displaying values of twelve auxiliary functions for 25 stars (see the extract in **Fig. 5**). The following list identifies the scientific Arabic terminology used in the manuscript, the modern equivalents, and the notation introduced above. The purpose of the auxiliary functions is, as we shall see, to determine for a specific latitude ϕ and observed altitude h the point of the ecliptic which is instantaneously rising over the eastern horizon, called the horoscopus (also known as the ascendant) and here labelled H with ecliptic longitude λ_H. The functions tabulated are:

al-ṭūl	ecliptic longitude	λ
al-ʾarḍ	latitude	β
al-mayl	second declination	$\delta_2(\lambda)$
al-buʿd ʿan muʿaddil al-nahār	declination	Δ
al-inḥirāf ʿan samt al-raʾs	zenith distance at culmination	$90°$-H
al-irtifāʿ	altitude when culminating	H
niṣf qaws al-nahār	half arc of visibility	D
maṭāliʿ ajzāʾ al-ṭulūʿ	ascensions of the co-ascendant	$\alpha_\phi(\rho)$
daraj al-ṭulūʿ	ecliptic co-ascendant	ρ
daraj al-ghurūb	ecliptic co-descendant	σ
wasaṭ al-samāʾ	ecliptic co-culminant	τ

Only two other tables of this kind are known, both for Baghdad, one from the early 10th century and the other from the early 11th.[9] The reader should be aware that here the last 10 of these quantities have been computed from the first two, and the first is a function of time. A note by the side of this table states:

> "(This is) a table for operations (of timekeeping) with the fixed stars according to the results of al-Balkhī in the year 1310. We recomputed (the entries for) several stars for the year 1424 and did not find any difference which would affect the determination of the (co)ascendants, because they (?) do not vary over long periods. We recomputed values for Sirius and Procyon, Betelgeuze, Spica and Arcturus, and Aldebaran, and for these there was no perceptible difference. We used what we derived in the table for finding the ascendants accurately from the fixed stars. God grants success."

The dates 1310 and 1424, which are written in the standard Arabic alphanumerical (*abjad*) notation in the text, must relate to the calendar of Alexander,[10] and thus correspond to 999

---

[9] **I-6.16.2-3**.

[10] On the calendars used by medieval Muslim astronomers see the article "Taʾrīkh, 2. Era chronology in astronomical handbooks" by Benno van Dalen in EI_2.

and 1113 in the Julian calendar. Also from the table itself (the longitude of Regulus is 136;39°) a date of *ca.* 1050 is suggested. We are clearly dealing essentially with al-Balkhī's original table. His successor asserts that he made some calculations to investigate the effect of precession on the values (particularly the co-ascendant is of interest) for some of the stars, but found that this could be neglected. However, the longitudes of the stars would have increased by almost 2° in the intervening period, and the co-ascendants would certainly have displayed a perceptible difference. In any case, the later astronomer seems to be saying that he used al-Balkhī's table to compile another set of tables of a different kind (see below).

Now who was the man whose family or whose good self hailed from Balkh in Khurasan, or what is now Northern Afghanistan? The only astronomer named al-Balkhī who was active *ca.* 1000 was a certain Nūr al-Dīn Abu 'l-Qāsim ʿAlī ibn Aḥmad al-Balkhī, known to us as the author of at least two treatises on astrology both extant in a few copies and both unstudied.[11] It seems that this is our al-Balkhī.

The latitude underlying this first set of tables is found by inspection to be 30;25°. This value is attested in the medieval sources only as the latitude of Qandahar. It was used for Qandahar in a recension of the geographical tables of the 9th-century Baghdad scholar al-Khwārizmī,[12] and it is certainly possible, in the light of present knowledge, that the table was intended for use there. Now if the stellar tables are for Qandahar, it is rather surprising that al-Balkhī used such a poor value, albeit on good authority, for the accurate value is 31°38′. We should bear in mind that al-Balkhī's contemporary, al-Bīrūnī,[13] was able to derive the latitude of nearby Ghazna correct to within 2 minutes.[14] But there was a difference of attitude between such scholars as al-Balkhī, interested only in astrology, and al-Bīrūnī, interested in all matters

---

[11] See Max Krause, "Stambuler Handschriften islamischer Mathematiker", *Quellen und Studien zur Geschichte der Mathematik, Astronomie und Physik*, B 3:4 (1936), pp. 437-532, esp. p. 514; Fuat Sezgin, *Geschichte des arabischen Schrifttums*, 12 vols. to date, VII: *Astrologie, Meteorologie und Verwandtes*, Leiden: E. J. Brill, 1979, pp. 176-177; and David King, *A Survey of the Scientific Manuscripts in the Egyptian National Library*, (American Research Center in Egypt, Catalogs, vol. 5), Winona Lake, In.: Eisenbrauns, 1987, no. B55. His *Sirr al-asrār fī ḥaqīqat al-tasyīr wa-'l-istimrār* deals with mathematical astrology and enables us to date the author to the 2nd half of the 10th century. His *Mudkhal fī ʿilm al-nujūm* in 60 chapters is dedicated to "*al-wazīr Abu 'l-Ḥasan ibn mawlānā al-kāmil dhi 'l-jalālatayn wazīr al-wuzarā*'", whom I have not identified, but he sounds more like a Būyid or Samānid minister rather than a Ghaznavid. The members of the Middle Eastern History discussion group on the Internet were unable to help me, but Prof. C. Edmund Bosworth kindly drew my attention to (1) the *wazīr* Abu 'l-Ḥusayn ʿAbdallāh ibn Aḥmad al-ʿUtbī, appointed in 977 by the son Nūḥ of the deceased Samanid ruler of Transoxania and Khurasan, al-Manṣūr, and (2) Abu 'l-Ḥasan al-Hāmulī, who became a *wazīr* of the Sāmanids in Bukhara *ca.* 1000. These are mentioned in Wilhelm Barthold, *Turkestan down to the Mongol Invasion*, 2nd edn., transl. from the original Russian, Oxford: University Press & London: Luzac and Co., 1928, pp. 252 and 265, but I am not equipped to pursue them further.

Inevitably this Nūr al-Dīn al-Balkhī in later times became confused with the better-known astrologer Abū Maʿshar al-Balkhī, who had lived and worked in Baghdad a century earlier: on him see the article by David Pingree in *Dictionary of Scientific Biography*, 14 vols. and 2 supp. vols., New York: Charles Scribner's Sons, 1970-80 (hereafter *DSB*). A third astrological treatise on conjunctions (*Kitāb al-Qirānāt*) attributed in several manuscripts and also by Fuat Sezgin to al-Balkhī is actually by Abū Maʿshar.

[12] Edward Stewart Kennedy and Mary Helen Kennedy, *Geographical Coordinates of Localities from Islamic Sources*, Frankfurt am Main: Institut für Geschichte der Arabisch-Islamischen Wissenschaften, 1987, p. 675, and also p. 261, entry for KHZ. No other Islamic tables for latitude 30;25° are known.

[13] On al-Bīrūnī see the article by E. S. Kennedy in *DSB* (n. 11).

[14] See Kennedy & Kennedy, *Geographical Coordinates* (n. 12), p. 125, and E. S. Kennedy, *A Commentary upon Bīrūnī's Kitāb Taḥdīd [nihāyāt] al-amākin, An 11th Century Treatise on Mathematical Geography*, Beirut: American University of Beirut, 1973, p. 50.

scientific, including accuracy for its own sake. Qandahar did flourish in the Ghaznavid and Ghurid periods (that is, from the end of the 10th century), but we have no other records of any scientific activity there.[15] Nevertheless, Qandahar is so far from both Iran and from Khurasan, known centres of astronomical activity, and al-Bīrūnī came to nearby Kabul then Ghazna about 20 years after al-Balkhī compiled his auxiliary tables for latitude 30;25°. What disturbs me more is the distance of Qandahar from the scene of Kūshyār's *zīj* and the additional tables in the Berlin manuscript.

Let us suppose that the parameter 30;25° was not intended for Qandahar but for somewhere else. Then we have a problem, because the parameter is not attested elsewhere and there is only locality of consequence in Greater Iran (assuming reasonably accurate measurements), namely, Kirman, with latitude 30°18′. But for Kirman all Eastern Islamic sources have precisely 30;0°, although one recension of Kūshyār's handbook uses 30;5° for Bardshir (Kirman).[16] In some tables for timekeeping in the *Ashrafī Zīj* of Sanjar al-Kamālī, compiled in Shiraz in 1302/03, the latitude 30;22° is used for an unspecified locality in Iran, possibly Kirman because 29;30° is used alongside for nearby Shiraz (accurately 29°38′).[17] I know of no other evidence of any astronomical activity in Kirman.

The next four pages in the Berlin manuscript (pp. 190-194) contain the only remaining part of these associated ascendant tables (see the extract in **Fig. 6**). The function tabulated is the longitude of the ascendant, $\lambda_H(h)$, for each degree of eastern and western altitudes for Sirius, Capella, Vega and Arcturus. Entries are given in zodiacal signs, degrees and minutes, and the underlying latitude appears to be again 30;25°. These tables were clearly compiled by al-Balkhī's successor using al-Balkhī's stellar data.

Just preceding these tables (on pp. 186-187 of the Berlin manuscript) is another anonymous set of auxiliary tables for finding the ascendant at night (*al-ṭāliʿ bi-'l-layl*) from stellar altitude (see the extract in **Fig. 7**). These tables appear at first sight not to be related to the tables of the ascendant at night that we have just discussed. For each of 34 stars, five quantities are tabulated to two sexagesimal digits, and at the head of the columns there are instructions on how to use each one:

1 *taḍrub jayb irtifāʿ al-waqt fīhi abad^{an}*
multiply the sine of the instantaneous altitude by this;
2 *taqsim mā yakhruj mina 'l-ḍarb ʿalayhi abad^{an}*
divide the product of the multiplication by this;
3 *tanquṣ al-maqsūm minhu abad^{an} wa-tajʿal al-bāqī qaws^{an} maʿkūsa^{tan}*
subtract the quotient from this and take the inverse versed sine;
4 *tanquṣ al-qaws al-maʿkūsa minhu aw tuzād ʿalayhi*
subtract or add the arc from or to this; and
5 *yuzād mā baqiya aw balugha mina 'l-qaws ʿalayhi abad^{an}*,
add the result to this.

[15] On the history of Qandahar, which is singularly obscure in the early Islamic period, see the article "Ḳandahār" by C. Edmund Bosworth in EI_2, esp. p. 536.
[16] Kennedy, "*Zīj* Survey", p. 156a, using the Leiden manuscript.
[17] **I-8.1.2**.

The quantities mentioned relate to the following functions:

$$1: \text{vers } D,\ \ 2: \sin H,\ \ 3: \text{vers } D,\ \ 4: D,\ \ \text{and}\ \ 5: \alpha_\phi(\rho)\ ,$$

and the instructions conform to the two formulae:

$$T(h) = D - \text{arc vers } \{ \text{ vers } D - \sin h \text{ vers } D / \sin H \} \text{ and}$$

$$\alpha_\phi(\lambda_H) = \alpha_\phi(\rho) + T(h)\ .$$

The first formula was known to Muslim astronomers from Indian sources, whereas the second uses ascensions in the Greek tradition: in other words, the combination came from Baghdad rather than, say, Sind.

The underlying latitude in these tables is about 30° (this needs to be investigated further) and they were probably intended for use in some city in the south of Greater Iran, possibly Qandahar, for which the main version of al-Khwārizmī's geographical tables gives a latitude of 30;0° (rather than 30;25°),[18] but also possibly Kirman.[19] I find it hard to accept that we should have two completely different sets of tables of the same kind of auxiliary functions from one and the same place, Qandahar or Kirman. I have not attempted the delicate task of trying to date the tables; the lack of basic stellar data would make this particularly difficult. There is no mention of an additional set of tables of $\lambda_H(h)$ for any of the individual stars. The only such table known to us from this region other than the one for latitude 30;25° described above is for the latitude of Baghdad and survives in a unique copy dated 1215.[20]

Now we have here two sets of tables that are virtually without precedents or successors in the history of astronomy. They are clearly useful from a theoretical point of view. They are also carefully computed. The use of "algebraic"-type notation for operations of mathematical astronomy in tabular form is unique in the known medieval literature. However, whichever, if any, of the tables are for Qandahar, they are flawed from a practical point of view by the fact that the underlying latitude is in error. Added to this is the fact that the basic data for the stars was not corrected for the precessional motion during some 125 years.

The purpose of these tables for astronomical timekeeping has nothing whatsoever to do with "astronomy in the service of Islam". For their purpose is purely astrological. One does not need to know the ascendant at night except to begin to determine horoscopes.[21] Now it is only by studying the original sources that one can hope to begin to establish what certain people did in history. We historians of science have the advantage over historians of other aspects of human activity that we have certain reliable tools for interpreting scientific texts. We can actually investigate the numbers in such texts and subject them to various tests. For example, from two numbers "30;25°" and "30;0°" we can, within the medieval context, derive "Qandahar", rightly or wrongly, as the case may be. From stated star positions we can check dates.

---

[18] Kennedy & Kennedy, *Geographical Coordinates*, entry under Qandahar for KHU. No other Iranian tables specifically for latitude 30° are known (most medieval tables for Cairo are computed for 30;0°). Basra is not a possibility, for its latitude was taken as 31;0° or 30;0° in virtually all medieval Islamic geographical lists (*ibid.*, pp. 71-72), although the correct value is 30°30′.

[19] See *ibid.*, pp. 187-188.

[20] See **I-3.2.2**.

[21] To compute a horoscope for a given time one needs to determine the ascendant, which defines the instantaneous configuration of the ecliptic relative to the local horizon and meridian, then determine the 12 astrological houses on the ecliptic as defined by that configuration, and then determine in which house the sun, moon and five planets are situated. One can then assess their influence according to the prevailing astrological doctrines.

And we can determine the context of scientific material by its very nature. In this case, tables for astronomical timekeeping by night turn out to be singularly sophisticated tables for practical astrology, with one of them involving a remarkable mathematical notation. These tables need to be investigated more thoroughly by someone from the next generation who is trained in the old methods associated with the "Quellen und Studien" tradition.[22]

But I have another point too. al-Balkhī's successor was **certain** (in other words, he convinced himself) that the tabulated values in the first set of auxiliary tables did not change enough with time to warrant preparing a new version. If they were intended for Qandahar, both al-Balkhī and his successor were oblivious to the fact that the tables, accurately computed though they may have been, were based on an incorrect value of the local latitude that they had clearly not measured themselves: they trusted their predecessors for the parameter and had no reason to **doubt** them. The fact that the underlying parameters were in **error**, and substantially so, for the latitudes were incorrect and the basic stellar coordinates no longer valid, was anyway immaterial to their purpose in presenting the tables. It is very, very rare that a medieval astronomer wrote by a table in a manuscript words to the effect: "this doesn't work" (Arabic, *bāṭil*). It was also actually immaterial for the purpose of the tables, since this was astrological. Mathematical astrology is one science where it does not really matter what data and procedures one uses, as long as one is satisfied with the result.[23] One could argue that astronomy in the service of Islam was another case in point, not because it can barely matter in the framework of the universe that one prays at a certain time or a few minutes before or after, but because the complicated apparatus of the medieval astronomers and their results were generally ignored altogether by the medieval legal scholars, who had their own apparatus based on folk astronomy, which inevitably produced different results.

The tabulation of the various auxiliary functions was a sign of considerable **innovation**, but the parameters underlying the tables were dictated by **tradition**, a force that tended to restrict scientific progress rather than to further it.[24] So whilst we are dealing here with two sets of tables of considerable mathematical sophistication, indeed without parallel in the history of astronomy, they could be judged useless from a practical point of view. But this does not lessen the interest of the tables from a historical point of view, or from a cultural point of view. If we are to seek a cultural context for these tables, we have to consider the relative security of the reign of Sultan Maḥmūd of nearby Ghazna (998-1030) as the only feature of the Eastern Islamic lands that would have allowed anyone the luxury to sit down in Qandahar and compile astronomical tables.[25] And one may wonder whether al-Bīrūnī, the greatest universal scholar in the exact sciences in Islamic history, who worked in Ghazna and dedicated his own astrono-

---

[22] See n. 3.

[23] In this connection see my obituary notice for Professor Frans Bruin (1922-2001), formerly of the American University of Beirut, in *Journal for History of Astronomy* 33:2 (2002), pp. 214-216, as well as the remarks at the beginning of **XIIa**. Always ready for a good laugh, Bruin prepared his own horoscope according to Ptolemaic procedures, albeit with accurate astronomical data, and predicted that he would die a peaceful death. In 1964 he purchased for a few dollars a computer-generated horoscope based on modern astrological procedures at Grand Central Station in New York City: this predicted he would die a violent death. Fortunately the first prediction was more accurate.

[24] See King, *Mecca-Centred World-Maps* (n. 5), p. 363, on this theme.

[25] See the article "Masʿūd b. Sebüktigin" by C. Edmund Bosworth in *EI*$_2$, esp. p. 66a.

mical handbook to Maḥmūd, had any contact with the al-Balkhī who compiled at least some of those tables for Qandahar. For the time being, it seems appropriate to be reminded that a thousand years ago, in what is now Afghanistan, there were people who had such serious interests.[26]

---

[26] And even if it can eventually be shown that the tables I have discussed were *not* intended for Qandahar, which would identify a serious **error** on my part, the reader of the present text need not **doubt** that Muslim astronomers in the Middle Ages from the Maghrib to India, such as the ones we have mentioned above, were pretty **certain** that they were doing the right thing and not wasting their time. I can hope the same for myself.

INDEXES

Note: Some of these indexes should be used in conjunction with the tables of contents of the different parts of this work. Sometimes references are given to parts of the text in which ample cross-references will guide the reader to other parts on the same topic that are not indexed here. A reference of the form **X**-1.2.3 is to Section 1.2.3 of Part **X**, and **X**-n. 1:2 to footnote 2 in Section 1 of Part **X**. References to footnotes are intended to locate more precisely the relevant part of the text. Titles are not indexed, though some anonymous ones are included in the index of personal names. Individual instruments are indexed in Volume 2.

INDEX OF TOPICS

Note: I have not included topics like sun, declination, prayers at the *ʿaṣr* and the *maghrib*, since they occur throughout this volume.

INDEX OF PROPER NAMES (AND TITLES)

Note: In the order of this index Abū, al-, and Ibn/ibn have been ignored. Anonymous titles are listed under "anonymous" at the beginning of the list. Occasional titles like *Calendar of Cordova* and *Libros del Saber* are included.

INDEX OF MODERN SCHOLARS (SELECTED)

Note: The selection emphasizes earlier contributions to the background of this study.

INDEX OF TECHNICAL TERMS

Note: All terms are Arabic, unless followed by (M) for Malay, (P) for Persian, (S) for Sanskrit, and (T) for Turkish.

INDEXES OF PARAMETERS
(1) TERRESTRIAL LATITUDE

Note: For latitudes of localities in Tunisia and Sicily see **I**-9.7 and **II**-13.4. For latitudes of localities in the Nile Valley see **II**-8.10. For al-Khalīlī's list of localities in Syria see **II**-10.8, and for Taqi 'l-Dīn's list of localities in the Ottoman Empire see **II**-14.9.

(2) OBLIQUITY OF THE ECLIPTIC

INDEX OF MANUSCRIPTS CONSULTED

Notes: An asterisk after a manuscript number indicates that I have not consulted the manuscript or a copy thereof. Section numbers that are underlined contain descriptions of the manuscripts concerned. For *Cairo* manuscripts see *Cairo ENL Catalogue* and *Survey*. For manuscript catalogues other than the *Berlin*, *Escorial*, and *Paris Catalogues*, see Sezgin, *GAS*, VI, pp. 295-466. The abbreviation PC indicates a private collection. More detailed references to manuscripts used in **III** and **IV** are given in the indexes *ad loc*. Manuscripts from the library of al-Ṭūlūnī listed in **V-10** are not included here.

DM 740,2	Prayer-tables of Ibrāhīm ibn Qāyitbāy	**II**-7.2
DM 758	al-Khalīlī's universal auxiliary tables	**I**-9.5; **II**-6.14, 10.2; **VIIb**-n. 5:1, Fig. 4.1
DM 776,1	Anonymous tables of time since sunrise for Cairo	**I**-2.1.1; **II**-4.0, <u>4.2</u>
DM 777	Anonymous timekeeping tables for Cairo	**I**-2.1.1; **II**-4.0, <u>4.2</u>, 4.5
DM 778	Anonymous hour-angle tables for Cairo	**I**-2.1.1; **II**-4.0, <u>4.2</u>
DM 786	al-Bakhāniqī's edition of the Cairo corpus	**I**-2.1.1, 5.1.1; **II**-4.0, <u>4.2</u>
DM 812	al-Kutubī's timekeeping tables for Cairo	**I**-2.1.8; **II**-7.14
DM 830,2	Anonymous notes on shadow-schemes	**III**-9.7
DM 832	Treatise on the use of the sine quadrant by ʿAlāʾ al-Dīn ibn Ṭaybughā	**V**-Fig. 12.1
DM 834	Yūsuf Kalārjī's prayer-tables for Crete	**I**-4.8.9, 5.6.13, 6.4.14, 6.10.12; **II**-8.8
DM 899,2	Yemeni shadow-scheme	**III**-4.7
DM 910	Coptic astronomical handbook by al-Asʿad ibn al-ʿAssāl	**III**-9.7
DM 948	al-Aṣbaḥī's treatise on folk astronomy and other Yemeni treatises on the same subject	**III**-2.2, 4.1, 4.8; **IV**-1, *etc.*
DM 954,2	Anonymous timekeeping tables for Cairo	**II**-<u>4.2</u>, 4.6
DM 970	al-Khujandī's treatise on his hemispherical instrument	**IV**-4
DM 1014	al-Maḥallī's prayer-tables for Damietta	**I**-6.4.12, 6.10.8; **II**-8.6
DM 1101	Anonymous solar azimuth tables for Cairo	**I**-2.1.1, 5.1.1; **II**-4.0, <u>4.2</u>
DM 1103	al-Kutubī's timekeeping tables for Cairo	**I**-2.1.8; **II**-7.14
DM 1104	al-Kutubī's timekeeping tables for Cairo	**I**-2.1.8; **II**-7.14
DM 1108,3	Tables for determining twilight for all latitudes by Ibn Abi ʾl-Khayr al-Ḥusnī	**II**-8.3, Fig. 8.3
DM 1108,9	Anonymous solar azimuth tables for Cairo (modified from those of Ibn Yūnus)	**I**-2.1.1, 5.2.1, Fig. 5.2.1; **II**-4.0, <u>4.2</u>, 4.4, 5.7
DM 1109	al-Bakhāniqī's edition of the Cairo corpus	**I**-2.1.1, 5.1.1; **II**-4.0, <u>4.2</u>
DM 1112	Contains anonymous shadow-scheme	**III**-9.7
DM 1169,6	Anonymous Maghribi treatise on the astrolabe	**III**-6.7; **IV**-2, 4
DM 1170,2	Anonymous prayer-tables for latitude 34° (due to Ibn al-Shāṭir)	**II**-9.3, Fig. 9.3a; **III**-8.2
DM 1211	Prayer-tables for Alexandria by Riḍwān Efendī	**II**-8.5
DM 1207,1	Anonymous prayer-tables for latitude 31°	**I**-4.8.8, 5.6.12, 6.4.13, 6.10.11; **II**-8.5
DM 1207,2	Anonymous prayer-tables for Rosetta	**II**-8.7, Fig. 8.7
DM 1218	Tables of the ascensions at nightfall by Muḥammad al-Ghazāwī	**II**-5.7
DM 1224	al-Bakhāniqī's edition of the Cairo corpus of tables for timekeeping, attributed to Ibn Yūnus	**I**-2.1.1, 5.1.1; **II**-4.0, <u>4.2</u>
DMF (*mīqāt fārisī*)		
DMF 1	*Īlkhānī Zīj* of al-Ṭūsī	**I**-2.3.2, Fig. 2.3.2a-b
DMF 8,1	Sexagesimal quotient table	**I**-Fig. 1.0c
DṢ (Dār al-Kutub *ṣināʿa*)		
DṢ 166,2	Anonymous tables for the lunar crescents during 1125-1130 H	**V**-Fig. 11.2
ḤJ (Ḥalīm *majāmīʿ*)		
ḤJ 47,1	*Kanz al-ʿulūm* by Ibn Tūmart al-Andalusī	**III**-6.1, Fig. 6.1
ḤM (Ḥalīm *mīqāt*)		
ḤM 19	Anonymous Maghribi treatise on sundial construction	**IV**-6.2

MS	Contents	References
MM 64	Anonymous solar azimuth tables for Cairo	**I**-2.1.1, 5.1.1; **II**-4.0, 4.2
MM 66	Tables of normed right ascensions by ʿAbd al-Qādir al-Minūfī	**II**-7.2
MM 67,1	Anonymous solar azimuth table with instructions in Turkish	**I**-5.5.2
MM 68	Anonymous prayer-tables for Mecca	**I**-6.4.4; **II**-5.3, 6.10, Fig. 6.10, 8.1
MM 81	al-Ikhṣāṣī' prayer-tables for Cairo	**I**-5.6.9; **II**-7.8
MM 71	Timekeeping tables for Damascus by al-Khalīlī, and solar azimuth tables for Damascus by al-Ḥalabī	**I**-2.1.4, 5.1.2, Fig. 5.1.2; **II**-8.5, 10.2, 10.5, 10.6, 11.2
MM 72	Anonymous solar altitude tables for Cairo	**I**-2.6.1, 4.4.1, Fig. 4.4.1, 5.5.1; **II**-4.1.1d, 5.5, 6.9
MM 98	al-Khalīlī's universal auxiliary tables	**I**-9.5; **II**-6.14, 6.15, 10.2
MM 100	Anonymous prayer-tables for Istanbul	**II**-14.8
MM 114	Treatise on the qibla by Ghars al-Dīn al-Ḥalabī	**VIIb**-n. 6:1
MM 123	Treatise on refraction at the horizon by ʿAbd al-Qādir al-Minūfī	**II**-8.2, Fig. 8.2
MM 132	Najm al-Dīn al-Miṣrī's universal auxiliary tables (for the first part see MS Oxford Marsh 676)	**I**-2.6.1, Fig. 2.6.1, 9.3a; **II**-6.5
MM 137	Solar azimuth tables for Cairo, attributed to Ibn Yūnus	**I**-2.1.1, 5.1.1; **II**-4.0, 4.2, 4.4, 5.2
MM 155,2	al-Wafāʾī's notes on refraction at the horizon (fol. 23v)	**II**-5.7
MM 167,4	Tables for timekeeping by the moon by Yūsuf ibn Ṭūghān	**II**-6.17, Fig. 6.17
MM 168,1	Egyptian shadow-scheme	**III**-9.4
MM 169,2	Anonymous notes on shadow-schemes	**III**-9.7
MM 171,6	Poem on shadows by Aḥmad al-Jālī	**III**-9.7
MM 181	Anonymous notes on folk astronomy	**II**-5.7, Fig. 5.7a; **III**-9.7
MM 184	Contains anonymous prayer-tables for Cairo	**I**-4.2; **II**-4.2
MM 188,1	Star catalogue of Ibn Yūnus	**I**-Fig. 1.2f
MM 201	al-Khalīlī's treatise on the quadrant (fols. 11v-12r)	**I**-6.5.1
MM 203,3	Anonymous auxiliary tables for timekeeping	**I**-6.14.2, 9.11, Fig. 9.11
MM 204	Anonymous spherical astronomical tables	**I**-6.9.3, Fig. 6.9.3; **II**-4.2
MM 209,1	Abū Bakr ibn al-Mushrif's tables of normed right ascensions	**I**-9.8; **II**-5.7
MM 221	al-Lādhiqī's prayer-tables for Cairo	**II**-7.8
MM 222	al-Lādhiqī's prayer-tables for Cairo	**II**-7.8
MM 223	al-Lādhiqī's prayer-tables for Cairo	**II**-7.8
MM 224	al-Lādhiqī's prayer-tables for Cairo	**II**-7.8
MM 225	al-Lādhiqī's prayer-tables for Cairo	**II**-7.8
MM 235	Treatise on refraction at the horizon by Muḥammad al-Minūfī	**II**-7.1
MM 241	Ibn al-Mushrif's universal auxiliary tables	**I**-6.4.3, 6.13.1, 7.3.1, 8.2.1, 9.8, Fig. 9.8; **II**-6.1
MR (Muṣṭafā Fāḍil *riyāḍa*)		
MR 40,2	Illustrations of astronomical instruments by Muṣafā Ṣidqī	**VIb**-Fig. 8.1, Fig. 8.3; **VIa**-Fig. 12.2
QM (Qawala *mīqāt*)		
QM 2	Anthology of treatises on unusual instruments	**V**-Fig. 6.1
Sh (magic)		
Sh 89,4	Ibn al-Rashīdī's introduction to his prayer-tables for Mecca (fols. 29v-32v)	**II**-6.11
Sh 89,11-15	Various treatises by al-Hattārī containing prayer-tables for Medina (fols. 157r-322v)	**II**-12.8, Fig. 12.8

Manuscript	Contents	References
TR 159,2	Treatise on the universal plate by Ibn Bāṣo	**V**-6
TR 191	Hour-angle tables for Cairo, attributed to Ibn Yūnus	**I**-2.1.1; **II**-4.0, 4.2, 4.5, 4.7, 4.9
TR 274	Anonymous Yemeni almanac for the year 808 H [= 1405/06]	**II**-12.5, Fig. 12.5
TR 275	*al-Lum'a* (*zīj*) by al-Kawm al-Rīshī	**I**-7.1.8, 7.2.3
TR 276	Riḍwān Efendī's prayer-tables	**II**-7.10
TR 303,2	Anonymous notes on refraction at the horizon	**II**-5.7, Fig. 5.7b
TR 338,2	Prayer-tables for the Western Maghrib by al-Jannād	**I**-6.4.11, 6.10.9; **II**-13.5, Fig. 13.5
TR 346	Ḥasan al-Jabartī's prayer-tables for Cairo	**II**-7.13
TR 354	Timekeeping tables for Alexandria, (falsely) attributed to Ibn Yūnus	**I**-2.1.1, 2.1.6, Fig. 2.1.6, 4.8.8, 5.1.3; **II**-4.0, 8.5
ZK (Zakiyya)		
ZK 154	Ibn Abi 'l-Khayr al-Ḥusnī's almanac for Cairo	**II**-7.4
ZK 287	Anonymous prayer-tables for Cairo (by al-ʿAjmāwī)	**II**-7.7
ZK 324	Anonymous prayer-tables for latitude 31°	**II**-8.5
ZK 782,5	Anonymous fragment of tables for timekeeping by the stars for Istanbul	**I**-2.8.1, Fig. 2.8.1
ZK 917,4	Notes on interpolation, some attributed to al-Khalīlī	**I**-n. 1:21; **II**-n. 10:41
Cambridge University Library		
Add. 3527	Contains miscellaneous timekeeping tables for Baghdad	**II**-3.12, 3.13
Arberry Supp. 110(a)	Treatise on folk astronomy by al-Aṣbaḥī	**IV**-1
Browne O.1	*Zīj-i mufrad* of Muḥammad ibn Ayyūb al-abarī	**I**-3.3.3; **II**-14.2; **IV**-5.3, Fig. 5.3
Gg. 3.27	*Muẓaffarī Zīj* of al-Fārisī, together with an anonymous recension	**I**-7.1.7, 7.2.2
Damascus, al-Maktaba ¯āhiriyya		
27	Prayer-tables for Medina by Tāj al-Dīn	**II**-12.10
3098	Contains Ibn al-Shāṭir's treatise on the complete quadrant	**II**-9.2, 9.3
3116	al-Khalīlī's hour-angle tables for Damascus	**I**-2.1.4, 7.1.8; **II**-10.2, 11.13
4893	Anonymous prayer-tables for Tripoli	**II**-8.1, 10.2, 11.4, 11.6
5579	shorter treatise on the qibla by al-Dimyāṭī	**VIIb**-n. 6:17, Fig. 9.3
5588	Prayer-tables for Medina by Tāj al-Dīn	**III**-3.4, 3.5; **IV**-8
7387	Recension of the *Zīj* of Ibn al-Shāṭir by al-Kawm al-Rīshī	**I**-2.1.4, 7.2.3; **II**-11.13
7388	*al-Lum'a* (*zīj*) by al-Kawm al-Rīshī (incomplete)	**II**-11.13
7564	Anonymous astronomical tables for Aleppo	**II**-4.10, 11.11, 12.10
9227	al-Ḥalabī's solar azimuth tables for Damascus	**I**-5.1.2; **II**-10.2, 11.1, 11.2, 11.3
9233	al-Khalīlī's hour-angle tables and prayer-tables for Damascus, attributed to al-Ṭanṭāwī	**I**-2.1.4, 6.1.2, 8.3.2; **II**-10.2, 10.6, 11.13
9353	Treatise on sundial construction by al-Ṭanṭāwī	(**II**-11.13)
10076	Contains tables of corrections for refraction at the horizon of Damascus	**II**-11.13
10387 (?)	Contains al-Khalīlī's prayer-tables	**II**-10.2
10732	Prayer-tables and shadow-schemes associated with Ibn Qudāma	**II**-6.2; **III**-8.1
Damascus, al-Maktaba al-ʿArabiyya bookshop (1973)		
unnumbered	Timekeeping tables for Damascus attributed to al-Ṭanṭāwī	**I**-2.2.5
Dublin, Chester Beatty Library		
3665	*Zīj* of Muḥyi 'l-Dīn al-Maghribī for Maragha	**I**-5.6.3; **II**-3.9
3673	Corpus of timekeeping tables for Cairo, attributed to Ibn Yūnus	**I**-2.1.1, Fig. 2.1.1a, 4.7.1, 4.8.4, 5.1.1, Fig. 5.1.1a, 6.14; **II**-4.0, 4.1.1d, 4.2,

		4.3, 4.4, 4.5, 4.6, 4.7, Fig. 4.7a, 4.8, 4.10, 4.11, Fig. 4.11c, 4.12, 5.2, 6.7
4067	Treatise on refraction at the horizon by ʿAbd al-Qādir al-Minūfī	**II**-8.2
4078	Anonymous timekeeping tables for Cairo	**I**-2.1.1; **II**-4.0, 4.2, 4.5
4087	*Zīj* of Ibn al-Bannāʾ	**I**-6.9.4
4090	al-Maqsī's treatise on sundial construction (without diagrams)	**I**-4.1.3; **II**-4.1.1b
4091	al-Khalīlī's solar azimuth tables for Damascus and auxiliary tables for timekeeping by the sun	**I**-5.3.1, 9.4, Fig. 9.4a-b; **II**-10.1, 10.2, 10.3, 10.10
4092	al-Bakhāniqī's tables for astrolabe construction	**II**-4.1.1e
4538	Treatise on the lunar mansions by al-Ṣiqillī	**III**-9.6, Fig. 9.6
4677	Recension of al-Kāshī's *Khāqānī Zīj* by Ibn al-Kayyāl	**II**-11.0
4833	Treatise on the sine quadrant by Abū Bakr ibn Shihāb al-Dawla al-Miṣrī al-Karakī	**VIIa**-Fig. 3.3
Persian 102,1	Treatise by [Najm al-Dīn al-Miṣrī] on his universal auxiliary tables	**I**-2.6.1
Persian 102,2	Treatise on instruments by [Najm al-Dīn al-Miṣrī]	**VIa**-Fig. 3.1; **VIIb**-5, Fig. 5.2
El Escorial, Bibliotheca de		
ár. 908*	*Zīj* of al-Battānī (published in Nallino, *al-Battānī*)	**I**-4.1.2, 5.4.2, 5.8.1, 6.16
ár. 909,1	*Zīj* of Ibn al-Bannāʾ	**I**-6.9.4
ár. 924,7	Solar azimuth tables for Cairo, attributed to Ibn Yūnus	**I**-2.1.1, 5.1.1; **II**-4.0, 4.2, 4.4, 5.2
ár. 927	Anonymous recension of the *Mumtaḥan Zīj* of Yaḥyā ibn Abī Manṣūr	**I**-4.1.1, 6.1, 6.16.3, Fig. 6.16.3, 9.2; **II**-6.3, 6.6
ár. 931,8	al-Khalīlī's universal auxiliary tables (fols. 171r-211v)	**I**-9.5; **II**-10.2, 10.7
ár. 941	Treatise on folk astronomy by al-Ḥasan ibn ʿAlī al-Umawī al-Quṭrubī	**III**-5.2
ár. 955	*al-Tabṣira fī ʾilm al-hayʾa* by al-Kharaqī	**IV**-3
ár. 961,1	Treatise by Ibn Bāṣo on his universal plate	V-6
ár. 961,3	Anonymous short treatise on timekeeping	**II**-2.5, 7.5
ár. 965	Treatise on instruments by Amīn al-Dīn al-Abharī	**II**-2.8
ár. 966	Almanach of Zacuto	**I**-4.3.7
Florence, Laurenziana		
24	al-Mawṣilī's recension of the *Īlkhānī Zīj* for Mosul and anonymous prayer-tables for Damascus (due to al-Mizzī)	**II**-9.2
Medici 269	*Īlkhānī Zīj* of Naṣīr al-Dīn al-Ṭūsī	**I**-2.3.2; **II**-3.8
Gotha, Thüringische Landsbibliothek		
A 1390	Sibṭ al-Māridīnī's treatise on sexagesimal arithmetic	**II**-5.8
A 1399	Prayer-tables of al-Lādhiqī (defective)	**II**-7.8
A 1403	*al-Rawḍ al-ʿāṭir* (*zīj*) by Ibn Zurayq	**II**-9.5
A 1410	Solar azimuth tables for Cairo, attributed to Ibn Yūnus	**I**-2.1.1, 5.1.1; **II**-4.0, 4.2, 4.4, 4.8, 5.2, 5.5
A 1411	Contains anonymous prayer-tables for Cairo	**II**-4.2, 4.7
A 1414	Amīn al-Dīn al-Abharī's treatise on instruments	**II**-2.8
Hartford, Conn., Theological Seminary		
621	Prayer-tables of al-Lādhiqī	**II**-4.3, 4.10, 5.6, 7.8
Hyderabad, Sndra Pradesh State Library		
Āṣafiyya 298	Syrian recension of the *Zīj* of Ibn Isḥāq	**I**-4.3.7, 6.9.1*, 6.9.5, 7.1.5; **II**-13.1; **VIa**-Fig. 7.2

Manuscript	Contents	References
Istanbul, Kandilli Rasathanesi (Observatory)		
176	[Taqi 'l-Dīn]'s treatise on the differences of opinion amongst the Cairo astronomers concerning horizon phenomena	**II**-14.9
196	Aḥmad Efendi's hour-angle tables for Istanbul	**I**-2.1.7; **II**-14.12
208	Anthology of treatises by Taqi 'l-Dīn (in his own hand)	
208,1	Taqi 'l-Dīn's treatise on the difference of opinion between Cairo astronomers (fols. 89v-90r)	**II**-14.9
208,2	Taqi 'l-Dīn's *ṭaylasān* table for Istanbul (fols. 96r-100v)	**I**-2.3.6, Fig. 2.3.6; **II**-14.7
219	Ṣāliḥ Efendī's corpus of timekeeping tables for Istanbul	**I**-2.2.3; **II**-14.13
220	Anonymous recension of Ṣāliḥ Efendī's tables for Istanbul	**I**-2.2.3; **II**-14.13
226	Saʿīd Beg Zāde's auxiliary tables for computing the prayer-times	**I**-9.12; **II**-14.16; **IV**-6
249	*Shāmil Zīj* of Ibn al-Raqqām	**I**-6.9.5, 7.1.5
424 (?)	Anonymous Ottoman-type prayer-tables for Cairo (fols. 5r-6v)	**II**-7.5, 7.11, 7.12
440,1	Anonymous recension of Ṣāliḥ Efendī's tables for Istanbul	**I**-2.2.3; **II**-14.13
440,2	*Ruznāme* of Darendelī	**II**-14.4
441,1	Anonymous recension of Ṣāliḥ Efendī's timekeeping tables for Istanbul	**I**-2.2.3, 6.4.18, 6.10.15; **II**-14.13
441,2	Geographical table of Taqi 'l-Dīn	**I**-6.4.8; **II**-14.9
522	Anonymous prayer-tables for Crete	**II**-14.14
Istanbul Köprülü		
Köprülü 949	Treatise on mechanical clocks by Riḍwān al-Khurasānī (see *Cairo ENL Survey*, no. C1)	**II**-n. 6:7
Köprülü 1619	Contains anonymous prayer-tables for Cairo and Mecca and al-Mizzī's treatise on the winged quadrant	**II**-8.1
Istanbul, Millet Genel Kütüphanesi		
Ali Emiri 2743	al-Lādhiqī's prayer-tables for Cairo	**II**-7.8
Istanbul Nurosmaniye (NO)		
NO 2781	Riḍwān Efendī's prayer-tables for Cairo	**II**-7.10
NO 2782	Anonymous Seljuq (?) astrological treatise containing a table for timekeeping	**I**-3.3.3; **II**-14.2; **IV**-5.3
NO 2903	Anonymous timekeeping tables for Cairo	**I**-2.1.1, 5.1.1; **II**-4.0, 4.2, 4.5, 5.2
NO 2904	Treatise on sundial construction by Ibn Abi 'l-Fatḥ al-Ṣūfī	**I**-5.5.2
NO 2914,1	Ruznāme of Shaykh Vefā	**II**-14.3
NO 2921,2	al-Wafāʿī's auxiliary tables	**I**-6.14.1, 9.9; **II**-6.15
NO 2925	Timekeeping tables for Cairo attributed to al-Maqsī	**I**-2.1.1, 4.7.1, 5.1.1, 6.2.4, 6.10.2; **II**-4.0, 4.1b, 4.2, 4.5, 4.8, 5.2, 5.6
NO 2929,4	ʿAbd al-Qādir al-Minūfī's tables for refraction at the horizon for all latitudes	**II**-8.2
NO 2930	*Zīj* (*Sidr al-muntahā*) of Taqi 'l-Dīn	**I**-6.1.4, 7.1.12
Istanbul, Süleymaniye Kütüphanesi		
Aşir Ef. 224	Ṣāliḥ Efendī's corpus of timekeeping tables for Istanbul	**I**-2.2.3; **II**-14.13
Aşir Ef. 470	Contains the *Ruznāme* of Darendelī	**II**-14.4
Ayasofya (AS)		
AS 2590	Recension of al-Khalīlī's universal auxiliary tables by Muḥammad ibn Kātib Sinān	**I**-2.7.2, 9.5; **II**-10.2, 14.6

Manuscript	Description	References
AIII 3342	Treatise on sundial construction by al-Sijzī, disordered (correctly, fols. 123r-153v then 114r-122v)	**I**-4.1.1, Fig. 4.1.1b
AIII 3343	al-Marrākushī's treatise on instruments	**II**-4.1.5
AIII 3515 (T3046)	Tables for timekeeping by the stars by Muḥammad ibn Kātib Sinān	**I**-2.7.2, Fig. 2.7.2; **II**-14.5
AIII 3515	Tables for timekeeping by the stars by Muḥammad Kātib Sinān	**I**-2.7.2; **II**-14.5
Bağdatli 411	Anonymous early-15th-century anthology containing astronomical tables for Isfahan and Shiraz	**I**-5.6.8; **II**-3.15
Revan Köşku 1638	Astrological treatise *al-Qānūn fī 'l-dunyā* by Ibn Zunbul	**II**-Fig. 4.1.1
unidentified	Contains anonymous prayer-tables for Tunis	**II**-13.3
Istanbul, UL = Istanbul University Library (A/P/T = Arabic, Persian, Turkish)		
A 4082,1	Prayer-tables for Cairo attributed to al-Lādhiqī (fols. 1r-20v)	**II**-7.8
P 1418,2	*Īlkhānī Zīj* of Naṣīr al-Dīn al-Ṭūsī (fols. 109v-245v)	**I**-2.3.2; **II**-3.8
T 1824,1	Anonymous prayer-tables for Istanbul (fols. 3r-9r)	**I**-6.4.9, 6.5.2, 6.10.7; **II**-14.8, 14.9
T 1963	Anonymous recension of Ṣāliḥ Efendī's tables for timekeeping for Istanbul	**I**-2.2.3; **II**-14.13
T 1964	Anonymous recension of Ṣāliḥ Efendī's tables	**I**-2.2.3; **II**-14.13
T 4203	Auxiliary tables for compuring the hour-angle by "Jamāl"	**I**-9.12
Leiden, Universiteitsbibliotheek		
Or. 65	Anonymous recension (?) of the *Zīj* of Ibn al-Shāṭir	**I**-4.3.5
Or. 98,1	Treatise by al-Samawʿal al-Maghribī	**V**-10
Or. 143	*Ḥākimī Zīj* of Ibn Yūnus (Chs. 1-22)	**I**-4.8.1, 5.1.1, 5.6.2, 6.1, 7.1.3, 8.1.1, 8.3.1, 9.1; **II**-2.5, 4.3, 4.12, 5.1, Fig. 5.1, 6.6, 6.10, 6.15
Or. 168,16	Treatise on the qibla by Ḥabash	**VIIa**-Fig. 3.5
Or. 199	Anonymous tables of time since sunrise for latitude 36° (fols. 21v-27v)	**I**-2.3.4; **II**-3.11
Or. 468	Anonymous Egyptian treatise on spherical astronomy entitled *Kanz al-yawāqīt*	**I**-9.1; **II**-13.6; **V**-6, Fig. 12.2
Or. 1001/1111 (?)	Ibn al-Shāṭir's treatise on his prayer-tables (fols. 108r-113r)	**II**-9.3; **III**-8.2; **IV**-2
Or. 2538	*al-Sirāj al-wahhāj* by al-Hattārī	**II**-12.8
Or. 2805	Possibly (?) contains Ibn al-Rashīdī's prayer-tables for Mecca	**II**-6.11
Or. 2808(2)	al-Dīsṭī's prayer-tables for Lattakia	**I**-4.8.7; **II**-11.8
Leipzig Universitätsbibliothek		
808	al-Karakī's timekeeping tables for Jerusalem	**I**-2.1.3, 2.2.1, Fig. 2.2.1a-b; **II**-9.4
814	Anonymous tables from the Damascus corpus	**II**-10.2
817	Anonymous timekeeping tables for Cairo	**I**-2.1.1; **II**-4.0, <u>4.2</u>, 4.5
London BL = British Library		
27261	*Rawḍat al-munajjimīn* by Shāhmardān Rāzī	**I**-6.4.1
Add. 7489	Treatise on astronomy by ʿAbd al-Raḥīm containing edition of Timurid geographical table	**VIa**-Fig. 6.1
Add. 9599,31	al-Khalīlī's universal auxiliary tables (Maghribi copy)	**I**-9.5; **II**-10.2, 10.7
Or. 411,2	Anonymous commentary on al-Jādarī's poem *Rawḍat al-azhār* (with tables, see also MS Cairo K 4311, without tables)	**I**-6.3.4, 6.4.15, 6.15.2, 8.1.4; **II**-2.0. 13.6; **IV**-5, 6

Manuscript	Contents	References
Or. 3624	*Mukhtār Zīj* of Abu 'l-ʿUqūl	**I**-2.1.2, 3.1.1, 6.14, 7.1.1, 7.1.3, 7.2.1; **II**-4.3, 12.1
Or. 5593	al-Bīrūnī's treatise on astrolabe construction	**I**-1.4
Or. 9116	*Taysīr al-maṭālib* (*zīj*) by al-Kawāshī	**I**-5.6.1
Madrid, Museo Naval		
unnumbered	Contains *inter alia* the *Zīj* of Ibn al-Bannāʾ and extracts from the *Qawīm Zīj* of Ibn al-Raqqām	**I**-6.9.4, 6.9.5
Manchester, Rylands		
361	Anthology of treatises mainly on instruments copied by Muṣṭafā Ṣidqī in 1154 H [= 1741], including *Khulāṣat al-durar* by al-Wafāʾī	I-10.11
Medina Aref Hikmet		
mīqāt 1	*Adwār al-anwār* (*zīj* for Maragha) by Muḥyi 'l-Dīn al-Maghribī (copied 691 H!)	**I**-4.3.1a; **II**-3.8
Meshed Shrine Library		
332(103)	*Zīj* of Muḥyi 'l-Dīn al-Maghribī	**I**-5.6.3; **II**-3.8
Milan, Biblioteca Ambrosiana		
227a (C49)	Treatise on spherical astronomy by Najm al-Dīn al-Miṣrī (fols. 85v-97r)	**I**-2.6.1, 4.4.1; **II**-2.5, 4.10
C46	Almanac attributed to Abu 'l-ʿUqūl ((fols. 52-58))	**II**-12.1
C82	Maghribī recension of the perpetual almanac of Zacuto (Yemeni copy?)	**I**-4.3.7, 6.9.5
C84	Anonymous fragments of the Taiz corpus of tables for timekeeping (by Abu 'l-ʿUqūl)	**I**-2.1.2, 3.1.1; **II**-12.1
C86	Fragment of the *Zīj* of al-Battānī (fols. 167r-168v) (see *RSO* 7 (1916-18), pp. 125-126)	**I**-5.8.1
Griffini 37 / 73 Sup.	al-Fārisī's treatise on folk astronomy	**III**-4.2; **IV**-1.5, *etc.*; **VIIa**-Fig. 2.1
281e	Miscellaneous notes attributed to Ibn Yūnus	**VIIb**-n. 5:3
Mosul, al-Muḥammadiyya		
129	Contains al-Khalīlī's tables of the hour-angle and time since sunrise	**I**-2.1.4; **II**-10.2
Munich, Bayerische Staatsbibliothek		
ar. 853	Almanach of al-Zarqāllu (edited in Millás Vallicrosa, *Azarquiel*)	**I**-6.9.1, 7.1.5
New Haven, Conn., Yale Medical School Historical Library		
unnumbered	*Ruznāme* of Darendelī	**II**-14.4
Oxford, Bodleian Library		
arab. e. 93	Anonymous timekeeping tables for Edirne (fols. 3v-25r)	**I**-2.2.2, Fig. 2.2.2; **II**-14.7
Bodley 133,2	Anonymous 13th-century Egyptian treatise on timekeeping	**II**-6.1, Fig. 6.1; **III**-9.3; **IV-4; V**-5, Fig. 5.1a-b;
Hunt. 143	*Īlkhānī Zīj* of Naṣīr al-Dīn al-Ṭūsī	**I**-2.3.2; **II**-3.8
Hunt. 233	al-Sulṭān al-Ashraf's treatise on astrology	**I**-3.3.1, Fig. 3.3.1a-b, 4.2.5; **II**-12.2; **III**-4.3
Hunt. 331	*Ḥākimī Zīj* of Ibn Yūnus (Chs. 21-44)	**I**-4.7.1, 5.1.1; **II**-5.1
Hunt. 566	al-Khujandī's treatise on his hemispherical instrument	**IV**-4.5
Hyde 32 (Nic. 284)	Anonymous prayer-tables for Cairo (fols. 69v-81r)	**II**-7.5
Marsh 39 (Uri 1042)	al-Khalīlī's hour-angle tables for Damascus (fols. 113v-142v)	**I**-2.1.4; **II**-10.2
Marsh 95 (Uri 961)	al-Khalīlī's universal auxiliary tables	**I**-9.5; **II**-10.2
Marsh 134	al-Aṣbaḥī's treatise on folk astronomy	**III**-4.9; **IV**-1
Marsh 139	Ibn al-Shāṭir's treatise *Nihāyat al-suʾl* on planetary theory	**II**-9.3

Manuscript	Description	References
Marsh 294	Navigational atlas of al-Ṣafāqusī	**III**-6.6
Marsh 592	Treatise on the qibla by al-Dimyāṭī	**VIIa**-Fig. 2.7; **VIIb**-n. 6:18, Fig. 9.5
Marsh 676 (Uri 944 = 995)	Najm al-Dīn al-Miṣrī's universal auxiliary tables (2nd part – see MS Cairo MM 132)	(**I**-2.1.6), 2.6.1, 9.3a; **II**-6.5, 8.1, 9.5
Pococke 249	Treatsie on folk astronomy by Abū ʿAbdallāh al-Muʿaddib	**III**-6.4. Fig. 6.4
Seld. A30	*Zīj* of Ibn al-Shāṭir	**I**-7.2.3; **II**-2.7
Seld. Supp. 100	al-Khalīlī's prayer-tables and hour-angle tables for Damascus	**I**-2.1.4, 6.5.1; **II**-10.2, 10.5, 10.6
Seld. Supp. 101 (Uri 1040)	Ibn Abi 'l-Fatḥ al-Ṣūfī's auxiliary tables	**I**-9.10; **II**-6.15
Seld. Supp. 99	al-Fawānīsī's prayer-tables for Cairo	**I**-4.8.4, 6.8.1, 6.14.2; **II**-7.5
Thurston 3, fols. 111r-114r	Abū Naṣr's auxiliary tables	**I**-6.2.3, 9.3, Fig. 9.3
Oxford, Museum of the History of Science		
MHS 84-33	Darendelī's tables for Istanbul	**II**-14.4, Fig. 14.4b-c
Paris BNF = Bibliothèque Nationale de France		
ar. 2186	Cosmography of Pseudo-Ibn al-Wardī	**V**-Fig. 12.3
ar. 2273	Navigational atlas of al-Ṣafāqusī	**III**-6.6
ar. 2486	*Zīj* of al-Baghdādī	**I**-2.3.1, Fig. 2.3.1, 6.4.1a, 7.1.6; **II**-3.2, 3.3, Fig. 3.3
ar. 2494	*al-Majisṭī* by Abu 'l-Wafāʾ (lacks tables)	**I**-5.8.2
ar. 2506,1	Treatise on sundial theory attributed on the title folio to Saʿīd ibn Khafīf al-Samarqandī or Ibn al-Ādamī (the copyist, Ibn Abi 'l-Fatḥ al-Ṣūfī, was not sure)	**I**-4.2.3
ar. 2507 and 2508	al-Marrākushī's *summa* on spherical astronomy and instruments	**I**-4.1.3, Fig. 4.1.3, 4.2.4, 5.6.4; **II**-2.7, 5.6
ar. 2513	Anonymous recension of the *Muṣṭalaḥ Zīj*	**I**-2.1.1, 6.7.1, 6.9.2, 6.10.3, 7.1.8, 7.4.1; **II**-2.5, 4.3, 5.3, 6.6, 6.7, 9.2
ar. 2514	Anonymous tables for timekeeping by the sun	**I**-2.5.2, Fig. 2.5.2, 3.2.2, Fig. 3.2.2; **II**-3.4
ar. 2516	*Tuḥfa shāhiyya* of Quṭb al-Dīn al-Shīrāzī	**II**-14-0
ar. 2520	Anonymous recension of the *Muṣṭalaḥ Zīj*	**I**-7.1.9, 9.1, Fig. 9.1a-b; **II**-6.6, 9.2, 10.7, 11.13
ar. 2521	Prayer-tables for Damascus based on those of al-Khalīlī	**II**-11.1
ar. 2523	Anonymous Yemeni *zīj*	**I**-2.5.3, 3.3.2, 4.3.3, Fig. 4.3.3, 4.3.4, 7.1.10; **II**-6.6, 12.1, 12.3
ar. 2525,1	al-Māridīnī's auxiliary tables	**I**-9.6; **II**-6.15
ar. 2525,2	Anonymous prayer-tables for latitudes 27° to 30°	**II**-4.2, 8.1
ar. 2540,2	Planetary tables by Nāṣir al-Dīn ibn ʿĪsā al-Ṭabīb al-Ḥiṣkafī	**III**-7.2, Fig. 7.2
ar. 2544,14	Anonymous prayer-tables for Istanbul	**II**-14.8
ar. 2545	Prayer-tables of al-Fawānīsī for Cairo	**II**-7.5, Fig. 7.5a-b
ar. 2552	Large sexagesimal multiplication table	**I**-Fig. 1.0b
ar. 2553	Prayer-tables of al-Lādhiqī	**II**-4.3, 4.10, 7.8, 7.8
ar. 2558	al-Khalīlī's prayer-tables and hour-angle tables for Damascus, universal auxiliary tables, and qibla-table	**I**-2.1.4, Fig. 2.1.4a-b, 4.8.3, 5.6.6, 6.5.1, 9.5, Fig. 9.5a-b; **II**-10.1, 10.2, Fig. 10.2, 10.4, 10.5, 10.6, 10.8, Fig. 10.8, Fig. 10.9, 10.10

www.ingramcontent.com/pod-product-compliance
Lightning Source LLC
Chambersburg PA
CBHW061148240926
51846CB00006B/951